LENGTH

scales for comparison of metric and U.S. units of measurement

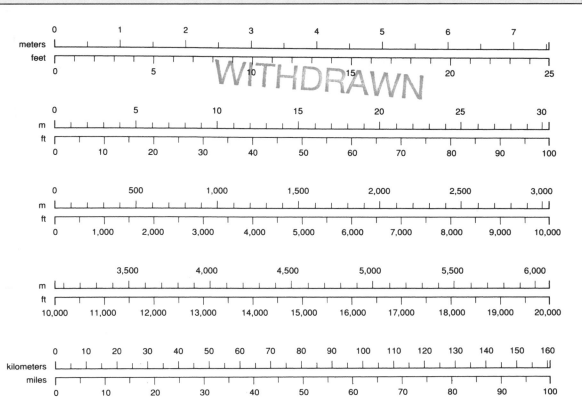

CONVERSION TABLES

	U.S. to Metric		Metric to U.S.	
	to convert	multiply by	to convert	multiply by
LENGTH	in. to mm.	25.4	mm. to in.	0.039
	in. to cm.	2.54	cm. to in.	0.394
	ft. to m.	0.305	m. to ft.	3.281
	yd. to m.	0.914	m. to yd.	1.094
	mi. to km.	1.609	km. to mi.	0.621
AREA	sq. in. to sq. cm.	6.452	sq. cm. to sq. in.	0.155
	sq. ft. to sq. m.	0.093	sq. m. to sq. ft.	10.764
	sq. yd. to sq. m.	0.836	sq. m. to sq. yd.	1.196
	sq. mi. to ha.	258.999	ha. to sq. mi.	0.004
VOLUME	cu. in. to cc.	16.387	cc. to cu. in.	0.061
	cu. ft. to cu. m.	0.028	cu. m. to cu. ft.	35.315
	cu. yd. to cu. m.	0.765	cu. m. to cu. yd.	1.308
CAPACITY (liquid)	fl. oz. to liter	0.03	liter to fl. oz.	33.815
	qt. to liter	0.946	liter to qt.	1.057
	gal. to liter	3.785	liter to gal.	0.264
MASS (weight)	oz. avdp. to g.	28.35	g. to oz. avdp.	0.035
	lb. avdp. to kg.	0.454	kg. to lb. avdp.	2.205
	ton to t.	0.907	t. to ton	1.102
	l. t. to t.	1.016	t. to l. t.	0.984

Abbreviations

avdp.	avoirdupois
cc.	cubic centimeter(s)
cm.	centimeter(s)
cu.	cubic
ft.	foot, feet
g.	gram(s)
gal.	gallon(s)
ha.	hectare(s)
in.	inch(es)
kg.	kilogram(s)
lb.	pound(s)
l. t.	long ton(s)
m.	meter(s)
mi.	mile(s)
mm.	millimeter(s)
oz.	ounce(s)
qt.	quart(s)
sq.	square
t.	metric ton(s)
yd.	yard(s)

Walker's **Mammals of the World**

Ernest P. Walker, 1891–1969

Walker's
Mammals of the World

Monotremes, Marsupials, Afrotherians, Xenarthrans, and Sundatherians

Ronald M. Nowak

JOHNS HOPKINS UNIVERSITY PRESS | BALTIMORE

© 2018 Johns Hopkins University Press
All rights reserved. Published 2018
Printed in China on acid-free paper
9   8   7   6   5   4   3   2   1

Johns Hopkins University Press
2715 North Charles Street
Baltimore, Maryland 21218-4363
www.press.jhu.edu

Library of Congress Cataloging-in-Publication Data
Names: Nowak, Ronald M.
Title: Walker's mammals of the world : Monotremes, marsupials,
    afrotherians, xenarthrans, and sundatherians / Ronald M. Nowak.
Description: Baltimore : Johns Hopkins University Press, 2018. |
    Updated accounts of 19 orders of mammals from Walker's mammals
    of the world. | Includes bibliographical references and index.
Identifiers: LCCN 2017022926 | ISBN 9781421424675 (hardcover :
    alk. paper) | ISBN 9781421424682 (electronic) | ISBN 1421424673
    (hardcover : alk. paper) | ISBN 1421424681 (electronic)
Subjects: LCSH: Mammals. | Mammals—Classification. | Monotremes. |
    Marsupials. | Afrotherians. | Sundatherians. | Xenarthrans.
Classification: LCC QL703 .W2222 2018 | DDC 599.01/2—dc23
LC record available at https://lccn.loc.gov/2017022926

A catalog record for this book is available from the British Library.

Special discounts are available for bulk purchases of this book.
For more information, please contact Special Sales at 410-516-6936
or specialsales@press.jhu.edu.

Johns Hopkins University Press uses environmentally friendly book
materials, including recycled text paper that is composed of at least
30 percent post-consumer waste, whenever possible.

To the MAMMALS, GREAT AND SMALL,
who contribute so much to the welfare and happiness
of man, another mammal, but receive so little in return,
except blame, abuse, and extermination.

Ernest P. Walker

Contents

Preface

More than half a century has passed since the first edition of this book was published by Ernest P. Walker (1891–1969), assistant director of the National Zoological Park in Washington, DC, and six coauthors. Yet this new updated volume retains Walker's objective of providing a physical description and a basic account of natural history for each genus of living mammal, in a manner that will be easily understood and appreciated by the general public but also serve the professional community. This has become more of a challenge, though. Since the first edition appeared in 1964, there has been an enormous increase in mammalian research and in the resulting literature. An especially large amount of new information has been obtained through the systematic study of the ecology and behavior of mammals under natural conditions and through modern field and taxonomic approaches such as radio telemetry, remote sensing, and genetic analysis. Many genera known during the preparation of the first edition only by casual observation or examination of a few specimens now have been studied in detail.

The second and third editions of *Mammals of the World*, edited by John L. Paradiso (1929–2007) and published in 1968 and 1975, respectively, included many new facts and several accounts of newly described genera, but made no attempt to extensively revise and expand the text. The fourth edition, authored by Ronald M. Nowak and Paradiso in 1983, incorporated major changes in format and content. The overall length of the text was increased by about 50 percent, and approximately 90 percent of the generic accounts received substantive modification, many being completely rewritten. There was a break with the original practice of restricting the information on each genus to a single page, and there was an expansion of the accounts of those genera for which more data were available. In the fifth edition, authored by Nowak in 1991, text length was increased by about 22 percent, substantive changes were made in about 80 percent of the previously existing generic accounts, and 106 new generic accounts were added (mainly because of elevation of former subgenera and synonyms to generic rank). Respective figures for the sixth edition, authored by Nowak in 1999, were a 25 percent increase in text length, 95 percent of the previous generic accounts substantively modified, and 81 new generic accounts.

This newest update, following the sequence and nomenclature of Wilson and Reeder (2005), covers Monotremata (one order), Marsupialia (seven orders), Afrotheria (seven orders), Xenarthra (two orders), and Sundatheria (two orders). In comparison to that portion of the text of the previous edition applicable to those 19 orders, there has been, in this update, a 97 percent increase in text length, substantive change to 100 percent of the previously existing generic accounts, and addition of 17 generic accounts. Updated volumes covering other orders may be issued in the future.

The last edition for which Walker is considered primary author was the third in 1975. Only about 10 percent of the written text of this update is the same as in the relevant portion of that third edition, but major aspects of Walker's format have been retained. Ordinal and familial accounts usually precede the generic accounts and give information applicable to all members of the group involved. When possible, the following topics are covered for each genus: scientific and common name, number and distribution of species, measurements, physical description, habitat, locomotion, daily and seasonal activity, diet, population dynamics, home range, social life, reproduction, longevity, and relationship with people. There is little emphasis on such subjects as internal morphology, physiology, laboratory experimentation, parasitology, pathology, and paleontology. Walker's charts showing the world distribution of mammals have been retained for the genera covered by this volume, but they have been revised to reflect changes in taxonomy and distributional knowledge. As in the earlier editions, the metric system is used throughout (conversion scales and tables appear on the endpapers).

Walker, an accomplished photographer, wished to include a satisfactory photograph of a living representative of every genus in *Mammals of the World*. Despite his efforts and those of Paradiso, it was judged that, of the approximately 1,000 accounts of living genera in the third edition, nearly 400 still were accompanied by only a photo of a museum specimen or dead individual, by a drawing, by an unsatisfactory photo of a live animal, or by no illustration at all. In the course of preparing the fourth, fifth, and sixth editions, acceptable photos of living representatives were found for more than 250 generic accounts. However, by the time of the sixth edition, the number of generic accounts in the book had grown to nearly 1,200. Moreover, all illustrations were in black-and-white.

For the present update, it was decided to essentially begin anew and seek color coverage of all genera. Color photos of living individuals have been obtained for 143 of the 157 genera in this updated volume. Paintings of another seven extinct genera, based on skeletal remains and available descriptive data, have been made for this volume. Of the remaining genera, one is represented by a colorized photo of a living individual, one by a color photo of a mount, one by a colorized photo of a mount, and one only by black-and-white photos of skin and skull; three genera are not illustrated. With supplementary material, the total number of images (all color) new to this updated volume is 503, and there is one colorized photo that had been used previously in black and white. Only four black-and-white photos from the sixth edition have been retained in original form, and it is not without some regret that many excellent old pictures, including about 30 taken by Walker himself, have been laid aside.

Beginning with the fourth edition, an effort was made to list the name and range of every species of every genus, and, for genera with more than one species, to arrange the list in systematic order. That objective continues in this update, with more closely related species grouped together and more basal species coming first. Admittedly, information still is far from complete for many genera, and taxonomic interpretations sometimes vary widely. Even when such problems have been resolved, it may be difficult to ad-

equately express relationships in a linear fashion. In cases where more than one reference is cited in support of the systematic aspects of a generic account, each of the references was consulted in making a decision regarding the names, number, arrangement, and distributions of the species listed. All references do not necessarily agree with the information given in the accounts here. Major points of systematic controversy are discussed after the accepted list of species.

Development of the species lists of previous editions relied heavily on published sources providing the names and distributions of the mammals of major parts of the world. Especially useful were Bannister et al. (1988), Cabrera (1957, 1961), Chasen (1940), Corbet (1978), Corbet and Hill (1991, 1992), Ellerman and Morrison-Scott (1966), Flannery (1995a), Hall (1981), Laurie and Hill (1954), Meester and Setzer (1977), Rice (1977), Ride (1970), and Taylor (1934). The most extensive compilation of the world's mammalian species, edited by Wilson and Reeder (2005), consists of sections prepared by 26 authorities; much original information is included. It was used both as a primary source and to check the systematic lists given here. All points of disagreement with Wilson and Reeder (2005), except those involving minor differences in species distribution and more recently published information, are discussed. Another important reference was the IUCN (International Union for Conservation of Nature) Red List of Threatened Species (http://www.iucnredlist.org), the assessments for which actually cover every mammal species, threatened or not, that has lived since 1500 AD. Although based primarily on Wilson and Reeder (2005), those assessments include updated taxonomic data and information on distribution, population status, and ecology, as well as conservation.

This update also largely follows Wilson and Reeder (2005) regarding the number and sequence of mammalian orders, which has resulted in major changes in the traditional arrangement accepted in early editions of *Mammals of the World*. Of particular note in this update is recognition of seven distinct orders of marsupials, dissolution of the former order Insectivora (see account of class Mammalia herein), grouping of some of the insectivores with six other placental orders in the clade Afrotheria, and division of Xenarthra into two orders. Also covered here, though not in Wilson and Reeder (2005), is the extinct but historical order

Bibymalagasia of Madagascar, which was described by MacPhee (1994). Substantive changes were made in the fourth, fifth, and sixth editions with respect to number and arrangement of families, and the present volume contains many additional changes at that level. The names and contents of suborders and subfamilies are given at the beginning of the respective ordinal and familial accounts. Accepted subgeneric designations appear at appropriate places within the lists of species in generic accounts.

Within the 19 orders covered by this update are 40 families, 157 genera, and 541 species. The genus remains the basic unit of treatment, though some generic texts contain separate species accounts. Full accounts of 17 genera are new to this volume, while two generic accounts in the applicable portion of the sixth edition have been deleted. Of the newly covered genera, eight were named subsequent to preparation of the sixth edition: *Hyladelphys* Voss, Lunde, and Simmons, 2001; *Tlacuatzin* Voss and Jansa, 2003; *Chacodelphys* Voss, Gardner, and Jansa, 2004; *Cryptonanus* Voss, Lunde, and Jansa, 2005; *Micromurexia* Van Dyck, 2002; *Phascomurexia* Van Dyck, 2002; *Paramurexia* Van Dyck, 2002; and *Murexechinus* Van Dyck, 2002. None of those genera represents a newly discovered mammal; all involve elevation of a previously known taxon to generic level. Two of the new accounts reflect an attempt, begun in the fifth edition, to resolve the inconsistent treatment in earlier editions of genera known only by subfossil material but thought to have lived in historical time. Here, a separate account is provided for every extinct genus known to have lived within approximately the last 5,000 years. Four such genera in this volume (pilosans) inhabited the West Indies, one (*Plesiorycteropus*) lived in Madagascar, and one (*Mammuthus*) survived historically on Wrangel Island in the Arctic; all may have disappeared through human agency. Reference is also made to extinct species that lived in historical time and to the former distribution of extant species.

Literature for work on the 19 orders included in this update was assembled primarily through consultation of *Zoological Record* from 1995 through 2014. Many references, some published as late as 2017, were obtained by other means. As with the fourth, fifth, and sixth editions, but unlike the first edition of *Mammals of the World*, there are textual references for all new information except minor changes in measurements.

The Literature Cited section of this update contains approximately 2,200 new entries, along with 1,000 retained from the sixth edition. Information lacking citation of specific references, much of it on physical description, generally dates from the first edition and, as explained in the preface thereto, was obtained largely through direct observation by the authors or taken from unpublished data received orally, through correspondence, or through addition to the manuscript by reviewers.

This update contains substantial information on the relationship between people and other mammals, especially with regard to economic importance and conservation. Each generic account discusses any species, subspecies, or population that at the time of preparation of the manuscript was classified in a category of concern by the IUCN, based on the individually authored and dated assessments provided at http://www.iucnredlist.org. The IUCN classifications and supporting assessments cited in this volume were the latest available as of mid-January 2016. At that time, 209 full species of mammals were classed as critically endangered, 481 as endangered, and 507 as vulnerable, for a total of 1,197 of the 5,502 modern mammal species recognized by the IUCN. Another 324 species were designated "near threatened," which is defined as coming close to meeting the criteria for classification as vulnerable or endangered. Still another 78 species were considered "extinct," and two were regarded as "extinct in the wild." In addition, the IUCN classified as critically endangered, endangered, vulnerable, and near threatened 92 mammal subspecies that are not components of any of the full species of mammals that also are in those categories. Recently, the IUCN has deemphasized assessment of subspecies; many that were covered earlier (Baillie and Groombridge 1996) no longer have formal classifications.

Also indicated here are species, subspecies, and populations of mammals that were on the List of Endangered and Threatened Wildlife (http://www.ecfr.gov/cgi-bin/text-idx?rgn=div8&node=50:2.0.1.1.1.2.1.1), as maintained by the USDI (United States Department of the Interior) as of mid-January 2016, and that were on appendix 1 or 2 of the CITES (Convention on International Trade in Endangered Species of Wild Fauna and Flora) as of 5 February 2015 (https://www.cites.org/eng/app/appendices.php). By definition, appendix 1 includes species threatened with extinction that are or may be affected by trade. Appendix 2 includes species that are not necessarily threatened with extinction now but may become threatened unless trade is subject to strict regulation. It must be emphasized that those designations are subject to constant change.

Readers also should be cautioned that the USDI classification process has become hopelessly subject to delay and manipulation by bureaucratic, political, and commercial interests. Its resulting List is now almost meaningless as an expression of the extent and diversity of the world's declining mammals. Until 1987 the USDI had an Office of Endangered Species that included a Listing Branch with personnel dedicated to authoritative development of the List of Endangered and Threatened Wildlife. That List and the IUCN Red List were then closely comparable, each containing just over 400 species and subspecies of mammals. However, in October 1987 the Office of Endangered Species was abolished. Subsequently, the IUCN Red List was considerably expanded, reaching a peak size with the publication of Baillie and Groombridge (1996). It since has been refined somewhat, but its total number of mammal species and subspecies in all categories of concern is now 1,693, which represents a net increase of 1,276 since 1987. In the same period the USDI List has had a net gain of only 49 mammals (57 added but 8 delisted). And of the 49, 13 represent aquatic species, subspecies, and populations listed through the efforts of the National Marine Fisheries Service, Department of Commerce, not the USDI.

The latter agency's listing program is a travesty, with practically all new classifications resulting from petitions and litigation by outside groups, and enormous resources devoted not to additions to the List but to delistings based on often spurious claims of recovery. For example, the USDI recently issued a rule (Rieck et al. 2016) delisting the Louisiana black bear (*Ursus americanus luteolus*), arguing that 700 animals scattered across about 7,000 sq km represents "recovery," even though that number includes an introduced population of another subspecies, which could hybridize with and dilute the genome of the native subspecies, and even though the original range of the native subspecies covered over 300,000 sq km of habitat, which, assuming normal black bear densities (Larivière 2001), could have supported about 80,000 animals.

The above notwithstanding, the United States remains the greatest hope for saving the world's threatened mammals and indeed for conservation in general. Such hope may be fading, however, as political trends that exacerbated deterioration of USDI listing activity continue to gain strength. Such trends also interfere with recognition of the crisis of climatic warming and development of appropriate countermeasures (Cook et al. 2016; Notz and Stroeve 2016; Urban 2015; USCGRP 2017). The scope of that problem is enormous, affecting not only arctic species, such as the polar bear (*Ursus maritimus*), which even the USDI (in conjunction with the National Marine Fisheries Service) has accordingly classified as threatened (Schliebe and Johnson 2008), but also mammals far from the north, such as the tiny mountain pygmy possum (*Burramys*) of southeastern Australia (see account thereof herein).

Even favorable political and climatic conditions may not be sufficient to overcome a second encompassing potential problem for wild mammals. World demographics are changing (Aiyar et al. 2016; Colby and Ortman 2015; United Nations 2015). Human population growth, once seen as the greatest threat to nature and wildlife, has actually ceased and even reversed in many areas. But that has occurred in the very countries, primarily in North America and western Europe, and to some extent in eastern Europe and northern Asia, that have provided the most financial, technical, and altruistic backing for conservation. In contrast, population continues to rise sharply in regions where there is little traditional support and minimal resources for wildlife preservation, and where, unfortunately, there is the greatest degree of mammalian diversity. Expanding human populations in those regions are spilling into much of Europe and North America and may alter the demographic structure that has provided the foundation of international wildlife protection. Appropriate education and elucidation of a conservation ethic could be helpful. That there is yet hope in this regard may be seen in the growing number of productive conservation-oriented biologists in Latin America, sub-Saharan Africa, and other relevant areas, many of whom are authors cited herein.

Many persons contributed to the preparation of this updated volume, but the one primarily responsible for its initiation, continued development through frequent delays and difficulties, and ultimate publication is Vincent J.

Burke, Editor Emeritus, Johns Hopkins University Press. He invited me to do the work and eventually overcame my reluctance to do so. Subsequently, he became a constant source of guidance, understanding, and, especially, patience. It was his determination and faith in the viability of the "Walker Books," far more than my own, that kept the project alive. Other valuable editorial assistance was provided by Patty Bolgiano, Tiffany Gasbarrini, Kyle Howard Kretzer, Juliana McCarthy, Lauren Straley, Meagan M. Szekely, and Gene A. Taft of Johns Hopkins University Press; Mikala Guyton and JodieAnne Sclafani of Westchester Publishing Services; and Tatiana Holway, who copyedited the manuscript.

The following authorities voluntarily reviewed portions of the 1999 edition: Steven M. Goodman, Tenrecidae; Galen B. Rathbun, Macroscelidea; Hendrick N. Hoeck, Hyracoidea; Daryl P. Domning, Sirenia; and Louise H. Emmons, Scandentia. Jeheskel Shoshani had begun a review of Proboscidea but, before it could be completed, died tragically in a terrorist attack in Ethiopia. While the reviewers made many corrections and suggestions for improvement, I retain full responsibility for all errors and other problems that may still be present. It should be added that the reviews were done early in the course of the project and that many of the comments were superseded by subsequently available published data. I also thank the many people who sent publications and other information, by email and postal mail, which were not otherwise accessible. The reference staff of the Fairfax County, Virginia, Tysons-Pimmit Regional Library was of great help in locating and securing very many books and papers necessary to my research through interlibrary loan.

Illustrations have always been an indispensable aspect of *Mammals of the World*. In such regard this volume simply could not have been completed without the cooperation of the following, who provided their own, often superb, photographs and/or arranged for the provision of photos taken by others, at either no cost or a cost far below a normal rate: Stacy Frank of Minden Pictures, Pavel German of Wildlife Images, Gerhard Körtner, Jiri Lochman of Lochman Transparencies, Alexander F. Meyer, and Patricia A. Woolley. Those six individuals, and, if applicable, their respective agencies, account for nearly half of all the photos in this volume.

Among the others who showed an unusual degree of kindness with respect to acquisition of images are Cathy A. Beck, Nancy Carrizales, Francois Catzeflis, Dave Davis, Fritz Geiser, Thomas E. Heinsohn, David G. Huckaby, Norman Lim, Rexford D. Lord, Link E. Olson, Johannes Pfleiderer, Fiona A. Reid, Klaus Rudloff, Richard D. Sage, Harald Schütz, Chris and Mathilde Stuart, Mariella Superina, and Tim Flannery. Special thanks are due to Martyn Colbeck for generously donating his magnificent shot of an African forest bull elephant. Stefanie L. Ervin worked with the project for a time, found numerous photographic leads, and pointed me to Shutterstock.com, which proved a surprisingly helpful resource. I also am grateful to the many others credited in the legends of the images in this volume and to all those who offered photos that, unfortunately, could not be published. It should be assumed in all cases that copyright of an image is held by the party credited in the legend thereof and that such party must be contacted with regard to any matter involving reproduction.

Helga Schulze of the Department for Neuroanatomy, Ruhr-University Bochum, did the paintings of extinct genera, based primarily on her own extensive research of the literature covering skeletal remains and other descriptive information on the involved animals. Given a rather short time frame, she persisted steadily through this work despite illness, fire, office moving, and the need to care for her personal collection of rescued slender lorises.

Finally, I thank the late John L. Paradiso, colleague of Walker, editor of the second and third editions, and my coauthor for the fourth edition. It was he who recommended my original participation in the project and introduced me to many of the procedures involved. I joined him in considering it an honor and challenge to try to improve on the work of Ernest P. Walker, a man of remarkable experience, energy, and devotion to the wild mammals of the world. I hope that both Walker and Paradiso would not be disappointed in this latest volume.

World Distribution of Mammals

For maximum usefulness, it has been necessary to devise the simplest practicable outline of the approximate distribution of the genera in the sequence used in the text. The tabulation should be regarded as an index guide to groups of mammals or to geographic regions. At the same time, it gives a good overall picture of the general distribution of the mammals covered by this volume.

The major geographic distribution of the genera of Recent mammals that appears in the tabulation is designed to show their natural distribution at the present time or within comparatively recent times. It should be noted that most of the animals occupy only a portion of the geographic region that appears at the head of the column. Some are limited to the tropical regions, others to temperate zones, and still others to the colder areas. Mountain ranges and streams have sometimes been natural barriers preventing the spread of animals on the lands that are a part of the same continent. Also, many restricted ranges cannot be designated either by letters to show the general area or by footnotes because of limited space on the tabulation. Therefore it should not be assumed that a mark indicating that an animal occurs within a geographic region implies that it inhabits that entire area. For more detailed outlines of the ranges of the respective genera, it is necessary to consult the generic texts.

Explanation of Geographic Column Headings

Europe and Asia constitute a single land mass, but this land mass consists of widely different types of zoogeographic areas created by high mountain ranges, plateaus, latitudes, and prevailing winds. The general distribution of Recent mammals can be shown much more accurately by two columns, headed "Europe" and "Asia," than by a single column headed "Eurasia."

Most islands are included with the major land masses nearby unless otherwise specified, though in many instances some of the mammals indicated for the continental mass do not occur on the islands.

With Europe are included the British Isles and other adjacent islands, including those in the Arctic.

With Asia are included the Japanese Islands, Taiwan, Hainan, Sri Lanka, and other adjacent islands, including those in the Arctic.

With North America are included Mexico and Central America south to Panama, adjacent islands, the Aleutian chain, the islands in the arctic region, and Greenland but not the West Indies.

With South America are included Trinidad, the Netherlands Antilles, and other small adjacent islands but not the Falkland and Galápagos Islands unless named in footnotes.

With Africa are included only Zanzibar Island and small islands close to the continent but not the Cape Verde or Canary Islands.

The island groups treated separately are:

Southeastern Asian islands, in which are included the Andamans, the Nicobars, the Mentawais, Sumatra, Java, the Lesser Sundas, Borneo, Sulawesi, the Moluccas, and the many other adjacent small islands;
New Guinea and small adjacent islands;
the Australian region, in which are included Australia, Tasmania, and adjacent small islands;
the Philippine Islands and small adjacent islands;
the West Indies;
Madagascar and small adjacent islands;
other islands that have only one or a few forms of mammals and are named in footnotes.

The dugong, Steller's sea cow, and manatees are native to water areas adjacent to the lands for which they are recorded. Manatees frequent rivers through some of those lands, and the dugong has a substantial oceanic range.

Footnotes indicate the major easily definable deviations from the distribution indicated in the tables.

Symbols

†	The mammals are extinct.
▪	The mammals occur on or adjacent to the land or in the water area.
N	Northern portion
S	Southern portion
E	Eastern portion
W	Western portion
Ne	Northeastern portion
Se	Southeastern portion
Sw	Southwestern portion
Nw	Northwestern portion
C	Central portion

Examples: "N, C" = northern and central; "Nc" = north-central. Numerals refer to footnotes indicating clearly defined limited ranges within the general area.

Genera of Recent Mammals	page	North America	West Indies	South America	Madagascar	Africa	Europe	Asia	Southeast Asia Islands	Philippine Islands	New Guinea	Australian Region	Antarctic Region	Arctic Region	Atlantic Ocean	Indian Ocean	Pacific Ocean
MONOTREMATA TACHYGLOSSIDAE																	
Tachyglossus	24										▪	▪1					
Zaglossus	29										▪	▪Nw					
MONOTREMATA ORNITHORHYNCHIDAE																	
Ornithorhynchus	32											▪E,1					
DIDELPHIMORPHIA DIDELPHIDAE																	
Glironia	44			▪Wc													
Caluromys	46	▪S		▪													
Caluromysiops	48			▪Nw													
Hyladelphys	50			▪N													
Monodelphis	51	▪S		▪													
Tlacuatzin	56	▪S															
Marmosa	57	▪S		▪													
Metachirus	63	▪S		▪													
Chironectes	65	▪S		▪													
Lutreolina	68			▪													
Philander	70	▪S		▪													
Didelphis	72	▪		▪													
Marmosops	77	▪S		▪													
Chacodelphys	80			▪S													
Cryptonanus	81			▪S													
Gracilinanus	83			▪													
Lestodelphys	85			▪S													
Thylamys	87			▪													
PAUCITUBERCULATA CAENOLESTIDAE																	
Caenolestes	93			▪Nw													
Lestoros	95			▪Wc													
Rhyncholestes	97			▪Sw													
MICROBIOTHERIA MICROBIOTHERIIDAE																	
Dromiciops	99			▪Sw													
NOTORYCTEMORPHIA NOTORYCTIDAE																	
Notoryctes	105											▪					
DASYUROMORPHIA DASYURIDAE																	
Planigale	116										▪S	▪N,E					
Antechinomys	118											▪					
Sminthopsis	120										▪S	▪1					
Ningaui	128											▪					
Phascogale	130											▪					
Micromurexia	133										▪						
Murexia	135										▪						
Phascomurexia	137										▪						

1. And Tasmania.

Genera of Recent Mammals	page	North America	West Indies	South America	Madagascar	Africa	Europe	Asia	Southeast Asia Islands	Philippine Islands	New Guinea	Australian Region	Antarctic Region	Arctic Region	Atlantic Ocean	Indian Ocean	Pacific Ocean
Paramurexia	137										▪Se						
Murexechinus	138										▪						
Antechinus	139											▪1					
Myoictis	147										▪						
Dasykaluta	149											▪Nw					
Parantechinus	150											▪Sw					
Pseudantechinus	152											▪N,W					
Dasycercus	154											▪					
Dasyuroides	156											▪Nc					
Phascolosorex	158										▪						
Neophascogale	160										▪						
Dasyurus	160										▪	▪1					
Sarcophilus	173											▪2					
DASYUROMORPHIA THYLACINIDAE																	
Thylacinus †	178											▪2					
DASYUROMORPHIA MYRMECOBIIDAE																	
Myrmecobius	183											▪S					
PERAMELEMORPHIA CHAEROPODIDAE																	
Chaeropus	191											▪					
PERAMELEMORPHIA THYLACOMYIDAE																	
Macrotis	193											▪					
PERAMELEMORPHIA PERAMELIDAE																	
Perameles	199											▪1					
Isoodon	203									▪S		▪1					
Peroryctes	208										▪						
Echymipera	209									▪3		▪					
Rhynchomeles	211								▪4								
Microperoryctes	213										▪						
DIPROTODONTIA PHASCOLARCTIDAE																	
Phascolarctos	218											▪E					
DIPROTODONTIA VOMBATIDAE																	
Vombatus	227											▪E,1					
Lasiorhinus	229											▪E,S					
DIPROTODONTIA BURRAMYIDAE																	
Cercartetus	233										▪	▪1					
Burramys	238											▪Se					
DIPROTODONTIA PHALANGERIDAE																	
Ailurops	243								▪5								
Strigocuscus	245								▪6								
Trichosurus	247											▪1					
Wyulda	253											▪Nw					
Spilocuscus	254								▪7		▪3	▪Ne					

1. And Tasmania. 2. Tasmania only (in modern times). 3. And Bismarck Archipelago. 4. Seram only. 5. Sulawesi and Talaud Islands only.
6. Sulawesi only. 7. And South Molucca Islands.

Genera of Recent Mammals	page	North America	West Indies	South America	Madagascar	Africa	Europe	Asia	Southeast Asia Islands	Philippine Islands	New Guinea	Australian Region	Antarctic Region	Arctic Region	Atlantic Ocean	Indian Ocean	Pacific Ocean
Phalanger	257								▪1		▪2	▪Ne					
DIPROTODONTIA PSEUDOCHEIRIDAE																	
Pseudochirops	262										▪	▪Ne					
Petropseudes	263											▪N					
Pseudocheirus	264											▪E,S,3					
Pseudochirulus	267										▪	▪Ne					
Hemibelideus	268											▪Ne					
Petauroides	269											▪E					
DIPROTODONTIA PETAURIDAE																	
Petaurus	273								▪4		▪5	▪					
Gymnobelideus	280											▪Se					
Dactylopsila	282										▪	▪Ne					
DIPROTODONTIA TARSIPEDIDAE																	
Tarsipes	285											▪Sw					
DIPROTODONTIA ACROBATIDAE																	
Distoechurus	289										▪						
Acrobates	290											▪E					
DIPROTODONTIA HYPSIPRYMNODONTIDAE																	
Hypsiprymnodon	292											▪Ne					
DIPROTODONTIA POTOROIDAE																	
Potorous	297											▪S,3					
Bettongia	301											▪3					
Caloprymnus †	306											▪C					
Aepyprymnus	308											▪E					
DIPROTODONTIA MACROPODIDAE																	
Lagostrophus	313											▪W,S					
Dorcopsis	315										▪						
Dorcopsulus	316										▪						
Dendrolagus	318										▪	▪Ne					
Setonix	322											▪Sw					
Thylogale	325										▪6	▪E,3					
Peradorcas	330											▪N					
Petrogale	332											▪					
Lagorchestes	340										▪Sc	▪					
Onychogalea	343											▪					
Wallabia	346											▪E					
Macropus	348										▪S	▪3					
AFROSORICIDA TENRECIDAE																	
Potamogale	387					▪W,C											
Micropotamogale	388					▪W,C											
Tenrec	390				▪												
Hemicentetes	393				▪												

1. Moluccas, Seram, and Timor. 2. And Bismarck and Louisiade Archipelagoes and Solomon Islands.

3. And Tasmania. 4. North Moluccas only 5. And Bismarck and Louisiade Archipelagoes.

6. And Bismarck Archipelago.

Genera of Recent Mammals	page	North America	West Indies	South America	Madagascar	Africa	Europe	Asia	Southeast Asia Islands	Philippine Islands	New Guinea	Australian Region	Antarctic Region	Arctic Region	Atlantic Ocean	Indian Ocean	Pacific Ocean
Echinops	395				■S												
Setifer	396				■												
Geogale	398				■												
Oryzorictes	399				■												
Microgale	401				■												
Limnogale	406				■												
AFROSORICIDA CHRYSOCHLORIDAE																	
Chrysochloris	410					■C,S											
Cryptochloris	412					■S											
Huetia	413					■C											
Eremitalpa	414					■Sw											
Chrysospalax	416					■S											
Chlorotalpa	417					■S											
Calcochloris	419					■Se											
Amblysomus	420					■S											
Neamblysomus	422					■Se											
Carpitalpa	424					■Se											
MACROSCELIDEA MACROSCELIDIDAE																	
Rhynchocyon	428					■E											
Petrodromus	431					■											
Macroscelides	434					■S											
Elephantulus	435					■											
BIBYMALAGASIA PLESIORYCTEROPODIDAE																	
Plesiorycteropus †	441				■												
TUBULIDENTATA ORYCTEROPODIDAE																	
Orycteropus	444					■											
HYRACOIDEA PROCAVIIDAE																	
Dendrohyrax	452					■											
Heterohyrax	455					■											
Procavia	458					■		■Sw									
PROBOSCIDEA ELEPHANTIDAE																	
Loxodonta	472					■											
Elephas	515						■1	■S	■2								
Mammuthus †	530							■3									
SIRENIA DUGONGIDAE																	
Dugong	541					■E		■S	■	■	■	■				■	■4
Hydrodamalis †	551	■5						■6									■7
SIRENIA TRICHECHIDAE																	
Trichechus	555	■Se	■	■		■W									■8		
CINGULATA DASYPODIDAE																	
Dasypus	579	■		■													
Euphractus	585			■													

1. Tilos Island only (questionably *Elephas* and questionably in historical time).   2. Sri Lanka, Sumatra, Java, and Borneo.
3. Wrangel Island only (in historical time).   4. East to Yap Island and Vanuatu.   5. Western Aleutians and St. Lawrence Island only.
6. Commander Islands only.   7. Bering Sea only.   8. Primarily Gulf of Mexico and Caribbean Sea.

Genera of Recent Mammals	page	North America	West Indies	South America	Madagascar	Africa	Europe	Asia	Southeast Asia Islands	Philippine Islands	New Guinea	Australian Region	Antarctic Region	Arctic Region	Atlantic Ocean	Indian Ocean	Pacific Ocean
Chaetophractus	587			■S													
Zaedyus	589			■S													
Calyptophractus	591			■S													
Chlamyphorus	593			■S													
Priodontes	594			■													
Cabassous	596	■S		■													
Tolypeutes	597			■													
PILOSA CYCLOPEDIDAE																	
Cyclopes	601	■S		■													
PILOSA MYRMECOPHAGIDAE																	
Tamandua	605	■S		■													
Myrmecophaga	608	■S		■													
PILOSA MEGALONYCHIDAE																	
Neocnus †	614		■1														
Choloepus	616	■S		■N													
Acratocnus †	620		■2														
Megalocnus †	622		■1														
Parocnus †	624		■1														
PILOSA BRADYPODIDAE																	
Bradypus	625	■S		■													
SCANDENTIA PTILOCERCIDAE																	
Ptilocercus	632							■3	■4								
SCANDENTIA TUPAIIDAE																	
Dendrogale	636							■Se	■5								
Anathana	638							■Sc									
Urogale	639									■S							
Tupaia	641							■Se	■	■W							
DERMOPTERA CYNOCEPHALIDAE																	
Galeopterus	651							■Se	■6								
Cynocephalus	654									■S							

1. Cuba and Hispaniola.   2. Cuba, Hispaniola, and Puerto Rico.   3. Malay Peninsula only.   4. Sumatra, Borneo, and nearby islands.
5. Borneo only.   6. Sumatra, Java, Borneo, and nearby islands.

Walker's **Mammals of the World**

Class Mammalia

Mammals

The class Mammalia is considered here to comprise 31 orders that lived during historical time (approximately the last 5,000 years) and, in turn, 156 families, 1,252 genera, and 5,511 species. Those figures were taken primarily from Wilson and Reeder (2005) but adjusted to account for newer information (particularly, numbers of taxa covered in this volume), certain points of disagreement on taxa covered by this volume, and inclusion of extinct taxa not listed by Wilson and Reeder, who dealt only with mammals that lived in the last 500 years. A representative of each historical order is illustrated in this class account.

Mammals traditionally are viewed as descendants of the class Reptilia, particularly of the order Therapsida of the subclass Synapsida, or mammal-like reptiles. Synapsida and the ancestral stock of modern reptiles, however, probably developed separately. About 320 million years ago in the late Carboniferous period of the Paleozoic era, both lineages arose from within Amniota, a group that had diverged from Amphibia in the early Carboniferous (Feldhamer et al. 2007). Synapsida sometimes is ranked as a full vertebrate class, with Therapsida a subclass thereof (Kemp 2005). The class Aves (birds) evolved somewhat later from the reptilian subclass Diapsida, which includes the lizards, snakes, crocodiles, and dinosaurs (Chatterjee 1997).

Of the three main groups of living mammals, monotremes lay eggs as did their synapsid ancestors, marsupials give birth to live but relatively undeveloped young,

Tree shrew (*Tupaia glis*), **Order Scandentia.** The tree shrews retain certain primitive features and perhaps superficially resemble some of the earliest mammals. Photo by Patipat Boonlae / Shutterstock.com.

and placentals retain the embryo until it reaches a more advanced stage. The three groups are thought to have developed during the Mesozoic era (comprising the Triassic, Jurassic, and Cretaceous periods), when dinosaurs and other reptiles dominated the earth's terrestrial fauna. Mammals then were mostly small, about the size of shrews and rats, and generally did not attain larger size until the extinction of the dinosaurs at the end of the Cretaceous (Feldhamer et al. 2007; Kemp 2005; Kielan-Jaworowska, Cifelli, and Luo 2004). However, a few became as big as foxes and beavers. Romiguier et al. (2012) suggested that some ancestral mammals were fairly large and long-lived, while small size was actually a derived state. *Repenomamus giganticus* of the early Cretaceous of China weighed 12–14 kg and apparently was large and powerful enough

Echidna, or spiny anteater (*Tachyglossus aculeatus*). **Order Monotremata.** This order forms one of the three main groups of living mammals and is reproductively characterized by egg laying. Photo by Eric Isselee/Shutterstock.com

to prey on and compete with small dinosaurs (Hu et al. 2005). *Castorocauda lutrasimilis* from the middle Jurassic of China shows adaptations for aquatic life, including a broad, flattened, partly scaly tail, analogous to that of modern beavers (Ji et al. 2006). Recent discoveries have revealed Mesozoic mammals to be far more diverse than once believed, some of them having been specialized for climbing, fossorial behavior, and even gliding (Luo 2007).

Mammalian Characters and Origins

Hall (1981) characterized the class Mammalia as follows:

Beings especially notable for possessing mammary glands (mammae) that permit the female to nourish the newborn young with milk; hair present although confined to early stages of development in most of the Cetacea; mandibular ramus of lower jaw made up of single bone (dentary); lower jaw articulating directly with skull without intervention of quadrate bone; two exoccipital condyles; differing from both Aves and Reptilia in possessing diaphragm and in having nonnucleated red blood corpuscles; resembling Aves and differing from Reptilia in having "warm blood," complete double circulation, and four-chambered heart; differing from Amphibia and Pisces in presence of amnion and allantois and in absence of gills.

Clemens (1989) noted other proposed definitions, some involving complex sets of skeletal characters. Those diagnoses serve mainly to distinguish Mammalia not from other living classes but rather from the ancestral synapsids. The morphological and temporal boundary between synapsids and the first mammals has been intensively debated. Fossil studies have revealed some specimens that do not clearly fall into either group. Rowe (1988, 1993) and Rowe and Gauthier (1992) reviewed the many proposed diagnoses and the differing views regarding the exact stage of the synapsid lineage from which mammals arose. They concluded that the class Mammalia originated with the most recent common ancestor of, on the one hand, the order Monotremata, and, on the other hand, the marsupials and placentals. Based on those criteria, the earliest known mammal was considered to be *Phascolotherium bucklandi* from the middle Jurassic of England, about 165 million years old.

McKenna and Bell (1997), using similar criteria, did not consider Mammalia to include Morganucodontidae. That family of mouse-sized creatures, on or near which most views of mammalian origin focus, developed in the late Triassic and persisted through much of the Jurassic, but did not fully meet the mammalian skeletal diagnoses indicated above. In particular, the quadrate bone of the upper jaw and the articular bone of the lower jaw retained some involvement with articulation; those bones were not fully incorporated into the middle ear, where in modern mammals they take the respective names incus and malleus and function solely for sound transmission (Feldhamer et al. 2007). McKenna and Bell (1997) utilized the term "Mammaliformes" for the overall group comprising the class Mammalia, together with Morganucodontidae and various related synapsids that approached the mammalian grade of evolution but diverged prior to the most recent common ancestor of monotremes, marsupials, and placentals.

Other authorities have considered the class Mammalia to include all lineages that achieved some particular morphological suite of characters, regardless of whether such lineages are in the direct ancestral line of modern mammals. Major assessments by Kemp (2005) and Kielan-Jaworowska, Cifelli, and Luo (2004) treated Mammalia as comprising the living mammals, morganucodontids, and *Sinoconodon*, a genus known from the early Jurassic of China, about 200 million years ago. Although available specimens are not as old as those of a few other mammalian genera, *Sinoconodon* appears to be the most basal form definitively attributable to the class. It was relatively large for a Mesozoic mammal, with the skull reaching a length of

Black-eared opossum (*Didelphis marsupialis*) of **Order Didelphimorphia**, one of the seven living orders of marsupials that form another main group of mammals, characterized by giving birth to live but relatively undeveloped young. Photo by Rexford D. Lord.

60 mm. Lucas and Luo (1993) had argued that the earliest known mammal is *Adelobasileus cromptoni*, from the late Triassic of Texas, about 225 million years old, but available material has not allowed assignment to any particular family or lineage.

Controversy surrounds the genus *Kuehneotherium* and a few related forms that date back to the late Triassic of Europe. Because the three main cusps of the molar teeth have the triangular arrangement characteristic of advanced mammals, *Kuehneotherium* sometimes is thought to have the same common ancestor as the modern monotremes, marsupials, and placentals. It was so allocated by McKenna and Bell (1997), who did not otherwise recognize any Triassic mammals. Kielan-Jaworowska, Cifelli, and Luo (2004) suggested that the dental pattern of the genus may represent early convergence and, based on its other, more primitive characters, tentatively placed *Kuehneotherium* outside the lineage of modern mammals. Kemp (2005) thought that more complete material was needed to resolve the issue.

Monophyly of Mammals

There now does seem to be a consensus that all currently living groups of mammals—monotremes, marsupials, and placentals—are monophyletic, having originated from a common ancestor, and that those groups probably are more closely related to one another than to Morganucodontidae and earlier mammal or mammal-like lineages. Until recently, a widely accepted view was that the monotremes had diverged at a very early stage of mammalian evolution or even emerged from a synapsid branch entirely separate from that giving rise to other living mammals and that they actually were more closely related to long-extinct lineages (see review by Musser 2003).

There has been disagreement regarding the affinity of the three living groups of mammals and Multituberculata, an order of herbivorous mammals with some superficial resemblance to rodents, known from middle Jurassic through late Eocene deposits in Europe, Asia, North Africa, and North America (Kemp 2005; Simmons 1993). McKenna and Bell (1997) recognized a multituberculate suborder, Gondwanatheria, but Kemp (2005) indicated that it is a separate group known from South America, Madagascar, India, and possibly Tanzania; Goin et al. (2006) reported a specimen from middle Eocene beds on Seymour Island, off the Antarctic Peninsula. New material examined by Gurovich and Beck (2009) suggests that Gondwanatheria and Multituberculata can be joined in a clade.

Some studies indicated that Multituberculata diverged from all other mammals even before the development of Morganucodontidae (Miao 1993), or branched off prior to monotremes but subsequent to morganucodontids (Wible and Hopson 1993). Other work suggests that monotremes diverged from the line leading to modern marsupials and placentals at an earlier stage than did multituberculates, but that all four groups are part of the same overall lineage (Archer et al. 1993; Cifelli 1993; Clemens 1989; Kemp 1982, 1983; Lucas and Luo 1993; Rowe 1988, 1993).

Recent reviews have tended toward the last position and to regard multituberculates as part of a Mesozoic

European hedgehog (*Erinaceus europaeus*). **Order Erinaceomorpha.** Although superficially showing some convergence with the echidna, *Erinaceus* represents Laurasiatheria, one of the four placental clades. The placentals form the third main group of living mammals and generally retain the embryo to an advanced stage. Photo by Zorandim/Shutterstock.com.

complex of mammals that developed after separation of the line leading to the monotremes but before development of the lineages of the modern marsupials and placentals (Kemp 2005; Kielan-Jaworowska, Cifelli, and Luo 2004; McKenna and Bell 1997). Those assessments, along with others (Baker et al. 2004; Belova and Hellman 2003; Hurtley 2009; Messer et al. 1998; Phillips and Penny 2003; Van Rheede et al. 2006), support the consequent view that monotremes and marsupials are not immediately related and do not form a group, sometimes called "Marsupionta," to the exclusion of placentals.

According to a new molecular analysis by Kullberg et al. (2008), the divergence of monotremes from marsupials and placentals occurred 168–178 million years ago in the middle Jurassic. The earliest known fossil evidence for the presence of distinct marsupial and placental lineages had been from the early Cretaceous of China, about 125 million years ago (Ji et al. 2002; Luo et al. 2003; Wible, Rougier, and Novacek 2005). However, Luo et al. (2011) reported a placental species, *Juramaia sinensis*, from the late Jurassic of China, 160 million years ago. Two recent estimates of the time of divergence of the marsupials and placentals, based on molecular study, are about 126–138 million years ago (Hallström et al. 2007) and 147 million years ago (Bininda-Emonds et al. 2007, 2008). Placentals are not known from South America prior to the early Tertiary, but may have been in Australia during the Cretaceous, 115–120 million years ago (Helgen 2003a).

Sunda pangolin (*Manis javanica*) of **Order Pholidota**, another unusual group within placental clade Laurasiatheria. Photo by Mooikum/Shutterstock.com.

Mammal Subclasses

The systematic division of Mammalia, above the ordinal level, has undergone much fluctuation. Recent authorities have tended to avoid subclass or infraclass assignment for the primitive mammals (including *Adelobasileus, Sinoconodon, Kuehneotherium*, and Morganucodontidae) that diverged prior to the line leading to modern monotremes, marsupials, and placentals. The lineage of the latter three groups, however, has been extensively divided between class and ordinal level (though not by Wilson and Reeder 2005). The monotremes were placed in a subclass, designated "Prototheria" by Feldhamer et al. (2007) and Kielan-Jaworowska, Cifelli, and Luo (2004) and (tentatively) "Australosphenida" by Kemp (2005). The multituberculates and certain other extinct groups were also placed in a subclass, the Allotheria, by Kemp (2005) and Kielan-Jaworowska, Cifelli, and Luo (2004). A third subclass, designated "Theria" by Feldhamer et al. (2007), "Tribosphenida" by Kemp (2005), and "Boreosphenida" by Kielan-Jaworowska, Cifelli, and Luo (2004), includes the infraclass Metatheria for the marsupials and Eutheria or Placentalia for the placentals. Dickman (2005) noted that the term "Metatheria" properly applies both to the lineage of existing marsupials and to several basal fossil taxa that appear most closely related to marsupials.

The term "Tribosphenida" refers to a characteristic triangular arrangement of the cusps of the cheek teeth, found in modern marsupials and placentals. The term "Australosphenida" indicates a clade of Mesozoic mammals in the Southern Hemisphere, which developed a dentition somewhat convergent with that of the Tribosphenida. Validity of such a lineage is controversial (Kemp 2005; Musser 2003; Pascual et al. 2002; Rauhut 2002; Rowe et al. 2008; Woodburne 2003); some authorities argue that Australosphenida comprises both the modern monotremes and a putative ancestral Gondwanan stock, while others regard the group as an early therian branch. Assessment of newly discovered specimens from the middle Jurassic of Argentina led Rougier et al. (2007) to support inclusion of Monotremata within Australosphenida. The specimens apparently retained some postdentary elements in the lower jaw, thus suggesting that incorporation of the articular bone into the mammalian middle ear (see above) occurred independently in Monotremata and

Indian flying fox (*Pteropus giganteus*), **Order Chiroptera** (bats), clade Laurasiatheria. This is the only mammalian order capable of true flight. Photo by Ondrej Prosicky / Shutterstock.com.

Theria. A review by Luo (2007), which considers Australosphenida basal to Monotremata, also indicates that development of both tribosphenic molars and a sophisticated middle ear occurred more than once in the evolution of mammals.

McKenna and Bell (1997) recognized only two subclasses, Prototheria for the monotremes and Theriiformes for all other members of the lineage of modern mammals. Theriiformes, in turn, was thought to comprise three infraclasses: Allotheria for the multituberculates, Triconodonta for certain extinct groups, and Holotheria for all others. Holotheria was broken down into numerous categories, but with the living components ranked under superlegion Trechnotheria, legion Cladotheria, sublegion Zatheria, infralegion Tribosphenida, supercohort Theria, and cohorts Marsupialia and Placentalia.

Placental Clades

There appears a general consensus that the placentals comprise 22 orders (or 23 if Pinnipedia is counted as an order separate from Carnivora) that have lived during historical time. The arrangement of those orders here follows that of Wilson and Reeder (2005), which in turn stemmed from McKenna and Bell (1997) and what the former authorities termed a subsequent "explosion of literature based on new techniques of molecular systematics." That the matter may be far from settled, and open to substantive revision through newer molecular methodology, is suggested by Dolgin (2012).

Both molecular and morphological evidence has supported recognition of four superordinal clades of the historical placental orders accepted here: (1) Afrotheria, for Afrosoricida (or Tenrecoidea), Macroscelidea, Tubulidentata, Bibymalagasia, Hyracoidea, Proboscidea, and Sirenia; (2) Xenarthra, for Cingulata and Pilosa; (3) Euarchontoglires (or Supraprimates), for Scandentia, Dermoptera, Primates, Rodentia, and Lagomorpha; and (4) Laurasiatheria, for Erinaceomorpha, Soricomorpha, Chiroptera, Pholidota, Carnivora (usually considered to include Pinnipedia), Perissodactyla, Artiodactyla, and Cetacea (Amrine-Madsen et al. 2003b; Archibald and Rose 2005; Asher 2007; Asher, Novacek, and Geisler 2003; Helgen 2003a; Madsen et al. 2001; Matthee et al. 2007; Murphy et al. 2001a; Scally et al. 2001; Springer and Murphy 2007; Springer et al. 2005; Waters et al. 2007; Wible, Rougier, and Novacek 2005). Laurasiatheria and Euarchontoglires sometimes are combined in the higher grouping Boreoeutheria, which in turn sometimes is united with Xenarthra to form Notolegia or Exafroplacentalia (Nikolaev et al. 2007).

A phylogenomic analysis by Hallström et al. (2007) showed a closer affinity of Xenarthra to Afrotheria, which together constitute the basal placental clade Atlantogenata. The analysis indicated that Atlantogenata and Boreoeutheria diverged 99–100 million years ago, with Xenarthra and Afrotheria splitting about 97–98 million years ago and Laurasiatheria and Euarchontoglires 95–96 million years ago. Approximately the same dates were derived by Bininda-Emonds et al. (2007,

Two-toed sloth (*Choloepus didactylus*), **Order Pilosa**, clade Xenarthra. This species, like most living pilosans, is arboreal. Photo by Seubsai / Shutterstock.com.

African savannah elephant (*Loxodonta africana*) of **Order Proboscidea**, which is part of Afrotheria, a placental clade comprising seven historical orders centered mainly in Africa. Although extremely diverse in size and other morphological features, most members of the clade are characterized by a mobile proboscis. Photo by Deborah Benbrook/Shutterstock.com.

Giant sengi (*Rhynchocyon cirnei*) of **Order Macroscelidea**, an afrotherian group comprising the so-called "elephant shrews." Photo by Klaus Rudloff at Prague Zoo.

2008), who reviewed nearly all extant mammals, employing both gene sequencing and cladistic assessment of fossils. However, they deemphasized the significance of the Cretaceous/Tertiary boundary, which marked the extinction of the dinosaurs, in the radiation of modern mammals. All placental clades and nearly all existing orders had originated by 85 million years ago, 20 million years before the Cretaceous ended; the remaining orders had appeared by 74 million years ago. Although the start of the Tertiary did see substantial development of new lineages of mammals, most became extinct; modern mammalian diversification seems not to have accelerated again until a favorable climatic period in the early Eocene, about 50–55 years ago.

In a contrasting molecular study, Kitazoe et al. (2007) concluded that the four supraordinal clades diverged from one another only about 75–85 million years ago. And based on morphological assessment of newly discovered fossils from the late Cretaceous of Mongolia, Wible et al. (2007) argued that placentals did not even originate until near the Cretaceous/Tertiary boundary, with Laurasiatheria and Euarchontoglires appearing shortly thereafter and Afrotheria and Xenarthra still later.

The name "Atlantogenata" implies that divergence of Xenarthra and Afrotheria resulted from opening of the Atlantic Ocean through plate tectonics, the underlying mechanism of continental drift. About 200 million years ago, when dinosaurs ruled and mammals were emerging, the earth's continents are thought to

have been conjoined in a single land mass, Pangaea, which later began to fragment (Feldhamer et al. 2007; Masters, de Wit, and Asher 2006; Murphy et al. 2007; Wildman et al. 2007). By about 150 million years ago, the Tethys Seaway had divided Pangaea into northern and southern sections. The northern continent, Laurasia, consisted of what would become North America and most of Eurasia, joined on their Atlantic sides. The southern continent, Gondwana, included South America, Africa, Madagascar, India, Antarctica, and Australia. By around 100 million years ago, North America and Eurasia had separated, Madagascar and India had drifted away from Africa, and South America and Africa were splitting, though there still were tenuous connections between South America, Antarctica, and Australia. Subsequently, those last three continents moved apart, India reached southern Eurasia, and land bridges alternately opened and closed between North and South America and between Eurasia and Africa.

The above geological sequence fits with some molecular studies in suggesting that the genetic divergence of Afrotheria and Xenarthra coincided with the split of Africa and South America in the Cretaceous (Hedges 2001; Kjer and Honeycutt 2007; Murphy et al. 2001b, 2007; Springer and Murphy 2007; Springer et al. 1997; Tabuce et al. 2007; Van Dijk et al. 2001). An analysis of DNA evolutionary rates (Wildman et al. 2007) supported the concept of Atlantogenata, as well as an overall subdivision of the placentals through plate tectonics. An initial split occurred when the Tethys Sea

Aardvark (*Orycteropus afer*), **Order Tubulidentata**, part of Afrotheria. Photo by Alexander F. Meyer at Brookfield Zoo.

separated the Boreoeutheria in Laurasia from the Atlantogenata in Gondwana. Later in the Cretaceous, disconnection of Africa and South America resulted in Afrotheria and Xenarthra. Divergence of Laurasiatheria and Euarchontoglires may have been associated with separation of North America and Eurasia. Based on both geological and molecular data, Nishihara, Maruyama, and Okada (2009) proposed that divergence of the eutherian ancestor into Boreoeutheria, Xenarthra, and Afrotheria occurred nearly simultaneously 120 million years ago, possibly with the almost concomitant cut-off of land bridges that had connected the three ancient continents, Laurasia, South America, and Africa.

Certain studies, while recognizing Afrotheria and Xenarthra, have suggested those clades did not originate through the division of Africa and South America and are not necessarily immediately related (Asher, Novacek, and Geisler 2003; Gheerbrant and Rage 2006; Hallström et al. 2007; Hunter and Janis 2006a, 2006b; Nishihara, Okada, and Hasegawa 2007; Zack et al. 2005). Instead, the two groups initially developed and diverged in Laurasia, with their ancestral stocks migrating south, when land bridges were available, and then radiating during the extensive isolation of the southern continents.

As explained by Teeling and Hedges (2013), there is considerable disagreement as to the basal phylogeny of placental mammals. Paleontologists generally favor a root anchored by Xenarthra, and molecular biologists

support either Afrotheria or the clade Atlantogenata (Afrotheria plus Xenarthra) as the basal lineage. Recent phylogenomic analysis have indeed shown Afrotheria (Romiguier et al. 2013) or Atlantogenata (Morgan et al. 2013) as the sister group to all other placentals (Boreoeutheria). In contrast, a study combining phenomic characters with molecular sequences (O'Leary et al. 2013) placed Xenarthra at the base, thereby making Afrotheria the sister taxon to Boreoeutheria (see also reviews by Asher, Bennett, and Lehmann 2009, and Tabuce, Asher, and Lehmann 2008).

Afrotheria

Afrotheria represents a relatively new concept, differing sharply from traditional taxonomic views. There long had been suggestions of affinities between the various constituent orders, especially Hyracoidea, Proboscidea, and Sirenia, which often were linked in the superorder Paenungulata (Lavergne et al. 1996; Novacek 1992, 1993; Prothero 1993). Mounting evidence indicated that Paenungulata also included Tubulidentata and Macroscelidea (Graur 1993; Madsen et al. 1997; Shoshani 1993). Prior to recognition of Afrotheria, however, Macroscelidea usually was associated with Lagomorpha. Afrosoricida (with the living families Tenrecidae and Chrysochloridae) traditionally was conjoined with what are here ranked as the orders Erinaceomorpha (family Erinaceidae) and Soricomorpha (families Nesophontidae, Solenodontidae, Soricidae, and Talpidae) to form what is now considered the

Bibymalagasy (*Plesiorycteropus madagascariensis*), **Order Bibymalagasia**. This extinct but historical order is also considered part of clade Afrotheria. Painting by Helga Schulze, based on sources cited in the account herein of *Plesiorycteropus*.

Rock hyrax (*Procavia capensis*), **Order Hyracoidea**. This afrotherian group is comparable to the elephants in having a single pair of upper incisors that grow continuously and are long and curved. Photo by Vladimir Wrangel / Shutterstock.com.

phylogenetically invalid order Insectivora (or Lipotyphla). Shoshani and McKenna (1998) presented a major summation of superordinal relationships among extant eutherians, based on data available prior to the widespread recognition of Afrotheria.

The erection of Afrotheria and dismemberment of Insectivora resulted initially from studies of mitochondrial DNA and RNA demonstrating a clade comprising not only Macroscelidea, Tubulidentata, Hyracoidea, Proboscidea, and Sirenia, but also the families Tenrecidae and Chrysochloridae, which consequently were placed in the new order Afrosoricida (Springer et al. 1997; Stanhope et al. 1998a, 1998b). There has been much debate as to whether the name "Tenrecoidea" should be used in place of "Afrosoricida" (Arnason et al. 2008; Asher and Helgen 2010, 2011; Bronner and Jenkins 2005; Hedges 2011; see also account of order Afrosoricida). Using both molecular and anatomical data, Asher, Novacek, and Geisler (2003) supported the validity of Afrotheria and also showed that it contained the order Bibymalagasia.

One feature that might seem to unite the members of Afrotheria is presence of a mobile, even prehensile, proboscis—notably the elephant's trunk, the manatee's movable muzzle, and the long, flexible snouts of the aardvark, tenrec, and elephant shrew. Whidden (2002), however, determined that the involved musculature is not homologous across the afrotherian orders and that it actually indicates that Tenrecidae and Chrysochloridae are closer to the other families of original Insectivora. Nonetheless, a torrent of research, mostly mo-

lecular but also morphological and paleontological, has supported the composition of Afrotheria, as shown above, and the requisite realignment of the insectivoran families (Asher 2005; Gheerbrant, Domning, and Tassy 2005; Holroyd and Mussell 2005; Kellogg et al. 2007; Kjer and Honeycutt 2007; Kuntner, May-Collado, and Agnarsson 2010; Malia, Adkins, and Allard 2002; Matthee et al. 2007; Mouchaty et al. 2000; Murata et al. 2003; Nishihara et al. 2005; Pardini et al. 2007; Redi et al. 2007; Robinson and Seiffert 2004; Sánchez-Villagra, Narita, and Kuratani 2007; Scally et al. 2001; Seiffert 2007; Wible, Rougier, and Novacek 2005).

Tabuce, Asher, and Lehmann (2008) reviewed the published data on the monophyly of Afrotheria, the opinions on its position within class Mammalia, and the alternatives proposed for the group's own systematic composition. Most analyses unite Tenrecidae and Chrysochloridae in a clade within Afrotheria, with Macroscelidae as their sister taxon; the resulting association is known as "Afroinsectivora." The latter plus Tubulidentata is called "Afroinsectiphilia." However, some authorities consider Tenrecidae and Chrysochloridae basal to the other afrotheres, or unite only Chrysochloridae or only Tenrecidae with Tubulidentata. Paenungulata (Hyracoidea, Proboscidea, Sirenia) is still consistently recognized and, in most recent

West Indian manatee (*Trichechus manatus*) of **Order Sirenia**, a fully aquatic component of Afrotheria. Photo by R. K. Bonde, US Geological Survey, through Cathy A. Beck.

assessments, considered the sister group of Afroinsectiphilia. Further suggestions have been to combine Proboscidea and Sirenia in a clade known as "Tethytheria," or instead to align either Hyracoidea with Sirenia or Hyracoidea with Proboscidea.

A recent comprehensive study of nine molecular markers, both mitochondrial loci and nuclear protein coding genes (Kuntner, May-Collado, and Agnarsson 2010), supported the monophyly of Afrotheria, with a basal division into Afroinsectiphilia and Paenungulata. Within Afroinsectiphilia, Afrosoricida plus Macroscelidea form a sister group to Tubulidentata. Within Paenungulata, Hyracoidea plus Proboscidea form a sister group to Sirenia. The study also supported a sister relationship of Afrotheria and Xenarthra.

Xenarthra
Although questions remain about the unity of Afrotheria, Rose et al. (2005) noted that there is little doubt regarding the monophyly of Xenarthra, based on either morphological or molecular evidence, and continued to treat the group as an order, with Cingulata and Pilosa suborders thereof. Such remains a common practice (Feldhamer et al. 2007), though Helgen (2003a) recognized three living xenarthran orders: Cingulata (armadillos), Vermilingua (anteaters), and Folivora (sloths). Based on albumin systematics, Sarich (1985) indicated that armadillos, anteaters, and sloths diverged at least 75–80 million years ago and that they

Nine-banded armadillo (*Dasypus novemcinctus*) of **Order Cingulata**, a generally terrestrial component of placental clade Xenarthra. Photo by Ron Kacmarcik / Shutterstock.com.

are at least as distinct from one another as are carnivores, bats, and primates. McKenna and Bell (1997) designated Xenarthra a magnorder, with Cingulata (armadillos) and Pilosa (anteaters and sloths) orders thereof; Gardner (2007a, 2007e) followed the same arrangement and Wilson and Reeder (2005) also accepted the two order division.

The name Xenarthra formerly was applied at subordinal level, with the name Edentata sometimes given to an order that included Xenarthra and the supposedly ancestral suborder Palaeanodonta. However, Glass (1985) explained that Palaeanodonta is ancestral to Pholidota, not to Xenarthra. In a review of fossil and morphological evidence, Rose and Emry (1993) also indicated a connection between Palaeanodonta and Pholidota and found no association between Pholidota and Xenarthra, though Novacek (1992) had continued to suggest possible affinity. Rose et al. (2005) found no compelling evidence, either morphological or molecular, to support relationship of Pholidota and Xenarthra, and suggested avoidance of the name Edentata. That term signifies toothlessness, though only the anteaters actually have no teeth. In the living sloths and armadillos, the teeth are reduced, are mostly undifferentiated, and lack enamel.

Xenarthrans are distinguished from all other mammals by what are known as xenarthrous vertebrae. One or more pairs of supplementary articulations are usually present between all lumbar and a variable number of posterior thoracic vertebrae. They lend support, particularly to the hips, which is especially valuable to the armadillos for digging. There are also ischial articulations with the spinal column, thus incorporating into the sacral vertebrae what in other mammals would be caudal elements (Glass 1985; Hall 1981). The supplementary intervertebral articulations, or xenarthrales, show some variation between genera and suborders, but span a wide range of locomotory habits, and there is no substantive and consistent difference between the two orders Cingulata and Pilosa (Gaudin 1999). Xenarthrans have a double posterior vena cava that returns blood from the posterior part of the body to the heart, whereas most other mammals have only a single vena cava. Females have a common urinary and genital duct. The testes are located in the abdominal cavity between the rectum and the urinary bladder.

California sea lion (*Zalophus californianus*), **Order Pinnipedia**, placental clade Laurasiatheria. The pinnipeds, now usually considered part of order Carnivora, are aquatic but, unlike Sirenia, bear and initially rear their young terrestrially. Photo by Roger de Montfort/Shutterstock.com.

Euarchontoglires

Euarchontoglires is not so well defined as is Xenarthra. For around a century, however, some authorities have considered Chiroptera, Dermoptera, Scandentia, and Primates to form a monophyletic group called Archonta. Indeed, Scandentia often was included in the order Primates, and Chiroptera and Dermoptera sometimes were coupled in the superordinal unit Volitantia. All living archontans are characterized by a pendulous penis suspended by a reduced sheath between the genital pouch and abdomen, and by a large distal sustentacular facet of the calcaneum articulating with a ventral extension of the navicular facet of the astragalus. Volitantians, in turn, are united by such features as elongation of the forelimbs, fusion of the distal part of the ulna to the radius, and presence of humeropatagalis muscles. Recent molecular analyses have consistently failed to show close relationship of Chiroptera to other archontans, but have supported a clade, designated Euarchonta, comprising Dermoptera, Scandentia, and Primates, and also have supported the taxonomic grouping of Euarchonta and Glires (Rodentia and Lagomorpha) to form the clade Euarchontoglires (Silcox et al. 2005). Multiple molecular data sets have strongly backed placement of Chiroptera in Laurasiatheria (Simmons 2005).

Three hypotheses have been proposed for relationships within Euarchonta: that Scandentia and Primates are most closely related, that Dermoptera and Primates are immediately related in a clade called Primatomorpha, and that Dermoptera and Scandentia form a group, Sundatheria, which is sister to Primates. Based on molecular and genomic data, Janečka et al. (2007) supported Primatomorpha; they estimated Euarchonta diverged from Glires 87.9 million years ago, Primatomorpha from Scandentia 86.2 million years ago, and Primates from Dermoptera 79.6 million years ago. However, using phenomic characters combined with molecular sequences, O'Leary et al. (2013) favored Sundatheria and argued that the placental orders did not originate until after the Cretaceous/Tertiary boundary.

The most characteristic feature of Glires is a pair of enlarged, evergrowing upper and lower incisor teeth that extend deep into the maxilla and dentary. The upper and lower canines, first upper and lower premolars, and second lower premolar have been lost, leaving a lengthy gap between the incisors and cheek teeth. Some biochemical investigation not only questioned the unity of Glires but suggested dividing Rodentia into several orders. However, a wave of subsequent molecular research has confirmed the monophyly of Rodentia, and many recent morphological and molecular studies have sustained the validity of Glires (Carleton and Musser 2005; Meng and Wyss 2005). Fossils, recently discovered in Mongolia, indicate that rodents and lagomorphs diverged from other placentals close to the Cretaceous-Tertiary boundary (Asher et al. 2005).

Laurasiatheria

Both Euarchontoglires and Laurasiatheria are thought to have evolved in Laurasia. Genetic association of the

Mountain gorilla (*Gorilla beringei*), **Order Primates**, placental clade Euarchontoglires. This is a generally terrestrial species within a mostly arboreal order. Photo by Bimserd/Shutterstock.com.

families Tenrecidae and Chrysochloridae with Afrotheria, rather than with the defunct order Insectivora (or Lipotyphla), implies an independent northern radiation of insectivoran families within the Laurasiatheria. The two most consistent characters uniting northern and southern insectivorans, a reduced pubic symphysis and a simplified intestinal tract, are also found in Chiroptera and certain other mammals (Asher 2005). Morphologically, however, insectivorans are among the more primitive eutherians. The abundant molecular evidence dividing the group between Afrotheria and Laurasiatheria is consistent with the view that insectivorans were the central stock that independently gave rise to diverse eutherian lineages in different geographic venues (Murphy et al. 2001b).

The molecular data that now suggest splitting insectivores into southern and northern groups, and splitting the latter into two orders, Erinaceomorpha and Soricomorpha (sometimes conjoined in the superordinal taxon Eulipotyphla), contrast strongly with evidence on the other Laurasiatherian orders. Numerous morphological and molecular data sets support the monophyly of Chiroptera, and multiple molecular data sets align Chiroptera with either Eulipotyphla or a group comprising Pholidota, Carnivora, Perissodac-

Bottlenose dolphin (*Tursiops truncatus*) of **Order Cetacea**, a fully aquatic group within clade Laurasiatheria. Photo by Natali Glados/Shutterstock.com.

tyla, Artiodactyla, and Cetacea (Simmons 2005); that group has been labeled either Ferungulata or Cetferungulata. Monophyly of Carnivora is undisputed, and incorporation of Pinnipedia therein is usually accepted (but see below). Molecular data suggest affinity of Carnivora and Pholidota (sometimes conjoined in the superordinal taxon Ostentoria or Ferae), Carnivora and Perissodactyla (Amrine-Madsen et al. 2003b; Flynn and Wesley-Hunt 2005; Hooker 2005; Matthee et al. 2007), or Perissodactyla and Artiodactyla, to the exclusion of Carnivora (Hou, Romero, and Wildman 2009).

Artiodactyla and Cetacea are separately monophyletic and very closely related, based on both molecular and morphological data, and there have been suggestions that the two could be united in a single order, Cetartiodactyla (Amrine-Madsen et al. 2003b; Boisserie et al. 2010; Gingerich 2005; Margot 2007; Springer et al. 2005; Theodor, Rose, and Erfurt 2005; Wible, Rougier, and Novacek 2005). Helgen (2003a) favored such an order but noted that its proper name would remain Artiodactyla; Geisler et al. (2007) followed that procedure. The case for union hinges on the apparent descent of Cetacea from a group within Artiodactyla (probably the lineage of Hippopotamidae), though there is disagreement as to the evolutionary sequence (Geisler and Theodor 2009; O'Leary et al. 2013; Thewissen et al. 2007, 2009). That position, hence, is similar to the argument for including Pinnipedia in the order Carnivora. Both matters involve cladistic principles that may not be fully acceptable to all taxonomists. Those issues, and additional questions of whether certain placental orders should be combined with others or split into further orders, will be discussed in appropriate ordinal accounts.

Marsupial Superordinal Groupings

Marsupials (Marsupialia or Metatheria), once lumped in a single order (Simpson 1945), are now commonly recognized as seven living orders, based on paleontological, morphological, and biochemical evidence (Amrine-Madsen et al. 2003a; Archer and Kirsch 2006; Asher, Horovitz, and Sánchez-Villagra 2004; Cardillo et al. 2004; Dickman 2005; Horovitz and Sánchez-Villagra 2003; Kemp 2005; Nilsson et al. 2003, 2004). The sequence of marsupial orders accepted here follows that given by Wilson and Reeder (2005): Didelphimorphia, Paucituberculata, and Microbiotheria of the New

Marsupial mole (*Notoryctes typhlops*), **Order Notoryctemorphia**. In body shape, claw structure, and lack of functional eyes, the marsupial mole shows remarkable convergence with the placental golden moles. Photo by Stanley Breeden / Lochman LT.

South African golden mole (*Amblysomus hottentotus*), **Order Afrosoricida**, placental clade Afrotheria. Photo by H. A. York / Mammal Images Library, American Society of Mammalogists.

World; Notoryctemorphia, Dasyuromorphia, Peramelemorphia (or Peramelia or Peramelina), and Diprotodontia of Australasia. Such an arrangement, along with acceptance of more marsupial families than formerly recognized, is reasonable in that the marsupials include species that seem just as diverse in morphology and ecology as are moles, rabbits, mice, flying squirrels, anteaters, wolves, and antelopes. Marsupials are united primarily by a common means of reproduction. Placentals also share a common reproductive method, but no one questions their division into many orders.

Nonetheless, based on detailed osteological analyses of fossil and modern specimens, Szalay (1982, 1993, 1994, 1999) considered the diversity of marsupials, in comparison with the placentals, sufficiently reflected by recognizing only three living orders: Didelphida (including Didelphimorphia and Paucituberculata), Gondwanadelphia (including Microbiotheria and Dasyuromorphia), and Syndactyla (including Notoryctemorphia, Peramelemorphia, and Diprotodontia). Szalay divided the living marsupials into two cohorts (ranked as magnorders by McKenna and Bell 1997): Ameridelphia (for Didelphida) and Australidelphia (for Gondwanadelphia and Syndactyla). An alternative view, stemming from molecular assessments (Burk et al. 1999; Kirsch, Lapointe, and Springer 1997), is that Peramelemorphia evolved separately from all other Australasian marsupials and actually is more closely related to the ameridelphian orders.

Subsequent morphological and molecular studies (see reviews by Archer and Kirsch 2006; Cardillo et al. 2004; Dickman 2005; Horovitz and Sánchez-Villagra 2003; Nilsson et al. 2010; Palma 2003; Phillips et al. 2001; Springer and Murphy 2007) generally have supported Australidelphia, with Microbiotheria and Peramelemorphia included, and also the basal position of Didelphimorphia in the overall sequence of marsupial divergence. However, there has been little recent support for Ameridelphia; Paucituberculata has not been grouped with Didelphimorphia but rather seen as the most basal part of a lineage that also includes the Australidelphia. The following are among the reported versions of the sequence of divergence within Australidelphia:

Microbiotheria, a clade of Notoryctemorphia + Dasyuromorphia + Peramelemorphia, Diprotodontia (Amrine-Madsen et al. 2003);
Microbiotheria, Diprotodontia, possibly Notoryctemorphia, a clade of Dasyuromorphia + Peramelemorphia (Phillips et al. 2006);
Microbiotheria, Diprotodontia, a clade of Notoryctemorphia + Dasyuromorphia + Peramelemorphia (Meredith et al. 2008);
Dasyuromorphia, a clade of Notoryctemorphia + Peramelemorphia, Microbiotheria, Diprotodontia (Horovitz and Sánchez-Villagra 2003);
Peramelemorphia, Microbiotheria, a clade of two groups—one of Notoryctemorphia

Monito del monte (*Dromiciops gliroides*), **Order Microbiotheria**, a marsupial showing some convergence with placental rodents. Photo by Mark Chappell / Animals Animals Earth Scenes.

Fat dormouse (*Glis glis*), **Order Rodentia**, placental clade Euarchontoglires. Photo by Robert Varndell / Shutterstock.com.

+ Dasyuromorphia and one of Diprotodontia (Asher, Horovitz, and Sánchez-Villagra 2004); Peramelemorphia, a clade of Notoryctemorphia + Dasyuromorphia, a clade of Microbiotheria + Diprotodontia (Cardillo et al. 2004); and Diprotodontia, Microbiotheria, a clade of Notoryctemorphia + Dasyuromorphia, Peramelemorphia (Nilsson et al. 2003, 2004).

Based primarily on morphological study of fossil and modern forms, Hershkovitz (1992a, 1992b, 1995, 1999) took a position strikingly different from the above. He put the order Microbiotheria in the cohort Microbiotheriomorphia, which he considered basal to Metatheria and to show some affinity to both monotremes and eutherians. He placed all other marsupials, living and extinct, in the cohort (not order) Didelphimorphia. Dickman (2005) noted that other studies

have provided little support for such arrangement, most authorities considering Microbiotheria part of the australidelphian radiation. Additional details are provided in the account of order Microbiotheria.

Marsupial Evolution

A traditional view holds that marsupials developed initially in the Northern Hemisphere and might formerly have been widespread but eventually were restricted to southerly regions. Some authorities came to support origin in Australia, now with the greatest diversity of marsupials, but subsequent discovery of primitive marsupial fossils in early Cretaceous deposits of North America and Asia gave credence to the earlier view (see Horovitz et al. 2009). All relevant fossils from South America and Australia have more derived characters. The oldest undoubted South American marsupials are

Greater bilby (*Macrotis lagotis*), **Order Peramelemorphia**. This marsupial order is often compared to the placental rabbits. Photo by Alexander F. Meyer at David Fleay Wildlife Park, Gold Coast, Australia.

European hare (*Lepus europaeus*), **Order Lagomorpha**, placental clade Euarchontoglires. Photo by Jenny Cottingham / Shutterstock.com.

from the latest Cretaceous to earliest Paleocene; the oldest in Australia are latest Paleocene or early Eocene (Archer and Hand 2006; Archer and Kirsch 2006; Beck et al. 2008). A thriving North American marsupial radiation was largely eliminated at the Cretaceous/Tertiary boundary, 65 million years ago, in the same wave of extinction that doomed the dinosaurs (Kemp 2005). However, in contrast to its minimal effect on placental evolution, that boundary seems to have coincided with diversification of the modern australidelphian orders; all five originated near or shortly after the boundary, though the Didelphimorphia and Paucituberculata had appeared earlier (Bininda-Emonds et al. 2007, 2008).

By the end of the Cretaceous, the southern continent of Gondwana had begun to drift apart, but final separation of Australia and Antarctica may not have been complete until 40–50 million years ago and that of Antarctica and South America not until about 36 million years ago in the late Eocene. Marsupials probably migrated from North to South America in the late Cretaceous before those two continents separated, not to rejoin until the late Pliocene. Marsupials then evidently spread across Antarctica and entered Australia in the Paleocene, though views vary as to the nature of

that movement and subsequent evolution (Archer and Hand 2006; Archer and Kirsch 2006; Dickman 2005; Kemp 2005; Nilsson et al. 2003, 2004; Palma and Spotorno 1999; Springer et al. 1998). If Microbiotheria is basal to the other australidelphian orders, as indicated in some of the above versions of the sequence of divergence, the development of Australasian marsupials would require only one dispersal of ancestral stock from South America. However, if Microbiotheria developed in South America, and if it is part of but not basal to the Australidelphia, as indicated in most of the above sequences, then there would have been a dispersal into Australia of the respective ancestors of each of the lineages more basal than Microbiotheria. Alternatively, Microbiotheria could have originated in Australia, spread to South America, then disappeared in Australia. Fossils of apparent microbiotheres have been found in Australia and on Seymour Island off the Antarctic Peninsula. Hershkovitz (1999), however, argued that Microbiotheria was always restricted to South America, where it actually gave rise to the entire marsupial lineage. Recently, Beck (2012) reported the first unequivocal non-australidelphian marsupial known from Australia, thus suggesting the possibility of dispersals even earlier than those of the lineages indicated above.

Five extinct orders and 37 extinct families (not including Thylacinidae) of marsupials and marsupial-like taxa are known. Recorded geological range is early Cretaceous to early Miocene and late Pliocene to Recent in North America, late Cretaceous to Recent in South America, late Cretaceous to middle Miocene in Europe, early Eocene to Oligocene in North Africa, early Cretaceous to Oligocene in Central Asia, middle Eocene to early Miocene in East Asia, early Eocene to early Oligocene in Southwest Asia, early Eocene in South Asia, middle Miocene in Southeast Asia, early Eocene to Recent in Australia, and middle Miocene in Antarctica (Seymour Island only) (Bajpai et al. 2005; Bown and Simons 1984; Dickman 2005; Ducrocq et al. 1992; Gabuniya, Shevyreva, and Gabuniya 1985; Goin et al. 1999; Martin et al. 2005; McKenna and Bell 1997; Woodburne and Zinsmeister 1982, 1984; Ziegler 1999). A reported fossil marsupial tooth from the late Cretaceous of Madagascar (Krause 2001) was identified as that of a placental by Averianov, Archibald, and Martin (2003). A report of a fossil marsupial on New Caledonia has been shown to be invalid (Rich et al. 1987).

Marsupial Reproduction

Renfree (1993) suggested that when mammals originated they simultaneously evolved lactation and certain other of their characteristic reproductive features. Lactation may have begun as a maternal sebaceous secretion with antibacterial functions that served to protect the eggs and hatchlings and then gradually evolved into a flow of nutrients from specialized glands. From the late Triassic through part of the Cretaceous the basic mammalian reproductive mode remained almost unchanged because it was the most appropriate for small, nocturnal, insectivorous creatures such as most early mammals are thought to have been. The major differences in living mammals evolved during the divergence of the marsupials and placentals and the subsequent adaptive radiations of the two groups in response to the metabolic requirements of increasing body size and the ecological constraints imposed by new ecological niches and modes of life. All three living groups of mammals share certain basic reproductive characters, including Graafian follicles, functional corpora lutea, bilaminar blastocysts, uterine secretion, yolk-sac placentae, mammary glands, and lactation. The monotremes are distinguished largely by factors associated with egg laying and incubation of the egg outside of the body. The specializations of the marsupials have to do mainly with extended lactation and endocrine controls thereof. The unique characters of the placentals are associated with extended embryonic gestation and include the universal precocious development of the chorioallantoic villous placenta and a tendency to develop a variety of luteotrophic and luteolytic controls of the corpora lutea.

Further details on Monotremata are given in the following ordinal account, but because there are seven metatherian orders to cover, a summary of the characters common to all marsupials and that distinguish them from placentals is provided at this point. For detailed discussion and comparison of the morphology,

Shrew opossum (*Caenolestes sangay*), **Order Paucituberculata**, a marsupial with morphological and behavioral characters comparable to those of placental shrews. Photo by Reed Ojala-Barbour.

Common shrew (*Sorex araneus*), **Order Soricomorpha**, placental clade Laurasiatheria. Photo by Erni / Shutterstock.com.

physiology, and natural history of the two groups, see Bryden (1989), Dawson et al. (1989), Jarman, Lee, and Hall (1989), and Russell, Lee, and Wilson (1989).

Most female marsupials possess an abdominal pouch, referred to as a marsupium, within which the young are carried. The pouch, however, is neither the only nor the most diagnostic characteristic of the group. The best-developed pouches are found in marsupials that climb (phalangers), hop (kangaroos), dig (bandicoots and wombats), or swim (the yapok), but some small terrestrial marsupials have no pouch (Kirsch 1977c). In certain didelphids and dasyurids, among others, the pouch consists merely of folds of skin around the mammae that help to protect the attached young. Many marsupials develop pouches only during the reproductive season. When well developed, a pouch may open either to the front or to the rear, depending on the genus. The mammae usually are abdominal and located within the pouch if one is present.

All marsupials lack a complete placenta, the membranous structure facilitating passage of nutrients from the mother's body to the embryo in the uterus. Marsupials form a yolk-sac placenta through which uterine secretions are absorbed by the amnion, an outer membrane of the embryo. In the order Pera-

melemorphia, and also evidently in the genus *Phascolarctos* (Lee and Carrick 1989), there is a more developed placenta consisting of additional embryonic membranes, the chorion and the allantois. In all other therian mammals, however, this chorioallantoic placenta develops villi that penetrate uterine tissue and provide for the most efficient means of nourishment. Mammals with such a completely developed placenta are known as placentals.

In marsupials the female reproductive tract is bifid; that is, the vagina and uterus are double. The two lateral vaginae spread sufficiently to allow the urinary ducts to pass between them (Tyndale-Biscoe 1973). During birth, however, the young typically are extruded through a third canal, the birth canal, or median vagina, which passes from a point of medial fusion between the two uteri to the urogenital sinus. In Didelphimorphia, Dasyuromorphia, Peramelemorphia, and most of Diprotodontia the median vagina is transitory and must be formed before each birth. In Macropodidae and Tarsipedidae, however, the median vagina becomes lined with epithelium at the first birth and forms a permanent birth canal (Sharman 1970). The passage of the ureters medially between the vaginae, rather than laterally as in placentals, is a critical expression of the dichotomy of the two groups

Tasmanian devil (*Sarcophilus harrisii*), **Order Dasyuromorphia**, a marsupial predator often considered to have been eliminated on the Australian mainland through competition from introduced placental canids. Photo by Patsy A. Jacks / Shutterstock.com.

Gray wolf (*Canis lupus*), **Order Carnivora**, placental clade Laurasiatheria. Photo by David Osborn / Shutterstock.com.

Eastern gray kangaroo (*Macropus giganteus*), **Order Diprotodontia**, a grazing marsupial showing some parallels with placental artiodactyls and often considered a competitor with sheep. Photo by Dennis Jacobsen / Shutterstock.com.

Argali sheep (*Ovis ammon*), **Order Artiodactyla**, placental clade Laurasiatheria. Photo by Karamysh / Shutterstock.com.

and is their essential reproductive distinction. Since the median vagina in marsupials is so constricted, the young must be very small at birth and hence must subsequently undergo an extended period of development and relatively longer lactation before independence (Renfree 1993). In placentals, the vagina opens to the outside, separate from the anus, but in most marsupials the vaginae open into a urogenital sinus that serves as a common passage for the reproductive, urinary, and excretory systems (Shaw 2006).

In male marsupials the scrotum is in front of the penis, except in Notoryctidae, and there is no baculum. The vas deferens (ducts that carry sperm from the testes) pass laterally to the ureters instead of mesially as in placentals. The males of most species have a bifid penis, the right and left prongs apparently being placed in the corresponding vaginal canals during mating (Sharman 1970). However, in marsupials with an undivided penis, such as the kangaroos and wallabies, large volumes of semen do pass into each of the lateral vaginae after mating. Sperm generated in the testis and released singly are stored and mature in the epididymis. In Didelphimorphia and Paucituberculata, but not in Microbiotheria or any other marsupials or vertebrates, the sperm pair up, head to head, during epididymal maturation, then again separate as they pass up the oviduct of the female (Shaw 2006).

The gestation period in marsupials is short compared with that of placentals of equivalent size, and the tiny young are born in a practically embryonic state.

The newborn crawl some distance anteriorly from the urogenital opening and take hold of nipples that expand in their mouths to ensure a firm attachment. Contrary to earlier belief, the mother apparently does not directly assist the young in moving to the nipples (Sharman 1970). However, as birth approaches in some species, the female increasingly licks the urogenital opening and she therefore is crouched over or curled up, bringing the opening closer to the pouch and/or mammary area (Shaw 2006).

The reproductive process of marsupials, often considered less advanced and less efficient than that of placentals, may actually have advantages. A female marsupial invests relatively few resources during the brief gestation period of her young. Her major commitment comes later, during lactation, a phase that is more environmentally sensitive than that of internal nutrition and more easily terminated by adverse factors. The marsupial that loses her young is, therefore, able to make a second attempt at reproduction more quickly and in better condition than a placental in a comparable situation. That means of reproduction perhaps should be seen as a dynamic and highly derived alternative befitting the particular evolutionary history of the marsupials rather than as the survival of a primitive developmental stage (Kirsch 1977c; Parker 1977; Renfree 1983, 1993).

Other Marsupial Characters

In addition to reproduction, other characteristics help to set marsupials apart. The number of teeth, usually 40–50, often exceeds that of placental mammals. The Vombatidae, however, have only 24 teeth. The seven or eight cheek teeth, usually present on both sides, are divided into three premolars and four or five molars, in contrast with the four premolars and three molars of typical placental mammals. The lower jaws do not have the same number of incisors as the upper jaws. Marsupials show a unique pattern of tooth eruption and replacement. Most mammals have two sets of teeth during their lifetime; the deciduous or milk teeth of young animals are replaced by permanent or secondary dentition. In placentals, all teeth except the molars have deciduous predecessors. In marsupials, however, only the third upper and lower premolars undergo deciduous and secondary stages; all other teeth erupt only once (Dickman 2005; Feldhamer et al. 2007).

The skull has a large facial area and a relatively small cranial cavity; the brain, especially the telencephalon, is small compared with that of a placental mammal. The brain of the marsupials, like that of the monotremes, lacks a corpus callosum, a mass of tissue of many functions that connects the hemispheres of the brain in the placental mammals. The convolutions of the cerebral hemispheres are generally simple. The nasal bones of the skull are large and expand posteriorly, and the zygomatic arches are complete, with the jugal bone extending backward below the zygomatic process of the squamosal bone as far as the glenoid fossa. The palate is usually imperfect, with spaces between the back molars. The angular process of the mandible is usually bent inward except in Tarsipes. Epipubic (marsupium) bones are associated with the pelvic girdle in both sexes of nearly all marsupials (and also monotremes and some multituberculates) but are small in Notoryctidae and absent in Thylacinidae (Dawson et al. 1989). Recent research suggests these bones form part of a complex muscular linkage between the pelvis and hind limbs that stiffens the trunk of the body during movement (Dickman 2005).

Marsupials traditionally have been viewed as primitive or second-class mammals. That line of thought now seems to be changing. Laboratory studies have demonstrated that the learning and problem-solving abilities of marsupials often equal or exceed those of some placental groups (Kirkby 1977). It is true that

Sunda colugo (*Galeopterus variegatus*), **Order Dermoptera**. Although the colugo is a placental (clade Euarchontoglires), its young is born in a marsupial-like undeveloped state, and the patagium of the mother can be folded in a manner somewhat analogous to the marsupial pouch. Photo by Norman Lim.

Burchell's zebra (*Equus burchellii*), **Order Perissodactyla**, placental clade Laurasiatheria. In marked contrast to those of the marsupials, the young of the zebra are born in a highly advanced state and ready to move about on their own almost immediately. Photo by Zachary Zirlin / Shutterstock.com.

some Australian marsupials lost ground when placentals were introduced by human agency. The Virginia opossum, however, continues to thrive and even to extend its range in North America despite a host of placental competitors. Following its introduction on the West Coast it rapidly expanded its range over a large area. Other marsupials also have had a successful introduction, notably brush-tailed possums in New Zealand and wallabies in New Zealand, Great Britain, Germany, and Hawaii (Gilmore 1977). If marsupials did evolve mainly on the southern continents, they would have had far less room to diversify than the placentals and much more difficulty in achieving a wide geographic distribution. Those factors, rather than any inferiority on a one-to-one basis, may explain their current restricted range.

Order Monotremata

Monotremes

This order, comprising two families, three genera, and five species of Australia and New Guinea, is the most distinctive of the 31 orders of historical mammals recognized here. McKenna and Bell (1997) divided Monotremata into two orders, listing them in the sequence: Platypoda for the family Ornithorhynchidae (platypuses) and Tachyglossa for the family Tachyglossidae (echidnas). Groves (2005a), however, recognized only the single order Monotremata, with the familial sequence Tachyglossidae and Ornithorhynchidae. Although platypuses appear more specialized than do echidnas, the former have an older fossil record, and there have been suggestions that they gave rise to the latter (see Musser 2003; Pettigrew 1999). As noted in the account of class Mammalia, most authorities now consider monotremes to have the same common ancestor as marsupials and placentals, but to have diverged from the lineage of the latter two groups prior to the divergence of the multituberculates. Other proposals have been: (1) that monotremes are closely related to multituberculates; (2) that they form a clade with the marsupials to the exclusion of the placentals; (3) that they belong in a subclass with long-extinct lineages, separate from all other extant mammals; and (4) that they are actually living therapsids (see Musser 2003).

Monotremes resemble reptiles and differ from all other mammals in that they lay shell-covered eggs that are incubated and hatched outside of the body of the mother. They also resemble reptiles in various anatomical details, such as the structure of the eye and the presence of certain bones in the skull. The pectoral or shoulder girdle of monotremes is reptilian, possessing distinct coracoid bones and an interclavicle, and some features of their ribs and vertebrae are reptile-like. Monotremes also exhibit similarities to reptiles in their digestive, reproductive, and excretory systems. In both sexes of all monotremes, as in all reptiles, the posterior end of the intestine, the ducts of the excretory system, and the genital ducts open jointly into a common chamber known as the "cloaca." Therefore, for the three systems there is only a single external opening, and that arrangement is the basis of the ordinal name, "Monotremata." Most marsupials also have a cloaca, but the urogenital and alimentary tracts open separately into it. In placentals the urogenital and intestinal canals open separately to the outside of the body.

According to Temple-Smith and Grant (2001), the reproductive system of female monotremes opens into the cloaca and features a urogenital sinus connected to separate left and right tracts, each comprising an ovary, oviduct (fallopian tube), uterus, and cervix. As in many bird species and some reptiles, only the left side of the female reproductive system in the platypus is functional, but both sides are functional in echidnas. In male monotremes the testes are abdominal. The penis is attached to the ventral wall of the cloaca, and the glans penis is bifid, divided into paired canals used only for the passage of sperm. In echidnas this division is symmetrical, complementing the functional symmetry of the female reproductive system. In the platypus, how-

Duck-billed platypus (*Ornithorhynchus anatinus*). Photo by D. Parer and E. Parer-Cook / Minden Pictures.

Long-beaked echidna (*Zaglossus bruijni*), which may have retained substantive electroreception in its bill, suggestive of an aquatic platypus-like ancestor. Photo by Klaus Rudloff at Moscow Zoo.

ever, the glans penis is asymmetrical, with the left side much larger than the right, therefore aligning with the functional (left) side of the female reproductive tract in the mating position. It is not certain whether the eggs are fertilized in the uterus, as suggested by Temple-Smith and Grant (2001) or in the oviducts, as stated by Feldhamer et al. (2007). Then, over a period of about 2 weeks, the eggs are covered with albumen and a flexible, sticky, somewhat rubbery shell. Unlike the case in birds, the shell of monotreme eggs is permeable, allowing absorption of nutrients. The eggs are small, about 16 mm long and 14 mm wide.

Although monotremes have many reptilian characteristics, in other ways they are typically mammalian. Like other mammals, they are furred, have a four-chambered heart (there is an incomplete right atrioventricular valve, however, and a single aortic arch on the left), nurse their young from milk secreted by specialized glands, and are warm-blooded (but the body temperature averages lower than that of other mammals, about 30–32° C). In addition, the skeleton has mammalian features, such as the single bone of each side of the lower jaw (not several, as in reptiles) and the three middle ear bones (rather than one, as in reptiles). In the brain of monotremes, as in that of other mammals, the "pallial" region is the most strongly developed part of the cerebral hemispheres, not the "striatal" part as in reptiles and birds. Along with marsupials and reptiles, however, monotremes lack the corpus callosum, a multifunctional bridge of nervous tissue that connects the two hemispheres in other mammals.

One striking feature that monotremes possess in common with marsupials is epipubic, or "marsupium," bones associated with the pelvis. These bones are said to aid in supporting a pouch, but that function is doubtful as they are equally developed in both sexes. It seems more likely that they are a heritage from reptile-like ancestors, associated with the attachment of abdominal muscles that support large hindquarters.

The monotreme skull has a smooth, rounded cranial portion terminating in a long rostrum covered by rubbery, sensitive skin. The sutures of the skull tend to become obscured. Young Ornithorhynchidae have teeth that do not cut through the gums, but true functional teeth are not present in adult monotremes.

The males of all monotremes bear a horny spur on each ankle. In *Ornithorhynchus* the spur is grooved to permit passage of a poisonous glandular secretion. According to Ligabue-Braun, Verli, and Carlini (2012), the female platypus is born with a spur but loses it during development; female echidnas have degenerate spurs. There is some question as to whether the spur in male echidnas has any role, defensive or otherwise.

"Uniquely among mammals, monotremes share with a number of fish and salamanders a capacity for electroreception. The skin of the bill of platypuses and echidnas is a mosaic of mechanoreceptors and electroreceptors that enable them to detect weak electrical fields, which they use for locating invertebrate prey" (Nicol 2003). However, Proske and Gregory (2003) observed that electroreception in echidnas is not as well understood as it is in the platypus. It is uncertain whether stimuli in the natural environment of echidnas are strong enough to have an effect, though perhaps, as an echidna pushes its snout through the soil, electrical activity of living prey in the immediate vicinity is sufficient to activate the receptors. Pettigrew (1999) noted that the number of electroreceptors in the bill of *Zaglossus* (2,000) is intermediate to the number in *Tachyglossus* (400) and *Ornithorhynchus* (40,000). Since *Zaglossus* forages in a moist rainforest environment, it may have retained more of the sensory capability of a presumed aquatic platypus-like ancestor, than has *Tachyglossus*, which lives in a drier habitat.

Fossils from the middle Jurassic of Madagascar, late Jurassic of Patagonia, and early Cretaceous of Australia suggest a widespread Mesozoic radiation of southern mammals (see account of class Mammalia) with possible affinity to a basal stock of Monotremata (Musser 2003). However, known geological range of the order is early Cretaceous to Recent in Australia, late Pleistocene to Recent in New Guinea, and early Paleocene in South America (Archer et al. 1985; McKenna and Bell 1997; Pascual et al. 1992). The order apparently originated in Australia long before the Cretaceous, when that continent still was joined to Antarctica, then dispersed through Antarctica to South America (Flannery et al. 1995a). Recent assessment of fossils and development of molecular clock models by Rowe et al. (2008) point to divergence of Tachyglossidae and Ornithorhynchidae during or before the early Cretaceous. Based on further morphological and genetic study, Phillips, Bennett, and Lee (2009) concluded that the two families did not diverge until

48–19 million years ago, and that platypus-like monotremes predate the separation, indicating that echidnas had aquatically foraging ancestors. Nicol (2015) favored an early Cretaceous divergence, more than 115 million years ago.

MONOTREMATA; Family TACHYGLOSSIDAE

Echidnas, or Spiny Anteaters

This family of two Recent genera and four species, is found in Australia, Tasmania, and New Guinea.

Echidnas have a body covering of fur intermixed with dorsal and lateral barbless spines. The average size of males is larger than that of females. The body is robust and muscular, yet supple. The limbs have broad, powerful feet modified for digging, each foot bearing three to five strong, curved claws. The second toe on the hind foot is elongated and is used for scratching and cleaning the fur and skin. The eyes are smaller than those of the Ornithorhynchidae. The pinna of the external ear is well developed but partly concealed in the pelage. The sense of smell is acute. The tail is vestigial.

The braincase of the skull is large and rounded, and the brain is relatively large, with convoluted cerebral hemispheres. The rostrum is long, tubular, and tapering; its surface contains electroreceptors that may function in finding prey (see above account of the order

Monotremata). The lower jaw bone, on each side, is slender. The mouth is small. Teeth are absent throughout life. The posterior part of the long, sticky tongue has horny serrations that work against hard ridges on the palate for grinding food.

In both genera, *Zaglossus* and *Tachyglossus*, the mammary glands open into the abdominal pocket of the breeding female. This pocket, or pouch, is not the same as the pouch of marsupials; it is a structure that develops only temporarily in the female at the time of breeding. Also, in the males of both genera a spur is present on each of the hind legs, on the inner surface near the foot. However, while there is a crural gland system, there appears to be no production of venom, as found in male *Ornithorhynchus*. Females have a degenerate spur; the role in each sex remains unknown (Ligabue-Braun, Verli, and Carlini 2012).

Based on blood samples from 13 echidnas, Bollinger and Backhouse (1960) indicated that the hemoglobin level is higher than in most mammals. Griffiths (1968), however, stated that wild marsupials and placentals exhibit higher hemoglobin levels than those found by Bollinger and Backhouse in echidnas. In other aspects, the blood of echidnas resembles that of other mammals.

Echidnas are powerful diggers. An individual in a horizontal position can burrow straight down with remarkable speed. Shelters include hollow logs, cavities under roots or rocks, and burrows. These animals use their spines and feet to wedge themselves into

Left, long-beaked echidna (*Zaglossus bruijni*); photo by Pavel German / Wildlife Images. *Right,* short-beaked echidna (*Tachyglossus aculeatus*); photo by Gerhard Körtner.

crevices and burrows in such manner that it is practically impossible to dislodge them. The spines can be erected and the limbs withdrawn, as in a hedgehog (*Erinaceus*). Contrary to what has often been written, echidnas (*Tachyglossus*, but probably not *Zaglossus*) do hibernate during the cold season; they may also enter a shallow state of torpor in response to lack of food. With their long, sticky tongue, provided with mucus from enlarged salivary glands, echidnas feed on termites, ants, other insects, and worms. Some individuals have thrived when fed milk, finely minced meat, hard-boiled eggs, mealworms, and bread. The quantity of milk consumed by captives is remarkable. Echidnas have shown themselves able to exist without nourishment of any kind for at least 1 month.

Echidnas are generally solitary. Brattstrom (1973) studied the social behavior of *Tachyglossus* and found that many complicated behavioral postures, such as grooming, aggression, courtship, and maternal behavior, were missing. He concluded that echidnas seem to have a behavior that is not only simpler than that found in most mammals but perhaps also simpler than that of many lizards. However, subsequent investigation has revealed more intricate social behavior (see especially the following account of *Tachyglossus*). According to Flannery (1995a), echidnas have "extraordinarily large and complex brains and relatively high intelligence."

During the breeding season, the single egg, or on rare occasions two or three, is transferred directly from the cloaca into the female's temporary pouch on the abdomen. When hatched, the young is naked and measures about 12 mm in length. It sucks the thick, yellowish milk that flows from the mammary glands, which open into the pouch. The young remains in the pouch for 6–8 weeks, or until its spines begin to develop. At this age it is about 90–100 mm long. It is then deposited in a sheltered spot by the mother, which continues to nurse it during periodic visits.

The known geological range of Tachyglossidae is middle Miocene to Recent in Australia and Pleistocene to Recent in New Guinea and Tasmania (Marshall 1984; Musser 2003). That range would be in keeping with the finding by Phillips, Bennett, and Lee (2009) that the family diverged from Ornithorhynchidae just 48–19 million years ago. However, other paleontological and molecular assessment suggested the split had occurred by the early Cretaceous (Nicol 2015; Rowe et al. 2008).

MONOTREMATA; TACHYGLOSSIDAE; **Genus**
TACHYGLOSSUS Illiger, 1811

Short-beaked Echidna

The single species, *T. aculeatus*, inhabits most of mainland Australia, many nearby islands, Tasmania, and eastern and south-central New Guinea (Collins 1973; Flannery 1995a; Griffiths 1968, 1978). Occurrences

Short-beaked echidna (*Tachyglossus aculeatus*). Photo by Eric Isselee / Shutterstock.com.

north of the central cordillera of New Guinea may possibly have resulted from early human translocation (Heinsohn 2003). A number of apparently well-marked subspecies have been described, based primarily on characters of the pelage and spines. The subspecies *T. a. setosus* of Tasmania is often considered a full species, and another distinctive subspecies, *T. a. multiaculeatus*, is found only on Kangaroo Island off South Australia. Aplin and Pasveer (2007) reported remains of *T. aculeatus* from a late Pleistocene to Holocene archeological site on Kobroor in the Aru Islands, where the species is not currently present. Nicol (2015) noted that a specimen had been recorded from Sulawesi.

Head and body length is 300–530 mm, tail length is about 90 mm, and adult weight in Australia is 2.0–7.0 kg. New Guinean specimens are smaller; a male weighed 1.4 kg, a female 1.7 kg (Flannery 1995a). The spines, which cover most of the body, are up to 60 mm long. They are usually yellow at the base and black at the tip or, less frequently, entirely yellow. These specialized hairs are generally large, hollow, and thin-walled. In some individuals the brown or black body fur is almost concealed by the spines, but in the Tasmanian subspecies the short spines are largely hidden by the fur. The underparts lack spines but are covered with fur and thick bristles.

Like that of the platypus, the body of the short-beaked echidna is compressed dorsoventrally, but it lacks the streamlining so evident in the platypus. The back of *Tachyglossus* is domed, and the belly is flat, even concave, in outline. The head is small and seems to emerge from the body without any indication of a neck. At the posterior end of the body is a short, stubby tail, which is naked on its undersurface. Each foot of *Tachyglossus* has five digits, with flat claws that are well adapted for digging. The eyes are small and situated well forward on the head, almost at the base of the snout; they look ahead more than sideways. After studying the vision of *Tachyglossus*, Gates (1978) concluded that its sight is somewhat better than "dismal." In fact, he stated that its visual discrimination is at least comparable to that of rats and that it compares favorably with that of "nonvisual" animals, such as bats.

The long and slender snout is about half the length of the entire head; it is nearly straight or curved slightly upward. According to Griffiths (1978), the snout is strong enough to be used as a tool to break open hollow logs and to plow up forest litter to get at the ants and termites the echidna catches with its long, sticky tongue. The snout also evidently contains about 400 electroreceptors, which may assist the animal in moving about or locating prey (Augee and Gooden 1992; Pettigrew 1999). However, Pettigrew, Manger, and Fine (1998) suggested that the short-beaked echidna's relatively few electroreceptors, compared to the number in *Zaglossus* and *Ornithorhynchus*, may not be of much use in its present relatively dry habitat and that it may be a vestige of a more aquatic and platypuslike ancestor (see above account of the order Monotremata).

There is no scrotum in males, the testes are internal, and in both sexes there is only a single opening for the passage of feces, urine, and reproductive products. During the breeding season females develop on the abdomen a crescentic fold of skin that forms a pocket in which the single egg is deposited and incubated.

The short-beaked echidna frequents a variety of habitats, including forests, rocky areas, hilly tracts, and sandy plains. It shelters either in burrows or in crevices among rocks, and it commonly emerges to forage in the afternoon or at night. It evidently retreats to its burrow when the air temperature exceeds 32° C and experiences severe heat stress if maintained above 35° C (Grant 1983). Abensperg-Traun and De Boer (1992) observed no foraging below 9° C or above 32° C, noting

Short-beaked echidna (*Tachyglossus aculeatus*). Photo by Fritz Geiser.

that winter activity generally began soon after midday, while during summer foraging usually started long after dark. On Kangaroo Island, females were found to greatly increase their activity time when lactating, occasionally to 18 hours per day, and to return to the nursery burrow (see below) at various times during the day or night (Rismiller and McKelvey 2009). In southeastern Queensland, Wilkinson, Grigg, and Beard (1998) found *Tachyglossus* to most often use hollow logs and depressions under the roots of fallen trees for daily shelter and rabbit burrows for hibernation.

When disturbed in its burrow, the echidna digs into the earth and clings with its claws and spines, or, if the substrate is too hard for digging, it rolls up into a spiny ball. The echidna walks with its legs fully extended, so that the stomach is relatively high off the ground, and with the hind toes directed outward and backward. It can run quite swiftly and can climb well. Augee (2008) noted that it is a surprisingly good swimmer.

Tachyglossus has a body temperature lower than that of most mammals, and it is controlled by shivering thermogenesis rather than nonshivering thermogenesis, which is found in mammals other than monotremes. When it is not hibernating or torpid, its body temperature is 28–35° C, with a modal of 31–32° C (Nicol and Andersen 2007). Augee (1978) thought that as long

Short-beaked echidna (*Tachyglossus aculeatus*), partially rolled up for self-protection. If it had been completely rolled up, as when danger threatens, the face and feet would have been entirely enclosed in the spiny ball. Photo by Gerhard Körtner.

as the echidna had sufficient food. it would not become torpid at any time of the year, regardless of temperature, but that without sufficient food it would enter torpor during any season. However, Nicol and Andersen (2007) reported that throughout its range *Tachyglossus* follows a pattern of maximal mass gain in late spring–early summer (December–February), followed by hibernation, regardless of food availability.

Although at one time scientists did not consider the echidna or any monotreme to be capable of true hibernation, recent investigations have demonstrated otherwise (Grigg, Beard, and Augee 1989). In Tasmania, hibernation occurs from about February or March to July, sometimes to September (Nicol and Andersen 2007). Studies in the Snowy Mountains of New South Wales (Beard, Grigg, and Augee 1992; Grigg, Augee, and Beard 1992) found the echidna to hibernate during the cold season, usually from about April to July. Physiologically, hibernation resembled that of eutherian mammals, with body temperature falling as low as 3.78° C and weight losses of 2–3 percent per month. However, there were periodic arousals, up to 24 hours long, during which normal temperature was regained and individuals sometimes moved to a different retreat. As in other areas, animals also were found to enter brief periods of shallow torpor during the nonhibernating season.

Griffiths (1968) found *Tachyglossus* to be almost exclusively an ant and termite eater. For unknown reasons, in cool, moist areas ants compose the major portion of the diet, while in hot, dry areas termites are most often consumed. It seems not to be a question of availability since there are plenty of ants in both kinds of habitat.

In a radio-tracking study on Kangaroo Island, South Australia, Augee, Ealey, and Price (1975) found the short-beaked echidna to have a definite home range averaging about 800 meters in diameter. They also reported that the ranges of different individuals may overlap, that there is no specific nest site, that individuals are solitary for most of the year, and that there is no evidence of territorial behavior. Other radio-tracking studies have also found extensive overlap, with home range size being 24–76 ha. in the Snowy Mountains of New South Wales (Augee, Beard, and Grigg 1992), 21–93 ha. on a sheep grazing property in southeastern Queensland (Wilkinson, Grigg, and

Beard 1998), 24–192 ha. for adults and 6–48 ha. for juveniles in the wheatbelt reserves of Western Australia (Abensperg-Traun 1991a), and 35–164 ha. for males and 17–104 ha. for females in Tasmania (Nicol et al. 2011). Mean size of areas used by females carrying young on Kangaroo Island was 24.8 (15.6–33.6 ha.); once young had been placed in a nursery burrow (see below), mean range size of females was 70.3 (55.8–90.2 ha.); those areas overlapped when females were both with and without young (Rismiller and McKelvey 2009).

Up to 33 individual *Tachyglossus* have lived together in captivity and generally ignored one another (Jackson 2003). Augee, Bergin, and Morris (1978) found captives to be mutually tolerant, but animals of the same sex formed a dominance hierarchy. On Kangaroo Island, South Australia, where hibernation is brief (less than 10 days), there is a courtship period of several weeks, when a female may be pursued by up to 10 males ("echidna trains"), which possibly form a dominance hierarchy (Augee 2008; Morrow, Andersen, and Nicol 2009; Rismiller and McKelvey 2000). In Tasmania, however, most matings occur 1 or 2 days following arousal from hibernation (Nicol and Andersen 2007). Most mating groups there have only one male and female, but 40 percent have multiple males, several of which may mate with the female (Morrow, Andersen, and Nicol 2009). Males in Tasmania may arouse in early winter and seek out still-hibernating females; the latter then re-enter hibernation between matings and sometimes when pregnant. Such timing ensures that maximum growth rate of the young coincides with the period of greatest ecosystem productivity, while female torpor through the mating period minimizes energy expenditure during the time of lowest food availability (Morrow and Nicol 2009; Nicol and Morrow 2012). In the Snowy Mountains, sexually mature females also mate immediately after hibernation (Beard, Grigg, and Augee 1992).

The height of the breeding season is July and August (winter) all over Australia (Griffiths 1968, 1978). In Tasmania, matings have been reported from early June to early September and egg-laying from July to late September (Morrow, Andersen, and Nicol 2009). Both sexes are promiscuous (Nicol et al. 2011). Although adult females are capable of producing young in consecutive years, most do so only once every 2 to 5 years (Nicol and Andersen 2007; Rismiller and McKelvey

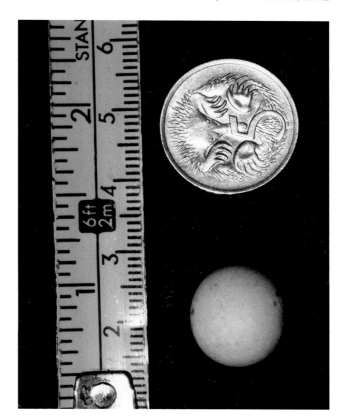

Short-beaked echidna (*Tachyglossus aculeatus*), egg. Photo by Fritz Geiser.

2000). A female breeding a second time during one season, after losing her first young, has been recorded (Beard and Grigg 2000). Estrous cycle has been estimated as 9 days to 1 month (Hayssen, Van Tienhoven, and Van Tienhoven 1993). The gestation period has been reported as 20–24 days (Morrow, Andersen, and Nicol 2009), 22–24 days (Rismiller and McKelvey 2000), and 19–27 days (Ferguson and Turner 2013).

When preparing to incubate young, the female constructs a short nursery burrow (Augee 2008). On Kangaroo Island all such burrows were found to be not the same ones previously used by the female for shelter; length was 1.25–2.2 meters (Rismiller and McKelvey 2009). In Tasmania the nursery burrow is just sufficient in size to admit the mother, and the young usually is found in a chamber 30–50 cm underground and up to 1.5 meters from an entrance that normally is closed with soil; females in Tasmania were found to remain in the burrow for about 6 weeks, though such may not be the case elsewhere (Morrow, Andersen, and Nicol 2009). The single egg is deposited into the pouch directly from the cloaca. This egg has a very large yolk and is enclosed in a flexible, leather-like shell through which the baby breaks at birth with the help of an egg tooth. Incubation in the pouch, during which the greater part of embryogenesis occurs, lasts about 10–13 days (Ferguson and Turner 2013; Griffiths 1968, 1978). Newly hatched young are about 24 mm long and weigh about 0.3 grams (Rismiller and McKelvey 2000, 2003).

There is considerable geographic variation in the process of maternal care. In eastern Australian populations the mother remains in the nursery burrow for 2–3 weeks after laying her egg, until the young is about a week old, then emerges to forage with the young in her pouch (Morrow, Andersen, and Nicol 2009; Nicol and Andersen 2007). On Kangaroo Island the female commonly carries the egg and then the young in the pouch as she forages, then places the young in a nursery burrow when it is 45–55 days old, and returns every 5, or occasionally 6, days to nurse it (Rismiller and McKelvey 2000, 2009). The young is then 15–21 cm long and weighs 180–270 grams (Griffiths 1978; Rismiller and McKelvey 2003). In Tasmania, the mother does not normally leave the nursery burrow until the young is 25–35 days old, after which she will leave it alone in the nursery burrow, typically for 3–5 days, while she forages (Nicol and Andersen 2007).

According to Griffiths (1978), the young does not lick up milk from the mammary glands, but sucks as does any other mammal. He observed that the sucking is quite vigorous, involves movement of practically the entire body, and is clearly audible. Age and average body mass at weaning vary geographically, being 139–152 days and 1.8 kg in Tasmania and 204–210 days and 1.4 kg on Kangaroo Island (Morrow, Andersen, and Nicol 2009; Rismiller and McKelvey 2009). Dispersal from the natal home range occurs at about 12 months (Abensperg-Traun 1991a). Adult mass is attained at 3–5 years (Nicol and Andersen 2007). Minimum age of sexual maturity in both males and females is

Short-beaked echidna (*Tachyglossus aculeatus*). Photo by Mark Higgins / Shutterstock.com.

5–12 years (Rismiller and McKelvey 2003). The short-beaked echidna may have an extremely long life span; wild individuals have been known to survive for as long as 49 years (Jackson 2003), and one captive lived to be more than 50 years old.

In contrast to many Australian mammals, *Tachyglossus* appears to have adapted well to the coming of European colonization, remaining common and widespread (Abensperg-Traun 1991b). Apparently, the echidna is not restricted by habitat and does not depend on vegetation for shelter. Its food source is abundant, readily available, and not subject to competition. Moreover, it is metabolically capable of tolerating low food supplies, as may result from fire or drought, and can go into torpor under extreme conditions. And because of its spines and other defensive strategies, the echidna is little affected by the introduced predators that have wiped out other native species. Reports compiled by Abbott (2008a), however, do suggest that in some areas populations have been suppressed by the red fox (*Vulpes vulpes*), which was introduced in Australia by people in the nineteenth century. A subspecies, *T. a. multiaculeatus*, endemic to Kangaroo Island, South Australia, is declining seriously because of predation, not by the fox but by two domestic mammals that have become feral, the house cat (*Felis catus*) and the pig (*Sus scrofa*). That factor, applied to IUCN criteria, recently led Woinarski, Burbidge, and Harrison (2014) to designate *multiaculeatus* endangered, though the current IUCN assessment itself (Aplin et al. 2008a) does not assign a category of concern. In parts of New Guinea *Tachyglossus* is threatened by overhunting by people for use as food, and there may be some localized declines; it also is widely used there for ceremonial purposes (Aplin et al. 2008a).

MONOTREMATA; TACHYGLOSSIDAE; Genus *ZAGLOSSUS* Gill, 1877

Long-beaked Echidnas

There are three species (Flannery 1995a, 1995b; Flannery and Groves 1998; Helgen et al. 2012):

Z. bruijni, New Guinea west of Paniai Lakes, Salawati and possibly Waigeo and Supiori islands, a single specimen collected in 1901 at Mount Anderson in the West Kimberley region of northern Western Australia;

Z. bartoni, Central Cordillera of New Guinea from Paniai Lakes to the Huon Peninsula and Nanneau Range;

Z. attenboroughi, known only by the type specimen from the Cyclops Mountains of north-central New Guinea.

Flannery and Groves (1998) recognized four subspecies of *Z. bartoni*: *diamondi*, Central Cordillera from Paniai Lakes to Kratke Range in the eastern highlands of Papua New Guinea; *bartoni*, Papua New Guinea Central Cordillera from the Efogi area to around Wau; *clunius*, mountains of Huon Peninsula; and *smeenki*, Nanneau Range. According to Leary et al. (2008a, citing K. Helgen), the subspecies *diamondi*, *bartoni* (including *clunius*), and *smeenki* may represent distinct species. Nicol (2015) noted that a new population of *Z. bartoni*, not clearly referable to any of the recognized subspecies, had been discovered in the Foja Mountains of northwestern New Guinea.

Pleistocene fossils referable to *Zaglossus* have been found in Australia, but the genus long was thought not to have survived there into historical time. Musser (2006) did note that the Pleistocene species *Zaglossus* (or *Megalibgwilia*) *ramsayi* appears to have survived into the Holocene in Australia, with the latest records being from Tasmania, and suggested that human hunting may have contributed to its extinction. Recently, Helgen et al. (2012) uncovered the well-documented but previously overlooked specimen of *Z. bruijni* from the West Kimberley region, as noted above. They also reported Aboriginal accounts of an echidna, other than *Tachyglossus* and perhaps referable to *Z. bruijni*, from the East Kimberley of Western Australia in the early to mid-twentieth century, and they considered the possibility that the species persists in that area. Woinarski, Burbidge, and Harrison (2014) accepted the historical occurrence of *Z. bruijni* in Australia but designated it as now extinct on the continent.

Head and body length is about 450–775 mm, and the tail, like that of *Tachyglossus*, is only a slight projection. Long-beaked echidnas usually weigh 4–10 kg. An individual in the London Zoo weighed more than 16 kg, but, like many zoo animals, it may have been excessively fat. Opiang (2009) found female *Z. bartoni* to

Long-beaked echidna (*Zaglossus bruijni*). Photo by Klaus Rudloff at Moscow Zoo.

average significantly longer and heavier than males. Flannery and Groves (1998) described *Z. bartoni* as the largest known living monotreme, noting that a male, after 12 years at the Taronga Zoo, weighed 16.5 kg. The hairs of *Zaglossus* are brownish or black and sometimes almost hide the spines on the back; the underparts are usually spineless. The spines range in color from white and light gray to solid black. The head is typically paler than the body and is almost white in one form that has a pale brown body.

Claw number varies geographically in *Zaglossus* (Flannery and Groves 1998). *Z. bruijni*, the westernmost species, has three (rarely four) claws on both the forefeet and hind feet; the claws are located on the three middle toes, while digits 1 and 5 are covered by a callosity. On the front feet, digits 1 and 5 are mere prominences in external aspect, but on the hind feet these digits are conspicuous. *Z. bartoni* has five claws on the forefeet and at least four (usually five) on the

hind feet. The only known specimen of *Z. attenboroughi* has five claws on each foot.

Zaglossus is larger than *Tachyglossus* and has spines that are shorter, less numerous, blunter, and with smaller central cavities. The tongue of *Zaglossus* is longer, and the anterior third has a very deep groove on the dorsal surface; in the groove are three longitudinal rows of backwardly directed, sharp, keratinous "teeth" or spines. In *Zaglossus* the long beak accounts for two-thirds of the length of the head and is curved downward, producing a pronounced convex profile, whereas the beak of *Tachyglossus* is shorter and straighter. Opiang (2009) recorded a beak length of 101–116 mm in three male *Z. bartoni* and 108–142 mm in eight females. The beak contains about 2,000 electroreceptors that may be used to detect prey (see above account of the order Monotremata).

Long-beaked echidnas generally are found in humid montane forests and may occur in alpine meadows at

Long-beaked echidna (*Zaglossus bartoni*). Photo by Klaus Rudloff at London Zoo.

an elevation of 4,000 meters (Griffiths 1978). Historically, *Z. bartoni* was recorded at sea level, but, because of human persecution, it now rarely occurs at lower elevations (Leary et al. 2008a). The habits of long-beaked echidnas are not well known but probably are comparable to those of *Tachyglossus*. They are primarily nocturnal and forage on the forest floor by probing with the beak and digging with the claws (Opiang 2009). They walk upright, more in the manner of therian mammals than that of lizards and salamanders (Gambaryan and Kuznetsov 2013).

In a five-year study of *Z. bartoni* in southern Simbu Province, Papua New Guinea, Opiang (2009) located 223 dens, 209 of which were underground. Most were on slopes, entrance diameter averaged about 10 cm, and burrows were 1.3–4.9 meters long. The resting site often was situated above the entrance. Most echidnas changed dens every night but occasionally reused a den or stayed inside over a period of days. Nightly move-

ments by individuals varied from 15 to 823 meters. Eight adults, radio-tracked for 1–12 months, had estimated home ranges of 2.2 to 75.5 ha., though the smallest estimate was based on only four radio fixes, the largest on 64; a dispersing juvenile moved over 168.2 ha. There was some overlap of home ranges.

Grigg et al. (2003) determined modal body temperature of two *Zaglossus bartoni* in the Taronga Zoo in Sydney, Australia, to be 31° C over 3.5 years. Temperature cycled a few degrees daily but not to an extent indicating long-term torpor. In contrast, two *Tachyglossus* monitored concurrently in the same pen underwent hibernation in the cooler months and short periods of torpor at any time of the year. The study thus suggested that torpor in wild adult *Zaglossus* is unlikely.

The diet of *Zaglossus* differs from that of *Tachyglossus* in that earthworms apparently make up a major portion (Griffiths 1978). In fact, Collins (1973) reported that some captive specimens of *Zaglossus* would not eat ants at all. *Zaglossus* ingests earthworms buried in the forest litter by hooking them with the "teeth" within the groove on its tongue; the tongue protrudes only 2–3 cm beyond the end of the snout, and the groove opens by muscle flexion (Griffiths, Wells, and Barrie 1991). Griffiths (1978) speculated that the reason *Tachyglossus* is more widely distributed than *Zaglossus* may be that its principal foods, ants and termites, are found over a larger area and in many more kinds of habitat than are earthworms. However, Helgen et al. (2012) observed that little definitive information on the diet of *Zaglossus* is available, and that most reports are based on anecdotes or extremely limited studies of *Z. bartoni*, which is thought to specialize on earthworms, though it also feeds on subterranean arthropods. There has been speculation that *Zaglossus* may have disappeared from Australia because post-Pleistocene climatic changes were unfavorable to the earthworms on which it supposedly preyed. However, Helgen et al. (2012), in support of their report of the Recent survival of *Z. bruijni* in northwestern Australia, suggested that the species may have depended largely on ants and termites. They also considered that the extinct *Zaglossus* (or *Megalibgwilia*) *ramsayi*, the other (and more southerly distributed) large echidna present in the Australian Quaternary, may also have fed mainly on insects, not earthworms.

Data on reproduction in *Zaglossus* are limited, but suggest that breeding may be centered in July (Griffiths 1978). Opiang (2009) captured a lactating female *Z. bartoni* during April and May. *Zaglossus* almost certainly lays eggs, probably carries the small offspring in the pouch, and is said by local hunters to have 4–6 young at a time (Flannery 1995a). One captive lived in the London Zoo for 30 years and 8 months, another lived at the Berlin Zoo for 30 years and 7 months, and both animals died within 6 days of each another in August 1943. An individual at the Taronga Zoo, captured in November 1963, was still living in January 2005 (Weigl 2005).

According to Thornback and Jenkins (1982), the average population density of *Zaglossus* has been estimated at 1.6/sq km, which would indicate a total of about 300,000 animals in the then-remaining suitable habitat. However, the IUCN classifies all species of *Zaglossus* as critically endangered due to a suspected continuing population decline of at least 80 percent over the last 45–50 years (Leary et al. 2008a, 2008b, 2008c). *Zaglossus* is now avidly hunted for food in New Guinea by natives using trained dogs. Because of such persecution, as well as loss of forest habitat to farming activity, *Zaglossus* has become rare in areas accessible to humans. *Z. bartoni* has largely disappeared in the western half of its range, and *Z. bruijni* has not been recorded since the 1980s. Flannery and Groves (1998) thought that *Z. attenboroughi* was confined to less than 50 sq km of montane habitat and might already be extinct. However, based on diagnostic field sign and information from local hunters, Baillie, Turvey, and Waterman (2009) concluded that the species still occurs in an extensive part of the Cyclops Mountains. All species of *Zaglossus* are on appendix 2 of the CITES.

MONOTREMATA; **Family ORNITHORHYNCHIDAE**; **Genus *ORNITHORHYNCHUS*** Blumenbach, 1800

Duck-billed Platypus

The single living genus and species, *Ornithorhynchus anatinus*, inhabits freshwater streams, lakes, and lagoons of eastern Australia, originally in Queensland, New South Wales, Victoria, southeastern South Australia, and Tasmania (Grant 1992; Griffiths 1978). The species occurs naturally on King Island north of Tasmania, and has been introduced to Kangaroo Island off the coast of South Australia, but no longer is present on the mainland of South Australia (Grant and Temple-Smith 1998). Genetic analyses have found a highly significant difference between the population of Tasmania and that on the nearby Victorian mainland (Furlan et al. 2010); they have also found that the population of western Victoria is distinct from other mainland populations (Furlan et al. 2013) and that the clade on Tasmania also includes the population of King Island (Gongora et al. 2012).

Head and body length is 300–500 mm, tail length is 100–150 mm, and adult weight in mainland Australia is usually 0.5–2.0 kg. Males are larger on average than females. In Tasmania, mean adult weight (and range) was 2.30 kg (1.40–3.00) for 42 adult males and 1.34 kg (1.05–1.75) for 29 females (Connolly and Obendorf 1998). Coloration is deep amber or blackish brown above and grayish white to yellowish chestnut below.

Grant and Carrick (1978) stated that the fur, which is short and very dense, consists of a woolly underfur overlain with bladelike guard hairs. This coat traps a layer of air in the kinks of the fibers of the underfur and between the underfur and the blades of the guard hairs and thus contributes substantially to the insulation of the body. Fur density, at about 600–900 hairs per square mm, is second only to that of the sea otter (*Enhydra*) (Pettigrew, Manger, and Fine 1998).

The body of the platypus is streamlined and compressed dorsoventrally. The limbs are short and stout, and the webbed feet are broad. Each of the four feet has five clawed digits. The tail looks something like that of a beaver. The snout, which resembles a duck's bill, is elongated and covered with moist, soft, naked, leathery skin; it is perforated over its entire surface by openings to sensitive nerve endings, many of them electroreceptive (see below). The nostrils also open through the snout, on its upper half, and are close together. The idea that the platypus has a horny bill like that of a bird arose from examination of dried skins only.

The brain is relatively small, with smooth cerebral hemispheres. The olfactory organs are not as well developed as in echidnas. This platypus has keen sight and hearing, but when it submerges, the eyes and ears, which lie in a common furrow on each side of the head, are covered by skin folds that form the edges of the furrow. There are no ear pinnae. The rarely heard voice is a low growl. For details on the senses and vocalization of *Ornithorhynchus*, see Pettigrew, Manger, and Fine (1998). Extensive physiological and biochemical coverage was provided by Pasitschniak-Arts and Marinelli (1998). For information on the anatomy of the reproductive system, see the above account of order Monotremata.

The young have small, calcified teeth with little enamel and numerous stubby roots; at least two pairs have replacement buds beneath. Adults lack functional teeth and instead have hornlike plates for most of the length of each jaw. Near the front of the mouth these plates are sharp ridges, but toward the back they are almost flat and function as crushing surfaces. They

Duck-billed platypus (*Ornithorhynchus anatinus*). Photo by Pavel German / Wildlife Images.

Duck-billed platypus (*Ornithorhynchus anatinus*). Photo by Alexander F. Meyer at Sydney Aquarium.

grow continuously and are rapidly worn by gritty food. The flattened tongue works against the palate and aids in mastication. The bones of the upper and lower jaws expand distally and support the bill. Grant and Temple-Smith (1998) noted that cheek pouches, used for food storage during foraging, lie beside the horny grinding pads.

The ankles of both hind limbs of the male have inwardly directed, hollow spurs that are connected with venom glands, which are components of a crural system. The spur is found on both sexes when they are young, but it degenerates in the female. The gland secretes venom that is passed to the spur and can be injected into other animals by erection of the spur (see Ligabue-Braun, Verli, and Carlini 2012). Griffiths (1978) stated that the venom causes agonizing pain in humans and can kill a dog. Most evidence supports an old idea that the crural system of the platypus is used primarily in intraspecific fighting between males during the breeding season and that it is a potentially effective mechanism for ensuring spatial separation of

males along riverine habitat. Both the crural glands and the testes increase in size before the mating season and decrease afterward (Grant and Temple-Smith 1998). While acknowledging that social function, Musser (1998) suggested the poisonous spur may also have developed as a defense against crocodiles, the range of which overlapped extensively with ancestral ornithorhynchids.

The distribution of the platypus indicates remarkable flexibility in habitat choice and adaptability to a wide range of temperature; it is able to cope successfully in both hot tropical forests and snow-covered mountainous areas, though presence of permanent fresh water is essential to its survival (Pasitschniak-Arts and Marinelli 1998). While it is normally found in and near water, genetic studies show it capable of overland dispersal between river systems (Kolomyjec et al. 2009). The platypus constructs two kinds of burrows in banks of streams and ponds. One, short and simple, provides a shelter for both sexes and is retained by the male during the breeding season. The other, usually much deeper and more elaborate, is made by the female and contains a nest for rearing the young; it may be up to 30 meters long (Carrick, Grant, and Temple-Smith 2008). Burrow entrances are normally close to or below water level (Grant and Temple-Smith 1998), but extend into the banks 1–7 meters above the water line. Upon leaving the water, a platypus may enter its burrow and emerge moments later dry and glossy; the tunnels apparently squeeze out the water, which the soil absorbs like a sponge.

The platypus swims, dives, and digs well, using the forefeet more than the hind feet. It swims smoothly with alternate thrusts of its webbed forefeet; the hind feet are employed in steering and breaking (Carrick, Grant, and Temple-Smith 2008). The part of the web of the forefoot that extends beyond the nails when the platypus is swimming is folded under the palm when the animal is on land. It has been said to be capable of staying underwater for as long as five minutes by holding onto some object. However, based on study at a Tasmanian lake, Bethge et al. (2003) estimated that oxygen reserves would be depleted after a maximum of 40 seconds. A few dives lasted up to 70 seconds, and previous reports indicated dives as long as 140 seconds; these would entail anaerobic metabolism and require extended recovery time at the surface. Mean dive duration in the study was 31.3 seconds and mean surface duration between dives was 10.1 seconds; there were up to 1,600 dives per foraging trip, with a mean of 75 dives per hour. Mean dive depth was 1.28 meters, with a maximum of 8.77 meters. Reporting on a population in New South Wales, Grant (2004b) wrote that the platypus prefers to forage in water deeper than 1.5 meters.

Early observations suggested the platypus to be crepuscular but recent radio-tracking studies have shown continuous or intermittent activity throughout the night, with diurnal activity in some individuals (Grant and Temple-Smith 1998). At a subalpine lake in Tasmania, mean daily foraging duration was 12.4 hours, with some animals foraging continuously for up to 29.8 hours (Bethge et al. 2009). Males have been documented to travel up to 10.4 km and females up to 4.0 km in a single overnight period (Serena and Williams 2013). After leaving its burrow, the platypus commences a unique quest for food. It swims along the bottom of a freshwater stream or lake, probing the mud and

gravel with the highly sensitive end of its rubbery bill. Until recently it was thought that edible material was located by the sense of touch alone, since the eyes and ears are closed when submerged, and that only those foodstuffs touched with the bill were snatched from the bottom. However, experiments by Scheich et al. (1986) indicate the bill of the platypus is an electroreceptive as well as mechanoreceptive organ that allows the detection of muscle activity in prey animals. Proske, Gregory, and Iggo (1992) determined that weak electrical stimulus of the bill evoked a reaction in the brain. Stimuli are detected by mucous glands innervated with electroreceptive terminals and transmitted to the brain via the trigeminal nerve system (Proske, Gregory, and Iggo 1998).

The bill has a high density of both mechanosensory and electrosensory receptors, the former numbering about 60,000, the latter about 40,000. Inputs from the two sensory arrays are integrated in a manner remarkably similar to that of the primate visual cortex. When foraging underwater, the platypus swings its head from side to side, seemingly scanning for prey. Experimentation has shown that the bill briefly moves vertically and horizontally, apparently when the electroreceptors detect the direction of electrical stimuli emitted through muscular activity of living prey. Bill mechanoreceptors are capable of detecting mechanical waves traveling through the water from moving prey. The mechanical waves arrive after the electrical signal from the same prey. The difference in time between receipt of the two stimuli may be used to determine distance, perhaps up to 50 cm. That information, combined with the directional input from the electroreceptors, would result in a three-dimensional fix on the prey (Pettigrew 1999; Pettigrew, Manger, and Fine 1998; Proske and Gregory 2003). In both its spoon-shaped bill, with its array of electroreceptors, and its method of hunting, the platypus has undergone parallel evolution with the North American paddlefish (*Polyodon spathula*) (Pettigrew and Wilkins 2003).

When sufficient prey has been gathered and stored in the cheek pouches, or when breathing becomes necessary, the platypus proceeds to the surface and planes out with limbs outstretched and flattened tail trailing behind. Mastication of the food is accomplished by crushing and grinding it between the horny plates that act as teeth. Grit taken up with foodstuffs probably

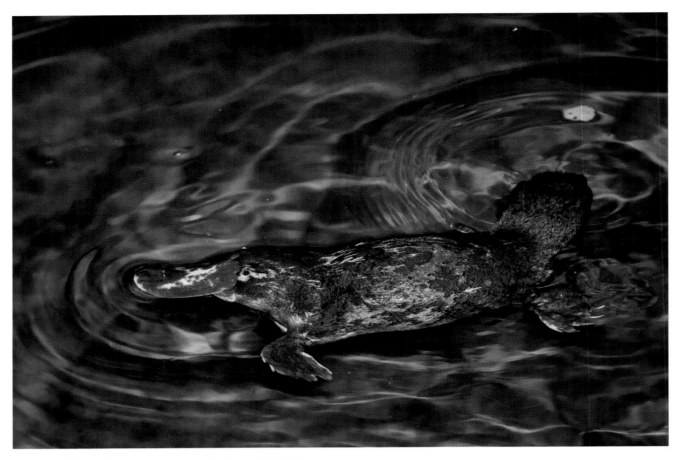

Duck-billed platypus (*Ornithorhynchus anatinus*). Photo by Fritz Geiser.

serves as an abrasive to aid in mastication. The diet consists of crayfish, shrimp, larvae of water insects, snails, tadpoles, worms, and small fish. The average daily intake of a 1.5-kg captive male included 450 grams of earthworms, 20–30 crayfish, 200 mealworms, 2 small frogs, and 2 coddled eggs. Data cited by Grant and Temple-Smith (2003) indicate that food consumption may be 15–28 percent of body weight per day during normal foraging and 90–100 percent in females during the later stages of lactation.

The platypus is active year-round, even in the very cold water temperatures of winter, and responds to the thermal stress of cold by markedly declining in weight and tail volume (Grant and Carrick 1978). The species is homeothermic, maintaining a low body temperature (32° C) even while foraging for hours in water below 5° C. Certain observations, both in the wild and in captivity, suggest that some individuals may undergo periods of winter dormancy or even hibernation, but the matter remains unresolved (Grant and Temple-Smith 1998).

Individuals may forage along several kilometers of a stream during a period of a few days (Grant et al. 1992). Apparently, a resident breeding population permanently occupies a particular area, though transients also may be present (Gemmell et al. 1992). Reported home range length is 0.2–15.1 km of channel, but young animals have dispersed up to 48 km (Grant 2004a; Serena and Williams 2013). The foraging area of three radio-tracked individuals in a Tasmanian alpine lake was 14–30 ha. (Grant and Temple-Smith 1998). Radio-tracking and marking/recapture studies in Victoria indicated a population density of 1.3–2.1 adults or subadults per kilometer of stream. Home range lengths of resident adult males were determined to be 6.0–13.9 km. Some of those ranges were mutually exclusive, and others overlapped substantially, but in the latter case the animals avoided one another,

Duck-billed platypus (*Ornithorhynchus anatinus*), male grasping female's tail during courtship. Photo by D. Parer and E. Parer-Cook / Minden Pictures.

spending most of their time in different parts of the shared area. All male ranges apparently overlapped those of two or more adult females, which in turn overlapped extensively. Female home ranges were up to 4.4 km of channel over 127 months (Gardner and Serena 1995; Serena 1994; Serena and Williams 2013). In captivity, two females can be housed together, as well as a male with one or two females, but two adult males will interact aggressively, and there are records of one killing the other with the venomous spur (Jackson 2003). Available information suggests intense competition between males during the breeding season, which is likely associated with defense of specific areas or changes in ranging behavior (Gust and Handasyde 1995). Monitoring of captives has shown adult males to be consistently dominant over females (Hawkins 1998).

The platypus mates from late July to November (winter and spring in Australia). Breeding begins ear-

lier in the northern than in the southern parts of the range (Connolly and Obendorf 1998; Temple-Smith and Grant 2001). A captive female had a receptive period of 4–6 days in the spring of each year, during which her behavior changed from avoidance to initiating contact with the male (Hawkins and Battaglia 2009). Mating takes place in the water and is preceded by an unusual and intricate courtship during which, among other maneuvers, the male grasps the female's tail and the two animals then swim slowly in circles. After mating, the female carries bundles of wet leaves under her forwardly folded tail into the nest chamber. The wet leaves provide moisture and prevent the eggs from becoming too dry. The burrow used for incubation and rearing is plugged with earth at one or more places by the female. Holland and Jackson (2002) reported that a captive female laid her eggs 15–21 days after mating, but noted that earlier observations in the wild suggested a gestation period of about 27 days.

Hawkins and Battaglia (2009) also estimated that a captive female had a gestation period of 15–21 days. One to three, usually two, eggs are laid in the nest. The incubating mother does not come out of the burrow for days at a time, and usually just briefly to defecate, wash, and wet her fur. However, observations cited by Temple-Smith and Grant (2001) indicate that she may occasionally exit the nest for about 30 minutes and enter the water. When leaving and returning to the nest chamber, she removes and then rebuilds the earth plugs in the tunnel. The eggs, 16–18 mm in length and 14–15 mm in diameter, are about the size of those of a house sparrow though slightly more spherical. They adhere together after they are laid and thus do not roll away. The female curls about the eggs while incubating. No pouch is present in the female at any time during the reproductive process. The young hatch after an incubation period of about 10 days. When hatched, the young are about 25.4 mm long, blind, and naked, and the female curls around them. After about 4 months, when fully furred and about 335 mm in length, the young emerge from the burrow; they apparently are weaned a few days later (Holland and Jackson 2002). Pasitschniak-Arts and Marinelli (1998) summarized reports that the platypus is playful; captive young have been observed rolling, wrestling, and playfully biting one another in shallow water.

Grant and Griffiths (1992) reported 512 platypuses captured during 18 years of study on the upper Shoalhaven River in New South Wales. Lactation was observed to last 3 to 4 months, with most lactating females being found from November to February. Females did not begin to breed until their second or third year of life. Further investigation in the same area (Grant, Griffiths, and Temple-Smith 2004) indicated that some females bred during at least two or three consecutive seasons but others failed to breed in successive years. After 30 years of study there, Grant (2004a) reported 700 captures, showing that females lived up to 21 years and males up to at least 7 years. Weigl (2005) listed a captive in the Melbourne Zoo at a probable age of 22 years and 7 months.

Until it was protected in all Australian states by 1912, the platypus was intensively hunted for its fur, which was used to make hats, slippers, and rugs; a single rug contained 40 to 60 pelts (Grant and Temple-Smith 1998). The platypus was also highly subject to accidental drowning in the nets of inland fisheries (Grant 1993). Those factors probably suppressed platypus numbers, and the species seems never to have regained its former abundance, though it has made a comeback through government conservation efforts. It now is considered generally common in its original range, but has disappeared on the mainland of South Australia and is rare or absent in the lower reaches of the Murrumbidgee and Murray River systems of New South Wales and Victoria. Populations in some other river systems have become fragmented and subject to rapid local extirpation. Immediate and severe threats include: interference with stream flow by dams, impoundments, and channelization; extraction of water for agricultural, domestic, and industrial supplies; and poor forestry and agricultural practices associated with stream bank erosion, loss of riparian vegetation, and channel sedimentation. Such factors tend to eliminate the large, deep pools that are favored by the platypus for foraging and are necessary for survival during droughts. Other problems include drought associated with climate change, water pollution, predation by the introduced red fox (*Vulpes vulpes*), and continued mortality in fish nets and traps (Grant and Temple-Smith 1998, 2003; Lunney et al. 2008a; Woinarski, Burbidge, and Harrison 2014). Of 124 mortality records from the 1980s to 2009, for which cause of death could be reliably assigned, 41 percent involved drowning in illegal nets or traps set to capture fish or freshwater crustaceans (Serena and Williams 2010). Although the current IUCN account for the platypus does not assign a category of concern (Lunney et al. 2008a), Australian authorities, applying IUCN criteria to the latest available data, have classified the species as near threatened (Woinarski, Burbidge, and Harrison 2014).

Using high resolution x-ray computed tomography, Rowe et al. (2008) proposed that *Teinolophos trusteri* from the early Cretaceous of Victoria, Australia, is an ornithorhynchid. X-rays revealed presence of a relatively large mandibular canal coursing the entire length of the dentary, a character unique to the family. That canal carries a branch of the trigeminal nerve, which transmits stimuli from the electroreceptive bill of the platypus to the somatosensory cortex. Further analysis showed presence of the same feature in the slightly younger *Steropodon galmani* from the early Cretaceous of New South Wales. That species thus was also

Duck-billed platypus (*Ornithorhynchus anatinus*). Photo by Dave Watts / Lochman LT.

considered an ornithorhynchid, as initially identified, though it later had been put into its own family, Steropodontidae (Archer et al. 1985; Flannery et al. 1995a; Musser 2003, 2006). A subsequent analysis of morphological and genetic data suggested *Teinolophos* and *Steropodon* actually predate the divergence of Ornithorhynchidae from Tachyglossidae, which did not take place until the Tertiary, 48–19 million years ago (Phillips, Bennett, and Lee 2009).

The next oldest known species in the family is the early Paleocene *Monotrematum sudamericanum*, which was discovered in southern Argentina, thereby providing evidence of a former land bridge that connected Australia, Antarctica, and South America and allowed dispersal of early mammals (Pascual et al. 1992, 2002). Subsequent ornithorhynchids are all Australian, the oldest being *Obdurodon insignis* and an undescribed species of *Obdurodon* of the late Oligocene of South Australia, and *Obdurodon dicksoni* of the early to middle Miocene of northwestern Queensland;

a well-preserved skull of the latter features a prominent bill and functional molar and premolar teeth (Archer et al. 1992, 1993; Musser 2003, 2006). The genus *Ornithorhynchus* has been reported from the Pliocene of New South Wales (Musser 1998, 2006). *Ornithorhynchus anatinus* is known from the Pleistocene and Recent of mainland Australia and Tasmania (Archer, Plane, and Pledge 1978; Musser 2003, 2006).

As related by Moyal (2004), platypus specimens began arriving in England in 1799 and shortly thereafter in Germany and France. In an initial popular publication, Thomas Bewick acknowledged inability to classify the creature as fish, bird, or mammal. Naturalist George Shaw wondered whether his dried specimen might have been cleverly stitched together from parts of various animals. He assigned the specific term *anatinus* or "ducklike," and Johann Blumenbach established the generic name *Ornithorhynchus* or "bird beak." Etienne Geoffroy Saint Hilaire placed platypus and echidna in a new group that he called "Monotremata," but believed he named a class intermediate to other vertebrates, not an order of mammals. There followed fervent debate on taxonomy and reproductive mode, and not until 1884 did the work of William Caldwell bring general recognition that the platypus is an egg-laying mammal. Nonetheless, controversy as to its position relative to class Mammalia (see account thereof) has continued to this day. A recent analysis led Warren et al. (2008) to conclude that the striking blend of reptilian and mammalian features, noted since the initial description of the platypus, penetrates to the level of its genome sequence.

Order Didelphimorphia

American Opossums

This marsupial order, containing the single Recent family Didelphidae with 18 genera and 112 species, is found naturally from southeastern Canada through the eastern and central United States and Mexico into South America to about 47° S in Argentina. Opossums also occur, possibly through introduction, on some islands in the Lesser Antilles, and introduced populations occupy much of the West Coast of North America. The vernacular name "opossum" generally applies to members of this order, which now is found only in the New World. The term "possum" is used mainly for certain marsupials in the Australasian region.

Dickman (2005), Gardner (2005a, 2007b), and McKenna and Bell (1997) used "Didelphimorphia" as the name of this order, considered Didelphidae the only full family therein, and accepted the subfamilies Caluromyinae for the genera *Glironia*, *Caluromys*, and *Caluromysiops*, and Didelphinae for all other genera. Voss and Jansa (2009) did the same with respect to order and family, but erected two new subfamilies, Glironiinae for *Glironia*, and Hyladelphinae for *Hyladelphys* (formerly placed in Didelphinae). Szalay (1982, 1993, 1994, 1999) used the name "Didelphida" for the order and included therein what is here regarded as the order Paucituberculata, as well as Didelphimorphia, which he reduced to subordinal rank, initially with the name "Didelphiformes." In contrast, Hershkovitz (1992a,

Short-tailed opossum (*Monodelphis brevicaudata*). Photo by Rexford D. Lord.

1992b, 1999), who also used the name Didelphida for the order, regarded Didelphimorphia as a cohort comprising all marsupials except the cohort Microbiotheriomorphia with the single order Microbiotheria (see account thereof). Hershkovitz and some other authorities (e.g., Palma 2003; Palma and Spotorno 1999) have treated Caluromyinae as a distinct family.

Voss and Jansa (2009) divided the subfamily Didelphinae into four tribes: Marmosini for the genera *Monodelphis*, *Tlacuatzin*, and *Marmosa*; Metachirini for *Metachirus*; Didelphini for *Chironectes*, *Lutreolina*, *Philander*, and *Didelphis*; and Thylamyini for *Marmosops*,

Chacodelphys, Cryptonanus, Gracilinanus, Lestodelphys, and *Thylamys.* Voss and Jansa's (2009) sequence of genera was presented alphabetically within tribes. Here, an attempt has been made to arrange the genera systematically, grouping those that are more closely related. A remarkable series of recent studies, using both morphological and molecular methodology, has contributed immensely to our understanding of didelphine relationships (Gruber, Voss, and Jansa 2007; Jansa, Forsman, and Voss 2006; Jansa and Voss 2000, 2005; Voss, Gardner, and Jansa 2004; Voss and Jansa 2003, 2009; Voss, Lunde, and Jansa 2005; Voss, Lunde, and Simmons 2001; Voss, Tarifa, and Yensen 2004). That work has indicated the new genus *Hyladelphys* to be intermediate to the subfamilies Caluromyinae and Didelphinae and best considered to represent its own subfamily, which is basal to Didelphinae. Also, the new genus *Tlacuatzin* evidently is associated with a group comprising *Marmosa* and *Monodelphis. Micoureus,* which here is listed as a subgenus of *Marmosa,* following Voss and Jansa (2009), sometimes has been treated as a separate genus. In the latter case, however, *Micoureus* would be nested within a paraphyletic *Marmosa* (see account thereof). *Metachirus* is associated with a group comprising *Philander, Didelphis, Lutreolina,* and *Chironectes; Philander* and *Didelphis* have close affinity to one another, somewhat less to *Lutreolina,* and still less to *Chironectes.* The new genus *Chacodelphys* apparently is closest to a group comprising *Lestodelphys*

Brown "four-eyed" opossum (*Metachirus nudicaudatus*). Photo by Louise H. Emmons.

and *Thylamys,* which are immediately related to one another and which in turn form a grouping with *Gracilinanus* and the new genus *Cryptonanus.* That grouping has affinity to *Marmosops.* Voss and Jansa (2009) pointed out that the relationships of *Tlacuatzin, Chacodelphys,* and *Cryptonanus* have not yet been conclusively resolved.

The recent studies listed above also support the affinity of *Caluromys* and *Caluromysiops* and suggest that *Glironia* is basal to the rest of the family Didelphidae. Those studies are reflected in the generic sequence presented here, as is Gardner's (2007a) tribal content for the Didelphinae and the interrelationships discussed by Hershkovitz (1992a, 1992b, 1997) and Palma (2003).

Hershkovitz (1992a, 1992b, 1997) had divided Didelphidae into four separate families (within a superfamily, the Didelphoidea), 10 subfamilies, and 16 genera, which, in his sequence, are: Marmosidae, with the subfamilies Marmosinae (for the genera *Gracilinanus, Marmosops,* and *Marmosa*), Thylamyinae (for *Thylamys*), Lestodelphyinae (for *Lestodelphys*), Metachirinae (for *Metachirus*), and Monodelphinae (for *Monodelphis*); Caluromyidae, with the subfamilies Caluromysinae (for *Caluromys*) and Caluromysiopsinae (for *Caluromysiops*); Glironiidae (for *Glironia*); and Didelphidae, with the subfamilies Didelphinae (for *Philander* and *Didelphis*), Chironectinae (for *Chironectes*), and Lutreolinae (for *Lutreolina*). The genera *Hyladelphys, Tlacuatzin, Chacodelphys,* and *Cryptonanus* had not been described at the time of Hershkovitz's work. Although his familial and subfamilial divisions have not been generally accepted for taxonomic purposes, his characterizations of the different generic groupings are utilized in part in this ordinal description.

Didelphimorphians are small to medium in size. Adult head and body length is about 68–500 mm, tail length is 45–535 mm, and weight is 10–7,000 grams. The tail of most genera is long, scaly, very scantily haired, and prehensile, but in some forms it is short and/or rather hairy. The tail is not prehensile or is only slightly so in *Lestodelphys, Metachirus, Monodelphis,* and *Lutreolina.* Some genera have long, projecting guard hairs in the pelage. The muzzle is elongate, and the ears have well-developed conches. The limbs are short in many species, with the hind limbs slightly longer. All four feet have five separate digits. The great toe is large, clawless, and opposable. Hershkovitz (1992b) reported that

Philander, Didelphis, Chironectes, and *Lutreolina* have stouter claws than do most other genera in what is here considered the subfamily Didelphinae, and that the claws are recurved and extend well beyond the digital tips (though not on the forefoot of *Lutreolina*). The claws are also stout in *Lestodelphys* and *Thylamys*. A distinct marsupial pouch is present in *Philander, Didelphis, Lutreolina, Chironectes, Caluromys* (*Mallodelphys*) *derbianus, C.* (*M.*) *lanatus*, and *Caluromysiops* (Gardner 2007b). In the other genera it is absent, or it may consist of only two longitudinal folds of skin, separate at both ends, near the median line of the body. Females have 4–27 mammae.

The facial part of the skull is long and pointed, but the cranial part is small. According to Hershkovitz (1992b), *Philander, Didelphis, Lutreolina, Chironectes, Caluromys*, and *Caluromysiops* generally have a more massive skull than do other didelphimorphians, with more prominent postorbital processes and sagittal crest. The skull of the first four of those genera has large maxillary-palatine vacuities, whereas in *Caluromys* and *Caluromysiops* the bony palate is nearly entirely ossified. Hershkovitz (1992b) also distinguished *Caluromys* and *Caluromysiops* by their auditory or tympanic bullae, the dome-shaped housings for the inner ears at the posterior base of the skull. In those two genera the bullae are bipartite, with the floor formed by the close junction (but not fusion) of the alisphenoid and petrous bones, thereby enclosing the ectotympanic bone. In all other living didelphimorphians, except *Thylamys*, the bullae are tripartite, with the ectotympanic forming a portion of the floor and there being a gap between the alisphenoid and petrous bones. The skull of *Thylamys* also is characterized by parallel-sided nasal bones. *Glironia* has tripartite bullae, lacks a sagittal crest, and has midpalatal and posterolateral vacuities. A sagittal crest is present in *Lestodelphys* and *Monodelphis*.

The dental formula for the order is: i 5/4, c 1/1, pm 3/3, m 4/4 = 50. The teeth are rooted and sharp. The upper incisors are conical, small, and unequal; the first is larger than, and separated from, the others. The canines are large, and the molars are tricuspidate. The third upper and lower deciduous premolars (the only marsupial teeth to have a deciduous stage) are multicuspidate and molariform, except in *Hyladelphys*, in which those teeth are nonmolariform and much

Virginia opossum (*Didelphis virginiana*). Photo by Cynthia Kidwell / Shutter stock.com.

smaller than in other didelphimorphians (Voss, Lunde, and Simmons 2001).

According to Hershkovitz (1992a), a critical distinction expressing the separate evolution of *Caluromys* and *Caluromysiops* is found in the anklebones—the astragalus and calcaneus. In *Philander, Didelphis, Chironectes*, and *Lutreolina*, this joint has a primitive pattern in which two separate facets on the calcaneal surface articulate with a corresponding pair of separate facets on the astragalar plantar surface. *Caluromys* and *Caluromysiops*, however, have a partially derived pattern in which there has been coalescence of the once dual facets of the calcaneus into a single continuous facet. Hershkovitz (1992b) noted that the anklebones are also specialized in *Gracilinanus, Marmosops, Marmosa*, and *Lestodelphys*, but are unspecialized in *Thylamys, Metachirus*, and *Monodelphis*. Reig, Kirsch, and Marshall (1987) gave extensive further details on the morphological characters of Didelphimorphia. Gardner (2007b) provided a key to the subfamilies, tribes, and most genera.

Opossums are active mainly in the evening and at night. Most forms are arboreal or terrestrial, and one genus (*Chironectes*) is semiaquatic. Opossums are insectivorous, carnivorous, or, more commonly, omnivorous.

Some people once believed that opossums copulate through the nose and that sometime later the tiny young are blown into the pouch. The idea may have originated with the observation that the pouch is often filled with newborn young soon after the female has

Slender mouse opossum (*Marmosops parvidens*). Photo by T. B. F. Semedo / Mammal Images Library, American Society of Mammalogists.

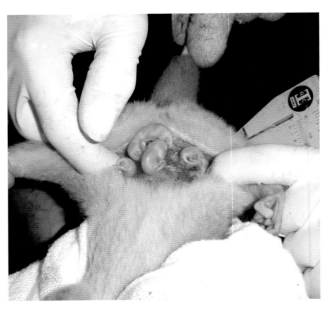

Woolly opossum (*Caluromys philander*), pouch young. Photo by Paula Ferreira through Diogo Loretto.

investigated the pouch with her nose. Or perhaps the tale stems from seeing the male's forked penis.

As in other marsupials, the gestation period is short, and the developmental period is long. The gestation period of *Didelphis*, about 12.5–13 days, is among the shortest found in mammals. The front legs of a newborn opossum are well developed, and the toes thereof are supplied with sharp, deciduous claws (at least in *Didelphis*); those claws drop off some time after the young reaches the pouch. The hind legs of opossums at birth are much smaller than the forelegs and practically useless. In young opossums the passage from the nasal chamber to the larynx is so separated from the passage to the esophagus that the baby can swallow and breathe at the same time. The pouch young of opossums and probably other pouch marsupials breathe and rebreathe air that contains 8–20 times the normal content of carbon dioxide; that may serve some unknown, useful role or may merely show unusual tolerance.

Contrary to an earlier opinion, the mother does not pump milk into the young; they suckle normally, though continuously, under their own power (Banfield 1974; Lowery 1974). The part of the brain that regulates body temperature is not functional in a young opossum; the baby is kept warm solely by the mother's body heat. After leaving the pouch the young of some species travel with the mother for awhile, usually riding by clinging to the fur on her back.

A number of carnivores prey on opossums, and people eat some of the larger species. Didelphid pelts are sometimes used for inexpensive trimmings and garments. Opossums, particularly *Monodelphis* and

Didelphis, are widely used in laboratory research. South American didelphids are susceptible to yellow fever virus.

The known geological range of this order is early Cretaceous to middle Miocene and middle Pleistocene to Recent in North America, late Cretaceous to Recent in South America, Eocene to Miocene in Europe, early Eocene and early Oligocene in North Africa, early Oligocene in Central Asia, middle or late Eocene in East Asia, early or middle Eocene to early Oligocene in Southwest Asia, and middle Miocene in Southeast Asia (McKenna and Bell 1997). Hershkovitz (1995) described the genus *Adinodon* from the latter part of the early Cretaceous of Texas; it is the oldest known member of the Didelphimorphia and among the earliest of marsupials. He thought it apparently a component of the family Marmosidae, which is here considered part of the family Didelphidae, but McKenna and Bell (1997) placed *Adinodon* basal to the Didelphidae.

DIDELPHIMORPHIA; DIDELPHIDAE; **Genus *GLIRONIA* Thomas, 1912**

Bushy-tailed Opossum

The single species, *Glironia venusta*, has been recorded from the Amazon lowlands of extreme southwestern

Colombia, eastern Ecuador, eastern Peru, northern and western Brazil, and northern Bolivia; fewer than 30 specimens are known (Ardente et al. 2013; Barkley 2007; Brown 2004; Calzada et al. 2008; Rossi et al. 2010; Santos-Filho et al. 2007). Gardner (2005a, 2007b) placed the genus in the subfamily Caluromyinae of the family Didelphidae (see account thereof), Voss and Jansa (2009) recognized *Glironia* as representing the subfamily Glironiinae of family Didelphidae, and Hershkovitz (1992a, 1992b) put both *Glironia* and the Caluromyinae in full and separate families.

Head and body length is 160–225 mm, and tail length is 195–226 mm. Rossi et al. (2010) gave weight as approximately 140 grams, and Ardente et al. (2013) reported that an adult male weighed 119 grams. The upper parts are fawn-colored or cinnamon brown; a dark brown to black stripe extends through each eye and gives the appearance of a mask; the tail is tipped with white or has only a sprinkle of white hairs, and the underparts are gray or buffy white. The texture of the fur varies from soft and velvety to dense and woolly.

This genus is much like *Marmosa* in general appearance, but the tail is well furred and bushy to the tip. The extent of the naked area on the tail of *Glironia* varies, but there is at least a trace of a ventral naked area on the terminal few centimeters. Voss and Jansa (2009) also diagnosed subfamily Glironiinae by its strongly recurved and laterally compressed manual claws, and by postorbital processes formed by the frontals and parietal (in other opossums postorbital processes are absent or are formed only by the frontals). Some other characters, as used by Hershkovitz (1992b) to distinguish *Glironia*, are given above in the account of the family Didelphidae. Astúa (2015) noted that females lack a pouch and have four mammae, two on each side; unlike most other opossums, there is no medial mamma.

Four of the known specimens of *Glironia* were collected by commercial animal dealers, and were taken in heavy, humid tropical forests. *Glironia* is presumed to be arboreal because of its large, opposable hallux (Marshall 1978c), though some specimens have been collected in traps on the ground or in the understory (Rossi et al. 2010). Characters of the postcranial skeleton are consistent with efficient and rapid arboreal locomotion (Flores and Diaz 2009). The diet is unknown but probably is comparable to that of *Marmosa*—insects, eggs, seeds, and fruits. Individuals have been observed licking exudates from tree branches (Astúa 2015).

Emmons (1997) reported seeing an individual at night in dense vegetation about 15 meters above the ground. It ran about the vines quickly, often jumping from one to another in a manner unlike that of other opossums, and seemed to be hunting insects. Another individual, seen in a dry forest, emerged from an eight-meter-high tree hole at nightfall and dashed up into the canopy. Da Silveira, de Melo, and Lima (2014) reported capture of an adult female and three cubs after the fall of a Brazil nut tree (*Bertholletia excelsa*) on 21 December 2011, and observation of another adult female with three cubs, just a few months old, foraging in a tree at the edge of a wooded area at twilight on 18 July 2012. Those records suggest that breeding is continuous or

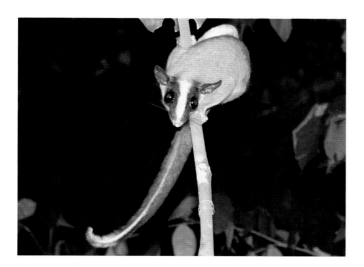 

Bushy tailed opossum (*Glironia venusta*); *left*, wild individual; *right*, captive mother and young. Photos by Jonatha Lima and Fabiano R. de Melo.

that there is more than one season per year. Although losing habitat to deforestation and apparently extremely rare, *Glironia* is thought to be widespread and to have a large global population; it no longer is classified as vulnerable by the IUCN (Patterson and Solari 2008).

DIDELPHIMORPHIA; DIDELPHIDAE; Genus *CALUROMYS* J. A. Allen, 1900

Woolly Opossums

There are two subgenera and four species (Brito et al. 2015; Brown 2004; Gardner 2007b; López-Fuster, Pérez-Hernández, and Ventura 2008):

subgenus *Caluromys* J. A. Allen, 1900

C. philander, the Guianas, northern and central Brazil, with the apparently isolated subspecies *C. p. dichurus* in southeastern Brazil;

C. trinitatis, Venezuela, including Margarita Island, Trinidad.

subgenus *Mallodelphys* Thomas, 1920

C. derbianus, southern Mexico to Colombia and Ecuador west of the Andes;

C. lanatus, northern and central Colombia, northwestern and southern Venezuela, eastern Ecuador and Peru, northern and eastern Bolivia, western and southern Brazil, southern Paraguay, extreme northeastern Argentina (Misiones).

Gardner (2005a, 2007b) treated *C. trinitatis*, originally described from Trinidad, as a subspecies of *C. philander* and assigned thereto *venezuelae* of northern Venezuela, including Margarita Island. Employing cranial morphometrics, López-Fuster, Pérez-Hernández, and Ventura (2008) distinguished *C. trinitatis*, *venezuelae* included, as a full species. The form *leucurus* of southern and eastern Venezuela, also placed in the synonymy

Woolly opossum (*Caluromys philander*). Photo by Rexford D. Lord.

of *C. philander* by Gardner (2007b), was considered by López-Fuster, Pérez-Hernández, and Ventura (2008) to group closely with *trinitatis* and *venezuelae*, based on skull shape. Astúa (2015) indicated that *C. p. dichurus* of southeastern Brazil shows little or no morphometric overlap with other forms of *C. philander* and may represent a separate taxon.

Head and body length is 160–300 mm, tail length is 250–450 mm, and weight (Bucher and Fritz 1977) is 200–500 grams. The pelage is long, fine, and woolly. Some forms are pale gray or otherwise not well marked, but woolly opossums usually have an ornate color pattern, including the diagnostic dark median stripe on the face, which extends between the ears and the eyes almost to the nose. Some individuals show a pale stripe extending backward from the shoulder region. The color pattern may also include indistinct dark patches around the eyes, a light creamy white to buffy white face, and a reddish and blackish body.

According to Gardner (2007b), these opossums are characterized by the striped face, woolly pelage, and large, naked ears. The tail is prehensile and longer than the head and body. In the subgenus *Caluromys* the tail is furred on only the basal 10–20 percent, and the naked portion is usually brown or brown mottled with paler markings, while in the subgenus *Mallodelphys* the tail is furred dorsally on the basal 30–70 percent and ventrally on the basal 20–40 percent, and the naked portion is usually predominantly white (but see a differing assessment by Cáceres and Carmignotto 2006). In the subgenus *Caluromys* the marsupial pouch consists of only two lateral abdominal folds, while in *Mallodelphys* the pouch is well developed.

Woolly opossums generally inhabit forested country and are more arboreal than the other large opossums. Nearly all specimens of *C. lanatus* and *C. philander* collected by Handley (1976) in Venezuela were taken in trees, usually near streams or other moist areas. These animals live in tree hollows or limbs and are active mainly during the evening, night, or early morning. Nests of *C. lanatus* have been found at heights of 12 meters, and *C. philander* is usually found at heights

Woolly opossum (*Caluromys philander*), female with young. Photo by Rexford D. Lord.

of 5–12 meters in the canopy (Astúa 2015). Woolly opossums are quite agile, presenting a sharp contrast to *Didelphis*. A nest of leaves made by *C. derbianus* was found in a vine tangle in a small tree. That species, like *Didelphis*, reportedly coils its tail to carry nesting materials (Hunsaker and Shupe 1977). Individual *C. philander* travel 500–1,000 meters per night (Astúa 2015). Woolly opossums are apparently fairly common throughout their wide range, but their numbers never seem to reach the population levels of *Philander* and *Didelphis*. Perhaps their arboreal habits make them less conspicuous than terrestrial didelphids. Reported population densities are 50–200 individuals per sq km for *C. philander* (Astúa 2015) and 13.3 per sq km for *C. lanatus* (Peres 1999).

Woolly opossums are omnivorous. Their diet in the wild consists of a variety of fruits, seeds, leaves, soft vegetables, insects, and small vertebrates. They reportedly feed on carrion as well. One captive group had a decided preference for meat (Collins 1973), while another colony had a special liking for fruit (Bucher and Fritz 1977). Cáceres and Carmignotto (2006) noted that *C. lanatus* is a nectar consumer and pollinator.

Information on *C. philander* compiled by Astúa (2015), Eisenberg (1989), and Eisenberg and Redford (1999) indicate that the nightly foraging area varies from 0.3 to 1.0 ha., depending on food availability. Overall estimated home range in southeastern Brazil is 2.5–7.0 ha. Individuals are solitary and adult interaction tends to be agonistic, but there seems to be no strong territorial defense and home ranges overlap. Vocalizations include agonistic hisses and grunts, clicks by suckling young, and a distress scream. Females can produce three litters a year but probably will not rear more than one litter if food is scarce. Breeding occurs from around October–December in southeastern Brazil but throughout the year in French Guiana; mean litter size in the latter country is 4.1 young. Studies of captive *C. philander* by Atramentowicz (1992) indicate that gestation lasts about 25 days, apparently the longest in the Didelphimorphia; litter size is 1–7; the young leave the pouch at three months and are weaned at four months. Sexual maturity is attained at about 10 months of age (Astúa 2015).

Captive females of both *C. derbianus* and *C. lanatus* have modal estrous cycle lengths of 27–29 days and are cyclic throughout the year (Bucher and Fritz 1977).

In the wild in Nicaragua *C. derbianus* also seems to be reproductively active all year (Phillips and Jones 1968). Litter size there averaged 3.3 (2–4). In Panama the breeding period reportedly begins with the onset of the dry season in February (Enders 1966), and litter size averages between three and four (with a range of one to six). Sexual maturity is said to be attained by *C. derbianus* at seven to nine months. Collections of wild *C. lanatus* indicate that there is reproductive activity throughout the year and that there are usually three pouch young (Gardner 2007b), though Cáceres and Carmignotto (2006) reported litter size to be just one or two in the Amazon Basin and three or four farther south. A number of captive *Caluromys* have survived for more than 3 years (Collins 1973); known record longevity, by a specimen of *C. lanatus* in the Toronto Zoo, is 9 years and 9 months (Weigl 2005).

All species are declining because of habitat destruction, but are still common over vast regions and have not been placed in a threatened category by the IUCN (Brito et al. 2015; Costa et al. 2015; Solari and Lew 2015). In the past woolly opossums were trapped extensively for their pelts, but at present their fur is not popular. They are said to damage fruit crops occasionally in South America, but otherwise they are of little economic importance. Recently they have been shown to have potential use in laboratory research (Bucher and Fritz 1977).

DIDELPHIMORPHIA; DIDELPHIDAE; **Genus CALUROMYSIOPS** Sanborn, 1951

Black-shouldered Opossum

The single species, *C. irrupta*, was for a long time known with certainty from only three localities in southern Amazonian Peru and one on the upper Jarú River in west-central Brazil; a reported specimen from extreme southern Colombia probably was a transported captive (Brown 2004). However, Emmons (2007) reported that three specimens now have been collected in extreme northern Peru, and Astúa (2015) indicated the range extends to western Brazil and probably northern Bolivia. The generic distinction of *Caluromysiops* from *Caluromys* has been questioned (Gewalt 1990; Izor and Pine 1987), but Hershkovitz

Black-shouldered opossum (*Caluromysiops irrupta*). Photo by Marcio Martins.

(1992b) put the two in separate subfamilies based in part on differences in the molar teeth (see also account of family Didelphidae).

Head and body length is 250–330 mm, tail length is 272–340 mm, and a wild-caught male weighed 445 grams (Emmons 2007). The upper parts are gray, with two separate black lines that begin on the forefeet and run onto the back, join on the shoulders, then separate again and run parallel to each other down the back and over the rump to the hind limbs. The face has faint, dusky lines running through the eyes. The underparts are gray with buffy tips to the hairs, and there is a faint dusky line along the middle of the belly. The upper side of the tail for the basal two-thirds to three-quarters of its length is slightly darker than the gray of the body; the remainder of the tail is creamy white. The tail is well furred, except the underside of the last three-fourths, which is naked. The fur is long, dense, and woolly. The skull is similar to that of *Caluromys*, but the molars are much larger and the rostrum is relatively shorter.

The black-shouldered opossum is thought to inhabit mature humid forests; all specimens have been taken at elevations below 700 meters (Astúa 2015). The genus is nocturnal and arboreal in habit and probably has a diet like that of *Caluromys* (Hunsaker 1977). It has been reported to use the upper levels of the forest, rarely descending even to the middle levels, and to move slowly and spend hours in the same flowering tree, feeding periodically on nectar (Emmons 1997). Females with one or two young have been received from animal dealers in July–August (Astúa 2015). The few specimens maintained in captivity thus far have reportedly been hardy in comparison with other didelphids; one survived for 7 years and 10 months (Collins 1973). *Caluromysiops* is generally rare and some populations are declining because of deforestation, but it is thought to occur over an extensive region and is no longer classified as vulnerable by the IUCN (Solari and Cáceres 2015).

Kalinowski's mouse opossum (*Hyladelphys kalinowskii*). Photo by Christian Marty through Francois Catzeflis.

DIDELPHIMORPHIA; DIDELPHIDAE; **Genus**
HYLADELPHYS Voss, Lunde, and Simmons, 2001

Kalinowski's Mouse Opossum

The single known living species, *H. kalinowskii*, occurs in northern French Guiana, southern Guyana, the vicinity of Manaus in Amazonian Brazil, and eastern Peru, and possibly in southeastern Colombia and northern Bolivia (Astúa 2006, 2015; Gardner 2007d). The species was first described by Hershkovitz (1992b) as *Gracilinanus kalinowskii*, but its distinctive morphology led Voss, Lunde, and Simmons (2001) to place it in its own new genus and even to question whether it is a member of the family Didelphidae. Subsequently, Jansa and Voss (2005) used more detailed morphological assessment and molecular analysis to demonstrate that *Hyladelphys* is a didelphid, possibly intermediate to the subfamilies Caluromyinae and Didelphinae but probably near the base of the didelphine radiation, and Voss and Jansa (2009) erected the new subfamily Hyladelphinae for the genus. Those authorities also sug-

gested that while fewer than a dozen individuals were known, further study might show as many as three species represented by that material. Hershkovitz (1992b) had reported the presence of an undescribed species in northwestern Ecuador, perhaps with some affinity to *Gracilinanus kalinowskii*, but Voss, Lunde, and Simmons (2001) examined the relevant specimen and referred it to *Marmosa*, indicating that it most closely resembles material assigned to *M. robinsoni*.

Oliveira et al. (2011) described *Sairadelphys tocantinensis* from fossils found in a cave in the central part of the state of Tocantins in east-central Brazil. They assigned the species to subfamily Hyladelphinae and observed that, while the remains are likely to be late Pleistocene in age, "there is a chance that the newly described taxon is still a living marsupial in the study area."

Head and body length of *Hyladelphys* is 67–91 mm, tail length is 96–117 mm, and weight is 8–18 grams. The upper parts are unpatterned dull reddish brown with dark gray hair bases; the underparts are pure white or cream. The pelage is smooth, not woolly. The face is boldly marked by a broad mask of black fur and by a prominent median streak of very pale orange extending from between the eyes to the rhinarium. The cheeks and throat are white. The eyes are large and the ears very large and apparently naked. The tail is longer than the head and body, slender, brownish, and apparently naked. The skull has a distinctively short rostrum, no postorbital frontal processes, a laterally inflated braincase, and very large orbits. There is no premaxillary rostral process, the palatal process of the premaxilla contacts the alveolus of the upper canine tooth on each side, and the nasals are long and conspicuously widened posteriorly. The alisphenoid bulla is well developed, smoothly globular, and without a well-developed anteromedial process or lamina.

Hyladelphys differs from all other didelphimorphians, and resembles the members of other orders of marsupials, in having an even-numbered mammary count. There are two pairs of inguinal-abdominal mammae but no unpaired median nipple. There is no pouch. *Hyladelphys* also differs from all other didelphids in having vestigial deciduous third upper and lower premolars (the only marsupial teeth to have a deciduous stage). The milk teeth are much smaller than those of other didelphids and uniquely nonmolariform in occlusal structure—again more like the condition seen in some Old World marsupial groups.

Almost nothing is known of the natural history of this new genus. Voss, Lunde, and Simmons (2001) reported taking two specimens on the ground near a small stream in well-drained primary forest in the coastal lowlands of French Guiana, and a third as it perched on a palm frond about 1 meter above the ground in swampy primary forest at 1835 hours. Two specimens were collected in Peru at elevations of 890 and 1,100 meters, respectively. Astúa (2006) referred to the apparent rarity of the genus, which probably is related to its elusiveness, diminutive size, and arboreal habits, though one specimen was hand-caught on the ground. Emmons (1997) indicated that it probably is mostly terrestrial and is nocturnal.

DIDELPHIMORPHIA; DIDELPHIDAE; **Genus *MONODELPHIS* Burnett, 1830**

Short-tailed Opossums

There are 26 named and valid species (Caramaschi et al. 2011; Carvalho et al. 2011; de la Sancha, Solari, and Owen 2007; Eisenberg and Redford 1999; Gettinger et al. 2011; Kirsch and Calaby 1977; Lemos, Weksler, and Bonvicino 2000; Lim et al. 2010; Pavan, Rossi, and Schneider 2012; Pine, Flores, and Bauer 2013; Pine and Handley 2007; Solari 2004, 2007; Solari et al. 2012; Ventura et al. 2005; Voss, Lunde, and Simmons 2001; Voss, Pine, and Solari 2012):

M. brevicaudata, Venezuela south of the Orinoco, northern Guyana, Brazil north of the Negro and west of the Branco River;

M. arlindoi, central and southern Guyana and adjacent Brazil north of the Amazon River;

M. touan, from French Guiana and the Brazilian state of Amapá south to the state of Pará south of the Amazon and east of the Xingu River (including Marajó Island);

M. glirina, west-central Brazil, southwestern Peru, northern Bolivia;

M. palliolata, northeastern Colombia, northern Venezuela;

M. sanctaerosae, known only from the type locality in the province of Santa Cruz in east-central Bolivia;

M. domestica, eastern and central Brazil, eastern Bolivia, northern Paraguay, extreme northern Argentina;

M. maraxina, Isla Marajó at the mouth of the Amazon in northeastern Brazil;

M. adusta, eastern Panama, lower to middle elevations in the Andes from northwestern Venezuela to northern Peru;

M. reigi, highlands of southeastern Venezuela and west-central Guyana;

M. peruviana, lower to middle elevations in the Andes of southern Peru and western Bolivia;

M. ronaldi, known only from the type locality in southeastern Peru;

M. handleyi, known only from the type locality in northeastern Peru;

M. osgoodi, southern Peru, western Bolivia;

M. kunsi, central and southwestern Brazil, eastern Bolivia, Paraguay, extreme northern Argentina;

M. emiliae, Amazonian Brazil, eastern Peru, northern Bolivia;

M. americana, Brazil east of 50° W;

M. iheringi, coastal region of southeastern Brazil;

M. unistriata, known only by one specimen from extreme southeastern Brazil and one from extreme northeastern Argentina (Misiones);

M. umbristriata, southeastern Brazil;

M. theresa, states of Rio de Janeiro and Minas Gerais in eastern Brazil;

M. gardneri, montane forests on eastern slopes of Andes in central and southern Peru;

M. rubida, known only from the type locality on the coast of Bahia in eastern Brazil;

M. scalops, coastal region of southeastern Brazil;

M. sorex, southern Brazil, southeastern Paraguay, northeastern Argentina;

M. dimidiata, extreme southeastern Brazil, Uruguay, northern Argentina.

According to Pine and Handley (2007), the subgenus *Minuania* Cabrera 1919, which sometimes has been used for *M. dimidiata* and a few related species, is not valid with respect to either nomenclature or phylogenetics. The diversity within *Monodelphis*, as shown by morphological and DNA analysis, does justify subgeneric division, but affinities between species currently are insufficiently understood to allow such arrangement. The genus as a whole has not been systematically revised since 1888. The sequence of species shown here is based on information taken from the sources cited above, suggesting: that *brevicaudata, arlindoi, touan, glirina, palliolata,* and *sanctaerosae* form a related group; that *domestica* and *maraxina* are immediately related and form a sister unit to the preceding group; that close affinity also is shown by *adusta, reigi, peruviana, ronaldi, handleyi,* and *osgoodi,* which in turn may form a sister group to *kunsi*; that *emiliae* may be the sister to all the previously mentioned species; that another grouping comprises *americana, iheringi, unistriata, umbristriata,* and *theresa,* and possibly *gardneri, rubida* and *scalops*; and that *sorex* and *dimidiata* have some affinity to one another but less to the other species listed. Molecular and morphometric

Short-tailed opossum (*Monodelphis glirina*). Photo by T. B. F. Semedo / Mammal Images Library, American Society of Mammalogists.

Short-tailed opossum (*Monodelphis emiliae*). Photo by T. B. F. Semedo / Mammal Images Library, American Society of Mammalogists.

analyses by Vilela, Russo, and Alves de Oliveria (2010) suggested an immediate relationship of *sorex* and *dimidiata*, even that the former is a synonym of the latter, and supported their distinction from the other species of *Monodelphis*.

Pine and Handley (2007) recorded the following three valid but unnamed species: "A," affinity to *brevicaudata*, north-central Venezuela (this is the species reported, incorrectly, as *M. orinoci* by Ventura, Perez-Hernandez, and Lopez-Fuster 1998, and several other authors); "D," possible affinity to *brevicaudata*, south of the Amazon and east of the Rio Xingu in eastern Brazil; "E," externally similar to *osgoodi*, known from a single locality near Rio de Janeiro. Pine and Handley (2007) had noted two other species: "B," possibly related to *domestica*, in east-central Bolivia, now described as *M. sanctaerosae* (Voss, Pine, and Solari 2012), and "C," most similar to *theresa*, on the eastern slope of the Andes in central Peru, now described as *M. gardneri* (Solari et al. 2012). Pine and Handley

(2007) also noted the existence of at least four other unnamed species. Caramaschi et al. (2011) identified two genetic clades of *M. domestica* in Brazil, which could represent different species.

Head and body length is 70–200 mm and tail length is 45–106 mm. Fadem and Rayve (1985) listed the weight of *M. domestica* as 90–150 grams in males and 80–100 grams in females. Other recorded weights are 46–150 grams for *M. brevicaudata*, 52–60 grams for *M. emiliae*, 15–35 grams for *M. adusta*, 23–35 grams for *M. americana*, and 40–84 grams for *M. dimidiata* (Eisenberg 1989; Emmons 1997; Redford and Eisenberg 1992). The tail is about half as long as the head and body, always shorter than the body alone, and that character immediately distinguishes *Monodelphis* from other opossums in its range (Eisenberg and Redford 1999). In most forms the tail is sparsely haired, with only a few millimeters at the base well furred, but in *M. brevicaudata* the pelage extends onto the top of the tail for a third or more of its length (Voss, Lunde, and

Simmons 2001). The tail is prehensile to a modest extent, at least in some species, but is not incrassate. No pouch is present. There are 11 to 17 or more mammae, depending on the species; most are arranged in a circle on the abdomen, but there also are some pectoral mammae (Hershkovitz 1992b). *M. sorex* has up to 27 mammae (Pine and Handley 2007), the largest number known in any mammal except *Tenrec ecaudatus* of Madagascar, which has up to 34 (Goodman, Ganzhorn, and Rakotondravony 2003). The fur is short, dense, and rather stiff.

Coloration varies greatly by species and sometimes within species, and pelage patterns do not necessarily conform to systematic affinities indicated by other characters. *M. brevicaudata* has both a bicolored form, with dark reddish upper parts and sides and abruptly paler underparts, and a tricolored form, with a broad brownish, grayish, or almost black middorsal stripe, distinctly reddish flanks from nose to rump, and ashy to buffy gray underparts. *M. glirina* and *M. palliolata*, which are related to *M. brevicaudata*, as well as *M.*

sorex and *M. dimidiata*, which are probably not, have reddish or orangish flanks that contrast with the dorsal color but not strikingly with the underparts. *M. sorex* also features a dark brown back, gray forequarters and head, and yellowish cheeks. In *M. domestica* the upper parts are entirely gray and the underparts are whitish or grayish; the related *M. maraxina* is distinguished by extensive orange suffusion on the flanks. In *M. emiliae* and *M. scalops* the back and anterior flanks are grizzled buff or olive-gray, but the rump and head are prominently rufous. The species *M. adusta*, *M. reigi*, *M. peruviana*, *M. ronaldi*, *M. handleyi*, *M. osgoodi*, and *M. kunsi* are generally characterized by a uniformly brownish, reddish, or grayish dorsum without contrastingly colored head, rump, or flanks. *M. rubida* has a distinctive reddish brown back; three very faint brown dorsal stripes have sometimes been reported. *M. unistrata* has a single chestnut stripe on an otherwise rusty gray back and yellowish orange flanks and underparts. *M. americana*, *M. iheringi*, and *M. theresa* are characterized by brownish to reddish upper

Short-tailed opossum (*Monodelphis domestica*). Photo by Alexander F. Meyer at Cincinnati Zoo.

parts with three blackish dorsal stripes. *M. umbristriata* has a reddish dorsum with three darker reddish stripes.

The habits of short-tailed opossums are not well known. Most species occur in forested habitat and are thought to be predominantly terrestrial and nocturnal (Emmons 1997; Pine and Handley 2007). They apparently are among the least well adapted members of the Didelphidae for arboreal life and are usually found on the ground, though they can climb fairly well. Most specimens of *M. brevicaudata* collected in Venezuela by Handley (1976) were taken on the ground in moist areas. About half were caught in evergreen forest and about half in open areas. Nests of *Monodelphis* usually are built in hollow logs, in fallen tree trunks that bridge streams, or among rocks. *M. americana* may be unusual in constructing nests in forks of trees or in bushes about 5 meters above ground (Pine and Handley 2007). The diet consists of small vertebrates, insects, carrion, seeds, and fruits. *M. americana* is said to readily kill mammals and birds nearly as big as itself (Pine and Handley 2007). *M. dimidiata* has been observed to kill mice larger than itself by biting the throat; morphometric analysis shows it to have relatively the largest canine teeth among extant marsupial carnivores and that it could be considered a living analogue of certain large extinct saber-toothed predators of the families Felidae and Nimravidae (Blanco, Jones, and Milne 2013).

M. domestica apparently occurs naturally in scrub forest and savannah but has followed humans into agricultural areas; it is both diurnal and nocturnal (Emmons 1997). It received its name because of its habit in Brazil of living in dwellings, where it destroys rodents and insects. It is welcomed by the householders. One opossum was seen running through the double cane walls of an Indian hut carrying a piece of paper by curling its tail downward around the paper. Streilein (1982a, 1982b) found that species to be an efficient predator and particularly adept at capturing scorpions. Individuals were highly intolerant of one another, though conflicts rarely resulted in serious injury. Population densities of up to 4 individuals per ha. have been found for *M. domestica* in the Caatinga region of Brazil, and home ranges there averaged about 1,209 sq meters for males and 1,788 sq meters for females (Macrini 2004).

In a study in Argentina, Pine, Dalby, and Matson (1985) found *M. dimidiata* to be diurnal, with most activity concentrated in the late afternoon. The species was considered predominantly insectivorous and not an efficient predator of rodents. Breeding occurred in the summer months of December and January, and the young dispersed from March to May. The species evidently is semelparous; that is, individuals breed but once in their life, and very few, if any, survive past their second summer. One captive female produced a litter of 16 young.

In the tropical part of their range short-tailed opossums apparently breed throughout the year. The number of young varies from 5 to 14 depending on the species, and the newborn cling to the nipples of the mother for about 2 weeks. Later the young ride on the back and flanks of the female. In *M. domestica*, according to Fadem and Rayve (1985), females have as many as four litters annually; the gestation period is 14–15 days, there are usually 5–12 offspring, postpartum dependence lasts about 50 days, sexual maturity is attained at 4–5 months, and breeding has occurred at up to 39 months of age in males and 28 months in females. The period of estrus in that species was found to be 3–12 days, and the estrous cycle showed a bimodal distribution, lasting about 2 weeks in one group of captive females and about 1 month in another group. A captive *M. domestica* lived to an age of 5 years and 1 month (Weigl 2005).

M. domestica has become an important laboratory animal since first being imported to the United States in 1978. It is the only marsupial that has been produced in captivity in very large numbers, more than 29,000 progeny having been weaned at a single colony. It now is a prototype species for basic research on marsupial biology, comparable to the role of *Mus musculus* for placental mammals, and is used extensively to study melanoma, corneal cancer, and hypercholesterolemia in humans (VandeBerg and Robinson 1997). It also serves as a model system for investigations in mechanisms of imprinting, immunogenetics, neurobiology, neoplasia, and developmental biology, and recently provided generation of the first complete genome sequence for a marsupial (Mikkelsen et al. 2007).

In the wild, many species of *Monodelphis* evidently are declining because of habitat destruction. The IUCN classifies *M. unistriata* as critically endangered; it is known only by one specimen collected before 1842 and one taken in 1899 (Flores and Teta 2011a). *M. reigi* and *M. umbristriata* are classified as vulnerable by the

Gray mouse opossum (*Tlacuatzin canescens*). Photo by Gerardo Ceballos.

IUCN; they are restricted to small areas subject to human disturbance, and the population of the latter, at least, is decreasing (Lew et al. 2011a; Weksler and Bonvicino 2008). *M. handleyi* is designated near threatened; while not rare it is known from only one locality and may be jeopardized by logging (Flores and Solari 2011).

DIDELPHIMORPHIA; DIDELPHIDAE; **Genus TLACUATZIN** Voss and Jansa, 2003

Gray Mouse Opossum

The single species, *T. canescens*, is found in western Mexico from southern Sonora to Oaxaca and Chiapas, on the Yucatan Peninsula, and on the Tres Marías Islands (Gardner 2005a; Hall 1981; Voss and Jansa 2003). A single specimen from an owl pellet has been reported for the southernmost tip of the Baja California Peninsula (Zarza, Ceballos, and Steele 2003), but that record

was not listed by Ceballos (2014). Long placed in *Marmosa, canescens* now has been found to differ widely from the other species of that genus on the basis of morphological, karyological, and molecular assessment. Voss and Jansa (2003) determined that the other species of *Marmosa* (see account thereof), including those of the subgenus *Micoureus*, form a monophyletic clade to the exclusion of *canescens* and that it thus was necessary to erect a new genus for the latter. *Tlacuatzin* still is thought to have some affinity to that clade and to the genus *Monodelphis* (Gruber, Voss, and Jansa 2007; Jansa and Voss 2005). The Yucatan subspecies, *Tlacuatzin canescens gaumeri*, is isolated from the more westerly populations and warrants study to determine if it may represent a separate species (Voss and Jansa 2003). The information for the remainder of this account was compiled from Armstrong and Jones (1971), Astúa (2015), Hall (1981), Reid (1997), Voss and Jansa (2003), and Zarza, Ceballos, and Steele 2003).

Head and body length is 85–149 mm, tail length is 93–167 mm, and weight is 30–70 grams. The upper parts are distinctly gray or reddish gray, the underparts are yellowish buff or cream white, and there are prominent black eye rings; the fur is slightly woolly. The tail is prehensile, relatively short, faintly bicolored, and not incrassate; the basal 10–15 mm are heavily furred and the remainder is naked. There is no pouch, the rhinarium has two ventrolateral grooves, and the plantar surface of the hind foot is naked from heel to toes. *Tlacuatzin* differs from all species currently referred to *Marmosa* in having caudal scales in unambiguously annular series, in lacking a premaxillary rostral process in the skull, in possessing an accessory opening on each side of the palate between the second upper molar and the normal maxillopalatine opening, and in having a diploid chromosome number of 22, rather than 14. Mature adults have very large, flattened, winglike postorbital processes that continue posteriorly as convergent temporal ridges. In having prominent postorbital processes, *Tlacuatzin canescens* resembles *Marmosa* (*Stegomarmosa*) *andersoni*, but differs from the latter in the other morphological characters that distinguish the two genera.

The gray mouse opossum inhabits seasonally dry (deciduous) tropical forest and desert scrub from the lowlands to an elevation of 2,100 meters. It is nocturnal and semiarboreal but may spend more time on the ground than do other mouse opossums. It builds ball nests of dry leaves and stems, lined with plant down, in small trees, bushes, or cacti. The diet consists of insects, figs, cultivated fruits, and probably cactus fruits. Most specimens taken in Sinaloa were found by day in their nests in hollows of dead cacti. In deciduous and semideciduous forests of Jalisco, population densities of 0.4 to 4.5 individuals per ha. have been estimated, and individuals moved an average of 35.2 meters between successive captures. Densities at different sites in Colima were 0.67–8.0 per ha.

Tlacuatzin is relatively common but solitary. Reproduction reportedly occurs throughout the year. Females with young have been found from July to September and males with scrotal testes from January to August. Juveniles have been captured from February to September. The number of young in 3 litters from Jalisco averaged 11 and ranged from 8 to 14. A female taken in Sinaloa on 5 September contained 13 embryos, 1 found on 4 September was accompanied by 8 nursing young, and females taken in October were lactating.

DIDELPHIMORPHIA; DIDELPHIDAE; **Genus** *MARMOSA* Gray, 1821

Murine, or Mouse, Opossums

There are 5 subgenera and 19 species (Brown 2004; Cabrera 1957; Creighton and Gardner 2007b; de la Sancha, D'Elía, and Teta 2012; Eisenberg 1989; Faria, Oliveira, and Bonvicino 2013; Gardner 2005a; Gardner and Creighton 2007b; Gutiérrez, Jansa, and Voss 2010; Gutiérrez et al. 2011; Hall 1981; Handley and Gordon 1979; Jansa, Forsman, and Voss 2006; Kirsch 1977a; Kirsch and Calaby 1977; Lambert et al. 2011; Pine 1972; Reig, Kirsch, and Marshall 1987; Rossi, Voss, and Lunde 2010; Solari and Pine 2008; Voss et al. 2014):

subgenus *Stegomarmosa* Pine, 1972

M. andersoni, Departamento Cusco in southern Peru;

M. lepida, eastern Ecuador to Surinam and central Bolivia;

subgenus *Micoureus* Lesson, 1842

M. constantiae, Mato Grosso of Brazil, Bolivia, Paraguay, extreme northern Argentina;

M. regina, Colombia, Ecuador, Peru, extreme western Brazil, northwestern Bolivia, also reported from Panama;

M. alstoni, Belize to northwestern Colombia;

M. demerarae, Colombia, Venezuela, the Guianas, northern and central Brazil, southwestern Peru, northern Bolivia;

M. paraguayana, southern Paraguay, extreme northeastern Argentina (Misiones), southeastern Brazil;

M. phaea, western slopes of the Andes in southwestern Colombia, western Ecuador, and probably extreme northwestern Peru (Tumbes);

subgenus *Marmosa* Gray, 1821

M. macrotarsus, eastern slope of the Andes in Peru, western Brazil, northern Bolivia;

M. tyleriana, highlands of southern Venezuela;

Mouse opossum (*Marmosa constantiae*). Photo by T. B. F. Semedo / Mammal Images Library, American Society of Mammalogists.

M. waterhousei, Colombia, Andes of western Venezuela, eastern Ecuador, northern Peru, extreme northwestern Brazil;

M. murina, Colombia, Venezuela, Trinidad and Tobago, the Guianas, Brazil;

subgenus *Eomarmosa* Voss, Gutiérrez, Solari, Rossi, and Jansa, 2014

M. rubra, lowlands of south-central Colombia, eastern Ecuador, and eastern Peru;

subgenus *Exulomarmosa* Voss, Gutiérrez, Solari, Rossi, and Jansa, 2014

M. simonsi, western Ecuador (including Isla Puná), northwestern Peru;

M. xerophila, extreme northeastern Colombia, northwestern Venezuela;

M. robinsoni, Panama (including San Miguel and Saboga islands), northeastern and central Colombia, northern Venezuela (including

Margarita Island), Trinidad and Tobago, Grenada and the Grenadines (possibly through introduction, see Long 2003);

M. isthmica, Panama, western Colombia, western Ecuador;

M. zeledoni, central Nicaragua to western Colombia;

M. mexicana, northeastern Mexico to Panama, Great Corn Island off Gulf Coast of Nicaragua, provisionally Ruatán Island off Gulf Coast of Honduras (see Rossi, Voss, and Lunde 2010).

Until recently *Marmosa* was often considered to include the species now assigned to the genera *Tlacuatzin*, *Chacodelphys*, *Thylamys*, *Gracilinanus*, *Cryptonanus*, and *Marmosops* (see accounts thereof). Moreover, the subgenus *Micoureus* was, for a time, generally recognized as a full genus (Corbet and Hill 1991; Gardner 2005a; Gardner and Creighton 1989, 2007b; Hershkovitz 1992a, 1992b). However, molecular analyses (Jansa, Forsman, and Voss 2006; Voss and Jansa 2003) indicated that, if *Micoureus* was considered a separate genus, *Marmosa* would be paraphyletic—meaning the genus contained only species descended from a common ancestor but not all the species descended from that ancestor. Of the five species sampled in those analyses, three (*mexicana*, *robinsoni*, and *rubra*) formed a related group, while two (*lepida* and *murina*) formed a clade with species then placed in the genus *Micoureus*. Thus, if the rules of cladistics were to be followed, it would be necessary either to synonymize *Micoureus* with *Marmosa* or to keep *Micoureus* as a genus or as a subgenus of *Marmosa* and consequently to recognize one or more additional genera or subgenera. As *murina* is the type species of *Marmosa*, the additional generic or subgeneric designations would apply to the group containing *mexicana*, *robinsoni*, *rubra*, and any related species. Voss et al. (2014) subsequently applied such an arrangement. The species *M. murina* itself was considered probably to consist of several distinct species (Creighton and Gardner 2007b; Patton and Costa 2003; Steiner and Catzeflis 2003). Again, such division was applied by Voss et al.

(2014), who also concluded that *M. lepida* should be placed in neither subgenus *Marmosa* nor subgenus *Micoureus*, but together with *M. andersoni* in subgenus *Stegomarmosa*. Reports of *M. andersoni* in Colombia and northern Peru were considered unverified by Solari and Pine (2008). *M. demerarae* formerly was known as *Marmosa cinerea*. *M. paraguayana* formerly was considered part of *M. demerarae* and sometimes has been referred to as *M. travassosi*. Gutiérrez et al. (2014) found *M. robinsoni* genetically separable into two distinctive groups, one west and one east of the Cordillera de Mérida in northwestern Venezuela, though as yet there is no corresponding morphological evidence that would support their recognition as separate species.

Head and body length is 85–220 mm and tail length is 125–270 mm. In any individual the tail is longer than the head and body. Weight in *M. robinsoni*, one of the larger, more common species, is 60–130 grams for males and 40–70 grams for females. *M. murina*, a smaller species, weighs 13–44 grams (Eisenberg 1989),

Mouse opossum (*Marmosa demerarae*). Photo by Rexford D. Lord.

and *M. lepida* only about 10 grams (Emmons 1997). In *M. mexicana* weight is 29–92 grams, but averages are 63.7 grams for males and only 35.2 grams for females (Alonso-Mejía and Medellín 1992). Some species of subgenus *Micoureus* are the largest of the Marmosini (Gardner and Creighton 2007b), with weights of up to 230 grams, though others are much smaller. The upper parts of *Marmosa* range from various shades of gray or brown to bright reddish brown and orange-red; the underparts are paler, varying from almost white through buff to yellowish. Almost all forms have dusky brown or black facial markings around the eyes. The fur is short, fine, and velvety in most species. The tail is strongly prehensile, has only a short basal brush of fur, and either is uniformly colored or has the ventral surface paler than the dorsal. There is no pouch.

According to Creighton and Gardner (2007b), mammae number 7–19, depending on the species, and are confined to the abdominal-inguinal region. *Marmosa* is distinguished from *Thylamys*, *Gracilinanus*, and *Marmosops* by its prominent superior postorbital process on the skull, in having the alisphenoid portion of its bulla lacking an anteromedial process, and in having its caudal scales arranged in a spiral, rather than annular, pattern.

The subgenus *Micoureus* traditionally has been distinguished by a number of characters (Emmons 1997; Gardner and Creighton 2007b; Hershkovitz 1992b; Redford and Eisenberg 1992; Reig, Kirsch, and Marshall 1987). Its pelage has been described as thick, lax, and crinkly, and longer and woollier than that of subgenus *Marmosa*. The fur has been reported to extend at least 20 mm onto the tail, whereas in *Marmosa* it usually extends less than 20 mm. *Micoureus* also has been said to differ from *Marmosa* in lacking a throat gland and in having the superior border of the frontal bone of the skull projected as a ledge. However, Voss et al. (2014), while noting that the species of *Micoureus* have commonly been known as "woolly mouse opossums," did not find pelage length or texture to be consistently useful as diagnostic traits. Some species, such as *M. demerarae*, do have longer and woollier pelts than sympatric species of *Marmosa*, but others, including lowland populations of *M. constantiae*, have short fur, and some highland forms of subgenus *Marmosa* (e.g., *M. tyleriana*) have long fur. "Woolliness" could not be defined by any satisfactory (nonsubjective) criterion. Extension of body fur onto the base of the tail, as well as presence of a white tail-tip, are characters often found in *Micoureus*, and not commonly in other subgenera, but are not consistently exhibited by all phenotypes that seem to belong in *Micoureus* based on molecular analyses. While *Micoureus* does lack a throat gland, the same is true for *Eomarmosa* and some forms of *Marmosa* and *Stegomarmosa*.

According to Hershkovitz (1992b), *Marmosa* differs from *Thylamys* in having the tail never incrassate, weaker claws on the forefoot, the nasal bones of the skull flared at the frontomaxillary suture, and the second premolar tooth usually larger than the third (rather than the opposite). *Marmosa* differs from *Gracilinanus* in lacking pectoral mammae, and from *Marmosops* in having a lambdoidal crest on the skull.

Most species of *Marmosa* are forest dwellers. The genus ranges vertically from sea level to about 3,400 meters. Most specimens of *Marmosa* collected by Handley (1976) in Venezuela were taken near streams or in other moist areas, but *M. xerophila* was almost always found in dry situations. Mouse opossums are nocturnal and usually arboreal, though some species are terrestrial. They are often found on banana plantations and among small trees and vine tangles. They build nests of leaves and twigs in trees or shelter in abandoned birds' nests. They have also been located in ground nests, in holes in hollow logs, and under rocks.

M. mexicana sometimes digs its own burrow but more commonly nests in trees (Alonso-Mejía and Medellín 1992). The species *M. robinsoni* does not seem to have a fixed abode but apparently spends the day in any suitable shelter when daylight overtakes it. That species reportedly is nomadic in Panama, occupying home ranges of about 0.22 ha. for two or three months and occurring at population densities of 0.31–2.25/ha. (Fleming 1972; Hunsaker 1977). In northern Venezuela *M. robinsoni* was estimated to occur at densities of 0.25–4.25 adults per hectare, and males were found to be more nomadic than females (O'Connell 1983). Mouse opossums are generally solitary, usually hunting and nesting alone. Fleming (1972) obtained data suggesting that while there was overlap between the home ranges of male *M. robinsoni* and between those of opposite sexes, the females had nonoverlapping home ranges and appeared intolerant of one another.

Mouse opossum (*Marmosa lepida*). Photo by Manuel Mejia.

Murine opossums are courageous fighters for their size. Like some of the other marsupials, they can lower their ears by crinkling them down much as a sail is furled. Their eyes reflect light as brilliant ruby red. When moving along tree limbs and vines, *M. robinsoni* curves its tail loosely around the branch, except when climbing vertically, and sometimes leaps across gaps. None of its motions, however, are particularly rapid. The diet of *Marmosa* consists mainly of insects and fruits, but also includes nectar, small rodents, lizards, and birds' eggs. Large grasshoppers are killed by a number of bites about the head and thorax; only the harder parts and lower legs are discarded. One mouse opossum in Mexico was noted hanging by its tail and eating a wild fig, which it held in its forefeet. Raw sugar is a particular favorite of *M. mexicana*. Banana and mango crops sometimes are damaged by these marsupials. Occasionally they are found in bunches of bananas in warehouses and stores in the United States, having remained in the bunch when it was shipped from the tropics. Occurrences in

the New Orleans area, apparently resulting from such journeys, were documented by Lowery (1974).

M. paraguayana has been intensively studied in the vicinity of Rio de Janeiro in southeastern Brazil (Barros, Crouzeilles, and Fernandez 2008; Brito and Fonseca 2006; Brito and Grelle 2004; Moraes Junior and Chiarello 2005; Pires and Fernandez 1999). It is found in evergreen and semideciduous forests, favoring dense vegetation with many vines and palm trees. It is nocturnal and arboreal, but often descends to the ground. Radio-tracking of seven individuals showed an average nightly movement of 583 meters for males, 335 meters for females. Population density was about 1–3 individuals per ha. Calculations indicate that 1,300 ha. of suitable habitat is necessary to maintain a genetically viable population of 2,000 individuals. Based on radio-tracking, the best estimate of home range size is 4.0–10.9 ha. for males and 1.3–5.9 ha. for females. Male ranges apparently become larger during the breeding season but do not overlap; female ranges may overlap one another and are partly or completely contained within a male's range. The species is polygynous and breeds during the wet season (September to May). Females were found to have at least two litters annually, one in October/November and one in January/February. Litter size was 6–11 young but median was also 11. Young males possibly disperse into new habitat. Both sexes can breed when 6 months old. The longest lifespan recorded in the wild is about 24 months.

M. demerarae, one of the better known species of subgenus *Micoureus*, has been recorded at densities as high as 1 per ha. The female builds an open nest of dead leaves in the crown of a palm tree or a vine tangle. She transports the leaves with her mouth or prehensile tail. Breeding in Venezuela is tied to rainfall and does not take place during the winter dry season. Reproductively active females have been found in western Brazil during both the rainy (February to April) and dry (September to November) seasons; litters of seven and nine young have been reported (Eisenberg 1989; Emmons 1997; Gardner and Creighton 2007b).

Some species of *Marmosa* may breed all year, while others are seasonal. Females of *M. murina* produce three litters annually and breed throughout the year (Hunsaker and Shupe 1977); the gestation period in that species is 13 days maximum, and litter size is 13 (Eisentraut 1970). Reproductively active female *M. regina*, some with six to eight attached young, have been taken in Peru and adjacent Brazil in February, July, August, September, October, and November (Gardner and Creighton 2007b). For female *M. constantiae* in Bolivia, Anderson (1997) reported one lactating in May, one with five young in August, and four with no embryos in July, August, and September. A study in the Panama Canal Zone found *M. robinsoni* to be a seasonal breeder, with reproduction occurring mainly from late March to September and litter size averaging 10 (Fleming 1973).

According to Hunsaker (1977), *M. robinsoni* has breeding peaks in February, June, and July through December, an estrous cycle averaging 23 days, and a gestation period of 14 days. Litter size is seven to nine in the wild and slightly smaller in captivity. Weaning

Mouse opossum (*Marmosa phaea*). Photo by Manuel Mejia.

occurs when the young are 60–70 days old, they leave the mother a few days later, and young females have their first estrus when 265–275 days old. Life expectancy probably is under 1 year in the wild but is about 3 years in captivity. Reproductive ability declines in the second year of life.

Some slightly different information compiled by Collins (1973) is that gestation is just under 14 days, average size of 70 laboratory litters was about seven, females reach sexual maturity at approximately six months, and average life span in captivity is 1–2 years. One male *M. robinsoni* obtained at an unknown age lived for almost 5 more years, an apparent record longevity. Godfrey (1975) found that litter size in a laboratory colony of *M. robinsoni* ranged from 1 to 13 and usually was 7–9. Litter sizes of up to 15 have been reported in Venezuela (O'Connell 1983).

As noted by Grimwood (1969), because of their small size and nocturnal habits, populations of mouse opossums are not directly threatened by people; however, some of the high-altitude species might be jeopardized by widespread clearing of brush in the Andes region. The species *M. phaea* is classified as vulnerable by the IUCN; its restricted range is severely fragmented, and continued decline is expected through destruction of forests for agriculture and human settlement (Solari and Patterson 2011). *M. andersoni*, known only from three montane localities, was formerly classified as critically endangered but now is termed "data deficient" because of the absence of recent information on occurrence and of status and ecological requirements (Solari 2015a; Solari and Pine 2008). *M. xerophila*, also restricted to a small area and losing its dry thorn shrub habitat to agriculture, is classed as vulnerable (Lew et al. 2011b). Dias et al. (2010) indicated that *M. paraguayana*, which occurs largely in the Atlantic forest of southeastern Brazil, may be endangered through habitat destruction.

DIDELPHIMORPHIA; DIDELPHIDAE; **Genus *METACHIRUS*** Burmeister, 1854

Brown "Four-eyed" Opossum

The single currently recognized species, *M. nudicaudatus*, is found from extreme southern Mexico to

southern Brazil and extreme northeastern Argentina (Brown 2004; Gardner and Dagosto 2007; Medellín et al. 1992). Because of technical problems of nomenclature, there has been argument regarding which generic name properly applies to that species. Hershkovitz (1976, 1981) supported *Metachirus*, which has been in more general use, while Pine (1973) favored *Philander* Tiedemann, 1808 (not *Philander* Brisson, 1762), at least for the time being. Hall (1981) followed Pine in applying *Philander* to the brown "four-eyed" opossum, but Corbet and Hill (1991), Gardner (1981), Kirsch and Calaby (1977), and most subsequent authorities have agreed with Hershkovitz. The generic designation *Philander* is actually applicable to the gray and black "four-eyed" opossums and is so employed here, again in accordance with Hershkovitz (1976, 1981) and his backers, rather than the name *Metachirops*, used by Pine (1973) and Hall (1981). For a detailed discussion of the matter, see Gardner and Dagosto (2007). Based on the high level of mitochondrial DNA sequence divergence within *Metachirus*, which is as large as or larger than that found between most well recognized species of all other didelphid genera, Patton and Costa (2003) suggested that *Metachirus* contains more than one distinct species.

Head and body length is 190–310 mm and tail length is 195–390 mm (Hershkovitz 1992b). Recorded weights in Argentina and Paraguay have ranged from 91 to 480 grams (Redford and Eisenberg 1992). A male and a female from Barro Colorado Island, Panama Canal Zone, each weighed 800 grams. The back and sides are brown, often dark cinnamon brown, and the rump may be washed with black. The face is dusky, almost black in some individuals, with a creamy white spot over each eye. Those spots, suggesting eyes, are usually smaller and more widely separated than those of *Philander*. The underparts are buff to gray. The tail is furred for a short distance basally. The pelage is short, dense, and silky.

Although the common names and general appearance of *Metachirus* and *Philander* are similar, the two are not immediately related. *Metachirus* may be distinguished externally by its brown coloration and longer tail. Unlike the gray and black "four-eyed" opossums, the females of *Metachirus* lack a pouch, having instead simple lateral folds of skin on the lower abdomen, in which are located the mammae. Females with five, seven, and nine mammae have been recorded

Brown "four-eyed" opossum (*Metachirus nudicaudatus*). Photo by Francois Catzeflis.

(Collins 1973). Kirsch (1977b) observed that whereas *Philander* is probably the most aggressive of didelphids, *Metachirus* is almost quiet when held in the hand. However, Loretto, Ramalho, and Vieira (2005) reported that female *Metachirus* carrying newborn young seemed quite aggressive when disturbed in the nest. They made a characteristic sound by clicking, gnashing their teeth, and hissing, while simultaneously bouncing against the top of the nest to create a pulsation of the surrounding litter surface, thereby suggesting presence of a much larger animal.

The brown "four-eyed" opossum seems to favor mature forest with little undergrowth, but is also present in dense habitat; it is often common but lives at low population densities (Emmons 1997). Loretto, Ramalho, and Vieira (2005) found 13 nests, all within the litter layer of the forest floor, immediately below the surface or between roots of trees; they were spherical with no apparent entrance, about 10 cm across, and made from dry ground leaves interlaced with pieces of roots. Handley (1976) reported that all 18 specimens taken in Venezuela were caught on the ground, mostly near streams or in other moist areas. Using spool-and-line tracking of 19 individuals over 18 months, Cunha and Vieira (2002) confirmed that *Metachirus* is specialized for terrestrial and cursorial life, never leaving the forest floor. It is completely nocturnal, rarely moving from the nest until dark. In three nights of radio-tracking an adult female, Moraes Junior (2004) found an average nightly movement of 550 meters and a home range of 8.4 ha., which encompassed two burrows in fallen tree trunks and two nests under foliage on the forest floor. The diet includes fruits, insects, mollusks, amphibians, reptiles, birds, eggs, and small mammals. *Metachirus* has been accused of damaging fruit crops in some areas.

Limited data indicate that this opossum is seasonally polyestrous (Fleming 1973). It reportedly breeds in November in Central America, has litters of one to nine young, and probably has a maximum life span of 3–4 years (Hunsaker 1977). The gestation period is 20–28 days, and the newborn remain attached to the nipples for 75–80 days, after which they spend 30–45 days in the nest until totally weaned; they disperse when about 130 days old and attain sexual maturity at about 10 months (Astúa 2015). Females lactating or with attached young were collected on 11 April in Colombia, 30 April and 1 May in Venezuela, 13 September and 28 October in Peru, and 20 April in Paraguay (Gardner and Dagosto 2007). At one Atlantic Forest site in southeastern Brazil, females with young were captured in February, April, June, November, and December, and lactating females in April, June, and October (Astúa 2015). The single, 51-mm young of a female obtained on 18 December was then already able to stand alone. It later rode on its mother's back or hips and was fully independent by early February (Collins 1973).

DIDELPHIMORPHIA; DIDELPHIDAE; **Genus** *CHIRONECTES* Illiger, 1811

Water Opossum, or Yapok

The single species, *C. minimus*, is known from Oaxaca and Tabasco in southern Mexico, Guatemala, Honduras, El Salvador, Nicaragua, Costa Rica, Panama, Colombia, Venezuela, Trinidad, the Guianas, Ecuador, Peru, northwestern and central Bolivia, southern Paraguay, extreme northeastern Argentina (Misiones), and northeastern, extreme western, and southern Brazil (Brown 2004). Stein and Patton (2007) referred to the distribution as disjunct but noted that the absence of records from the central Amazon Basin might represent a collecting bias. Recently, Ardente et al. (2013) reported a specimen from Parauapebas in the southwest of the Brazilian state of Pará, which fills part of the distributional gap.

According to Marshall (1978d), head and body length is 270–400 mm, tail length is 310–430 mm, and

Water opossum, or yapok (*Chironectes minimus*). Photo by George Smiley.

weight is 604 to 790 grams. Collins (1973), however, stated that the weight of a captive female stabilized at about 1,200 grams. In contrast, living wild adults weighed by Galliez et al. (2009) had a range of 405–595 grams for males and 395–420 grams for females. The pelage is relatively short, fine, and dense. The back is marbled gray and black, the rounded black areas coming together along the midline. The muzzle, a band through the eye to below the ear, and the crown are blackish; a prominent grayish white, crescent-like band passes from the front of one ear to the other, just above the eyes. The chin, chest, and belly are white. The striking color pattern is unique among marsupials and may serve as camouflage while the yapok is swimming, the dorsal bands blending with ripples and presenting a disruptive appearance to aerial predators (Brosset 1989). The long, ratlike tail is well furred only at the base; it is black near the body and yellowish or whitish toward the end. The facial bristles are stout, long, and placed in tufts. One of the wrist bones is enlarged and simulates, in some respects, a sixth digit on the forefoot. The ears are moderately large, naked, and rounded.

This opossum is the only marsupial well adapted for a semiaquatic life. It has dense, water-repellent pelage, webbed hind feet, a streamlined body, and, in females, a rear-opening, waterproof pouch. Both sexes actually have a pouch. In the male it cannot be fully closed, as in the female, but the scrotum is pulled up into it when the animal is swimming or moving swiftly. The female is able to swim with the young in her pouch. A well-developed sphincter muscle closes the pouch, creating a watertight compartment, and the young can

Water opossum, or yapok (*Chironectes minimus*), hind foot. Photo by Alexander F. Meyer at Dallas World Aquarium.

tolerate low oxygen levels for many minutes (Marshall 1978d; Rosenthal 1975b). Four or five mammae are contained within the female's pouch (Stein and Patton 2007). Male and female yapoks, unlike most marsupials, lack a cloaca and have separate anal and urogenital openings (Nogueira et al. 2004).

The yapok is confined mainly to tropical and subtropical habitats, where it frequents freshwater streams and lakes. In some areas it is found at considerable elevations along mountain rivers. Grimwood (1969) reported that one was killed in Peru at an elevation of 900 meters, and Handley (1976) collected specimens at up to 1,860 meters in the mountains of Venezuela. This opossum generally is considered rare throughout its range, but that view may stem from the nocturnal habits of the animal and the inaccessibility of its habitat. In a lowland rainforest of French Guiana, Voss, Lunde, and Simmons (2001) readily observed *Chironectes* at night because of its bright eyeshine and noisy splashing, even though its presence in the area had been unknown to local foresters and hunters. All individuals seen during a study by Galliez et al. (2009) near Rio de Janeiro used stream stretches with preserved riparian forest, fast-flowing water, and stony substrate. Most dens in that area were in holes formed by stones and tree roots by the river margin; others were underground in riparian forest near the streams. If in a subterranean cavity, a main den may be reached through a hole in the stream bank just above water level. A ground nest of leaves or grasses in a dimly lit area may be used as a place of rest during the day. Like *Didelphis*, the yapok collects nesting material with the forepaws, pushing it under the body into a bundle that is then held by the tail (Marshall 1978d).

The water opossum is an excellent swimmer and diver, the broadly webbed hind feet serving as effective paddles. Only the rear feet are used in swimming, thereby avoiding the reduced propulsion efficiency that might result from interference between the fore- and hind limbs. The pelage does not soak up water, so the nose can remain above the surface without assistance from the front feet (Fish 1993). *Chironectes* can climb but rarely does so; its tail, though somewhat prehensile, is too thick to be of effective use. It is largely carnivorous, feeding on such aquatic life as crayfish, shrimp, and fish. Prey is located in the water by contact with the forefeet, which are not used for propulsion but are held out in front while the animal is swimming (Marshall 1978d). The yapok also may eat some aquatic vegetation and fruits (Hunsaker 1977).

In southeastern Brazil, Galliez et al. (2009) found home range lengths of 2,616, 3,388, and 3,724 meters of stream for three males and 844 and 1,216 meters for two females. The ranges of males overlapped with one another and with those of females. Males used dens close to those of females and shared the same den when the latter had offspring, thus suggesting they participated in parental care or were guarding their mates. Females with three pouch young each were captured in August, September, and October. However, juveniles were taken throughout the year in both wet and dry seasons, there thus being no evidence of reproductive seasonality. Anecdotal reports for the same region indicated that females with offspring had been captured in all seasons.

Galliez et al. (2009) also cited records of females with young in January, July, and November in Venezuela and of offspring in August in northern Argentina.

Eisenberg and Redford (1999) stated that litter size is one to five young, with two or three most common. However, Rosenthal (1975b) reported that three captive-born litters at the Lincoln Park Zoo numbered four, five, and five. Experience with those litters indicated that the young of *Chironectes* develop more rapidly than those of any other didelphid. One captive female had her first estrous cycle at 10 months of age (Eisenberg and Redford 1999). Weigl (2005) listed an individual that was caught in the wild and then lived at the Bronx Zoo for 3 years and 2 months.

Grimwood (1969) observed that yapok skins had been considered worthless but were beginning to command a price in Peru. He thought that intensive commercial hunting could readily reduce populations and that protective measures were necessary. Emmons (1997) noted that the genus seems common where it occurs but appears to be absent from many regions. McLean and Ubico (1993), who reported the first record for Guatemala, suggested that it is very rare in that country. Galliez et al. (2009) indicated that in Brazil the

Water opossum, or yapok (*Chironectes minimus*). Photo by Fiona A. Reid.

water opossum may be jeopardized by degradation of areas of river confluence, thereby blocking gene flow and isolating populations; they also cited reports that the genus is rare and vulnerable in Mexico, Peru, Uruguay, and Argentina, and that fur hunting remains a concern. The IUCN does not consider *Chironectes* eligible for listing in a formal threatened category, because of its wide distribution and presumed large population, but notes that is decreasing and is threatened by deforestation and contamination and deterioration of freshwater ecosystems (Cuarón et al. 2008).

DIDELPHIMORPHIA; DIDELPHIDAE; **Genus** *LUTREOLINA* Thomas, 1910

Thick-tailed, or Lutrine, Opossums

There are two species (Martínez-Lanfranco et al. 2014; Stein and Patton 2007):

L. massoia, montane forests from southern Bolivia to northwestern Argentina;

L. crassicaudata, a greatly disjunct range, with subspecies *L. c. turneri* known from east-central Colombia, eastern Venezuela, and northwestern Guyana, and subspecies *L. c. crassicaudata* occurring east of the Andes in northern and eastern Bolivia, southern Brazil, Paraguay, Uruguay, and northeastern Argentina.

Martínez-Lanfranco et al. (2014) pointed out that the two species appear to be geographically separated by the arid Chaco region. Lew et al. (2011b) noted that *L. crassicaudata* also has been recorded from southeastern Peru on the border of Bolivia and is suspected to occur continuously down the Orinoco River in Venezuela.

Head and body length is 210–445 mm and tail length is 209–310 mm. Regidor, Gorostiague, and Sühring (1999) recorded average weights of 791 grams for male

Thick-tailed, or lutrine, opossum (*Lutreolina crassicaudata*). Photo by Jorge La Grotteria.

and 522 grams for female *L. crassicaudata*. *L. massoia* is considerably smaller; Martínez-Lanfranco et al. (2014) reported a body mass of just 284 grams for the holotype, an adult female. The fur is short, dense, and soft but not water-repellent. The upper parts are a rich, soft yellow, buffy brown, or dark brown, and the underparts are reddish ochraceous or pale to dark brown. The pelage of some live individuals has a peculiar purplish tinge. There may be faint markings on the face, but there are no eye spots or other prominent markings.

The body form is long and low, almost weasel-like. The ears are short and rounded and barely project above the fur. The limbs and feet are short and stout, and the pads are small and narrow. The tail is characteristic in its extremely thick, heavily furred base; only about 5 cm of the undersurface of the tip is naked, though in some individuals the terminal half is thinly haired, showing the scales. The tail is not as prehensile as in other didelphids. The thumbs and great toe are not fully opposable. Although some reports have indicated otherwise, specimens examined by Lemke et al. (1982) did have a well-developed pouch. Reig, Kirsch, and Marshall (1987) also confirmed the presence of a pouch. The mammae, at least in one specimen, number nine.

L. crassicaudata is restricted mostly to grassland, savannahs, and gallery woodland, often near the shores of streams and lakes. It is considered to be the didelphid most adapted to life on the pampas. *L. massoia* is restricted to Yungas forest, from premontane areas, at 450 meters, to the upper forested belts, at 2,000 meters (Martínez-Lanfranco et al. 2014). These opossums climb well, are agile on the ground, and seem suited for wetland habitat. Although one captive reportedly was clumsy in the water, most information indicates that *Lutreolina* is a good swimmer under natural conditions (Gewalt 1990; Hunsaker 1977; Marshall 1978a; Martínez-Lanfranco et al. 2014). After studying videotapes of a laboratory animal, Santori et al. (2005) concluded that *Lutreolina* is a strong swimmer, with specialized stroke frequency and buoyancy ability similar to those of *Chironectes* and not shared by other didelphids. However, unlike *Chironectes*, which uses only its hind feet for propulsion in the water, *Lutreolina* employs a less efficient quadrupedal gait; its swimming speed is equivalent to that of more terrestrial didelphids.

In wooded areas these opossums often shelter in tree holes, but in wetlands they construct a snug round nest of grasses and rushes among the reeds, and on the pampas they often utilize abandoned armadillo and viscacha burrows. The young of *Lutreolina* are raised in a nest of dry grass. Thick-tailed opossums are nocturnal, emerging after dark to prey on small mammals, birds, reptiles, fishes, and insects. Cáceres, Ghizoni-Jr, and Graipel (2002) found them to eat mainly crabs and beetles but also to consume considerable fruit during the warm season. They occasionally raid chicken houses and pigeon lofts. They are said to be sometimes savage in temperament but can be tamed. They actually appear somewhat more sociable than other didelphids; one male and two females were maintained together successfully (Collins 1973). Home ranges of two individual *L. massoia* were 650 and 950 sq meters (Martínez-Lanfranco et al. 2014).

Lutreolina breeds in the spring and again later in the year after the young of the first litter have become independent. The gestation period is thought to be about 2 weeks. Regidor, Gorostiague, and Sühring (1999) live-trapped and tagged 36 individuals near Buenos Aires. Breeding was found to commence in September, after an anestrous of 5 months, and was finished by early April. The first litters were born in late September (early spring), the second in late December or early January (summer). Litter size ranged from 7 to 11 and averaged 8.6. The young apparently were weaned at about 3 months of age and reached sexual maturity after 6 months or more. A study by Muschetto, Cueto, and Suarez (2011) also showed reproductive seasonality in east-central Argentina, with a continuous breeding period from September to February (spring to summer). A female taken in Colombia on 10 August carried seven attached young in the pouch (Stein and Patton 2007). Collins (1973) cited a record of a specimen living at least 3 years in captivity.

Roig (1991) reported that *Lutreolina* had disappeared from a large part of central and northern Argentina because of human habitat disruption. However, Lew et al. (2011c) observed that, while *L. crassicaudata* is generally rare and may be adversely affected by drainage of wetlands for agriculture in the southern part of its range, it does not qualify for a formal IUCN threatened category. Martínez-Lanfranco et al. (2014) noted that the Yungas habitat of *L. massoia* is extremely

Gray "four-eyed" opossum (*Philander opossum*). Photo by Johannes Pfleiderer.

fragmented and under great anthropogenic pressure, but that the species is frequently captured.

DIDELPHIMORPHIA; DIDELPHIDAE; **Genus** *PHILANDER* Brisson, 1762

Gray and Black "Four-eyed" Opossums

There are seven species (Brown 2004; Chemisquy and Flores 2012; de la Sancha and D'Elía 2015; Flores, Barquez, and Díaz 2008; Lew, Pérez-Hernández, and Ventura 2006; Patton and Da Silva 1997, 2007):

P. frenatus, southern Brazil, southern Paraguay, extreme northeastern Argentina (Misiones Province);

P. andersoni, southern Venezuela, northwestern Brazil, southern Colombia, eastern Ecuador, northeastern and central Peru;

P. mcilhennyi, Amazon Basin of east-central Peru and adjacent western Brazil;

P. opossum, northeastern and southern Mexico, Central America, western Colombia and Ecuador, the Guianas, central and eastern Peru, the Amazon and upper Paraná basins of Brazil, Bolivia and immediately adjacent Brazil west of Paraguay River, Paraguay, northeastern Argentina (Chaco and Formosa Provinces);

P. olrogi, known from one locality in northeastern Peru and two in eastern Bolivia;

P. deltae, Orinoco Delta of eastern Venezuela;

P. mondolfii, central Colombia to eastern Venezuela, possibly Guyana and northern Brazil.

Tiedemann, 1808, long the accepted authority for the name *Philander*, was invalidated by the International Commission on Zoological Nomenclature in 1998. Pine (1973) had argued that the generic name *Metachirops* Matschie, 1916, is the correct designation for the above species and that *Philander* actually applies to the brown "four-eyed" opossum (here called *Metachirus*). Both Husson (1978) and Hall (1981) agreed, but most others have not. For further comment see the account of *Metachirus* Burmeister, 1854, and for a more detailed discussion see Patton and Da Silva (2007). *Philander* generally was thought to contain only the single living species *P. opossum* until Gardner and Patton (1972) described *P. mcilhennyi*, which subsequently was designated a synonym of *P. andersoni*. *P. frenatus*, meanwhile, usually was treated as a subspecies of *P. opossum*. Assessing mitochondrial DNA, however, Patton and Da Silva (1997) found *P. frenatus* to be highly distinct from all other components of the genus, and *P. mcilhennyi* to be separable from *P. andersoni*. Later, based mainly on characters of the skull and pelage, Lew, Pérez-Hernández, and Ventura (2006) distinguished *P. deltae* and *P. mondolfii* from *P. opossum*. Flores, Barquez, and Díaz (2008) described *P. olrogi* as a close relative of *P. opossum*—externally much the same but with distinctive cranial features. Chemisquy and Flores (2012) suggested that *P. andersoni* and *P. mcilhennyi* might be only subspecies of *P. opossum* and, if specific status for *P. andersoni* is maintained, that *P. opossum fuscogriseus*, found from Honduras to southern Ecuador, should also be treated as a full species. Likewise, de la Sancha and D'Elía (2015) noted that if *andersoni* and *mcilhennyi* are retained as distinct species, *P. opossum canus*, which occurs from eastern Peru and western Brazil to Paraguay and northeastern Argentina, should be elevated to species status.

According to Hershkovitz (1997), head and body length is 200–380 mm, tail length is 195–355 mm, and natural weight is 230–675 grams. Collins (1973), however, stated that healthy, well-fed captive males weighed 800–1,500 grams and females, 600–1,000 grams. The fur is rather straight and short in *P. opossum*, *P. frenatus*, and *P. olrogi*, short and velvety in *P. deltae*, short and woolly in *P. mondolfii*, longer in *P. andersoni*, and especially long and shaggy in *P. mcilhennyi*. The upper parts are gray or brownish gray in *P. opossum* and *P. olrogi*, brown in *P. deltae*, pale gray in *P. mondolfi*, dark gray in *P. frenatus*, dark with a blackish mid-dorsal stripe in *P. andersoni*, and black in *P. mcilhennyi*. The underparts vary from yellowish or buffy white to dark gray. A white spot above each eye accounts for the vernacular name "four-eyed" opossum.

The tail, which is furred for about 50–80 mm from the rump and naked toward the tip, is black, brown, or grayish on its basal half and white toward the end, with a pink tip. The ears are bicolored or black but appear naked. The body is slim and usually lean, and the head is large, with an elongate, conical muzzle. The tail is slender, tapering, and prehensile. Females have a distinct pouch. The number of mammae is usually seven but varies from five to nine.

These opossums inhabit forested areas and are often found near swamps and rivers. They are smaller and more agile than *Didelphis* and quick in their actions. Although good climbers and swimmers, they are mainly terrestrial. All 46 specimens taken in Venezuela by Handley (1976) were caught in moist areas, nearly always on the ground. Globular nests, about 30 cm in diameter, are built in the lower branches of trees or in bushes; Emmons (1997) noted that nests may be up to 8–10 meters above the forest floor. These opossums may also inhabit ground nests and burrows. They are thought to be mainly nocturnal, though Husson (1978) stated that in Surinam they are as active in the day as at night. Their diet includes small mammals, birds and their eggs, reptiles, amphibians, insects, freshwater crustaceans, snails, earthworms, fruits, and probably carrion. A captive female described by Hershkovitz (1997) would eat anything served and readily attack and devour live mice. Occasional damage to fruit crops and cornfields has given *Philander* a bad reputation in certain areas.

In Panama, Fleming (1972) found maximum population densities of 0.55–0.65 individuals per ha. Hershkovitz (1997) cited reports of about 1–2 per ha. and that home ranges overlap and there are no defended territories. Radio-tracking of *P. frenatus* in southeastern Brazil indicated home ranges of 0.6–7.4 ha., with no significant difference between sexes (Lira and Fernandez 2009). Unlike *Didelphis*, these opossums are not known to feign death when danger threatens, but will open their mouths wide, hiss loudly, and fight savagely. When disturbed they may utter a long, chattering cry.

Philander breeds year-round in some areas, possibly including the state of Veracruz in Mexico, French Guiana, and Peru, but is seasonal in others; reproductive activity generally increases during the rainy season (Barros, Crouzeilles, and Fernandez 2008;

Castro-Arellano, Zarza, and Medellín 2000; Fleck and Harder 1995; Hershkovitz 1997; Hunsaker and Shupe 1977). Hingst et al. (1998) reported a gestation period of 13–14 days and a post-lactation estrous. Collins (1973) cited records of females with pouch young being taken in Nicaragua from February to October, in Panama from April to July, and in Colombia in June, September, and October. Jones, Genoways, and Smith (1974) caught a female with six nursing young in March on the Yucatan Peninsula. Phillips and Jones (1969) collected data indicating that the main reproductive season in Nicaragua extends from March through July. Litter size there averaged 6.05 (3–7). According to Fleming (1973), *P. opossum* is seasonally polyestrous in Panama, with two or more litters probably being produced by each female from January to November. Litter size averaged 4.6 (2–7). Husson (1978) found females with young in Surinam during January, March, and April. Litter size there ranged from 1 to 7 but averaged only 3.4. Those records support the statement by Phillips and Jones (1969) that litter size tends to be larger in the north. However, size may increase again moving south of the equator, as Eisenberg and Redford (1999) reported an average of 2.8 in Surinam but 4.5 in southeastern Brazil. At a captive colony in Rio de Janeiro, births occurred from August to February and mean litter size was 5.5 (1–10) at birth and 4.4 at weaning, which occurred when the young were 70–80 days old and 24–49 grams in weight (Hingst et al. 1998). In French Guiana, female *P. opossum* have been observed to bear 2–4 litters per breeding season, with an interbirth interval of about 90 days (Astúa 2015). Records cited by Hershkovitz (1997) indicate that young attain sexual maturity when 5 to 6 months old. Longevity in the wild probably does not exceed 2.5 years (Castro-Arellano, Zarza, and Medellín 2000), but Weigl (2005) listed a captive that lived to an age of 4 years and 5 months.

DIDELPHIMORPHIA; DIDELPHIDAE; **Genus *DIDELPHIS*** Linnaeus, 1758

Large American Opossums

There are six species (Brown 2004; Cabrera 1957; Cerqueira and Tribe 2007; Gardner 1973; Gardner and Sunquist 2003; Hall 1981; Lemos and Cerqueira 2002; Ventura et al. 2002):

Virginia opossum (*Didelphis virginiana*). Photo by Ivan Kuzmin / Shutterstock.com.

D. albiventris, eastern and southern Brazil, southeastern Bolivia, Paraguay, Uruguay, northern and eastern Argentina;

D. pernigra, forested slopes of the Andes from northwestern Venezuela to central Bolivia and possibly northern Argentina;

D. imperfecta, south of the Orinoco in Venezuela and adjacent Brazil, the Guianas, possibly eastern Ecuador and northeastern Peru;

D. virginiana (Virginia opossum), naturally from New Hampshire to Colorado and New Mexico, and from southern Ontario to northern Costa Rica, also widely introduced (see below);

D. marsupialis, northeastern and central Mexico to central Bolivia and northeastern Brazil, Trinidad and Tobago, also in the Lesser Antilles from Dominica southward where it may have been introduced on some islands;

D. aurita, southeastern Brazil, southeastern Paraguay, extreme northeastern Argentina (Misiones).

According to Cerqueira and Tribe (2007), South American *Didelphis* comprises two groups of species, the white-eared opossums *D. albiventris*, *D. pernigra*, and *D. imperfecta*, and the black-eared opossums *D. marsupialis* and *D. aurita*. *D. albiventris*, often under the name *D. azarae*, long was considered the only representative of the white-eared group, but morphological and craniometric assessment by Lemos and Cerqueira (2002) and Ventura et al. (2002) demonstrated that *D. pernigra* and *D. imperfecta* warrant specific status. Likewise, *D. marsupialis* once was thought to be the only species in the black-eared group, but Cerqueira (1985) and Cerqueira and Lemos (2000) distinguished *D. aurita* therefrom. *D. marsupialis* also was long regarded as the only living North American opossum, but Gardner (1973) showed *D. virginiana* to be a separate species. He hypothesized that *D. virginiana* had evolved from *D. marsupialis* relatively late in the Pleistocene, but molecular studies have suggested that *D. virginiana*, though it has black ears, is more closely related to the white-eared group of opossums (Gardner and Sunquist 2003).

Head and body length is 305–500 mm, and tail length is 255–535 mm. Gardner and Sunquist (2003) reported weights of about 0.5–7.0 kg for *D. virginiana*,

noting that males average significantly larger than females in any given population, that growth continues throughout life, and that the species is the only didelphid capable of putting on large stores of body fat, so that animals are heaviest in fall and early winter. That species is the largest living New World marsupial. Emmons (1997) gave a weight range of 0.5–2.0 kg for the other species. The pelage, consisting of underfur and white-tipped guard hairs, is unique in the family Didelphidae. Guard hairs are lacking or few in number in the other opossums. Coloration is gray, black, reddish, or, rarely, white. Head markings are sometimes present in the form of three dark streaks, one running through each eye and another running along the midline of the crown of the head. The basal tenth of the tail is furred, and the remainder is almost naked. Cerqueira and Tribe (2007) wrote that the ears are large, oval, naked, and leathery. At the northern limits of the range many individuals that survive the winter lose part of the tail and ears from frostbite.

As in the other didelphids, all four feet have five digits, and all digits have sharp claws, except that the first toe of the hind foot is clawless, thumblike, and opposable to the other digits in grasping. The female has a well-developed pouch, commonly having 13 mammae arranged in an open circle with one in the center. Number and arrangement of mammae, however, are variable.

These opossums usually inhabit forested or brushy areas but also have been found in open country near wooded watercourses. *D. virginiana* is reported to favor moist woodlands or thick brush near streams or swamps (Banfield 1974; Jackson 1961; McManus 1974; Schwartz and Schwartz 1959). However, it may occur in almost any habitat from sea level to elevations over 3,000 meters, while *D. marsupialis*, in Mexico, is restricted to humid tropical and subtropical areas, generally below 1,000 meters (Gardner and Sunquist 2003). In Venezuela, Handley (1976) collected most specimens of *D. marsupialis* on the ground in moist areas but found most *D. imperfecta* in dry situations. These opossums are largely nocturnal, spending the day in rocky crevices, hollow tree trunks, vine tangles, piles of dead brush, culverts, attics and foundations of buildings, or burrows dug by other animals. They may use numerous dens within a home range, each for only a couple days; recent radio-tracking studies of

White-eared opossum (*Didelphis albiventris*). Photo by Fotos593 / Shutterstock.com.

D. virginiana have found nightly movements to commonly be about 1–2 km (Gardner and Sunquist 2003). Large American opossums construct rough nests of leaves and grasses, using their curled-up tail and their mouth to transport dry vegetation. They are mainly terrestrial, moving with a slow, ambling gait, and are strong swimmers. They can climb well, with the help of the prehensile tail, and can even hang by the tail (Lowery 1974). In the northern part of its range, *D. virginiana* accumulates fat in the autumn and may become inactive during severe winter weather, remaining in its nest for several days, but it does not hibernate. Females have a greater tendency toward winter inactivity and also are more sedentary in their movements at other times. Opossums have an extremely varied diet that includes small vertebrates, carrion, invertebrates, and many kinds of vegetable matter. An extensive review by Gardner and Sunquist (2003) indicates that cottontail rabbits (*Sylvilagus*) are an important food for *D. virginiana*. Emmons (1997) stated that *D. marsupialis* will climb to the treetops to feed on fruit and nectar.

Records compiled by Hunsaker (1977) show that for *D. virginiana* in the United States, population density averaged 0.26/ha. (0.02–1.16/ha.), home range averaged about 20 ha. (4.7–254.0 ha.), and nightly foraging distance was 1.6–2.4 km; and that for *D. marsupialis* in Panama, density was 0.09–1.32/ha., with the animals being nomadic and remaining in an area for only 2–3 months. According to Hunsaker and Shupe (1977), *D. virginiana* also is nomadic and stays in a particular area for 6 months to a year; there is no territoriality, but individuals will defend the space occupied at a given time. In a study in Venezuela, Sunquist, Austad, and Sunquist (1987) found the home range size of *D. marsupialis* to average 11.3 ha. in the dry season and 13.2 ha. in the wet season. Male home ranges overlapped one another extensively, and each overlapped the ranges of several females. The latter occupied exclusive home ranges for at least part of the year.

Opossums usually are thought to be solitary and antisocial, either avoiding one another or acting aggressively. However, Holmes (1991) found that most interaction among unrelated individuals kept in large enclosures was neutral or affiliative, with females sometimes nesting together. The animals readily formed stable dominance hierarchies, with females usually dominant. There seems agreement that there almost always is extreme agonistic behavior when two males meet, but if opposite sexes meet during the breeding season, initial aggressive displays turn to courtship and the two animals may spend several days together. If a male is placed with a female that is not in estrus, she becomes aggressive, but the male does not return her attacks. The vocal repertoire of *D. virginiana* consists of a hiss, a growl, and a screech, which are used in agonistic and defensive situations, and a metallic lip-clicking, which is heard under a variety of conditions, including mating (McManus 1970, 1974).

Death feigning, referred to popularly as "playing possum" and technically known as "catatonia," is a passive defensive tactic of *D. virginiana* employed occasionally, but not always, in the face of danger. In this state the opossum becomes immobile, lies with the body and tail curled ventrally, usually opens the mouth, and apparently becomes largely insensitive to tactile stimuli. The condition may last less than a minute or as long as 6 hours (McManus 1974). Although catatonia seems partly under the conscious control of the animal, physiological changes suggest a state analogous to fainting in humans (Lowery 1974). Death feigning may cause a pursuing predator to lose the visual cue of motion or to become less cautious in its approach, thereby giving the opossum a better chance of escape. McManus (1970) thought that anal secretions, sometimes accompanying catatonia, might contribute to deterring a predator. He also suggested that catatonia may have evolved primarily as a means of reducing intraspecific aggression. If so, the condition may be vaguely comparable to submission in certain other animals. According to Hunsaker and Shupe (1977), death feigning has been reported for, but is rare in, *D. albiventris*. Emmons (1997) noted that *D. marsupialis*

Southern black-eared opossum (*Didelphis aurita*). Photo by Leonardo Mercon / Shutterstock.com.

Virginia opossum (*Didelphis virginiana*) feigning death. Photo by Sari ONeal/Shutterstock.com.

does not play dead but will threaten with open mouth, lunge, and bite when cornered or grasped.

Records compiled by Rademaker and Cerqueira (2006) indicate that length of the breeding season in *Didelphis* varied from 12 months near the equator to just 6 months in southern Brazil and Argentina; litter size also tended to vary inversely with latitude. For *D. marsupialis* in Panama, Fleming (1973) found that females were polyestrous, breeding began in January, and two or possibly three litters were born between then and October. Litter size averaged 6 (2–9). Tyndale-Biscoe and Mackenzie (1976) reported that reproduction in that species also began in January on the llanos of eastern Colombia. A second litter was produced there in April or May, and there was no breeding from September to December. Mean litter size was 6.5 (1–11). In western Colombia the breeding period was the same, but a third litter was known to be produced in August. Mean litter size there was 4.5 (1–7). No reproductive season could be determined for *D. pernigra* in Colombia, but litter size averaged only 4.2 (2–7). Eisenberg and Redford (1999) cited average litter sizes of 9.4 young for *D. albiventris* in Uruguay and 10.7 for *D. aurita* in southeastern Brazil.

Except as noted, the following life history data on *D. virginiana* were derived from Collins (1973), Hunsaker (1977), Lowery (1974), Schwartz and Schwartz (1959), and Tyndale-Biscoe and Mackenzie (1976); an extensive review of the topic was provided by Gardner and Sunquist (2003). Females are polyestrous, have an estrous cycle of about 28 days, and are receptive for 1 or 2 days.

As many as seven cycles are possible during a breeding season in the absence of pregnancy. First matings usually occur in January or February in most of the United States but can be as early as mid-December in Louisiana (Edmunds, Goertz, and Linscombe 1978) or as late as March in Wisconsin (Jackson 1961). There are two litters per year in most areas, though Jackson (1961) thought there usually was only one in Wisconsin and a third occurs rarely in the South. The second peak of mating takes place in the spring or early summer, and an average of 110 days separates the two litters. The gestation period is 12.5–13 days. The tiny young are remarkably undeveloped but have pronounced claws, which are used in the scramble from the birth canal to the mother's pouch. The young measure only 14 mm in length and weigh 0.13 grams. Twenty could fit in a teaspoon, and it would take 217 to make one ounce. There usually are about 21 young, but there is a report of 56 being born at once. Since there normally are only 13 mammae, and since the young must continually grasp the mammae to suckle, some newborn often cannot be accommodated and quickly perish. Additional mortality usually occurs later. The number of pouch young generally observed is 5–13 and reportedly averages about 8–9 in northern areas and 6–7 in the South. The young first release their grip on the mammae when they are about 50 days old, begin to leave the pouch temporarily at 70 days, and are completely weaned and independent at 3–4 months. After the pouch becomes too small to hold all the young, but before weaning, some of them ride on the mother's back. Sexual maturity is attained by about 6–8 months, and females apparently

Virginia opossum (*Didelphis virginiana*), newborn. Photo by C. H. Tyndale-Biscoe/Mammal Images Library, American Society of Mammalogists.

have only 2 years of reproductive activity. Very few opossums survive their third year of life, though some laboratories have maintained captives for 3–5 years. Weigl (2005) listed a wild-born individual still living after 6 years and 7 months in the Little Rock Zoo.

The Virginia opossum has adapted better to the presence of people than have most mammals. When European colonists first arrived in North America, *D. virginiana* apparently did not occur north of Pennsylvania. Subsequently it extended its range, the first being taken in Ontario in 1858 (Hunsaker 1977) and in New England during the early twentieth century (Godin 1977). Introductions in New York contributed to this expansion (Long 2003). The species also moved westward on the Great Plains, its progress facilitated by human agricultural development (Armstrong 1972; Jones 1964; Genoways, and Smith. 1974). In 1890 captives were released in California and, boosted by other introductions there in and in Washington and Oregon, the Virginia opossum eventually spread all along the West Coast, becoming well established from southwestern British Columbia to the vicinity of San Diego and entering northern Baja California. Introduced populations also occupy parts of Arizona, western Colorado, and Idaho (Gardner and Sunquist 2003; Gwinn, Palmer, and Koprowski 2011; Hall 1981; Long 2003).

Opossums are sometimes hunted or trapped by people for food and sport and as predators of poultry. Leopold (1959) wrote that in Mexico they are considered chicken and egg thieves, they are hunted for food and local use of their fur, and certain of their parts are believed to have medicinal value. In Peru both *D. marsupialis* and *D. pernigra* have a bad reputation for poultry killing, but populations have not been adversely affected by human settlement (Grimwood 1969). All serious studies indicate that opossums have little impact on poultry and wildlife (Gardner and Sunquist 2003). The value of opossum fur and the consequent commercial harvest have varied widely. During a single year in the 1920s, when prices were high, 518,295 opossum skins were sold in Louisiana alone (Lowery 1974) and 350,286 from Kansas were sold (Hall 1955). In the 1970–71 annual season in the United States 101,278 pelts of *D. virginiana* were reported sold at an average price of $0.85. The corresponding figures for the 1976–77 season were 1,069,725 and $2.50 (Deems and Pursley 1978). A high harvest of 1,750,338

pelts, with an estimated value of $5,715,928, was reported for the United States in 1979–80 (Gardner and Sunquist 2003), a period of general economic distress but when certain commodities, such as silver and gold, also commanded especially high prices. The take for 1983–84 in Canada and the United States was 515,832 skins, and the average price in Canada was $1.75 (Novak et al. 1987). The number of skins taken in the 1991–92 season in the United States fell to 145,290, and the average price to $1.26 (Linscombe 1994). More recently, better skins have sold for $2.00 to $3.00, but trappers commonly discard the others or release the animals (Gardner and Sunquist 2003). In addition to its other uses, *Didelphis* is finding increasing employment in laboratory research.

DIDELPHIMORPHIA; DIDELPHIDAE; **Genus** *MARMOSOPS* Matschie, 1916

Slender Mouse Opossums

There are 17 species (Brown 2004; del Carmen Peralta and Pacheco 2014; Díaz-N. 2012; Díaz-N., Gómez-Laverde, and Sánchez-Giraldoas 2011; Gardner and Creighton 1989, 2007a; Hall 1981; Handley and Gordon 1979; Mustrangi and Patton 1997; Pine 1981; Rocha et al. 2012; Voss and Jansa 2003; Voss, Lunde, and Simmons 2001; Voss, Tarifa, and Yensen 2004; Voss et al. 2013):

M. noctivagus, Amazon Basin of western Brazil, southern Colombia, Ecuador, Peru, and Bolivia;

M. creightoni, Valle de Zongo in west-central Bolivia;

M. ocellatus, eastern Bolivia and adjacent Brazil;

M. impavidus, eastern Panama, Colombia, northwestern Venezuela, Ecuador, Peru, northwestern Bolivia, western Brazil;

M. neblina, known from eastern Ecuador, extreme southern Venezuela, and western Brazil;

M. paulensis, coastal mountains of southeastern Brazil;

M. incanus, eastern and southeastern Brazil;

M. parvidens, Colombia, Venezuela, the Guianas, northern Brazil;

M. pakaraimae, highlands of southeastern Venezuela and western Guyana;

M. invictus, Panama, possibly Colombia;

M. bishopi, southern Colombia, southeastern Peru, Bolivia, western Brazil;

M. handleyi, now known from several localities in the Department of Antioquia in northwestern Colombia;

M. juninensis, Junín Department of central Peru;

M. pinheiroi, southeastern Venezuela, the Guianas, northeastern Brazil;

M. caucae, northwestern Colombia;

M. fuscatus, mountains from central Colombia to northern Venezuela, Trinidad;

M. cracens, known only from the type locality in northwestern Venezuela.

Marmosops was long considered a synonym or subgenus of *Marmosa* but was recognized as a full genus by Corbet and Hill (1991), Gardner and Creighton (1989),

Hershkovitz (1992a, 1992b), and all subsequent authorities cited herein. Much uncertainty remains regarding the number and affinities of the inclusive species (Gardner and Creighton 2007a). However, going by recent studies, notably those of Mustrangi and Patton (1997), Voss and Jansa (2003), Voss, Lunde, and Simmons (2001), and Voss, Tarifa, and Yensen (2004), there appear to be two major clades within *Marmosops*. One, centered on *noctivagus*, also comprises *creightoni*, *impavidus*, and *ocellatus*, with the last two having particular affinity to one other. The second clade, centered on *parvidens*, may include all other species in the genus, though *neblina* sometimes has been associated with *parvidens*, sometimes with *impavidus*. *M. dorothea* of Bolivia often has been considered a distinct species, as by Brown (2004) and Gardner (2005a), but was treated as a synonym of *M. noctivagus* by Gardner and Creighton (2007a) and Voss, Tarifa, and Yensen

Slender mouse opossum (*Marmosops incanus*). Photo by Guilherme Grazzini.

(2004). *M. caucae*, usually placed in the synonymy of *M. impavidus*, was treated by Díaz-N., Gómez-Laverde, and Sánchez-Giraldoas (2011) as a full species with some affinity to *M. parvidens*, *M. pinheiroi*, and *M. handleyi*. Except as noted, the information for the remainder of this account was compiled from Eisenberg (1989), Eisenberg and Redford (1999), Emmons (1997), Gardner and Creighton (2007a), Hershkovitz (1992b), and Voss, Tarifa, and Yensen (2004).

Head and body length is 90–162 mm, tail length is 105–220 mm, and weight is 15–140 grams. In at least some species males are much larger than females. The upper parts are pale to dark brown or gray, the underparts are usually pale gray to white but darker gray or yellowish in some species, and there is a dark eye ring. The pelage is not as long and woolly as in the subgenus *Micoureus* of genus *Marmosa*. The tail is substantially longer than the head and body, prehensile, and, unlike that of *Thylamys*, never incrassate (seasonally thickened by fat deposition). The caudal scales are ar-

ranged in a spiral pattern, not in an annular pattern as in *Gracilinanus*. The skull is proportionally longer and narrower than in *Hyladelphys*, *Marmosa* (including subgenus *Micoureus*), *Gracilinanus*, and *Cryptonanus*. Unlike that of *Marmosa*, the skull lacks strong postorbital processes and a lambdoidal crest. The upper canine teeth are relatively short and often straight, laterally compressed, and bladelike. There is no pouch, and mammae number 13 or fewer.

Notwithstanding the local abundance and relatively accessible habitats of some species, the natural history of *Marmosops* remains little studied. Slender mouse opossums are found in lowland rain forests, lowland dry forests, and montane cloud forests to an elevation of about 3,000 meters. They seem to live predominantly in the understory and are often collected on the ground or a few meters above, on logs, branches, and lianas, but apparently never in the canopy. Both terrestrial and arboreal activity has been reported. They are nocturnal and said to feed on insects and

Slender mouse opossum (*Marmosops ocellatus*). Photo by T. B. F. Semedo / Mammal Images Library, American Society of Mammalogists.

fruit. A population density of 14.4 individuals per sq km has been recorded for *M. parvidens*. Leiner and Silva (2009) found home range of *M. paulensis* to average about 0.40 ha.; male ranges overlapped one another by an average of 30 percent, male ranges overlapped those of females by an average of 72 percent, and ranges of females did not overlap one another, indicating the latter are territorial. It was suggested that males moved between female territories during the breeding season.

M. impavidus apparently breeds year-round. The breeding season of *M. fuscatus* probably extends from May to January or early February; the estimated number of young is seven to nine. Pregnant female *M. parvidens* have been found with six or seven embryos. *M. handleyi* apparently is reproductively active during the rainy season, September–November (Díaz-N., Gómez-Laverde, and Sánchez-Giraldoas 2011). *M. incanus* is said to be nearly semelparous (see account of *Gracilinanus*), there being a single breeding season during the rains from September to December, after which all adult males die. The adult females survive to May and then die, so there is a complete annual turnover of the population. *M. paulensis* also has been found to be semelparous, with adult males disappearing from the population after the September–March reproductive period and adult females disappearing after weaning their young in March and April (Leiner, Setz, and Silva 2008). Notwithstanding the evident short lifespans of those two species, a captive *M. noctivagus* lived for 4 years and 6 months (Weigl 2005).

M. handleyi is classified as critically endangered by the IUCN (Weksler, Patterson, and Bonvicino 2008), as it was thought to occupy an area of less than 10 sq km of forest habitat, which is threatened by conversion to agriculture and pasture. Recently, however, Díaz-N., Gómez-Laverde, and Sánchez-Giraldoas (2011) found it to occupy a somewhat larger area and suggested that it be reclassified to endangered. *M. juninensis* is classed as vulnerable by the IUCN (Pacheco, Solari, and Patterson 2008), though del Carmen Peralta and Pacheco (2014) recommended it be designated endangered. It is known by only six specimens, and its restricted montane forest habitat is being lost to agriculture and illicit coca crops. *M. fuscatus* also is known to be declining because of expanding farms and settlements, but is not in a formal threatened category.

DIDELPHIMORPHIA; DIDELPHIDAE; Genus *CHACODELPHYS* Voss, Gardner, and Jansa, 2004

Pygmy Opossum

The single species, *C. formosa*, is known from Formosa, Chaco, and Misiones Provinces in northern Argentina (Gardner 2007c; Teta and Pardiñas 2007). The species originally was placed in *Marmosa*, then in the genus *Thylamys*, and finally in the genus *Gracilinanus*, sometimes being treated only as a subspecies or synonym. However, on the basis of morphological and molecular studies, Voss, Gardner, and Jansa (2004) erected for it the new genus *Chacodelphys*. They concluded that *formosa* clearly does not belong in *Marmosa* or *Gracilinanus*. It evidently is a sister taxon either to *Monodelphis* or to a clade comprising *Thylamys* and *Lestodelphys*, but to include it in either of those groups would compromise their diagnosability. The type specimen was collected by renowned naturalist Alexander Wetmore in 1920 at Linda Vista in Formosa Province. It remained the only known specimen until Teta, Pardiñas, and D'Elía (2006) and Teta and Pardiñas (2007) reported fragmentary material from owl pellets at four localities and the skull of an individual trapped in another place.

This may be the smallest known American marsupial. The type has a head and body length of 68 mm and a tail length of 55 mm, and it probably weighed no more than 10 grams. The upper parts are brownish with gray-based fur, somewhat darker middorsally, the underparts are gray-based but superficially washed with buff-yellow, and the eyes are narrowly surrounded by a mask of dark fur. The body pelage does not extend onto the tail, which, however, is densely covered to the tip with short hairs and is distinctly bicolored, dark above and pale below; its scales are arranged in annular series, and it is not incrassate. On the forefoot, the third digit is longer than the second and fourth, the claws are shorter than the apical digital pads, and the central palmar surface is densely covered with small convex tubercles. On the hind foot, the fourth digit is slightly longer than the adjacent digits, and the plantar epithelium is naked from heel to toes. In the skull, the rostral process of the premaxillae is absent, the nasals are very narrow and have subparallel lateral margins, the supraorbital ridges are smoothly rounded, and the maxillopalatine openings are very large. The upper molar teeth are strongly dilambdodont (cusps in W-shaped pattern) and highly carnassialized (adapted for holding and cutting) and increase in width from front to back.

Chacodelphys differs from *Marmosa* (including subgenus *Micoureus*), and *Gracilinanus* by the long third digit and densely tuberculate central palmar surface of its forefoot, its very short and apparently nonprehensile tail, its absence of a rostral process of the premaxillae, and its narrow nasals. It differs from *Monodelphis* by the dark mask surrounding its eyes, the short claws and densely tuberculate central palmar surface of its forefoot, the long fourth digit on its hind foot, and its narrow nasals. And *Chacodelphys* differs from *Thylamys* and *Lestodelphys* by not having distinctly tricolored pelage, its non-incrassate tail, and having the plantar epithelium on its hind foot naked from heel to toes.

The pygmy opossum is known from the humid Chaco of Formosa and Chaco Provinces, a mosaic of moist and seasonally flooded grassy savannas and riparian forests, and from the Northern Campos, an area of grasslands and plantations. One specimen was caught in a pit-fall trap set in tall grasses. The short and nonprehensile tail indicates terrestrial habits, and the carnassialized molars suggest an almost exclusive diet of insects and other arthropods. Voss, Gardner, and Jansa (2004) thought the apparent rarity of *Chacodelphys* might stem in part from it being too light to depress the triggers of traps set by mammalogical collectors. However, Teta, Pardiñas, and D'Elía (2006), noting that it also occurs at very low frequencies in owl pellet samples and considering its restricted known range of under 10,000 sq km, believed it to be truly rare. For that reason, and because populations are decreasing and becoming fragmented by encroaching agriculture, *Chacodelphys* now is classified as vulnerable by the IUCN (Teta and de la Sancha 2008).

DIDELPHIMORPHIA; DIDELPHIDAE; **Genus**
CRYPTONANUS Voss, Lunde, and Jansa, 2005

Savannah Gracile Mouse Opossums

Five species are currently recognized (de la Sancha and D'Elía 2015; Gardner 2005a, 2007c; Martinelli, Ferraz, and Teixeira 2011; Quintela et al. 2011; Voss, Lunde, and Jansa 2005):

C. chacoensis, southern Bolivia, Paraguay, northern Argentina, Uruguay, southern Brazil (as far north as states of Mato Grosso do Sul and Minas Gerais);
C. ignitus, known only from the type locality in Jujuy Province in northwestern Argentina;
C. guahybae, state of Rio Grande do Sul in extreme southern Brazil;
C. unduaviensis, northern and central Bolivia, northern Paraguay;
C. agricolai, states of Ceará, Pernambuco, Tocantins, Goiás, Minas Gerais, and Mato Grosso do Sul in eastern and southern Brazil.

The species *chacoensis, guahybae, unduaviensis,* and *agricolai* were described in 1931 or 1943 and at various times considered species of *Marmosa* and/or subspecies or synonyms of species of that genus. Eventually, *chacoensis* and *unduaviensis* came to be treated as synonyms of *Gracilinanus agilis, guahybae* as a subspecies of *Gracilinanus microtarsus,* and *agricolai,* with some question, as a full species of *Gracilinanus* (Gardner 2005a, 2007; Voss, Lunde, and Jansa 2005). Meanwhile, Díaz, Flores, and Barquez (2002) described *ignitus* as a new species of *Gracilinanus*. After analyzing morphological and molecular data, Voss, Lunde, and Jansa (2005) assigned all five taxa to a new genus, *Cryptonanus,* which they considered closely related to but distinct from *Gracilinanus* and part of a clade also containing *Lestodelphys* and *Thylamys.* Although the five taxa were provisionally recognized as full species, it was observed that the series examined were small, that there was an absence of unambiguously diagnostic characters, and that further study might show the five to be populations of a single geographically variable species. It also was noted that, in the event that *chacoensis, guahybae,* and *unduaviensis* proved to be conspecific, the name *chacoensis* would have priority as the name of the resulting species. Regarding the new species *ignitus,* it was suggested that the only known specimen could represent a very old male *chacoensis.*

According to Voss, Lunde, and Jansa (2005), head and body length is 82–111 mm, tail length is 108–135 mm, and weight is 14–40 grams. The upper parts are brownish, reddish brown, or grayish brown; the underparts are whitish, buffy, or orange. The eye is narrowly surrounded by a mask of dark fur, contrasting in color with the paler fur of the cheeks and crown.

The tail appears naked but is sparsely covered by short hairs; it is dark above and paler below, its scales are in distinctly annular series, and it is not incrassate and is prehensile—that is, the ventral surface is naked distally, with an apical pad bearing dermatoglyphs. On the fore-foot, the third and fourth digits are longer than the second and fifth, the claws are shorter than the fleshy digital pads, and the central palmar surface is sparsely covered with tubercles. On the hind foot, the fourth digit is longer than the third and fifth and the plantar epithelium is naked from heel to toes. There is no pouch, and there are 9 to 15 mammae. In the skull the rostral process of the premaxillae is absent, the nasals are conspicuously wider posteriorly than anteriorly, the supraorbital ridges are rounded, and the maxillopalatine openings are very large, but maxillary fenestrae are absent.

Savannah gracile mouse opossum (*Cryptonanus* sp.). Photo by Antoine Baglan through Francois Catzeflis.

Cryptonanus and *Gracilinanus* are similar in size and external characters and might be indistinguishable in the field, though generally *Gracilinanus* tends to have a broader circumocular mask, larger ears, longer mystacial vibrissae, and a relatively longer tail. In the skull, the rostral process of the premaxillae is present in *Gracilinanus* but absent in *Cryptonanus*, a secondary foraman ovale is uniformly present in *Gracilinanus* but absent in *Cryptonanus*, maxillary fenestrae are consistently present in most species of *Gracilinanus* but are absent or very small in *Cryptonanus*, and the second and third upper premolar teeth are equal or nearly equal in height in *Gracilinanus* but the second upper molar is distinctly shorter than the third in *Cryptonanus*.

Most ecological regions in which *Cryptonanus* has been collected are predominantly covered with savannahs or other kinds of open vegetation; it apparently has not been found in Amazonian rainforest or in the wet tropical coastal zone of the Atlantic rainforest. All explicit collection descriptions mention terrestrial capture, and some refer to presence in marshes or seasonally flooded grassland. *C. unduaviensis* may depend on shrub-covered termite mounds for refuge during floods.

According to Gardner (2007c), there is a report of *C. chacoensis* constructing nests in such places as tree holes and among clusters of bromeliads within wet riparian habitat. Specimens were caught there at night or in their nests by day. Seven individuals, interpreted as a multi-generational family unit, were found in one nest 1.6 meters above the ground, and it was believed that females might have as many as 12 young. However, since *C. chacoensis* has only nine mammae and because of other discrepancies in the report, it is possible that the information refers to another genus.

According to IUCN accounts, *C. chacoensis* is considered common over a large region and may even be increasing in some areas (Carmignotto et al. 2011), but prospects for the other four species are dim. Despite extensive searches, *C. ignitus* has not been found since the type specimen was collected in 1962 and presumably was driven to extinction when its habitat was lost to agricultural and industrial development, cattle ranching, and indiscriminate deforestation (Diaz and Barquez 2008a). *C. guahybae* is known from only three localities, which have been highly altered (Costa et al. 2008). *C. unduaviensis* and *C. agricolai* also are suspected to be threatened by loss of habitat to agriculture and deforestation (Brito et al. 2008; Pires Costa and Patterson 2008a).

DIDELPHIMORPHIA; DIDELPHIDAE; **Genus** *GRACILINANUS* Gardner and Creighton, 1989

Gracile Mouse Opossums

There are six species (Brown 2004; Creighton and Gardner 2007a; Gardner 2005a; Tate 1933; Teta et al. 2007; Voss, Lunde, and Jansa 2005; Voss, Lunde, and Simmons 2001):

> *G. agilis*, Peru, Bolivia, eastern and south-central Brazil, Paraguay, northeastern and extreme northern Argentina, western Uruguay;
> *G. microtarsus*, southeastern Brazil, extreme northeastern Argentina (Misiones Province);
> *G. marica*, central Colombia, northern Venezuela;
> *G. emiliae*, central Colombia, northeastern Peru, Venezuela, the Guianas, northeastern Brazil;
> *G. aceramarcae*, southeastern Peru, west-central Bolivia;
> *G. dryas*, the Andes of Colombia and northwestern Venezuela.

The species of *Gracilinanus* long were placed in *Marmosa*, but Creighton (1985) suggested their distinction, and most subsequent authorities have agreed. The species formerly known as *Gracilinanus kalinowskii* has now been placed in the new genus *Hyladelphys* (see account thereof). Based on analysis of mitochondrial DNA, Faria et al. (2013) suggested *G. aceramarcae* is basal to a clade of *G. emiliae* and *G. microtarsus*, which in turn forms a sister group to *G. agilis*. In contrast, morphological and molecular data analyzed by Voss, Fleck, and Jansa (2009) indicated *G. emiliae* to be sister to a clade comprising all other species and that *G. agilis* is sister to a clade of *G. aceramarcae* and *G. microtarsus*; *G. marica*, *G. dryas*, and an as-yet undescribed species from northwestern Colombia and probably eastern Panama were considered less closely related to *G. emiliae* than to its sister clade. Although Costa, Leite, and Patton (2003) suggested *G. microtarsus* might be separable into two species, Loss, Costa, and Leite (2011) subsequently concluded that such division was not warranted. Except as noted, the information for the remainder of this account was taken from Gardner and Creighton (1989) and Hershkovitz (1992b).

Head and body length is 70–135 mm, and tail length is 90–167 mm. Pires Costa, Leite, and Patton (2003) listed weights of 15–40 grams for male and 13–25 grams for female *G. agilis*, and 17–52 grams for male and 12–37 grams for female *G. microtarsus*. Martins et al. (2006) found one population of *G. microtarsus* to be even more sexually dimorphic in size, with males weighing 30–45 grams and females 20–30 grams, but Pires et al. (2010) did not report such dimorphism. An adult male *G. emiliae*

Gracile mouse opossum (*Gracilinanus agilis*). Photo by T. B. F. Semedo / Mammal Images Library, American Society of Mammalogists.

weighed 10 grams (Voss, Lunde, and Simmons 2001). The upper parts range from dull brownish gray to bright reddish brown; the underparts are paler, varying from almost white through cream to orange. There is a dark brown or blackish eye ring, the upper surface of the muzzle is pale, and the tail is scaled and weakly bicolored or unicolored fuscous. There is no pouch. Mammae number about 11–15 and are mostly abdominal, but usually a few are pectoral. The closely related genera *Marmosops* and *Marmosa* lack pectoral mammae.

The skull of *Gracilinanus*, unlike that of *Marmosa*, lacks postorbital processes, the rostrum is slender, and the hard palate is highly fenestrated. The auditory bullae are tripartite and are relatively large compared with those of *Marmosa* and *Marmosops*, but relatively smaller than those of *Thylamys*. The second upper premolar is always larger than the third, and the lower canine teeth are relatively shorter than those of *Marmosa* and *Thylamys* but not as short or premolariform as in Marmosops. The caudal scales of *Gracilinanus*

and *Thylamys* are arranged in an annular pattern, whereas those of *Marmosa* and *Marmosops* have a spiral pattern. Unlike that of *Thylamys*, the tail of *Gracilinanus* does not become seasonally thickened (incrassate) by fat deposition.

These tiny opossums inhabit forests and woodlands from coastal areas to elevations of about 4,500 meters in the Andes. They dwell in trees or shrubs but frequently forage on the ground. *G. microtarsus* often uses tree hollows; its nests are composed mainly of dry leaves and have a central chamber where individuals rest (Pires et al. 2010). Locomotion is by a short, swift quadrupedal gait interchanged with overhand climbing and headfirst descents, assisted by use of the prehensile tail. Individuals have been observed to enter torpor in response to cool weather. The diet consists of fruit, plant exudates, insects, and other small invertebrates. Breeding in some species may continue throughout the year, at least when food is abundant.

According to Pires et al. (2010), population density of *G. microtarsus* in a cerrado remnant ranged from 6.5 to 23.4 individuals per ha. and was higher from December to March and lower from September to November. Average home-range size in that area, as estimated during a 12-month period, was 0.12 ha. for females and 0.14 ha. for males. Home-range size was larger for males than for females and was not affected by season.

Martins et al. (2006) described *G. microtarsus* as an arboreal, nocturnal, insectivorous, solitary, and short-lived (1–2 years) marsupial inhabiting the Atlantic rainforest and highly seasonal cerrado biomes of southeastern Brazil. It begins to breed at the end of the cool, dry season (August–September), and the young are raised and weaned in the first half of the warm, wet season (October–December). Females seem to produce a single litter with 6–14 offspring. Weaning likely occurs when young are 2–3 months old (Pires et al. 2010). Individuals reproduce for the first time when they are ap-

proximately a year old. Martins et al. (2006) attempted to determine whether *G. microtarsus* exhibited semelparity, a condition in which reproduction occurs only once in a lifetime, followed by death of adults, at least of the males, and thus which leads to discrete, nonoverlapping generations. A sharp decrease of males was found in the study area after the beginning of the breeding season, but mortality was not complete, and a small percentage of males apparently survived to breed in a second season. It was concluded that the species is partially semelparous. Lopes and Leiner (2015) found *G. agilis* also to exhibit semelparity, with a breeding season from July to January/February, during which females sometimes produced two litters but mostly disappeared after weaning their young, along with a post-mating die-off of males. Weigl (2005) reported that a zoo specimen of *G. microtarsus* lived only 5 months, but that a captive *G. agilis* was still living at an age of 6 years and 1 month.

The IUCN classifies *G. dryas* as near threatened (Pérez-Hernandez et al. 2011). Its habitat is being severely fragmented as forests are replaced by human settlement and agriculture, especially coffee and cattle production. *G. aceramarcae*, formerly listed as critically endangered, now is considered too poorly known to qualify for a formal threatened category.

DIDELPHIMORPHIA; DIDELPHIDAE; **Genus** *LESTODELPHYS* Tate, 1934

Patagonian Opossum

The single species, *L. halli*, is endemic to Argentina, where it has been recorded from northern Mendoza Province, in the west-central part of the country, south to central Santa Cruz Province in Patagonia. Marshall (1977) reported that the species was known from only nine specimens, but hundreds more have since been found, mostly in owl pellets (Birney et al. 1996; Formoso et al. 2011; Pearson 2007; Sauthier, Carrera, and Pardiñas 2007). *L. halli* occurs farther south than any other living marsupial. Martin, De Santis, and Moreira (2008) reported the southernmost locality to be Estancia "La Primavera" at 48°25'14"S 69°33'41"W. However, Formoso et al. (2011) reported that locality to be at 47°51'10"S 68°56'46"W, and the

Gracile mouse opossum (*Gracilinanus agilis*). Photo by Helder Faria.

Patagonian opossum (*Lestodelphys halli*). Photo by Fritz Geiser.

actual southernmost recorded specimen to be from Estancia La María, about 65 km farther south, just past 48°S.

Head and body length is 132–144 mm and tail length is 81–99 mm. Weight is 60–90 grams (Pearson 2007). The fur is not particularly long, but it is dense, fine, and soft. The back is dark gray, the sides of the body are clear gray, and the forearms, hands, ankles, feet, and underparts are white. The top of the head and an area between the eyes are dark, and there is a black eye ring, but the cheeks, a patch over each eye, and a patch at the posterior base of the ear are whitish. There are dark shoulder and hip patches. The tail, which is furred like the body for about 20 mm at the base, then thickly covered with short, fine hairs, is dark grayish brown above and whitish below and at the tip.

In general appearance individuals of this genus resemble certain species of *Marmosa* and related genera, but they differ in characters of the feet and in distinctive cranial and dental features. The feet are stronger than those of *Marmosa*; the Patagonian opossum probably is more terrestrial than the murine opossums. In *Lestodelphys* the claw of the thumb and other digits extends considerably beyond the soft terminal pad; in *Marmosa* it is markedly shorter than in those of the others and does not extend beyond the pad. The Patagonian opossum has a short, broad skull, small incisors, long, sharp, straight canines, and large molars. In *Lestodelphys* the bullar floor of the skull is complete and does not have a gap between the petrous and alisphenoid components, such as is found in *Gracilinanus*, *Marmosops*, *Marmosa*, *Thylamys*, *Metachirus*, and *Monodelphis* (Hershkovitz 1992b). As in *Thylamys*, the tail occasionally becomes thickened near the base due to a seasonal accumulation of fat. The ears are short, rounded, and flesh-colored.

Southern locality records for *Lestodelphys* generally are in semidesert Patagonian shrubland or steppe, while those in the north are in areas of Monte scrub vegetation (Birney et al. 1996). According to Pearson (2007), some specimens have been taken in traps set on the ground or in the burrows of the rodent *Ctenomys*. Two individuals became torpid while in traps, but hibernation may be erratic, because the southernmost specimen was trapped during winter. That view was supported by Geiser and Martin (2013), who found two captives to exhibit strong daily fluctuations of body temperature, and to enter spontaneous and induced torpor for periods of up to 42.5 hours, but not to demonstrate prolonged or predictable hibernation. Martin and Udrizar Sauthier (2011) reported captives to enter torpor, sometimes for at least 4 days, but only when deprived of food and exposed to lowered ambient temperature. They also observed the captives to climb, jump, and dig well, and to have powerful grasping hind feet and a prehensile tail-tip that could hold the body suspended for several seconds. Both males and females were seen to construct subterranean galleries and nests of grasses in a spiral arrangement, opened at the top.

Thomas (1921) commented: "This interesting little opossum . . . appears, from the structure of its skull, to be of a more carnivorous and predaceous nature than any of the other small members of the family. Ordinarily *Marmosa* feeds mainly on insects and fruit, and as insects are rare and fruit almost non-existent in its far-southern habitat, this opossum has had to acquire peculiar habits, and no doubt lives largely on mice and small birds." Redford and Eisenberg (1992) cited Oliver P. Pearson as reporting that captive *Lestodelphys* "killed live mice at lightning speed, eating everything—bones, teeth, and fur. A 70 gram animal will eat an entire 35 gram mouse in one night." Nonetheless, Zapata et al. (2013) suggested that the Patagonian opossum is not an efficient rodent predator in the wild, and that its diet consists largely of invertebrates, with some birds, reptiles, and fruit.

Birney et al. (1996) videotaped another captive that quickly killed and ate a mouse (*Abrothrix*) placed in its enclosure; they noted also that at least some localities where *Lestodelphys* has been found appear to have a high diversity and abundance of rodents. They collected two probably young but independent *Lestodelphys* in early April (early autumn) and speculated that reproduction most likely is in late summer and early autumn, when prey is most abundant. Considering that factor, and the high number of mammae (at least 19 in one specimen), it was hypothesized that breeding occurs just once a year and that litters are large. Weigl (2005) reported a captive still living after 3 years in the Parque de Animales Silvestres, Temaiken, Argentina.

Populations of *L. halli* are thought to be declining, and some are losing habitat to agriculture. However, considering the recent recovery of hundreds of specimens from owl pellets, the species now is believed to occur in large numbers over a large region. It thus is no longer classified as vulnerable by the IUCN (Martin, Flores, and Teta 2015a).

DIDELPHIMORPHIA; DIDELPHIDAE; **Genus THYLAMYS** Gray, 1843

Fat-tailed Mouse Opossums

The following 2 subgenera and 12 species are tentatively accepted (Braun et al. 2005; Carmignotto and Monfort 2006; Creighton and Gardner 2007c; Formoso et al. 2011; Gardner 2005a; Giarla and Jansa 2014; Giarla, Voss, and Jansa 2010; Palma et al. 2002, 2014; Solari 2003; Teta et al. 2009):

subgenus *Xerodelphys* Giarla, Voss, and Jansa, 2010

> *T. velutinus*, southeastern Brazil;
> *T. karimii*, central and east-central Brazil;

subgenus *Thylamys* Gray, 1843

> *T. macrurus*, southern Paraguay and adjacent Brazil;
> *T. venustus*, southern Bolivia, northwestern Argentina;
> *T. cinderella*, extreme southern Bolivia, northwestern Argentina;
> *T. sponsorius*, extreme southern Bolivia, northwestern Argentina;
> *T. pusillus*, south-central Bolivia, Paraguay, northern Argentina (eastern Formosa Province);
> *T. citellus*, northeastern Argentina (Corrientes and Entre Rios Provinces);
> *T. pulchellus*, dry Chaco region of north-central Argentina;
> *T. elegans*, central Chile;
> *T. tatei*, west-central Peru;
> *T. pallidior*, southern Peru, southwestern Bolivia, northern Chile, northwestern and central Argentina as far south as south-central Chubut Province.

Thylamys was long treated as a synonym or subgenus of *Marmosa*, but since the 1980s both morphological and molecular assessments have overwhelmingly supported its recognition as a full and monophyletic genus (see reviews by Braun et al. 2005; Carmignotto and Monfort 2006; Gardner and Creighton 1989; and Solari 2003). Recent analyses, based on mitochondrial DNA and morphology (Giarla and Jansa 2014; Giarla, Voss, and Jansa 2010), led to recognition of the basal subgenus *Xerodelphys* for *T. velutinus* and *T. karimii*, and to two species groups within the subgenus *Thylamys*, the *venustus* group, with *T. venustus* and *T. sponsorius*, and the *elegans* group, with *T. tatei*, *T. elegans*, and *T. pallidior*; the species *T. macrurus* and *T. pusillus* were also recognized and were considered basal to the *elegans* group, but were not themselves assigned to a group. Another recent study, employing both mitochondrial and nuclear DNA (Palma et al. 2014), sug-

gested a similar phylogenetic sequence, with *T. karimii* being the first species to diverge, followed closely by *T. velutinus*, and then a clade with *T. venustus* and *T. sponsorius*. *T. macrurus* was thought to have diverged subsequently and to be sister to all remaining species. It was followed by *T. pusillus*, which was in a clade with the additionally recognized species *T. citellus* and *T. pulchellus*. Next came *T. tatei* and an undescribed sister species represented by specimens from Arequipa and Lima in the middle and high elevations of the Peruvian Andes. Last to diverge were the sister species *T. elegans* and *T. pallidior*.

The one other species listed above, *T. cinderella*, was not accepted by Giarla and Jansa (2014), Giarla, Voss, and Jansa (2010), or Palma et al. (2014), and the name sometimes is considered a synonym of *T. sponsorius* or *T. venustus*. However, Flores, Diaz, and Barquez (2000) found cranial structure of *T. cinderella* and *T. sponso-*

Fat-tailed mouse opossum (*Thylamys pulchellus*). Photo by Richard D. Sage.

rius to be very different, and Braun et al. (2005) recognized *T. cinderella* and *T. venustus* to be separate but closely related species. Creighton and Gardner (2007c) accepted *T. cinderella* as a distinct species and also noted that additional undescribed species of *Thylamys* might exist in Argentina and Uruguay. Voss et al. (2009) suggested that *T. bruchi*, described from San Luis Province in west-central Ecuador, is a species distinct from *T. pallidior* and more closely related to *T. pusillus*. Martin (2009) distinguished *T. fenestrae*, found in western Buenos Aires, eastern La Pampa, and southern Cordoba and San Luis Provinces of Argentina, as a species separate from *T. pallidior*. However, neither Formoso et al. (2011) nor Palma et al. (2014) believed the morphological differences between the two taxa to warrant more than subspecific distinction.

Some morphological and molecular work, prior to that of Giarla and Jansa (2014), Giarla, Voss, and Jansa (2010), and Palma et al. (2014), resulted in quite different phylogenies of *Thylamys* (again, see reviews by Braun et al. 2005; Carmignotto and Monfort 2006; Gardner and Creighton 1989; and Solari 2003). In particular, both methodologies indicated that *T. macrurus* is basal to the other living species. Morphological and some molecular work (Carmignotto and Monfort 2006; Palma et al. 2002; Solari 2003) suggested further that *T. macrurus* and *T. pusillus* are associated in a "Paraguayan" species group, that *T. velutinus* and *T. karimii* form a "Brazilian" group, which may have descended directly from the Paraguayan group, and that *T. venustus*, *T. elegans*, *T. tatei*, and *T. pallidior* form an "Andean" group, which appears to be a later derivative. However, molecular analysis by Braun et al. (2005) showed *T. macrurus* strongly differentiated from all other species, with *T. venustus* and the related *T. cinderella* as the earliest derivatives, followed by *T. pusillus* and then a group comprising *T. elegans*, *T. tatei*, and *T. pallidior*. Braun et al. (2005) were unable to analyze *T. velutinus* and *T. karimii*. Carvalho, Oliveira, and Mattevi (2009) also suggested affinity of *T. pusillus*

Fat-tailed mouse opossum (*Thylamys karimii*). Photo by T. B. F. Semedo / Mammal Images Library, American Society of Mammalogists.

with *T. elegans*, *T. tatei*, and *T. pallidior*, and suggested those four species form a sister grouping to *T. karimii*.

Head and body length is 68–150 mm, tail length is 65–161 mm, and weight is about 12–55 grams. The tail is slightly longer than the head and body in *T. pallidior* but is shorter in *T. velutinus* and *T. karimii* (Braun, Pratt, and Mares 2010). The upper parts are various shades of gray or brown, the underparts are yellowish or white, and there is a dark eye ring. The fur is very dense and soft. The ears are large and naked. The tail is usually bicolored and is naked or only finely haired, except at the base (Redford and Eisenberg 1992). With the possible exception of *T. macrurus*, the tail is seasonally incrassate—that is, it becomes thickened through storage of fat (Creighton and Gardner 2007c). In the last regard *Thylamys* differs from *Marmosa* (including subgenus *Micoureus*), *Gracilinanus*, and *Marmosops*. It also differs in having stout and well-projecting claws on the forefoot, parallel-sided nasal bones, and the third upper premolar larger than the second (Hershkovitz 1992b). The auditory bullae are relatively large, rounded, and close together (Creighton and Gardner 2007c). There are up to 19 mammae arranged in two bisymmetrical lines on the venter, and there is little or no pouch development (Eisenberg and Redford 1999).

Carmignotto and Monfort (2006) stated that *Thylamys* is found mainly in open and semiarid formations, whereas most other mouse opossums prefer forested habitats. *T. velutinus* and *T. karimii* have characters (such as a short tail lacking a conspicuously prehensile tip, and very small feet and toes with granules covering the whole palmar and plantar surface and with very small or no dermatoglyph-bearing plantar pads) that show a tendency toward an exclusively terrestrial way of life. In contrast, Cáceres et al. (2007) reported *T. macrurus* to be scansorial and capable of climbing trees. And Flores, Díaz, and Barquez (2007) reported that *T. sponsorius* and *T. venustus* occur primarily in humid forests and are arboreal. *T. pallidior* was found mainly in rocky areas and at elevations as great as 4,500 meters. That species is terrestrial but can climb well and sometimes nests in tree holes; it may be active, at least sporadically, throughout the year, but has been found to enter torpor at temperatures of less than 15° C (Braun, Pratt, and Mares 2010). According to Redford and Eisenberg (1992), *T. elegans* lives in wet forest, brush, and scrub and may nest in trees, rocky embank-

ments, or holes in the ground made by Guinea pigs (*Cavia*); it stores fat in its tail, like other species, and hibernates during the winter. *T. pusillus* frequently occurs in much drier areas, such as thorn forest, and is active even when there is snow on the ground.

Thylamys is nocturnal. Its diet consists primarily of insects but also includes fruit and small vertebrates. A 13-year study in Chile found population density of *T. elegans* to fluctuate from about 2 to 22 individuals per ha. (Lima et al. 2001). *T. velutinus* has been reported to reach densities of 0.55 per ha. in the Brazilian cerrado; home range has been estimated at 2.28 ha. for males and 1.7 ha. for a female (Eisenberg and Redford 1999).

T. elegans breeds from September to March, during which time a female can have two litters, each with as many as 17 young, usually 8–12 (Palma 1997; Redford and Eisenberg 1992). Lactating female *T. pallidior* have been reported in Argentina in December and February, and juveniles and young were collected there in February, March, April, May, and June; that species may have around 15 young in a litter (Braun, Pratt, and Mares 2010). Young *T. karimii* were found by Carmignotto and Monfort (2006) in both the wet and dry seasons, suggesting that the species breeds throughout the year. In contrast, Cáceres et al. (2007) found most young of *T. macrurus* in southwestern Brazil during the wet season from October to March. Flores, Diaz, and Barquez (2000) collected breeding females and/or young individuals of several species in northwestern Argentina during January, February, March, April, July, and December. Weigl (2005) reported a captive *T. elegans* that lived 1 year and 7 months in captivity.

Thylamys has a vast distribution but one coinciding extensively with growing human disturbance. *T. karimii* is listed as vulnerable by the IUCN; it is rapidly declining as its habitat is converted to soy bean production (Pires Costa and Patterson 2008b). *T. macrurus* is classified as near threatened in light of widespread population declines resulting from loss of habitat to agriculture and logging; it may be more threatened than currently suspected (de la Sancha and Teta 2015). For the other species, the IUCN accounts report: *T. velutinus*, "intense habitat destruction over much of its range" (Vieira, Astua de Moraes, and Brito 2008), *T. venustus*, "occurs in an area which is being developed and could be threatened" (Diaz and Barquez 2008b), *T. cinderella*, "some populations are declining

Fat-tailed mouse opossum (*Thylamys pusillus*). Photo by Rexford D. Lord.

due to deforestation" (Diaz and Barquez 2008c), *T. sponsorius*, "some populations are threatened by deforestation" (Diaz and Barquez 2008d), *T. pusillus*, "not abundant . . . decreasing" (de la Sancha et al. 2015), *T. citellus*, "deforestation in Entre Ríos Province and conversion of large areas to rice plantations in Corrientes Province could threaten the species" (Flores and Teta 2011b), *T. pulchellus*, "some populations are declining due to deforestation, conversion of most areas from native grassland, shrubland and xerophytic chacoan forests to agricultural lands" (Flores and Teta 2011c), *T. elegans*, "numerous populations are in decline . . . central Chile is a highly perturbed ecosystem" (Solari and Teta 2008), *T. tatei*, "occurs in a region which is being developed in Peru . . . suspected to be threatened" (Solari 2015b), *T. pallidior*, "some populations are threatened by habitat conversion" (Albanese et al. 2015).

Order Paucituberculata

PAUCITUBERCULATA; **Family CAENOLESTIDAE**

"Shrew" Opossums

This order, containing the single Recent family Caenolestidae, with three genera and seven species, is found in western South America. The group sometimes has been placed at only the level of a superfamily (Marshall 1987) or infraorder (Szalay 1994), but full ordinal status was supported by Aplin and Archer (1987), Bublitz (1987), Dickman (2005), Gardner (2005b), Kirsch (1977a), Marshall, Case, and Woodburne (1990), McKenna and Bell (1997), and Patterson (2007a). Some authorities have suggested there may be only a single valid genus and as few as three species (Marshall 1980), but Gardner (2005b) and Patterson (2007a) recognized three genera and six species. Ojala-Barbour et al. (2013) described one more species but also recognized three genera, with a sister relationship between *Lestoros* and *Rhyncholestes*.

These marsupials are somewhat shrewlike in appearance. Head and body length is 90–135 mm, and tail length is 65–135 mm. The head is elongate and conical in shape. The eyes are small, and vision is poor, but the sense of smell is well developed. Sensory vibrissae are present on the snout and cheeks. The lips have well-developed flaps of flesh. The tail is entirely covered with stiff, short hairs. The hind limbs are slightly longer than the forelimbs, and all have five separate, clawed digits. The humerus is large and heavy in comparison with the slender forearm; the hind foot is relatively long and narrow. Females of *Caenolestes* and *Lestoros* have four mammae, and females of *Rhyncholestes* have seven. A marsupium is absent in adult females, though a rudimentary pouch may possibly be present in the young. Patterson (2007a) noted that the penis is bifid and shaped like a corkscrew.

According to Marshall (1984), the mandibular rami are unfused along the symphysis. There is an antorbital vacuity between the nasal, maxilla, and frontal bones. The teeth are rooted, sharp, and cutting. The dental formula is: i 4/3–4, c 1/1, pm 2–3/3, m 4/4=44, 46, or 48. There is one large, laterally compressed, procumbent incisor in each lower jaw, followed by six or seven tiny, spaced, vestigial teeth, among them a vestigial lower canine. The molar teeth are four-sided or almost triangular in outline, and there is a sharp reduction in size from the first to the fourth. Szalay (1994) stated that the Paucituberculata are diagnosed by emphasized vertical shearing between the third upper and the first lower molar and by a carpus in which the lunate and magnum are in contact but are indented by a slight lateral wedge of the scaphoid.

All species prefer densely vegetated, humid habitat, typically in scrub adjacent to meadows of high, moist Andean paramo (Kirsch 1977a). Comparatively few specimens have been collected, probably more because of the inhospitable nature of the habitat than because of the rarity of the animals (Kirsch and Waller 1979). Habits are not well known but may be like those of large forest shrews. Caenolestids use runways on the surface of the ground and are mainly terrestrial but

Common "shrew" opossum (*Caenolestes fuliginosus*). Photo by Hugo Loaiza.

climb well. They are active during the evening and night. Patterson (2015) described them as opportunistic feeders with diets typically including a wide variety of invertebrates (mainly arthropods and annelids), fruits, fungi, and small vertebrates. Kirsch and Waller's (1979) observations suggest that the lip flaps help prevent the sensory vibrissae and fur at the side of the mouth from becoming clogged with blood and dirt during feeding.

The living Paucituberculata are highly relictual, the last survivors of a formerly widespread group, and most species are continuing to decline in the face of human environmental disruption. McKenna and Bell (1997) listed 12 extinct families, which taken together are known from the late Cretaceous of North America, the early Paleocene to the late Pliocene of South America, and the middle Eocene of Antarctica. Aplin and Archer (1987) indicated those families had a morphological and ecological diversity comparable to that of the Australian order Diprotodontia. The geological range of the family

Caenolestidae is late Oligocene to Recent in South America. At least one genus of that family and three of the order were described from Tertiary fossils prior to the formal description of the living caenolestids.

PAUCITUBERCULATA; CAENOLESTIDAE; **Genus** ***CAENOLESTES*** Thomas, 1895

Common "Shrew" Opossums

There are five species (Albuja V. and Patterson 1996; Barkley and Whitaker 1984; Bublitz 1987; Cabrera 1957; Kirsch and Calaby 1977; Ojala-Barbour et al. 2013; Timm and Patterson 2007):

C. convelatus, the Andes of western Colombia and north-central Ecuador;
C. fuliginosus, the Andes of western Colombia, extreme western Venezuela, and Ecuador;

93

Common "shrew" opossum (*Caenolestes sangay*). Photo by Reed Ojala-Barbour.

C. condorensis, known only from the type locality on the eastern versant of the Andes in southern Ecuador;

C. caniventer, the Andes of western Ecuador and extreme northwestern Peru;

C. sangay, the eastern slope of the Andes in southern Ecuador.

The genus *Lestoros* (see account thereof) sometimes has been included in *Caenolestes*. Timm and Patterson (2007) stated that four of the above species could be differentiated into two groups that are well defined morphologically and ecologically. One consists of only *C. fuliginosus*, which is smaller than the other species, has a more slender build, has darker and glossier pelage, and tends to inhabit higher elevation temperate zone forests and paramos. The other group comprises *C. caniventer*, *C. convelatus*, and *C. condorensis*, which

are large and robust, have coarse pelage with pronounced counter shading, and inhabit mid-elevation subtropical forests. However, Ojala-Barbour et al. (2013) found that grouping to be unnatural, based on morphological and molecular analysis, and that *C. convelatus* is the most basal species and a sister to a group formed by the other species, including the newly described *C. sangay*. Albuja V. and Patterson (1996) suggested that *C. condorensis* is more closely related to *C. convelatus* than to *C. caniventer*. Eisenberg and Redford (1999) treated *C. tatei* of southern Ecuador as a species distinct from *C. fuliginosus*, noting that the two might be sympatric, but Ojala-Barbour et al. (2013) confirmed that *tatei* falls within the synonymy of *C. fuliginosus*.

Head and body length is 90–146 mm, and tail length is 93–150 mm. Tail length is approximately the same as head and body length in all species. Kirsch and Waller (1979) recorded weights of 25.0–40.8 grams for adult males and 16.5–25.4 grams for females. The type specimen of *C. condorensis*, the largest known living caenolestid, weighed 48 grams (Albuja V. and Patterson 1996). The pelage is soft and thick over the entire body, but it appears loose and uneven because of the different textures of the hairs. Coloration varies somewhat with the species, but all have dark upper parts, generally deep sooty brown, blackish brown, fuscous black, or plumbeous black. Some species are uniformly colored, or nearly so, whereas some are markedly lighter below. The tail is scantily haired and about the same color as the back, except that in some specimens the tip is white.

These marsupials could with equal justification be described as shrewlike or ratlike in general form. The head is elongate, the eyes are very small, and the ears are rounded and project above the pelage. The tail tapers gradually and is nonprehensile. The forefoot has five digits; the outer toes are small and bear blunt nails, and the other three digits have sharp, curved claws. On the hind foot the great toe is small and bears a small nail, and the other four digits have well-developed, curved claws. Females have four mammae but no pouch. Albuja V. and Patterson (1996) characterized *Caenolestes* by its conical single-rooted incisor teeth, with the fourth incisor mostly filling the space between the third incisor and the canine, the first and second premolars being similar in size, and the infraorbital foramen opening to anterior (not lateral) view. Based on 19 selected craniodental characters, Martin (2013) found *Caenolestes* to have a much greater resemblance to *Rhyncholestes* than to *Lestoros*.

Common "shrew" opossums occur in the alpine forest and meadow zone of the Andes at altitudes of about 1,000–4,300 meters. They prefer cool, wet areas covered by thick vegetation. They are nocturnal and terrestrial and move from one favored area to another by means of runways through the surface vegetation. Kirsch and Waller (1979) found that *C. fuliginosus* may bound at high speed but is not saltatorial. It is primarily terrestrial, but is an agile climber and uses its tail as a prop as it moves up a vertical surface. That species was heard to hiss, squeak, click, and apparently make a sound by drawing air in and out between the lower incisors. *C. fuliginosus* reportedly has well-developed senses of hearing and smell but poor vision. It had long been considered mainly insectivorous, but Kirsch and

Waller (1979) found it readily trapped by baits of meat and that it efficiently killed newborn rats. It also feeds on earthworms and fruit (Patterson 2015). Stomach contents of a series of *C. caniventer* from Peru suggest the diet consists largely of invertebrate larvae but also includes small vertebrates, fruit, and other vegetation (Barkley and Whitaker 1984).

Of the six female *C. fuliginosus* collected by Kirsch and Waller (1979) in southern Colombia from 25 August to 4 September 1969, one was small, four were lactating, and one had enlarged but empty uteri. None had attached young, so apparently the breeding season must begin several weeks before August. Three of the lactating females had all four mammae enlarged, and one had three of the mammae enlarged, suggesting a litter size as great or greater than the mother's capacity to suckle. Barnett (1991) collected a pregnant female *C. caniventer* with two fetuses in southern Ecuador on 9 September.

The IUCN classifies *C. convelatus* as vulnerable (Patterson and Gomez-Laverde 2008). The species is known from a habitat of under 20,000 sq km, one part of which is in Colombia and the other in Ecuador, which is decreasing and becoming severely fragmented because of deforestation. *C. condorensis*, also classed as vulnerable, is known from only one locality with very specific habitat (Solari and Martínez-Cerón 2015a). *C. caniventer* is designated near threatened in light of an ongoing population decline and fragmentation inferred from conversion of forest to agriculture and from logging; it may be more threatened than currently suspected (Solari and Martínez-Cerón 2015b). The newly described *C. sangay* appears to be uncommon, and its known habitat is threatened by road construction and land conversion (Ojala-Barbour et al. 2013).

PAUCITUBERCULATA; CAENOLESTIDAE; **Genus *LESTOROS* Oehser, 1934**

Incan "Shrew" Opossum

The single species, *L. inca*, occurs on the eastern slopes of the Andes in the departments of Cusco and Puno in southeastern Peru and the department of La Paz in extreme western Bolivia. Various authorities long argued that *Lestoros* (synonyms of which are

Incan "shrew" opossum (*Lestoros inca*). Photo by Bruce D. Patterson.

Orolestes Thomas, 1917, and *Cryptolestes* Tate, 1934) would be shown invalid once a sufficient number of specimens had been examined (Marshall 1980). Bublitz (1987) did study a large collection, did place *Lestoros* in the synonymy of *Caenolestes*, and also described *Caenolestes gracilis* of Cusco as a species distinct from what he regarded as *Caenolestes inca*. Corbet and Hill (1991) and some other workers followed suit, but Albuja V. and Patterson (1996), Gardner (2005b), Martin (2013), Myers and Patton (2007), and Ojala-Barbour et al. (2013) maintained generic status for *Lestoros* and treated *Caenolestes gracilis* as a synonym of *Lestoros inca*.

Head and body length is 86–120 mm, and tail length is 96–135 mm. Three adult males collected by Kirsch and Waller (1979) weighed 25.8–31.3 grams, while a female weighed only 13.6 grams. A male listed by Anderson (1997) weighed 31 grams. The pelage is thick and loose, and is uniformly dark brown or slate gray-black or is dark brown above and paler below. In size and general appearance, *Lestoros* resembles *Caenolestes*, particularly *C. fuliginosus*, but is distinguished by various features of the skull and teeth, notably by a low, double-rooted canine, a greatly reduced or complete loss of the first upper premolar, and a very small fourth upper molar. Martin (2013) listed 13 well-marked craniomandibular and dental differences between *Lestoros* and *Caenolestes*. Adult females lack a pouch; they typically have four inguinal mammae but sometimes have a median fifth on the abdomen (Patterson 2015).

The Incan "shrew" opossum inhabits densely vegetated country at elevations of about 2,800 to 4,000 meters. Myers and Patton (2007) collected it in habitats ranging from wet, mossy elfin forest to highly disturbed scrub and second growth, but Kirsch and Waller (1979) found preferred habitat considerably drier than that of *Caenolestes*. Like the latter, *Lestoros* is nocturnal and primarily terrestrial, may bound along

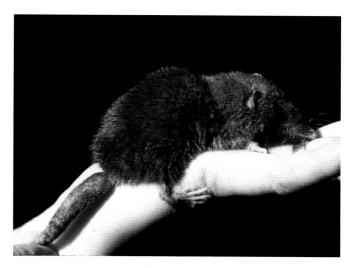 

Chilean "shrew" opossum (*Rhyncholestes raphanurus*). Photos by Peter L. Meserve.

the ground at high speed, and is an agile climber. Captives ate insects, worms, and live newborn rats; the natural diet probably is much like that of *Caenolestes*. Patterson (2015) wrote that females in Manu, Peru, were lactating in November and March.

PAUCITUBERCULATA; CAENOLESTIDAE; Genus *RHYNCHOLESTES* Osgood, 1924

Chilean "Shrew" Opossum

The single species, *R. raphanurus*, is found in southern Chile within the Región de Los Lagos on both the mainland and Chiloé Island, as well as in immediately adjacent Argentina at one locality in the extreme western part of Rio Negro Province. It also sometimes has been reported to occur in the neighboring Argentine province of Neuquen, but this was not mentioned in Martin's (2011) compilation of records. Bublitz (1987) restricted *R. raphanurus* to Chiloé Island and described a second species, *R. continentalis*, for the mainland population. However, Flores, Díaz, and Barquez (2007), Gardner (2005b), and Patterson, Meserve, and Lang (1990) considered *R. raphanurus* the only species in the genus. Patterson (2007b, 2015) provisionally retained *continentalis* as a subspecies of *R. raphanurus*.

Head and body length is 97–128 mm, tail length is 65–88 mm, and weight is 20–40 grams (Patterson 2007b; Pine, Miller, and Schamberger 1979; Redford and Eisenberg 1992). The pelage is loose and soft. The coloration is dark brown or brownish gray above and below with no markings, and the tail is blackish. The external appearance is much like that of *Caenolestes* except that the tail is shorter and, at least periodically, thickened at the base. The skull is narrow and elongate, especially the facial part. The lateral upper incisors are unique among living marsupials in that they have two cusps. Females have seven mammae, the seventh being in a medial position; there is no pouch (Patterson and Gallardo 1987).

According to Patterson and Gallardo (1987), *Rhyncholestes* apparently is restricted to temperate rainforests at elevations from sea level to 1,135 meters. Most specimens have been taken on the ground, alongside logs, and in dense cover. Some have been collected near burrow entrances at the bases of trees or under fallen logs. The genus appears to be mainly terrestrial and nocturnal. Patterson (2007b) suggested it is slower, less agile, and less inclined to prey on rodents than are other caenolestids. It is insectivorous but also eats earthworms and plant matter. Patterson, Meserve, and Lang (1990) described it as semifossorial, feeding chiefly on soil-inhabiting invertebrates and fungi. Kelt and Martínez (1989) found that the tails of animals caught during the autumn were thicker than those of specimens taken in the summer, thus suggesting that fat was being deposited in preparation for winter torpor.

However, some individuals were taken on packed snow during mid-winter, suggesting the absence of extended hibernation. Kelt and Martínez (1989) also found lactating females in February, March, May, October, November, and December.

These marsupials appear to be rare; only three specimens were known through the 1970s. Larger numbers were subsequently collected, but Martin (2011) noted that none had been taken in the previous 14–15 years. Patterson, Meserve, and Lang (1990) stated that *Rhyncholestes* tends to occupy tall, wet forest but exhibits broad elevational range and wide habitat tolerance, and becomes locally abundant. Miller et al. (1983) warned that the dense forest habitat is now shrinking because of logging. Such declines have led the IUCN to classify *R. raphanurus* as near threatened (Diaz and Teta 2008), but the species may be in greater jeopardy than currently suspected (Patterson 2015).

Order Microbiotheria

MICROBIOTHERIA; **Family MICROBIOTHERIIDAE; Genus *DROMICIOPS*** Thomas, 1894

Monito del Monte

This order contains one living family, Microbiotheriidae, with the single living genus and species *Dromiciops gliroides*, which occurs on the mainland of southern Chile between about 36° S and 42° S, on Chiloé Island south to about 43° S, and in adjacent Argentina in the western parts of Neuquen, Rio Negro, and Chubut provinces, south to about 42°30'S (Flores, Díaz, and Barquez 2007; Hershkovitz 1999; Lobos et al. 2005; Martin 2010; Patterson and Rogers 2007). Except perhaps for the extinct Bibymalagasia, Microbiotheria has the most restricted Recent distribution of

Monito del monte (*Dromiciops gliroides*). Photo by Richard D. Sage.

any mammalian order. However, the range may have contracted somewhat southward in historical time (Martin 2010). Until the 1990s, *D. gliroides* commonly was called *D. australis* and divided into two subspecies, *D. a. australis* from the mainland and *D. a. gliroides* from Chiloé Island. However, the name *australis* is preoccupied and subspecific validity is questionable (Gardner 2005c; Hershkovitz 1999; Patterson and Rogers 2007). Until about the 1970s, *Dromiciops* generally was placed in the family Didelphidae. Subsequent investigation indicated it to be distinct from that group and more appropriately referred to Microbiotheriidae, an otherwise extinct family of New World marsupials, which, in turn, was considered part of the group referred to here as the order Didelphimorphia (Archer 1984; Kirsch 1977b; Kirsch and Calaby 1977; Marshall 1982). Even more recently the characters of *Dromiciops* have suggested that it represents a full order distinct from all other living groups of marsupials (Aplin and Archer 1987; Gardner 2005c; Hershkovitz 1992a, 1999; Patterson and Rogers 2007).

The affinities of *Dromiciops* have become a central point of interest and contention with respect to marsupial phylogeny. On the basis of both morphological and biochemical evidence, Reig, Kirsch, and Marshall (1987) regarded the genus as very distinctive but clearly part of Didelphimorphia and having no special similarity to Australian marsupials. Szalay (1982), followed by Aplin and Archer (1987), divided the marsupials into two great cohorts, Ameridelphia, which includes most living New World genera, and Australidelphia, comprising all living Australasian genera. Considering anatomy, serology, and cytology, they assigned *Dromiciops* and the order Microbiotheria (initially with the name Dromiciopsia) to Australidelphia, as did Marshall, Case, and Woodburne (1990). Based to a considerable extent on the analysis of limb bones, Szalay (1994) recognized approximately the same cohort division but designated a new order, Gondwanadelphia, with the New World Microbiotheria and the Australasian Dasyuromorphia as suborders. That group, it was suggested, arose in South America, spread to Antarctica and then Australia when the southern continents were joined during the late Cretaceous, and became the founding stock of the entire great Australasian marsupial radiation.

Hershkovitz (1992a, 1992b, 1995, 1999) disagreed with Szalay's assessment of limb bones and with the assignment of *Dromiciops* and Microbiotheria to an otherwise Australasian cohort. He argued that *Dromiciops* is the sole living member of an entirely separate cohort, Microbiotheriomorphia, which diverged from the other marsupials before development of the existing orders. All living marsupials besides *Dromiciops* can be united in the cohort (not order) Didelphimorphia; an early migration of the latter group from South America led to the Australasian radiation. The development of Microbiotheriomorphia may have occurred as early as the Jurassic, apparently in South America in a cool, humid biome dominated by a beech-bamboo (*Nothofagus-Chusquea*) plant association. That environment and the microbiotheres shifted south to Patagonia and Antarctica during an Eocene warming trend, retreated back north when conditions cooled in the Oligocene and Miocene, and finally became restricted to the current range; microbiotheres never reached Australia. The retention today by *Dromiciops* of a basicaudal cloaca such as found in monotremes but in no other marsupials is suggestive of a very primitive and independent origin. Moreover, *Dromiciops* has a number of key characters found regularly in placental mammals but lost in other marsupials, such as an entotympanic bone at the base of the auditory bullae, a sagittally keeled presphenoid of the mesopterygoid fossa, and a nonstaggered third lower incisor tooth. All other marsupials with comparable dentition are seen to have undergone a reduction in muzzle length, with associated crowding of teeth, so that the third lower incisor is not in line with the other teeth and is usually supported by a bony buttress.

Although Hershkovitz's (1999) review has been cited as a comprehensive examination of the morphology, natural history, and biogeography of Microbiotheria (Patterson and Rogers 2007), his interpretation of the relationships of the order with other marsupials has not been generally accepted (Dickman 2005; Giannini, Abdala, and Flores 2004; Palma 2003; Spotorno et al. 1997). There has been little support for his idea that Microbiotheria represents a cohort separate from and basal to all other existing orders. In most recent studies (e.g., Nilsson et al. 2010), *Dromiciops* appears more closely related to australidelphians than to didelphi-

Monito del monte (*Dromiciops gliroides*). According to the photographer, the individual shown actually represents a separate species (*Dromiciops bozinovici*), as described by a newly published paper (D'Elía, Hurtado, and D'Anatro 2016). Photo by Richard D. Sage.

morphians. According to Palma and Spotorno (1999), chromosomal comparisons, sperm morphology, studies of mitochondrial and nuclear DNA, and biogeographic evidence unequivocally support Szalay's (1982) original proposal that *Dromiciops* is phylogenetically related to Australasian marsupials. However, as shown in the account of class Mammalia (under "Marsupial Superordinal Groupings" and "Marsupial Evolution"), there is considerable disagreement as to whether Microbiotheria is the sister taxon of the other australidelphian orders or nested at some level within the Australidelphia, and as to how Microbiotheria and related groups dispersed through the Southern Hemisphere.

Head and body length of *Dromiciops* is 83–130 mm, tail length is 90–132 mm, and weight is 16–49 grams; the sexes appear identical in size and other nonreproductive characters (Hershkovitz 1999; Marshall 1978b; Redford and Eisenberg 1992). Females evidently have a higher body mass than males in most months, especially in late summer and early fall, suggesting they in-

crease body fat storage before hibernation, probably in order to have energy reserves for the subsequent reproductive season—gestation, lactation, and rearing of young (Franco, Quijano, and Soto-Gamboa 2011). The fur of *Dromiciops* is silky, short, and dense. The upper parts are agouti or brown, with several ashy white patches on the shoulders and rump arranged in vague whorls in a manner somewhat like the color pattern of *Chironectes*. The underparts are buffy to nearly white. Other than pronounced black rings around the eyes, there are no well-defined markings on the face.

Dromiciops may be recognized by its short, furry ears and thick, hairy tail, and by features of the skull and dentition. The dental formula is the same as in the Didelphidae, but the canines, last molars, and first incisors are relatively smaller than those of typical didelphids. The upper incisors are spatulate and relatively large and form a complete series with no gaps (Patterson and Rogers 2007). The tympanic bullae of

Dromiciops are greatly inflated and occupy a relatively large area on the base of the skull. The face is quite short. The basal third of the moderately prehensile tail is furred like the body; the terminal two-thirds is also well furred, but the hairs are straight and slightly different in color, being nearly all dark brown. The only naked part of the tail is a narrow strip 25–30 mm long on the underside of the tip. Females have four mammae in a small but distinct pouch.

The monito del monte has usually been reported to inhabit dense, humid forests, especially areas with thickets of Chilean bamboo (*Chusquea* sp.), but recent records indicate tolerance to a broader range of habitats, including other forest types, drier areas of forest-steppe transition, and human-disturbed areas (Fontúrbel et al. 2012; Martin 2010). It has been found from sea level to an elevation of 1,460 meters (Patterson and Rogers 2007). It sometimes is said to be arboreal, but Pearson (1983) considered it scansorial, noting that while it is a good climber, it may take refuge underground when released from a trap. Patterson, Meserve, and Lang (1990) stated that the highly scansorial habits of this genus are underscored by its prehensile tail, opposable toes, and unqualified aversion to entering enclosed live traps. It makes small round nests, about 200 mm in diameter, out of sticks and the water-repellent leaves of *Chusquea*, lined with grasses and mosses. The nests may be under rocks or fallen tree trunks, in hollow trees, on branches, or suspended in lianas or patches of *Chusquea* 1–2 meters above ground (Hershkovitz 1999; Marshall 1978b; Rageot 1978). *Dromiciops* is almost entirely nocturnal (see Fontúrbel, Candia, and Botto-Mahan 2014). In some areas, at least, it hibernates for lengthy periods when cold temperatures prevail and food is scarce, and thus it has been considered one of the few true hibernators among marsupials (see Bozinovic, Ruiz, and Rosenmann 2004). Hibernation is preceded by an accumulation of fat in the basal part of the tail, which may more than double the weight of an individual in one week (Kelt and Martínez 1989; Rageot 1978). *Dromiciops* also has been observed to enter periods of daily torpor, even when food was readily available (Grant and Temple-Smith 1987). Nespolo et al. (2010) thought further study necessary to determine a clear demarcation between daily and long-term torpor in *Dromiciops*, and whether the genus exhibits the long-term physiologi-

Monito del monte (*Dromiciops gliroides*) in torpor. Photo by Richard D. Sage.

cal and molecular changes that characterize true mammalian hibernation.

The natural diet was once thought to consist mainly of insects and other invertebrates, though Rageot (1978) observed that captives could be maintained on vegetable matter, and Pearson (1983) reported that captives ate large quantities of apple, as well as grubs, flies, and small lizards. Subsequent study of natural populations has shown *Dromiciops* to be omnivorous, commonly feeding on arthropods but also on a diverse array of fleshy fruits, the seeds of which it is a significant disperser, and occasionally the eggs and young of birds (Celis-Diez et al. 2012).

Reported average long-term population densities are about 6 individuals per ha. (Celis-Diez et al. 2012), 20 per ha. (Fontúrbel et al. 2010), 21 per ha. (Fontúrbel and Jiménez 2009), 26 per ha. (Franco, Quijano, and Soto-Gamboa 2011), and 27 per ha. (Rodríguez-Cabal et al. 2008). Considerable variation also has been reported for home range size, Celis-Diez et al. (2012) finding it to be 0.12 to 0.35 ha. for 80 individuals, with those of males being twice as large as those of females, and Fontúrbel et al. (2010) finding it to be 0.7 to 2.2 ha. for 14 individuals, with a spatial overlap of about 50 percent and no difference between sexes.

The monito del monte is considered a sociable marsupial. Its voice has been recorded as a chirring trill with a slight coughing sound at the end. Recent field studies have found communal nesting, which may occur during summer and early fall, but also may involve several (up to nine) individuals in torpor, especially during winter. Celis-Diez et al. (2012) inferred that such groups might be genetically unrelated, because some had five or more juveniles (a single litter never contains more than four surviving individuals), and thus that the arrangement appears an efficient means of conserving energy during times of low temperature. However, Franco, Quijano, and Soto-Gamboa (2011), noted that since communal nesting frequently occurs in warm seasons and involves groups composed of post-reproductive females and juveniles, it could be related more to parental care associated with kin selection than to thermoregulation.

Dromiciops reportedly lives in pairs, at least during the mating season in the austral spring (September–December), though there is no evidence that the male stays to help care for the offspring. Females with suckling young have been recorded from November to January, lactating females have been found as late as March, and litters of one to five have been reported, though no more than four could survive, as females have only four mammae (Collins 1973; Hershkovitz

1999; Marshall 1978b). Although *Dromiciops* has usually been reported to breed once annually, data gathered by Celis-Diez et al. (2012) indicate that females may sometimes bear a second litter. The young release the nipples and leave the pouch when about 10 weeks old. When they outgrow the nest they may ride on the mother's back, then maintain a loose association. In contrast to earlier reports that males and females become sexually mature during their second year of life, observations by Celis-Diez et al. (2012) suggest that individuals of both sexes born in mid-spring (November) could reach sexual maturity by the next breeding season at an approximate age of 10 months, if they reach a minimum body mass of 25 grams in males and 27 grams in females. Rageot (1978) reported catching a female, probably at least a year old, which then lived in captivity for 2 years and 2 months.

Captives may be docile and even willing to remain in human presence (Hershkovitz 1999), but some natives of Chile's Región de Los Lagos, who call this genus "colocolo," believe it is very bad luck to see it or to

Monito del monte (*Dromiciops gliroides*). Photo by Mark Chappell / Animals Animals Earth Scenes.

have one living about the house. Some even have been known to burn their houses to the ground after seeing one of these inoffensive little animals running around inside. *Dromiciops* apparently is common in some areas but now is declining as its restricted habitat is destroyed and fragmented by agriculture, livestock, and logging. Fontúrbel et al. (2012) considered the genus more tolerant of habitat disturbance than previously thought, but observed that its incapability to disperse across nonforested areas suggests that ongoing disruption likely poses a serious threat. The IUCN classifies it as near threatened but notes that it almost qualifies for a threatened category (Martin, Flores, and Teta 2015b).

Reig, Kirsch, and Marshall (1987) considered Microbiotheriidae to include extinct genera from the late Cretaceous of North and South America. Archer et al. (1999) referred to "discovery of at least two microbiotheriid-like marsupials in the early Eocene Tingamarra Local Fauna" of Queensland, Australia. However, most other authorities (e.g., Hershkovitz 1999; McKenna and Bell 1997; Patterson and Rogers 2007) restrict the geological range of the family to early Paleocene to early Miocene in Argentina, Bolivia, and Brazil, middle or late Eocene on Seymour Island off the Antarctic Peninsula, and Recent in Chile and adjacent Argentina. Goin et al. (2007) described a new family of order Microbiotheria, Woodburnodontidae, based on an Eocene specimen from Seymour Island; it is thought to represent an animal that weighed 900–1,300 grams and that would have been the largest known microbiotherian.

Order Notoryctemorphia

NOTORYCTEMORPHIA; **Family NOTORYCTIDAE;**
Genus *NOTORYCTES* Stirling, 1891

Marsupial "Moles"

The single family, Notoryctidae, contains one known
genus, *Notoryctes*, and two species (Benshemesh 2004,
2008; Benshemesh and Aplin 2008; Benshemesh and
Johnson 2003; Woinarski, Burbidge, and Harrison
2014):

  N. typhlops, inland deserts of eastern Western
    Australia, southern Northern Territory, western
    and northern South Australia, and extreme
    southwestern Queensland;
  N. caurinus, deserts of Western Australia from
    the northwestern coast to near the border with
    Northern Territory and south to the Warburton
    area, probably also southwestern Northern
    Territory.

Over the years *Notoryctes* has been placed in various
orders of other Australian mammals either at the familial or the subordinal level. However, studies of morphology, serology, and karyology (Archer 1984; Calaby
et al. 1974; Kirsch 1977b) did not provide a clear idea of
the affinities of this unusual genus with the other marsupial groups. Eventually it was recognized that *Notoryctes* is not closely related to those other groups and that it
belongs in its own order (Aplin and Archer 1987; Marshall, Case, and Woodburne 1990; McKenna and Bell
1997; Westerman 1991). Most recent workers, including

those employing molecular methodology, have ranked
Notoryctemorphia as a full order, though with widely
varying views as to its position relative to other Australidelphian orders (see account of Class Mammalia, especially under "Marsupial Superordinal Groupings"; see
also reviews by Benshemesh and Johnson 2003; Dickman 2005; and Warburton 2006). Part of the disagreement centers on whether the digits of the hind feet of
Notoryctes show signs of syndactyly as found in orders
Peramelemorphia and Diprotodontia (see accounts
thereof); Szalay (1982, 1993, 1994, 1999) thought they
did and combined *Notoryctes* with the members of those
other two groups in the single order Syndactyla.

The species *N. caurinus*, described from the coast of
Western Australia in 1920, often was treated as a synonym of *N. typhlops* (e.g., Corbet and Hill 1991; Ride
1970). However, Johnson and Walton (1989) believed
there to be no substantive basis for synonymy. *N. caurinus* has been accepted as a species by Groves (2005b),
Walton (1988), and most other recent authorities. Unpublished morphological and molecular studies cited
by Benshemesh (2004), Benshemesh and Aplin (2008),
Benshemesh and Johnson (2003), Maxwell, Burbidge,
and Morris (1996), Pearson and Turner (2000), and
Warburton (2006) support specific recognition of *N.
caurinus* and indicate that *N. typhlops* is divisible into a
northern and a southern subspecies.

Head and body length is about 70–140 mm, tail
length is 12–26 mm, and weight is 30–70 grams. Coloration reportedly varies from almost white through
pinkish cinnamon to rich golden red, though Howe

Marsupial mole (*Notoryctes typhlops*). Photo by Stanley Breeden / Lochman LT.

(1975) observed that the reddish color probably results from iron staining. A live specimen lost this coloration after a week in captivity, then again became red after its terrarium was filled with red sand. In life, the pelage, which consists almost entirely of underfur, is remarkably iridescent, fine, and silky.

Notoryctes has about the size and general proportions of placental moles (Talpidae), with a thick, powerful, and somewhat elongate body. The short, cylindrical, stumpy tail is hard and leathery, is marked by a series of distinct rings, and terminates in a horny knob. There is a horny shield on the nose. The vertebrae in the neck are fused, enhancing the rigidity of the body as the animal digs. The third and fourth digits of the forefoot are greatly enlarged and bear large, triangular claws that form a cleft spade or scoop. The remaining three digits of the forefoot are small, but the first and second have claws and are opposed to the third and fourth. The middle three digits of the hind foot also possess enlarged claws. The only external indication of ears

is a small opening beneath the fur on either side of the head. The epipubic (marsupium) bones are reduced. The eyes of *Notoryctes* are vestigial, measure 1 mm in diameter, and are hidden under the skin. There is no lens or pupil, and the optic nerve to the brain is reduced. The female has a distinct pouch that opens posteriorly and contains two mammae. The testes of the male are situated between the skin and the abdominal wall (Benshemesh 2008), and there is no scrotum.

The skull is conical in shape with an expanded occipital region and overhanging nasal bones. The sutures between the bones of the posterior half of the braincase are obliterated. The variable dental formula is: i 3–4/3–4, c 1/1, pm 2–3/2–3, m 4/4 = 40–48, but other arrangements have been reported (see Johnson and Walton 1989; Warburton 2006). The incisors, canines, and premolars are simple and often blunt, though the last upper premolar is bicuspidate; the upper molars are tritubercular. The teeth are well separated from each other. According to Archer et al. (2011),

Notoryctes is unique among living marsupials in having zalambdodont dentition, characterized by upper molars with a single, central cusp homologous with either the paracone or the metacone. That condition, however, is shared with the placental families Tenrecidae, Chrysochloridae, and Solenodontidae (see account of order Afrosoricida).

The discovery of *Notoryctes* in 1888 created a stir in the scientific community comparable to the sensation over discovery of the duck-billed platypus. Marsupial "moles" afford an interesting example of evolutionary parallel between marsupial and placental mammals. They especially resemble the golden moles (order Afrosoricida, family Chrysochloridae) in general body form, burrowing habits, texture of fur, and even external features of the brain (Kirsch 1977b). In an exhaustive study of the functional morphology of *Notoryctes*, Warburton (2006) noted particular convergence in subterranean locomotion with *Eremitalpa* (see account thereof), though her work involved primarily dissec-

tion and X-rays. She stated that collection of marsupial moles has been purely opportunistic and attempts at systematic surveys have invariably failed, thus making statistically significant experimental morphometry impossible. Weigl's (2005) only record on longevity in captivity is of an individual that lived just 1 month. In consequence, the natural history and behavior of Notoryctemorphia may be the most poorly known of the living orders of mammals, though Benshemesh (2004) did compile extensive information and report that about 300 specimens have been recorded.

The habitat of *Notoryctes* is sandridge desert with acacias and shrubs, especially on the crests and slopes of sand dunes, but sandplain and sandy river flats are also used (Benshemesh 2004, 2008; Benshemesh and Johnson 2003; Corbett 1975; Ride 1970). Marsupial "moles" are the most fossorial marsupials and among the most fossorial mammals. They are highly specialized for an almost completely subterranean existence and rarely venture to the surface (Benshemesh 2008;

Marsupial mole (*Notoryctes typhlops*) eating a centipede. Photo by Mike Gillam / Auscape II / Minden Pictures.

Warburton 2006). When they do appear above ground, they shuffle about with frantic haste but very little speed, the body propelled forward mainly by the hind limbs and simultaneously hauled by the powerful forelimbs (Benshemesh and Johnson 2003; Ride 1970). *N. typhlops* leaves a peculiar triple track, with outer marks from the feet and a sinuous central line produced by the downward held tail, but the track of *N. caurinus* shows no obvious tail marks (Benshemesh and Aplin 2008). Surfacing is most common during cooler months and after rains and typically involves just a few meters of movement, but individuals occasionally emerge repeatedly over a period of hours, perhaps in search of conspecifics (Benshemesh 2008; Benshemesh and Johnson 2003).

The actual burrowing process of *Notoryctes* remains unknown, but Warburton's (2006) morphological analysis indicates the foreclaws initially extend anteriorly and function to pull the soil back and down, while the hind feet push and provide support. Then the hind limbs come forward under the body, receive the excavated soil, and thrust it back to fall in behind the animal. The thrust also provides force to propel the animal forward through the soil, the shielded snout perhaps acting as a bore. The tail effectively becomes a fifth limb, serving as a prop to support the body while the hind limbs are drawn under the body to remove the soil.

Marsupial moles sometimes have been said to practically "swim" through the ground and generally to not leave a long-lasting tunnel as do true moles. However, remnants of their excavations persist for months, possibly years, and studies thereof suggest that *N. typhlops* spends most of its time 20–60 cm below the surface (Benshemesh 2008). A recent investigation showed that, rather than sand "swim," *N. typhlops* carves circular tunnels of remarkable consistency and backfills them with sand as it moves forward, leaving clearly distinguishable and long-lasting traces of its passage (Benshemesh 2014). Tunnels are twice as common in the top meter as in the second meter, though some also occur below 2 meters (Benshemesh and Johnson 2003). It is possible that females construct a deep, permanent burrow in which to bear young, but none has been found. As there are data indicating a low metabolic rate and highly variable body temperature in *Notoryctes* (see Withers, Thompson, and Seymour 2000), vertical subterranean movement may be associated with thermoregulation (Benshemesh and Aplin 2008; Benshemesh and Johnson 2003).

Notoryctes is reported to be active, timorous, apparently solitary, and both diurnal and nocturnal (Collins 1973). One captive proved to be extremely active and moved continuously about its enclosure in search of food. Its nose was always held downward. It fell asleep suddenly on several occasions and awoke just as suddenly to resume its feverish activity. Despite the appearance of being highly nervous, it did not seem to resent handling; it would even consume milk rapidly while being held and then would suddenly fall asleep again. This individual fed ravenously on earthworms, but some other captives have refused them (Collins 1973). Actually, it is unlikely they are part of the natural diet since only one rare species of earthworm occurs in the range of *Notoryctes*. Normal prey consists of a variety of small subterranean creatures, especially colonial insects, such as ants and termites, and the larvae and eggs of beetles (Benshemesh 2008). Examination of the digestive tracts of 10 specimens revealed mostly ants, some other insect remains, and some seeds (Winkel and Humphrey-Smith 1987). Another 16 digestive tracts contained mainly ants, termites, and beetle larvae, though termites were present in a significantly smaller proportion than their availability in the areas where the specimens of *Notoryctes* were collected (Pavey, Burwell, and Benshemesh 2012). Captive *N. typhlops* have proved capable of killing and consuming a wide variety of prey, including gekkonid lizards, centipedes, earthworms, and larvae and pupae of moths (Aplin 2015a).

One captive made low-intensity, sharp squeaking noises when restrained (Howe 1975). Such vocal capability is probably used in communication, though the parts of the brain associated with olfaction are highly developed and suggest that social contact is mostly through scent (Benshemesh and Johnson 2003). The only information on reproduction is that in the Northern Territory females give birth about November, that the pouch of one contained a single young, and that there is a museum specimen with two sucklings, but that one pregnant female contained six sub-terminal embryos (Benshemesh and Johnson 2003; Collins 1973; Johnson and Walton 1989; Ride 1970).

Surprisingly, there reportedly was once an active trade in the skin of these small and poorly known crea-

Marsupial mole (*Notoryctes typhlops*), tail visible at left end. Photo by Mike Gillam / Auscape II / Minden Pictures.

tures. Whether such activity involved commercial or scientific demand is not clear (see Aplin 2015a). Nonetheless, several thousand pelts are said to have been brought by Aborigines to trading posts in the Northern Territory from about 1900 to 1920 and later sold on the market for around two British pounds each (then about US$9). For some time afterward, the collection of specimens reportedly remained relatively steady at about 5–15 every 10 years (Johnson and Walton 1989). However, data compiled by Pearson and Turner (2000) suggest a general downward trend in collection, from about 28 in the 1930s to 6 in the 1990s. Anecdotal reports from long-established pastoralists also indicate a decline in abundance (Benshemesh and Johnson 2003). One factor could be predation by the introduced red fox (*Vulpes vulpes*) and domestic cat (*Felis catus*); by the late nineteenth century the latter had established feral populations even in remote desert regions of Australia (Abbott 2002). Recent studies of the scats of the fox and cat, as well as the dingo (*Canis familiaris dingo*), suggest *Notoryctes* is common prey (Woinarski, Burbidge, and Harrison 2014). Other threats include habitat disturbance by livestock and disruption of the Aborig-

inal winter burning that had tended to benefit wildlife (see account of genus *Lagorchestes*). Nonetheless, Benshemesh (2004) questioned the evidence for a substantive decline, and the IUCN no longer classifies *N. typhlops* or *N. caurinus* as endangered, considering they still occur over a wide region and there is insufficient information about population status and threats (Benshemesh and Burbidge 2008; Dickman et al. 2008). Recent surveys have shown both species to be widespread and relatively common (Woinarski, Burbidge, and Harrison 2014).

Although molecular analyses indicated a Paleocene or Eocene origin for Notoryctemorphia, there long were no descriptions of fossils referable to the order (Benshemesh and Johnson 2003). Nonetheless, for some time there had been relevant speculation about certain remains from early Miocene deposits in the Riversleigh World Heritage Area of northwestern Queensland. Archer et al. (1999) thought the remains were from at least one distinct notoryctid and saw no reason why it could not be ancestral to Recent *Notoryctes*. Aplin and Archer (1987) and Warburton (2006) thought the fossils suggested distant relationship between Notoryctemorphia and Peramelemorphia. Such affinity was also supported by some morphological assessments of modern specimens, though most molecular comparisons aligned Notoryctemorphia with Dasyuromorphia (Dickman 2005). Recently, Archer et al. (2011) described the specimens as *Naraboryctes philcreaseri*, and considered that species to represent a transitional stage leading to modern *Notoryctes*. The morphology of the fossils exhibits numerous adaptations for fossorial life, very similar to but less well developed than those of *Notoryctes*, and demonstrate that notoryctids were incipiently zalambdodont and at least semi-fossorial by the early Miocene. However, the faunal assemblages containing *N. philcreaseri* appear to have been adapted to wet forest palaeoenvironments, which suggests that notoryctids, despite being confined today to Australia's sandy deserts, may actually have evolved their burrowing adaptations in soft rainforest floors.

Order Dasyuromorphia

Australasian Carnivorous Marsupials

This order of 3 Recent families, 23 genera, and 79 species is found in Australia, Tasmania, and New Guinea and on some nearby islands. Essentially this order has the same living content as the family Dasyuridae as the latter was accepted by Simpson (1945) and most other authorities until about the 1960s. Subsequent investigation led to the distinction of Myrmecobiidae and Thylacinidae as separate families but eventually to their union with the remaining Dasyuridae at ordinal level (Aplin and Archer 1987; Marshall, Case, and Woodburne 1990). Szalay (1994) regarded Dasyuromorphia as a suborder of his new order Gondwanadelphia, within which he also placed Microbiotheria as a suborder, but Hershkovitz (1992a, 1995) saw no close relationship between those groups. McKenna and Bell (1997) ranked Dasyuromorphia as a grandorder to comprise the families Thylacinidae and Dasyuridae, the latter including Myrmecobiidae as a subfamily. Groves (2005c), while recognizing Myrmecobiidae as a full family, also implied its affinity to Dasyuridae by arranging the dasyuromorphian families in the sequence Thylacinidae, Myrmecobiidae, Dasyuridae. However, recent molecular, morphological, and paleontological work suggests the lineage of Dasyuridae and Thylacinidae diverged from the lineage of Myrmecobiidae prior to the separation of Dasyuridae and Thylacinidae from one another (Archer and Hand 2006; Krajewski, Blacket, and Westerman 2000; Krajewski and Westerman 2003; Krajewski, Wroe, and Wes-

terman 2000; Wroe and Muirhead 1999; Wroe et al. 2000). Hence, the familial sequence followed here corresponds to that recognized by Wroe (1997): Dasyuridae, Thylacinidae, Myrmecobiidae. A new assessment of the mitochondrial genome sequence of *Thylacinus* supports a basal position of Thylacinidae within Dasyuromorphia, with Myrmecobiidae as the sister clade to Dasyuridae (Miller et al. 2009).

The living members of Dasyuromorphia are small to medium in size and vary widely in appearance; head and body length is about 50–1,300 mm, weight about 3 grams to 35 kg. They have a general superficial resemblance to the didelphimorphians and indeed sometimes were combined with the latter group in an order designated Marsupicarnivora or Polyprotodontia. However, dasyuromorphians have only four upper and three lower incisor teeth on each side of the jaw, compared with five upper and four lower in Didelphimorphia. In both orders the incisors are polyprotodont, being relatively numerous, small, and sharp. In that regard the two orders differ from the diprotodont marsupials, which have only one or two lower incisor teeth on each side of the jaw, including a very strongly developed pair. The two orders also share a didactylous condition (having all their digits separate), whereas in the Australasian orders Diprotodontia and Peramelemorphia the second and third digits of the hind foot are syndactylous (joined together by integument). Unlike Didelphimorphia, dasyuromorphians lack a caecum and never have a prehensile tail. According to Szalay (1994), the talonids of the molar teeth of

Red-cheeked dunnart (*Sminthopsis virginiae*). Photo by P. A. Woolley and D. Walsh.

Dasyuromorphia, compared to those of Didelphimorphia, are reduced relative to the size of the trigonids, and there is a lack of magnum contact with the lunate in the carpus. Clemens, Richardson, and Baverstock (1989) indicated that Dasyuromorphia is also characterized by a large neocortex, loss of the calcaneal vibrissae, and a number of adaptations of tarsal morphology for terrestrial existence. Various cranial features that had been proposed to define the order were considered invalid by Wroe (1997), but Wroe et al. (2000) listed nine detailed cranial and dental characters that support dasyuromorphian monophyly.

The known geological range of the order is late Oligocene to Recent in Australasia, though the oldest recorded Australian marsupial carnivore is from the early Eocene (Wroe 2003). Morton, Dickman, and Fletcher (1989) suggested that Dasyuromorphia differentiated from the syndactylous Australian marsupials during the later Eocene, at about the time when the last connection between Antarctica and Australia was broken.

That view, however, is not consonant with the finding of Bininda-Emonds et al. (2007, 2008) that all australidelphian orders originated near or shortly after the Cretaceous/Tertiary boundary.

DASYUROMORPHIA; Family DASYURIDAE

Marsupial "Mice" and "Cats," and Tasmanian Devil

This family of 21 Recent genera and 77 species occurs in Australia, Tasmania, and New Guinea and on some adjacent islands. Reflecting affinities within the family, Archer (1982a, 1984) recognized five subfamilies: Murexinae (sometimes designated by the preoccupied name Muricinae), for the genus *Murexia*; Phascolosoricinae (originally named Phascolosorexinae but emended by Archer 1989), for *Phascolosorex* and *Neophascogale*; Phascogalinae, for *Phascogale* and

False antechinus (*Pseudantechinus ningbing*). Photo by P. A. Woolley and D. Walsh.

Antechinus; Sminthopsinae, with the tribe Planigalini, for *Planigale*, and the tribe Sminthopsini for *Ningaui, Sminthopsis,* and *Antechinomys*; and Dasyurinae, with the tribe Parantechini for *Parantechinus, Dasykaluta,* and *Pseudantechinus,* and the tribe Dasyurini for *Myoictis, Dasyuroides, Dasycercus, Dasyurus,* and *Sarcophilus.*

Analysis of mitochondrial DNA led Krajewski et al. (1994) to recommend abolishing the subfamilies Murexinae and Phascolosoricinae and transferring their constituent genera to, respectively, the Phascogalinae and Dasyurinae. Krajewski et al. (1994) also suggested that the New Guinean species of *Antechinus* were more closely related to the New Guinean genus *Murexia* than to Australian *Antechinus.* Flannery (1995a) observed that the New Guinean species of *Antechinus* would have to be assigned to a separate genus. After analyzing DNA sequence variation and also considering phallic morphology, Krajewski et al. (1996) recom-

mended transferring those species to *Murexia,* hence leaving *Antechinus* endemic to Australia. Armstrong, Krajewski, and Westerman (1998) formerly took that step, while at the same time recognizing the subfamily Phascogalinae as comprising *Phascogale, Murexia,* and *Antechinus,* in that order. Archer and Hand (2006) accepted the three-subfamily arrangement.

However, based on a DNA hybridization study, Kirsch, Lapointe and Springer (1997) had reduced Phascogalinae to a tribe, Phascogalini, within Dasyurinae. The resulting arrangement, as adopted and augmented by Blacket et al. (1999), Krajewski, Blacket, and Westerman (2000), Krajewski and Westerman (2003), Krajewski, Woolley, and Westerman (2000), Krajewski, Wroe, and Westerman (2000), and Krajewski et al. (1997, 2004), has just two subfamilies: Sminthopsinae, with the tribe Planigalini for *Planigale,* and the tribe Sminthopsini for *Antechinomys, Sminthopsis,* and *Ningaui*; and Dasyurinae, with the tribe Phascogalini for

113

Phascogale, Murexia, and *Antechinus,* and the tribe Dasyurini for *Myoictis, Dasykaluta, Parantechinus, Pseudantechinus, Dasycercus, Dasyuroides, Phascolosorex, Neophascogale, Dasyurus,* and *Sarcophilus.* All of that work also involved primarily molecular methodology, though in some cases with consideration of morphological characters.

Van Dyck (2002) made a detailed morphological assessment of the New Guinean species originally placed in *Antechinus* and subsequently transferred to *Murexia* (see above). He concluded that three species were represented and erected a new genus for each. He also assigned one of the two remaining species of *Murexia* (a genus endemic to New Guinea) to a new genus, thus leaving only one species in *Murexia.* The five New Guinean genera so recognized, in order of their apparent evolutionary derivation, are *Micromurexia, Murexia, Phascomurexia, Paramurexia,* and *Murexechinus.* While those genera were considered to form a clade that is primitive relative to Australian *Antechinus* and other dasyurid genera, they lacked clear signs of close relationship to each other and the study did not support resurrection of the subfamily Murexinae.

Van Dyck (2002), however, did treat Planigalinae, Sminthopsinae, Phascogalinae, and Dasyurinae as subfamilies, not merely tribes, for the Australian taxa. Citing largely biochemical criteria, Woolley (2008a) recognized the same four subfamilies for the Australian genera. Her generic content was the same as that of the corresponding four tribes recognized by Krajewski and Westerman (2003) and other authorities cited above, except that she assigned *Ningaui* to Planigalinae.

Groves (2005c) followed Krajewski and Westerman's (2003) two-subfamily/four-tribe arrangement. He accepted Van Dyck's (2002) four new genera but placed them in subfamily Dasyurinae, tribe Phascogalini. Extensive new analysis of DNA sequences and reassessment of morphological data by Krajewski et al. (2007) supported relegating those four genera to the synonymy of *Murexia,* hence keeping their species in that genus, where they had been put by Armstrong, Krajewski, and Westerman (1998), and

Kowari (*Dasyuroides byrnei*). Photo by Alexander F. Meyer at Frankfurt Zoo, Germany.

maintaining *Murexia* as part of the Phascogalini. The order of divergence implied by Krajewski et al. (2007) for those four genera and *Murexia* is: *Paramurexia, Micromurexia, Murexechinus, Phascomurexia,* and *Murexia.*

For the sequence of genera that follows herein, all of the above authorities have been considered, but primary reliance is on the time-scale depiction shown by Krajewski and Westerman (2003) and the modifications suggested by Krajewski et al. (2004). Placement of the genus *Murexia* corresponds to Krajewski et al. (2007), though in keeping with Groves (2005c), the four new genera described by Van Dyck (2002) are tentatively maintained.

Dasyurids are small to medium in size. The genus *Planigale* includes the world's smallest marsupials, some measuring only 95 mm in total length and weighing under 4 grams. The largest living member of the family is the Tasmanian devil (*Sarcophilus*), which may weigh up to 14 kg. The tail of a dasyurid is long, hairy, and nonprehensile. The limbs usually are about equal in length, and the digits are separate. The forefoot has five digits, the hind foot four or five. The longest front toe is the third; digits 2 through 5 of the hind foot are well developed, but the great toe, which lacks a claw, is small or absent. Morton, Dickman, and Fletcher (1989) noted that arboreal species have broad hind feet and a mobile hallux, while species that are both arboreal and terrestrial have longer hind feet and a reduced hallux. Some genera walk plantigrade, that is, on the soles of the feet, and others walk digitigrade, on the toes. The marsupium is absent in some; when present, it usually opens posteriorly and often is poorly developed. The pouches of some genera become conspicuous only during the breeding season. Females have 2–12 mammae, the usual number being 6, 8, or 10.

The basic pattern of dasyurid dentition resembles that of *Didelphis*, though there has been a characteristic reduction (see account of order Dasyuromorphia). The incisors are small, pointed, or bladelike, and the canines are well developed and large, with a sharp cutting edge. The molars have three sharp cusps. The teeth in this family are specialized for an insectivorous or carnivorous diet—rooted, sharp, and cutting. The basic dental formula is: i 4/3, c 1/1, pm 2–3/2–3, m 4/4 = 42 or 46, though Archer and Hand (2006) indicated variations thereof.

Dasyurids have acute senses and are considered alert and intelligent. They are active animals and move rapidly. Most of the members of this family are terrestrial, though some marsupial "mice" are primarily arboreal. The usual shelter is a hollow log, a hole in the ground, or a cave. The mouselike forms are usually silent, but the Tasmanian devil growls and screeches loudly. Dasyurids are active mainly at night and prey upon almost any living thing they can overpower. According to Geiser (2003), it is likely that most dasyurids are heterothermic; all those that have been investigated have displayed daily torpor, usually lasting for several hours.

Morton, Dickman, and Fletcher (1989) provided an extensive summary of the physiology and natural history of the Dasyuridae. They identified six different life-history strategies based on variation in age of sexual maturity, frequency of estrus, seasonality, and duration of male reproductive effort. Krajewski, Woolley, and Westerman (2000) found that two of those strategies characterize a basal separation of the lineages of the two subfamilies recognized through molecular phylogeny: Dasyurinae (11 months to sexual maturity, monestrous females, restricted breeding season, perennially reproductive males) and Sminthopsinae (8–11 months to maturity, polyestrous females, extended breeding season, perennial males). The basic dasyurine strategy appears to have evolved into a third strategy in much of the tribe Phascogalini (males breeding only once and then dying off). A fourth strategy (facultative polyestry, in which females may attempt a second litter if they fail with a first) arose twice in the tribe Dasyurini and once in Sminthopsini. A fifth strategy (aseasonal breeding) appears to have arisen independently in certain species of each of the four recognized dasyurid tribes. A sixth strategy (maturity reduced to 6 months or less) originated in *Sminthopsis*.

As explained by Schmitt et al. (1989), although male dasyurids generally tend to die in large numbers at the end of their very first breeding season, such mortality is nearly complete in certain genera, notably *Antechinus* and *Phascogale*. The phenomenon is thought to result from stress mediated through an increase in the concentration of plasma glucocorticoid and to be exacerbated by a reduction in the concentration of plasma-corticosteroid-binding globulin associated with a progressive rise in plasma androgen. Increased glucocorticoids, which inhibit most stages of the inflammatory and immune responses of animals, results in debilitating effects manifested by anemia, lymphocytopenia and neutrophilia, splenic hypertrophy, gastrointestinal hemorrhage and disease, immune suppression and disease, degeneration of major organs, and negative nitrogen balance.

The geological range of the family is early or middle Miocene to Recent in Australia and middle Pliocene to Recent in New Guinea (Archer 1982a; Archer and Hand 2006; Marshall 1984). Until at least the late Miocene, dasyurids appear to have been rare and to have lacked diversity, while a wide variety of Oligocene and Miocene carnivorous niches were dominated by different species of Thylacinidae. Subsequently, Dasyuridae began a spectacular radiation at the expense of the thylacinids, which underwent a dramatic decline until their probable extinction in modern times (Wroe 1999).

DASYUROMORPHIA; DASYURIDAE; **Genus PLANIGALE** Troughton, 1928

Planigales, or Flat-skulled Marsupial "Mice"

There are five species (Archer 1976; Blacket, Kemper, and Brandle 2008; Blacket et al. 2000; Burnett 2008b; Fisher 2008; Flannery 1995a; Read 2008a, 2008b):

> *P. maculata*, mainly coastal and sub-coastal areas of northern Western Australia, Northern Territory, New South Wales, and northeastern New South Wales;
>
> *P. novaeguineae*, southern Papua New Guinea;
>
> *P. tenuirostris*, inland parts of southeastern Northern Territory, southern Queensland, western New South Wales, and eastern South Australia;
>
> *P. ingrami*, mainly interior plains of northeastern Western Australia, Northern Territory, Queensland, and northeastern South Australia;
>
> *P. gilesi*, inland parts of eastern South Australia, southwestern Queensland, northern and western New South Wales, and northwestern Victoria.

According to Archer (1976), *P. maculata* and *P. novaeguineae* are very similar and might prove conspecific, *P.*

ingrami and *P. tenuirostris* also might not be separate species, and *P. gilesi* is distinct from all other species in possessing only two, rather than three, premolars in both the upper and the lower jaw. Analysis of mitochondrial DNA (Painter, Krajewski, and Westerman 1995) initially indicated that *P. gilesi* is immediately related to *P. novaeguineae* and *P. ingrami* and that *P. maculata* is the most divergent member of the genus, possibly with closer ties to *Sminthopsis*. A subsequent molecular study (Blacket et al. 2000) supported the very close association of *P. maculata* and *P. novaeguineae* and suggested that *P. gilesi* is the most divergent species, but indicated affinity of *P. maculata* and *P. tenuirostris* and a more distant position for *P. ingrami*. The study also raised the possibility that *P. maculata* actually comprises two or three related species and provided evidence that two specimens, collected years earlier in the Pilbara region of northwestern Western Australia, represent two undescribed species of *Planigale*.

Head and body length for the genus is about 50–100 mm, and tail length is about 45–90 mm. The smallest species of *Planigale* and the world's smallest marsupial, *P. ingrami*, weighs 2.6–6.6 grams (Fisher 2008). A series of *P. maculata*, one of the largest species, averaged 15.3 grams for males and 10.9 grams for females (Morton and Lee 1978). The upper parts are pale tawny olive, darker tawny, or brownish gray; the underparts are olive buff, fuscous, or light tan. The feet are light grayish olive or pale brown, and the tail is grayish or brownish. The fur is soft and dense on the body; the tail is short-haired and nontufted. The tail of *P. gilesi* becomes swollen (incrassate) with stored fat when food is abundant, but is thin and bony when conditions are poor (Read 2008a). The central areas of the foot pads are usually smooth but sometimes are striated, the latter condition perhaps being most common in *P. novaeguineae*. In appearance and behavior planigales resemble the true shrews, *Sorex*.

These marsupials are remarkable for their extremely flat skull, which has an almost straight upper profile and a depth of as little as 6 mm. The same condition, however, is closely approached in some species of *Antechinus* (Archer 1976; Ride 1970). The flattened head serves as a wedge to probe through litter for prey and allows the animals to squeeze into narrow crevices (Read 2008a, 2008b). The female's pouch becomes fairly well developed during the breeding season and opens to the

Long-tailed planigale (*Planigale ingrami*). Photo by Jiri Lochman / Lochman LT.

rear. The known number of mammae is 5–10 (possibly up to 15) in *P. maculata*, 6–12 in *P. ingrami* and *P. tenuirostris*, and 12 in *P. gilesi* (Archer 1976).

Planigales occur mainly in savannah woodland and grassland, though *P. maculata* also has been reported from rainforest. They may shelter in rocky areas, clumps of grass, the bases of trees, or hollow logs (Archer 1976). *P. tenuirostris*, *P. ingrami*, and *P. gilesi* depend on deep cracks in clay soils for shelter and may also hunt therein (Fisher 2008; Read 2008a, 2008b). Although primarily terrestrial, planigales climb fairly well (Collins 1973); they are generally nocturnal. *P. gilesi* and *P. tenuirostris* are most active in the 3 hours after sunset and in the hours before sunrise, but during winter they may bask in the sun or become torpid for short periods; except for females with young, individuals have shifting home ranges and may travel more than a kilometer within a few days (Read 2008a, 2008b). Planigales are avid predators, feeding on insects, spiders, small lizards, and small mammals such as *Leggadina* (Collins 1973). They are capable of catching and eating grasshoppers almost as large as themselves. Even tiny *P. ingrami* is a ferocious nocturnal

hunter, both down in the soil cracks and on the surface, and takes centipedes several times its own length (Fisher 2008). Captive female *P. ingrami* ate six to eight grasshoppers 50 mm long every day and appeared to thrive on this diet. Webb et al. (2008) found *P. maculata* to readily attack the introduced and highly toxic cane toad (see account of *Dasyurus hallucatus*) but to usually survive the encounter and then learn to avoid the toad.

Even in the best habitat, the population density of *P. tenuirostris* is probably less than 1 individual per ha., but in winter several huddle together in a nest to conserve energy (Read 2008b). When displaying aggression to another individual, *P. gilesi* has a loud, sharp vocalization; in the breeding season males and females call to one another with high-pitched clicking sounds (Read 2008a). In all Australian species, females are polyestrous, males may breed for more than one season, and the young attain sexual maturity when 8–11 months old (Krajewski, Woolley, and Westerman 2000). Captive females of *P. maculata* can breed throughout the year and produce several litters annually. Gestation is 19–20 days, and litter size is 5–11 young. Males are

Planigale (*Planigale gilesi*). Photo by Gerhard Körtner.

capable of breeding until they are at least 24 months old (Aslin 1975). In a study of wild populations of that species in the Northern Territory, Taylor, Calaby, and Redhead (1982) found breeding to occur in all months; litter size ranged from 4 to 12 young and averaged 8. The young are raised in a spherical nest of thin-bladed grasses (Burnett 2008b). Read (1984) reported that *P. gilesi* has an estrus of 3 days and an estrous cycle of 21 days, while *P. tenuirostris* has an estrus of 1 day, an estrous cycle of 33 days, and a gestation period of 19 days; the breeding season of each species extends from July–August to mid-January, and some females produce two litters per season. A female *P. novaeguineae* with two half-grown young was found in September, near the end of the dry season (Flannery 1995a). Three reports on *P. ingrami* are that it breeds from February to April and has litters of 4–6, that in northeastern Queensland it breeds during the wet season (December–March) and has litters of

4–12, and that females collected in the northern part of Western Australia in December–January had pouch young (Archer 1976). The young of that species are capable of an independent existence at approximately 3 months of age and are mature within their first year (Collins 1973). A captive *P. maculata* lived to an age of 5 years and 9 months (Weigl 2005).

The subspecies *P. ingrami subtilissima*, found in the Kimberley Division in the northern part of Western Australia, and the species *P. tenuirostris* are classified as endangered by the USDI. Those designations seem to have been applied mainly because, at the time (1970), relatively few specimens of either taxon had been collected. Both forms subsequently were found to occur over a more extensive region than previously thought (Thornback and Jenkins 1982). The IUCN does not now assign a threatened classification to any species or subspecies.

DASYUROMORPHIA; DASYURIDAE; Genus *ANTECHINOMYS* Krefft, 1866

Kultarr

The single species, *A. laniger*, originally was found over much of arid and semi-arid Western Australia, Northern Territory, South Australia, western and southern Queensland, central and western New South Wales, and northwestern Victoria (Molsher 2002; Woinarski, Burbidge, and Harrison 2014). Archer (1977) recognized only that one species but observed that there are two distinctive allopatric forms. Lidicker and Marlow (1970) considered those two forms full species: *A. laniger* in New South Wales and Queensland and *A. spenceri* in more westerly areas. Molsher (2002) and Valente (2008) referred to the two forms as subspecies, *A. l. laniger* and *A. l. spenceri*. On the basis of skeletal and dental characters, Archer (1981) treated *Antechinomys* only as a subgenus of *Sminthopsis*, but Woolley (1984) pointed out that biochemical analysis and phallic morphology indicate the two genera are distinct. The former arrangement was used by Corbet and Hill (1991) and McKenna and Bell (1997), the latter by Groves (2005c) and Mahoney and Ride (1988). Molecular assessment has supported generic status for *Antechinomys* and suggested it diverged from *Smin-*

thopsis and *Ningaui* prior to the evolutionary separation of the latter two genera (Blacket et al. 1999; Krajewski and Westerman 2003; Krajewski et al. 1997). There have been suggestions that the species *Sminthopsis longicaudata* belongs in *Antechinomys* (Krajewski et al. 2012).

Head and body length is 70–100 mm and tail length is 100–150 mm; males average about 30 grams in weight, females about 20 grams (Valente 2008). The upper parts are grizzled fawn-gray to sandy brown; the underparts usually are whitish, with gray hairs at the base. There is a dark ring around the eye and a dark patch in the middle of the forehead. The ears are relatively large. The fur is long, soft, and fine, with few guard hairs except on the back and rump. More than the basal half of the very long tail is fawn-colored, then the hairs increase in length, and the terminal third is covered with long brown or black hairs. The upper third of the limbs is furred like the body; the remaining two-thirds is covered by short, fine, white fur. The face is well provided with vibrissae, and usually there are long vibrissae arising from the carpal pads on the wrists.

The body form is well adapted for a bounding locomotion. The feet are narrow and have granular pads. Each hind foot is elongated and has one large, well-haired, cushion-like pad on the sole and only four toes, the first toe of a normal succession of five being absent. The forelimbs are also graceful and elongated. It was once thought that the long hind feet betokened a bipedal, hopping means of progression comparable to that of such rodents as *Notomys*, *Dipodomys*, and *Jaculus*; however, studies have demonstrated that the kultarr actually moves in a graceful gallop, springing rapidly from its hind feet and landing on its forefeet (Ride 1970).

During the breeding season the female's pouch becomes fairly well developed. It consists of folds of skin that enlarge from the sides to partially cover the nipples. The fold is least developed on the posterior side, but unlike in most dasyurids, the pouch does not open to the rear (Lidicker and Marlow 1970). The number of

Kultarr (*Antechinomys laniger*). Photo by P. A. Woolley and D. Walsh.

mammae has been recorded at 4, 6, 8, and 10, there usually being 8 in *A. l. laniger* and 6 in *A. l. spenceri* (Archer 1977).

The kultarr occurs in a wide variety of mostly dry habitats—savannah, grassland, and desert associations. It appears to be strictly terrestrial and nocturnal, sheltering by day in logs, stumps, vegetation, deep cracks in the soil, and the burrows of other small animals, such as trapdoor spiders and hopping mice (Valente 2008). Although it has not been observed to construct its own burrow under natural conditions, captives have dug shallow holes and covered the entrances with grass, and wild individuals have been found in shallow burrows that may have been used for shelter and the rearing of young (Molsher 2002). Radio-tracking has shown that different individuals may use the same burrow on different days, daily home ranges overlap little, and individuals move up to 500 meters a night (Woinarski, Burbidge, and Harrison 2014). The natural diet consists mainly of insects and other small invertebrates. It is not definitely known whether vertebrates, such as lizards and mice, are taken under normal conditions (Archer 1977). Torpor has been induced experimentally through the withholding of food and may be an adaptive mechanism to deal with deteriorating environmental conditions in the wild (Geiser 1986).

Field and laboratory investigations (Stannard and Old 2010; Woolley 1984) show there is a long breeding season. Animals in southwestern Queensland are in reproductive condition from mid-winter to mid-summer (July–January). Females are polyestrous, being able to enter estrus up to six times during the season. The estrous cycle lasts about 35 days. The gestation period, once thought to be less than 12 days, now is thought to be longer, possibly in association with embryonic diapauses or sperm storage. A female may rear up to 8 young per litter. Weaning occurs after about 3 months and sexual maturity at 11.5 months. Both sexes may survive to breed in more than one season, and captive females up to 3 years of age have given birth. Individual females are known to have lived for 48 and 67 months in captivity. Weigl (2005) listed a captive male that lived to an age of 3 years and 6 months.

Although the kultarr is widespread, it is rare over much of its range, has declined to a critically low level in New South Wales, and appears to have disappeared from Victoria, southeastern South Australia, and some parts of Queensland. The main problems seem to be habitat degradation through overgrazing by domestic livestock, predation by feral domestic cats (*Felis catus*) and introduced foxes (*Vulpes vulpes*), large-scale fires, and possibly the ingestion of poisons used to kill insects and predators (Molsher 2002). While acknowledging that populations are decreasing, the IUCN does not assign a formal threatened classification (Morris et al. 2008a). The USDI designates *A. laniger* as endangered, but that classification was intended to apply only to the subspecies *A. l. laniger* of New South Wales and Queensland.

DASYUROMORPHIA; DASYURIDAE; **Genus**
SMINTHOPSIS Thomas, 1887

Dunnarts, or Narrow-footed Marsupial "Mice"

There are 20 species in 7 species groups (Archer 1979b, 1981; Baverstock, Adams, and Archer 1984; Blacket et al. 1999, 2001, 2006; Burbidge, McKenzie, and Fuller 2008; Churchill 2001; Cooper, Adams, and Labrinidis 2000; Crowther, Dickman, and Lynam 1999; Dickman 2008c; Flannery 1995a; Groves 2005c; Kemper et al. 2011; Kitchener, Stoddart, and Henry 1984; Kutt, Van Dyck, and Christie 2005; Mahoney and Ride 1988; McKenzie and Archer 1982; Morton, Wainer, and Thwaites 1980; Pearson and McKenzie 2008; Pearson and Robinson 1990; Van Dyck 1985, 1986; Van Dyck, Woinarski, and Press 1994; Woinarski 2008a; Woinarski, Burbidge, and Harrison 2014; Woolley 2008h, 2008i, 2008j; Woolley, Westerman, and Krajewski 2007):

crassicaudata **group**

> S. *crassicaudata*, southern Western Australia, southern Northern Territory, southwestern Queensland, South Australia, western and central New South Wales, Victoria;

macroura **group**

> S. *bindi*, northern Northern Territory;
> S. *douglasi*, inland parts of northwestern Queensland;

Fat-tailed dunnart (*Sminthopsis crassicaudata*). Photo by Fritz Geiser.

S. *virginiae*, southern lowlands of New Guinea, Aru Islands, northeastern Western Australia, northern Northern Territory, Melville Island, northeastern Queensland;

S. *macroura*, mostly inland parts of northern and central Western Australia, Northern Territory, Queensland, South Australia, and northern New South Wales;

longicaudata group

S. *longicaudata*, central Western Australia, southwestern Northern Territory;

murina group

S. *archeri*, south-central New Guinea, Cape York Peninsula and other parts of northeastern Queensland;

S. *dolichura*, southern Western Australia, southern South Australia;

S. *butleri*, known only from the type locality in extreme northeastern Western Australia and from Bathurst and Melville Islands off the northwest coast of Northern Australia;

S. *leucopus*, known from one isolated area of rainforest in northeastern Queensland, from the coast of extreme southeastern New South Wales and southern Victoria, and from Tasmania (including Bruny, Cape Barren, Clarke, West Sister, and East Sister Islands);

S. *murina*, eastern Queensland, New South Wales, Victoria, southeastern South Australia;

S. *gilberti*, southwestern and extreme southeastern Western Australia, extreme southwestern South Australia;

psammophila group

S. *hirtipes*, southern Western Australia, southern Northern Territory, western South Australia;

S. *psammophila*, isolated spots in southwestern Northern Territory, south-central Western Australia, and southern South Australia;

S. *ooldea*, central and southeastern Western Australia, southern Northern Territory, western South Australia;

S. *youngsoni*, northern Western Australia, southern Northern Territory;

granulipes group

S. *granulipes*, inland parts of southwestern Western Australia;

griseoventer group

S. *griseoventer*, coastal parts of southwestern Western Australia;

S. *boullangerensis*, Boullanger Island in Jurien Bay on the southwest coast of Western Australia and on the nearby mainland;

S. *aitkeni*, Eyre Peninsula of South Australia and nearby Kangaroo Island, known also by subfos-

sils from South Australia in the southwest and southeast and on the Yorke and Eyre Peninsulas.

The above species groups are based largely on the morphological work of Archer (1981) and Van Dyck, Woinarski, and Press (1994), and the molecular analyses of Blacket et al. (1999, 2001, 2006). While there is disagreement regarding the affinities of certain species, Groves (2005c) noted that the groups are strongly distinct and that some may ultimately be given generic rank. Woolley, Westerman, and Krajewski (2007) found that the various forms of penis in the different species of *Sminthopsis* show greatest congruence with the groups delineated by molecular data, but that the six species in the *murina* group display four different penile forms.

A recent analysis of both nuclear and mitochondrial DNA (Krajewski et al. 2012) supported the position that *Sminthopsis* is not monophyletic, and specifically that the species S. *longicaudata* is more closely aligned with the genus *Antechinomys*. That analysis also indicated that the remaining species of *Sminthopsis* are di-

Stripe-faced dunnart (*Sminthopsis macroura*). Photo by Gerhard Körtner.

vided into two major clades: (1) what above are the *crassicaudata* and *macroura* groups, and (2) what above are the *granulipes*, *psammophila*, and *murina* groups. The species *S. ooldea* was found to have more affinity to the *murina* group than to the *psammophila* group. There also were indications that the species of the *crassicaudata* and *macroura* groups may have somewhat more affinity to the genus *Ningaui* than to other species groups of *Sminthopsis*.

Mitochondrial DNA analysis by Cooper, Adams, and Labrinidis (2000) indicates that *S. crassicaudata* comprises two distinctive clades, one in Victoria and southeastern South Australia and the other in the rest of the range of the species, which do not correspond to the two subspecies previously recognized. Blacket et al. (2001) indicated that *S. froggatti* of northeastern Western Australia and *S. stalkeri* of Northern Territory and western Queensland are species separate from *S. macroura*. However, *froggatti* and *stalkeri* were treated as being not more than subspecifically distinct by Groves (2005c), Morton and Dickman (2008b), and Woolley, Westerman, and Krajewski (2007). The form *fuliginosus* of southwestern Western Australia was considered a subspecies of *S. murina* by Kitchener, Stoddart, and Henry (1984), treated as a full species by Mahoney and Ride (1988) and Groves (2005c), discussed in detail but left in question by Crowther, Dickman, and Lynam (1999), referred to as a synonym of *S. dolichura* by Friend and Pearson (2008), and not recognized by Blacket et al. (2006).

S. boullangerensis was found to be completely distinct morphologically from *S. griseoventer* by Crowther, Dickman, and Lynam (1999), but they described it only as a subspecies of the latter because the two differed very little in mitochondrial DNA; Groves (2005c) treated it as a full species. However, Start et al. (2006) argued that neither molecular nor morphological data support taxonomic differentiation and recommended that *boullangerensis* be relegated to the synonymy of the subspecies *S. griseoventer griseoventer*. Kemper et al. (2011) suggested that *S. aitkeni* may also be a subspecies of *S. griseoventer*.

Head and body length is about 40–135 mm, and tail length is about 40–130 mm except in *S. longicaudata*, in which tail length is approximately 200 mm or about twice head and body length. Adult *S. crassicaudata* average about 75 mm in head and body length and about

15 grams in weight. Average weight of *S. macroura* is 19 grams for males, 16 grams for females (Godfrey 1969). In *S. douglasi*, one of the largest species, males weigh 50–70 grams, females 40–60 grams (Woolley 2008i). Respective figures for *S. virginiae* are 31–58 and 18–34 grams (Woolley 2008j) and for the smaller *S. ooldea*, 9–17 and 8–15 grams (Foulkes 2008). The fur is soft, fine, and dense. The back and sides are buffy to grayish, the underparts are white or grayish white, the feet are usually white, and the tail is brownish or grayish. Some species have a median facial stripe.

This genus is differentiated from other marsupial "mice" largely by features of the skull and dentition. The feet are slender, and the pads are striated or granulated. The hind part of the soles lacks pads. Most species have 8–12 mammae, though some have only 6. The pouch is relatively better developed than in most other marsupial "mice." In some species the tail accumulates fat and becomes carrot-shaped (incrassate) during times of abundant food. The fat reserve may be utilized when food is scarce, and the tail then will become thin. In *S. murina*, *S. leucopus*, *S. virginiae*, *S. longicaudata*, and *S. psammophila* the tail never becomes fat, not even under the best of conditions (Ride 1970).

Most of the permanently thin-tailed species live in moist forest or savannah, but *S. longicaudata* and *S. psammophila* occupy arid grassland and desert. As would be expected, species capable of storing food in the tail are mainly inhabitants of dry country (Ride 1970), though *S. crassicaudata* is sometimes found in moist areas (Morton 1978a). Most species are known to be strictly nocturnal. Dunnarts rest by day in burrows, often made by other animals, or they construct nests of grasses and leaves in hollow logs or clumps of grass or under bushes or stumps. During the summer *S. crassicaudata* reportedly shelters among rocks. That species uses bipedal bounds when traveling at top speed, but over short distances it has a peculiar quadrupedal ramble and holds its tail above the ground in a stiff upward curve. It and the other species are mainly terrestrial, but some are agile climbers. Ewer (1968) found that *S. crassicaudata* has alternating periods of activity and rest during the night. Individuals of that species, along with *S. douglasi*, *S. macroura* and *S. murina*, also have been reported occasionally to enter a state of torpor during times of low food supplies (Geiser 2003; Geiser et al. 1984; Morton 1978d; Ride 1970).

Long-tailed dunnart (*Sminthopsis longicaudata*). Photo by P. A. Woolley and D. Walsh.

Torpid periods of around 8–20 hours have been reported (Geiser 2003). Dunnarts are mainly insectivorous, but small vertebrates, such as lizards and mice, also are fiercely killed and eaten. Most insects are caught on the ground, but *S. murina* sometimes leaps high into the air to catch moths in flight. Because *S. crassicaudata* and some other species obtain enough water from juicy prey, such as cockroaches and spiders, they do not need to drink (Morton and Dickman 2008a).

Using radio-telemetry in the Simpson Desert of western Queensland, Haythornthwaite and Dickman (2006) found *S. youngsoni* to be serially nomadic and able to move a long way to areas of greater food availability when local resources are depleted. Average nightly foraging distance was 412 meters, but males moved about three times as far as females, sometimes traveling more than 2 km per night. Males generally remained in valleys between sand dunes, while pregnant or lactating females apparently moved to dune sides and tops in response to the spring abundance there of invertebrate prey. By day, individuals usually sheltered in scorpion or spider burrows, rarely returning to the same one twice. In a contrasting study in the coastal heathlands of Victoria, Laidlaw, Hutchings, and Newell (1996) found *S. leucopus* to occupy overlapping home ranges of only about 1 ha. and no difference between sexes in distances moved. The animals sheltered in hollow trees or underground burrows, but it was unclear as to whether they dug the burrows themselves.

According to Churchill (2001), a radio-tracking study of 15 *S. psammophila* showed that individuals generally moved 200 to 300 meters per foraging period but were capable of going farther; one moved 1,940 meters in 2 hours. Home range averaged 7.8 ha. (1.8–19.0 ha). Male ranges overlapped those of other males and females, but female ranges may have been exclusive.

Limited data indicate individuals remain within an area for at least 8 months, but the boundaries of the home range may drift over time. At one site, five adults occupied an area of 20 ha., implying a potential density, in suitable habitat, of up to 25 per sq km. Fox (2008) reported densities of up to 6/ha. for *S. murina* in habitat regenerating after burning.

For *S. crassicaudata* in two parts of Victoria, Morton (1978b) reported population densities of 0.5/ha. and 1.3/ha. Individuals occupied home ranges, but precise size could not be determined as the areas utilized shifted over a period of months. Home ranges overlapped among the same sex and between sexes even during the breeding season, and no territorial behavior by males was observed. Breeding females, however, tended to be sedentary and possibly defended a small territory around the nest. Both sexes usually nested alone during the breeding season, but at other times up to 70 percent of the population shared nests in groups of 2–8 that included members of either or both sexes. Such groups apparently were nonpermanent, random aggregations. If nest sharing did occur during the breeding season, it generally involved pairs of a male and an estrous female. Males tolerated one another to some extent during the breeding season but may have attempted to monopolize estrous females.

Collins (1973) reported that captive dunnarts could be maintained in pairs or small groups if sufficient space and multiple nesting facilities were provided, but that an adult male might be attacked by a female with a litter. Fox (2008) stated that males of *S. murina* become increasingly hostile to each other at the onset of breeding, but are attracted by the vocalization of females. The latter have an estrous cycle of 24 days and can produce two litters of up to 10 young each from August to March. Aslin (1983) found captive adult *S. ooldea* to be generally intolerant of one another, even of the opposite sex, and to fight to the death when caged together. That species attains sexual maturity at around 10 months and gives birth to 5 to 8 young from September to January. According to Foulkes (2008), the young of *S. ooldea* release the teats after about 30 days, open their eyes and cling to the mother's back at around 45 days, and are largely independent by 70 days.

White-tailed dunnart (*Sminthopsis granulipes*). Photo by P. A. Woolley and D. Walsh.

In most species of *Sminthopsis* reproductive activity is seasonal but extended, females are polyestrous, males may survive to breed in more than one season, and sexual maturity is attained by 6 months of age or less. Exceptions include *S. griseoventer* and *S. leucopus*, in which females are, respectively, monestrous and facultatively polyestrous, and sexual maturity comes at 11 months (Krajewski, Woolley, and Westerman 2000). Female *S. psammophila* also may exhibit facultative polyestry; they probably usually produce a single litter each year in September or October and wean the young in December or January, but they may be able to extend the time of reproduction and produce a second litter during good seasons (Churchill 2001). *S. virginiae* is capable of breeding year-round, at least in captivity; its estrous cycle is about 30 days, gestation about 15 days. The 6–8 young suckle for 65–70 days and are mature when 4–6 months old (Woolley 2008j).

The reproductive traits of *S. crassicaudata* are fairly well known (Collins 1973; Ewer 1968; Godfrey and Crowcroft 1971; Jackson 2003; Morton 1978c). Females are able to produce litters continuously in captivity. Estrus occurs in repeated cycles of 25–37 days, extending through a season of at least 6 months and perhaps the entire year. Estrus lasts 1–3 days. In the wild, breeding appears to be restricted to a period of 6–8 months, starting in June or July, during which each female is thought to give birth to two litters. The gestation period is about 13 days. Litter size is 3–10, usually 7–8 in the wild and slightly smaller in captivity. The young are carried in the mother's pouch until they are about 42 days old and then are left in the nest until they are about 63 days old. They are then practically self-sufficient, and the family soon breaks up. Minimum age of sexual maturity is 115 days in females and 159 days in males.

The breeding season of *S. macroura* was found to last from July to February in a captive colony in South Australia (Godfrey 1969), and reproductive activity in the wild occurs at the same time (Morton and Dickman 2008b). Females are polyestrous, with an average cycle length of 26.2 days. Gestation is about 11 days, the shortest known for any marsupial. The usual litter size of eight is equal to the number of mammae; litters of only one or two sometimes are produced but are not reared. The young are carried in the pouch for 40 days, then suckled in the nest for 30 more days. Females mature at about 4 months of age, much earlier than

males, which mature at 9 months. The same phenomenon has been observed in *S. douglasi* (Lundie-Jenkins and Payne 2000; Woolley 2008i), with females maturing at 17–27 weeks, males at 28–31 weeks. Maturation of females, always prior to males of the same litter, may be an important mechanism for avoiding inbreeding.

Morton (1978c) found that in the wild few female *S. crassicaudata* lived past 18 months, and few males lived past 16 months. In captivity dunnarts seem to survive and reproduce over a longer period (Collins 1973). The longevity record is held by a specimen of *S. youngsoni* that lived for 5 years and 4 months (Weigl 2005).

The IUCN classifies *S. aitkeni* as critically endangered (Van Weenen 2008), *S. psammophila* as endangered (Robinson et al. 2008), *S. butleri* and *S. leucopus* as vulnerable (Lunney, Menkhorst, and Burnett 2008; McKnight 2008a), and *S. douglasi* as near threatened (Burnett and Winter 2008a). All are highly restricted in distribution and/or population size, and further declines are expected. Although once found on both sides of Kanagaroo Island, *S. aitkeni* now is restricted to an area of around 100 sq km on the western end, where fewer than 500 individuals are thought to survive. Numbers are continuing to fall because of wildfires and other human-induced habitat disturbance. Status of two recently discovered, isolated groups on the Eyre Peninsula is not clear. Woinarski, Burbidge, and Harrison (2014) treated *aitkeni* as a subspecies of *S. griseoventer* (see above) and classified it only as endangered.

S. psammophila has been reduced to fewer than 2,500 mature individuals in remnant patches of suitable habitat, and still is decreasing through predation

Common dunnart (*Sminthopsis murina*), mother with young in pouch. Photo by Fritz Geiser.

by the introduced red fox (*Vulpes vulpes*) and domestic cat (*Felis catus*), environmental disturbance by domestic livestock and the introduced rabbit (*Oryctolagus cuniculus*), land clearance for agriculture, and a change in fire patterns through removal of traditional Aboriginal burning practices. According to Churchill (2001), the Aborigines burned old vegetation to encourage new growth, a process that over time established a complex mosaic of plant communities providing more cover and food. Termination of the process reduced habitat diversity and allowed build up of vegetation and thus an increase in large-scale wildfires. Aitken (1971a) reported that *S. psammophila* had been known from only a single specimen collected in 1894 in the Northern Territory but that a colony was discovered on the Eyre Peninsula of South Australia in 1969. Additional specimens were subsequently found in both South and Western Australia (Pearson and Robinson 1990).

S. butleri long was known from only a couple of specimens discovered in the Kimberley region of Western Australia in the 1960s, but later was found in larger numbers on Bathurst and Melville Islands (Woolley 2008h). It now may have disappeared on the mainland, because of human habitat disruption, and is threatened on the islands by the same factor, as well as introduced predators and changes in burning practices.

S. leucopus apparently has a patchy distribution, with fewer than 1,000 mature individuals in the New South Wales population, fewer than 2,000 in Victoria, and probably fewer than 5,000 in Tasmania. Status of the Queensland population is not well known, but only a few specimens have been collected. The species evidently depends on early to mid-successional forest stages and is unable to adapt to severe disturbances that promote dense regrowth. Based on reapplication of IUCN criteria, Woinarski, Burbidge, and Harrison (2014) did not place the overall species in a category of concern, but did classify the subspecies *S. l. ferruginifrons* of southern Victoria and southeastern New South Wales as near threatened.

Julia Creek dunnart (*Sminthopsis douglasi*), mother with three 60-day-old young. Photo by P. A. Woolley and D. Walsh.

S. douglasi occurs at scattered sites within a small area used predominantly for livestock. Grazing by sheep and cattle and the introduction of prickly acacia trees (*Acacia nilotica*) to provide shade for livestock has eliminated ground cover necessary to *S. douglasi* (Lundie-Jenkins and Payne 2000). It is also declining because of predation by cats and foxes. It had been known by only four specimens, the last collected in 1972, but was rediscovered by Woolley (1992).

Another species, *S. murina*, is not in a formal IUCN threatened category but is known to be declining, probably on account of the same factors threatening the other species. The IUCN (Dickman, Burnett, and McKenzie 2008) noted a need to reassess the taxonomic status of the subspecies *S. murina tatei* of northeastern Queensland, which it formerly designated as near threatened. Blacket et al. (2006) did determine the genetic distinction of *tatei*, but Woinarski, Burbidge, and Harrison (2014) did not place the subspecies in a category of concern. Using IUCN criteria, however, the latter authorities did classify the species *S. archeri* and *S. bindi* as near threatened; both occur in small areas and are thought to be declining because of habitat degradation by people and predation by feral cats.

The IUCN also formerly applied a critically endangered classification to what is listed above as the species *S. boullangerensis*, though at the time considered it an unnamed subspecies of *S. griseoventer*. The current IUCN account (McKenzie and Kemper 2008) indicates that the taxonomic status of the involved population, on Boullanger Island, needs clarification (see above), but that if it is a species, it would almost certainly qualify for a threatened category. Woinarski, Burbidge, and Harrison (2014), citing Start (2006), did not consider *boullangerensis* a distinct taxon.

The species *S. longicaudata* and *S. psammophila* are listed as endangered by the USDI and are on appendix 1 of the CITES. *S. longicaudata* had been known by only five specimens, the last found in 1975 (Thornback and Jenkins 1982), and its apparent rarity was the basis for the USDI listing (in 1970). However, the species later was rediscovered alive in the Northern Territory and Western Australia (Burbidge, McKenzie, and Fuller 2008).

DASYUROMORPHIA; DASYURIDAE; Genus *NINGAUI* Archer, 1975

Ningauis

There are three species (Archer 1975; Baverstock et al. 1983; Carthew and Bos 2008; Dunlop, Cooper, and Teale 2008; Kitchener, Stoddart, and Henry 1983; Krajewski, Wroe, and Westerman 2000; McKenzie and Dickman 2008):

> *N. timealeyi*, Pilbara and adjacent parts of northwestern Western Australia;
> *N. yvonneae*, south-central Western Australia, southern South Australia, southwestern New South Wales, northwestern Victoria;
> *N. ridei*, interior parts of Western Australia, Northern Territory, southwestern Queensland, and northern South Australia.

Molecular assessment (Krajewski, Wore, and Westerman 2000; Krajewski et al. 1997) indicates that *N. timealeyi* was the first species to evolutionarily diverge and that *N. yvonneae* and *N. ridei* are sister species.

Head and body length is 46–75 mm, tail length is 59–95 mm, and weight is 3.5–14.0 grams (Carthew and Bos 2008; Dunlop, Cooper, and Teale 2008; McKenzie and Dickman 2008). The upper parts are gray or brown to black, the underparts are usually yellowish, and the sides of the face are salmon to buffy brown. The tail is thin, without a brush or crest. The pouch is a simple depression; *N. timealeyi* has 4–6 mammae, *N. yvonneae* has 7, and *N. ridei* has 6–8 (Dunlop, Cooper, and Teale 2008; Kitchener, Stoddart, and Henry 1983; McKenzie and Dickman 2008). *Ningaui* is considered most similar to *Sminthopsis* but to differ in smaller size, longer hair, broader hind feet with enlarged apical granules, and various cranial and dental characters. Archer (1981) stated that *Ningaui* resembles *Sminthopsis* in having wide molars and narrow nasals and in lacking posterior cingula on the upper molars, but differs from *Sminthopsis* in lacking squamosal-frontal contact in the skull.

Ningauis occur in dry grassland, brush country, and savannah, frequently in association with spinifex or hummock grass (*Triodia*), and appear adapted to life under arid conditions (see Bos and Carthew 2003). They are nocturnal, sheltering by day in dense hum-

Ningaui (*Ningaui ridei*). Photo by Jiri Lochman / Lochman LT.

mocks of spinifex, hollow logs, or small burrows, most likely dug by other animals, such as lizards and large spiders. At night they hunt on the ground and in shrubs. *N. ridei* often climbs along spinifex leaves, using the tail in a semiprehensile manner (McKenzie and Dickman 2008). Captive *N. yvonneae* often enters daily torpor, lasting up to about 12 hours, especially in response to low temperature and withdrawal of food (Geiser 2003; Geiser and Baudinette 1988).

Ningauis prey on insects, other invertebrates, and possibly small vertebrates. *N. timealeyi* may attack and overcome desert centipedes and cockroaches much larger than itself (Dunlop, Cooper, and Teale 2008). According to Joan M. Dixon (National Museum of Victoria, pers. comm., c. 1981), a female ningaui captured in Victoria on 28 December 1977 lived until 6 April 1980. It was fed on a wide variety of insects as well as oranges and apples. The animal was vocal throughout its period in captivity and used a high-pitched rasping noise to demand food.

Radio-tracking of *N. ridei* in the Simpson Desert suggests individuals have no fixed home ranges and can move at least 1.5 km within 3 days (McKenzie and Dickman 2008). *N. yvonneae* occasionally moves up to 2 km, but usually females are fairly sedentary, the average distance between recaptures being less than 100 meters. Male *N. yvonneae* regularly travel more than 200 meters between recaptures, especially during the breeding season (Carthew and Bos 2008). In the breeding season of *N. timealeyi*, males become aggressive toward each other, and females with pouch young drive other adults away (Dunlop, Cooper, and Teale 2008).

Ningaui generally is characterized by seasonal but extended breeding, polyestrous females, males that can potentially breed in a second year, and sexual maturity attained at 8–11 months of age (Krajewski, Woolley, and Westerman 2000). In years with good rainfall the reproductive period of *N. timealeyi* may extend from September to March (spring and summer). Usually five or six young are carried to weaning. By March in most years the population of *N. timealeyi* consists predominantly of the now independent young. They attain sexual maturity in late winter and both sexes can survive into a second breeding season (Dunlop, Cooper, and Teale 2008).

In a study of a group of *N. ridei* captured in the Northern Territory, Fanning (1982) found the breeding season to extend from early September to late February. Males uttered a distinctive mating call, "tsitt," and females responded similarly. Gestation was 13–21 days and most litters contained 7 young. They remained attached to the nipples until about 42–44 days, then were left in a nest constructed by the mother, and became independent at 76–81 days. Bos and Carthew (2001) reported that on the Eyre Peninsula of South Australia *N. yvonneae* mates from September to February but mostly in mid-October; litter size is usually 5–7 and longevity averages 14 months, although one female was known to have lived 23 months.

DASYUROMORPHIA; DASYURIDAE; **Genus** *PHASCOGALE* Temminck, 1824

Tuans, Phascogales, or Brush-tailed Marsupial "Mice"

There are three species (Bradley, Foster, and Taggart 2008; Rhind, Woinarski, and Aplin 2008; Ride 1970; Soderquist and Rhind 2008):

P. tapoatafa, southwestern and extreme northern Western Australia, northern and eastern Queensland, eastern New South Wales, eastern and southern Victoria, southeastern South Australia;

P. pirata, northern Northern Territory, Melville Island, Sir Edward Pellew Islands;

P. calura, originally known from central and southern Western Australia, southern Northern Territory, southeastern South Australia, extreme

Brush-tailed marsupial "mouse" (*Phascogale tapoatafa*), subadult. Photo by Todd Soderquist.

southwestern New South Wales, and northwestern Victoria.

Groves (2005c) and, until recently, most other authorities treated *P. pirata* as a subspecies or synonym of *P. tapoatafa*. However, both morphological and molecular studies (Rhind, Bradley, and Cooper 2001; Spencer, Rhind, and Eldridge 2001) have revealed distinctions suggesting that specific status is appropriate. Further investigation may show the same for the population of *P. tapoatafa* in southwestern Western Australia.

In *P. tapoatafa* head and body length is 148–261 mm, tail length is 160–234 mm, and weight is 106–311 grams (Soderquist and Rhind 2008). In *P. pirata* head and body length is 150–210 mm, tail length is 180–208 mm, and weight is 108–234 grams (Rhind, Woinarski, and Aplin 2008). *P. calura* is considerably smaller, with a head and body length of 93–122 mm, tail length of 119–145 mm, and weight of 38–68 grams (Bradley, Foster, and Taggart 2008). Males are larger on average than females. Each species is grayish above and whitish below. The tail in *P. tapoatafa* and *P. pirata* is black except for the gray base; in *P. calura* the proximal part is reddish brown and the distal half is black. The upper surface of the hind feet is pure white in *P. pirata* but grizzled in *P. tapoatafa*.

Phascogale is distinguished from the other broad-footed marsupial "mice" in having the terminal portion of the tail covered with a silky brush of long black hairs. In *P. tapoatafa* and *P. pirata* those hairs are capable of erection, producing a striking "bottle brush" effect, which is normal when the animal is active. Soderquist (1994) pointed out that the action is intended to distract predators and deflect their strike away from the body. When the animal is at rest the hairs are pressed along the tail and are not conspicuous. The ears are relatively large, thin, and almost naked. Strong, curved claws are present on all digits except the innermost digit of each hind foot, which is clawless. The hind foot can be rotated 180° at the ankle to facilitate climbing (Soderquist and Rhind 2008). Females have 8 mammae (Cuttle 1982), although those in the southwestern population of *P. tapoatafa* may have 6, 7, or 8 (Rhind 2002; Rhind, Bradley, and Cooper 2001). Although there is no true pouch, there is a pouch area, marked by light-tipped brown hairs coarser in texture than the body fur, and that begins to develop protective folds of skin about 2 months before parturition and will completely enclose the newborn.

These marsupial "mice" are primarily arboreal. *P. calura* once occupied a large region of mostly arid country but depends on isolated patches of forest, preferably with wandoo (*Eucalyptus wandoo*), the hollows of which provide nest sites (Bradley, Foster, and Taggart 2008). According to Soderquist and Rhind (2008), *P. tapoatafa* formerly occurred throughout the dry sclerophyll forests and woodlands of Australia, preferring open forest with sparse groundcover but also living in rainforest. An individual of that species may use more than 40 nest sites in a year, including hollow trees, rotted stumps, and house ceilings. Lactating females prefer large tree cavities with small entrances, where they build spherical nests, sometimes 30 times their own size, of bark strips, feathers, and fur.

Tuans are mostly nocturnal and as agile as squirrels, spiraling up tree trunks and leaping as much as 2 meters between trees. They feed mainly on insects and spiders but occasionally capture and eat small mammals, birds, and lizards. *P. tapoatafa* finds prey by tearing away bark with its incisor teeth and probing into crevices with its dexterous forepaws. It also relishes nectar and may spend much of a night foraging in a heavily flowering eucalypt. It sometimes destroys poultry, but that is outweighed by its extermination of insects and mice.

When disturbed, *P. tapoatafa* utters a low, rasping hiss, which apparently is an alarm note. When angered, tuans emit a series of staccato "chit-chit" sounds. Sometimes, when excited, tuans slap the pads of their forefeet down together while holding an alert, rigid pose, thus producing a sharp rapping sound. At times they also make a rapid drumming noise by quick vibrations of the tail.

Radio-tracking studies by Soderquist (1995) of *P. tapoatafa* in Victoria showed that female home ranges averaged 41 ha. and had little or no overlap with one another. Adult females generally were highly agonistic, but a mother sometimes relinquished a portion of her range to a newly independent daughter. Male ranges averaged 106 ha., expanded during the breeding season, and overlapped extensively with one another and with those of females. Observations in the wild (Soderquist and Ealey 1994) indicate that females are able to dominate males and to deter mating. Young animals disperse at an average age of 162 days; juvenile males may move many kilometers, but females either stay

Red-tailed marsupial "mouse" (*Phascogale calura*). Photos by Alexander F. Meyer at Sydney Zoo.

within their mother's range or settle adjacent to it (Soderquist and Lill 1995; Soderquist and Rhind 2008). Although tuans are usually solitary, groups of 2–4 individuals, of any age and sex combination and not known to be related, have been found sharing nests during winter and prolonged drought. Such communal nesting seems a response to thermoregulatory difficulties posed by declining temperatures, reduced food, and lowered body mass (Rhind 2003).

P. tapoatafa is semelparous, with a reproductive biology very much like that of *Antechinus* (Cuttle 1982; Krajewski, Woolley, and Westerman 2000; Millis et al. 1999; Rhind 2002; Soderquist 1993; Soderquist and Rhind 2008). The breeding season is restricted to a few weeks, females are monestrous, and ovulation is synchronized. Estrous cycle length averages about 41 days, and mating occurs from mid-May to early July (fall and winter), depending on locality. The sperm are stored in the oviduct for 8–20 days and subsequent gestation averages 27 (range 22–35) days. Litters commonly consist of eight young but sometimes contain as few as one. They remain attached to the nipples for about 40–50 days and are then left in the nest while the mother forages. They begin to emerge just before weaning at an age of around 5 months, reach adult size at about 8 months, and attain sexual maturity by 11 months. The males, evidently weakened by stress-related diseases and susceptible to predators, disappear entirely from the wild population just after mating and before they are a year old (see also account of family Dasyuridae). Captive males have lived for more than three years but are not reproductively viable after their first breeding season. Females may survive to breed for a second year in the wild. Weigl (2005) listed one captive female *P. tapoatafa* that lived to an age of about 5 years and 11 months.

The reproductive pattern of *P. calura* is similar (Bradley 1997; Bradley, Foster, and Taggart 2008; Foster et al. 2008; Kitchener 1981). Mating in the wild is synchronized, occurring in a 21- to 24-day period in July (winter), and parturition follows during a 21- to 24-day period in August. Births at the Adelaide Zoo have occurred from early June to late August. Females mate with multiple males. Apparent gestation length is 28–30 days. Litters contain up to 13 young, though only 8 can be reared; they weigh about 15 mg each at birth, first detach from the mammae at around 45 days, and are weaned before the end of October. Wild male *P. calura* live for only 11.5 months and die just after mating. Apparently they are able to adapt stress-related endocrine changes to provide energy for the reproductive period, but that metabolic process results in gastric ulcers and other problems that soon prove fatal. Captive males have survived for up to five years. Stannard et al. (2013) reported that females can live at least as long in captivity and that some have bred at an age of four years. According to Baker (2015), very limited data indicate that the life history of *P. pirata* is similar to that of *P. calura* and *P. tapoatafa*: annual die-off of males, births mid-year, and litters of up to eight young.

The IUCN designates *P. pirata* as vulnerable (Woinarski, Rhind, and Oakwood 2008) and *P. calura* and *P. tapoatafa* as near threatened (Friend 2008a; Menkhorst, Rhind, and Ellis 2008). *P. pirata* is estimated to number fewer than 10,000 mature individuals; there may be less than 2,500. Rhind, Woinarski, and Aplin (2008) noted it was common a century ago but that fewer than 10 records had come from the mainland in the prior decade, despite intensive survey work. There also are two recent records from Melville Island. *P. pirata* was found on West Pellew Island in 1988 but a more recent survey there was unsuccessful. Continued decline is likely as habitat is degraded through pastoralism and altered human burning practices (see account of *Sminthopsis*).

P. calura has disappeared from more than 99 percent of its range and now survives only in southwestern Western Australia (Kennedy 1992; Woinarski, Burbidge, and Harrison 2014). Its decline has been attributed to fires, forest clearance, and livestock grazing, which eliminated critical nesting trees and dense cover. Predation by introduced placental foxes and domestic cats also may have been a factor. The decline seems to have

been arrested in the last decade, and the species now occurs at 40–50 sites, many of which are in protected reserves. Interestingly, survival in its current range may be related to presence of *Gastrolobium* and *Oxylobium*, leguminous plants that produce the toxin monosodium fluoroacetate ("1080"). Native animals have evolved tolerance to that toxin, so that their bodies can build up levels sufficient to be lethal to introduced predators (Bradley, Foster, and Taggart 2008). The range of *P. tapoatafa* has been reduced by about half since European settlement, and declines are continuing due to the same factors that threaten the other two species, notably the loss of trees with nesting hollows.

DASYUROMORPHIA; DASYURIDAE; **Genus** *MICROMUREXIA* Van Dyck, 2002

Habbema Dasyure

The single species, *M. habbema*, is known from mid- to upper montane areas of the central cordillera of New Guinea. The species long was placed in the genus *Antechinus*, sometimes as a synonym of *A. naso* (here referred to *Phascomurexia*), but Krajewski et al. (1994) suggested that it and the other New Guinean species assigned to *Antechinus* are more closely related to the New Guinean genus *Murexia* than to the Australian *Antechinus*. Flannery (1995a) treated *habbema* as a distinct species and provisionally kept it in *Antechinus* but realized it eventually would have to be put in a separate genus. Based on analysis of DNA sequence variation and also considering phallic morphology, Krajewski et al. (1996) recommended transferring *habbema* and the other New Guinean *Antechinus* to *Murexia*, and Armstrong, Krajewski, and Westerman (1998) formerly took that step. However, morphological assessment by Van Dyck (2002) led him to erect an entirely new genus, *Micromurexia*, for *habbema*. Groves (2005c) followed that arrangement. Further analysis of DNA sequences and reassessment of morphological data by Krajewski et al. (2007) supported relegating *Micromurexia* to the synonymy of *Murexia* (see also account of family Dasyuridae), and this was done by Baker (2015).

Van Dyck (2002) listed the following measurements: total length (including tail), 224–279 mm; tail length, 109–157 mm; weight, 22.68–45.36 grams. Males

Habbema dasyure (*Micromurexia habbema*). Photo by P. A. Woolley and D. Walsh.

average larger than females. The tail is normally substantially longer than the head and body. The upper parts are reddish brown to blackish, and the underparts are grayish olive. The tail varies in color from light reddish brown in the east to black in the west and has a distinctive crest of hairs along the entire ventral edge. Woolley (2003) reported that females have four mammae and a pouch consisting only of folds of skin that develop when breeding occurs.

The first upper molar tooth is very broad, with a wide protocone and complete anterior cingulum, and its anterior margin is straight or anteriorly convex but never indented or concave. The first upper incisor is needlelike and minutely crowned, and the upper canine is long, extremely thin, and needlelike. *Micromurexia* is also characterized by a generally uniform pelage, a skull markedly domed posterior to the frontals, and fluted nasal bones. By comparison, *Antechinus* has a tail shorter than the head and body, no ventral crest on the tail, and little or no nasal fluting; its first upper molar has a reduced

protocone and an anterior margin indented posteriorly and never anteriorly convex (Van Dyck 2002).

The Habbema dasyure occurs at elevations of 1,600–3,330 meters in rainforest, mid-mountain forest, beech forest, mossy forest, and subalpine grassland (Van Dyck 2002). It apparently is nocturnal and largely terrestrial. The diet evidently consists mostly of insects and also includes worms and small vertebrates (Baker 2015). Two females were tracked to burrow systems with nesting chambers 80–100 cm below ground level; nests were composed of interwoven leaves and ferns. One nest contained an adult female, two adult males, an adult of unknown sex, a subadult female, and four juveniles (Flannery 1995a). Reproductively, *Micromurexia* is characterized by extended and aseasonal breeding, polyestrous females, males that can potentially breed in more than one season, and sexual maturity attained at 8–11 months of age (Krajewski, Woolley, and Westerman 2000). Lactating females have been collected in every month of the year except April (Van

Short-haired dasyure (*Murexia longicaudata*). Photo by P. A. Woolley and D. Walsh.

Dyck 2002; Woolley 1994, 2003). Lactation lasts over three months, litters usually contain four young but may have only two or three, and if pregnancy is not successful, the mother will return to estrus 146–166 days after the prior estrus (Woolley 2003).

DASYUROMORPHIA; DASYURIDAE; **Genus MUREXIA** Tate and Archbold, 1937

Short-haired Dasyure, or Marsupial "Mouse"

Van Dyck (2002) and Groves (2005c) restricted this genus to the single species *M. longicaudata*, which is

found on New Guinea, Japen (or Yapen) Island to the northwest, and the Aru Islands to the southwest. Another New Guinean species, *rothschildi*, traditionally is placed in *Murexia*, but was transferred to the new genus *Paramurexia* (see account thereof) by Van Dyck (2002). Three more species, *habbema, naso,* and *melanurus* (including *wilhelmina*) long were placed in *Antechinus* but were transferred to *Murexia* by Armstrong, Krajewski, and Westerman (1998) and then moved to the respective new genera *Micromurexia, Phascomurexia,* and *Murexechinus* (see accounts thereof) by Van Dyck (2002). More recently, Krajewski et al. (2007) supported restoring *habbema, naso, rothschildi,* and *melanurus* to *Murexia* (see also account of family Dasyuridae), and this was done by Baker (2015). A specimen from Normanby Island was originally assigned to *M. longicaudata*, but then was placed provisionally in *Antechinus* by Flannery (1995b) and considered by Van Dyck (2002) to represent *Murexechinus*.

Head and body length is 136–284 mm, and tail length is 147–283 mm. Flannery (1995a) listed weights of 114–434 grams for males and 57–88 grams for the much smaller females. The upper parts are dull grayish brown, and the underparts are white. The fur is short and dense. The tail is long and sparsely haired with a few longer hairs at its tip. The skull is heavy and strong with deep zygomatic arches. The foot pads are striated, indicating that the animals probably are scansorial (Collins 1973). Females have four mammae and a pouch consisting only of folds of skin that develop when breeding occurs (Woolley 2003). *Murexia* is similar to *Micromurexia* in dentition and some other characters, but is much larger and has stronger, thicker, and more curved claws; its fur is shorter and harsher and its tail is almost naked dorsally, rather than well haired (Van Dyck 2002).

The short-haired dasyure is found in all lowland and midmountain forests of New Guinea, from sea level to 2,200 meters. Woolley (1989) found it to utilize a spherical nest about 20 cm wide, composed of leaves and located in a tree about 4 meters above the ground. Baker (2015) noted that it also uses subterranean nests. It is at least partly diurnal, arboreal, and insectivorous, but its morphology suggests it could be a powerful predator (Flannery 1995a; Ziegler 1977). Reproduction is characterized by extended and aseasonal breeding, polyestrous females, males that can potentially breed in more than one season, and sexual maturity attained at 8–11 months of age (Krajewski, Woolley, and Westerman 2000). Estrous females have been recorded in all months, and lactating females in all months except May and July (Van Dyck 2002; Woolley 2003). Time between estruses averages 68 days, and two gestation periods were 19 and 20 days. In 23 births, the number of young averaged 3.8 (range 2–4), and lactation lasted about 3.5 months. Wild-caught individuals were maintained in captivity for up to 28 months (Woolley 2003).

Long-nosed dasyure (*Phascomurexia naso*). Photo by P. A. Woolley and D. Walsh.

DASYUROMORPHIA; DASYURIDAE; **Genus**
PHASCOMUREXIA Van Dyck, 2002

Long-nosed Dasyure

The single species, *P. naso*, is known from mid- to lower montane areas of the central cordillera of New Guinea. Although Flannery (1995a) indicated its occurrence on the Vogelkop Peninsula, Van Dyck (2002) assigned the relevant specimens from that area to *Murexechinus melanurus*. *P. naso* long was placed in the genus *Antechinus*. Its subsequent referral either to *Murexia* (Armstrong, Krajewski, and Westerman 1998; Baker 2015; Krajewski et al. 2007) or to the new genus *Phascomurexia* (Groves 2005c; Van Dyck 2002) corresponds to the taxonomic history discussed above in the accounts of the family Dasyuridae and the genus *Micromurexia*.

Van Dyck (2002) listed a total length (including tail) of 231–307 mm and a tail length of 109–175 mm. Woolley (2003) listed weights of 29–74 grams, males being on average heavier than females. According to Van Dyck (2002), the upper parts are uniformly grayish brown, and the underparts are pale olive-buff. The tail is somewhat longer than the head and body and is seminaked dorsally, with a weak ventral crest developing towards the tip; some specimens have white tail tips. *Phascomurexia* has teeth generally similar to those of *Micromurexia* but has absolutely longer upper and lower molar tooth rows. *Phascomurexia* is distinguished from *Murexia* by a shorter lower tooth row, having long and luxurious rather than short and harsh pelage, and not having a post-metatarsal granule on the hind foot. Woolley (2003) reported that females have four mammae and a pouch consisting only of folds of skin that develop when breeding occurs.

The long-nosed dasyure occurs in a narrow elevational range of 1,400–2,800 meters and has been collected in rainforest, mid-montane forest, beech forest, pandanus forest, and mossy forest (Van Dyck 2002). It seems to be crepuscular and partly arboreal, individuals having been found in trees, on the ground, and in burrows; one nesting chamber was 65 cm beneath the surface, about 20 cm wide, and heavily lined with dry leaves (Flannery 1995a). The diet consists mainly of insects, spiders, and worms (Baker 2015). Reproduction is characterized by extended and aseasonal breeding, polyestrous females, and males that can potentially breed in more than one season (Krajewski, Woolley, and Westerman 2000). Lactating females have been recorded in all months except March and July, litter size is 2–4 young, recorded interval between estrus averages 60 days, and age at sexual maturity is estimated at 14 months for males (Woolley 2003).

DASYUROMORPHIA; DASYURIDAE; **Genus**
PARAMUREXIA Van Dyck, 2002

Broad-striped Dasyure

The single species, *P. rothschildi*, is known from middle elevations in southeastern Papua New Guinea (Flannery 1995a). The species traditionally is placed in *Murexia* but morphological assessment led Van Dyck (2002) to

Broad-striped dasyure (*Paramurexia rothschildi*). Photo by P. A. Woolley and D. Walsh.

erect the new genus *Paramurexia* for *rothschildi*. Analysis of DNA sequences and reassessment of morphological data by Krajewski et al. (2007) supported restoring *rothschildi* to *Murexia* (see also account of family Dasyuridae), and this was done by Baker (2015).

Head and body length is 124–170 mm, tail length is 140–184 mm, and weight is 32–102 grams (Flannery 1995a). The most striking external feature is a broad black dorsal stripe that begins at the nose and ends at the base of the tail. There are also dark eye rings. Otherwise the upper parts are grayish brown, and the underparts are cinnamon buff. The feet are thinly covered with buffy brown hairs. The tail is blackish and covered with short hairs; it is thin and tapers towards the tip. *Paramurexia* is close in size and external appearance to *Phascolosorex dorsalis* but has a much broader dorsal stripe and paler and generally sleeker pelage. The first upper molar tooth is similar to that of *Murexia* and *Micromurexia* (see account thereof) and different from that of *Antechinus*. *Paramurexia* differs from *Murexia* in having a narrow skull, a shorter upper tooth row, broad rather than needlelike first upper incisors, shorter canine teeth, and a large auxillary granule outside the third interdigital pad of the hind foot (Van Dyck 2002). Females have four mammae and a pouch consisting only of folds of skin that develop when breeding occurs (Woolley 2003).

The broad-striped dasyure is found in forests at elevations of 600–1,400 meters. It apparently is partly diurnal and partly arboreal and is known to prey on birds (Flannery 1995a). However, according to Leary et al. (2008d, citing P. Woolley), it is terrestrial and scansorial, nests underground, and is active mostly at night. Reproduction is characterized by extended and aseasonal breeding, polyestrous females, males that can potentially breed in more than one season, and sexual maturity attained at 8–11 months of age (Krajewski, Woolley, and Westerman 2000). Females in estrus have been noted in all months, recorded interval between estrus averaged 71 days, one gestation period was 20 days, lactation lasted 3.5 months, and individuals were maintained in captivity for up to 34 months. One female was taken from an underground nest in January 1986 with four suckling young, which were weaned soon thereafter; the female gave birth to a second litter of four young in March 1987, which it then reared in captivity (Woolley 2003).

The IUCN classifies *P. rothschildi* (placing it in the genus *Murexia*) as vulnerable (Leary et al. 2008d). The species occurs in an area of under 20,000 sq km and is known from fewer than 10 sites and approximately 23 wild-caught specimens. Numbers and habitat are declining due to conversion of forest habitat for cultivation and predation by hunting dogs.

DASYUROMORPHIA; DASYURIDAE; Genus *MUREXECHINUS* Van Dyck, 2002

Black-tailed Dasyure

The single species, *M. melanurus*, occurs throughout New Guinea, including the Vogelkop Peninsula, and on Normanby Island just off the southeastern tip (Flannery 1995a; Van Dyck 2002). The single specimen from

Black-tailed dasyure (*Murexechinus melanurus*). Photo by P. A. Woolley and D. Walsh.

Normanby Island originally was assigned to *Murexia longicaudata*. *M. melanurus* long was placed in the genus *Antechinus*. Its subsequent referral either to *Murexia* (Armstrong, Krajewski, and Westerman 1998; Baker 2015; Krajewski et al. 2007) or to the new genus *Murexechinus* (Groves 2005c; Van Dyck 2002) corresponds to the taxonomic history discussed above in the accounts of the family Dasyuridae and the genus *Micromurexia*. Another taxon, *wilhelmina*, of the central mountains of New Guinea, was also long designated a distinct species of *Antechinus*. It was treated as a species of *Murexia* by Armstrong, Krajewski, and Westerman (1998) and Krajewski, Woolley, and Westerman (2000). However, Van Dyck (2002) showed that *Antechinus wilhelmina* is a synonym of *Murexechinus melanurus*. That arrangement was followed by Groves (2005c), and neither Krajewski and Westerman (2003) nor Krajewski et al. (2007) recognized *A. wilhelmina*.

Head and body length is about 90–160 mm, tail length is 101–165 mm, and weight is 26–70 grams (Flannery 1995a; Van Dyck 2002; Woolley 2003). The tail is usually longer than the head and body. Males' average size is larger than females'. The upper parts show a change in color from head to rump, usually agouti gray-brown to russet, but there is much variation; the underparts are olive-buff. The two most distinguishing external features are rich rufous to light fawn patches on the backs of the ears and a uniformly black or dark brown tail, thickly haired and with a ventral crest. The claws are thick and strongly curved. The first upper molar tooth is similar to that of *Micromurexia* (see account thereof) and different from that of *Antechinus*. Unlike those of *Micromurexia* and *Murexia*, the first upper incisor is broad and heavily crowned, and the upper canine is short and thick. *Murexechinus* also differs from *Murexia* in having shorter tooth rows and a generally narrower skull (Van Dyck 2002). Females have four mammae and a pouch consisting only of folds of skin that develop when breeding occurs (Woolley 2003).

The black-tailed dasyure is found in forests from sea level to 2,800 meters. Arboreal and nocturnal activity has been reported (Flannery 1995a). The diet seems to consist mainly of insects and spiders, and also includes worms and small vertebrates (Baker 2015). Reproduction is characterized by extended and aseasonal breeding, polyestrous females, males that can potentially breed in more than one season, and sexual maturity attained at 8–12 months of age. Estrous and/or lactating females have been noted in all months, recorded interval between estrus averaged 69.5 days, one gestation period was estimated to be 22 days, litter size is usually four young, lactation lasts 3.5 months, and individuals were maintained in captivity for up to 25.5 months (Krajewski, Woolley, and Westerman 2000; Woolley 2003).

DASYUROMORPHIA; DASYURIDAE; **Genus *ANTECHINUS*** Macleay, 1841

Antechinuses, or Broad-footed Marsupial "Mice"

There are 15 described species (Baker, Mutton, and Hines 2013; Baker, Mutton, and Van Dyck 2012; Baker and Van Dyck 2013a, 2013b, 2013c; Baker et al. 2014, 2015; Baverstock et al. 1982; Crowther 2002a, 2002b, 2008; Crowther, Sumner, and Dickman 2003; Crowther et al. 2002; Dickman 2008b; Dickman et al. 1988, 1998; Groves 2005c; Kirsch and Calaby 1977; Krajewski and Westerman 2003; Krajewski et al. 2007; Leung 2008; Mahoney and Ride 1988; McNee and Cockburn 1992; Ride 1970; Van Dyck 1980, 1982a, 1982b; Van Dyck and Crowther 2000; Watson and Calaby 2008; Woolley 1982b):

> *A. swainsonii*, extreme southeastern Queensland, eastern New South Wales, southern Victoria, Tasmania;
> *A. vandycki*, known only from the eastern Tasman Peninsula in extreme southeastern Tasmania;
> *A. minimus*, coastal parts of southeastern South Australia and Victoria, Tasmania, islands of Bass Strait;
> *A. mimetes*, northeastern New South Wales to southern and western Victoria;
> *A. arktos*, known only from the Tweed Volcano caldera of far southeastern Queensland and northeastern New South Wales;
> *A. godmani*, Atherton Tableland between the towns of Cardwell and Cairns in northeastern Queensland;
> *A. adustus*, rainforests of northeastern Queensland;

Dusky antechinus (*Antechinus swainsonii*). Photo by Gerhard Körtner.

A. stuartii, extreme southeastern Queensland, eastern New South Wales;

A. subtropicus, extreme southeastern Queensland and extreme northeastern New South Wales;

A. agilis, southeastern New South Wales, eastern and southern Victoria;

A. flavipes, eastern Queensland, central and eastern New South Wales, Victoria, southeastern South Australia, southwestern Western Australia;

A. argentus, known from the plateau on the eastern escarpment of Kroombit Tops National Park in southeastern Queensland;

A. mysticus, east coast of Queensland from the town of Mackay to just north of the border with New South Wales;

A. leo, rainforests on east coast of Cape York Peninsula in northern Queensland;

A. bellus, northern Northern Territory, Melville Island.

The genera *Micromurexia, Phascomurexia, Murexechinus, Dasykaluta, Parantechinus,* and *Pseudantechinus* (see accounts thereof) sometimes have been included in *Antechinus*. The species once known as *Antechinus maculatus* has been transferred to the genus *Planigale* (Archer 1975, 1976). The above sequence of the remaining species is based largely on recent molecular studies (Armstrong, Krajewski, and Westerman 1998; Krajewski and Westerman 2003; Krajewski, Wroe, and Westerman 2000; Krajewski et al. 2007; but see also Van Dyck 2002), which generally support earlier assessments of relationship (as listed above) and indicate that *A. swainsonii* and *A. minimus* have affinity to one another and are the most basally divergent species; that *A. godmani* is also primitive and is sister to the group comprising all subsequently diverging species; that *A. adustus*, *A. subtropicus*, and *A. agilis*, which until recently were considered part of *A. stuartii*, form a related complex of species with the latter; that *A. flavipes* was the next to diverge; and that *A. leo* and *A. bellus* are immediately related. A more recent series of papers describing additional species and assessing their phylogenetic relationships (Baker, Mutton, and Hines 2013; Baker, Mutton, and Van Dyck 2012; Baker et al. 2014, 2015) has indicated that *A. vandycki* is a sister species to *A. swainsonii*; that *A. mimetes*, formerly considered a subspecies of *A. swainsonii*, remains part of a complex with the latter; that *A. arktos* is also closely related to *A. swainsonii* and *A. minimus*; and that *A. argentus* and *A. mysticus* are sister species forming a clade that in turn is sister to *A. flavipes*. Dickman et al (1998) had noted that *A. flavipes* probably also is a complex of separate species. Crowther et al. (2002) reported that the subspecies *A. flavipes flavipes* of southeastern Australia and *A. flavipes leucogaster* of southwestern Western Australia are very distinctive, in terms of morphological and molecular data and reproductive seasonality, but are not so divergent as to warrant classification as independent species. *A. flavipes rubeculus* of northeastern Queensland also seems very different morphologically, but molecular assessment is still needed.

Within any species males are generally larger than females. In *A. agilis*, the smallest species, males have a head and body length of 65–124 mm, a tail length of 63–116 mm, and a weight of 20–40 grams; respective figures for females are 62–128 mm, 60–103 mm, and 16–25 grams (Dickman 2008a). In *A. flavipes*, the most widely distributed species, males have a head and body length of 93–165 mm, a tail length of 70–151 mm, and a weight of 26–79 grams; respective figures for females are 86–127 mm, 65–107 mm, and 21–52 grams (Crowther 2008). In *A. swainsonii*, the largest species, males have a head and body length of 103–188 mm, a tail length of 80–121 mm, and a weight of 43–178 grams; respective figures for females are 89–140 mm,

Rusty antechinus (*Antechinus adustus*). Photo by Jiri Lochman / Lochman LT.

75–100 mm, and 37–100 grams (Dickman 2008b). The fur of *Antechinus* is short, dense, and rather coarse. Coloration of the upper parts varies from pale pinkish fawn through gray to slate and dark or coppery brown; the underparts are buffy, creamy, or whitish. The tail, usually the same color as the back, is short-haired in most species. *A. flavipes* features a distinct change in pelage, from slate-gray head to warm rufous rump, feet, belly, and sides, along with yellow-brown feet and black-tipped tail (Crowther 2008). *A. bellus* is strikingly pale gray and has white feet and large ears (Baker and Van Dyck 2013a). *A. arktos* is distinguished from other species by its evenly black tail (Baker et al. 2014).

The short, broad feet are distinctive. The great toe is present but is small and clawless. All species have transversely striated pads on the bottom of the feet. In the arboreal species, such as *A. flavipes*, the pads are prominent and strongly striated; in *A. swainsonii* and *A. minimus*, which are poor climbers, the pads are small and the striations faint (especially so in *A. minimus*, which occurs in treeless habitat). Those two closely related species have long and strong foreclaws modified for digging. *A. flavipes* has short, hooked claws, is much more active, and is a great climber.

A pouch is known to develop during the breeding season in most species but in some may consist only of prominent skin folds around the mammae. Female *A. stuartii* lack a pouch (Crowther and Braithwaite 2008). The number of mammae varies between and within species, having been reported as 6–8 in *A. swainsonii* (Collins 1973), 6 (Tasmania), 8 (mainland) in *A. minimus* (Wilson and Bachman 2008), 6 in *A. adustus* (Burnett and Crowther 2008a), 6–10 in *A. stuartii* (Crowther and Braithwaite 2008), 10–12 in *A. flavipes* (Collins 1973), 6 in *A. godmani* (Burnett 2008a), 8 in *A. subtropicus* (Burnett and Crowther 2008b), 6–10 in *A. agilis* (Dickman 2008a), 10 in *A. leo* (Van Dyck 1980), and 10 in *A. bellus* (Watson and Calaby 2008).

With regard to habitat, some species are generalists, others are specialists, but all require cover and nest sites. *A. agilis*, *A. stuartii*, and *A. minimus* have been recorded from forest, woodland, grassland, and heath, with the last species preferring damp areas (Crowther and Braithwaite 2008; Dickman 2008a; Wilson and Bachman 2008). *A. flavipes* occupies an even broader range of habitats, including swamps, dry country, and suburban gardens (Crowther 2008). *A. agilis* has been

observed swimming, and rivers have not historically restricted gene flow in *A. flavipes* (Lada, Mac Nally, and Taylor 2008). *A. swainsonii* is found in alpine heath or tall open forest with a dense understory of shrubs or ferns (Dickman 2008b). *A. godmani*, *A. adustus*, *A. subtropicus*, and *A. leo* are restricted to rainforest (Burnett 2008; Burnett and Crowther 2008a, 2008b; Leung 2008), and *A. bellus* dwells on tropical savannahs (Watson and Calaby 2008).

Antechinuses construct nests of vegetation for shelter and to rear young. Nests may be located in hollow trees, logs, rock crevices, abandoned birds' nests, or pockets in cave ceilings (Collins 1973). The arboreal *A. godmani*, *A. adustus*, and *A. leo* depend on trees with large hollows, but *A. subtropicus* sometimes enters houses and builds its leafy nests in walls, ceilings, and furniture (Burnett 2008; Burnett and Crowther 2008a, 2008b; Leung 2008). *A. bellus* shelters in standing or fallen trees, entering either through burrows at the base or through holes more than 10 meters above ground (Watson and Calaby 2008). Females of the largely terrestrial *A. swainsonii* excavate nests during winter in creek banks or below the snow and under decaying logs, making a roughly spherical nest chamber about 10 cm wide and lined with dry grass and leaves (Dickman 2008b).

Broad-footed marsupial "mice" are secretive, active animals characterized by rapid movements. The claws and ridged foot pads of some species enable them to run upside down on the ceilings of rock caverns. Most species are nocturnal and at least partly arboreal,

Brown antechinus (*Antechinus stuartii*). Photo by Fritz Geiser.

though there is considerable variation. *A. stuartii* is usually terrestrial but becomes quite arboreal in dry forests with little ground cover and in wetter areas where *A. swainsonii* occupies the forest floor. It is primarily nocturnal but sometimes diurnal, especially in winter, and it may then also enter torpor for a few hours at a time to conserve energy (Crowther and Braithwaite 2008). Torpor also has been reported in *A. stuartii* and *A. flavipes* (Geiser 2003). *A. leo* is an excellent climber, darting up and down trees with great speed and agility. When on the forest floor it tends to run silently along logs and buttresses rather than in leaf litter (Leung 2008), while *A. flavipes* bulldozes through the leaf litter in search of prey (Crowther 2008). *A. adustus* forages, mostly during daylight, from ground level to canopy (Burnett and Crowther 2008), but *A. subtropicus* prefers to hunt in the dense understory of vine tangles and rotting logs (Burnett and Crowther 2008). *A. agilis* pursues its prey on branches and trunks of trees (Dickman 2008a), but the terrestrial *A. minimus* nests at ground level and uses its long claws to forage in leaf litter; it is crepuscular with some daytime activity (Wilson and Bachman 2008). *A. bellus* is also mainly crepuscular (Watson and Calaby 2008).

The diet consists mostly of invertebrates. Small vertebrates, including amphibians and introduced house mice, are also taken. Green (1972) reported captive *A. swainsonii* to accept items ranging in size from mosquitoes to house mice and domestic sparrows. In the wild, *A. swainsonii* digs up soil invertebrates and occasionally eats blackberries, holding food in the forepaws (Dickman 2008b). Nagy et al. (1978) found *A. agilis* in winter to consume about 60 percent of its weight in arthropods each day. That species also occasionally eats soft berries (Dickman 2008a). *A. adustus* and *A. godmani* sometimes feed on carrion (Burnett 2008; Burnett and Crowther 2008). *A. flavipes* takes mostly insects but also nectar, mice, and, frequently, birds (Crowther 2008).

Population density of *A. minimus* varies from a maximum of 18 individuals per ha. on the mainland to 90/ha. on islands off the Victoria coast (Wilson and Bachman 2008). Density of *A. agilis* was found to be 3/ha. and 18/ha. in two sections of wet sclerophyll forest (Nagy et al. 1978). Home range size is about 1 ha. in *A. godmani* (Burnett 2008) and 1–2 ha. in *A. adustus* (Burnett and Crowther 2008a). Ranges of three

female *A. subtropicus* near Brisbane were 0.80, 0.98, and 1.15 ha.; males utilized larger but uncalculated areas (Wood 1970). Male *A. agilis* forage over about 1 ha., almost three times the area of females, but both sexes use much larger areas for social interaction (Dickman 2008a).

Studies by Cockburn and Lazenby-Cohen (1992), Lazenby-Cohen (1991), and Lazenby-Cohen and Cockburn (1988) indicated that outside of the breeding season both male and female *A. agilis* forage in clearly defined individual home ranges but that neither sex is territorial (at the time of those studies, *A. agilis*, as well as *A. adustus* and *A. subtropicus*, were considered taxonomically to be components of the species *A. stuartii*). As temperatures dropped during the winter, individuals might leave their nocturnal foraging range to spend the day in a communal nest that held up to 18 animals simultaneously. The total number of individuals visiting a single nest during the winter could be more than 50. When the mating season began, males were found to abandon their foraging range and aggregate in a few of the nest trees, where they spent most of the night. Males always nested communally and always in groups that included females. Such mating aggregations were considered comparable to the "leks" that have been described for various large hoofed mammals. Prior to the mating season females nested alone or in small unisexual groups. Even during the breeding season females continued to use their foraging range and sometimes to nest alone, but they made excursions to the aggregations, thus determining the time and place of mating.

More recent studies of *A. agilis*, as well as *A. adustus*, *A. subtropicus*, and *A. stuartii*, confirmed communal nesting but yielded no evidence of a lek mating system (Fisher et al. 2011). The communal nesting of males during winter mating was thought to have a thermoregulatory function, resulting in an energy savings and improved breeding success. Males were found to change nesting sites at the same rate as females during winter and to be more mobile than females throughout the year, with larger home ranges. Females had a stable social structure, sharing ranges with close female relatives and nesting regularly in groups that included close kin. Males apparently moved continually between nests containing different groups of females, while increasing their ranges to locate new females as

Agile antechinus (*Antechinus agilis*). Photo by Jiri Lochman / Lochman LT.

the year progressed. A pattern of mobile males, overlapping with more females during the mating season, together with promiscuous mating and mixed paternity, minimal mate choice in females, and greater reproductive success in larger and more dominant males was consistent with a mating system of scramble polygyny, not lek promiscuity.

Males become increasingly aggressive to one another as the breeding season approaches. The staccato "chee" calls of male *A. stuartii* then are heard more frequently (Crowther and Braithwaite 2008). *A godmani* may stand on its hind legs, hissing and lunging with mouth agape (Burnett 2008). Fighting male *A. swainsonii* utter a shrill "siss" and also wheeze, cackle, and smear objects with secretions from a chest gland (Dickman 2008b). Intromission is vigorous and prolonged, generally lasting about 6 hours in *A. stuartii* (Crother and Braithwaite 2008), up to 9.5 hours in *A. swainsonii* (Dickman 2008b), and up to 12 hours in *A. agilis* and *A. flavipes* (Crowther 2008; Shimmin, Taggart, and Temple-Smith 2002). Holleley et al. (2006) found that 92 percent of

wild female *A. stuartii* mated with multiple males, producing litters of eight that had up to four fathers; paternal success was related most strongly to body mass and scrotal size.

All species now are known to be monestrous, to produce only one litter per year, and to have a restricted breeding season. The females of some species store sperm in the oviducts for up to 2 weeks prior to ovulation, hence extending the time between mating and birth well beyond the true gestation period. In most species reproduction occurs during part of the Australian winter (June–September), but in some may start in the spring and/or extend into autumn. Although the season for an entire species may cover several months, mating within any one population takes place within 3, usually for 1–2, weeks, and births are correspondingly highly synchronized. Timing varies little from year to year at any given locality. In at least some species the process apparently is stimulated by increasing length of daylight (photoperiod); this factor contributed to the systematic discovery that *A. stuartii* actually comprises several species, as each species is sensitive to a different rate of increase. Such distinctions may serve as a reproductive barrier where species of *Antechinus* are sympatric, with each species breeding at a different time. The foregoing information and the following data on mating, birth, and gestation periods, number of young per litter, and age at weaning were compiled from Baker and Van Dyck (2013b), Burnett and Crowther (2008a), Crowther and Braithwaite (2008), Dickman (2008a, 2008b), Krajewski, Woolley, and Westerman (2000), Leung (2008), McAllan (2003), McAllan, Dickman, and Crowther (2006), McAllan and Geiser (2006), Taggart et al. (1999), Wainer (1976), Watson and Calaby (2008), and Wilson and Bachman (2008):

A. swainsonii—mating May–September, births June–October, gestation about 30 days, litter size 6–10, weaning 90–95 days;

A. minimus—mating late May–early August, births July–September, gestation 28–32 days, litter size 6–8, weaning 80–90 days;

A godmani—mating late May–early August, births late June–mid-September, gestation 30 days, litter size 4–6, weaning about 140 days;

A. adustus—mating early July–early August, births early August–early September, gestation 25–31 days, litter size 6, weaning about 120–140 days;

A. stuartii—mating late July–September, births late August–October, gestation 27 days, litter size 6–8, weaning 90–95 days;

A. subtropicus—mating September, births October, gestation about 25 days, litter size 8, weaning about 90 days;

A. agilis—mating mid-July–late August, births mid-August–mid-September, gestation 27 days, litter size 6–10, weaning 90–110 days;

Yellow-footed antechinus (*Antechinus flavipes*). Photo by Jiri Lochman / Lochman LT.

A. flavipes—mating early June–mid-September, births early July–late October, gestation about 30 days, litter size 8–12, weaning 90–120 days;

A. leo—mating mid-September–mid-October, births mid-October–early November, gestation about 30 days, litter size up to 10, weaning about 100–120 days;

A. bellus—mating late August, births late September–early October , gestation 28–35 days, litter size up to 10, weaning about 120 days.

The newborn measure 4–5 mm in length and weigh about 0.016 gram. They remain attached to the teats and are carried around by the mother for 1–2 months. The young then are left in the nest for about another month, after which they accompany the mother until weaning. Young males are forced by the mother to disperse, and thus the subsequent winter aggregations consist largely of unrelated males, but young females remain in their natal vicinity and inherit the mother's range and nest sites. All species evidently attain sexual maturity when about 9–11 months of age (Cockburn and Lazenby-Cohen 1992; Crowther 2008; Crowther and Braithwaite 2008; Krajewski, Woolley, and Westerman 2000; Leung 2008; Marlow 1961; McAllan 2003).

A remarkable feature of the biology of *Antechinus* is the abrupt and total mortality of males following mating, when they are 11–12 months old, a form of the life history process called "semelparity" (Arundel, Barker, and Beveridge 1977; Barker et al. 1978; Bradley 2003; Emison et al. 1978; Lazenby-Cohen and Cockburn 1988; Lee, Bradley, and Braithwaite 1977; Morton, Dickman, and Fletcher 1989; Nagy et al. 1978; Naylor, Richardson, and McAllan 2008; Scott 1987; Wainer 1976; Wilson and Bourne 1984; Wood 1970). Although the period of mortality varies geographically, it occurs at the same time each year in any given population. Males become more active and aggressive during the breeding season. They stop feeding and live on their reserves. As the season approaches a climax, they move about considerably, even in daylight and mainly from one male aggregation to another. They apparently are subject to intensive physiological stress resulting from continuous competition for females, as well as from gluconeogenic mobilization of body protein, a process that sustains them temporarily. Increasing levels of testosterone and other androgens evidently depress plasma-corticosteroid-binding globulin and result in a rise in corticosteroid concentration. The trauma associated with stress and endocrine changes causes suppression of the immune system, major ulceration of the gastric mucosa, increased susceptibility to parasites and pathological conditions, anemia, and death. Even males captured during the mating season die in the laboratory at the same time as those remaining in the wild.

Males taken before sexual maturity have been maintained until they are at least 2 years and 8 months old (Rigby 1972). Males castrated four months prior to the breeding season and released, as well as males naturally lacking a scrotum, have occasionally been captured after the other males have died (Bradley 2003). Many wild females die after rearing their first litter, but some survive for at least another year (Lee, Bradley, and Braithwaite 1977). Females of several species have lived for over 3 years in captivity, the record being an *A. stuartii* that attained an age of 5 years and 5 months (Weigl 2005). The physiological processes culminating

Yellow-footed antechinus (*Antechinus flavipes*), mother and young. Photo by Gerhard Körtner.

in deterioration and death in *Antechinus*, notably the sexual and stress-related factors affecting males, have been suggested as models for studies of aging in humans (McAllan 2006). Evolutionary advantages of semelparity may involve a concentration of physiological resources towards maximizing reproductive potential, rather than longevity, and ensuring that adult females and young are relatively free from competition for food and shelter (Naylor, Richardson, and McAllan 2008).

Antechinuses generally are not under serious human pressure, though some populations have been reduced by predation from the domestic cat (*Felis catus*), feral populations of which had become established throughout Australia by 1890 (Abbott 2002). Even the most widespread species, *A. flavipes*, has been adversely affected by massive alteration of its habitat, and careful management of woodland and floodplains may be needed to prevent extensive local extinctions (Lada and Mac Nally 2008). *A. godmani*, classified as near threatened, is the only species now in a formal IUCN threatened category (Burnett and Winter 2008a). Laurance (1990) determined that deforestation had reduced the historical range of that species by about 30 percent, resulting in extensive fragmentation of remaining populations. According to Burnett (2008a), fragmentation is continuing, and *A. godmani* cannot survive in forest segments under about 190 ha. That factor plus anticipated climatic changes (reduced rainfall, rising temperatures) could severely jeopardize the species within a few decades. Applying IUCN criteria to updated information, Woinarski, Burbidge, and Harrison (2014) classified another species, *A. bellus*, as vulnerable; it reportedly declined by more than 30 percent in the previous years, because of the same factors mentioned above.

Populations of the mainland subspecies *A. minimus maritimus* also are being fragmented by drainage, grazing, burning, and clearing of wetlands. The IUCN regards that subspecies as rare and decreasing in numbers (Van Weenen and Menkhorst 2008), while Woinarski, Burbidge, and Harrison designate it vulnerable. Green (1972) considered the Tasmanian subspecies *A. minimus minimus* "rare and endangered" because of its limited habitat and the threats posed by mineral development, grazing of domestic livestock, and flooding by hydroelectric dams. Baker et al. (2015) indicated that another Tasmanian species, the

newly described *A. vandycki*, should be placed in an IUCN threatened category, as it is known from a small patch of forest subject to lumbering and fires. Baker, Mutton, and Hines (2013) recommended that the newly described *A. argentus*, be classified as endangered, because its very restricted habitat is jeopardized by fires, grazing, and feral pigs. The distribution of the newly described *A. arktos* appears to have recently contracted to the very highest parts of the Tweed Volcano caldera, in association with global warming, and the species may warrant classification as endangered or critically endangered (Baker et al. 2014). The subspecies *A. swainsonii insulanus*, which is restricted to the Grampians Range of western Victoria, and the habitat of which is jeopardized by fires and clearing, was designated near threatened by Woinarski, Burbidge, and Harrison (2014).

DASYUROMORPHIA; DASYURIDAE; **Genus *MYOICTIS*** Gray, 1858

Three-striped Dasyures, or Marsupial "Mice"

There are four species (Westerman et al. 2006; Woolley 2005a):

M. melas, northern and western New Guinea, including the Vogelkop Peninsula, and on Japen (Yapen), Waigeo, and Salawati Islands to the northwest;

M. wallacei, southern New Guinea, from Merauke in the west to Avera in the east, and on the Aru Islands to the southwest;

M. leucura, southern side of the central cordillera in Papua New Guinea;

M. wavicus, vicinity of Wau in east-central Papua New Guinea.

Flannery (1995a, 1995b), like many authorities up to that time, recognized only the single species *M. melas*, though he noted that *wallacei*, then usually treated as a subspecies, is very distinctive in appearance and might be a full species. Subsequent molecular studies (Krajewski and Westerman 2003; Krajewski, Wroe, and Westerman 2000; Krajewski et al. 1997) supported specific status for *wallacei* and suggested *M. melas* might include more than one species. Groves (2005c) listed *wallacei* (using the spelling *wallacii*) as a separate species. Based on morphological assessment, Woolley (2005a) recognized all four of the above species, noting that they showed no overlap in range, considering available specimens, but that interbreeding would be unlikely even if they were sympatric. Molecular analysis by Westerman et al. (2006) generally supported Woolley's (2005a) assessment, but indicated *M. melas* could warrant further taxonomic division.

Head and body length is about 165–250 mm, and tail length is approximately 150–230 mm. Based on four specimens each, weight is 172–255 grams in *M. melas*, 206–245 grams in *M. wallacei*, and 200–230 grams in *M. leucura*, but two specimens of *M. wavicus* weighed only 110 and 134 grams (Woolley 2005a). The specific name *melas* means "black" and was applied because the first specimen known to science was melanistic. Flannery (1995a) wrote that melanistic and partly melanistic individuals seem fairly common, though Woolley (2005a) indicated there is only a small number. Normal individuals of *Myoictis* are among the

most colorful of all marsupials. The upper parts are rusty or grayish brown, or richly variegated chestnut, mixed with black, yellow, and white. There are three dark longitudinal stripes extending most of the length of the back. The crown may be tawny brown, and *M. melas* has bright reddish patches on the nape and behind the ears. The central dorsal stripe often extends to the tip of the nose. The underparts are yellowish gray or whitish, and the chin and chest may be pale rufous. The evenly tapered tail serves to distinguish the four species. In *M. melas* it is well haired, has a dorsal crest, and is black above and rufous brown below; in *M. wallacei* it is very bushy and reddish; in *M. leucura* it is white-tipped but otherwise has long, dark hairs on the top and sides; and in *M. wavicus* it has short hair and is darker above (Woolley 2005a).

The general body form is similar to that of a small mongoose. The footpads are striated; in some specimens the first interdigital and thenar pads are fused. The pouch is only slightly developed. There are six mammae in *M. melas* and *M. wallacei*, and four in *M. leucura* and *M. wavicus*. Cranial characters that help

Three-striped dasyure (*Myoictis wallacei*). Photo by P. A. Woolley and D. Walsh.

define the genus include the blunt rostrum and wide interorbital region, a diastema between the first and second incisor teeth, the second premolar tooth large in comparison with the first and third, large and well rounded bullae, and nasal bones markedly broadened at their bases (Woolley 2005a).

Ziegler (1977) wrote that three-striped dasyures occur in most rainforests of the lowlands and midmountains of New Guinea but are relatively uncommon. He observed that their external similarities to the largely terrestrial, open forest chipmunks and other sciurids suggest comparable habits and parallel activity patterns. Information compiled by Woolley (2005a) indicates that all four species are diurnal. One of her informants said they live on the ground, eat insects and lizards, and seek refuge in rock piles and tree hollows; one specimen was caught as it moved up the trunk of a tree. Flannery (1995a) wrote that three-striped dasyures are probably diurnal and primarily terrestrial, but noted that native hunters report them to enter villages at night in pursuit of murid rodents. A. R. Wallace stated in 1858 that in the Aru Islands these animals "were as destructive as rats to everything eatable in houses."

Reproductively, *M. melas* is characterized by polyestrous females, extended and aseasonal breeding, sexual maturity attained at 8–11 months of age, and males that may live to breed in more than one season (Krajewski, Woolley, and Westerman 2000). Lactating and estrous females have been taken in nearly all months of the year. Males caught in the wild have been subsequently maintained for up to 25 months in captivity, females for up to 31 months (Woolley 1994, 2003).

DASYUROMORPHIA; DASYURIDAE; Genus *DASYKALUTA* Archer, 1982

Little Red Kaluta

The single species, *D. rosamondae*, is found in the Pilbara region of northwestern Western Australia (Woolley 2008c); it also occurs on nearby Potter Island (Burbidge 2008a). The species was described in 1964

Little red kaluta (*Dasykaluta rosamondae*). Photo by P. A. Woolley and D. Walsh.

and placed in the genus *Antechinus*, but Archer (1982a) considered it to represent a separate and not closely related genus. A recent molecular study (Westerman, Young, and Krajewski 2008) suggests *Dasykaluta* is distinct from but has affinity to *Parantechinus*. *Dasykaluta* was treated as a synonym of *Parantechinus* by Mahoney and Ride (1988) and of *Antechinus* by McKenna and Bell (1997).

According to Woolley (2008c), head and body length is 90–110 mm, tail length is 55–70 mm, and weight is 20–40 grams. The overall coloration is russet brown to coppery. The fur is rather rough, the head and ears are short, and the forepaws are strong and well haired on the back. *Dasykaluta* has a general form similar to that of *Dasycercus* but is distinguished by its small size, coloration, and lack of black hair on the tail. Females have eight mammae. Among the characters that Archer (1982a) used to describe the genus are: second upper incisor tooth larger than the fourth, third upper premolar absent, protocone of fourth upper molar compressed, rapid posterior flaring of nasal bones, and ability to develop an incrassate tail.

The kaluta dwells in hummock grassland among the mazes formed by tussocks of spinifex (*Triodia lanigera*). Although Woolley (2008c) referred to it as mainly nocturnal, Baker (2015) wrote that, unlike most dasyurids, the kaluta, in both the wild and captivity, is almost exclusively diurnal and retreats into underground burrows during the night. It feeds on insects and small vertebrates, such as lizards (Woolley 2008c). Its unusually robust morphology and aggressive behavior, as well as feeding observations, suggest the kaluta is the most carnivorous of the small dasyurids and that its diet includes mice. Laboratory investigation showed the kaluta to have a high propensity to enter torpor at ambient temperatures of 11–21° C, which resulted in significant absolute energy and water savings (Withers and Cooper 2009).

Observations on both wild-caught and laboratory-maintained animals (Woolley 1991a) show that females are monestrous and that there is a short annual breeding season. Mating occurs in September, and the young are born in November. The total period of pregnancy averages 50 days but ranges from 38 to 62 days, the great variation perhaps being associated with arrested development in some cases. Normally there are eight young, but there may be as few as one. They open their eyes after 58–60 days and are weaned at 90–120 days. Both sexes attain sexual maturity at about 10 months and thus are able to participate in the first breeding season following their birth. However, as in *Antechinus* (see account thereof), all mature males perish shortly thereafter. Females produce only one litter annually, but some live to breed for a second and possibly a third season. The kaluta is considered widespread and abundant, though population trend is unknown (Burbidge 2008a).

DASYUROMORPHIA; DASYURIDAE; **Genus *PARANTECHINUS*** Tate, 1947

Dibbler

At the time of European settlement, the single species, *P. apicalis*, was found along the coast of southwestern Western Australia from the Moore River to King George Sound; geologically Recent fossils indicate the species once occurred from Shark Bay to Bremer Bay in Western Australia and on the Eyre Peninsula of South Australia (Woolley 2008d). Fossil material also has been reported from Rockhampton on the coast of southeastern Queensland (Baynes 1982; Mahoney and Ride 1988). Neither Kirsch and Calaby (1977) nor Ride (1970) recognized *Parantechinus* as generically distinct from *Antechinus*, but cranial and dental features, biochemical and molecular analysis, and phallic morphology suggest the two genera are not closely related (Archer 1982a; Baverstock et al. 1982; Krajewski and Westerman 2003; Woolley 1982b). A second species, *P. bilarni*, was transferred from *Antechinus* to *Parantechinus* by Archer (1982a), but recent morphological and molecular assessment (Cooper, Aplin, and Adams 2000; Westerman, Young, and Krajewski 2008) show it more appropriately referred to the genus *Pseudantechinus*.

Head and body length is about 140–145 mm, tail length is 95–115 mm, and weight is 30–125 grams. The upper parts are brownish gray speckled with white, and the underparts are grayish white tinged with yellow. *Parantechinus* can be distinguished from *Antechinus* by its tapering and hairy tail, a white ring around the eye, and the freckled appearance of its rather coarse fur (Woolley 1991c, 2008d). The pouch of females con-

Dibbler (*Parantechinus apicalis*). Photo by P. A. Woolley and D. Walsh.

sists merely of folds of skin on the lower abdomen enclosing the eight mammae. According to Cooper, Aplin, and Adams (2000), *Parantechinus* differs from *Pseudantechinus* in having a high vaulted braincase, broader nasals, a small but distinct postorbital process, less extensive fenestration of the maxillary and palatine portions of palate, lesser inflation of the auditory bulla, and noticeable enlargement of the second upper premolar tooth.

The dibbler occurs in heathland, scrub, and thickets, seeming to prefer dense, unburned vegetation. It was said to make nests in dead logs or stumps. Friend (2003) noted that such nests had not been found in recent field studies but that the dibbler does often spend time in seabird burrows. According to Woolley (1983), captives tend to be nocturnal but may emerge from cover during the day to bask in the warmth of the sun. They burrow through leaf litter, which suggests they do the same in the wild to search for insects. They also eat chopped meat, honey, and nectar. The latter food, plus the climbing ability of captives, suggests that nectar of flowers is sought in the wild. Friend (2003) indi-

cated that the natural diet consists mostly of insects and also includes small vertebrates and plant material.

Radio-tracking showed that home ranges overlap in summer, those of males being on average larger than those of females (Friend 2003). Mating occurs early in the year, in March or April; females are monestrous, the estrous cycle is 10–20 days, estrus lasts 1–3 days, and the gestation period, at least on the mainland, is 41–48 days (Jackson 2003; Woolley 1971b, 1991c). On Boullanger and Whitlock Islands gestation has been found to last only about 38 days, possibly because the animals there are significantly smaller than on the mainland (Mills et al. 2012). There are up to eight young per litter; they remain dependent for 3–4 months and reach sexual maturity at 10–11 months. Dickman and Braithwaite (1992) found some populations to experience a synchronized male die-off following the mating season, as occurs in *Antechinus* (see account thereof). However, observations of other populations in the wild and captivity indicate that such is not inevitable (Woolley 1991c). The process in *Parantechinus* appears to be a facultative response to adverse environmental conditions, rather than the physiological, stress-related breakdown seen in *Antechinus* (Woolley 2008d). A captive female lived for 5 years and 6 months (Weigl 2005).

Parantechinus is one of the rarest of mammalian genera. It is classified as endangered by the IUCN (Friend, Burbidge, and Morris 2008) and the USDI. It had not been collected for 63 years when, in 1967, two specimens were taken alive at Cheyne Beach in the extreme southwestern corner of Western Australia (Morcombe 1967). Individuals subsequently were collected at a few more locations and, starting in 1984, at numerous places in Fitzgerald River National Park on the south coast of Western Australia. Captive bred animals from there have been used for reintroduction to two other mainland sites. Meanwhile, in 1985 the species was discovered on Boullanger and Whitlock Islands off the west coast of the state, and captive-bred stock from there has been released on nearby Escape Island (Friend 2003; Woolley 2008d). However, *P. apicalis* continues to decline, mainly because of burning, habitat degradation, and predation by feral house cats. Fewer than 1,000 individuals likely survive on the mainland and only about 180 in the two island populations (Woinarski, Burbidge, and Harrison 2014).

DASYUROMORPHIA; DASYURIDAE; **Genus *PSEUDANTECHINUS* Tate, 1947**

False Antechinuses

There are six species (Cooper 2008; Cooper, Aplin, and Adams 2000; Johnson, Woinarski, and Langford 2008; Kitchener 1988, 1991; Kitchener and Caputi 1988; Westerman, Young, and Krajewski 2008; Woolley 2008e, 2008f, 2008g, 2011):

P. bilarni, Table Top Range to Wollogorang Station in northern Northern Territory, Marchinbar Island;

False antechinus (*Pseudantechinus macdonnellensis*). Photo by P. A. Woolley and D. Walsh.

P. woolleyae, Pilbara, Ashburton, Murchison, and Little Sandy Desert regions of Western Australia;

P. ningbing, northeastern Western Australia, extreme northwestern Northern Territory, Augustus and Heyward Islands;

P. macdonnellensis, east-central Western Australia, central and southern Northern Territory;

P. roryi, Cape Range to Clutterbuck Hills in central Western Australia, Barrow Island;

P. mimulus, Mittiebah Range northeast of Alexandria Station in northeastern North Australia, vicinity of Mount Isa in west-central Queensland, Sir Edward Pellew Islands.

Neither Kirsch and Calaby (1977) nor Ride (1970) recognized *Pseudantechinus* as generically distinct from *Antechinus*. Subsequent systematic work involving cranial and dental features, biochemical analysis, and phallic morphology suggested the two genera are not closely related (Archer 1982a; Baverstock et al. 1982; Morton, Dickman, and Fletcher 1989; Woolley 1982b). McKenna and Bell (1997) continued to treat *Pseudantechinus* as a synonym of *Antechinus*, but recent molecular studies have consistently supported their distinction, even placing the two genera in separate tribes (see account of family Dasyuridae). Based primarily on morphological characters, Cooper, Aplin, and Adams (2000) recognized *P. roryi* as a species separate from *P. macdonnellensis*, but molecular assessment led Johnson, Woinarski, and Langford (2008) to recommend synonymy. Burbidge, Cooper, and Morris (2008, citing M. Westerman) noted that isolated populations of *P. roryi* on Barrow Island and in the Cape Range are genetically distinct and might be separate species. Except as noted, the information for the remainder of this account was taken from Cooper (2008), Johnson, Woinarski, and Langford (2008), and Woolley (2008e, 2008f, 2008g).

Head and body length is 63–115 mm, tail length is 56–125 mm, and weight is 14–45 grams. The upper parts are brown, grayish brown, or reddish brown, the underparts are grayish or whitish, and there are chestnut or orange patches behind the ears in most species. The tail usually tapers and, when food is plentiful, becomes very thick (incrassate) at the base for purposes of fat storage. *P. bilarni* is the only species with a consistently thin tail. There are six mammae, except in *P. ningbing*, which has four. According to Cooper, Aplin, and Adams (2000), *Pseudantechinus* differs from all other genera of Dasyurinae in having a broad and flattened braincase, minimal development of the sagittal crest, and reduction in size of the upper and lower canine teeth. It further differs from *Parantechinus* in having a shorter facial skeleton, more extensive fenestration of the maxillary and palatine portions of palate; no postorbital process on the frontal bone, greater inflation of the middle ear cavity, a less procumbent first upper incisor, no enlargement of the second upper premolar, and greater reduction of the talonids and no protoconules on the second and third upper molars.

False antechinuses are found mostly in dry habitats and are commonly associated with rugged, rocky country, covered with shrubs, grasses, or open woodland. The diet consists mostly of insects. The best known species, *P. macdonnellensis*, is found mainly on rocky hills and breakaways but also lives in termite mounds in some areas. It is predominantly nocturnal but may emerge from shelter among the rocks to sunbathe. Morning basking and a shift towards diurnal foraging, mostly within 3 hours of sunset, have been reported during the winter (Pavey and Geiser 2008). Individuals radio-tracked for up to 26 days had overlapping home ranges of 0.15–1.77 ha.; those of males were up to three times larger than those of females.

Reproduction is seasonal and restricted, females are monestrous, males may live to breed in more than one season, and sexual maturity is attained at about 11 months of age (Krajewski, Woolley, and Westerman 2000). *P. ningbing* mates in June and gives birth during a period of just two to three weeks in late July and early August after a gestation of 45–52 days (Woolley 1988). Maximum litter size is four. The young are weaned at about 16 weeks. *P. bilarni* mates from late June to early July, the interval between mating and parturition is 38 days, litters contain four to five young, and weaning occurs late in the year.

Observations by Woolley (1991b) show that *P. macdonnellensis* has a short annual breeding season during the austral winter. Mating occurs in June and early July in the eastern part of the range and in August and early September farther west. The gestation period is 45–55 days. There commonly are five or six young, they open their eyes after 60–65 days, they are weaned at about 14 weeks, and they are able to breed in the first season following their birth. Unlike male *Antechinus*, male *Pseudantechinus* do not experience mass die-off after reproduction; indeed, males are potentially capable of breeding for at least 3 years. Females may breed in at least four seasons. Weigl (2005) listed a captive female *P. macdonnellensis* that lived to an age of about 7 years.

The IUCN classifies *P. mimulus* as endangered (Woinarski and Dickman 2008). It has declined because of habitat degradation by fires and livestock and predation by introduced cats. However, applying IUCN criteria to updated information indicating continued presence on the mainland of Northern Territory and Queensland, as well as the Sir Edward Pellew Islands, Woinarski, Burbidge, and Harrison (2014) categorized the species as near threatened. Those authorities, as well as the IUCN (Woinarski 2008b), also classify *P. bilarni* as near threatened. Not discovered until 1948, that species contains fewer than 10,000 mature individuals, and numbers are decreasing.

DASYUROMORPHIA; DASYURIDAE; **Genus** *DASYCERCUS* Peters, 1875

Mulgaras

There are two species (Masters 2008; Woinarski, Burbidge, and Harrison 2014; Woolley 2005b, 2006, 2008b):

> *D. cristicauda*, historically found in the arid region from central Western Australia to southwestern Queensland and coastal South Australia;
> *D. blythi*, arid region from west-central Western Australia to southwestern Queensland.

Although *D. blythi* long was treated as a synonym of *D. cristicauda*, as by Groves (2005c), Woolley (2005b, 2008b) pointed out the clear morphological distinctions of the two. She noted, however, that precise past and present distribution of each species cannot be clarified until available specimens have been properly identified. She also explained that *D. hillieri*, described from a specimen collected east of Lake Eyre and sometimes treated as a distinct species, is synonymous with

Mulgara (*Dasycercus blythi*). Photo by Jiri Lochman / Lochman LT.

D. cristicauda. Ellis (1992) stated that remains found in owl pellets indicate the range of *Dasycercus* may have extended as far as western New South Wales in the nineteenth or early twentieth century. Some authorities include *Dasyuroides* (see account thereof) within *Dasycercus*. Recent molecular studies have supported separation of the two genera and of the two species of *Dasycercus* (Krajewski et al. 2004; Westerman, Young, and Krajewski 2008).

Head and body length is 120–230 mm, tail length is 60–125 mm, and weight is 60–185 grams; males are larger on average than are females (Masters 2008; Woolley 2008b). *Dasycercus* sometimes exceeds *Dasyuroides* in size but usually is smaller. The upper parts vary from buffy to bright red brown, and the underparts are white or creamy. The close and soft pelage consists principally of underfur with few guard hairs. *Dasycercus* is compactly built, with short limbs, ears, and muzzle. The tail is usually thickened (incrassate) for about two-thirds of its length and is densely covered near the body with coarse, chestnut-colored hairs. Beginning at about the middle of its length the tail is covered with coarse, black hairs. Woolley (2005b) explained that in *D. cristicauda* those hairs increase in length toward the tip to form a distinct dorsal crest but that in *D. blythi* there is no crest. *D. cristicauda* also is characterized by having three upper premolars (the third very small and sometimes present on one side only) and eight mammae, whereas *D. blythi* has only two upper premolars (and a diastema between the second upper premolar and the first molar) and six mammae. The pouch area consists of only slightly developed lateral skin folds.

Mulgaras are terrestrial but capable of climbing (Collins 1973). Although both species are desert dwellers, there is some difference in habitat. *D. cristicauda* occupies sparsely vegetated sand dunes or areas around salt lakes, while *D. blythi* inhabits spinifex grassland with medium to dense cover (Masters 2008; Woolley 2008b). The burrows of *D. cristicauda* tend to be higher on the dunes, while those of *D. blythi* are on lower slopes or in swales (Masters 2008). The latter species excavates two kinds of burrows (Woolley 1990, 2008b). One has a grass-lined nest area at a depth of about 0.5 meter, connected to the surface by a single large tunnel, and several side tunnels and near vertical popholes.

That kind of burrow is occupied by a female with pouch young. The other kind, occupied by a male, has up to five entrance holes, a complex system of interconnecting tunnels, and one grass-lined nest at a depth of up to 1 meter.

Mulgaras hunt by night but are not strictly nocturnal (Woolley 2008b). Ride (1970) noted that they avoid exposure to heat by remaining in their burrows during the hot part of the day. Collins (1973), however, observed that captives bask in the sun whenever opportunity arises and even make use of sun lamps. When sunning themselves, they flatten the body against the substrate and the tail twitches sporadically. Mulgaras can withstand considerable exposure to both heat and cold. However, captives were found to enter torpor for up to 12 hours, during which metabolic rate and body temperature were greatly reduced (Geiser and Masters 1994). Studies of captives indicate that the physiological adaptations are such that they are able to subsist without drinking water or even eating succulent plants, because they can extract sufficient water from a diet of just lean meat or mice (Ride 1970).

This genus is relatively uncommon but reportedly increases in numbers when a house mouse plague occurs within its range. *Dasycercus* attacks a mouse with lightning action and then devours it methodically from head to tail, inverting the skin of the mouse in a remarkably neat fashion as it does so. Mulgaras also prey on other small vertebrates, arachnids, and insects. They reportedly can skillfully dislodge insects from crevices by means of their tiny forepaws.

Masters (2003) found *D. cristicauda* to occupy moderately overlapping home ranges of 1.0 to 14.4 ha. Körtner, Pavey, and Geiser (2007) reported that six male *D. blythi* occupied an average home range of 25.5 ha. and that three females had an average range of 10.8 ha.; there was substantial overlap of ranges both between and within sexes. Mulgaras can be maintained in captivity in pairs except when the females have young (Collins 1973). They generally do not fight among themselves and appear quite solicitous of each other. Although in the wild only a single adult is usually found in a burrow, Körtner, Pavey, and Geiser (2007) observed burrows to be shared occasionally by a male and female, by two females, and even by two males. Masters and Dickman

Mulgara (*Dasycercus blythi*), mother and young. Photo by Jiri Lochman / Lochman LT.

(2012) found that young female *D. blythi* remained near their mother's home range but that males moved to other areas.

Females are facultatively polyestrous and may attempt a second litter during a season if they fail with a first (Krajewski, Woolley, and Westerman 2000). *D. cristicauda* gives birth to up to eight young in the winter and early spring (Masters 2008). In *D. blythi* (Woolley 2008b), females with up to six young have been captured in September. Captives of that species have been observed mating from mid-May to mid-June, and young have been born in late June, July, and August after a gestation period of 5–6 weeks. The young suckle for 3–4 months and become sexually mature at 10–11 months. Individuals of both sexes have been known to come into breeding condition each year for 6 years, suggesting a fairly long life span. Weigl (2005) listed a captive male that lived for 7 years and 5 months.

Archer (1979a) considered mulgaras to be vulnerable or possibly endangered, noting they apparently had disappeared or become very rare in Queensland and South Australia. Kennedy (1992) suggested their decline occurred because of habitat disturbances and predation caused by the European introduction of livestock, domestic cats, foxes, and rabbits. Maxwell, Burbidge, and Morris (1996) listed them generally as vulnerable and recognized *D. hillieri* as a distinct and endangered species. The IUCN, citing their wide distribution, presumed large population, and lack of continued substantive decline, does not now place *D. cristicauda* or *D. blythi* in a category of concern (Woolley 2008k, 2008l). However, applying IUCN criteria to recent information indicating that *D. cristicauda* consists of not substantially more than 10,000 individuals and is continuing to decline due to the above factors, Woinarski, Burbidge, and Harrison (2014) designated the species vulnerable; its current range may be restricted to a small area of desert in southeastern Northern Territory, northeastern South Australia, and southwestern Queensland.

DASYUROMORPHIA; DASYURIDAE; Genus *DASYUROIDES* Spencer, 1896

Kowari

The single species, *D. byrnei*, historically occurred in the south of the Northern Territory of Australia, the southwest of Queensland, and the northeast of South Australia; subfossil remains indicate that its range also extended into western New South Wales (Lim 2008).

Kowari (*Dasyuroides byrnei*). Photo by Alexander F. Meyer at Poznan Zoo.

Dasyuroides was considered a synonym of *Dasycercus* by Corbet and Hill (1991) and Mahoney and Ride (1988) but was maintained by Morton, Dickman, and Fletcher (1989), McKenna and Bell (1997), and Groves (2005c). Recent molecular studies have supported the distinction between the two genera, as well as their close affinity (Krajewski et al. 2004, 2007; Westerman, Young, and Krajewski 2008).

Head and body length is 135–180 mm, tail length is 110–160 mm, and weight is 70–175 grams, with males being on average heavier than females (Lim 2008). The back and sides are grayish with a faint rufous tinge. The underparts are pure creamy white, the feet are white, and there are fawn eye rings. Less than the basal half of the tail is rufous; the remainder is densely covered on both the upper and lower surfaces with long black hairs that form a distinct brush, heavier than that of *Dasycercus blythi*. The soft and dense pelage is composed mainly of underfur with few guard hairs. The body form is strong and stout. The hind feet are very narrow and lack a first toe; the bottoms of the feet are hairy. The tail does not become thick with fat deposits (incrassate). Pouch development is sufficient to conceal the young completely, and there usually are six (five to seven) mammae (Aslin 1974).

The kowari inhabits desert surfaces covered with closely packed pebbles and rock fragments and with less than 25 percent vegetative cover (Lim 2008). It may shelter in the hole of another mammal or dig its own burrow in sandy places, and both sexes construct therein a nest of soft materials (Aslin 1974; Ride 1970). It is primarily terrestrial but climbs well and is capable of vertical leaps of at least 45.7 cm (Collins 1973). In the field, Aslin (1974) observed *Dasyuroides* only after dark; in captivity one individual was largely nocturnal, while another was sporadically active both day and night. Collins (1973) stated that the kowari enjoys sun- and sandbathing. He also reported that the diet consists of insects, arachnids, and probably small vertebrates such as birds, rodents, and lizards. Aslin (1974) reported that one wild individual had fed on *Rattus villosissimus* and some insects. Torpor has been induced experimentally in *Dasyuroides* through a moderate reduction in diet, and it is likely that individuals also enter torpor in the wild in response to declining food supplies (Geiser, Matwiejczyk, and Baudinette 1986). Daily torpor, lasting up to 7.5 hours, has been reported (Geiser 2003).

Aslin (1974) estimated that a population of at least 14 adults occupied an area of less than 750 ha. and suggested that some kind of social aggregation might occur in the wild. Lim (2008) stated that the kowari is solitary except during the mating and post-natal periods. Collins (1973) observed that a captive pair or small group could be maintained together but that serious fighting might result if the enclosure were too small. Aslin (1974) noted, however, that aggressive behavior is stylized in such a way as to prevent serious injury in intraspecific conflict. She also reported that each sound made by adults—hissing, chattering, and snorting—is restricted to threat situations and that both sexes use scent produced by sternal and cloacal glands to mark parts of their home range.

The following reproductive information is based on studies of both wild and captive individuals (Aslin 1974, 1980; Collins 1973; Krajewski, Woolley, and Westerman 2000; Woolley 1971a). Females are polyestrous and may have up to four estrous periods a year, which recur at two-month intervals if young are not being suckled. Mating takes place from April to December, though mostly from May to July, and females may produce two litters in a season. Gestation is 30–36 days. Litter size is 3–7, having averaged 5.1 in captivity and 5.8 in the wild. The young are 4 mm long at birth, first detach from the nipples at 56 days, are weaned and practically independent at about 100 days, reach sexual maturity at 8–11 months, and attain adult weight at 1 year. The young may ride on the mother's back or sides when they are 2–3 months old, and she remains tolerant of them for a while after weaning. Both sexes are capable of breeding through their fourth year of life. Weigl (2005) listed a captive kowari that lived to an age of just over 8 years.

The IUCN classifies *D. byrnei* as vulnerable (McKnight et al. 2008). Total numbers are estimated at fewer than 10,000 mature individuals, and they are subject to extreme fluctuations. Declines have occurred where livestock grazing is intense, especially near water holes used by stock. Continued decline is likely, mainly because of future impacts of global warming and subsequent range contraction. There were 24 known populations but some are thought to have recently been lost. Kennedy (1992) reported an over-all decline of 50 to 90 percent. Lim (2008) indicated that *Dasyuroides* has disappeared from the type locality at Charlotte Waters and from the rest of the original range in the Northern Territory.

DASYUROMORPHIA; DASYURIDAE; Genus *PHASCOLOSOREX* Matschie, 1916

Marsupial "Shrews"

There are two species (Flannery 1995a):

P. dorsalis, highlands of eastern and extreme western New Guinea;
P. doriae, western New Guinea, including the Vogelkop Peninsula.

There is some question whether *P. doriae* and *P. dorsalis* are distinct species, given that they are remarkably alike in basic characters and differ markedly only in coloration and body size. The two were treated as separate species by Flannery (1995a), Groves (2005c), and Krajewski et al. (2004). *P. doriae* is generally the larger animal, with a head and body length of 117–226 mm and a tail length of 116–191 mm. Its general body color is dark grizzled orange brown with bright rufous on its underside, whereas *P. dorsalis* is a grizzled gray brown with chestnut red underneath. The head and body length of *P. dorsalis* is 134–167 mm, and tail length is 110–160 mm. Flannery (1995a) noted that specimens of *P. dorsalis* from the eastern segment of its range are relatively small, with adult males weighing only 50–60 grams, while those from the west weigh 130–140 grams.

A distinctive character of *Phascolosorex* is the thin black stripe that runs from the head all the way down the middle of the back to the base of the tail. This dorsal stripe is much narrower than that of *Paramurexia rothschildi*. The tail is black and short-haired

Marsupial "shrew" (*Phascolosorex dorsalis*). Photo by P. A. Woolley and D. Walsh.

except at the base, where it is covered by thick, short, soft fur like that covering the body. The tail may or may not terminate in white. The ears are relatively short and more sparsely haired than the body. The feet, which are black in *P. doriae* and brown in *P. dorsalis*, possess striated pads and relatively short claws. There are four mammae in the pouch area, which Woolley (2003) noted is covered by a circular skin fold.

These marsupials occur in mountain forests. *P. dorsalis* is more abundant at high altitudes but has been collected at 1,210–3,600 meters; *P. doriae*, the more geographically restricted species, has been recorded only in the lower to middle elevations, at 900–1,900 meters. The differences in locality of the two species when they both occur in the same general area are more strikingly vertical than horizontal. They usually are thought to be nocturnal and scansorial in habit (Collins 1973). However, extended laboratory observa-tions by Woolley et al. (1991) demonstrate that *P. dorsalis* is predominantly diurnal; in that regard it is unusual among the Dasyuridae. Dwyer (1983) described it as "a rainforest species said to be terrestrial and of-ten diurnal in activity."

Reproduction of *P. dorsalis* is characterized by extended and aseasonal breeding, polyestrous fe-males, males that can potentially breed in more than one season, and sexual maturity attained at 8–12 months of age (Krajewski, Woolley, and Westerman 2000; Woolley 2003). Estrous females have been re-corded in all months and lactating females in January, April, May, August, September, October, and Decem-ber. Time between estruses of captive females was 58–427 days and four gestation periods were 18–21 days. The young numbered 3–4 in 4 births and lactation lasted 4.0–4.5 months. Wild-caught individuals were maintained in captivity for up to 39 months (Woolley 2003).

DASYUROMORPHIA; DASYURIDAE; **Genus**
NEOPHASCOGALE Stein, 1933

Speckled Dasyure, or Long-clawed Marsupial "Mouse"

The single species, *N. lorentzi*, is found in the western and central mountains of New Guinea (Flannery 1995a).

This marsupial resembles a tree shrew in general appearance. Head and body length is 166–230 mm, and tail length is 170–215 mm. A single male weighed 212 grams (Flannery 1995a), and a captive female weighed about 130 grams when it first entered estrus (Woolley 2001). The color ranges from deep rufous to dull, pale cinnamon on the upper parts, and the fur is profusely speckled with white hairs or white-tipped hairs. The underparts are rich rufous with whitish, subterminal bands to the hairs. The head is deep rufous brown, and the backs of the ears are white. The limbs are rusty in color and blend to brown on the feet. The tail is rufous, except for the last third, which is white. The fur is long, soft, and dense. A few specimens taken have been melanistic. There are long claws on all the feet, and the pads are striated. Structural differences that support separation from *Phascogale* are mainly in the dentition: there is no gap between incisors 1 and 2, and the last premolar is considerably smaller than the first two.

The long-clawed marsupial "mouse" inhabits humid moss forests at altitudes of 1,500–3,400 meters. It is partly diurnal and probably is largely arboreal (Ziegler 1977). Two specimens that survived several months in captivity were fed strips of beef, cicadas, and cockchafer beetles; live insects were consumed greedily (Collins 1973). Females are polyestrous (Krajewski, Woolley, and Westerman 2000). Lactating females have been collected only in November, though sampling has been insufficient to determine whether reproduction is seasonal (Woolley 1994). One female was captured as a juvenile and then lived for 34 months; estrus occurred at intervals of 93–126 days and eventually during most months, thus suggesting a potential to breed throughout the year (Woolley 2001).

Long-clawed marsupial "mouse" (*Neophascogale lorentzi*). Photo by P. A. Woolley and D. Walsh.

DASYUROMORPHIA; DASYURIDAE; **Genus**
DASYURUS E. Geoffroy St.-Hilaire, 1796

Quolls, or Native or Tiger "Cats"

There are six species (Archer 1979a; Belcher, Burnett, and Jones 2008; Firestone 2000; Flannery 1995a; How, Spencer, and Schmidt 2009; Jones 2008a; Jones and Rose 2001; Kirsch and Calaby 1977; Krajewski et al. 2004; Morris et al. 2003; Oakwood 2008; Orell and Morris 1994; Serena and Soderquist 2008; Van Dyck 1987):

- *D. hallucatus* (northern quoll), Pilbara and Kimberley areas of northern Western Australia and numerous offshore islands, northern Northern Territory and several nearby islands, northern and eastern Queensland;
- *D. viverrinus* (eastern quoll), originally southeastern South Australia, Victoria, eastern New South Wales, Tasmania, Kangaroo Island, King Island, and Flinders Island;

Tiger quoll (*Dasyurus maculatus*). Photo by Alexander F. Meyer at Columbus Zoo.

D. geoffroii (western quoll, or chuditch), originally throughout Australia except the extreme north, eastern and southeastern coasts, and Tasmania;

D. spartacus (bronze quoll), south-central New Guinea;

D. albopunctatus (New Guinea quoll), throughout New Guinea except the south-central region;

D. maculatus (tiger or spotted-tailed quoll), originally eastern Queensland, eastern and southern New South Wales, Victoria, southeastern South Australia, Tasmania, Kangaroo Island, King Island, and Flinders Island.

These species sometimes were placed in separate genera or subgenera as follows: *D. hallucatus* and *D. albopunctatus* in *Satanellus* Pocock, 1926; *D. viverrinus* in *Dasyurus* E. Geoffroy St.-Hilaire, 1796; *D. geoffroii* in *Dasyurinus* Matschie, 1916; and *D. maculatus* in *Dasyurops* Matschie, 1916. Based on paleontological data,

Archer (1982a) again recognized *Satanellus* as a full genus. That view was not accepted by Van Dyck (1987), who considered *D. albopunctatus* to be closely related not to *D. hallucatus* but to his newly described *D. spartacus*. The assignment of all the above species to *Dasyurus* was accepted by Corbet and Hill (1991), Groves (2005c), Mahoney and Ride (1988), and Morton, Dickman, and Fletcher (1989). Recent molecular studies have supported the monophyly of *Dasyurus* and its affinity to *Sarcophilus* (Krajewski and Westerman 2003; Krajewski et al. 2004), though various phylogenetic sequences within *Dasyurus* have been proposed (see review by Firestone 2000). Assessment of mitochondrial DNA led Firestone (2000) to question the specific distinction of *D. spartacus* from *D. geoffroii*. A recent comprehensive analysis of mitochondrial and nuclear DNA, plus penis morphology, showed *D. hallucatus* to be sister to all other species and *D. viverrinus*, *D. geoffroii*, and *D. albopunctatus* to form a clade that is sister to *D. maculatus*; a single specimen of *D. spartacus* fell within the range of variation of *D. geoffroii* but more study of their relationship was thought necessary (Woolley, Krajewski, and Westerman 2015). Two subspecies of *D. maculatus* are usually recognized, *D. m. gracilis* in northeastern Queensland and *D. m. maculatus* in southeastern Australia and Tasmania. However, Firestone et al. (1999) suggested that, while the two are valid, the Tasmanian population is highly distinct from those of the mainland and probably warrants its own taxonomic designation. Aplin and Pasveer (2007) reported remains of *D. albopunctatus* from a late Pleistocene to Holocene archeological site on Kobroor in the Aru Islands, where *Dasyurus* is not now known to occur.

As with many other Australian mammals, the common names applied to *Dasyurus* stem from European terminology. After extensive research on Aboriginal names, Abbott (2013) recommended the following applications: "digul" for *D. hallucatus*, "luaner" for *D. viverrinus*, "chuditch" for *D. geoffroii*, and "bindjulang" for *D. maculatus*. Pronunciation is with stress on the first syllable. Continued use of "quoll" was endorsed as a suitable common equivalency for the genus *Dasyurus*.

The species of *Dasyurus* are the largest living members of the Dasyuridae, except for *Sarcophilus*. While the pelage of the latter genus may have a few white patches, *Dasyurus* is distinguished from all other dasyurids by numerous prominent white spots or blotches on the back and sides, regardless of basic coloration. It also is characterized by relatively short fur, short legs, and massive skull. The pads of the feet are granulated in *D. viverrinus*, *D. geoffroii*, and *D. spartacus* and striated in the other species. All species have five toes on the hind foot, except *D. viverrinus*, which has only four. In *D. hallucatus*, *D. spartacus*, and *D. albopunctatus* pouch development consists only of lateral folds of skin; in the other species there is a shallow pouch formed by a flap of skin

All species are primarily terrestrial but able to climb well, all are nocturnal but are occasionally seen by day, all are solitary but can be maintained in captivity as pairs except when the female has young, and all are predators but will eat vegetable matter (Collins 1973; Ride 1970). They occasionally raid poultry yards and are therefore disliked by farmers, but they probably also benefit human interests by destroying many mice and insect pests (Ride 1970).

All species have suffered since the settlement of Australia by Europeans and evidently are continuing to decline. Besides direct human persecution and clearing of cover, native "cats" have been adversely affected by introduced placental predators and competitors, such as the domestic cat (*Felis catus*) and dog (*Canis familiaris*) and the red fox (*Vulpes vulpes*). The placental cat was introduced to Australia by Europeans in the early nineteenth century, and feral populations thereof had become established throughout the continent by 1890 (Abbott 2002, 2008b). During the late nineteenth and first decade of the twentieth century a severe epidemic seems to have greatly reduced the populations of native "cats," as well as certain other marsupials, in southeastern Australia and Tasmania, and in some areas there apparently never was a substantial recovery. Abbott (2006) hypothesized that a major epizootic disease began near Shark Bay in Western Australia around 1875 and spread through most of the state by 1920, severely reducing populations of the species of mammals therein by about a third. The disease may have been transmitted by deliberate or incidental introduction of affected domestic animals and rodents; a similar process may have taken place in eastern Australia. Additional information is provided separately for each species.

Dasyurus hallucatus (Northern Quoll)

The account of this species was adapted from that of Oakwood (2008); other literary sources are noted. Head and body length is 249–370 mm, tail length is 202–345 mm, and weight is 240–1,120 grams. Males on average weigh 760 grams, females 460 grams (Oakwood 2002). The upper parts are brown or drab gray, with white spots on the back, rump, and head; the underparts are creamy white or yellowish. The tail is well haired and colored like the back for most of its length; the base is brown with occasional white spots, and the tip is dark brown or black with no spots. Burnett and Mott (2004) pointed out that, while *D. maculatus* is often reported to be the only species of *Dasyurus* characterized by a spotted tail, some specimens of at least some populations of *D. hallucatus* also exhibit spotting as far back as the middle of the tail. The coat is short and coarse and has little underfur. There

Northern quoll (*Dasyurus hallucatus*). Photo by Jiri Lochman / Lochman LT.

is a prominent first toe and the foot pads are striated. Females have 5–9 mammae, usually 8.

The northern quoll is most common on dissected rocky escarpment, though it is also found in eucalypt forest and savannah woodland and occasionally in rainforest or on beaches. While predominantly nocturnal, it is sometimes active by day, particularly during the mating season or in overcast weather. It is both terrestrial and arboreal, is quick and agile, and may carry its tail erect when walking or running. Dens include tree hollows, rock crevices, logs, termite mounds, roofs of buildings, old stoves in abandoned houses, and burrows of the large monitor lizards (*Varanus*). Dens usually are changed every day. Mean distance between successive dens varies from around 0.1 to 0.8 km, depending on season and sex (Oakwood 2002). *D. hallucatus* eats arthropods, fruit, and small vertebrates, including bandicoots and fish. It drinks water, when available, but can obtain sufficient moisture from its food.

Density of resident females in a savannah population varied from 4 per sq km in the wet season to 1/sq km in the late dry season; male density also reached a maximum of about 4/sq km (Oakwood 2002). The adult males had home ranges of over 100 ha., which overlapped several female territories, the latter averaging 35 ha. Both sexes were solitary, and den sharing was rare. Dissected escarpment supports higher densities, and females there appear less territorial, with greater overlap and smaller ranges.

Female *D. hallucatus* are monestrous (Krajewski, Woolley, and Westerman 2000). They breed annually, exhibiting synchronous reproduction every year at each site. According to McAllan (2003), mating takes place in May–June and births in June–July, though Schmitt et al. (1989) found females in Western Australia to produce a single litter in July or August. In the Pilbara, mating reportedly occurs in July–August, and young are born in late August–September (Woinarski, Burbidge, and Harrison 2014). The pouch becomes deep and moist, and gestation lasts 21–26 days. Average litter size is 7 young, but McAllan (2003) reported a range of 6–10; they measure only 3 mm in length (Collins 1973). When 8–9 weeks old, the young are deposited in a series of nursery dens, the mother returning regularly to allow suckling. Juveniles begin to eat insects at 4 months but are not fully weaned until 6 months old. The mother's pouch region then reverts to resemble an udder. Only 2–3 of the young are weaned. Sexual maturity is attained at about 11 months of age. Oakwood (2000) found that only 27 percent of females survived to wean a second litter and none bore a third. The oldest female recorded in the wild was 3 years of age.

The northern quoll is the largest species of mammal known to undergo male die-off after mating (see account of *Antechinus* for further information on that phenomenon). The process appears to be associated with the intense physical effort made by the males of *D. hallucatus* during the mating period to regularly visit numerous females in rapid succession to monitor the onset of estrus. Oakwood (2000) reported complete post-mating mortality, with most males dying within 2 weeks of mating. Schmitt et al. (1989) found that most males died during the post-mating period from July to September, evidently because of high testosterone levels and resulting trauma; males with lower testosterone levels, apparently the less dominant individuals, had increased prospects of survival. Dickman and Braithwaite (1992) determined that a given population of *D. hallucatus* may experience a complete male die-off in some years but not in others. Weigl (2005) listed a male, captured at an age of about 2 months, which subsequently lived in captivity for 5 years and 9 months.

Populations of *D. hallucatus* are still scattered across northern Australia, but the species has disappeared from most of its range, particularly areas of lower rainfall. It is classified as endangered by the IUCN, because of a serious population decline, estimated to exceed 50 percent over the last 10 years and projected to continue at a similar rate for the next 10 years (Oakwood, Woinarski, and Burnett 2008). Precise reasons for the decline are uncertain, but may include predation by and passage of disease from the domestic cat and dog and the burning and agricultural destruction of habitat. The decline evidently corresponds in large part with the spread of the cane toad (*Bufo marinus*), a native of the Western Hemisphere originally brought to Australia to control beetles feeding on sugarcane. The large (ca. 1 kg) amphibian has salivary glands toxic to mammals. *D. hallucatus* seems particularly susceptible to poisoning when it attempts to capture and eat the toad. *D. hallucatus* remains relatively common in the Kimberley and Pilbara Regions of Western Australia and on nearby islands, where the cane toad has not yet spread. DNA analysis indicates the populations there are demographically intact and that the islands may serve as refuge, even when the toad reaches the mainland (How, Spencer, and Schmitt 2009). Nonetheless, Woinarski, Burbidge, and Harrison (2014) considered the effects of the toad "catastrophic," having caused the virtual extirpation of *D. hallucatus* in Arnhem Land, including Kakadu National Park, and also on the Sir Edward Pellew Islands; they projected a similar decline in the Kimberley over the next 10 years.

Dasyurus viverrinus (Eastern Quoll)

Except as noted, the information for the account of this species was taken from Jones (2008a) and Jones and Rose (2001). Head and body length is 280–450 mm, tail length is 170–280 mm, and weight is 900–1,900 grams in males and 700–1,100 grams in females. In about 75 percent of the animals the upper parts are fawn colored, with black guard hairs, and the underparts and feet are pure white. The remaining individuals are jet black above and below, except where grading to brownish on the venter and underside of the tail. In both color morphs, 60–80 irregularly shaped white spots, 5–20 mm in diameter, cover the dorsal surface of the body from the top of the head and cheeks to the rump and tops of the legs. The black phase appears unrelated to sex and often is found in the same litter with young of the more common color. The fur is soft, thick, and fairly short. *D. viverrinus* differs from all other species of *Dasyurus* in lacking a hallux (first toe) on the hind foot. The foot pads are granulated. Females usually have six, occasionally eight, mammae in a shallow pouch formed by enlargement of lateral folds.

The eastern quoll is found in most habitats of Tasmania, from sea level to 1,500 meters, but is most abundant in the drier eastern half and prefers grassland, open forest and woodland, and alpine heaths. It is nocturnal, commonly sheltering by day within a wooded area and foraging in adjoining pasture, sometimes moving over 1 km in a night. Dens include hollow logs, rock piles, buildings, and underground burrows that range from simple tunnels with no nests to complex interconnected passages with one or more nests. Den sites frequently are changed on successive

Eastern quoll (*Dasyurus viverrinus*) in the two color phases. Photos by Johannes Pfleiderer.

days, except when a female has nursing young. *D. viverrinus* is known to enter torpor occasionally in extremely bleak weather. It moves with a bounding gait and goes onto fallen branches but does not regularly climb trees. Prey is pinned with the forepaws and killed by biting. The diet includes small vertebrates, insects, and soft fruits. Woinarski, Burbidge, and Harrison (2014) referred to *D. viverrinus* as "an impressive hunter, taking mammals such as rabbits, mice, and rats." It regularly scavenges and may have a commensal relationship with *Sarcophilus harrisii*. It darts about while the latter species is feeding and tries to steal pieces of flesh; it also depends on the more powerful *S. harrisii* to rip open tough carcasses.

The eastern quoll is solitary. Home ranges overlap extensively but each individual has a core area of intensive use that is avoided by neighboring residents. Mean range size is 44 ha. for males and 35 ha. for females. Male ranges increase significantly in the mating season. Most males and some females disperse from their natal ranges. Olfactory communication is done through urination, defecation, sternal and ventral rubbing, and face washing. Vocalizations are infrequent but include agonistic hisses and grunts, an alarm shriek, and softer calls between mother and young.

Females are facultatively polyestrous (Krajewski, Woolley, and Westerman 2000), breeding synchronously each winter but able to return to estrus if they fail to conceive or lose a litter early in the season. The estrous cycle is 34–37 days, and estrus lasts 5 days (Jackson 2003). Mating occurs from mid-May to early June and most births from May to July. Gestation lasts 20–24 days (Fletcher 1985). A female may give birth to as many as 30 young but only 6–8 can be accommodated in the pouch. Each is about 4 mm long and weighs 12.5 mg. They remain attached to the mammae for 60–65 days and then are left in a grass-lined den. The eyes open at about 79 days. After 85 days they forage with the mother, often clinging to her back. They are completely weaned at 150–165 days and reach sexual maturity when about a year old. Longevity in the wild is probably 3–4 years. Weigl (2005) listed a captive female that lived 6 years and 10 months.

D. viverrinus is known to survive only on Tasmania and, possibly through introduction, on nearby Bruny Island. It is classified as endangered by the USDI and as near threatened by the IUCN (McKnight 2008b).

Beginning around the middle of the nineteenth century, some mainland populations reportedly reached very high numbers, but there were drastic declines in the 1850s and 1860s, evidently caused in large part by parasites and disease, such as canine distemper, spread by introduced carnivores (Peacock and Abbott 2014). That factor was critical in the initial and continued suppression of the eastern quoll, though during the 1930s the species was still common on the southeastern mainland of Australia and was found even in the suburbs of large cities such as Adelaide and Melbourne. It subsequently became very rare on the mainland (Emison et al. 1975), but remained common and even increased in some parts of Tasmania (Green 1967). The last known mainland population, in Nielson Park in the Sydney suburb of Vancluse, disappeared around 1964 because of brush clearance, fire, and drought. Other factors that may have contributed to extirpation on the mainland are excessive hunting for fur and to protect poultry, as well as predation by and competition from the red fox (*Vulpes vulpes*). Recent introduction of the fox to Tasmania may pose a severe threat. Fancourt, Hawkins, and Nicol (2013) found evidence of a rapid and severe decline; for example, spotlight surveys from 150 sites across Tasmania indicated a 52 percent reduction in the number of eastern quoll sightings over 10 years up through 2009.

Dasyurus geoffroii (Western Quoll or Chuditch)

Head and body length is 260–400 mm, tail length is 210–350 mm, and weight is 710–2,185 grams in males and 615–1,130 grams in females (Serena and Soderquist 2008). The pelage is thick and soft. The back and sides are olive-gray tinged with rufous; the underparts are creamy white. The face is paler and grayer than the upper parts. The head, back, and sides are spotted with white, the spots being more numerous and smaller than in *D. viverrinus*. The basal half of the bushy tail is olive-gray, without spots, and the terminal half is black. A first toe is present on the hind foot but the associated pad is virtually obsolete. The foot pads are granular, except at their center, which is smooth. The female's pouch consists of a crescentic fold of skin enclosing the front and sides of the mammary area; there are six mammae arranged in line with the folds of the pouch.

According to Serena and Soderquist (2008), the chuditch originally occupied a diverse array of forest, shrub, and desert habitats across most of mainland

Western quoll or chuditch (*Dasyurus geoffroii*). Photo by Martin Withers / FLPA / Minden Pictures.

Australia, but now survives only in parts of southwestern Western Australia dominated by sclerophyll forest or drier woodland, heath, and mallee shrubland. It is primarily nocturnal but sometimes is active by day, especially at the height of the breeding season and during cold or wet weather. Morris et al. (2003) noted that it is a swift runner and efficient climber, is capable of moving several kilometers in a 24-hour period, and dens in hollow logs, tree limbs, rock outcrops, termite mounds, and earth burrows. Until at least the 1930s, it also commonly nested in the roofs of suburban houses around Perth (Serena and Soderquist 2008). Females excavate burrows in which to give birth, the main tunnel descending to a nursery chamber nearly 1 meter below the ground (Serena and Soderquist 1989a). The diet includes mammals up to at least the size of a rabbit, birds up to the size of a parrot, lizards, frogs, crustaceans, and insects (Serena and Soderquist 2008).

Radio-tracking studies in the Murray River Valley (Serena and Soderquist 1989b) showed females to use home ranges averaging about 337 ha., including a core area of about 90 ha. with numerous dens, where most activity is concentrated. There was usually little or no overlap of those core areas, suggesting that females are intrasexually territorial, but sometimes a daughter shared her mother's core area. Males used much larger ranges, including core areas of at least 400 ha., and there was broad overlap of their core areas with those of both other males and females. Both sexes were found to be essentially solitary and to scent-mark with feces. Another study, in the Batalling Forest, recorded home range sizes of 278–314 ha. for females and 509–791 ha. for males (Morris et al. 2003).

Females are facultatively polyestous; breeding extends from May to September and peaks in May and June (Krajewski, Woolley, and Westerman 2000; Serena and Soderquist 1990). The gestation period is 16–23 days (Collins 1973). According to Serena and Soderquist (2008), most litters appear in June and July, and up to six young are accommodated in the pouch. By 9 weeks of age they have outgrown the pouch and are left in a den while the mother forages. They are weaned by 22–24 weeks and typically disperse shortly thereafter, in the austral summer. Both sexes are capable of breeding when 1 year old. Wild individuals seldom survive more than 3 years, but Weigl (2005) listed a female that was born in the Perth Zoo and released at an age of 6 years and 3 months.

D. geoffroii once was the most widely distributed species of *Dasyurus*, but has declined drastically since European settlement (Glen et al. 2009; Morris et al. 2003; Orell and Morris 1994; Serena and Soderquist 2008) and is classified as near threatened by the IUCN (Morris, Burbidge, and Hamilton 2008). It was last collected in New South Wales in 1841, in Victoria in 1857, in Queensland between 1884 and 1907, and in South Australia in 1931. The most recent reliable reports from the central deserts of Australia date from the mid-1950s. The decline seems associated with loss of cover due to clearing, grazing, and frequent fires. *D. geoffroii* also has been hunted and poisoned by people who considered it destructive to poultry. Another major factor is competition for food with and direct predation by the introduced feral cat (*Felis catus*) and red fox (*Vulpes vulpes*), which at up to 6.5 kg and 8.0 kg, respectively, enjoy a considerable size advantage over the chuditch and also have considerable dietary similarity. Since the 1970s, *D. geoffroii* has been confined to southwestern Western Australia, about 5 percent of its original range, though there are a few unconfirmed reports from elsewhere. By the end of the 1980s, fewer than 6,000 individuals were thought to survive. Subsequent efforts to control foxes and reintroduce *D. geoffroii* appear to have allowed numbers of the latter to stabilize or moderately increase, though there is concern that feral cats may also become more abundant in the absence of foxes. Interestingly, *D. geoffroii*, like some other mammals native to southwestern Australia, has a relatively high tolerance to the toxin sodium fluoroacetate ("1080") and is not affected by amounts used in baits to poison foxes. Woinarski, Burbidge, and Harrison (2014) cited recent estimates of about 14,500 individuals remaining.

Dasyurus spartacus (Bronze Quoll)

This species is known by only 12 specimens; it originally was thought to be conspecific with *D. geoffroii* of Australia and appears immediately related to the extinct eastern subspecies *D. geoffroii geoffroii* (Leary et al. 2008e). It was described as a separate species by Van Dyck (1987). Head and body length was 345 mm and 380 mm in two males and 305 mm in one female; tail length was 285 mm in one of the males and 250 mm in the female (Flannery 1995a). A sexually mature female weighed 684 grams; its pouch area had very thin lateral skin folds and eight small mammae (Woolley 2001). The body color is deep bronze to tan brown with small white spots, and the tail is black with no spots. As in *D. geoffroii*, the foot pads are granulated, not striated. However, in *D. spartacus* the hallux is vestigial (3.5 mm or less), the rostrum is narrow between the lachrymal canals (20.5 mm or less), the body spots are small (8.8 mm or less), and the ears are small (45 mm or less). In *D. geoffroii* the hallux is well developed (more than 3.5 mm), the rostrum is broad between the lachrymal canals (more than 20.5 mm), the body spots are large (more than 8.8 mm), and the ears are large (more than 45 mm) (Jones, Rose, and Burnett 2001).

The bronze quoll appears restricted to the low, mixed savannah woodland of the monsoonal trans-Fly ecoregion of southeastern Irian Jaya and southwestern Papua New Guinea (Flannery 1995a). It is thought to range in elevation from sea level to 200 meters but has been recorded only up to 60 meters. During the wet season, much of its lower lying range is inundated, confining both predator and prey species to smaller islands of habitat. It is nocturnal and has been caught raiding chicken houses (Leary et al. 2008e).

A female captured by Woolley (2001) in October 1989 and held until August 1990 entered estrus only in February and again 55 days later in April. That sequence, together with the age and reproductive status of other individuals collected, suggests that *D. spartacus* has a restricted breeding period during the wet season and is facultatively polyestrous—i.e., females can return to estrus if they lose a litter early in the season. A gestation period of 20 days also was indicated. If further study confirms that *D. spartacus* is a seasonal breeder, it would be in contrast to all other studied dasyurid species of New Guinea.

The IUCN classifies *D. spartacus* as near threatened (Leary et al. 2008e). The species has a restricted distribution, about 26,600 sq km, and almost certainly numbers fewer than 10,000 mature individuals. It may be jeopardized by predation from dogs, both feral and

Bronze quoll (*Dasyurus spartacus*). Photo by P. A. Woolley and D. Walsh.

those used by people for tracking game. Its habitat is predicted to decline due to invasive weeds (e.g., *Mimosa*) and fires. A serious potential problem is introduction of the toxic cane toad, which has devastated *Dasyurus hallucatus* in the nearby Northern Territory of Australia.

Dasyurus albopunctatus (New Guinea Quoll)

Head and body length is approximately 228–350 mm, and tail length is 210–310 mm. Three males weighed 580–710 grams (Flannery 1995a). The upper parts are drab gray to dark rufous brown and spotted with white; the underparts are drab, yellowish, or white. The tail is well haired, not spotted, and colored like the back for most of its length on the dorsal surface; the tip and entire ventral surface are dark brown or black. The coat is short and coarse and has little underfur. *D. albopunctatus* is distinguished from *D. spartacus* by its smaller

size, larger hallux, striated (not granulated) foot pads, and less hairy tail, and in having 6 (rather than 8) mammae (Flannery 1995a; Jones, Rose, and Burnett 2001; Woolley 2001).

The New Guinea quoll is widespread in forested areas above about 1,000 meters, though it has an overall elevational range from sea level to 3,500 meters. It is respected as a fierce predator by the native people, who say it will attack animals larger than itself and will approach human habitations to steal food and prey on rats. Some reports suggest it may be partly diurnal and has a call not unlike that of a small dog (Flannery 1995a). Collection of reproductively active females and young show that breeding occurs throughout the year (Woolley 2001). There reportedly are 4–6 young, which spend about 2 months in the pouch (Baker 2015). A captive lived for 4 years and 7 months (Weigl 2005). *D. albopunctatus* is declining, probably because

New Guinea quoll (*Dasyurus albopunctatus*). Photo by P. A. Woolley and D. Walsh.

of habitat disruption and predation by dogs, and is classified as near threatened by the IUCN (Woolley et al. 2008).

***Dasyurus maculatus* (Tiger or Spotted-Tailed Quoll)**
The information for this account was taken primarily from Belcher, Burnett, and Jones (2008) and Jones, Rose, and Burnett (2001); substantive exceptions are noted. *D. maculatus* is the largest carnivorous marsupial still present on the mainland of Australia. Head and body length is 310–759 mm, tail length is 285–550 mm, and weight is usually 0.85–5.0 kg in males and 0.8–2.5 kg in females; sexual size dimorphism is pronounced. Maximum weights of 8.85 kg for a male and 4.0 kg for a female have been recorded (Glen and Dickman 2006). The southeastern subspecies *D. m. maculatus* is on average substantively larger than *D. m. gracilis* of northeastern Queensland. The general coloration of the back and sides is rich rufous brown to dark brown; the underparts are pale yellow to sandy. White spots, varying in size, are distributed over the back, sides, and tail. The crown of the head and shoulders are rarely spotted in *D. m. maculatus* but are usually spotted in *D. m. gracilis*. The fur is thick and short. The jaws are strong, and adult males develop a noticeable heaviness about the jowls. The ears are relatively small. The first toe of the hind foot (hallux) is well developed, and the foot pads are striated. The female's pouch is a crescentic flap of skin enclosing six mammae in two rows; the pouch opens to the rear.

The spotted-tailed quoll is found in a wide range of treed habitats including wet and dry sclerophyll forest, vine thickets, woodland, coastal scrub, and tropical, subtropical, and temperate rainforest; in Tasmania it also occupies heathland. It is predominantly nocturnal but frequently active in daylight. It is usually found on the forest floor or moving atop fallen timber, but is moderately arboreal and adept at climbing high into trees to capture possums and sleeping birds at night. It can climb down head first and can move through the canopy from tree to tree. During the day it dens in a hollow tree or log, cave, rock crevice, or subterranean burrow; it sometimes enters or digs beneath a human-made structure. Captive females excavate underground chambers, where they construct nests. Glen and Dickman (2006, 2011) found logs particularly important for shelter and traveling in northeastern New South Wales and argued for retention of fallen timber for conservation of *D. m. gracilis*. Belcher and Darrant (2006) recommended the same for *D. m. maculatus* in southeastern Australia, but found rock dens more important in some areas; they reported that females used up to 15 dens each during a 7–9 month period.

The tiger "cat" may cover 3–5 km during its daily activity and has been recorded moving up to 8 km overnight and up to 17.5 km in a 2-year period. It stalks its prey, then seizes with the forepaws and kills by biting, or it may pounce on the back of a larger animal and hold on with all four paws while biting the neck. The diet includes insects, crayfish, reptiles, birds, and small mammals, but adults depend mostly on medium-sized mammals, such as wallabies, bandicoots, possums, and rabbits. Larger mammals—kangaroos, feral pigs, cattle—may be consumed as carrion. Assessment of skull structure and mechanics indicated *D. maculatus* to have a relatively more powerful bite, at least in some aspects, than the other two modern large carnivores of Tasmania, *Sarcophilus harrisii* and *Thylacinus cynocephalus*, and to be more capable of hunting animals considerably larger than itself (Attard et al. 2011).

A population density of one individual per 20 sq km has been estimated for Tasmania (Jones et al. 2004), and a density of approximately 0.3 per sq km was calculated for a 55 sq km area of northeastern Queensland (Glen 2008). *D. maculatus* is solitary and occupies a large home range. Females are intrasexually territorial, maintaining virtually exclusive home ranges (Belcher and Darrant 2004; Glen 2008), but tolerate their own female offspring within their ranges, which vary in size from 90 to 1,000 ha. Male ranges of 580–5,000 ha. overlap those of other males and a number of females, but may contain a small core area of exclusive use. Contrary to some earlier observations of captives, radio-tracking in the wild by Belcher and Darrant (2004)

showed that males do not to participate in the rearing of young. Males threaten one another with open mouths and may fight viciously. Vocalizations are infrequent but are given in all social interactions. They include agonistic huffs, coughs, sharp hisses by exhaling forcibly through the nose, and piercing screams likened to short blasts of a circular saw. Softer sounds accompany adult greetings and contacts between mother and young. Olfactory communication includes mutual sniffing and self marking by washing the face with mouth and ear secretions. Communal latrines are used by some populations and may provide a means by which males monitor the reproductive status of females. Both sexes mate with several individuals in succession; females exhibit strong mate choice but subsequently become aggressive towards males. There is a lengthy courtship, during which the female may be bitten severely about the head and shoulders. Copulation may last up to 17 hours.

Females are facultatively polyestrous (Krajewski, Woolley, and Westerman 2000), reproducing synchronously within a population each winter but recycling immediately if conception fails or a litter is lost. The estrous cycle is 28 days, during which receptivity is 3–5 days. The pouch develops as the breeding season approaches, whether or not young are born. Most matings occur from late May to early August, births from June to August. Gestation lasts 21 days. Litter size is 2–6, averaging 5.3. The young are 7 mm long in their curled up position at birth; 4 weeks later their length is about 38 mm. At 7 weeks of age they are no longer permanently attached to the nipples, their eyes begin to open, and spotted fur is apparent. At 18–20 weeks they are fully independent, though only one-third grown. The young are able to kill a mouse when 95 days old and a rabbit at 18 months. Sexual maturity comes at 1 year of age, but most females do not breed until they are 2 years old, when they reach full size (Belcher 2003); males are not full grown until 3 years. Individuals may live for up to 5 years in the wild. Weigl (2005) recorded a captive female that lived for 6 years and 10 months.

The IUCN classifies *D. maculatus* as near threatened and estimates that about 20,000 mature individuals survive (Burnett and Dickman 2008). However, that assessment is based on the status of the entire species

Tiger quoll (*Dasyurus maculatus*). Photo by Gerhard Körtner.

and may not fully consider that there are probably three distinctive subspecies, fully isolated from one another (see also Firestone 2000, Firestone et al. 1999), and that *D. m. gracilis* and mainland *D. m. maculatus*, at least, are dangerously fragmented. Remnant areas of habitat are often not sufficiently large or protected to insure survival of populations. A more precise assessment, applying IUCN criteria to updated information, was provided by Woinarski, Burbidge, and Harrison (2014). They also designated the overall species as near threatened, but distinguished *D. m. gracilis* as endangered, mainland *D. m. maculatus* as vulnerable, and the unnamed Tasmanian subspecies as vulnerable.

The mainland range has been reduced by 50–90 percent since European settlement and numbers are continuing to fall. Glen and Dickman (2011) reported the species to be still abundant in northeastern New South Wales, possibly because of a large expanse of relatively undisturbed habitat and high availability of prey, particularly arboreal mammals. Glen and Dickman (2008) did caution that there was considerable dietary overlap, and hence strong potential for competition, between *D. maculatus* and the introduced red fox (*Vulpes vulpes*), though the latter was not numerous in the area. Woinarski, Burbidge, and Harrison (2014) cited estimates of 2,000–10,000 mature individuals for mainland *D. m. maculatus* and fewer than 1,000 individuals for *D. m. gracilis*, which is restricted to a small area of the wet tropics in northeastern Queensland. Jones et al. (2004) cited a 1996 estimate of 3,500 individuals in Tasmania, but the species has been extirpated on Flinders and King Islands and on all Tasmanian offshore islands. Past and present threats include logging and clearing of forests, inappropriate burning practices, predation by introduced carnivores, deliberate trapping and poisoning by people seeking to protect poultry, ingestion of the toxic cane toad (see account of *Dasyurus hallucatus*), and incidental consumption of poison aimed at other species. Unlike *D. geoffroii* (see account thereof), *D. maculatus* seems highly susceptible to poisoning by compound 1080, used to kill foxes and other animals considered pests, and some populations have been devastated in such manner (Belcher 1998, 2003).

DASYUROMORPHIA; DASYURIDAE; Genus *SARCOPHILUS* E. Geoffroy St.-Hilaire and F. Cuvier, 1837

Tasmanian Devil

The single living species, *S. harrisii*, is now found only in Tasmania. *Sarcophilus* also once occupied much of the Australian mainland (Calaby and White 1967), but it probably disappeared there before European settlement, through competition with the introduced dingo (*Canis familiaris dingo*). Johnson and Wroe (2003) argued that another factor was increased human population and more advanced hunting technology, which intensified about 4,000 years ago on the mainland but not on Tasmania. Owen and Pemberton (2005) suggested that climatic deterioration and more aridity also had a role. Subfossil remains indicate that *S. harrisii* or a closely related species was still present as recently as 3,120 years ago in the Northern Territory and 430 years ago in southwestern Western Australia (Dawson 1982a). Subfossils, thought to date from the early 1800s, also are known from Flinders Island, where *S. harrisii* does not presently occur (Jones et al. 2007). The species was also present on Bruny Island, off southeastern Tasmania, in the early 1800s, but there are no confirmed records from after 1900; it still occurs on Robbins Island to the northwest of mainland Tasmania (Hawkins et al. 2008). Remains from as late as 600 years ago have been found in Victoria (Guiler 1983); Groves (2005c) considered those specimens to represent a subspecies, *S. harrisii dixonae*. Living specimens collected in Victoria in 1912 and 1971 are generally thought to represent escaped captives (Guiler 1982; Ride 1970), as are subsequent reports and road kills in that state (see White and Austin 2017).

Head and body length is about 525–800 mm, and tail length is about 230–300 mm. Weight is about 6–14 kg in males and 4–9 kg in females (Green 1967; Jones 2008b). The coloration is blackish brown or black except for a white chest patch, often one or two white patches on the rump and sides, and the pinkish white snout. The general form, except for the tail, resembles that of a small bear. The head is short, broad, and covered with masses of muscle; the skull and teeth are extremely massive and rugged. The canines are large, and the molar teeth are developed into heavy bone

Tasmanian devil (*Sarcophilus harrisii*). Photo by Johannes Pfleiderer.

crushers in a manner reminiscent of the hyenas. The first toe is not present, and the granular pads of the feet lack striations. The pouch during the breeding season forms a complete enclosure, unlike that of many dasyurids, and opens to the rear. There are four mammae.

Sarcophilus is found throughout Tasmania, except in areas where cover has been extensively cleared, and is most numerous in coastal heath and sclerophyll forest (Guiler 1970a). It occurs at lower densities in the rainforests and moorlands of southwestern Tasmania (Jones 2008b). It is nocturnal, sheltering by day in any available cover, such as caves, hollow logs, wombat holes, or dense bushes. Jones (2008b) wrote that the usual site is an underground burrow and that *Sarcophilus* maintains a primary den but has several others within its home range. Owen and Pemberton (2005) reported that once established in dens, adults tend to use them for life, that old wombat burrows are favored maternity dens, and that dens are commonly found under human houses. Both sexes make nests of bark, button grass, or leaves.

The Tasmanian devil is terrestrial but capable of climbing and occasionally emerges from its lair to bask in the sun (Collins 1973). A radio-tracking study showed that it does not enter torpor, even under prolonged severe weather conditions, and that time spent active does not differ between summer and winter or between moderate and severe winter weather (Jones, Grigg, and Beard 1997). Movements may appear slow and clumsy, but *Sarcophilus* has great stamina and can travel a long way at a good pace; speeds of 10 km per hour are common for extended periods and one individual was clocked at 35 km per hour for 300 meters (Owen and Pemberton 2005). Its habit of continually nosing the ground suggests a well-developed sense of smell (Ride 1970). It forages about 8 hours per night, starting at dusk, and covers about 9 km on average, but one adult male moved 50 km in 48 hours and juvenile dispersal distance probably exceeds 30 km (Jones 2008b).

The diet consists of a wide variety of invertebrates and vertebrates, including poisonous snakes, and a small amount of plant material. Guiler (1970a) found the main foods to be wallabies (*Macropus* and *Thylogale*), wombats, sheep, and rabbits, most of which were taken as carrion. Pemberton et al. (2008) reported the species contributing most significantly to the diet, both in terms of prey composition and frequency of occurrence, were the ringtailed possum *Pseudocheirus peregrinus* and the wallabies *Thylogale billardierii* and *Macropus rufogriseus*; remains of birds also were commonly found in scats, and their presence, along with that of *Pseudocheirus*, indicates that *Sarcophilus* is capable of active arboreal predation. Some laboratory and field observations indicate that *Sarcophilus* is an inefficient killer that does not usually hunt large, active prey (Buchmann and Guiler 1977). However, Jones (2008b) stated that it is an effective predator of medium-large mammals, persistently pursuing or ambushing its prey, primarily macropods, wombats, and possums. Wroe, McHenry, and Thomason (2005) calculated that the Tasmanian devil's biting force, relative to its size, is the highest of any living mammal. All sources agree that it is an efficient scavenger, consuming all parts of a carcass, including the fur and bones. It is said to have formerly fed on the remains of animals killed by the larger Tasmanian wolf (*Thylacinus*).

Population densities in two parts of Tasmania were about 3/sq km and 25/sq km, the latter figure being considered abnormally high (Buchmann and Guiler 1977; Guiler 1970a). Known densities in unmodified habitats are 0.3–0.7 per sq km (Jones et al. 2004). Home ranges, which are larger for males, are about 5–27 sq km (Jones 2008b). In an area of abundant food, ranges were found to be small, with individuals traveling about 3.2 km during a night. In an area of less food, ranges

Tasmanian devil (*Sarcophilus harrisii*). Photo by Gerhard Körtner.

Tasmanian devil (*Sarcophilus harrisii*). Photo by Pavel German / Wildlife Images.

were larger, and nocturnal movement covered about 16 km. There was considerable overlap in home ranges, both between and among sexes, and no evidence of territorial behavior (Guiler 1970a).

As in so many animals, disposition of *Sarcophilus* is individually variable; some animals seem quite vicious, others more tractable. Eric R. Guiler (University of Tasmania, pers. comm., 1972) reported that most of the more than 7,000 Tasmanian devils he handled were docile to the point of being lethargic and could be handled with ease; he said that ferocity has been greatly exaggerated. Captives, however, generally are highly aggressive toward their own kind, and there is severe fighting over food (Buchmann and Guiler 1977). Agonistic encounters are accompanied by much vocalization, starting with low growls and proceeding to a rising and falling vibrato or even loud screeching. Eventually two individuals may form a stable dominant-subordinant relationship with reduced aggression. Observations of wild animals congregated to feed on car-

casses indicated considerable agonistic interaction but few serious physical clashes (Pemberton and Renouf 1993). Collins (1973) reported that pairs could be maintained in captivity if the animals were gradually introduced to each other. Turner (1970) observed that adult males would assist females in cleaning the offspring and that after the young were too big to fit in the mother's pouch, they might cling to the back of either parent. Guiler (1971c) cautioned, however, that at a later stage the father might devour the young.

Sarcophilus usually has been considered monestrous, as by Krajewski, Woolley, and Westerman (2000), but Jones (2008b) wrote that females are seasonally polyestrous and can recycle a second time, though not a third, if they lose a litter or do not conceive. Keeley et al. (2012) confirmed that three estrous cycles are not only possible but common in captive females, though the chances of producing young may decline through those cycles. Length of each cycle is about 33 days, during which females are receptive for up to 5 days. Jones

(2008b) noted that mating involves conflict, blood-curdling screams, and real affection. Females aggressively assert their preference for older and larger males, and the latter intensely guard their mates. Most matings occur in February and March, but female reproductive cycles have been recorded from January to June, and successful reproduction reportedly has occurred as late as August (Keely et al. 2012). Jones (2008b) stated that the gestation period is 21 days, but most earlier sources (e.g., Collins 1973; Guiler 1970b, 1971c; McAllan 2003) give it as about 31 days, while Phillips and Jackson (2003) indicated 21–31 days. Keely et al. (2012) cited an unpublished estimate of only 13–14 days, as calculated by examining developmental stages of the fetuses through to parturition. The variation in those reports may reflect a period in which the female stores sperm in the oviducts prior to ovulation, and thus an extension of the time between mating and birth beyond the true gestation period. Some 15–20 young are born from about 40 ovulated eggs, but only 4 young can attach to the teats and have a chance at survival. They weigh 0.18–0.29 grams at birth and are 6 mm long. They first release the nipples and open their eyes at 90–121 days and leave the pouch shortly thereafter. They roam, at increasing distances from the den, from October until January, when they are weaned and the males disperse. Most females do not become sexually mature until they are 2 years old, and males may not mate until they are 3 or even 4 years old. Longevity is around 6 years in the wild, but Weigl (2005) listed captive males that lived to ages of about 12 and 13 years.

Sarcophilus sometimes is considered a nuisance and has been subject to official control measures. Through the 1980s and 1990s, systematic poisoning in sheep-growing areas probably killed more than 5,000 Tasmanian devils annually (Hawkins et al. 2008, citing N. Mooney). Buchmann and Guiler (1977), however, stated that its reputation as a killer of livestock is unmerited, that sheep are probably eaten only as carrion, and that money spent on control is wasted. Apparently because of persecution by European settlers, and also destruction of forest habitat and a severe epidemic in the early twentieth century, the Tasmanian devil had become rare by about 100 years ago. There was concern that it, along with the Tasmanian wolf, might be facing extinction. *Sarcophilus* subsequently recovered, even becoming abundant in many areas, perhaps partly because of newly available food supplies in the form of carrion from livestock and commercial trapping operations (Green 1967; Guiler 1982; Ride 1970). Jones et al. (2004) cited a 1996 estimate of 130,000 individuals in Tasmania, but believed that prior to European settlement numbers actually were much lower, roughly 45,000. They suggested that pastoral development created a patchy mosaic of grazing land and forest remnants conducive to a growing abundance of grazing prey species. Hence, *Sarcophilus* also increased but became more susceptible to epidemic.

In 1996 came the first reports of Devil Facial Tumor Disease, an emerging and invariably fatal infectious cancer that subsequently caused population collapses across much of the range of *Sarcophilus* (Hawkins et al. 2006, 2008; Jones 2008b; McCallum et al. 2007). It is transmitted during contact in the breeding season and continues to spread. As a result, the IUCN now classifies *S. harrisii* as endangered and projects a general population decline of at least 90 percent over the next 10 years in the 60 percent of the Tasmanian devil's distribution currently affected, with at least a 100 km extension of the disease (Hawkins et al. 2008). Based on recent surveys, the rate of decline, and a total population estimate of 130,000–150,000 individuals in the mid-1990s, the number remaining in 2007 was estimated at 50,000 animals, of which half or fewer were mature. The disease first affects adults and then younger individuals, so there is a progressive lowering of the age of populations and a reduced likelihood of success in rearing litters. An estimate of 17,000–42,000, most likely about 30,000, surviving individuals was cited by Woinarski, Burbidge, and Harrison (2014). As pointed out by McCallum et al. (2009), transmission of the disease does not appear to be density dependent, hence restricting the prospect of recovery after populations fall to a low point. This is the first case of a host-specific disease threatening to cause extinction of its host. Epstein et al. (2016) recently reported hopeful evidence that populations are evolving immune-modulated resistance that could aid the long-term survival of *Sarcophilus*.

Declines, however, could be exacerbated if the introduced red fox (*Vulpes vulpes*) increases in response to the food surplus created as numbers of *S. harrisii* fall, then competes with the survivors of the latter and even preys on their young at the den while the female forages. Bradshaw and Brook (2005) suggested that the current epidemic may be part of a natural cycle, as there is evidence of a major population peak and crash in the early to mid-nineteenth century, as well as at the start of the twentieth century. They noted, however, that the current situation could be more severe because of the threat posed by the fox.

In response to the crisis, the Tasmanian state government launched the "Save the Tasmanian Devil Program," its top priority being establishment of insurance populations of healthy devils in places isolated from the disease, which would serve as a source for eventual reintroduction to the wild. Jones et al. (2007) recommended that such populations be placed on large offshore islands, where they would be the most secure and could live under relatively natural selective conditions and retain normal behaviors. Johnson and Wroe (2003) proposed that *Sarcophilus* be reintroduced to mainland Australia, where competition for resources with the Aboriginal people, which may have contributed to its original extirpation there, no longer would be a factor.

DASYUROMORPHIA; **Family THYLACINIDAE;** **Genus *THYLACINUS*** Temminck, 1824

Thylacine, or Tasmanian "Wolf" or "Tiger"

The single Recent genus and species, *Thylacinus cynocephalus*, apparently was found only in Tasmania within modern time and is now likely to be extinct. There were many reports of animals bearing some resemblance to the thylacine from the Australian mainland during the eighteenth, nineteenth, and twentieth centuries (Heuvelmans 1958; Ride 1970); none has been confirmed, though Archer (1984) noted some hints that a population may have survived in the north until the early twentieth century, Helgen and Veatch (2015) cited reputable accounts from South Australia at around the same time, and Mooney and Rounsevell (2008) mentioned a bone from northwestern Western Australia with an estimated age of not more than 80 years. During the late Pleistocene and early Recent periods the genus was widespread in Australia and New Guinea, with the latest definite subfossil mainland records dating from just over 3,000 years ago. Helgen

and Veatch (2015) indicated that *T. cynocephalus* was the species of eastern and far southwestern Australia, but that the forms found in the more arid inland parts of Australia and in New Guinea were morphologically distinctive.

The thylacine may have disappeared from the mainland and New Guinea through competition with semi-domestic dogs introduced by humans starting several thousand years ago (Archer 1974; Dawson 1982b; Partridge 1967). Those dogs, which in Australia have the name "dingo" (*Canis familiaris dingo*), became fully feral and spread over a large region but did not enter Tasmania. Flannery (1995a) pointed out that those dogs were introduced in New Guinea about 2,000 years ago, somewhat later than in Australia; thus, the thylacine's extirpation on the island may have been around that time. It has been suggested that competition with the dingo could not have been the primary factor in the disappearance of the thylacine, as the latter had very different hunting adaptations and niches (see below) and the modern thylacine of Tasmania averaged much larger than the dingo and could have prevailed in confrontations over food. However, based on subfossil specimens, Letnic, Fillios, and Crowther (2012) found the dingo to average somewhat heavier than the Holocene thylacine of Western Australia. Females were considerably smaller, and direct killing thereof by the dingo could have driven *Thylacinus* to extinction.

Thylacinus often was placed in the family Dasyuridae but subsequently came to be considered distinct from that group and to have closer affinity with several extinct South American families of large predatory marsupials (Kirsch 1977b). Those families, plus Thylacinidae, were then grouped in the superfamily Borhyaenoidea, which was thought to have originated before the process of continental drift had separated South America, Antarctica, and Australia (see Argot 2004 for detailed information on the Borhyaenoidea). More recent studies of morphology, serology, and mitochondrial DNA (Archer 1982b, 1984a; Krajewski et al. 1992; Krajewski, Buckley, and Westerman 1997; McKenna and Bell 1997; Sarich, Lowenstein, and Richardson 1982; Szalay 1982; Wroe 2008) have indicated once again that Thylacinidae is most closely related to Dasyuridae, though the two are considered very distinct families within the order Dasyuromorphia (see account

Thylacine, or Tasmanian wolf or tiger (*Thylacinus cynocephalus*). This originally was a black-and-white photo of an individual in the Hobart Zoo, Australia. Photo colorized by Dave Davis based on examination of old drawings, skins, and other references; see https://neitshade5.wordpress.com/category/nature/.

thereof for more information on family relationships; see also Miller et al. 2009).

Head and body length of *Thylacinus* is 850–1,300 mm, tail length is 380–650 mm, shoulder height is 350–600 mm, and weight is 15–45 kg; males apparently were larger, on average, than females (Helgen and Veatch 2015; Moeller 1990a; Mooney and Rounsevell 2008; Owen 2004). The upper parts are tawny gray or yellowish brown with 13–19 blackish brown transverse bands across the back, rump, and base of the tail; the underparts are paler. The face is gray with some indistinct white markings around the eyes and ears. The fur is short, dense, and coarse. The dental formula is: i 4/3, c 1/1, pm 3/3, m 4/4=46. The unusual dental features,

as noted below, distinguish *Thylacinus* from the Dasyuridae. In addition, the epipubic bones of *Thylacinus* are vestigial and cartilaginous, the clavicles are reduced, and the foramen pseudovale is lacking. Unlike that of the Myrmecobiidae, the palate of *Thylacinus* has large vacuities (Marshall 1984).

With its general external resemblance to a canid, the thylacine represents one of the most remarkable examples of convergent evolution found in mammals. Its scientific name means "pouched dog with a wolf head." The jaws have a remarkably wide gape, though Attard et al. (2011) doubted reports that they could be opened much more than 75–80°. Long canine teeth, shearing premolars, and grinding molars show further similarity

Thylacine, or Tasmanian wolf or tiger (*Thylacinus cynocephalus*), mounted specimen in Melbourne Museum. Photo by Alexander F. Meyer.

with the dog family. The overall build is doglike, but the legs are relatively short and the tail is rigid and tapers into the hindquarters. The feet also resemble those of dogs, and like them the thylacine was digitigrade, walking on its toes, though it sometimes stood on its soles (Moeller 1990a). The pads of the feet are granulated, not striated, and the forefoot leaves a five-toed print. In females there is a rear-opening pouch consisting of a crescent-shaped flap of skin enclosing the four mammae.

Despite the superficial appearance of the thylacine, an analysis of its skeletal proportions by Keast (1982) indicated it is essentially like a large dasyurid and does not have the specialized pursuit adaptations of the placental wolf (*Canis lupus*). Also, Jones and Stoddart (1998) found the thylacine's canine teeth to be more ovoid in cross section than those of the wolf, and suggested the thylacine killed with a puncturing, crushing bite in a manner similar to that of the quolls (*Dasyurus*) and smaller canids, rather than with a slashing bite like

that of the wolf. Its teeth were not adapted for bone crushing and consumption, as are those of the Hyaenidae. Wroe et al. (2003) noted that brain mass in *C. lupus* (132 grams) exceeds that in *T. cynocephalus* (53.4 grams) by a factor of 2.45.

The preferred habitat of the thylacine probably was open forest or grassland, but its last populations may have occupied dense rainforest in southwestern Tasmania. Its lair reportedly was in a rocky outcrop or a hollow log, and it apparently brought in nesting material. Although nocturnal, it was sometimes seen moving about or basking in the sun. Early observers reported the thylacine feeding on kangaroos, wallabies, smaller mammals, and birds. It was said to trot relentlessly after its prey until the victim was exhausted and then to close in with a rush. Modern studies have offered several alternatives. Mooney and Rounsevell (2008) stated that the thylacine had a rather stiff gait and could not run very fast, but Owen (2004) wrote that recent anatomical studies indicate it was capable of genuine speed and could turn swiftly and precisely.

Examination of elbow morphology suggests it was more of an ambush predator, comparable to a large cat (Felidae), than a pack-hunting pursuit predator like a large canid, and hence that the name "Tasmanian tiger" is more appropriate than "Tasmanian wolf" (Figueirido and Janis 2011). Based on a comparison of its canine tooth strength and limb bone length ratios with those of extant marsupial and placental carnivores, Jones and Stoddart (1998) concluded that it was probably a pounce-pursuit predator of fairly open habitats, which killed medium-sized prey (1–5 kg) that was small relative to its body size; its trophic niche was more similar to that of the coyote (*Canis latrans*) than to that of the wolf (*C. lupus*). Considering skull structure and mechanics, Attard et al. (2011) concluded the bite force of *Thylacinus* was relatively weaker than that of the other two large carnivores of Tasmania, *Dasyurus maculatus* and *Sarcophilus harrisii*, and that it likely preyed on animals smaller than itself.

The thylacine seems to have been mostly solitary but sometimes reportedly hunted in pairs or small family groups. Various calls were reported, including a whine (which may have been for communication), a low growl expressing irritation, and a coughing bark when hunting. Based on a study of bounty records, Guiler (1961) found that births occurred throughout the year, with a pronounced peak in the summer (December–March). However, Sleightholme and Campbell (2014) argued that the timing and process of bounty submission would have distorted a reproductive assessment and that more precise information can be derived from historical newspaper, museum, and zoo accounts of females with young. Accordingly, mating apparently occurred from the late austral autumn to early spring (about April–September) and females with pouch-dependent young could normally be found from May to December. The gestation period was estimated to be 28 days. The two to four young were thought to leave the pouch after around 16 weeks but not to be fully weaned until they were about 8 months old and to remain with the mother for at least 12 months. Although many individuals were successfully maintained in zoos, the thylacine was never bred in captivity. Record longevity was nearly 13 years (Moeller 1990a).

Owen (2004) thought that prior to European colonization thylacine numbers may have been around 5,000, while Mooney and Rounsevell (2008) wrote that there may never have been more than 2,000 adults in Tasmania. As with most large predatory mammals, there is controversy regarding the extent of damage the thylacine caused to domestic livestock. Nonetheless, it was generally considered to be a sheep killer in Tasmania, and intensive private bounty hunting began about 1840. By 1863 the thylacine already had been restricted to the mountains and more inaccessible parts of the island. From 1888 to 1909, 2,184 thylacines were destroyed for payment of a government bounty then in effect, and many more were killed for the private bounties. Information compiled by Owen (2004:42, 115) indicates the number killed was even greater; from 1878 to 1896, 3,482 skins were sent from a tannery to London, where they are said to have been made into waistcoats, though no such garment is now known to exist. In any case, a rapid decline in the population was noticed about 1905; by 1914 the thylacine was a rare species, and some conservationists were already calling for protection (Owen 2004). In addition to human hunting, trapping, and poisoning, the demise of the thylacine has been blamed on disease, habitat modification, and increased competition from settlers' domestic dogs. Paddle (2012) presented detailed information indicating that a distemper-like disease was causing serious problems for both captive and wild thylacines around the time the species was in sharp decline. However, a population viability analysis by Prowse et al. (2013) indicated disease was not a major factor and that extinction resulted mainly from the effects of European colonization, including a loss of native prey populations through competition with introduced sheep.

There is much uncertainty about the thylacine in the late nineteenth and early twentieth centuries, even as to whether specimens photographed at the time were alive or mounted (see Paddle 2008; http://www.natural worlds.org/thylacine/captivity/Burrell/Burrell_1 .htm, accessed 20 December 2015). The last definite record of a wild individual being killed was in 1930 (Ride 1970). It is questionable whether the last animals brought into captivity were taken in 1924 or 1933, but it is clear that the last of them died in 1936 (see Mooney and Rounsevell 2008; Owen 2004; Paddle 1993). Several

Thylacine, or Tasmanian wolf or tiger (*Thylacinus cynocephalus*). Photo from US National Zoological Park.

organized field searches over the next 40 years were unsuccessful. A number of reports, including the supposed killing of a thylacine in western Tasmania in 1961, were doubted by Brown (1973). *T. cynocephalus* has received complete legal protection since 1938 and a 647,000-ha. game reserve was established in the southwestern part of the island in 1966, partly to protect the species, should it exist. It is classified as endangered by the USDI, is on appendix 1 of the CITES (but with the notation "possibly extinct"), and is designated extinct by the IUCN (McKnight 2008c).

Some persons still believe the Tasmanian wolf survives. Many alleged sightings both in Tasmania and on the mainland of Australia were discussed by Rounsevell and Smith (1982) and M. Smith (1982). According to Mooney and Rounsevell (2008), some reports from the mainland seem of equal veracity to the most noteworthy of those from Tasmania. However, despite investigations and offered rewards, no irrefutable evidence has been presented; some accounts have been shown to be hoaxes or errors but most remain unresolved (see also http://www.naturalworlds.org/thylacine/index .htm, accessed 20 December 2015). A recent attempt to clone the species was unsuccessful (for earlier details on that project see Owen 2004; http://www.naturalworlds .org/thylacine/mrp/mrp.htm and http://cubits.org/the extinctioncubit/db/extinctmammals/view/18756/, accessed 20 December 2015). Based on a mathematical and economic assessment of known mortality data, Bulte, Horan, and Shogren (2003) concluded that bounty hunting did not result in extinction and that a viable population might persist in peripheral habitat.

The geological range of the family Thylacinidae is late Oligocene to Recent in Australasia. The group once was more diverse than in historical time, with 9 known genera and at least 13 species varying in size from that of a small house cat to that of a large wolf, but by the Pliocene only *Thylacinus cynocephalus* remained (Archer and Hand 2006; Long et al. 2002; Wroe 2003, 2008).

Numbat (*Myrmecobius fasciatus*). Photo by Jiri Lochman / Lochman LT.

DASYUROMORPHIA; **Family MYRMECOBIIDAE;**
Genus *MYRMECOBIUS* Waterhouse, 1836

Numbat, or Banded Anteater

The single known genus and species, *Myrmecobius fasciatus*, occurred in suitable areas from southwestern Western Australia, through South Australia and extreme southwestern Northern Territory, to southwestern New South Wales (Friend 1989, 2008a; Ride 1970). *Myrmecobius* sometimes has been placed in the family Dasyuridae, but the sum of available information indicates it represents a distinct family (Archer and Kirsch 1977; Groves 2005c; see also account of order Dasyuromorphia). There are Pleistocene records of *M. fasciatus* from Australia (Archer 1984); otherwise Myrmecobiidae is known only from the Recent.

Head and body length is 200–290 mm, tail length is 125–213 mm, and weight is 305–752 grams (Friend 2008a). In *M. fasciatus fasciatus*, the subspecies that in-

habits southwestern Western Australia, the foreparts of the body are grayish brown with some white hairs. In *M. fasciatus rufus*, which occurred farther to the east, the main body color is rich brick red. Both subspecies usually have six or seven white bars between the midback and the base of the tail, producing an effect of transverse light and dark stripes. Cooper (2011) indicated there may be as few as 4 and as many as 11 such stripes. There is a dark cheek stripe running through the eye, with a white line above and below. The body pelage is short and coarse, and both the head and hindquarters are remarkably flattened above. The body is larger posteriorly than anteriorly. The tail is long, somewhat bushy, and nonprehensile. It is covered with long, stiff hairs that are sometimes bristly and form a brush. The snout is long and tapering, the mouth small; the slender tongue can be extended at least 100 mm. The tip of the nose is naked. The front legs are comparatively thick and widely spaced; the forefoot has five toes and the hind foot has four, all bearing strong claws. A complex gland on the chest opens onto the skin through a number of conspicuous pores. The female has four mammae but lacks any suggestion of a pouch. When the young are attached to her nipples, they are protected only by the long hair on her underparts. She supplies them with milk and warmth, but they are dragged beneath her as she travels.

There may be four or five upper molar teeth and five or six lower molars (Archer and Kirsch 1977). Therefore, the total number of teeth may be as high as 52, more than that of any other land mammal. The overall dental formula is: i 4/3, c 1/1, pm 3/3, m 4–5/5–6 = 48– 52. Most of the teeth are small, delicate, and separated from one another. The size is not constant; the molars on the right and left sides of the skull often vary in length and width. Cooper (2011) noted that the molar teeth lack the typical tribosphenic structure of dasyurids, and are instead simple conical cusps that in some instances barely extend above the gum. The numbat's bony palate extends farther back than in many mammals and may be associated with the long, extensible tongue. This type of palate is also present in the armadillos (Cingulata, Dasypodidae) and the pangolins, or scaly anteaters (Pholidota, Manidae). Friend (1989) wrote that the palate has no vacuities, the alisphenoid tympanic wing forms virtually the entire floor of the middle ear, processes from the frontal and jugal bones form an incomplete bony bar behind the orbit of the skull, and the lachrymal bone is very large and extends a long way out onto the face.

The numbat inhabits open scrub woodland, mostly in arid or semiarid country and generally where eucalyptus trees predominate and provide hollow logs, trunks, and branches for shelter. According to Friend (2008a), some hollows are used to avoid predators, but nests are built in others and the numbat sleeps therein at night. It also digs burrows, in which it constructs nests and sleeps, particularly in cold winter weather. The burrow usually consists of a single narrow shaft, sloping downward at a gentle angle for 1–2 meters, then opening into a spherical chamber about 250 mm in diameter, which is packed with grass,

Numbat (*Myrmecobius fasciatus*), group. Photo by Jiri Lochman / Lochman LT.

shredded bark, leaves, flowers, and other soft material. Christensen, Maisey, and Perry (1984) reported one nest chamber about 100 mm below ground level. The numbat also is known to bed down in the burrow of some other animal.

Unlike most marsupials, *Myrmecobius* is exclusively diurnal. Winter activity peaks in the middle of the day, while in summer the numbat is active in the morning and late afternoon, resting during the heat of the day; the pattern corresponds to the activity of termites in the upper soil layers (Cooper 2011; Friend 2008a). As captives do not cease activity in the middle of the day during summer, the inactivity of wild individuals at that time apparently is a response to food availability rather than a thermoregulatory measure (Cooper and Withers 2004a). The numbat often undergoes torpor during rest periods (Friend 2008a). Using radiotelemetry, Cooper and Withers (2004b) found that it entered spontaneous, shallow, daily torpor, which reduced energy expenditure by 13–42 percent over 12.5 hours.

The numbat is nimble and climbs readily. It can run quickly when threatened; one was timed at close to 35 km/hour (Friend 2015). It trots and leaps about with jerky movements, carrying its tail in line with its body, but with a slight upward curve. When startled or frightened it may sit bolt upright, flatten its entire body and fluff its tail, or run into a hollow log. When caught, it reportedly emits snuffling and hissing sounds but does not attempt to bite. During cooler periods it often basks in the sun.

Myrmecobius searches by scent for underground termite galleries in the woodland floor, making small excavations and turning over sticks and small branches (Friend 2008a). Termites and ants form the regular diet, and other invertebrates are eaten only occasionally. Termites actually are the preferred food; most of the ants consumed are of small predatory species that rush in when the numbat uncovers a termite nest and are lapped up along with the termites. Captive females require a pure termite diet during the breeding season for successful reproduction (Cooper 2011). One captive numbat consumed 10,000–20,000 termites of the smaller species daily. Those are swallowed whole, whereas those of the larger species are masticated. Termites are obtained from rotten logs, dead trees, and subsurface soil. The strong foreclaws are used in scratching into soil and decayed wood, and the long snout sometimes functions as a lever. The extensible, cylindrical tongue extracts the termites from crevices in the wood and is manipulated with great speed and dexterity. According to Cooper (2011), the soft palate has 13 or 14 transverse ridges that scrape termites from the tongue when it is withdrawn into the oral cavity. In southwestern Australia, wandoos (*Eucalyptus reduca elata*) are often attacked by a species of termite, so that the woodland floor is littered with hollow branches from the infested trees. Those logs provide both food and shelter for the numbat.

Friend (2008a) wrote that home ranges of animals of the same sex are exclusive but that those of males and females overlap to some degree. Range size is 25–50 ha., though males search for females over a much wider area during the breeding season and may fight one another at that time. Radio-tracking studies by Christensen, Maisey, and Perry (1984) showed that some animals used over 100 ha. during a two-month period; there were many shelters within a range, with individuals using hollow logs during warm weather and changing to burrows in the cold. Except during the breeding season the numbat is usually solitary, though Collins (1973) observed that captives could be maintained in pairs. A courting male and female vocalize to one another with a soft series of clicks (Jackson 2003).

Krajewski, Woolley, and Westerman (2000) stated that *Myrmecobius* is monestrous. However, Friend (2008a) wrote that females first come into estrus in January, with mating occurring up to 48 hours after the onset, and that a second estrus can occur in the absence of conception. The gestation period is 14 days (Friend 1989). The female digs a burrow and constructs a nest, as described above, and gives birth therein (Christensen 1975). In southwestern Australia the young are born from January to April or May. Litters of 2–4 have been recorded but 4 seems the usual number. The young are carried by the female, attached to her nipples, for about 4 months and then are suckled in a nest for about 2 more months. At that time a female may carry her young between nests on her back (Cooper 2011). Juveniles are weaned and have been seen foraging for themselves by late October. Friend (2008a) wrote that by the end of November most spend the night apart from the mother, and by mid-December they have dispersed and established their own home ranges;

Numbat (*Myrmecobius fasciatus*). Photo by Chris and Mathilde Stuart.

dispersal distances of more than 15 km have been recorded. Females breed in their first year, but males are not sexually mature until their second year. Longevity rarely exceeds 5 years, but Weigl (2005) listed a wild-born male that lived in the Perth Zoo to an age of about 11 years, and a captive female that lived for 9 years and 6 months.

According to J. A. Friend (1990a), a decline of *M. fasciatus* began in the east shortly after European settlement around 1800 and moved progressively westward. Native populations disappeared entirely from New South Wales, South Australia, and the Northern Territory and survived only in a few suitable areas in extreme southwestern Western Australia. The eastern subspecies, *M. f. rufus*, is apparently extinct. There had been speculation that the main cause of the decline was clearing of land for agriculture, which eliminates the dead and fallen trees from which termites can be

obtained; however, the continued abundance of the echidna (*Tachyglossus*) in the same region suggests no shortage of food for a mammal that depends on termites and ants. Probably an overriding factor in the decline of the numbat has been predation by introduced carnivores, especially the red fox (*Vulpes vulpes*), exacerbated by the loss of vegetative cover to fire and clearing. Unlike the echidna, which can dig into the ground and is protected by its spiny pelage, the numbat depends on cover to escape predators. The numbat actually remained common in many of the arid parts of South Australia and Western Australia until about 1930, but then underwent a drastic population crash in the 1940s and 1950s. That decline coincided both with the establishment of the red fox in the region and with the departure of the Aboriginal people from the deserts. While those people had traditionally burned off small patches of vegetation throughout the year, the cessation of this practice led to a buildup of scrub and eventually to huge summer wildfires that destroyed all cover. In addition, introduced rabbits (*Oryctolagus cuniculus*) increased in the same region and further encouraged numbers of fox and domestic cat (*Felis catus*). Even in the 1970s the numbat was locally common on reserves in southwestern Western Australia, but populations there underwent another crash at that time and disappeared entirely from some reserves, again apparently because of fox predation.

By the late 1970s the numbat numbered only in the low hundreds, and by 1985 only two isolated populations survived, at Dryandra and Perup to the southeast of Perth (Friend 2008a). Subsequent control of the fox allowed a modest increase and successful re-introduction of the numbat at seven natural sites in Western Australia and at two fenced sanctuaries in South Australia and New South Wales (Friend and Thomas 2003). Unfortunately, the population at Dryandra has since fallen precipitously, from a peak of about 600 in 1992 to just 50. The reason is not well understood, but the feral cat is now reportedly the main predator of the numbat at Dryandra (Woinarski, Burbidge, and Harrison 2014). There has been no decline at Perup, and there are 500–600 animals in the reintroduced populations but none is yet considered secure. The IUCN estimates total numbers at fewer than 1,000, which continue to decrease, and classifies *M. fasciatus* as endangered (Friend and Burbidge 2008a), as does the USDI.

Order Peramelemorphia

Bandicoots

This order of 3 Recent families with 8 genera and 23 species occurs in Australia, Tasmania, and New Guinea and on certain nearby islands. The Recent families are placed in the superfamily Perameloidea, and there also is an extinct superfamily, Yaraloidea, known from Oligocene and Miocene deposits (Archer and Hand 2006; Paull 2008a). For many years the entire group was generally considered to be no more than a suborder within a larger assemblage of syndactylous marsupials that also included Diprotodontia. However, Aplin and Archer (1987) explained that there evidently is closer affinity to Dasyuromorphia and that, in any case, full ordinal status is warranted. The latter procedure was followed by Groves (2005d), though not by Szalay (1994). Marshall, Case, and Woodburne (1990) and McKenna and Bell (1997) accepted this order but used the respective names Peramelina and Peramelia for it. There also have been suggestions of a direct relationship between Peramelemorphia and the New World Didelphimorphia (Gordon and Hulbert 1989). For some of the various proposals of affinity between Peramelemorphia and other australadelphian orders, see the account of class Mammalia, especially under "Marsupial Superordinal Groupings" (see also Kirsch, Lapointe, and Springer 1997; Paull 2008a).

Regardless of their ordinal status, bandicoots once were usually placed in the single family Peramelidae. Groves and Flannery (1990), however, erected the separate family Peroryctidae for the rainforest genera *Echymipera*, *Rhynchomeles*, *Microperoryctes*, and *Peroryctes*. In addition, Archer and Kirsch (1977) had placed the genus *Macrotis* in still another family, Thylacomyidae, but such an arrangement was not considered appropriate by Groves and Flannery (1990) or Baverstock et al. (1990). Szalay (1994) did not accept Peroryctidae as a full family and treated Thylacomyidae as a subfamily of Peramelidae. McKenna and Bell (1997) incorporated Thylacomyidae and the genus *Chaeropus* into the subfamily Chaeropodinae of the family Peramelidae, and also accepted the family Peroryctidae (erecting the subfamily Echymiperinae for *Echymipera*, *Rhynchomeles*, and *Microperoryctes*). Kirsch et al. (1990) had supported recognition of the members of Peroryctidae as a monophyletic group but also indicated that Thylacomyidae was even more distinct from Peramelidae. Kirsch, Lapointe, and Springer (1997) questioned the anatomical evidence for placing *Macrotis* in a separate family, but their DNA-hybridization studies showed that genus to be distinct from all other bandicoots; they treated Thylacomyidae and Peramelidae as full families, while again reducing Peroryctidae to a subfamily of the latter.

More recent molecular assessments (Westerman, Springer, and Krajewski 2001; Westerman et al. 1999) show *Chaeropus* basal to and far removed from all other bandicoot genera; *Macrotis* diverged subsequently, then *Peroryctes*, and then a group comprising all other genera. Based on that sequence and the affinities indicated within the last group, the order Peramelemorphia can be divided into three families:

Eastern barred bandicoot (*Perameles gunnii*). Photo by Alexander F. Meyer, Healesville Sanctuary, Australia.

Chaeropodidae for the genus *Chaeropus*; Thylacomyidae for *Macrotis*; and Peramelidae, with the subfamilies Peroryctinae for *Peroryctes*, Peramelinae for *Perameles* and *Isoodon*, and Echymiperinae for *Echymipera*, *Rhynchomeles*, and *Microperoryctes* (Groves 2005d; Paull 2008a). Such an arrangement of families and subfamilies was supported by a study of nuclear DNA (Meredith, Westerman, and Springer 2008), but there were indications that *Macrotis* is basal to all other living genera and that *Chaeropus*, though highly distinct from the others, is sister to Peramelidae. An extensive new analysis of both nuclear and mitochondrial DNA of all Recent genera (Westerman et al. 2012) again indicated *Chaeropus* as most divergent, but that Peroryctinae is sister to Echymiperinae; the generic sequence suggested by that analysis is followed here.

The term "bandicoot" is a corruption of a word in the Telugu language of the people of the eastern Deccan Plateau of India meaning "pig rat," which was originally applied to a large species of rodent (*Bandicota indica*) from India and Sri Lanka. It seems likely that the name was first applied to the Australian marsupials by the explorer Bass in 1799, and use of the name is of questionable desirability because of confusion with *Bandicota*. It has been suggested that the semantic association of marsupial bandicoots with placental rats has reduced the popularity of the former as subjects of natural history investigation. That situation now seems to have changed; detailed summaries of studies of the ecology, behavior, physiology, morphology, and systematics of the Peramelemorphia were provided by Gordon and Hulbert (1989) and Lyne (1990).

The species expressing both extremes of size in this order are endemic to New Guinea. *Microperoryctes murina* probably weighs less than 100 grams, while *Peroryctes broadbenti* probably exceeds 5 kg (Flannery 1995a). The muzzle is elongate and pointed. The hindquarters are not as elongated as in the kangaroos (Macropodidae), but the hind limbs are larger than the forelimbs. In the family Chaeropodidae all limbs are long. The tail is hairy, of variable length, and nonprehensile.

The second and third digits of the hind foot are syndactylous, that is, bound together by integument; only the tops of the joints and the nails are separate. The combined toes function somewhat like a single digit and are useful in grooming. The main digit of the hind foot is the fourth; the fifth is well developed in most species but is shorter than the fourth; the first toe of the hind foot lacks a nail and is generally poorly de-

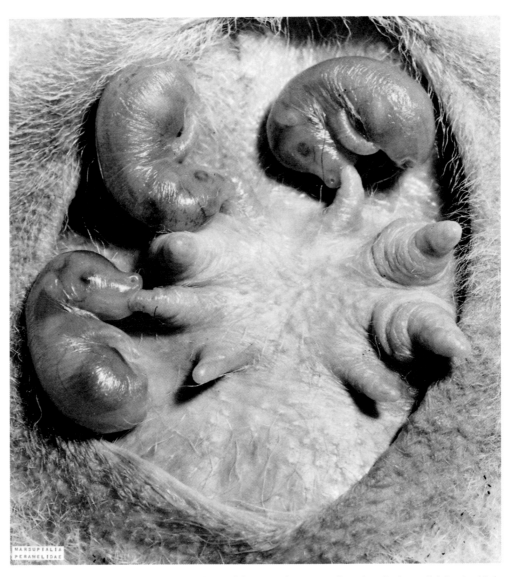

Long-nosed bandicoot (*Perameles nasuta*), litter of three pouch young at four days (estimated) following birth. Three of the unoccupied teats are still enlarged and were suckled by a previous litter. Photo by A. Gordon Lynne through Australian Commonwealth Scientific and Industrial Research Organization.

veloped, and in the genera *Macrotis* and *Chaeropus* it is absent. Except in *Chaeropus* the forefoot has five digits: the third is the longest, the second is slightly shorter, the fourth is somewhat shorter than the second, and the first and fifth are vestigial. In *Chaeropus* the second and third digits of the forefoot are about the same size, the fourth is vestigial, and the first and fifth are absent. Those digits of the forefoot that are well developed have sharp nails for digging.

Although the Peramelemorphia are syndactylous, like the Diprotodontia, their dentition is polyprotodont, as in the Dasyuromorphia. The incisors are numerous and small; the uppers are flattened and unequal, and the lowers slope forward; the crown of the last lower incisor has two lobes. The canines are slender and pointed. The premolars are narrow and pointed, and the molars are four-sided or triangular in outline with four external cusps. The dental formula is: i 4–5/3, c 1/1, pm 3/3, m 4/4 = 46 or 48.

The members of this order are alert and sprightly, sometimes traveling rapidly on all four feet with a galloping action. Paull (2008a) also described a slow gait, in which the weight of the body is taken by the forelimbs while the hind limbs are brought forward together, and an ungainly quadrupedal run, in which the hind legs move alternately. Bandicoots are terrestrial and mainly nocturnal, sheltering by day in grassy nests on the surface of the ground. Most species have mixed feeding habits, taking mainly arthropods but also small vertebrates, fruit, and fungi. They drink water in captivity, but dew seems sufficient for their needs in the wild when water is not present.

The Peramelemorphia are the only marsupials in which a chorioallantoic placenta develops after the transient yolk-sac placenta. This chorioallantoic placenta, however, unlike that of placental mammals, lacks villi. It evolved independently from the placenta of the latter group and is probably functionally correlated with a high rate of reproduction. The bandicoot placenta is retained in the uterus after parturition, and the young remain attached thereto by long umbilical cords until some time after they have secured themselves to teats. The Peramelemorphia are characterized by very short gestation periods, rapid litter succession in polyestrous females, rapid development of pouch young, minimal parental care, and early sexual maturity (Gordon and Hulbert 1989). Females have 6, 8, or 10 mammae (usually 8) and generally have 2–5 young per litter. The pouch opens downward and backward. Newly born young of some bandicoots have deciduous claws that are shed soon after the young reach the pouch.

Bandicoots have suffered to varying degrees since the settlement of Australia by Europeans. Several kinds have been adversely affected by such factors as habitat modification and introduced predators. In general, those that live in the coastal forests and woodlands of Australia and in Tasmania have survived well. In contrast, those that live on the inland plains and deserts have declined severely (Aitken 1979).

Based on fossil evidence, the geological range of the Peramelemorphia is early Eocene to middle Miocene and early Pliocene to Recent in Australia (Archer and Hand 2006; Travouillon et al. 2014) and Pleistocene to Recent in New Guinea (Flannery 1995a). Molecular assessment suggests the bandicoots diverged from other Australian marsupials in the late Paleocene or early Eocene (Meredith, Westerman, and Springer 2008; Westerman et al. 1999).

PERAMELEMORPHIA; Family CHAEROPODIDAE; Genus *CHAEROPUS* Ogilby, 1838

Pig-footed Bandicoot

The single genus and species, *Chaeropus ecaudatus*, formerly occurred in suitable localities from southwestern and west-central Western Australia, through South Australia and the southern part of the Northern Territory, to southwestern New South Wales and western Victoria (Baynes 1984; Ride 1970). Late Pleistocene fossils of the species have been collected in northeastern and east-central Queensland (Hocknull 2005). *Chaeropus* sometimes has been placed in the family Peramelidae, sometimes in its own subfamily, but now is thought to be highly divergent from all other bandicoots (see account of order Peramelemorphia). The name *ecaudatus* is an unfortunate misnomer as the original description was based on a mutilated specimen from which the tail was missing.

Head and body length is 230–260 mm, tail length is 100–150 mm, and estimated weight is 200 grams (Johnson and Burbidge 2008a). The hair, though coarse, is not spiny. The upper parts are grizzled gray tinged

Pig-footed bandicoot (*Chaeropus ecaudatus*), mounted specimen. Photo by Eric Woods / FLPA / Minden Pictures.

with fawn, or almost orange brown, and the underparts are white or pale fawn. The tail is gray or fawn below and on the sides and black above. There is an inconspicuous crest above, with a few white hairs towards the tip.

The body form is light and slender. The head is broad with a long, sharply pointed snout. The ears are long and narrow. Characters of the limbs and feet distinguish Chaeropodidae from the rest of the Peramelemorphia (see account thereof). The limbs are long, slender, and peculiarly developed. The first and fifth digits of the forefeet are absent, the fourth is minute, and the second and third are well developed with long, sharp nails and conspicuous pads under their tips. The forefeet thus resemble those of Artiodactyla, though in pigs the functional digits are the third and fourth rather than the second and third. On the hind feet the first digit is absent, the fifth is barely evident, the fourth is large and long and bears a short, stout nail and a well-developed pad, and the second and third are small and united. The hind feet show some resemblance to those of Perissodactyla, except that in the horses the functional digit is the third, not the fourth.

The pig-footed bandicoot inhabited sand dunes and sand plains with hummocks and tussock grass, grassy plains, and open woodland with a shrub and grass understory; it was said to dig a short, straight burrow with a nest at the end (Johnson and Burbidge 2008a). The nest was constructed of grass, sticks, and leaves and was lined with soft grasses (Ride 1970). *Chaeropus* was cursorial; the gait was described by G. Krefft as resembling that of a "broken-down hack in a canter, apparently dragging the hind quarters after it." However, it apparently was capable of considerable speed, running on its toes (Paull 2008a). Only the fourth toe of the hind foot was employed for locomotion; the tiny fused second and third toes may have been used, as in other bandicoots, for grooming (Johnson and Burbidge 2008a). *Chaeropus* was said to squat in the open with its ears laid back and to run for shelter in hollow logs and trees when chased. It appears to have been primarily nocturnal but on several occasions individuals were seen out in daylight. The diet was omnivorous but tooth and gut structure indicate herbivory, even a degree of grazing (Johnson and Burbidge 2008a). Breeding reportedly occurred in May or June. The maximum recorded litter size was only two, but females possess eight mammae (Collins 1973).

This family may well be extinct. Its decline is thought to have been caused by introduced predators, such as the red fox (*Vulpes vulpes*) and house cat (*Felis catus*), destruction of habitat by domestic livestock, and competition with introduced rabbits (Thornback and Jenkins 1982). Another major factor may have been an epizootic disease transmitted from domestic animals and rodents introduced in the vicinity of Shark Bay, Western Australia, around 1875 (Abbott 2006). The last reliably dated specimen was taken at Lake Eyre in South Australia in either 1901 (Johnson and Burbidge 2008a) or 1907 (Ride 1970). However, there is extensive evidence that *Chaeropus* persisted in the interior of Western Australia well into the twentieth century (Friend 1990b), and the native people of the Great Sandy and northern Gibson Deserts recall it surviving there until the 1950s (Johnson and Burbidge 2008a). Its final disappearance may have been associated with the same factors that affected *Isoodon* (see account thereof). The pig-footed bandicoot is designated as extinct by the IUCN (Burbidge, Dickman, and Johnson 2008) and endangered by the USDI and is on appendix 1 of the CITES (but with the notation "possibly extinct").

PERAMELEMORPHIA; **Family THYLACOMYIDAE; Genus *MACROTIS* Reid**, 1837

Bilbies, or Rabbit-eared Bandicoots

The single living genus, *Macrotis*, contains two species (K. A. Johnson 1989, 2008a, 2008b; Woinarski, Burbidge, and Harrison 2014):

M. lagotis (greater bilby), originally Western Australia, central and southern Northern Territory, South Australia, southwestern and south-central Queensland, and western and central New South Wales;

M. leucura (lesser bilby), formerly east-central Western Australia, southern Northern Territory, and northern South Australia.

The generic name *Thylacomys* Owen, 1840, has sometimes been applied to those species, but for technical reasons is no longer considered valid. Based on cranial and dental morphology, serology, and karyology, Archer and Kirsch (1977) considered *Macrotis* to represent a distinct family. In accordance with the rules of zoological nomenclature, that family is properly called "Thylacomyidae" even though the name is based on an incorrect generic name. Aplin and Archer (1987) suggested that assignment of *Macrotis* to a separate family may have been premature, and Groves and Flannery (1990) restored the genus to Peramelidae, noting a particularly close relationship to *Chaeropus*. McKenna and Bell (1997) conjoined *Macrotis* and *Chaeropus* in the subfamily Chaeropodinae of the family Peramelidae. However, analyses of DNA and RNA (Kirsch, Lapointe, and Springer 1997; Kirsch et al. 1990; Westerman, Springer, and Krajewski 2001; Westerman et al. 1999) support maintenance of the Thylacomyidae (see also account of order Peramelemorphia). A subfossil of *Macrotis* of late Pleistocene–Holocene age has been reported from northeastern Queensland; additional remains from the late Pleistocene of Australia are known (Hocknull 2005; Long et al. 2002). Thylacomyidae also is represented by one genus, *Ischnodon*, from the Pliocene of South Australia (Archer and Hand 2006; Long et al. 2002), and by the newly described species, *Liyamayi dayi*, from the middle Miocene of northwestern Queensland (Travouillon et al. 2014).

According to K. A. Johnson (1989, 2008a, 2008b), head and body length of *M. lagotis* is 290–550 mm, tail length is 200–290 mm, and weight is 1,100–2,500 grams in males and 800–1,100 grams in the much smaller females. Respective figures for *M. leucura* are 200–270 mm, 120–170 mm, and 361–435 grams in males and 311 grams in females. The pelage is long, silky, and soft. The upper parts are fawn gray, blue gray, or ash gray, often pale vinaceous along the sides, and the underparts are white. The color of the body fur extends onto the base of the tail. In *M. lagotis* the remainder of the proximal half of the tail is black, and the terminal half is white, with a conspicuous dorsal crest of long hairs. The tail of *M. leucura* is entirely white except for a gray line extending from the body onto the upper surface of the base. The end of the tail bears a prominent dorsal crest of hairs, as in *M. lagotis*, and in both species the extreme tip of the tail is naked.

Greater bilby (*Macrotis lagotis*). Photo by Pavel German / Wildlife Images.

The general form is light and delicate with a long, tapered muzzle. The ears are very long, pointed, and finely furred. As in all the Peramelemorphia except *Chaeropus*, the forefeet have three functional toes that bear stout, curved claws; the remaining toes are small. On the hind foot the first toe is not present, and in this characteristic both *Macrotis* and *Chaeropus* differ from the other genera of Peramelemorphia. In *Macrotis*, as in the rest of the order, the second and third hind toes are partially united, the fourth toe is the largest, and the fifth is of moderate size. The pouch opens downward and slightly backward, enclosing eight mammae.

Groves and Flannery (1990) pointed out that *Macrotis* shares many characters with *Chaeropus*. However, the skull of *Macrotis* has an extremely broadened braincase and narrowed snout, the most excessively flattened cranium in Peramelemorphia, and enormous, pear-shaped auditory bullae. The dental formula is identical to that of Peramelidae. Archer and Kirsch (1977) observed that the teeth of *Macrotis* differed

from those of Peramelidae in several ways, particularly with regard to the structures involved in the squaring of the molars. Groves and Flannery (1990) noted that while the molars of *Macrotis* have undergone an enormous expansion of the metacone, those of *Chaeropus* show a less modified version of the same condition.

Bilbies live in a variety of mainly dry habitats, including woodland, savannah, shrub grassland, and sparsely vegetated desert. Their distribution seems to have been associated with the presence of suitable soils for burrowing. Unlike most peramelids, bilbies are powerful diggers and live mainly in burrows of their own making. They dig with the forelimbs, kicking out the soil with the hind limbs (Paull 2008a). The burrows are characteristic in that they usually descend from a single opening as a fairly steep, ever-widening spiral to a depth of 1–2 meters. According to Ride (1970), *M. leucura* burrows only in sand hills, never on flats, and the burrows of *M. lagotis* in Western Australia were found to be most common in shrub grassland. An area

Greater bilby (*Macrotis lagotis*), mother and young. Photo by Roland Seitre / Minden Pictures.

occupied by the latter species had 58 burrows, none more than 168 meters from any other. It seems likely that each animal has a number of burrows within its range. K. A. Johnson (2008a, 2008b) wrote that the burrows of both species are similar, containing no vent shafts or nesting material, but that *M. leucura* closes the entrance when in residence and *M. lagotis* does not.

Bilbies are nocturnal, spending the day in their burrows, and terrestrial. They progress on all four feet by means of a shuffling gait; hopping is facilitated by the relatively long hind feet. These animals usually do not lie down to sleep; instead, they squat on their hind legs and tuck their muzzle between their forelegs; the long ears are laid back and then folded forward over the eyes and along the side of the face. They feed on insects, small vertebrates, and some vegetable matter. Southgate and Carthew (2006) found seeds and bulbs to be most important for *M. lagotis* in some areas and invertebrates, especially termites and beetles, in others. Conical depressions up to 25 cm deep, scratched in the earth around trees and bushes, give evidence of foraging for subterranean insect larvae. *M. lagotis* licks seeds from the ground with its long, slender tongue, ingesting a great deal of sand in the process (K. A. Johnson 2008a). Raw or cooked meat, insects, snails, birds, mice, bread, and cake are readily accepted by captives of that species, but roots and fruit usually are rejected. Captives of *M. leucura* reportedly were fierce, hissing and snapping at handlers; stomach contents from a limited sample of that species included extensive remains of rodents, as well as seeds, but no insects (K. A. Johnson 2008b).

According to studies cited by Woinarski, Burbidge, and Harrison (2014), mature males range more widely than females and have been recorded occupying burrows over 5 km apart on consecutive days. Estimates of short-term home range sizes in the Northern Territory varied from 1.1 to 3.0 sq km, and mean home ranges in Queensland were 51 ha. for males and 20 ha. for females. These bandicoots are found alone or in small colonies usually consisting of a single adult male, a female, and an independent young (Johnson 1989). Burrows generally seem to have only one occupant (Ride 1970) but reportedly sometimes contain a pair or a female with young. Groups with more than one male or with several females can be maintained in captivity if the enclosure is sufficiently large (Collins 1973). In that situation the males form a rigid dominance hierarchy without severe fighting; the dominant male chases its subordinates away from the burrows but freely shares these places with females (Johnson and Johnson 1983).

According to McCracken (1990), *M. lagotis* is physiologically capable of breeding throughout the year and captives have been observed to do so. The former wild populations in the temperate zone of South Australia had a distinct breeding season from about March to May, but the existing populations in arid regions may breed at any time of year, perhaps depending on rainfall and food availability. Ovulation occurs toward the end of lactation; thus births occur as soon as the young of the previous litter are weaned. The average estrous cycle is 20.6 days, and the average gestation is 14 days. Litter size is usually one or two young; triplets are rare. The young develop rapidly and have a pouch life averaging about 80 days. They then are left in a burrow for about 2 more weeks and suckled at regular intervals. Johnson (1989) added that sexual maturity occurs at about 175–220 days in females and 270–420 days in males. Females may produce four litters per year under ideal conditions and commonly continue to breed past the age of 4 years (Southgate 2015). A captive male and female each lived for 9 years and 7 months (Weigl 2005).

Both *M. leucura* and *M. lagotis* are listed as endangered by the USDI and are on appendix 1 of the CITES. The IUCN now designates *M. leucura* as extinct and *M. lagotis* as vulnerable (Burbidge, Johnson, and Dickman 2008; Friend, Morris, and van Weenen 2008). *M. leucura* was last collected alive in 1931 in northeastern South Australia, though Aboriginal reports indicate survival in the Great Sandy and Gibson Deserts of Western Australia until the 1960s; a skull of unknown age was found near Alice Springs in the Northern Territory in 1967 (K. A. Johnson 2008b). *M. leucura* formerly was common but was reduced drastically in the early twentieth century through such factors as trapping for its pelt, predation by introduced foxes, wildfires, and competition with introduced rabbits for burrows.

The same problems beset the once even more abundant *M. lagotis*. It may even have occupied Victoria (Scarlett 1969) as well as suitable habitat all across Australia south of about 18° S. However, Southgate (1990) reported that it had disappeared from New South Wales by 1912 and from South Australia by about 1970 and was restricted to a few isolated colonies in Western Australia, the Northern Territory, and southwestern Queensland; the progression of the decline seems to have coincided with the spread of foxes and rabbits. Abbott (2001) pointed out that the last definite specimens of *M. lagotis* from southwestern Western Australia were collected in 1935, shortly after foxes entered the region, but that numerous reports indicate survival of some populations there as late as the 1980s; hence, caution is advised when declaring a species extinct. Johnson (1989) suggested the decline of the greater bilby in arid areas was associated with the disappearance of the traditional Aboriginal practice of burning off vegetation gradually, thereby producing a patchwork of habitat types in different stages of regeneration. The current regime of large wildfires results in a homogeneity of habitats that is less favorable to the species. Thornback and Jenkins (1982) noted that its numbers may be kept at very low densities by the depletion of food supplies by grazing cattle. *M. lagotis* still has a wide distribution, and there now are reintroduced colonies in southern Queensland, New South Wales, South Australia, and southwestern Western Australia, but total numbers may be less than 10,000 and are decreasing (Friend, Morris, and van Weenen 2008).

PERAMELEMORPHIA; Family PERAMELIDAE

Typical Bandicoots

This family of 6 Recent genera and 20 species is found in Australia and New Guinea and on a number of nearby islands. Three subfamilies are currently recognized: Peroryctinae for *Peroryctes*, Peramelinae for *Perameles* and *Isoodon*, and Echymiperinae for *Echymipera*, *Rhynchomeles*, and *Microperoryctes* (Groves 2005d; Paull 2008a; Westerman, Springer, and Krajewski 2001). Peramelidae sometimes has been considered to comprise all living bandicoots and sometimes has been divided into two families, Peroryctidae for

New Guinean spiny bandicoot (*Echymipera rufescens*). Photo by Pavel German / Wildlife Images.

the rainforest genera—*Peroryctes, Echymipera, Rhyn-chomeles,* and *Microperoryctes*—and Peramelidae for the dry-country genera—*Perameles* and *Isoodon*—and also for *Chaeropus* and *Macrotis*. However, recent molecular studies indicated that both *Chaeropus* and *Macrotis* belong in their own families and that Peroryctidae is polyphyletic, *Peroryctes* being more divergent from the other three genera than are *Perameles* and *Isoodon* (Westerman, Springer, and Krajewski 2001; Westerman et al. 1999). It therefore became necessary to designate the three subfamilies now recognized (Groves 2005d; Paull 2008a), arranged in the sequence: Peroryctinae, Peramelinae, Echymiperinae (see also account of order Peramelemorphia). However, a new analysis of both nuclear and mitochondrial DNA of all genera indicated that Peroryctinae is sister to Echymiperinae (Westerman et al. 2012); therefore, it becomes

necessary here to group the genera of those two sub-families. Some authorities, such as Cuthbert and Denny (2013), continue to treat Peroryctidae as a separate family.

Head and body length is 150–558 mm, tail length is 50–335 mm, and weight is around 100–5,000 grams. The pelage may be soft, coarse, or spiny, depending on genus, and coloration varies considerably. The tail and ears vary in length, depending on genus, and the hind limbs are larger than the forelimbs. The tail and limbs are shorter and the ears much shorter than in Thylaco-myidae. The tail and ears are usually shorter and the limbs much shorter than in the Chaeropodidae. The first toe of the hind foot is generally poorly developed but is present, whereas it is absent in the other two bandicoot families. The dental formula of most genera is i 5/3, c 1/1, pm 3/3, m 4/4=48, but in *Echymipera* and

Rhynchomeles there are only four upper incisors on each side.

Groves and Flannery (1990) described the genera assigned here to the subfamily Peramelinae as flat-skulled. The genera assigned here to Peroryctinae and Echymiperinae, however, were characterized as being "cylindrical-skulled," the crania being much higher and subcylindrical in cross section in both the rostral and the neurocranial portion. In the Peramelinae the anteromedial orbital margin slopes diagonally outward, forming a remarkable ridge, or crest, along its edge; the facial extension of the lacrimal bone is very narrow, its suture with the maxilla parallel to the antorbital ridge; the lower margin of the mandible is convex; the auditory bullae (bones housing the inner ear) are complete; and the molar teeth are squared. In the Peroryctinae and Echymiperinae the foramen rotundum is prolonged into a tube; the alveolar plate in the molar region is reduced, so that the zygomatic arch swings forward low over the posterior molars; the infraorbital fossa is low and compressed dorsoventrally; the main palatal vacuities are long and narrow and do not tend to fuse across the midline; the mesopterygoid fossa is broad and parallel-sided; the antorbital surface is flattened; the gonial hooks are directed posteriad; the superior temporal lines are nearly parallel throughout, not converging farther back; and the auditory bullae are incomplete.

Although the above description may suggest immediate affinity between Peroryctinae and Echymiperinae, the involved characters are thought to represent ancestral retention and thus not to offer good evidence of relatedness (Westerman, Springer, and Krajewski 2001). Moreover, the morphological characters distinguishing the rainforest Peroryctinae and Echymiperinae from the dry-country Peramelinae are not thought to express that difference in habitat (Groves and Flannery 1990).

The geological range of Peramelidae is middle Miocene to Recent in Australia (Gordon and Hulbert 1989; Travouillon et al. 2014) and Pleistocene to Recent in New Guinea (Flannery 1995a). Molecular assessment shows that Peramelidae had diverged from Thylacomyidae and Chaeropodidae in the later Oligocene (Westerman et al. 2012).

Eastern barred bandicoot (*Perameles gunnii*). Photo by Alexander F. Meyer, Healesville Sanctuary, Australia.

PERAMELEMORPHIA; PERAMELIDAE; **Genus**
PERAMELES E. Geoffroy St.-Hilaire, 1804

Barred or Long-nosed Bandicoots

The four species originally occurred as follows (Dickman and Stodart 2008; Friend 2008b; Gordon 2008c; Kirsch and Calaby 1977; Ride 1970; Seebeck 2001; Seebeck et al. 1990; Woinarski, Burbidge, and Harrison 2014):

> *P. bougainville*, western and southern Western Australia and Bernier and Dorre Islands in Shark Bay off west coast, South Australia, western New South Wales, northwestern Victoria;
>
> *P. eremiana*, formerly found in deserts from northwestern Western Australia to southern Northern Territory and northern South Australia;
>
> *P. gunnii*, extreme southeastern South Australia, southern Victoria, Tasmania and nearby Bruny Island;
>
> *P. nasuta*, eastern Queensland and New South Wales, southern and eastern Victoria.

Morphological and molecular data indicate that *P. bougainville* is basal to and divergent from the lineage of *P. gunnii* and *P. nasuta*, though the skull of *P. bougainville* closely resembles that of *P. gunnii*; the apparently extinct *P. eremiana* is thought to be immediately related to and possibly conspecific with *P. bougainville* (Gordon 2008c; Seebeck 2001; Westerman and Krajewski 2000). Recent molecular and craniodental assessment suggests the subspecies *P. nasuta pallescens* of the Cape York Peninsula is a separate species, possibly with more affinity to *P. bougainville* and *P. eremiana* than to *P. nasuta* and *P. gunnii* (Westerman et al. 2012). *P. pallescens* was accepted as a full species by Woinarski, Burbidge, and Harrison (2014).

Perameles is characterized by a long and tapered snout, forefeet with strong claws, relatively elongate hind feet, and conspicuous pointed ears. The tail is relatively short and usually thin, but Gordon and Hall (1995) discovered that the tails of *P. bougainville* and *P. eremiana* occasionally are swollen for fat storage. The sleek-looking pelage is composed chiefly of coarse, distinct hairs. In all species except *P. nasuta* there are transverse or diagonal dark and light bars on the back and rump, forming in some cases an elaborate pattern. Females have a backward-opening pouch and eight mammae. Seebeck (2001) noted that length of skull is more than twice maximum width, the auditory bullae are small and nearly hemispherical, and the dentary does not have a posterobuccal process.

In *P. bougainville* head and body length is 173–280 mm, tail length is 75–106 mm, and weight is 165–302 grams; the upper parts are light gray to brownish gray, with two or three alternating paler and darker bars, pronounced in some populations but muted in others; the underparts and feet are white (Friend 2008b). In *P. eremiana* head and body length is 180–285 mm, and tail length is 77–135 mm; the upper parts are dull orange, rufescent, or brown, with the dark mid-dorsal area extending in one or two dark bands on either side of the rump; the underparts are white and the tail is very dark brown above and white laterally and ventrally (Gordon 2008c). In *P. gunnii* head and body length is 270–350 mm, tail length is 70–110 mm, and weight is 500–1,450 grams; the upper parts are grizzled yellowish brown with three or four pale bars on the hind quarters, the underparts are slaty gray, and the tail is white above except at the base (Seebeck and Menkhorst 2008). In *P. nasuta* head and body length is 310–445 mm, tail length is 120–160 mm, and weight is 520–1,330 grams; the upper parts are drab grayish brown, only a few individuals show a faint barred pattern, and the underparts, forefeet, and upper surface of the hind feet are creamy white (Dickman and Stodart 2008).

Habitats include semiarid country with thickets, hummock grassland, heaths, or dune vegetation for *P. bougainville*, sand plains and ridges with spinifex grassland for *P. eremiana*, woodland and open country with good ground cover for *P. gunnii*, and rainforest and sclerophyll forest for *P. nasuta* (Friend 2008b; Gordon 2008c; Ride 1970). Collins (1973) wrote that in heavy vegetation long-nosed bandicoots construct a nest consisting of an oval mound of twigs, leaves, and humus on the surface of the ground. In more open areas they excavate a nest chamber, line it with plant fibers, and then cover it with a mound of twigs and leaves. Abandoned rabbit burrows, rock piles, and hollow logs are also used as nest sites. Most individuals of *P. bougainville* use the same nest over a week but some

Western barred bandicoot (*Perameles bougainville*). Photo by Jiri Lochman / Lochman LT.

use a different one each night (Friend 2008b). Long-nosed bandicoots are nocturnal, terrestrial, and highly active. Their rapid running has been described as a kind of gallop, and they have been seen to jump straight up into the air—as high as 1.5 meters (Seebeck 2001)—and then to take off immediately in another direction. Nightly movements of *P. bougainville* averaged 226 meters for males and 185 meters for females over 11 nights in a study on Bernier and Dorre Islands, but overnight movements of over 1 km have been recorded (Dickman 2015).

Long-nosed bandicoots are omnivorous, often concentrating on insects, earthworms, and other invertebrates, but also taking lizards, mice, fruits, bulbs, and other plant material. *P. nasuta* scratches and digs in the ground for insects and grubs, including weevils found on fruit tree roots and the larvae of the scarabaeid beetle, which feeds on the grass roots of lawns and pastures. Stodart (1977) observed that the claws are used to make a hole big enough for the snout to enter.

Reported population densities of *P. gunnii* in Victoria and Tasmania have varied from about 0.35 to 4.80 individuals per ha. (Mallick, Driessen, and Hocking 2000). Male *P. gunnii* have much larger home ranges (25.6 ha.) than females (3.2 ha.), but core ranges are smaller, 4–5 ha. for males, 1.5–2.5 ha. for females (Seebeck 2001). There may be considerable overlap in female ranges, and presumably several are included within the home range of each male. The home ranges of female *P. bougainville* on Dorre Island average 6.2 ha. when population density is low and 1.4 ha. at high density; respective figures for males are 14.2 and 2.5 ha. (Friend 2008b). In *P. nasuta* mean home range size over a year was found to be 4.4 ha. for males and 1.7 ha. for females; females frequently had overlapping ranges, while males appeared to avoid each other during nonbreeding months but to expand and overlap their ranges in the breeding season (Scott, Hume, and Dickman 1999).

P. nasuta has been reported to be extremely solitary, with males acting aggressively when coming into contact with one another (Ride 1970). Stodart (1977) found that species to be not gregarious in enclosures; individuals tended to ignore each other, and there was no territorial defense. However, Scott, Hume, and Dickman (1999) thought that in the wild males probably defend territories that include parts of the home ranges of several females. Such was suggested by evidence of wounds from fighting during the breeding season and by territorial displays, in which a male assumed a bipedal stance while hissing at another male. Collins (1973) wrote that pairs or family groups could be kept together if the enclosure was sufficiently large but that two or more males should not be placed in the same enclosure. Seebeck (2001) reported similar behavior for *P. gunnii.*

Long-nosed bandicoots are polyestrous, with an average estrous cycle of 21 days in *P. nasuta* (Lyne 1976) and about 26 days in *P. gunnii* (Seebeck 2001). If fertilization occurs, there is a 62- to 63-day interval from one period of female receptivity, through pregnancy, to the next period of receptivity (Stodart 1977). *P. nasuta* and *P. gunnii,* at least, are physiologically capable of reproducing throughout the year. However, female *P. nasuta* usually stop breeding by early winter, when food resources decline (Dickman and Stodart 2008). Breeding of *P. gunnii* in Tasmania extends through most of the winter, spring, and early summer (around May to February), with each female producing three or four litters at intervals of 60 days (Ride 1970; Stodart 1977), but may cease in autumn and early winter. Breeding of *P. gunnii* in Victoria slows or stops during hot, rainfall-deficient summers (Seebeck 2001). *P. bougainville* produces young continuously from April to October on Bernier and Dorre Islands; litter size is 1–3, usually 2 (Friend 2008b). Gestation in *P. nasuta* and *P. gunnii* has been timed at about 12.5 days and litter size is 1–5, usually 2–3 (Dickman and Stodart 2008; Seebeck 2001). Female *P. nasuta* and *P. gunnii* may become pregnant again while lactating and give birth to a new litter immediately after the last one has been weaned (Dickman and Stodart 2008; Seebeck and Menkhorst 2008).

Since females have eight mammae, the number of young seems rather low, but actually there is an advantage, considering the short interval between births. A new litter is born at the same time that the previous one is weaned. The mammae used by the previous litter have greatly expanded and are too large for the newborn to grasp, but because usually at least four other mammae have not been used, the newborn are assured nourishment (Collins 1973).

Newborn *P. gunnii* are about 13.5 mm long and weigh about 200 mg; their eyes open when they are around 30–35 days old (Seebeck 2001). Pouch life is about 55 days, after which the young are left in a grass nest in a scraped depression until weaning at 70–80 days; they disperse at 3–5 months (Seebeck and Menkhorst 2008). Females are sexually mature at around 3 months and can bear one or two litters in the same breeding season in which they were born. Adult size and male sexual maturity are attained at 4–6 months (Collins 1973; Ride 1970). In *P. nasuta* the young are carried in the pouch 50–54 days, then remain briefly in the nest, are weaned at 60 days, and begin to forage with the female at 62–63 days; sexual maturity comes at 5 months (Dickman and Stodart 2008; Stodart 1977). Wild *P. bougainville* have lived over 4 years (Friend 2008b), and a captive *P. gunnii* lived for 6 years and 1 month (Weigl 2005).

The IUCN classifies *P. bougainville* as endangered, *P. eremiana* as extinct, and *P. gunnii* as near threatened (Burbidge, Johnson, and Aplin 2008; Friend and Richards 2008; Menkhorst and Richards 2008). The USDI lists *P. bougainville* and *P. eremiana* as endangered. *P. bougainville* is also on appendix 1 of the CITES. Those species declined severely following European settlement, apparently because of clearing of natural vegetation and other habitat modification, inadvertent destruction in poisoning and trapping efforts to control introduced rabbits, and the spread of introduced predators, particularly the red fox (*Vulpes vulpes*) and domestic cat (*Felis catus*) (Aitken 1979; Menkhorst and Seebeck 1990). Apparently the most drastic losses in the inland arid country were coincident with the large-scale departure of Aboriginal people from the 1930s to the 1950s. The gradual, patchwork burning practiced by these people has been replaced by intensive, lightning-caused wildfires that destroy habitat diversity (Johnson and Southgate 1990).

P. eremiana was still common in the 1930s and may have survived through the 1950s, but has not been collected since a specimen was taken in Western Australia in 1943 (Burbidge, Johnson, and Aplin 2008; Kemper 1990). *P. bougainville,* formerly occurring all

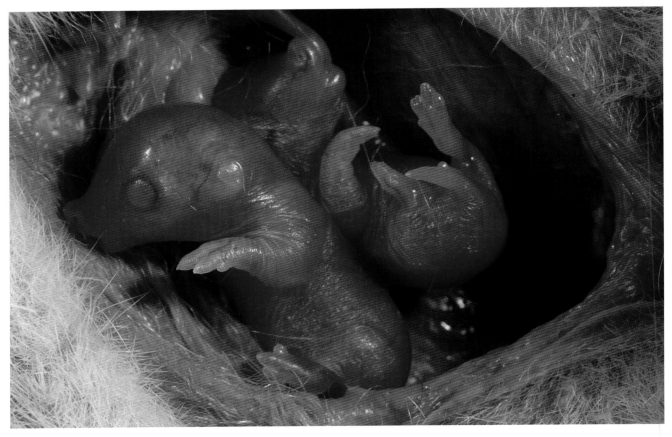

Western barred bandicoot (*Perameles bougainville*), newborn. Photo by Jiri Lochman / Lochman LT.

across southern Australia, had disappeared from New South Wales by the 1860s and from South Australia and the mainland of Western Australia by about 1930 (Friend 1990b; Friend and Richards 2008). The decline of both species may have been associated with a hypothesized major epizootic disease that spread from affected domestic animals and rodents introduced in the vicinity of Shark Bay, Western Australia, around 1875 (Abbott 2006). The several mainland subspecies of *P. bougainville*, or possibly closely related species, are extinct: *myosuros* in southern Western Australia, *notina* in South Australia, and *fasciata* in New South Wales (Friend 1990b; Kemper 1990; Woinarski, Burbidge, and Harrison 2014). Natural populations of *P. bougainville* are known to survive only on Bernier and Dorre Islands in Shark Bay. Numbers recently have been estimated to fluctuate from about 120 to 900 individuals on Bernier, depending on rainfall, and from 140 to 1,500 on Dorre. A population on the nearby, larger island of Dirk Hartog has disappeared. There are

small reintroduced colonies at two other Shark Bay sites and at Roxby Downs, South Australia. All populations are highly vulnerable to the potential introduction of predators, fires, and disease (Friend and Richards 2008; Richards and Short 2003; Woinarski, Burbidge, and Harrison 2014).

P. nasuta seems to have fared better than the other species. However, *P. gunnii* has been declining in Tasmania (Robinson, Sherwin, and Brown 1991) and is all but extinct on the mainland. Applying IUCN criteria to updated information, Woinarski, Burbidge, and Harrison (2014) reclassified the mainland populations of *P. gunnii* to endangered and the Tasmanian populations to vulnerable. Mallick et al. (1997) indicated that *P. gunnii* is now absent from most of its original range in Tasmania, principally because of removal of native vegetation by agriculture and livestock grazing. Recent introduction of the fox to Tasmania poses a major additional threat (Menkhorst and Richards 2008). *P. gunnii* formerly occupied a large part of Victoria, but by the 1960s was essentially restricted to suburban gardens and lightly farmed areas around the city of Hamilton in the southwestern part of the state (Emison et al. 1978; Minta, Clark, and Goldstraw 1989). Survival in that area depended partly on maintenance of gardens and the willingness of gardeners to accept digging activity by the bandicoot. However, notwithstanding intensive conservation efforts and community cooperation, the population underwent an uncontrollable decline. In the 1970s over 1,000 individuals were present in 3,000 ha., but by 1988 there were only 200 in about 600 ha. (Watson and Halley 1999); fewer than 100 animals remained in 1992 (Robinson, Murray, and Sherwin 1993). It had been suggested the decline was associated with inbreeding, but Robinson (1995) found genetic variability in Hamilton *P. gunnii* to be relatively high and thought that bringing in animals from Tasmania would be of no benefit. The original Hamilton population seems to have disappeared entirely around 2002. Individuals captured there were used to reintroduce other colonies in Victoria, but most, including one near Melbourne that had grown to over 700 animals, have disappeared; about 350 individuals survive at two other sites (Menkhorst and Richards 2008; Woinarski, Burbidge, and Harrison 2014).

PERAMELEMORPHIA; PERAMELIDAE; **Genus** *ISOODON* Desmarest, 1817

Short-nosed Bandicoots

Three species are currently recognized (Ellis, Wilson, and Hamilton 1991; Flannery 1995a; Gordon 2008b; Lyne and Mort 1981; McKenzie, Morris, and Dickman 2008; Paull 2008b; Pope et al. 2001; Ride 1970; Seebeck et al. 1990; Westerman and Krajewski 2000):

I. macrourus, south-central and southeastern New Guinea, northeastern Western Australia, northern Northern Territory and nearby islands, northern and eastern Queensland and nearby islands, northeastern New South Wales;

I. obesulus (here including *peninsulae*, see below), northern Queensland, southeastern New South Wales, southern Victoria, southeastern South Australia and nearby Kangaroo Island and Nuyts Archipelago, southwestern Western Australia and nearby Recherche Archipelago, Tasmania and several nearby islands, West Sister Island in Bass Strait;

I. auratus, originally northern and eastern Western Australia and nearby Barrow and Augustus islands, Northern Territory and nearby Marchinbar Island, inland South Australia, western New South Wales.

The name *Thylacis* Illiger, 1811, sometimes used instead of *Isoodon* for these species, is considered here to be a synonym of *Perameles*. Although *I. auratus* was treated as a full species by Groves (2005d) and McKenzie, Morris, and Dickman (2008), there is mounting genetic evidence that the various populations of *auratus* should be included within the species *I. obesulus* (Pope et al. 2001; Westerman and Krajewski 2000; Zenger, Eldridge, and Johnston 2005). *I. peninsulae* of the Cape York Peninsula and Lamb Range in northern Queensland was treated as an extremely disjunct population of *I. obesulus* by Ride (1970) and Seebeck et al. (1990), but Close, Murray, and Briscoe (1990) considered it the most allozymically distinctive member of *Isoodon*. Molecular assessment by Westerman and Krajewski (2000) found little support for *peninsulae* even at subspecific level, but molecular and morphological analysis by Pope et al. (2001) strongly supported recognition of *peninsulae*, at least as a subspecies. Subsequently, *peninsulae* has usually been treated as a subspecies or synonym of *I. obesulus* (Groves 2005d; Paull 2008b; Zenger, Eldridge, and Johnston 2005). However, a new assessment of its mitochondrial DNA (Westerman et al. 2012) suggests it is a full species, more closely related to *I. auratus* than to *I. obesulus*. Aplin and Pasveer (2007) reported remains of *I. macrourus* from late Pleistocene to Holocene archeological sites on Kobroor in the Aru Islands southwest of New Guinea, where the species no longer seems to occur.

In *I. macrourus* head and body length is 300–470 mm, tail length is 80–215 mm, and weight is 500–3,100 grams. In *I. obesulus* head and body length is 280–360 mm, tail length is 90–145 mm, and weight is 400–1,850 grams. In *I. auratus* head and body length is 190–295 mm, tail length is 84–121 mm, and weight is 250–670 grams. Males are on average substantially larger than females. The pelage of *Isoodon* is coarse and

Short-nosed bandicoot (*Isoodon macrourus*). Photo by Gerhard Körtner.

glossy with distinct hairs. The upper parts generally are a fine mixture of blackish brown and orange or yellow, and the underparts are yellowish gray, yellowish brown, or white. An elongated snout is characteristic of all bandicoots, but in *Isoodon* the head is relatively broader than in *Perameles* and the jaws are slightly shorter and stouter. The ears are short and rounded. The well-developed pouch opens backward, as in other bandicoots, and the pouch contains eight mammae.

Short-nosed bandicoots inhabit open woodland, thick grass along the edges of swamps and rivers, arid grassland, and thick scrub on dry ridges. They are active and inoffensive animals and are often mistaken for large rats or rabbits from a distance. When walking, they move the forelegs and hind legs separately; they do not bound when running, though Bennett and Garden (2004) reported them to commonly use a half-bound gait, in which a pair of feet contacts and leaves the ground together, while ground contact of the other

pair is asynchronous. Short-nosed bandicoots often make long tunnels through the grass. Their nests of sticks, leaves, and grass, sometimes mixed with earth, are located in a scraped out depression on the ground or in a hollow log or burrow of some other animal. These bandicoots level off the surrounding vegetation in making their nests. They are terrestrial and often nocturnal, but *I. obesulus* is active both day and night (Paull 2008b). *I. auratus* has peaks of activity 3–4 hours after dusk and again before dawn, during which it covers up to 10 ha. (McKenzie, Morris, and Dickman 2008). *I. obesulus* seems to detect the approach of bad weather, as it will enlarge its nest before heavy rains. After entering or leaving its nest, it closes the opening behind it. That species also has been known to construct a short burrow (Kirsch 1968).

Isoodon feeds in definite areas, with the location changing from time to time, as evidenced by scratch marks in the ground. *I. obesulus* uses a rapid movement

Short-nosed bandicoot (*Isoodon obesulus*). Photo by Israel Didham.

of its forefeet to crush its living prey. That species feeds on a variety of items, at least in captivity, but prefers insects and worms. Individuals that were not fed for several days refused potatoes, carrots, and turnips despite their hunger. The diet of *I. macrourus* does include significant amounts of seeds, fruits, and fungi, as well as earthworms and other invertebrates (Mallick, Driessen, and Hocking 1998a). *I. auratus* is known to eat insects, centipedes, turtle eggs, small reptiles, and rodents (McKenzie, Morris, and Dickman 2008).

I. auratus is most common in coastal areas of Barrow Island, where there may be 10 adults per ha. (McKenzie, Morris, and Dickman 2008). On Marchinbar Island, home ranges of that species vary from 4.4 ha. to 35.0 ha. for males and 1.7 to 12.7 ha. for females (Palmer and Burbidge 2003). Population densities of *I. obesulus* have been reported to be as high as 0.35 per ha. in Tasmania (Mallick, Driessen, and Hocking 1998b) and about 1.5 per ha. in the Franklin Islands

off South Australia (Copley et al. 1990). Home ranges in that species were found to be 4.05–6.48 ha. for four males and 2.31 ha. for one female. Home ranges of *I. macrourus* in Queensland were 1.7–5.2 ha. for males and 0.9–2.1 ha. for females (Stodart 1977). The species *I. obesulus* has been reported to be highly aggressive and territorial, with probably very little overlap in home ranges of individuals of the same sex. Mallick, Driessen, and Hocking (1998b) found total separation in the ranges of two male *I. obesulus*, averaging 6.95 ha., and those of five females, averaging 3.28 ha. However, Ride (1970) indicated that the larger territories of males might overlap those of several females. Southgate et al. (1996) reported individuals of *I. auratus* to maintain overlapping home ranges of about 10-35 ha. on Marchinbar Island. Broughton and Dickman (1991) suggested the degree of range exclusion may be a function of food availability and population density. *Isoodon* is solitary in the wild, apparently coming together only

Golden bandicoot (*Isoodon auratus*). Photo by Elizabeth Tasker / Wildlife Images.

to mate (Stodart 1977). Two individuals can rarely be kept together in captivity as they are intolerant of each other and pugnacious. Two males should never be placed in the same enclosure (Collins 1973). They fight with open mouths, but their long claws usually inflict greater injury than their teeth. Gordon (2008b) noted that both sexes of *I. macrourus* have a gland behind each ear, which is used to mark the ground and vegetation during aggressive encounters and which enlarges during the breeding season.

Short-nosed bandicoots are polyestrous, with the estrous cycle averaging 22 days in *I. macrourus* (Gemmell 1988; Lyne 1976). Female *I. auratus* give birth throughout the year on Barrow Island, though the proportion with pouch young increases after substantial rains; normal litter size is two (McKenzie, Morris, and Dickman 2008). Breeding reportedly can take place year-round for *I. macrourus* in temperate and subtrop-

ical parts of Queensland and New South Wales and in May–February for *I. obesulus* in Tasmania (Ride 1970). In the wet tropics of northeastern Queensland, almost all births of *I. macrourus* occur during the late dry (August–October) and early wet (November–January) seasons (Vernes and Pope 2009). Several litters can be produced by each female during one season, there usually being two or three in *I. obesulus* in Tasmania (Ride 1970) and two to four in *I. macrourus* (Friend 1990; Vernes and Pope 2009). In a population of *I. obesulus* in the Franklin Islands, off South Australia, breeding occurred throughout the year, and females produced up to five litters during the period (Copley et al. 1990). A captive *I. macrourus* bore eight litters totaling at least 32 young in 17 months (Collins 1973).

The gestation period in *I. macrourus* has been calculated as from 12 days and 8 hours to 12 days and 11 hours—among the shortest recorded for any mammal

(Lyne 1974). In *I. macrourus* there are up to 7 young per litter, with reported averages being 2.7 (Friend 1990), 3.1 (Vernes and Pope 2009), and 3.4 (Stodart 1977). In *I. obesulus* there are up to 6 young (Gordon 2008b), though Copley et al. (1990) reported an average ranging from 1.6 in the summer to 2.4 in the spring. According to Collins (1973), probably no more than four young of any one litter survive, so that four small, unused mammae are immediately available for the next litter, and reproduction can proceed at a rapid rate, just as in *Perameles*. The young of *I. macrourus* leave the pouch after 7–8 weeks, are weaned at 8–10 weeks, and reach adult size at about 1 year (Collins 1973; Friend 1990). Females of that species may attain sexual maturity at only 3–4 months but more commonly at about 8 months (Gordon 2008b). Wild *I. obesulus* live up to 4 years (Lobert and Lee 1990). A specimen of *I. macrourus* lived for 6 years and 9 months in captivity, then was released (Weigl 2005).

I. obesulus and *I. auratus*, which may actually represent a single species (see above), had allopatric ranges that together covered nearly all of Australia, but their distribution has been reduced drastically since European settlement began (Ashley et al. 1990; Friend 1990b; Gordon, Hall, and Atherton 1990; Johnson and Southgate 1990; Kemper 1990; Kennedy 1992; McKenzie, Morris, and Dickman 2008; Palmer, Taylor, and Burbidge 2003; Paull 2008b; Woinarski, Burbidge, and Harrison 2014). *I. obesulus* has been eliminated from 50–90 percent of its original range because of habitat loss and introduced predators, though it is still found at scattered localities across southern Australia and Tasmania and on Kangaroo Island, in the Nuyts Archipelago off South Australia, on Daw Island off southern Western Australia, and on Bruny, Three Hummock, and West Sister Islands off Tasmania; subfossils show that it also once occurred on Flinders Island. *I. auratus* is known to survive only in remote parts of the Kimberley area of northern Western Australia, on nearby Augustus, Storr, Lachlan, and Uwins Islands, on Barrow and Middle Islands farther to the west off the Pilbara coast (it is extinct on Hermite Island), and on Marchinbar Island off the northeastern tip of Arnhem Land. It has been introduced recently to several other islands off Arnhem Land. The largest population is on Barrow Island, where at least 20,000 individuals occur.

The decline of short-nosed bandicoots may be associated with habitat destruction by introduced grazing animals, competition from the introduced black rat (*Rattus rattus*), and predation by introduced cats, foxes, dogs, and pigs. The greatest current threat to island populations is considered to be deliberate or inadvertent introduction of cats. However, the main period of decline took place in the arid interior from about the 1930s to the 1950s, when there was a large-scale movement of Aboriginal people from their native lands to permanent settlements and cattle stations. In their former nomadic movements those people had burned off vegetation gradually, thereby producing a mosaic of habitats that provided abundant food and cover to small mammals. Disappearance of those conditions led to massive destruction of habitat by wildfires started by lightning and the subsequent loss of habitat diversity.

Both *I. obesulus* and *I. auratus* are still decreasing, though the IUCN now puts only *I. auratus* in a threatened category, as vulnerable (Burbidge, Woinarski, and Morris 2008). Formerly, when the IUCN gave more attention to subspecies, it classified *I. obesulus nauticus* of South Australia as vulnerable and *I. o. obesulus* of eastern mainland Australia and *I. o. fusciventer* of Western Australia as near threatened (Baillie and Groombridge 1996). Using IUCN criteria, Woinarski, Burbidge, and Harrison (2014) maintained *I. o. obesulus* (in which they included *I. o. nauticus*) as near threatened, but did not place *I. o. fusciventer* in a category of concern. They considered the latter locally abundant in Western Australia, where fox control is ongoing, but noted that the distribution of *I. o. obesulus* is now badly fragmented and restricted mostly to coastal areas and islands.

I. macrourus occurs predominantly in zones of higher rainfall and greater vegetative cover than those occupied by the other two species; it has disappeared from much of its inland habitat, because of the same factors, but is well adapted to unstable environments and is still widely distributed on the mainland and islands (Gordon 2008b). *I. macrourus* is one of relatively few native Australian ground-dwelling mammals able to survive within urbanized landscapes and occurs at a density of about 1 individual per ha. in parts of Brisbane (Fitzgibbon, Wilson, and Goldizen 2011).

Giant bandicoot (*Peroryctes broadbenti*). Photo by Roy MacKay, from Flannery (1995a).

PERAMELEMORPHIA; PERAMELIDAE; **Genus** *PERORYCTES* Thomas, 1906

New Guinean Bandicoots

There are two species (Aplin, Helgen, and Lunde 2010; Flannery 1995a; Westerman et al. 2012; Ziegler 1977):

> *P. raffrayana*, New Guinea and Japen (Yapen) Island to northwest;
> *P. broadbenti* (giant bandicoot), southeastern New Guinea.

Kirsch and Calaby (1977) did not separate *P. broadbenti* from *P. raffrayana*, but Ziegler (1977) argued that the two are distinct and seem to occur together in southeastern Papua New Guinea. The latter arrange-

ment was accepted by Groves and Flannery (1990), who also transferred the species *P. longicauda* and *P. papuensis* to the genus *Microperoryctes* (see account thereof). Aplin and Pasveer (2007) reported fossil remains of a new and unnamed species of *Peroryctes* from late Pleistocene to Holocene archeological sites on Kobroor in the Aru Islands southwest of New Guinea.

Measurements (based in part on Aplin, Helgen, and Lunde 2010, and Lidicker and Ziegler 1968) are: *P. raffrayana*, head and body length 175–384 mm, tail length 110–230 mm; and *P. broadbenti*, head and body 340–558 mm, tail 95–335 mm. Flannery (1995a) listed weights of 650–1,000 grams for *P. raffrayana* and 4.8 kg for *P. broadbenti* and indicated that the latter could exceed 5 kg. Aplin, Helgen, and Lunde (2010) wrote: "The few available body weights for *P. broadbenti* indicate a striking level of sexual dimorphism, with males attaining body weights to 4.9 kg (n = 5), more than three times the weight of females (to 1.4 kg; n = 2)." Cuthbert and Denny (2013) reported average weights of 1,038 grams for five male *P. raffrayana* and 550 grams for two females. In *P. raffrayana* the upper parts are dark brown with a slight mixture of black, and the underparts are white, brownish yellow, or buff. *P. broadbenti* also is dark brown dorsally but has reddish buff on the flanks and is near white ventrally. The fur of *Peroryctes* is comparatively soft and long, not spinous as in *Echymipera*. Flannery (1995a) noted that *Peroryctes* has longer hind feet than do the other bandicoots of New Guinea and can be further differentiated from *Microperoryctes* by having a single pair of posterior palatal vacuities rather than two.

New Guinean bandicoots dwell mainly in forests. Ziegler (1977) gave elevation ranges of 60–3,900 meters for *P. raffrayana* and 0–2,700 meters for *P. broadbenti*. He also noted that *P. raffrayana* is rare below 500 meters, occurs primarily in dense forest, and avoids grassland. Flannery (1995a) found *P. raffrayana* most common in undisturbed forest at around 1,000 meters, but thought *P. broadbenti* to be essentially a lowland species. Available information suggests the latter species occurs in dense rainforest habitat and is most often encountered along creeks or rivers (Aplin, Helgen, and Lunde 2010). The hind foot of *P. raffrayana* seems adapted for leaping. Collins (1973) wrote that New Guinean bandicoots are terrestrial, nocturnal, and apparently solitary. According to Flannery (1995a), *P. raffrayana* feeds on figs, is usually encountered in the early evening, and sometimes has been taken from nests; females of that species with one pouch young each were found in June, August or September, November, and December, and a female with two pouch young was examined in March; a female *P. broadbenti* reportedly had two young. Cuthbert and Denny (2013) radio-tracked a male *P. raffrayana* for 19 days, finding an average daily movement of 87 meters and an estimated home range of 2.7 ha. A specimen of *P. raffrayana* lived in captivity for 3 years and 3 months (Jones 1982).

The IUCN classifies *P. broadbenti* as endangered; it is restricted to the peninsula that forms the southeastern extension of Papua New Guinea and is declining seriously because of excessive hunting and loss of habitat to agriculture (Leary et al. 2008f). George and Maynes (1990) had observed that the great size of the species renders it vulnerable to hunting pressure. They thought it might already be extinct, though noted that there remain extensive tracts of suitable habitat with a sparse human population. Aplin, Helgen, and Lunde (2010) were unaware of any new specimens of *P. broadbenti* collected in more than three decades, aside from one recently obtained ear clip that was assigned to the species based on sampling of mitochondrial DNA. *P. raffrayana* is more widespread and common but also is subject to hunting (Leary et al. 2008g).

PERAMELEMORPHIA; PERAMELIDAE; **Genus *ECHYMIPERA* Lesson, 1842**

New Guinean Spiny Bandicoots

There are five species (Flannery 1990, 1995a, 1995b; George and Maynes 1990; Heinsohn 2003; Menzies 1990; Westerman, Springer, and Krajewski 2001; Ziegler 1977):

> *E. clara*, northern New Guinea, Japen (Yapen) Island;
>
> *E. rufescens*, New Guinea, Misool Island, Japen Island, Kai Islands (possibly introduced), Aru Islands, Goodenough Island, Fergusson Island, Normanby Island, Cape York Peninsula of Queensland;
>
> *E. kalubu*, New Guinea and many nearby islands including Waigeo, Salawati, Misool, Supiori, Biak, Japen, Manus (perhaps introduced), and New Britain in the Bismarck Archipelago;
>
> *E. echinista*, known only from Western Province in southwestern Papua New Guinea;
>
> *E. davidi*, known only from Kiriwina Island in the Trobriand Islands off southeastern Papua New Guinea.

E. rufescens and *E. kalubu* appear to be closely related and possibly conspecific; *E. clara* is genetically divergent from them but shows some affinity to *E. rufescens*, as does *E. echinista*; *E. oriomo* of the Fly River Plateau in south-central New Guinea could be specifically distinct from *E. kalubu*; *E. davidi* does not seem closely related to any other species, though it shares some

New Guinean spiny bandicoot (*Echymipera davidi*). Photo by Daniel Heuclin / npl / Minden Pictures.

similarities with populations of *E. rufescens* on Goodenough, Fergusson, and Normanby Islands (Flannery 1990, 1995b; Groves 2005d; Menzies 1990; Westerman, Springer, and Krajewski 2001). *E. rufescens* long had been known in Australia by only a single specimen collected in 1932, but in 1970 five more individuals were captured (Hulbert, Gordon, and Dawson 1971), and subsequently the species was found over a large part of the Cape York Peninsula (Gordon and Lawrie 1977). Probable remains of *E. kalubu* (in addition to *E. rufescens*) have been reported from late Pleistocene to Holocene archeological sites on Kobroor in the Aru Islands (Aplin and Pasveer 2007). Remains, possibly of *E. rufescens* and dating as late as 1,870 years ago, have been found on Halmahera Island west of New Guinea (Flannery 1995b; Flannery et al. 1995b).

Head and body length is about 200–500 mm, and tail length is 50–125 mm. Often the tail is missing, perhaps bitten off, or it may detach easily. Weight is 525–2,225 grams for male and 300–1,200 grams for female *E. rufescens* (Gordon 2008a), 1,140–1,700 grams for male and 825–1,140 grams for female *E. clara*, and 840–1,500 grams for male and 450–820 grams for female *E. kalubu* (Flannery 1995a). The upper parts are bright reddish brown, dark coppery brown, black mixed with yellow, or black interspersed with tawny; the underparts usually are buffy or brownish. The entire pelage is stiff and spiny. The snout is comparatively long and sharp. There are only four upper incisor teeth on each side of the jaw. There are three pairs of mammae in *E. kalubu* and four pairs in *E. rufescens* (Lidicker and Ziegler 1968). Flannery (1995a) noted that the sexes of *E. clara* differ so greatly that they could easily be mistaken for separate species; males are much larger and distinguished by a very large head, massive skull with extremely developed sagittal crest, and huge canines and premolars, while females actually more closely resemble *E. rufescens* externally and *E. kalubu* in skull morphology.

These bandicoots generally inhabit rainforest. Flannery (1995a) listed the following altitudinal ranges: *E. clara*, 300–1,700 meters; *E. rufescens*, 0–1,200 meters; *E. kalubu*, 0–2,000 meters; and *E. echinista*, 40–80 meters. They are apparently terrestrial, nocturnal, and omnivorous (Collins 1973). Van Deusen and Keith (1966) reported that *E. clara* feeds in part on the fruit of *Ficus* and *Pandanus* and in turn forms an important food staple for the native people within its range. Gordon (2008a) wrote that *E. rufescens* usually shelters in shallow burrow complexes, 50–80 cm deep and occupying 2–3 sq meters, with two or more entrances. It forages on the forest floor and in leaf litter, or digs conical holes in the soil with its forefeet, seeking invertebrates, fruit, fungi, and seeds. On the Cape York Peninsula, Shevill and Johnson (2008) found *E. rufescens* to feed mostly on fruits and seeds; approximately 48 individuals there occupied an area of just 2.25 ha.

In a spool-and-line tracking study, Anderson et al. (1988) found *E. kalubu* to forage for fallen fruit and to root in the forest floor for earthworms and grubs. Average nightly movement was 344 meters. Individuals maintained regular home ranges; two males utilized 1.0 ha. and 2.1 ha., respectively, over a period of three nights. There were several nests within each range, located in hollow logs, leaf piles, and shallow burrows. Ranges overlapped, though there were probably exclusive core areas. Cuthbert and Denny (2013) radio-tracked individuals of *E. kalabu* an average of 11 days. Movements between daytime nest sites averaged only about 93 meters. A population density of about 85 individuals per sq km was estimated. Home range averaged 2.79 ha. and varied from 0.9 to 7.3 ha. There was little overlap between the ranges of males, but extensive overlap between the ranges of individuals of opposite sex.

Available evidence indicates these bandicoots are solitary and highly intolerant of their own kind. According to Flannery (1995a), breeding of *E. kalubu* occurs throughout the year, average birth interval is probably about 120 days, 1–3 pouch young are usually present, and females begin breeding at a very early age; pouch young of *E. rufescens* were recorded in New Guinea in May, October, March, and August. In Australia, *E. rufescens* breeds from November to May, a female has up to three litters in that period, litter size is 1–4, pouch life lasts about 65 days, weaning occurs at about 70 days of age, and sexual maturity comes at 4–5 months (Gordon 2008a). A specimen of *E. rufescens* lived in captivity for 3 years and 1 month (Weigl 2005).

New Guinean spiny bandicoot (*Echymipera kalubu*). Photo by P. A. Woolley and D. Walsh.

George (1979) considered *E. clara* to be threatened in Papua New Guinea because of constant hunting and expanded agricultural activity. The IUCN places neither *E. clara* nor *E. rufescens* in a threatened category, yet states that in New Guinea the latter species is "threatened" by overhunting for meat and by disturbance of forest habitats and that it is uncommon in much of its range and populations are decreasing (Leary et al. 2008h). The IUCN reports that *E. echinista* is still definitely known only by the two specimens used for its original description (Menzies 1990), which were collected in the Fly-Strickland river drainage, but that there is an unconfirmed specimen from Mount Menawi well to the north (Leary et al. 2008i). The IUCN classifies *E. davidi* as endangered, because there almost certainly is a continuing decline in extent and quality of its very restricted habitat due to presence of high numbers of people and extensive agricultural activity (Leary et al. 2008j).

PERAMELEMORPHIA; PERAMELIDAE; **Genus *RHYNCHOMELES*** Thomas, 1920

Seram Island Long-nosed Bandicoot

The single species, *R. prattorum*, is known only by seven specimens from Seram Island, located in the South Moluccas between New Guinea and Sulawesi. Those specimens are in the British Museum of Natural History (Flannery 1995b). The species is named after the collectors, Felix Pratt and his sons Charles and Joseph.

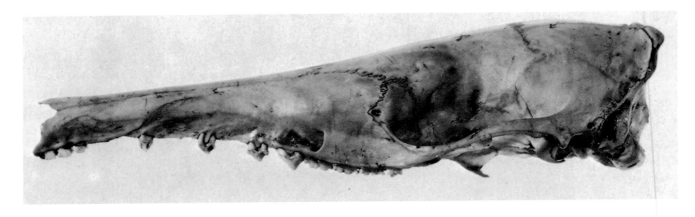

Seram Island long-nosed bandicoot (*Rhynchomeles prattorum*). Photos from British Museum (Natural History).

Head and body length is 245–330 mm, and tail length is 105–130 mm (Flannery 1995b). Weight is probably more than 500 grams (Dickman 2015). The fur is crisp, glossy, and completely nonspinous. There is little underfur. This bandicoot is dark chocolate brown above and below with a patch of white on the chest. The head is somewhat lighter than the back, and there may be a whitish area on the forelimbs. The tail is blackish brown and almost naked.

As the generic name suggests, the muzzle is long because of the elongation of the nasal bones of the skull. The ears are small and oval in shape. As in *Echymipera*, there is no fifth upper incisor. Groves and Flannery (1990) noted that *Rhynchomeles* differs from *Echymipera* in only a few features but is characterized by the extreme length and slenderness of its muzzle, its very small last molar, and its generally well-spaced cheek teeth.

All known specimens were obtained during February 1920 in heavy jungle in very precipitous limestone country on Mount Manusela, at an elevation of 1,800 meters (Flannery 1995b). Archer (1984) stated that

the genus was probably extinct, but Flannery (1995b) doubted that and even suggested it may be locally common. Kitchener et al. (1993) were unable to locate it during an expedition in 1991, but conversations with local villagers indicated it might survive in certain areas of undisturbed montane forest.

The IUCN now classifies *R. prattorum* as endangered, because of continued decline in extent and quality of its potential habitat (Leary et al. 2008k). The lower altitude forests of Seram are being cleared for timber and cultivation. That affects upper montane forest by increasing desiccation and susceptibility to fire. Introduced pigs, rats, dogs, and possibly cats may be jeopardizing *R. prattorum* through both habitat degradation and predation. However, the collection locality is within Manusela National Park and is protected. Surveys are needed to determine whether the species still exists and if it occurs elsewhere on Seram or the island of Buru.

PERAMELEMORPHIA; PERAMELIDAE; **Genus**
MICROPERORYCTES Stein, 1932

New Guinean Mouse Bandicoots

There are two subgenera and five species (Flannery
1995a; Groves and Flannery 1990; Helgen and Flannery 2004a; Ziegler 1977):

subgenus *Microperoryctes* Stein, 1932

> *M. murina*, known only by three specimens taken
> in Weyland Range of western New Guinea;

subgenus *Ornoryctes* Tate and Archbold, 1937

> *M. aplini*, Arfak Mountains of Vogelkop Peninsula
> in western New Guinea;
>
> *M. papuensis*, southeastern Papua New Guinea;
>
> *M. longicauda*, Arfak Mountains of Vogelkop
> Peninsula, Central Cordillera of New Guinea
> from Weyland Range to Star Mountains and
> Strickland River;
>
> *M. ornata*, from Star Mountains in central New
> Guinea, along the eastern Central Cordillera, to
> Mount Simpson in far southeastern Papua.

Microperoryctes was long considered to comprise
only the species *M. murina*, but recent investigations
prompted the transfer of *M. papuensis* and *M. longicauda* to this genus from *Peroryctes* (Flannery 1995a;
Groves 2005d; Groves and Flannery 1990). Kirsch
et al. (1990) agreed that *M. longicauda* is not immediately related to *Peroryctes* but suggested that it has closer
affinity to *Echymipera*. Molecular analysis by Westerman, Springer, and Krajewski (2001) supported a sister
relationship of *M. papuensis* and *M. longicauda*, with
Echymipera being more divergent. Helgen and Flannery
(2004a) recognized *M. ornata* as a species distinct from
M. longicauda and described *M. aplini* from material
previously assigned to *M. murina*; they also suggested
the systematic arrangement shown above. Westerman
et al. (2012) reported that a morphologically and genetically distinct specimen taken from Tembagapura in
the Snow Mountains of West Papua might represent an
undescribed species with close affinity to *M. aplini*.

In *M. murina* head and body length is 152–174 mm,
tail length is 105–111 mm, and weight is probably under

100 grams; the pelage is an unpatterned smoky gray
above and below and is very soft and woolly (reminiscent of many shrews), the scrotum is furry and grayish white, the feet have scattered white hairs, and the
short-haired tail is dark fuscous above and below
and lacks a white tip (Flannery 1995a; Helgen and
Flannery 2004a). In *M. aplini* head and body length is
142–160 mm, tail length is 115–120 mm, and weight is
probably under 100 grams; the upper parts are sleek
and brown, there is a pronounced black dorsal stripe but
no rump stripes, the underparts are gray-brown and
have a creamy white mid-ventral stripe, and the tail is
brown above and white below and has a white tip (Helgen and Flannery 2004a). In *M. papuensis* head and
body length is 175–205 mm, tail length is 135–158 mm,
and weight is 137–184 grams (Aplin and Woolley 1993;
Helgen and Flannery 2004a); the upper parts are dark
with a prominent black middorsal stripe, there are
paired dark lateral rump stripes and dark eye stripes,
the underparts are rich orange buff, and the tail has a
white tip,.

M. longicauda and *M. ornata* do not differ substantially in size and proportions across most of their
largely allopatric ranges, and coloration varies geographically from relatively dull and unpatterned in
the west to a brighter and boldly marked pelage in the
east (Flannery 1995a; Helgen and Flannery 2004a;
Lidicker and Ziegler 1968). Head and body lengths in
large series of specimens are 250–290 mm for *M. longicauda* and 239–303 mm for *M. ornata*; respective tail
lengths are 160–205 mm and 161–258 mm. The overall weight range for the two species is 350–670 grams.
The upper parts are generally pale brown to reddish
brown speckled with black, the underparts are grayish, buff, or white, and there is a white tail tip. The
subspecies *M. longicauda longicauda* of the Vogelkop
Peninsula lacks a dorsal stripe and rump stripes. *M. l.
dorsalis* of the western Central Cordillera sometimes
has a moderately pronounced dorsal stripe but rump
stripes are absent or faint. *M. ornata ornata* and *M. o.
magna* of east-central and southeastern New Guinea
have a prominent black dorsal stripe, paired lateral
rump stripes, and sometimes dark facial stripes. In the
Star Mountains of central New Guinea, where the ranges
of the two species overlap, *M. ornata* occurs in rainforest and moss forest at elevations of about 1,400–2,600
meters, while *M. longicauda* occupies moss forest

New Guinean mouse bandicoot (*Microperoryctes longicauda*). Photo by Harold G. Cogger, from Flannery (1995a).

and subalpine grassland above 3,100 meters. In that region, *M. longicauda* has a much shorter tail and longer ears than *M. ornata*, as well as a more elongate rostrum, thicker pelage, darker dorsal coloration, a less well-defined dorsal stripe, no lateral rump stripes, and gray-based rather than pure white belly fur.

Microperoryctes is characterized by a broadened braincase but a very narrow snout, a highly fenestrated palate with two pairs of vacuities that are shifted forward, straight posterior palatal margins, a deeply wavy coronal suture, small and incomplete auditory bullae lacking anterior spurs, trituberculate molars equal in size to or slightly larger than premolars, granular soles of feet, small ears, a soft and dense pelage, and only three pairs of mammae (Gordon and Hulbert 1989; Groves and Flannery 1990; Helgen and Flannery 2004a).

Helgen and Flannery (2004a) gave the elevation range of both *M. longicauda* and *M. ornata* as approximately 1,000–4,000 meters, except in their zone of sympatry in the Star Mountains (see above). Ziegler (1977) stated that *M. murina* lives in moss forest at an altitude of 1,900–2,500 meters. He observed that its short, soft fur suggests a semifossorial existence. Helgen and Flannery (2004a) thought that its pelage also suggests a lifestyle convergent with that of insectivorous or venomous shrews and shrew-mice. The hind foot seems to be adapted for running in *M. longicauda* and *M. ornata*. Dickman (2015) noted that *M. longicauda* constructs nests of leaves, moss, and twigs in shallow burrows or on the ground among grass tussocks or tree roots; it rests by day in its nest and appears to be active mostly or entirely at night. Dwyer (1983) collected a 170-gram immature male *M. longicauda* in June. According to Flannery (1995a), *M. ornata* inhabits primary forests, has been taken in nests among roots and at the base of a tree, and is primarily insectivorous but will eat some fruit; a female taken in March had four furred young, a female taken in October had three unfurred young, and there is evidence that mothers leave their young in nests while foraging and have a lengthy association with their female offspring. Aplin and Woolley (1993) reported that most specimens of *M. papuensis* have been taken in lower montane forest or associated secondary growth at 1,200–1,450 meters but that the species also has been found at elevations of up to 2,650 meters. Limited data suggest that *M. papuensis* breeds year-round and litter size is one; young have been collected in burrows (Flannery 1995a).

The IUCN does not place any species of *Micropero-ryctes* in a threatened category, but does note (Leary et al. 2008l) that *M. longicauda* is "threatened by hunting for food, especially with dogs" and is "also threatened by predation by feral dogs." Helgen and Flannery (2004a) thought that *M. murina*, given its extremely limited range, may be vulnerable to extinction; it is known by only three specimens collected in 1931 on Gunung Sumuri in the Weyland Mountains.

Order Diprotodontia

Koala, Wombats, Possums, Wallabies, and Kangaroos

This order of 11 Recent families, 40 genera, and 149 species is found throughout Australia, Tasmania, and New Guinea and on many islands of the East Indies from Sulawesi to the Solomons. Although this group was at one time generally considered to form only a marsupial suborder or superfamily, there has long been recognition of its unity and increasing acceptance of its ordinal status (Aplin and Archer 1987; Archer and Hand 2006; Flannery 1989; Hume et al. 1989; Kavanagh et al. 2004; Marshall, Case, and Woodburne 1990; McKenna and Bell 1997; see also sources cited under "Marsupial Superordinal Groupings" in account of class Mammalia). However, Szalay (1994) considered Diprotodontia a semiorder of Syndactyla, which he regarded as also including what are here designated Peramelemorphia and Notoryctemorphia.

Groves (2005e), following mainly the molecular assessment of Kirsch, Lapointe, and Springer (1997), recognized three suborders with the following sequence and content: Vombatiformes, with families Phascolarctidae and Vombatidae; Phalangeriformes, with superfamily Phalangeroidea for families Burramyidae and Phalangeridae, and superfamily Petauroidea for families Pseudocheiridae, Petauridae, Tarsipedidae, and Acrobatidae; and Macropodiformes with families Hypsiprymnodontidae, Potoroidae, and Macropodidae. Van Dyck and Strahan (2008), and their included accounts by Aplin (2008) and Eldridge (2008a), fol-

"Dormouse" possum (*Cercartetus concinnus*). Photo by Gerhard Körtner.

lowed a comparable arrangement but recognized infraorder Phascolarctomorphia for Phascolarctidae and infraorder Vombatomorphia for Vombatidae, used the subordinal name "Phalangerida" in place of "Phalangeriformes," placed Burramyidae in the separate superfamily Burramyoidea, and ranked Macropodiformes as only a superfamily, Macropodoidea, of suborder Phalangerida. A comprehensive molecular analysis by Phillips and Pratt (2008) also accepted suborder Vombatiformes for families Phascolarctidae and Vombatidae and suborder Phalangerida for all other families of Diprotodontia; however, the families of Phalangeroidea, Macropodoidea, and probably Burramyoidea formed a clade, distinct from those of Petauroidea. A similar basic division resulted from a study of the nuclear DNA of Diprotodontia (Meredith, Westerman, and Springer 2009b), which suggests an early Eocene divergence of Petauroidea from a clade comprising Burramyoidea, Phalangeroidea, and Macropodoidea, and then a middle Eocene divergence of Macropodoidea from a clade of Burramyoidea and Phalangeroidea.

Kavanagh et al. (2004) generally supported the foregoing arrangement, including placement of Tarsipedidae within Petauroidea, but indicated that Burramyidae does not belong in Phalangeriformes and might have more affinity to Vombatiformes. Osborne, Christidis, and Norman (2002) suggested that Burramyidae is basal to all other living components of Diprotodontia. Burk and Springer (2000) considered Macropodiformes (or Macropodoidea) limited to two living families, Hypsiprymnodontidae and Macropodidae, with Potoroidae only a subfamily of the latter. McKenna and Bell (1997), did not associate Tarsipedidae with Phalangeriformes, nor Phascolarctidae with Vombatiformes.

Diprotodontia is the largest and most diverse marsupial order, with extensive variation in size and shape, but there are several critical characters common to the group. As the name of the order implies, the dentition is diprotodont; the two middle incisor teeth of the lower jaw are greatly enlarged and project forward. The condition probably evolved as a carnivorous adaptation, the procumbent incisors being used to stab prey. Usually there are no other remaining lower incisors or canine, but if any are present, they are very small, and there is a gap between the incisors and the lower cheek teeth. A second major unifying character is syndactyly; the second and third digits of the hind foot are joined together by integument, are relatively small, and retain claws (see Hall 1987 for a detailed discussion of the condition and its phylogenetic signifi-

Sugar glider (*Petaurus breviceps*). Photo by Gerhard Körtner.

cance). No other marsupial order is both diprotodont and syndactylous. Other diagnostic characters of Diprotodontia include reduction of the upper incisors to three or fewer, selenodont (having crescent-shaped cusps) upper molar teeth, a fasciculus aberrans (connection between the cerebral hemispheres) and large neocortex in the brain, a superficial thymus gland, an expanded squamosal epitympanic wing in the roof of the tympanic bullae, and an unusually complex morphology of the glenoid fossa in the basicranial region of the skull (Aplin 1987; Aplin and Archer 1987; Archer 1984; Archer and Hand 2006). Serological investigations also support the content of the Diprotodontia as set forth herein (Baverstock 1984; Baverstock, Birrell, and Krieg 1987).

The order now is predominantly herbivorous, though some members consume invertebrates and even small vertebrates. The smallest species, *Acrobates pygmaeus*, may weigh less than 15 grams, is adapted for

Eastern gray kangaroo (*Macropus giganteus*), mother and young. Photo by Gerhard Körtner.

gliding like a flying squirrel, and eats nectar, pollen, and insects. The largest living species, the red kangaroo (*Macropus rufus*), weighs up to 100 kg, is structured for a leaping mode of progression, and like many placental bovids is a grazer. The largest extinct species, and the largest marsupial ever to exist, was *Diprotodon optatum*, a member of the suborder Vombatiformes, which had the size and general appearance of a modern hippopotamus; Wroe et al. (2004) estimated its average weight at 2,786 kg, and Price (2008a) concluded that males were substantially larger than females. The largest Australian carnivore was *Thylacoleo carnifex*, also a vombatiform, which may have been comparable to a modern leopard (*Panthera pardus*) in predatory habits (Long et al. 2002); Wroe et al. (2003) estimated its weight at 87–130 kg, and Wroe, McHenry, and Thomason (2005) calculated that its biting power approached that of a modern lion (*Panthera leo*). *Diprotodon* and *Thylacoleo* probably became extinct in the late Pleistocene, perhaps by 30,000 years ago, ap-

parently because of overhunting and environmental disruption by invading humans (see Johnson 2005; Johnson and Prideaux 2004; Turney et al. 2008). There are, however, reports suggesting the possibility that *Thylacoleo* survived in the tropical zone of northern Queensland until the early twentieth century (Helgen and Veatch 2015). Other fossils extend the geological range of Diprotodontia back to the late Oligocene and show that the order was highly diverse even then.

DIPROTODONTIA; Family PHASCOLARCTIDAE; Genus *PHASCOLARCTOS* De Blainville, 1816

Koala

The single living genus and species, *Phascolarctos cinereus*, has a modern range extending from east-central Queensland through eastern New South Wales to Victoria and southeastern South Australia (Martin, Handasyde, and Krockenberger 2008; Ride 1970). During the late Pleistocene the koala also occurred in southern Western Australia, where suitable habitat still seems to exist. The possibility that a population survived in that region into colonial times is suggested by an 1841 report cited by Martin and Handasyde (1999), though the latest fossil record there dates from around 31,000–43,000 years ago (Black et al. 2014). Animals translocated from eastern Australia in 1938, 1948, and 1992, have been used to establish and maintain a breeding colony in Yanchep National Park, near Perth, Western Australia (http://www.everytrail.com/guide/yanchep-national-park, accessed 27 November 2015). Introduced populations also occur on a number of islands and in other places in eastern Australia, beyond the original modern distribution (see below). *Phascolarctos* was often placed in the family Phalangeridae but now is considered to represent a distinct family, apparently the most primitive extant group of Diprotodontia, with some affinity to the Vombatidae (Archer 1984; Archer and Hand 2006; Harding, Carrick, and Shorey 1987; Kirsch 1977b).

Head and body length is 600–850 mm, the tail is vestigial, and weight is 4–15 kg. Males are on average substantially larger than females, and individuals from the southern part of the range are significantly larger than those from the northern part. Martin, Handasyde,

Koala (*Phascolarctos cinereus*). Photo by Pavel German / Wildlife Images.

and Krockenberger (2008) listed average weight in Victoria as 12.0 kg for males, 8.5 kg for females, and in Queensland as 6.5 kg for males, 5.1 kg for females. The dense, woolly fur is relatively short and pale gray in the north and longer and gray-brown in the south (Martin, Handasyde, and Krockenberger 2008). The underparts are paler or whitish, the rump is often dappled, and the ears are fringed with white.

The compact body, large head and nose, and big, hairy ears produce a comical, appealing appearance. Cheek pouches and a 1.8- to 2.5-meter caecum, proportionally the most capacious of any mammal (Martin, Handasyde, and Krockenberger 2008), aid in digesting the bulky, fibrous eucalyptus leaves that form the main diet. The koala develops a characteristic eucalyptus-like odor. Both the forefoot and the hind foot have five digits, all strongly clawed except the first digit of the hind foot, which is short and greatly broadened. The second and third digits of the hind foot are relatively small and partly syndactylous but have separate claws. The first and second digits of the forefoot are opposable to the other three. The palms and soles are granular. Females have two mammae; their marsupium opens in the rear and extends upward and forward.

The skull is massive and flattened on the sides. The tympanic bullae are elongated and flattened from side to side. The dental formula is: i 3/1, c 1/0, pm 1/1, m 4/4=30. The lower incisor and first upper incisor are large, the canine is small, and the molars are blunt and tubercular. Unlike those of wombats, the koala's teeth are not ever-growing; they may be worn flat in old age and unable to break up food adequately for efficient digestion, hence leading to malnutrition and death (Martin and Handasyde 1999).

The koala is confined to eucalyptus forests, which provide it food and shelter. It is largely nocturnal and arboreal, spending up to 20 hours a day resting or sleeping in a low fork, then climbing into the canopy around dusk to feed (Martin, Handasyde, and Krockenberger 2008). It tends to feed in a series of 6–10 bouts averaging 20 minutes each, mostly during the first half of the night. In cold or wet weather it curls into a tight ball to conserve energy (Martin and Handasyde 1999). It apparently does not enter torpor, use a den, or make a nest. It comes to the ground only to move between food trees or to lick up soil or gravel, which serves as a

digestive aid. However, while often viewed as a sluggish animal and usually moving at a sedate pace, its relatively long legs can propel it rapidly over the ground or up the trunk of a tree (Lee and Carrick 1989). The koala seems to prefer the leaves and young bark of about 12 species of *Eucalyptus* ("gum" trees), but also sometimes eats from many other species of eucalyptus and other trees, such as *Acacia*, *Banksia*, and *Tristania* (box). According to Martin and Handasyde (1999), a 10-kg animal eats about 600–800 grams of leaf per day. The food is masticated by the powerful jaws and digested mostly in the small intestine. However, remnant finely crushed food is retained in the huge caecum and proximal colon for hindgut fermentation, a process apparently functioning to retain and recycle nitrogen. Larger leaf particles proceed through the distal colon, where water is reabsorbed. The koala rarely drinks in the wild, obtaining required moisture from its food leaves.

Population density varies with food abundance, having been reported as over 800 koalas per sq km in the fertile Strathbogie Ranges of northeastern Victoria, 70–160 per sq km in the drier Brisbane Ranges of central Victoria, 40 per sq km in the open forests of southeastern Queensland, and about 0.5–1.5 per sq km in a semiarid zone of central Queensland (Martin and Handasyde 1999). Home range size also varies, from as small as 1–2 ha. in high quality habitat to 100 ha. in semiarid country (Martin, Handasyde, and Krockenberger 2008). Mitchell (1990a) found the koala to be usually solitary and to occupy a restricted home range, centered on a few large food trees, for periods of years. Range size averaged 1.70 ha. for adult males and 1.18 ha. for females. There was extensive overlap between male and female ranges, with pairs tending to use the same trees in their common areas. In a study of the introduced koala population on Kangaroo Island, South Australia, Eberhard (1978) found each individual to remain in an area of about 1.0–2.5 ha. containing a few favorite food trees. Ranges of adults were largely separate, but there was some overlap between sexes. Of 943 sightings only 11 percent were of pairs and only 1 percent of three animals. Trespass into an occupied tree led to a savage attack by the resident.

Although antagonistic encounters are common, and males form a dominance hierarchy in high density populations, there is evidently no defense of a substantive

Koala (*Phascolarctos cinereus*). Photo by Alexander F. Meyer at San Diego Zoo.

territory (Martin, Handasyde, and Krockenberger 2008). Gordon, McGreevy, and Lawrie (1990) suggested that the males in any given established population tend to be mostly more than 5 years or less than 2 years of age, with the older group comprising dominant residents and the younger group made up of immature animals that will soon disperse to seek openings in other populations. In contrast, Mitchell and Martin (1990) indicated that young females tend to remain in the vicinity of their mothers and establish adjoining home ranges. Mitchell (1990b) distinguished a series of sounds, most frequently emitted at night, including the bellows of males, wails, and screams; he also observed urine marking as well as rubbing of trees with powerful scent from the male's large sternal gland. Martin and Handasyde (1999) wrote that the male's bellow sounds vaguely like the braying of a donkey and can be heard up to a kilometer away; it is given throughout the year but most often prior to and during the breeding season. Females produce a high-pitched

scream in response to males attempting to mate with them. In a study of captive animals, Bercovitch et al. (2006) concluded that the male's sound and scent advertisements are courtship signals providing indications of age and that female preference is for similarly aged, or slightly older, males. Ellis et al. (2011) had similar findings in a study of the introduced koala population on St. Bees Island off Queensland.

The following natural history data were compiled from Eberhard (1978), Handasyde et al. (1990), Johnston et al. 2000, Martin and Handasyde (1990, 1999), Martin, Handasyde, and Krockenberger (2008), and Martin and Lee (1984). The breeding season is in the Australian spring and summer, with births occurring from October to May, mostly from November to March, though the peak period may vary in different populations. Females are seasonally polyestrous, with a non-mated estrous cycle of about 35 days and an estrus of about 10 days (if inclusive of a luteal phase, the estrous cycle lasts about 50 days). They usually breed once every year but loss of a pouch young is followed by further ovulations and sometimes another birth. The gestation period is variable but averages about 35 days, and the usual litter size is 1, though twins have been recorded regularly. The young weighs as little as 0.36 grams at birth, has a pouch life of 5–7 months, and is weaned at 6–12 months. Toward the end of pouch life, for a period of up to 6 weeks, the young feeds regularly on material passed through the mother's digestive tract. It continues to associate with the mother, traveling around on her back, until fully weaned. Both sexes may attain sexual maturity at 2 years but may not begin to breed that soon. Full physical maturity is reached during or after the fourth year in females and the fifth year in males, and the latter probably do little mating before that time. Wild females have bred when as old as 13 years and commonly live to an age of 13–18 years; longevity in wild males is usually shorter. Menkhorst (2008a) reported that a few tagged and translocated animals in Victoria are known to have lived for over 20 years. Weigl (2005) listed a captive male that lived to 22 years and 1 month and a captive female that lived for more than 21 years and 8 months.

At one time, there were probably millions of koalas in Australia, and populations may even have grown in response to the great reduction of Aboriginal people, and of their hunting and burning practices, which followed European settlement. Despite grievous losses from forest fires and epidemics, koala numbers remained high in the late nineteenth century. Subsequent clearing of wooded habitat and increased access to the koala's range, combined with a demand for its beautiful, warm, durable fur, resulted in a precipitous decline (Eberhard 1978; Martin and Handasyde 1999; Melzer et al. 2000; Ride 1970). The commercial kill increased until in 1924 more than 2 million koala skins were exported. By the end of that year the species appeared to have been extirpated in South Australia and nearly wiped out in Victoria and New South Wales. There still was a huge population in Queensland, but in 1927 the state government, yielding to pressure from economic interests, allowed an open season, and nearly 600,000 more pelts were exported. Consequent public outcry both in Australia and abroad led to permanent bans on commercial hunting. Some accounts suggest that only a few thousand scattered animals then survived throughout Australia, though compilations by

Gordon and Hrdina (2005) and Hrdina and Gordon (2004) indicate that populations remained high and even expanded in parts of Queensland in the 1920s and 1930s. Meanwhile, protective measures had been initiated, with koalas introduced to several small islands off eastern Queensland (Brampton, Magnetic, St. Bees) and southern Victoria (French, Phillip, Raymond, Snake) and the larger Kangaroo Island, Eyre Peninsula, and Adelaide Hills area of southeastern South Australia (Woinarski, Burbidge, and Harrison 2014). The resulting island populations began growing to high levels (Martin and Handasyde 1999).

Since the 1920s, intensive conservation efforts, including the breeding and transplantation of thousands of individuals by the Victoria Fisheries and Wildlife Division, have allowed partial recovery of the species in some parts of southeastern Australia. Menkhorst (2008a) reported that beginning in the late nineteenth century, populations were established on six large coastal islands and several smaller islands of Victoria, and that since 1923 more than 24,000 of the animals have been translocated to about 250 release sites in the state. However, even though the koala is now legally protected from direct killing by people, its habitat is being reduced and fragmented by agriculture, settlement, road construction, fires, logging, and massive cutting of forests for production of woodchips (mostly for export to Japanese paper mills). More than half of the tall to medium-sized trees, on which the koala depends for survival, have already been destroyed (Phillips 1990). Habitat destruction is now the most significant threat to the koala, especially in Queensland and New South Wales (Melzer et al. 2000). Another problem is the widespread prevalence of the bacterial pathogen *Chlamydia*, which infects the reproductive tract of females, reducing fertility and thus restricting the size of some populations, but it is not considered an overriding threat to the survival of the species as a whole (Martin and Handasyde 1990, 1999).

There is much controversy regarding the precise status of this highly popular mammal and what conservation measures should be taken (Clark et al. 2000; Cork, Clark, and Mazur 2000a, 2000b). Estimates of numbers remaining in the wild range from about 40,000 to more than 400,000 (Black et al. 2014; Melzer et al. 2000; Payne 1995; Thompson 1995). What is perhaps the most precise recent assessment indicates

Koala (*Phascolarctos cinereus*), mother and young. Photo by Benjamint / Shutterstock.com.

approximately 300,000 mature individuals across Australia and that there has been a decline of about 29 percent since 1990 (Woinarski, Burbidge, and Harrison 2014). As indicated above, relatively high koala densities have been reported in some areas. Based on such figures, it has been suggested that as many as 200,000 koalas might be present in the Strathbogie Ranges alone (Martin and Handasyde 1999). However, Phillips (2000) argued that those numbers were extrapolated from studies at sites not random or stratified and relatively close to one another and that alternative methodology produces an estimate of only 2,000–18,000 koalas in the Strathbogies. Regardless, it seems likely that with natural predators extirpated and human hunting ended in parts of southeastern Aus-

tralia, the koala will increase to the point at which it will deplete its food supply, devastate its ecosystem, and become subject to mass starvation; this already has occurred on some of the islands where it was introduced (Martin and Handasyde 1999). The estimated 28,000 koalas now on Kangaroo Island are highly inbred, all being descended from 10 animals brought there in the 1920s, and have caused severe environmental damage (Gordon et al. 2008). There is also concern that the small number of individuals that were used to found the island populations off Victoria may have resulted in low genetic diversity and increased vulnerability of the current mainland populations of the state, which are descended in large part from animals translocated from the islands (Menkhorst 2008a).

Koala (*Phascolarctos cinereus*). Photo by Nattanan726 / Shutterstock.com.

About a quarter century ago, the Australasian Marsupial and Monotreme Specialist Group of the IUCN Species Survival Commission designated the koala as "potentially vulnerable" and indicated that the species had lost 50–90 percent of its habitat (Kennedy 1992). The Australian Department of the Environment, Water, Heritage, and the Arts used the same figure and referred to the koala as "near threatened" (Maxwell, Burbidge, and Morris 1996). Gordon, Hrdina, and Patterson (2005) calculated the loss in Queensland at only 31 percent and did not consider the species threatened. However, Martin and Handasyde (1999) wrote: "The current situation in Queensland is alarming. More than 500,000 hectares of forest is still being cleared each year, most of it for agricultural pursuits of dubious economic benefit."

Lee et al. (2010) reported that habitat destruction and fragmentation, together with mortality caused by dogs, motor vehicles, and disease, had led to sharp recent declines in koala numbers in southeastern Queensland, and that the species was further threatened by resulting restriction of gene flow in the area. Black et al. (2014) cautioned that the entire family Phascolarctidae has been undergoing a decline in distribution and diversity since the Miocene and that the group is particularly sensitive to climate change. Predicted hotter and drier conditions in Australia may result in further range contractions eastward and southward into areas where the koala is currently threatened by the effects of urbanization. Climate change has already been identified as a factor exacerbating a drastic reduction of koala populations in part of southeastern New South Wales (Lunney et al. 2014). Although the recent decline in koala numbers across Australia has sometimes been questioned and sometimes attributed simply to severe but temporary drought, Woinarski, Burbidge, and Harrison (2014) stated that evidence for overall reduction is compelling and that many factors, aside from drought, are involved.

The IUCN, at least in its latest published assessment, no longer places the koala in a threatened category, "in view of its wide distribution, presumed large population, and because it is unlikely to be declining at nearly the rate required to qualify for listing" (Gordon et al. 2008). However, the USDI has classified the koala as "threatened" (Clark 2000), a measure demonstrating the foresightedness and now seldom implemented potential of the U.S. Endangered Species Act of 1973. The first stated purpose of the Act is "to provide a means whereby the ecosystems upon which endangered species and threatened species depend may be conserved." The Act defines a "threatened species" as one "likely to become an endangered species within the foreseeable future throughout all or a significant portion of its range."

The USDI did not initiate the measure but acted in response to a 1994 petition by nongovernment Australian and American conservation groups. The content of the petition was such that the USDI was obligated by law to issue a finding that the petition had "presented substantial information indicating that the requested action may be warranted." Having done so, the USDI had to evaluate the status of the koala pursuant to the mandates of the Endangered Species Act, as stated above. Accordingly, a proposal to classify the koala as threatened was issued by the USDI in 1998. The proposal was opposed by all commenting Australian federal and state authorities, who pointed out that the koala was common and well protected. And yet the most telling points in the USDI's case for classification (Clark 2000) were taken from official Australian government documents, specifically data showing that 54 to 63 percent of the koala's forest habitat had already been destroyed and that the process was continuing, even intensifying. Notwithstanding reports of high and even overabundant populations in some areas, the USDI was legally required to give precedence to the long-term prospects for the koala's ecosystem and had no choice but to declare the species threatened. Remarkably, 12 years later an assessment by Australian authorities, utilizing IUCN category standards, recognized the koala as "vulnerable," a classification expressing a degree of concern at least as great as that of "threatened" by the USDI (Woinarski, Burbidge, and Harrison 2014).

Except for the populations introduced to islands in modern times, the family Phascolarctidae seems al-

ways to have been restricted to the mainland of Australia. It was represented in the late Oligocene by eight known genera, in the early to middle Miocene by six, in the Pliocene by three, and in the middle Pleistocene by two. By the late Pleistocene, only the single genus *Phascolarctos* was present, and it contained three species, including *P. yorkensis*, which was three times the size of *P. cinereus*; only the latter species survived into the Recent (Archer and Hand 2006; Black, Archer, and Hand 2012; Black et al. 2014; Long et al. 2002; Price and Hocknull 2011). *P. cinereus* evidently is not a derived dwarfed form of a giant Pleistocene koala but has always been a distinct and smaller species (Price 2008b).

DIPROTODONTIA; Family VOMBATIDAE

Wombats

This family of two Recent genera and three species is known from eastern and southern Australia, Tasmania, and islands of Bass Strait. The sequence of genera presented here follows that suggested by Archer and Hand (2006) and Murray (1998). The name "Phascolomyidae" has sometimes been used for this family.

Wombats resemble small bears in general appearance. The body is thick and heavy; adult head and body length is about 700–1,200 mm and weight is 15–35 kg. The muzzle is naked and the pelage is coarse in *Vombatus*, but in *Lasiorhinus* the muzzle is haired and the pelage is soft. Underfur is almost lacking. The eyes are small. The limbs are short, equal in length or nearly so, and extremely strong. There are five digits on all the limbs, and all have claws except the first toe, which is vestigial. The second and third digits of the hind foot are partly united by skin. The marsupium opens posteriorly and encloses a single pair of mammae. Wombats have traces of cheek pouches. They also have a group of glands of unusual structure inside the stomach, which may be associated with digestion of a special type of plant food. The skull is massive and flattened.

The dentition is remarkably similar to that of rodents. Wombats also resemble rodents in their manner of feeding and in the rapid side-to-side movements of their jaws. All the teeth of modern wombats are

Left, common wombat (*Vombatus ursinus*); photo by Nicholas Toh/Shutterstock.com. *Right,* hairy-nosed wombat (*Lasiorhinus latifrons*); photo by Pavel German/Wildlife Images.

rootless and ever-growing, which compensates for wear. The two rootless incisors in each jaw are large and strong, have enamel on their front and lateral surfaces only, and are separated by a wide space from the cheek teeth. The premolars are small, single-lobed, and close to the molars; the molar teeth have two lobes and are rather high-crowned. The dental formula is: i 1/1, c 0/0, pm 1/1, m 4/4 = 24. According to Archer and Hand (2006), wombats are the only animals known to deliberately reshape their teeth to suit their diet. While still in the pouch, the young grind their molars together, wearing away the crowns and carving transverse ridges that can efficiently process grass.

Wombats are shy, timid, and difficult to observe in the wild; they are sometimes active during the day but are considered nocturnal. They live in burrows and are rapid, powerful diggers. Under certain conditions they may construct burrow systems more than 30 meters in length. Wombats dig with their front feet, thrusting the soil out with the hind feet, and use their strong incisors to cut such obstructions as roots. The burrow entrance normally is a low arch that fits the body. A grass or bark nest is located near the end of the burrow. A shallow resting place is usually excavated against a tree or log near the mouth of the burrow as a site for sunbathing.

Paths often lead from burrows to feeding areas. Wombats eat mainly grasses, roots, bark, and fungi.

Wombats do well in captivity and often become interesting and affectionate pets. They have suffered serious reduction in numbers and range. People are their chief enemy. Colonies have been exterminated near settled areas because of damage to crops and because domestic livestock may injure a leg by breaking through into the burrows. Wombats also have been destroyed in the campaign against rabbits, as these introduced pests often shelter in wombat burrows. There are many regions in Australia where these unusual mammals could reside without disturbing human developments. Adequate protection would probably enable wombats to maintain stable populations, but it is surprising how long it has taken to gain any detailed understanding of their biology and conservation needs. Wells (1998) noted that the first attempts at in-depth studies in the wild did not begin until the 1970s and that wombats remain the most enigmatic of Australian marsupials.

The geological range of this family is late Oligocene to Recent in Australia (Murray 1998). *Phascolonus gigas*, a giant wombat that lived from the Pliocene to late Pleistocene, would have been about 1.6–1.8 meters long, a meter tall, and up to 250 kg in weight (Long et al. 2002; Murray 1998). Although *P. gigas* is sometimes considered a burrower, like modern wombats, an analysis by Woolnough and Steele (2001) suggested that the energetic costs of constructing and maintaining a burrow for such a large mammal would have outweighed its need to range extensively for grazing. C. N. Johnson (1998) thought the same, pointing out that the modern wombats' extraordinary combination of large size, herbivory, and burrowing occurs in no other radiation of mammals and would have been very unlikely in the still larger wombats of the past.

DIPROTODONTIA; VOMBATIDAE; **Genus**
VOMBATUS E. Geoffroy St.-Hilaire, 1803

Common, or Coarse-haired, Wombat

The single species, *V. ursinus*, originally occupied extreme southeastern Queensland, eastern New South Wales, eastern and southern Victoria, extreme southeastern South Australia, Tasmania, and the islands of Bass Strait (McIlroy 2008; Ride 1970). The mainland populations have sometimes been designated a separate species with the name *V. hirsutus*.

Head and body length is 700–1,200 mm, tail length is about 25 mm, and adult weight is about 15 to 35 kg. General coloration is yellowish buff, silver gray, light gray, dark brown, or black. In *Vombatus* the fur is coarse and harsh, the nose is naked, the ears are short and somewhat rounded, and the nasal bones are relatively narrow. In the other living wombat genus, *Lasiorhinus*, the fur is soft and silky, the nose is finely haired, the ears are longer and relatively pointed, and the nasal bones are broader. In both genera the teeth are rootless, that is, they grow continuously from pulpy bases. Wombats are the only living marsupials that possess two rootless incisors in each jaw, an arrangement also seen in rodents. In the lesser curvature of the stomach is a peculiar gland patch that is also found

Common wombat (*Vombatus ursinus*). Photo by Marco Tomasini / Shutterstock.com.

in the koala (*Phascolarctos*) and the beaver (*Castor*). The pouch of the female opens posteriorly and contains two mammae.

According to McIlroy (2008), in Queensland and northern New South Wales the common wombat lives only in sclerophyll forest above 600 meters; at night it moves into clearings and forest edges to graze. Farther south it also occurs at lower elevations and in more open country, such as woodland, coastal scrub, and heathland. It is one of the few species of marsupials active above the snowline during winter. Some of its burrows, 2–5 meters long, are used mainly for temporary refuge, but some are eventually extended up to 20 meters for diurnal shelter. They often branch into a complex network of tunnels with several entrances and nesting chambers lined with vegetation. An individual visits 1–6 burrows each night and up to 20 over several weeks. In one study area, Evans (2008) found density of active burrows, 0.25 per ha., to far exceed estimated density of wombats, 0.13 per ha. The animals there typically spent 1–4 days sleeping in the same burrow and then moved to another; on average, each active burrow was used by 2.2 different individuals.

The common wombat is primarily nocturnal except for occasional bouts of basking in the sun (Collins 1973). It is quick in its movements and can run rapidly, at least for short distances. When touched, especially near the

Common wombat (*Vombatus ursinus*). Photo by Gerhard Körtner.

hindquarters, it kicks backward with both hind feet, and when annoyed, it may emit a hissing growl. Often, however, it becomes a playful and affectionate pet. Individuals formerly occurring on King Island in Bass Strait were reportedly domesticated by fishermen. Those animals, in a reversal of normal routine, would feed in forests during the day and return in the evening to the cabins that served as their retreat. *Vombatus* eats mainly fibrous native grasses, apparently preferring fresh seed stems, but also takes sedges, rushes, roots, and fungi. It uses its forefeet to tear and grasp pieces of vegetation. It has also been seen foraging among sea refuse along the shore.

Population densities of 0.3–0.6 individuals per ha. occur in native forests containing, or adjoining, open grassy areas, and can reach 1.9 per ha. in pastoral areas. Home ranges measure 2–82 ha. and contain a number of burrow systems. Each animal generally maintains a discrete feeding area, but in zones of high density ranges overlap extensively, and individuals use one another's burrows, either at separate times or simultaneously (McIlroy 2008). Evans (2008) reported ranges to be almost circular and to average 17.7 ha., with a core area averaging 2.9 ha. Collins (1973) wrote that the common wombat is solitary except during the breeding season and that keeping more than one individual in an enclosure often results in fighting and injury. He added, however, that pairs and groups of compatible individuals have been maintained together successfully. Ride (1970) cited a study of a wild population in Victoria in which usually only a single wombat occupied each burrow, but individuals were sociable and would visit one another's burrows. Wells (1989) indicated that territories are maintained by rubbing scent on logs and branches and by depositing feces along trails. There are a number of vocalizations, the most common being a harsh cough.

Vombatus is polyestrous and may breed at any time of the year (McIlroy 2008). In Tasmania births were found to occur throughout the year, though about 48 percent took place from October to January (Green and Rainbird 1987). The estrous cycle in three captive females averaged 47.2 days, with a range of 35–60 days (West et al. 2004). One captive female exhibited estrous periods of 24–81 hours (Böer 1998). Gestation lasts approximately 21 days, and the young are fully weaned at 12 months (Wells 1989). A single young is the usual case, but twins are known to occur. The young remains in the pouch for 6–10 months, stays with the mother for another 11 months, and attains sexual maturity after 2 years (McIlroy 2008). A genetic assessment by Banks, Skerratt, and Taylor (2002) provided strong evidence that dispersal of the young is predominantly by females, a rare finding among mammalian species. *Vombatus* appears capable of long life in captivity; record longevity is about 30 years (Weigl 2005).

Cooke (1998) concluded that the introduced European rabbit (*Oryctolagus*), which eats many of the same grasses as *Vombatus*, displaced the latter from much of its range in South Australia. Roger, Laffan, and Ramp (2007) found the genus much less abundant in southern New South Wales than usually thought; many burrows had been abandoned, and population size in the study area was about the same as average annual mortality from road kill. The common wombat has also declined as a result of human persecution, and its distribution has become fragmented, though some mainland and Tasmanian populations probably are secure in their mountain forest habitat (Ride 1970). By the late nineteenth century, however, the species had been exterminated from all islands of Bass Strait except Flinders Island, where it still occurs. The IUCN formerly classified the subspecies there, *V. ursinus ursinus*, as vulnerable, but no longer assigns any threatened classification to *Vombatus* as a whole (Taggart, Martin, and Menkhorst 2008). *V. ursinus ursinus* was designated "near threatened" by Woinarski, Burbidge, and Harrison (2014) because of its restricted range and relatively low numbers (estimated at 4,000 individuals in 1996).

DIPROTODONTIA; VOMBATIDAE; Genus *LASIORHINUS* Gray, 1863

Hairy-nosed, or Soft-furred, Wombats

There are two species (Horsup 2004; Horsup and Johnson 2008; Kirsch and Calaby 1977; Ride 1970; Woinarski, Burbidge, and Harrison 2014):

L. latifrons, originally from southeastern Western Australia to southwestern New South Wales and western Victoria;

L. krefftii, historically known only from Epping Forest in east-central Queensland, along the Moonie River near Saint George in southeastern Queensland, and near Jerilderie and Deniliquin in south-central New South Wales.

L. krefftii was first described from a Pleistocene fossil found in the Wellington Caves of east-central New South Wales; other fossils were subsequently found in southwestern New South Wales, Victoria, and Queensland (Horsup and Johnson 2008). The historical populations in east-central Queensland and south-central New South Wales sometimes have been considered a distinct species with the name *L. barnardi*, and the historical population in southeastern Queensland sometimes has been called *L. gillespiei* (Ride 1970). Kirsch and Calaby (1977) included *barnardi* and *gillespiei* in *L. krefftii*. Groves (2005e) accepted that arrangement with question, noting that the historical populations might not be referable to *L. krefftii*, in which case the latter species would be restricted to the Pleistocene.

Head and body length is 840–1,110 mm, tail length is 25–60 mm, and weight is 19–36 kg (Horsup and Johnson 2008; Taggart and Temple-Smith 2008). Females are about equal in size to or slightly larger than males. The upper parts are usually dappled with gray and black or brown; the underparts are gray, except that the cheeks, neck, and chest are often white. The specific name *latifrons* refers to the great width of the anterior part of the skull. Externally, *Lasiorhinus* can be differentiated from *Vombatus* by its haired nose and relatively longer, more pointed ears. The fur in *Lasiorhinus* is soft and silky, not coarse and harsh as in *Vombatus*. The marsupial pouch is well developed and opens to the rear; there are two mammae.

These wombats occupy relatively dry country—savannah woodland, grassland, and steppe with low shrubs. They construct complex tunnel systems consisting of a large number of separate burrows that join together to form a warren; the entrances are connected above ground by a network of trails, and additional trails radiate out to feeding areas and other warrens (Ride 1970). Burrows are elliptical in shape and 0.6–3.5 meters in depth. A warren may have 1–80 entrances. It may be used by generations of wombats (Taggart and Temple-Smith 2008). Warrens examined by Shimmin,

Hairy-nosed wombat (*Lasiorhinus latifrons*). Photo by Pavel German / Wildlife Images.

Skinner, and Baudinette (2002) consisted of systems up to 89 meters long, with deep tunnel temperatures showing a daily range of less than 1° C, while surface temperatures varied up to 24° C. The burrows of one colony in the early twentieth century covered an area 800 meters long and about 80 meters wide. Smaller, single-entrance burrows for temporary refuge may be scattered around the major systems (Horsup and Johnson 2008).

While these wombats may appear to be slow and bumbling and to have poor eyesight, Taggart and Temple-Smith (2008) noted they have good hearing and a keen sense of smell and can run as fast as 40 km/hr for short distances. They are primarily nocturnal, resting by day in their burrows, but in the winter and spring, especially on warm days, they sometimes emerge in the late afternoon to bask or graze nearby. They feed mainly on native perennial grasses and forbs. In a field and laboratory study, Wells and Green (1998) determined that *L. latifrons* rarely drinks, obtaining necessary water from its diet. *Lasiorhinus* is said to be more docile in captivity than *Vombatus*.

Wells (1978) found density of a population of *L. latifrons* in South Australia to be 1 animal per 4.8 ha., while McGregor and Wells (1998) calculated that 33,871 animals occupied an area of 3,530 sq km. Taggart and Temple-Smith (2008) observed that, for herbivorous animals of their size, hairy-nosed wombats have exceptionally small home ranges, around 1.3–4.8 ha. The ranges are centered on preferred warrens, are similar in size for both sexes, and overlap substantially, at least intersexually. Several individuals share each warren but they rarely share burrows, which may be defended and marked near the entrance with dung and urine. Johnson and Crossman (1991) reported individual *L. krefftii* of the same sex to keep to separate ranges when leaving the burrow system to feed. They also found that, in contrast to the situation in *Phascolarctos*

Hairy-nosed wombat (*Lasiorhinus latifrons*), mother and young. Photo by Jiri Lochman / Lochman LT.

(see account thereof) and most other mammals that have been studied, dispersal from the natal colony was common for young females but rare for young males.

Taggart and Temple-Smith (2008) stated that the breeding season of *L. latifrons* falls between July and December and varies in length from year to year depending on rainfall and pasture growth; most young are born from mid-August to October. Estrus lasts around 36 days, and females appear to be monogamous, though may mate again and produce a second litter in the same season if pouch young are lost early in lactation. Crowcroft (1977) found gestation for two births of *L. latifrons* to be 20–21 and 21–22 days. There usually is one young, but twins have been reported. The young first leave the pouch at approximately 8 months, vacate it permanently at 9 months, are weaned at 12 months, may reach full size by 2 years, and are usually sexually mature at 3 years (Wells 1989). A wild-born female *L. krefftii* was captured at an age of about 5 years and

lived for another 25 years, and a captive-born male *L. latifrons* was still living after 29 years and 11 months (Weigl 2005).

Hairy-nosed wombats declined in numbers and distribution following persecution by European settlers. Ride (1970) expressed serious concern about the steadily contracting range of *L. latifrons* in South Australia. Aitken (1971b), however, thought the species was still abundant over large areas, though density had been reduced in the late nineteenth century, probably because of competition from introduced rabbits; human take was considered negligible but there was some killing based on the unfounded belief that the wombat damaged fences, competed with livestock for grazing, and dug burrows that allowed rabbits (considered pests) to find shelter. St. John (1998) reported that about 1,000 wombats are destroyed each year under permit in South Australia. According to the IUCN (Taggart and Robinson 2008), estimated numbers are

50,000–100,000 on the Nullarbor Plain in southwestern South Australia and 10,000–15,000 in the Murray Lands to the southeast, though the population in the latter area has declined by about 70 percent since 2002, probably because of drought and sarcoptic mange. Smaller and highly fragmented groups survive on the Yorke and Eyre Peninsulas. Outside of South Australia, *L. latifrons* is known to survive only at a few sites in Western Australia, New South Wales, and Victoria (Woinarski, Burbidge, and Harrison 2014). The IUCN does not currently place *L. latifrons* in a category of concern, but Woinarski, Burbidge, and Harrison (2014), using IUCN criteria, designated the species as near threatened.

The status of *L. krefftii* is worse (Horsup 1998, 2004; Horsup and Johnson 2008; Ride 1970). The population in southeastern Queensland, sometimes called *L. gillespiei* (and here treated as subspecies *L. krefftii gillespiei*), and the once plentiful population in south-central New South Wales disappeared by 1908 because of habitat disruption. The existing population in east-central Queensland (here treated as subspecies *L. krefftii barnardi*) was first scientifically documented in 1937 and has subsequently declined, mainly because of competition with cattle for forage, especially during drought. A second population in the same vicinity may have existed until the late 1930s. In 1971 the area was designated Epping Forest National Park but habitat continued to deteriorate, as cattle were allowed in the park. The population fell to 30–40 individuals by 1982, when cattle were finally excluded. Numbers subsequently increased to 115, including perhaps 25 breeding females. The population occupies only about a sixth of the 3,300-ha. park. After 15–20 wombats were lost to dingo (*Canis familiaris dingo*) predation in 2000–2001, a 20-km-long dingo-proof fence was built around the entire area inhabited by *L. krefftii*. A small population recently was reintroduced near St. George in southern Queensland (Woinarski, Burbidge, and Harrison 2014). Loss of genetic diversity is a concern, though an assessment by Taylor, Sherwin, and Wayne (1994) found the population to remain genetically viable. *L. krefftii* is classified as critically endangered by the IUCN (Taggart, Martin, and Horsup 2008) and as endangered by the USDI and is on appendix 1 of the CITES. Tisdell and Nantha (2007) raised the legitimate question of why conservation of *L. krefftii* receives only about a third of the funding allotted to *Phascolarctos cinereus* in Queensland, even though the former is by far the more seriously threatened species.

DIPROTODONTIA; Family BURRAMYIDAE

Pygmy Possums

This family of two Recent genera and five species is found in Australia, Tasmania, and New Guinea. The members of the Burramyidae sometimes have been placed in the family Phalangeridae but now are considered to represent a distinct group (Kirsch 1977b). The sequence of genera presented here follows that of Kirsch and Calaby (1977). The genera *Distoechurus* and *Acrobates* were also placed in the Burramyidae when that family was first resurrected in the 1970s, but were later moved to a new family, the Acrobatidae (Aplin and Archer 1987; Archer 1984). Although Westerman, Sinclair, and Woolley (1984) found the karyotype of *Distoechurus* to be very similar to that of the burramyids, Aplin (2008) explained that the two true burramyid genera have no particularly close relatives among the other members of suborder Phalageriformes (Phalangerida). The separate position of Burramyidae has been supported by comprehensive molecular analyses (Meredith, Westerman, and Springer 2009b; Phillips and Pratt 2008).

Head and body length is 50–120 mm, tail length is 70–160 mm, and weight is 6–82 grams. The pelage is soft. Digital structure is much like that of the Phalangeridae, but neither genus has opposable digits on the

"Dormouse" possum (*Cercartetus nanus*). Photo by Gerhard Körtner.

forefoot. The dental formula is: i 3/2, c 1/0, pm 2–3/3, m 3–4/3–4=34–40. Archer (1984) stated that the Burramyidae are characterized by reduction of the first upper and second lower premolars, a bicuspid tip on the third lower premolar, and a well-developed posteromesial expansion of the tympanic wing of the alisphenoid. Aplin (2008) noted that the molar teeth of all species have low, smooth cusps, and that the tail is long, slender, lightly haired, and strongly prehensile. Both genera are good climbers.

According to Aplin (2008), pygmy possums occupy a wide range of environments, including cool and semiarid heath and savannah, alpine regions, temperate wet sclerophyll forest, and tropical rainforest. *Burramys* is terrestrial, and the four species of *Cercartetus* are arboreal. They are primarily insectivorous, but *Cercartetus nanus* includes much nectar in its diet, *C. lepidus* takes small lizards, and *Burramys parvus* eats hard seeds. All species construct nests, usually in a tree hollow, but that of *B. parvus* is on the ground, typically among boulders, and in winter lies under snow. All species can enter torpor, and *B. parvus* hibernates for up to seven months. The females of most species have four mammae; those of *C. concinnus* have six. The pouch opens anteriorly and is least well developed in *C. nanus*.

The known geological range of the Burramyidae is late Oligocene to Recent on the mainland of Australia; a possible specimen has been reported from the early Miocene of Tasmania, where otherwise the family is known only from the late Pleistocene to Recent (Archer and Hand 2006; Harris 2009b). In New Guinea the family is known only from the Recent (Flannery 1995a).

DIPROTODONTIA; BURRAMYIDAE; **Genus *CERCARTETUS*** Gloger, 1841

"Dormouse" Possums

There are four species (Carthew and Cadzow 2008; Carthew, Cadzow, and Foulkes 2008; Flannery 1995a; Haffenden and Atherton 2008; Kirsch and Calaby

"Dormouse" possum (*Cercartetus nanus*). Photo by Jiri Lochman/Lochman LT.

1977; Osborne and Christidis 2002a; Ride 1970; Ward and Turner 2008):

> *C. caudatus*, Central Cordillera of New Guinea from Vogelkop Peninsula to extreme southeast, northeastern Queensland between Townsville and Cooktown;
>
> *C. lepidus*, southeastern South Australia, northwestern Victoria, Kangaroo Island, Tasmania;
>
> *C. nanus*, extreme southeastern Queensland, eastern New South Wales, Victoria, extreme southeastern South Australia, King and Flinders Islands, Tasmania;
>
> *C. concinnus*, southern Western Australia, southern South Australia, western Victoria, southwestern New South Wales, Kangaroo Island.

Molecular and morphological evidence indicates that *C. caudatus* is the most primitive species, *C. lepidus* was next to diverge, and *C. nanus* and *C. concinnus* have a sister relationship (Osborne and Christidis 2002a). *C. caudatus* and *C. lepidus* have sometimes been placed in a separate genus, *Eudromicia* Mjöberg, 1916. Groves (2005e), but not Flannery (1995a, 1995b), recorded *C. caudatus* from Fergusson Island off the southeastern coast of Papua New Guinea. Except for fossil and subfossil remains on the mainland, *C. lepidus* long was known only from Tasmania. In 1964, however, it was found alive on Kangaroo Island, and in the 1970s living specimens were collected on the mainland (Aitken 1977; Barritt 1978; Dixon 1978; Wakefield 1970a). *C. nanus* was also initially discovered in Tasmania, in 1802, and then, in 1896, reported as a subfossil in New South Wales; presence of a natural, living population on the mainland was not confirmed until 1904 (Harris 2006). It has been suggested that *macrurus*, the Australian subspecies of *C. caudatus*, may be a separate species, and that the mainland and Tasmanian populations of *C. lepidus* may be distinct species (Carthew and Cadzow 2008; Flannery 1995a; Osborne and Christidis 2002a). Analysis of the mitochondrial DNA of *C. concinnus* indicated no taxonomic differentiation across the range of the species (Pestell et al. 2008).

In *C. concinnus* head and body length is 64–106 mm, tail length is 53–101 mm, weight is 8–21 grams, the upper parts are fawn, reddish brown, or gray, and the underparts and feet are white (Carthew, Cadzow, and

Foulkes 2008). In *C. nanus* head and body length is 70–110 mm, tail length is 75–105 mm, weight is 15–43 grams, the upper parts are grayish or fawn-colored, and the underparts are whitish (Ward and Turner 2008). In *C. lepidus*, smallest of all possums, head and body length is 50–73 mm, tail length is 60–75 mm, weight is 6–10 grams, the upper parts are pale fawn or gray, and the underparts are gray (Carthew and Cadzow 2008; Harris 2009b). In *C. caudatus* head and body length is 86–109 mm, tail length is 128–156 mm, weight is 17–40 grams, the upper parts are brownish gray, and the underparts are pale gray (Flannery 1995a; Haffenden and Atherton 2008). *C. caudatus* also features well-marked, broad black bands passing from the nose through the eyes, not quite reaching the ears. The pelage is dense and soft. The prehensile, cylindrical tail is well furred at the base and scantily haired for the remainder of its length. With the approach of winter *C. nanus* becomes very fat and its tail becomes greatly enlarged, especially at the base, so that stored food will be available during periods of torpor. Geiser (2007) reported that captive *C. nanus*, provided a surplus of food, attained an average weight of 53.8 grams but then dropped to 20.9 grams after a lengthy period of induced torpor. Perhaps the same process is reflected in the above weight ranges for the different species.

The common name, "dormouse" possum, reflects the superficial resemblance of this marsupial genus to the dormouse (*Glis*) so well known in Europe. In addition, both genera are nocturnal and undergo torpor in cold weather. "Dormouse" possums have large, thin, almost naked ears. The forefoot looks somewhat like a human hand, and the great toe of the hind foot is thumblike and widely opposable. The claws are small, but the pad of each digit is expanded into two lobes. The pouch is well defined in females and opens anteriorly. The normal number of mammae is four except in *C. concinnus*, which has six (Collins 1973). *C. nanus* usually has four functional and two nonfunctional teats (Ward 1990a).

Habitats include rainforest and eucalyptus forest for *C. caudatus*, mallee scrub to wet and dry sclerophyll forest for *C. lepidus*, rainforest and sclerophyll forest for *C. nanus*, and mallee heath and dry sclerophyll forest for *C. concinnus* (Archer and Hand 2006; Osborne and Christidis 2002a). *C. caudatus* also is common in undisturbed subalpine grassland above 3,000 meters

"Dormouse" possum (*Cercartetus concinnus*). Photo by Jiri Lochman / Lochman LT.

throughout the Central Cordillera of New Guinea (Flannery 1995a). All species construct a small, spherical or dome-shaped nest of grass, leaves, or bark in a hollow limb, hollow stump, crevice of a tree trunk, or thick clump of vegetation. They sometimes reside in abandoned birds' nests. *C. nanus* usually makes a nest of soft bark and may travel as much as 500 meters to secure the desired kind of bark. Its nest is about 6 cm in diameter, and each individual may use several nests (Ward and Turner 2008). The nest of *C. caudatus* may be up to 15 cm across (Flannery 1995a).

These tiny, arboreal possums hang by the tail when reaching from one branch to another and also skillfully use the tail when climbing or descending. They are nocturnal, foraging through trees and bushes and leaping and running freely at night. They generally sleep soundly during the day but sometimes become active in cloudy weather. Morrant and Petit (2012) reported the mean linear distance traveled per night by radio-tracked *C. concinnus* to be 969 meters when *Eucalyp-*

tus rugosa, the preferred food source, was flowering (October–March), and 297 meters when it was not (April–August); some individuals moved up to 4.7 km in a single night in search of resources. Harris et al. (2007) found average overnight distance to be 44 meters for one male *C. nanus* and 19 meters for one female.

Hickman and Hickman (1960) found *C. nanus* and *C. lepidus* to undergo alternate periods of activity and dormancy throughout the year in Tasmania. Neither experienced prolonged hibernation; the longest period of torpor was 6 days for *C. lepidus* and 12 days for *C. nanus*. Body temperature during dormancy was about equal to air temperature. Experimental inducement of deep hibernation in *C. nanus* for periods of up to 35 days was achieved by Geiser (1993). Subsequently, Geiser (2007) experimentally induced hibernation in five individuals of *C. nanus* for periods averaging 310 days and up to 367 days, though there were numerous intervals of brief arousal, when water was available but

"Dormouse" possum (*Cercartetus caudatus*). Photo by Pavel German/ Wildlife Images.

no food; all of the animals survived. *C. concinnus* regularly enters short-term torpor during inclement weather (Carhew, Cadzow, and Foulkes 2008); its longest recorded bout of torpor is 11 days (Harris 2009a). In a study in South Australia during winter, Turner et al. (2012) found *C. concinnus* to use torpor on 63 percent of days; individuals employed both short and prolonged torpor bouts, depending on temperature.

All species seem to be omnivorous, feeding on leaves, nectar, pollen, fruit, nuts, insects and their larvae, spiders, scorpions, and small lizards. One captive specimen of *C. nanus*, however, was a strict vegetarian, refusing to eat meat or insects, and lived 8 years (Perrers 1965). In a study during a particularly dry year in South Australia, Morrant and Petit (2012) found *C. concinnus* to feed almost exclusively on nectar and pollen, but noted that insectivory would likely increase in times of greater rainfall.

Recorded population densities of *C. nanus* are 14–20 individuals per ha. in coastal banksias woodlands, 3.1 per ha. in heathlands, and 2.5 per ha. in rainforest, but densities probably are lower elsewhere (Ward and Turner 2008). Home ranges in that species were found by Laidlaw and Wilson (1996) to average 0.71 ha., to be about the same for both sexes, and to not be exclusive. However, Harris et al. (2007) reported home range to average 0.85 ha. for one male *C. nanus* and 0.19 ha. for one female. The ranges of individual *C. concinnus* radio-tracked for at least 5 nights varied from 0.4 to 300.5 ha.; the larger areas were used when the favorite food, *Eucalyptus rugosa*, was flowering, perhaps re-

flecting increased energy provided by the nectar (Morrant and Petit 2012). *Cercartetus* generally seems to be solitary, though has been maintained in pairs in captivity (Collins 1973); three males and one female were found hibernating together, and other small groups have been observed nesting or feeding together. The vocalization of *C. concinnus* has been described as a rapid chattering "chi-chi-chi" (Harris 2009a). *C. lepidus* may make a hissing noise when alarmed or threatened (Harris 2009b).

Captive *C. caudatus* have bred twice yearly, in January and February and from late August to early November (Atherton and Haffenden 1982). A wide range of birth dates has been recorded for *C. caudatus* in New Guinea, but there is a breeding recession during the dry season from mid-May to early August (Dwyer 1977). *C. lepidus* gives birth throughout the year in South Australia and Victoria, mainly from late winter to spring, but only from spring to summer in Tasmania (Harris 2009b). Births of *C. nanus* occur all year on the mainland, mostly from late spring to early autumn (November to March), but in Tasmania breeding appears confined to late winter and spring. Most female *C. nanus* in Victoria produce two litters per season and some have three within a 12-month period, but one per season is typical in northern New South Wales (Ward 1990a, 1990c; Ward and Turner 2008).

C. concinnus can breed any time of year and can have up to three consecutive litters of six young in ideal conditions, though normally only 2–4 young are reared at a time (Carthew, Cadzow, and Foulkes 2008). Minimum gestation in *C. concinnus* is 51 days (Hayssen, Van Tienhoven, and Van Tienhoven 1993). Young of that species leave the pouch when around 25 days old, before their eyes open, but then remain in the nest until they are weaned at about 50 days of age. Female *C. concinnus* are thought to reach sexual maturity at 12–15 months (Harris 2009a). Litters of *C. caudatus* have 1–4 young, which leave the pouch at 45 days of age, became independent at 92 days, and first breed at 15 months (Atherton and Haffenden 1982; Haffenden and Atherton 2008). *C. lepidus* has up to 4 young, which emerge after about 42 days and then are either left in the nest while the mother forages or travel clinging to her fur; they become independent at about 90 days (Harris 2009b). In *C. nanus* there normally are 4 young, which remain in the pouch about 30 days and subsequently

"Dormouse" Possum (*Cercartetus nanus*), mother and young. Photo by Gerhard Körtner.

are left in a nest. They are weaned when 65 days old and become independent immediately thereafter but may continue to associate with one another and other juveniles; sexual maturity comes as early as 4.5 to 5 months (Ward 1990a; Ward and Turner 2008). *C. nanus* appears to do well in captivity (Collins 1973); one female lived to an age of about 10 years and 4 months (Weigl 2005). Maximum longevity in the wild is at least 4 years (Ward 1990a).

Female *Cercartetus* may return to estrus in late lactation, mate, and give birth to a litter immediately after the previous one is weaned. It has been suggested that a period of embryonic diapause (see account of *Macropus*) subsequently occurs. Apparently such is not the case, but only about 2 days are needed between the weaning of one litter and giving birth and commencing to suckle the next; in that short period the mammary secretion changes from milk to colostrum and the enlarged teats regress to a size small enough to be taken into the mouth of a neonate (Harris 2009a, 2009b; Ward and Turner 2008).

The subspecies *C. caudatus macrurus* of Queensland was formerly designated near threatened by the IUCN. Currently, the IUCN reports no major threats to any subspecies or species of *Cercartetus*, but does note that populations of *C. nanus* are decreasing because of clearing and grazing of its habitat (Dickman, Lunney, and Menkhorst 2008). Surveys have indicated *C. nanus* to be rare and threatened in New South Wales, Victoria, and possibly throughout its range (Bowen and Goldingay 2000; Harris and Goldingay 2005). Harris (2009a, 2009b) expressed concern for both *C. nanus* and *C. lepidus*; the latter species declined historically through loss of habitat to clearing and also underwent

Mountain pygmy possum (*Burramys parvus*). Photo by Linda S. Broome.

substantial range contraction since the late Pleistocene, when it occurred as far as eastern Victoria and southern New South Wales.

DIPROTODONTIA; BURRAMYIDAE; **Genus** *BURRAMYS* Broom, 1896

Mountain Pygmy Possum

During the period 1966–2010, the single species, *B. parvus*, came to be known from three living, isolated populations: one at Mount Buller, one between Mount Bogong and Mount Higginbotham (including the Bogong High Plains) in northeastern Victoria, and one in the southern part of Kosciuszko National Park in extreme southeastern New South Wales (Broome 2008; Broome et al. 2013). Molecular clock estimates suggest those three populations diverged from one another in the mid-Pleistocene and have been isolated ever since (Heinze, Broome, and Mansergh 2004). In 2010 and 2011, three new populations were discovered farther north in Kosciuszko National Park—at Happy Jacks Valley, the Rough Creek headwaters, and Snow Ridge (Broome et al. 2012; Schulz, Wilks, and Broome 2012).

Prior to 1966, the species had been known only from late Pleistocene cranial and dental fossils found at the Wombeyan Caves, near Goulburn in New South Wales, and described in 1895. The specimens were controversial, some scientists thinking they represented a kind of miniature kangaroo. In the early 1950s they were examined by W. D. L. Ride, who showed that *Burramys* actually was a small possum related to *Cercartetus*. In August 1966 a live possum of an unknown kind was found in a ski hut at Mount Hotham (an area that includes Mount Higginbotham), Victoria. Upon examination it was identified as *B. parvus*. In the words of Ride (1970:16), "The dream dreamed by every paleontologist had come true. The dry bones of the fossil had come together and were covered with sinews, flesh and skin."

From 1970 to 1972, three more specimens were taken at Mount Hotham, one was trapped in the Falls Creek area of the Bogong High Plains, and nineteen were collected in Kosciuszko National Park (Dimpel and Calaby 1972; Dixon 1971). Subsequently, many hundreds were live-trapped in the course of field investigations (Mansergh and Scotts 1990). Meanwhile, more fossils of *B. parvus* were found, at Buchan Caves in Victoria and Jenolan Caves in New South Wales (Broome 2008; Mansergh and Broome 1994). That material, like the original specimens, dates from 25,000–30,000 years ago. The fossils indicate that the distribution of *B. parvus* has been declining with the receding snowline since the end of the last Ice Age. It is likely that *B. parvus* once occupied a still larger region, including Tasmania (Harris and Goldingay 2005; Mansergh and Broome 1994). At least three extinct species of *Burramys* are known from the late Oligocene to Pliocene of the eastern Australian mainland (Archer and Hand 2006). It has been suggested that assessment of their paleoecology could help in the conservation of existing *B. parvus*, which, unfortunately, has apparently declined rapidly in the last 15 years and could be approaching true extinction in association with predicted climate change (Broome et al. 2012). The remainder of this account was compiled primarily from Broome (2008), Heinze, Broome, and Mansergh (2004), and Mansergh and Broome (1994); other literary sources are noted.

Head and body length is 100–120 mm, tail length is 130–160 mm, and weight is 30–82 grams. Males and females are about equal in size. The upper parts are gray-brown, sometimes darker from the mid-dorsal area to the top of the head, and there is a dark ring around each eye. The underparts are paler gray-brown to cream, but with increasing age the ventral fur and flanks become bright fawn-orange, especially in males during the breeding season. The tail is thin, prehensile, scaly, and almost naked, except for about 1 cm at the base where the body fur continues. The feet are pink with sparse white hairs. The pouch is well developed, opens lengthwise, and contains two pairs of mammae. The upper and lower third premolar teeth are greatly enlarged and have crowns that are rounded in profile and have serrated edges; they are used for cutting and for cracking open hard seeds and exoskeletons of arthropods.

The mountain pygmy possum is the only Australian mammal limited to alpine and subalpine country, where there is continuous snow cover for up to 6 months. Until 2010, all recorded specimens had been taken in mountainous areas at elevations of about 1,300–2,228 meters. Habitat then was reported to consist primarily of fields of large boulders and associated patches of shrubby heathland. The boulders ameliorate temperature extremes and provide deep hibernacula and sheltered nesting sites. Areas of suitable habitat range from less than 1 to more than 5 ha. and often are over 1 km apart, but there apparently is substantial movement between them, especially by males and dispersing juveniles. Specimens collected at Happy Jacks Valley in northern Kosciuszko in 2010 were taken at elevations of 1,200–1,310 meters at sites modified by humans and lacking typical boulder field characteristics and associated alpine heath (Schulz, Wilks, and Broome 2012).

Burramys is primarily terrestrial. Individuals released under controlled conditions ran rapidly on the ground and over rocks in a ratlike manner and did not hesitate to dive into holes. They climbed shrubs rapidly by grasping the stems with the forefeet and hind feet (Calaby, Dimpel, and Cowan 1971). In captivity, at least, individuals use their large premolars to cut grass, from which they construct a nest over a depression in the ground. Only a single nest has been discovered in the wild; it was among granite boulders, consisted of moss, was spherical and 15–20 cm in diameter, and contained a mother and four young (Heinze and Olejniczak 2000).

Most activity is nocturnal. Radio-tracked males frequently have traveled 1–3 km per night, and even females with large pouch young have moved up to 1 km from nest sites to the highest peaks, where the Bogong moth (*Agrotis infusa*) is most abundant. That insect migrates in vast numbers to the high country in the warmer time of year (late October to mid-April), sheltering and aestivating in the boulder fields. Along with other arthropods, it is then heavily exploited by *B. parvus*. In late summer and autumn the seeds and fruits of heathland shrubs become increasingly important in the possum's diet.

This is the only marsupial definitely known to undergo long periods of winter hibernation under natural conditions (though hibernation may also occur to

Mountain pygmy possum (*Burramys parvus*) eating a Bogong moth. Photo by Jiri Lochman / Lochman LT.

some extent in the related *Cercartetus* and in the New World *Thylamys* and *Dromiciops*). The climate in Kosciuszko National Park is severe, with an average annual precipitation of about 125–200 cm, much of it winter snow. Temperatures under the snow in winter are 0–2° C. After breeding, the possums fatten, about doubling their weight before winter snowfall. Adult females enter hibernation as early as February at Mount Higginbotham but as late as April in New South Wales. Juveniles and males are active for 1 or 2 more months. Hibernation lasts up to 7 months in adults, 5–6 months in juveniles. As shown by studies of captives, there are prolonged bouts of deep torpor lasting up to 20 days, during which body temperature is regulated at around 2° C, interspersed by periods of less than a day of normal temperature. Studies in the wild by Körtner and Geiser (1998), using temperature-sensitive radio transmitters, showed that hibernation was interrupted by periodic arousals, during which both males and females were often active and left their hibernacula for up to 5 days.

According to Broome et al. (2012), population viability is closely associated with snow, and recent declines appear partly related to diminishing snow cover and early snow melt. A good cover of snow (more than 100 cm) provides insulation during the coldest part of the winter and promotes prolonged torpor in hibernating possums. Shallow cover lowers temperature in hibernacula and leads to more frequent arousals, thus depleting energy stores and decreasing winter survival. Early spring melt may leave the possums with little food during the critical early spring period of final arousal from hibernation and the beginning of breeding.

Burramys, at least in captivity, is the only marsupial known to make caches of durable foods. Storage usually is not in the nest but under leaf litter around the cage. Whether wild populations depend on such activity during winter hibernation and arousal is not known. Körtner and Geiser (1996) suggested that food caches or foraging during that period are probably used only when an animal's body fat reserves run short.

In the highly productive basalt boulder heaths at Mount Higginbotham, population density is as high as 94 individuals per ha.; adult females there are sedentary and have overlapping home ranges averaging 0.06 ha. In the granodiorite boulder heaths of Kosciuszko and Mount Buller, densities are 8–28 per ha., and home ranges are 0.2–7.7 ha. During the breeding season, males expand their ranges into female habitats. Otherwise, sexual segregation is evident in all populations. After mating, aggression from resident females apparently forces males to disperse into less favorable habitat at lower elevations. Such behavior presumably reduces competition for resources and enhances development and prehibernation fattening of juveniles. Male dispersal may also be motivated by need for additional food or warmer hibernacula, which allow earlier emergence for spring mating. Juvenile males disperse, like their fathers, into suboptimal habitat, while most juvenile females remain in their natal habitat, and some hibernate together with their mothers. Nonbreeding captives are tolerant of one another, thus suggesting that huddling, another winter survival strategy, occurs in the wild.

Burramys is polygynous, with no lasting pair bonds and males mating with multiple females. However, unlike most such arrangements in mammals, mating apparently is dominated by matriarchal lineages of females that permit temporary presence of males. Females are polyestrous, with an overall reproductive period of 35–40 days and an estrous cycle of about 20 days; hence, if one litter is lost, a second can be produced. The short alpine summer requires rapid development of young and usually just one litter can be successfully reared. Mating begins as early as September but depends on the time of spring snowmelt and may vary by site. Gestation lasts 14–16 days. In the wild, births usually occur from October to early December. Litters contain up to nine young, each about 1 cm long, but only four actually survive to be carried in the pouch. They emerge after 4–5 weeks, then spend another 4–5 weeks in the nest, the mother returning to suckle them. Their eyes open at 5–6 weeks, and they are fully weaned and independent at 9–10 weeks, when each weighs about 16 grams. They can breed the following spring, when about a year old, but do not attain adult weight until the end of their second summer. In the wild, females have been known to live up to 12 years, males only up to 5 years.

Mountain pygmy possum (*Burramys parvus*) in torpor. Photo by Fritz Geiser through Linda S. Broome.

The mountain pygmy possum is listed as endangered by the USDI and as critically endangered by the IUCN (Menkhorst, Broome, and Driessen 2008). The species occurs in several genetically isolated populations separated by low-lying river valleys that do not provide suitable habitat. The total area of habitat actually occupied by the populations at Mount Buller, Mount Bogong–Mount Higginbotham, and southern Kosciuszko National Park is less than 6 sq km, and total numbers there in 2011 were approximately 1,655 adult females and 620 adult males; there are an estimated 350 more animals in the recently discovered populations farther north in Kosciuszko and another 80 in a captive breeding colony at Healesville Sanctuary (Woinarski, Burbidge, and Harrison 2014). According to Menkhorst, Broome, and Driessen (2008), the extremely restricted habitat has been fragmented or destroyed by construction of roads, dams, and aqueducts, as well as by environmental modifications associated with development of ski resorts. About 50 percent of the habitat on Mount Bogong and Mount Higginbotham and 20 percent of the Kosciuszko habitat was burned in bushfires in January 2003. Predation by the house cat (*Felis catus*) and red fox (*Vulpes vulpes*) is also a threat; the former was introduced to Australia in the early nineteenth century, partly in an effort to control harmful rodents, and feral populations thereof had become established throughout the continent by the late nineteenth century (Abbott 2002, 2008b), while the fox was introduced in Victoria in the mid-nineteenth century for sporting purposes (Abbott 2011). Global climate warming is predicted to lessen snow cover, thereby promoting invasion of predators and competitors and, more significantly, reducing insulation necessary for thermoregulation and sustained hibernation. A recent trend of second litters following early snowmelt has been observed at Mount Buller; neither second litters nor their mothers are known to survive the winter, as they are unable to accumulate the fat reserves necessary for successful hibernation. Only 40 individuals now remain at Mount Buller, down from 300 in 1996 (Woinarski, Burbidge, and Harrison 2014). The associated loss of genetic diversity in that population is the most rapid ever documented in a mammal (Mitrovski et al. 2008).

DIPROTODONTIA; Family PHALANGERIDAE

Possums and Cuscuses

This family of 6 living genera and 26 species occurs in Australia, Tasmania, and New Guinea and on islands from Sulawesi to the Solomons. The family sometimes has been considered much larger and to include all the genera that here, in accordance with Aplin and Archer (1987) and Kirsch and Calaby (1977), are allocated to the families Phascolarctidae, Burramyidae, Pseudocheiridae, Petauridae, Tarsipedidae, and Acrobatidae. The sequence of genera presented here follows that of Flannery, Archer, and Maynes (1987), who recognized two subfamilies: the basal Ailuropinae, with the single genus *Ailurops*, and Phalangerinae, with the tribe Trichosurini for the genera *Strigocuscus* and *Trichosurus* and the tribe Phalangerini for *Spilocuscus* and *Phalanger*. *Wyulda*, considered only a subgenus of *Trichosurus* by Flannery, Archer, and Maynes (1987), has been given generic rank by most recent authorities, including Groves (2005e), who otherwise accepted the same division of the family. Assessment of the morphology of the periotic bone (Crosby and Norris 2003) reinforced the case for a monophyletic tribe Trichosurini consisting of the three genera indicated above. However, analyses of mitochondrial (Ruedas and Morales 2005) and nuclear (Raterman et al. 2006) DNA supported removing *Strigocuscus* from Trichosurini and placing it in Ailuropinae. Those analyses suggested elevating

Cuscus (*Phalanger orientalis*), adult male. Photo by Thomas E. Heinsohn.

Trichosurini (restricted to *Trichosurus* and *Wyulda*) to the level of a subfamily, Trichosurinae, and treating it, rather than Ailuropinae, as the most primitive phalangerid grouping. A DNA-hybridization experiment (Kirsch and Wolman 2001) had also indicated a basal position for Trichosurinae and had reduced Ailuropinae to tribal level within Phalangerinae. An analysis of the nuclear DNA of Diprotodontia (Meredith, Westerman, and Springer 2009a) supported the affinity of *Strigocuscus* and *Ailurops*.

Head and body length is 320–650 mm and tail length is 240–610 mm. The soft, dense pelage is woolly in all genera except *Wyulda*. According to Aplin (2008), phalangerids have a rather short face with eyes directed forward, the rhinarium is prominent and moist, the ears vary from relatively large in brushtailed possums to small and hidden by fur in many cuscuses, and the tail is relatively long and strongly prehensile, though there is considerable variation in its morphology. The

Brushtailed Possum (*Trichosurus vulpecula*). Photo by Pavel German / Wildlife Images.

well-developed marsupium opens to the front; there are two or four mammae.

All phalangerids are excellent climbers, usually moving slowly and deliberately (Aplin 2008). All four limbs have five digits, and all the digits except the first toe of the hind foot have strong claws; this first toe is clawless but opposable and provides for a firm grip on branches. The second and third digits of the hind foot are partly syndactylous, being united by skin at the top joint, but the nails are divided and serve as hair combs. The largest digits of the hind foot are the fourth and fifth. In the genera other than *Trichosurus* and *Wyulda* the first and second digits of the forefoot are opposable to the other three.

Flannery, Archer, and Maynes (1987) listed the following diagnostic characters of the Phalangeridae: the mastoid wing on the rear face of the cranium is reduced in size and the rear face of the cranium above the mastoid is composed of an extension of the squamosal; al-

most the entire plantar surface of the hind foot is covered by a large striated pad; the distal portion of the tail is nearly naked, and at least part is covered in scales or tubercles; and the upper second premolar is greatly reduced or lost but the upper first premolar is large. The dental formula of the family is: i 3/1–2, c 1/0–1, pm 2–3/3, m 5/5=40–46 (Archer 1984). The first incisors are long and stout. The second and third incisors, if present, are vestigial, as are the lower canine and some of the premolars. The molars have sharp cutting edges. The skull is broad and flattened.

The known geological range of this family is late Oligocene to Recent in Australia (Archer 1984; Crosby et al. 2004) and Pleistocene to Recent in New Guinea (Flannery 1995a).

DIPROTODONTIA; PHALANGERIDAE; Genus *AILUROPS* Wagler, 1830

Large Sulawesi, or Bear, Cuscus

Most recent authorities (e.g., Aplin 2008; Corbet and Hill 1992; Flannery 1995a, 1995b; Heinsohn 2001; Kirsch and Wolman 2001; Raterman et al. 2006; Ruedas and Morales 2005) recognize a single species, *A. ursinus*, which occurs on Sulawesi and the nearby islands of Muna, Butung (Buton), Peleng, Malenge (Togian Islands), and Lembeh. A single preserved specimen is also known from Lirung on Salebabu (Talaud Islands), about 300 km northeast of the northeastern tip of Sulawesi, and it usually is thought to represent a subspecies, *A. ursinus melanotis*. However, Groves (2005e) listed *melanotis* as a separate species, observing that measurements of the type "fall strongly outside those from Sulawesi and offshore islands, and the distinctive colouration and patterning of the type is seen in a second (living) specimen, whose photograph I examined." In contrast, Flannery (1995b) noted the possibility that the population on Salebabu was introduced by prehistoric humans (see also Heinsohn 2004a). Riley (2002) reported a captive specimen on Sangihe Island, midway between Sulawesi and Salebabu, and indicated that it represented a natural population on the island; he stated that the specimen's "ash-grey coat resembles skins of the subspecies *A. ursinus melanotis* from Salebabu" and that "the Sangihe population is

Large Sulawesi, or bear, cuscus (*Ailurops ursinus*). Photo by Thomas E. Heinsohn.

610 mm, a tail length of 580 mm, and a weight of 10 kg. The upper parts vary greatly in color, being black, grayish, or brown, and the underparts are usually whitish. The limbs are relatively long, the feet and claws very large, the rostrum unusually short and broad, the rhinarium large and naked, the canine tooth short, and the third upper molar large. George (1987) noted that the pelage consists of fine but wiry underfur and coarse, bristly guard hairs and that the ears are short and well furred internally. Flannery (1995b) pointed out that the pupil of the eye is round, not ovoid or catlike as in all other phalangerids.

Flannery, Archer, and Maynes (1987) listed 10 cranial characters that distinguish *Ailurops* from all other phalangerids. In the others, the mastoid and ectotympanic are usually broadly continuous, but in *Ailurops* those bones are separated by a deep and continuous groove. In the others the entire basicranial region, particularly the squamosal and mastoid, are much more pneumatized than in *Ailurops*. In the others the mastoid is restricted to a thin ventral band on the posterior face of the cranium, but in *Ailurops* the mastoid has an extensive wing on the rear face of the cranium. The third upper incisor of the others is reduced in size, but in Ailurops this tooth is larger than the second upper incisor. The first upper premolar is double-rooted in the others but single-rooted in *Ailurops*.

Although this cuscus is a distinctive and relatively large animal, up to 1,200 mm in total length, and has the most northwesterly distribution of any Old World marsupial, little is known of its natural history. Available information was provided by Dwiyahreni et al. (1999), Flannery (1995a), and Heinsohn (2001). *Ailurops* occurs mainly in rainforests and has been recorded from sea level to 2,400 meters, but seems most common at around 400 meters. It is arboreal, living in the upper canopy and crossing gaps between trees by bracing itself with its tail and hind feet or hanging from its tail, then reaching out with extended forepaws to pull in an adjacent branch. As suggested by the shape of its pupil, it is apparently largely diurnal, not nocturnal like all other phalangerids, but it reportedly does move by night, sometimes on the ground. It has been found to spend about 63 percent of the day resting and only 5.6 percent feeding. Its diet seems to consist mostly of leaves, supplemented by fruit. It is most often seen in pairs but sometimes alone or in groups of two or three

probably best classified with this subspecies." There have been suggestions that *A. u. furvus* of the highlands of Sulawesi (Flannery 1995b) and *A. u. tongianus* of Malenge Island (Flannery, Archer, and Maynes 1987) also might be separate species. Helgen and Jackson (2015) recognized both *melanotis* and *furvus* as full species. *Ailurops* long was treated as a synonym of *Phalanger*, but Flannery, Archer, and Maynes (1987) considered it a highly distinctive genus and representative of its own subfamily. Based on morphology, they regarded *Ailurops* as the most primitive genus in Phalangeridae, but that view has been challenged by subsequent studies of DNA (see account of family Phalangeridae).

According to Tate (1945), one specimen had a head and body length of 564 mm and a tail length of 542 mm. Miller and Hollister (1922) indicated that weight is about 7 kg. For another specimen, a male, Flannery and Schouten (1994) listed a head and body length of

Large Sulawesi, or bear, cuscus (*Ailurops ursinus*). Photo by C. Smith / Mammal Images Library, American Society of Mammalogists.

animals. In 1979 it was estimated to occur at a density of one pair per 4 ha. in the lowland rainforest at Tangkoko-DuaSudara Nature Reserve on the northeastern tip of Sulawesi. Frequently, a single dependent juvenile is seen riding on its mother's back, its tail wrapped around the base of hers. Weigl (2005) listed a captive female that lived 5 years and 11 months.

The IUCN treats *A. melanotis* as a separate species and classifies it as critically endangered (Flannery and Helgen 2008). Surveys in the late 1990s did not locate it on Salebabu Island and indicated that on Sangihe it depends on a remnant block of 4,268 ha. of primary forest, which continues to decrease. The IUCN designates *A. ursinus* as vulnerable, because of widespread habitat loss to logging and agriculture (Salas et al. 2008a). A survey in the Tangkoko-DuaSudara Nature Reserve in 1993–94 found that population density had fallen to only one pair per 100 ha. (O'Brien and Kinnaird 1996). Both species reportedly are heavily hunted by people for food.

DIPROTODONTIA; PHALANGERIDAE; **Genus** *STRIGOCUSCUS* Gray, 1862

Dwarf Cuscuses

There are two species (Corbet and Hill 1992; Flannery 1995b; Flannery and Schouten 1994):

> *S. celebensis*, Sulawesi and the islands of Sangihe and Siau to the northeast and Muna to the southeast;
>
> *S. pelengensis*, Peleng, Taliabu, and Mangole Islands to the east of Sulawesi.

Strigocuscus was at one time commonly placed in *Phalanger*, but Flannery, Archer, and Maynes (1987) regarded the two as generically distinct. Those authorities also tentatively considered *Strigocuscus* to include the species here designated *Phalanger ornatus*, *Phalanger gymnotis*, and *Phalanger mimicus* (see Corbet and Hill 1992; Flannery 1995b; Flannery and Schouten 1994;

Dwarf cuscus (*Strigocuscus celebensis*). Photo by Chien Lee / Minden Pictures.

George 1987; Groves 2005e; Kirsch and Wolman 2001; Menzies and Pernetta 1986; Raterman et al. 2006; Springer et al. 1990). Menzies and Pernetta (1986) regarded *S. pelengensis* as a subspecies of *S. celebensis*, and Corbet and Hill (1992) and Flannery, Archer, and Maynes (1987) included it in the genus *Phalanger*, but the species was reassigned to *Strigocuscus* by Flannery and Schouten (1994). A comprehensive analysis of the nuclear DNA of Diprotodontia indicated that *Strigocuscus*, as delineated above, is paraphyletic, with *S. pelengensis* actually showing closer affinity to *Phalanger* and *Spilocuscus* than to *S. celebensis* (Meredith, Westerman, and Springer 2009a). Helgen and Jackson (2015) did place *S. pelengensis* in the genus *Phalanger*, but also recognized *Strigocuscus sangirensis* of Sangihe and Siau islands as a species distinct from *S. celebensis*.

In *S. celebensis* head and body length is 294–380 mm, and tail length is 270–373 mm (Groves 1987b). Weight is about 500–1,000 grams (Helgen and Jackson 2015). In *S. pelengensis* head and body length is 350–370 mm, tail length is 245–300 mm, and weight is 1,070–1,150 grams (Flannery and Schouten 1994). *S. celebensis* is pale buff in color, while *S. pelengensis* is clear tawny brown to reddish above and paler brown or yellowish below.

According to Flannery, Archer, and Maynes (1987), characters of the skull indicate that *Strigocuscus* is more closely related to *Trichosurus* and *Wyulda* than to *Phalanger*. In *Strigocuscus*, *Trichosurus*, and *Wyulda* the rostrum is relatively narrower than in other phalangerids, the lachrymal is retracted from the face, the ectotympanic is almost totally excluded from the anterior face of the postglenoid process, and the third upper premolar tooth is set at a more oblique angle relative to the molar row than it is in other phalangerids. George (1987) indicated that *Strigocuscus* is characterized by the very large size of the third upper premolar, a widening of the zygomatic arches at the orbits, and short paroccipital processes. Flannery (1995b) noted that *Strigocuscus* also shows affinity to *Trichosurus* by the lack of a dorsal stripe and the presence of hairs on the portion of the tail that is usually naked in *Phalanger*.

The little available information about *Strigocuscus* was summarized by Flannery (1995b) and in accounts of the IUCN (Helgen et al. 2008a; Leary et al. 2008m), which classify *S. celebensis* as vulnerable but regard *S. pelengensis* as abundant, despite the latter's more restricted range. Dwarf cuscuses are arboreal and found in moist, lowland forests, sometimes in secondary forests, coconut plantations, and gardens. *S. celebensis* reportedly is nocturnal and frugivorous and usually occurs in pairs. It has declined sharply because of hunting by humans for food and destruction of its habitat by logging and agriculture. *S. pelengensis* has an elevational range from sea level to over 1,000 meters. It evidently breeds throughout the year and normally rears a single young; individuals have been noted sleeping in the crowns of coconut palms. It is not threatened by hunting for food, as its range is inhabited by Muslim people, whose religion proscribes eating animals of this kind. Weigl (2005) listed a captive female *S. celebensis* that lived 3 years and 5 months.

Brushtailed possum (*Trichosurus vulpecula*). Photo by Chris and Mathilde Stuart.

DIPROTODONTIA; PHALANGERIDAE; **Genus**
TRICHOSURUS Lesson, 1828

Brushtailed Possums

There are three species (Flannery and Schouten 1994; Kerle and How 2008; Kerle, McKay, and Sharman 1991; Lindenmayer, Dubach, and Viggers 2002; Ride 1970; Viggers and Lindenmayer 2004; Woinarski, Burbidge, and Harrison 2014):

> *T. vulpecula*, originally throughout mainland Australia and Tasmania and on many nearby islands including Barrow, Bathurst, Melville, Groote Eylandt, the Sir Edward Pellew Group, Magnetic, Prudhoe, Kangaroo, Thistle, King, and the Furneaux Group;
> *T. caninus*, southeastern Queensland, eastern New South Wales south to near Sydney;

> *T. cunninghami*, mountainous districts of southeastern New South Wales and central and eastern Victoria.

Flannery, Archer, and Maynes (1987) considered *Wyulda* (see account thereof) a subgenus of *Trichosurus*. *T. arnhemensis*, found in northern parts of Western Australia and Northern Territory and on Barrow, Bathurst, and Melville Islands, long was considered a separate species and was maintained as such by Flannery and Schouten (1994) and Groves (2005e). However, morphological, karyological, electrophoretic, and ecological studies by Kerle, McKay, and Sharman (1991) indicate that it is best treated as a subspecies of *T. vulpecula*; that position was followed by Kerle and How (2008) and Woinarski (2004). *T. johnstonii*, found between Koomboolomba and Kuranda on the Atherton Tableland in northeastern Queensland, also was recognized as a separate species by Flannery and Schouten (1994), Groves (2005e), and Helgen and Jackson (2015), but not by Kerle and How (2008) or Winter et al. (2004).

Head and body length is 320–580 mm, tail length is 240–400 mm, and weight is 1.2–5.0 kg. The coloration of *T. vulpecula* is extremely variable, with gradations within the main color phases of gray, brown, black and white. In the extensive natural range of that species there is considerable geographic variation (Kerle and How 2008; Ride 1970; Wayne et al. 2005a). It is often silver gray above and white to pale gray below, has a brushy tail and long oval ears, and weighs 2–3 kg. Males of any given population are usually on average somewhat heavier than females. The subspecies *fuliginosus* of Tasmania is much larger than the other races, very woolly, and very commonly black. The subspecies *johnstonii* of the Atherton Tableland is smaller and copper colored.

The western subspecies *hypoleucus* is also smaller (males average about 1.6 kg, females about 1.5 kg) and has longer fur. The subspecies *arnhemensis* of the northern tropics is gray and short haired, has a scrawny tail, and weighs as little as 1 kg. *T. vulpecula* also exhibits sexual dimorphism in color; adult males usually blend to reddish across the shoulders. *T. caninus* generally weighs 3–4 kg and is steely gray above and whitish below, though occasionally black; the ears are short and rounded, and the bushy tail tapers to the tip (How 2008; Viggers and Lindenmayer 2004); *T. cunninghami* is distinguished by longer ears and hindfeet and shorter tail (Martin 2008). The coat of *Trichosurus* is usually thick, woolly, and soft; the tail is usually well furred and prehensile, and at least the terminal portion is naked on the lower side. Females of the genus have a well-developed pouch that opens forward and encloses two mammae.

Brushtailed possum (*Trichosurus vulpecula*). Photo by Pavel German / Wildlife Images.

Brushtailed possums have a well-developed glandular area on the chest. This brown-pigmented, sternal patch is more prominent in the male than in the female. Observations of animals in the field and in captivity have revealed that these areas are rubbed against stumps or fallen logs to form "scent posts." Both sexes possess two pairs of anal glands. One pair secretes copious amounts of oil with a strong musky odor that also apparently functions in demarcating territory. The other pair does not liquefy the secretion, which eventually appears in the urine as cells; this phenomenon has not been reported for placental mammals. Viggers and Lindenmayer (2004) noted that in *T. caninus* and *T. cunninghami* the sternal gland is less conspicuous than in *T. vulpecula*. Other scent glands are located in the labial, chin, and pouch regions (Cowan 2005).

All species are nocturnal and arboreal and usually nest in tree hollows. They generally occupy forest habitat, but *T. vulpecula* may also be found in areas devoid of trees, where it shelters by day in caves, rock cavities, or burrows of other animals. That species is even found, though in greatly reduced numbers, in arid central Australia, where it takes shelter in eucalyptus trees along watercourses. Resident populations of *T. vulpecula* occur in most cities in parks and suburban gardens, where the animals may shelter in attics and recesses of roofs. *T. caninus* and *T. cunninghami* are more dependent on wet forests than is *T. vulpecula* and possibly exclude the latter species from such habitat in zones of potential sympatry (Viggers and Lindenmayer 2004).

The diet of *Trichosurus* consists mostly of leaves, young shoots, flowers, fruits, and seeds. Because *T. vulpecula* has long been established in New Zealand and is considered of much economic importance, it has been studied there at least as much as in Australia. According to Cowan (2005), New Zealand animals feed mainly on leaves but readily take cultivated grains and vegetables, insects, and native birds and their eggs, and scavenge on deer and pig carcasses. There are 2–3 feeding sessions per night, separated by 2–3 hours of inactivity in dens; an individual may use 5–10 dens at any one time. Dens are not usually shared in forests, where there are many suitable sites, but in farmland up to five individuals may share a den.

Reported population densities in New Zealand have varied from about 0.5 individuals per ha. in beech forests and 1 per ha. in scrubby farmland to about 10–12 per ha. in podocarp-broadleaf forests and 25 per ha. in forest-pasture margins; numbers are usually highest February–May with seasonal influx of newly independent young and lowest September–October due to winter mortality (Cowan 2005). Densities in Australia vary from 0.2 to 4.0 per ha. (Kerle and How 2008). Based on studies in New Zealand, home range size is 0.7–3.4 ha. for males, 0.6–2.7 ha. for females; larger areas, averaging 24.6 ha. for males and 18.3 ha. for females, were reported when animals left their dens in native forest and moved downhill up to 1.5 km to feed on pasture, and some individuals have used annual ranges as large as 60 ha. (Cowan 2005). However, four individuals in New Zealand had nightly ranges usually under 1,000 sq meters and annual ranges of only 0.28–3.21 ha. (Ward 1978). Home range in *T. vulpecula* averaged 3 ha. for males and 1 ha. for females in the Canberra vicinity (Crawley 1973). For the species *T. caninus* in southeastern Australia, density was 0.3 per ha.; home range was 7.67 ha. for males and 4.85 ha. for females (How 1978). For *T. cunninghami*, reported densities vary from as low as 0.01 per ha. to more than 2/ha. and home ranges from about 1 to 4 ha. (Viggers and Lindenmayer 2004).

According to Collins (1973), two captive males cannot be kept in the same enclosure, but one male and several females can coexist. There is strong evidence that wild *T. caninus* are paired, with an adult male and female sharing a home range (How 1978), and some evidence that *T. cunninghami* is polygynous (Viggers and Lindenmayer 2004). The social organization of *T. vulpecula* is not fully understand but exhibits a degree of complexity (Cowan 2005; Crawley 1973; How 1978). The species seems basically solitary. In its low-density populations a territorial system may function, with individuals of the same sex maintaining discrete ranges. In its high-density populations, as in parts of New Zealand, there is considerable overlap of home range for both same sex and different sexes, and territorial behavior is not evident. However, individuals avoid one another and actively defend the area around dens. Young females tend to remain on or near their mothers' ranges, giving rise to groupings of related females. During the breeding season an older female may be accompanied by one or two males for 30–40 days before estrus. Dominant males may mate with several females. Spacing between individuals is maintained through agonistic encounters and olfactory and vocal

Mountain brushtailed possum (*Trichosurus cunninghami*). Photo by Gerhard Körtner.

communication. Scent-marking includes rubbing trees with the chin or sternal glands and depositing secretions from the anal glands

At least 22 different sounds have been reported for *T. vulpecula* (Cowan 2005). Its mating call is a loud, rolling, guttural sound terminating in a series of staccato "ka-ka-ka" syllables. When mildly disturbed, all species make a metallic clicking sound and then, if further aroused, a series of harsh exhalations with the mouth open. Further disturbance causes the animal to rise on its hind limbs with the forelimbs raised and outstretched. The possum then extends itself to full height and screams. All species are powerful fighters.

Females are polyestrous; the estrous cycle averages about 25 days in *T. vulpecula* (Van Deusen and Jones 1967) and about 26.4 days in *T. caninus* (Smith and How 1973). Breeding may extend throughout the year, but generally for *T. vulpecula* in Australia there are two distinct seasons or peaks, one in spring and one in autumn.

In some areas a female may produce a litter in both seasons, but usually most females give birth during the autumn (How 1978; Kerle and How 2008). The subspecies *T. vulpecula arnhemensis* in the tropics of extreme northern Australia has been found to breed continuously throughout the year, with females evidently conceiving before weaning the pouch young or returning to estrus in about 10 days if the pouch young is lost (Kerle and Howe 1992). For *T. vulpecula* in New South Wales, Smith, Brown, and Frith (1969) reported 89.6 percent of births in autumn (late March–June) and 9.1 percent in spring (September–November). In the Tasmanian subspecies *fuliginosus* and at least some populations of the southwestern subspecies *hypoleucus*, there is no spring breeding (Wayne et al. 2005a). Births occur throughout the year in the introduced population of *T. vulpecula* in New Zealand, but are concentrated in the autumn from April to June; it is less common in New Zealand than in Australia for a female to give birth in the autumn and/or to have two litters in 1 year (Cowan 2005). For *T. caninus* in New South Wales, How (1976) found 87.5 percent of births to take place from February to May, mostly in March and April, with a few scattered over the rest of the year. *T. cunninghami* produces one young per year in April or May (Martin 2008).

Gestation averages about 17.5 days in *T. vulpecula* (Van Deusen and Jones 1967) and 16.2 days in *T. caninus* (Smith and How 1973); it is 15–17 days in *T. cunninghami* (Martin 2008). Usually a single young is produced; of 60 births of *T. vulpecula* recorded at the London Zoo, six resulted in twins and one in triplets (Collins 1973). Of the three species, development appears to take place most rapidly in *T. vulpecula* (Cowan 2005; How 1976, 1978, 2008; Smith, Brown, and Frith 1969); the young weighs 0.2 grams at birth, develops fur by 90–100 days of age, and opens its eyes about 10 days later. It leaves the pouch at about 4–5 months, is weaned by 6–7 months, and separates from the mother at 8–18 months. Females generally attain sexual maturity and begin to breed at 9–12 months, though in some populations this phase is delayed by 1 or even 2 years. In *T. caninus* the young emerges from the pouch at 5–6 months, weaning occurs at 8–11 months, dispersion occurs at 18–36 months, and females do not reach sexual maturity until 24–36 months. In *T. vulpecula* mortality is high among dispersing young; only 25 percent reach the age of 1 year (How 1978). Adult mortality has

Brushtailed possum (*Trichosurus vulpecula*). Photo by Gerhard Körtner.

been found to be relatively low—20 percent in one study—and life expectancy is 6–7 years. Longevity in wild populations of *T. vulpecula* is usually less than 11 years but one animal is known to have lived for 13 years (Kerle and How 2008). Wild female *T. caninus* and *T. cunninghami* have survived up to 17 years, males up to 15 (How 2008; Martin 2008). *Trichosurus* also does well in captivity; a male *T. vulpecula* lived for 15 years and 11 months (Weigl 2005).

Because of its fecundity and adaptability to a variety of conditions, *T. vulpecula* has been compared to the Virginia opossum (*Didelphis virginiana*) in North America. Both species seem able to live near people and even to expand their range in the face of human development. Ride (1970) wrote that Australians had more contact with *T. vulpecula* than with any other native mammal. That species can find shelter in artificial structures and can subsist on garden vegetation.

Although declining drastically in many regions, it is sometimes considered a major pest. In Australia it is said not only to damage flowers, fruit trees, and buildings but also to adversely affect regenerating eucalyptus forests and introduced pine plantations and to carry diseases that are potentially harmful to humans and livestock (How 1976; McKay and Winter 1989; Ride 1970).

T. vulpecula has been used extensively in Australia for the fur trade. In the late nineteenth and early twentieth centuries, that species and several other kinds of possums were taken in association with the hunting of *Phascolarctos* (see account thereof) and in even larger numbers. In 1906, about 4.7 million Australian possum skins were sold in London. The harvest in Queensland alone during six open seasons from 1923 to 1936 averaged about 1.8 million possums, at least 80 percent of which were probably *T. vulpecula* (Gordon and Hrdina 2005; Hrdina and Gordon 2004). At present,

Short-eared possum (*Trichosurus caninus*). Photo by Alexander F. Meyer at Healesville Sanctuary, Australia.

commercial hunting is restricted to Tasmania and serves to suppress populations that allegedly damage forestry and agriculture. As recently as 1980 more than 250,000 skins were exported with a value of about US$6 each, but subsequently prices fell and the market declined. In 1990 and 1991 exports totaled fewer than 15,000 skins at a value of about $2 each (Callister 1991). *T. caninus* also causes considerable forestry damage, stripping off bark in pine plantations, and has been taken extensively for its pelt (How 2008).

A desire to share in the fur market led to the introduction of *T. vulpecula* in New Zealand, first in 1840 and more extensively in the 1890s. It adapted well, spread throughout both main islands of New Zealand, and, at least in part because of absence of natural predators and competitors, achieved far higher densities than in Australia; it is now often treated more as a pest than as a fur bearer (Cowan 2005; Crawley 1973; Fitzgerald 1976, 1978; Gilmore 1977). It reportedly damages gardens, orchards, crops, pasture, exotic tree plantations, and remaining native forests. All legal restrictions on

killing it in New Zealand were removed in 1947, a bounty was in effect from 1951 to 1962, and the species has been subject to intensive programs of poisoning, trapping, and shooting. Its overall numbers, however, do not seem to have been greatly affected. From 1962 to 1974 the annual number of skins exported from New Zealand ranged from 346,000 to 1,605,000. According to Cowan (2005), during a period of strong demand, 1978–1982, more than 2 million skins were exported each year. The average annual harvest fell to about 250,000 in 1996–1998. Since 2000, development of mixed possum fur/wool fibers has stimulated the market and pushed the price of fur to around US$30 per kg.

Although *T. vulpecula* may thrive in close proximity to people and in areas of introduction, its natural populations have undergone significant reduction across Australia (Kerle and How 2008). It has disappeared from almost the entire vast arid and semiarid zones of the western and central mainland of the continent, where it was still common in the nineteenth and early twentieth centuries. Now only a few scattered colonies survive there, all in southern Northern Territory and southwestern Queensland; those groups appear most closely related to populations of the subspecies *T. v. vulpecula* in extreme southeastern South Australia (Taylor and Foulkes 2004). The inland decline of *T. vulpecula* began about 1880, because of an epizootic (see Abbott 2006), and continued during 1920–1960 as a result of predation by the introduced red fox (*Vulpes vulpes*) (Abbott 2012).

The species is also declining where it does persist in the forests and woodlands of the far north, west, and east. Based on IUCN criteria, Woinarski, Burbidge, and Harrison (2014) designated the subspecies *T. v. arnhemensis* vulnerable, noting that in the area of most intensive monitoring, Kakadu National Park in central Arnhem Land, numbers fell about 80 percent from 2001 to 2009. There are very few recent records from the Kimberley area of northeastern Western Australia, and the subspecies has disappeared entirely from the Sir Edward Pellew Islands. The largest remaining populations are thought to be on Bathurst and Melville Islands and, surprisingly, in the city of Darwin and vicinity. The subspecies *T. v. hypoleucus* survives in the forests of extreme southwestern Western Australia but is jeopardized by fox predation and the destruction by the woodchip industry of eucalyptus trees, which have the large hollows needed for denning (Jones 2004). That subspecies also persists in the Pilbara of northwestern Western Australia and on nearby Barrow Island and was recently introduced at a site in central Western Australia (Woinarski, Burbidge, and Harrison 2014). Populations of *T. vulpecula* remain high on Tasmania (Munks et al. 2004) and Kangaroo Island (Carthew 2004). A substantial population also exists on Magnetic Island, off the northeast coast of Queensland, where absence of predators and competitors has allowed possum densities (5 per ha.) comparable to those of New Zealand (Isaac 2005).

DIPROTODONTIA; PHALANGERIDAE; **Genus *WYULDA*** Alexander, 1918

Scaly-tailed Possum

The single species, *W. squamicaudata*, is restricted to the Kimberley Division in extreme northern Western Australia (Ride 1970). It was known from only four specimens until 1965, when eight more were collected (Fry 1971). It subsequently has been observed with more regularity. *Wyulda* was considered a subgenus of *Trichosurus* by Flannery, Archer, and Maynes (1987), but restored to generic rank by Flannery and Schouten (1994).

Head and body length is 290–470 mm, tail length is about 250–325 mm, and adult weight range is 0.9–2.0 kg (Helgen and Jackson 2015; Humphreys et al. 1984). The pelage is short, soft, fine, and dense. The general dorsal color is pale or dark ashy gray. A dark stripe, obscure or distinct, runs along the middorsal line from the shoulders to the rump. The nape, shoulders, and rump of one specimen were mottled with buff, and its throat and chest were gray. The sides and back are usually the same color, though the sides are somewhat paler. The underparts are creamy white.

The prehensile tail is densely furred at the base and has nonoverlapping, thick scales for the remainder of its length. Short, bristly hairs are present around the scales. This is the only member of the family Phalangeridae with a tail of this kind. The head is short and wide. The claws are short and not strongly curved.

Information on the natural history of wild populations was provided by Burbidge and Webb (2008), Collins (1973), Humphreys et al. (1984), and Runcie (1999). The scaly-tailed possum inhabits rugged, rocky country covered by open woodland or closed forest. It is nocturnal and scansorial, sheltering by day in rock piles, under rock slabs, and in underground crevices. An individual uses several dens, spaced up to 500 meters apart. After leaving the rocks at night it forages mainly in trees but sometimes on the ground in open areas. It is very agile and uses its tail to hang vertically from a tree branch while it feeds. The diet includes leaves, blossoms, fruit, seeds, insects, and possibly small vertebrates. This possum is solitary, foraging alone and with neither sex sharing a den. However, radio-tracking showed home ranges to overlap and to average 1.0 ha. during an 8-day period. Population densities have been estimated at 1.0 to 4.6 possums per ha. Females give birth mainly in the dry season from March to August. Only one young is carried; it remains in the pouch for 150–200 days and is weaned sometime after 8 months. Males do not reach sexual maturity until past 18 months, and females not until their third year. A captive individual was said to be gentle and affectionate and to make a chittering noise like a bird (Fry 1971). It was still living after being maintained in a private home for 4 years and 4 months (Collins 1973).

The genus has been designated as endangered by the USDI since 1970, though Winter (1979) stated that it was probably more plentiful than previously thought, and Thornback and Jenkins (1982) reported no evidence of any major threat. Later, however, Flannery and Schouten (1994) reaffirmed that *Wyulda* should be regarded as endangered; while seemingly abundant at one site, it has a very limited and patchy distribution and may be declining as its habitat is disrupted by the expanding pastoral industry. The IUCN notes that the population is decreasing but applies the designation "data deficient" (McKnight 2008d). The genus was described from a specimen that reportedly originated well inland near Turkey Creek in the eastern Kimberley in 1917, but subsequent records, until recently, all came from coastal areas with much higher rainfall. This suggested a decline in more arid country or that the type was actually transported inland by human agency. A 2003 survey indicated survival in the Prince Regent Nature Reserve and on Bigge and Boongaree Islands (Burbidge and Webb 2008; Jones 2004). Doody et al. (2012) rediscovered a population of *Wyulda* in the inland part of the eastern Kimberley, and suggested a broader environmental adaptability than previously thought. Potter et al. (2014a) reported low genetic differentiation between the eastern and western populations and indicated that more survey work is needed to determine if the genus also is present in the intervening 200–300 km.

DIPROTODONTIA; PHALANGERIDAE; Genus *SPILOCUSCUS* Gray, 1862

Spotted Cuscuses

There are six species (Aplin and Pasveer 2007; Feiler 1978a, 1978b; Flannery 1995a, 1995b; Flannery, Archer,

Scaly-tailed possum (*Wyulda squamicaudata*). Photo by Pavel German / Wildlife Images.

Spotted cuscus (*Spilocuscus maculatus*). Photo by Pavel German / Wildlife Images.

and Maynes 1987; Flannery and Calaby 1987; Flannery and Schouten 1994; George 1979, 1987; Heinsohn 2002a, 2004a; Helgen and Flannery 2004b; Helgen and Jackson 2015):

S. rufoniger, mainly northern New Guinea with a few records from the south in the vicinity of the Lorentz River and the headwaters of the Fly River, probably also on Japen (Yapen) Island;

S. maculatus, New Guinea and some nearby small islands (Batanta, Japen, Karkar, Misool, Numfoor, Pulau Num, Roon, Salawati, Su Mios, Walis, Wammer), Aru Islands (Kobroor, Maikoor, Trangan, Warmar, Wokam, Wonoembai), Cape York Peninsula of Queensland, and likely through introduction on South Molucca Islands (Ambon, Banda Islands, Buru, Pulau Panjang, Seram, Tioor), Kai Islands, Selayar (Salayer) Island (south of Sulawesi), and Bismarck Archipelago (northeastern New Ireland, Mussau, Eloaua);

S. nudicaudatus, Cape York Peninsula of northern Queensland to north of Stewart River;

S. wilsoni, Numfoor, Supiori, and Biak Islands off northwestern New Guinea;

S. kraemeri, Manus and some small nearby islands (Baluan, Los Negros, Lou, Luf, Pak, Ponam, Rambutyo) in the Admiralty Group northeast of New Guinea, Luf in the Hermit Islands and Ninigo and Wuvulu Islands in the Ninigo Group north of New Guinea, Bali Island and possibly nearby New Britain in the Bismarck Archipelago;

S. papuensis, Waigeo Island and nearby Gam Island off the western tip of New Guinea, possibly also nearby Batanta Island.

Spilocuscus had been placed in *Phalanger* by most authorities, but Flannery, Archer, and Maynes (1987) and George (1987) regarded the two as generically distinct. Helgen and Flannery (2004b) noted the possibility that *Spilocuscus* might comprise only a single, highly variable species. *S. nudicaudatus* was considered a subspecies

Spotted cuscus (*Spilocuscus maculatus*), adult male. Photo by Thomas E. Heinsohn.

of *S. maculatus* by Groves (2005e) and most of the authorities cited above, but was treated as a distinct species by Helgen and Jackson (2015).

Head and body length is 338–640 mm, tail length is 310–590 mm, and weight is 1.5–7.0 kg (Flannery 1995b; Flannery and Calaby 1987; Flannery and Schouten 1994; Winter and Leung 2008a). On average females are larger than males. The sexes also are colored differently in some species. In *S. maculatus* the adult males may be brilliant white all over, mottled ginger, or gray spotted with white above and white below. The females are uniformly gray and usually unspotted. In both sexes of *S. nudicaudatus* the upper parts are generally steely gray but males have cream colored spots and blotches; the underparts are cream (Helgen and Jackson 2015). The single available adult specimen of *S. wilsoni* is a male and has a pure white coat; that species also has blue-green eyes, whereas in all other species the iris is brown or hazel (Helgen and Flannery 2004b). In *S. rufoniger* and the much smaller *S. kraemeri* there is a dark saddle on the back of females but only an area of mottling or spots on the males. In *S. papuensis* both sexes are marked with very small spots. The young go through a sequence of color changes. The fur is dense and woolly, the snout is short, and the ears are almost invisible. Females have four mammae.

Flannery, Archer, and Maynes (1987) listed the following unique characters that distinguish *Spilocuscus* from related genera: there is sexual dichromatism as noted above; in both sexes the frontal bones of the skull are markedly convex and accommodate a large sinus that does not open into the nasal cavity; there is a well-developed protocone on the first upper molar; and the alisphenoid and basoccipital consistently form a more extensive suture that is developed earlier in life than it is in other phalangerids.

Although relatively common and widespread, *S. maculatus* and *S. nudicaudatus* have not been extensively studied; available information has been summarized by Flannery (1995a), Heinsohn (2004b), and Winter and Leung (2008a). On Cape York *S. nudicaudatus* occurs mainly in rainforest from sea level to an elevation of 820 meters, but in New Guinea, where it competes with other possums, it ranges up to about 1,200 meters, and on mountainous islands, such as Seram, it may be found still higher. It has also been seen in mangroves and open forests and adapts well to coconut plantations, groves of shade trees around villages, and certain other modified habitats. It may even adopt a semi-commensal niche, particularly in areas where it is not hunted for food, such as Muslim parts of Indonesia and Seventh-Day Adventist parts of Melanesia. It shelters amidst dense foliage on a branch or in the canopy of a tree, sometimes building a small sleeping platform of twigs and leaves. It is mainly nocturnal and arboreal, climbs slowly and de-

Spotted cuscus (*Spilocuscus maculatus*), juvenile male. Photo by Thomas E. Heinsohn.

liberately, maintains a strong grip with its feet and tail, and bounds at the speed of a fast human walk when on the ground. The natural diet seems to be primarily folivorous but may also include a substantial amount of coconuts and other fruit; captives have readily taken dog food, beef, chicken, and eggs. A radio-tracking study found short-term home ranges of about 0.9 ha. for a male and 0.6 ha. for a female. Males are aggressive, possibly defend small territories in the wild, and cannot be housed together in captivity. Females make a high frequency hiss, thought to be a mating call, and reportedly vocalize almost continuously through the night at the peak of estrus. The estrous cycle lasts about 28 days. Apparently there is an extended breeding season in Australia, New Guinea, and New Ireland. Although three young have been recorded from a pouch, it is likely that only one is reared to weaning. The pouch may be vacated at 6–7 months of age. Juvenile females possibly maintain an association with the mother longer than do males. Weigl (2005) listed a wild-caught specimen thought to have been over 17 years old when it died in the Perth Zoo.

The other four species are much more limited in distribution, more dependent on natural forests, and of greater conservation concern. The IUCN designates *S. rufoniger* as critically endangered; a decline of more than 80 percent is anticipated in the period 2005–2020 because of intensive hunting for food and symbolic use and loss of habitat to logging, oil palm concessions, and agriculture (Leary et al. 2008n). *S. wilsoni*, also designated critically endangered, is threatened by rapid deforestation and hunting by local people for use as food and pets (Aplin and Helgen 2008). *S. papuensis* is listed as vulnerable, because it occurs only on one island that is susceptible to increased hunting pressure and large-scale deforestation (Helgen, Aplin, and Dickman 2008). *S. kraemeri*, classed as near threatened, is declining as its restricted range becomes subject to increased human population pressures (Helgen et al. 2008b). *S. kraemeri*, *S. papuensis* and *S. maculatus* are on appendix 2 of the CITES.

S. maculatus and *S. kraemeri* are widely taken for food and their attractive pelts; they are often maintained alive for future utilization or kept as pets, and have long been transported and traded across considerable distances by local people. It seems likely that *S. maculatus*, at least, reached most of its current range beyond New Guinea and the Cape York Peninsula through introduction (Heinsohn 2004a). It has been suggested that *S. kraemeri* was brought to the Admiralty, Ninigo, and Hermit Islands only 1,000–2,000 years ago, and since it probably could not have evolved its distinctive characters over so short a period, that there may be an as-yet undiscovered parent population (Flannery 1995b; Flannery and Schouten 1994). Although Heinsohn (2004a) pointed out that recent archeological excavation showed the presence of *S. kraemeri* on Manus Island about 13,000 years ago, he supported the idea of an early introduction. In contrast, Helgen and Flannery (2004b) believed the subfossil evidence indicates *S. kraemeri* may be a natural part of the Manus fauna, though they also noted possible records of the species from New Britain, which could be its place of origin.

DIPROTODONTIA; PHALANGERIDAE; **Genus PHALANGER** Storr, 1780

Cuscuses

There are 13 species (Aplin and Pasveer 2007; Colgan et al. 1993; Feiler 1978a, 1978b; Flannery 1987, 1995a, 1995b; Flannery, Archer, and Maynes 1987; Flannery

Southern common cuscus (*Phalanger intercastellanus*). Photo by Pavel German / Wildlife Images.

and Boeadi 1995; Flannery and Schouten 1994; George 1979, 1987; Groves 1976, 1987a, 1987b, 2005e; Heinsohn 2002b, 2004a; Kirsch and Calaby 1977; Laurie and Hill 1954; Menzies and Pernetta 1986; Norris and Musser 2001; Osborne and Christidis 2002; Ride 1970; Springer et al. 1990; Ziegler 1977):

P. gymnotis, mainland New Guinea and Aru Islands (Wokam, Kobroor) to southwest;

P. lullulae, Woodlark Island east of New Guinea and (possibly through introduction) nearby Alcester and Madau Islands;

P. rothschildi, Obi and Bisa Islands in the North Moluccas;

P. ornatus, Halmahera, Bacan (Batjan), and possibly Morotai Islands in the North Moluccas;

P. matabiru, Ternate and Tidore Islands off western Halmahera Island;

P. alexandrae, Gebe and possibly other islands in the North Moluccas;

P. matanim, known only by five specimens from the Telefomin area in the mountains of western Papua New Guinea;

P. orientalis, northern New Guinea and some nearby small islands (Waigeo, Batanta, Salawati, Misool, Biak-Supiori, Numfoor, Japen, Su Mios, Koil, Vokeo, Mioko) and probably through introduction on Halmahera Island, South Molucca Islands (Sanana, Buru, Ambon, Saparua, Seram, Banda Islands, Gorong), Timor and nearby Wettar and Letti Islands, Kai Islands, Bismarck Archipelago (Karkar, Bagabag, Umboi, Long, New Britain, New Ireland, Tabar, Lihir, Boang), and Solomon Islands (Nissan, Buka, Bougainville, Choiseul, Vella Lavella, New Georgia, Santa Isabel, Russell Islands, Guadalcanal, Malaita, San Cristobal, possibly Santa Ana);

P. intercastellanus, southeastern Papua New Guinea, Trobriand Islands (Kiriwina), D'Entrecasteaux Islands (Goodenough, Fergusson, Normanby, Ito, Sariba, Sidela), Louisiade Archipelago (Misima, Sudest, Rossel);

P. mimicus, southern New Guinea from Mimika River to Mount Bosavi and Oriomo River, Kobroor and probably other of the Aru Islands, eastern Cape York Peninsula of Queensland;

P. sericeus, highlands of central and eastern New Guinea;

P. vestitus, highlands of western, central, and southeastern New Guinea;

P. carmelitae, highlands of central and eastern New Guinea.

Phalanger was often considered to include the genera *Ailurops*, *Strigocuscus*, and *Spilocuscus* (see accounts thereof). Flannery, Archer, and Maynes (1987) referred *P. ornatus* and *P. gymnotis* to *Strigocuscus*, but Flannery and Schouten (1994) reassigned both to *Phalanger* based in part on the molecular analyses of Springer et al. (1990); this was supported by Hamilton and Springer's (1999) assessment of mitochondrial DNA. George (1987) regarded *P. leucippus* of mainland New Guinea a species separate from *P. gymnotis*. *P. intercastellanus*, usually included within *P. orientalis*, was recognized as a full species by Colgan et al. (1993) and Flannery and Schouten (1994). *P. mimicus* formerly was treated as part of *P. intercastellanus* but considered a distinctive species by Norris and Musser (2001); they did not discuss the population of *P. intercastellanus* reported from the Aru Islands (Flannery 1995a, 1995b), but on geographic grounds that population would most likely represent *P. mimicus*. *P. matabiru* was described as a subspecies of *P. ornatus* by Flannery and Boeadi (1995) and retained as such by Helgen and Jackson (2015), but listed as a full species by Groves (2005e). Other factors contributing to the above sequence of species are: *P. gymnotis* and *P. lullulae* are the most basal species in *Phalanger* (Flannery 1995a; Osborne and Christidis 2002); there is affinity between *P. gymnotis*, *P. rothschildi*, and *P. ornatus* (Norris 1999); *P. alexandrae* is related to *P. rothschildi* and *P. ornatus* (Groves 2005e); *P. matanim* does not appear closely related to any other species but is considered primitive (Flannery 1995a); *P. orientalis*, *P. intercastellanus*, and *P. mimicus* may form a coherent group (Norris and Musser 2001); and *P. orientalis*, *P. sericeus*, *P. vestitus*, and *P. carmelitae* are closely related (Flannery 1995a; Osborne and Christidis 2002).

Head and body length is 325–600 mm, tail length is 240–610 mm, and weight is 1,045–4,850 grams (Flannery 1995a, 1995b; Flannery and Schouten 1994; Lidicker and Ziegler 1968). In some species, females average slightly larger than males. The fur in most

Ground cuscus (*Phalanger gymnotis*). Photo by Alexander F. Meyer at Plzen Zoo, Czech Republic.

forms is thick and woolly. Coloration in the genus ranges from white, reds, and buffs through various shades of brown to light grays and different intensities of black. Some members of this genus have suffusions of yellow or tawny over the shoulder region, and others have dark dorsal stripes that extend from the head to the rump. In *P. vestitus* some individuals are pale silvery brown and short-haired, others are dark brown and long-haired, and still others represent intermediate forms.

These are heavy and rather powerfully built animals. Their yellow-rimmed protruding eyes, bright yellow noses, inconspicuous ears, and prehensile tails give them a somewhat monkey-like appearance. The terminal portion of the tail is covered with scales and lacks hair. The digits of the forefeet are not of equal length; the claws are long, stout, and curved; and the soles are naked and striated. Females have four mammae in a well-developed pouch.

Cuscuses inhabit mainly tropical forests and thick scrub. *P. gymnotis* is common in many parts of New Guinea from sea level to 2,700 meters but avoids swampy areas and floodplains; *P. sericeus* lives in pri-

mary montane forest at elevations of 1,500–3,900 meters (Flannery 1995a). Cuscuses are arboreal animals with strongly prehensile tails but sometimes descend to the ground. They are mostly nocturnal, feeding in trees at night and resting by day curled up in the thick foliage of a vine tangle, in a tree or tree hollow, in burrows under tree roots, or among rocks. They are somewhat sluggish, resembling the slow loris (*Nycticebus*) in their movements, though Winter and Leung (2008b) noted that *P. mimicus* moves through the canopy faster than does *Spilocuscus maculatus* and likely jumps across gaps. The diet consists mainly of fruits and leaves but also includes insects, small vertebrates, and birds' eggs. Flannery (1995a) cited two reports from native people that females carry fruit back to the den in their pouch.

In dense lush rainforest, with ample foliage, an individual *P. orientalis* can subsist in a home range of less than 1 ha., where it may maintain several den sites (Heinsohn 2005a). Cuscuses are solitary but can be kept in pairs in roomy enclosures (Collins 1973). Their sounds include snarls, barks, and hisses. They also make noise by hindfoot thumping (Flannery 1995a).

They scent-mark the vicinities of dens with urine and glandular secretions; even when handled gently, they usually emit a penetrating musk odor. Severe fighting seems frequent in *P. gymnotis*, and probably occurs in other species, even between opposite sexes when females are anestrous (Flannery 1995a). The breeding season of *Phalanger* appears extensive, perhaps lasting throughout the year, at least in some species; females have up to three young, though they probably rear only one or two; the young of *P. gymnotis* begin to leave the pouch when 108 days old, stay out regularly at about 138 days, when they may ride on the mother's back, and remain permanently out at 160–200 days (Flannery 1995a, 1995b; Helgen and Jackson 2015). Captive *P. gymnotis* have lived for more than 18 years and 11 months (Weigl 2005).

The IUCN classifies *P. matanim* as critically endangered; its only known habitat was completely destroyed by fire in 1998 (Leary et al. 2008o). *P. lullulae* and *P. alexandrae* are classed as endangered and *P. matabiru* as vulnerable; all are restricted to small islands and jeopardized by hunting, conversion of natural forest to agriculture, and potential introduction of *P. orientalis*, which would be a competitor for limited resources (Leary et al. 2008p, 2008q, 2008r). In addition, the IUCN reports that *P. gymnotis*, *P. intercastellanus*, *P. mimicus*, and *P. vestitus* are decreasing in numbers, apparently because of widespread hunting for food, pelts, and commercial trade (Leary et al. 2008s, 2008t, 2008u; Salas, Dickman, and Helgen 2008). *P. orientalis*, *P. intercastellanus*, and *P. mimicus* are on appendix 2 of the CITES.

P. orientalis has been extensively translocated by native people because it is valued for ceremonial purposes and used as food or pets. Archeological, anthropological, zoogeographical, and genetic research indicates that many island populations were introduced through human agency in prehistoric time (Heinsohn 2004a, 2005a). Based on electrophoretic and morphological analyses, Colgan et al. (1993) suggested the species was introduced to New Ireland from New Britain between 10,000 and 19,000 years ago and then spread to the Solomon Islands between 2,000 and 6,000 years ago. Flannery, Archer, and Maynes (1987) noted that *P. orientalis* probably was introduced on Timor about 4,000–5,000 years ago. Although lack of morphological differentiation between the Woodlark and Alcester populations of *P. lullulae* might suggest introduction to the latter island (Flannery 1995a), the great genetic divergence of the two populations indicates they may be separate natural subspecies (Osborne and Chistidis 2002b).

DIPROTODONTIA; **Family PSEUDOCHEIRIDAE**

Ring-tailed and Greater Gliding Possums

This family of 6 Recent genera and 18 species inhabits Australia, Tasmania, New Guinea, and certain nearby islands. The members of the Pseudocheiridae were long regarded as components of the family Phalangeridae. Kirsch (1977b), however, explained that they differ in serology, karyology, and other characteristics, and he placed them in the subfamily Pseudocheirinae of the family Petauridae. On the basis of additional serological studies, Baverstock (1984) elevated the Pseudocheirinae to familial rank but indicated that this group is closely related to the Petauridae. That view has been supported by Aplin and Archer (1987), Baverstock, Birrell, and Krieg (1987), and most subsequent authorities. However, the group was regarded as only a tribe by Szalay (1994) and as a subfamily by McKenna and Bell (1997) within the family Petauridae. Based on the molecular assessment of Kirsch, Lapointe, and Springer (1997), Groves (2005e) recognized three subfamilies: Pseudochiropsinae, for the genus *Pseudochirops*; Pseudocheirinae, for *Petropseudes*, *Pseudocheirus*, and *Pseudochirulus*; and Hemibelideinae, for *Hemibelideus* and *Petauroides*. The sequence of genera presented here considers that arrangement and also the affinities discussed by Aplin (2008), Archer and Hand (2006), and Flannery and Schouten (1994). A comprehensive recent assessment of nuclear DNA by Meredith et al. (2010) suggests about the same sequence but with a major proviso. *Petropseudes* was found to group more closely with the species of *Pseudochirops* in New Guinea than with the Australian species *Pseudochirops archeri*, thereby rendering *Pseudochirops* paraphyletic and also necessitating the transfer of *Petropseudes* to subfamily Pseudochiropsinae. Jackson (2015a) did transfer *Petropseudes* accordingly, while continuing to treat it as a full genus.

Greater gliding possum (*Petauroides volans*). Photo by Pavel German / Wildlife Images.

Head and body length is 163–480 mm, and tail length is 170–550 mm. The pelage may be woolly or silky. According to McKay (1989), the dentition of the Pseudocheiridae is characterized as follows: three upper incisors on each side of the jaw, only the first pair prominent, and also a reduced upper canine; three upper premolars, the third elongated and bearing two or three cusps; four strongly selenodont upper molars, with crescent-shaped ridges connecting the cusps; the large, procumbent first lower incisor compressed and bladelike; a second, vestigial lower incisor may be present in some species but lost in others; three large lower premolars; and four lower molars, all approximately equal in size. The skull is characterized by a robust zygomatic arch and posterior palatal vacuities. The forefoot has the first two digits at least partly opposable to the other three, an adaptation for grasping small branches. The tail is prehensile and may have a ventral friction pad of naked calloused skin or only a small naked area at the tip. In the digestive tract the caecum is greatly enlarged to act as a fermentation chamber. Females have a forward-opening pouch with two or four mammae.

All members of the Pseudocheiridae are arboreal. One genus has attained a gliding ability through development of a membrane uniting the front and hind limbs. That genus, *Petauroides*, has a phylogenetic position analogous to that of *Petaurus* in the family Petauridae (Kirsch 1977b). Pseudocheirids are generally folivorous; their crescentic molars enable leaves to be finely shredded, a large caecum breaks down the leaf fiber, and, in at least some species, coprophagy assists in full digestion (Aplin 2008).

The known geological range of Pseudocheiridae is late Oligocene to Recent in Australia (Archer and Hand 2006) and Pleistocene to Recent in New Guinea (Flannery 1995a). Details on Miocene pseudocheirid fossils and their significance to the phylogeny of the family were presented by Woodburne, Tedford, and Archer (1987).

DIPROTODONTIA; PSEUDOCHEIRIDAE; **Genus** *PSEUDOCHIROPS* Matschie, 1915

Coppery and Silvery Ring-tailed Possums

There are five species (Flannery 1995a; Flannery and Schouten 1994; Kirsch and Calaby 1977; Ride 1970; Winter, Krockenberger, and Moore 2008; Ziegler 1977):

P. cupreus, Central Cordillera of New Guinea;

P. albertisii, highlands of northern and western New Guinea and nearby Japen Island;

P. coronatus, Arfak Mountains of Vogelkop Peninsula in far western New Guinea;

P. corinnae, highlands of Central Cordillera of New Guinea and Huon Peninsula;

P. archeri, highlands from Paluma to Mount Windsor Tableland in northeastern Queensland.

Kirsch and Calaby (1977) and most other authorities once considered *Pseudochirops* to be a subgenus of *Pseudocheirus* and also to include the species *P. dahli*. Some later serological and morphological studies suggested that *Pseudochirops* should be elevated to generic rank (Baverstock 1984; Kirsch, Lapointe, and Springer 1997) and that *P. dahli* should be placed in the separate genus *Petropseudes* (see account thereof). Recent comprehensive analyses of nuclear DNA indicate that *Petropseudes* has greater affinity to the New Guinean species of *Pseudochirops* than to the Australian species *Pseudochirops archeri*. Hence, if Petropseudes is regarded as a full genus, *Pseudochirops* would be paraphyletic, and so it was recommended that *Petropseudes dahli* now be recognized as *Pseudochirops (Petropseudes) dahli* (Meredith, Westerman, and Springer 2009b; Meredith et al. 2010). That recommendation was not followed by Jackson (2015a). While McKenna and Bell (1997) continued to include *Pseudochirops* in *Pseudocheirus*, and Corbet and Hill (1991) restricted *Pseudochirops* to the single species *P. archeri*, Groves (2005e) accepted the general arrangement given above. *P. coronatus* usually was considered a synonym of *P. albertisii* until restored to specific rank by Flannery and Schouten (1994).

Head and body length is 289–410 mm, tail length is 258–371 mm, and weight is 640–2,250 grams (Flan-

Coppery ring-tailed possum (*Pseudochirops cupreus*). Photo by Pavel German / Wildlife Images.

nery 1995a). The upper parts are coppery or silvery green, the underparts are paler, and there are dark dorsal stripes and pale facial markings on some of the species. The tail is prehensile and has distally bare areas similar to those of *Pseudocheirus*. McKay (1989) characterized *Pseudochirops* as follows: fur short, dense, and fine; tail shorter than head and body, tapering rapidly from a thickly furred base to sparsely furred tip, friction pad long; pupil a vertical slit.

The following natural history information was taken from Flannery (1995a) and Winter, Krockenberger, and Moore (2008). The New Guinea species are found mainly in undisturbed montane forests at elevations of 1,000–4,000 meters. *P. archeri*, of Australia, lives in upland rainforest and seems able to survive in areas partially logged over. Both *P. archeri* and *P. corinnae* are unusual in that they sleep on exposed branches rather than in a nest. In contrast, *P. cupreus* nests both in tree hollows and in burrows under tree roots. All species presumably are nocturnal but sometimes move about by day. *P. archeri* is usually a slow climber but can run rapidly along branches. All species are primarily folivorous but may also eat fruit. *P. archeri* is largely solitary; both it and *P. cupreus* appear to be silent and to breed throughout the year. The young of *P. archeri* permanently exit the pouch at around 200 days of age, but lactation continues for another 100 days, and the young remain close to the mother for about 2 months after weaning. The resulting 300–360 days of mother-young association is longer than in most pseudocheirids and may reflect a low-energy diet. Reproductively active female *P. albertisii* have been taken in March, July, and September. *Pseudochirops* normally produces a single young.

The IUCN classifies *P. coronatus* as vulnerable and *P. albertisii* and *P. corinnae* as near threatened (Flannery et al. 2008; Helgen, Dickman, and Salas 2008a; Leary et al. 2008v). All are in decline because of hunting for food by local people and conversion of forest habitat to cultivated land. Using IUCN criteria, Woinarski, Burbidge, and Harrison (2014) designated *P. archeri* near threatened; a total population estimate of 100,000 individuals was cited, but the species was considered to have undergone a substantial historical decline because of clearing of its rainforest habitat.

Rock possum (*Petropseudes dahlia*). Photo by Pavel German / Wildlife Images.

DIPROTODONTIA; PSEUDOCHEIRIDAE; **Genus *PETROPSEUDES*** Thomas, 1923

Rock Possum

The single species, *P. dahli*, is found from the Kimberley region of northeastern Western Australia, through areas of suitable habitat in northern Northern Territory, to Lawn Hill National Park in extreme northwestern Queensland, and on the island of Groote Eylandt (Flannery and Schouten 1994; Webb, Kerle, and Winter 2008). Although *Petropseudes* was long considered a full genus or subgenus, Kirsch and Calaby (1977) treated it as a synonym of *Pseudochirops*, and McKenna and Bell (1997) included both taxa within *Pseudocheirus*. Some other recent systematic accounts restored *Petropseudes* to generic rank (Aplin 2008; Groves 2005e; McKay 1988). However, detailed molecular assessments (Meredith, Westerman, and Springer 2009b; Meredith et al. 2010) indicate that *Petropseudes* is phylogenetically closer to the species of *Pseudochirops* in New Guinea than to the Australian species *Pseudochirops archeri* and call for treatment of *Petropseudes* as a subgenus of *Pseudochirops*. Jackson (2015a) continued to recognize *Petropseudes* as a full genus.

According to Webb, Kerle, and Winter (2008), head and body length is 334–375 mm, tail length is 200–266 mm, and weight is 1,280–2,000 grams; mean measurements of females are slightly larger than those of males. The fur is long and woolly. The upper parts are gray to reddish gray, the underparts are lighter gray, and there is a mid-dorsal stripe from the crown of the head to about the middle of the back. The short but prehensile tail is thickly furred at the base; the terminal half is almost naked and often held at a sharp angle to the basal half. The tail is naked on the terminal two-thirds of the undersurface. The claws are reduced and blunt. Adults have sternal and caudal scent glands, stained ginger, which leave markings with a sweet musky aroma. Aplin (2008) indicated that the female's well-developed pouch opens anteriorly and contains two mammae.

This possum inhabits rocky outcrops on tropical savannahs. It is nocturnal, denning by day in caves or crevices among the rocks; it usually selects the darkest recesses and is not known to make a nest. When at rest, it often lays squeezed flat in a crevice or sits in a bent-over position in a corner. It moves out at night to forage in trees and shrubs growing up to 100 meters from the edge of the outcrops (Webb, Kerle, and Winter 2008). It is a good climber and enters trees to feed on leaves, fruits, and flowers.

Petropseudes lives in cohesive social groups of 2–9 members, which den and forage together (Webb, Kerle, and Winter 2008). Groups may include a permanently mated pair and the young of 1 or 2 years. Nocturnal observations and radio-telemetric data from 3 years of study in Kakadu National Park (Runcie 2000, 2004) showed six possum groups to maintain nonoverlapping home ranges. Frequency of scent-marking by both sexes indicated that the rocky area containing the dens of a group is a defended territory. Ten different vocalizations were reported, the most common being "quark," "hoar," and "sheellrr," used to maintain contact when group members are separated. A hissing alarm call may be accompanied by repeated beating of the underside of the tail against a branch or rock, resulting in a loud slapping noise. Males were found to contribute more than females to maintenance of the pair bond, but both sexes contribute equally to parental care. The study revealed several parental behaviors not previously reported in marsupials, including embracing by the father, marshaling of young, taking turns as a sentinel while the group forages, and bridge formation, when adults hang between two treetops while young run across their backs. The study suggested a mating system of obligate social monogamy, in which male care is essential for survival of the young. *Petropseudes* apparently breeds throughout the year, and there is normally a single offspring, which may ride on an adult's back after leaving the pouch. Weigl (2005) listed two specimens that were living after 3 years and 1 month in captivity.

The IUCN reported no major threats but noted concern that there might be ongoing population declines (Burbidge et al. 2008c). *Petropseudes* is locally uncommon, with many populations isolated in restricted patches of suitable habitat. Many such areas are now jeopardized by dry season fires, which have been exacerbated through the rapid spread of introduced pasture grasses (Woinarski 2004).

DIPROTODONTIA; PSEUDOCHEIRIDAE; Genus *PSEUDOCHEIRUS* Ogilby, 1837

Common Ring-tailed Possums

Two species are now recognized (De Tores 2008a; McKay and Ong 2008):

> *P. peregrinus*, eastern coast of mainland Australia from the Cape York Peninsula of northeastern Queensland to southeastern South Australia, Kangaroo Island, King and Flinders Islands in Bass Strait, Tasmania;
> *P. occidentalis*, historically found in southwestern Western Australia but fossil remains suggest a former distribution east along the coast to the border of South Australia.

McKay and Ong (2008) indicated four subspecies of *P. peregrinus* and that some probably warrant specific status. Groves (2005e) included *occidentalis* in *P. peregrinus* but noted that the latter is probably a species complex. The genera *Pseudochirulus*, *Pseudochirops*, *Petropseudes*, and *Hemibelideus* (see accounts thereof) sometimes have been included in *Pseudocheirus*.

Head and body length is 287–400 mm, tail length is 287–360 mm, and weight is around 700–1,400 grams (De Tores 2008a; Flannery and Schouten 1994; McKay and Ong 2008). The fur is short, dense, and soft. The coloration is highly variable, ranging from predominantly gray to rich red, and the underparts are paler. There are always distinctive white ear tufts and a white tail tip. The tail is tapered, with the fur progressively shorter distally and a long, bare friction pad on the undersurface of the distal portion. This prehensile end of the tail is commonly curled into a ring. The first two digits of the forefoot are opposable to the other three digits. Females have four functional mammae.

Common ring-tailed possum (*Pseudocheirus peregrinus*). Photo by Pavel German / Wildlife Images.

Common ring-tailed possums are scansorial and nocturnal. They may be found in rainforest, sclerophyll forest, woodland, or brush. In the northern part of Australia they shelter by day in a tree hollow, sometimes lined with dry leaves (McKay and Ong 2008). In the south and west they generally sleep in a spherical or dome-shaped nest of interwoven leaves, bark, and ferns located in the branches of a shrub or tree or in dense undergrowth. The height of the nest varies from only a few centimeters above the ground to about 25 meters in mistletoe clusters (Collins 1973). *P. occidentalis* sometimes nests in hollow logs on the ground and in roof spaces of human buildings (Woinarski, Burbidge, and Harrison 2014). In the wild these possums seem strictly herbivorous, feeding mostly on a variety of leaves and also on fruits, flowers, bark, and sap. In major Australian cities, *P. peregrinus* lives in close association with humans and feeds in gardens (McKay and Ong 2008).

Movements are generally slow, and the quiet, retiring manner may make these possums uninteresting to human observers. When moving from one limb to another they usually keep hold of the old resting place with the tail until they have grasped the new branch with the forefeet. They sleep with the head beneath the body and between the hind feet. Although they are agile and graceful, their usual reaction to a sudden encounter is to remain motionless with a vacant stare. If the intruder remains, the possums may slowly creep away. On the ground they travel at fair speed with a waddling gait.

The scarce western species, *P. occidentalis*, has been reported to occur at densities of around 1 to 13 per ha. (Finlayson et al. 2010) and to have an average home range of 1.0–2.5 ha. (Jones, How, and Kitchener 1994). Density of *P. peregrinus* is commonly 12–16 per ha., up to 19 per ha. in favorable habitat, and home range averages 0.37 ha. The home range of a male of the latter

Common ring-tailed possum (*Pseudocheirus peregrinus*). Photo by Pavel German / Wildlife Images.

species usually includes the ranges of two females, while the ranges of adjacent males have little overlap. In areas where density is high, individuals are very aggressive toward one another. Otherwise *P. peregrinus* is semisocial and even gregarious, with reports of up to three adults per nest, up to eight individuals constructing their nests in close proximity, and males and females remaining together all year and into subsequent breeding seasons. The species can often be detected by its soft, high-pitched, twittering call (Collins 1973; How 1978; How et al. 1984; McKay and Ong 2008; Ride 1970).

P. peregrinus is polyestrous, with an estrous cycle of about 28 days and gestation period of 14–16 days (Jackson 2003). According to McKay and Ong (2008), *P. peregrinus* breeds from April (autumn) to November (spring) throughout its range. Most females in Victoria and northern Queensland give birth at the begin-

ning of the season but in the vicinity of Sydney there is a high incidence of breeding in the spring. In the suburbs of Melbourne, a first litter is born in May or early June. The pair may then mate again and produce a second litter in mid- to late November. After the young permanently exit the pouch when about four years old, both parents care for them and carry them about on their backs. Litter size is one to four, usually two. The young are weaned at 6–7 months, disperse at 8–12 months, and reach sexual maturity at about 1 year; longevity in the wild is commonly 4–6 years (Collins 1973; How 1978; How et al. 1984; Jackson 2003; Ride 1970; Tyndale-Biscoe 1973). Weigl (2005) listed a captive female *P. peregrinus* that lived for 10 years and 1 month and a captive male that lived for 9 years and 7 months. *P. occidentalis* breeds year-round, with birth peaks in April–June and September–November, but no female gives birth twice in the same year, and there is usually only a single young; pouch life is about 104 days, and young males disperse when about 7 months old, while young females tend to remain within their mother's home range (De Tores 2008a; Jones, How, and Kitchener 1994; Wayne et al. 2006).

The relatively low reproductive rate of *P. occidentalis* may be one factor in the recent disappearance of the species from much of its historic range. The decline also is associated with predation, especially by the introduced red fox (*Vulpes vulpes*) but also by the house cat (*Felis catus*) and python (*Morelia spilota imbricata*), as well as loss and fragmentation of forest habitat because of logging and land clearing (De Tores 2008a; Wayne et al. 2005b). The current IUCN classification is vulnerable, though that is based on outdated information, including the assessment simply that the range is now less than 20,000 sq km (Morris, Burbidge, and Friend 2008). Using IUCN criteria, Woinarski, Burbidge, and Harrison (2014) reclassified *P. occidentalis* as critically endangered, noting that the remaining area of occupancy is less than 500 sq km and is severely fragmented. In addition to continuation of the above threats, there has been an acute decline in association with a drying climate that is causing heat waves and droughts. Overall numbers were thought to have fallen by more than 80 percent in the last 10 years.

P. peregrinus seems to have fared better, though populations in parts of Victoria and on the Cape York Peninsula have become uncommon or restricted in distribution, apparently because of agricultural development and predation by the fox and house cat (Emison et al. 1978; McKay and Ong 2008; Winter 1979). In Tasmania, where presumably the cold winters result in desirable pelts, *P. peregrinus* was an important fur bearer; approximately 7.5 million were taken from 1923 to 1955 (Ride 1970). Such hunting contributed to a rapid decline in Tasmania after 1940, though populations began to recover in the 1960s (Munks et al. 2004).

DIPROTODONTIA; PSEUDOCHEIRIDAE; Genus *PSEUDOCHIRULUS* Matschie, 1915

New Guinean and Queensland Ring-tailed Possums

There are eight species (Flannery 1995a, 1995b; Flannery and Schouten 1994; Groves 2005e; Kirsch and Calaby 1977; Musser and Sommer 1992; Ride 1970; Winter and Trenerry 2008; Ziegler 1977):

> *P. canescens*, New Guinea and Japen and Salawati Islands to the west;
> *P. mayeri*, Central Cordillera of New Guinea;
> *P. caroli*, west-central New Guinea;
> *P. schlegeli*, Arfak Mountains of the Vogelkop Peninsula in western New Guinea;
> *P. forbesi*, mountains of southeastern Papua New Guinea;
> *P. larvatus*, eastern Central Cordillera, Huon Peninsula, and North Coast Range of northeastern New Guinea;
> *P. herbertensis*, Ingham to Atherton Tablelands in northeastern Queensland;
> *P. cinereus*, Thornton Peak massif, Mount Windsor Tableland, and Mount Carbine Tableland in northeastern Queensland.

Pseudochirulus was long considered part of the genus *Pseudocheirus* and was still treated as such by McKenna and Bell (1997). However, Flannery and Schouten (1994), interpreting the molecular analyses of Baverstock (1984) and Baverstock et al. (1990), regarded *Pseudochirulus* as a separate genus. *P. larvatus* usually has been considered a subspecies of *P. forbesi*, but Musser and Sommer (1992) suggested that it and

Queensland ring-tailed possum (*Pseudochirulus herbertensis*). Photo by Jiri Lochman / Lochman LT.

also *P. forbesi lewisi* of the Vogelkop Peninsula might be distinct species. Groves (2005e) formally listed *P. larvatus* as a species but followed Flannery (1995a) in recognizing *lewisi* as a synonym of *P. schlegeli*.

Head and body length is about 167–400 mm, tail length is 151–470 mm, and weight is 105–1,450 grams (Flannery and Schouten 1994; Winter and Moore 2008). In most species the fur is dense, soft, and woolly. The upper parts of the body are gray or brown, often very dark, and the underparts are white, yellowish, or almost as dark as the back. Some species have dark and light markings on the head, a median stripe on the back, or stripes around the thighs. The distal end of the tail is usually bare for some length on the undersurface and only sparsely haired on the upper surface. In most species the prehensile end of the tapered tail is usually curled into a ring, hence the common name of these possums. The first two digits of the forefoot are opposable to the other three digits. Females have four mammae, but normally only two are functional.

All species are arboreal and mainly nocturnal forest dwellers. Their habits are not well known but are thought to be comparable to those of *Pseudocheirus*. Some species build nests like those of *Pseudocheirus* or shelter in hollow trees. The diet is largely folivorous (leaf-eating) and also includes fruit. *P. mayeri* eats lichens, mosses, pollen, and fungi, as well as leaves (Flannery 1995a). According to Winter and Moore (2008), *P. herbertensis* attains highest densities, about 9 individuals per ha. at elevations above 800 meters. Its home range is 0.5–1.0 ha., and those of males and females overlap. Adults of that species are solitary except when males are attracted to estrous females. They have a variety of screeches and grunts in antagonistic situations. Breeding extends through much of the year but most births occur from May to July. Usually two young

New Guinean ring-tailed possum (*Pseudochirulus forbesi*). Photo by Pavel German / Wildlife Images.

are reared. They emerge from the pouch after 4 to 5 months and may then be carried on the mother's back for a short period; thereafter they are left in the nest and make increasingly longer forays alone. *P. cinereus*, *P. forbesi* and *P. canescens* also appear to breed year-round; *P. cinereus* rears two young, while *P. forbesi* usually has only a single young and litter size in *P. canescens* is 1–3 (Flannery 1995a; Winter and Trenerry 2008).

The IUCN classifies *P. schlegeli* as vulnerable because it occurs in less than 20,000 sq km and at fewer than five known locations, and it continues to decline through human encroachment and hunting (Helgen, Dickman, and Salas 2008b). No other New Guinea species is in a threatened category, though populations of some are decreasing and subject to excessive hunting. Winter (1984) noted that *P. herbertensis* may be in jeopardy because of logging of its tropical forest habitat. However, Winter and Moore (2008) reported that, while 20 percent of its habitat had been cleared, the species is secure as most remaining habitat is now within the Wet Tropics World Heritage Area. Using IUCN criteria, Woinarski, Burbidge, and Harrison (2014) classified *P. cinereus* as near threatened; the species is restricted to three discontinuous localities within an area of less than 1,000 sq km, and a severe decline is projected because of clearing and fragmentation of its habitat.

DIPROTODONTIA; PSEUDOCHEIRIDAE; Genus
***HEMIBELIDEUS* Collett, 1884**

Brush-tipped Ring-tailed Possum

The single species, *H. lemuroides*, is found only in rainforest above an elevation of 450 meters between Ingham and Cairns in northeastern Queensland, with a small isolated population above 1,100 meters on the Mount Carbine Tableland west of Mossman (Winter, Moore, and Wilson 2008). *Hemibelideus* sometimes is considered a subgenus of *Pseudocheirus*, and was retained in the latter genus by McKenna and Bell (1997), but serological and morphological studies suggest the two are generically distinct (Baverstock 1984; Groves 2005e; Johnson-Murray 1987). Molecular studies point to close affinity between *Hemibelideus* and *Petauroides* (Kirsch, Lapointe, and Springer 1997; McQuade 1984).

Brush-tipped ring-tailed possum (*Hemibelideus lemuroides*). Photo by Martin Withers / FLPA / Minden Pictures.

Head and body length is 313–400 mm, tail length is 314–384 mm, and weight is 810–1,170 grams (Winter, Moore, and Wilson 2008). The upper parts vary in color from light fawn to dark blackish gray, and the underparts are yellowish gray. The head is dark brown with a tinge of red, and the limbs are dark brown with black near the ends. The ears project only a short distance beyond the thick fur of the head. The pelage is soft and woolly, even on the feet. The prehensile tail is black, bushy, thickly covered with fur for the whole length on the upper surface, and slightly tapering; the naked underside of the tip is very short.

Hemibelideus possesses suggestions of gliding membranes, in that it has small folds of skin (less than 25 mm in width) along the side of the body. It has been observed making long jumps from tree to tree and from tree to ground. Those factors, along with molecular study (see above), suggest the genus is transitional between the other ring-tailed possums and the greater gliding possum (*Petauroides*). It is said to be very agile and active in the trees, and some of its jumps almost have the appearance of true glides. It seems to use its furry tail as a rudder during those jumps.

According to Winter, Moore, and Wilson (2008), this possum is strictly arboreal and nocturnal, spends the day in a tree hollow, and emerges just after dark to forage. Leaps from branch to branch often cover 2–3 meters. The diet consists almost exclusively of leaves. Population density is up to 7 per ha. and home range is no more than 1.2 ha. *Hemibelideus* is frequently seen in groups of two individuals (mother and young or male and female) and in family groups of three. Up to three may share a den, and feeding aggregations of as many as eight have been found in a single tree. Females have two mammae in the pouch, but usually only one young is reared. Young have been recorded in the pouch from August to November and riding on the mother's back from October to April. Weigl (2005) listed a specimen that lived in captivity for 5 years and 5 months.

Winter (1984) cautioned that *Hemibelideus* may be in jeopardy because of logging of its tropical forest habitat. Laurance (1990) found that the genus could be relatively abundant on large tracts of primary forest but that it declined by more than 97 percent when such areas were fragmented into small areas, thereby preventing normal movement through trees. Woinarski, Burbidge, and Harrison (2014) cited a total population estimate of 186,000 individuals. The IUCN designates *Hemibelideus* as near threatened, noting that it occurs in less than 3,000 sq km, is very sensitive to habitat disturbance, and is "a prime candidate to be affected by global warming" (Burnett and Winter 2008c).

DIPROTODONTIA; PSEUDOCHEIRIDAE; **Genus PETAUROIDES** Thomas, 1888

Greater Gliding Possum, or Greater Glider

The single species recognized here, *P. volans*, occurs in eastern Queensland, eastern New South Wales, and eastern Victoria (Ride 1970). The use of the name *Pet-*

auroides in place of the former *Schoinobates* Lesson, 1842, was explained by McKay (1982). There are two subspecies, *P. v. volans* in that part of the range from about the Burnett River and Rockhampton in central eastern Queensland south to Victoria, and *P. v. minor* from the Windsor Tableland and Barron River in northeastern Queensland south to about the Dawson River in central eastern Queensland (Woinarski, Burbidge, and Harrison 2014). Regarding the latter and referring to Kenneth P. Aplin, Jackson and Thorington (2012:6, 118) stated: "Recently, Aplin (pers. comm.) recognized its distinctiveness, both morphologically and genetically, as being distinct from *P. volans*, so it is likely to be elevated to species rank in the future." Jackson (2015a) did treat *P. v. minor* as a full species and also accepted specific distinction for *P. armillatus*, which hitherto has usually been considered a synonym of *P. v. minor* and which occurs from just north of Townsville, south to Eungella Range and the vicinity of Roma. Still another species that could be referable to *Petauroides*, *P. ayamaruensis*, was described from mid-Holocene remains from Kria Cave in the central Vogelkop Peninsula of western New Guinea by Aplin, Pasveer, and Boles (1999) and was considered by Helgen and Veatch (2015) likely to be extant in isolated mountains of New Guinea.

This is the largest gliding marsupial. The thick, shaggy coat obscures the basic body form, making the animal appear much larger than it actually is (Harris and Maloney 2010). Head and body length is 300–480 mm and tail length is 450–600 mm. Adult weight is 900–1,700 grams (McKay 2008). The fur is soft and silky. Coloration is variable, ranging from black to smoky gray or creamy white on the upper parts and sooty gray, grayish white, or pure white on the underparts. All-white and white-headed individuals are common. The long tail is prehensile and evenly furred except the underpart of the tip, which is naked. Females have a well-developed pouch and two mammae.

This attractive marsupial glides with the use of a patagium, consisting of a fold of skin extending from the elbow to the leg. The position of the arms of *Petauroi-*

Greater gliding possum (*Petauroides volans*). Photo by Jiri Lochman / Lochman LT.

Greater gliding possum (*Petauroides volans*). Photo by Pavel German/ Wildlife Images.

des when gliding is entirely different from that of gliding squirrels and of the lesser gliding possums (*Petaurus*): they are bent at the elbows, so that the forearms are directed in toward the head and the hands almost meet on the front part of the chest. Grzimek and Ganslosser (1990) noted that the gliding membrane apparently has a secondary use as a blanket: the animal wraps it around its body as a protection against loss of heat.

Wakefield (1970b) wrote that attributes of the greater gliding possum have been confused with those of the largest of the lesser gliding possums, *Petaurus australis*. Whereas the latter species can glide up to about 114 meters, is highly maneuverable in the air, and calls loudly while gliding, *Petauroides* is a sedentary, slow-moving, silent animal of "minor gliding ability." McKay (2008), however, wrote that glides of *Petauroides* may cover a horizontal distance of up

to 100 meters and involve changes in direction of as much as 90 degrees. One individual was observed moving more than 0.5 km in six successive glides from tree to tree. After each landing it ran straight up the vertical trunk to the top of the tree to begin the next glide; it emitted a squeaking sound as it climbed (Grzimek and Ganslosser 1990). Just before reaching the target tree, *Petauroides* directs its flight upward, so that it slows and lands with all four feet against the trunk (McKay 2008).

The habitat is sclerophyll forest and tall woodland (Ride 1970). In many parts of its range it appears to be most common in high-elevation, tall, wet eucalypt forests (Harris and Maloney 2010). Although this possum apparently travels across open ground at times, it spends nearly its entire life in trees. It is nocturnal, sheltering by day in large hollows high up in old trees. Typically, 4–20 different dens are used by an individual within its home range (Harris and Maloney 2010). Some individuals make a nest of stripped bark or leaves in their den, but often no material is added. There is some indication that this possum transports nesting material in its rolled-up tail. The musty eucalyptus smell of this animal permeates the shelter. The diet is very specialized, consisting mainly of the leaves, bark, and bud debris of certain species of *Eucalyptus* (Marples 1973). Like *Phascolarctos* (koala), *Petauroides* has a greatly enlarged caecum, in which much of the cellulose of leaves is broken down to assimilable substances though bacterial fermentation (McKay 2008).

The following information on population structure and life history was compiled from Collins (1973), Harris and Maloney (2010), Henry (1984), How (1978), Kavanagh and Wheeler (2004), Kerle and Borsboom (1984), Pope, Lindenmayer, and Cunningham (2004), Smith (1969), and Tyndale-Biscoe and Smith (1969). Population densities of 0.24–3.80 individuals per ha. have been reported, the variation reflecting availability of suitable tree hollows, foliar nutrient levels, and other ecological factors. Captive pairs or groups can be maintained in a sufficiently large enclosure. Wild adult males utilize home ranges that usually are largely separate from those of other males but overlap with those of females. The ranges of females overlap to some extent, but the females tend to avoid one another. Home range size is commonly about 0.5 to 4.0 ha., averaging around 2 ha. for males and 1 ha. for females, but in southern Queensland males have been recorded using ranges of up to 19.3 ha. Some males are monogamous and others are bigamous. There is some interaction between the sexes all year, but contact peaks just prior to and during the mating season. Females are polyestrous and have one litter per season, though up to 50 percent of females do not breed in any given year. In New South Wales mating occurs from March to May and births occur from April to June. In Victoria the young usually are born in July and August. Litters contain a single young. It first releases the nipple at about 6 weeks but usually spends 4, sometimes up to 6, months in the pouch and then another 4 months as a dependent nestling. It may sometimes be carried on the back of the mother, though not when gliding. There evidently is no paternal care, and adult males are believed to be antagonistic to their male offspring, but young females presumably inherit a part of their mother's range. Full

independence comes at about 10–13 months, and sexual maturity is attained in the second year of life. Longevity may be 15 years in the wild.

Kennedy (1992) regarded *Petauroides* as "potentially vulnerable" because of the continued fragmentation and disturbance of the coastal forests on which it depends. The IUCN assigns no threatened category but does note that population size is decreasing (Lunney et al. 2008b). McKay (2008) observed that conservation of *Petauroides* is utterly dependent on the maintenance of sufficient amounts of old-growth forest and that its abundance in undisturbed areas contrasts strongly with its absence from pine plantations and its scarcity in regenerated forest lacking old trees with suitable large hollows for nesting. McCarthy and Lindenmayer (1999) suggested that the greater gliding possum could persist if suitable old-growth habitat, in patches of at least 3 ha., covered more than 10 percent of the landscape. Using IUCN criteria, Woinarski, Burbidge, and Harrison (2014) classified the species *P. volans* and each of its subspecies (or possibly species; see above) as vulnerable. Long-term monitoring at several sites indicated substantive declines in Victoria and Queensland, and the total population of *P. v. minor* was suspected to be fewer than 10,000 individuals.

Greater gliding possum (*Petauroides volans*). Photo by Pavel German / Wildlife Images.

DIPROTODONTIA; **Family PETAURIDAE**

Gliding and Striped Possums

This family of 3 Recent genera and 11 species inhabits Australia, Tasmania (through introduction), New Guinea, and certain nearby islands. The members of the Petauridae sometimes have been placed in the family Phalangeridae. Kirsch (1977b), however, explained that the Petauridae differ from the Phalangeridae in serology, karyology, and other characteristics. He regarded the Pseudocheiridae (see account thereof) as one subfamily of the Petauridae and also recognized two other subfamilies: Petaurinae, for the genera *Gymnobelideus* and *Petaurus*, and Dactylopsilinae, for the genus *Dactylopsila*. Flannery and Schouten (1994) and Aplin (2008) maintained the latter two subfamilies, though DNA analyses by Edwards and Westerman (1992) and Osborne and Christidis (2001) indicate that *Gymnobelideus* actually is more closely related

Sugar glider (*Petaurus breviceps*). Photo by Gerhard Körtner.

to *Dactylopsila* than to *Petaurus*. Recently, Jackson (2015b) placed *Gymnobelideus* in subfamily Dactylopsilinae, but that procedure is not followed herein. On the basis of serological studies, Baverstock (1984) and Baverstock, Birrell, and Krieg (1987) suggested that *Distoechurus* and *Acrobates* have affinity to Petauridae, but those genera now are placed in a separate family, Acrobatidae (see account thereof).

Head and body length is 120–320 mm, and tail length is 150–480 mm. The pelage may be woolly or silky. There is a prominent dark dorsal stripe that extends onto the forehead (Aplin 2008). All members of the Petauridae are arboreal, and one genus, *Petaurus*, has attained a gliding ability through development of a membrane uniting the front and hind limbs.

According to McKay (1989), the dentition of the Petauridae is characterized as follows: three upper incisors on each side of the jaw, the first pair normally longer than the succeeding pairs and projecting anteriorly; the upper canine small in the Dactylopsilinae but larger and laterally compressed in the Petaurinae; three single-cusped upper premolars, the first and sec-

ond laterally compressed and the third conical; four bunodont upper molars, decreasing in size from first to last, each having four low pointed cusps; the procumbent first lower incisor slightly curved in the Petaurinae and strongly curved and greatly enlarged in the Dactylopsilinae; a second, vestigial lower incisor; three small lower premolars; and four lower molars, decreasing in size from first to fourth. The skull is characterized by a slender and rounded zygomatic arch and a palate without posterior vacuities. The forefoot shows no specialization for climbing other than enlarged claws; in the Petaurinae the second to fifth digits are all subequal in length, but in the Dactylopsilinae the fourth digit is considerably elongated as an adaptation for feeding. The tail is semiprehensile; in the Petaurinae it is entirely furred, but in the Dactylopsilinae there is a small ventral naked patch at the tip. The caecum of the digestive tract is large in the Petaurinae but very small in the Dactylopsilinae.

Goldingay and Jackson (2004) reviewed the ecology of the Australian Petauridae. The various species live in forest and woodland and feed primarily on plant and insect exudates, except for *Dactylopsila trivirgata*, which takes mostly arthropods. Most petaurids are highly social, with groups defending territories that include the entire home range or a core area with den and feeding trees. There is a preponderance of breeding in the winter.

The geological range of the Petauridae is late Oligocene to Recent in Australia (Long et al. 2002) and Recent in New Guinea (Flannery 1995a). Although Flannery (1995a) suggested a New Guinean origin for the Dactylopsilinae and that the single Australian species is a recent immigrant from New Guinea, Aplin (2008) noted that the subfamily is known from Oligo-Miocene deposits in Queensland.

DIPROTODONTIA; PETAURIDAE; **Genus** *PETAURUS* Shaw and Nodder, 1791

Lesser Gliding Possums

Six species are currently recognized (Aplin and Pasveer 2007; Colgan and Flannery 1992; Flannery 1995a, 1995b; Flannery and Schouten 1994; Goldingay 2008; Jackson 2008; McKay 1988; Ride 1970; Smith

Sugar glider (*Petaurus breviceps*). Photo by McDonald Wildlife Photog. / AnimalsAnimals Earth Scenes.

1973; Suckling 2008; Van Der Ree and Suckling 2008; Ziegler 1981):

P. breviceps (sugar glider), Halmahera Islands (Halmahera, Batjan, Ternate, Gebe) in the North Moluccas, New Guinea and many nearby islands (Salawati, Misool, Numfoor, Japen, Koil, Vokeo, Wei, Kadovar, Bam, Blup Blup, Karkar, Bagabag, Adi, Kai Besar, Wamar), Bismarck Archipelago (New Britain, Duke of York), D'Entrecasteaux Islands (Goodenough, Fergusson, Normanby), Woodlark Island, Louisiade Archipelago (Misima, Sudest), northeastern Western Australia, northern Northern Territory and nearby Tiwi and Groote Eylandt Islands, northern and eastern Queensland and nearby Normanby Island, eastern New South Wales, Victoria, extreme southeastern South Australia, introduced in Tasmania;

P. biacensis (Biak glider), Biak, Supiori, and Owi Islands off northwestern New Guinea;

P. norfolcensis (squirrel glider), central Cape York Peninsula through eastern Queensland and eastern New South Wales to Victoria and extreme southeastern South Australia, several islands off coast of southeastern Queensland (Ricketts, Fraser, Moreton, North and South Stradbroke);

P. gracilis (mahogany glider), vicinity of Tully to Ingham on coast of northeastern Queensland;

P. abidi (northern glider), North Coast Range in northwestern Papua New Guinea;

P. australis (fluffy or yellow-bellied glider), coastal parts of eastern Queensland and New South Wales, eastern and southern Victoria, extreme southeastern South Australia.

Populations, now assigned to *P. breviceps*, in the Tifalmin area of western Papua New Guinea, in the D'Entrecasteaux Islands off southeastern New Guinea, and on Karkar Island off northeastern New Guinea show certain distinctive features and may possibly represent separate species (Colgan and Flannery 1992; Flannery 1995a; Flannery and Schouten 1994; Groves 2005e). Although Helgen (2007) agreed that *P. biacensis* is "morphologically distinctive," he considered it not specifically distinct from *P. breviceps*, since it differs mainly in its higher incidence of melanism. There were no formal records of *P. gracilis* from 1886 to its rediscovery in 1989, and during that period it had come to be treated as a synonym of *P. norfolcensis*, but it has since been generally recognized as a full species (Jackson 2011; Jackson and Thorington 2012). Analysis of mitochondrial and nuclear DNA by Malekian, Cooper, and Carthew (2010) and Malekian et al. (2010) provided strong support for a sister relationship of *P. australis* to all other species of *Petaurus*, and for a sister relationship of *P. abidi* to a clade containing *P. norfolcensis*, *P. gracilis*, and *P. breviceps* (*P. biacensis* was not assessed). It was also found that *P. breviceps* showed considerable diversity, with two divergent clades in Australia and five in New Guinea; levels of divergence among those clades was similar to that observed between *P. breviceps* and *P. norfolcensis*, thus suggesting the possible existence of additional species.

Head and body length is 120–320 mm and tail length is 150–480 mm. Weights are 69–160 grams in *P. breviceps* (Flannery 1995a; Suckling 2008), 79–100 grams in *P. biacensis* (Flannery 1995b), 190–300 grams in *P. norfolcensis* (Van Der Ree and Suckling 2008), 310–500 grams in *P. gracilis* (Jackson 2008), 228–332 grams in *P. abidi* (Flannery 1995a), and 435–725 grams in the larger *P. australis* (Goldingay 2008). The fur is fine and silky. In *P. breviceps*, *P. biacensis*, *P. norfolcensis*, and *P. abidi* the upper parts are generally grayish and the underparts are paler. The color of *P. gracilis* varies from overall mahogany brown (dorsal and ventral) to gray-brown dorsally with buff to apricot belly (Jackson 2008). In all those species a dark dorsal stripe runs from the nose to the rump, and there are stripes on each side of the face from the nose through the eye to the ear. In *P. australis* the upper parts usually are dusky brown, markings are less conspicuous, the feet are black, and the underparts are orange yellow.

Lesser gliding possums resemble flying squirrels (*Glaucomys*) in form and in having a large gliding membrane, but the tail of the former is furred all around and not as flattened as that of *Glaucomys*. As in *Glaucomys*, but not the greater gliding possum (*Petauroi-des*), the gliding membrane extends all the way from the outer side of the forefoot to the ankle and is opened by spreading the limbs straight out.

Females have a well-developed pouch during the breeding season; mammae number four, occasionally two. Pouch morphology of *P. australis* is unique among marsupials in having two compartments separated by a well-furred septum (Craig 1986). Other petaurids have developed the same feature to some extent; it serves to position the young on either side of the center of the body, so they are not subject to the full force of impact upon landing after a glide (Jackson 1999).

These possums are extremely active. *P. breviceps* can glide up to 50 meters and has been observed to leap at and catch moths in flight, and *P. australis* can glide for 140 meters; vertical and lateral angle can be changed in the air (Goldingay 2008; Ride 1970; Smith 1973; Suckling 2008). Gliding distance increases with body size; mean distances are 20 meters for *P. breviceps*, 30 meters for *P. gracilis*, and 40 meters for *P. australis* (Goldingay and Jackson 2004). Jackson (1999) reported that *P. gracilis* appears to carefully assess direction and landing point before a long glide. It climbs to the top of a tree and generally sways from side to side and bobs its head up and down as if determining whether to glide or not; this may be a method of triangulation to measure distance, as has been suggested for flying squirrels. It may hesitate or back away from a glide, frequently finding an alternative route along tree branches. Both *P. gracilis* and *P. breviceps* have a remarkable ability to steer by changing position of limbs and the tension of the muscular gliding membrane. A long glide may include several turns around nontarget trees and branches. One individual was observed to make a full U-turn and land on the same tree from which it had departed. Both species swoop upward before landing so that speed is reduced and landing is made with all four feet.

All species inhabit wooded areas, preferably open forest, and are arboreal and largely nocturnal. They generally shelter by day in a spherical or bowl-shaped leaf nest in a tree hollow. Crane, Lindenmayer, and Cunningham (2010) found that each individual *P. norfolcensis* used an average of seven hollow trees as den sites; average distance between dens used on successive days was 218 meters. *P. breviceps* usually collects leaves by hanging by its hind feet and passing the leaves

Fluffy or yellow-bellied glider (*Petaurus australis*). Photo by Gerhard Körtner.

via the forefeet to the hind feet and then to the tail, which then coils around the nest material. The tail cannot be used for gliding when it is employed in this manner, so the possum transports its load of leaves by running along branches to its hollow. All species are omnivorous, feeding on sap, blossoms, nectar, pollen, insects and their larvae, arachnids, and small vertebrates (Collins 1973; Smith 1973; Wakefield 1970b). *P. australis* is an obligate sap-feeder (Goldingay 1990, 2008; Goldingay and Jackson 2004). It removes bark from various eucalyptus trees to get the sugary sap. It makes characteristic V-shaped cuts in the bark that channel the sap to its mouth, placed at the bottom of the V. *P. australis* devotes 90 percent of the time outside of

its den to foraging, 70 percent of that time to feeding on eucalyptus nectar. It probably assists in cross-pollinating eucalyptus trees. It may travel over 2 km in a night while foraging. Nightly movement in *P. gracilis* is about 600–3,400 meters (Jackson 2000b, 2008).

Fleming (1980) reported that captives of *P. breviceps* huddle together to conserve energy during cold weather and that they may simultaneously enter daily torpor when winter food supplies are low. Studies of that species in the wild by Körtner and Geiser (2000) showed that from autumn to spring (May–October) individuals frequently entered torpor on cold or rainy nights for an average period of 13 (2–23) hours, during which body temperature fell from a normal 32–34° C to a minimum of 10.4° C. The process apparently served to lower energy loss at times when foraging was not feasible. Torpor also has been observed in *P. australis* (Geiser and Körtner 2004).

In a study in remnant natural vegetation in Victoria, Suckling (1984) found the average population density of *P. breviceps* to vary from 2.9 per ha. in summer to 6.1 per ha. in autumn. Average home range was about 0.5 ha., though males extended their range slightly during the breeding season. In a study of the same species in New South Wales, Quin et al. (1992) found home ranges nearly 10 times as large. In an area of continuous habitat in northern Queensland, average density was 0.27 per ha for *P. breviceps* and 0.24 per ha. for *P. gracilis*, but densities in fragmented habitat were 0.46 per ha. for *P. breviceps* and only 0.16 per ha. for *P. gracilis* (Jackson 2000a). Monogamous pairs of *P. gracilis* have a home range of approximately 20 ha., within which they have 6 to 13 dens. Ranges in fragmented habitat are barely half as large. Territorial defense ap-

pears strong, with both sexes traversing the edge of the home ranges at least once every 2–3 nights (Jackson 2000b, 2008, 2011). Average range size for six groups of *P. norfolcensis* was 6.7 ha.; there was about 50 percent overlap of those ranges but only 12 percent overlap of core areas containing den trees (Sharpe and Goldingay 2007).

All species of *Petaurus* appear highly social. Collins (1973) reported that captive *P. australis* can be maintained in pairs and that *P. breviceps* and *P. norfolcensis* can be kept in groups but that established groups might attack newly introduced individuals. Wild groups of *P. norfolcensis* typically include one or two adult males, one or two adult females, and their offspring; one group occupied 19 different tree hollows over 40 days of radio-tracking (Van Der Ree and Suckling 2008).

P. breviceps nests in groups of up to seven adult males and females and their young (Ride 1970; Smith 1973; Suckling 1984). All members probably are related and descended from an original colonizing pair. Groups

Sugar glider (*Petaurus breviceps*). Photo by Pavel German / Wildlife Images.

appear mutually exclusive, territorial, and agonistic toward one another. One or two dominant, usually older males are responsible for most territorial maintenance, aggression against intruders, and fathering of young. There are a few lone adults, and the young generally leave their natal group at 10–12 months. *P. breviceps* has a complex chemical communication system based on scents produced by frontal, sternal, and urogenital glands of males and by pouch and urogenital glands of females. Each animal of a group has its own characteristic smell, which identifies it to other individuals and which is passively spread around the group's territory. In addition, a dominant male actively marks the other members of the group with his scent. Although some studies of *P. breviceps* in captivity indicate formation of male coalitions and dominance hierarchies, Sadler and Ward (1999) found little evidence for such in the wild; males generally dispersed, while females tended to remain in the natal group.

These possums produce a variety of sounds. *P. breviceps* has an alarm call resembling the yapping of a small dog and a high-pitched cry of anger (Smith 1973). It also makes buzzing, droning, hissing, and clicking sounds (Flannery 1995a). The most frequent sound of *P. norfolcensis* is a nasal grunt that appears to regulate individual spacing (Sharpe and Goldingay 2009). According to Goldingay (1994, 2008), *P. australis* is the most vocal of all marsupials, with loud shrieking calls audible 500 meters away and a moan or gurgle usually emitted during glides. Individuals call as often as 10–15 times per hour when foraging. The loud vocalizations serve to define territories of a group. Trespassers are readily attacked and even killed.

Studies of *P. australis* (Brown, Carthew, and Cooper 2007; Craig 1985, 1986; Goldingay 1992, 2008; Goldingay and Kavanagh 1990, 1993; Goldingay, Quin, and Churchill 2001; Henry and Craig 1984) show that the species is thinly distributed but has a variable social organization. Most estimates of population density range from about 1 per 25 ha. to 1 per 6 ha. A population is divided into pairs or small groups of adults, with or without dependent young. Pairs may form stable, monogamous units; one unit existed throughout a 41-month study period. A male and female may spend most of their waking time in close proximity and generally sleep together in a den. Polygynous units consisting of one adult male and several adult females also

Mahogany glider (*Petaurus gracilis*). Photo by Alexander F. Meyer at David Fleay Wildlife Park, Gold Coast, Australia.

may be found, especially where food resources are abundant. Some groups include two or three adult males and/or two or three adult females, but membership may change through the year. Groups occupy exclusive home ranges, or territories, of 7–100 ha. Those areas include a variety of habitats, especially eucalyptus forest, favored for denning and feeding. The ranges of a monogamous male and female overlap closely, forming a joint territory, which shows little overlap with the ranges of other individuals. Breeding occurs mainly from about June to December in Victoria and New South Wales, but at one site in the latter state females were observed to give birth predominantly from February to April, presumably in association with availability of food resources. Breeding is year-round in Queensland. Almost always a single young per pair is raised in a year, though there is a record of twins. Both parents provide care for the young. Dispersal and independence has been observed at various ages from 9 to 24 months.

The species *P. breviceps* is polyestrous, with an estrous cycle averaging about 29 days and a gestation period of about 16 days; litter size is commonly two young, occasionally three (Collins 1973; Smith 1971, 1973). In captivity there seems to be no definite breeding season for *P. breviceps* and *P. norfolcensis* (Collins 1973). In the wild, *P. norfolcensis* usually gives birth from April to November, with a peak during winter; occasionally a second litter is raised (Van Der Ree and Suckling 2008). Gestation is slightly less than 3 weeks in *P. norfolcensis* (Smith 1973); litter size is one or two, averaging 1.73 in one study in southeastern Queensland (Millis and Bradley 2001). Female *P. breviceps* breed throughout the year in northern Queensland and may produce a second litter after the first is weaned (Jackson 2000a). The same seems true for *P. breviceps* in the wild in New Guinea and Arnhem Land in the Northern Territory, but in southeastern Australia the young of that species are born only from June to November (Flannery 1995a; Smith 1973). *P. gracilis* has a distinct breeding season from April to October, during which a litter of one or two young is produced; the young are weaned after 4 to 5 months. A second litter is produced only if the first is lost (Jackson 2000a, 2008). The young of *P. breviceps* weigh about 0.19 grams at birth, first release the nipple at about 40 days, first leave the pouch at about 70 days, first leave the nest at 111 days, and are independent shortly thereafter (Collins 1973; Smith 1973). Observation of a wild adult male *P. breviceps* remaining at a nest with the young during the night is demonstrative of direct male parental care and is consistent with the hypothesis that, while the mother forages, care from other family members is required to prevent the young becoming hypothermic (Goldingay 2010). Sexual maturity in both *P. breviceps* and *P. gracilis* appears to come at approximately 12–18 months (Jackson 2000a).

Squirrel glider (*Petaurus norfolcensis*). Photo by Pavel German / Wildlife Images.

Longevity in the wild is about 3 to 5 years in *P. nor-folcensis* (Van Der Ree, Harper, and Crane 2006), 5 or 6 years in *P. gracilis* (Jackson 2008), and usually 4 or 5 years for *P. breviceps*, though ages up to 9 years have been recorded (Suckling 2008). Wild *P. australis* are known to have lived at least 6 years (Goldingay and Kavanagh 1990). Lesser gliding possums may have a long life span in captivity, the recorded maximums being 17 years and 9 months for *P. breviceps*, 14 years and 10 months for *P. norfolcensis*, and 15 years and 3 months for *P. australis* (Weigl 2005).

P. breviceps was introduced into Tasmania in 1835 and has since spread over the island (Smith 1973). It was probably also introduced to parts of its Melanesian distribution (Flannery 1995b). It is comparatively common in mainland Australia and New Guinea, though it and the other species are sensitive to habitat disturbance, especially because of their dependence on patches of natural forest with old trees containing hollows suitable for nesting.

The largest species, *P. australis*, appears to have become rare in some areas (Emison et al. 1975, 1978; Mackowski 1986). Although smaller than *Petauroides volans*, it has a much greater home range and requires a much larger area of old-growth habitat (McCarthy and Lindenmayer 1999). Assessment by Goldingay and Possingham (1995) indicated that *P. australis* is vulnerable due to widespread habitat loss and low population densities resulting from social groups using exclusive home ranges of 25–85 ha. At least 150 groups, in a forested area of 18,000 to 35,000 ha., were considered necessary to support a viable population. Based on analysis of molecular, genetic, morphological, and ecological characters, Brown et al. (2006) suggested that the disjunct population of *P. australis* in northeastern Queensland is distinct and should be given special priority for conservation. The IUCN considers *P. australis* to be declining in numbers and distribution and to be seriously threatened by degradation of habitat due to timber harvesting and agriculture, yet does not place the species in a formal threatened category (Menkhorst et al. 2008).

However, using IUCN criteria, Woinarski, Burbidge, and Harrison (2014) designated the species *P. australis* and the subspecies *P. a. australis*, which comprises the entire species except for the disjunct northeastern population, near threatened. They designated the lat-ter population, which evidently represents a distinct but unnamed subspecies (see also Jackson and Thorington 2012), endangered. It is restricted to a narrow and discontinuous range along the western edge of the Great Dividing Range in northeastern Queensland, from the Windsor Tablelands in the north to Kirrima and the Herbert River in the south. In addition to "catastrophic" loss and fragmentation of habitat through clearing by people, the population is severely threatened by the spread of unnatural rainforest vegetation into the tall open forest, on which it depends for nesting, feeding, and gliding.

The IUCN also currently assigns no threatened classification to *P. norfolcensis*, yet notes the species is decreasing because of steady attrition of quality and extent of habitat by removal of timber, a lack of suitable den hollows in most remaining habitat, and hindrance of forest regeneration by inappropriate fire management and excessive grazing by livestock, introduced rabbits, and macropodids (Winter et al. 2008). Claridge and Van Der Ree (2004) did consider the species endangered in New South Wales.

The IUCN classifies *P. abidi* as critically endangered and *P. gracilis* as endangered (Burnett, Winter, and Martin 2008; Leary et al. 2008w). Both species are narrowly distributed and decreasing in numbers because of deforestation and human encroachment. *P. abidi* occurs in less than 100 sq km; all individuals are in a single location in the Torricelli Mountains and also are subject to local meat hunting. Only 20 percent of the former habitat of *P. gracilis* remains, and that is severely fragmented. Woinarski, Burbidge, and Harrison (2014) also noted that, because of an unnatural reduction in fire frequency and intensity, rainforest vegetation is encroaching on the species' preferred habitat of tall open forest.

DIPROTODONTIA; PETAURIDAE; **Genus *GYMNOBELIDEUS* McCoy, 1867**

Leadbeater's Possum

The single species, *G. leadbeateri*, historically occurred in eastern Victoria from coastal swamp forests on the margins of Westernport Bay, near Melbourne, to alpine ash forests in the Australian Alps (Smith and Harley

Leadbeater's possum (*Gymnobelideus leadbeateri*). Photo by Brent Huffman.

2008). There are historical records from as far as Mount Willis (Harley 2004), and the range may have extended into southeastern New South Wales (Ride 1970). Late Pleistocene fossils have been collected at Wombeyan Cave and Marble Arch in the latter region, and unconfirmed reports indicate that small populations may persist near the border with New South Wales and at other localities beyond the present known range (Harley 2004). *Gymnobelideus* had been known from only six specimens taken between 1867 and 1909 but was rediscovered in 1961 (Lindenmayer and Dixon 1992).

Head and body length is 150–170 mm, tail length is 145–180 mm, and weight is 100–166 grams (Smith and Harley 2008). The sexes are equal in size. The fur is soft but neither as long nor as silky as in *Petaurus*. The upper parts are gray or brownish gray with a dark mid-dorsal stripe extending from the forehead to the base of the tail. The fur of the ventral surface and the inner surface of the limbs is a dull creamy yellow at the tips and light gray beneath. The markings about the ears and the eyes are very similar to those of the sugar glider, *Petaurus breviceps*.

In many respects this possum closely resembles the sugar glider, but it lacks the gliding membrane. Unlike in other possums, the tail is flattened laterally, narrow at the base, and bushes out evenly to the tip. It is not

prehensile to any marked degree and appears to be used for balance when the animal is climbing or jumping. The digits are very wide or spatulate at the tip and bear short, strong claws. *Gymnobelideus* is very active and appears to rely on the grip of the toe pads rather than the claws when climbing. The well-developed pouch of the female contains four mammae.

This possum is found in dense, wet sclerophyll forest at elevations up to about 1,200 meters. It is arboreal and nocturnal and constructs a nest of loosely matted bark about 1–30 meters up in a large hollow tree. Most individuals have dens in two or more trees. The diet consists mainly of plant and insect exudates and also includes a variety of arthropods. A notch may be gnawed into a tree in order to obtain gums (Smith 1984b).

Detailed studies of *Gymnobelideus* have been made in mountain ash forest at Cambarville and in lowland swamp forest at Yellingbo (Harley and Lill 2007; Lindenmayer and Meggs 1996; Smith 1984a, 1984b; Smith and Harley 2008). Populations have a density of 1.6 to 2.9 individuals per ha. and are grouped into colonies of 2 to 12 individuals. Each colony occupies a den tree centered in a defended territory of 1–2 ha. There are prolonged territorial disputes that involve chasing and grappling. Colonies usually consist of a monogamous breeding pair and one or more generations of their offspring. One or two unrelated adult males may also be present. Adult females are more socially aggressive than males, readily attacking and pursuing females from neighboring colonies. Males appear to move freely between colonies. Births occur throughout the year at Yellingbo, while at Cambarville most young are born

Leadbeater's possum (*Gymnobelideus leadbeateri*). Photo by Dave Watts/ Lochman LT.

May–June (autumn) and October–November (spring). Females are polyestrous and can give birth within 30 days of losing a litter. At Yellingbo females typically produce two litters annually. Gestation probably lasts 15–17 days. Litters usually contain one or two young. They remain in the pouch for about 90 days and then spend another 5–40 days in the nest. Female offspring disperse at 7–14 months, and males at 11–26 months. Since the young females are excluded from established colonies, they suffer a high rate of mortality. A captive female, however, was reproductively active until the age of 9 years. A wild male was still alive at 7.5 years. Weigl (2005) listed a captive female that lived for about 14 years and 1 month.

The first four known specimens of this species apparently came from the Bass River Valley and Koo-wee-rup Swamp, lowland areas southeast of Melbourne. Each was taken long before *Gymnobelideus* actually was described, and little or no information was available regarding the living population represented. That lack of information, plus clearing of the involved habitat, led to the assumption that the genus was extinct. In 1931, however, it was learned that another specimen had been taken in 1909 in a highland area farther inland. Initial searches for a population in that area failed, but in 1961 individuals were observed in the Cumberland Valley of eastern Victoria. Subsequent investigation showed the genus to occur in numerous localities over a large area of mountain forest and that it apparently increased in numbers and range following severe brush fires in 1939 (Ride 1970). The fires left many large, dead trees with hollows required for nesting and stimulated regrowth with abundant feeding habitat. An estimated peak population of around 7,500 individuals was reached in the early 1980s (Smith and Harley 2008).

More recently, however, forest clearing destroyed a substantial area of habitat and initiated a decline. Kennedy (1992) reported a population estimate of 5,000 and expected further decreases and habitat fragmentation. Smith and Harley (2008) wrote that a population decline of approximately 90 percent, caused by timber harvesting and the collapse of the dead hollow trees left by the 1939 fires, was anticipated over the next 30 years. Hollows will remain scarce until the year 2075, as those suitable for *Gymnobelideus* do not develop until trees are typically over 190 years old. The genus and its remnant habitat are also closely tied to a narrow set of climatic conditions that could be severely affected by global warming (Lindenmayer et al. 1991). Leadbeater's possum is classified as endangered by the USDI and the IUCN (Menkhorst 2008b), which cited an estimate of 2,000 mature individuals in the main mountain population and about 200 animals in the separate lowland population at Yellingbo.

Hansen et al. (2009) showed that the Yellingbo group is highly genetically differentiated from the more extensive highland population, having been isolated from the rest of the species' range since before European-induced changes to the montane landscape, and that it formed part of a larger genetic unit that is now otherwise extinct. According to Woinarski, Burbidge, and Harrison (2014), the Yellingbo population has fallen to just 60 individuals and has less than 20 ha. of high quality habitat. The highland population is now estimated to contain only about 1,000 animals, having lost 45 percent of its habitat to severe fires in 2009. Further declines in both populations are anticipated, and the species now is considered critically endangered.

DIPROTODONTIA; PETAURIDAE; **Genus *DACTYLOPSILA* Gray, 1858**

Striped Possums

Two subgenera and four species have been described as living taxa (Aplin and Pasveer 2007; Collins 1973; Flannery 1995a, 1995b; Handasyde 2008; Kirsch and Calaby 1977; Ride 1970):

subgenus *Dactylopsila* Gray, 1858

> *D. trivirgata*, New Guinea and Waigeo and Japen Islands to northwest, Aru Islands (Wokam, Kobroor), Cape York Peninsula south to Townsville in northeastern Queensland;
> *D. tatei*, Fergusson Island off southeastern Papua New Guinea;
> *D. megalura*, central and western highlands of New Guinea;

subgenus *Dactylonax* Thomas, 1910

> *D. palpator*, highlands of New Guinea, including Vogelkop Peninsula.

Striped possum (*Dactylopsila trivirgata*). Photo by Pavel German / Wildlife Images.

Head and body length is 170–320 mm, and tail length is 165–400 mm. Weight is 280–545 grams in *D. trivirgata* and 260–550 grams in *D. palpator* (Flannery 1995a; Handasyde 2008). Average size of males, especially in *D. palpator*, is larger than that of females. In the subgenus *Dactylopsila* the fur is thick, close, woolly, and rather coarse; in *Dactylonax* it is silky and dense. All species have three parallel, dark stripes on the back; in the subgenus *Dactylopsila* these stripes are black on a basal color of white or gray, and in *Dactylonax* they are smoky brown on a basal color of grayish tawny. In *D. trivirgata* and *D. megalura* there is a black chin spot. The bushy tail is well haired except at the undersurface of the tip and is mostly dark-colored except that the tip is usually whitish. Females have two mammae and a well-developed pouch. Flannery (1995a) noted that the pouch of *D. palpator* is divided into two subunits, each with a single nipple.

This genus is characterized by large first incisor teeth and a slender, elongated fourth digit with a hooked nail on the forefoot. Both features are more pronounced in the subgenus *Dactylonax* and are shared by the aye aye (*Daubentonia*), a primate found in Madagascar. The claw of the fourth front digit of *Dactylonax* is much smaller than those of the other digits. The brain of *D. trivirgata* shows expanded cortical representation for the elongated fourth digit (Sanderson 2004). *D. trivirgata* has the largest brain, relative to body weight, of any marsupial (Handasyde 2008).

Striped possums live in rainforest or sclerophyll forest and are nocturnal and arboreal. The members of the subgenus *Dactylopsila*, at least, are superb climbers and seldom if ever descend to the ground. They shelter by day in dry leaf nests in tree hollows, where they lie curled into a ball, flat on their side and not rolled up in a sitting position like most possums. *D. trivirgata* is very active, racing and crashing noisily through the canopy with huge leaps and rapidly covering distances of more than 1 km (Handasyde 2000). Adults of that species use a number of dens spread across their home range and change dens frequently (Handasyde 2008). The subgenus *Dactylonax* seems less adept at climbing, may spend considerable time hunting insects in rotting logs on the forest floor, and nests on the ground among tree roots as well as in tree hollows (Flannery 1995a). The extremely unpleasant and penetrating odor of striped possums is of glandular

An additional species, *D. kambuayai*, was described from mid-Holocene remains from Kria Cave in the central Vogelkop Peninsula of western New Guinea by Aplin, Pasveer, and Boles (1999), who noted there is no reason why that species should not persist as a living animal somewhere in the region. Jackson (2015) indicated that a living population of *D. kambuayai* was discovered recently in the Arfak Mountains and that the species is probably most closely related to *D. palpator*, but that no descriptive or natural history information is available. *Dactylonax* was considered a full genus by Helgen (2007) and McKay (1988) but not by Jackson (2015b). Ziegler (1977) did not consider *D. tatei* and *D. megalura* to be specifically distinct from *D. trivirgata*. However, the above arrangement was followed by Corbet and Hill (1991) and Groves (2005e), and Menzies and Singadan (2005) confirmed the specific status of *D. megalura*.

Striped possum (*Dactylopsila trivirgata*). Photo by Pavel German / Wildlife Images.

origin, but it cannot be ejected in "skunklike" manner. Nonetheless, the odor, together with the black and white markings, provide a striking case of convergence with North American skunks (Flannery and Schouten 1994). Both sexes use scent to mark territories (McKenna 2005).

A captive striped possum cleaned itself elaborately upon emerging at night and then began to hunt for food with long-legged striding movements, so that it seemed to flow rather than jump from branch to branch. Striped possums sniff loudly around likely food sources, use their incisors to tear wood and gnaw out wood-boring grubs, and use their long finger to hook grubs from deep and narrow holes. They also tap their forefeet rapidly on loose bark, presumably to disturb insects beneath it. Their natural diet consists

mainly of insects but also includes fruits and leaves. A captive occasionally attacked and consumed mice. Smith (1982) reported that *D. trivirgata* evidently feeds mainly on ants and other social insects, which it obtains by breaking into nests with its incisors. Rawlins and Handasyde (2002) found the most frequent components of the diet of *D. trivirgata* to be wood-boring coleopteran and lepidopteran larvae, and adult hymenopterans and isopterans.

According to Handasyde (2008), *D. trivirgata* has a large home range for its body size, males using around 100–150 ha., females 20–40 ha. Ranges of adult males overlap those of several adult females but apparently overlap little with ranges of other males; limited data suggest that ranges of adult females do not overlap substantially. In Australia that species appears to breed throughout the year, but there may be a seasonal birth peak from March to June. The one or two young are carried in the pouch until fully furred, then left in the den while the mother forages. Flannery (1995a) wrote that *D. trivirgata* and *D. palpator* have been found in female pairs, while males have been found alone or paired with a female. Two males competing for an estrous female were heard to emit loud guttural shrieks; other vocalizations include growls, rasping coughs, and screams. Females with single young have been collected in New Guinea in January, April, July, August, September, and October. A captive-born *D. trivirgata* vacated the pouch when 121 days old and was then carried on the mother's back; it began showing independence and interest in solid food when 171 days old (McKenna 2005). One specimen of *D. trivirgata* lived in captivity for 9 years and 7 months (Weigl 2005).

D. tatei of Fergusson Island long was known by only nine specimens collected in 1935 and could not be located by several expeditions from the 1950s to the 1980s, but another specimen was taken in 1992 (Flannery and Schouten 1994). The IUCN classifies *D. tatei* as endangered, noting that it occurs in less than 200 sq km, its entire population is in a single location, and there is continuing decline in the extent and quality of its habitat, but a new specimen was collected in 2003 (Leary et al. 2008x). *D. trivirgata* is not placed in a threatened category but is decreasing as forest habitat is lost to subsistence agriculture (Salas et al. 2008b).

DIPROTODONTIA; **Family TARSIPEDIDAE;**
Genus TARSIPES Gray, 1842

Honey Possum

The single known genus and species, *Tarsipes rostratus*, is found in the southwestern part of Western Australia (Ride 1970). This genus was at one time usually placed in the family Phalangeridae but eventually came to be recognized morphologically and serologically as one of the most divergent of Australian marsupials and was put in its own superfamily (Kirsch 1977b; Kirsch and Calaby 1977). Later serological and morphological studies suggested that this superfamily should also include the genera *Acrobates* and *Distoechurus* (Aplin

and Archer 1987; Baverstock et al. 1987). Recent assessments have united Tarsipedidae, Acrobatidae, Pseudocheiridae, and Petauridae in the superfamily Petauroidea (see above account of order Diprotodontia). The use of the name *T. rostratus* in place of the previously used *T. spenserae* was explained by Mahoney (1981).

Head and body length is 65–90 mm, tail length is 70–105 mm, and weight is 7–11 grams for males and 8–16 grams for females (Russell and Renfree 1989). The head is pale brown, and the body is grayish brown with three dark stripes along the back. The central stripe, which is almost black, extends from the head to the base of the tail; the two outer stripes are fainter and do not reach the tail. The underparts are pale yellowish or white, the limbs are pale rufous, and the feet are white. Some individuals are more grayish, but the stripes are always present. The fur is rather coarse, short, and close. The long whiplike tail is almost hairless and has a naked prehensile undertip.

This possum can be distinguished from the other small Australian marsupials by its coloration and its

Honey possum (*Tarsipes rostratus*). Photo by P. A. Woolley and D. Walsh.

Honey possum (*Tarsipes rostratus*). Photo by Jiri Lochman / Lochman LT.

extremely long snout, which is about two-thirds of the length of the rest of the head. The long tongue, bristled at the tip, can be extended about 25 mm beyond the nose and deep into flowers. Ridges on the hard palate scrape honey and pollen off the tongue when it is withdrawn through a channel formed by flanges on the upper and lower lips, assisted by the long, slender lower incisors. The pair of procumbent lower incisors are the only well-developed teeth; the at-most 20 other teeth are reduced to small pegs, reflecting the soft diet of the animal. All the digits except the united second and third of the hind feet, which are equipped with functional claws, are expanded at the tip and bear a short

nail much like those of the tarsier (*Tarsius*), hence the generic name *Tarsipes* for the honey possum. The gut is unusual, both in lacking a caecum and in possessing a bilobed stomach with a marked diverticulum, which is associated with a very rapid passage of pollen particles (Bradshaw and Bradshaw 2012). Females have a well-developed pouch and four mammae.

The honey possum dwells on tree and shrub heaths and is an active, nimble climber, though most trails are on the ground (Wooller et al. 2004). It moves short distances to trees and shrubs that are producing its favorite flowers, and several animals may assemble in one desirable place. It often hangs upside down, especially when feeding on flowers. Sometimes it shelters in deserted birds' nests or hollow trees. Vose (1973) reported three peaks of feeding activity in each 24-hour period: between 0600 and 0800, 1700 and 1900, and 2330 and 0130. More recent observations and examination of eye structure suggest largely crepuscular and some diurnal activity (Arrese and Runham 2002; Wooller et al. 2004). Nightly movements across straight-line distances of up to 368.3 meters were recorded by Bradshaw et al. (2007). The honey possum may enter torpor in response to cold temperature or reduced food, but the torpor is both short-term and deep, whereas other heterothermic marsupials are characterized by torpor that is either long-term and deep or short-term and shallow (Bradshaw and Bradshaw 2012). The diet consists exclusively of nectar and pollen; *Tarsipes* eats no seeds, fruits, or insects (Wooller et al. 2004). Like some bats and hummingbirds, the honey possum is well adapted for probing into flowers and licking up its food. Nearly every floret in a favorite flower is explored. *Tarsipes* has a high rate of nectar and pollen intake and is thought to be a significant pollinator (Bradshaw and Bradshaw 2012; Renfree 2008). It apparently has developed finely tuned, trichromatic color vision that assists in detecting food-bearing flowers; this possession of three types of cone photoreceptors is found in only a few other mammals, such as certain primates, including humans (Sumner, Arrese, and Partridge 2005).

The following information on socialization and life history was compiled from Bradshaw and Bradshaw (2012), Bradshaw et al. (2007), Renfree (1980, 2008), Renfree, Russell, and Wooller (1984), Russell and Renfree (1989), and Wooller et al. (2000, 2004). Individuals usually occupy permanent, overlapping home ranges averaging 700 sq meters for females and 1,280 sq meters for males, though ranges of up to 9 ha. have been estimated. *Tarsipes* is mostly solitary in the wild, though may seem gregarious in the laboratory, huddling together in groups of two or more. Animals are often caught in a torpid state during cold or wet weather, and presumably huddling helps to minimize body heat loss. Large adult females interact very little with one another and are dominant to males; when accompanied by young, they may exclude other individuals from their range. Females with pouch young have been found throughout the year but predominantly in

early autumn (February–March), winter (May–July), and spring (September–October), when pollen and nectar are most abundant. Few young seem to be born in the summer months of December and January. Males have the largest known sperm of any mammal. Females are polyestrous and apparently polyandrous, and some have up to four litters annually, though two or three are usual. Embryonic diapause such as in the Macropodidae also occurs in *Tarsipes*. Dormant embryos, carried in the uterus while the female is lactating, resume development after the young have left the pouch. Although the active gestation period is not definitely known, it has been estimated to range from 21 to 28 days but 19–21 days may be more precise; since a postpartum estrus occurs approximately 2–4 days following birth, it is likely that the gestation period is 2–4 days shorter than the estrous cycle, which has been estimated to range from 21 to 28 days, with an average of 24. Two or three young are usually reared but litter sizes of from one to four have been recorded. With a weight of under 5 milligrams each, the young are the smallest known mammalian neonates. Lactation lasts about 10 weeks; for the first 8 weeks the young remain in the pouch, and for the last 2 weeks they follow the mother while she forages and sometimes ride about on her back. No nest is constructed. Adult size is attained at about 8 months, and both sexes breed in their first year of life, before they are fully grown. Following the spring breeding peak there is a drastic population decline; most individuals do not survive more than a year, though field recapture data indicate a maximum longevity of 1.6 years for males and 4 years for females.

Vose (1973) cautioned that the existence of the honey possum in the wild is directly dependent on a continued supply of blossoms from eucalyptus, *Banksia* and *Callistemon*. Large-scale urbanization and habitat destruction in southwestern Australia is affecting the feeding of this species. Kennedy (1992) indicated that it had already lost up to 50 percent of its habitat and designated it "potentially vulnerable." Bradshaw et al. (2007) observed that it had been drastically reduced since European arrival and that research on its unique physiology and habitat linkage was needed to insure long-term survival. However, the IUCN assigns no threatened classification, noting that population size is stable and likely to number in the hundreds of thousands (Friend et al. 2008).

The geological range of the family Tarsipedidae is late Pleistocene to Recent in Australia (Marshall 1984).

DIPROTODONTIA; Family **ACROBATIDAE**

Feather-tailed Possums

This family of two genera and two species occurs only in New Guinea and eastern Australia. The two genera, *Distoechurus* and *Acrobates*, were at one time usually placed in the family Phalangeridae, but serological data led to their transfer to the Burramyidae (Kirsch 1977b). For a time they were considered to have affinity to the Petauridae, but new serological and morphological studies indicated a closer relationship to the Tarsipedidae and, in any case, that they warranted inclusion in an entirely separate family (Aplin and Archer 1987; Ar-

cher 1984; Baverstock, Birrell, and Krieg 1987; Kirsch, Lapointe, and Springer 1997). The latter arrangement was followed and/or supported by Aplin (2008, 2015b), Groves (2005e), McKenna and Bell (1997), and Meredith, Westerman, and Springer (2009b), though Corbet and Hill (1991) continued to include the two genera in the Burramyidae, and Szalay (1994) considered them to constitute only a tribe of the Petauridae.

Head and body length is 60–120 mm, and tail length is 65–155 mm. The distichous or "pen" tail is distinctive in having paired lateral fringes of long stiff hairs that give a feather-like appearance. The genus *Acrobates* has a narrow patagium, or gliding membrane, that extends from the forelimbs to the hind limbs. The dental formula is: i 3/2, c 1/0, pm 3/3, m 4/4 = 40. Aplin and Archer (1987) provided a detailed technical diagnosis of the family based in part on the structures of both the external and the internal ear. The external ear is uniquely complex, with a prominent anterior helical process, paired antitragal processes, and a well-defined

Pygmy gliding possum, or feathertail glider (*Acrobates pygmaeus*). Photo by Pavel German / Wildlife Images.

Feather-tailed possum (*Distoechurus pennatus*). Photo by P. A. Woolley and D. Walsh.

bursa. In the internal ear the auditory bulla lacks an alisphenoid component and instead is formed from the petrosal, squamosal, and ectotympanic bones; the bulla is under-run medially and posteriorly by secondary tympanic processes from basi- and exoccipital bones; the primary tympanic cavity is complexly compartmentalized by numerous septa. Other diagnostic characters include a distinct lingual eminence at the rear of the tongue, a long upper canine that projects well below the level of the upper incisor teeth, greatly enlarged posterior palatal vacuities, and an intestinal caecum moderately to very elongate. The Burramyidae and Petauridae lack these and other key features, but the Tarsipedidae show some similarity in the structure of the internal ear and associated basicranial components.

In a study of the organogenesis and fetal membranes of *Distoechurus*, Hughes et al. (1987) found that the bilaminar yolk sac is totally invasive and that the maternal endometrial glands exhibit total degeneration. These two features are without parallel in marsupials so far

studied and may provide further evidence of the distinctiveness of the Acrobatidae. Aplin (2008) observed that *Acrobates* is one of the few possums to have the reproductive capacity of embryonic diapause, a feature it shares with *Tarsipes* and Macropodiformes.

The geological range of the family Acrobatidae is early Miocene to Recent in Australia (Crosby et al. 2004) and Recent in New Guinea (Flannery 1995a). *Acrobates* is reportedly the oldest extant genus of small mammal, with representatives dating back 29 million years (Ward and Woodside 2008).

DIPROTODONTIA; ACROBATIDAE; **Genus**
DISTOECHURUS Peters, 1874

Feather-tailed Possum

The single species currently recognized, *D. pennatus*, is found in suitable habitat throughout mainland New

Guinea (Ziegler 1977). It has not been recorded from any nearby islands (Flannery 1995a). There are diverse forms, and preliminary genetic and morphological assessments indicate that they represent three or more distinct species (Aplin 2015b).

Head and body length is 103–132 mm, tail length is 126–155 mm, and weight is 38–62 grams; females are on average larger than males (Flannery 1995a). The general body coloration is dull buff, light brown, slightly darker and more olivaceous, or slightly darker and grayer. In contrast to the dull, plain body, the head is strikingly ornamented. The face is white, with two broad well-defined dark brown or black bands that pass from the sides of the muzzle through the eyes to the top of the head just between the ears, and there is a conspicuous black patch just below each ear. The basal part of the tail is well furred, and the remainder is nearly naked but fringed laterally with long, relatively stiff hairs. The pelage is soft, thick, and woolly.

Gliding membranes are not present. The claws are sharp and curved; the terminal pads of the digits are not expanded. The eyes are large, and the ears are small and naked. Although the tail lacks a true prehensile tip, it is wrapped around branches as extra support while the animal is climbing (Aplin 2015b). Females have two, occasionally four, teats in a well-developed pouch, which opens anteriorly.

Relatively little is known about this possum; the following was compiled from Aplin (2015b), Collins (1973), Flannery (1995a), Woolley (1982a), and Ziegler (1977). *Distoechurus* has been collected in undisturbed climax forest across its entire elevational distribution, from sea level to about 2,600 meters. It also is found in young regrowth and disturbed forest, lowland swamp and peat forests, and old gardens. It nests in tree hollows and leaf nests within the foliage of trees, including banana and palm trees. It is active, arboreal, and nocturnal and has been reported to eat blossoms, fruit, insects, and other invertebrates. Individuals nest alone or in groups of 2–3; an adult male and an adult female with a pouch young were collected together in a leafy nest in a small tree in January. Most breeding data come from the Kariumi Plateau of central Papua New Guinea and suggest an extended reproductive season, coinciding with the dry season, from at least April to August. However, without comparative data from the wetter period, it cannot be concluded that high intensity breeding occurs year-round, at least during favorable years. Females in other areas that were lactating or that had pouch young have been collected from June to October. Observation of one female with an advanced young riding on her back, a single embryo in the uterus, and evidence of recent release of a single ovum confirms polyestry, as well as sequential production of two litters. The number of young per litter seems to be usually one, sometimes two. Individuals have been kept in captivity for as long as 20 months

DIPROTODONTIA; ACROBATIDAE; **Genus *ACROBATES*** Desmarest, 1817

Pygmy Gliding Possum, or Feathertail Glider

The single species usually recognized, *A. pygmaeus*, occurs along the eastern coast of Australia, from the extreme northern tip of the Cape York Peninsula to southeastern South Australia (Ride 1970). According to Aplin (2015b), however, another species, *A. frontalis*, occupies that entire distribution, including the Great Dividing Range and its seaward draining catchments and southern tableland, while *A. pygmaeus* is restricted to a region from the southeastern corner of Queensland to the southeastern corner of South Australia, not including the coastal drainages of the Great Dividing Range. Aplin (2015b) stated that the two species are extensively sympatric and well differentiated, and that their existence was revealed by a recent morphological and molecular study, but did not precisely identify that study. Another named species, *A. pulchellus*, was based on a single specimen from a small island to the north of New Guinea and is now believed to represent an introduced pet (Groves 2005e; Jackson and Thorington 2012). However, Helgen (2003b) discussed another specimen supposedly taken in New Guinea, and Harris (2015) thought it possibly premature to discard the records. Flannery (1995a) suggested *Acrobates* may still occur in New Guinea but has not yet been discovered because the most likely habitat has not been well explored.

Head and body length is 65–80 mm, tail length is 70–80 mm, and weight is 10–15 grams; males average slightly larger in size (Ward and Woodside 2008). The

Pygmy gliding possum, or feathertail glider (*Acrobates pygmaeus*). Photo by Eric Isselee / Shutterstock.com.

pelage is soft and silky. The upper parts are grayish brown, and the underparts and inner sides of the limbs are white. The ears are sparsely haired and bear tufts of hair at their base. The tail is light brown through-out, the ventral surface being somewhat paler than the dorsal surface. Aplin (2015b) reported that in *A. pygmaeus* the tail is furred ventrally to within 1–2 mm of the tip, lacking a prehensile section, but that in the nominal *A. frontalis* the ventral fur of the tail stops about 2–4 mm short of the tip, leaving a naked and prehensile pad. Otherwise, the two forms are remarkably similar in body size, general proportions, fur texture, coloration, and patterning. The pads at the tips of the digits (see discussion below) are moderately large and round in *pygmaeus* but larger and heart-shaped in *frontalis*.

This beautiful, delicate creature is the smallest mammal capable of gliding flight and is indeed, as the name *Acrobates pygmaeus* indicates, a pygmy acrobat. The gliding membrane is a very narrow fold of skin, fringed with long hairs along its margin and extending from each limb along the sides of the body. The structure of the tail is such that it provides an additional plane: it is flattened by a fringe of hairs along the sides, a characteristic giving *Acrobates* the common name "feathertail glider." The fringe of hairs spreads from

each side to a total width of 8 mm. The tip of the tail is without hair below, and the tail is somewhat prehensile. The tips of the digits on both the forefeet and hind feet have expanded, deeply scored or striated pads, which assist the animal in clinging to surfaces. A histological study by Rosenberg and Rose (1999) showed the pads to consist of epidermal ridge–sweat gland complexes that allow running on vertical panes of glass. The fourth digit on all four limbs is the longest, and the claws of the digits are sharp. The well-developed pouch encloses four mammae and is lined with yellow hair.

This active, agile marsupial inhabits sclerophyll forest and woodland and is much like a flying squirrel (*Glaucomys*) in its actions. It can glide from one tree to another as well as travel short distances, from limb to limb, by air. In studies conducted in Victoria, it was found to average 3 to 5 glides per hour, with an average distance of 14 meters and a maximum of 28 meters (Ward and Woodside 2008). Because of its small size and nocturnal habits, *Acrobates* often is overlooked. Some individuals are caught by domestic cats or are discovered when trees are felled. *Acrobates* makes small, spherical nests of dry leaves and other vegetation in hollow branches and knotholes of trees at a height of 15 meters or more. During cold days it may

Pygmy gliding possum, or feathertail glider (*Acrobates pygmaeus*). Photo by Fritz Geiser.

become torpid in the nest. Jones and Geiser (1992) experimentally induced deep torpor, during which body temperature fell to 28° C for periods of up to 5.5 days. Apparently, however, multi-day torpor bouts are used only in emergency situations of food scarcity, and there is no prolonged hibernation (Geiser and Körtner 2004). The diet consists of nectar, pollen, and insects (Ward and Woodside 2008).

A 10-year study on 7 hectares of forest indicated a maximum of 70 and a minimum of 12 individuals (Ward and Woodside 2008). Other estimates of density have been in excess of 50 per 100 ha. and only 1.7–4.4 per ha. (Harris 2015). Home ranges of 0.15 to 2.1 ha. have been reported during periods of up to 10 days (Ward 2004). The feathertail glider does not appear to be territorial, shows remarkable conspecific tolerance, and has an extended breeding season (Collins 1973; Fleming and Frey 1984; Ward 1990b; Ward and Renfree 1988a, 1988b; Ward and Woodside 2008). Family units consisting of one or both parents plus the offspring of one or two litters are present for much of the year, and those units freely share their nests with other individuals. Aggregations of up to 25 have been found, including several adults of both sexes, but such are not stable units. Groups more commonly have 2–5 individuals. Breeding appears to be year-round in the northern parts of the range. In Victoria adult males are usually solitary from January to April and then form pairs with the females from May to August. Births occur there from July to February, with a peak from August to November. Females undergo a postpartum estrus and a period of embryonic diapause and normally produce two litters during the season. A female may carry as many as four young in her pouch, but litter size averages 2.5 at weaning. When they are well furred, the young ride on the mother's back. The young are left in the nest at 60 days and weaned at 95–105 days. A second litter may then be born immediately. The young from the first litter remain with the mother while she raises the second litter. Sexual maturity is attained at 8 months in females and by 12–18 months in males. A DNA assessment by Parrott, Ward, and Taggart (2005) indicates that most litters result from matings with more than one male within a single estrous period, and also that mothers suckle and care for young other than their own. Females have been found to be reproductively active in their fifth annual breeding season, males in their fourth (Harris 2015). A captive specimen lived for 8 years and 9 months (Weigl 2005).

DIPROTODONTIA; **Family**
HYPSIPRYMNODONTIDAE; Genus
HYPSIPRYMNODON Ramsay, 1876

Musky "Rat"-kangaroo

The single living genus and species, *Hypsiprymnodon moschatus*, occupies approximately 320 km of the coast of northeastern Queensland, from Mount Amos, near Cooktown, south to Mount Lee, near Ingham (Calaby 1971; Dennis and Johnson 2008a). That distribution is what remains after a "severe range reduction" caused by clearing of tropical forest for agriculture and grazing (Claridge, Seebeck, and Rose 2007). *Hypsiprymnodon* was for a long time usually placed with the other rat-kangaroos in the family Potoroidae or the subfamily Potoroinae of the family Macropodidae. The latter procedure was still followed by McKenna and Bell

Musky "rat"-kangaroo (*Hypsiprymnodon moschatus*). Photo by Lorraine Harris.

(1997). However, most authorities, such as Flannery (1989), had come to recognize *Hypsiprymnodon* as the sole living member of another subfamily, Hypsiprymnodontinae, of the family Potoroidae. Subsequently, both morphological and molecular studies showed *Hypsiprymnodon* to represent a lineage with an early divergence from the line leading to Potoroidae and Macropodidae and thus that it warrants inclusion in its own family (Burk and Springer 2000; Burk, Westerman, and Springer 1998; Cooke 2006; Kear and Cooke 2001; Kirsch, Lapointe, and Springer 1997; Meredith, Westerman, and Springer 2009a, 2009b; Prideaux and Warburton 2010; Szalay 1994). Such a procedure was followed by Eldridge (2008a) and Groves (2005e).

This is the smallest "rat"-kangaroo and indeed the smallest living member of suborder Macropodiformes and/or superfamily Macropodoidea. Head and body length is 153–341 mm, tail length is 123–165 mm, and weight is 337–680 grams (Dennis and Johnson 2008a; Johnson and Strahan 1982). The pelage is close, crisp, and velvety and consists mainly of underfur. The general coloration is rich brown or rusty gray, brightest on the back and palest on the underparts. Some individuals have a white area on the throat that extends as a narrow line to the chest.

This genus is unique among Macropodiformes in that the hind foot has a well-developed first digit that is movable and clawless but nonopposable to the other four digits. The limbs are more equally proportioned than those of the Potoroidae and Macropodidae. The claws are quite small, weak, and unequal in length. The tail, which is used to gather nesting material, also differs from that of all other members of the suborder in being almost completely naked and scaly; only the extreme base is hairy. The muzzle is naked, and the ears are rounded, thin, and naked except at their posterior base. The dental formula is: i 3/2, c 1/0, pm 1/1, m 4/4 = 32. Females have four mammae and a well-developed pouch. The stomach is relatively simple, not divided into compartments as in the Potoroidae and Macropodidae. The specific name refers to the musky scent emitted by both sexes.

The musky "rat"-kangaroo occurs in rainforests, often in dense vegetation bordering rivers and lakes. It seems to be fairly common but is shy, quick in its movements, and difficult to observe in its dense habitat. It differs from most members of the Potoroidae in being completely diurnal (Seebeck and Rose 1989). One individual was seen sunbathing, lying spread-eagle on a fallen tree trunk. *Hypsiprymnodon* does not hop on its hind feet, as do all other members of Macropodiformes, but runs on all four limbs. Johnson and Strahan (1982) noted that adults have been seen to climb on fallen branches and horizontal trees, and juveniles have ascended a thin branch inclined at about 45°. At night and in the middle of the day *Hypsiprymnodon* sleeps in a nest in a clump of vines or between the plank buttresses of a large tree.

Nests are regularly made and well constructed; leaves are collected with the mouth and forepaws, placed on the ground, and transferred to the prehensile tail with a kick of the hind legs before being carried to the new nest site (Dennis and Johnson 2008a). *Hypsiprymnodon* sits on its haunches while eating. The diet differs from that of other "rat"-kangaroos and consists of fruits, invertebrates, and fungi in varying proportions through the year (Dennis and Johnson 2008a). By far the largest component is fruits and seeds, especially large, fleshy drupes

and seeds with a moderate to soft coat (Dennis 2002). Fallen fruit is important and may be consumed at the site of the fall, carried off and eaten at another site if the individual is challenged by a conspecific, or "scatter hoarded" in holes in the leaf litter for later use; the caches are up to 68 meters from the source, and the musky "rat"-kangaroo is thus a primary dispersal agent for rainforest seeds (Dennis 2003). Animals occasionally fight over fruit, engaging in dramatic chases and emitting guttural grunts (Dennis and Johnson 2008a). *Hypsiprymnodon* appears to be solitary, but feeding aggregations of up to three individuals have been observed (Johnson and Strahan 1982). Population densities of 1.4–4.5 individuals per ha. and overlapping home ranges of 0.8–4.2 ha. have been reported (Claridge, Seebeck, and Rose 2007).

In a captive breeding colony, Lloyd (2001) found the estrous cycle to be 25–27 days, the gestation period to be 19 days (the shortest of any macropodiform), and no evidence of a postpartum estrus or embryonic diapause; of the litters recorded, four had three young each, one had two young, and one had a single young. In a wild population, Dennis and Marsh (1997) found

Musky "rat"-kangaroo (*Hypsiprymnodon moschatus*). Photo by Lorraine Harris.

breeding to correspond generally with the rainy season. The testes of males expanded greatly in October and remained enlarged until April. Births occurred from February to April; of the litters examined, 18 had 2 young, 5 had 3 young, and 3 had a single young. Pouch eviction occurred in October, when fruits were abundant, and the young then were left in the maternal nest and periodically suckled by the mother. For the next few months the young began to explore and feed themselves, with weaning in January. Most subadults attained sexual maturity during the subsequent breeding season, in their second year of life. Females apparently could breed in consecutive years and live at least 4 years. *Hypsiprymnodon* was considered the most fecund of any macropodiform. Johnson et al. (2005) reported discovery of a female carrying four pouch young.

Kennedy (1992) considered the musky "rat"-kangaroo "potentially vulnerable" because its rainforest habitat had become restricted through clearing for agricultural development, though Seebeck, Bennett, and Scotts (1989) observed that it seemed relatively common where it survived. Dennis and Johnson (2008a) wrote that most remaining habitat is well protected in national parks, other forest reserves, and the Wet Tropics World Heritage Area.

Hypsiprymnodontidae is usually considered to include an extinct subfamily, Propleopinae, which has a known geological range from the late Oligocene to late Pleistocene of Australia, while subfamily Hypsiprymnodontinae ranges from the early Miocene to Recent; however, certain fossil evidence suggests alignment of Propleopinae with the extinct kangaroo family Balbaridae, rather than with Hypsiprymnodontidae (Cooke 2006; Kear and Cooke 2001; Long et al. 2002). Some species of Propleopinae were very different from modern *Hypsiprymnodon moschatus*. *Propleopus oscillans*, with a weight of about 70 kg, dagger-like lower incisor teeth, and premolars shaped like circular saws, may have been highly predatory and somewhat like the Australian dingo (*Canis familiaris dingo*) in feeding habits, though some evidence suggests it was omnivorous. There has been speculation that it survived into geologically Recent times, not disappearing until the dingo entered Australia, about 4,000 years ago (Ganslosser 2003).

DIPROTODONTIA; Family POTOROIDAE

"Rat"-Kangaroos

This family of 4 Recent genera and 12 species occurs in Australia, especially the southern and eastern parts, and on a few nearby islands and in Tasmania. In the past the Potoroidae were often regarded only as a subfamily of the Macropodidae and some authorities have continued to follow that arrangement (Burk and Springer 2000; Kirsch, Lapointe, and Springer 1997; McKenna and Bell 1997; Szalay 1994). However, there has been increasing recognition of the two as distinct families (Calaby and Richardson 1988a, 1988b; Claridge, Seebeck, and Rose 2007; Corbet and Hill 1991; Eldridge 2008a; Groves 2005e; Kear and Cooke 2001). The sequence of genera presented herein follows that suggested by Burk, Westerman, and Springer (1998), Flannery (1989), and Kear and Cooke (2001). Potoroidae sometimes is divided into two tribes, Potoroini (potoroos) with the genus *Potorous*, and Bettongini (bettongs) with *Bettongia*, *Caloprymnus*, and *Aepyprymnus*.

Descriptions of the Potoroidae were provided by Claridge, Seebeck, and Rose (2007) and Seebeck and Rose (1989). Head and body length is 243–415 mm, tail length is 183–387 mm, and weight is 637–3,500 grams. The body is covered with dense fur. The upper parts range in color from browns and grays to rufous and sandy; the underparts are usually pale. Unlike in the Macropodidae, coat color is usually uniform with no stripes or other markings. All living genera are compact animals with variously elongate muzzles, short and rounded ears, and short but muscular forelegs bearing small paws with short, forward-pointing spatulate claws. The hind limbs are well developed and heavily muscled, and the hind feet are elongate. Unlike in *Hypsiprymnodon*, all genera have lost the first digit of the hind foot. The second and third digits of the hind foot are syndactylous (united by integument). The fourth toe is the largest. The long tail is furred; it is markedly prehensile in *Bettongia* but only weakly so in the other genera. The stomach is complex, being divided into compartments, in all genera that have been studied. The female reproductive tract of potoroids differs from that of most macropodids in the presence of an anterior vaginal expansion.

Rufous "rat"-kangaroo (*Aepyprymnus rufescens*). Photo by Jiri Lochman / Lochman LT.

The skull is short and broad in most genera but elongate and narrow in some species of *Potorous*. The squamosal bone has wide contact with the frontal bone, thereby separating the parietal and alisphenoid bones. In the Macropodidae parietal-alisphenoid contact is usual. In the Potoroidae the masseteric canal of the mandible is confluent with the inferior dental canal and extends farther anteriorly than it does in the Macropodidae. The dental formula is: i 3/1, c 1/0, pm 1/1, m 4/4×2=30. The dentition is typically diprotodont, the pair of lower incisors being enlarged and projecting forward. There are two deciduous premolars on each side of both jaws, but these are shed near maturity and replaced by a single large, permanent tooth. This premolar is bladelike and has vertical corrugations. Unlike in the Macropodidae, there is no forward progression of molar teeth. The molars are mostly quadrituberculate, and they decrease in size from the anterior to the posterior.

Habitats range from hummock grassland to tropical rainforest, but most species are found in drier country in southern Australia and are mainly nocturnal. Most species now are known to feed primarily on fungi, though other vegetation and animal matter may be consumed. There is little evidence of territoriality and mating seems essentially promiscuous. The reproductive biology of the Potoroidae is much like that of the Macropodidae, but the "rat"-kangaroos generally have more than one young per year and have a relatively shorter pouch life. According to Claridge, Seebeck, and Rose (2007), the length of estrous cycle and gestation are nearly the same. Immediately after the female has given birth, she can mate again. If the pouch is occupied at that time, the new fertilized egg develops only slightly (about 100 cells) before entering what is termed "embryonic diapause," a state of suspended animation when little growth occurs. It remains in that state until near the end of pouch life of the previous young or after that young is lost. In most species, the end of pouch life is accompanied, often on the same night, by a new birth and subsequent postpartum estrus. Unlike many of the larger wallabies and kan-

Long-nosed potoroo (*Potorous tridactylus*). Photo by Pavel German / Wildlife Images.

garoos, most potoroids are able to breed continuously. This may allow them to maintain populations in challenging habitats. Despite their fecundity, potoroids have declined sharply because of the effects of European settlement, especially predation by the introduced red fox (*Vulpes vulpes*) and habitat disturbance. Only 2 of the 11 Recent species remain relatively common over most of their original ranges (Claridge, Seebeck, and Rose 2007; Finlayson, Finlayson, and Dickman 2010; Rose 1989; Seebeck, Bennett, and Scotts 1989).

The known geological range of the Potoroidae is late Oligocene to Recent in Australia (Long et al. 2002).

DIPROTODONTIA; POTOROIDAE; **Genus POTOROUS** Desmarest, 1804

Potoroos

Four species now are recognized (Friend 2008c; Johnston 2008; Johnston and Sharman 1976, 1977; Kirsch and Calaby 1977; Kitchener and Friend 2008; Menkhorst and Seebeck 2008; Seebeck and Johnston 1980; Sinclair and Westerman 1997; Westerman, Loke, and Springer 2004):

P. tridactylus (long-nosed potoroo), southeastern Queensland, coastal New South Wales, Victoria, extreme southeastern South Australia, islands in Bass Strait (King, Flinders, Cape Barren, Clarke), Tasmania;

P. gilbertii, extreme southwestern Western Australia;

P. longipes, eastern Victoria, extreme southeastern New South Wales;

P. platyops (broad-faced potoroo), occurred historically in southwestern Western Australia.

Analysis of mitochondrial DNA by Frankham, Handasyde, and Eldridge (2012) showed the populations of *P. tridactylus* in Queensland and northern New South Wales to have a level of divergence from the extant populations in southern New South Wales, Victoria, and Tasmania equivalent to that seen between full mammalian species. However, until additional assessment could be made of nuclear DNA and morphology, it was recommended that *P. tridactylus* be recognized as comprising three distinctive subspecies: *P. t. tridactylus* in the northern part of the mainland range, *P. t. trisulcatus* on the southern mainland, and *P. t. apicalis* on Tasmania. Subfossil specimens indicate that the geologically Recent range of *P. platyops* extended eastward along the southern coast of Western Australia to southeastern South Australia and Kangaroo Island, but it is not clear whether those records date from historical time (Kitchener and Friend 2008; Seebeck 1992b; Seebeck and Rose 1989; Woinarski, Burbidge, and Harrison 2014).

Head and body length is 243–415 mm, tail length is 183–325 mm, and weight is 660–2,200 grams (Friend 2008c; Johnston 2008; Kitchener and Friend 2008; Menkhorst and Seebeck 2008). The pelage, at least in *P. tridactylus*, is straight, soft, and loose. The upper parts are grayish or brownish, and the underparts are grayish or whitish. The tail of *P. tridactylus* is often tipped with white. The muzzle is elongated in *P. tridactylus* and shortened in *P. platyops*. The hind feet of *Potorous* are shorter than the head, except in *P. longipes*. Females have a well-developed, anteriorly opening pouch, within which are four mammae, though normally only a single young is reared.

Following three years of field and laboratory observations, Buchmann and Guiler (1974) were able to clear up the uncertainty regarding potoroos' means of locomotion. There are three principal methods: (1) the "quadrupedal crawl," a slow (20–23 meters per minute) plantigrade movement in which the hind feet are employed for the main forward propulsion and then the weight is shifted to the forefeet, which is used in leisurely feeding and foraging; (2) the "bipedal hop," a series of synchronous digitigrade thrusts by the hind limbs, moving the animal 30–56 meters per minute, used for escape and chasing in aggression; and (3) "jumping," a single bound 2.5 meters long and 1.5 meters high resulting from a powerful thrust of the hind legs, used in initial escape or aggression.

Potoroos inhabit rainforest, riparian forest, damp or wet sclerophyll forest, dense grassland, or low, thick scrub especially in damp places (Johnston 2008; Menkhorst and Seebeck 2008; Ride 1970). Most species are nocturnal but occasionally may engage in early morn-

ing basking in the sun (Collins 1973). *P. longipes* is frequently active during daylight (Claridge, Seebeck, and Rose 2007). When not active, potoroos shelter in shallow "squats," usually excavated at the base of a tussock or under dense shrubs, and do not construct complex nests (Seebeck, Bennett, and Scotts 1989). *P. tridactylus* digs small holes in the ground when feeding. These holes are not round like those made by bandicoots. Well-defined trails lead from one feeding site to another. A study of *P. tridactylus* in Tasmania found fungi of substantial importance from May to December and to constitute more than 70 percent of food in May and June (Guiler 1971a). However, that species depends on insects more than do most potoroids, especially in summer (Guiler 1971b). Its diet also includes grass, roots, seeds, and other forms of vegetation. In a study of *P. longipes*, Green et al. (1999) determined that fecal material comprised about 91 percent fungal remains, with no significant seasonal variation. Friend (2008c) described *P. longipes* and *P. gilbertii* as the most fungi-dependent mammals known, as the diet of both species includes over 90 percent fungi at most times of the year.

Potoroo home ranges are sometimes rather large, perhaps because of the need to procure sufficient fungi, and populations usually occur at relatively low densities. *P. tridactylus* has been found at 0.2–2.55 individuals per ha. (Seebeck, Bennett, and Scotts 1989). Its home range in Tasmania was reported to be at least 12–34 ha. for males and 2–11 ha. for females (Claridge, Seebeck, and Rose 2007). In Victoria, however, range averaged only about 2.0 ha. for males and 1.5 ha. for females. Male ranges may overlap those of several females, but female ranges are often exclusive (Seebeck and Rose 1989). Observations indicate that males are territorial and defend a small part of their home range but tend to avoid conflict under natural conditions. They may be kept in captivity together with other males and several females, but when a female is in estrus the males will fight fiercely, with one eventually establishing dominance (Collins 1973; Ride 1970; Russell 1974b). Individuals usually forage alone but may tolerate others nearby if food is abundant (Claridge, Seebeck, and Rose 2007).

A density of 0.06 per ha. has been recorded for *P. longipes* (Claridge, Seebeck, and Rose 2007). It has a reported home range of about 9 to 60 ha. and appears

Gilbert's potoroo (*Potorous gilbertii*). Photo by Jiri Lochman / Lochman LT.

territorial in some areas, with ranges of males almost exclusive of one another but overlapping extensively with those of females; female ranges show moderate overlap (Green, Mitchell, and Tennant 1998; Menkhorst and Seebeck 2008). *P. gilbertii* lives in colonies of 3–8 animals; home ranges are at least 15–25 ha. for adult males and around 3–6 ha. for females and subadults, with extensive overlap between, but little within, the sexes (Claridge, Seebeck, and Rose 2007; Friend 2008c).

Potoroos are polyestrous, with an estrous cycle of about 42 days in *P. tridactylus* and 25 days in *P. longipes* (Claridge, Seebeck, and Rose 2007). They apparently lack a well-defined breeding season. In Tasmania *P. tridactylus* bears young throughout most of the year, with peaks from July (winter) to January (summer). Each female is able to breed twice a year. Nondelayed gestation is 38 days, the longest in the Macropodiformes, and normal litter size is one. Four days after the season's first young is born, the female mates again, but because of embryonic diapause, development is arrested and parturition does not occur for about 4.5

months unless the first young is lost. The newborn measures 14.7–16.1 mm in length, is able to detach from the nipple at 55 days, and leaves the pouch at about 130 days. Females are able to breed at 1 year (Bryant 1989; Collins 1973; Ride 1970). Reproduction in *P. longipes* is similar, but pouch life is 140–150 days and sexual maturity may not be attained until 2 years (Green and Mitchell 1997; Seebeck 1992a). *P. gilbertii* breeds all year, the single young remaining in the pouch 3–4 months, then staying in the mother's home range until well after weaning and finally leaving when 7–18 months old; females reach sexual maturity at around 9 months, and both sexes have a maximum longevity of over 10 years (Friend 2008c). A wild specimen of *P. tridactylus* is known to have lived at least 7 years and 4 months (Guiler and Kitchener 1967), and captives of that species have attained ages of over 15 years (Weigl 2005).

All species of *Potorous* have been seriously affected by modern human activity. *P. platyops* apparently has been extinct since about 1875, when five specimens were sold to the National Museum in Victoria, and the

Long-footed potoroo (*Potorous longipes*). Photo by Dave Watts / Lochman LT.

species may have declined considerably even prior to the coming of European settlers; it is designated extinct by the IUCN (Australasian Mammal Assessment Workshop 2008a). Calaby (1971) indicated that *P. gilbertii* also might be extinct, none having been collected in more than 80 years, though Thornback and Jenkins (1982) reported unconfirmed sightings. In 1994 *P. gilbertii* was rediscovered on Mount Gardner promontory in Two Peoples Bay Nature Reserve on the southern coast of Western Australia. Friend (2008c) referred to the species as Australia's rarest mammal, being known by only a single population of 30–40 individuals. Since that information was compiled, however, there have been translocations to Bald Island, just off the southern coast of Western Australia, and the population there had grown to around 60 animals by 2012 (Woinarski, Burbidge, and Harrison 2014). The IUCN classifies *P. gilbertii* as critically endangered; it is jeopardized by potential fires and predation by the introduced red fox (*Vulpes*

vulpes) and house cat (*Felis catus*) (Friend and Burbidge 2008b). The IUCN classifies *P. longipes* as endangered (McKnight 2008e). It is known from only three disjunct, fragmented populations, one in southeastern New South Wales, and two in northeastern Victoria. Total numbers may be from a few hundred to a few thousand. Threats include predation from the fox and domestic dog (*Canis familiaris*), competition for the species' specialized food requirements from the introduced pig (*Sus scrofa*), and loss of habitat and food sources because of fires and logging.

On the mainland of southeastern Australia *P. tridactylus* apparently has declined because of its dependence on dense vegetation and resultant vulnerability to clearing operations and brush fires, but it is not as rare as once thought. It is widespread from Victoria to southern Queensland and remains common in much of Tasmania. It still is present, though rare, on King and Flinders Islands in Bass Strait, but seems to have disappeared from Clarke Island, which is heavily grazed by livestock and introduced rabbits (Calaby 1971; Johnston 2008; Poole 1979; Seebeck 1981) and from Cape Barren Island (Woinarski, Burbidge, and Harrison 2014). The IUCN formerly recognized a single mainland subspecies, *P. t. tridactylus*, and classified it as vulnerable, noting that total numbers were under 10,000. The IUCN does not currently assign any threatened classification, even though the over-all species population is thought to be decreasing and the subspecies on Tasmania (*P. t. apicalis*) is now threatened by the fox, which was recently introduced to the island (Menkhorst and Lunney 2008). Recently, using IUCN criteria and the taxonomic arrangement suggested by Frankham, Handasyde, and Eldridge (2012, see above), Woinarski, Burbidge, and Harrison (2014) classified the entire species *P. tridactylus* as near threatened, the subspecies *P. t. trisulcatus* and *P. t. apicalis* also as near threatened, and the subspecies *P. t. tridactylus* as vulnerable. It was noted that, while numbers of *P. t. trisulcatus* may recently have increased to somewhat

more than 10,000, *P. t. tridactylus* was in a substantial decline because of habitat loss and fragmentation and predation by introduced carnivores.

DIPROTODONTIA; POTOROIDAE; **Genus BETTONGIA** Gray, 1837

Short-nosed "Rat"-kangaroos, or Bettongs

There are six species (Burbidge and Short 2008; Claridge, Seebeck, and Rose 2007; De Tores and Start 2008; Kirsch and Calaby 1977; McDowell et al. 2015; McNamara 1997; Ride 1970; Sharman et al. 1980; Winter, Johnson, and Vernes 2008):

B. penicillata (woylie), originally from central and southwestern Western Australia to central New South Wales and northern Victoria;

Short-nosed "rat"-kangaroo (*Bettongia gaimardi*). Photo by Dave Watts / Lochman LT.

B. pusilla, known only by subfossil but late Holocene remains from the Nullarbor Plain in
southeastern Western Australia and southwestern South Australia;

B. tropica, originally from Mount Windsor Tableland to Dawson Valley in eastern Queensland;

B. gaimardi, originally southeastern Queensland,
coastal New South Wales, southern Victoria,
extreme southeastern South Australia, and
Tasmania;

B. lesueur (boodie or burrowing bettong), origi-
nally from Western Australia, including certain
coastal islands (Barrow, Boodie, Bernier, Dorre,
Dirk Hartog, Faure), through southern North
Australia and South Australia, to southwestern
Queensland and western New South Wales.

B. anhydra, known by only one specimen collected in 1933 in the McEwin Hills area of
southwestern Northern Territory and a
subfossil Holocene specimen from Stegamite
Cave in southeastern Western Australia.

Woinarski, Burbidge, and Harrison (2014) referred to
additional subfossil material of *B. anhydra* from Meer-
kanoota in western Western Australia and suggested
the species may once have had a wide distribution. *B.
cuniculus* of Tasmania was formerly recognized as a
distinct species but now is included in *B. gaimardi*.
Sharman et al. (1980) concluded there is no chromosomal basis for the specific distinction of *B. tropica*
from *B. penicillata*. Other factors, however, have led to
general acceptance of *B. tropica* as a full species (Calaby and Richardson 1988a; Claridge, Seebeck, and
Rose 2007; Corbet and Hill 1991; Flannery 1989; Groves
2005e; Winter, Johnson, and Vernes 2008). The animals
of the populations of *B. lesueur* on Bernier and Dorre
Islands in Shark Bay are about twice the size of those in
the populations on Barrow and Boodie Islands farther
north (Claridge, Seebeck, and Rose 2007). According to
Richards (2007), preliminary findings suggest that the
differences between those two groups may warrant specific distinction.

Head and body length is 280–450 mm and tail length
is 250–360 mm. Weight is 750–1,850 grams in *B. penicillata*, 1,200–2,250 grams in *B. gaimardi*, and 900–1,400
grams in *B. tropica* (De Tores and Start 2008; Rose and
Johnson 2008; Winter, Johnson, and Vernes 2008). The

upper parts are buffy gray to grayish brown; the underparts are paler. The tail is crested in all extant species
except *B. lesueur* and usually is white-tipped in *B.
gaimardi* and *B. lesueur*. The unworn adult pelage in *B.
penicillata* is often crisp or even coarse; in *B. lesueur* it is
soft and dense. The tip of the muzzle is naked and flesh-
colored, and the ears are short and rounded. The hind
feet are longer than the head. Females have four mammae and a well-developed pouch.

Habitats are mainly in drier regions and include
grassland, heath, and sclerophyll woodland. These
"rat"-kangaroos are nocturnal, sheltering in a nest by
day. They carry nesting material in the curled-up tip
of their tail. Rose and Johnson (2008) stated that *B.
gaimardi* constructs a densely woven nest of dry grass
and bark in a depression about 150 mm in diameter and
50 mm deep, commonly sited under a fallen limb or
among short bushes or tussocks. The nest is ovoid in
shape and about 300 mm long and 200 mm wide, with
walls 20–30 mm thick and a small opening at one end.
It may be used regularly for at least a month, vacated
for the same amount of time, then reoccupied. An individual may travel up to 1.5 km between its nest and
a feeding area. *B. lesueur* constructs a large burrow for
community dwelling and may modify a rabbit warren
for its own use. Burbidge and Short (2008) wrote that
its burrows typically have two or three entrances in
sand soils but up to 120 in other soil types. That species is entirely bipedal in locomotion, never using its
forefeet, even when moving slowly (Ride 1970), but *B.
gaimardi* walks on all four feet when probing and feeding (Rose and Rose 1998). The claws of all species are
used for digging. When fleeing, *B. penicillata* travels
with head held low, back arched, and the tail brush displayed conspicuously.

The diet of most species consists predominantly of
the fruiting bodies (sporocarps) of underground fungi
("truffles"), which exist in symbiotic relationship with
trees (De Tores and Start 2008; Johnson and McIlwee
1997; Rose and Johnson 2008; Seebeck, Bennett, and
Scotts 1989; Taylor 1992, 1993; Winter, Johnson, and
Vernes 2008). Other vegetative material is taken, including roots, tubers, bulbs, seeds, and legume pods,
and there are some reports of feeding on marine refuse,
carrion, and meat (Collins 1973; Ride 1970); *B. lesueur*
reportedly eats arthropods, as well as fungi and plant
matter (Burbidge and Short 2008).

Boodie or burrowing bettong (*Bettongia lesueur*). Photo by Jiri Lochman / Lochman LT.

Bettongs have relatively low population densities and large home ranges for animals of their size. Density of *B. gaimardi* was calculated as 19 individuals per sq km (Rose and Rose 1998), male range was 47–85 ha., and female range 38–63 ha. (Claridge, Seebeck, and Rose 2007). For *B. penicillata*, density was 7–45 per sq km, male range 28–43 ha., and female range 15–28 ha. (Claridge, Seebeck, and Rose 2007; Seebeck, Bennett, and Scotts 1989). Home ranges of both species overlap. Winter, Johnson, and Vernes (2008) wrote that a density of 4–7 *B. tropica* per sq km occurs on the Lamb Range, where the species is known from an area 25 km long and 3 km wide; home range may be as large as 120 ha., but is typically about 50–70 ha. Although two individual *B. gaimardi* may occasionally occupy the same nest, the species is normally solitary, and in captivity males are relatively aggressive (Rose and Johnson 2008). *B. lesueur* is very vocal, making a variety of grunts, hisses, and squeals (Burbidge and Short 2008).

A male and several females of *B. lesueur* may form a social group and occupy a burrow system. Large warrens may have 40–50 occupants (Seebeck, Bennett, and Scotts 1989). Males are aggressive toward one another and seem to defend groups of females but not a particular territory. Females generally are amicable but sometimes will establish a territory and exclude other females. They are polyestrous, with modal length of estrous cycle being 23 days, and can produce up to three litters annually. Breeding may go on throughout the year in some areas, but in the Bernier Island population of *B. lesueur* most births occur between February and September. The modal length of undelayed gestation is 21 days, and the usual litter size is one, though twins occasionally occur. Just after one young is born the female mates again, but because of embryonic diapause development is delayed, and parturition of the second young does not take place for about 4 months unless the first young is lost. The young weigh about 0.317 grams at birth, leave the pouch permanently at about 116 days, are weaned at 165 days, and attain adult size at 280 days. Females apparently are capable of giving birth at about 200 days, and males are sexually mature at 14 months (Ride 1970; Tyndale-Biscoe 1968).

In *B. penicillata* the estrous cycle, gestation period, and embryonic diapause are nearly the same as in *B. lesueur* (Rose 1978); pouch life is 90–110 days and females breed continuously (De Tores and Start 2008). *B. tropica* also has been found to breed continuously, with no significant difference in number of births from month to month (Vernes and Pope 2002); means for that species are: estrous cycle, 22.5 days; gestation, 21 days; pouch life, 106.4 days; birth to weaning, 177 days; sexual maturity, 11.5 months in females, 14.6 months in males (Johnson and Delean 2001). *B. gaimardi* has an estrous cycle of 22.6 days, a gestation period of 21.6 days, and continuous breeding. The young permanently vacates the pouch at 15 weeks, at which time the pouch sphincter tightens, preventing reentry. Simultaneously there is often a new birth and subsequent postpartum estrus and mating. Hence, the mother can have one young outside and still nursing, another in the pouch, and a third in the embryonic stage. Females mature at 8 to 11 months and have produced up to 15 young in 5 years of breeding in captivity (Rose 1987; Rose and Rose 1998). Record longevity for the genus is held by a captive specimen of *B. penicillata* that lived 18 years and 10 months (Weigl 2005).

Considering especially reduction of range, modern human agency seems to have harmed *Bettongia* more than any other polytypic genus of marsupials. All four extant species are listed as endangered by the USDI and are on appendix 1 of the CITES. The IUCN designates *B. pusilla* as extinct, *B. penicillata* as critically endangered, *B. tropica* as endangered, and *B. gaimardi* and *B. lesueur* as near threatened (Burbidge 2008b; Burnett and Winter 2008d; Menkhorst 2008c; Richards, Morris, and Burbidge 2008; Wayne et al. 2008). In

Woylie (*Bettongia penicillata*). Photo by Israel Didham.

addition, the following subspecies are probably extinct, formerly were classified as such by the IUCN, and more recently were classified as such by Woinarski, Burbidge, and Harrison (2014): *B. penicillata penicillata* of eastern Australia (*B. p. ogilbyi* of Western Australia survives), *B. gaimardi gaimardi* of mainland Australia (*B. p. cuniculus* of Tasmania survives), and *B. lesueur graii* of mainland Australia (*B. l. lesueur* survives on a few islands off Western Australia, though it may represent more than one subspecies or species; see above).

B. penicillata once occurred all across the southern part of the continent, including certain islands, as far as the northwestern corner of Victoria. Its range extended well into southwestern Northern Territory and may have reached Queensland (Seebeck, Bennett, and Scotts 1989). It was described as "very abundant" in New South Wales in 1839–1840 but disappeared from that state shortly thereafter. It was common in South Australia at the end of the nineteenth century but was gone from there by 1923. By the 1970s it was restricted to four tracts of woodland at the extreme southwestern tip of the continent—Dryandra, Tutan-ning Nature Reserve, Kingston, and Perup Forest. The primary causes of its decline seem to have been predation by the introduced red fox (*Vulpes vulpes*) and house cat (*Felis catus*), clearing of brush for agricultural development and other habitat disruption, changes in fire management, competition with domestic and feral introduced herbivores, and possibly disease (Australian National Parks and Wildlife Service 1978; Calaby 1971; Poole 1979; Ride 1970; Wayne et al. 2008; Woinarski, Burbidge, and Harrison 2014).

Intensive efforts to save *B. penicillata*, featuring reintroduction and predator control, began in the mid-1970s and initially seemed highly successful (Claridge, Seebeck, and Rose 2007; Delroy et al. 1986; Finlayson, Finlayson, and Dickman 2010). The species was translocated to 46 sites in Western Australia, 3 in New South Wales, and 12 in South Australia, including several offshore islands. In 1996 the species was even removed from threatened classifications by the state of Western Australia, the national government of Australia, and the IUCN (Groom 2010). However, since 2001 most of the original and translocated populations have

Short-nosed "rat"-kangaroo (*Bettongia tropica*). Photo by Hans and Judy Beste / Lochman LT.

declined at a rate of 25–95 percent annually, and some have disappeared. Woinarski, Burbidge, and Harrison (2014) cited estimates that a total population of around 250,000 individuals in 1999 had fallen to just 18,000 in 2010. The cause is unknown but numbers are continuing to fall and within a year or two a total of only 1,000 individuals may be left in the three surviving natural populations (the one at Tutanning has disappeared). Only a few hundred animals are present at the mainland reintroduction sites, though the South Australian island populations contain several thousand and are apparently stable (Wayne et al. 2008). In 2008 *B. penicillata* was restored to an IUCN threatened category, this time as critically endangered (Wayne et al. 2008).

B. pusilla is known only by subfossils but the condition of the remains, plus Aboriginal accounts, suggests the species persisted into the early colonial period (McNamara 1997). *B. anhydra* may have survived until the 1950s or 1960s but is now considered extinct (Woinarski, Burbidge, and Harrison 2014). *B. tropica* had been known from only six specimens, the latest collected in 1932, and was thought to be possibly extinct, but a population was discovered in the Davies Creek National Park in northeastern Queensland (Poole 1979). Three populations now are known but they occupy less than 500 sq km and are declining because of habitat degradation. The species *B. gaimardi* has not been recorded from the mainland since 1910; it still is found in reasonable numbers in many localities of Tasmania but may be jeopardized there by logging and poisoning (Calaby 1971; Ride 1970; Rose 1986; Woinarski, Burbidge, and Harrison 2014). Burbidge (2008b) noted that recent introduction of the fox to Tasmania is a potential major threat to *B. gaimardi*, even bringing the species close to qualifying for an IUCN vulnerable classification. Using IUCN criteria, Woinarski, Burbidge, and Harrison (2014) did designate *B. gaimardi* vulnerable because of the threat of predation by the fox and also the feral domestic cat, which evidently is increasing in Tasmania, possibly in response to the decline of the Tasmanian devil (see account of genus *Sarcophilus*).

B. lesueur was once among the most widely distributed of native Australian mammals. Although the last specimen in New South Wales was collected in 1892, *B. lesueur* remained common in parts of central and southwestern Australia until the 1930s. By the early 1960s it had disappeared completely from the mainland and become restricted to Barrow, Boodie, Bernier, and Dorre Islands off the west coast. Its drastic decline apparently resulted, at least in part, from competition with introduced rabbits for burrows and food, habitat disruption and disturbance by domestic livestock, predation by the fox, and direct killing by people. Abbott (2006) hypothesized that a major epizootic began near Shark Bay in Western Australia around 1875 and spread through most of the state by 1920, severely reducing populations of many mammals, *B. lesueur* being one of those particularly affected. The disease may have been transmitted by deliberate or incidental introduction of affected domestic animals and rodents; a similar process may have taken place in eastern Australia. Populations on islands may have avoided infection and once were thought to be viable and well protected (Calaby 1971; Poole 1979). However, surveys by Short and Turner (1993) revealed that the species had been wiped out on Boodie Island, apparently as an inadvertent consequence of a poisoning campaign against *Rattus rattus*. It was reintroduced to Boodie from Barrow Island in 1992 (Richards 2007). It has also been introduced to Faure Island and to three fenced sites on the nearby mainland of Western Australia. There are about 7,500 individuals in the three natural island populations and about 7,800 at the reintroduction sites. Populations are currently stable or increasing but are jeopardized by potential introduction of the fox and house cat, fire, disease, and other factors (Richards, Morris, and Burbidge 2008; Woinarski, Burbidge, and Harrison 2014; see also Finlayson, Finlayson, and Dickman 2010).

DIPROTODONTIA; POTOROIDAE; **Genus *CALOPRYMNUS* Thomas, 1888**

Desert "Rat"-kangaroo

The single species, *C. campestris*, was described in 1843 on the basis of three specimens from an unknown locality in South Australia. There were no further records until 1931, when a specimen was collected in the northeastern corner of South Australia. A subsequent survey indicated that in the early 1930s the species occurred in an area about 650 km north to south and 250 km wide in the Lake Eyre Basin in northeastern South Australia and southwestern Queensland. A later

Desert "rat"-kangaroo (*Caloprymnus campestris*). Painting by Helga Schulze, based on descriptive information in this account and a black-and-white photo in Finlayson (1943).

investigation suggested the historical distribution could have extended farther to the south and more than 600 km east to west. In addition, apparently Recent cave remains were found in extreme southeastern Western Australia, and late Pleistocene fossils were located in southwestern New South Wales (Calaby 1971; Carr and Robinson 1997; Claridge, Seebeck, and Rose 1997; Ride 1970; Smith and Johnson 2008).

Head and body length is 254–282 mm, tail length is 297–377 mm, and weight is 637–1,060 grams (Smith and Johnson 2008). The pelage is soft and dense. The coloration of the upper parts is clear pale yellowish ochre, which matches that of clay pans and plains; the underparts are whitish. The ears are longer and narrower than those of any other "rat"-kangaroo, and the muzzle is naked. The long, cylindrical tail is evenly short-haired without a trace of a crest. There is a well-defined neck gland, at least in the skin of the type specimen. The most conspicuous feature, however, is the relatively enormous hind foot. The forelimb is small and delicate; the bones of its three segments weigh only 1 gram, whereas the bones of the hind limb weigh 12 grams.

A peculiar feature of the hopping gait of *Caloprymnus* is that the feet are not brought down in line with one another; rather, the right toe mark registers well in front of the left toe mark. This "rat"-kangaroo seldom dodges or doubles back when moving rapidly and is noted for endurance rather than speed. A young adult male exhausted two galloping horses in a 20-km run. The gait, when the animal moves on all four limbs and uses the tail as a support, is normal for the Potoroidae.

The area known to have been inhabited consists mainly of gibber plains, clay pans, and sandridges. There is a sparse cover of saltbush and other shrubs. *Caloprymnus* is a nest builder, not a burrower, and shelters in simple leaf and grass nests in scratched-out excavations despite the glaring heat of its habitat. It has the unusual habit of protruding its head through a gap in the roof of the nest to observe its surroundings. It has been reported to feed mainly on the foliage and stems of plants and to feed less on roots than do the other "rat"-kangaroos. Dixon (1988) examined the stomach and colon contents of a preserved specimen and found extensive remains of beetles. Smith and Johnson (2008) noted that *Caloprymnus* seems quite independent of surface water, to shun succulent plants, and to be nocturnal and solitary.

Females with a single pouch young have been taken in June, August, and December. The breeding season is apparently irregular: one female had a small, naked young in her pouch at the same time that two other females were carrying well-furred and almost independent "joeys." Weigl (2005) listed a specimen caught in 1853 that subsequently lived in captivity for about 12 years and 11 months.

The desert "rat"-kangaroo apparently was rare for many years but then became fairly common in its restricted range when severely dry conditions abated about 1931. It soon seemed rare again, the last specimens being taken in 1933 (Smith and Johnson 2008). However, extensive interviews by Carr and Robinson (1997) indicate *Caloprymnus* remained numerous until 1943, was less common but present until 1954, and may have persisted in northeastern South Australia and

southwestern Queensland into the 1980s; reported sightings occurred most often during exceptionally wet seasons, as in 1956–57 and 1974–75. Woinarski, Burbidge, and Harrison (2014) cited unconfirmed reports in 1993 and 2011. *Caloprymnus* is classified as extinct by the IUCN (Australasian Mammal Assessment Workshop 2008), which notes that habitat alteration by humans and predation by the introduced house cat and fox contributed to its demise; it is listed as endangered by the USDI and is on appendix 1 of the CITES.

DIPROTODONTIA; POTOROIDAE; **Genus** *AEPYPRYMNUS* Garrod, 1875

Rufous "Rat"-kangaroo

The single species, *A. rufescens*, occurred historically from Cooktown in northeastern Queensland to north-central Victoria, with possible distributional gaps in

Rufous "rat"-kangaroo (*Aepyprymnus rufescens*). Photo by Jiri Lochman / Lochman LT.

New South Wales (Claridge, Seebeck, and Rose 2007; Dennis and Johnson 2008b; Ride 1970). Remains have been found in cave deposits in southwestern Victoria and on Flinders Island in Bass Strait (Calaby 1971).

This is the largest of the "rat"-kangaroos. Head and body length is 345–520 mm, and tail length is 314–407 mm. Johnson (1978a, 1978b) reported that adults stand 350 mm tall, and weight in a series of specimens was 2.27–2.72 kg for males and 1.36–3.60 kg for females. The pelage is crisp and often coarse. The upper parts, grizzled in appearance, are rufescent gray, and the underparts are whitish. The tail is thick, evenly haired, and not crested. This genus can be distinguished by its ruddy color, black-backed ears, whitish hip stripe (usually not distinct), and hairy muzzle. Females have four mammae in a well-developed pouch.

At present this "rat"-kangaroo is found from sea level to the tops of plateaus, in open forest and woodland with a dense grass floor. It is nocturnal and constructs dome-shaped nests in which it shelters during the day. Nests are built in a shallow excavation and lined and covered with grass or bark (Seebeck, Bennett, and Scotts 1989). An individual constructs one or two clusters of nests that provide protection from the sun and concealment from predators (Wallis et al. 1989). Although not particularly fast, *Aepyprymnus* is wonderfully agile, adept at dodging, and difficult to approach on foot. On horseback, however, a person may get quite close. When startled it usually seeks shelter in a hollow log, if available. It has little fear of people at night, is often attracted by a bush camp, and, if not molested by dogs, can be enticed up to a tent door to receive scraps of food. When taken young and treated kindly, it becomes tame and responsive. According to Dennis and Johnson (2008b), *Aepyprymnus* emerges from its nest 40 minutes after sunset to browse on herbs and grasses or dig for roots and tubers with its long foreclaws; it may consume entire plants, underground fungi, bones of dead animals, and exudates from trees. To access sufficient resources, an individual may cover 2.0–4.5 km in a normal night's foraging. Johnson (1978a) noted that in captivity it accepts a variety of foods and apparently thrives on them. Like most coastal species, it has little resistance to drought; during dry periods it excavates holes in creek beds to reach the water level.

Pope, Blair, and Johnson (2005) reported population densities of about 3 to 37 individuals per sq km, the difference being associated with soil fertility. Dennis and Johnson (2008b) noted that densities are even more variable, showing a 50-fold change in some places, and that home ranges are around 75–110 ha. for males and 45–60 ha. for females. Frederick and Johnson (1996) found *Aepyprymnus* to be predominantly solitary but to occasionally form groups of up to six individuals. Some males seemed to maintain close associations with one to three females, even when the latter were not in estrus; those males would attempt to prevent the approach of other males but were not always successful, so that mating was considered "promiscuous, but with a hint of monogamy." P. M. Johnson (1980) found captive males extremely aggressive toward one another. Vocalizations include loud guttural growls and grunts, usually given by a nonestrous female in repelling a male (Dennis and Johnson 2008b).

Aepyprymnus is polyestrous, with an estrous cycle of about 21–25 days and a gestation period of 22–24 days (Johnson 1978a). Breeding occurs throughout the year (Dennis and Johnson 2008b). In the Dawson Valley of central coastal Queensland females with pouch young were taken from January to March. Near Killarney in southeastern Queensland three females with pouch young were caught in June, and local residents reported finding pouch young all year. Females in a captive colony at Canberra had large pouch young in February, July, and August and smaller ones in July, August, and December (Moors 1975). Litters usually consist of a single offspring, but twins occasionally occur. Within a few hours of giving birth, the mother mates again, producing a quiescent blastocyst. The pouch young releases the nipple at 7–8 weeks, vacates the pouch permanently at about 16 weeks, and then remains with the mother for another 7 weeks. Sexual maturity is attained at about 11 months in females and 12–13 months in males (Dennis and Johnson 2008b; Rose 1989). Female young settle in ranges close to their mother's, while male young usually disperse, sometimes for distances of up to 6.5 km (Pope, Blair, and Johnson 2005). A captive specimen lived for 12 years and 10 months (Weigl 2005).

The rufous "rat"-kangaroo is still widely distributed along the coast from northeastern Queensland to northeastern New South Wales. It is fairly common in much of that region and seems able to coexist with grazing beef cattle (Calaby 1971). Farther south the species has disappeared, the last record in Victoria being dated 1905. Its decline may have resulted from predation by the introduced red fox and competition from introduced rabbits; those problems and climate change may potentially threaten the surviving populations (Dennis and Johnson 2008b).

DIPROTODONTIA; Family MACROPODIDAE

Wallabies and Kangaroos

This family of 12 Recent genera and 68 species is native to Australia, Tasmania, and New Guinea and some nearby islands. Potoroidae, the "rat"-kangaroos (see account thereof), sometimes has been considered a subfamily of Macropodidae but now generally is treated as a separate family (as is Hypsiprymnodontidae). Flannery (1983, 1989), whose phylogenetic sequence is followed here, employed morphological criteria in recognizing two living macropodid subfamilies: Sthe-

nurinae, with the single living genus *Lagostrophus*, and Macropodinae, with three tribes: an unnamed tribe for the genera *Dorcopsis* and *Dorcopsulus* (the name "Dorcopsini" was designated for this tribe by Prideaux and Warburton 2010); Dendrolagini for *Dendrolagus*; and Macropodini for *Setonix*, *Thylogale*, *Peradorcas* (considered part of *Petrogale* by Flannery [1983, 1989]), *Petrogale*, *Lagorchestes*, *Onychogalea*, *Wallabia*, and *Macropus*. However, Flannery (1989) indicated that similarity of the genera *Lagorchestes* and *Onychogalea* might be due to convergence and not true phylogenetic relationship. Other authorities have also suggested those two genera represent separate lineages (Burk and Springer 2000; Eldridge 2008a; Kirsch 1977b).

Some investigations involving DNA hybridization, mitochondrial DNA, and albumin immunologic relationships support immediate affinity of *Petrogale* and *Dendrolagus* and a more distant position for *Setonix* (Baverstock et al. 1989; Burk and Springer 2000; Eldridge 2008a; Kear and Cooke 2001; Kirsch, Lapointe, and Springer 1997). An exhaustive paleontological and

Red kangaroo (*Macropus rufus*). Photo by Gerhard Körtner.

morphological assessment by Prideaux (2004) found no support for placing *Lagostrophus* in the otherwise extinct Sthenurinae and indicated closer affinity to Macropodinae. Analysis of molecular sequences for three mitochondrial genes and one nuclear gene (Westerman et al. 2002) showed *Lagostrophus* to represent a distinct lineage outside of Macropodinae but to have a sister-group relationship with the latter; the analysis neither corroborated nor refuted placing *Lagostrophus* in the Sthenurinae. A comprehensive analysis of the nuclear DNA of the suborder Macropodiformes (Meredith, Westerman, and Springer 2009a) showed the same sister-group relationship of *Lagostrophus*, and that *Onychogalea* was the next genus to diverge, followed by a clade comprising *Dorcopsis, Dorcopsulus, Thylogale, Dendrolagus, Petrogale,* and *Peradorcas*; the remainder of the sequence was *Setonix, Lagorchestes,* and *Macropus* (including *Wallabia*). Based on an osteological appraisal, Prideaux and Warburton (2010) placed *Lagostrophus* in a new subfamily, Lagostrophinae, separate from and basal to Sthenurinae and Macropodinae; within the latter subfamily the suggested phylogenetic sequence was tribe Dorcopsini for the genera *Dorcopsis* and *Dorcopsulus*, the genus *Setonix* in an undetermined tribe, tribe Dendrolagini for *Dendrolagus* and *Petrogale*, and tribe Macropodini for *Thylogale, Wallabia, Lagorchestes,*

Red-bellied pademelon (*Thylogale billardierii*), mother and young. Photo by Gerhard Körtner.

Onychogalea, and *Macropus.* Subfamily Lagostrophinae was accepted by Eldridge and Coulson (2015).

Adults in this family vary in head and body length from less than 300 mm to as much as 1,600 mm in *Macropus*. Full-grown individuals of the latter genus weigh as much as 100 kg (Hume et al. 1989). The head is rather small in relation to the body, and the ears are relatively large. The tail is usually long, thick at the base, hairy, and nonprehensile. It is used as a prop or additional leg, as a balancing organ when leaping, and sometimes for thrust. In all Recent members of this family except *Dendrolagus* the hind limbs are markedly larger and stronger than the forelimbs. The forelimbs are small with five unequal digits. The hind foot is lengthened and narrowed in all genera—hence the family name Macropodidae, meaning "large foot." Digit 1 of the hind foot is lacking in all genera, the small digits 2 and 3 are united by skin (the syndactylous condition), digit 4 is long and strong, and digit 5 is moderately long. The well-developed marsupial pouch opens forward and encloses four mammae.

In kangaroos and wallabies the dentition is suitable for a grazing or browsing diet. The first upper incisors are prominent and with the other incisors form a U- or V-shaped arcade; the two remaining lower incisors are very large and forward-projecting (the diprotodont condition), and except in the Sthenurinae their tips fit within the upper incisor arcade and press on a pad on the front of the palate; there are no lower canines, and the upper canines are small or absent, thus leaving a space between the incisors and cheek teeth; the premolars are narrow and bladelike; and the molars are broad, generally high-crowned, rectangular teeth with two transverse ridges separated by a deep trough crossed by a longitudinal ridge that is weakly formed in the browsing species but strong in grazers (Hume et al. 1989). In many macropodid genera, as in elephants, manatees, and certain pigs, an anterior migration of the molariform teeth occurs throughout life, making room for the late-erupting rear molars. The fourth molar may not erupt until well after adulthood is attained. The dental formula in this family is: i 3/1, c 0–1/0, pm 2/2, m 4/4=32 or 34.

Most species are nocturnal, but they may sunbathe on warm afternoons and some are active at intervals during the day. All the members of this family except *Dendrolagus* progress rapidly by leaps and bounds, using only the hind limbs. Representatives of the genus

Macropus illustrate the peak of development of the jumping mode of progression. Macropodids commonly move on all four limbs and often use the tail as a support when feeding, progressing slowly, or standing erect.

The members of this family are mainly grazers or browsers, feeding on many kinds of plant material. The occurrence of ruminantlike bacterial digestion in the kangaroos and wallabies enables them to colonize areas that would be nutritionally unfavorable to most other large mammals. In this kind of digestion the food is fermented by a dense bacterial population in the esophagus, stomach, and upper portion of the small intestine, thus providing the available energy for chemical breakdown of foods over a longer period of time and enhancing the uptake of nitrogen and other nutrients.

Females of most macropodid genera usually give birth to one young at a time. A phenomenon known as embryonic diapause, or delayed birth, approximately equivalent to delayed implantation in placental mammals, occurs in some, perhaps most, members of this family. Following the birth of one young, the female often mates again. Usually this mating occurs only a day or two after the birth, but in some species it is late in the pouch life of the young. In one species, *Wallabia bicolor*, the mating takes place just before the young is born (Kaufmann 1974). The development of the embryo resulting from this mating is arrested at about the 100-cell stage. The embryo remains in this state of diapause until the first young nears the end of its pouch life, dies, or is abandoned. At such a time the embryo resumes development, and a second young is soon produced. This reproductive approach seems well adapted to the variable, often severe climate of inland Austra-

Quokka (*Setonix brachyurus*). Photo from Pixabay

lia. Should adverse conditions lead to the death of a pouch young or force the female to discard it, a successor is in reserve for another try at rearing during the season. If conditions are consistently favorable, both offspring can be raised, the second being born shortly after the first permanently vacates the pouch. Other views regarding embryonic diapause are that it functions simply to prevent two young from crowding the pouch at the same time or that it helps to ensure synchrony between the embryo and the uterus (Russell 1974a). A feature unique to the macropodids and potoroids is the capability of females to produce milk in one mammary gland that has a very different nutritional composition from that of the other gland when young at different stages of development are being reared (Merchant 1989).

Modern humans have greatly affected many species of this family (Calaby and Grigg 1989; Maxwell, Burbidge, and Morris 1996; Pople and Grigg 1999). Perhaps the single most important factor was introduction of vast herds of domestic livestock, which cropped grasslands, thereby eliminating the cover needed by smaller macropodids. Much devastation has also resulted from introduction of the red fox (*Vulpes vulpes*), which has wiped out entire wallaby populations and species. Surprisingly, perhaps, the effects of settlement thus far have seemed less detrimental to the larger species of *Macropus*, even though most now are subject to large-scale government-sanctioned culling and commercial hunting. Adult kangaroos and wallaroos, at 20–70 kg, are not bothered by foxes weighing 4–8 kg. The dingo (*Canis familiaris dingo*), which weighs 10–20 kg and hunts in packs, is dangerous to the large macropodids. Again, however, the latter have benefited, because humans have greatly reduced dingo numbers to protect sheep. Some large species also appear to have at least temporarily prospered through elimination of tall, dry grass by livestock and the resultant production of more favorable vegetation (Newsome 1975).

The known geological range of the family Macropodidae is Miocene to Recent in Australaia (Kirsch 1977a) and Pliocene to Recent in New Guinea (Flannery 1995a). The largest extinct kangaroo was *Procoptodon goliah* of the late Pleistocene, a member of the subfamily Sthenurinae, which weighed around 232 kg and stood 2–3 meters tall (Cooke 2006; Helgen et al. 2006; Long et al. 2002). Although sometimes considered the largest hopping animal that ever lived, Janis, Buttrill, and Figueirido (2014) concluded that *P. goliah* was anatomically very different from large *Macropus*, lacking the specialized features for fast hopping, and apparently relied primarily on a bipedal striding gait. Closer to *Macropus*, both phylogenetically and behaviorally, was *Protemnodon anak*, which stood about 2 meters tall and weighed 100–150 kg; it is thought to have disappeared on the Australian mainland about 46,000 calendar years ago, but to have survived on Tasmania for another few thousand years, until the arrival of human hunters (Diamond 2008; Long et al. 2002; Turney et al. 2008)

DIPROTODONTIA; MACROPODIDAE; Genus *LAGOSTROPHUS* Thomas, 1887

Banded Hare Wallaby, or Munning

Subsequent to European colonization, the single species, *L. fasciatus*, is known to have occupied Bernier, Dorre, Dirk Hartog, and Faure Islands in Shark Bay, as well as parts of the mainland of extreme southwestern Western Australia and the lower Murray River region of South Australia. Subfossil remains, dating from 3,000–5,000 years ago, have been found on the mainland just south of Shark Bay, on the Nullarbor Plain of southeastern Western Australia, in coastal and southeastern South Australia, and in southwestern New South Wales and western Victoria (Calaby 1971; Helgen and Flannery 2003; Prince and Richards 2008; Richards 2007).

Head and body length is 400–450 mm, tail length is 230–360 mm, and weight is 1.0–2.3 kg (Prince and Richards 2008). The fur is thick, soft, and long. *Lagostrophus* can be readily distinguished by its banded color pattern. The dark transverse bands on the posterior half of the body contrast sharply with the general grayish coloration. The underparts are buffy white with gray hair bases, and the hands, feet, and tail are gray. This genus, like *Lagorchestes*, is referred to as a "hare wallaby" because of its harelike speed, its jumping

Banded hare wallaby (*Lagostrophus fasciatus*). Photo by Jiri Lochman / Lochman LT.

ability, and its habit of crouching in a "form." The muzzle in *Lagostrophus* is rather long and pointed, and the nose is naked rather than hairy. The tail is evenly haired throughout, except for an inconspicuous pencil of longer hairs at the tip. The claws of the hind feet are hidden by the fur. Unlike the condition in all other genera of the Macropodidae, the lower incisors of *Lagostrophus* occlude with the upper incisors, not with a pad between the upper incisors (Hume et al. 1989). Additional technical characters of the dentition that distinguish *Lagostrophus* were listed by Flannery (1983, 1989).

On the mainland the banded hare wallaby inhabited prickly thickets on the flats and the edges of swamps. On the islands it lives in dense thickets of thorny species of acacia bushes. Runs and "forms" are made in these tangled masses under low-hanging limbs. This wallaby is nocturnal, emerging at night from its retreats in the scrub to feed on grasses, shrubs, and other plant matter. Average home range in one small sampling was 11 ha. (Richards 2007). Ride (1970) referred to *Lagostrophus* as gregarious. However, Prince and Richards (2008) indicated that, while several individuals are often found in one patch of scrub, adults appear to live in separate, well-defined ranges. Each adult male probably defends a territory, which overlaps those of several females. Relations between adult females seem largely peaceful but interactions of males show a high level of aggression, the intensity of fighting apparently related to competition for food.

Females are polyestrous, have a gestation period of about 30 days, and normally have a single young, though twins have occurred (Collins 1973; Richards 2007). According to Prince and Richards (2008), breeding on Bernier and Dorre Islands appears to be continuous, with some indication of a peak in autumn and a decrease in the latter half of the year. A postpartum estrus may occur, followed by embryonic diapause while the first young still is suckling. The young spends about 6 months in the pouch and is weaned after another 3 months; both sexes are sexually mature in their first year of life but usually do not breed until their second year. Although females appear capable of producing two young annually, they normally are able to rear only one. Individuals have been known to survive at least 6 years in the wild. Richards (2007) reported that captives have lived for 10 years.

The last mainland specimens of *Lagostrophus* were collected in the Pingelly area of Western Australia in 1906 (Prince and Richards 2008). Disappearance of the genus may have been associated with clearing of vegetation for agriculture, competition for food with the introduced rabbits and livestock, and predation by the introduced red fox (*Vulpes vulpes*) and house cat (*Felis catus*). Natural populations survive only on Bernier and Dorre Islands and are well protected, though numbers fluctuate between relative abundance and scarcity (Ride 1970; Thornback and Jenkins 1982). Surveys by Short and Turner (1992) indicated a minimum of about 3,900 individuals on Bernier and 3,800 on Dorre Island. Earlier, when domestic sheep were brought temporarily to Bernier Island, numbers of *Lagostrophus* were reduced severely. A natural population on nearby Dirk Hartog Island disappeared entirely in the 1920s following establishment of sheep. Attempts were made to reintroduce *Lagostrophus* on that island in 1974 and at a nearby mainland site on the Peron Peninsula in 2001, but both efforts failed due to predation by the house cat and various other factors. Reintroduction to Faure Island, begun in 2004, appears successful so far (Richards 2007).

The banded hare wallaby is classified as endangered by the IUCN (Richards et al. 2008a) and USDI and is on appendix 1 of the CITES. The IUCN formerly assigned an extinct designation to *Lagostrophus fasciatus albipilis*, a subspecific name then applied to all mainland populations. However, Helgen and Flannery (2003) placed that name in the synonymy of *L. f. fasciatus* and considered the latter name to apply to both the mainland and the island populations of Western Australia. Those authorities also described the new subspecies *L. f. baudinettei* for the populations of South Australia, New South Wales, and Victoria; that subspecies apparently became extinct shortly after a specimen was collected in 1863. Woinarski, Burbidge, and Harrison (2014) recognized the latter subspecies and designated it extinct; they also considered the appropriate classification for the species *L. fasciatus* as a whole to be vulnerable.

New Guinean forest wallaby (*Dorcopsis hageni*). Photo by Klaus Rudloff at Prague Zoo.

DIPROTODONTIA; MACROPODIDAE; **Genus**
DORCOPSIS Schlegel and Müller, 1842

New Guinean Forest Wallabies

Four species now are recognized (Groves and Flannery 1989):

> *D. luctuosa*, coastal lowlands of southeastern New
> Guinea;
> *D. muelleri*, lowlands of western New Guinea and
> nearby Misool, Salawati, and Japen (Yapen)
> Islands;
> *D. hageni*, lowlands of northern New Guinea;
> *D. atrata*, Goodenough Island off southeastern
> New Guinea.

The genus *Dorcopsulus* (see account thereof) has sometimes been considered a subgenus or synonym of *Dor-*

copsis. The name *D. veterum* has sometimes been used
in place of *D. muelleri*, but the latter designation was
recommended by George and Schürer (1978) and accepted by Groves and Flannery (1989). Helgen (2007)
noted that *D. muelleri* might occur on Batanta and
Waigeo Islands but that there were no confirmed records. Groves (2005e) reported the range of *D. muelleri*
to extend to the Aru Islands off southwestern New
Guinea, though Flannery (1995a, 1995b) did not mention its presence there. Aplin and Pasveer (2007) also
did not list *Dorcopsis* as a living component of the fauna
of the Aru group, but did report late Pleistocene to Holocene specimens from an archeological site on Kobroor Island, which might be referable to *D. muelleri*
or to *D. luctuosa*. Skeletal remains indicate that *D.
muelleri* or a closely related species occurred on
Halmahera Island in the North Moluccas until at least
1,870 years ago (Flannery et al. 1995b) and on nearby

Gebe Island until about 2,000 years ago (Flannery et al. 1998), though presence there may possibly have resulted from early human introduction (Heinsohn 2003).

Head and body length is 340–970 mm, and tail length is 270–550 mm. Flannery (1995a, 1995b) listed weights of 3.6–11.6 kg for *D. luctuosa*, 5.0–6.8 kg for *D. muelleri*, 5.0–6.0 kg for *D. hageni*, and 4.0–7.5 kg for *D. atrata*. The pelage is short and sparse in *D. hageni* but long and thick in the other species. In *D. muelleri* the coloration of the upper parts is dull brown or blackish brown tipped with light buff, and the underparts are gray or white. *D. hageni* is light brown or fuscous above with a narrow white dorsal stripe and grayish white below. *D. luctuosa* is dark gray above and drab gray to creamy orange below. In *D. atrata* the upper parts are black or blackish brown and the underparts are also blackish brown. In all species the nose is large, broad, and naked; the ears are small and rounded; and the hairs are reversed on the nape. The tail is evenly haired except for a fifth or less of the terminal half, which is naked. Females have four mammae and a well-developed pouch that opens forward.

On New Guinea these wallabies generally inhabit lowland rainforests up to 400 meters in elevation. *D. atrata* of Goodenough Island lives in oak forest at elevations of 1,000–1,800 meters in the forested mountains (Flannery 1995b). These animals do not appear to be as adapted for hopping as most other wallabies. They are presumed to be mainly nocturnal, but there is some evidence that they move about in daytime in dense forest. Observations of a captive colony revealed a crepuscular pattern of activity (Bourke 1989). The diet of most species consists of roots, leaves, grasses, and fruit, but Flannery (1995a) noted that native hunters report encountering *D. hageni* mostly on river banks, where it turns over flat stones to search for cockroaches and other invertebrates. Ganslosser (1990) pointed out a number of unusual behaviors of *Dorcopsis*. When it is sitting or hopping slowly, only the end of the tail is in contact with the ground. During mating the male bites the female's neck, a primitive habit also seen in the Didelphidae and Dasyuridae.

Observations of *D. luctuosa* in captivity indicated little agonistic behavior and a tendency to form social groups (Bourke 1989). Females usually give birth to one young at a time (Collins 1973). According to Flannery (1995a, 1995b), *D. muelleri* and *D. atrata* may breed throughout the year, naked pouch young of *D. hageni* were found in January and April, a captive male *D. luctuosa* first emerged from the pouch on 22 May 1983 and reached sexual maturity by June 1985, and a specimen of *D. muelleri* was born in captivity on 16 October, first left the pouch on 13 April, and permanently left on 17 May. A captive *D. luctuosa* lived for 13 years and 11 months (Weigl 2005).

These wallabies are valued as food and are sometimes intensively hunted by native people with dogs. The IUCN classifies *D. atrata* as critically endangered, because it occurs in only one area of less than 100 sq km and is continuing to decline because of hunting and destruction of its forest habitat by slash-and-burn agriculture (Leary et al. 2008y). *D. luctuosa* is classed as threatened and is projected to decline by at least 30 percent in a 15-year period (Leary et al. 2008z). It is killed in large numbers for local markets and is losing habitat to agriculture, logging, and road building (which will lead to more hunting).

DIPROTODONTIA; MACROPODIDAE; **Genus**
DORCOPSULUS Matschie, 1916

New Guinean Forest Mountain Wallabies

There are two species (Baker 2016; Flannery 1989, 1995a):

D. vanheurni, Central Cordillera and Huon Peninsula of New Guinea;
D. macleayi, mountains of central and southeastern Papua New Guinea.

Kirsch and Calaby (1977) suggested it would be reasonable to put these two species in the genus *Dorcopsis*, and that was done by Ziegler (1977) and Corbet and Hill (1991). Kirsch and Calaby (1977) also stated that *D. vanheurni* probably is conspecific with *D. macleayi*. Nonetheless, *Dorcopsulus* was accepted as a distinct genus, and *D. vanheurni* as a separate species, by Eldridge and Coulson (2015), Flannery (1989, 1995a), and Groves (2005e). Subfossil remains show that *D. vanheurni* originally occurred in the Vogelkop Peninsula and the North Coast Ranges of Papua New Guinea (Aplin, Pasveer, and Boles 1999; Helgen 2007).

New Guinean forest mountain wallaby (*Dorcopsulus macleayi*). Photo by Nick Baker of a wild individual, 10 km southwest of Moro Airfield, Southern Highlands Province, Papua New Guinea (see Baker 2016).

Head and body length is 315–460 mm, tail length is 225–402 mm, and weight is 1,500–3,400 grams (Flannery 1995a). Both species are deep gray-brown above with a darker mark above the hips and light brownish gray below. The chin, lips, and throat are whitish. The pelage is long, thick, soft, and fine. The tail is evenly haired, the terminal quarter to half being bare with a small white tip. Females have four mammae and a well-developed pouch that opens forward. *Dorcopsulus* is distinguished from *Dorcopsis* in being smaller and more densely furred and in having more of the tail naked.

These wallabies inhabit mountain forests, the known altitudinal ranges being 800–3,100 meters for *D. vanheurni* and 1,000–1,800 meters for *D. macleayi*. Although they are presumed nocturnal, Lidicker and Ziegler (1968) observed that *D. vanheurni* apparently is active by day. *D. macleayi* is reported to be fond of the fruit and leaves of *Ficus* and various other trees (Flannery 1995a). According to Leary et al. (2008aa), *D. vanheurni* occurs in primary and secondary forests and small garden clearings, is often associated with small streams, uses holes in the ground, and has a home range of about 1.0–1.5 ha.; females give birth to one or two young. Two adult female *D. vanheurni* collected by Lidicker and Ziegler (1968) in October were each carrying a single pouch young and did not have visible embryos. A female *D. macleayi* taken in January had two pouch young, and one taken in March had a single pouch young (Flannery 1995a). A captive female *D. macleayi* lived 7 years and 11 months (Weigl 2005).

The IUCN classifies *D. vanheurni* as near threatened (Leary et al. 2008aa). The species is declining rapidly throughout its range because of overhunting for food

Tree kangaroo (*Dendrolagus inustus*). Photo by Pavel German / Wildlife Images.

and predation by the naturalized New Guinea singing dog (*Canis familiaris dingo*). Flannery (1995a) indicated that it is hunted by large-scale drives and burning of the forest, and Lidicker and Ziegler (1968) expressed concern about the ease with which it could be caught by small domestic dogs. *D. macleayi* is locally threatened by subsistence hunting but the IUCN no longer assigns a threatened classification (Aplin, Dickman, and Salas 2008).

DIPROTODONTIA; MACROPODIDAE; **Genus** *DENDROLAGUS* Schlegel and Miller, 1839

Tree Kangaroos

There are 14 species (Bowyer et al 2003; Flannery 1989, 1993, 1995a, 1995b; Flannery, Boeadi, and Szalay 1995; Flannery and Seri 1990; Flannery and Szalay 1982; Groves 1982, 2005e; Helgen 2007; Johnson and Newell 2008; Long 2003; R. Martin 2005; McGreevy, Dabek, and Husband 2012):

> *D. inustus*, Vogelkop and Fak Fak Peninsulas and east along north coast of New Guinea to Wewak area, also nearby Japen and possibly Waigeo and Salawati Islands;
>
> *D. lumholtzi*, Mount Carbine Tablelands to Cardwell Range in northeastern Queensland;
>
> *D. bennettianus*, Mount Amos to Daintree River in northeastern Queensland;
>
> *D. ursinus*, Vogelkop and Fak Fak Peninsulas in far western New Guinea;
>
> *D. matschiei*, mountains of Huon Peninsula of eastern Papua New Guinea and (probably through human introduction) nearby Umboi and western New Britain Islands;
>
> *D. spadix*, Fly River to Purari River in south-central Papua New Guinea;
>
> *D. goodfellowi*, Central Cordillera and adjacent highlands from the western border to the southeastern tip of Papua New Guinea;
>
> *D. pulcherrimus*, Torricelli and possibly Foja mountains in north-central New Guinea;
>
> *D. mbaiso*, Maokop (Sudirman Range) of west-central New Guinea;
>
> *D. dorianus*, mountains of southeastern Papua New Guinea;
>
> *D. notatus*, Central Cordillera and adjacent highlands from Strickland River to Wau in Papua New Guinea;
>
> *D. stellarum*, Wissel Lakes to Victor Emmanuel Range in Central Cordillera of New Guinea;
>
> *D. mayri*, known only by a single specimen from the mountains of the Wondiwoi Peninsula in far western New Guinea;
>
> *D. scottae*, Toricelli Mountains in northwestern Papua New Guinea.

Review of the sources cited above suggests that *D. inustus* is the most basal species, that *D. lumholtzi* and *D. bennettianus* are also very primitive, that the evolutionary sequence then runs through *D. ursinus* to *D. matschiei*, *D. spadix*, *D. goodfellowi*, and *D. pulcherrimus*, which form a closely related group, and that *D. mbaiso*, *D. dorianus*, *D. stellarum*, *D. mayri*, *D. notatus*, and *D. scottae* are also a species complex. Eldridge

Tree kangaroo (*Dendrolagus goodfellowi*). Photo by Johannes Pfleiderer.

and Coulson (2015) treated *D. pulcherrimus* as a subspecies of *D. goodfellowi*, and *D. notatus, D. stellarum,* and *D. mayri* as subspecies of *D. dorianus*. Eldridge and Coulson (2015) also referred to a population of *D. scottae* in the Bewani Mountains of northwestern Papua New Guinea, which may represent an undescribed subspecies.

Head and body length is 412–810 mm, and tail length is 408–935 mm. Weight is 5.1–13.7 kg in the Australian species (Johnson and Newell 2008; Martin and Johnson 2008). Flannery (1995a) listed weights of 6.5–17.0 kg for the New Guinea species, and Flannery and Seri (1990) noted that the maximum recorded weight for a tree kangaroo is 20.0 kg in a wild-caught *D. dorianus*. The sexes are about equal in size, except

for *D. inustus* and *D. bennettianus*, the males of which are much larger. The pelage is usually fairly long; in some species it is soft and silky, in others coarse and rough. Coloration in *D. ursinus* is blackish, brown, or gray above and white or buff below. *D. scottae* is uniformly blackish (Flannery and Seri 1990), *D. spadix* uniformly brownish, and *D. inustus* grizzled in color, with distinct black ears (Flannery 1995a). *D. dorianus* and most of its related complex of species are some shade of brown over most of the body. *D. mbaiso* is generally dark above but has distinctly white facial markings and underparts (Flannery, Boeadi, and Szalay 1995). *D. bennettianus* is mostly brown with a grayish tinge, its feet are black, and its tail is light fawn above and black below (Martin and Johnson 2008). In

319

D. lumholtzi the upper parts are grayish or olive buff, the underparts are white, the feet are blackish, and the terminal part of the tail is blackish brown. *D. matschiei* is among the most brilliantly colored of marsupials: its back is red or mahogany brown, its face, belly, and feet are bright yellow, and its tail is mostly yellow. *D. goodfellowi* is similar but also has yellow lines along each side of the spine, and the tail is mostly the same color as the dorsum with some yellow markings (Lidicker and Ziegler 1968). *D. pulcherrimus* is distinguished from the former by its orange shoulders and head, pinkish face and limbs, and pale yellow to white tail rings (Flannery 1993).

The forelimbs and hind limbs are of nearly equal proportions. The cushionlike pads on the large feet are covered with roughened skin. The claws on the forefeet, which are used for climbing, are much larger than those of terrestrial kangaroos and are strongly curved. The long, well-furred tail is of nearly uniform thickness and acts as a balancing organ; it is not prehensile but is often used to brace the animal when climbing. The thick fur on the nape and sometimes on the back grows in a reverse direction and apparently acts as a natural water-shedding device as the animals sit with the head lower than the shoulders. Females have a well-developed pouch and four mammae.

Tree kangaroos dwell mainly in mountainous rainforests, though *D. spadix* is a lowland species, and Johnson and Newell (2008) noted that *D. lumholtzi* may formerly have occupied coastal lowland forests. Reported altitudinal ranges (in meters) are: *D. inustus*, 100–1,400; *D. lumholtzi*, 800–1,600; *D. bennettianus*, 450–760; *D. ursinus*, 1,000–2,500; *D. matschiei*, 1,000–3,300; *D. spadix*, 0–800; *D. goodfellowi*, 1,000–2,865; *D. pulcherrimus*, 680–1,120; *D. mbaiso*, 3,250–4,200; *D. dorianus*, 600–4,000; *D. notatus*, 900–3,100; *D. stellarum*, 2,600–3,200; *D. mayri*, 1,600; and *D. scottae*, 900–1,520 (Eldridge and Coulson 2015; Flannery 1995a; Flannery, Boeadi, and Szalay 1995; Flannery and Seri 1990; Ride 1970; Ziegler 1977).

They are very agile, often traveling rapidly from tree to tree and leaping as much as 9 meters downward to

Tree kangaroo (*Dendrolagus lumholtzi*). Photo by Alexander F. Meyer at David Fleay Wildlife Park, Gold Coast, Australia.

an adjoining tree. They also jump to the ground from remarkable heights, up to 18 meters or perhaps even more, without injury. When descending trees, they usually back down, unlike the possums (Phalangeriformes). When on the ground, they may progress by means of relatively small leaps, leaning well forward to counterbalance the long tail, which is arched upward. Ganslosser (1992) observed that *D. inustus* and *D. lumholtzi* use predominantly bipedal hops on the ground, as well as in trees, and that they use their forepaws more or less simultaneously in feeding and grooming. In contrast, *D. matschiei* and *D. dorianus* rarely hop bipedally but often walk quadrupedally and use their forelimbs individually in foraging and grooming. *D. ursinus* and *D. goodfellowi* seem to use both methods of movement equally.

Studies suggest that *D. dorianus*, the heaviest arboreal marsupial, may actually have become mainly a ground-dweller (Moeller 1990b), and that *D. mbaiso* also is largely terrestrial (Flannery 1995a), but that *D. lumholtzi* spends only 2 percent of its time on the ground, the remainder in the middle and upper layers of the canopy (Hutchins and Smith 1990). Time of activity varies: *D. bennettianus* and *D. lumholtzi* are primarily nocturnal (Johnson and Newell 2008; Martin and Johnson 2008), while *D. stellarum* and *D. goodfellowi* may be crepuscular, and *D. matschiei* and *D. scottae* are principally diurnal (Flannery 1995a). The diet consists mainly of leaves and fruit, obtained either in trees or on the ground. Ferns reportedly are important on the Huon Peninsula, and captives have been fed chickens and have caught and eaten other birds (R. Martin 2005).

Population structure is variable. *D. goodfellowi* is often found in male-female pairs, while *D. matschiei* appears solitary (Flannery 1995a); a density of 0.6–1.4 individuals per ha. has been recorded for the latter species (R. Martin 2005). Female *D. dorianus* live in groups, and related individuals may form coalitions that interact in a friendly manner and cooperate in agonistic displays toward unfamiliar males; a single male may also be present in a group (Flannery 1995a; Ganslosser 1984). Groups of female *D. lumholtzi* with one male can be maintained amicably in captivity, but two males in the presence of females will fight savagely. Under natural conditions, *D. lumholtzi* is mostly solitary, nongregarious, and highly territorial (Hutchins and Smith

1990). Females maintain home ranges of 0.7–2.0 ha., which are almost entirely discrete but overlap the larger ranges (1.9–2.1 ha.) of several males; the latter can be extremely aggressive to each other and inflict severe injuries (Johnson and Newell 2008). Male *D. bennettianus* are also belligerent and defend territories as large as 20 ha.; those areas encompass several adult female ranges, which are about 6–12 ha. in size and usually discrete (R. Martin 2005; Martin and Johnson 2008).

Mating is polygynous in *D. bennettianus*; females breed annually and have a single young, mostly at the start of the wet season. Pouch life is around 9 months, and the young remains with the mother around 2 years (Martin and Johnson 2008). In a study of captive *D. lumholtzi*, Johnson and Delean (2003) recorded births throughout the year. Estrous cycle was 47–64 days, gestation was 42–48 days, and interval between loss of pouch young and new mating was 22 days; postpartum estrus and embryonic diapause were not observed. The single young first emerged from the pouch at 164–192 days, permanently left at 246–275 days, and was weaned 87–240 days later. Sexual maturity was attained at 2.04–3.71 years of age in females and at 4.6 years in one male.

In New Guinea, breeding appears aseasonal; females of various species there have been found with pouch young in March, April, May, June, September, October, November, and December (Flannery 1995a). Breeding may also occur at any time of the year in captive *D. matschiei*. Its estrous cycle is about 54–56 days, gestation is approximately 44 days, there is no embryonic diapause, the young permanently leaves the pouch at 10 months of age, females first give birth when 2.0–2.5 years old and are subsequently capable of producing young at 12-month intervals, and one captive female continued breeding until age 14 (R. Martin 2005). Tree kangaroos seem capable of long life in captivity; the record is held by a female *D. matschiei* that lived 26 years and 11 months, and specimens of *D. dorianus*, *D. goodfellowi*, *D. inustus*, and *D. ursinus* have lived for about 20 years or more (Weigl 2005).

Considering especially number and percentage of species in jeopardy, *Dendrolagus* is among the world's most seriously threatened mammalian genera. The IUCN classifies *D. mayri*, *D. pulcherrimus*, and *D. scottae* as critically endangered (Leary et al. 2008bb, 2008cc, 2008dd); *D. matschiei*, *D. goodfellowi*, *D. mbaiso*, and

Tree kangaroo (*Dendrolagus matschiei*). Photo by Alexander F. Meyer at Bronx Zoo.

Martin 2005; Thornback and Jenkins 1982). Another factor cited by the IUCN for the New Guinea species is habitat loss through conversion of forest habitat to subsistence agriculture. All species have a relatively low reproduction rate and most have restricted ranges. *D. mayri* may already be extinct, but if it still exists the population must have fewer than 50 mature individuals. *D. pulcherrimus* is undergoing a drastic population decline suspected to have exceeded 80 percent over the past 30 years; it has been extirpated from 99 percent of its historic range, and total numbers are probably fewer than 250. *D. scottae* is thought to have declined by more than 80 percent in just the last 10 years; even before that, Flannery and Seri (1990) were concerned that it might number only in the low hundreds and occupy only 25–40 sq km of suitable habitat.

There are questions about the status of the two Australian species, *D. lumholtzi* and *D. bennettianus*, some reports suggesting that clearing of rainforest in northeastern Queensland considerably reduced their historical ranges (Calaby 1971; Hutchins and Smith 1990; Poole 1979). Woinarski, Burbidge, and Harrison (2014) indicated that while about two thirds of the original habitat of *D. lumholtzi* had been cleared, there had been little change in that of *D. bennettianus*, and there was little ongoing decline for either species, though both occupied very restricted areas of less than 2,000 sq km. They are also subject to predation by the domestic dog (*Canis familiaris*), but still appear to be common in protected reserves. Heise-Pavlov, Jackrel, and Meeks (2011) believed *D. lumholtzi* should be placed in an IUCN threatened category.

DIPROTODONTIA; MACROPODIDAE; **Genus SETONIX** Lesson, 1842

Quokka

The single species, *S. brachyurus*, originally inhabited southwestern Western Australia from Jurien south and east at least to Bremer Bay; the range includes Rottnest Island, near Perth, and Bald Island and possibly Breaksea Island, near Albany (De Tores et al. 2007). Bones of this species have also been found in surface deposits at several sites along the south coast as far east as Fitzgerald River National Park (De Tores et al. 2008).

D. notatus as endangered (Leary et al. 2008ee, 2008ff, 2008gg, 2010); *D. inustus*, *D. ursinus*, *D. dorianus*, and *D. stellarum* as vulnerable (Leary et al. 2008hh, 2008ii, 2008jj, 2008kk); and *D. bennettianus* as near threatened (Winter, Burnett, and Martin 2008a). Recently, using IUCN criteria, Woinarski, Burbidge, and Harrison (2014) also classified *D. lumholtzi* as near threatened. Therefore, of the 14 species in the genus, only *D. spadix* is not placed in a category of concern, though even it is reported to be decreasing in numbers (Leary et al. 2008ll). *D. inustus* and *D. ursinus* are on appendix 2 of the CITES.

The most devastatingly consistent problem in New Guinea is intensive hunting by people for use as food. This has already caused widespread declines of all species there and problems are intensifying with the spread of modern weapons and other technologies into remote areas (Flannery 1995a; George 1979; R.

Quokka (*Setonix brachyurus*). Photo by Kitch Bain/Shutterstock.com.

Head and body length is about 390–600 mm, tail length is 235–350 mm, and adult weight is 1.6–5.0 kg. The hair is short and fairly coarse, and the general coloration is brownish gray, sometimes tinged with rufous. There are no definite markings on the face. The ears are short and rounded. The tail, which is only about twice as long as the head, is sparsely furred and short. Females have four mammae and a well-developed pouch.

On islands the quokka occurs in a variety of habitats with sufficient cover, but on the mainland it seems to be restricted to dense vegetation in swamps amidst dry sclerophyll forest (Ride 1970). An important factor in the quokka's ecology, at least on Rottnest Island, is the diurnal shelter in a thicket or some other shady location where the animal can avoid the heat (Nicholls 1971). An individual returns to the same shelter every day through most of the year but may change sites in May or June. At night the quokka emerges from its shelter to feed. It makes runways and tunnels through dense grass and undergrowth. When moving quickly,

it hops on its hind legs; when moving slowly, it does not use its tail as a third prop for the rear end of the body as do kangaroos and the larger wallabies. Like most macropodids, it is terrestrial, but it can climb to reach twigs up to about 1.5 meters above the ground (Ride 1970). The quokka is herbivorous, feeding on a variety of plants.

Ruminant-like digestion in the Macropodidae was first demonstrated in *Setonix*. The pregastric bacterial digestion in the quokka is much like that in sheep: most of the 15 or so morphological types of bacteria present in the large stomach region of the quokka are comparable to those in the rumen of sheep. Most of the bacterial fermentation in the stomach of the quokka takes place in the sacculated part. This wallaby seems to occupy a position intermediate to the ruminant and the nonruminant herbivores in its efficient digestion of fiber and the rate of food passage.

During the wet season the feeding area or home range used by a quokka on Rottnest Island covers about

1.0–12.5 ha. In the dry season (November–April) this area increases to 2.0–17.0 ha. (Nicholls 1971). In addition, some individuals move up to 1.8 km to soaks or fresh-water seepages during the summer since there is almost no free surface water on Rottnest Island at that time. Other individuals do without water (Ride 1970). Population density on the island has been reported as 1 individual per 0.4 ha. to 1 per 1.2 ha. (Main and Yadav 1971) and 10.5 to 25.3 individuals per ha. (McLean et al. 2009). Kitchener (1972) reported that on Rottnest the quokka population is organized into family groups. Adult males dominate the other members of the family and also form a linear hierarchy among themselves, which is usually stable. On hot summer days, however, the adult males may fight intensively for possession of the best shelter sites. Apparently the availability of such sites, rather than food, is the main factor in limiting the population. McLean et al. (2009) found the heaviest males among a human-acclimated population on Rottnest to be the most dominant and to sometimes violently drive a smaller male away from an estrous female. Males routinely attempted to form liaisons with females; most pairings lasted less than 10 minutes but the involved two individuals tended to come together repeatedly over at least two breeding seasons, thus suggesting a long-term relationship. Females formed such "consorts" with 1–3 males and males with 1–5 females. Females clearly indicated preferences for certain males. Females rarely appeared to avoid associating with other females or forming any kind of identifiable social group. Nicholls (1971) observed that other than the conflict for shelters, there is little evidence of territoriality on Rottnest, and groups of 25–150 individuals may have overlapping home ranges. During the summer some parts of the population concentrate around available fresh water (Ride 1970).

In a study of five remnant mainland populations, Hayward et al. (2003, 2004) found population densities of 0.07 to 4.30 per ha. and home ranges averaging 6.39 ha., including a core range of 1.21 ha. where an animal

Quokka (*Setonix brachyurus*), mother and young. Photo by Elizabeth Tasker / Wildlife Images.

spent about 50 percent of its time. Ranges tended to contract and shifted to the edge of swamps in winter, as the swamps became inundated following rain, and expanded and shifted toward the center in autumn as the swamps dried. Hayward (2008) reported that 52 percent of the average quokka home range is overlapped by the ranges of other quokkas; there is considerable intersex and intrasex overlap, but female ranges overlapped other ranges more than male ranges did. That, combined with the male weight-based dominance hierarchy, suggests uni-male polygyny, in which one male defends the ranges of a small number of females.

Reproduction in the quokka has been studied in some detail (Collins 1973; De Tores 2008b; Hayward et al. 2003; Ride 1970; Rose 1978; Shield 1968). Females are polyestrous, with an average estrous cycle of 28 days. They are capable of breeding throughout the year in captivity and in some wild mainland populations. On the islands anestrus occurs from about August through January, and most births take place from February to April. Nondelayed gestation is 26–28 days. On the mainland, litter size is one but a female can produce two young annually. Twins have been observed occasionally on Rottnest Island but usually only a single young can be successfully reared each year. One day following birth the female mates again. Embryonic diapause, which was first demonstrated in *Setonix*, then occurs. If the young already in the pouch should die within a period of about 5 months, the embryo resumes development and is born 24–27 days later. If the first young lives, the embryo degenerates when the female enters anestrus. In captivity or under unusually good conditions in the wild the second embryo can resume development even if the first young is successfully raised. The young initially leaves the pouch at about 175–195 days but will return if alarmed or cold and will continue to suckle for 3 to 4 more months. Earliest recorded sexual maturity is 389 days for males and 252 days for females. In the wild on Rottnest Island, however, it is unlikely that females give birth until well into their second year. According to Shield (1968), a marked 10-year-old wild female carried a young, and a wild male lived more than 10 years. Weigl (2005) listed a captive female that attained an age of 13 years and 10 months.

At the time of European settlement and through the early twentieth century the quokka was very common in coastal parts of southwestern Western Australia. In the 1930s there was a precipitous decline on the mainland, though populations on Bald and Rottnest Islands remained relatively numerous (De Tores et al. 2007). The quokka now is classified as endangered by the USDI and as vulnerable by the IUCN (De Tores et al. 2008). Mainland populations continue to decrease, and the range is severely fragmented. The two island populations fluctuate widely, but are considered stable over the long term. Habitat clearing and predation by the introduced fox (*Vulpes vulpes*) and feral house cat (*Felis catus*) caused the past decline on the mainland and are still serious problems. The feral pig is now degrading habitat and excluding the quokka from swampy areas. Total numbers of the quokka are estimated at 7,850–17,150 mature individuals, of which 4,000–8,000 are on Rottnest and 500–2,000 on Bald Island. There have been suggestions to augment the mainland populations with animals from Rottnest Island, but Alacs et al. (2011) found strong genetic differentiation between the two groups and cautioned that intermixing should be avoided to maintain natural patterns of genetic structuring and prevent potential outbreeding depression.

DIPROTODONTIA; MACROPODIDAE; **Genus *THYLOGALE*** Gray, 1837

Pademelons

Seven species are reported to exist (Flannery 1992, 1995a, 1995b; Groves 2005e; Heinsohn 2005b; Helgen 2007; Johnson and Rose 2008; MacQueen et al. 2011; Ride 1970; Ziegler 1977):

> *T. billardierii* (red-bellied pademelon), southeastern South Australia, Victoria, Tasmania, King Island and Furneaux Group in Bass Strait;
>
> *T. thetis* (red-necked pademelon), eastern Queensland, eastern New South Wales;
>
> *T. stigmatica* (red-legged pademelon), south of Fly River in southwestern Papua New Guinea, eastern Queensland, eastern New South Wales;
>
> *T. brunii* (dusky pademelon), southern lowlands of New Guinea, Aru Islands, Kai Besar Island;

Red-legged pademelon (*Thylogale stigmatica*). Photo by Alexander F. Meyer at Sydney Wildlife World.

T. browni, northern and eastern New Guinea, possibly Japen Island, at least in part through human introduction on the Bismarck Archipelago (Bagabag, Umboi, New Britain, New Ireland, Lavongai, Tabar, Lihir, Malendok, Ambitle, Buka, and possibly Emirau, Dajul, and Eloaue);

T. lanatus, mountains of Huon Peninsula in northeastern Papua New Guinea;

T. calabyi, Mount Giluwe and Pureni and Kaijende Highlands in central New Guinea, and Mount Albert Edward in southeastern Papua New Guinea.

According to MacQueen et al. (2010, 2011), mitochondrial DNA analysis supports recognition of the three Australian species of *Thylogale* in the above list. However, for New Guinea the analysis suggests a phylogeny not conforming precisely to the above designation of, on the one hand, a group of related species—*T. browni*, *T. lanatus*, and *T. calabyi*—in the north and along the Central Cordillera, and, on the other hand, the species *T. brunii* in the southern lowlands and on the Aru Islands. Instead, an "eastern clade" would comprise those populations of what above are *T. browni*, *T. lanatus* (treated as a subspecies of *T. browni*), and *T. calabyi* in the Bismarck Archipelago and on the mainland east of the Ramu-Markham and Watut-Tauri valleys, and also the population of *T. brunii* at Varirata in the extreme southeast. A "western clade" would comprise those populations of *T. browni* and *T. calabyi* west of the Ramu-Markham and Watut-Tauri valleys, as well as the populations of *T. brunii* on that part of the mainland and on the Aru Islands. However, there was no definite assignment of New Guinea populations to particular species, and the remainder of this account reflects the names and distributions in the above list. Eldridge and Coulson (2015) also treated *T. lanatus* (using the spelling *lanata*) as a subspecies of *T. browni* and indicated that *T. brunii* and *T. calabyi* might not be specifically separable from *T. browni*.

Hope (1981) described an additional species, *T. christenseni*, from remains at an archeological deposit in the western Maokop (Sudirman Range) of west-central New Guinea and thought that both that species and *T. brunii* survived there until less than 5,000 years ago. Flannery (1995a) reported that specimens of *T. christenseni* had been found at two other sites in the western Maokop, that the remains attributed to *T. brunii* actually represented another extinct but undescribed species, and that both species were extant until at least 3,000 years ago. Helgen (2007) suggested the latter species is closely related to *T. calabyi* and that both it and *T. christenseni* may yet survive. Helgen (2007) also predicted the range of *T. stigmatica* would be shown to extend into western New Guinea. Aplin and Pasveer (2007) reported abundant remains of *T. stigmatica* from two late Pleistocene to Holocene sites on Kobroor in the Aru Islands. Maynes (1989) indicated that presence of *T. brunii* on Kai Besar Island and of *T. browni* on Umboi and New Britain might be attributable to introduction by the Melanesians. Flannery (1995a) wrote that the island distribution of *T. browni* is largely or entirely the result of human introduction. Flannery (1995b) stated that *T. browni* was brought by people to New Ireland in the Bismarck Archipelago about 7,000 years ago and then carried from there to smaller islands to the north and east. Heinsohn (2005b) postulated that food shortages during World War II led to intensive hunting of *T. browni* in the Bismarck Archipelago and a consequent sharp contraction of its range on New Ireland and Lavongai and its complete dis-

Red-necked pademelon (*Thylogale thetis*). Photo by Gerhard Körtner.

Red-bellied pademelon (*Thylogale billardierii*). Photo by Gerhard Körtner.

appearance from all the smaller islands to the north and east (though records from some of those islands may be attributable to animals held for use as food, not to established populations). Van Gelder (1977), on the basis of captive hybridization between *T. thetis* and *Macropus rufogriseus*, recommended that *Thylogale* be considered a synonym of *Macropus*.

Head and body length is about 290–670 mm, tail length is 246–570 mm, and weight is 1.8–12.0 kg; males are larger on average than females (Flannery 1995a; K. A. Johnson 2008c; Johnson and Rose 2008; Johnson and Vernes 2008). The fur is soft and thick. *T. thetis* is grizzled gray above, becoming rufous on the shoulders and neck, and often has a light hip stripe. In *T. stigmatica* the upper parts are mixed gray and russet and the yellowish hip stripe is conspicuous. *T. brunii* is gray brown to chocolate brown above with a dark cheek stripe from behind the eye to the corner of the mouth, above which is a white area; it also has a prominent hip stripe. The underparts in those three species are considerably paler than the dorsal parts. *T. billardierii* is grayish brown above tinged with olive; a yellowish hip stripe is often evident, and the undersurface is rufous or orange. *T. browni* is dark brown above and grayish below, *T. lanatus* differs in having a pale mantle over the shoulders, and both species consistently lack a pale hip stripe. The body of *Thylogale* is compact, and the comparatively short tail is sparsely haired and thickly rounded. Females have four mammae and a well-developed pouch.

Habitats include rainforest, sclerophyll forest, savannah, thick scrub, and grassland. At least some species tend to concentrate where forest or other dense cover abuts open fields. Pademelons may move about in undergrowth through tunnel-like runways. They are inoffensive and apparently quite curious, as they sometimes will allow a close approach before bounding away. They have a habit, comparable to that of rabbits, of thumping on the ground with their hind feet, presumably as a signal or warning device (see Rose et al. 2006 for a discussion of this behavior in Macropodiformes). In the absence of predators or other disturbing factors they may graze or browse in the dusk and early morning near thickets, into which they hop quickly when alarmed. The diet includes grass, leaves, shoots, seeds, and fruit.

Very little is known of the natural history of the New Guinea species, though *T. browni* is found in many places, including disturbed areas, from sea level to 2,100 meters; *T. brunii* and *T. stigmatica* occur in rainforest near sea level, and *T. calabyi* and *T. lanatus* are montane species, the latter ranging from 3,000 to 3,800 meters (Flannery 1995a). The Australian species have been studied more extensively (K. A. Johnson 2008c; Johnson and Rose 2008; Johnson and Vernes 2008). They live mainly in dense rainforest and eucalypt forest but move along well-defined runways to clearings to feed from dusk to dawn. They rarely venture more than 70–100 meters from the forest edge. *T. billardierii* may travel more than 2 km from a daytime rest site to a regularly used feeding area; its home range may be as large as 170 ha. Between periods of sleep during the day, *T. thetis* moves through the forest grazing or browsing; on chilly winter mornings it often basks where the sun penetrates to the forest floor. Including forest and open feeding grounds, its home range is 5–30 ha. *T. stigmatica* is both nocturnal and diurnal but activity is reduced in the early afternoon and around midnight. By day it stays within the forest to forage, typically moving in a slow quadrupedal gait with its nose close to the ground. After dusk it proceeds rapidly to the forest edge to graze. The home range of *T. stigmatica* is 1–4 ha. and is temporally partitioned, the diurnal range located well within the forest and often more than twice the size of the nocturnal range, which is primarily pasture.

Captive pademelons can be maintained in groups (Collins 1973). *T. billardierii* has been reported to be gregarious in the wild, but Johnson and Rose (2008) stated that, while feeding aggregations of 10 or more individuals may form at night, there is no evidence of colonial structure or persistent bonds. The same is true for *T. thetis* and *T. stigmatica*, though both make a number of socially motivated sounds, including harsh rasps or guttural growls during hostile interaction and soft clucks by courting males (K. A. Johnson 2008c; Johnson and Vernes 2008). Captive male *T. billardierii* were observed to make numerous aggressive displays but not actually to fight; females also were aggressive toward one another, sometimes fought, and seemed to form a linear dominance hierarchy (Morton and Burton 1973).

Very limited data suggest that breeding is continuous in New Guinea (Flannery 1995a). In Tasmania *T. billardierii* is a seasonal breeder, with most births in April, May, and June (Rose and McCartney 1982). Female *T. thetis* breed continuously, but, at least in northeastern New South Wales, there is a birth peak in autumn and spring (K. A. Johnson 2008c); there is a single young, which spends about 6 months in the pouch and attains sexual maturity at 11 months in captivity and 17 months in the wild (Eldridge and Coulson 2015). Births in captivity have been recorded in January, February, and July for *T. billardierii* and in September for *T. thetis*. Embryonic diapause is known to occur in those two species. In *T. billardierii* the estrous cycle averages 30 days, nondelayed gestation is 29.6 days, and the period from the end of embryonic diapause to birth is 28.5 days (Rose 1978). Litter size usually is one, but twins have been recorded (Collins 1973). Pouch life of *T. billardierii* is 202 days, weaning occurs about 4 months after the young permanently leave the pouch, and sexual maturity comes at about 14–15 months (Morton and Burton 1973; Rose and Mc-

Red-necked pademelon (*Thylogale thetis*). Photo by Pavel German / Wildlife Images.

Cartney 1982). A specimen of *T. billardierii* lived in captivity for 8 years and 10 months (Collins 1973).

Johnson and Vernes (1994) maintained captive colonies of one adult male and eight adult female *T. stigmatica* and found births to occur therein throughout the year. Supplementary studies indicated that young were produced in the wild in Queensland from October to June. A postpartum estrus and mating generally followed birth by 2–12 hours. The estrous cycle was 29–32 days, gestation was 28–30 days, the pouch was permanently vacated at 184 days, weaning occurred an average of 66 days later, and mean age at sexual maturity was 341 days for females and 466 days for males. Weigl (2005) listed a captive female *T. stigmatica* that lived for 9 years and 8 months.

T. billardierii disappeared from the mainland of Australia by the 1920s, mainly because of predation by the introduced fox (*Vulpes vulpes*), and it has also been exterminated on the smaller islands of Bass Strait. It still survives on the larger of those islands and is common in many parts of Tasmania (Calaby 1971; Johnson and Rose 2008; Menkhorst and Denny 2008), where it is sometimes considered a pest to agriculture and silviculture and is taken in large numbers for its skin (Ride 1970). Population levels are thought to be high enough to sustain the regulated hunting and control operations that are carried out (Poole 1978). The distribution of *T. thetis* and *T. stigmatica* has been reduced in Australia by land clearance for agriculture, dairying, and forestry, but those two species still are common in some areas (Calaby 1971; K. A. Johnson 2008c; Johnson and Vernes 2008; Poole 1979). There is some genetic evidence of hybridization between *T. thetis* and *T. stigmatica*, where they are sympatric in southeastern Queensland, and between recognized subspecies of the latter (Eldridge et al. 2011). The status of the New Guinea population of *T. stigmatica* seems to be worse, with widespread disappearances and range fragmentation, apparently caused by excessive human hunting (Flannery 1992). The IUCN acknowledges that *T. stigmatica* is decreasing but does not assign a threatened classification (Burnett and Ellis 2008).

The IUCN now classifies *T. brunii* and *T. browni* as vulnerable and *T. lanatus* and *T. calabyi* as endangered (Leary et al. 2008mm, 2008nn, 2008oo, 2008pp). All are threatened primarily by local people hunting for food with their dogs. *T. calabyi*, which is restricted to less than 500 sq km of alpine grasslands, is also jeopardized by habitat disturbance caused by the feral pig (*Sus scrofa*). *T. brunii* has been exterminated in southeastern New Guinea, because of hunting, and *T. browni* may have been extirpated on Japen Island (see also previous comments on status of *T. browni* in the Bismarck Archipelago).

DIPROTODONTIA; MACROPODIDAE; **Genus** *PERADORCAS* Thomas, 1904

Little Rock Wallaby, or Nabarlek

The single species, *P. concinna*, is found in the Kimberley area of northeastern Western Australia, including several offshore islands (Long, Hidden, Augustus, Border), and in Arnhem Land in the northern Northern Territory (Churchill 1997; Pearson and Kinnear 1997). Historically, *Peradorcas* sometimes has been regarded as a distinct genus and sometimes considered part of *Petrogale* (see Eldridge 1997). The latter procedure was suggested by Poole (1979) based on serological and karyological study and has been followed in most recent systematic treatments (Calaby and Richardson 1988b; Corbet and Hill 1991; Eldridge and Close 1997; Eldridge and Coulson 2015; Flannery 1989; Groves 2005e; Potter et al. 2012a, 2012b, 2014b; Sanson and Churchill 2008; Sharman, Close, and Maynes 1990). Nonetheless, *Peradorcas* was retained as a full genus by Ganslosser (1990), McKenna and Bell (1997), and Nelson and Goldstone (1986) based on morphological and behavioral data. Chromosomal analyses have indicated that if *Peradorcas* is a separate genus, it should include the evidently closely related *Petrogale brachyotis* and *P. burbidgei*, even though the latter two species do not possess continually erupting molar teeth, the character on which

Little rock wallaby (*Peradorcas concinna*). Photo by Brad and Lynn Weinert.

Peradorcas was originally founded (Eldridge, Johnston, and Lowry 1992). However, a DNA/DNA hybridization analysis of all species of *Petrogale*, while confirming the immediate relationship of *P. brachyotis* and *P. burbidgei*, did not cover *P. concinna* (Campeau-Péloquin et al. 2001). An analysis of mitochondrial DNA did not strongly support an association of *Petrogale* and *Peradorcas* and treated the latter as a distinct genus (Burk and Springer 2000). In a comprehensive study of the nuclear DNA of suborder Macropodiformes, Meredith, Westerman, and Springer (2009a) recognized *Peradorcas* as a separate genus closely aligned with *Petrogale*.

Head and body length is 290–350 mm, tail length is 220–310 mm, and weight is 1.05–1.70 kg (Sanson and Churchill 2008). The fur is short, soft, and silky. The upper back is rusty red to grayish, the rump is reddish to orange, the underparts are white or grayish white, and the tail becomes darker toward the tip. The hind foot is well padded and the sole is roughly granulated. Females have four mammae in the pouch.

Peradorcas resembles *Petrogale burbidgei* but is distinguished by its longer ears and dental characters. Its molar dentition is unique among marsupials in that there are supplementary replacement molars behind the last regular molar. The actual number of molar teeth is not known, but study suggests that as many as nine molars may erupt successively and that there are seldom more than five molars in place at any one time.

According to Sanson and Churchill (2008), the little rock wallaby uses a variety of rocky environments, from steep fissured sandstone cliffs to isolated granite boulder piles. Associated vegetation may be monsoon rainforest, open woodland, vine thickets, or hummock grassland. The wallaby commonly rests by day in a cave or on a ledge and emerges at night to forage. Individuals may move several hundred meters, much farther than the sympatric *Petrogale brachyotis*, with which it sometimes forms mixed groups. *P. concinna* also is extremely fast and agile, traversing areas of cliff beyond the ability of *P. brachyotis*. In the wet season activity is partly diurnal, with individuals basking on rocks for up to 3 hours after dawn and feeding for several hours before dusk. Study of a floodplain population indicates a diet of grasses and ferns with an unusually high silica content. Those abrasive foods are probably associated with development of the continual molar replacement system. Breeding probably occurs through-

Little rock wallaby (*Peradorcas concinna*). Photo by Brad and Lynn Weinert.

out the year, though a greater number of large pouch young and young accompanying the mother have been observed in February, during the wet season.

In studies of a captive colony Nelson and Goldstone (1986) found *Peradorcas* to have a postpartum estrus and embryonic diapause. The estrous cycle lasted 31–36 days, and gestation about 30 days. Dominant females had shorter cycles than did subordinates. The single young opens its eyes at about 110 days, leaves the pouch at about 160 days, and is independent at about 175 days. Weaning is much more sudden than in *Petrogale*. Sexual maturity is attained in the second year. Weigl (2005) reported a male that lived in captivity for 11 years and 8 months.

Churchill (1997) reported finding *Peradorcas* more common than previously believed and observed that its range is not occupied by introduced rabbits, foxes, goats, and sheep, which have been largely responsible for the decline of various species of *Petrogale* (see account thereof). The IUCN considers *P. concinna* a species of *Petrogale* and refers to it as "data deficient," in view of absence of recent status information, but believes its habitat is decreasing due to changes in the fire regime and that there have been localized extinctions over the last 30–40 years in Northern Territory (Woinarski et al. 2008). Applying IUCN criteria to updated information, Woinarski, Burbidge, and Harrison (2014) classified *P. concinna* and two of its three subspecies, *P. c. monastria* of Kimberley and *P. c. canescens* of Arnhem Land, as near threatened, mainly because of historic and ongoing declines due to habitat disturbance and predation, not by foxes but by feral house cats. Such problems were thought likely to have almost or entirely extirpated the third subspecies, *P. c. concinna* of the Victoria River district in extreme northwestern Northern Territory, which has not been recorded for at least 170 years. It was classified formally as "critically endangered (possibly extinct)." Potter et al. (2014b) noted: "Urgent attention is required to clarify the taxonomic status of . . . *P. concinna* since it appears to be declining . . . and may represent multiple threatened species."

DIPROTODONTIA; MACROPODIDAE; **Genus** *PETROGALE* Gray, 1837

Rock Wallabies

There are 16 species (Campeau-Péloquin et al. 2001; Eldridge 2008c; Eldridge and Close 1992, 1993, 1997, 2008a, 2008b, 2008c, 2008d, 2008e, 2008f, 2008g; Eldridge, Johnston, and Lowry 1992; Eldridge, Moore, and Close 2008; Eldridge and Pearson 2008; Eldridge and Telfer 2008; Eldridge et al. 2001, 2008; Flannery 1989; Johnson and Eldridge 2008a, 2008b; Johnson et al. 2001; Kitchener and Sanson 1978; Maynes 1982; Pearson and Eldridge 2008; Pearson and Kinnear 1997; Pearson et al. 2008; Potter et al. 2012a, 2014b):

P. burbidgei, northwestern Kimberley area of northeastern Western Australia, and several nearby islands (Boongaree, Bigge, Katers, and possibly Darcy, Uwins, and Wollaston);

Little rock wallaby (*Petrogale burbidgei*), the species also called "monjon." Photo by Jiri Lochman / Lochman LT.

P. brachyotis, Kimberley area of northeastern Western Australia and nearby Koolan Island, northwestern Northern Territory;

P. wilkinsi, east of Daly River in northern Northern Territory, and several nearby islands (Koolan, Groote Eylandt, and the Wessel, Sir Edward Pellew, and English Companys groups);

P. persephone, Whitsunday area of east-central coast of Queensland, including Gloucester Island, and introduced to nearby Hayman Island;

P. xanthopus, eastern South Australia, northwestern New South Wales, southwestern Queensland;

P. rothschildi, Pilbara and Ashburton areas of northwestern Western Australia, islands of Dampier Archipelago (Rosemary, Dolphin, Enderby, West Lewis, East Lewis);

P. lateralis, suitable areas of mainland Western Australia, central and southern Northern Territory, and South Australia, and also various islands off the western and southern coasts (Depuch, Barrow, Wilson, Mondrain, Westall, Salisbury, North Pearson, and introduced to South Pearson, Middle Pearson, Thistle, and Wedge) but not Tasmania;

P. purpureicollis, western Queensland from Lawn Hill National Park to Bladensburg National Park;

P. godmani, northeastern Queensland from Bathurst Head south to Mount Carbine and west to "Pinnacles";

P. assimilis, central Queensland from Croydon east to Townsville and south to Hughenden and Burdekin and Bowen Rivers, also nearby Palm and Magnetic Islands;

P. coenensis, from the Pascoe River south to Musgrave on the northeastern Cape York Peninsula of Queensland;

P. mareeba, northeastern Queensland from near Mount Carbine south to the upper Burdekin River and west to Bullaringa National Park;

P. sharmani, Seaview and Coane ranges near Ingham on east-central coast of Queensland;

P. penicillata, extreme southeastern Queensland, central and eastern New South Wales, Victoria;

P. herberti, southeastern Queensland from Clermont east to Rockhampton and south to Nanango;

P. inornata, east-central Queensland from Home Hill south to Rockhampton, also nearby Whitsunday Island.

Petrogale is often considered to include the genus and species *Peradorcas* (see account thereof) *concinna.* If that species is so included, it would be placed in a systematic group also including *P. brachyotis* and *P. burbidgei.* Molecular evidence indicates that group diverged first in the evolution of *Petrogale,* that the lineage of *P. persephone* also is primitive, that *P. xanthopus* was next to diverge, and that all other species represent a more recent radiation, with *P. rothschildi* at its base (Campeau-Péloquin et al. 2001; Eldridge 2008b; Eldridge and Close 1997; Groves 2005e; Potter et al. 2012a). Much taxonomic work on *Petrogale* has been based on evaluation of chromosomes, with certain populations that appeared morphologically separable being found to have identical karyotypes, and with populations that appeared morphologically indistinguishable being designated as distinct, though cryptic, species based on karyological differences. Over the past several decades at least eight alternatives to the above phylogenetic arrangement have been suggested, and the issue likely is far from settled (Eldridge 1997, 2008b; Eldridge and Close 1993; Sharman, Close, and Maynes 1990). Introgression of genetic material from and to certain karyotypically and physically distinct species has been reported in the wild, and hybrids have also been produced in captive experiments (Close and Bell 1997; Kirsch et al. 2010). Narrow hybrid zones have been formed between the natural ranges of *P. godmani* and *P. mareeba* (Eldridge and Close 2008b) and of *P. herberti* and *P. penicillata* (Eldridge and Close 2008c). Assessment of the mitochondrial DNA of *P. brachyotis* suggests that it is divided into several distinct lineages that have been separated by major biogeographic barriers for millions of years and that are distinguishable at the specific level; one of the lineages is more closely related to *P. burbidgei* than to the other lineages, hence making *P. brachyotis* paraphyletic as presently construed (Potter et al. 2012b).

Excluding the very small *P. burbidgei,* the genus is characterized as follows: head and body length about

Proserpine rock wallaby (*Petrogale persephone*). Photo by Hans and Judy Beste / Lochman LT.

400–700 mm; tail length, 300–700 mm; adult weight, 2–12 kg; and thick, long, dense fur. Eldridge (2008b) noted that males are generally larger than females, by up to 40 percent in some species, but that there are no pronounced sexual differences in coloration or markings. In *P. burbidgei* head and body length is 290–353 mm, tail length is 252–322 mm, and weight is 0.96–1.43 kg (Kitchener and Sanson 1978).

The general coloration of the upper parts varies from pale sandy or drab gray to rich dark vinaceous brown; the underparts are paler, usually buffy, yellowish gray, yellowish brown, or white. Several species have stripes, patches, or other striking markings. *P. xanthopus*, the ring-tailed or yellow-footed rock wallaby, is one of the most brightly colored members of the kangaroo family. Eldridge (2008c) described it as fawn-gray above and white below and having a white cheek stripe, a dark mid-dorsal stripe from between the ears to the middle of the back, orange to bright yellow limbs and feet, orange or gray-brown ears, a reddish brown axillary patch, a buff-white side stripe, sometimes a

white hip stripe, and an orange-brown tail with irregular dark brown rings. *P. purpureicollis* has a striking purple-mauve tinting over the face, head, and neck, caused by a pigmented secretion and varying in intensity through the year (Johnson and Eldridge 2008b). The upper back of *P. rothschildi* also becomes a brilliant purple at certain times. *P. lateralis* is dark gray or brown above and has a dark dorsal stripe, a distinct white to yellowish cheek stripe, a white side stripe, and ears with a pale base; other markings vary widely by subspecies (Eldridge and Pearson 2008). In *P. burbidgei* the upper parts are mostly ochraceous tawny, and the underparts are ivory yellow (Kitchener and Sanson 1978).

The tails of rock wallabies are long, cylindrical, bushy, and thickly haired at the tip. They are less thickened at the base than in *Thylogale*, *Macropus*, and *Wallabia* and are used primarily for balancing rather than as props for sitting. The hind foot of *Petrogale* is well padded; the sole is roughly granulated, permitting a secure grip on rock, and edged with a fringe of stiff hair that extends onto the digits. The central hind claws are short, exceeding the toe pads by only 2–3 mm. Females have a well-developed, forward-opening pouch and four mammae.

These wallabies usually inhabit rocky slopes, gorges, and boulder-strewn outcrops with an associated cover of forest, woodland, heath, or grassland. They are nocturnal but in winter may emerge in the afternoon to warm up in the sun before foraging and also may bask in the sun in the morning before retiring. Eldridge (2008b) observed that while they commonly spend the day deep within rock piles, crevices, and caves, they also sometimes shelter in hollow logs, dense vegetation, and branches of large trees. According to Warburton (2005e), introduced *P. penicillata* in New Zealand excavate dens, usually with more than one entrance, under the roots of pohutukawa trees growing on cliffs. Telfer and Griffiths (2006) found each individual *P. brachyotis* to use an average of four dens.

The agility of rock wallabies in difficult terrain is astonishing; some of their leaps measure up to 4 meters horizontally. The friction of their fur and feet on regular paths, over many generations, imparts a glasslike sheen to limestone. They can scramble up cliff faces and leaning tree trunks with relative ease. Eldridge (2008b) noted that they ascend trees to feed or escape danger and pointed out that genetic studies have shown *Petrogale* to be most closely related to *Dendrolagus* (tree kangaroos). Rock wallabies usually progress in a series of short or long leaps. They travel awkwardly in open country. Eldridge (2008b) considered them generalist herbivores, eating a wide range of grasses, leaves, forbs, fruit, flowers, and seeds. In dry seasons they can exist for long periods without water by eating the juicy bark and roots of various trees. *P. xanthopus* travels up to several km from its daytime shelter to feeding and drinking sites (Eldridge 2008c), while average movement by *P. brachyotis* is only about 200 meters (Telfer and Griffiths 2006).

Home ranges of *P. wilkinsi* (originally reported as *P. brachyotis*) average about 18 ha. in the dry season, overlap by about 39 percent, and show no difference

Yellow-footed rock wallaby (*Petrogale xanthopus*). Photo by Gerhard Körtner.

between sexes (Telfer and Griffiths 2006). Other reported data on home ranges: *P. persephone*, averaging around 20 ha., averaging larger for males (Johnson and Eldridge 2008a); *P. xanthopus*, 24–200 ha., overlapping with those of other members of a colony (Eldridge 2008c); *P. lateralis*, 171 ha. for one female (Eldridge and Coulson 2015); *P. assimilis*, around 8 ha. in the wet season, doubling in size in the dry season when food is scarce (Eldridge and Close 2008a); *P. penicillata*, 2–30 ha., averaging larger for males (Eldridge and Close 2008f).

According to Eldridge (2008b), individuals show high site fidelity, defend preferred resting and sunning spots, and rarely move between suitable patches of habitat. They typically live in colonies varying in size from fewer than 10 to well over 100 animals. *P. xanthopus* has the largest reported colony size (Eldridge 2008c), and *P. sharmani* also is quite social, forming groups of over 40 at favorable sites (Eldridge and Close 2008g). *P. penicillata* lives in groups that have a defended territory, and there is a linear dominance hierarchy among males established through fighting (Russell 1974b). Colony size in that species reportedly varies from 2 to as many as 43 individuals, though Jarman and Bayne (1997) found an average of fewer than 6 adults in colonies on the tablelands of northern New South Wales. Each colony there contained one or several groups with 1–3 females and 1 male; the latter defended the females and their ranges. Young female *P. penicillata* settle in or close to their mother's range, thereby establishing matrilineal kin groups within a colony; males move farther away but might not actually leave the colony (Eldridge and Close 2008f; Hazlitt, Eldridge, and Goldizen 2010). Laws and Goldizen (2003) found that a population of about 37 *P. penicillata* in Queensland consisted of three groups, the males and females of which had nocturnal ranges overlapping extensively with those of both sexes of their own group. Observations of wild *P. assimilis* indicate that within a colony an adult male may form a long-term stable relationship with one or two females. Such pair bonding involves regular mutual grooming, foraging together at night, and sharing and defense of daytime resting places and other parts of an exclusive home range. However, genetic studies show that a female's social partner does not necessarily father all of her offspring (Barker 1990; Eldridge and Close 2008a).

Females are polyestrous (Rose 1978). Litter size usually is one, but twins occur on occasion (Collins 1973; Delaney 1997; Spencer and Marsh 1997). A postpartum estrus and embryonic diapause are widespread characteristics, with most, if not all, species able to breed continuously, though in southern Australia births often have a seasonal peak and in arid areas births increase following a period of good rainfall (Eldridge 2008b). The peak for *P. penicillata* comes in February–May, so most young emerge from the pouch and are weaned during spring and early summer when food is abundant (Eldridge and Close 2008f). In a study of captive *P. penicillata*, Johnson (1979) found an estrous cycle of 30.2–32.0 days and a gestation period of 30–32 days. There was usually a postpartum mating, and the resultant embryo developed and was born 28–30 days after premature removal of the original young. Normal pouch life was 189–227 days, and sexual maturity was attained at 590 days by males and 540 days by females.

The following additional reproductive data have been reported (Delaney 1997; Eldridge 2008b, 2008c; Eldridge and Close 2008a, 2008d; Eldridge and Coulson 2015; Eldridge and Pearson 2008; Johnson and Delean 1999, 2002a; Johnson and Eldridge 2008a, 2008b). Estrous cycle is about 33–38 days in *P. persephone*, 32–37 days in *P. xanthopus*, 30 in *P. lateralis*, 36–38 in *P. purpureicollis*, 29–34 in *P. assimilis*, and 30–32 in *P. inornata*. Gestation period is 30–34 days in *P. persephone*, 32 in *P. xanthopus*, 30 in *P. lateralis*, 33–35 in *P. purpureicollis*, 29–34 in *P. assimilis*, about 32 days in *P. penicillata*, and 30–32 in *P. inornata*. Female *P. persephone*, *P. purpureicollis*, and *P. assimilis* are known to enter estrus and mate within 48 hours of giving birth, and most other species probably have similar capability. If the young of that birth undergoes normal development, the new embryo is held in diapause and will be born within a day of permanent emergence of the previous young from the pouch. If the first young is lost or experimentally removed, birth of the new embryo occurs 30–32 days later. *P. xanthopus* weighs less than 0.5 gram at birth and grows to about 1 kg before permanent emergence. The young of *P. assimilis* is continuously attached to a teat for 110–143 days. Normal pouch life is around 6.0–7.5 months in all species investigated, and weaning comes 3–6 months after permanent emergence from the pouch. Unlike those of other macropodids, a mother rock wallaby often initially

Black-footed rock wallaby (*Petrogale lateralis*). Photo by Pavel German / Wildlife Images.

leaves her young secreted in a small rock crevice, returning regularly to suckle and groom it. Sexual maturity is attained in *P. persephone* at 20.5 months for females and about 24 months for males, in *P. xanthopus* at 18 months for both sexes, in *P. purpureicollis* as early as 18 months for females and about 22 months for males, in *P. assimilis* at around 19 months for females and 24 months for males, in *P. mareeba* at 20–24 months for males, in *P. penicillata* at about 18 months for females and 23 months for males, and in *P. inornata* at about 18 months for both sexes. *P. assimilis* is known to live up to 13 years in the wild. Weigl (2005) reported a male *P. xanthopus* still living after 14 years and 5 months in captivity.

While rock wallabies have disappeared in places, and their numbers have declined considerably in the sheep country of southern Australia, they seem to have maintained themselves better than other small macropodids (Calaby 1971; Poole 1979). Portions of their rocky habitat are inaccessible to sheep and rabbits, which may compete for forage, though feral goats are a problem in some areas. Several species are common, especially those of the complex along the coast of Queensland (Eldridge and Close 2008a, 2008b, 2008c, 2008d, 2008e; Johnson and Eldridge 2008b). There, *P. assimilis* has colonized boulders used to construct dams, and *P. inornata* is regularly sighted atop buildings and stationary cars.

In contrast, the more primitive *P. persephone*, occurring in the same region, is classified as endangered by the IUCN (Burnett and Winter 2008e), which reports numbers decreasing because of road kills and attacks by the domestic dog. Moreover, Johnson and Eldridge (2008a) noted that *P. persephone* is restricted to just 14,500 ha. of habitat, which is jeopardized by tourist development and urbanization. *P. sharmani* and *P. coenensis*, which also occupy the Queensland coast, are classified as near threatened by the IUCN, partly because their relatively restricted habitat may be adversely affected by livestock (Winter, Burnett, and Martin 2008b). Applying IUCN criteria to updated information, Woinarski, Burbidge, and Harrison (2014)

337

Brush-tailed rock wallaby (*Petrogale penicillata*). Photo by Gerhard Körtner.

lis, and the Victoria River population of *P. brachyotis*, one of the distinctive and geographically restricted lineages identified by Potter et al. (2012b), near threatened. The status of *P. burbidgei* is not well known, though suitable habitat is very limited, and predation by the house cat (*Felis catus*) is likely a severe problem.

P. xanthopus, listed as endangered by the USDI, declined severely after European settlement because of intensive hunting for its beautiful pelt, habitat degradation by livestock, and predation by the red fox (*Vulpes vulpes*). It disappeared in some areas and is still declining in the Flinders Range of South Australia, but control of the fox and goat has led to recovery of some populations (Copley 1983; Eldridge 2008c; Gordon, McGreevy, and Lawrie 1978). Gordon et al. (1993) estimated numbers in Queensland at 5,000–10,000, though the IUCN now indicates fewer than 10,000 mature individuals exist throughout the range of the species and that about 6,000 of those are in South Australia (Copley, Ellis, and van Weenen 2008). Woinarski, Burbidge, and Harrison (2014) generally agreed with that assessment but distinguished the subspecies *P. xanthopus celeris* of Queensland as vulnerable, mainly because of an ongoing decline, probably resulting from predation by the fox and habitat degradation by livestock.

The most widely distributed species, *P. lateralis* and *P. penicillata*, have been greatly reduced in numbers and range. Despised as agricultural pests and valued for their skins, hundreds of thousands were killed in the nineteenth and early twentieth centuries (Pearson 1992; Short and Milkovits 1990). Remaining populations continued to decline because of predation by the fox and other introduced carnivores, destruction of vegetative cover by introduced rabbits and hares, competition with sheep and goats for food and with the latter for rock shelters, and brushfires and land clearance. The latest IUCN assessment places the total number of *P. lateralis* at over 10,000 mature individuals but notes that it is decreasing (Burbidge et al. 2008b). More recently, Woinarski, Burbidge, and Harrison (2014) indicated there were probably fewer than 10,000 mature individuals.

The species *P. lateralis* is divided into a number of subspecies, which are of particular conservation concern. According to Eldridge and Pearson (2008), by the 1970s *P. l. lateralis* of Western Australia (including Depuch, Barrow, and Salisbury Islands) was found only

designated *P. sharmani* vulnerable and *P. coenensis* endangered. The total population of the former now numbers fewer than 800 individuals, while that of the latter is under 2,000, badly fragmented, and declining. *P. godmani* and *P. mareeba* have comparable problems and were designated near threatened by Woinarski, Burbidge, and Harrison (2014), but they are locally common, and one population has become habituated to tourists.

Of species farther west and south, the IUCN classifies *P. burbidgei*, *P. xanthopus*, *P. lateralis*, and *P. penicillata* as near threatened (Burbidge, McKenzie, and Start 2008; Burbidge et al. 2008b; Copley, Ellis, and van Weenen 2008; Taggart, Menkhorst, and Lunney 2008). However, in a more recent classification based on IUCN criteria, Woinarski, Burbidge, and Harrison (2014) designated *P. lateralis* and *P. penicillata* vulnerable. They also designated *P. burbidgei*, *P. purpureicol-*

Unadorned rock wallaby (*Petrogale inornata*). Photo by Chris and Mathilde Stuart.

in small numbers in several widely scattered localities, but subsequent control of the fox allowed substantial recovery and recolonization. More recently, Woinarski, Burbidge, and Harrison (2014) designated *P. l. lateralis* as endangered, noting that some populations were declining even with fox control, and that the Depuch Island population had become extinct. *P. l. hacketti*, restricted to Westall, Wilson, and Mondrain Islands in the Recherche Archipelago off the south coast of Western Australia, is classed as vulnerable, because of small population sizes (500 or fewer on each island) and the constant threat of introduced predators, disease, and fires (Woinarski, Burbidge, and Harrison 2014). *P. l. pearsoni*, originally found only on North Pearson Island but introduced to South Pearson, Middle Pearson, Thistle, and Wedge Islands, is considered near threatened because of small population sizes and low genetic diversity (Jones Lennon et al. 2011; Woinarski, Burbidge, and Harrison 2014). An as-yet unnamed subspecies of *P. lateralis* in the MacDonnell Ranges of central Australia began a steep decline in the

1930s, which continues today because of habitat degradation and predation by the fox and domestic dog and cat. That subspecies was designated vulnerable by Woinarski, Burbidge, and Harrison (2014), who noted that the great majority of the total population, perhaps 4,500–6,000 individuals, is present in southern Northern Territory, while only about 200 remain in northern South Australia and evidently very few in east-central Western Australia. Another unnamed subspecies, restricted to a rugged area of about 70 sq km in the western Kimberley area of northern Western Australia was classed as endangered by Woinarski, Burbidge, and Harrison (2014), because of an ongoing decline resulting from fox predation and habitat disturbance by livestock.

P. penicillata had become uncommon in most of its range by 1920. The major initial cause was hunting for the commercial fur trade. At least 144,000 skins of the species were sold in the 1890s; 92,590 skins were marketed by a single company in 1908 alone. Many other animals were killed for government bounties until

1914. As populations fell, they became less able to withstand fox predation and the other human-induced problems listed above (Lunney, Law, and Rummery 1997). Eldridge and Close (2008f) stated that *P. penicillata* is continuing to decline and is severely threatened in most of its remaining range. The latest IUCN assessment estimates total numbers at 15,000–30,000 and cites predation by the fox and competition with goats as the greatest threats (Taggart, Menkhorst, and Lunney 2008). That assessment was accepted by Woinarski, Burbidge, and Harrison (2014), but they designated the species vulnerable, because declines are continuing at many or most sites where it survives. *P. penicillata* was thought to have disappeared in Victoria by 1905 (Harper 1945), but two small colonies were found to have persisted for a time (Emison et al. 1978; Wakefield 1971). Only one now remains, in the Little River Gorge in East Gippsland, where 12 individuals were counted in 2008 (Bluff et al. 2011).

P. penicillata was introduced to Kawau, Rangitoto, and Motutapu Islands off northern New Zealand in 1870–1873 and attained high numbers. Population density on Motutapu was 12 individuals per ha. in the 1970s. The species was considered a pest, because it competed with livestock for pasture and accelerated erosion of cliffs by excavating its dens. An intensive government eradication program eliminated the wallaby from Rangitoto and Motutapu by 1999 (Warburton 2005e), and apparently it is also now extirpated on Kawau (Woinarski, Burbidge, and Harrison 2014). In 1916 a male, female, and young *P. penicillata*, apparently from Queensland, were brought to the island of Oahu, Hawaii, and escaped. A free-roaming population resulted and has numbered around 100 individuals in recent years (Eldridge and Browning 2002; Lazell, Sutterfield, and Giezentanner 1984; Long 2003; Maynes 1989).

DIPROTODONTIA; MACROPODIDAE; **Genus**
LAGORCHESTES Gould, 1841

Hare Wallabies

There are four species (Burbidge and Johnson 2008a; Burbidge et al. 2008a; Flannery 1989; Hitchcock 1997; Johnson and Burbidge 2008b; Ride 1970; Strahan 2008):

L. conspicillatus, northern Western Australia and nearby islands (Barrow, Hermite, Trimouille), Northern Territory, northern Queensland, known also by a single live specimen collected in 1997 on the upper Bensbach River in southwestern Papua New Guinea;

L. leporides, formerly eastern South Australia, western and central New South Wales, northwestern Victoria;

L. hirsutus, formerly Western Australia and nearby islands (Bernier and Dorre in Shark Bay, introduced to Trimouille), central and southern Northern Territory, western South Australia;

L. asomatus, known only by a single adult skull collected between Mount Farewell and Lake Mackay in southwestern Northern Territory but likely once found in much of the surrounding desert country.

Woinarski, Burbidge, and Harrison (2014) showed a single pre-1993 record of *L. hirsutus* from a site in west-central Queensland.

Head and body length of *L. conspicillatus* is 390–490 mm, tail length is 370–530 mm, and weight is 1,600–4,750 grams (Burbidge and Johnson 2008a); respective figures for *L. hirsutus* are 310–390 mm, 245–305 mm, and 780–1,960 grams (Johnson and Burbidge 2008b). Females are larger on average than males. *L. leporides* is usually gray brown above with reddish sides and grayish white underparts; the coloration of *L. hirsutus* is much the same except that long reddish hairs are present on the lower back, imparting a shaggy appearance. *L. conspicillatus* is gray brown or yellowish gray above and whitish or reddish below with reddish patches encircling the eyes and white hip marks. The guard hairs are long and coarse, and all species have long, soft, thick underfur.

Lagorchestes is stocky, thickset, and short-necked (Burbidge and Johnson 2008a). Its generic name means "dancing hare." It is not much larger than a hare and resembles the latter in movements and, to some extent, in habits. Its nose is wholly or partially covered by hair; the central hind claw is long and strong and is not hidden by the fur of the foot, and the tail is evenly haired throughout. *Lagorchestes* has a small canine on each side of the upper jaw, thus giving it a total of 34 teeth, 2 more than most other macropodids.

Hare wallaby (*Lagorchestes conspicillatus*). Photo by Hans and Judy Beste/Lochman LT.

Hare wallabies live in open forests and grassy or spinifex plains with or without shrubs or trees (Burbidge and Johnson 2008a; Ride 1970). They generally are nocturnal and rest by day in a "hide" or "form" lightly scratched in the ground in the shade of a bush or by a tuft of grass. *L. hirsutus* may also use a burrow more than 70 cm deep, especially in the heat of summer; when flushed, it escapes in a zigzag burst of speed, often uttering a high-pitched nasal squeak (Johnson and Burbidge 2008). Jumps of 2–3 meters have been credited to *L. leporides* when pressed, and one individual chased by dogs for 0.4 km doubled back on its track and leaped over the head of a man standing in its path.

According to Burbidge and Johnson (2008a), on the mainland about 65 percent of the diet of *L. conspicillatus* is grasses, the remainder mostly forbs or herbs. On Barrow Island that species is a selective feeder, browsing mainly on colonizing shrubs and also eating the tips of spinifex leaves; it does not drink even when water is available. It constructs several hides within a home range of about 8–10 ha. However, average home range in central Queensland was found to be 177 ha. The species breeds throughout the year, but on Barrow Island there are birth peaks in March and September. Birth is followed very closely by mating and then by embryonic diapause. If the first young is lost prematurely, the quiescent embryo resumes development to be born 28–30 days later. Young vacate the pouch at about 150 days, and both sexes are reproductively mature at about one year. Johnson (1993) noted an estrous cycle of 30 days and a gestation period of 29–31 days in *L. conspicillatus*. Johnson and Burbidge (2008b) wrote that reproduction of *L. hirsutus* is continuous under favorable conditions, pouch life is about 124 days, and sexual maturity is attained at 5–18 months by females and about 14 months by males. Weigl (2005) listed a captive *L. hirsutus* that lived for 13 years and 3 months.

Lagorchestes is among those genera most severely affected by the European colonization of Australia. Ride (1970) suggested that an important factor in the decline on the mainland may have been alteration of grassland habitat through trampling and grazing by sheep and cattle. Woinarski, Burbidge, and Harrison (2014) listed the most damaging problems as predation by the introduced red fox (*Vulpes vulpes*) and domestic cat (*Felis catus*). The species *L. asomatus* is known only from one specimen taken in 1932, though Aboriginal accounts indicate it survived in the Tanami Desert of Northern Territory until the late 1940s and was still present near Kiwirrkura southwest of Lake Mackay in 1960 (Burbidge et al. 2008a). *L. leporides*, common in South Australia and New South Wales until the mid-nineteenth century, has not been recorded since 1890 (Calaby 1971). The IUCN designates both species as extinct (Australasian Mammal Assessment Workshop 2008c; Burbidge and Johnson 2008b).

The IUCN also formerly classified *L. conspicillatus* generally as near threatened and the subspecies *L. c. conspicillatus* of Barrow, Hermite, and Trimouille Islands as vulnerable. Populations on the latter two islands disappeared by the early twentieth century,

probably because of predation by the house cat (Burbidge and Johnson 2008a), but Short and Turner (1991) reported a population of about 10,000 individuals on Barrow Island. The IUCN (Winter, Woinarski, and Burbidge 2008) no longer assigns formal threatened categories but does note that the mainland subspecies, *L. c. leichardti*, which formerly occupied almost half the Australian continent, now has an extremely patchy distribution and is decreasing. In Western Australia it is rare and reduced to a few isolated populations in the Pilbara and Kimberley regions. In Northern Territory it now rarely occurs south of 21° S—a range contraction of more than 200 km. In Queensland it is still widespread, though in the central highlands its area of occupancy may have fallen 20–30 percent due to broadscale clearing and development and fox predation. Using IUCN criteria, Woinarski, Burbidge, and Harrison (2014) did designate the entire species *L. conspicillatus*, including *L. c. leichardti*, and the subspecies *L. c. conspicillatus* of Barrow, Hermite, and Trimouille Islands,

Hare wallaby (*Lagorchestes hirsutus*). Photo by Gerhard Körtner.

as near threatened, and reported that a reintroduction to Hermite Island was undertaken in 2010.

L. hirsutus disappeared from most of its mainland range by the 1950s; its initial decline may have resulted from an epizootic disease transmitted by deliberate or incidental introduction of affected domestic animals and rodents to the vicinity of Shark Bay, Western Australia, around 1875 (Abbott 2006). However, presence of two small colonies in the Tanami Desert Sanctuary in Northern Territory, each containing only 6–10 animals, was later confirmed. One increased to about 20 individuals and was studied in some detail (Lundie-Jenkins, Corbett, and Phillips 1993), but it was destroyed by fire in 1991; the other was destroyed by fox predation in 1987 (Johnson and Burbidge 2008b). According to Richards et al. (2008b), those two colonies represented an unnamed central Australian subspecies. Animals from those colonies were used to establish captive breeding groups, which in turn served as the source for a successful introduction to Trimouille Island in 1998. The IUCN recognizes three other subspecies: *L. h. hirsutus* of southwestern Western Australia and now extinct, *L. h. bernieri* of Bernier Island, and *L. h. dorreae* of Dorre Island. Woinarski, Burbidge, and Harrison (2014) cited unpublished molecular data showing *L. h. dorreae* should be made a synonym of *L. h. bernieri*. Populations of the two are known to fluctuate, but surveys carried out by Short and Turner (1992) during a drought and thus probably when numbers were minimal indicated the presence of 2,600 individuals on Bernier Island and 1,700 on Dorre. The IUCN now classifies the entire species *L. hirsutus* as vulnerable, notes that it is decreasing, and lists major threats as habitat alteration by the introduced rabbit (*Oryctolagus cuniculus*), grazing by livestock, wildfire, and predation by the fox and house cat (Richards et al. 2008b). *L. hirsutus* also is listed as endangered by the USDI and is on appendix 1 of the CITES.

Although the Aborigines of Australia avidly hunted hare wallabies for food, they actually may have benefited the animals by regularly setting winter fires in order to clear areas for easier hunting and thereby producing a mosaic of different regenerating vegetative stages. That process not only provided food for *Lagorchestes* but also prevented the buildup of brush and the devastation of the habitat by lightning-caused fires during the summer. The decline of hare wallabies coincided with the removal of the Aborigines from large areas and the reduction of winter fires. The two mainland colonies of *L. hirsutus* that persisted until 1987 and 1991 were in localities where regular winter burning was still practiced (Australian National Parks and Wildlife Service 1978; Bolton and Latz 1978; Ingleby 1991a).

DIPROTODONTIA; MACROPODIDAE; **Genus** *ONYCHOGALEA* Gray, 1841

Nail-tailed Wallabies

There are three species (Burbidge 2008c; Calaby 1971; Gordon and Lawrie 1980; Ride 1970):

> *O. unguifera*, northern Western Australia to northeastern Queensland;
> *O. lunata*, formerly east-central and southern Western Australia, southern Northern Territory, South Australia, and southwestern New South Wales;
> *O. fraenata*, formerly inland parts of eastern Queensland, New South Wales, and northern Victoria.

In the two living species, *O. unguifera* and *O. fraenata*, head and body length is 430–700 mm, tail length is 360–730 mm, and weight is 4–9 kg; *O. lunata* was smaller, with a weight of approximately 3.5 kg (Burbidge 2008c; Evans and Gordon 2008; Ingleby and Gordon 2008). The fur is soft, thick, and silky. In *O. unguifera* the upper parts are fawn-colored, and a dark brown mid-dorsal stripe extends from neck to base of tail. The general color in the other two species is gray. In *O. fraenata*, the bridled wallaby, the white shoulder stripes run from beneath the ears along the back of the neck and down around the posterior part of each shoulder to the white undersurface of the body; the center of the neck is black or gray. In *O. lunata* the white shoulder stripes form a crescent around the posterior region of each shoulder and do not extend onto the neck, which is dark rufous. In all three species the underparts are white, and white hip stripes, often indistinct, are present.

The tail is tapering, short-haired, not bushy, more or less crested toward the tip, and, at its extreme end, provided with an unusual horny excrescence, forming a spur or nail at the tip. The function of the tail spur has never been satisfactorily explained. The tail operates as a lever during leaps, as it does in larger wallabies and kangaroos. The central hind claws of *Onychogalea* are long, narrow, and sharp. Females have four mammae and a well-developed pouch.

Nail-tailed wallabies inhabit woodland, savannah, brushland, or steppe with thickets. They seem to depend on dense vegetation for cover and emerge at night to feed. Although they occasionally move about in daylight, they spend most of the day in a shallow nest, scratched out beneath a tussock of grass or a bush, or in some other shelter, such as a hollow tree or log. When disturbed, *O. unguifera* sometimes emits a quickly repeated "u-u-u" and then flees. All species are remarkably swift when startled from their shelter and often escape into new cover. Unlike other wallabies, they carry their arms outward almost at a right angle to the body. When hopping they move their arms in a rotary motion, which has led to the common name "organ grinders."

Evans and Gordon (2008) explained that *O. fraenata* starts toward the edge of the scrub about an hour before dusk, often grazing in small clearings along the way, then rarely venturing more than 200 meters from the edge, except during drier months; the diet of that species consists of mixed forbs, grass, and browse. *O. lunata* reportedly ate grass (Burbidge 2008c) and *O. unguifera* is a selective feeder, choosing mostly dicotyledonous herb foliage, succulents, fruits, and green grass shoots (Ingleby and Gordon 2008).

Fisher, Hoyle, and Blomberg (2000) reported a maximum population density of 32 individual *O. fraenata* per sq km. Evans and Gordon (2008) stated that home ranges of *O. fraenata* vary in size from 20 to 90 ha. and overlap within and between sexes. That species is usually solitary, but feeding aggregations of four or five animals sometimes form. In central Queensland, Fisher and Lara (1999) found long-term home ranges of female *O. fraenata* to average about 23 ha., significantly smaller than those of males, which averaged

Nail-tailed wallaby (*Onychogalea unguifera*). Photo by Jiri Lochman / Lochman LT.

about 53 ha. Ranges of tagged males overlapped those of 4–21 tagged females. Most observations were of lone individuals, but estrous females were usually accompanied by a single male, sometimes up to six males at once. Males were aggressive to one another in such situations but, unlike the males of some other macropodids, did not establish a dominance hierarchy. Both sexes were promiscuous. Sigg and Goldizen (2006) reported more than half of observed estrous females to associate with more than one male (2–6); fighting among males was rare and restricted to individuals of similar size, and large males were most successful in mating.

O. fraenata breeds throughout the year, though wild females with accompanying young and pouch young are seen more frequently in late spring and summer (Evans and Gordon 2008). In a study of captive *O. fraenata*, Johnson (1997) found estrous cycle to average 36.2 days, gestation period 23.6 days. Postpartum estrus was not observed, and mating occurred either when pouch young were 80–92 days old or 8–21 days after premature loss of a pouch young. Final departure from the pouch came at 119–126 days, and weaning came 70–118 days later. Fisher and Goldizen (2001) found an identical pattern in wild *O. fraenata* except for age at weaning, which was much earlier, just 7–8 weeks after permanent exit from the pouch. Also, after that final exit, the young were found to always spend the day concealed in dense cover, generally more than 200 meters from their mothers. Johnson's (1997) study showed minimum interbirth interval to be 119 days and thus that a female could have up to 3 young per

Nail-tailed wallaby (*Onychogalea fraenata*). Photo by Alexander F. Meyer at David Fleay Wildlife Park, Gold Coast, Australia.

year. Females became sexually mature when 136–277 days old, males when 270–419 days old. The relatively young age at maturity and high annual number of young apparently give *O. fraenata* a fecundity comparable to that of potoroids and much higher than that of most macropodids. Fisher and Lara (1999) reported female *O. fraenata* to have an average longevity of 6 years. Weigl (2005) listed a captive-born female that lived for 7 years and 5 months.

Ingleby and Gordon (2008) wrote that *O. unguifera* also breeds all year. It has an estrous cycle of 20–23 days and a gestation period of 17–24 days. Birth is followed in 1–3 days by mating and then by embryonic diapause. Pouch life is 19–22 weeks, and the young are weaned 13–21 weeks after permanent emergence. Sexual maturity is attained at 45–47 weeks by males and 54–62 weeks by females.

Both *O. fraenata* and *O. lunata* are listed as endangered by the USDI and are on appendix 1 of the CITES. The IUCN designates the former as endangered but the latter as extinct (Burbidge and Johnson 2008c; McKnight 2008f). Both have suffered drastic declines largely through habitat alteration by farming and grazing and because of predation by introduced foxes, dogs, and cats. Newsome (1971a) suggested that livestock grazing and deliberate clearing destroyed the thickets where the wallabies sheltered, leaving them homeless, insecure, and greatly vulnerable to predation. *O. lunata* was still common in some parts of its range until the 1930s, but the last reliable records are from central Australia in the 1950s (Burbidge 2008c).

O. fraenata was common in parts of Queensland and New South Wales in the mid-nineteenth century but subsequently became rare. There were no confirmed sightings from 1937 to 1973, and the species was considered possibly extinct. In the latter year, however, a population of a few hundred animals was discovered in central Queensland. About half the occupied area was purchased by the state government to form a reservation for the species (Gordon and Lawrie 1980; Poole 1979). The reserved area, now known as Taunton National Park, covers 11,470 ha. (Evans and Gordon 2008). After cattle were removed from the area, the wallaby population increased to a high of 1,400 individuals in 1991, but drought subsequently reduced numbers by about half. Some of the Taunton animals were used to establish another population at Idalia National Park in

central Queensland, and there also are translocated colonies at Avocet Nature Refuge in Queensland and at Scotia Sanctuary in New South Wales (McKnight 2008f). Woinarski, Burbidge, and Harrison (2014) reported a total of only a few hundred individuals remaining at Taunton, Idalia, and Avocet, which are unfenced and exposed to predators, but about 2,000 at the fenced Scotia Sanctuary and another 1,500 in a captive breeding program. Considering those figures and IUCN criteria, those authorities reclassified *O. fraenata* as vulnerable.

The third species of the genus, *O. unguifera*, is still widespread in northern Australia and reportedly is common at several localities (Calaby 1971; Ingleby and Gordon 2008). It may have survived through an association with wetter habitats than those used by the other species of *Onychogalea* and also by many other now endangered or extinct Australian marsupials. It thus would be less dependent on the core areas of rich vegetation that were needed by the others during droughts but that were degraded by livestock, introduced animals, and other problems caused by humans (Ingleby 1991b).

DIPROTODONTIA; MACROPODIDAE; **Genus *WALLABIA*** Trouessart, 1905

Swamp Wallaby

The single species, *W. bicolor*, occurs from the Cape York Peninsula, through eastern Queensland (including nearby Fraser, Bribie, and North and South Stradbroke Islands), eastern New South Wales, and Victoria, to extreme southeastern South Australia (Merchant 2008b; Ride 1970). Some species here assigned to the genus *Macropus* have often been put in *Wallabia*. Calaby (1966), however, argued that because of its karyology, reproductive physiology, behavior, and dental morphology, *W. bicolor* should be placed in a monotypic genus. Calaby and Richardson (1988b), Flannery (1989), Groves (2005e), and most other recent systematic accounts have followed that arrangement. However, Kirsch (1977b) thought, on the basis of serological investigation, that *W. bicolor* probably could be included in *Macropus*, and DNA-hybridization studies by Kirsch, Lapointe, and Springer (1997) associated

Swamp wallaby (*Wallabia bicolor*). Photo by Alexander F. Meyer at Lone Pine Koala Sanctuary, Brisbane.

Wallabia with *Macropus rufus* and *M. robustus* in the subgenus *Osphranter*. Results of a comprehensive analysis of the nuclear DNA of suborder Macropodiformes (Meredith, Westerman, and Springer 2009a) showed *Wallabia* to nest within the genus *Macropus*, hence making the latter paraphyletic; *Wallabia* was thought possibly to be sister to the subgenus *Notamacropus*. Such a result was consistent with Van Gelder (1977), who recommended that *Wallabia* be considered a synonym of *Macropus* since hybridization in captivity had occurred between *Wallabia bicolor* and *Macropus* (*Notamacropus*) *agilis*. Nonetheless, an osteological assessment by Prideaux and Warburton (2010) suggested that *Wallabia* is actually well removed from *Macropus* and is a phylogenetic sister to a clade consisting of *Lagorchestes*, *Onychogalea*, *Macropus*, and a number of extinct genera.

Head and body length is 665–847 mm, tail length is 640–862 mm, and weight is 10.3–15.4 kg in females and 12.3–20.5 kg in males (Merchant 2008b). The fur is long, thick, and coarse. In the north the back and head are reddish brown and the belly is orange; in the south the general coloration is brownish or grayish black with grayish sides and sometimes brownish red underparts. The paws, toes, and terminal part of the tail are usually black. There may be a distinct light-colored stripe extending from the upper lip to the ear. Females have four mammae and a well-developed, forward-opening pouch.

Despite its common name, *Wallabia* is not restricted to swamps. It does inhabit moist thickets in gullies and even mangroves but is also found in open forest in upland areas as long as there are patches of dense cover. It hops heavily with the body well bent over and the head held low (Ride 1970). In Queensland the diet includes pasture, brush, and agricultural crops (Kirkpatrick 1970b). In southeastern Australia *Wallabia* also eats a variety of underground-fruiting fungi (Claridge, Trappe, and Claridge 2001). A study in Victoria found this genus to be mainly a browser and to occupy elongated home ranges, with long axes measuring up to 600 meters (Edwards and Ealey 1975). However, Troy and

Coulson (1993) reported individuals of both sexes to maintain small, overlapping home ranges of about 16 ha. Di Stefano et al. (2010) radio-tracked *Wallabia* at all times of the day and night, finding home ranges of 16.0–82.6 ha. for males and 3.4–37.5 ha. for females. Genetic analysis indicates that maturing females tend to remain near the ranges of their mothers but that males disperse to new areas (Paplinska et al. 2009).

The swamp wallaby is usually solitary, but unrelated animals may gather at an attractive food source (Kirkpatrick 1970b). There does not appear to be any territorial defense (Edwards and Ealey 1975). Females are polyestrous, with an estrous cycle averaging 31 (29–34) days (Kaufmann 1974). Births in captivity have been recorded from January to May and from October to December (Collins 1973), but there appears to be no sharply demarcated breeding season in the wild. Under

Swamp wallaby (*Wallabia bicolor*). Photo by Gerhard Körtner.

good conditions females can breed continuously and give birth about every 8 months (Edwards and Ealey 1975; Kirkpatrick 1970b; Paplinska et al. 2006). Average undelayed gestation is 36.8 (35–38) days, and litter size usually is one, but twins have been reported (Collins 1973; Kaufmann 1974). This is the only marsupial with a gestation period longer than its estrous cycle. There thus would seem to be a prepartum estrus and mating during the last 3–7 days of pregnancy. However, evidence is based on small sample sizes from captive animals; presence of a prepartum estrus, especially in the wild, requires further study for confirmation (Paplinska et al. 2006). Available evidence indicates that after mating there is a near-term embryo in one uterus and a segmenting egg in the other. After the first embryo is born, its suckling induces diapause in the other embryo. Should the first young die or be removed from the pouch, the second embryo will resume development and another birth will occur in about 30 days. Otherwise the second embryo will remain dormant until late in the pouch life of the first young and then will resume development in time to be born when the first young vacates the pouch at an age of about 250 days (Collins 1973; Kaufmann 1974; Russell 1974a). Sexual maturity is attained by both sexes at about 15 months, and maximum life span in the wild may be 15 years (Kirkpatrick 1970b). A captive individual lived 16 years and 10 months (Weigl 2005).

Although *Wallabia* has declined in numbers and distribution because of clearing of its habitat, it still is widespread and common (Calaby 1971; Merchant 2008b). Kirkpatrick (1970b) stated that approximately 1,500 skins are marketed each year in Queensland, and a larger number of animals are shot as agricultural pests, but such killing is not considered a threat to survival of the genus. The swamp wallaby was introduced on Kawau Island, New Zealand, about 1870 and still occurs there (Maynes 1977a; Warburton 2005f).

DIPROTODONTIA; MACROPODIDAE; **Genus**
MACROPUS Shaw, 1790

Wallabies, Wallaroos, and Kangaroos

There are 3 subgenera and 14 species (Caughley 1984; Clancy and Croft 2008; Collins 1973; Coulson 2008a,

Agile wallaby (*Macropus agilis*). Photo by Alexander F. Meyer at Munich Zoo.

2008b; Croft and Clancy 2008; Dawson and Flannery 1985; Flannery 1989, 1995a, 1995b; Hinds 2008; Jarman and Calaby 2008; P. M. Johnson 2008a; Kirsch and Calaby 1977; Kirsch and Poole 1972; Merchant 2008a; Richardson and Sharman 1976; Ride 1970; Ritchie 2005; Smith and Robinson 2008):

subgenus *Notamacropus* Dawson and Flannery, 1985

M. irma (western brush wallaby), southwestern Western Australia;

M. greyi (toolache wallaby), formerly southeastern South Australia and southwestern Victoria;

M. parma (parma wallaby), eastern New South Wales;

M. dorsalis (black-striped wallaby), from near Chillagoe in northeastern Queensland south to Coonabarabran in northeastern New South Wales;

M. agilis (agile wallaby), south central and southeastern New Guinea and—perhaps at least in part through introduction—several islands to the east (Kiriwana, Goodenough, Fergusson, possibly New Ireland and Normanby), northeastern Western Australia, northern Northern Territory and several nearby islands (Tiwi Islands, Groote Eylandt), northern and eastern Queensland and several nearby islands (Mornington, Friday, Peel, Russell, North and South Stradbroke, Woogoompah);

M. rufogriseus (red-necked wallaby), coastal areas from southeastern Queensland to southeastern South Australia, Tasmania, islands in Bass Strait (King, Furneaux Group);

M. eugenii (tammar wallaby), southwestern Western Australia and several nearby islands (East and West Wallabi, Garden, North Twin Peaks, Middle), southern South Australia and several nearby islands (St. Peter, St. Francis, Flinders, Kangaroo);

M. parryi (whiptail wallaby), eastern Queensland, northeastern New South Wales;

subgenus *Macropus* Shaw, 1790

M. giganteus (eastern gray kangaroo), Cape York Peninsula, eastern and central Queensland and several nearby islands (Fraser, Moreton, North Stradbroke), New South Wales, Victoria, extreme eastern South Australia, eastern Tasmania;

M. fuliginosus (western gray kangaroo), southern Western Australia, southern and northwestern South Australia, western and central New South Wales, south-central Queensland, western Victoria, Kangaroo Island;

subgenus *Osphranter* Gould, 1842

M. bernardus (black wallaroo), Arnhem Land of northern Northern Territory;

M. robustus (common wallaroo, euro), nearly throughout mainland Australia, Barrow Island off west coast of Western Australia;

M. antilopinus (antilopine wallaroo), northeastern Western Australia, northern Northern Territory, northern Queensland;

M. rufus (red kangaroo), throughout mainland Australia except northeastern and southwestern Western Australia, northern Northern Territory, Cape York Peninsula, east coast of Queensland and New South Wales, most of Victoria, and extreme southeastern South Australia.

McKenna and Bell (1997) retained *M. rufus* in the separate genus *Megaleia* Gistel, 1848. Groves (2005e) noted that subgenus *Notamacropus*, as constituted above, is almost certainly paraphyletic. Such a view is supported in part by recent molecular studies indicating that the genus *Wallabia* (see account thereof) is more closely related to subgenus *Notamacropus* than are subgenera *Macropus* and *Osphranter* (Meredith, Westerman, and Springer 2009a, 2009b; see also Kear and Cooke 2001). Those studies suggest resolving the problem by recognizing *Wallabia* as a fourth subgenus of *Macropus*, which might appropriately be placed above after subgenus *Notamacropus*. However, Eldridge and Coulson (2015) considered that the divergency of *Wallabia* warranted retention as a separate genus and that the best solution would be to elevate the above three subgenera

Red-necked wallaby (*Macropus rufogriseus*). Photo by Gerhard Körtner.

to full generic status; their sequence was *Wallabia, Macropus, Osphranter, Notamacropus*.

McKenzie and Cooper (1997) pointed out that the population of *M. eugenii* in Western Australia has probably been isolated from the population in South Australia for 50,000–100,000 years; captive individuals from the two areas have mated and produced fertile offspring, though at a rate lower than within the parental populations. *M. fuliginosus* was sometimes treated as a part of *M. giganteus* but was confirmed as a separate, closely related species by Kirsch and Poole (1972). In captivity, male *M. fuliginosus* have hybridized with female *M. giganteus* but never the reverse. Only the female offspring of those crosses are fertile and they readily backcross with males of either parental species. There is evidence that hybridization also occurs occasionally in the wild and involves introgression into both species (Coulson 2008a; Neaves et al. 2010; Poole 1975).

Kangaroos and wallaroos (subgenera *Macropus* and *Osphranter*) include the largest living marsupials.

Common wallaroo, or euro (*Macropus robustus*). Photo by Gerhard Körtner.

When standing in a normal, plantigrade position, male gray and red kangaroos are usually about 1.5 meters tall and sometimes reach nearly 1.8 meters. Reports of kangaroos 2.1 meters (7 feet) tall seem unfounded (W. E. Poole, Commonwealth Scientific and Industrial Research Organization, Division of Wildlife Research, pers. comm., 1977). When standing on their hind toes, however, as when in an aggressive position, males may reach or slightly exceed 2.1 meters. Wallaroos are shorter on average than kangaroos but more heavily built and may weigh just as much. The fur of kangaroos and wallaroos is generally thick and coarse, being especially long and shaggy in the wallaroos. The muzzle is completely hairless in wallaroos, partly haired in the red kangaroo, and fully haired in the gray kangaroos. Wallabies (subgenus *Notamacropus*) are generally smaller than kangaroos and wallaroos. The enlarged hindquarters of *Macropus* are powerfully muscled, and the tapered tail acts as a balance and rudder when the animal is leaping and as a third leg when it is sitting.

The tail is strong enough to support the weight of the entire animal. Females have a well-developed, forward-opening pouch and four mammae.

Habitat varies widely in this genus (see following individual species accounts). Most wallabies rely on dense vegetation for cover but usually move into open forest or savannah to feed. The gray kangaroos, *M. giganteus* and *M. fuliginosus*, are found in forests and woodland and also seem to require heavy cover. *M. robustus* and *M. bernardus* occur mainly in mountains or rough country. *M. antilopinus* and *M. rufus* dwell on grass-covered plains or savannah but, again, depend on places with denser vegetation for shelter. Nearly all species are primarily crepuscular or nocturnal, feeding from late afternoon to early morning and resting during the day, but they do sometimes move about in daylight. When moving slowly, as in grazing or browsing, these animals exhibit an unusual "five-footed" gait, balancing on their tail and forelimbs while swinging their hind legs forward, then bringing their forelimbs and tail upward.

They are remarkably developed in their bipedal leaping mode of progression, which involves only the hind limbs. At a slow pace the leaps of wallaroos and kangaroos usually measure 1.2–1.9 meters; at increased speeds they may leap 9 or more meters. One gray kangaroo jumped nearly 13.5 meters on a flat. Normally they do not jump higher than 1.5 meters. Speeds of about 48 km/hr are probably attained for short distances when animals are pressed in relatively open country.

All species of the genus *Macropus* are herbivorous, and most are mainly grazers. In some areas well over 90 percent of the food eaten by kangaroos and wallaroos is grass (Russell 1974a). Some species can go for long periods without water. Where they shelter in cool caves, they can exist indefinitely without water, even when outside temperatures exceed 45° C. Some species dig in the ground for water or eat succulent roots. Nonetheless, availability of water, particularly through rainfall, is a critical factor in the survival and reproduction of the larger species of kangaroos, which live primarily in semiarid regions. Population size may fluctuate dramatically in response to drought (Caughley, Grigg, and Smith 1985), though other factors may also be involved (see Pople et al. 2010).

Social structure is variable in *Macropus* and does not seem fully related to the size, taxonomy, or ecology of the animals (see following individual species accounts). The small *M. parma* and *M. eugenii*, the medium-sized *M. rufogriseus* and *M. agilis*, and the large *M. robustus* are often solitary but may aggregate temporarily in the vicinity of favored resources (food, water, shelter). The medium- and large-sized *M. dorsalis, M. parryi, M. giganteus, M. fuliginosus, M. antilopinus,* and *M. rufus* occur in organized groups or "mobs" (Croft 1989; Russell 1974a). Notwithstanding any anthropomorphic connotation of the term, a mob is technically considered a relatively stable set of individuals that interact commonly and have overlapping ranges. It is the long-term pool of associates amongst which the individual leads its social life (Jarman and Coulson 1989).

Eastern gray kangaroo (*Macropus giganteus*). Photo by Dennis Jacobsen / Shutterstock.com.

Extensive information is available on reproduction in *Macropus*, though only a brief compilation can be provided here. A detailed summary also was given by Hume et al. (1989). The females of all species that have been investigated are polyestrous, with lengthy reproductive seasons but relatively brief intervals of receptivity lasting perhaps only a few hours (Kaufmann 1974). Embryonic diapause occurs in most species of *Macropus*. In those species, unlike in most mammals, conception alone does not affect the estrous cycle, and the next period of receptivity and mating come at the same time as they would have if the female had not become pregnant (Russell 1974a). After the second mating, estrus finally is suppressed through the suckling stimulus of the newly born first young. The embryo resulting from the second mating develops only to the blastocyst stage and then becomes quiescent until the first young is approaching the end of pouch life or perishes or is experimentally removed. Subsequently the second embryo resumes development, and birth occurs at a time when the pouch is vacant. Most individuals probably do not live to maturity but life span is potentially great.

The effect of people on *Macropus* has varied (see above account of family Macropodidae and following individual species accounts). Some species have declined in response to environmental disruption, while others, especially the red and gray kangaroos, have maintained high numbers and come into direct confrontation with human interests. A detailed summary of the involved interrelationships, along with natural history data and maps showing overall population density estimates for the three species of kangaroos, is available online (Pople and Grigg 1999, http://www .environment.gov.au/biodiversity/trade-use/wild -harvest/kangaroo/harvesting/roobg-01.html).

Kangaroos have long been heavily hunted, first by the Aborigines, then by European settlers for meat and skins, and later by stockmen because of alleged damage to crops, pasture, and fences. In the nineteenth century the eastern Australian states enacted measures to encourage eradication of the species. Available records show that the number of bounty payments was approximately 20 million in Queensland from 1877 to 1906. In New South Wales about 8 million bounties were paid from 1881 to 1889, and the officially estimated number of kangaroos in the state reportedly

fell from 6.1 million to 1 million during that period. Bounties in both states were discontinued just after 1900 (Robertshaw and Harden 1989).

Subsequently an industry developed based on the harvesting and processing of about 300,000–400,000 skins per year (Poole 1978). In the late 1950s, improved refrigerating equipment allowed a great increase in the take of kangaroos for use as pet food. Such hunting, along with continued killing for skins and pest control, was not well regulated and coincided with severe droughts in parts of Australia. The number of kangaroos commercially harvested in Queensland rose from about 200,000 in 1954 to over 1 million in 1959. Kangaroo populations dropped alarmingly over large areas (Ride 1970; Robertshaw and Harden 1989; Russell 1974a).

During the 1970s the Australian state governments increased their control of kangaroo hunting. From 1973 to 1975 the Australian federal government prohibited the general export of kangaroo products but subsequently ended the ban as the states established acceptable management programs (Poole 1978). The states now develop such programs that are then subject to review and approval by the federal government, which sets annual harvest quotas. Management programs and exportation have been approved for six species of *Macropus*: *M. rufus*, *M. giganteus*, *M. fuliginosus*, *M. robustus*, *M. parryi*, and *M. rufogriseus*. The first three species, the red and gray kangaroos, make up over 95 percent of the commercial harvest (Pople and Grigg 1999).

Most kangaroo skins had been going to the United States, which imported an annual average of more than 1 million from 1963 to 1966 (Gaski 1988). In 1974, however, the USDI listed *M. giganteus*, *M. fuliginosus*, and *M. rufus* as threatened, pursuant to the Endangered Species Act of 1973 (Short 1995). Accompanying regulations prohibited the importation of those species and products thereof until there was satisfactory certification that wild populations would not be adversely affected. Such certification subsequently was received, and importation was opened in 1981. Surveys in 1980–1982 indicated there were then about 19.1 million red, western gray, and eastern gray kangaroos in Australia (Grigg et al. 1985). The USDI proposed completely removing the three species from threatened status in 1983 (Short 1995). However, mainly because of severe drought, estimated kangaroo numbers fell to 13.3

Western gray kangaroo (*Macropus fuliginosus*), mother and young. Photo by Gerhard Körtner.

million by 1984, and the USDI proposal was withdrawn that year. Importation to the United States continued under the USDI regulations.

Fletcher et al. (1990) reported new surveys indicating that populations had recovered to approximately their 1980 levels. In 1993 the USDI again proposed delisting and announced the finding that kangaroo populations were high and that the four states with a commercial harvest (New South Wales, Queensland, South Australia, Western Australia) had effective conservation programs ensuring protection of the species (Short 1995). In 1993 the total estimate of red and gray kangaroos in those areas where commercial harvesting occurs, which includes most of the range of the affected species, was 21.4 million, the commercial take was 2.8 million, and an additional 153,000 were reported killed for damage mitigation purposes. Many

conservationists formally opposed the USDI proposal, indicating kangaroo populations were not safely managed in Australia and subject to precipitous decline. Over such opposition, and despite the role of its regulations in improving such management, the USDI proceeded to cancel the regulations and remove the three species from threatened status in 1995 (Short 1995).

Subsequent figures have continued to fluctuate dramatically. The 2001 official population estimates for the commercial harvest areas were about 50.6 million red, western gray, and eastern gray kangaroos plus 6.8 million common wallaroos (*M. robustus*). Surveys of the same areas in 2010 indicated a total of 22.7 million for the three kangaroos and 2.4 million for *M. robustus* (Australian Department of Sustainability, Environment, Water, Population and Communities 2011). Notwithstanding the drop from 2001 to 2010, reportedly because of further drought, the total number of those four species probably exceeds the number of humans (about 22 million) in Australia. No other terrestrial region in the world of comparative size still has more large wild mammals than people.

The major question, however, is not whether there are plenty of kangaroos but whether populations can sustain present intensive levels of hunting, especially considering the additional factors of human habitat modification, competition with sheep for grazing land, and drought. There has long been controversy as to how many kangaroos can be safely harvested and whether commercial utilization jeopardizes the species or helps by giving them a value that discourages their mass destruction as pests (Australian National Parks and Wildlife Service 1988; Grigg 2002; Hercock 2005; Lunney 2010; Poole 1984; Pople and Grigg 1999; Short 1995). There also are animal welfare issues, especially regarding the fate of young kangaroos after their mothers are shot. Although regulations call for humane dispatch, Witte (2005) estimated that 4.5 million young, no longer in the pouch but still suckling and fully dependent, were left to a lingering death from 1994 to 2003. Ben-Ami et al. (2014) concluded that, on balance, commercial killing at present levels does more harm than good, because of loss of dependent young, inhumane killing of adults, intensified mortality during drought, and a disruption of social stability and evolutionary potential.

Red kangaroo (*Macropus rufus*). Photo by Gerhard Körtner.

Red and gray kangaroos often are said to have benefited and even greatly increased in numbers because of European settlement of Australia (Ellis et al. 2008b; Newsome 1971a; Poole 1979; Pople and Grigg 1999; Russell 1974a; Short 1995). They seem to have been aided by establishment of artificial water holes for livestock, as well as by the grazing of stock, which crops the long, dry grass and browse avoided by the kangaroos and causes the sprouting of soft, green shoots. Favorable conditions for the kangaroos thus were presumably created over vast areas. However, based on numerous recorded observations from the eighteenth and early nineteenth centuries, plus assessment of the capacity of native habitat and available water to support herbivores, Auty (2004) estimated that red and gray kangaroos were far more abundant prior to settlement than they are today, probably numbering 100 to 200 million. Croft (2005) suggested even higher numbers could have been supported by the amount of water and for-

age available prior to any manipulation for or by livestock.

At present, kangaroo populations do appear high, perhaps as high as at any time since the mid-nineteenth century, and official survey data indicate no marked decline in response to harvesting programs in the last several decades. As explained by Pople and Grigg (1999), however, the average proportion of the overall quota taken each year has fluctuated at around 80 percent, and a sustained meeting of the quota could potentially reduce kangaroo numbers by 40 percent. Such reduction, in conjunction with severe drought and habitat disruption, might be at least as damaging as the intensive eradication programs of the late nineteenth and early twentieth centuries (Croft 2005; Newsome 1975). Additional information on the kangaroos and each of the other species of *Macropus* is provided separately in the following accounts.

Macropus irma (Western Brush Wallaby)

Except as noted, the information for the account of this species was taken from Morris and Christensen (2008). Head and body length is about 1,200 mm, tail length is 540–970 mm, and weight is 7–9 kg. Coloration is pale gray with a distinct white facial stripe, black and white ears, black feet, and a dorsal crest of black hair on the long tail, particularly toward its extremity. The sexes are similar in size and appearance.

Optimum habitat for this species is open forest or woodland, particularly seasonally wet flats with low grasses and scrubby thickets. Unlike the sympatric *M. fuliginosus*, it does not appear to venture into open pasture areas adjacent to its bushland refuges, but it does resemble larger kangaroos in some of its habits. Clearly adapted for life in the open, it is speedy and able to weave or sidestep with ease as it moves low to the ground with its long tail extended. Activity is greatest in early morning and late afternoon. It rests during the hotter part of the day, singly or in pairs, in the shade of a bush or in small thickets. It is a grazer, not a browser, and seems able to manage without free water. Young appear to be born in April or May and to emerge from the pouch in October and November.

M. irma was very common in the early days of settlement, and large numbers of its skins were traded commercially—for example, 122,000 in 1923 and 105,000 in 1924 (Morris, Friend, and Burbidge

Western brush wallaby (*Macropus irma*). Photo by Jiri Lochman / Lochman LT.

2008). Extensive clearing and fragmentation of bushland in the wheatbelt from the 1930s to the 1960s reduced suitable habitat. In the 1970s a dramatic increase in the fox (*Vulpes vulpes*) population in southwestern Western Australia apparently led to a decline of around 80 percent in the population of *M. irma*. Fox control was initiated in much of the region in 1996 and allowed the wallaby's population to increase. The IUCN estimates its numbers at around 100,000 individuals and no longer assigns a threatened classification (Morris, Friend, and Burbidge 2008).

Macropus greyi (Toolache Wallaby)

According to Smith and Robinson (2008), head and body length is around 810 mm in males and 840 mm in females; tail length is 730 mm in males and 710 mm in females. The upper parts are pale grayish fawn, the back is banded with 10–12 light gray bars, the underparts are fawn, there is a dark cheek stripe from muzzle to ear, and the tail is almost white and has a crest on the distal third. In life this species was characterized by a slim, graceful body, high speed, endurance, and erratic gait. Its preferred habitat was the edge of stringybark heath in flat or gently undulating grassland, which became swampy in winter and was interspersed with depressions filled by a matted growth of black rush among tussocks of tall grass. By day it sheltered on little islands of higher ground with sparse stands of drooping sheoak. It fed in shorter grassland between heath and tussocks, where it could also best display its swiftness and agility. It could outrun all but the best dogs and was adroit at changing direction or length of stride among obstacles; its gait gave the appearance of two short hops, then a longer one. In poorer country it seemed to be solitary but in richer grassland it was gregarious, grazing and resting in groups.

The species was still common in the first years of the twentieth century but subsequently declined rapidly. In 1923–1924 an effort was made to capture the last known wild individuals and transfer them to a sanctuary on Kangaroo Island, but the project was mishandled (Harper 1945). The only animal taken alive died in 1939 (Smith and Robinson 2008). Competition from livestock, bounty and sport hunting, and killing for its beautiful pelt may all have contributed to the demise of the toolache wallaby (Calaby 1971), but the major problem was the almost complete draining and clearing of its specialized wetland-edge habitat for agriculture. Reliable reports indicate that small remnant populations may have persisted until the early 1970s (Smith and Robinson 2008), but the IUCN now considers the species extinct (Australasian Mammal Assessment Workshop 2008d).

Macropus parma (Parma Wallaby)

Except as noted, the information for the account of this species was taken from Maynes (1973, 1974, 1977a, 1977b, 2008). Head and body length is 424–527 mm, tail length is 405–544 mm, and weight is 2.6–5.9 kg. Males may average somewhat larger than females and have a more robust chest and forelimbs. The back and shoulders are uniform grayish brown, with a dark dorsal stripe extending from neck to shoulders. The throat and chest are white and about half the animals have a white tip to the tail.

Optimum habitat appears to be wet sclerophyll forest with a thick, shrubby understory associated with grassy patches. One of the most nocturnal species of *Macropus*, it takes cover among the shrubs during the day and emerges at dusk or shortly before, to feed on grasses and herbs. When hopping, it remains close to the ground in an almost horizontal position; at medium pace the tail is curved upward.

M. parma is normally solitary under natural conditions, but temporary feeding aggregations are sometimes observed. A series of sightings in Australia included 52 of lone animals, 14 of pairs, and 5 of groups of 3. Larger aggregations occur within the introduced population on Kawau Island in New Zealand, where population density is much higher.

Evidence from wild and captive populations indicates that breeding occurs throughout the year in Australia, but most births take place from February to July. Environmental factors may influence continuity of reproduction. Hence, the population on Kawau Island bred throughout 1970 and 1971 but only from

Parma wallaby (*Macropus parma*). Photo by Alexander F. Meyer at San Antonio Zoo.

March to July in 1972, when less food was available. Calculated average length of estrous cycle is 41.8 days, average nondelayed gestation is 34.5 days, and litter size is usually one. Some females have an estrus 4–13 days after giving birth, but most do not become receptive until the pouch young is 45–105 days old, and a few do not enter estrus at all while carrying a pouch young. If fertilization does take place when a pouch young is present, embryonic diapause occurs. Natural loss or experimental removal of the pouch young is followed by birth in 30.5–32.0 days. If the pouch young survives and vacates the pouch normally, the next young is born 6–11 days later. First departure from the pouch occurs at an age of 161–175 days, permanent emergence is at 207–218 days, and weaning is at 290–320 days. Sexual maturity has been observed at 11.5–16.0 months of age in captivity and as early as 12 months in wild females in Queensland. On Kawau Island, females may attain maturity when only 19 months old but usually do not do so until 2, occasionally 3, years old. Males appear to reach maturity at 20–24 months in both places. Weigl (2005) listed a captive male that lived to an age of 15 years and 11 months.

M. parma seemed to have disappeared through clearing of its natural habitat in New South Wales, with the last known specimens taken in 1932. About 1870, however, *M. parma*, along with several other species of wallaby, had been introduced on Kawau Island off northern New Zealand. There, a relatively dense population built up, but because it was confused with the other wallaby species present, it remained unknown to science until 1965. The wallabies on Kawau were considered highly detrimental to forestry and were subject to shooting and poisoning; thousands of *M. parma* are thought to have been killed. Knowledge that *M. parma* was present on Kawau brought protection in 1969 and intensive efforts to capture animals to establish breeding colonies elsewhere. Some animals were also used in reintroduction attempts in Australia, but most failed because of fox predation (Long 2003; Ride 1970; Warburton 2005c).

In 1967 *M. parma* was rediscovered in its original range in New South Wales. Later investigation showed that the species still occurred along several hundred km of the coast but at low densities. The IUCN now classifies the species as near threatened and estimates there are fewer than 10,000 mature individuals (Lunney and McKenzie 2008); it is listed as endangered by the USDI.

Word of a viable population in Australia led to the end of protection on Kawau Island in 1984 and a renewed effort to totally eradicate the species there (Warburton 2005c).

Macropus dorsalis (Black-striped Wallaby)

P. M. Johnson (2008a) indicated that this species is highly sexually dimorphic. Head and body length is 616–820 mm in males, 481–617 mm in females; tail length is 656–832 in males, 521–619 mm in females; weight is 8.7–21.0 kg in males, 5.2–7.6 kg in females. The upper parts are brown and the sides are paler, grading almost to white below. There is a dark brown mid-dorsal stripe from neck to rump, a white cheek stripe above the upper lip, a white spot on the cheek, and a horizontal white stripe on the thigh.

P. M. Johnson (2008a) wrote that this species is shy and prefers forested country with a dense shrub layer. By day it rests in cover then moves along established pathways to more open areas where it forages from dusk to dawn. It has a distinctive gait, involving a short hop with the head held low, the body strongly curved and the rump tucked under the body, the forelimbs usually extended forward and outward from the body. It is considered a gregarious species, sometimes living in groups of 20 or more individuals, though old males usually live alone.

Based on their study in Taunton National Park, central Queensland, Hoolihan and Goldizen (1998) suggested that *M. dorsalis* forms open-membership groups. Animals did not associate consistently with particular individuals for long periods of time and were seen in groups that varied in size and membership. Nearly half the sightings were of a single individual. Home range averages 91 ha. for both sexes; ranges overlap extensively with those of other individuals of both sexes (Eldridge and Coulson 2015).

Johnson and Delean's (2002b) study of two captive colonies showed *M. dorsalis* to be capable of breeding throughout the year. The gestation period was 33–36 days. Birth was usually followed by a postpartum estrus and mating, then a period of embryonic diapause. Pouch life was 192–225 days, and weaning was 81–159 days after permanent emergence. Sexual maturity came at 11.3 months in females and 15.7 months in males. Weigl (2005) listed a captive that lived in the Bronx Zoo for 12 years and 5 months.

Black-striped wallaby (*Macropus dorsalis*). Photo by Alexander F. Meyer at Australia Zoo, Beerwah, Queensland.

Much of the natural habitat of this species has been completely removed or heavily modified (P. M. Johnson 2008a). The IUCN indicates that deterioration may be continuing and that numbers are decreasing, but does not assign a threatened category (Winter, Burnett, and Menkhorst 2008). *M. dorsalis*, together with several other species of wallaby, may have been introduced on Kawau Island off northern New Zealand about 1870. However, the only evidence for its presence is a questionable and now lost specimen collected in 1954. If a breeding population was established, it is now extinct or extremely rare (Long 2003; Warburton 2005d).

Macropus agilis (Agile Wallaby)

Except as noted, the information for the account of this species was taken from Merchant (2008a). Head and body length is 717–850 mm in males and 593–722 mm in females, tail length is 692–840 mm in males and 587–700 mm in females, and weight is 16–27 kg in males and 9–15 kg in females. Males are considerably larger than

females and have much stouter forelimbs. The upper parts are sandy brown, and the underparts are whitish. The head may have a median dark brown stripe between the eyes and ears and a faint light buff cheek stripe. There is a distinct light stripe on the thigh, and the edges of the ears and tip of the tail are black.

The agile wallaby is the most common macropodid in tropical coastal Australia and the southern and eastern lowlands of New Guinea. Its preferred habitat is along streams in open forests and adjacent grasslands, but in Northern Territory it is abundant from coastal sand dunes to the base of rugged inland hills. It is a comparatively nervous macropodid, as demonstrated by its foot-thumping. It appears to eat most native grasses and may dig 30 cm or more to obtain the roots. In the dry season it also browses on leaf litter, fruits, and flowers of various shrubs and trees.

Home range size averages 17 ha. for males and 11 ha. for females in the wet season and 25 ha. for males and 15 ha. for females in the dry season (Eldridge and Coulson 2015). The agile wallaby is considered gregarious,

Agile wallaby (*Macropus agilis*). Photo by Chris and Mathilde Stuart.

living in groups of up to 10 individuals, which may form much larger aggregations in feeding areas. Kirkpatrick (1970a) indicated that the basic groups are made up mainly of females sharing the same resting and feeding areas. Such groups are not necessarily related, however, as it is thought that young animals quickly separate from the females. Several studies have indicated that, aside from the small groupings, *M. agilis* is most frequently seen alone or in consorting units (Dressen 1993). There apparently is intense competition among males for access to breeding females; they establish dominance hierarchies through repetitive ritualized bouts of display, sparring, wrestling, and kicking (Eldridge and Coulson 2015).

Breeding continues through the year under suitable conditions, and litter size is one (Merchant 1976). Calculated average length of estrous cycle is 30.6 days, and nondelayed average gestation period is 29.2 days (Rose 1978). There is a postpartum estrus before the pouch young is 2 days old, then a period of embryonic diapause. If the pouch young is lost prematurely, the next birth occurs an average of 26.5 days later. If the pouch young survives, its age when the second young normally is born is about 7 months. The young first leaves the pouch at an age of 176–211 days and permanently emerges at 207–237 days (Merchant 1976). It remains with the mother until weaned, when 10–12 months old. In captivity, females reach sexual maturity at an average age of 12 months, males at 14 months, but in the wild maturity probably comes several months later. Weigl (2005) listed a captive specimen that lived to an age of at least 16 years and 11 months.

In southeastern New Guinea *M. agilis* has declined because of excessive hunting for use as food (Flannery 1995a). In parts of Australia it is considered an agricultural pest and has been subject to shooting and poisoning. The IUCN (Aplin et al. 2008b) stated that it remains common in both countries, though overall numbers are decreasing; it was introduced on Vanderlin Island off northeastern Northern Territory. It was also introduced to Long Island off east-central Queensland, where it no longer occurs, and to Baniara Island off southeastern New Guinea (Long 2003). A small population, likely originating from individuals that escaped or were released from a wildlife park, recently became established on the central east coast of Tasmania (Pauza et al. 2013).

Macropus rufogriseus (Red-necked Wallaby)

Except as noted, the information for the account of this species was taken from Jarman and Calaby (2008). Head and body length is 770–888 mm in males and 708–837 mm in females, tail length is 703–876 mm in males and 664–790 mm in females, and weight is 15.0–23.7 kg in males and 12.0–15.5 kg in females. In the subspecies *M. r. banksianus* of mainland Australia, males are grizzled medium gray to reddish above, with a pronounced reddish brown neck, and white or pale gray below. Females are somewhat paler. In both sexes the muzzle, paws, and largest toe are black and there is a white stripe on the upper lip. *M. r. rufogriseus* of Tasmania and the Bass Strait islands is darker gray and has a more brownish neck.

This wallaby inhabits eucalypt forests with a good shrub stratum and open vegetation nearby, as well as hedge, sedge, and buttongrass communities. It rests in dense cover by day and emerges in the late afternoon, or earlier on wet, dull, or cool days. It eats largely grasses and herbs and occasionally browses shrubs or tree seedlings.

A maximum density of 2–3 individuals per ha. has been reported for the introduced population in New Zealand (Warburton 2005b). In an agricultural area of northern New South Wales, average home range was 32 ha. for males and 12 ha. for females. Seasonal averages there were 10.9 ha. in winter and 5.9 ha. in summer. Ranges were larger in a eucalypt plantation, averaging 72 ha. for males and 33 ha. for females (Eldridge and Coulson 2015). An individual's range shifts little from year to year and overlaps widely with those of others of both sexes. Daughters usually settle in a range overlapping their mother's, while males disperse when about 2 years old. *M. rufogriseus* is essentially solitary but may aggregate, especially in winter, to feed. Females associate mostly with related females, but males associ-

Red-necked wallaby (*Macropus rufogriseus*), male. Photo by Gerhard Körtner.

ate mostly with unrelated males of similar size. Males establish a dominance hierarchy based on relative size, less strongly on age. As a female's estrus approaches, she is followed by an increasing number of males. Eventually the largest and locally dominant male drives the others away and mates with the female.

Breeding in northern New South Wales and southern Queensland occurs year round and is continuous. In Tasmania births are strongly seasonal, occurring from January to July, the great majority in February and March. Hinds (2008) noted that the Tasmanian subspecies, *M. r. rufogriseus*, is the only macropodid, other than *M. eugenii*, to have a strictly seasonal pattern of breeding. Calculated average length of estrous cycle is 32.4 days, and average nondelayed gestation period is 29.6 days (Rose 1978). Litter size usually is one, but twins have been recorded and also one set of triplets (Collins 1973). There is a postpartum estrus, with mating 2 hours after the previous young's birth. Embryonic diapause may follow conception. If the pouch young is

Red-necked wallaby (*Macropus rufogriseus*), mother and young. Photo by Gerhard Körtner.

lost prematurely, the next birth occurs on average 26.7 days later (McEvoy 1970). Normal pouch life averages 9.7 months for female young and 9.2 months for males. Captive females may give birth when as young as 14 months. Wild females reach sexual maturity when 2 years old and may live to 10 years, bearing on average 9 offspring. Weigl (2005) listed a captive specimen that lived for 15 years and 2 months. However, one wild individual was estimated to have reached an age of 18.6 years (Kirkpatrick 1965).

In Tasmania *M. rufogriseus* is probably more common than when settlement began (Calaby 1971). It is also especially abundant in southeastern Queensland and northeastern New South Wales. In the late nineteenth and early twentieth centuries it was heavily hunted as an agricultural pest and for pelts. Now it is covered by game laws that permit killing the species when it is detrimental to crops, pasture, or plantation forestry. Tasmania allows sport and commercial hunting. In 1984 the total kill there was estimated at 71,000 individuals. In fiscal 2007 the recorded kill, 26,380, was limited to King and Flinders Islands. In 2008 Tasmania withdrew its harvest quota submission because of a lack of export market demand for wallaby products (Australian Department of Sustainability, Environment, Water, Population and Communities 2011).

M. rufogriseus was brought to the South Island of New Zealand in the 1870s, eventually increased to an estimated 750,000, became a pest to the sheep industry, and was reduced by shooting and poisoning in the 1960s to only about 3,500 (Wodzicki and Flux 1971). However, Warburton (2005b) indicated that by the 1980s numbers had increased to around 15,000, with 2,500–3,000 being shot each year. *M. rufogriseus* was also introduced to several European countries, but the only surviving feral populations are two small groups in England (Long 2003).

Macropus eugenii (Tammar Wallaby)

Except as noted, the information for the account of this species was taken from Hinds (2008) and applies to the well-known population on Kangaroo Island off South Australia. Head and body length is 590–680 mm in males and 520–630 mm in females, tail length is 380–450 mm in males and 330–440 mm in females, and weight is 6–10 kg in males and 4–6 kg in females. Animals from other populations are smaller on average. The upper parts are dark, grizzled gray-brown, becoming rufous on the sides of the body and on the limbs, especially in males; the underparts are pale gray-buff.

The tammar wallaby inhabits coastal scrub, heath, dry sclerophyll forest, and thickets in mallee and woodland. It requires dense low vegetation for shelter and open grassy areas for feeding. By day it rests in scrub, usually not leaving until after dark and then returning before dawn. Warburton (2005a) stated that *M. eugenii* feeds an average of 6.5 hours a night in pasture; it is primarily a grazer but also eats leaves. It is constantly alert when feeding, frequently looking and listening. A disturbed individual will alert others nearby by posture or hind foot thumping and if necessary fleeing, accompanied by all the others. In certain areas where there is almost no fresh water, *M. eugenii* obtains what it needs from salty plant juices and even is able to drink sea water (Ride 1970).

A population density of one individual per 1.6 ha. has been recorded (Main and Yadav 1971). Each individual has a defined home range that overlaps the ranges of others. Home ranges of 10–39 ha. have been recorded for the introduced population on the North Island of New Zealand, with diurnal forested sections and nocturnal open areas connected by clear trails (Warburton 2005a). Eldridge and Coulson (2015) cited home range

Tammar wallaby (*Macropus eugenii*). Photo by Jiri Lochman / Lochman LT.

sizes averaging 42 ha. in summer and 16 ha. in winter, there being no difference between the sexes, and referred to the species as moderately gregarious. However, other than loose feeding aggregations, sometimes with over 20 individuals, and females with young, no regular social grouping has been observed. Warburton (2005a) noted that opposite sexes may associate, resting and emerging in the evening together, and males threaten and, rarely, fight one another; females sometimes emit a deep screeching call of unknown function.

There is a rigid breeding season. In the natural population of Kangaroo Island most young are born in late January or early February, very few later than March, and none from July to December. The season is the same in the introduced population on Kawau Island off northern New Zealand and in captivity in Australia; the season is reversed when animals are taken to the Northern Hemisphere (Maynes 1977a). This evidently fixed reproductive pattern is associated with increasing and decreasing daylight from solstice to solstice (Tyndale-Biscoe 1989). Calculated average

length of estrous cycle is 28.4 days, and average non-delayed gestation period is 28.3 days (Rose 1978). Litter size usually is one, but twins have been recorded (Collins 1973). Within a few hours of giving birth, the female has a postpartum estrus and mates again, but then enters embryonic diapause. If the previous young is prematurely lost or experimentally removed before the winter solstice in mid-June, embryonic development resumes. The new young is born 26.5 days later (Kirkpatrick 1970a). From late June to November the female is in an anestrous period (Russell 1974a), and the embryo remains quiescent even if the pouch young is lost and lactation ceases. Young females reach sexual maturity at about 9 months of age while still suckling; they may mate at that time but their embryos also become quiescent. Typically, all diapausing embryos are reactivated soon after the summer solstice on 22 December and young are born up to 12 months after the mating at which they were conceived. The young first releases the teat when about 100 days old, opens its eyes at about 150 days, and first exits the pouch at about 200 days (Janssens and Rogers 1989). Total pouch life is 8–9 months, ending in September or October. Young males do not mature until nearly 2 years old. Wild males may live to at least 11 years, females to 14 years. Weigl (2005) listed captives of each sex that lived around 15 years.

Because of its unusual breeding seasonality, which may produce large groups of synchronously developing individuals, as well as its small size and ease of handling, *M. eugenii* has become the macropodid most used in research (Janssens and Rogers 1989). Captive colonies derived from the Kangaroo Island population have been established by several Australian institutions. The species was also introduced to Kawau Island about 1870 and to the North Island of New Zealand in 1912. It is considered a threat to natural vegetation in both areas, and efforts are being made to eradicate it, though in the latter area its range is still expanding (Warburton 2005a).

The tammar wallaby was common until about 1920 but subsequently vanished in mainland South Australia and became restricted to a few isolated colonies in Western Australia, because of habitat loss, competition with domestic sheep, and predation by the introduced red fox (Calaby 1971; Harper 1945; Poole 1979; Poole, Wood, and Simms 1991; Woinarski, Burbidge, and Harrison 2014). It remains abundant on Kangaroo Island but has been extirpated on St. Francis, St. Peter, Thistle, and Flinders Islands. The populations on St. Francis and St. Peter in the Nuyt's Archipelago off South Australia formed a subspecies, *M. e. eugenii*, which would be extinct except that it was the source of the New Zealand introductions (Warburton 2005a). Introduced populations also are present on North Island off Western Australia and Greenly and Boston Islands off South Australia (Morris et al. 2008b). Animals from Kawau Island are being used for reintroduction on the mainland of South Australia; reintroduction projects and control of the fox have also recently allowed the tammar wallaby to increase at several sites in Western Australia.

Macropus parryi (Whiptail Wallaby)

P. M. Johnson (2008b) listed a head and body length of 736–1,003 mm in males and 610–879 mm in females, a

Whiptail wallaby (*Macropus parryi*). Photo by Alexander F. Meyer at Lone Pine Koala Sanctuary, Brisbane.

tail length of 861–1,045 mm in males and 728–858 mm in females, and a weight of 14–26 kg in males and 7–15 kg in females. The upper parts are light gray in winter and brownish gray in summer; the underparts are white. The forehead and bases of the ears are dark brown. There is a white stripe on the upper lip, a light brown stripe down the neck to the shoulder, and a white hip stripe. The tail is long and slender and has a dark tip.

P. M. Johnson (2008b) wrote that the whiptail wallaby inhabits undulating or hilly country with open forest and a grassy understory. It is unusually diurnal for a macropodid, grazing mostly at dawn and continuing into the early morning, with increased periods of rest during the middle of the day, when it may seek patches of shade. It feeds primarily on grasses and other herbaceous plants, including ferns. Except during drought it seldom drinks, apparently obtaining sufficient water from its food and dew.

M. parryi is perhaps the most social of all marsupials. It was studied intensively in northeastern Queensland by Kaufmann (1974), who found a population density of 1 individual per 2 to 4 ha. In the study area were three loosely organized but discrete "mobs," each having a year-round membership of 30–50 individuals. The animals in any one mob did not keep together in a single group but usually split into varying subgroups of fewer than 10 individuals. All the animals in a particular mob, however, used the same home range, measuring 71–110 ha., which had little overlap with the home ranges of other mobs. Members of adjacent mobs mingled peaceably in the zones of overlap but kept mostly to their own undefended ranges. Within each mob, males established a dominance hierarchy through ritualized bouts of pawing, which did not cause injury. Larger males were dominant over smaller ones. The hierarchy functioned only to determine access to estrous females and ensured that fathering of offspring was limited to higher-ranking adult males. Courtship involved wild chasing of females by males. As they approached maturity, some subadult males left their natal mobs and joined other mobs, but no females were observed to do that.

P. M. Johnson (2008b) noted that a dominant male keeps others at bay through a ritual involving pulling up clumps of grass with his forepaws while directing his head toward his rival. Vocalizations include a soft cough, indicating fear or submission, a sound intermediate to a hiss and a growl, used by females as a defensive threat, and a soft clucking noise made by courting males. An alarmed individual thumps the ground with its hind feet, and all animals in its group hop away, taking a zigzag course that probably confuses a predator.

Kaufmann (1974) reported breeding to occur throughout the year in the wild in northeastern Queensland, and P. M. Johnson (1998) found females to give birth in all months of the year in a captive colony in the same area. Both studies, and other investigations cited therein, indicated that mean length of estrous cycle is 42 days and mean length of gestation period is about 38 days. There is no postpartum estrus, and mating does not occur again until the pouch young is 118–210 days old or unless the pouch young is prematurely lost; in the latter case, return to estrus and mating has been observed within 31 days. Embryonic diapause may follow mating. The young normally first leaves the pouch at about 240 days of age, permanently emerges at 256–295 days, and is weaned 104–215 days later. If the pouch young dies, the next young is born about a month later; otherwise the next young is born just after the first young vacates the pouch. Sexual maturity occurs at 18–29 months, but because of social factors males are prevented from mating until they are 2–3 years old. Weigl (2005) listed a female that lived in captivity for 9 years and 8 months.

The whiptail wallaby is common and may have benefited from limited logging that allowed more areas of grass (P. M. Johnson 2008b). In 1976, 45,259 *M. parryi* were killed for commercial purposes in Queensland (Poole 1978). The take subsequently fell to 14,954 in 1990 and only 357 in 2000 (Australian Department of Sustainability, Environment, Water, Population and Communities 2011), apparently because of a lack of demand.

Macropus giganteus (Eastern Gray Kangaroo)

Many millions of *M. giganteus* and *M. fuliginosus* (western gray kangaroo) have been killed in government regulated programs in the past several decades (see above generic discussion). Yet apparently no set of standard morphometrics, based on substantial series of fully mature animals, has been compiled and published. The oft-cited "head and body length" measurements listed by Coulson (2008a, 2008b) and Eldridge and Coulson (2015) seem actually to be the same as the

Eastern gray kangaroo (*Macropus giganteus*). Photo by Pavel German / Wildlife Images.

"head to tail" (or total length) measurements listed by Poole (1983a, 1983b, 1995a, 1995b). Those dimensions, along with tail length, were derived from animals as young as 1 year old, but Poole (1982) observed that young kangaroos grow rapidly for about 2 years before growth slows. Jarman (1989) noted that the larger kangaroos continue to grow well past the age of 1 year, perhaps throughout life, and that females attain sexual maturity (when 20–36 months old, see below) well before reaching their maximum weight.

With respect to adults of both *M. giganteus* and *M. fuliginosus*, Grzimek and Heinemann (1975) gave head and body length as 1,050–1,400 mm in males and 850–1,200 mm in females and tail length as approximately 950–1,000 mm in males and 750 mm in females. Based on live individuals of *M. giganteus* aged 2 years and above, Poole (1982) gave tail length as 750–1,000 mm for males and 700–840 mm for females. Specifically for adult *M. giganteus*, Coulson (2008b) listed weight as

19–85 kg for males and 17–42 kg for females. Poole (1982) wrote that the general coloration of *M. giganteus* varies from light silver gray to dark gray and may have brownish overtones on the back and flanks. The pelage along the mid-dorsal line is often dark, the dorsal surface of the ears may be darker than the body, and the face may be paler. The information for the remainder of this account was taken from Coulson (2008b), except for those specific sentences with a citation.

Natural habitats include sclerophyll forest, woodlands, shrubland, and heathland, and the species also occurs in highly modified areas, such as pine plantations. Both thick cover and open grassy patches for feeding are essential. Individuals typically emerge from cover 2–3 hours before sunset, graze through the night in the open, and return to cover 1–2 hours before sunrise. Clarke, Jones, and Jarman (1989) reported daily movements of 400–5,600 meters, averaging 2,280 for males and 1,467 for females. A "five-footed" gait while feeding and bipedal locomotion for speed are typical for the genus (see above). McCullough and Mc-Cullough (2000) clocked a sustained hopping pace of about 40 km per hour and noted that greater speeds can be attained for short distances. If alarmed, *M. giganteus* thumps the ground with its feet to alert others in the vicinity, then all hop rapidly to cover. Adult males may turn to fight or enter water and then use their height and strong forelimbs to drown any predator that follows. Grasses may comprise up to 99 percent of the diet, but seedlings of trees and shrubs, soft herbs, ferns, seeds, bark, and crops are also consumed.

At a time when kangaroo populations throughout Australia had seemingly reached a modern high, maps updated by Grigg (2002) indicated densities of over 40 individual *M. giganteus* per sq km in about 25,000 sq km of the sheep ranges of south-central Queensland, 20–40 per sq km in a surrounding area of about 100,000 sq km, and lower densities in other parts of the range of the species. Caughley, Sinclair, and Wilson (1977) estimated that in 496,000 sq km of New South Wales,

overall population densities were 3.18 individuals per sq km for the combined populations of *M. giganteus* and *M. fuliginosus*. For *M. giganteus* alone in an area of 29,500 sq km of agricultural and grazing lands in southeastern Queensland, Pople et al. (2006) calculated a density of approximately 10 per sq km. Other reported densities for *M. giganteus* across substantial areas have been about 1 per sq km in marginal habitat and 5 per sq km in favorable habitat (McCullough and McCullough 2000). However, in the wetter and more predictable parts of its range the species can at times achieve very high local densities, exceeding 3 per ha.

M. giganteus is generally considered a relatively sedentary species, not prone to long-distance movements and preferring to remain in a restricted home range even when food and water become scarce there and even when noticeable thunderstorms with heavy rains yield more favorable areas nearby (McCullough and McCullough 2000; Poole 1982). Individuals occupy stable home ranges that overlap those of other individuals. Ranges vary in size from 30–160 ha. in southern

Eastern gray kangaroo (*Macropus giganteus*). Photo by Gerhard Körtner.

Victoria to roughly 10 times larger in semiarid western New South Wales. At Yathong Nature Reserve in central New South Wales, McCullough and McCullough (2000) found home ranges to average about 12 sq km for males and 5 sq km for females.

M. giganteus may be the most social of the larger species of *Macropus* (Kirkpatrick 1967). In at least some areas there are stable social units known as "mobs," which contain up to 50 individuals that regularly meet each other. Within a mob, temporary subsets of 2–20 or more animals sometimes form at feeding sites; such groups are quite fluid, with individuals joining and leaving freely. The adult females of a mob associate frequently, but not exclusively, with their female kin. Adult males are more often alone, moving regularly among groups of females to check their stage in the estrous cycle. Kaufmann (1975) studied two mobs, each consisting of 20–25 animals and divided into subgroups averaging 3.7 individuals; he found a persistent group structure but no territorial defense or permanent association between males and females. McCullough and McCullough (2000) found average group size to be only two individuals in an area of marginal habitat and suggested that the social dynamics of *M. giganteus* there were more like those of *M. fuliginosus* and *M. rufus* than like those reported for *M. giganteus* in more favorable habitat. *M. giganteus* has several vocalizations and, like most other macropodids, thumps the ground with its hind feet in response to potential danger (see Bender 2006).

Adult males form dominance hierarchies to determine access to females, and females also form them (Grant 1973; Russell 1974b). Establishment of the male hierarchy may involve aggressive interaction but is mainly limited to ritualized display. A high-ranking male rubs his chest on a bush, then growls and clucks while standing on his toes and tail tip to emphasize his height, or approaches a subordinate with his back arched in an exaggerated walk. The subordinate gives a hoarse cough and retreats. If two males are evenly matched, they may lean back onto their tails and kick at each other and also use their muscular forearms and long claws in wrestling bouts. The dominant male is the most likely to mate when a female is in estrus.

M. giganteus is capable of breeding throughout the year (Collins 1973; Poole 1975; Russell 1974a). Breeding in the wild is mainly from September to March, a period that follows winter rainfall and coincides with the time of maximum growth of vegetation; in New South Wales most births occur from November to January (Poole 1973, 1975). The interval between successive births is approximately 1 year (Poole 1977). Calculated average length of estrous cycle is 45.6 days, average nondelayed gestation period is 36.4 days, and litter size usually is one, but twins have been recorded (Poole 1975; Poole and Catling 1974; Rose 1978). Embryonic diapause is known to occur in *M. giganteus* but appears to be rare, having been found in only about 5 percent of wild females with pouch young, which have been shot for study (Clark and Poole 1967; Poole 1973), and in only seven captive females during 10 years of observation (Poole and Catling 1974); in the latter animals, diapause followed matings that took place 160–209 days after the first young was born. Females with pouch young may become receptive as early as 112 days after the first young is born (Eldridge and Coulson 2015), but conception does not generally occur until there will be time before birth for the first young to vacate the pouch (Poole 1977). If the pouch young dies, the female returns to estrus within an average of 10.92 days (Poole and Catling 1974). The neonate weighs 800 mg but grows rapidly and begins to leave the pouch for short excursions after about 8 months. Total pouch life averages around 319–320 days (Croft 1989; Poole 1977). Lactation continues until the young is at least 18 months old (Poole 1975). Sexual maturity has been reported to come at an average age of 42.5 months for captive males, at 20–72 months for wild males, at an average of 21 months for captive females, and at 20–36 months for wild females (Poole 1973; Poole and Catling 1974). Longevity has been reported to be as long as 19.9 years in the wild and 25 years in captivity (Kirkpatrick 1965; Poole 1982).

The eastern gray kangaroo has disappeared from many densely settled localities, and there is concern for some remaining populations, such as that in southeastern South Australia (Poole 1977). The subspecies *M. giganteus tasmaniensis* of Tasmania was greatly reduced in numbers and distribution by the early twentieth century by excessive sport and commercial hunting and occurs in just a few small parts of its original range (Barker and Caughley 1990; Harper 1945). It is listed as endangered by the USDI. The IUCN (Munny, Menkhorst, and Winter 2008) cites an estimate of

Eastern gray kangaroo (*Macropus giganteus*). Photo by John Carnemolla/Shutterstock.com.

10,000–20,000 individuals for that subspecies and states that it is "threatened by loss of habitat through agricultural clearing," but no longer assigns a formal threatened classification. Otherwise, *M. giganteus* still occurs over most of its original range and generally is not thought to be in any immediate jeopardy (Calaby 1971; Calaby and Grigg 1989; Poole 1979). There is conflicting evidence as to whether the species was originally far more abundant than it is today or increased in numbers following European settlement and the introduction of livestock (see above generic account for *Macropus*). However, it is likely that *M. giganteus* has been able to expand into the arid zone of western New South Wales because of establishment of artificial water holes for livestock and the latter's cropping of long grass and browse, which stimulates growth of short grasses favored by the kangaroo (see Dawson, McTavish, and Ellis 2004; Dawson et al. 2006).

As with other large kangaroos, the survival and reproduction of *M. giganteus* are strongly influenced by rainfall. Population size may fluctuate dramatically in response to drought (Caughley, Grigg, and Smith

1985). Surveys from 1980 to 1982 indicated that there were then about 8,978,000 eastern gray kangaroos in Australia (Grigg et al. 1985). Mainly because of a severe drought, by 1984 numbers had fallen to 5,791,000. Later surveys reported by Fletcher et al. (1990) suggested that populations had recovered to approximately their 1980 levels. In 2001 the official estimate was 29,721,271 for those areas where commercial harvesting occurs, which includes most of the range of *M. giganteus*, but surveys of the same areas in 2010 indicated a figure of 11,409,235 (Australian Department of Sustainability, Environment, Water, Population and Communities 2011). Like the other large kangaroos, *M. giganteus* is subject to large-scale killing to reduce its impact on agriculture and to provide commercial products, particularly skins and meat (Robertshaw and Harden 1989). From 1987 to 2009, the annual number of eastern gray kangaroos killed ranged from 704,137 in 1997 to 1,810,426 in 2002 (Australian Department of Sustainability, Environment, Water, Population and Communities 2011). Additional discussion of the commercial harvest is provided above in the generic account for *Macropus*.

Macropus fuliginosus (Western Gray Kangaroo)

For information on head and body and tail lengths, see the above account of *M. giganteus*. Except as noted, the information for the remainder of this account was taken from Coulson (2008a). Adult weight is 18–72 kg for males and 17–39 kg for females. *M. fuliginosus* resembles *M. giganteus*, but its overall color is brown, its face is dark, and its ears are longer, black in color, and almost hairless behind.

M. fuliginosus occupies open woodland and shrub habitats that are generally more xeric than those of *M. giganteus*, with less free water (McCullough and McCullough 2000). Rainfall there occurs mostly in winter, and the annual average (303 mm) is half that for the range of *M. giganteus*. Where the two species are sympatric, *M. fuliginosus* shows more preference for shrubby cover but is also seen together with *M. giganteus* in the open. McCullough and McCullough (2000) found both species to be most active from about 1800 to 0600 hours, considering all seasons of the year, with a sharp drop-off toward mid-day. However, the western gray kangaroo was found to move significantly less than the eastern gray and red kangaroos; mean distance between successive locations of radio-tracked individuals of *M. fuliginosus* (made every third day on average) was 0.60 km for females and 0.93 km for males. *M. fuliginosus* is well adapted to dry summer conditions. During the hottest part of the day it retreats into shade and rests in deep hip holes scooped out at the base of trees. Both species are predominantly grazers, but *M. fuliginosus* appears to browse shrubs more frequently.

M. fuliginosus is very similar to *M. giganteus* (see account thereof) in most aspects of its biology, but there are some clear differences. Grigg (2002) indicated population densities of over 20 individuals per sq km in a small part of the sheep country of central New South Wales and 2–20 per sq km through about half of the total range of the species. Individuals occupy overlapping home ranges, varying from 40–70 ha. in the temperate southwest of Western Australia to 10 times larger in semiarid western New South Wales. Males usually have larger ranges than females. At Yathong Nature Reserve

Western gray kangaroo (*Macropus fuliginosus*). Photo by Hugh Lansdown / Shutterstock.com.

Western gray kangaroo (*Macropus fuliginosus*), mother and young. Photo by Pavel German / Wildlife Images.

in central New South Wales, McCullough and McCullough (2000) found home ranges to average about 9 sq km for males and 3 sq km for females. There is little evidence of persistent "mobs," as in *M. giganteus*; instead *M. fuliginosus* forms labile groups of 2–25 individuals. Groups usually contain both sexes during the mating season, which is concentrated in spring and early summer, while single-sex groups are more common at other times. MacFarlane and Coulson (2009) reported that adult males, in particular, seem to have a need to form groups of their own.

M. fuliginosus is capable of breeding throughout the year (Poole 1975), though Mayberry et al. (2010) reported that in the vicinity of Perth, Western Australia, nearly all young of the subspecies *M. f. ocydromus* are conceived from November to February. The interval between successive births is approximately 1 year (Poole 1977). Calculated average length of estrous cycle is 34.9 days, average gestation period is 30.6 days, and litter size usually is one (Poole 1975; Poole and Catling 1974). Embryonic diapause is not known to occur in *M.*

fuliginosus. Average age at first emergence from the pouch is 298.4 days and at final departure is 323.1 days; as in *M. giganteus*, lactation usually exceeds 18 months, longer than in any other marsupials (Poole 1975). If the pouch young dies, the female returns to estrus within an average of 8.25 days (Poole and Catling 1974). Sexual maturity in captivity has been reported to come at an average age of 29.0 months for males and 22 months for females (Poole 1973; Poole and Catling 1974), though Eldridge and Coulson (2015) cited respective figures of 20 months and 14 months. A captive female lived to an age of 23 years and 3 months (Weigl 2005).

M. fuliginosus fuliginosus of Kangaroo Island off South Australia, confirmed as a highly distinctive subspecies by Poole, Carpenter, and Simms (1990), may be subject to loss of habitat as human activity increases in its limited range (Poole 1976, 1979). Otherwise, the western gray kangaroo is thought to be abundant and even increasing in numbers (Burbidge et al. 2008d), though some evidence suggests the species originally was more numerous than it is today (see above generic account for *Macropus*). The same patterns of population fluctuation described above for *M. giganteus* apply to *M. fuliginosus*. The 1980–1982 surveys reported by Grigg et al. (1985) indicated about 1,774,000 western gray kangaroos in Australia. Numbers fell to 1,162,000 by 1984 because of drought, but subsequently recovered (Fletcher et al. 1990). In 2001 the official estimate was 3,424,992 for those areas where commercial harvesting occurs, and surveys of the same areas in 2010 indicated a figure of 2,790,358. From 1987 to 2009, the annual number of western gray kangaroos killed in government-sanctioned culling and commercial operations ranged from 107,154 in 1989 to 353,650 in 2002 (Australian Department of Sustainability, Environment, Water, Population and Communities 2011). Additional discussion of population history and the commercial harvest is provided above in the generic account for *Macropus*.

Macropus bernardus (Black Wallaroo)

Except as noted, the information for the account of this species was taken from Telfer and Calaby (2008). Four males had a head and body length of 595–725 mm, a tail length of 545–640 mm, and a weight of 19–22 kg. Respective measurements for one female were 646 mm, 575 mm, and 13 kg, but it was noted that females probably grow larger. There is striking sexual dimorphism in color. Males are dark sooty brown to black, with black feet and tail tip. Females are pale gray to gray-brown, with dark brown feet and tail tip. There are no face markings, the fur is long and shaggy, the ears are relatively short, and the body is stocky.

The black wallaroo is restricted to the steep, rocky escarpments and tops of the deeply dissected sandstone plateau and outliers of western Arnhem Land. It uses habitats dominated by spinifex grassland, sandstone heath, eucalypt woodland, and patches of rain-

Black wallaroo (*Macropus bernardus*). Photo by Stephen Zozaya.

forest. It may come down from the escarpments to drink at springs or waterholes or to graze close to the base of the higher country. It is extremely wary and, when disturbed, rapidly ascends the escarpment until out of sight. It is largely nocturnal and is usually not active in the daytime except on overcast days during the wet season. Daylight hours are generally spent resting in caves or in the shade of large rocks and trees. Its diet apparently includes grass and the leaves, fruits, and flowers of a variety of trees and shrubs. Adults are usually solitary. Large, furred young tend to be seen in pouches in the middle of the dry season, June–September. Weigl (2005) listed a male that was born in the wild about January 1993, brought into captivity in July 1993, and still living in October 2004.

M. bernardus was once thought to be extinct. It had not been collected since 1914 but was rediscovered in 1969 (Goodwin and Goodwin 1973; Parker 1971). Aerial surveys now suggest it is abundant but its distribution is very limited, thus increasing susceptibility to habitat disruption, especially changing fire conditions. It is designated as near threatened by the IUCN (Woinarski 2008c). Woinarski, Burbidge, and Harrison (2014) estimated the total population at fewer than 10,000 individuals in an overall range of about 30,000 sq km.

Macropus robustus (Common Wallaroo, or Euro)

As with *M. giganteus* (see account thereof), there appears to be no complete and published set of standard body measurements for *M. robustus* based on substantial series of fully mature individuals. The figures listed by Clancy and Croft (2008) and Eldridge and Coulson (2015) date back to Poole (1983c) and seem to be derived in part from animals that had not reached adult size. Grzimek and Heinemann (1975) gave head and body length as 1,000–1,400 mm in adult males and 750–1,000 mm in females and tail length as 800–900 mm in males and 600–700 mm in females. However, they treated *M. antilopinus* as a subspecies of *M. robustus*, not a distinct species as considered here, and thus their measurements for the latter include those of the former. The dimensions of *M. antilopinus* listed by Ritchie (2008) suggest a slightly smaller median value, though Russell (1974a) indicated that species to be the largest wallaroo and that males might weigh just as much as male gray and red kangaroos. Maximum

Common wallaroo, or euro (*Macropus robustus*). Photo by Alexander F. Meyer at Featherdale Wildlife Park, Sydney.

weights of *M. robustus*, as listed by Clancy and Croft (2008), are 60 kg for males and 28 kg for females; except as noted, the information for the remainder of this account was taken from those authorities.

Coloration varies widely in *M. robustus*. In the east, males are usually dark gray to almost black, females generally bluish gray above and paler below. Elsewhere, males are light gray to red or almost black, females generally light gray to light brown. The fur is longer and shaggier than that of *M. giganteus* and *M. rufus*, the build is much stockier, and the rhinarium is completely bare.

The common wallaroo is found over a vast region in varied habitats that usually feature steep escarpments, rocky hills, or stony rises. Home ranges may extend to lower slopes and plains but then are associated with drainage lines or dense scrub. The species does well in arid country, relying just on shade from trees and scrub for cover and on waterholes, soaks, or stock-watering points. In contrast to the relatively mobile *M. rufus*, but like *M. giganteus*, *M. robustus* is sedentary, remaining in a restricted home range even when food and water become scarce. Over a period of years in a semiarid part of New South Wales, no individual was found to move

Common wallaroo, or euro (*Macropus robustus*). Photo by Gerhard Körtner.

more than 7 km (Croft 1991a). *M. robustus* is assisted in such a way of life by its ability to withstand dehydration better than the red and gray kangaroos, to survive on food containing less nitrogen, and to minimize water loss by sheltering in caves during the daytime (Russell 1974a). Individuals can live up to 3 months with no access to free water. The diet consists of more than 90 percent grass, except during relatively wet periods.

Populations fluctuate markedly, especially in arid country, in response to rainfall-driven changes in food resources. Densities have reached over 80 individuals per sq km in some areas, but averages of 10–20 per sq km are more typical in good habitat. Home ranges are relatively small and stable over time. In a semiarid part of New South Wales, weekly ranges were 116–283 ha. in summer and winter (Croft 1991a). In an arid part of that province, weekly winter ranges averaged 77.2 ha. for males and 30.5 ha. for females, but weekly summer ranges averaged 30.2 ha. for males and 27.6 ha. for fe-

males; on a yearly basis males ranged over an area approximately three times the size of that used by females (Clancy and Croft 1990). Average home range sizes over 2 years at an arid site were 309 ha. for males and 150 ha. for females, then 215–295 ha. for males and 75–85 ha. for females over the subsequent 3 years; ranges overlap extensively within and between sexes (Eldridge and Coulson 2015).

A young animal may remain with its mother for some months after weaning, but other animals seen together usually represent only chance aggregations at some favorable site (Dawson 1995; Kirkpatrick 1968). Those groups are small and highly labile, though mothers and independent daughters may loosely associate in their permanently overlapping home ranges; such association has been observed to last for up to 3.5 years. Young males disperse from their natal ranges, with movements of up to 18 km reported (Eldridge and Coulson 2015).

The mating system is polygamous. The largest male consorts with an estrous female for about a week, mating with her and defending her from other males that may follow. *M. robustus* is an opportunistic breeder, with both sexes remaining potentially fertile throughout the year; young are produced continuously when there is adequate rainfall and forage, but breeding may cease altogether during prolonged drought (Collins 1973; Newsome 1975; Tyndale-Biscoe 1973). The estrous cycle averages 33.6 days for the eastern subspecies *M. r. robustus* and 36.7 days for *M. r. erubescens* of western and central Australia, respective gestation periods are 32.7 and 33.4 days, and litter size usually is one (Dawson 1995). There is a postpartum estrus before the pouch young is 2 days old and embryonic diapause (see above generic account). The mother gives birth again at an average of 30.8 days after any premature loss or experimental removal of the pouch young; if the pouch young survives, it normally vacates the pouch when 8–9 months old, and the second young is then born (Kirkpatrick 1968). Average age at weaning is 409 days (Dawson 1995) but may be several months above that. Males are capable of reproduction when about 2 years old but normally do not attain a significant breeding role within a wild population until reaching maximum size at around 6–7 years of age; females have bred at just over 14 months in captivity but may not become fully mature until at least 21 months old (Dawson 1995). A wild individual is known to have reached an age of 18.6 years (Kirkpatrick 1965), and a captive lived for 22 years (Weigl 2005).

According to Ellis et al. (2008a), the subspecies *M. robustus isabellinus*, restricted to Barrow Island off Western Australia, has been found to suffer from anemia and poor condition, which may be related to nutritional stress. Woinarski, Burbidge, and Harrison (2014) cited total population estimates of 1,800 individuals (in 1991) and 528–914 (in 2003) for *isabellinus* and, using IUCN criteria, designated it vulnerable. Aside from that subspecies, the IUCN considers *M. robustus* to be widespread and relatively common, to have a stable population trend, and to face no major threats (Ellis et al. 2008a). The species, like the red and gray kangaroos, is hunted commercially for skins and meat pursuant to government management plans that are based on regular population surveys. In 2001 *M. robustus* was estimated to number 6,849,250 individuals in those areas where commercial harvesting occurs; surveys of the same areas in 2010 indicated a figure of 2,416,285. From 1984 to 2009, the recorded annual kill ranged from a low of 74,608 in 1987 to a high of 347,914 in 2003 (Australian Department of Sustainability, Environment, Water, Population and Communities 2011). Additional discussion of the commercial harvest is provided above in the generic account for *Macropus*.

Macropus antilopinus (Antilopine Wallaroo)

Except as noted, the account of this species was taken from Ritchie (2008, 2010). Head and body length is 830–1,200 mm in males and 733–935 mm in females, tail length is 745–960 mm in males and 664–813 mm in females, and weight is 18.6–51.0 kg in males and 14.0–24.5 kg in females. Males are reddish tan above and cream below, with the tips of the feet dark black and no distinctive facial marks. Females usually have a pale gray head and forequarters, but may be all gray, and have a distinctive white fringing of the ears. Both sexes have a large, "swollen," black nose, very pronounced in adult males. Longer limbed and more slender than the other wallaroos, *M. antilopinus* resembles the gray and red kangaroos in general appearance and behavior.

The antilopine wallaroo occurs in various kinds of tropical savannah vegetation, from scattered to dense eucalypt woodland to tall open forest, with grass-dominated understorys and typically in flat to gently undulating terrain. In the dry season when temperature is high, this species is inactive during the day, resting in scrapes under trees, bushes, or boulders, usually close to water. It begins grazing in late afternoon and continues through the night, returning to cover in early morning. On overcast or wet days during the rainy season, it may be active at any time. The diet consists largely of grasses but also includes some forbs.

A study at 30 sites in Queensland, differing in vegetation, rainfall, and temperature, found an average population density of 4.5 individuals per square km. A study at two sites in Northern Territory (Croft 1987) found densities of 8.8–30.9 per sq km and home range sizes of about 76 and 102 ha. for two adult males and 14 ha. for a female. *M. antilopinus* is considered gregarious, with an average group size of about three

Antilopine wallaroo (*Macropus antilopinus*). Photo by ChameleonsEye/Shutterstock.com.

individuals and groups of three to eight being common, though medium- and large-sized males are often solitary. Aggregations of more than 20 form in response to predators or localized patches of fresh grass. Groups tend to include both sexes in the wet season but to be all-male or all-female in the dry season.

Births have been recorded in all months but most occur toward the end of the wet season (February–April). Young vacate the pouch at the start of the next wet season, when grasses are most abundant and nutritious. Observations of a male and two females in captivity indicated a long estrous cycle, averaging 40 days, and an average gestation period of 34 days. Females did not appear to have a postpartum estrus. Available evidence suggests that there is no embryonic diapause. According to Dawson (1995), the young weighs 0.66 gram at birth, first leaves the pouch at an estimated age of 210 days, permanently exits at 269 days, and is weaned at about 380 days; no data on sexual maturity are available. Wild individuals have been known to live up to 10 years, and Weigl (2005) listed a captive-born female that was still living at an age of 19 years and 9 months.

The antilopine wallaroo is restricted to a relatively narrow band of suitable savannah habitat at the northern extremity of the Australian continent. If global temperatures increase to the extent predicted by some authorities, the species could undergo a severe range contraction, with possible extinction by 2070. For now, however, the IUCN sees no major threats; populations are decreasing, but the decline is minor and localized (Woinarski, Ritchie, and Winter 2008).

Macropus rufus (Red Kangaroo)

Croft and Clancy (2008) listed head and body length as 935–1,400 mm in males and 745–1,100 mm in fe-

Red kangaroo (*Macropus rufus*). Photo by Alexander F. Meyer at Chapultapec Zoo.

males, tail length as 710–1,000 mm in males and 645–900 mm in females, and weight as 22–92 kg (averaging 55 kg) in males (but see discussion of growth, below) and 17–39 (averaging 25 kg) in females. Grzimek and Heinemann (1975) indicated a maximum of 1,600 mm for head and body length in males. The upper parts are usually rich reddish brown in males and bluish gray in females. However, both sexes can be either color or an intermediate shade, and the proportion of animals of each color varies from place to place (Dawson 1995). Adults are distinguished by white to light gray underparts, a black and white patch at the side of the muzzle, a broad white stripe from the corner of the mouth to the base of the ear, dark feet, and a pale tail tip (Croft and Clancy 2008; Dawson 1995).

The red kangaroo dwells on grass-covered plains, preferring more open habitat than that of the gray kangaroos, but needs places with denser vegetation and trees for shelter and shade. It sometimes becomes relatively mobile, covering considerable ground to seek food and adjusting, or even completely changing, its range in response to environmental fluctuation. It once was believed to be truly nomadic, following the rains and showing only temporary attachment to any one site. Studies over the last several decades have indicated that long-distance movement usually involves temporary range modification, extensive positional shifts within a large but stable home range, or dispersal of young males. Most individuals show fidelity to established ranges and may not leave even during harsh dry periods or when thunderstorms produce more favorable conditions nearby. Animals sometimes do move off to congregate at sites with abundant forage, but most eventually return to their original ranges. A number of cases suggest that some individuals are indeed nomadic, though more data are needed to settle the entire issue (Dawson 1995; McCullough and McCullough 2000).

According to Croft and Clancy (2008), the red kangaroo's preferred diet is short green grass in warmer

Red kangaroo (*Macropus rufus*). Photo by Gerhard Körtner.

months and germinating winter forbs. It forages from dusk to dawn but may be more diurnal in cooler months, with an extended rest period in the middle of the night. It may make opportunistic forays of around 50 km to feed on vegetation arising from local rains. Adult females typically return to their core ranges after such travel, but some animals, especially younger males, go on to another storm site, sometimes moving over 200 km over several months. In summer the red kangaroo drinks only once every 4–10 days, having adaptations that allow it to move 10–20 km from water and to cope with blistering heat and sparse shade. Its coat is highly reflective to heat, its camel-like nostrils are narrow slits, and it tucks its long tail between its legs to assume a spheroid shape of minimal surface area to reduce convective heat gain.

Dawson (1995) observed that the red kangaroo is without equal among mammals in the dissipation of heat. It sweats during extreme exertion, thereby removing heat from the blood, but stops as soon as exercise ceases, thus conserving water. It pants while resting, greatly increasing air flow over, and blood flow to, the upper airways for evaporative heat loss. Panting also produces excess saliva that is wiped or licked onto the forearms, which have a dense superficial network of fine blood vessels that bring heat to the surface. Dawson (1995) added that the red kangaroo hops comfortably at 20–25 km per hour, can maintain 40 km per hour for a couple of kilometers, has been clocked at up to 50 km per hour in the wild, and has been reported to reach speeds of 65–70 km per hour in emergencies.

The greatest population densities, currently 10–20 per sq km, occur in the sheep rangelands of New South Wales and Queensland, though temporary aggregations can give the impression of much higher numbers (Croft and Clancy 2008). Maps published by Grigg (2002) at a time when kangaroo populations throughout Australia had seemingly reached a modern high indicated densities of over 20 per sq km in about 50,000 sq km of northwestern New South Wales. Other reported long-term densities vary from about 1 to 10 per sq km (Main and Yadav 1971; McCullough and McCullough 2000). In a semiarid part of New South Wales, Croft (1991b) found weekly home ranges of about 2.59–5.16 sq km, but during a prolonged dry spell there was a general movement to a more favorable area 20–30 km away. At Yathong Nature Reserve in central New South Wales, McCullough and McCullough (2000) found red kangaroo home ranges to overlap extensively and to average about 20 sq km for males and 10 sq km for females—significantly larger than the ranges of gray kangaroos. According to Dawson (1995), daily home ranges at Fowlers Gap in western New South Wales averaged around 1.5 sq km, and weekly ranges were two or three times larger; during the hot summer individuals usually left their ranges every 3–10 days to go to water. Otherwise there was little seasonal difference in area used. Ranges were about the same size for both sexes and were not exclusive. However, in Western Australia males were found to use ranges twice as large as those of females, and the ranges of individuals of each sex did not overlap those of animals of the same sex.

M. rufus appears to be less social than the other large macropodids, but it does commonly occur in organized groups or "mobs" that appear to be more than just random, temporary assemblies (Croft 1989; Dawson 1995; McCullough and McCullough 2000; Russell 1974a, 1974b). The typical number of individuals seen together is around 2–5, though more may be part of a social grouping in a given area. Those groups are based primarily on related adult females and are in a different category from the large but unorganized aggregations of kangaroos that sometimes collect at a favorable feeding site. Newsome (1971b) observed a gathering of 1,500 M. rufus in central Australia and stated that aggregations of 50–200 were common during periods of drought. Such assemblies do not have permanent bonds, except those between mothers and dependent young, but they do facilitate contact between potential mates and between males that may form a dominance hierarchy based on size (Croft and Clancy 2008; McCullough and McCullough 2000). The largest male in an area consorts with and defends any estrous female in his home range. He may gain temporary control of several females, along with their young, but there usually is no permanent association between adults of opposite sexes (Russell 1974b, 1979). As in other species of Macropus, there appears to be no territorial defense, but the male may fight fiercely with other males that challenge his possession of the females, and there may also be agonistic behavior relative to competition for food and resting sites. Croft (1980) stated that in such situations male M. rufus might suddenly rear up and begin hitting each other from an upright position (boxing) and also kick with the hind feet. Collins (1973) reported that captive M. rufus, as well as other species of Macropus, could be kept in groups but that the presence of more than one adult male would lead to trouble.

The red kangaroo is an opportunistic breeder, with both sexes remaining potentially fertile throughout the year. Young are produced continuously when there is adequate rainfall and forage, but breeding may cease altogether during prolonged drought (Caughley, Grigg, and Smith 1985; Newsome 1975; Tyndale-Biscoe 1973). Average length of estrous cycle is 34.8 days and average gestation period is 33.2 days (Dawson 1995). Litter size usually is one, but twins have been recorded (Collins 1973).

Embryonic diapause is perhaps best known in M. rufus (Dawson 1995; Russell 1974a; Tyndale-Biscoe 1973). The female has a postpartum estrus and mates within 2 days after giving birth. The resultant embryo develops into a blastocyst of approximately 85 cells and then becomes dormant, provided the original young still is suckling in the pouch. When the pouch young is about 204 days old, or at any time sooner if the pouch young dies or is removed, the embryo resumes development. In 31 days, and within a day after the first

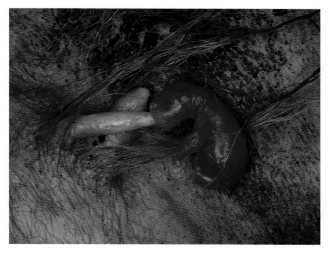

Red kangaroo (*Macropus rufus*), newborn. Photo by Gerhard Körtner.

Red kangaroo (*Macropus rufus*), mother and young. Photo by Vladislav T. Jirousek / Shutterstock.com.

young permanently leaves the pouch, a second birth occurs. Immediately thereafter the female again becomes receptive, and another mating can take place. If the pouch young is lost before it is 204 days old, the female becomes receptive 33–35 days after such loss. As a result of this process, a female red kangaroo can produce one young approximately every 240 days as long as favorable conditions hold. Under such conditions most adult females examined in the field are found to have one quiescent embryo, one pouch young, and one accompanying young outside of the pouch. When food supplies are low, a pouch young may die from inadequate lactation, but the diapausing embryo will then resume growth and may reach term by the time conditions are better. If there has been no environmental improvement, the second young may also die, but another fertilization would have taken place and a third embryo would be on its way. In severe droughts, which are common in much of Australia, females become anestrous and breeding ceases.

The red kangaroo, largest of marsupials, weighs only 0.75 grams at birth (Sharman and Pilton 1964). At that time it has a large tongue and well-developed nostrils, forelimbs, and digits but is embryonic in other external features. Like other newborn marsupials, it scrambles from the birth canal to the pouch without

the assistance of the mother and grasps one of the mammae. It first releases the nipple at about 70 days, first protrudes its head from the pouch at 150 days, temporarily emerges at 190 days, and permanently vacates the pouch at 235 days. It then continues to suckle by placing its head in the pouch, and it is finally weaned at about 1 year. Survival of young to weaning is dependent on rainfall and reportedly has varied from 20 to 85 percent, though Bilton and Croft (2004) found females at Fowlers Gap to wean an average of 3.7 young during their reproductive life.

According to Dawson (1995), kangaroos, especially males, grow through most of their lives. Male *M. rufus* usually begin to produce sperm when they are about 3 years old and weigh 20–25 kg, but in the wild they are then still too small to compete for a mate. At 5 years and 40 kg, males may be able to breed successfully in areas where populations are low. Elsewhere, they might have to live another 5–7 years and grow to 70 kg before making a significant contribution to breeding. Females may reach sexual maturity at an age of 15–20 months under good conditions but sometimes take much longer, especially during periods of drought. Once a red kangaroo is fully mature, it can cope well with its harsh environment. Much of a natural breeding population is well over 12 years old. Bailey (1992) reported a male that lived at least 27 years in the wild. Weigl (2005) listed a female that lived 25 years in captivity.

The red kangaroo still occupies most of its original range and is considered abundant, though abundance is highly variable depending on climatic conditions from year to year (Croft and Clancy 2008). The IUCN (Ellis et al. 2008b) considers populations to be stable but higher where the dingo (*Canis familiaris dingo*), the red kangaroo's main natural predator, has been eliminated or suppressed. Population densities of *M. rufus* have been found to be much higher inside, as compared to outside, a dingo-proof fence extending 5,531 km through southeastern Australia, built to protect sheep rangelands (Fillios et al. 2010; Newsome et al. 2001). *M. rufus* seems to have benefited from livestock grazing and creation of artificial water sources, though some evidence suggests the species originally was far more numerous than it is today (see above generic account for *Macropus*). The 1980–1982 surveys reported by Grigg et al. (1985) indicated about 8,351,000 red kangaroos in Australia. Numbers fell to 6,330,000 by 1984 because

Red kangaroo (*Macropus rufus*). Photo by Rafael Ramirez Lee / Shutterstock.com.

of drought, but subsequently recovered (Fletcher et al. 1990). In 2001 the official estimate was 17,434,513 for those areas where commercial harvesting occurs, and surveys of the same areas in 2010, after further drought, indicated a figure of 8,542,148. From 1984 to 2009, the annual number of red kangaroos killed in government-sanctioned culling and commercial operations ranged from 605,630 in 1984 to 1,500,588 in 2002 (Australian Department of Sustainability, Environment, Water, Population and Communities 2011). Additional discussion of the commercial harvest is provided above in the generic account for *Macropus*.

Much of the range of the red kangaroo is overlapped by herds of domestic sheep. The fate of the former may hinge on the degree to which sheep compete for resources and degrade habitat, as well as on the extent that landholders perceive the kangaroo as a competitor. Newsome (1975) warned that simultaneous competition with sheep, cattle, or rabbits, especially during drought, could eventually bring about conditions

under which *M. rufus* would become rare. He also observed that in northwestern Australia *M. rufus* initially increased in numbers following introduction of sheep but then lost in competition with the sheep and *M. robustus*. The latter species had invaded the plains because overgrazing by sheep had eliminated the luxuriant grass cover and allowed the spread of highland vegetation favored by *M. robustus*.

Edwards (1989) indicated that the ecological relationship between kangaroos and sheep is complex and not fully understood but that competition does occur under certain circumstances. Pople and Grigg (1999) reviewed various studies that had yielded diverse views on the amount of competition and the amount of harm caused and concluded that much more experimental research was needed to resolve the issue. However, there was no question that most landholders considered kangaroos detrimental and favored their reduction. Grigg (2002) doubted that such reduction would benefit sheep production. He observed that

381

sheep had degraded their ranges and that major land management changes were needed to avoid further desertification. He supported expanding the market for kangaroo products, thereby providing an economic incentive for landholders to reduce sheep numbers and encourage kangaroo populations. Munn et al. (2009) presented data showing that *M. rufus* has a field energy requirement of 35 percent and a water turnover rate of only 13 percent of that of a domestic sheep. Therefore, removal of kangaroos would not markedly improve rangeland capacity for domestic stock, and, given the low resource requirements of kangaroos, their use in consumptive and nonconsumptive enterprises would benefit the rangelands. Croft (2005), citing Newsome (1975) and other studies, suggested that grazing pressure from sheep is unsustainable. That factor, combined in unpropitious circumstance with continued intensive harvest and drought, could reduce *M. rufus* to a level from which it could not recover. Jonzén et al. (2010) calculated that if average rainfall drops by more than about 10 percent, any level of harvesting may be unsustainable.

Order Afrosoricida

Tenrecs and Golden Moles

This order of 2 families, 20 genera, and 58 species occurs naturally in Africa, from Guinea to Somalia and south to the Cape of Good Hope, and in Madagascar. The name "Afrosoricida" was proposed by Stanhope et al. (1998b) in conjunction with their molecular analysis indicating that families Tenrecidae (tenrecs) and Chrysochloridae (golden moles) form a related group that in turn is part of the placental clade Afrotheria (which also comprises orders Macroscelidea, Tubulidentata, Bibymalagasia, Hyracoidea, Proboscidea, and Sirenia; see account of class Mammalia). Bronner and Jenkins (2005) "reluctantly" accepted the name "Af-

rosoricida," observing that, while it had come to be used widely for the tenrec-golden mole group, some authorities argued that a more proper name is "Tenrecoidea" or "Tenrecomorpha." Feldhamer et al. (2007), Hedges (2011), Kuntner, May-Collado, and Agnarsson (2010), and Wilson (2009) continued to use Afrosoricida, though Arnason et al. (2008) considered it technically invalid, and Asher and Helgen (2010, 2011) argued in favor of Tenrecoidea.

Prior to the work of Stanhope et al. (1998b), Tenrecidae and Chrysochloridae had commonly been put in an order called "Insectivora," together with various other kinds of small fossil and living mammals with primitive characters, especially groups that could not

Lesser hedgehog tenrec (*Echinops telfairi*). Photo by Eric Isselee / Shutterstock.com.

readily be assigned to some other order. A common arrangement had been to divide Insectivora into two suborders: Lipotyphla, with the living families Tenrecidae, Chrysochloridae, Erinaceidae, Nesophontidae, Solenodontidae, Soricidae, and Talpidae; and Menotyphla, with the families Macroscelididae and Tupaiidae. As explained by Yates (1984), however, the two menotyphlan families eventually were placed in separate orders (Macroscelidea and Scandentia, respectively) and not considered immediately related to the seven lipotyphlan families. Therefore, the order Insectivora, sometimes with the name "Lipotyphla," was usually restricted to those seven families. A modified version of this arrangement was suggested by Novacek (1986), who ranked the Insectivora as a superorder, with Lipotyphla being the only living order and containing two suborders: Erinaceomorpha for Erinaceidae, and Soricomorpha for the other six families. Subsequently, MacPhee and Novacek (1993) designated a third suborder, Chrysochloromorpha, for the family Chrysochloridae.

Even before the molecular work of Stanhope et al. (1998b), some authorities recognized the affinity of Tenrecidae and Chrysochloridae, based on morphological assessment. McDowell (1958) placed the two in their own superfamily, for which he used the name "Tenrecoidea." Van Valen (1967) put Tenrecidae, Chrysochloridae, and Solenodontidae in a suborder, Zalambdodonta, of an otherwise extinct order, Deltatheridia, while retaining the other four lipotyphlan families in the order Insectivora. However, he acknowledged that, although the relationship of Tenrecidae and Chrysochloridae was "reasonably well established," it was "very possible" that Solenodontidae would be removed to the group comprising the lipotyphlan families. Eisenberg (1981) did realign Solenodontidae with the lipotyphlans, while uniting Tenrecidae and Chrysochloridae in a separate order, Tenrecomorpha, and suggesting that both families might warrant ordinal rank.

More recently, a detailed mitogenomic study by Arnason et al. (2008) supported placing Tenrecidae and Chrysochloridae in separate orders, with the respective names "Tenrecidea" and "Chrysochloridea." The study also indicated that Chrysochloridea is more closely related to Macroscelidea than to Tenrecidea; such an arrangement would effectively eliminate the

Yellow golden mole (*Calcochloris obtusirostris*). Photo by Johnny Wilson.

basis for order Afrosoricida. For now, however, Bronner and Jenkins (2005) and Wilson (2009) are followed here with respect to the name, ordinal status, and content of Afrosoricida, and in recognizing the suborders Tenrecomorpha for Tenrecidae and Chrysochloridea for Chrysochloridae.

Stanhope et al. (1998b) observed that, while their study was molecular, there are morphological characters supporting common ancestry of Tenrecidae and Chrysochloridae and hence proposal of order Afrosoricida. Those characters include zalambdodont cheekteeth (also in Solenodontidae), a basisphenoid contribution to the bulla (also in Erinaceidae), and orthomesometrial implantation of the blastocyst; data from albumin immunodiffusion comparisons also conjoin tenrecs and golden moles in a distinct clade, separate from all other insectivores.

Seiffert (2010) stated that the molars of all extant afrosoricids have the zalambdomorph occlusal pattern, a condition in which upper molar metacones and lower molar talonids are greatly reduced or absent altogether, but that the condition apparently evolved independently in Tenrecidae and Chrysochloridae. McDowell (1958) noted that zalambdodonty in Solenodontidae differs in detail and seems to represent parallel evolution rather than true relationship.

McDowell (1958:202–203) provided the following technical diagnosis of his superfamily Tenrecoidea, which, with regard to living members, is equivalent to order Afrosoricida of Stanhope et al. (1998b):

basisphenoid with well-developed tympanic wing, in contact with tympanic; epitympanic recess large to very large; squamosal making extensive contribution to roof of tympanic cavity; pyriform fenestra absent or incipient, the alisphenoid forming a large portion of the tympanic roof and providing a large bony area of origin for the tensor tympani; tympanohyal ventral in plane to the posterior limb of the tympanic; sinus canal foramen opening within the foramen lacerum anterius; inferior ramus of stapedial artery passing out of ear through a foramen in the alisphenoid, or a foramen that is converted into a sharply incised notch; glenoid capitular facet simple; nuchal crest, when produced (e.g., *Tenrec, Potamogale*), extended upward; maxilla produced backward along the lateral border of lamina defining the interpterygoid (choanal) fossa; no postpalatine torus; hamular process well developed (except in *Geogale*); levator labii superioris proprius arising within temporal fossa and partially covered by the temporalis; well-developed anal glands, but no lateral glands; allantoic sac large to very large; male reproductive tract without conus abdominalis. Milk dentition well developed, persisting nearly or quite until the animal reaches full size; premolars gradually becoming more complex from front to back of series.

Asher, Novacek, and Geisler (2003) used a number of McDowell's (1958) characters, together with much other morphological and fossil evidence, combined with molecular data, to demonstrate strong support for order Afrosoricida within Afrotheria. Tenrecidae and Chrysochloridae also share various external characters generally found in mammals formerly placed in order Insectivora—small size, narrow snout, small eyes and ears (*Geogale* is an exception), nonopposable digits, flattened head, a common urogenital opening or cloaca—but the two families differ strikingly in appearance. Seiffert (2010) indicated that they diverged in the late Cretaceous, around 70 million years ago, but that they do form a clade and can be united in order Afrosoricida, which forms a sister group to all other afrotherians. He reported afrosoricid fossils, apparently representing lineages that diverged prior to the tenrecid-chrysochlorid split, from the late Paleocene, late Eocene, and early Oligocene of North Africa.

AFROSORICIDA; Family TENRECIDAE

Tenrecs, or Madagascar "Hedgehogs"

This family of 10 Recent genera and 37 species is found in western and central equatorial Africa and Madagascar (one genus, *Tenrec*, has been introduced to several islands in the western Indian Ocean). With one exception, the sequence of genera presented here follows that suggested by the molecular and morphological analyses of Olson and Goodman (2003), who recognized four subfamilies: Potamogalinae, with the genera *Potamogale* and *Micropotamogale*; Geogalinae, with *Geogale*; Tenrecinae, with *Tenrec, Hemicentetes, Echinops,* and *Setifer*; and Oryzorictinae, with *Oryzorictes, Microgale,* and *Limnogale*. The same four subfamilies were listed by Bronner and Jenkins (2005), McKenna and Bell (1997), and Poux et al. (2008). However, using DNA assessment, Poux et al. (2008), as well as Kuntner, May-Collado, and Agnarsson (2010), found *Geogale* to have much closer affinity to Oryzorictinae than to Tenrecinae. That position is followed in the generic sequence herein. Potamogalinae sometimes has been accorded familial rank, as by Eisenberg (1981), but most authorities have treated it as a subfamily, especially since description in 1954 of *Micropotamogale*, a genus that seems intermediate to *Potamogale* and the other tenrecids (Corbet 1977a). A recent study of hind limb characters did support family level distinction for Potamogalinae (Salton and Szalay 2004) but that position was refuted by Asher and Hofreiter (2006).

An alternative systematic arrangement of Tenrecidae, in which *Limnogale* is grouped with Potamogalinae in a "semi-aquatic" clade, has been suggested over the years, most recently (with some question) by Asher (1999, 2002). However, subsequent molecular and morphological analyses have indicated that association to be the result of convergence and that the true phylogenetic affinity of *Limnogale* is with *Microgale* (Asher and Hofreiter 2006; Olson and Goodman 2003; Salton and Szalay 2004).

Head and body length is 40 to approximately 400 mm. In Tenrecinae there is only a rudimentary tail; in the other subfamilies tail length is 30–290 mm. *Microgale longicaudata* and possibly some closely related species have 47 tail vertebrae, more than any other mammal except some of the tree pangolins (*Manis*); the

Streaked tenrec (*Hemicentetes nigriceps*). Photo by Jonas Livet.

tip of the tail in those species is naked, modified for prehension. In Tenrecinae the body is covered by spines or bristly hairs. Other subfamilies have soft fur. The two lower bones of the hind leg can be united or free. The mammae are often numerous; *Tenrec* has up to 17 pairs (Goodman, Ganzhorn, and Rakotondravony 2003), the most of any mammal (Garbutt 2007). The testes are located in a scrotum in Potamogalinae, in the pelvic region in *Microgale*, and against the kidneys in the other genera. The penis is retractile, but Bedford, Mock, and Goodman (2004) reported that, when extended, it is much the longest, relative to body length, among mammals so far examined; that of *Tenrec ecaudatus* extends to approximately 70 percent of nose-rump length.

Feldhamer (2007) noted that the morphology of Tenrecidae defies a general description. The otter shrews (Potamogalinae) of mainland Africa resemble true otters (Lutrinae) in general appearance and aquatic habits. The other subfamilies probably became established in Madagascar in the Cretaceous and subsequently developed a number of specialized body forms. *Tenrec* resembles *Didelphis*, an American genus of opossums, in certain external, skeletal, and dental features. The Madagascar "hedgehogs," *Setifer* and *Echinops*, resemble true hedgehogs (Erinaceidae) in several dental features and in the spiny pelage. The rice tenrecs (*Oryzorictes*) look like moles (Talpidae) and have acquired definite fossorial characters. *Limnogale* resembles the semiaquatic mole *Desmana* of Eurasia. The shrew tenrecs (*Microgale*) have some resemblance to the shrews of the family Soricidae and, like the latter, are primarily terrestrial, though several long-tailed species are thought to be at least partly arboreal (Garbutt 2007).

The dental formula in Potamogalinae is: i 3/3, c 1/1, pm 3/3, m 3/3 = 40. The first upper incisor and the second lower incisor are caninelike, and the first upper incisors are separated by a greater space than are any other teeth. The canines are low and resemble the premolars. The upper molars of the otter shrews have V-shaped or weak W-shaped cusps.

In the other subfamilies the dental formula is: i 2/3, c 1/1, pm 3/3, m 4/3=40. The last molar, however, does not erupt until after the first molar has been shed, so that the complete dentition is not present at any one time. The incisors and canines in these genera are variable; in *Tenrec* the canines are enlarged. The first premolar is absent. The crowns of the molars are V-shaped, occasionally W-shaped.

The skull has long, narrow nasal bones and a small braincase that is not constricted between the orbits. Those lobes of the brain associated with the sense of smell are well developed. Clavicles are present except in Potamogalinae.

The known geological range of Tenrecidae is early Miocene and Recent in Africa and Recent in Madagascar (McKenna and Bell 1997). The family also is represented by subfossil material from the latest Pleistocene or Holocene of Madagascar (Goodman, Ganzhorn, and Rakotondravony 2003).

AFROSORICIDA; TENRECIDAE; Genus *POTAMOGALE* Du Chaillu, 1860

Giant African Water Shrew, or Giant Otter Shrew

The single species, *P. velox*, is found in the rainforest zone of central Africa, from southeastern Nigeria to western Kenya, northern Zambia, and central Angola (Aggundey 1977; Corbet 1977a; Happold 1987). Stephenson (2015) listed Sudan as part of the range, though possibly meant the new nation of South Sudan.

Head and body length is 290–350 mm, and tail length is 235–290 mm. Weight was reported as 340–397 grams by Nicoll (1985) and 517–780 grams by Vogel (2013). The fur is short, soft, and dense with a protective coat of coarse guard hairs. The upper parts are dark brown or blackish, and the underparts are whitish or yellowish.

This genus, among the largest of the many formerly grouped in the obsolete order Insectivora, resembles an otter in appearance—hence one of its vernacular names. The flattened muzzle has long, stiff, white whiskers. External ears are present, the eyes are minute, and the nostrils are covered by flaps that act as valves when the animal is submerged. The body is cylindrical, and the thick, powerful tail is strongly compressed laterally. The short, rather weak limbs have five non-webbed digits. Digits 2 and 3 of the hind foot are fused to form a comb, used to groom the pelage (Vogel 2013). A longitudinal flange of skin is present along the inner border of the hind foot, so that it may be pressed smoothly against the body and tail in swimming. The two lower bones of the hind leg are united distally. Females have two mammae located in the groin region.

Habitats include both sluggish lowland streams and cold, clear mountain torrents, from sea level to about 1,800 meters. In some areas the giant otter shrew frequents small forest pools during the rainy season and

Giant African water shrew (*Potamogale velox*). This originally was a black-and-white photo of a mounted specimen at the Bristol City Museum, England. Photo colorized by Dave Davis based on descriptive references; see https://neitshade5.wordpress.com/category/nature/.

migrates overland to streams at the beginning of the dry season. It normally does not occur in large rivers, though one was found in the Ivindo in Gabon, where that river was several hundred meters wide (Stephenson 2015). It becomes active in the late afternoon, after sheltering during the day in holes and tunnels in stream banks. Although the entrance to its burrow reportedly is often below water level, Fons (1990) wrote that it is above mean water level, under or near a tree. The entry branches into two tunnels, one terminating below water level and the other leading to a small living chamber floored with leaves.

Potamogale is an extremely agile and rapid swimmer, propelling itself through the water entirely by lateral undulations of the back and very muscular tail (Kingdon 1974; Vogel 2013). That method of propulsion is shared with fish and crocodiles, but not with fully aquatic mammals—pinnipeds, cetaceans, sirenians—which swim by undulating the tail in a vertical plane, or with other semi-aquatic mammals, such as the water opossum (*Chironectes*), web-footed tenrec (*Limnogale*), water shrews (*Neomys*), fish-eating rats (*Ichthyomys*), and beavers (*Castor*), which paddle with their feet. On land the giant otter shrew walks on the soles of its feet with the heels touching the ground. Its terrestrial movements are rather clumsy, but it can move at considerable speed.

Potamogale hunts by dives, each lasting for only a few seconds (Stephenson 2015). As it is commonly underwater in the dark, it apparently does not use its eyes to seek prey but depends mainly on the vibrissae and scent. It reportedly feeds on crabs, fish, and amphibians. In some areas it eats freshwater crabs almost exclusively. It turns a crab over and tears out the flesh from the body and claws. A captive adult consumed 15–20 crabs per night and left those he could not eat. *Potamogale* comes out on land to eat and apparently to void excrement. Vogel (2013) referred to the genus as territorial and indicated that latrines are probably used to mark territorial boundaries.

According to Happold (1987), *Potamogale* is solitary, each adult utilizes 500–1,000 meters of a stream, females probably have two litters annually, and litter size is one or two young. Fons (1990) suggested that there is no definite breeding season, as pregnant females have been found both in late February at the beginning of the rainy season and in June and August when it is ex-

tremely dry. Kuhn (1971) reported that all pregnant females he examined contained two embryos.

Potamogale is widely hunted for its skin and is also trapped accidentally, but if its forest habitat remains intact, even as a narrow strip along a riverbank, a viable population can apparently be maintained. Unfortunately, logging and subsequent soil erosion, with increased opaqueness of watercourses, is leading to local disappearances (Nicoll and Rathbun 1990). In 1935 it was reported to be widely distributed in the Niger Delta of Nigeria, but by 1952 there was no evidence of its occurrence there; instead, it was apparently restricted to the area east of the Cross River (Happold 1987). The IUCN considers the giant otter shrew to be declining but not at a rate sufficient to qualify for a threatened category (Stephenson 2015).

AFROSORICIDA; TENRECIDAE; **Genus**
MICROPOTAMOGALE Heim de Balsac, 1954

Dwarf African Water Shrews, or Dwarf Otter Shrews

There are two species (Corbet 1977a; Kingdon 1974; Vogel 2008a, 2008b):

> *M. lamottei*, Nimba Mountains of southeastern Guinea and adjacent parts of Liberia and Ivory Coast, Putu Range of eastern Liberia;
> *M. ruwenzorii*, known only from the Rwenzori region of southwestern Uganda and eastern Democratic Republic of Congo and in the mountainous region of the latter country west of Lake Edward and Lake Kivu (as indicated in the legend of the accompanying photo, a specimen also recently was collected in Rwanda).

The latter species sometimes is placed in the genus or subgenus *Mesopotamogale* Heim de Balsac, 1956. Vogel (2008b) suggested that such a distinction might be warranted for *M. ruwenzorii* because of its webbed feet and other morphological adaptations for aquatic life.

Head and body length is 120–200 mm, and tail length is 100–150 mm; weight is about 60–80 grams in *M. lamottei* and 135 grams in *M. ruwenzorii* (Fons 1990; Kingdon 1974; Kuhn 1971; Rahm 1966). In *M. ruwenzorii* the coloration is brownish gray above and gray be-

Dwarf African water shrew (*Micropotamogale lamottei*). Photo by Ara Monadjem through Jan Decher.

low, the feet are webbed between the digits, and the tail is short-haired throughout and roughly oval in cross section with a slight keel along the upper and lower surfaces. *M. lamottei* has unwebbed feet and a round tail. From *Potamogale*, *Micropotamogale* is distinguished by its smaller size, zalambdodont (V-shaped) rather than dilambdodont (W-shaped) upper cheek teeth, its fleshy rather than hornlike or leathery rhinarium, and the remarkable evolution of the middle ear, which is transformed into an almost perfect bulla with a carotid canal rather than only a groove. A female *M. lamottei* had four pairs of mammae (Kuhn 1971).

Dwarf otter shrews live along upland forest streams. *M. ruwenzorii* is found at elevations of 800–2,200 meters (Vogel 2008b). *M. lamottei* also occurs in agricultural areas where dense vegetation remains along streams (Vogel 2008a). Both species are nocturnal and semiaquatic, foraging in water and propelling themselves with their feet, not the tail as in *Potamogale velox*. Although *M. lamottei* lacks the aquatic specializations of the other two species of otter shrews, it is an efficient swimmer and diver; it can remain underwater for over 10 minutes when escaping danger by lowering its metabolic rate (Vogel 2008a). *M. ruwenzorii* is known to dig tunnels and make sleeping chambers with hay or grass. Its dives are brief, and it frequently comes out of the water, bringing larger prey to eat; it sometimes hunts on dry land (Kingdon 1974). Foods recorded for the genus include worms, insects and their larvae, crabs, fish, and small frogs.

According to Vogel (2008a), available evidence suggests *M. lamottei* is rather solitary and territorial, though a pair kept over several months in the same enclosure did not exhibit aggressive behavior. Births and embryo counts for 11 females showed a mean litter size of 2.6 (range 1–4) and a gestation period of more than 50 days. One young opened its eyes when 23 days old and first took solid food on day 40. Kuhn (1971) reported that a female *M. lamottei* taken on 5 December in Liberia had four embryos, each 18.5 mm long. Kingdon (1974) noted that a female *M. ruwenzorii* collected in late September contained two embryos

These animals reportedly have considerable strength. They do damage in fish traps, generally killing all the fish in the trap, and then, being unable to escape, they drown. Many of them have been found in the holes of the diamond diggers along small creeks in the Sanniquellie District of Liberia. Two specimens of *M. ruwenzorii* were caught in fishing nets.

Dwarf African water shrew (*Micropotamogale ruwenzorii*); note the webbing in the feet of this species. Photo by Nicolas Ntare in Nyungwe National Park, Rwanda.

Accidental killing in such traps and nets is one problem, but a more serious concern is the loss of the very restricted habitat in which the two species occur. The future of *M. lamottei*, known from an area of less than 1,500 sq km, seems particularly bleak (Nicoll and Rathbun 1990). Mining activities have devastated the Liberian part of its range and may be spreading into Guinea. Wetland rice farming, introduced to the Nimba region in 1980, has also caused large-scale habitat destruction. The range of *M. lamottei* is now fragmented, and the species is rare and declining; it is classified as endangered by the IUCN (Vogel 2008a). *M. ruwenzorii* is also losing habitat as human numbers and activity increase, but part of its range remains intact and not severely fragmented. That species is now designated near threatened by the IUCN but is close to qualifying for classification as vulnerable (Vogel 2008b).

AFROSORICIDA; TENRECIDAE; **Genus *TENREC***
Lacépède, 1799

Tenrec

The single species, *T. ecaudatus*, occurs naturally in most habitats throughout Madagascar and is also found on Mayotte in the Comoro Islands, Réunion and Mauritius in the Mascarene Islands, and Mahé in the Seychelles (Fons 1990; Nicoll 2003). Bronner and Jenkins (2005) indicated that the population in the Comoros is natural, while Nicoll (2003) referred to it as a presumed introduction, though the time it took place and the reason are unknown. He noted that the other island populations were introduced to provide food for plantation workers. In the United States National Museum of Natural History, Washington, DC, there is a male tenrec that was captured in July 1892 by Dr. W. L. Abbott on Mahé. The label reads: "Introduced from Madagascar via Bourbon (Mauritius [actually Reunion (Nicoll 2003)]). Introduced about 10 years ago and now very abundant."

Head and body length is approximately 265–390 mm, and tail length is 10–16 mm. Weight averages 832 grams under natural conditions (Goodman, Ganzhorn, and Rakotondravony 2003) but in captivity is usually 1.6–2.4 kg (Eisenberg and Gould 1970). General coloration is grayish brown or reddish brown, but some individuals are dark brown on the back and rump. The pelage, which is not dense, consists of hairs and spines. Young animals have strong white spines arranged in longitudinal rows along the back. Nicoll (2003) stated that those specialized spines make up a stridulating organ that vibrates to produce sounds, by

Tenrec (*Tenrec ecaudatus*). Photo by Reptiles4all/Shutterstock.com.

which the young communicates with family groups. In adults those dorsal spines are replaced by a crest of long, rigid hairs. The adult pelage is coarse and spiny but not sharp-tipped. When angry, alarmed, or excited, the tenrec is capable of erecting the mane of the upper back, which has longer hairs than the rest of the body. Nicoll (2003) noted there is a second set of specialized spines on the nape that are associated with scent glands; females rub their forefeet over that area and use them to mark sites already marked by other females.

The body is stout, the snout long and pointed, the forelimbs shorter than the hind limbs, the skull cylindrical and elongated, and the digits all well developed. Nicoll (2003) observed that the general form is like that of a North American opossum (*Didelphis*) but without the long tail. The head is slender in females but broadens and domes in older males. The gape spans up to 10 cm, the canine teeth are up to 15 mm log, and the snapping defensive bite can inflict serious wounds. *Tenrec* emits a low hissing noise when annoyed and also squeaks or squeals. It is muscular and powerful and does not curl into a protective ball as does *Setifer*.

Females have up to 17 pairs of mammae (Goodman, Ganzhorn, and Rakotondravony 2003).

The tenrec is common in most parts of Madagascar, from sea level to montane regions, though it is probably absent from the highest peaks (Garbutt 2007). It occurs in all natural forest formations, secondary open wooded grasslands, farmlands, and even urban centers (Nicoll 2003). For daily shelter it may use a nest in a hollow log or under a rock. It is generally nocturnal, but during late pregnancy and lactation females (and young as soon as old enough to leave the burrow) are active night and day (Nicoll 2003). Observations of wild and captive individuals demonstrated a bimodal pattern of activity with peaks between the hours of 1800 and 2100 and between 0100 and 0500 (Eisenberg and Gould 1970). Body temperature is variable during the summer active season (Nicoll 2003). It averages 28°–29° C at that time but during physical activity may reach 35.5° C. During inactivity or sleep, temperature may fall below 24° C, resulting in temporary torpor.

In the winter dry season (May–October) there is a much longer period of torpor or hibernation. The tenrec then uses a ground burrow 1–2 meters long, which it plugs from the inside with soil before going into its long sleep. Hibernation lasts 5–6 months, with males emerging in the spring 1 month, on average, ahead of females (Nicoll 2003). Torpid individuals are cold to the touch and have a breathing rate of about 30 respirations per minute. Apparently the sleep is continuous, as all the animals examined in the dormant state have been fat and have not had any trace of food or feces in the intestine. Prior to hibernation is a period of fattening, during which body mass may double; in the Seychelles a considerable volume of fruit is eaten at that time (Nicoll 2003). However, the diet consists mostly of a variety of invertebrates, which the tenrec probes and digs for in the soil with its snout. The tenrec also occasionally captures small vertebrates.

According to Nicoll (2003), in the Seychelles, where there is little hunting by people, tenrec population densities may reach 15 adults per ha. Home ranges there overlap, with individuals typically covering about 1 ha. in a night's activity and at least 2 or 3 ha. in a season.

After emergence from hibernation in the austral spring (October–November), females tend to restrict their movements, sometimes to less than 0.1 ha., and males search for them. A large, dominant male may remain with a female for the 2–4 days she is receptive, fighting with or chasing other males that approach. He may leave the female in her burrow for an hour or two to seek other mates, but if another male comes at that time, the female will accept him until he is driven off when the original male returns. Fons (1990) noted that pairs brought together in captivity form family groups and that adult males are very tolerant of their offspring.

In Madagascar the young are born in December and January following a gestation period of 56–64 days. It is unlikely that there is more than one litter per year, though it is possible that a second litter could be produced if the first is lost immediately after being born in December. Litter size is often around 25 young in dry western Madagascar, about 20 in the humid eastern forests, and seldom more than 15–16 in the tropical Seychelles; overall range is 1 to 32, making *Tenrec* one of

Tenrec (*Tenrec ecaudatus*), mother and young. Photo by Harald Schütz.

the most prolific of mammals (Eisenberg 1975; Louwman 1973; Nicoll 2003). The young weigh 10–18 grams at birth (Fons 1990). They emerge from the nest when 18–20 days old and forage as a group with the mother. At about 35 days they are fully weaned and may forage without her, sometimes in small, loose groups until they are solitary. Molt to adult pelage begins at around 35 days and is complete at 60–70 days (Nicoll 2003). The young are full grown by March or April (Eisenberg and Gould 1970). Females can first conceive at 6 months (Hayssen, Van Tienhoven, and Van Tienhoven 1993). In the Seychelles, where the tenrec population is not extensively hunted by people, individuals survive up to 4 or 5 years in the wild (Nicoll 2003). A captive male and female lived to respective ages of 8 years, 8 months and 8 years, 7 months (Weigl 2005).

Aside from the Seychelles population, *Tenrec* is widely hunted for use as human food. Organized collection, involving many people and dogs, is common in Madagascar, and some villages specialize in tenrec hunting. Consequently, numbers have been greatly reduced around some larger cities, but the genus remains common elsewhere (Nicoll 2003).

AFROSORICIDA; TENRECIDAE; **Genus**
HEMICENTETES Mivart, 1871

Streaked Tenrecs

There are two species (Goodman et al. 2000):

> *H. semispinosus*, rainforests of eastern Madagascar;
> *H. nigriceps*, eastern edge of central highlands of Madagascar from around Antananarivo south to Fianarantsoa.

Although Genest and Petter (1977) treated *H. nigriceps* as a subspecies of *H. semispinosus*, Goodman et al. (2000) considered the two distinct species based on morphological characters and sympatric occurrence in at least one area. Subsequent authorities have followed suit (Bronner and Jenkins 2005; Garbutt 2007; Goodman, Stephenson, and Soarimalala 2015a, 2015b; Olson and Goodman 2003; Stephenson 2003b).

Head and body length is 120–190 mm, the tail is vestigial, and weight is 70–220 grams (Stephenson 2003b). The pelage is mostly spiny and sharply pointed. *H. semispinosus* has blackish brown upper parts, a yellowish central stripe from the muzzle to behind the ears, a prominent crest of erectile spines on the nape, and a number of yellowish to orange longitudinal stripes on the back and sides. Coloration of *H. nigriceps* is similar, but the stripes are dull white to pale yellow, broader, and not as distinct as in *H. semispinosus*; the crest is equally pronounced. The underparts are chestnut brown in *H. semispinosus* and whitish or buffy in *H. nigriceps*. The hairs covering the underparts are spiny in *H. semispinosus* and soft and woolly in *H. nigriceps*.

Hemicentetes is easily recognized by its streaked color pattern. Its skull is relatively long and tapered, compared to that of *Tenrec*, and its canine teeth are less prominent; they are no longer than the anterior premolars, whereas in *Tenrec* they are much longer (Genest and Petter 1977). Female *H. semispinosus* have up to 16 mammae; female *H. nigriceps* have up to 10 (Goodman, Ganzhorn, and Rakotondravony 2003).

Elevational range is sea level to 1,550 meters for *H. semispinosus* and 1,200 to 2,350 meters for *H. nigriceps* (Goodman, Stephenson, and Soarimalala 2015a, 2015b). The latter species is restricted to montane and sclerophyllous forests and adjacent habitats (Garbutt 2007), while *H. semispinosus* is found in lowland rainforest and brushland and has extended its range into cultivated rice fields. *H. semispinosus* excavates a burrow consisting of a nest chamber at the end of a tunnel up to 150 cm long and about 15 cm below the surface of the ground, or there may be an extensive burrow system for a group of animals, with tunnels several meters in length (Stephenson 2003b); burrows of *H. nigriceps* are smaller and less complex (Marshall and Eisenberg 1996). Available evidence indicates that *H. nigriceps* is strictly nocturnal but *H. semispinosus* may be active at any time of the day (Stephenson 2003b). In locomotion *Hemicentetes* uses a diagonal coordination pattern of limb movement. The heel may be lifted off the ground in running, but generally the animals are plantigrade. They are capable of climbing but do not often do so. When disturbed or touched, an individual spreads the crest of spines on its nape, lowers its head, and then quickly thrusts its head upward. A small area of heavy spines in the middle of the back is often vibrated rapidly, independently of the rest of the spines

Streaked tenrec (*Hemicentetes semispinosus*). Photo by Ryan M. Bolton / Shutterstock.com.

on the back. *Hemicentetes* forages by inserting its nose at the roots of grasses or under leaves (Eisenberg and Gould 1970). The diet consists largely of earthworms and other soft-bodied invertebrates, as reflected in the reduced dentition of the genus (Stephenson 2003b).

A state of torpor, in which body temperature is less than 1° C above ambient temperature, may be attained during the winter by both species. Stephenson and Racey (1994) found that *H. semispinosus*, which lives at lower, warmer elevations, may be able to avoid torpor when favorable environmental conditions prevail but that the upland *H. nigriceps* must hibernate throughout the austral winter (May–October).

Streaked tenrecs produce a number of sounds that are audible to humans. During agonistic behavior there may be a noisy "crunch" and a "putt-putt." Gould (1965) presented evidence that *Hemicentetes* uses tongue clicks for echolocation. Sound also results from the vibration of the specialized quills in the middorsal region. The latter process, known as stridulation, seems to be a means

of communication and apparently is especially important in helping the mother and young locate one another (Eisenberg and Gould 1970).

Except as indicated, the following life history information applies only to *H. semispinosus* (Eisenberg 1975; Eisenberg and Gould 1970; Marshall and Eisenberg 1996; Stephenson 2003b). The social system is the most complex of any species of the obsolete order Insectivora (including Deltatheridia and Lipotyphla). Because of the speed of maturation and relatively large litter size, a family group may consist of more than 20 individuals from three related generations. They forage together, maintaining contact by stridulation. One observed group consisted of an extended colony of 18 individuals, including 2 adult males, 2 adult females, and 14 juveniles. Most other groups are smaller, usually having 1 or 2 adults and several young. *H. nigriceps* is not known to form large family groups or colonies. Apparently there is a lengthy period of association between the sexes. Mating begins in September or October and continues through December. The young are born from November to March, following a gestation period of about 55–63 days. Litter size averages 2.8 (range 2–4) in *H. nigriceps* and 6.6 (2–11) in *H. semispinosus*. Neonates weigh about 11 grams each. Development is extraordinarily fast. The young open their eyes when 7–12 days old and are weaned after 17–20 days. Females attain sexual maturity within only 35–40 days of their birth but become senescent after their second year. *H. nigriceps* has lived more than 3 years in captivity. Weigl (2005) reported a captive *H. semispinosus* that lived for 2 years and 9 months.

AFROSORICIDA; TENRECIDAE; **Genus *ECHINOPS***
Martin, 1838

Lesser Hedgehog Tenrec, or Small Madagascar "Hedgehog"

The single species, *E. telfairi*, occurs in the arid part of southwestern and southern Madagascar, at least as far as Morombé in the north and Andohahela National Park in the southeast (Garbutt 2007; Genest and Petter 1977; Goodman, Jenkins, and Pidgeon 1999).

Head and body length is 130–180 mm, tail length is around 10 mm (Garbutt 2007), and weight in captivity is 110–250 grams. Mean natural weight of adults is 140.6 grams (Goodman, Ganzhorn, and Rakotondravony 2003). Coloration is variable in a given population, ranging from pale, almost white to a very dark color resulting from more intense melanin deposition in the annular bands of the quills (Eisenberg and Gould 1970). *Echinops* closely resembles *Setifer* but is slightly smaller, has shorter claws, and has only 32 teeth (*Setifer* has 36).

The lesser hedgehog tenrec is confined to dry deciduous forests, xerophytic spiny forest, and gallery forest (Garbutt 2007). It dens in tree cavities, and the location of its nests indicates considerable arboreal ability. Garbutt (2007) stated that it is surprisingly agile and capable of climbing along thin branches, where its short tail is used as a brace; its antipredator response is to roll into a tight ball and emit hisses and grind the teeth. It is nocturnal and hibernates for 3–5 months during the cold season (Godfrey and Oliver 1978). It forages both in trees and on the ground, and captives take a variety of animal food, including insects, baby mice, and chopped meat (Eisenberg and Gould 1970).

Echinops exhibits considerable social tolerance: captive mated pairs or several adult females have shared a

Lesser hedgehog tenrec (*Echinops telfairi*); right image shows animal in antipredator response. Photos by Link E. Olson.

small cage even when a litter was being raised, and occasionally two adults have been found hibernating together. Males may be highly aggressive toward one another. Communication is by tactile, chemical, and a few auditory signals. Mating probably begins in October, during the austral spring, shortly after the adults emerge from torpor, and births take place in December or January. Females appear to be polyestrous and to exhibit several cycles of 6 days each during the mating season. Gestation evidently lasts 61–64 days. Litters contain 1–10 young, usually 5–7. The young weigh about 8 grams at birth, open their eyes by 9 days, and are weaned and functionally independent at 30–35 days. They are sexually mature after they complete their first winter's hibernation (Eisenberg 1975; Eisenberg and Gould 1967, 1970; Godfrey and Oliver 1978; Goodman, Ganzhorn, and Rakotondravony 2003; Mallinson 1974). A captive male and female lived to respective ages of 19 years and 18 years, 2 months (Weigl 2005).

AFROSORICIDA; TENRECIDAE; **Genus *SETIFER***
Froriep, 1806

Greater Hedgehog Tenrec, or Large Madagascar "Hedgehog"

The single species, *S. setosus*, is found throughout Madagascar, from sea level to around 2,250 meters (Goodman, Stephenson, and Soarimalala 2015c). *Dasogale* Grandidier, 1928, with the species *D. fontoynonti*, is now considered to be part of *Setifer* and probably to represent an immature *S. setosus* (MacPhee 1987b). *Ericulus* I. Geoffroy St. Hilaire, 1837, is also a synonym of *Setifer*.

Head and body length is 150–225 mm, tail length is 15–16 mm, and weight is 180–300 grams, averaging 282 grams for adults (Eisenberg and Gould 1970; Garbutt 2007; Goodman, Ganzhorn, and Rakotondravony 2003). In one population, individuals were lightest after emergence from hibernation and heaviest immediately

Greater hedgehog tenrec (*Setifer setosus*). Photo by Thomas Marent / Minden Pictures.

prior to hibernation or during pregnancy, with weights ranging from around 100 to over 350 grams (Levesque et al. 2013). The upper parts are covered with close-set, sharp, and bristly white-tipped spines that extend from the forehead along the back and sides and even cover the short tail projection. The muzzle, limbs, and underparts are covered with soft, scanty hair. As in *Tenrec*, there are a number of vibrissae, about 50–75 mm long, growing from the face. A finely mottled color effect results from the white-tipped spines; some individuals are comparatively pale with broadly white-tipped spines, whereas others are blackish with narrowly white-tipped spines. The underparts are buffy, dark brown, or whitish.

The general appearance is that of a small species of hedgehog (*Erinaceus*), but the body is relatively longer and the muzzle is less pointed. The tip of the muzzle projects beyond the lower lip. The ear is not longer than the muzzle and is much like the ear of *Tenrec* in general form. The hind limbs are only slightly longer than the forelimbs, and all the digits are present and well developed. Females have five pairs of mammae.

The greater hedgehog tenrec dwells in rainforest, deciduous forest, dry spiny forest, grasslands, agricultural land, and even some urban areas; it seems to prefer rainforests below 900 meters (Garbutt 2007). It appears to be strictly nocturnal. It sometimes constructs short tunnels leading to a leaf-lined nest, though in a radio-tracking study in the dry deciduous woodland of western Madagascar, Levesque, Rakotondravony, and Lovegrove (2012) found the most common resting sites to be hollows of either live or dead trees. In that area, all individuals, except lactating females, changed rest sites daily, rarely retuning to the same area. Four lactating females, however, used one site for 20–25 days, returning there each morning after nightly foraging, until the young were old enough to move. Although an overall population of *Setifer* may be active throughout the year, individuals have the capacity to become torpid and may spend most of the winter within a restricted foraging radius, passing much time in a torpid state. In their study area, Levesque, Rakotondravony, and Lovegrove (2012) confirmed that *Setifer* does undergo true hibernation during the dry season, which covers a substantial part of the year. Males entered hibernation from as early as 18 February up to 2 April, while females began from April to May after weaning their last litter of the season (Levesque et al. 2013). Hibernation subsequently lasted 5–7 months until September or October.

For defense the greater hedgehog tenrec rolls itself into a protective ball of spines that point in every direction. For general locomotion it employs a crossed extension coordination pattern. When running it rises up on its toes, keeping its heels off the ground. It climbs very well but slowly. Captives have eaten earthworms, grasshoppers, raw ground meat, and carcasses of mice but have seemed incapable of killing larger prey. Wild individuals consume a variety of soil invertebrates (Soarimalala and Goodman, 2003) and have been observed to take carrion and, in metropolitan areas, to scavenge around garbage dumps (Eisenberg and Gould 1970).

Levesque, Rakotondravony, and Lovegrove (2012) found home range to be surprisingly large for such a small animal, averaging 13.7 ha. for five males and 6.7 ha. for five females. High overlap between the ranges of multiple individuals of the opposite sex indicated a promiscuous mating system. Adults appear to be generally solitary and to avoid one another, but there may be temporary associations between two adults of the opposite or the same sex. Several sounds have been recorded during agonistic or sexual encounters.

In one investigated population mating took place from late September to mid-October (Eisenberg and Gould 1970). However, Levesque, Rakotondravony, and Lovegrove (2012) reported that in their study area all females were in a reproductive state nearly throughout the non-hibernating portion of the year, October–April; females in late lactation (30–40 days) were also gestating. There evidently is thus a postpartum estrus, and females may have up to three litters annually (Levesque et al. 2013). Gestation is variable, being shorter when ambient temperature is higher. Gestation in captivity was reported as 65–69 days by Eisenberg (1975) and 51–61 days by Mallinson (1974). Eisenberg (1975) reported litter size to average 3 young and to range from 1 to 5, but Goodman, Ganzhorn and Rakotondravony (2003) listed a maximum litter size of 10 or more. The young weigh about 25 grams at birth, they are weaned after just 15–44 days, and females can conceive by the time they are 6 months old (Hayssen, Van Tienhoven, and Van Tienhoven 1993; Levesque et al. 2013). A captive lived 14 years and 1 month (Weigl 2005).

Large-eared tenrec (*Geogale aurita*). Photo by Link E. Olson, Field Museum of Natural History number 159732.

AFROSORICIDA; TENRECIDAE; **Genus *GEOGALE***
Milne-Edwards and Grandidier, 1872

Large-eared Tenrec

The single species, *G. aurita*, is found in the arid south and southwest of Madagascar, from the Tsiribihina River on the west coast to Andohahela National Park in the southeast (Garbutt 2007; Stephenson 2003a). It was also reported recently from Ankarafantsika National Park in the northwest (Goodman, Stephenson, and Soarimalala 2015d). A subspecies, *G. a. orientalis*, was described from a single specimen taken in the northeast near Fénérive-Est, but recent surveys of that area did not obtain evidence of its continued presence (Stephenson 2003a). The genus and species *Cryptogale australis* Grandidier, 1928, described from cranial remains found in the Grotte d'Androhomana, south of Fort Dauphin, Madagascar, is included in *Geogale aurita* (Genest and Petter 1977).

Head and body length is 60–76 mm, tail length is 34–41 mm, and weight is 5.0–8.5 grams (Garbutt 2007; Stephenson 2003a). The coloration of the type is reddish brown above and soiled yellowish white below. Usually, however, the upper parts are pale gray and the underparts creamy white, sometimes with orange flecks on the flanks; the pelage is soft, short, and dense (Garbutt 2007; Stephenson 2003a). The muzzle projects beyond the lower jaw. The tail is cylindrical, scaly, and covered with fine hairs. All limbs have five digits. The ears are relatively large and more prominent than those of other afrosoricids. Females have eight mammae (Goodman, Ganzhorn, and Rakotondravony 2003).

According to Stephenson (1993, 2003a), the large-eared tenrec inhabits dry deciduous forest, gallery forest, and spiny bush. It is terrestrial and nocturnal. It is among the most heterothermic tenrecs, its body temperature paralleling ambient temperature at all times of the year. It appears to enter daily torpor in all seasons and then can be found in crevices within decomposing logs on the forest floor. It probably becomes torpid more frequently during the dry season (winter), though captive individuals eat and drink throughout winter. The ears are extended during feeding and appear to be used in searching for prey. Nicoll and Rathbun (1990) reported finding numerous individuals within fallen logs, all torpid by day but active by night. They noted that *Geogale* takes a wide variety of invertebrate prey but shows a marked preference for termites. Gould and Eisenberg (1966) reported a single specimen taken from beneath a fallen tree, where it was sleeping in a cavity of sand. Individuals are usually found alone, though occasionally a male and female are found in close proximity (Stephenson 2003a).

Stephenson (1993, 2003a) established a captive colony with 21 individuals. Mating occurred from September to February, births from November to March during the austral summer. Females exhibited a postpartum estrus, unusual for the Tenrecidae, enabling one litter to be suckled while a second was developing in the uterus. Average gestation was 57 days, though in some cases it lasted as long as 69 days, probably because development was arrested when the females entered torpor. Litter size averaged 3.9 and ranged from 2 to 5 young. They weighed about 0.7 grams at birth, opened their eyes at about 24 days, and were weaned after about 5 weeks. Nicoll and Rathbun (1990) reported that a male caught as an adult lived 25 months in captivity.

According to Goodman, Stephenson, and Soarimalala (2015d), although *Geogale* is rarely found, it is common in owl pellets, suggesting it may be more widespread and abundant than once thought. It seems adaptable to some habitat disturbance and is not dependent on forests, having been found in savannahs and grassland. A more important factor in its distribution may be presence of termite mounds.

AFROSORICIDA; TENRECIDAE; **Genus**
ORYZORICTES Grandidier, 1870

Rice Tenrecs, or Mole Tenrecs

There are two species (Goodman 2003; Goodman, Jenkins, and Pidgeon 1999):

Rice tenrec (*Oryzorictes hova*). Photo by Jonas Tonboe.

O. hova, northern and eastern Madagascar from Manongarivo region in northwest to Masoala Peninsula and south to Andohahela National Park, Nosy Mangabe Island;

O. tetradactylus, eastern central highlands of Madagascar from Antsirabe south to Andringitra National Park and Vinanitelo.

Genest and Petter (1977) retained *O. tetradactylus* in the separate genus *Nesoryctes* Thomas, 1918. Bronner and Jenkins (2005) recognized *Nesoryctes* as a subgenus.

Head and body length is 99–124 mm, tail length is 38–62 mm, and weight is 28–48 grams (Fons 1990; Goodman 2003; Stephenson 1994). According to Goodman (2003), the upper parts of *O. hova* are dark blackish brown to light tan brown, and the underparts are dark brown to buffy white; albinism apparently is not rare on Nosy Mangabe Island. *O. tetradactylus* has largely dark brown fur that often is mottled with lighter brown, particularly on the dorsum. Both species have very soft, dense, and silky pelage with a slight iridescence. The hair texture of *O. tetradactylus* is coarser than that of *O. hova*; its tail is bicolored like the body and is relatively short, while that of *O. hova* is naked and about half the length of the head and body.

Rice tenrecs are modified for a burrowing life and externally resemble true moles (Talpidae) and golden moles (Chrysochloridae) more than they do true shrews (Soricidae) and shrew tenrecs (*Microgale*). The body is robust, the nose broad and naked, and the eyes and ears small. The forefeet are stout and larger than the hind feet, and they have strong digging claws. *O. tetradactylus* has four digits on the forepaws, *O. hova* has five. Goodman (2003) observed that, when held in the hand, both species of *Oryzorictes* show surprising forelimb strength and can exert considerable lateral force. The skull differs from that of *Microgale* in being more robust, with a shorter and wider braincase and distinctly longer canine teeth.

O. hova inhabits marshy areas, rice fields, and a wide range of natural formations from rainforest at sea level to sclerophyllous forest near the tree line at an elevation of 1,990 meters. *O. tetradactylus* is known from only a few sites, most at high elevations, up to 2,450 meters, and some above the tree line (Goodman 2003). Rice tenrecs are said to be mainly nocturnal, but because of their burrowing habits, it is possible they

Rice tenrec (*Oryzorictes hova*). Photo by Link E. Olson, Field Museum of Natural History number 159466.

move about at all hours without being observed. Goodman (2003) noted that they often are captured after heavy rains, presumably because flooding of their burrows forces them to the surface. They have been collected in the banks of rice paddies in tunnels that resemble those of moles. In some parts of Madagascar their burrowing activities in the water-retaining walls of the fields has resulted in damage to the dikes. The diet is assumed to consist mainly of insects and other invertebrates, such as mollusks. Observations by Stephenson (1994) indicate that *O. hova* sometimes forages above ground on the forest floor, using its muzzle to probe beneath the leaf litter and humus, and may feed on earthworms. Very little life history information is available for either species; female *O. hova* have six mammae, and, based on embryo counts, the maximum litter size in that species is four young (Goodman 2003).

The IUCN considers *O. hova* to be uncommon, declining in numbers, and threatened by drainage of natural wetland habitat but does not assign a formal threatened category (Goodman, Stephenson, and Soarimalala 2015e). Most specimens of *O. tetradactylus* were collected over 100 years ago, and its current status is unknown (Goodman, Stephenson, and Soarimalala 2015f).

Shrew tenrec (*Microgale drouhardi*). Photo by Link E. Olson.

AFROSORICIDA; TENRECIDAE; **Genus** *MICROGALE* Thomas, 1882

Shrew Tenrecs

There are currently 23 recognized species (Garbutt 2007; Goodman and Jenkins 1998; Goodman, Jenkins, and Pidgeon 1999; Goodman and Soarimalala 2004; Goodman, Vasey, and Burney 2007; Goodman et al. 2006; Jenkins 1988, 1992, 1993; Jenkins and Goodman 1999; Jenkins, Goodman, and Raxworthy 1996; Jenkins, Raxworthy, and Nussbaum 1997; MacPhee 1987a; Muldoon et al. 2009; Olson and Goodman 2003; Olson, Goodman, and Yoder 2004; Olson et al. 2009; Soarimalala and Goodman 2008; Soarimalala, Raheriarisena, and Goodman 2010; Stephenson 1995; Stephenson, Goodman, and Soarimalala 2015a):

M. brevicaudata, lower elevations of western and northeastern Madagascar from Montagne d'Ambre in north to Masoala on east coast and Soahany River on west coast, also known by subfossil remains, about 500 years old, from Ankilitelo Cave in southwestern Madagascar;

M. grandidieri, lowland forest of western Madagascar from Namoroka in north to Onilahy River in south;

M. macpheei, known only by subfossil remains, about 2,000 years old, from Andrahomana Cave in southeastern Madagascar, but possibly still extant in nearby forests;

M. drouhardi, low to mid-elevation rainforest in northern and eastern Madagascar from Montagne d'Ambre in north to Manongarivo near west coast and Andohahela in southeast;

M. monticola, mountains of Andapa region in northeastern Madagascar;

M. soricoides, mostly in highlands of eastern Madagascar from Anjanaharibe-Sud and possibly Manongarivo in north to Andohahela in south;

M. fotsifotsy, mostly low to mid-elevation rainforest in eastern Madagascar from Montagne d'Ambre in north to Andohahela in south;

M. nasoloi, known only from forests of Menabe region, Vohibasia, and Analavelona Massif in southwestern Madagascar and by subfossil remains, about 500 years old, from nearby Ankilitelo Cave;

M. dryas, known only from the Ambatovaky Special Reserve and a few other sites in northeastern Madagascar;

M. gymnorhyncha, mid- to high elevation forests of eastern Madagascar from Anjanaharibe-Sud and Marojejy in north to Andohahela in south;

M. gracilis, mostly highland rainforest of eastern Madagascar from Marojejy Massif in north to Andohahela in south;

M. taiva, eastern rainforest belt of Madagascar from Tsaratanana in north to Andringitra in south;

M. cowani, mid- to high elevations in northern and eastern Madagascar from Marojejy in north to Andohahela in south, also known from several sites farther west;

M. jobihely, known only from two sites on the southwestern slopes of the Tsaratanana Massif in northern Madagascar, and from another site in the Ambatovy Forest in central-eastern Madagascar;

M. thomasi, central and southern parts of eastern rainforest zone of Madagascar from Zahamena in north to Andohahela in south;

M. pusilla, eastern rainforest and central highlands of Madagascar from north of Antanarivo to Tolagnaro in southeast;

M. jenkinsae, known only by two specimens collected in 2003 at Forêt des Mikea in southwestern Madagascar but probably also represented by subfossil material from Lelia and Anjohimpaty, about 200 km farther south;

M. principula, low to mid-elevation rainforest in eastern Madagascar from Marojejy and Anjanaharibe-Sud in north to Andohahela in south;

M. longicaudata, mostly highland rainforest of northern and eastern Madagascar from Montagne d'Ambre in north to Andohahela in south, possibly also in some western areas;

Shrew tenrec (*Microgale taiva*). Photo by Link E. Olson.

M. majori, most of northern and eastern rainforest zone of Madagascar, also known from several sites in west;

M. parvula, northern and eastern forest zone of Madagascar from Montagne d'Ambre in north to Andohahela in south;

M. talazaci, northern and eastern humid forests of Madagascar from Montagne d'Ambre in north to Vondrozo in south;

M. dobsoni, eastern rainforest and central highlands of Madagascar from Marojejy and Anjanaharibe-Sud in north to Andohahela in south.

The above distributional data were taken mostly from Garbutt (2007). The sequence of species generally follows the arrangement shown by Olson and Goodman (2003) but also reflects additional information on affinity as provided by the authorities cited above. *M. jenkinsae* is the only species for which no phylogenetic relationship has been suggested, but in describing it Goodman and Soarimalala (2004) compared it most closely to *M. pusilla*. Olson, Goodman, and Yoder (2004) suggested that *M. prolixacaudata* of the extreme northern tip of Madagascar may be a species separate from *M. longidaudata*. Recent recognition of the distinction of various other species of *Microgale* formerly in synonymy, along with description of new species, has resulted from extensive survey work and development of improved taxonomic techniques, and is likely to continue (Garbutt 2007; Jenkins 2003; Olson and Goodman 2003; Olson et al. 2009).

Several species were formerly placed, as by Genest and Petter (1977), in the separate genera *Leptogale* Thomas, 1918, *Paramicrogale* Grandidier and Petit, 1931, and *Nesogale* Thomas, 1918. Those names generally have come to be regarded as synonyms of *Microgale* (Bronner and Jenkins 2005; MacPhee 1987a). *Nesogale*, however, might warrant resurrection, as the species it comprised, *M. talazaci* and *M. dobsoni*, now are thought to be more closely related to *Limnogale mergulus* than to the other currently recognized species of *Microgale* (see Olson and Goodman 2003). Therefore, *Microgale*, as delineated above, is paraphyletic, meaning that the genus contains only species descended from a common ancestor but not all the species descended from that ancestor. If the rules of cladistics were to be followed, it would be necessary

either to synonymize *Limnogale* with *Microgale* or to treat *Limnogale* as a genus or as a subgenus of *Microgale* and consequently to recognize *Nesogale* as a genus or subgenus for *M. talazaci* and *M. dobsoni* and possibly to divide *Microgale* into several more genera or subgenera. Additional molecular evidence for *Microgale* being paraphyletic and for the inclusion of *Limnogale* therein was provided by Kuntner, May-Collado, and Agnarsson (2010) and Poux et al. (2008).

According to Jenkins (2003), there is no obvious sexual dimorphism in external appearance or size. In many species the tail is not substantially longer or shorter than the head and body, but the long-tailed species, *M. principula*, *M. longicaudata*, and *M. majori*, have tails one and a half to more than twice the length of the head and body, and the short-tailed species, *M. brevicaudata* and *M. grandidieri*, have tails about half the head and body length (see also Olson, Goodman, and Yoder 2004; Olson et al. 2009). In *M. parvula*, the smallest species and among the world's smallest mam-

Shrew tenrec (*Microgale majori*). Photo by Link E. Olson.

mals, head and body length is 50–64 mm, tail length is 46–66 mm, and weight is 2–4 grams. Garbutt (2007) listed size for most other species, including the small *M. pusilla*, head and body length 47–56 mm, tail length 65–77 mm, weight 3.1–4.2 grams; the mid-sized *M. soricoides*, head and body length 79–103 mm, tail length 84–104 mm, weight 14–22 grams; the long-tailed *M. principula*, head and body length 70–80 mm, tail length 146–171 mm, weight 8.5–12.5 grams; the short-tailed *M. brevicaudata*, head and body length 63–74 mm, tail length 35–41 mm, weight 6.3–12.0 grams; and the large *M. talazaci*, head and body length 105–138 mm, tail length 120–150 mm, weight 31.5–47.0 grams. Eisenberg and Gould (1970) noted that *M. dobsoni*, another large species, usually weighs 34–45 grams but that with the onset of winter it accumulates fat reserves in its body and tail, and weight may then reach 84.7 grams in captivity. The tail of *M. thomasi* also is known to accumulate fat.

The pelage is short, dense, and soft and lacks spines (Jenkins 2003). The upper parts are buffy, dark rufous brown, olive brown, or nearly black; the underparts are usually buffy, buff gray, or lead-colored. *M. drouhardi* has a dark mid-dorsal stripe (Jenkins, Raxworthy, and Nussbaum 1997). The usual number of mammae is six (Goodman, Ganzhorn, and Rakotondravony 2003).

These animals are shrewlike and apparently have become adapted to the same sort of habitat that on the African mainland would be occupied by true shrews (Soricidae). Jenkins (2003) wrote that in general the body is long; the head is elongated and has a tapering

Dobson's shrew tenrec (*Microgale dobsoni*). Photo by J. F. Eisenberg / Mammal Images Library, American Society of Mammalogists.

rostrum, numerous vibrissae, and small eyes; the limbs are short and have five digits; and both sexes have a cloaca. In the long-tailed species, which evidently is semi-arboreal, the tip of the tail is modified for prehension, and the hind limbs are lengthened. *M. gracilis* and *M. gymnorhyncha*, which are probably semifossorial, have a velvety pelage, reduced eyes and ears, an elongated rostrum and rhinarium, and short forelimbs with stout claws. Garbutt (2007) indicated that in most other species the ears are prominent and project well above the fur. Most species have a thin, papery skull, with light and delicate teeth, but in *M. talazaci* and *M. dobsoni* the skull is heavily built and has stout, strong teeth.

According to Jenkins (2003), most species of *Microgale* occur exclusively in the eastern humid forests of Madagascar, though there are exceptions (see above list). In parts of that region, up to 11 species have been found in sympatry. Some species have a broad elevational range; *M. parvula* and *M. talazaci* are found from lowland to sclerophyllous montane forest, *M. cowani* from lowlands to sites above the treeline in ericoid shrubland at 2,450 meters. In contrast, *M. monticola* apparently is restricted to a narrow belt of montane forest at 1,550–1,950 meters, *M. nasoloi* to an isolated transitional zone between slightly humid and dry forest. Several species, including *M. brevicaudata*, *M. cowani*, *M. pusilla*, and *M. dobsoni*, have also been recorded from nonforested habitats. Shrew tenrecs generally forage among leaf litter, fallen branches, and roots on the forest floor, but most probably are also scansorial, running along inclined lianas and up into low underbrush. *M. principula* showed good climbing ability when released from traps, leaping short distances between branches and using its long tail to balance and coil around thin branches. Gould (1965) presented evidence that *M. dobsoni* employs echolocation to assist in its movements.

Eisenberg and Gould (1970) proposed a functional classification based partly on tail length. A modified version follows, covering most currently recognized species of *Microgale* and reflecting available behavioral information (Garbutt 2007; Jenkins 2003): (1) semi-fossorial forms (*M. brevicaudata*, *M. grandidieri*, *M. gymnorhyncha*, *M. gracilis*), in which the tail is relatively short and/or the forefeet have strong claws; (2) surface-foraging forms (*M. drouhardi*, *M. cowani*, *M.*

thomasi, M. parvula), in which tail length approaches the head and body length; (3) surface foragers and climbers (*M. soricoides, M. fotsifotsy, M. pusilla, M. talazaci, M. dobsoni*), in which the tail is up to 1.5 times as long as the head and body; and (4) partly or largely arboreal forms (*M. principula, M. longicaudata, M. majori*), in which the tail is 1.5–2.6 times as long as head and body.

Garbutt (2007) noted that activity is presumed to be both diurnal and nocturnal. Seasonal variation in activity has not been well studied. Olson et al. (2009) reported no evidence that *M. grandidieri* is active during the cold-dry season; presumably it aestivates or hibernates for extended periods when food availability is drastically reduced, but specimens showed no indication of tail incrassation, such as observed in *M. dobsoni*. According to Eisenberg and Gould (1970), a captive *M. dobsoni* did not enter deep torpor, but when it had accumulated fat reserves it showed a tendency to become inactive, sleep a great deal, eat less, and decline in body temperature. It constructed a nest when a suitable box was provided. *M. talazaci*

did not accumulate extensive amounts of fat and showed no tendency toward inactivity. In the wild, *M. talazaci* lives in a more stable habitat than does *M. dobsoni* and may not have evolved the latter's adaptations for an extended dry season. *M. talazaci* appeared to make use of an extensive tunnel system, as well as regular runways on the surface. *M. dobsoni* was observed to move about sniffing the substrate, occasionally to insert its nose under litter, and to rush at insects when it detected them by hearing. Both species, and probably all other species of *Microgale*, feed primarily on insects. A study of six species in Ranomafana National Park found Orthoptera and Hymenoptera to be the most commonly consumed prey (Soarimalala and Goodman 2003).

The large *M. talazaci* and *M. dobsoni*, *M. soricoides* with its enlarged incisors, and several other species apparently take some vertebrate prey, including frogs and even smaller species of *Microgale* (Jenkins 2003).

Observations by Eisenberg and Gould (1970) indicate that *M. dobsoni* is basically solitary and well spaced

Least shrew tenrec (*Microgale pusilla*). Photo by Harald Schütz.

in the wild. Strange individuals are antagonistic toward one another and sometimes fight, but a male and a female may establish a stable relationship. Several sounds, including squeals, are produced during agonistic encounters. The behavior of *M. talazaci* is much the same, but opposite sexes may maintain an association throughout the year. Both sexes scent-mark by dragging the cloacal area over a surface.

In the eastern humid forests of Madagascar, many species of *Microgale* generally give birth from late November to early December, coinciding with the start of the rainy season; litter size is 1–4 young (Jenkins 2003). Field studies indicate that *M. dobsoni* breeds in the spring and summer and that *M. talazaci* may have a more extended season. In *M. dobsoni* gestation is 62–64 days, maximum litter size is 6, the young open their eyes at 22–27 days, minimum age of conception is 22 months, and maximum known longevity in captivity is 5 years and 7 months. In *M. talazaci* gestation is 58–63 days, litter size averages 2 (1–5), birth weight is 3.6 grams, weaning takes place at 28–30 days, sexual maturity is reached at about 21 months, and maximum longevity is 5 years and 10 months (Eisenberg 1975; Eisenberg and Maliniak 1974; Goodman, Ganzhorn, and Rakotondravony 2003; Hayssen, Van Tienhoven, and Van Tienhoven 1993).

Of the 23 species of *Microgale* listed above, there is an IUCN assessment for all except the apparently extinct *M. macpheei*. Of those assessed, 17 have not been put in a formal threatened category, mainly because of their widespread distributions, but only one of the 17, *M. pusilla*, is considered to have a stable population; it appears better suited for survival outside of forests than are most other species (Goodman, Stephenson, and Soarimalala 2015g). The population trend of one other species, the newly described *M. grandidieri*, is unknown (Soarimalala, Stephenson, and Goodman 2015). The remaining 15 species without formal threatened classifications are nonetheless thought to be decreasing in numbers, usually through loss and fragmentation of natural habitat due to agricultural activity and logging (Goodman, Soarimalala, and Stephenson 2015a, 2015b, 2015c, 2015d; Goodman, Stephenson, and Soarimalala 2015h, 2015i, 2015j, 2015k, 2015l, 2015m; Soarimalala, Goodman, and Stephenson 2015; Stephenson, Goodman, and Soarimalala 2015a, 2015b, 2015c, 2015d). The same problems confront the five species that are in IUCN threatened categories but each of those has a very restricted range and is much more susceptible to human activity. Of those five, *M. jobihely* and *M. jenkinsae* are classified as endangered (Goodman, Soarimalala, and Stephenson 2015e; Hoffmann 2008) and *M. monticola*, *M. nasoloi*, and *M. dryas* are designated vulnerable (Goodman, Soarimalala, and Stephenson 2015f; Goodman, Stephenson, and Soarimalala 2015n, 2015o).

AFROSORICIDA; TENRECIDAE; **Genus *LIMNOGALE* Forsyth Major, 1896**

Web-footed Tenrec

The single species, *L. mergulus*, is known reliably from 10 localities in eastern Madagascar, from the Sihanaka Forest (south of Lake Alaotra) south to the Iantara River in Andringitra National Park; the range may extend as far north as Mananara-Nord near the east coast and Tsaratanana in the northwest (Benstead and Olson 2003). Recent systematic studies suggest *Limnogale* should be included in the genus *Microgale* (see account thereof).

Head and body length is 121–170 mm, tail length is 119–161 mm, and weight is 60–107 grams (Benstead and Olson 2003; Eisenberg and Gould 1970). The fur is close, dense, and soft and resembles that of an otter in texture. The upper parts are brownish with an admixture of reddish and blackish hairs. The underparts are pale yellowish gray. Females have six mammae.

Web-footed tenrec (*Limnogale mergulus*). Photo by P. J. Stephenson.

Web-footed tenrec (*Limnogale mergulus*). Photo by Kevin H. Barnes through Jonathan P. Benstead.

According to Benstead and Olson (2003), the skull of *Limnogale* is relatively broad but otherwise similar to that of *Microgale dobsoni* and *M. talazaci*. The eyes and ears are small and nearly hidden by the dense surrounding fur. The muzzle appears swollen, owing to enlarged mystacial pads underlying and supporting the vibrissae (whiskers) above the lips. The digits of all four feet are webbed to the base of the claws and there is a stiff fringe of hairs along the margins of the hind feet. The powerful tail, approximately equal in length to the head and body, is thick and almost square in the proximal half and laterally compressed in the distal half. It is covered with dense hairs that form a distinct keel along the underside.

Except as noted, the remainder of this account is based on the investigations of Benstead, Barnes, and Pringle (2001), which involved radio-tracking of two individuals, analysis of fecal droppings, observations in captivity, and assessment of distributional records of *Limnogale*. This semiaquatic tenrec seems restricted to fast flowing streams at elevations of 450–2,000 meters.

It is apparently active throughout the year (Benstead and Olson 2003). It is strictly nocturnal and remains in streamside burrows during daylight. A single burrow excavated in a previous study was dug horizontally into the bank about 50 cm above water level; it was 10 cm wide, 17 cm deep, and lined with grass and twigs. Nocturnal movement of the two tracked animals was restricted solely to stream channels and consisted of active foraging by swimming and diving. Mean distances traveled per night were 860 and 1,067 meters, with an overall range of 200–1,550 meters along the stream channel. Total lengths of stream channel used by the two animals were 505 and 1,160 meters, though it is not certain if those areas constituted entire home ranges. Activity typically began at, or just after, sunset and ended 60–90 minutes before sunrise. The animals sometimes remained active throughout the night and sometimes returned to their burrow for rest periods of up to 4 hours.

Limnogale is a tactile and exclusively aquatic predator (Benstead and Olson 2003). Observation of a cap-

407

tive showed that swimming is accomplished primarily by the hind feet, with the tail used as a rudder. Foraging involves dives of 10–15 seconds and sweeping the stream bed with the vibrissae. Once contacted, prey is seized in the mouth, brought to the surface and subdued, then taken to the land, where it is held in the forefeet and eaten. The diet consists mainly of larval and adult aquatic insects, larval anurans, and crayfish. Previous investigations indicated that small fish and frogs are also sometimes taken. Although *Limnogale* is often found in the same area as the aquatic lace plant, *Aponogeton*, the base of which harbors abundant aquatic invertebrates, earlier reports of a dependence on the plant seem unfounded (Benstead and Olson 2003).

The radio-tracked individuals did not show a regular and predictable pattern of movement and activity. This may indicate lack of territorial behavior and occurrence at naturally low densities. In contrast, *Limnogale* does use regular latrine sites for depositing feces, usually located on prominent boulders. It is not certain if the function is territorial. Nothing else is known of social life, very little of reproduction. Young are believed to be born during December and January, one pregnant female contained two embryos, and two juveniles were found in a burrow (Benstead and Olson 2003). However, Eisenberg and Gould (1970) thought average litter size probably is about three.

Nicoll and Rathbun (1990) regarded *Limnogale* as the most severely threatened tenrecid. The fast-flowing streams on which it depends are becoming increasingly isolated as agricultural expansion fragments the remnant upland forests. Although it evidently can survive without forest cover (see Benstead, Barnes, and Pringle 2001), it is jeopardized by siltation and soil erosion caused by deforestation. It also is subject to accidental capture in fish traps. It is known only from a few sites and has disappeared or become extremely rare at several of those. The IUCN currently classifies *Limnogale* as vulnerable, noting that less than 2,000 sq km of its habitat is estimated to remain and that the area is continuing to decrease (Goodman, Soarimalala, and Stephenson 2015g).

Golden Moles

This family of 10 genera and 21 species inhabits Africa from Cameroon and Somalia southward to the Cape of Good Hope. The sequence of genera presented here follows that suggested by Asher et al. (2010), who, after an extensive molecular and morphological assessment, tentatively recognized two subfamilies: Chrysochlorinae, with the genera *Chrysochloris, Cryptochloris, Huetia, Eremitalpa, Chrysospalax,* and *Calcochloris*; and Amblysominae, with *Amblysomus, Neamblysomus,* and *Carpitalpa*. The systematic position of the genus *Chlorotalpa* was considered unresolved but apparently intermediate to the two subfamilies. Previously, partly on the basis of dental formula and development of the malleus of the middle ear, Bronner and Bennett (2005) had used the following systematic sequence: subfamily Chrysochlorinae for the genera *Chrysospalax, Cryptochloris, Chrysochloris, Eremitalpa, Carpitalpa,* and *Chlorotalpa*; and subfamily Amblysominae for *Calcochloris* (including *Huetia*), *Neamblysomus,* and *Amblysomus*; the same subfamilial division and genera were listed by Bronner and Jenkins (2005). Still earlier, other classifications of the family were proposed. Simonetta (1968) designated another subfamily, Eremitalpinae, for *Chrysospalax* and *Eremitalpa*, and considered *Cryptochloris* a subgenus of *Chrysochloris*. McKenna and Bell (1997) did not recognize the subfamily Amblysominae, included *Carpitalpa* in *Chlorotalpa*, and included *Calcochloris* and *Neamblysomus* in *Amblysomus*.

In contrast to tenrecids, chrysochlorids display a high degree of morphological conformity (Bronner and Bennett 2005). Head and body length is 60–235 mm; the tail is not visible externally. The skin is tough and loosely attached to the body; the fur is thick and has dense woolly underfur. Notwithstanding the common name, not all species are gold in color. The pelage has a unique metallic luster or iridescence of red, yellow, green, bronze, or violet. Microscopic study of nanostructure (Snyder et al. 2012) indicates the hairs are flattened and have highly reduced cuticular scales forming multiple layers of light and dark materials, strikingly similar to those in the elytra of iridescent beetles. The layers produce color through thin-film

South African golden mole (*Amblysomus hottentotus*). Photo by Chris and Mathilde Stuart.

interference, and sensitivity of that mechanism to slight changes in layer thickness and number explains color variability. As chrysochlorids are blind, the structural basis of the color seems unrelated to sexual ornamentation and may be associated with other selective factors, including the ability to move and keep clean in dirt and sand.

Golden moles somewhat resemble true moles (Talpidae) in appearance. The muzzle terminates in a smooth, leathery pad used in working in the soil, and the nostrils are located under a fold of skin at the front of the long snout. The eyes are vestigial and covered with hairy skin; in the genus *Eremitalpa* the eyelids fuse at an early age, and the skin covering the eye increases in thickness. The ears are small and concealed in the pelage. Golden moles have short forelimbs with four clawed digits; the third digit is the longest and has a powerful claw, and the other three vary in relative length according to the species. The anterior walls of the chest cavity are deeply hollowed to provide space for the thick, muscular forelimbs. There are only four digits on the front feet of chrysochlorids, whereas there are five in talpids (Bronner and Bennett 2005). The hind limbs are also short, the two lower bones (the tibia and fibula) being fused near the ankle. The hind feet have naked soles and five sharp-clawed toes connected by membranous skin. Females have one pair of abdominal mammae and one pair of inguinal mammae. The urogenital system has a single external opening.

In most genera the dental formula is: i 3/3, c 1/1, pm 3/3, m 3/3 = 40. *Calcochloris, Amblysomus*, and some specimens of *Neamblysomus* have the same number of incisors, canines, and premolars but only 2 molars above and 2 below, making a total of 36 teeth. Presence or absence of the third molar is not considered a reliable feature in subfamilial diagnosis (Asher et al. 2010). The teeth are slightly separated from each other. The incisors are enlarged, the first premolar is much like the canine, and the two posterior incisors and the last two premolars are like the molars—narrow and wedge-shaped with a bladelike medial half. The molars usually have two lateral cusps and one medial cusp, with an enamel ridge on each side of the medial cusp. Golden moles have high-crowned teeth, the crowns of the lower teeth being twice as high as those of the upper teeth. The skull is conical in shape, broad at the braincase, and not constricted at the orbits. The olfactory region and the premaxillary bones form a narrow, elongate snout that widens at the end into two processes on which the nose pad is attached. Bronner and Bennett (2005) indicated certain unique cranial features, such as a hyoid-dentary articulation, which apparently supports the tongue during prey-handling and mastication, and a hypertrophied malleus of the middle ear in most genera, which permits extreme sensitivity to underground sounds and vibrations. Mallear enlargement probably occurred independently in each subfamily (Asher et al. 2010).

Golden moles inhabit sandy areas, plains, forests, and cultivated areas. They burrow by means of powerful thrusts, as do the true moles (Talpidae), but use the armored nose more than do the talpids. Some forms burrow just below the surface of the ground; others generally burrow deeper. Local soil conditions probably determine the depth of the tunnels. Members of the genus *Chrysospalax* (and probably some others) rest in chambers and passages in mounds of earth reached by a system of tunnels made in part by them and in part by certain rodents. Golden moles are active day or night. They are believed to be inactive for varying periods during the colder months. The majority of forms seem to find food underground and come to the surface only after a rain brings worms and other invertebrates to the surface. *Chrysospalax* and *Eremitalpa*, however, may seek food on the surface at night. Golden moles have an extraordinary sense of orientation and can return directly to the burrow entrance with rapid speed. Most forms feed on invertebrates; *Eremitalpa* includes species of sand-burrowing skinks in its diet, and *Cryptochloris* feeds on legless lizards in addition to various invertebrates. At least one genus (*Cryptochloris*) has the unusual habit of feigning death when picked up or even when turned over with a spade. This may be similar to the involuntary "fainting" of *Didelphis*. Golden moles seldom bite when handled. Very little is known about their social life and reproduction.

The geological range of this family is early Miocene, late Pliocene, Pleistocene, and Recent in Africa (McKenna and Bell 1997). Molecular dating suggests that golden moles diverged from a common ancestor approximately 28.5 million years ago (Gilbert et al. 2006),

within the later Oligocene. The Miocene records are from Kenya. *Proamblysomus*, an extinct genus from the late Pliocene and/or Pleistocene in South Africa, known from skull fragments found in the Sterkfontein caves, is a small golden mole closely resembling *Amblysomus*.

AFROSORICIDA; CHRYSOCHLORIDAE; **Genus** *CHRYSOCHLORIS* Lacépède, 1799

Cape Golden Moles

There are two subgenera and three species (Aggundey and Schlitter 1986; Asher 2010; Bronner 1995a, 2015a; Bronner and Bennett 2005; Kingdon 1974; Lamotte and Petter 1981; Meester 1977a):

subgenus *Chrysochloris* Lacépède, 1799

> *C. asiatica*, western parts of Northern Cape and Western Cape provinces in South Africa, Robben Island;
> *C. visagiei*, known only from the type specimen collected at Gouna in Northern Cape Province in western South Africa;

subgenus *Kilimatalpa* Lundholm, 1955

> *C. stuhlmanni*, scattered mountainous parts of western Cameroon, northern and northeastern Democratic Republic of Congo, Uganda, western Rwanda, western Kenya, and central Tanzania.

Unpublished phylogenetic analyses support elevation of *Kilimatalpa* to generic rank (Bronner 2015a). The form in Cameroon, *C. stuhlmanni balsaci*, known only from Mount Oku, may be a full species (Bronner and Jenkins 2005). Several field trips to the type locality of *C. visagiei*, which was described in 1950, have yielded no specimens or signs of golden moles, suggesting either an error in recording provenance of the original specimen or that it was an aberrant individual of *C. asiatica* transported to the collection locality by people or floodwaters (Bronner 2015b).

Head and body length is 90–140 mm. Reported weight is 23–46 grams in *C. asiatica* (Bronner and Bennett 2005) and 26–53 grams in *C. stuhlmanni* (Kingdon

Cape golden mole (*Chrysochloris asiatica*). Photo by Hamish G. Robertson, Iziko Museums of South Africa.

1974). The ground color is tawny olive, brownish, or grayish. Depending on the angle from which the animal is viewed, the fur has an iridescent luster of greenish, bronze, or violet. The face and nose pad are often paler in color than the back. The underparts are light, almost white or creamy in some individuals. The fur is soft, fine, and dense. The Bakiga people of the Kigezi area of Uganda were said to use the skins of *C. stuhlmanni* as charms at one time.

Chrysochloris is distinguished by prominent temporal bullae that form bulges in the braincase toward the rear of the eye orbits (Bronner and Bennett 2005). There are two well-developed foreclaws; the one on the second digit is shorter than that of the third. The first claw is still shorter, and the fourth digit is represented by a small tubercle. Females have four mammae.

In the Rwenzori Mountains, *C. stuhlmanni* lives at elevations as high as 3,500 meters (Kingdon 1974), whereas *C. asiatica* is found in coastal and other lowland areas (Bronner and Bennett 2005). These golden moles usually burrow just below the surface of the ground, so that fresh workings may be traced by the cracked and ridged ground surfaces above the shallow tunnels. These tunnels sometimes radiate from the base of a bush and sometimes penetrate the large mounds of mole-rats (*Bathyergus*). Occasionally, however, golden moles burrow downward to such depths that they must remove soil from the burrow, which they push to the surface of the ground as mounds of fresh soil. They are common in certain areas and are often found in gardens, where they can burrow in the loose, cultivated soil. They become active on the surface in rainy weather, when they root like little pigs for earthworms in the damp earth. Sometimes they make short overland journeys at night. Most activity of *C. asiatica* is at night, with a minor peak in the late afternoon (Bronner and Bennett 2005). In addition to worms, the diet includes other invertebrates, such as beetles and grubs.

The breeding season of *C. asiatica* in Western Cape Province seems to be during the rainy months, April–

July. In that area the animals make a round nest of grass in which the young are born. Litter size is one to three young (Bronner and Bennett 2005). At birth they are about 47 mm in length when stretched straight. They suckle until nearly full grown, that is, for 2–3 months, as the teeth do not cut through the gums until the young are almost mature. Kingdon (1974) wrote that reproductive details for *C. stuhlmanni* are unknown, though small, probably subadult individuals have been collected in July in Uganda and in March in Democratic Republic of Congo, and no females in breeding condition were found in the Rungwe district of Tanzania in January–February. Weigl (2005) listed an individual *C. asiatica* that was born about July 1979 in the wild and died 23 December 1981 in captivity.

AFROSORICIDA; CHRYSOCHLORIDAE; Genus *CRYPTOCHLORIS* Shortridge and Carter, 1938

De Winton's and Van Zyl's Golden Moles

There are two species (Bronner 2015c, 2015d; Bronner and Bennett 2005; Helgen and Wilson 2001; Meester 1977a):

> *C. wintoni,* known only from Port Nolloth on coast of northwestern Northern Cape Province of South Africa;
> *C. zyli,* known only from Compagnies Drift, 16 km inland from Lamberts Bay on coast of Western Cape Province, and from Groenriviermond on coast of western Northern Cape Province of South Africa.

Unpublished phylogenetic analyses based on both morphological and genetic data support the distinction of these two species and justify making *Cryptochloris* a subgenus of *Chrysochloris* (Bronner 2015c).

Head and body length is 80–92 mm. The fur is short, soft, and dense. In *C. wintoni* the upper parts are pale fawn with a tinge of yellow, which becomes more intense on the face, and a silvery sheen. In *C. zyli* the upper parts are drab lead or brown with a violet iridescence, and there are whitish buff facial markings. In both species the underparts are somewhat paler than the upper parts. The colors of both are remarkably like those of *Eremitalpa granti.*

Compared to *Chrysochloris*, *Cryptochloris* has greater frontal expansion of the cranium, smaller temporal bullae, and a better developed first digit of the forefoot. There are three well-developed foreclaws. The flattened and roundly oval body shape is most pronounced in *C. zyli.* As in *Eremitalpa,* this shape may be an adaptation for burrowing through loose sand.

Cryptochloris is endemic to the same general region as *Eremitalpa.* In the Lamberts Bay area both genera occupied the white coastal sand dunes. Their habits also appear to resemble those of one another. Based on early observations, *C. wintoni* generally burrows just below the surface. Occasionally it tunnels deeper, often to the base of a bush, where presumably it secures shelter and rears its young. In those areas where the sand is extremely powdery the roofs of its surface tunnels collapse almost immediately and form shallow furrows. After a rain the fresh workings of this golden mole are indicated on the surface by minute cracks that disappear as soon as the ground dries. Unsuccessful attempts to dig the animals out usually result in a desertion of the disturbed ground the following night, when the moles travel overland to a new area, often a distance of several hundred meters. *C. zyli* is said to utter

Van Zyl's golden mole (*Cryptochloris zyli*). Photo by Jenny Jarvis.

Congo golden mole (*Huetia leucorhina*). Photo by Jabruson / npl / Minden Pictures.

a fairly sharp squeak when handled; when first picked up or when turned over with a spade, it feigns death.

The diet consists of various invertebrates and legless lizards (*Typhlosaurus*). The lizards, which grow to a length of nearly 200 mm, shelter in the sand under bushes. It has been conjectured that *Cryptochloris* uses its long foreclaws to kill and devour them.

Both species are now extremely rare, apparently because of human disruption of their restricted habitat. Mining of coastal sands for alluvial diamonds could be a factor in the decline. *C. wintoni* is represented by only three museum specimens, has not been collected for over 50 years, and may already be extinct. *C. zyli* is represented by only four museum specimens; three, from Compagnies Drift were taken in 1937–1938, but the one from Groenriviermond, 150 km farther north, was collected in 2003, suggesting the species may still occur over a substantial area. The IUCN classifies *C. zyli* as endangered and *C. wintoni* as critically endangered (Bronner 2015c, 2015d).

AFROSORICIDA; CHRYSOCHLORIDAE; **Genus** *HUETIA* Forcart, 1942

Congo and Somali Golden Moles

There are two species (Bronner 1995b; Kitchener et al. 2008):

H. leucorhina, known from scattered localities in southern Cameroon, southwestern Central African Republic, Gabon, Congo, Democratic Republic of Congo, and northeastern Angola;

H. tytonis, known only from the type specimen from an owl pellet collected at Giohar in southern Somalia.

Except as noted, the information for the remainder of this account was taken from Bronner (1995b), who treated *Huetia* as a subgenus of *Calcochloris* and noted that the few characters by which *Calcochloris tytonis* could be assessed suggest it may be closely related to *C. leucorhina*, though he favored not making a definite subgeneric assignment until more specimens were available. The type specimen of *tytonis* is from an owl pellet, is incomplete, and may have been lost, though descriptive data, measurements, and a photograph are available. Asher et al. (2010) elevated *Huetia* to generic level but did not discuss the status of *tytonis*. Maree (2015a, 2015b) accepted *Huetia* as a full genus for the species *H. leucorhina*, but tentatively treated *tytonis* as a species of *Calcochloris*.

H. leucorhina is similar in size to *Calcochloris obtusirostris*, with a head and body length of 65–126 mm. The pelage is dark brown to slate- or mouse-gray, the underparts being slightly paler. A creamy white mask occurs on the face, extending laterally almost to the ear. A third molar is invariably present, the palate is long and narrow, and the malleus is not inflated, so that there is no distinct sub-temporal bulla. *C. obtusirostris* differs in having yellow-orange underparts, a broad palate, and no third molar.

Measurements of the mandible of *H. tytonis* show that it is considerably larger than *H. leucorhina* and approximately the size of that of the larger species of *Amblysomus*. However, *H. tytonis* has a third molar, and its stylohyal bone is not nearly as robust as that in *Amblysomus*. The malleus of *H. tytonis* is not inflated and resembles that of *H. leucorhina*.

H. leucorhina is restricted primarily to lowland equatorial forests of western Africa and montane forests of central Africa. It has also been found in forest-savannah mosaics, and signs of activity have been observed in pastoral and cultivated lands as well as rural and urban gardens (Maree 2015a). The habitat where *H. tytonis* was collected consists of dense bush and savannah; the IUCN formerly classified the species as critically endangered but now indicates that data are insufficient to ascertain status (Maree 2015b).

AFROSORICIDA; CHRYSOCHLORIDAE; **Genus
EREMITALPA Roberts, 1924**

Grant's Golden Mole

Currently, a single species is recognized, *E. granti*, which occurs from the Kuiseb River in western Namibia, south through the coastal sand dunes of the Namib Desert and Northern Cape Province, to Port Nolloth in Western Cape Province of South Africa (Bronner and Bennett 2005). Unpublished phylogenetic analyses, based on molecular, cytogenetic, and morphological characters, support full species status for *E. g. namibensis*, found north of the Orange River, and *E. g. granti*, found to the south (Maree 2015c).

This is the smallest golden mole. Head and body length is 60–88 mm. Bronner and Bennett (2005) listed weights of 17–30 grams for males and 15–23 grams for females. The fur is long and silky. The upper parts of subadults are a beautiful aluminum gray but change with age to sandy buff or gray. The face is whitish or buffy in adults, with indications of pale cheek markings in the immature. The underparts vary from buffy white to pale rufous. The tips of the hairs, especially on the lower back, shine like spun glass, giving a silvery sheen.

Eremitalpa resembles *Chrysochloris* but is distinguished by lacking temporal bullae and in having well-developed claws on the first three digits of the forefoot (not just on two digits). The claws are broad, flat, and leaflike. *Eremitalpa* is further distinguished from other genera by having a prominent thickened pad on the hind foot that probably takes the place of a heel, and in bearing a well-developed, scraperlike claw on the vestigial fourth digit of the forefoot (Bronner and Bennett 2005; Perrin and Fielden 1999). The body shape of *Eremitalpa*, like that of *Cryptochloris*, is flattened and roundly oval, like a small tortoise, and may be an adaptation for tunneling through loose sand. Fons (1990) stated that *Eremitalpa* moves by "literally swimming" in the sand. It evidently is not a good swimmer in water (Bronner and Bennett 2005).

Grant's golden mole is confined largely to sand dunes. It probably does not range inland because of the relatively firm and level sandveld. It makes shallow, winding tunnels in the dunes. The tunnels often can be traced for 45 meters or more and occasionally connect with deeper excavations that descend several meters

Grant's golden mole (*Eremitalpa granti*). Photo by Galen Rathbun / California Academy of Sciences.

below the ground. It has been speculated that the young are borne in these shafts. Mounds of sand are not thrown up in the manner of some moles that burrow in harder soil. Surface openings occur irregularly wherever excavations are present. While foraging, Eremitalpa often runs on the surface, periodically dipping its head below the surface, and then plunging into the sand when abundant food is sensed. The head-dipping behavior may involve use of seismic cues for navigation and prey detection, which may be facilitated by the extraordinary development of the malleus in the middle ear (Bronner and Bennett 2005; see also Lewis et al. 2006).

Fielden (1991), Fielden, Perrin, and Hickman (1990), and Perrin and Fielden (1999) have reported on the ecology and behavior of *Eremitalpa*. Unlike most other subterranean mammals, it lacks a permanent burrow system and is a nocturnal surface forager. Individuals cover different parts of their range each night, usually forage along a path of several hundred meters, and seldom return to the rest site occupied the previous day. Termites are the major prey, but the diet includes a variety of other invertebrates and also small lizards. Most rest sites are shallow excavations under vegetative cover. No deep underground nesting chambers have yet been discovered, though they presumably would be needed for the rearing of young. Population densities are estimated at 0.014–1.19 per ha., the large variation resulting from differences in prey abundance. Home ranges are stable and average 4.63 ha., with those of males (3.10–12.30 ha.) usually being larger than those of females (1.80–4.59 ha.). Ranges overlap considerably, but individuals are solitary and avoid one another while foraging. Two pregnant females, each with a single fetus near full-term, were taken in October. Two individuals were found again, within 100 meters of the initial capture points, after two years. Rathbun and Rathbun (2006a) followed six individuals for up to 21 days by radio-tracking; the animals moved an average straight-line distance of 8.8 meters during daylight and 5.6 meters at night, though some did not move at all for periods of 2–11 days, when they apparently were torpid. Those distances were much shorter than those found in other studies—for example, a mean track length of 1,412 meters reported by Seymour, Withers, and Weathers (1998)—but may have been associated with cold winter conditions. Bronner and Bennett (2005) suggested that *Eremitalpa* may be more social than other chrysochlorids and noted that pregnant females with one or two young had been recorded in October and November.

Maree (2015c) stated that *Eremitalpa* is probably more widespread than current records suggest and does not appear to be declining to the extent that would warrant assignment to an IUCN threatened category. Although mining of coastal sands for alluvial diamonds, as well as tourist development and agricultural practices have altered and fragmented some habitat, most of the range of the species coincides with deserts where human influence is not substantial.

AFROSORICIDA; CHRYSOCHLORIDAE; **Genus** *CHRYSOSPALAX* Gill, 1883

Large Golden Moles

There are two species (Bronner and Bennett 2005; Meester 1977a):

> *C. villosus*, Gauteng, Mpumalanga, KwaZulu-Natal, and Eastern Cape Provinces of South Africa;
>
> *C. trevelyani*, Eastern Cape Province of South Africa from King William's Town and East London districts east to Port St. Johns.

In *C. villosus* head and body length is 127–175 mm, and weight is 108–142 grams; in *C. trevelyani*, which is the largest golden mole, head and body length is 208–235 mm, and weight is 410–500 grams (Bronner and Bennett 2005). The fur is harsher in texture than that of the other genera of golden moles and lacks a pronounced sheen. *C. villosus* has rufous, yellowish brown, dark brown, or dark slaty upper parts and gray-ish underparts. In *C. trevelyani* the upper parts are dark glossy brown to yellowish or grayish brown, the underparts are slightly paler, and there are yellow patches on the head and forequarters.

Distinguishing features of this genus are the large body size and the shape of the zygomatic arch of the skull, which is produced upward posteriorly and meets the lambdoid crest at the back. The claw of the fourth finger is usually much more developed in *C. trevelyani* than in *C. villosus*. Females of *C. villosus*, at least, have four mammae.

C. villosus has been found in grassy areas, especially in meadowlike ground bordering marshes. *C. trevelyani* is associated with indigenous forests and is known to be predominantly nocturnal (Bronner and Bennett 2005). Large golden moles live in burrow systems consisting of chambers and passageways. They can excavate the burrows themselves but are said to sometimes use tunnels made by blesmols (*Cryptomys*) or mole-rats (*Bathyergus*). Poduschka (1980) reported that *C. trevelyani* makes numerous hills 40–60 cm in diameter and about 25 cm high, in which there are plugged openings to the tunnels below. In another study of that species, however, Maddock (1986) found no molehills and that surface entrance holes are left open; he reported that burrows are up to 13.6 meters long and are connected by runways through the leaf litter. *C. villosus* also makes burrows with open entrance holes, but apparently no subsurface runways (Bronner and Bennett 2005).

Poduschka (1980) questioned earlier reports that *C. trevelyani* actively seeks prey above ground, that it hibernates in winter, and that it can swim across a stream. Hickman (1986) confirmed that *C. trevelyani* is a poor

Large golden mole (*Chrysospalax villosus*). Photo by Gary Bronner.

Large golden mole (*Chrysospalax villosus*). Photo by Graham C. Hickman / Mammal Images Library, American Society of Mammalogists.

swimmer. However, Maddock (1986) determined that *C. trevelyani* does emerge nightly to forage or to move up to 95 meters across the surface to another tunnel and that it does occasionally enter nonforested areas. One male was found to move an average of 36.6 meters per night over 7 nights (Bronner and Bennett 2005). Whereas Poduschka (1980) indicated that *C. trevelyani* feeds mainly on giant worms (*Microchaetus*), Maddock (1986) found a wider dietary range. *C. villosus* is known to emerge from its burrow, normally after a rain, and to search for insects and earthworms (Bronner and Bennett 2005). At such times it is said to root about like a little pig. It has a remarkable sense of direction in regard to the exact location of its burrow; when threatened, it dashes rapidly and surely for the entrance. It is not known if it also forages in underground tunnels, but that seems likely given the long periods without rainfall during the dry winter months throughout most of its range (Maree, Bennett, and Bronner 2005). Although

evidently not herbivorous, *Chrysospalax* and other golden moles have been blamed for damage to vegetables and crops that was done by rodents, particularly the blesmol (*Cryptomys*). One female of each species of *Chrysospalax* has been found to contain two embryos, and a female *C. trevelyani* captured in October gave birth to a single young (Bronner and Bennett 2005).

Both species require very specific habitat, which is being lost, degraded, and fragmented in association with agricultural expansion, forest clearance, logging, livestock overgrazing, burning of grasslands, mining, and urbanization. Total remaining areas of occupancy are estimated to be only 128 sq km for *C. villosus* and 272 sq km for *C. trevelyani*. *C. villosus* no longer occurs at 3 of the 11 sites from which it is known historically. *C. trevelyani* also may have vanished from many localities and appears able to survive only in larger patches of indigenous montane forest with soft soils, well-developed undergrowth, and deep leaf-litter layers. The IUCN now classifies *C. villosus* as vulnerable and *C. trevelyani* as endangered (Bronner 2015e, 2015f).

AFROSORICIDA; CHRYSOCHLORIDAE; Genus *CHLOROTALPA* Roberts, 1924

Duthie's and Sclater's Golden Moles

There are two species (Bronner and Bennett 2005):

C. duthieae, coastal belt between George in Western Cape Province and Port Elizabeth in Eastern Cape Province of South Africa;

C. sclateri, Sutherland area of south-central Northern Cape Province of South Africa, northeastern Western Cape Province, northern Eastern Cape Province, Lesotho and adjacent areas of Free State and KwaZulu-Natal, Wakkerstroom area of Mpumalanga.

Chlorotalpa was treated as a full genus by Simonetta (1968) and Meester (1977a) but considered part of *Amblysomus* by Petter (1981) and Yates (1984). Morphological and karyological studies by Bronner (1991, 1995a, 1995b) support the generic distinction of *Chlorotalpa*. Except as noted, the information for the remainder of this account was taken from Bronner and Bennett (2005).

Duthie's golden mole (*Chlorotalpa duthiae*). Photo by Peter Dawson and Pauline Hawkins.

In *C. duthieae* head and body length is 95–130 mm, and mean weight is 33.6 grams in males, 26.5 grams in females. The upper parts are reddish black to brownish black and have a very distinct green sheen. The sides of the face are yellowish, with that color extending upward on each side as a triangular patch with the apex where the eye would have been. In *C. sclateri* head and body length is 82–135 mm, and weight is 22–54 grams in males, 22–27 grams in females. The upper parts are glossy reddish brown to dark brown, the flanks have a reddish tinge, and the underparts are a dull gray. Both species have long, slender foreclaws. However, the first digit of the forefoot is reduced, the tip of the claw barely reaching the halfway point of the second digit. The third foreclaw is the longest (7–9 mm). Females have two pairs of mammae.

Like the genera of subfamily Amblysominae, *Chlorotalpa* does not have prominent temporal bullae. However, *Chlorotalpa* is distinguished from those genera by having the malleus of the inner ear enlarged, with a spherical or clublike shape, and in consistently having a third molar plus well-developed talonids on the lower molars (see account of family Chrysochloridae).

Little is known of the natural history of these golden moles. *C. duthieae* occurs in alluvial sand and sandy loam within coastal forests and sometimes savannahs, cultivated areas, and gardens. *C. sclateri* is restricted to montane grassland, scrub, and forest at elevations up to 3,000 meters or more, often near streams. It constructs shallow subsurface tunnels for foraging, and those radiate out from deeper tunnels, in which nesting chambers are located, usually under rocks or trees. It is active mainly at night but also for short spells during the day after rain. Stomach contents of one specimen consisted exclusively of earthworms. Adults of *C. duthieae* are solitary, but up to 4 individuals per hectare have been trapped on the same night, suggesting population densities are relatively high in areas of suitable habitat (Bronner 2015g). Two male *C. duthieae* with enlarged testes and a female of the same species, pregnant with two fetuses, were trapped in November, suggesting that breeding occurs mainly in the wetter summer months. Bronner (2013b) cited information indicating that pregnant female *C. sclateri* have been recorded in the wet summer months of December and January, that litters of that species contain two young and are born in grass-lined nests, and that females have a postpartum estrus.

The IUCN classifies *C. duthieae* as vulnerable. It is known from only nine localities and is jeopardized by disruption of its specialized habitat by extensive coastal housing and tourism development (Bronner 2015g). In contrast, *C. sclateri* is relatively widespread, coexisting and often thriving in close proximity to people, pro-

vided that habitat disturbance is not too great; much of its range coincides with mountains where human influence is not substantial (Bronner 2015h).

AFROSORICIDA; CHRYSOCHLORIDAE; **Genus** *CALCOCHLORIS* Mivart, 1867

Yellow Golden Mole

The single species, *C. obtusirostris*, occurs in extreme southeastern Zimbabwe, southern Mozambique, and the extreme northeastern parts of Limpopo and KwaZulu-Natal provinces of South Africa (Bronner and Bennett 2005). *Calcochloris* was considered a synonym of *Amblysomus* by Petter (1981), Simonetta (1968), and Yates (1984), but a full genus by Meester (1977a) and Meester et al. (1986). Morphological and karyological studies by Bronner (1991, 1995a, 1995b) support the generic distinction of *Calcochloris*. Bronner and Jenkins

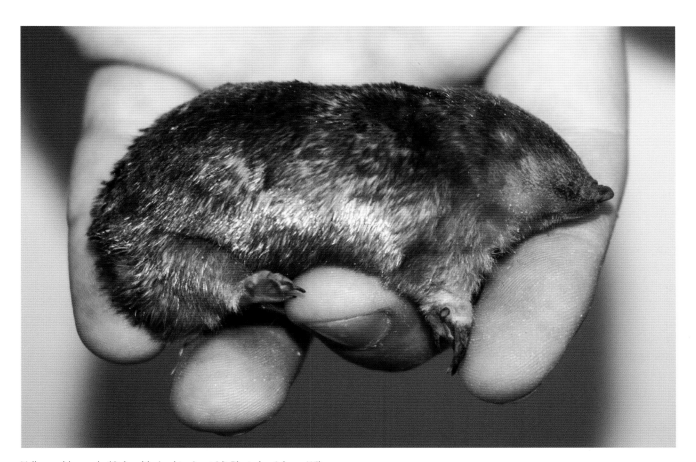

Yellow golden mole (*Calcochloris obtusirostris*). Photo by Johnny Wilson.

(2005) treated *Huetia* (see account thereof) as a subgenus of *Calcochloris*. What is here treated as the species *Huetia tytonis* was tentatively listed as a species of *Calcochloris* by Maree (2015b).

Head and body length is 82–110 mm, and weight is 15–37 grams (Bronner and Bennett 2005). The coloration of the upper parts varies geographically from glossy reddish brown (with yellow underfur) to bright golden yellow. The chin, throat, limbs, and belly are yellow. The sides of the face are yellow, and there is a broad yellow or buffy band across the top of the snout. The claw on the third digit of the front foot is about 10 mm long, and the fourth digit has a claw about 2 mm long (Bronner and Bennett 2005). From *Amblysomus*, *Calcochloris* is distinguished by its striking coloration, broader skull, and more molariform first upper and lower premolar teeth (Meester et al. 1986).

According to Bronner and Bennett (2005), the yellow golden mole is confined to light sandy soils, sandy alluvium, and coastal sand dunes, with a cover of forest, savannah, or thornveld. In captivity it tunnels to a depth of 20 cm in loose, sandy soils but not in heavy clay or schist soils. In the wild it moves in subsurface runs leading from chambers around the bases of trees, and if disturbed it may burrow deeply into the substrate. It is highly sensitive to the movements of prey dropped onto the surface of the soil and quickly burrows to the vicinity. At times it may emerge and move porpoise-like across the surface at surprising speed. The diet consists of earthworms and insects. Pregnant females, some found to contain one or two fetuses, were collected in September, October, and January, and lactating females were taken in January, thereby suggesting that breeding is restricted to the wet summer months. Maree (2015d) noted that the yellow golden mole occurs close to human settlements, thrives in gardens, and is unlikely to be in decline.

AFROSORICIDA; CHRYSOCHLORIDAE; **Genus** *AMBLYSOMUS* Pomel, 1848

South African Golden Moles

There are five species (Bronner 1995a, 1996, 2000; Bronner and Bennett 2005; Bronner and Mynhardt 2015a, 2015b, 2015c; Rampartab 2015a; 2015b):

A. corriae, from vicinity of Porterville and Stellenbosch to Riversdale and from George to Humansdorp in Western Cape and Eastern Cape Provinces of South Africa;

A. septentrionalis, from Parys to Heilbron and around Harrismith in eastern Free State, Barberton and Ermelo to Wakkerstroom in southeastern Mpumulanga Province of South Africa, possibly northwestern Swaziland;

A. hottentotus, coastally from Van Stadens River in Eastern Cape Province to St. Lucia district in KwaZulu-Natal Province of South Africa, inland to Maclear-Ugie in the south to Van Reenen in the north, possibly also in Lesotho and Swaziland, an apparently isolated subspecies (*A. h. meesteri*) in the Barberton-Graskop region of Mpumalanga Province;

A. marleyi, Ubombo and Ingwavuma on eastern slopes of Lebombo Mountains in northeastern KwaZulu-Natal Province of South Africa, apparently also owl pellet remains at Weenen to southwest;

A. robustus, Steenkamps Mountains in Belfast and Dullstroom districts of eastern Mpumalanga Province of South Africa, extending eastward to Lydenberg and possibly southward toward Ermelo.

Meester (1977a), and various other authorities, recognized an additional species, *A. iris*, that comprised the subspecies *A. i. iris*, *A. i. corriae*, and *A. i. septentrionalis*. However, Bronner (1996) showed that *A. i. iris* is actually a subspecies of *A. hottentotus* and that *corriae* and *septentrionalis* are separate species. *Amblysomus* is sometimes considered to include *Chlorotalpa*, *Calcochloris*, and *Neamblysomus* (see accounts thereof). Karyological analysis by Gilbert, Maree, and Robinson (2008) found *A. hottentotus meesteri* (see above) to be distinct from other of the named subspecies of *A. hottentotus*, which in turn group more closely with *A. robustus*, and hence that *meesteri* should be elevated to specific status.

Head and body length is 90–145 mm, and weight is 37–101 grams (Bronner and Bennett 2005; Kuyper 1985). The fur is short, fine, and dense. The upper parts are usually dark reddish brown, but sometimes blackish in *A. corriae*. The flanks are generally lighter and

South African golden mole (*Amblysomus hottentotus*). Photo by Sarita Maree.

are bright reddish brown in *A. robustus*. The underparts also are paler than the upper parts and are orangish in some species. All species have pale markings on the head and usually yellowish cheek patches. Distinguishing features of this genus are the absence of temporal bullae, lack of a third molar, and the presence of two well-developed foreclaws. The four mammae are in two pairs, one inguinal and one on the flanks.

These golden moles vary geographically in South Africa because of local isolation in much the same manner as the American pocket gophers (Geomyidae). Some species inhabit areas of peaty soil in the sheltered ravines of mountains or in forests; others occur in escarpment forests, and certain forms are present at times in forests but are usually more common in open areas where there is a good grass cover. Some forms reportedly usually burrow just below the surface of the ground, so that raised ridges mark the course of the tunnel. Others are said to burrow deeper and to throw up

mounds of fresh earth. Local soil conditions may be the major factor in determining the depth of the tunnel. These animals are sometimes reported to be nocturnal, though at least some species are often active in the daytime and during the rainy season. They make special, grass-lined nesting chambers in which they bear and nurse their young. Tunnels are often located near the nest chamber so that if an enemy, such as a mole snake, appears, the golden mole can burrow rapidly through the walls and escape. Hickman (1986) found *A. hottentotus* to be a competent swimmer. The diet consists of worms, larvae, pupae, and insects.

According to Kuyper (1985), *A. hottentotus* occupies a wide range of habitats from sea level to an elevation of 3,300 meters. It is active at intervals throughout the day and night. When the temperature is above about 30° C or below 15° C, it enters daily torpor to conserve energy. Excavated burrows have varied from 9.5 to 240 meters in length. The more permanent burrows are

29–94 cm below the surface and have several vertical passages that extend to a greater depth where the animal can sleep or escape danger. It burrows through about 4–12 meters of soil per day; during the winter, when rainfall is low, its movements are more restricted and at greater depths. It is solitary, fighting sometimes occurs, and a dominant individual may take over a neighbor's burrow. The young weigh 4.5 grams at birth and are evicted from the maternal burrow when they reach 35–45 grams. In a study of *A. hottentotus* from coastal KwaZulu-Natal, Schoeman et al. (2004) found breeding to be aseasonal; reproductive organs of both sexes exhibited no regression at any time of the year. However, there was evidence of enhanced fecundity during the warm, wet summer months. Litter size averaged two young and ranged from one to three. For *A. septentrionalis*, limited data indicate pregnancies occur throughout the year, mostly in the wet summer months from November to March; litters contain one or two young (Bronner 2013b). Pregnant female *A. corriae*, each with two embryos, have been captured in August, May, and December, suggesting aseasonal polyestry (Bronner 2013a).

According to Retief et al. (2013), some evidence suggests that females communicate their ovulatory status to males acoustically and that scent-marking may play a role in finding mates. Males attempt to reach and mate with females, and possibly several males may converge on a female during her receptive period, as territories appear to be small (0.02 ha) and are likely to overlap. However, there is no evidence that males sequester or control females. Sexual selection may be based on penis length; a longer penis would enable a male to deposit an ejaculate closer to the site of fertilization within the female reproductive tract.

A. hottentotus occurs over a relatively large region and is common in farmland, gardens, golf courses, and other areas in close proximity to people (Bronner and Mynhardt 2015b). *A. corriae* and *A. septentrionalis* also do well in such areas and have fairly widespread distributions but may be jeopardized as their habitat is degraded, that of the former species primarily by tourism development and intensive agricultural practices, that of the latter by coal mining; they are both designated near threatened by the IUCN and are considered close to the vulnerable level (Bronner and Mynhardt 2015a; Rampartab 2015a). *A. marleyi* and *A. robustus* face the same kinds of problems but have much more restricted ranges. *A. marleyi*, increasingly imperiled by livestock overgrazing and other poor agricultural practices, is classified as endangered (Bronner and Mynhardt 2015c). *A. robustus*, losing habitat to mining of shallow coal deposits to fuel power stations, is designated vulnerable (Rampartab 2015b).

AFROSORICIDA; CHRYSOCHLORIDAE; **Genus NEAMBLYSOMUS** Roberts, 1924

Gunning's and Juliana's Golden Moles

There are two species (Bronner and Bennett 2005; Maree 2015e, 2015f):

> *N. gunningi*, known only from the far northern Drakensberg escarpment between Haenertsburg, New Agatha, and Tzaneen in Limpopo Province of South Africa;
>
> *N. julianae*, three isolated populations known from the Bronberg Ridge near Pretoria in Gauteng Province, on the Nyl Floodplain in Limpopo Province, and in and near southern Kruger National Park in Mpumalanga Province of South Africa.

Neamblysomus was included in *Amblysomus* by Meester (1977a), Meester et al. (1986), Petter (1981), and Simonetta (1968), but restored to generic status by Bronner (1995b) on the basis of cytogenetic and cranial divergence. Ongoing molecular research suggests that the population of *N. julianae* in Kruger National Park may be a species distinct from that represented by the other two populations (Maree 2015g). Except as noted, the information for the remainder of this account was taken from Bronner and Bennett (2005).

In *N. gunningi* head and body length is 111–132 mm, and weight is 56–70 grams in males, 39–56 grams in females. The upper parts are dark golden brown, and there is a rich brown sheen tinged with bronze. The cheeks and anterior muzzle are paler and tinged with yellow. The underparts are fawn. In *N. julianae* head and body length is 92–111 mm, and weight is 21–35 grams in males, 23–29 grams in females. The upper parts are brown or cinnamon brown, the flanks are paler, the underparts are fawn, and there are whitish

Juliana's golden mole (*Neamblysomus julianae*). Photo by Craig R. Jackson.

markings on the head and upper chest. *Neamblysomus* is distinguished from *Amblysomus* by lacking talonids on the molars and in having a broader palate; the third molar may be present or absent.

N. gunningi is found in montane forest and grassland, where it makes lengthy and winding subsurface runways and deeper burrows. *N. julianae* is endemic to the savannah biome, where it is confined to sourish bushveld on sandy soils. It appears to depend on softer soil substrates that offer significantly less resistance to burrowing (Jackson et al. 2008). It forages in shallow subsurface runways and also constructs deeper tunnels, the entrances to which are sometimes marked by small mounds. Subsurface activity increases dramatically after rain and appears to cease during the dry winter months, when periods of daily torpor are extended and foraging is probably restricted to deeper tunnels. Using an implanted radio transmitter, Jackson et al. (2009) determined body temperature of a free-ranging individual *N. julianae* to be remarkably vari-

able, ranging from 27° to 33° C. Body temperature declined during periods of low ambient temperature; such a state of shallow torpor could result in a daily energy saving of about 20 percent. During high ambient temperature, the animal underwent thermoregulation by moving into deeper, cooler tunnels, but it could warm itself passively when soil temperatures rose. The study also revealed that, contrary to anecdotal reports of nocturnal behavior, activity was bimodal and centered around 1,100 and 2,100 hours. The diet of *Neamblysomus* consists of earthworms and insects.

Population density of *N. julianae* in prime habitat is 2–3 individuals per ha. (Maree 2015f). Adults of that species are solitary, and intraspecific aggression is well developed. Breeding individuals have been taken in July and August and from November to February, suggesting aseasonal polyestry and sustained but low levels of reproduction throughout the year (Bronner 2013d). Breeding specimens of *N. gunningi* have been collected in February, April, and May.

The overall species *N. julianae* is designated endangered by the IUCN, because it has very specific habitat requirements, its range is severely fragmented, and its total area of occupancy is only 160 sq km. Although two of its three isolated populations occur partly within protected areas, the remainder of its habitat is being degraded by agricultural activity and urbanization (Maree 2015f). The habitat of the Bronberg Ridge population near Pretoria is completely unprotected and is further jeopardized by quartzite mining, road construction, and high-density housing developments nearly throughout its extent of less than 10 sq km. Therefore, in an unusual procedure, the IUCN now classifies that population separately as critically endangered (Maree 2015g). *N. gunningi* is classified as endangered, as it is known from just six localities with an occupied area estimated at only 96 sq km, where its preferred habitat is jeopardized by commercial forestry operations and other human activity (Maree 2015e).

AFROSORICIDA; CHRYSOCHLORIDAE; Genus *CARPITALPA* Lundholm, 1955

Arend's Golden Mole

The single species, *C. arendsi*, occurs in the Inyanga Highlands of eastern Zimbabwe, from about 18° S to 20° S and at elevations of 850–2,000 meters, and also marginally in the adjacent Vila Perey district of western Mozambique (Bronner 2015i; Bronner and Bennett 2005). *Carpitalpa* was originally described as a subgenus of *Chlorotalpa*, made a full genus by Simonetta (1968), placed in the synonymy of *Chlorotalpa* by Meester (1977a), and again raised to generic rank by

Bronner (1995b). Except as noted, the information for the remainder of this account was taken from Bronner and Bennett (2005).

Carpitalpa is a medium-sized golden mole. Head and body length is 115–141 mm, and mean weight is 52.7 grams in males and 41.6 grams in females. The dorsal pelage is glossy black or very dark brown with a bronze sheen; the limbs, throat, and sides of the face are paler. The underparts are only a shade lighter than the upper parts and have a faint green sheen. The claw on the first digit of the forefoot is short and narrow but stoutly built, the second digit is about three-fourths the length of the third digit, and the fourth digit is vestigial, its claw appearing as a round knob. From *Chlorotalpa*, *Carpitalpa* is distinguished by its more inflated head of the malleus, more robust stylohyal bone, relatively longer palate, and lack of well-developed molar talonids.

Arend's golden mole has been found in montane grasslands and in riverine forests with a dense undercover and a deep substrate of leaf litter. It favors loamy soils where it constructs subsurface runs between clumps of tussock grass. After rains, it often moves about on the surface of the ground. Its diet probably consists principally of earthworms and insects. No information is available on reproduction.

Bronner (2015i) stated that *Carpitalpa* is common in cultivated lands and gardens and probably is preyed on by domestic cats and dogs. However, a more serious threat is posed by alteration of the restricted habitat of the genus through uncontrolled timber harvesting and overgrazing by cattle, in association with political instability and land transformation initiatives in Zimbabwe, and a breakdown of conservation management in protected areas. The IUCN now classifies *Carpitalpa* as vulnerable.

Order Macroscelidea

MACROSCELIDEA; **Family MACROSCELIDIDAE**

Sengis, or Elephant Shrews

This order, containing the single family Macroscelididae with 4 Recent genera and 19 species, inhabits Africa, including the islands of Zanzibar and Mafia. Although long assigned to the obsolete order Insectivora by many mammalogists, Macroscelididae is now generally considered to represent an entirely separate order (see account of order Afrosoricida). For a time, most authorities placed the group phylogenetically near Lagomorpha (see McKenna and Bell 1997; Meester et al. 1986), though that view was questioned by Hartenberger (1986), Simons, Holroyd, and Bown (1991), and Woodall (1995). More recently, there has been growing acceptance that Macroscelidea is part of the mammalian clade Afrotheria, now centered in Africa, which also comprises orders Afrosoricida, Bibymalagasia, Tubulidentata, Hyracoidea, Proboscidea, and Sirenia (see account of class Mammalia). Evidently, elephant shrews are much more closely related to elephants than to shrews.

The sequence of genera presented here follows that of Corbet and Hanks (1968), who recognized two subfamilies: Rhynchocyoninae for the genus *Rhynchocyon*; and Macroscelidinae for *Petrodromus*, *Macroscelides*, and *Elephantulus*. Rathbun (2005) followed the same arrangement, but Schlitter (2005a) did not indicate recognition of subfamilies. A molecular analysis by Douady et al. (2003) suggested an alternative arrangement and an immediate relationship of *Petrodromus* and the species *Elephantulus rozeti*. Subsequent comprehensive assessments of both nuclear and mitochondrial DNA (Kuntner, May-Collado, and Agnarsson 2010; Smit et al. 2011) supported recognition of the above subfamilial division, with Rhynchocyoninae and Macroscelidinae, but also that *Elephantulus* is paraphyletic and that *E. rozeti* should be grouped with *Petrodromus* and *Macroscelides*. Smit et al. (2011) suggested subsuming *Petrodromus* and *Macroscelides* in *Elephantulus*, though without delineating a precise arrangement of taxa. Meanwhile, two geometric morphometric analyses of elephant shrews, while supporting the same basic subfamilial division, yielded somewhat differing results with respect to the content of *Elephantulus*. Panchetti et al. (2008) found *E. rozeti* to fall within the range of variation of a monophyletic *Elephantulus*. Scalici and Panchetti (2011), however, concluded that *Elephantulus* is not monophyletic, with the species *E. rozeti*, *E. revoilii*, and *E. rufescens* belonging in a clade together with *Petrodromus* and *Macroscelides*. Subsequent investigations of the petrosal and bony capsule of the inner ear provided further evidence that *Elephantulus* is paraphyletic and that *E. rozeti* is aligned with *Petrodromus* and *Macroscelides* (Benoit, Maeva, and Tabuce 2013; Benoit et al. 2014). Notwithstanding the preponderance of data supporting such alignment, and as pointed out by Holroyd (2010) and Rathbun (2009), there has been no actual taxonomic revision that would allow assignment of *E. rozeti* to another genus or to a new genus, or that has provided a

Giant sengi, or checkered elephant shrew (*Rhynchocyon cirnei*). Photo by Alexander F. Meyer at Prague Zoo.

relevant arrangement of genera and/or subgenera. Therefore, *rozeti* is treated herein within the account of *Elephantulus*. Nyári, Peterson, and Rathbun (2010) proposed that a hypothetical common ancestor of *rozeti* and *Petrodromus* dispersed northward during the Last Glacial Maximum of the Pleistocene, when what is now the Sahara Desert region was much more moist and probably traversed by corridors with habitat conditions favorable to elephant shrews.

Head and body length is 95–318 mm, and tail length is 80–265 mm. The long and slender tail is sparsely covered with short hair, which in some species forms a small terminal tuft (Perrin and Rathbun 2013a). The pelage, which is slightly coarse in Rhynchocyoninae and soft in Macroscelidinae, is lacking in the rump region. In some forms of the genus *Petrodromus* the distal third of the underside of the tail has stiff bristles, which probably function to spread sweat and sebaceous gland products on the substrate during scent marking (Jennings and Rathbun 2001; Rathbun 2013a). In that genus,

such material is first exuded by a naked black perianal gland in the form of small beads of moisture that have a musky odor, especially in the males. Some other genera have subcaudal or sternal glands for scent marking.

The long, narrow snout, broadest at the base, is extremely sensitive; it is not retractile but is very flexible and is movable in a circular manner at the base. The nostrils are located at the end of the snout. The tongue is pink, thinly tapering, used to flick small prey from the substrate, and so long that it can curl around the top of the muzzle to lick the fur after eating (Rathbun 2005). The eyes are relatively large, and there is a pale eye ring in some species. The broad, upright ears are as mobile as the snout (Rathbun 2005). The hind leg is much longer than the forelimb. The two lower bones of the hind leg are united. In *Rhynchocyon* each of the four feet has only four digits; in *Petrodromus* there are five digits on the forefoot and four on the hind foot; and in *Macroscelides* and *Elephantulus* each of the four feet has five digits (Corbet 1977b). Females

Rock sengi, or long-eared elephant shrew (*Elephantulus myurus*). Photo by Chris and Mathilde Stuart.

have two or three pairs of mammae. Rathbun (2005) noted that the phylogenetic association of Macroscelidea with Afrotheria is confirmed by presence of a large functional caecum, abdominal testes, and a long ventral penis. In some species the penis is divided into three forks (see Douady et al. 2003).

The dental formula is: i 1–3/3, c 1/1, pm 4/4, m 2/2–3 = 36–42. When the formula is complete, the first upper incisors are larger than the others, and the lower incisors are about equal in size. The front incisors are rather widely separated from each other. The upper canines are premolar-like except in *Rhynchocyon*; the upper premolars increase in size and complexity from front to back, the posterior premolar being the largest of the molariform teeth. The molars usually have four cusps, arranged in a W; this has been called "dilambdodont" dentition. The hard palate extends behind the molars and has large openings through it near the median line, in the muzzle region, and between the back teeth. The braincase is relatively large.

Habitats include thornbush country, grassy plains, thickets, the undergrowth of forests, gravel plains, and rocky outcrops. *Rhynchocyon* is diurnal, while most Macroscelidinae are crepuscular, with some activity during the day and night; solar basking is a prominent behavior of the latter subfamily (Rathbun 2009). Sengis may take refuge under any available shelter when alarmed, but ordinarily they reside in burrows, ground depressions, rock crevices, and termite mounds or under logs in forests. Their feet are not well adapted for burrowing, but they sometimes construct their own burrows in the wild. The burrows of rodents are occasionally used. Sengi burrows often have an emergency exit in addition to the usual entrance and exit hole. *Rhynchocyon* builds and uses leaf nests on the forest floor, whereas none of the Macroscelidinae builds or uses nests (Rathbun 2009). Sengis sometimes have surface runways through grass or leaf litter, which can be distinguished by their broken appearance because of the jumping locomotion of the animals. Their jumps have been compared to the bounces of a rubber ball. When moving slowly they progress on all four limbs, but when alarmed or pursued, some hop on their hind limbs with the tail extended upward. Sengis often sunbathe. No sengis are known to be bipedal or ricochetal, but all are highly saltatorial and capable of exceedingly fast quadrupedal gaits, including half-bounds and a peculiar footfall pattern in the Macroscelidinae, in which one hind foot is suspended and out of action when walking (Rathbun 2009). The smaller species feed mainly on ants and termites, but sometimes take fruits and other plant material. *Rhynchocyon* and other of the larger forms feed mostly on beetles.

Sengis produce acoustic signals by rapidly tapping their hind feet on the substrate; such foot drumming occurs during agonistic encounters, when an animal is agitated, and when it encounters a predator (Rathbun 2005). In addition, *Rhynchocyon* makes a slapping noise by beating its tail on the ground. Squeaks have been reported for *Elephantulus* and *Rhynchocyon,* and a cricketlike call has been noted in *Petrodromus.*

Sengis usually live in monogamous pairs or, during the breeding season, in small family groups. Gestation periods of around 42 to 75 days have been reported for the different species. There usually are just one or two young, which, particularly in the Macroscelidinae, are well developed and comparatively large at birth. They are fully covered with hair, and their eyes are open at birth or soon thereafter. Some walk almost immediately after birth. It appears that nursing bouts are relatively short and weaning may take place within a month. In some instances the nursing young are dragged about by the mother as they cling to her nipples. However, this is probably rare as the young are mobile soon after birth. Sexual maturity is probably attained at 5–6 weeks.

The geological range of Macroscelididae is early Eocene to Recent in Africa (Butler 1995; Hartenberger 1986; Simons, Holroyd, and Bown 1991). Recent analysis of fossil material suggests that Macroscelidea originated in Eurasia (Tabuce et al. 2001) or North America (Zack et al. 2005) and spread to Africa in the late Paleocene to early Eocene. Such a movement would be in keeping with those views that support a non-African origin for the clade Afrotheria (see account of class Mammalia). Molecular clock estimates by Douady et al. (2003) suggest the two living subfamilies of Macroscelididae diverged around 43 million years ago, in

the later Eocene, and that the radiation of subfamily Macroscelidinae took place in the Miocene, partly through development of the Sahara Desert.

MACROSCELIDEA; MACROSCELIDIDAE; **Genus *RHYNCHOCYON* Peters, 1847**

Giant Sengis, or Checkered Elephant Shrews

There are four species (Corbet 1977b; Rathbun 2008; Rovero et al. 2008):

R. chrysopygus, coast of southeastern Kenya;

R. cirnei, northern and eastern Democratic Republic of Congo between Congo and Ubangi Rivers, Uganda, southern Tanzania, northeastern Zambia, Malawi, Mozambique north of Zambezi River;

R. udzungwensis, Udzungwa Mountains of central Tanzania;

R. petersi, southeastern Kenya, northeastern Tanzania, Zanzibar and Mafia Islands.

Kingdon (1974) listed *chrysopygus* and *petersi* as subspecies of *R. cirnei,* but also referred to the former two as incipient species. He indicated that a large hybrid zone between *petersi* and another subspecies, *R. cirnei reichardi,* had formed in southern Tanzania and northern Mozambique, where there are intergrading variations of the specific pelage patterns described below, but that *chrysopygus* apparently had not been affected by hybridization. Lawson et al. (2013) reported historical mitochondrial introgression between *R. cirnei* and *R. udzungwensis,* but considered them separate species. Andanje et al. (2010) reported an apparently new form of *Rhynchocyon* in the coastal forests of Kenya between the Tana River and the Somali border. Lawson et al. (2013) noted that its coloration and patterning are very similar to those of *R. udzungwensis* but that it has not yet been adequately sampled or described.

Head and body length is about 218–324 mm, and tail length is 190–270 mm. Reported weights are 320–420 grams for *R. cirnei* (Ansell and Ansell 1973), about 540 grams for *R. petersi* (Rathbun 2009), 410–690 grams for *R. chrysopygus,* and 658–750 grams for *R. udzungwensis* (Rovero et al. 2008). In *R. chrysopygus,* the golden-rumped sengi, the rump is straw-colored, the remainder of the upper parts is mostly maroon and black, the underparts are slightly paler, and there is a white band near the end of the tail. In *R. cirnei*, the checkered sengi, there is a series of up to three stripes on each side of the midline of the back, with an associated series of chestnut or pale buffy, or sometimes whitish or dark brown, somewhat rectangular spots (or checkers). The ground color is grizzled yellow to dark brown with a tendency to melanism in the eastern part of the range. There is a white band toward the distal portion of the tail. In *R. petersi*, the black and rufous sengi, the rump and center of the back are black, the remainder of the body is mostly orange rufous, and the tail and ears are pale orange-brown. *R. udzungwensis* is distinguished by its maroon back, jet-black lower rump and thighs, grizzled gray forehead and face, orange-rufous sides, and pale yellow to cream chest and chin; the tail is slightly bicolored, nearly black dorsally and dark brown ventrally, with a subterminal white band 4–6 cm long.

Rhynchocyon, the largest genus of the family, has only four digits on each foot; neither pollex nor hallux is present. The feet have long claws. The dentition differs from that of the other genera of elephant shrews in having only one vestigial upper incisor on each side, which is often absent, so that there are no functional teeth in front of the enlarged canines. Rathbun (2013b) noted the genus is characterized by a narrow, ungulate-like body; a long, rodent-like tail; and long, spindly legs.

These elephant shrews are confined mainly to forest (lowland and montane), thick riverine bush, and coastal scrub, though they have been taken in clearings amid grass and cane growth. According to Rathbun (1979), *R. chrysopygus* occupies a narrow strip of moist, dense coastal scrub amid lowland semideciduous forest and woodland along the Kenyan coast from sea level to 30 meters. *Rhynchocyon* is restricted to the forest floor and never burrows or climbs. It is diurnal and spends the night in one of several nests on the forest floor. Nests, used for 1–3 nights, are composed of a

Giant sengi, or black and rufous elephant shrew (*Rhynchocyon petersi*). Photo by Alexander F. Meyer at Brookfield Zoo.

shallow cup excavated in the soil and lined with dry leaves. This chamber is then covered with a pile of leaves, resulting in a cryptic dome about 10 cm high and 30 cm across. FitzGibbon (1997) noted further that each male and female *R. chrysopygus* makes about six separate nests and never shares its nest. *R. udzungwensis* constructs similar nests and is strictly diurnal (Rovero et al. 2008).

Giant sengis have been observed both alone and in small groups (most likely adult pairs with associated young) hunting for food on the forest floor in the daytime. They are very nervous animals; their noses, ears, and vibrissae twitch constantly, and they continually utter little squealing noises and squeaks. They do not bound or leap as much as the other sengis. Kingdon (1974) wrote that, while feeding, *Rhynchocyon* explores and turns over the litter with its nose but uses its forefeet when it senses something, scuffing the surface with several rapid scratches and immediately returning the

nose and mouth to the spot, darting the very long tongue in and out to sweep up the insects together with the earth. The process often results in small conical or saucer-shaped excavations in the soil of the forest floor.

The diet is composed mostly of insects and other arthropods, which are located by probing the leaf litter with the long flexible snout and flicked up with the long extensible tongue, but earthworms are excavated from the soil with the strong claws of the forefeet (Rathbun 2013c). There are reports that small mammals, birds, eggs, mollusks, and other animal foods are taken on occasion. There have been claims that *R. cirnei* feeds on elephant and human dung, but the sengi probably finds that material a rich source of dung beetles.

FitzGibbon and Rathbun (1994) determined population density of *R. chrysopygus* in coastal forest of Kenya to be 23–75 per sq km. Rovero et al. (2008) assumed a density of 50–80 *R. udzungwensis* per sq km. Rathbun (1973) found that in Kenya *R. chrysopygus* is

Giant sengi, or golden rumped elephant shrew (*Rhynchocyon chrysopygus*). Photo by Klaus Rudloff at Frankfurt Zoo, Germany.

distributed in pairs on contiguous territories. No permanent territorial boundary changes occurred in the more than 21 months of study. Both sexes scent-marked and defended their territories. Chases, however, involved mostly individuals of the same sex. Only territorial animals were observed to breed. According to Rathbun (2013c), territories of *R. chrysopygus* are 1.5–5.0 ha. in size, those of males being slightly larger than those of females. Territories of a monogamous pair overlap to a large extent, but there is little overlap with territories of neighboring pairs. Young remain on the parental territory indefinitely, eventually finding a vacant territory of their own. Vocalizations include very soft chattering between individuals and a loud distress scream. FitzGibbon's (1997) study in the Arabuko-Sokoke Forest of Kenya found that males occasionally court neighboring females, and incorporate their territories if that female's original mate disappears, but that defense of multiple territories is costly in terms of energy and vulnerability to intrusion by other males, thus making monogamy the superior option. Males provide no direct paternal care to young.

Rathbun (1979) reported that *R. chrysopygus* breeds throughout the year, gestation is approximately 42 days, and there was a mean interbirth interval of 81 days in one area. That species usually has a single young (occasionally two), which is confined to a nest for about 2 weeks. For about 5 days after emergence, it stays close to its mother, but subsequently the two are rarely seen together. From 5 to 20 weeks after emerging from the nest, the young may secure its own home range; once it does, it may live 4–5 years. Weigl (2005) listed a captive *R. chrysopygus* that lived to be over 11 years old.

Of two female *R. cirnei* taken in May one was pregnant with two embryos and the other was nursing. Pregnant females of that species were collected in Malawi during September (Ansell 1960). Three pregnant females, each with two small embryos, were taken in Zambia in August (Ansell and Ansell 1973). Observations of a group of captive *R. petersi* showed that the young are born naked, with closed eyes, and are kept in a nest constructed by the adults for 3–4 weeks. Interbirth interval was typically over 80 days following a successful litter, but shorter—somewhat over 40 days—if no infants survived from the first litter. Unre-

lated females were relatively tolerant of one another, more so than were males, and the latter did not appear to be involved with infant care (Baker et al., 2005). Litter size in *R. udzungwensis* is 1–2 (Rathbun 2009).

The IUCN classifies *R. chrysopygus* as endangered, *R. petersi* and *R. udzungwensis* as vulnerable, and *R. cirnei* as near threatened. The only substantial remaining population of *R. chrysopygus* is in the Arabuko-Sokoke Forest, an area of 372 sq km in Kenya. That population fell from an estimated 20,000 individuals in 1993 to 14,000 in 1996, mainly because of logging, which eliminates the leaf litter, canopy cover, and hollow trunks that the species requires. Another 9 percent reduction of that population may have occurred from 1996 to 2008. Other populations are smaller, scattered, and threatened by clearing of habitat for agriculture and by fires; the species is not deliberately hunted, as its meat is not good, but it is caught in traps and snares set for other animals (FitzGibbon and Rathbun 2015). Although *R. petersi* occurs over a larger region, its restricted forest habitat, some on coasts and islands, is being fragmented and degraded by urban and agricultural expansion (Rathbun and Butynski 2008). The newly discovered *R. udzungwensis* is known only from a forested area of about 390 sq km, where there are an estimated 15,000–24,000 individuals; the area is protected but subject to drought-driven fires and increasing human-induced fires (Rovero and Rathbun 2015; Rovero et al. 2008). *R. cirnei* is widely distributed but its habitat is naturally fragmented and vulnerable to the same threats confronting the other species; of particular concern are the subspecies *R. c. hendersoni*, restricted to an isolated montane forest in Malawi that is being cleared, and *R. c. cirnei*, known only by the type specimen from coastal Mozambique (Rathbun 2008).

MACROSCELIDEA; MACROSCELIDIDAE; **Genus** *PETRODROMUS* Peters, 1846

Four-toed Sengi, or Four-toed Elephant Shrew

The single species, *P. tetradactylus*, occurs in Democratic Republic of Congo, southern Uganda, Rwanda,

Four-toed sengi (*Petrodromus tetradactylus*). Photo by Galen Rathbun / California Academy of Sciences.

southeastern Kenya, Tanzania (including Zanzibar and Mafia Islands), northeastern Angola, Zambia, Malawi, Mozambique, southeastern Zimbabwe, and extreme northeastern Limpopo and KwaZulu-Natal provinces of South Africa; the species has also been reported on occasion from Congo and the Caprivi Strip of Namibia, but such occurrence is unconfirmed (Rathbun 2005, 2013a; Rathbun and FitzGibbon 2015). There have been suggestions that *Petrodromus* should be placed in a single genus together with *Macroscelides* and *Elephantulus* and that the species *Elephantulus rozeti* is more closely aligned with *Petrodromus* than with the other species of *Elephantulus* (see above account of order Macroscelidea).

Head and body length is 165–220 mm, and tail length is about 130–180 mm. Rathbun (2005) listed weight as 160–280 grams. Like other members of the family, *Petrodromus* has an attractive pelt, though it is not as colorful as that of *Rhynchocyon*. The fur is rather long and soft. The coloration above is buffy, with an orange or yellow tinge, or brown with a rufous wash; some forms have a grayish or brownish streak on the sides and flanks. There is a conspicuous white eye ring that is interrupted posteriorly by an irregular dark brown to black spot that extends back to below the ear (Jennings and Rathbun 2001). The underparts are white or ochraceous. This relatively large elephant shrew has four toes on the hind foot; the first digit is not present. There are three upper incisors on each side, and the upper canine is not large. Females have four mammae.

Although sometimes found in rocky districts, the four-toed elephant shrew seems to prefer thickets and dense forest undergrowth. It has much the same habits as the smaller species of the family; it is terrestrial and cursorial and makes runways through brush, grass, and leaf litter. Its mode of progression when not alarmed is a scurrying walk, but when disturbed, it progresses by fairly long jumps. It runs with its tail pointed upward. Brown (1964) stated that all natives agree that this elephant shrew makes no nest but sleeps under dense vegetation or in chance holes; he found no evidence to support the previous suspicion that it sleeps in deserted termite holes. Jennings and Rathbun (2001) noted that it is active mainly at dawn and dusk, with lesser periods of activity during the night and least of all during midday. In foraging along trails, it uses the hind feet to scuff the leaf litter, the long mobile nose as a probe, and the extendible tongue, in anteater fashion, to glean small invertebrates from the surface of the debris. The diet consists mainly of insects, particularly termites and beetles, but some fruits, seeds, and green plant matter are also eaten.

Jennings and Rathbun (2001) indicated that population densities of 2.1 individuals per ha. are attained in areas of the Arabuko-Sokoke Forest of Kenya, which is dominated by the tree genus *Afzelia*, but that densities can be about twice or half that high in other areas, perhaps depending on availability of prey and dense cover. Radio-tracked individuals were found to have an average home range of 1.2 ha., with no difference between the sexes and a high degree of overlap between males and females. Another radio-tracking study, in sand forest at Tembe Elephant Park, South Africa, showed *Petrodromus* to live in monogamous male-female pairs that defended partially overlapping home ranges; there was virtually no overlap between the ranges of adjoining pairs (Oxenham and Perrin 2009). Observations in captivity also suggest *Petrodromus* is territorial; male-female pairs fight less than same-sex pairs and larger groups. Animals may rise on their hind feet and "box" with the forepaws. When alarmed they often drum loudly with their hind feet on the substrate; this may have a communication function, though an early suggestion was that the sound was to disturb ants, which respond with a noise of their own, enabling the elephant shrew to locate them. A few vocalizations have been reported, including a shrill cricket-like call or squeak.

Available information from southern Africa indicates the young are born just before and at the commencement of the rains from about August to October (Rathbun 2005). In East Africa breeding probably occurs throughout the year (Kingdon 1974). Three females, each with a single embryo, were collected in that region in December, and Brown (1964) reported that of 14 females taken there throughout June, 8 were

pregnant, with the pregnancies ranging from early embryos to full-term fetuses. A birth was recorded in Zambia in July, another in April, and a small fetus in September. Sheppe (1973) reported that a pregnant female with two embryos was taken in Zambia in October. Pregnant females have been recorded in Katanga (southern Democratic Republic of Congo) in January and July. According to Jennings and Rathbun (2001), seasonality of reproduction may increase from tropical to temperate regions. A review of data from throughout the range of *Petrodromus* showed pregnant females in all months except March, May, September, and November, and that females typically carry single embryos, sometimes two. Neonates weigh about 31.5 grams and are highly precocial; they are fully furred with adult color, their eyes are open, and they are able to walk within hours. Weigl (2005) listed a captive female that lived to an age of about 7 years and 2 months.

Rathbun and FitzGibbon (2015) observed that *P. tetradactylus* occurs across a large region in a variety of habitats and that overall the species faces minimal threats. However, the two subspecies found in South Africa, *P. t. beirae* and *P. t. warreni*, were designated rare by Smithers (1986). And the subspecies *P. t. sangi* is known only from the Taita Hills of Kenya, where its isolated forest habitat is being increasingly lost to agriculture (Nicoll and Rathbun 1990).

Four-toed sengi (*Petrodromus tetradactylus*). Photo by Klaus Rudloff at Tierpark Berlin.

Round-eared sengi (*Macroscelides proboscideus*). Photo by Eric Isselee / Shutterstock.com.

MACROSCELIDEA; MACROSCELIDIDAE; **Genus** *MACROSCELIDES* A. Smith, 1829

Round-eared Sengis, "Short-eared" Elephant Shrews

There are three species (Corbet 1977b; Dumbacher et al. 2012, 2014; Rathbun 2005; Rathbun and Smit-Robinson 2015a; Smithers 1971):

M. micus, gravel plains of the Etendeka Plateau in northwestern Namibia;

M. proboscideus, mainly in the Karoo biomes of southern Namibia, extreme southwestern Botswana, and the South African provinces of Northern Cape, Western Cape, and Eastern Cape;

M. flavicaudatus, desert areas of western Namibia.

There have been suggestions that *Macroscelides* belongs in the same genus as *Elephantulus* and *Petrodromus* (see above account of order Macroscelidea).

Head and body length is 85–104 mm, tail length is 83–137 mm, and adult weight is 22–50 grams (Dumbacher et al. 2012, 2014; Rosenthal 1975a). The fur is long and soft. The upper parts are sandy brown, pale buff, pale grayish buff (sometimes with a tinge of rufous), orange buff, or buffy with a brownish or blackish suffusion. This genus lacks the pale rings around the eyes that are characteristic of nearly all other soft-furred sengis (subfamily Macroscelidinae). The underparts are white or grayish. The newly described *M. micus* is distinguished by its small size, rusty coloration, lack of dark pigmentation, and prominent subcaudal gland.

Notwithstanding one of the common names, Rathbun (2005) observed that the ears of *Macroscelides* are not shorter when compared to those of other small forms of Macroscelididae. Indeed, they are very broad and expanded, almost circular in shape, and that is what distinguishes *Macroscelides*. The shape of the supratragus of the ear is also characteristic, being square at the end and sparsely haired. Distinguishing features of the skull and teeth are the enormously enlarged bullae and the presence of three upper incisors on each side; the upper canine is not extremely dominant. The first digit of the hind limb is small and bears a claw. Females reportedly have six mammae.

Round-eared sengis dwell in open country, either on loose sandy soils with thornbush for cover or on hard gravelly plains with scattered boulders for cover. Elevational range is sea level to 1,400 meters (Dumbacher et al. 2012). They sometimes excavate a shallow burrow under a bush. The burrow has an ordinary entrance and an emergency exit, the latter being inconspicuous and almost perpendicular to the ground. In rocky areas *Macroscelides* may use a crevice, and it sometimes shelters in the warrens of the murid rodents of the subfamily Otomyinae. In Namibia, at least near Berseba, *Macroscelides* and *Elephantulus rupestris* often occur together in about equal numbers. Dumbacher et al. (2014) observed that *M. flavicaudatus* uses trails across the gravel desert, whereas *M. proboscideus* and *M. micus* are not known to use trails. *Macroscelides* is somewhat diurnal; it may be seen during the day hopping from one bush to another, its tail raised at an upward slant. During the warmest part of the day it often sits on its hind legs and basks in the sun. However, Rathbun (2005) wrote that it is active at dawn, dusk, and night. A laboratory study found that it enters daily torpor when deprived of food; it did not become torpid in response to cold, if provided ample food (Lovegrove, Lawes, and Roxburgh 1999). The diet of *Macroscelides* includes insects, particularly ants and termites, but also tender shoots, roots, and berries. Kerley (1995) referred to *Macroscelides* as an omnivore because a study revealed that 45 percent of its food consists of plant matter.

Round-eared sengis are usually seen alone in the wild, but an adult male and female usually share the same area. Breeding pairs may be maintained together in captivity. Sauer (1973) reported that in Namibia individuals occupied largely undefended home ranges of about 1 sq km, but portions of them are apparently territories. In the Goegap Nature Reserve of western Northern Cape Province, population density was found to range from 0.35 to 1.59 individuals per ha. during a 2.5-year period, and social organization was monogamous (Schubert 2011; Schubert et al. 2009a, 2009b). Overall home range size averaged about 2.0 ha., but during the breeding season the means were 1.7 ha. for males and 0.8 ha. for females. Generally, female areas had little overlap with those of neighboring females and most male areas overlapped the area of a single female. The pairings and territories thus formed were perennial and maintained year-round. Male territories tended to contract in the breeding season, evidently because the males were then guarding their mates from other males, though some males temporarily intruded into the ranges of "widows."

Reproduction in the Karoo occurs throughout the year, with a decline in pregnancy in early winter (Rathbun 2005). There is a postpartum estrus, and one captive female underwent six pregnancies during two seasons (Rosenthal 1975a). Assessment of records from European zoos indicates an interbirth interval of around 75 days and a gestation period of 60–61 days; of 570 litters, 29 percent contained a single young, 70 percent had two, and 0.9 percent had three (Olbricht, Kern, and Vakhrusheva 2006). Although no nest is constructed, birth does take place in a sheltered spot. The female then leaves, but she returns during the night at about 24-hour intervals to nurse the young (Eisenberg 1975; Sauer 1973). At birth the young are well developed, covered with hair, and able to move about. They leave their initial shelter after 18–36 days, by which time they are eating solid food. They attain adult weight at approximately 46 days (Rosenthal 1975a; Sauer 1973). Weigl (2005) listed a captive male and female, each of which lived to an age of 8 years and 8 months.

Although *Macroscelides* apparently occurs at low densities, it is widespread in suitable habitat over an area considerably greater than 500,000 sq km. Most of that habitat is very arid, not practical for development, and not known to be threatened (Rathbun and Dumbacher 2015; Rathbun and Eiseb 2015; Rathbun and Smit-Robinson 2015a).

MACROSCELIDEA; MACROSCELIDIDAE; **Genus *ELEPHANTULUS*** Thomas and Schwann, 1906

Rock Sengis, or Long-eared Elephant Shrews

There are 11 species (Corbet 1977b, 1995; Corbet and Hanks 1968; Kuntner, May-Collado, and Agnarsson 2010; Meester et al. 1986; Rathbun 2005, 2015a, 2015b, 2015c; Rathbun, Agnelli, and Innocenti 2014; Rathbun, and Smit-Robinson 2015b; Smit et al. 2008, 2011; Smithers 1971):

E. rozeti, northern Western Sahara, Morocco, northern Algeria, Tunisia, northwestern Libya;

E. revoili, northern and central Somalia;

E. fuscipes, southern South Sudan, northeastern Democratic Republic of Congo, Uganda;

E. fuscus, southeastern Zambia, southern Malawi, central Mozambique;

E. brachyrhynchus, savannah zone of Uganda, Kenya, Tanzania, southern Democratic Republic of Congo, Angola, Zambia, Malawi, northeastern Namibia, northern and eastern Botswana, western and southern Mozambique, and the South African provinces of Limpopo, North West, Gauteng, and Mpumalanga;

E. rufescens, steppe and savannah zone of southern South Sudan, central and southern Ethiopia, northwestern and southern Somalia, eastern Uganda, Kenya, and northern and central Tanzania;

E. rupestris, western and southern Namibia, northern and southeastern Northern Cape

Rock sengi, or Cape elephant shrew (*Elephantulus edwardii*). Photo by Fritz Geiser.

Province, eastern Western Cape Province, western Eastern Cape Province of South Africa;

E. intufi, southwestern Angola, Namibia, central and southern Botswana, northwestern Limpopo Province, northern Northern Cape Province, and extreme western North West Province of South Africa;

E. myurus, Zimbabwe, western Mozambique, eastern Botswana, parts of Lesotho, and eastern South Africa in the provinces of Limpopo, Gauteng, Mpumalanga, North West, Free State, Northern Cape, and Eastern Cape;

E. edwardii, western and southern Northern Cape Province, Western Cape Province, and western Eastern Cape Province of South Africa;

E. pilicaudus, Karoo region of south-central Northern Cape Province and northern Western Cape Province of South Africa.

The species *E. fuscipes*, *E. fuscus*, and *E. brachyrhynchus* were formerly placed in the separate genus *Nasilio* Thomas and Schwann, 1906 (see Kingdon 1974; Meester et al. 1986; Rathbun 2005). However, that name was considered a synonym of *Elephantulus* by Corbet and Hanks (1968). A number of recent molecular and morphological studies have indicated that the isolated northern species *E. rozeti* is more closely related to the genus *Petrodromus* than to the other species of *Elephantulus*, but there has not yet been a formal taxonomic revision that would allow relevant reassignment of *rozeti* or appropriate generic or subgeneric rearrangement (see account of order Macroscelidea). Placement of *E. rozeti* and certain of the other species in the above list attempts to account for the DNA assessments of Kuntner, May-Collado, and Agnarsson (2010) and Smit et al. (2011). The latter study also indicated the existence of several undescribed species of *Elephantulus*.

Head and body length is 90–145 mm, tail length is 80–165 mm, and adult weight is 25–100 grams. The fur is soft. The upper parts are grayish yellow, gray, buffy white, buffy brown, reddish, or dark brown. The underparts are white, buffy, or grayish. The snout has been reported to be comparatively short and tapering in *E. brachyrhynchus* and *E. fuscus*, and to be longer and more slender in most other species, though this

North African rock sengi (*Elephantulus rozeti*). As explained in the account of order Macroscelidea, there is extensive evidence that this species is more closely related to the genus *Petrodromus* than to the other species of *Elephantulus*. This was recently confirmed by Dumbacher, Carlen, and Rathbun (2016), who erected the new genus *Petrosaltator* for *rozeti*. Photo by Galen Rathbun / California Academy of Sciences.

feature may be difficult to ascertain in a live animal. The first digit of the hind foot is small and bears a claw. There are 10 lower teeth on each side, except in *E. brachyrhynchus*, *E. fuscipes*, and *E. fuscus*, which have 11. Females have three pairs of mammae.

These sengis occur in a variety of habitats, including savannah, steppe, and semidesert. They are commonly found in rocky areas, and several species are closely associated with kopjes, or rocky outcrops. *E. brachyrhynchus*, however, depends on a dense cover of grass and scrub bush; in some areas where its range overlaps that of *E. myurus*, individuals of the two species may be found just a few meters apart, but the former in vegetative cover, the latter among rocks (Rathbun 2005). *E. brachyrhynchus* and some other species use burrows, though it is not certain if they dig their own or occupy ones originally excavated or started by small rodents. In flatlands their burrows may have a vertical escape hole located a meter or so from the main entrance. Most species commonly shelter in rock crevices, though when pursued they hide in any available refuge. Most make a series of runways, visible where there is surface litter, radiating out from their retreats to feeding areas. *E. myurus*, however, does not travel along a network of trails; instead, it uses its swift cursorial gait to travel from rock to rock (Ribble and Perrin 2005). Some species tend to be more active during daylight, while others may not emerge until dusk during the hot season, or when harassed by hawks and other diurnal predators. *E. intufi* is active day or night, with peaks at dawn and dusk. *E. edwardii* has been reported to be predominantly nocturnal (Rathbun

2005), but was seen regularly during daylight hours by Stuart, Stuart, and Pereboom (2003). Free-ranging *E. myurus* has been observed to become torpid when ambient temperatures are low; captives of both that species and *E. rozeti* were found to enter torpor when deprived of food (Lovegrove, Ramin, and Perrin 2001; Mzilikazi and Lovegrove 2004). The natural diet of *Elephantulus* consists mostly of insects, especially ants, termites, and beetles, but also includes plant material. Captive animals accept various foods, including fruits and vegetables.

Rathbun (2005) wrote that the southern African species that have been studied—including *E. brachy-rhynchus*, *E. intufi*, *E. myurus*, and *E. edwardii*—are solitary or occur in monogamous pairs. In *E. intufi* and *E. myurus*, at least, the male and female of a pair share greatly overlapping home ranges that are essentially exclusive of other pairs. Each of those species uses its hind feet on occasion to "drum" the ground, and each shows a different pattern of beats. The exact function of that behavior remains unknown but may involve communication of warning or threat to conspecific individuals. Vocalizations include a high-pitched squeak. Employing radio-tracking and direct observation of *E. intufi* in the foothills of the Erongo Mountains of

Namibia, Rathbun and Rathbun (2006b) found home ranges to average 0.47 ha. for males and 0.32 ha. for congruent females. Occasionally a male expanded its range to incorporate the territory of a neighboring widow. However, such activity was considered costly in terms of energy and vulnerability to predators, and the would-be polygynous male generally retreated to his original territory and mate when another male began to associate with the widow.

In another radio-tracking study, of *E. myurus* in KwaZulu-Natal, Ribble and Perrin (2005) estimated home range size to average 9,901 sq meters for males and 3,623 sq meters for females. The ranges of males overlapped one another by an average of 67 percent, whereas female intrasex overlap was only 18 percent. The data indicated that individuals are spatially associated in lasting monogamous pairs and that female ranges tend to be overlapped by only one male. There was no evidence that males engaged in the direct care of young, but it was suggested that male defense of a large territory increased resource availability for the female and offspring, hence contributing to their survival.

The species *E. rufescens*, of East Africa, has also been studied in some detail (Koontz and Roeper 1983). A mated pair share a territory of about 0.34 ha. The

Rock sengi, or bushveld elephant shrew (*Elephantulus intufi*), showing flexibility of snout. Photo by Arab/Shutterstock.com.

Rock sengi, or rufous elephant shrew (*Elephantulus rufescens*). Photo by Johannes Pfleiderer.

animals make trails through that area and rest at established points but not in a hole or burrow. They spend little time together, but when they are, the female usually dominates the male. Each individual defends the territory from conspecifics of the same sex. Boundary encounters are characterized by drumming of one or both hind feet, ritualized gestures, and finally high-speed chases. There also is scent-marking by rubbing a sternal gland on the substrate and probably by urination and defecation. No reproductive seasonality has been observed in the wild in Kenya or in captivity. Females have a modal estrous cycle of 13 days, with estrus lasting about 12 hours. They may have several litters annually, and recorded interbirth intervals range from 56 to 145 days. Gestation has been estimated to last 57–65 days. All known litters have contained only one or two young. The young are precocious, weigh 10.6 grams at birth, and are weaned by 25 days. At 50 days they reach adult size, are sexually mature, and are driven from the parental territory.

Generally in *Elephantulus*, breeding appears to be continuous close to the equator, to be reduced during the cooler season at intermediate latitudes (15°–20°), and to cease during the period of declining photoperiod at higher latitudes (Neal 1995). *E. rozeti*, of North Africa, breeds from January to August, with births beginning in March or April after a gestation period of 75 days; females may have two litters of 1–4 young annually (Perrin and Rathbun 2013b; Séguignes 1989). *E. brachyrhynchus* was found to breed year-round in Tanzania, though there was a higher incidence of pregnancy in January–February (Leirs et al. 1995). Births and near-term embryos of *E. brachyrhynchus* were recorded in January in central Mozambique, young were born in a burrow in April in eastern Angola, and embryos were recorded in October and births in February and October in Zambia.

In Zimbabwe *E. brachyrhynchus* breeds year-round, though at a reduced rate during the cool season from mid-May to mid-August; females can produce five or six litters annually, with an average of 1.6 young per litter (Neal 1995). The breeding season of *E. myurus* in South Africa is in the warmer, wetter months. Gravid females of that species have been taken in Botswana in March, April, and September (Smithers 1971), in Zimbabwe in January and May (Wilson 1975), and in Limpopo Province of South Africa from August to January (Medger, Chimimba, and Bennett 2012). Lactating females were caught in Limpopo in September and December–March. Pregnant female *E. edwardii* were found from September to December (Stuart, Stuart, and Pereboom 2003). The young of *E. intufi* evidently are born during the warmer, wetter months of the year in Botswana, from about August to March, but throughout the year in Namibia (Rathbun 2005; Smithers 1971). Litter size in the southern African species is usually two, occasionally one or three. In *E. myurus* gestation lasts 8 weeks, and the precocious young are born from September to March; like those of the other species that have been studied, they are fully furred, their eyes are open, and they are able to walk very soon. Females attain sexual maturity at 5–6 weeks, and available data suggest that most do not live long enough to produce more than one litter, though natural life span may have been underestimated (Rathbun 2005). Record longevity for *Elephantulus*, held by a captive specimen of *E. intufi*, is 9 years and 3 months (Weigl 2005).

E. rufescens is a carrier of a type of malaria that apparently is not contracted by humans. That species and other members of the genus have been used in malarial research. *E. rozeti* sometimes is thought to have been the animal model used for the characteristic head of the ancient Egyptian god Set (Koontz and Roeper 1983). However, the Recent presence of that species in northeastern Africa was questioned by Douady et al. (2003), and the depiction of Set actually may have been taken from *Orycteropus*, the aardvark (Kingdon 1971). The IUCN formerly classified *E. revoili* as endangered and *E. edwardii* and *E. rupestris* as vulnerable, but now indicates no known major threats to their habitat, which is largely arid, rocky, and unsuitable for human development (Rathbun 2015b; Rathbun and Smit-Robinson 2015b, 2015c); much the same is reported for the other species of *Elephantulus*.

Order Bibymalagasia

BIBYMALAGASIA; **Family**
PLESIORYCTEROPODIDAE; Genus
PLESIORYCTEROPUS Filhol, 1895

Bibymalagasy

This Recently extinct order contains one family, Plesio-rycteropodidae, with the single genus, *Plesiorycteropus*, and two currently recognized species (MacPhee 1994):

P. madagascariensis, known by skeletal remains from 12 sites in Madagascar;

P.germainepetterae, known only by skeletal remains from a site in central Madagascar, probably Ampasambazimba.

The ordinal and common names mean simply "animals of Madagascar." *Plesiorycteropus* was commonly treated as a relative of *Orycteropus*, the aardvark of Africa, and placed in the order Tubulidentata, and there were also suggestions of affinity with Cingulata, Pilosa, or Pholidota. However, an intensive study by MacPhee (1994) indicated that, while *Plesiorycteropus* did seem to be part of a systematic assemblage comprising Tubulidentata and the other ungulates (Sirenia, Proboscidea, Perissodactyla, Hyracoidea, and Artiodactyla), it is as morphologically distinctive as is any eutherian group currently granted ordinal status. More recently, there has been general recognition that orders Afrosoricida, Macroscelidea, Tubulidentata, Hyracoidea, Proboscidea, and Sirenia are phylogenetically united in the clade Afrotheria (see account of class Mammalia). Helgen (2003a) thought it highly possible that *Plesiorycteropus* was part of Afrotheria. Asher, Novacek, and Geisler (2003) found *Plesiorycteropus* to fall within Afrotheria and to group most closely with *Orycteropus*. A molecular analysis by Buckley (2013) suggested *Plesiorycteropus* is likely a form of giant tenrec within what is here recognized as the order Afrosoricida. A recent assessment of the structure of the bony labyrinth (osseous inner ear) supported maintaining Bibymalagasia as a distinct order, phylogenetically intermediate to, on the one hand, a clade comprising Afrosoricida and Macroscelidea, and, on the other hand, Tubulidentata, but also suggested that *P. madagascariensis* and *P. germainepetterae* might not be specifically distinct (Benoit et al. 2015). Except as noted, the information for the remainder of this account was taken from MacPhee (1994).

Plesiorycteropus was much smaller than *Orycteropus*, and there is no evidence to indicate that its general shape and proportions were very similar. Its skull is estimated to have averaged about 100 mm in length, less than half the length of the skull of the latter genus, and to have had a comparatively long neurocranium and short face. It probably was edentulous, but neither teeth nor mandibles have been found. *P. madagascariensis* was the larger of the two species and has an especially more prominent braincase. Various estimates put its weight at 9–18 kg, whereas *P. germainepetterae* would have weighed about 6–10 kg.

Plesiorycteropus is distinguished from all other known eutherian mammals (including all recognized

Bibymalagasy (*Plesiorycteropus madagascariensis*). Painting by Helga Schulze, based on descriptive information in sources cited in this account, especially MacPhee (1994), and on a sketch provided through Ross D. E. MacPhee, American Museum of Natural History.

tubulidentates) by the following combination of skeletal characters: (1) a large, flat mandibular fossa (the area of the cranium articulating with the mandible) restricted to the facies articularis and not involving the zygomatic process of the squamosal (in *Orycteropus* this fossa is moderately deep and does impinge on the zygomatic process); (2) nasal bones markedly widened anteriorly (the nasals of *Orycteropus* are narrow anteriorly and widened posteriorly); (3) the neural arches of the posterior thoracic vertebrae and all lumbar vertebrae pierced by large longitudinal channels, or "transarcual canals" (*Orycteropus* also shows some evidence of such channels, but they are tiny and not perforating); (4) ischial tuberosities of the pelvis highly modified, expanded, and caudally flattened (such expansion is absent in *Orycteropus*); and (5) posteromedial process of the astragalus (anklebone) present, very large, and possessing a deep groove on the underside, probably for passage of the flexor tendons (the astragalus of *Orycteropus* also has a posteromedial process, but it lacks the distinct ventral groove).

Remains of *Plesiorycteropus* have been found both in forested areas along the coast of Madagascar and in open country in the interior of the island. In the latter areas it may have been associated with narrow belts of forest, now vanished, in and around major wetlands. Nothing definite is known about its natural history, though MacPhee's detailed assessment of its skeletal morphology offers some suggestions. Small eyes and poor vision are indicated by the tiny size of the optic canal; a highly developed olfactory sense seems likely based on the broad nasals; a soft, possibly insectivorous diet is suggested by the shallow jaw articulation and the apparent absence of teeth; and strong, sharp claws for digging are evidenced by transversely narrow and pointed, not hooflike, phalanges.

Plesiorycteropus may have been a rather cumbersome animal, and it had many of the attributes typifying mammals specialized for scratch-digging. This is shown especially by the forelimb, in which the radius is shorter than the humerus and the metapodials are short and wide. Also, the long hind limbs and large tail, with a massive base, probably served to anchor the body during digging. Certain features suggest a frequent sitting posture, notably a remarkable transformation of the ischial tuberosities of the pelvis into smooth-surfaced triangular plateaus. The same occurs in various anthropoid primates for support of ischial callosities, or "sitting pads." Some characters of *Plesiorycteropus* indicate roles other than digging, particularly a limited climbing ability. Aside from the likelihood of strong claws, the limb structure resembles that of known arboreal mammals; the head of the humerus rises above the tuberosities, facilitating the raising of the arms to or above shoulder height. Although probable olfactory and digging abilities, together with evident characters of the jaws, point to myrmecophagy, the overall small size of *Plesiorycteropus* suggests that it would not have broken into hardened termite mounds as modern *Orycteropus* does. It more likely took larvae and adult insects while digging in loose soil, tearing nests apart, or stripping off bark, perhaps while in trees.

The burning of the forests on which it depended or other environmental disruptions wrought by the early human invaders of Madagascar probably led to the extinction of *Plesiorycteropus*. According to Burney et al. (2004), the first evidence of people on the island dates from about 2,300 years ago, and a pelvic fragment of *Plesiorycteropus* from Masinandriana was dated at 2,154 years ago. However, radiocarbon dating of bones of another genus, found in association with those of *Plesiorycteropus* in the Anjohibe Caverns of northwestern Madagascar, suggest *Plesiorycteropus* may have survived until less than 500 years ago (Burney et al. 1997).

Order Tubulidentata

TUBULIDENTATA; **Family ORYCTEROPODIDAE;**
Genus ORYCTEROPUS E. Geoffroy St.-Hilaire, 1796

Aardvark, or Ant Bear

This order contains one family, Orycteropodidae, with the single living genus and species, *Orycteropus afer*, which occurs throughout Africa south of the Sahara, wherever suitable habitat is available (Meester 1977b). Taylor and Lehmann's (2015) map indicates occurrence as far north as southern Mauritania on the west coast and the Hala'ib Triangle (southeastern Egypt / northeastern Sudan border area) on the east coast. Early Egyptian paintings suggest the species ranged as far north as the Mediterranean in historical time (Kingdon 1971). The above use of "E. Geoffroy St.-Hilaire, 1796" as the authority for the generic name is in keeping with Lehmann (2007), but Schlitter (2005b) used "G. Cuvier, 1798." The extinct but Recent Malagasy genus *Plesiorycteropus* (see account thereof) sometimes has been placed in Tubulidentata but now is thought to represent an entirely separate order.

This animal resembles a medium-sized to large pig. Head and body length is 1,000–1,580 mm, tail length is 443–710 mm, and shoulder height is approximately 600–650 mm. Adult weight is 40–100 kg, though most individuals weigh from about 50 to 70 kg. Males are slightly larger than females (Shoshani, Goldman, and Thewissen 1988). The thick skin is scantily covered with bristly hair that varies in color from dull brownish gray to dull yellowish gray. The hair on the legs is often darker than that on the body. Numerous vibrissae occur on the face around the muzzle and about the eyes. The dull pinkish gray skin is so tough that it sometimes saves the aardvark from the attacks of other animals.

The aardvark has a massive body with a long head and snout. The round, blunt, piglike muzzle is pierced by circular nostrils, from which grow many curved whitish hairs 25–50 mm long. The tubular ears are 150–210 mm in length; they fold back to exclude dirt when the animal is burrowing and can be moved independently of one another. They are waxy and smooth, like scalded pig's skin. The tapering tongue often hangs out of the mouth, with the end coiled like a clock spring. The neck is short, the forequarters are low, and the back is arched. The strong and muscular tail is thick at the base and tapers to a point. The legs are short and stocky; the forefoot has four digits and the hind foot has five, the digits being webbed at the base. The long, straight, strong, blunt claws are suited to burrowing. The females have one pair of inguinal and one pair of abdominal mammae. In the males the penis has a fold of skin that covers scent glands at its base; females also have scent glands in the genital area.

The teeth in the embryo are numerous and traversed by a number of parallel vertical pulp canals. The milk teeth do not break through the gums. In the adult the teeth are located only in the posterior part of the jaw. The usual dental formula is: i 0/0, c 0/0, pm 2/2, m 3/3 = 20; additional vestigial premolars are sometimes present. The teeth do not grow simultaneously. Those

Aardvark (*Orycteropus afer*). Photo by Eric Isselee / Shutterstock.com.

nearest the front of the jaw develop first and fall out about the time the animal reaches maturity; they are succeeded by others farther back. The cheek teeth, which are covered externally by a layer of cement, each resemble a flat-crowned column and are composed of numerous hexagonal prisms of dentine surrounding tubular pulp cavities, hence the ordinal name "Tubulidentata" ("tubule-toothed"). The teeth of aardvarks grow continuously and lack enamel. The skull is elongate, and the lower jaw is straight, bladelike anteriorly, and swollen at the molars.

The aardvark occurs in a wide variety of habitats, including grassy plains, bush country, woodland, and savannah (Smithers 1971). Its range includes the rainforests of the Congo Basin (Taylor and Lehmann 2015). It appears to prefer sandy soils. The presence of sufficient quantities of termites and ants is apparently the main factor in distribution. The word "aardvark" means "earth pig" in the Afrikaans language—an appropriate name, as *Orycteropus* looks somewhat like a pig and is an extraordinarily active burrower. If overtaken away from its den, it digs into the ground with amazing speed. It can dig faster in soft earth than sev-

eral persons with shovels, and even the hardest sun-baked ground is no obstacle to its powerful forefeet. When digging, it pushes the ground backward under its body while resting on its hind legs and tail. When a sufficient amount of soil has accumulated, the aardvark shoves it back or to one side with the hind feet, sometimes with the aid of the tail.

According to Taylor (2005), *Orycteropus* makes excavations to find food and for shelter. Feeding digs vary from a simple surface scratch to holes over 1.5 meters deep, depending on nesting habits of the prey. An animal may make numerous burrows for shelter, there being one tally of 101 burrows within 1.5 ha., but usually each burrow is used a number of times. Three individuals used burrows for averages of 4.9, 6.9, and 8.6 consecutive days. Burrows consist of entrances, tunnels, and chambers. There is usually a single entrance but as many as five have been found. Tunnels are just wide enough for the animal to pass through and may be up to 13 meters long and 3 meters below the surface of the ground. At least one chamber is present, and it is about 0.75–1.0 meter wide and high. It must be large enough for the aardvark to turn in, as the animal

Aardvark (*Orycteropus afer*). Photo by Chris and Mathilde Stuart.

generally enters and leaves its burrow headfirst. When reentering a burrow at the end of an activity period, the aardvark displaces soil, effectively closing the entrance and, during colder months, providing protection from low ambient temperatures.

Orycteropus sometimes occupies a termite nest as a temporary shelter; its thick skin seems to be impervious to insect bites. It occasionally digs its holes in areas that seasonally flood and thus may have to evacuate at certain times of the year. Rahm (1990), however, cited reports that the aardvark is a good swimmer. Abandoned aardvark burrows are used by dozens of other kinds of animals. Also, the aardwolf (*Proteles cristatus*) has been observed to feed on termites from mounds that have been torn open by *Orycteropus* (Taylor and Skinner 2000).

The aardvark tends to walk on its claws, and its tail often leaves a track on soft ground. It is an extremely powerful animal. In one case, a man with a firm grip on the tail of an aardvark in its den was slowly drawn into the burrow as far as his waist and finally had to re-linquish his hold, despite the additional leverage afforded by two other persons holding onto his legs. The aardvark's movements are awkward and slow, but the animal is said to escape with surprising rapidity if alarmed. Its hearing is acute, and at the least alarm it seeks a burrow. Its eyesight does not appear to be good, since the aardvark frequently crashes into bushes, tree trunks, and other obstructions when running (Smithers 1971). It avoids enemies by digging or running, but if cornered, it will fight by striking with its tail or shoulders, by rearing on its hind legs and slashing with its forefeet, or by rolling on its back and slashing with all four feet.

The aardvark is mainly nocturnal but sometimes goes abroad by day and occasionally suns itself in the early morning at the burrow entrance. It usually sleeps inside during the day, curled up in a tight circle, with the snout protected by the hind limbs and tail. By night it moves from one termite nest to another, following a route that it regularly covers about once a week. When foraging, it travels in a zigzag path with its snout to the ground inspecting a strip about 30 meters wide.

Shoshani, Goldman, and Thewissen (1988) wrote that individuals move an average of about 10 km a night and that an estimated maximum of 30 km has been reported. Taylor and Skinner (2003) reported nightly movements of 2.7–4.4 km. Taylor (2005) indicated that, while above ground, the aardvark remains active 5–7 hours during winter and up to 9 hours in summer.

The diet consists principally of ants and termites, which are obtained by digging, tearing into nests, or seeking out the insects when they are on the march. A moving column of termites may contain tens of thousands of individuals and be about 40 meters long and can be detected by the aardvark by sound and smell. *Orycteropus* gathers ants and termites with its long, sticky tongue, which can be extended up to 300 mm. Other insects are taken occasionally, and Smithers (1971) reported apparent predation on the fat mouse (*Steatomys pratensis*). In captivity the aardvark accepts mealworms, boiled rice, meat, eggs, and milk and seems to thrive when some carbohydrate is included in the diet. Shoshani, Goldman, and Thewissen (1988) indicated that *Orycteropus* must have a source of water and sometimes digs up the fruits of a wild cucumber to obtain moisture. However, Taylor (2005) questioned that view, noting that, while the arrdvark has been observed to drink occasionally, water is not an essential requirement and the genus occurs in terrain where water may be available only seasonally.

Dorst and Dandelot (1969) believed that females are attached to a particular place, to which they return regularly, whereas males are more vagabond. However, in South Africa both sexes have been reported to utilize definite home ranges, the average size of which was determined to be 3.5 sq km in one study (Taylor 2005). In another study there, Taylor and Skinner (2003) found home ranges of 133–302 ha.; ranges of individuals of each sex overlapped those of the same sex and of the other sex; population density was about 8 aardvarks per 1,000 ha. Little is known of intraspecific interaction, but individuals do not share burrows with one another; the only reported noise is a vigorous sniffing with the nose pressed close to the ground (Taylor 2005). Both sexes use potent secretions from their scent glands to mark sites within their home ranges (Taylor 2011).

Orycteropus is generally solitary, but the young accompanies its mother for a long time. Births in Central Africa reportedly occur at the beginning of the second rainy season, in October or November. In southern Africa neonates or near-term fetuses have been recorded in February, April, May, June, July, September, and November; births in captivity have occurred in all months of the year (Taylor 2005). The gestation period has been variously cited as about 7–9 months (Hayssen, Van Tienhoven, and Van Tienhoven 1993); recent compilations of records indicate a possible range of 210–268 days and an average of 243 days (Reason, Gierhahn, and Schollhamer 2005; Taylor 2005). There is usually a single offspring, occasionally two. It weighs about 1.3–2.0 kg and is naked and flesh-colored. It remains in the burrow for about 2 weeks, and then begins to accompany its mother on nightly excursions. For the next several months the mother and young occupy a series of burrows, moving from one to another. The young can dig for itself after about 6 months. Shoshani, Goldman, and Thewissen (1988) stated that adult size is reached at about 12 months, sexual maturity is attained at about 2 years, a female gave birth to 11 young during 16 years of captivity, and a male had sired 18 offspring by the time he was 24 years old. Record longevity in captivity is 29 years and 9 months (Weigl 2005).

The flesh of *Orycteropus* has the appearance of coarse beef and is prized by some persons, but others say it is strong-smelling and tough as leather. In some areas the hide is made into straps and bracelets, and the claws are worn as good-luck charms. The burrows sometimes cause serious damage to farming equipment and earthen dams and thus make the aardvark subject to persecution by agricultural interests. However, in certain areas where the aardvark and other insectivorous animals have been exterminated, pasture and cereal crops have suffered enormous damage from termites (Shoshani, Goldman, and Thewissen 1988). The IUCN does not assign a threatened classification but does note that populations in eastern, central, and western Africa may be declining because of increasing human numbers, destruction of habitat, and hunting for meat (Taylor and Lehmann 2015). Lehmann (2007) listed 18 subspecies that have been described over the years. While many are probably not valid, some occur within restricted regions where the indicated problems are prevalent, and they should not be written off without assessment, particularly by molecular technology to determine presence of distinct lineages. Morphological

Aardvark (*Orycteropus afer*). Photos by Alexander F. Meyer at Brookfield Zoo (*left*) and at Faunia, Spain (*right*).

differences between rainforest and savannah populations were noted long ago and deserve further investigation (Taylor 2011).

The geological range of Tubulidentata is Oligocene to Pliocene in Europe, middle Miocene to late Pliocene in Asia, and early Miocene to Recent in Africa (McKenna and Bell 1997). Relationships with other orders, living and extinct, were not well understood based on morphological assessment (Thewissen 1985), but recent molecular analysis has indicated that Tubulidentata is part of the clade Afrotheria, which also comprises orders Afrosoricida, Macroscelidea, Bibymalagasia, Hyracoidea, Proboscidea, and Sirenia (see account of class Mammalia).

Order Hyracoidea

HYRACOIDEA; **Family PROCAVIIDAE**

Hyraxes, or Dassies

This order contains the single Recent family Procaviidae, which comprises three Recent genera and five species and occurs in southwestern Asia and most of Africa. The sequence of genera presented herein follows the phylogenetic arrangement of Pickford (2005). There is considerable controversy regarding the systematic affinities of the group. For most of the twentieth century a common practice was to place Hyracoidea, together with Proboscidea and Sirenia, in the superordinal taxon Paenungulata (or Uranotheria), and that position was supported by Shoshani (1992a, 1992b, 1993). However, there was also increasing support for a return to the older view that Hyracoidea is most closely related to Perissodactyla (Fischer 1989; Fischer and Tassy 1993). Indeed, based mainly on morphology of modern and fossil specimens, Prothero and Schoch (1989a, 1989b) proposed that Hyracoidea be regarded as a suborder of Perissodactyla. More recently, there has been growing acceptance that Hyracoidea is part of the clade Afrotheria, which also comprises orders Afrosoricida, Macroscelidea, Tubulidentata, Bibymalagasia, Proboscidea, and Sirenia. However, Paenungulata (Hyracoidea, Proboscidea, Sirenia) is still often considered valid and to be the sister group of Afroinsectiphilia, which comprises the other afrotherian orders (see account of class Mammalia).

The living members of the order are comparable in size and external appearance to large rodents and lagomorphs. Head and body length is 300–600 mm; the tail is 10–30 mm long or lacking. Adults sometimes weigh as much as 5.4 kg. The pelage consists of fine underhairs and coarser guard hairs. Scattered bristles, presumably tactile, are located mainly on the snout. A gland on the back is surrounded by hair usually of a different color from that on the rest of the body. According to Hoeck (2011), the odiferous gland lies beneath a raised patch of skin about 1.5 cm long. The surrounding hairs are erected when the animal is aroused during courtship, and probably the glandular secretions provide olfactory stimulation and disseminate scent that communicates identity and status. Another function of the gland may involve recognition of mother by young. There is also marking by secretions from the mouth and eyes and by urination and defecation (Hoeck 1990).

Hyraxes have a short snout, a cleft upper lip, short ears, and short, sturdy legs. The eye is unique in that a portion of the iris above the pupil bulges slightly into the aqueous humor, thus cutting off light from almost directly above the animal. The vertebral column is convex from the neck to the tip of the tail. The forefoot has four digits, with flattened nails resembling hooves. The hind foot has three digits; the inner toe, the second digit, has a long curved claw, while the other digits have short, flattened, hooflike nails. The soles have special naked pads for traction. The pads are kept continually

Bush hyrax, or yellow-spotted rock hyrax (*Heterohyrax brucei*). Photo by Chris and Mathilde Stuart.

moist by a glandular secretion and have a muscle arrangement that retracts the middle of the sole. This retraction forms a hollow that is a suction cup of considerable clinging power.

The dental formula of the deciduous teeth is: i 2/2, c 1/1, pm 4/4 = 28; the formula for the permanent teeth is: i 1/2, c 0/0, pm 4/4, m 3/3 = 34. The single pair of upper incisors grows continuously and is long and curved. The upper incisors are triangular in cross section and semicircular in form; the flattened back surfaces are without enamel and so produce pointed cutting edges. The lower incisors are chisel-shaped; the first pair has three cusps, the second pair only one. There is a wide space between the incisors and the cheek teeth. The milk canines are rarely persistent. From front to rear, the premolars become more like the molars in structure. The molars vary from low-crowned to high-crowned. Each has four roots, and the lower ones bear two crescents, as in the corresponding teeth

of rhinos and horses. The skull is stout and the roof is flattened.

The terrestrial genera (*Procavia* and *Heterohyrax*) inhabit rocky areas, arid scrub, and open grassland, whereas tree hyraxes (*Dendrohyrax*) are usually arboreal and found in forested areas, though in eastern Africa they inhabit lava flows. The elevational range of the family is sea level to 4,500 meters.

Although only *Dendrohyrax* is arboreal, all genera apparently can climb well; *Heterohyrax*, for example, sometimes suns itself in a tree. Hyraxes move quickly and are extremely agile on rugged and steep surfaces, running and jumping with skill and gaining traction by means of specialized foot pads and probably the inner claw on the foot. This claw is apparently also used to groom the hair. Hyraxes travel on the sole of the foot, with the heel touching the ground, or partly on the digits. In their habits the terrestrial forms are similar to the pikas, *Ochotona*, sheltering in colonies of 5 to about

Rock hyrax (*Procavia capensis*). Photo by Klaus Rudloff at Erfurt Zoo.

50 individuals, usually among rocks. Bush and rock hyraxes are active mainly during the daylight hours and are fond of basking in the sun and rolling in the dust. Tree hyraxes shelter singly or in family groups, using tree hollows and dense foliage; they are active during the night. They are less gregarious than the terrestrial forms and usually more tractable. The terrestrial hyraxes whistle, scream, and chatter; the tree hyraxes utter a series of loud cries.

Hyraxes have acute sight and hearing. They feed mainly on vegetation and are both grazers and browsers. Tree hyraxes feed on the ground as well as in trees. All species habitually work their jaws in a manner reminiscent of cud chewing. Distribution of hyraxes is not limited by lack of water to drink. They sometimes travel more than 1.3 km for food. Hyraxes are preyed on mainly by rock pythons, eagles, and leopards.

Hyraxes, referred to in the Bible as "conies," are in an order that once was far more diverse than at present and had more than 20 genera ranging in size from that of a rabbit to at least that of the Sumatran rhinoceros (*Dicerorhinus*). Hyracoidea is usually reported to comprise two families, the living Procaviidae and the extinct Pliohyracidae. Gheerbrant, Domning, and Tassy (2005) listed three subfamilies of the latter—Saghatheriinae, Geniohyinae, and Pliohyracinae—but considered the group paraphyletic, containing some members more closely related to other families. Pickford (2004, 2005) treated those three subfamilies as full families and also recognized a fourth family, Titanohyracidae.

The pliohyracid group apparently arose in the early Eocene of Africa, underwent considerable diversification by the Oligocene, and during that epoch formed the dominant small- to medium-sized herbivores of Africa. Communities of hyracoids then demonstrated a range in body mass (3 kg to 2,000 kg) and structure comparable to that of modern ungulate communities (Rasmussen 1989; Rasmussen et al. 1996; Schwartz,

Rasmussen, and Smith 1995). Distinctive lineages are known, showing convergence with such ungulate families as Equidae, Suidae, Tapiridae, and Bovidae (Gheerbrant, Domning, and Tassy 2005). The pliohyracids subsequently became much less diverse in response to competition from more advanced mammals. They persisted into the Miocene of Africa and also were present in the Miocene and Pliocene of Europe and Asia; the last known representatives lived about 2 million years ago in eastern China.

Pickford (2005) concluded that the most likely ancestor of the family Procaviidae is the family Saghatheriidae, in particular the genus *Meroehyrax* of the early Miocene of East Africa. Procaviidae subsequently underwent some diversification during the Pliocene and Pleistocene, though not to the extent of the pliohyracids. One extinct procaviid genus, *Gigantohyrax*, was an immediate relative of the living *Dendrohyrax* (Pickford 2005) but about three times the size. Outside of Africa Procaviidae is known only from the late Pleis-

tocene and Recent of southwestern Asia (McKenna and Bell 1997).

HYRACOIDEA; PROCAVIIDAE; **Genus** *DENDROHYRAX* Gray, 1868

Tree Hyraxes

There are three species (Bothma 1977; Butynski, Dowsett-Lemaire, and Hoeck 2015; Butynski, Hoeck, and de Jong 2015a; Hoeck et al. 2015; Shultz and Roberts 2013):

D. dorsalis, forest zone from Sierra Leone to the Volta River in Ghana and from Benin to Uganda and Cabinda in extreme northwestern Angola, island of Bioko (Fernando Poo);

D. arboreus, known from forested parts of eastern and southern Democratic Republic of Congo,

Tree hyrax (*Dendrohyrax validus*). Photo by Chris and Mathilde Stuart.

central and southern Uganda and Kenya, Rwanda, Burundi, Tanzania, northeastern Angola, Zambia, Malawi, northwestern and central Mozambique, and Eastern Cape and KwaZulu-Natal provinces of South Africa.

D. validus, restricted to montane and coastal forests in central and northeastern Tanzania and southeastern Kenya, and on the islands of Pemba, Zanzibar and Tumbatu.

Based on examination of available specimens, Shoshani (2005a) considered *D. validus* a synonym of *D. arboreus* and did not recognize any subspecies of the latter. However, Hoeck (2011) treated *D. validus* as a full species and listed seven subspecies of *D. arboreus*.

Head and body length is 400–600 mm, tail length is 10–30 mm, and weight is usually 1.5–4.5 kg. The hair on the upper parts is brown tipped with gray or yellow; black hairs are also present on the back. Some forms have two color phases, the darker coat and a yellow pelage. A white patch of hair marking the location of a gland is present on the back (see above generic account), and the ears are edged with white hairs. The underparts are usually brownish. The fur is longer and slightly more silky than that of other hyraxes. Gaylard (2005) noted that the number of mammae is variable; usually there is an inguinal pair but sometimes also a pectoral pair, and some females have two inguinal pairs and one pectoral pair.

According to Jones (1978), *D. dorsalis* occurs most frequently from sea level to an elevation of about 1,500 meters but has been recorded at up to 3,000 meters. It inhabits upland and riverine areas within tropical rainforest or tropical closed forest and is more numerous in areas containing lianas than in other habitats. It is arboreal but frequently descends and moves about on the ground. This species is strictly herbivorous, feeding on leaves, fruits, twigs, and bark, usually from the upper canopy of the forest. It is nocturnal, with most activity taking place soon after dark and just before dawn.

Studies of captive *D. arboreus* by Rudnai (1984) indicate the species is arboreal, sometimes nocturnal, and solitary. Milner and Gaylard (2013) noted, however, that in some areas it is active by day and also undergoes thermoregulation through sun basking. Milner and Harris (1999) found that species to spend 90 percent of its time above ground level and more time in the upper canopy at night. Kingdon (1971) indicated that some populations of *D. arboreus* inhabit subalpine areas at elevations of up to 4,500 meters. It is partly diurnal at high elevations, where it lives among rocks and is gregarious. It can be very noisy, and the calls often seem to follow a frequency pattern; most vocal activity seems to occur following intensive feeding periods, between 2000 and 2300 hours and between 0300 and 0500 hours. The cry is made by both sexes but is much more powerful in the male, and it may have both sexual and territorial functions. Gaylard (2005) described the vocalization of *D. arboreus* as "an unearthly noise, starting off with a series of cackling barks followed by piercing tremulous screams that rise to a high crescendo. When a few are vocalizing at the same time the forest resounds with what sounds like an orgy of mass murder." Other sounds include growling, barking, and teeth grinding. Based on counts of calls, estimated population densities were roughly 1–2 individuals per sq km for *D. dorsalis* in Taï National Park, Ivory Coast (Shultz and Roberts 2013) and up to 17 per ha. for *D. validus* in undisturbed, closed-canopy forest in the Udzungwa Mountains (Hoeck et al. 2015). Milner and Harris (1999) made a radio-tracking study of 35 individuals of *D. arboreus* in 2.6 ha. of the Virunga Mountains of Rwanda. Home ranges were 590–2,630 sq meters, with considerable overlap. Mature males had smaller core areas that overlapped those of several females.

Hoeck (2011) wrote that *D. dorsalis* is primarily solitary, but groups of two and three can be found, most likely a mother and subadult young. Each defended male territory overlaps several smaller female ranges. According to Jones (1978), each animal uses a small area in the forest, usually centered on a single tree. Both mating and birth of *D. dorsalis* reportedly occur during the dry season. The young are born primarily during March and April in Gabon and Cameroon and from May to August in the western and southern parts of Democratic Republic of Congo. In the eastern parts of the range of the species, young are born throughout the year. A single young, occasionally two, is produced after a gestation period of about 8 months. Birth weight is 130–180 grams (Shultz and Roberts 2013). The young are precocious; they reach adult length at about 120 days and grow little thereafter. In *D. arboreus*

Tree hyrax (*Dendrohyrax arboreus*). Photo by Mike Wilkes / npl / Minden Pictures.

breeding seems to continue throughout the year, the gestation period is 7–8 months, and the young number 1–3 (Gaylard 2005; Kingdon 1971). The young of *Dendrohyrax* weigh about 200 grams at birth, weaning occurs after about 5 months, and both sexes usually reach sexual maturity at around 16 months (Hayssen, Van Tienhoven, and Van Tienhoven 1993). A captive *D. arboreus* lived to an age of about 13 years and 7 months (Weigl 2005).

Predators of *D. dorsalis* include hawk eagles, leopards, golden cats, genets, servals, and pythons (Jones 1978). Tree hyraxes often assume a characteristic defensive position when threatened by a predator: the back and rump are turned toward the enemy, and the hairs around the dorsal gland are spread out and separated from each other, so that the naked glandular area is exposed. They are said to be less irritable than the

other dassies and to tame readily. They are trapped and hunted by people, commonly using dogs, for food and pelts throughout their range, and Hoeck (1984) suggested that they may be endangered in some areas through forest destruction. The subspecies *D. arboreus arboreus* is restricted to an isolated area of southeastern South Africa, in Eastern Cape and KwaZulu-Natal Provinces, and appears to be declining seriously in association with the disappearance of the forests on which it depends (Gaylard 1992, 2005). That subspecies was formerly classified as vulnerable by the IUCN. While such designation is no longer active, it is recognized that the forest habitats of the entire species *D. arboreus* are severely threatened, that the species is in decline, and that it may warrant a near threatened classification (Butynski, Hoeck, and de Jong 2015a). *D. validus* was recently classified as near threatened by the IUCN, as its total area of occupancy may be as small as 3,078 sq km, its viable habitat is severely fragmented, it is subject to intensive hunting, and its numbers apparently have fallen substantially in the last one or two decades (Hoeck et al. 2015).

HYRACOIDEA; PROCAVIIDAE; **Genus** *HETEROHYRAX* Gray, 1868

Bush Hyrax, or Yellow-spotted Rock Hyrax

The single currently recognized species, *H. brucei*, occurs in northeastern Sudan, Eritrea, Ethiopia, Somalia, Kenya, Uganda, Rwanda, Burundi, Tanzania, eastern Democratic Republic of Congo, Zambia, Malawi, northern Mozambique, Zimbabwe, eastern Botswana, and Limpopo and Mpumalanga Provinces of South Africa. In addition, the disjunct subspecies *H. b. bocagei* is found in west-central Angola and possibly

northern Namibia, and the apparently isolated subspecies *H. b. chapini* occurs at Matadi, near the mouth of the Congo River in western Democratic Republic of Congo. Bothma (1977) and Hoeck (1984) listed *H. b. chapini* as a full species, but most recent authorities have treated it as a subspecies of *H. brucei* (Barry and Shoshani 2000; Hoeck 2011; Shoshani 2005a). *H. antineae* from the Ahaggar Massif in southern Algeria was formerly considered a full species of *Heterohyrax* (Bothma 1977; Hoeck 1984) or an isolated subspecies of *H. brucei* (Barry and Shoshani 2000; Corbet and Hill 1991; Roche 1972; Shoshani 2005a), but it is now thought to represent the genus and species *Procavia capensis* (Butynski, Hoeck, and de Jong 2015b; Hoeck 2011; Hoffmann, Hoeck, and De Smet 2008). *H. brucei* has frequently been reported to occur in southeastern Egypt but such records also stem from confusion with *Procavia capensis* (Butynski, Hoeck, and de Jong 2015b; Hoffmann, Hoeck, and De Smet 2008). Hoeck (2011) listed 24 subspecies of *H. brucei*, suggesting that many are based only on variation in coat color, but noting also that mitochondrial DNA sequences indicate that at

Bush hyrax, or yellow-spotted rock hyrax (*Heterohyrax brucei*). Photo by Klaus Rudloff at Tierpark Berlin.

least *H. b. hindei* of Kenya, *H. b. ruddi* of Zimbabwe and Mozambique, and *H. b. granti* of South Africa are highly distinct and may represent cryptic species. Bloomer (2009) noted that genetically *ruddi* is more closely related to *hindei* than to *granti*. Ellerman and Morrison-Scott (1966) and Roche (1972) regarded *Heterohyrax* as only a subgenus of *Dendrohyrax*, but Hoeck (1978) showed that generic distinction is warranted.

Head and body length is 320–560 mm, an external tail is absent, shoulder height is about 305 mm, and weight is around 1.3–4.5 kg. The body is covered with thick, short, rather coarse hair. The upper parts are generally brownish or grayish, sometimes suffused with black, and the underparts are white. In general appearance it is much like a large guinea pig (*Cavia*). *Heterohyrax* is distinguished by a patch of yellowish, reddish, or whitish hair in the middle of the back. This patch covers a gland that is exposed when the surrounding hairs are erected (see above generic account). Females usually have one pair of pectoral mammae and two pairs of inguinal mammae.

Sclater (1900) wrote: "The soles, which are naked, are covered by a very thick epithelium which is kept constantly moist by the secretion of the sudorific glands there present in extraordinary abundance; furthermore, a special arrangement of muscles enables the sole to be contracted so as to form a hollow air-tight cup which, when in contact with the rock, gives the animal great clinging power, so much so that even when shot dead it remains attached to almost perpendicular surfaces as if fixed there."

Smithers (1971) reported that in Botswana *H. brucei* is confined to areas of rocky kopjes, rocky hillsides, krantzes, and piles of loose boulders, particularly where there is a cover of trees and bushes on which it can feed. This species is more strictly confined to the larger areas of such a habitat and is less prone than *Procavia* to colonize smaller outlying kopjes and piles of boulders. The elevational distribution of *Heterohyrax* is from sea level to at least 3,800 meters in mountains. *Heterohyrax* is sharp-sighted, keen of hearing, and quite aggressive, prepared to bite anything that attacks. It is noted for its wariness, being ever alert and seeking shelter upon the slightest alarm, but is extremely curious and will soon show itself again. It enjoys basking in the sun and lying on rocky ledges during the afternoon, and individuals like to play and chase one another among the rocks. Its presence often is detected by its calls, some of which resemble shrill screams.

Heterohyrax is primarily diurnal. Its habits and thermoregulatory needs are very much like those described below for *Procavia*, with which it may occur in close association. The two genera sometimes inhabit the same kopjes, they shelter in the same crevices and huddle together when outside, and their young play together. Such heterospecific association tends to increase during the birth season, thereby allowing larger group size, more efficient predator detection, and greater survival of young (Barry and Mundy 2002). *Heterohyrax* is more of a browser than *Procavia*, some 80 percent of its food being obtained in this manner, and it is also known to climb trees in order to feed. The diet consists of many kinds of bushes and trees, even those that are poisonous to most other mammals. Dobroruka (1973) reported that at the Numba Caves in Zambia *H. brucei* fed almost exclusively on the leaves of wild bitter yam (*Dioscorea dumetorum*), a plant often used by natives of the area to prepare poisoned arrows. According to Hoeck (1975), in Serengeti National Park *H. brucei* had feeding peaks in the morning and evening, and though primarily a browser, it grazed more in the wet season than in the dry season.

Except as noted, the following information on social life, demography, and reproduction was taken from the reports of Hoeck (1975, 1982, 1984, 1989, 2011) and Hoeck, Klein, and Hoeck (1982) on their studies in Serengeti National Park, Tanzania. In that area, where there is concentrated favorable habitat, *Heterohyrax* lives in colonies of sometimes hundreds of animals, with population densities of 20–81 per ha. Where habitat is more evenly distributed, densities may be much lower; in Matobo National Park, Tanzania, densities over a 5-year period were 0.51–1.10 animals per ha. (Barry and Mundy 1998). The large colonies are divided into cohesive, stable, polygynous family units of 5–34 animals each. On small kopjes (4,000 sq meters or less) there may be only a single family. One group observed for several years occupied an area of 3,600 sq meters and had an average of 16.3 members. Groups typically contain a single dominant, territorial, adult male, several peripheral males, 3–17 related adult females, and the young of 1 or 2 years. If several groups inhabit the same kopje, their ranges overlap, but the females of each

Bush hyrax, or yellow-spotted rock hyrax (*Heterohyrax brucei*). Photo by Klaus Rudloff at Tierpark Berlin.

family have their own core area, averaging about 2,100 sq meters, which is defended by the dominant male. He repulses and may fight all intruding adult males. He often stands guard on a high rock while the other animals feed or bask and is the first to give the shrill alarm call in the event of danger. He also emits a distinctive long and shrill territorial call. The females do not defend the home range and sometimes allow an outside female to join their group. On a large kopje there may be a number of peripheral or young dispersing males that live alone and occupy ranges overlapping a number of group ranges. They may harass the territorial males, attempt to rush in and mate with their females, and await their chance to completely take over a family unit. On a small kopje a young male must either quickly challenge the dominant male or depart the area.

The territorial male seeks to monopolize all females more than 28 months old and shows little interest in younger ones. This behavior may serve to prevent inbreeding. In the Serengeti there are discrete mating seasons of about 7 weeks, during which adult females enter estrus several times for several days each. The two corresponding birth seasons are May–July, following the long rainy season, and December–January, after the short rains. Females of a given family group become receptive in only one of the annual breeding seasons. All the pregnant females of a group then give birth within 3 weeks of one another at one time of the year. Gestation lasts about 230 days, and there are 1–3 young. The average number of young is 1.6 in the Serengeti. Gaylard (2005) stated that the average is 2.1 in Matobo National Park of Zimbabwe and birth weight is 220–230 grams. Fully developed at birth, the young are soon able to scamper about, and they play together in their own nursery groups. They are weaned at 1–5 months and reach sexual maturity at 16–17 months. Females usually remain in their mother's family, but young males disperse at 12–30 months. Average longevity of a group of 17 females was 50.6 months, but one female is known to have lived at least

11 years. Males do not seem to live as long, though Weigl (2005) listed one captive male that died at an age of 11 years and 9 months.

HYRACOIDEA; PROCAVIIDAE; **Genus** *PROCAVIA* Storr, 1780

Rock Hyrax

Most authorities tentatively recognize only a single species, *P. capensis*, occurring in Lebanon, Israel, Jordan, Sinai, the Arabian Peninsula (mainly the west and south but also scattered localities elsewhere), Egypt east of the Nile, isolated mountains of the Sahara (central and southern Algeria, southern Libya, northern Chad), and most of Africa farther south but with a large apparent hiatus, where the species is absent, including Gabon, most of Congo, central and southern Democratic Republic of Congo, central and southern Tanzania, Angola except the southwest, most of Zambia, western and central Botswana, and much of Mozambique (Butynski et al. 2015; Corbet 1979; Ellerman and Morrison-Scott 1966; Gaylard 2005; Harrison and Bates 1991; Hoeck 2011; Meester et al. 1986; Olds and Shoshani 1982; Roche 1972; Shoshani 2005a; Yalden, Largen, and Kock 1986). Reported occurrence in Turkey is in error, and reported presence in Syria has not been confirmed (Butynski et al. 2015). Although Hoeck (2011) indicated that the range of the subspecies *P. c. syriaca* includes Syria, Harrison and Bates (1991) showed that the actual type locality of *syriaca* is well inside the current borders of Lebanon and did not mention its presence in Syria. Some authorities, including Bothma (1977), consider *P. capensis* restricted to South Africa, central and southern Namibia, eastern Botswana, and Zimbabwe and treat the following as distinct species: *P. welwitschii*, southwestern Angola, northwestern Namibia; *P. johnstoni*, northeastern Democratic Republic of Congo to central Kenya,

Rock hyrax (*Procavia capensis*). Photo by Chris and Mathilde Stuart.

northern Tanzania, and possibly Malawi and western Mozambique; *P. ruficeps*, southern Algeria and Senegal to Central African Republic; and *P. syriaca*, Egypt to Kenya, and the southwest Asian portion of the range of the genus. Corbet and Hill (1991) recognized three of those species but included *P. johnstoni* in *P. capensis* and *P. ruficeps* in *P. syriaca* and also regarded *P. habessinica* of Ethiopia as a full species. Recent studies of mitochondrial DNA (Bloomer 2009) suggest there are two distinct and allopatric genetic groups of the subspecies *P. c. capensis*, one in northeastern South Africa and adjacent Zimbabwe, the other in the rest of South Africa and Namibia.

Head and body length is 305–580 mm, an external tail is lacking, shoulder height is 202–305 mm, and average weight is about 4 kg in males and about 3.6 kg in females. Hoeck (2011) gave the overall weight range as 1.8–5.4 kg. The hair pelage is dense and rather coarse, with short, thick underfur and long, scattered guard hairs. The general coloration of the upper parts is brownish gray, the flanks are somewhat lighter, and the underparts are creamy; however, there is considerable variation in color and intensity. The black whiskers may be as long as 180 mm. In general appearance *Procavia* is similar to a large pika (*Ochotona*) or a tail-less woodchuck (*Marmota*). However, the similarities go no further than the superficial appearance. The soles are moist and rubberlike, which gives the animals traction on smooth surfaces and steep slopes. *Procavia* is distinguished by a black (in *P. c. capensis*) or white or pale yellow (in *P. c. welwitschii*) dorsal patch covering a gland that is exposed if the surrounding hairs are erected; erection of those hairs is a demonstration of aggression (Gaylard 2005). Also, in *Procavia* the premolar series of teeth is much shorter than the molar series; in *Heterohyrax* the two series are about equal in length, and in *Dendrohyrax* the premolar series is longer (Gaylard 2005). Female *Procavia* have six mammae.

Procavia frequents rocky, scrub-covered habitat where there are suitable shelters in, between, or under rocks or where it can dig burrows of its own. In such habitat, it often is found living in close association with *Heterohyrax* (see account thereof). Butynski et al. (2015) gave elevational range of *Procavia* as sea level to 4,300 meters. In spite of its rather heavy build, *Procavia* is active and agile, running up steep, smooth rock surfaces with ease. Its senses are keen, and it becomes quite shy when persecuted. When alarmed, it quickly dashes into a rocky cranny or its own burrow. Although not a large animal, it can put up a vigorous fight in self-defense, biting savagely. Much of its lifestyle seems to be associated with its relatively poor thermoregulatory ability (Gaylard 2005; Kingdon 1971; Olds and Shoshani 1982). Its body temperature fluctuates relative to ambient temperature, tending to increase during warmer parts of the day and after its extensive sessions of sun-basking. It is predominantly diurnal, emerging in cool weather well after sunrise and retiring to its rock shelter before sunset. In warm weather it may remain out if there is a half- to full moon. It is reluctant to emerge if the weather is overcast or cold and does not do so at all during rainy periods, but it will seek shade during very hot weather. Individuals regularly crowd together in holes to conserve heat.

Procavia may be found feeding at any time of the day provided it is warm and sunny, with peaks of activity in the early morning and late afternoon. However, the actual time spent feeding is remarkably short for a vegetarian, amounting to less than 1 hour per day. Animals tend to stay near shelter but may forage 50–100 meters away. The diet includes a great variety of grasses, shrubs, and forbs. *Procavia* is predominantly a browser in some areas and a grazer in others. It feeds on practically any plant, including some that are poisonous to most other animals, such as plants of the families Solanaceae and Euphorbiaceae. It is adept at climbing trees and going out on branches to obtain fresh leaves. It can secure sufficient moisture from its food but will drink regularly if water is available (Gaylard 2005; Olds and Shoshani 1982).

Overall population density on Mount Kenya has been estimated at 20–100 animals per sq km (Butynski et al. 2015). Densities can be much higher at prime sites, but numbers are subject to great fluctuation, in part because of effects of drought. Population density in Serengeti National Park varied from 5 to 56 individuals per ha. (Hoeck 1984, 2011), but in Matobo National Park of Zimbabwe, where favorable habitat is less concentrated, density was only 0.73–0.94 per ha. over a 5-year period (Barry and Mundy 1998). Group home range averages 4,250 sq meters. The dominant male of a group is territorial and repels other males. The females do not defend their range and may allow an outside female to join the group. The gregarious

Rock hyrax (*Procavia capensis*). Photo of an individual in northern Israel by Klaus Rudloff.

nature of *Procavia* partially reflects its needs with respect to heat conservation and shelter (Gaylard 2005; Hoeck 1975; Olds and Shoshani 1982). Observations of captives indicate that as many as 25 animals will crowd together in a shelter to conserve warmth, either huddling on one level or heaping on top of one another. Depending on the amount of suitable habitat in a given location, especially the extent of shelter among rocks, a colony may be limited to a single family or may consist of hundreds of animals. The basic family unit consists of one adult territorial male, sometimes a subordinate male, 3–17 adult and subadult females, and young. Some males live alone after leaving their family group at 16–30 months. Group size in the wild is 2–26 individuals, usually about 4–10. One of the older females stays on a high rock or branch while the others feed or sun themselves, and if danger threatens, that sentry gives an alarm call—a sharp bark—upon which the others take to cover (Gaylard 2005).

Fourie (1977) distinguished 21 adult vocalizations that are apparently important for information transfer. Hoeck (2011) stated that the calls convey accurate information about the size, condition, social status, and hormonal state of the singer; the same attributes can be signaled chemically by scent-marking. However, Kershenbaum et al. (2012) observed that while males produce long, complex songs, lasting up to several minutes, the purpose is unclear. A song typically consists of a series of bouts, each being a sequence of "syllables," which can be classed as "wails, chucks, snorts, squeaks, and tweets." There are syntactic dialects of such songs, and syntax differs significantly from one area to another.

Reproductive activity is highly seasonal and seems to vary geographically (Gaylard 2005; Gombe 1983; Kingdon 1971; Mendelssohn 1965; Millar 1971; Smithers 1971). In South Africa, mating occurs in February–March, parturition in September–November in Western Cape Province; respective times are March–April and October–November in Eastern Cape, April and November–December in Free State, May and December–January in Limpopo, and September–November and June–July in Northern Cape. Mating also has been reported in Israel from August to September, in Kenya from August to November, in Zimbabwe in August, and in Tanzania at the end of the wet season, from April to June. Births peak in Israel from mid-March to early May and in Zimbabwe from March to April. Females in a colony give birth at around the same time. The estrous cycle lasts about 13 days. The gestation period is 202–245 days. Births usually occur in a crevice. The overall number of young per litter is 1–6, with reported averages of 3.2 in Israel, 2.4 in Zimbabwe, 2.7 in KwaZulu-Natal Province, and 3.3 in Western Cape. The newborn weigh 170–240 grams each, are fully covered with hair, have their eyes open, and can move about with agility after a day. They climb onto any adult's back within hours of birth and suckle from any female in the group. They are capable of taking solid food within a day of birth but may continue to nurse for 1–5 months. Females reach sexual maturity at about 16–17 months and then usually remain in their family group. Although females are commonly reported to give birth annually, Barry and Mundy (1998) found them to breed in alternate years in Matobo National Park, Zimbabwe. According to Hoeck (2011), males may mature physiologically at about the same age as females do but may not be able to mate until 28–29 months old, because of intimidation by the dominant male of a group. Most males disperse from their natal group when 16–24 months old, and the rest do so by 30 months. Dispersing animals may move at least 2 km, sometimes through inhospitable habitat, where they are highly vulnerable to predation. Males tend to have shorter lives than do females, which have survived over 10 years in the wild. Weigl (2005) listed a captive male and female, which lived to respective ages of 14 years, 3 months, and 14 years, 10 months.

The chief natural enemies of *Procavia* are leopards and eagles, but it also is preyed upon by foxes, weasels,

Rock hyrax (*Procavia capensis*), group on rocks. Photo by Graeme Shannon / Shutterstock.com.

and mongooses. Its flesh is prized by some people. Wanton killing for use as food and destruction of natural vegetation by humans have led to a drastic reduction of the formerly abundant populations of *Procavia* in Egypt (Osborn and Helmy 1980). In some areas it takes up residence in culverts or stone walls and becomes a pest. There had been suggestions that it sometimes competes with sheep for forage in South Africa and should be reduced in numbers, but Lensing (1983) showed that only about 1 percent of the potential agricultural yield of the involved area is lost each year to hyraxes. Populations in KwaZulu-Natal Province became locally extinct about 2000 and reintroduction attempts failed (Hoeck 2011). In Israel, where *Procavia* is protected and has few natural enemies, its numbers increased so much that crops were damaged and countermeasures, such as erection of electric fences, were taken (Butynski et al. 2015). Young rock hyraxes make interesting pets and will eat practically any vegetable matter.

Order Proboscidea

PROBOSCIDEA; **Family ELEPHANTIDAE**

Elephants

This order contains one living family, Elephantidae, with two living genera, *Loxodonta* and *Elephas*. *Loxodonta* herein is considered to contain two living species, *L. africana* in most of sub-Saharan Africa, and *L. cyclotis* in the forested region of western and central Africa and perhaps originally in northern Africa. Some authorities do not recognize *L. cyclotis* as a species distinct from *L. africana*. *Elephas* contains one living species, *E. maximus* of southern Asia. An additional elephantid genus and species, *Mammuthus primigenius*, evidently survived into early historical time on Wrangel Island off northeastern Siberia. Proboscidea also contains many extinct genera, at least four of which are sometimes reported to have persisted into early historical time—*Mammut* in North America, *Cuvieronius* or *Notiomastodon* in South America, *Stegodon* in eastern Asia, and *Palaeoloxodon* (sometimes treated as a subgenus of *Elephas*) on Tilos Island off southwestern Asia Minor—but definitive evidence is lacking (see below).

McKenna and Bell (1997) listed Proboscidea as a parvorder of infraorder Behemota (Proboscidea + the extinct aquatic parvorder Desmostylia) of suborder Tethytheria (Behemota + infraorder Sirenia) of order Uranotheria (=Paenungulata: Tethytheria + suborder Hyracoidea). They considered family Elephantidae to include (among other members) subtribe Loxodontina, comprising *Loxodonta*, and subtribe Elephantina, comprising *Mammuthus* and *Elephas* (including *Palaeoloxodon*). However, a new analysis of mitochondrial and nuclear DNA suggests that the Pleistocene *Palaeoloxodon antiquus* of Europe and western Asia is more closely related to *Loxodonta cyclotis* than to *Elephas maximus* (Meyer et al. 2017).

While most authorities continue to recognize Proboscidea as a full order (e.g., Sanders et al. 2010; Shoshani 2005b; Whyte 2005; Wittemyer 2011), McKenna and Bell's (1997) suggestion that *Elephas* and *Mammuthus* are more closely related to one another than either is to *Loxodonta*, has received substantial authoritative acceptance (e.g., Shoshani and Tassy 2005; Wittemyer 2011). Originally based on morphological characters, the position has been supported by abundant molecular assessment (Capelli et al. 2006; Greenwood et al. 1999; Kuntner, May-Collado, and Agnarsson 2010; Miller et al. 2008; Ozawa, Hayashi, and Mikhelson 1997; Roca 2007; Roca et al. 2015; Rogaev et al. 2006; Rohland et al. 2007, 2010; Vidya, Sukumar, and Melnick 2009; Yang, Golenberg, and Shoshani 1996). Some of that work has involved extracting ancient DNA from remains of *Mammuthus* and even from *Mammut* (see below), the latter being used as an outgroup. It has been indicated that phylogenetic divergence of *Mammut* from the other three genera occurred about 30 million years ago in the late Oligocene, that *Loxodonta* separated from the lineage of *Elephas* and *Mammuthus* about 7 million years ago in the late Miocene, that *Elephas* and *Mammuthus* diverged from

African savannah elephant (*Loxodonta africana*). Photo by Johan Swanepoel/Shutterstock.com.

one another about 5–6 million years ago in the late Miocene or early Pliocene, and that *Loxodonta cyclotis* and *L. africana* also diverged about 5–6 million years ago. That chronology has been considered in the sequence of elephantid genera presented herein. However, some other studies support the view that the lineage of *Elephas* diverged first and that *Loxodonta* and *Mammuthus* are immediately related (Barriel, Thuet, and Tassy 1999; Debruyne 2001; Debruyne, Barriel, and Tassy 2003; Thomas et al. 2000). Also, a recent analysis of cranial and dental morphology indicated that *Loxodonta* and *Elephas* have a common lineage, separate from that of *Mammuthus* (Todd 2010).

The most conspicuous external feature, at least in *Loxodonta*, *Elephas*, and *Mammuthus*, is the long, tubular, muscular, and very sensitive trunk. The trunk is actually a great elongation of the nose, the nostrils being located at the tip. The fingerlike extremity of the trunk is skillfully used to pick up food, such as peanuts,

or other small articles and to manipulate or examine objects. Lister and Bahn (2007) noted that *Loxodonta* has two "fingers" of equal length, *Elephas* has only one, and *Mammuthus* has one short and one long. The trunk also is used for breathing (along with the mouth) and to suck up water, which the animal then squirts into its mouth for drinking or sprays over its body for cooling. Sometimes, when no external water is available on a hot day, an elephant is seen to insert its trunk into its mouth and withdraw water, with which it then sprays itself. Shoshani (1998) showed that the source of the water is the pharyngeal pouch, just behind the tongue. When the elephant drinks, the pouch stretches and fills with water, which then is stored for future use.

Living members of Proboscidea have a maximum height of nearly 400 cm, and males weigh as much as 10,000 kg. The head is huge, the neck is short, the body is long and massive, and the tail is of moderate length. The highest point of *Loxodonta* is the shoulder; that of *Elephas* is the head, which is double domed (Laursen

Asian elephant (*Elephas maximus*). Photo by Andaman / Shutterstock.com.

and Bekoff 1978). The forehead of *Loxodonta* is rounded, that of *Elephas* is dished in. *Loxodonta* may appear "swaybacked" in profile, with a mid-dorsal depression (concave), while the back of *Elephas* tends to be rounded upward (convex). The ears are very large in *Loxodonta*, less so in *Elephas*, and relatively small in *Mammuthus*. The limbs are long, massive, and columnar. All the limb bones are well developed and separate; lacking marrow cavities, they are filled with spongy bone. The feet are short and broad; the weight of the animal rests on a pad of elastic tissue. There are 5 toes on each foot, but the outer pair may be vestigial and lack hooves (nails), and the hooves on some other toes may be worn away during life. Commonly, *Elephas* shows 4 hooves on the hind foot and 5 on the forefoot, *Loxodonta africana* 3–5 on the hind foot and 4–5 on the forefoot, and *L. cyclotis* 4–5 on the hind foot and 5 on the forefoot (see also account of genus *Loxodonta*). Females have two nipples just behind the front legs; the young nurse with the mouth, as do young of other mammals. Males retain the testes permanently within the abdomen.

Although *Mammuthus* had dense fur, the skin of living adult elephants is sparsely haired. The glands associated with the hair follicles in most mammals (sebaceous glands), which soften and lubricate the hair and skin, are not present. Studies by Myhrvold, Stone, and Bou-Zeid (2012) suggest that the low density hair covering of living elephants (about 1,500 hairs per sq meter as compared to about 2 million per sq meter for the human head) does not function as an insulator, as does the fur of most mammals, but instead serves to facilitate heat loss. Elephant hairs may be analogous to cactus spines and the leaf "hairs" of plants: they increase the contact area for thermal exchange and allow the transfer of heat into the surrounding air. Ear flapping also has such a function but is thought to be responsible for only a small percentage of necessary heat radiation.

The dental formula for Recent species of Elephantidae (including *Mammuthus primigenius*) is: i 1/0, c 0/0, pm 3/3, m 3/3 = 26. The single upper incisor grows throughout life into a large tusk (up to 330 cm long in *Loxodonta*). The tusk, which is usually absent in female *Elephas*, has enamel only on the tip, where it is soon worn away (some extinct proboscideans had lower tusks, and others had longitudinal bands of enamel on their tusks). The grinding teeth, which technically are the molars and premolars, are all commonly referred to simply as molars, as they look and function as such. Those teeth are generally large, averaging about 30 cm long and 10 cm across. They are high-crowned in the Recent species and have a complex structure. Each tooth is composed of a large number of transverse plates of dentine covered with enamel; the spaces between the ridges of enamel are filled with cement. Ridges do not show on an unworn tooth since it is covered with cement. The grinding teeth increase in size and in number of ridges from front to back. Those teeth do not succeed one another vertically in the usual mammalian pattern but come in successively from behind, the series thus moving obliquely forward (see Kingdon 1979:18–19). When the foremost tooth is so worn down as to be of no further use, it is pushed out, mostly in pieces. As those teeth are relatively large and the jaws are fairly short, on each side only one tooth above and one below are commonly in use at the same time (part of a second tooth may also be in use).

The skull is huge and shortened. The premaxillary bones have been converted into sheaths for the tusks, and the nasal bones are extremely shortened. All the bones forming the braincase are greatly thickened and, at the same time, lightened by the development of an extensive system of communicating air cells and cavities. The brain chamber is hidden within the middle of the huge mass of the skull. The skeleton of a proboscidean is massive, accounting for 12–15 percent of the body weight in Recent forms. Elephants have the largest absolute brain size among land animals, weighing up to 5.5 kg in *Elephas maximus* and 6.5 kg in *Loxodonta africana* (Byrne and Bates 2011). Elephants also have, on average, the largest relative cerebellum size of all mammals studied to date, including cetaceans and humans (Maseko et al. 2012).

The living proboscideans occupy a variety of habitats but generally live in forests, savannahs, and river valleys within more arid country. Vegetarians, they may consume more than 225 kg of forage a day. Recent elephants are gregarious animals, commonly found in organized social units, and have a potential life span of about 80 years. The Asian elephant is commonly used as a beast of burden. Ivory is obtained from the tusks of both species, and that factor has led to the widespread disappearance of wild populations in modern times.

Already over 175 species and subspecies of Proboscidea are known to have become extinct, some likely through human agency. The order has one of the most extensive and studied paleontological records of any group of mammals. The following account of its evolution was compiled largely from McKenna and Bell (1997), Prothero and Schoch (2002), Sanders et al. (2010), Shoshani (1998), Shoshani and Tassy (2005), Shoshani et al. (2007), and Sukumar (2003); substantive exceptions are noted.

Order Proboscidea and the fully aquatic order Sirenia are united in Tethytheria (herein considered a superorder). Divergence of the two probably occurred in the warm bays and marshes lining the Tethys Seaway, which had formed when the northern and southern land masses of the world split apart in the Mesozoic (see account of class Mammalia). Certain paleontological and embryological evidence suggests the ancestral stock of elephants was semiaquatic (Gaeth, Short, and Renfree 1999).

Proboscidea is divided into a primitive group, Plesielephantiformes, and a more evolutionarily advanced clade, Elephantiformes. Those two groups are sometimes treated as suborders. The earliest known plesielephantiform is *Eritherium azzouzorum* from the late Paleocene of Morocco, about 60 million years ago (Gheerbrant 2009). It is also the smallest definitely known species of Proboscidea, with a weight estimated at just 3–8 kg, about that of a large modern rock hyrax (*Procavia*). The next earliest, and much better known, species is *Phosphatherium escuilliei* from the Paleocene/Eocene transition of Morocco, about 55 million years ago (Gheerbrant 1998; Gheerbrant, Sudre, and Cappetta 1996; Gheerbrant et al. 1998, 2005). It is placed in its own family, Phosphatheriidae. Diagnosis as a proboscidean is based on cranial and dental characters, though it retained canine teeth, its only hint of tusks is the moderately enlarged lower central incisor, and its estimated weight is only 10–15 kg. The first proboscidean to be considered a large animal (i.e., larger

African savannah elephant (*Loxodonta africana*), with trunk raised and showing the two equal length "fingers" at tip. Photo by John Michael Evan Potter / Shutterstock.com.

Asian elephant (*Elephas maximus*), with trunk raised and showing the single "finger" at tip. Photo by Pigprox / Shutterstock.com.

than an average modern human) is *Daouitherium rebouli* from the earliest Eocene of Morocco, which weighed 80–170 kg; it was tentatively assigned to the family Numidotheriidae (Gheerbrant et al. 2002). The only other genus now assigned to that family is *Numidotherium* of the early Eocene of Algeria. About a meter high at the shoulder, with a very high forehead and some development of upper tusks, it appears to be in the ancestral lineage leading to Elephantiformes.

However, several other families branched off prior to development of advanced elephants. Moeritheriidae is known from the middle Eocene to early Oligocene of North Africa. The only described genus, *Moeritherium*, weighed about 225 kg and probably resembled a modern pygmy hippopotamus (*Hexaprotodon*) in size, general appearance, and mode of life. Its nasal bones suggest a fleshy proboscis, and it had short but prominent upper and lower tusks. Barytheriidae, from the late Eocene to early Oligocene of Egypt and Libya, also contains a single genus, *Barytherium*, which had a shoulder height of 2–3 meters, a meter-long skull, and the beginnings of a trunk. Its feet resembled those of modern *Hippopotamus* and it may have occupied a similar amphibious ecological niche. Still another side branch was Deinotheriidae, which lacked upper tusks but had a pair of lower tusks that curved downward and back; they may have been used as a point of leverage for the trunk, for stripping bark, or for digging. The skull was long and flattened, the first molar was trilophodont, the second bilophodont. One species, *Deinotherium giganteum*, reached 4 meters in height and, with large bulls perhaps approaching 20,000 kg, it may have been among the two or three largest land mammals ever to have lived (Christiansen 2004). Deinotheres probably diverged in the later Eocene but are known by fossils only from the early to middle Miocene of Europe and western Asia, from the late Miocene of central China (Qiu et al. 2007), and from the late Oligocene to early Pleistocene of Africa. They were the last survivors of Plesielephantiformes, coexisting for millions of years with many members of Elephantiformes.

Palaeomastodontidae, with the genera *Palaeomastodon* and *Phioma*, from Oligocene deposits of Africa and Oman, is the earliest known elephantiform family (*Phioma* sometimes is placed in its own family, Phiomidae). *Palaeomastodon* stood about 2 meters high at the shoulder and had a moderate trunk, upper and lower tusks that were oval in cross section, and bilophodont molar teeth with cusps arranged in transverse pairs. The family appears to form a phylogenetic stage subsequent to *Numidotherium* and prior to the more evolved members of Elephantiformes. A recently described genus and species from the late Oligocene of Eritrea, *Eritreum melakeghebrekristosi* (Shoshani et al. 2006) may represent populations intermediate to Palaeomastodontidae and two advanced lineages.

One, Mammutidae (mastodons), is known by fossils from the late Oligocene to early Pleistocene of Africa and Eurasia and from the mid-Miocene to very late Pleistocene or early Recent of North America. Mastodons resembled modern elephants in size and overall body shape, but their back was straighter, legs shorter, and skull relatively long and low. They had only upper tusks, which were very long, extending up to 3 meters from the skull. As in living elephants, the tusks of males were much larger than those of females (Smith and Fisher 2013). The grinding teeth of mastodons were fairly small and low-crowned and had three or four prominent transverse ridges of enamel formed into pairs of conical cusps; on each side only two such teeth above and two below were in use at the same time.

The mastodon species *Mammut americanum*, about the size of modern *Elephas maximus*, occurred across much of North America in the late Pleistocene. Numerous remains of that species have been radiocarbon dated to as recently as 5,000 years ago (Fiedel 2009), and many Native American legends tell of a creature resembling a mastodon. However, most authorities do not accept the more recent datings and consider *M. americanum* to have disappeared at the very end of the Pleistocene (Fisher 2009; Haynes 2002; P. S. Martin 2005; Martin and Steadman 1999; Prothero and Schoch 2002; Sukumar 2003). Fiedel (2009) listed the "youngest published credible dates" as about 10,200–11,000 radiocarbon years ago and indicated that later reported dates may involve contamination of the involved materials. Woodman and Athfield (2009) obtained a radiocarbon date of about 10,032 years ago (equivalent to around 11,500 calendar years ago) from remains found in northern Indiana. In any case, the mastodon lineage became extinct and is not ancestral to modern elephants.

The second advanced lineage to follow Palaeomastodontidae led initially to Gomphotheriidae, a family

African forest elephant (*Loxodonta cyclotis*). Photo by Brent Huffman.

that was highly diverse but generally characterized by trilophodont molars, the low blunt cusps being arranged in staggered rows and covered with thick enamel. As many as three molars, above and below, were simultaneously functional on each side of the mouth. Gomphotheres were up to 3 meters tall at the shoulder and had a short trunk and a long, low skull. The most conservative genus, *Gomphotherium*, had relatively straight upper and lower tusks of about equal length, but the lower two became flattened and spatula-shaped. Several advanced genera lost the lower tusks. In others, however, the upper tusks were reduced or absent, and the lower tusks and jaw joined to form a large "shovel," possibly used to slice through vegetation that then was consumed. Gomphotheres evidently arose in Africa in the Oligocene and are known there by fossils from the early Miocene to late Pliocene. The group spread through Eurasia, where it persisted until the mid-Pleistocene, and into North America, where it occurred from the mid-Miocene to late Pleistocene.

It entered South America by the late Pliocene or early Pleistocene, when a land bridge was available, and two genera, *Cuvieronius* and *Notiomastodon*, existed there at least to the end of the Pleistocene (Mothé, Avila, and Cozzuol 2012). Some radiocarbon dates and stone artifacts suggest possible presence as late as 5,000 years ago in Colombia (Correal U. and Van der Hammen 2003) and 6,700 years ago in the Bolivian Chaco (Coltorti et al. 2012). However, Lima-Ribeiro et al. (2013) set the extinction date of *Cuvieronius* and *Notiomastodon* at around 10,500 years ago, and Turvey (2009) noted that recent radiometric investigations have not found evidence that any of the South American megafauna survived long beyond the Pleistocene-Holocene boundary.

From within Gomphotheriidae there apparently arose two more main lineages. One family, Stegodontidae, comprises two genera, the more primitive of which, *Stegolophodon*, featured a pair of tusks on both the upper and lower jaws. The more advanced *Stegodon* had only two upper tusks, but they were very long, had J-shaped curves toward the ends, and were so close together in the middle that the trunk apparently could not fit between them and had to drape over one side. Other dental characters approached those of modern elephants. Replacement of the molars was horizontal, and only one or two, above and below, were in use at the same time on each side. Each tooth had up to 14 transverse ridges or lophs composed of a row of thickly enameled cusps. However, the molars were relatively low crowned. Stegodontidae is known from the late Miocene to late Pliocene of Africa, the Pliocene of Europe, and the early Miocene to late Pleistocene or early Holocene of Asia.

Several stegodontid species, about the size of modern elephants, were widely distributed in the late Pleistocene forests of mainland southeastern Asia and the East Indies (Louys, Curnoe, and Tong 2007). However, *Stegodon florensis insularis* of Flores Island was a dwarf form that may have been regularly hunted by the diminutive *Homo floresiensis* until both seemingly succumbed to effects of a volcanic eruption around 17,000 years ago (Morwood et al. 2004; Van Den Bergh et al. 2008, 2009). The species *S. orientalis* was found approximately in that part of China south of the Cháng Jiāng (Yangtze River), where it coexisted with and apparently was more common than *Elephas maximus* (Saegusa 2001; Tong and Patou-Mathis 2003). Remains discovered at two localities have been dated to the mid-Holocene: Jinhua Shuanglong Cave, Zhejiang Province, about 7,800 years old, and Xiaohe Cave, Yunnan Province, 4,100–5,000 years old (Saegusa 2001; Tong and Liu 2004; Tong and Patou-Mathis 2003). Once again, however, Turvey (2009) considered such dates questionable and in need of clarification by critical radiometric study. Turvey et al. (2013) noted that the relevant specimens had never been properly dated and are now lost.

Stegodontidae is a phylogenetic sister family to Elephantidae and is sometimes included as a subfamily thereof, but evidently is not a direct ancestor. The latter role probably belongs to a second lineage that emerged from Gomphotheriidae and led to the genus *Stegotetrabelodon* of the late Miocene to early Pliocene of Africa and the late Miocene of Italy (Ferretti, Rook, and Torre 2003). In that genus, the earliest known elephantid, the skull became shorter and more vertical, the lower jaw shortened and turned down, and the

ridged surface of the molar teeth developed into plates consisting of enamel outside and dentine inside, with deep, V-shaped valleys between the plates. The modifications of teeth and jaws allowed for improved grinding capability and exploitation of grasslands.

Stegotetrabelodon may be the ancestor of the most advanced elephants—*Loxodonta, Elephas, Palaeoloxodon,* and *Mammuthus*—though a contemporary African genus, *Primelephas,* has sometimes been assigned that role. *Loxodonta* is known only from Africa, initially in the late Miocene. *Elephas* developed in Africa in the late Miocene and spread through Asia by the late Pliocene. *Palaeoloxodon* diverged from *Elephas* or *Loxodonta* (see Meyer et al. 2017) in Africa during the mid-Pliocene, moved across Asia as far as Japan, and entered Europe. Mainland species probably resembled modern elephants externally, but populations spread to many Mediterranean islands, where dwarf forms often developed. *Palaeoloxodon* (sometimes treated as a subgenus of *Elephas*) was the dominant proboscidean of Africa for several million years, evidently restricting the range of *Loxodonta* to forest zones, but it vanished from that continent by the late Pleistocene. Based on dental remains and depictions on bronze wares found at archeological sites, Li et al. (2012) suggested that *Palaeoloxodon* still was present in northeastern China about 3,000 years ago. However, Turvey et al. (2013), who examined and dated one of the teeth, found it Pleistocene in age and referable to *Elephas maximus*; they also indicated that identification of the animals depicted on the bronzes was questionable.

Mammuthus arose in sub-Saharan Africa during the early Pliocene; it disappeared there by the mid-Pleistocene but meanwhile evolved into a number of species, starting in the late Pliocene of Eurasia and the early Pleistocene of North America. *Mammuthus* was extirpated from most of its range by the end of the Pleistocene but persisted into the early Holocene in northern Siberia (Mol et al. 2006) and Alaska (Haile et al. 2009) and into the mid-Holocene on Wrangel Island off northeastern Siberia (Arslanov et al. 1998) and St. Paul Island off southwestern Alaska (Enk et al. 2009). The well-studied Wrangel Island population, originally thought to be a dwarf form, was subsequently shown to consist of small but normal sized (shoulder height, 180–230 cm) members of the species *M. primigenius* (Reumer, Mol, and De Vos 2002). A true dwarf

Woolly mammoth (*Mammuthus primigenius*). © Jon Baldur Hlidberg / Wildlife Art Co. / Minden Pictures.

species, *M. exilis* (shoulder height, 120–180 cm), evidently a descendant of the North American *M. columbi*, developed in the late Pleistocene on the Channel Islands off southern California (Roth 1996; Tikhonov, Agenbroad, and Vartanyan 2003). Long before, by the early Pleistocene, a European lineage of *Mammuthus* apparently reached the island of Crete, where it evolved into the smallest known mammoth, *M. creticus*, with an estimated shoulder height of just 113 cm (Herridge and Lister 2012).

M. creticus was sometimes placed in the genus *Palaeoloxodon*, as were most other Pleistocene elephants of the Mediterranean islands, but Poulakakis et al. (2006) referred a sample of mitochondrial DNA, retrieved from an 800,000-year-old bone fragment, to *Mammuthus*. Their methodology was challenged (Binladen, Gilbert, and Willerslev 2007), but a morphological assessment of Cretan specimens supported assignment to *Mammuthus* (Herridge and Lister 2012). Poulakakis et al. (2006) also suggested that the elephants of Sicily and Malta might represent *Mammuthus*, but Herridge and Lister (2012) continued to refer those populations to *Palaeoloxodon*. Accordingly, while *Mammuthus* is known from Sardinia, as well as Crete, all other Mediterranean elephants represent *Palaeoloxodon*. That genus occupied Favignana, Sicily, Malta, Crete (along with *Mammuthus*), Rhodes, Cyprus, and many Aegean islands; in some cases there were repeated invasions of an island, resulting in species of different sizes (Caloi et al. 1996; Masseti 2001). *P. falconeri* of Sicily and Malta, the smallest known elephantiform, had an adult shoulder height of only 90–140 cm and an estimated average weight of 130 kg in males and 80 kg in females (E. Anderson 1984; P. S. Martin 1984; Raia, Barbera, and Conte 2003; Roth 1992). There seems general agreement that all island populations disappeared by the late Pleistocene or very early Holocene, with one possible exception.

According to Masseti (2001), remains of a dwarf elephant, slightly larger than *P. falconeri*, are known from Charkadio Cave on the island of Tilos, west of Rhodes and about 20 km off southwestern Asia Minor. Some of the remains have been dated to the late Pleistocene, but two radiocarbon dates are 7,090 +/− 680 and 4,390 +/− 600 years before the present. The latter date, equivalent to about 4,800 calendar years ago, is well within the time of Dynastic Egypt. A wall painting in an eighteenth-Dynasty tomb, from about 3,470–3,445 years ago, shows what appears to be an adult elephant with fully developed tusks. However, presence of adult humans in the painting indicates the elephant probably is less than 100 cm in height. Rosen (1994) suggested the animal is a dwarf mammoth, perhaps from Wrangel Island (see account of *Mammuthus* for discussion of whether the Wrangel mammoths could be considered dwarfs), but Masseti (2001) argued it is more likely an elephant from a Mediterranean island, possibly Tilos. Other explanations, including that the image simply represents artistic license involving *Elephas maximus*, have been offered. Poulakakis et al. (2002) reported that mitochondrial DNA from Tilos elephant bones showed closer affinity to *Elephas* than to *Mammuthus*.

Theodorou, Symeonidis, and Stathopoulou (2007) named the Tilos elephant *Elephas tiliensis*; they noted that it had morphological adaptations for agility and movement on rough terrain and that it became extinct around 4,000–3,500 years ago. However, such recent survival is questionable. Poulakakis et al. (2006) indicated that all dwarf elephants of the Mediterranean disappeared about 10,000 years ago. Turvey (2009) called the radiocarbon dates for *tiliensis* dubious; they were reported in 1975–1976, and it since has been impossible to verify them independently. Finally, it must be noted that *Elephas maximus* evidently did occur in Asia Minor until about 3,000 years ago (see account of *Elephas*), and presence of it on the mainland and a dwarf related species on a nearby island would seem comparable to occurrence of *Mammuthus columbi* in North America and *M. exilis* on the California Channel Islands in the late Pleistocene (see above).

PROBOSCIDEA; ELEPHANTIDAE; **Genus** *LOXODONTA* F. Cuvier, 1827

African Elephants

Two species are recognized herein (Allen 1936; Ansell 1977; Gautier et al. 1994; Groves and Grubb 2000a; Grubb et al. 2000; Ishida et al. 2011b; Kingdon 1979;

Laursen and Bekoff 1978; Le Quellec 1999; Roca, Georgiadis, and O'Brien 2005, 2007; Roca et al. 2001; Rohland et al. 2010):

L. cyclotis (African forest elephant), known from the tropical forest zone and adjacent forest/savannah zone of western and central Africa from Senegal to western Uganda and northeastern and central Democratic Republic of Congo, also possibly the elephant historically found along the northern rim of Africa from western Mauritania to Egypt;

L. africana (African savannah or bush elephant), known from throughout sub-Saharan Africa except the western and central tropical forest zone and parts of the coastal deserts of Somalia and Namibia, also probably the elephant historically found in much of the Sahara region before desertification.

There long has been disagreement as to whether the above two species are distinct. In the early twentieth century, all African elephants were commonly assigned to *L. africana*, but Allen (1936, 1939) distinguished *L. cyclotis* as a species separate from and more primitive than *L. africana*. Harper (1945) supported the distinction, and Ansell (1977) cited studies up to 1966 that continued to favor specific status for *L. cyclotis*. For about the next 30 years, most authorities again included *cyclotis* within *L. africana* (e.g., Ansell 1977; Baillie and Groombridge 1996; Corbet 1978; Corbet and Hill 1991; Douglas-Hamilton and Michelmore 1996; Eisenberg 1981; Grzimek 1990; Jones 1984; Kingdon 1979; Laursen and Bekoff 1978; Martin and Guilday 1967; Shoshani and Tassy 1996; Sikes 1971). Nonetheless, there was general agreement that the elephant of the tropical rainforest zone of western and central Africa was a smaller, darker kind as compared to the larger, paler bush or savannah elephant in the remainder of the range. Some authorities (e.g., Kingdon 1979) regarded those two kinds as being equivalent to subspecies, *L. africana cyclotis* for the forest elephant and *L. a. africana* for the savannah elephant. Others (e.g., Ansell 1977) listed additional named subspecies, but still recognized a basic division between the forest and savannah kinds of elephant. Rejection of both being full species may have stemmed from indications that *cyclotis* and *africana* intergrade where their ranges

Left, African savannah elephant (*Loxodonta africana*); photo by Albie Venter / Shutterstock.com. *Right,* African forest elephant (*Loxodonta cyclotis*); photo by Brent Huffman.

meet, as is normal for subspecies. Several reports of considerable intergradation in the region along the eastern side of the central forest zone were cited by Ansell (1977), and populations intermediate to *africana* and *cyclotis* were reported to occupy a wide band of transitional vegetation in western Africa (Roth and Douglas-Hamilton 1991).

To determine whether *africana* and *cyclotis* do interbreed, Groves and Grubb (2000a) carried out a discriminant function analysis of 117 skulls taken throughout the modern range of *Loxodonta*. Other morphometric studies were done as well, and Grubb et al. (2000) made a general assessment of 295 skulls and numerous observations of living animals. Results indicated that limited hybridization had occurred at two of the three sites evaluated, where the eastern edge of the range of *cyclotis* meets the range of *africana*, but that there was

no extensive intergradation. Discriminant function analysis showed all skulls from western Africa to be *cyclotis*, though two were from somewhat north of the forest zone (northern Ghana, central Togo); reported field observations from that same region indicated overlapping populations of *cyclotis* and *africana* (Groves 2000). It was concluded that the two forms are full species, morphometrically highly distinctive and instantly recognizable in the field over vast areas.

Tentative support for that view had already come from a preliminary, unpublished analysis of mitochondrial DNA (see Tangley 1997). A proliferation of mitochondrial DNA studies followed but with a disparity of results: *cyclotis* is as divergent from *africana* as *Loxodonta* is divergent from *Elephas*, but it cannot be concluded that the African forest and savannah elephants are separate species (Barriel, Thuet, and

Tassy 1999); *cyclotis* and *africana* diverged millions of years ago and are distinct species (Miller et al. 2008; Rohland et al. 2007); *cyclotis* and *africana* interbreed extensively and cannot be called distinct species (Debruyne 2005); the forest elephant of central Africa, the savannah elephant to the east and south, and the forest elephant plus the savannah elephant of western Africa represent separate lineages that may be three full species (Eggert, Rasner, and Woodruff 2002); west African savannah and forest elephants are indistinguishable, but central African forest elephants represent at least two lineages (M. B. Johnson et al. 2007).

Studies of nuclear DNA, perhaps more reliable for assessing true genetic composition and affinity of populations, have yielded relatively straightforward results strongly supporting recognition that *L. cyclotis* and *L. africana* are highly distinct, that they are the only two living species of *Loxodonta*, and that their distribution and content conforms closely with earlier phenotypic assessment and field observation (Capelli et al. 2006; Comstock et al. 2002; Ishida et al. 2011a, 2011b; Murata et al. 2009; Roca, Georgiadis, and O'Brien 2005, 2007; Roca et al. 2001; Rohland et al. 2010). The most comprehensive study (Ishida et al. 2011b) genotyped 461 elephants from 17 localities across the savannah regions of Africa, 75 elephants from 5 localities in the tropical forest zone, and 19 elephants from Garamba National Park in northeastern Democratic Republic of Congo, where there is some admixture of savannah and forest habitat. All savannah elephants were found to have the same kind of nuclear DNA, with very little variation. The forest elephants all had a different genotype, with somewhat more variation. Of the Garamba specimens, 17 were of the forest type and two were indicative of hybridization. Not a single savannah genotype was found within the forest zone. There was no sign of a hybrid swarm or intergradation of forest and savannah elephants.

Ishida et al. (2011b) also reassessed previously published work on the mitochondrial DNA of over 1,000

African savannah elephant (*Loxodonta africana*), near Mount Mt. Kilimanjaro. Photo by Volodymyr Burdiak / Shutterstock.com.

African elephants. The material was divisible into savannah and forest clades. No elephant in the 30 forest zone localities had the savannah clade of mitochondrial DNA. However, the forest DNA clade was present at 17 of the 51 localities outside the forest zone, even at some sites as far away as Zimbabwe and Botswana. That factor had been used in earlier studies of mitochondrial DNA to argue that forest and savannah elephants were undergoing extensive interbreeding over a vast region and thus could not be separate species. And yet no forest nuclear DNA was found at any of the sites outside the forest zone, and no field observations suggested interbreeding there. Most likely the mitochondrial DNA at those sites is a relict of ancient crossings when the forest zone and forest elephant had a greater distribution. Even with that evidence of former widespread hybridization, and with the apparent existing hybrid zone, isolating mechanisms appear to be functioning and preventing gene flow from *L. africana* to *L. cyclotis*, thereby demonstrating the specific status of the latter.

Roca, Georgiadis, and O'Brien (2005, 2007) explained that genomic dissociation in savannah locations implies that the ancient hybridization was between forest females and savannah males, which are larger and reproductively dominant to forest or hybrid males. Recurrent backcrossing of female hybrids to savannah bulls then would have steadily diluted the forest type nuclear DNA until the populations had overwhelmingly savannah type nuclear DNA but retained the maternally inherited forest type mitochondrial DNA. To the degree detectable, those savannah populations are completely savannah-like in nuclear DNA genotype and in morphology. Meanwhile, there would have been selection against and a lack of reproductive success among hybrid males and hence no passage of savannah genes to forest populations.

Possibly the most questionable region with respect to the interrelationship of *cyclotis* and *africana* is western Africa, where the tropical forest and forest/savannah zones continue along the coast from southern Cameroon to western Senegal, and a parallel band of open woodland runs just to the north (Barnes 1999; Ishida 2011b). Because of the confusion there, several conservation bodies have deferred a decision on recognizing the specific status of *L. cyclotis* (Blanc 2008; CITES 2012; Dublin 2011). However, Roca, Georgiadis, and O'Brien (2007) considered it unwise to delay con-

servation decisions for the rest of Africa based on lack of data for a region containing less than 2 percent of the continent's remaining elephants. Moreover, widespread deforestation and hunting likely disrupted natural genetic patterns in western Africa long ago, leaving elephants in extremely fragmented habitats within only 7 percent of their original range.

Ishida et al. (2011b) found mitochondrial DNA from southern Mali to carry the savannah clade, as did that from northern Cameroon, thus indicating that the range of *L. africana* continues westward, north of the forest zone, in accord with morphological studies and field observations (Groves 2000). Elephant distribution in western Africa was further clarified by Wasser et al. (2004), who identified specimens of nuclear DNA from Ivory Coast and Ghana as forest elephant. Their work, aimed primarily at finding means to determine the provenance of ivory, also covered much of Africa—399 samples of nuclear DNA from 28 sites—and was always able to distinguish *cyclotis* and *africana*, though they did not definitely refer to the two forms as species. Part of their study involved testing methodology that might pinpoint the location where a sample originated. Not all samples were accurately placed, but a review by Ishida et al. (2011b) showed that no savannah elephants were assigned to forest locations and no forest elephants to savannah locations. There was mis-assignment between the savannahs of eastern Africa and those of northern Cameroon but no mis-assignment between the latter region and the central forests just to the south. And there was substantial mis-assignment between the central and western forest regions. Such mis-assignment actually demonstrates genetic similarity and thus that in western Africa the forest and savannah elephant populations represent continuity of the respective species *L. cyclotis* and *L. africana*, and should not be combined into a third species, as suggested by some previous studies.

And yet there have long been reports that another species of elephant lives in the dense lowland jungles from Sierra Leone to Democratic Republic of Congo (Ansell 1977; Meester et al. 1986). That so-called pygmy elephant (*L. pumilio*) is said to be very small, more solitary than *L. cyclotis* and *L. africana*, and possibly semi-aquatic. There are old corroborative specimens, photographs, and measurement data. Newer information suggesting that a pygmy elephant may actually exist was cited by Greenwell (1993). Most authorities, however,

African forest elephant (*Loxodonta cyclotis*). Photo by Gudkov Andrey/Shutterstock.com.

consider such reports to reflect unusually small adult or immature *L. cyclotis*; subadult males of that species, sometimes with precocious tusk development, commonly leave their familial groups and circulate alone (Cousins 1993; Roth and Douglas-Hamilton 1991; Turkalo and Fay 2001). Groves and Grubb (2000b) reviewed the entire issue but could find no evidence that *L. pumilio* is a separate species; in particular, recorded measurements fell within the size range of *L. cyclotis*, photographs were not definitive, and the four available skulls were referable to *L. cyclotis*. Debruyne et al. (2003) analyzed mitochondrial DNA of nine museum bone or tooth specimens previously recorded as *L. pumilio*. They concluded that the specimens represented several different lineages within the overall content of *cyclotis* and that the specific taxon *L. pumilio* is not valid.

Although disagreement on the systematics of *Loxodonta* likely will continue, recognition of *L. cyclotis* and *L. africana* as distinct species, and as the only living species in the genus, has now received substantial authoritative support (Kingdon et al. 2013; Sanders et al. 2010; Shoshani 2005b; Shoshani and Tassy 2005; Whyte 2005; Wittemyer 2011). However, the IUCN continues to treat all African elephants under the species name *L. africana*, partly because more sampling is thought necessary, especially in western Africa, and partly because allocation to more than one species might leave hybrids in an uncertain position for conservation purposes (Blanc 2008; Dublin 2011).

Regardless of whether *L. cyclotis* and *L. africana* are considered distinct species, there remain questions as to where each is and originally was distributed. At the start of historical time, about 5,000 years ago, *Loxodonta* apparently occurred throughout Africa except for the more arid parts of the Sahara region—then in transition from relatively moist savannah to desert—and the coastal deserts of what are now Somalia and Namibia (Ansell 1977; Gautier et al. 1994; Grubb et al.

2000; Kingdon 1979; Klein 1984; Kröpelin et al. 2008; Laursen and Bekoff 1978; Le Quellec 1999). By 3,000–4,000 years ago the genus had disappeared from most of the Sahara because of continued desiccation and from Egypt because of human hunting, though it persisted in northwestern Africa until about 1,500 years ago (Lobban and De Liedekerke 2000; Osborn and Osbornova 1998). *Loxodonta* historically had occupied the entire Nile Valley and all other suitable parts of Egypt, and Sikes (1971) suggested it also was present in adjacent southwestern Asia until less than 2,000 years ago. Limited available information indicates the ancient elephant populations of the latter region represented *Elephas* (see account thereof), though Shalmon (2000) reported that "strange ancient rock carvings" at gebel Khazali in southwestern Jordan, associated with archeological sites dating from less than 7,000 years ago, may depict *Loxodonta*.

Through analysis of nuclear DNA, Ishida et al. (2011b) demonstrated that *L. cyclotis* is present in the tropical forest zone of Sierra Leone, Gabon, Congo, southwestern Central African Republic, and northern Democratic Republic of Congo. By reviewing previous studies of mitochondrial DNA, Ishida et al. (2011b) indicated the range of *L. cyclotis* also to include the forests of Liberia, southern Ivory Coast, southern Ghana, southern Cameroon, and central Democratic Republic of Congo. Recently, the organization Fauna and Flora International announced that *L. cyclotis* had been discovered in Western Equatoria State in extreme southwestern South Sudan (http://www.fauna-flora.org/news/remote-cameras-offer-glimpse-into-the-forgotten-forests-of-south-sudan/, accessed 10 December 2015). The two assessments by Ishida et al. (2011a, 2011b) showed that *L. africana* does not occur anywhere within the range of *L. cyclotis* but that it is found in the surrounding, more open country of southern Mali, northern Ghana, northern Cameroon, southern Chad, South Sudan, Uganda, Kenya, Tanzania, southern Democratic Republic of Congo, northeastern Angola, northern Namibia, Botswana, Zimbabwe, Mozambique, and South Africa.

African savannah elephant (*Loxodonta africana*), group. Photo by Artush / Shutterstock.com.

Based on measurements of 295 elephant skulls from sub-Saharan Africa, plus extensive observations of living animals, Grubb et al. (2000) presented a more generalized distribution, with *L. africana* occupying the savannah zone from about eastern Senegal to Eritrea and Somalia, then south to eastern South Africa, and the range of *L. cyclotis* extending west through the coastal forest zone as far as western Senegal. However, review of Grubb et al. (2000), together with Kingdon (1979) and Laursen and Bekoff (1978), also indicates that the range of *Loxodonta* once continued northward along both west and east coasts and then along the entire Mediterranean coast to completely surround the Sahara. The northern populations, which were still present in early historical time (see above), have sometimes been recognized as a subspecies, *pharaohensis*, with supposed affinity to *L. cyclotis* (Ansell 1977). If such designation is accepted, and if *L. cyclotis* is considered a distinct species, then there would be two subspecies thereof: *L. c. cyclotis* and *L. c. pharaohensis*. The latter may still have been present in modern times in Mauritania (Roth and Douglas-Hamilton 1991) and also reportedly on the coast of Sudan and Eritrea (Yalden, Largen, and Kock 1986). Another subspecies, *orleansi*, was historically found just to the south in northern Somalia and adjacent Ethiopia; a small isolated population still exists in the vicinity of Harar in east-central Ethiopia (Demeke, Renfree, and Short 2012; Largen and Yalden 1987). *Orleansi* has usually been placed in *L. africana* but reportedly is morphologically distinctive and quite small (Ansell 1977), as are *L. cyclotis* and *pharaohensis*.

Regardless of the status of *orleansi*, there would have been a 1,500-km gap in the range of *L. cyclotis*, between the populations of the central African forests and those said to occur along the Red Sea. There have been attempts to explain such an unusual distribution based on biogeographical history. Possibly, when *Palaeoloxodon* dominated a relatively cool, dry Africa during much of the Pliocene and Pleistocene (see account of family Elephantidae), *Loxodonta* was able to maintain itself only in remnant forest (Sukumar 2003). The presence of *Palaeoloxodon* may have been a barrier that isolated the precursors of *L. cyclotis* and *L. africana* in separate forest zones (Rohland et al. 2010). Then, when *Palaeoloxodon* vanished at the end of the Pleistocene, *Loxodonta* would have been able to ex-

pand its range (Kingdon 1979). During the early Holocene the Sahara region contained a series of linked lakes, rivers, and inland deltas that provided favorable conditions, exclusive of the Nile Basin, for many animals to occupy and disperse across the entire Sahara (Drake et al. 2011). Such conditions allowed *Loxodonta* to reach the Mediterranean Sea, though which species is not known. It may have been *L. cyclotis*, which then slowly adapted to the Mediterranean habitat zone, where vegetation is relatively heavy compared to that of the Sahara to the south. Meanwhile, *L. africana* may have occupied much of the Sahara region (see Gautier et al. 1994).

Later, as drier conditions again prevailed in the Holocene, *L. africana* retreated from the Sahara but spread through most of the rest of Africa, both north and east of the central forest zone, effectively creating the gaps between what were presumed to be the different populations of *L. cyclotis* (Ansell 1977). The type of mitochondrial DNA characteristic of *L. cyclotis* is carried by many elephants well outside the current forest zone, even though those animals are *L. africana* with respect to nuclear DNA and phenotype; that condition apparently represents a former much greater range of *L. cyclotis* and occasional hybridization of the two species over many generations (Ishida et al. 2011b; Roca, Georgiadis, and O'Brien 2005, 2007). The ancient widespread occurrence of *L. cyclotis*, as compared to a more constricted origin for existing populations of *L. africana*, may also be reflected by the significantly lower genetic diversity of the latter (Comstock et al. 2002).

Some modification of the above biogeographic scenario for *Loxodonta* is suggested by a recent molecular assessment of the mitochondrial and nuclear DNA of a small population of elephants in the Gash-Barka area, which is in the extreme western, inland part of northern Eritrea (Brandt et al. 2014). The population in that area of dry savannah contains only about 100 animals and is isolated by more than 400 km from other elephants. The study indicated that the population represents the species *L. africana* and has closer affinity to the savannah elephants in East Africa than to those of the north-central Sudanian/Sahelian region. Assuming the population is a remnant of the original kind of elephant found in Eritrea and the general Red Sea region, the study contradicts the view, mentioned

African forest elephant (*Loxodonta cyclotis*), group. Photo by Sergey Uryadnikov / Shutterstock.com.

above, that the species there was *L. cyclotis*. Building on this apparent refutation of claims that the latter species once had a widely disjunct distribution, Roca et al. (2015) suggested that further study might show that *L. cyclotis* also may not have been the elephant of North Africa.

The lineage of the genus *Loxodonta* is thought to have diverged from that of *Elephas* in Africa during the late Miocene, around 7 million years ago (Rohland et al. 2007, 2010; Todd 2010). Compared to *Elephas*, *Loxodonta* is morphologically conservative (Sanders et al. 2010). It has a globular cranium with a biconvex frontoparietal surface, unexpanded parietals, and a nearly vertical occipital region. In *Elephas* the skull is high and anteroposteriorly compressed, the frontoparietal surface is flat to concave, and there are usually distinct parieto-occipital bosses. The molars of *Loxodonta* are not as high crowned as are those of the more advanced species of *Elephas* or of the genera *Palaeoloxodon* and *Mammuthus*, and they are indicative of a dietary emphasis on browsing. Anterior and posterior central accessory conules are incorporated in the molar crowns of *Loxodonta* but are usually absent or reduced in *Elephas*. Laursen and Bekoff (1978) noted that the molars of *Loxodonta* are larger and have fewer transverse plates.

The lineages of *L. cyclotis* and *L. africana* evidently diverged about 4–7 million years ago (Roca et al. 2015; Rohland et al. 2007, 2010). *L. cyclotis* is considered the more primitive, retaining cranial and dental characters that are found in ancestral species of *Loxodonta* and that seem associated with forest habitat and a diet dominated by browse (Allen 1936; Groves and Grubb 2000a; Grubb et al. 2000; Kingdon 1979; Sanders et al. 2010). Its skull is much less pneumatized than that of *L. africana*, dorsally flattened, and relatively broader across the tusk bases and the roof, with more widely separated temporal ridges and a less flared rostrum. Its mandible is longer and lower, the condyles are transverse-oval, not spherical as in *L. africana*, and the anteroposterior diameter of the mandibular symphysis is relatively and absolutely greater. Its tusks are thinner and tend to grow straight and downward, rather than curve out and forward as in *L. africana*; this may be an adaptation for passing through dense forest. Its molar teeth are lower crowned, indicative of dependence more on softer vegetation and fruit than on coarse grasses. *L. cyclotis* usually is considerably smaller than *L. africana*. The males of both species grow throughout life but those of *L. africana* grow faster, experiencing a growth spurt at around 10 years of age, and end up much larger. Growth also is continuous in female *L. africana* but ceases at maturity in female *L. cyclotis*. Of the two species, *L. cyclotis* has a more compact body build, a straighter back, and smaller, more rounded ears. It also reportedly has darker skin and more hair, especially on the head, trunk, and tail.

For *Loxodonta* in general, head and body length (including the trunk, which is really an elongated nose) is roughly 500–750 cm and tail length is 90–150 cm. The skin has sparsely scattered black bristly hairs (see discussion of hair in account of family Elephantidae) and generally is dull brownish gray in color. The flat-tened end of the tail has a tuft of coarse, crooked hairs 38–76 cm long. Since the elephant wallows in streams and pools and tosses dirt or mud onto its back, its coloration usually appears similar to that of the soil it frequents.

In both sexes one incisor tooth on each side of the upper jaw is greatly developed to form a tusk. A few tuskless elephants, mostly females, exist naturally. The largest known tusk measures about 350 cm and weighs about 107 kg; the largest female tusk weighs about 18 kg (the average is about 7 kg). The tusks are used for fighting, digging, feeding, and marking (Laursen and Bekoff 1978). Adult males use them to knock over trees, sometimes to obtain succulent leaves or fruit for themselves but more often in a seeming display of dominance, though the available food commonly attracts other elephants. Interestingly, the largest tusks generally develop, not on the largest animals, but rather on those of mild temperament, which tend to avoid the combat and tree-ramming that may break the teeth

African savannah elephant (*Loxodonta africana*). Photo by Jez Bennett / Shutterstock.com.

(Kingdon 1979). However, broken tusks have been observed to grow back rapidly (Poole, Kahumbu, and Whyte 2013). The tusks of males grow in both length and bulk throughout life; growth accelerates with age (Pool, Kahumbu, and Whyte 2013). The tusks of females also grow continuously, though only slowly past the age of 30, and after age 15 there is a considerable reduction in the rate of development of diameter in relation to length, so that the result is a rather slender tusk (Layser and Buss 1985). Tusks are the source of ivory, which can be carved into decorative objects and jewelry. Except when found on an elephant that died of natural causes, they must be obtained by killing the animal and cutting them away from the body. That factor has led to the slaughter of countless millions of elephants and the disappearance of most wild populations. Intensive ivory poaching is associated with a marked increase in the proportion of tuskless elephants in some areas (Poole, Kahumbu, and Whyte 2013), though lack of tusks in 98 percent of the females in Addo Elephant National Park, South Africa, seems a result of nonselective genetic change, rather than selective ivory hunting (Whitehouse 2002).

Aside from the tusks, the most conspicuous external features of *Loxodonta* are the long trunk and large ears. As noted in the family account, the genus has two finger-like projections at the distal end of its trunk. Chevalier-Skolnikoff and Liska (1993) reported that *Loxodonta* sometimes handles "tools" with its trunk and throws objects in the same manner described in the account of *Elephas*. *Loxodonta* has much larger ears than *Elephas* and often has a hole through the lower portion of the earlobe caused by injury. The ears are sometimes nearly 200 cm in length from top to bottom in *L. africana*, but only about half that in *L. cyclotis*. The undersides of the ears have an extensive supply of blood vessels. Ear flapping, characteristically seen when elephants stand in the shade on hot days, creates air currents over the blood vessels and promotes the radiation of excess body heat (Whyte 2005; Wright 1984). Both species originally have five well-formed digits on both the forefeet and hind feet, but the nails (or hooves) tend to wear down more quickly in *L. africana*, perhaps because of the terrain it inhabits (Whyte 2005).

The very size and conspicuous nature of African elephants have allowed observations of their ecology and behavior for many years (Grzimek 1990; Happold 1987; Kingdon 1979; Whyte 2005). They have adapted to a great variety of habitats from dense tropical rainforest to riverine valleys of the coastal Namib Desert. They seem equally at home submerged up to their backs in broad rivers and climbing to elevations of up to 5,000 meters in the mountains. Populations in some areas migrate seasonally in response to availability of food and water, while other groups remain in favorable areas throughout the year. Their most critical requirement is a source of water, though they can go several days without drinking. They are probably the only animals in Africa that dig holes in search of water, which they do by loosening soil with their tusks and then excavating with their trunks, often to a depth of several meters. The waterholes enable many other species to survive. Elephants also need shade during the hottest part of the day and an average daily supply of food equivalent to about 5 percent of their body weight. However, they are remarkably nonselective in diet. No other animals are able to exploit such a wide range of plant life. Their great strength, powerful tusks, and long trunk give them an unmatched ability to utilize all parts of a tree. They smash through the thickest vegetation, strip bark, and freely knock trees down, not only for food but as part of social display by bulls. Habitats are devastated, but if the elephants are not constrained by human activity, effects are only temporary and ultimately beneficial (see below). Aside from people, no other animals have had such a role in shaping the environment of the continent. As a result of their abilities and versatility, their distribution and numbers under natural conditions potentially far exceed those of most other ungulate herbivores, whether large or small. Only human interference prevents tens of millions of elephants from occupying nearly all of Africa. Additional information is provided separately for each species.

Loxodonta cyclotis (African Forest Elephant)

Kingdon (1979) listed normal shoulder height as 160–286 cm (mean 250 cm) in males and 160–240 cm (mean 210 cm) in females, and weight as 2,700–6,000 kg. Grubb et al. (2000) listed shoulder height as 240–300 cm in males and 180–240 cm in females, and weight as 2,000–4,000 kg. Cousins (1994), however, argued that the true forest elephant does not exceed

African forest elephant (*Loxodonta cyclotis*), adult male. Photo by Martyn Colbeck.

218 cm in height and that larger reported individuals are attributable to *L. africana* that enter forested areas or to hybrids of *L. cyclotis* and *L. africana*. Happold (1987) noted that the ears measure about 90 cm from top to base. For additional descriptive information and comparison with *L. africana*, see the above generic account for *Loxodonta*.

The modern range of *L. cyclotis* corresponds largely to the tropical rainforests that originally extended across about 4.5 million sq km of western and central Africa (Naughton-Treves and Weber 2001). Those forests now cover an overall area of approximately 3.3 million sq km, though probably less than half of the area is primary forest (FAO 2010). Current occupied range within the original forest zone now amounts to perhaps 600,000 sq km (IUCN African Elephant Specialist Group 2013), though Blake and Hedges (2004) indicated a potential remaining elephant range of around 2 million sq km. Barnes, Blom, and Alers (1995) re-

ported elephant densities to be highest in mosaics of primary and secondary forest left by shifting human settlement and cultivation in the past. Individuals also are found in adjoining country with savannah or a mix of savannah and tropical forest, especially in western Africa and northern Cameroon (Groves 2000; Groves and Grubb 2000a; Ishida 2011b). In southeastern Nigeria, herds formerly lived in the Niger Delta during the dry season and moved north to the Anambra River or east toward the Cross River when the Delta flooded in the wet season; farther west, herds in the forests south of Ondo, Akure, and Owo migrated seasonally southward towards Okitipupa (Happold 1987).

In the Ndoki Forest of northern Congo, Blake (2002) found *L. cyclotis* to aggregate around large swamps and watercourses, especially in the dry season, probably because of high availability of browse. Such aggregation decreased as rainfall and fruit availability increased. The animals then moved in an essentially nomadic pattern, tracking the irregular distribution of ripe fruit. It seems likely that the elephants develop a cognitive map detailing the precise spatial and temporal location of zones of high fruit availability.

L. cyclotis has been studied far less than *L. africana*, largely because of the former's inaccessible habitat and the difficulty of extensive observations through thick vegetation (Tangley 1997). Recently, investigators have found that saline clearings ("bais") along watercourses offer unimpeded viewing, especially in the dry season when elephants emerge from the forest to dig for mineral-rich water. Such clearings are present in the forests from Nigeria to Democratic Republic of Congo. Turkalo and Fay (2001) reported that elephants typically begin to appear at Dzanga Bai in the Central African Republic at about 1400 hours, after feeding in the forest, and continue to arrive through the night. In addition to excavating, the animals have been observed to graze, bathe, fight, and sleep in the clearing. In the early morning they return to the forest to feed again. Using acoustic monitoring equipment, Wrege et al. (2011) also determined most activity in clearings to be nocturnal. However, using GPS and radio telemetry, Blake, Douglas-Hamilton, and Karesh (2001) found activity of an adult female to be greatest from 1200 to 2100 hours and least from 0000 to 0900 hours. That animal utilized a minimum area of 880 sq km during 11 months and made several journeys, each reaching a straight-line distance of up to 60 km from the bai where she was initially captured. Blake (2002) reported individuals to move an average of 0.33 km per hour and about 7.8 km per day, but to travel up to 57 km in 48 hours. Movements clearly showed that the elephants had learned to avoid areas with logging activity, a recent history of hunting, or other human disturbance.

Permanent trails up to several meters across, created and maintained by *L. cyclotis*, are conspicuous features of Africa's equatorial forest (Blake and Inkamba-Nkulu 2004). In Odzala National Park, Congo, Vanleeuwe and Gautier-Hion (1998) found these permanent paths or "boulevards" to extend as far as 34 km and to follow a relatively straight course between the mineral-rich clearings and other favorable sites. There are also more sinuous and narrow foraging paths, up to 3 km long, through adjacent forest where the elephants feed. In the immediate vicinity of the clearings are additional systems of trails, evidently used by the elephants to search for humans and natural predators before venturing into the open. Blake (2004) confirmed that leopards (*Panthera pardus*) prey on young *L. cyclotis*.

According to Campos-Arceiz and Blake (2011), in the Ndoki Forest of Congo, *L. cyclotis* eats at least 500 plant species, the highest known dietary diversity of any mammal. *L. cyclotis* depends to a large extent on fruit and is the most frugivorous of the three living proboscidean species, though Blake (2002) found leaves to be the most commonly consumed plant part, especially in swamps. Sukumar (2003) indicated that some forest populations might be almost entirely browsers, with different kinds of fruit varying in importance depending on season. One study in Ivory Coast found seeds to constitute up to 35 percent of the dry weight of elephant droppings during the fruiting season. African forest elephants have been called "megagardeners of the forest" (Campos-Arceiz and Blake 2011) and "tree planters of the Congo" (Blake et al. 2009). They are the exclusive or near-exclusive dispersers of more plants than any comparable mammal, defecating the seeds over vast distances and into nutrient-rich and protective dung. Ingestion has neutral or positive effects on seeds and accelerates the germination process. Large numbers of forest elephants ranging over large areas may be essential for the function of entire ecosystems. Their loss, because of human hunting for ivory and meat, could bring about a depauperate tree community. Beaune et al. (2013) found that, of 18 tree species studied at a site in Democratic Republic of Congo, 12 were unable to recruit sufficient replacement without elephant dispersal and would likely disappear, along with dependent primate species.

Fishlock, Lee, and Breuer (2008) indicated a potential individual home range of 1,000–2,000 sq km. Blake et al. (2008) determined mean home range to be 547 (26–2,226) sq km for 28 elephants that wore GPS telemetry collars for an average of about a year in wilderness areas of Gabon, Congo, and Central African Republic. Those ranges, however, were constrained to some extent by roadways. Territoriality has not been reported, though dominant males defend waterholes in clearings (Turkalo and Fay 2001). A somewhat different situation is suggested by a radio-tracking study of six adult females for 5–18 months in Loango National Park, Gabon (Schuttler, Blake, and Eggert 2012).

African forest elephant (*Loxodonta cyclotis*) digging in "bai." Photo by Gudkov Andrey / Shutterstock.com.

Those females maintained adjacent home ranges with minimal overlap throughout the year. The ranges were relatively stable, with no evidence of migration or long-distance movements, and at just 11–105 sq km were among the smallest recorded for any elephant species or population.

Population densities of 0.08 to 2.48 forest elephants per sq km have been reported for various sites in Gabon, Cameroon, Ivory Coast, and Democratic Republic of Congo (Morgan 2007). Variation may reflect presence or absence in response to fruiting or other favorable resources in an area, as well as depletion because of human hunting. An idea of natural, long-term density over a large region was provided by Blake et al. (2007), who estimated 22,000 individuals in the 7,592-sq km Minkébé National Park in Gabon (2.9 per sq km) and 14,000 in the 13,545-sq km Odzala-Koukoua National Park in Congo (1.0 per sq km). Much higher densities are attained temporarily in forest clearings. According to Turkalo and Fay (2001), more than 100 individuals have been present at one time within Dzanga Bai, an area of about 10 ha. Such aggregations do not represent organized units. True social groups of *L. cyclotis* usually have 2–5 individuals and are smaller than those of *L. africana*, perhaps because of the patchy nature of food resources (see also summary tables in Maréchal, Maurois, and Chamberlan 1998; Morgan and Lee 2007; Querouil, Magliocca, and Gautier-Hion 1999; Theuerkauf, Ellenberg, and Guiro 2000).

African forest elephant (*Loxodonta cyclotis*), family group. Photo by Sergey Uryadnikov / Shutterstock.com.

The most common grouping at Dzanga Bai is an adult female with one or two offspring, though such units may be components of larger groups of up to 18 animals, which periodically come together and then split back into separate families (Turkalo and Fay 2001). Analysis of DNA extracted from dung along elephant trails in Gabon (Munshi-South 2011) suggested groups may contain an adult female, as well as and her sister or half-sister, there occasionally being more than one reproductive female per group. Additional studies in Gabon, involving both assessment of mitochondrial DNA and observations in the wild, revealed social networks of up to 22 individuals, mostly of the same matriline (Schuttler et al. 2014a, 2014b). Those observations, and others at Mbeli Bai and Maya Nord clearing in Congo (Fishlock and Lee 2013; Fishlock, Lee, and Breuer 2008), showed the fission-fusion sociality of *L. africana* to extend to *L. cyclotis*, though with smaller group size and a much higher proportion of lone individuals, especially females. The open spaces function as social arenas that allow younger individu-

als to join larger groups, often with older elephants, and males to join females.

Studies of *L. cyclotis* at Dzanga Bai (Turkalo and Fay 1996, 2001) indicated that disassociation of young males and females from their family groups begins at an earlier age than in *L. africana* and that males leave at an earlier age than females. Juvenile and subadult males there do not appear to form permanent associations, though they might come together for a few hours. Adult forest males were never observed to form bachelor groups, like those of *L. africana*, but at Dzanga Bai they did confront one another in ritual tests of strength, apparently to establish and maintain a dominance hierarchy. Munshi-South (2011) reported some evidence of limited associations of males and of groups containing juveniles from multiple related females.

Acoustic monitoring in Gabon and Ghana showed that *L. cyclotis* has a series of low-frequency rumblings and calls comparable to the vocalizations of *L. africana* (Thompson, Schwager, and Payne 2009; Wrege et al. 2011). The reported lower range of 5 hertz (Turkalo and Barnes 2013) is below the minimum of 8 hertz recorded for *L. africana* (Poole 2011). Such calls may be important in maintaining group contact in the forest and coordinating arrival times at clearings (Tangley 1997). Garstang (2004) suggested that low-frequency elephant communication, which suffers little or no attenuation by trees or other vegetation, may have evolved first in forests and then found use on the savannahs. However, atmospheric conditions in the latter region favor such communication by night, while in a closed canopy forest more low-frequency elephant calls might be expected during the day.

Allen (2006) found a "complete paucity of published data on reproduction" in *L. cyclotis*, but postulated that, with the likely exception of the stable equatorial ambient temperature and rainforest food supply ensuring a more uniform age at puberty, and intercalving interval, no significant differences exist in basic repro-

ductive processes between *L. cyclotis* and *L. africana*. Theuerkauf, Ellenberg, and Guiro (2000) observed that females in Ivory Coast can calve at 8 years of age, somewhat younger than female *L. africana*. Turkalo (1996) and Turkalo and Fay (1996, 2001) reported that males in "musth" (see account of *L. africana*) are commonly observed at Dzanga Bai, sometimes entering the clearing in search of estrous females. In that area musth appears to be synchronized, occurring at about the same time each year in any given male, and to occur most frequently in the two dry seasons, November–February and July–August. Wittemyer (2011) mentioned speculation that the females of one population of *L. cyclotis* ovulate in the dry season, when underground mineral resources are accessible. According to Turkalo and Barnes (2011), a single young is born after a gestation period of approximately 660 days, and some calves have been observed suckling until an age of

African forest elephant (*Loxodonta cyclotis*), interaction. Photo by Gudkov Andrey / Shutterstock.com.

5 years. Interbirth interval at Dzanga Bai is 3.5–4.0 years (Turkalo and Fay 2001). A captive male *L. cyclotis* lived to an age of about 41 years (Weigl 2005), but potential longevity in the species likely approaches that of *L. africana*.

African elephants are intelligent and not particularly difficult to tame, but have not been employed by people for labor and transportation to the same extent as has the Asian elephant. Most efforts to utilize African elephants have involved *L. cyclotis*, which is considerably smaller than *L. africana* and presumably easier to handle. Barnes (1996) suggested it could be useful in the logging industry, while causing less collateral damage to the environment than does machinery, but acknowledged that its small size and other factors might make it less efficient than *Elephas* in such a role. Since the 1920s a training station for work elephants has been operating at Gangala na Bodio in what is now Garamba National Park in Democratic Republic of Congo. Civil disturbance and military activity nearly closed the station down in the 1960s, but it was revived in the 1990s with a successful program to train elephants to carry tourists (Iversen 1995). Subsequently, renewed unrest again stopped most work, though reportedly four old trained elephants are still maintained at the site (UNEP 2011).

Long before, African elephants were subject to far more extensive utilization, primarily for war (Kistler 2006; Lobban and De Liedekerke 2000; Sukumar 2003). Controlled by mahouts and sometimes armored and carrying troops, elephants could smash through enemy lines and spread terror among conventional infantry and cavalry. They were used widely by ancient kingdoms in what are now northeastern Sudan, Eritrea, and northern Ethiopia. Ptolemaic Egypt (305–30 BC) had a large-scale operation to capture and trade for elephants in that region and established facilities along the Red Sea, where the animals were assembled and tamed. They then were shipped north to Memphis to undergo further combat training. The operation's scope and duration reflect its military value. However, in 217 BC at Raphia in Palestine, the Ptolemies fought the Seleucids of Syria. The former used *Loxodonta*, and the latter deployed a force of *Elephas*, thereby occasioning the greatest known confrontation between the two genera. Although the Ptolemies were ultimately victorious, accounts of the battle specify that their elephants were routed by the much larger Asian animals. As *Elephas* usually is smaller than *Loxodonta africana*, there has been speculation that the Ptolemaic elephants, brought from lands southeast of Egypt, represented *L. cyclotis*, particularly the subspecies *L. c. pharaohensis*. However, recent assessment of the DNA of a small surviving population of elephants in inland northwestern Eritrea suggests the species in the Red Sea region is *L. africana*, hence casting doubt on the notion that the Ptolemaic elephants represented *L. cyclotis*, and indeed on the resulting complex biogeography of that species that has been conjectured (Brandt et al. 2014; Roca et al. 2015; see also above generic account for *Loxodonta*). In any case importation from the Red Sea region eventually ceased, possibly because of changing military tactics by the Mediterranean kingdoms or because elephant populations were depleted, but the elephants that were designated *pharaohensis* may have survived in coastal Sudan and Eritrea until the mid-nineteenth century (Yalden, Largen, and Kock 1986).

Loxodonta, possibly *L. cyclotis*, from what are now Morocco, Algeria, and Tunisia was employed by ancient Carthage in its wars against Rome. It was instrumental in a major victory at the Battle of Tunis in 255 BC, and some of the animals were brought over the Alps by Hannibal in 218 BC. The Roman legions soon developed effective countermeasures but for a time used elephants themselves. Rome also imported many live elephants for circus and gladiatorial performances (commonly ending in the death of the animals) and developed a capacious trade in ivory. Such exploitation soon made the elephant rare in northern Africa, and by the fifth or sixth century AD it no longer occurred north of the Sahara. A population of small elephants related to the northern group seems to have persisted until modern times in Mauritania (Roth and Douglas-Hamilton 1991), though it became extinct by 1991 (Barnes 1999).

European and Mediterranean societies may have been largely cut off from further contact with *L. cyclotis* until Portuguese mariners began to trade for ivory along the coast of western Africa in the fifteenth century. The trade flourished through the seventeenth century and came to involve the British, French, and Dutch as well, but then declined as elephant numbers fell in the eighteenth and nineteenth centuries (Barnes 1999; Sukumar 2003). Commerce in ivory, at least with respect to *L. cyclotis*, then tended to shift toward central Africa.

African forest elephant (*Loxodonta cyclotis*), apparent agonistic interaction. Photo by C. Smith / Mammal Images Library, American Society of Mammalogists.

Milner-Gulland and Beddington (1993) calculated that in 1814 the carrying capacity of 2.8 million sq km of tropical rainforest in western and central Africa was approximately 1.4 million elephants (assuming an average density of 0.5 per sq km). Considering also the above higher estimate of the original extent of the forest zone, the higher densities recently reported for large protected areas, and the apparent former extent of the range of *L. cyclotis* for some distance into the western savannah zone, it is possible that the natural population of the species in equatorial Africa reached several million individuals. In any case, numbers declined precipitously during the nineteenth century, mainly because of continued exploitation for ivory but also from killing of elephants for sport and to prevent interference with expanding agriculture.

According to Barnes (1999), intense hunting to meet the demand for ivory from Europe and North America led to the collapse of elephant populations in both the forest and savannah zones of western Africa (the region from what is now Senegal to Nigeria) by about 1910. The ivory trade subsequently slackened, but exponential human population growth and habitat usurpation prevented any substantive elephant recovery. For a time *L. cyclotis* seems to have fared better in the forests than *L. africana* did on the savannahs, remaining locally common in some of the west African countries until the 1950s. However, renewed ivory exploitation and conflict with human interests, compounded by civil unrest and breakdown of legal protection, led to further decreases. By around 1990 there were estimated to be fewer than 15,000 elephants of both species in western Africa (Cumming and Du Toit 1989; Roth and Douglas-Hamilton 1991; Said 1995).

The most recent report by the IUCN African Elephant Specialist Group (2013) indicates the maximum number of all elephants in western Africa to be around 12,000; of that figure only about 4,200 are *L. cyclotis*. The largest national number is that of Liberia, which may have about 1,600 individuals in six populations, four of which have more than 200 each. However, the figures for three of those groups are completely speculative and date from 1989–1990, when the country was entering an extended period of civil war and political instability. Only one other population in western Africa was thought to possibly have over 200 *L. cyclotis*, Digya National Park in Ghana, but it is in a forest-savannah transition area and its taxonomic status is indefinite (see Kumordzi et al. 2008). In any case, Danquah and Oppong (2014) recently observed that an isolated population of *L. cyclotis* in the Bia Conservation Area, with an estimated 146 individuals, is probably the largest remaining in Ghana. At the extreme western edge of the current range of *Loxodonta*, a group of perhaps 20 elephants occupies a small area of forest, partly cleared for agriculture, in southern Guinea Bissau, on the border with Guinea (Brugière et al. 2006). Given the likelihood of increased habitat disruption and hunting, as well as the loss of genetic variability, fragmentary populations of fewer than 200 elephants are very vulnerable and have a low probability of surviving the twenty-first century (Barnes 1999).

In the central African forest zone (what is now southern Cameroon, southwestern Central African Republic, Gabon, Equatorial Guinea, Congo, and northern and central Democratic Republic of Congo) human populations were not so dense, and habitat was less accessible. There was heavy hunting for ivory in that region in the late nineteenth and early twentieth centuries, with an estimated 550,000 elephants killed in the Belgian (now Democratic Republic of) Congo alone between 1889 and 1950 (Sukumar 2003), but numbers do not appear to have declined as sharply as in western Africa. In the 1970s, the price of ivory rose dramatically, bringing on a surge of illegal killing throughout Africa (see account of *L. africana*). For the central African forests, Michelmore et al. (1994) estimated that from the mid-1970s to 1989 the number of elephants dropped from about 306,000 to 170,000. What had been among the largest remaining elephant populations of Africa, that of the forests of what is now Democratic Republic of Congo, fell from about 172,000–208,000 to 64,000–72,000 (Alers et al. 1992; Barnes, Blom, and Alers 1995).

African forest elephant (*Loxodonta cyclotis*), mother with young of two ages. Photo by Sergey Uryadnikov / Shutterstock.com.

By 1989 the only large population of either species of African elephant that had not been extensively exploited or unnaturally constrained was that of *L. cyclotis* in Gabon. Although distribution in that country was influenced to some extent by human activity, habitat remained the least disturbed in Africa and had not yet been hit by the intensive poaching that had ravaged the elephants of Democratic Republic of Congo (Barnes et al. 1991, 1993). There were various estimates for the number of *L. cyclotis* in Gabon, which has an area of 267,667 sq km, but the most refined figure was 76,470 (Barnes et al., 1997). Unfortunately, waves of illegal hunting were moving toward Gabon from all directions. Michelmore et al. (1994) observed that if the kill rate of the 1980s continued, elephants would be eliminated throughout central Africa shortly after the turn of the century.

In June 1989 the United States completely banned the importation of ivory, and in October 1989 African elephants were transferred to appendix 1 of the CITES, which largely ended legal international trade in ivory. Those measures seem to have substantively slowed the decline of elephant populations in some parts of Africa, though shortly afterward poaching was reported to be continuing unabated in Cameroon, Congo, and Gabon (Dublin, Milliken, and Barnes 1995; Fay and Agnagna 1991). Positive effects of the trade bans were partly negated by subsequent events, including the breakdown of legal controls in much of the strife-torn continent and perhaps the limited reopening of commerce by CITES in 1997 (see account of *L. africana*, below). Hunter, Martin, and Milliken (2004) gathered evidence that the central African forests were under stronger poaching pressure than any other part of the continent and were the main source of illegal ivory. About 4,862 to 12,249 elephants were being killed annually, which is 2.5 percent to 6.3 percent of the 195,753 elephants that were estimated by the IUCN (Blanc et al. 2003) to occur in central Africa. Such figures were of much concern, since even under favorable conditions an elephant population will decline if mortality from all causes (natural and human-induced) is greater than 4–5 percent (Douglas-Hamilton 1987; Sukumar 2003).

In 2003–2005, Blake et al. (2007) carried out the first comprehensive assessment of central African elephants since 1989. Results indicated *L. cyclotis* to be severely threatened by ivory hunting, especially where there was road access. While guarded reserves were helpful in some areas, there had been severe declines even in supposedly protected areas of Democratic Republic of Congo. In that country's Salonga National Park, which at 36,000 sq km is the largest tropical rainforest reserve in Africa, as few as 1,900 elephants remained, there thus being an unnaturally low mean density of 0.05 elephant per sq km. In the wake of military battles in the Ituri Forest, an estimated 2,500 elephants were killed for ivory in 2003. A follow-up survey in 2011 found that in the previous 5 years the number of forest elephants in the area around Nouabalé-Ndoki National Park in Congo had declined by more than 50 percent (Wildlife Conservation Society 2012). An identical decline, associated largely with warfare and anarchy from 1996 to 2009, was reported for the Okapi Faunal Reserve in northeastern Democratic Republic of Congo (Beyers et al. 2011).

Subsequent intensive field surveys, together with assessment of mortality monitoring sites and demographic data, showed the overall extent of the disaster to be even worse than had been thought (Maisels et al. 2013; Wittemyer et al. 2014). From about 2002 to 2012, *L. cyclotis* had declined substantially (perhaps by around 63 percent in some areas) and lost 30 percent of its geographical range. Its population now is less than 10 percent of potential size and occupies less than 25 percent of potential range. Blanc et al. (2007) indicated there is intense poaching for both ivory and meat, and that such activity is exacerbated by new roads for logging operations and mineral and oil extraction, which provide both access to deep forest and routes for the transport of ivory and meat. Extraction of data from the most recent report by the IUCN African Elephant Specialist Group (2013) suggests the following maximum numbers of *L. cyclotis* in the central African countries: 13,278 in Cameroon, 2,104 in Central African Republic, 1,630 in Equatorial Guinea, 77,252 in Gabon, 49,248 in Congo, and 11,707 in Democratic Republic of Congo (see account of *L. africana* for information on numerical derivation). While the figures for some countries suggest little change from earlier assessments, Blanc et al. (2003) had provided data indicating a total of 5,987 *L. cyclotis* in Central African Republic and 54,224 in Democratic Republic of Congo.

African forest elephant (*Loxodonta cyclotis*), mother with young of two ages. Photo by Brent Huffman.

Although CITES implemented new restrictions on ivory trade in 2007, Douglas-Hamilton (2009) considered them ineffective. Poaching still increased, apparently fueled by demand in Asia, a rise in the price of ivory, and proliferation of illegal, uncontrolled markets. Successive investigations by regional experts in central Africa indicated continued ongoing declines in elephant numbers, densities, and distribution. Nowhere has the situation been worse than in Democratic Republic of Congo, once believed to have more elephants than any other country. There, amidst social disorder, political turmoil, and civil war, elephant numbers have declined consistently for three decades and now are in the order of just 14,000, including both species. That figure is well below the total number of elephants represented by the 418,000 kg of ivory (about 10 kg for an average tusk and 2 tusks per animal) exported from the country in the single peak year of 1919 (Blanc et al. 2003).

Recent reports from CITES (CITES Secretariat 2012; CITES Secretariat, IUCN/SSC African Elephant Specialist Group, and TRAFFIC International 2013) show that poaching to supply the international ivory trade has been increasing since 2005 and is leading to major declines in some elephant populations, particularly in central Africa. Monitoring at various sites indicates that the mortality rate far exceeds what those populations can sustain. There has also been a steady increase in illicit trade since 2004 and a major upsurge since 2009, with 2013 being the worst year on record for large ivory seizures. Available data show China to be the paramount destination for an escalating number of large consignments of illegal ivory leaving Africa. Consumer demand in that country is mirrored by a steady increase in the wholesale price paid for illegal ivory by carvers and processors, while legal control mechanisms there have been inadequate (for further details, see account of *L. africana*). According to Nishihara (2012), an active but poorly controlled market also continues in Japan, where most demand is specifically for the so-called "hard ivory" of the African forest elephant, which is considered superior to the "soft ivory" of the savannah elephant in production of the *bachi*, the plectrum used in playing the *shamisen*, a traditional stringed musical instrument. A single *bachi* requires a complete tusk weighing over 15 kg. Ivory is also still used in Japan for the manufacture of *hanko* (personal name seals), though not as extensively as in the late twentieth century, when such demand was a critical factor in large-scale illegal exploitation.

Thus, while restrictions on the ivory trade may have helped slow the decline of *L. cyclotis*, the dire warnings of two decades ago were basically valid. The poaching cyclone swept through central Africa, nearly wiping out the vast elephant aggregations of Democratic Republic of Congo and cutting down numbers elsewhere, even in parks and reserves. Gabon now stands alone in having a large and relatively intact forest elephant population. Can it be maintained if elephant numbers dwindle in surrounding countries but the demand for illegal ivory continues unabated? The sad answer may already be unfolding. In February 2013, the Gabon government reported that an estimated two-thirds of the more than 11,000 elephants in Minkébé National Park, in the northeastern corner of the country, had been killed by poachers since 2004 (Nellemann et al. 2013).

Notwithstanding the abundant relevant data, there has been little official recognition of either the conservation or systematic status of *L. cyclotis*. It is not even formally listed, as a species or subspecies, on the IUCN Red List of Threatened Species, the United States Department of the Interior (USDI) List of Endangered and Threatened Wildlife, or the CITES appendices (though all populations of *L. cyclotis* are covered by those promulgations, under the name *L. africana*, as, respectively, vulnerable, threatened, and appendix 1 and 2). The USDI is currently considering a petition from a major conservation organization, the Center for Biological Diversity, to recognize *L. africana* and *L. cyclotis* as separate species and to classify each as endangered (Easter et al. 2015).

African forest elephant (*Loxodonta cyclotis*), mother and young. Photo by Sergey Uryadnikov / Shutterstock.com.

The assessment of *L. africana* in the IUCN Red List (Blanc 2008) states:

> Although elephant populations may at present be declining in parts of their range, major populations in Eastern and Southern Africa, accounting for over two thirds of all known elephants on the continent, have been surveyed, and are currently increasing. . . . If current rates of increase continue, the number of elephants born in these populations between 2005 and 2010 will be larger than the currently estimated total number of elephants in Central and West Africa combined. In other words, the magnitude of ongoing increases in Southern and Eastern Africa are likely to outweigh the magnitude of any likely declines in the other two regions.

While probably not intended as detrimental to *L. cyclotis*, which is restricted to central and western Africa, the above may be among the most unfortunately worded statements ever issued by a major conservation body. It seems to repudiate responsibility—to accept the extermination of one species because a related species is (perhaps temporarily) increasing in a relatively small part of its original range. Douglas-Hamilton (2009) suggested that the IUCN treatment of African elephants focuses on the growing southern populations (and the strong lobbying of some countries to allow the sale of their ivory stockpiles) while ignoring the enormous contraction in numbers and range elsewhere.

Such official disregard is contributing to what, in terms of sheer biomass of a striking and distinct species, may be our greatest ongoing mammalian conservation disaster. Citation herein of reports and statistics can relate only a small part of the tragedy. We, who merely read and write about elephants, cannot share the feelings of authorities who have studied the animals in the field for decades and witnessed firsthand the destruction of most of their subjects. Nor can we comprehend the degree of suffering of countless huge and highly social creatures, which, in their way, may be at least as sensitive to physical and emotional distress as ourselves. We cannot see the butchering of adults or the starvation of orphaned calves. We may denounce the poachers partaking directly in the slaughter or the loggers destroying habitat, but we should remember they are driven by the supposedly highly cultured societies that demand ivory and the other resources of the African forests.

Loxodonta africana (African Savannah Elephant)

This is the largest living terrestrial mammal, some reports indicating an overall weight range of up to 10,000 kg (Wittemyer 2011). Kingdon (1979) listed the following normal size ranges (means in parentheses): shoulder height 300–400 cm (320 cm) in males and 240–340 cm (250 cm) in females, weight 4,000–6,300 kg (5,000 kg) in males and 2,400–3,500 kg (2,800 kg) in females. Happold (1987) noted that the ears measure about 150 cm from top to base, though Poole, Kahumbu, and Whyte (2013) reported size as great as 200 cm vertical and 120 cm across. For additional descriptive information and comparison with *L. cyclotis*, see the above generic account for *Loxodonta*.

L. africana lives in many kinds of habitats, including open savannahs, wet marshes, thornbush, and semi-desert scrub. It has been recorded from sea level to elevations of more than 5,000 meters. According to Whyte (2005), its requirements are a supply of fresh water, plentiful food in the form of grass or browse, and some shade. To find these conditions, the elephant may have to make annual migrations of several hundred kilometers. If food, water, and shade all remain available, elephants or elephant units will not venture far; when those items are scarce, large-scale movements may occur. Rodgers and Elder (1977) stated that nearly all seasonal movements have the same general pattern, that is, a migration from permanent water sources at the start of the rainy season, followed by a movement back to permanent water when the rains end and water holes dry up. Often there is a concentration of animals in suitable areas during the dry season or in times of drought and a dispersal in the wet season. Human activity and agricultural development have now forced permanent, unnatural concentrations in many areas (Kingdon 1979). Shorter, nonseasonal movements of elephants may involve travel between water and feeding areas.

Notwithstanding the importance of water to elephants, they evidently do not interfere with other herbivores around waterholes (Valeix et al. 2009). However, *L. africana* is sometimes aggressive toward other species. A group of young males, introduced to Pilanesberg National Park in South Africa, matured rapidly in the absence of mature bulls and experienced

African savannah elephant (*Loxodonta africana*), old male. Photo by J. F. Jacobsz / Shutterstock.com.

longer than normal periods of "musth" (see below). In that excitable state, they killed over 40 southern white rhinoceros (*Ceratotherium simum simum*) from 1992 to 1997. The killing stopped only after six older bulls from Kruger National Park were introduced and apparently suppressed musth in the younger males (Slotow et al. 2000).

In their movements, elephants follow regular routes and may create well-worn paths. In the ecosystem comprising Amboseli National Park and surrounding parts of Kenya, trail use was found to vary between elephant groups. Large aggregations consistently preferred heavily used "trunk-trails" or "boulevards," while resident families preferred minor, less traveled "family trails" (Mutinda, Poole, and Moss 2011). Although movements are determined primarily by availability of resources, local distribution also seems to depend on relationships of groups, particularly among the dominant females or "matriarchs" of a family (Croze and Moss 2011). Older matriarchs, apparently using their knowl-

edge of sources of permanent water and food, lead their groups over large areas during droughts, thus enabling them to survive such periods better than do groups with smaller ranges (Byrne and Bates 2011).

Activity is both diurnal and nocturnal but drops during the hottest hours of the day. The animals sleep either at noon or after midnight, lying down or standing and leaning against one another or against trees. An elephant usually seeks water once a day, or at least every few days, and then may bathe, wallow, or even submerge with only the nostrils at the tip of the trunk showing above the surface. In this manner an animal may also swim or walk underwater for several kilometers. A normal fast pace on land is 10–16 km/hr, but when charging, the animal may move at a rate of 35–40 km/hr (Kingdon 1979). Daily average movements of 12 km have been reported (Merz 1986). *L. africana* may consume 200–300 kg of food and up to 200 liters of water per day (Kingdon 1979; Whyte 2005). Individuals in the Amboseli ecosystem were

African savannah elephant (*Loxodonta africana*), family group at waterhole. Photo by Andre Klopper / Shutterstock.com.

observed to feed from about 42 to 79 percent of the time during daylight hours, with such time increasing during the dry season and dropping in the wet season (Lindsay 2011). Although primarily a browser, the savannah elephant is a mixed feeder and can switch to a grass- or browse-dominated diet, depending on which provides the highest ratio of protein to fiber (Poole, Kahumbu, and Whyte 2013). The diet includes grass, tree foliage, bark, twigs, herbs, shrubs, roots, and fruit. There seems to be pronounced seasonal variation, with grass intake increasing during the rainy season but falling to low levels during the dry season (Hanks 1979). The elephant plucks bundles of grass by curling the end of the trunk around the stems and pulling the plant up by the roots; it then beats the grass against its body to get rid of the soil (Whyte 2005).

Because of its enormous size, the elephant can be destructive to vegetation, pushing over trees to obtain edible twigs and leaves and sometimes modifying the habitat over large areas. Bulls also seem to knock trees down in response to social pressures or other excitement (Kingdon 1979). In the past, when the species could roam more freely over the African continent, such environmental modification presented few problems and in fact may have been necessary for the maintenance of ecosystems that could support large and diversified populations of animals. As human populations increased in Africa, however, and as large areas were cleared for agricultural purposes, the range of the elephant became fragmented and more confined to restricted sites such as parks and reserves. Within these areas, elephants, being artificially concentrated, sometimes gave the impression of becoming very numerous. An elephant population can increase at an annual rate of 4–5 percent under favorable conditions; the maximum known is 7 percent (Calef 1988; Cumming 1981; Douglas-Hamilton 1987).

Unable to migrate or disperse naturally, a growing population can begin to damage its habitat, changing savannah woodland to grassland and reducing biotic diversity (Poole, Kahumbu, and Whyte 2013). Therefore, some governments have carried out deliberate culling programs for the stated purposes of protecting habitat and preventing the elephants from destroying their own food supply. Those operations are questioned by some authorities. Guldemond and Van Aarde (2008) reviewed 238 published studies on the issue and found that a disproportionate few, at least some of which were conducted in fenced areas where short-term influences could be readily determined, may have introduced a bias toward concluding that elephants have a negative impact on woody vegetation. Skarpe et al. (2004) pointed out that the elephant population of northern Botswana had grown to become the largest on earth and had contributed to substantial habitat changes. However, such modification was considered a return to the natural situation, before severe ivory hunting had left the elephant rare in Botswana by the late nineteenth century, and no ecological reason for culling was seen. The northern Botswana elephants are part of a "metapopulation," consisting of about 250,000 animals in a region of approximately 520,000 sq km extending into Namibia, Angola, Zambia, and Zimbabwe. The region includes 36 parks and reserves, which, together with interspersed lands, were conjoined by a 2012 international treaty into the Kavango-Zambezi Transfrontier Conservation Area (Peace Parks Foundation 2014). Van Aarde and Jackson (2007) suggested not culling or otherwise restricting the elephant population, but instead allowing, pursuant to ecological conditions, movement of excess animals out of the protected zones into "sinks," where they might be hunted at a sustainable level for local use.

Poole et al. (2011b) considered culling an unjustified and unethical destruction of a highly social and intelligent animal, which leaves survivors individually disturbed and socially disrupted, while being ineffective beyond the short term and perpetuating a need for constant intervention. The sudden reduction of numbers removes natural density-related suppression of reproduction, thereby creating conditions in which females give birth at an earlier than normal age (see below). An alternative to lethal culling, immunocontraception through a vaccine injected by dart gun, has been tested successfully on small reserves in South Africa, with reportedly minimal disturbance to the animals, though practicality in controlling large populations is debat-

able (Delsink et al. 2007; Druce et al. 2013; Fayrer-Hosken et al. 2001; Whyte and Fayrer-Hosken 2008).

Food quantity and quality and availability of water are the most important natural factors determining the spatial distribution of the elephant. Differences in those factors have resulted in considerable variation in typical population densities, 0.26–5.00/sq km, and home ranges, 14–3,120 sq km (Kingdon 1979). Milner-Gulland and Beddington (1993) calculated that pristine carrying capacities in most habitats of Africa were about 2/sq km. Armbruster and Lande (1993) estimated that an area of approximately 2,600 sq km, with a stable density of about 1.2/sq km, would be necessary to ensure a population's permanent survival. Western and Lindsay (1984) reported a relatively undisturbed population of about 500 elephants occupying about 3,500 sq km in and around the Amboseli National Park of southern Kenya. Overall density thus would have been 0.14/sq km, but increasing during the dry season to 0.4–0.9/sq km and even 10/sq km when elephants concentrated in certain habitats retaining sufficient water and forage. Croze and Lindsay (2011) suggested that the above was an underestimate and that population size in the Amboseli ecosystem probably increased from about 850 elephants in the 1970s to just over 1,500 in 2008. Croze and Moss (2011) noted further that the actual range of the Amboseli elephant population covers 11,500 sq km and that overall density is thus only 0.13/sq km, even though the population is well protected.

In the Lake Manyara area of Tanzania, Douglas-Hamilton (1973) found home ranges of family units and of individual bulls to be 15–52 sq km and to overlap widely. Hanks (1979) found that on the whole, elephants in southern Africa had small home ranges, but some moved considerable distances for short periods of time. In Tsavo National Park, Kenya, Leuthold and Sale (1973) recorded mean home ranges of 530 sq km in Tsavo West and 1,580 sq km in Tsavo East. Long-term studies in the Amboseli of Kenya determined family home ranges to average about 570 (95–1,690) sq km (Mutinda, Poole, and Moss 2011). In northwestern Namibia, where scarce vegetation and water cause the elephant to make seasonal movements of about 650 km, home range is up to 12,600 sq km (Poole, Kahumbu, and Whyte 2013). The Gourma elephants of Mali make a counterclockwise annual migration of about 450 km

African savannah elephant (*Loxodonta africana*), mother and young at water source. Photo by Four Oaks/Shutterstock.com.

through near-desert conditions, moving between widely dispersed waterholes in an overall home range of around 25,000 sq km (Blake et al. 2003). *L. africana* is not territorial, and there may be extensive overlap of the home ranges of different family units, but females are wary in their use of areas occupied by unrelated groups (Poole, Kahumbu and Whyte 2013). In the rare instance of an aggressive encounter, larger families, with older matriarchs, are dominant to smaller families, with younger matriarchs, and tend to win (Mutinda, Poole, and Moss 2011). Dominance of one group matriarch over another group matriarch is apparently a function of age, not physical size, and may result from social interaction early in the animals' lives (Wittemyer and Getz 2007). Dominant families have been found to take control of favorable areas offering safety and proximity to water during the dry season in northern Kenya (Wittemyer et al. 2007).

The African savannah elephant is gregarious, sometimes being found in aggregations of hundreds or even more than 1,000 individuals. Those large groups, which have been observed mainly in East Africa, are thought to be temporary associations of organized family units, consisting of females and young, and individuals that come together in times of drought, human interference, or other disruptions of the normal pattern of social life. In the Amboseli, Western and Lindsay (1984) found that groups seldom exceeded 20 individuals in the dry season, but that aggregations of more than 400 sometimes formed during the rains. Mutinda, Poole, and Moss (2011) confirmed such aggregations at Amboseli, noting also that individual families, particularly those led by older females, are typically discernible as discrete socio-spatial units within the larger grouping. Douglas-Hamilton (1973) suggested that grouping could provide a collective defense against predators in times of social stress or disturbance. Wittemyer (2011), however, suggested that the large aggregations of female-based families, which typically form in the wet season when reproduction is most common, serve to attract dominant breeding males and thus facilitate mating. There are also temporary associations of males, the members joining and leaving at will (see below). Those bachelor groups usually are small but have been known to contain as many as 144 animals (Kingdon 1979). Solitary males are common, and very old bulls are often found alone. Solitary females are extremely rare. Both male and female groups spend most of their time in particular areas, which Mutinda, Poole, and Moss (2011) designated, respectively, as "bull areas" and "clan areas."

Savannah elephant society is essentially matriarchal, organized around a stable family unit of cows and their calves (Hanks 1979). There may be only a single adult female, the matriarch, with one or more of her offspring, but usually the units contain about 10 individuals (Kingdon 1979). Nearly all females are part of a family unit. Based on long-term genetic studies at Amboseli, Archie, Moss, and Alberts (2011) confirmed earlier observations that females are "matrilocal," tending to remain in their mother's family unit, though a few do leave and are accepted into other families. Some reports have suggested that the dominance of a family matriarch is undisputed, that she rebuffs outsiders at-tempting to enter the group, and that she usually maintains her position until death, when she is succeeded by her eldest daughter (Laursen and Bekoff 1978). While such observations may on occasion be valid, studies at Amboseli and at Tarangire National Park, Tanzania, have provided a much more refined assessment, depicting considerable moderation and flexibility. Females do tend to form and maintain their closest social relationships with maternal kin, but do not exclude non-kin from cooperation and affiliation. Aggression is rare, and matriarchs may associate freely. While there are female dominance hierarchies, they are based not on kinship but on age and size. Older, larger females consistently dominate smaller, younger ones, as can be determined by observations of aggressive and submissive movements. There is no evidence of nepotism or that a matriarch's daughter would take precedence over another female that is larger or similar-sized (Archie, Moss, and Alberts 2005, 2011; Lee 2011). Genetic investigation in northern Kenya showed approximately 20 percent of the members of family units to not be immediately related, demonstrating

African savannah elephant (*Loxodonta africana*), partly submerged. Photo by Travel Stock / Shutterstock.com.

that kinship is not a prerequisite for social affiliation at that social level (Wittemyer et al. 2009). Matriarchs determine interaction with other groups, decide which trails to follow, and initiate and guide family movements; in all observed cases when an entire multifamily aggregation moved in response to an initiation, the oldest matriarch present was responsible (Mutinda, Poole, and Moss 2011).

According to Poole, Kahumbu, and Whyte (2013), social units are subject to "fission and fusion," with regular forming and splitting based on kinship, home range, and season. The animals tend to divide into smaller groups during drier periods, when food is scarce, but amalgamate when food becomes more plentiful. Douglas-Hamilton (1973) found that sometimes two to four family units, jointly including as many as 50 animals, may come together into slightly less stable associations that he termed "kinship groups." Moss and Lee (2011b) defined the equivalent "bond groups" as families that frequently associate and whose members regularly and repeatedly exhibit affiliative behavior such as greeting and touching. A bond group

may form when a family grows large and a portion splits off to form a new family, but the two spend more time with each other than with other families in the overall population. During 30 years of study at Amboseli, the average number of animals in known family units increased from 7 to over 18, though some families had as many as 52 members. Bond groups consisted of up to 5 families and 76 members. The genetic studies at Amboseli found that adult females remained with their maternal relatives when family units underwent temporary fission, and that different family units were more likely to undergo fusion to form a bond group when the oldest females in each group were related. It was concluded that associations between those social groups probably persist for decades after the original maternal kin have died (Archie, Moss, and Alberts 2005). However, the genetic studies in northern Kenya, where elephant populations have been disrupted by poaching, indicated that relatedness at the level of the bond group and clan is less than in the relatively undisturbed population at Amboseli (Wittemyer et al. 2009).

African savannah elephant (*Loxodonta africana*), group seeking shade. Photo by Tr3gin / Shutterstock.com.

The term "clan" has been used for the social grouping comprising all families that share the same home range during the dry season (Archie et al. 2011; Poole, Kahumbu, and Whyte 2013). However, based on observations in northern Kenya, Wittemyer, Douglas-Hamilton, and Getz (2005) believed that, while there is a fourth-tier social level—above mother-calf association, family unit, and bond group—it does not correspond precisely to the definition of clan, as it is a function of social rather than spatial processes. Although that social grouping was discernible throughout the year, cohesion decreased significantly during dry periods. Genetic studies in northern Zimbabwe found that different families within a clan are not necessarily matrilineally related (Charif et al. 2005). In summarizing the social structure of *L. africana*, Wittemyer (2011) again indicated a correspondence of clan and fourth-tier social level, but noted that bonds are much weaker than at lower levels, with groups closely affiliating just 10–20 percent of the time. He noted that clans may not come together fully until the rainy season, when they are seen within aggregations, which constitute a still higher level of social organization, as they show coordinated behavior and movements.

It was long assumed that as a young male approached sexual maturity, when around 8–20 years old, he was forced out of his natal family unit by the older females, and that he then joined with others in a bachelor group. However, long-term studies at Amboseli have shown that dispersal depends mainly on male initiative and is a gradual process that does not involve female aggression (Lee et al. 2011; Moss, Croze, and Lee 2011; Poole et al. 2011a). Confusion may have resulted from observations of female coalitions driving away troublesome but unrelated young males. In contrast, there is typically a period of several months to several years when males freely leave and rejoin their natal families. Once they have left permanently, they are considered adult, but even then they sometimes associate with their original families, particularly when large aggregations form. Mean age of independence at Amboseli is 14 (range 8–19) years. Younger independent males, up to about 25 years of age, which are sexually active but not yet in musth, actually do not spend most of their time in all-male groups. Instead, they associate with family groups containing females, not necessarily their own natal units, but over time they gradually develop a pref-

erence for a particular bull area. Males 15–20 years old do seek mating opportunities, but they are subordinate to older bulls, which weigh twice as much. Most older males spend over 50 percent of their non-musth time in a bull area, though they are more successful at guarding and mating with females than are the younger males. Some older males readily form associations with other males, while some are less sociable. In northern Kenya, Goldenberg et al. (2014) found a surprising degree of affiliation between males, even when sexually active. In a study in the Okavango Delta of Botswana, Evans and Harris (2008) found adolescent males, 10–20 years old, to be highly social, maintaining close proximity to other elephants and frequently greeting and sparring with peers. However, they preferred the company of a fully mature bull, at least 36 years old, from which they apparently could gain social and ecological knowledge.

Elephant society normally is peaceful, but females with young are unpredictable, and males compete with one another for dominance in their groups and for access to estrous females (Kingdon 1979; Laursen and Bekoff 1978; Whyte 2005). Aggression is shown by raising the head and trunk, extending the ears perpendicular to the body, kicking dust, swaying the head, and making either a mock or a serious charge. Most dominance struggles are resolved after some pushing and light tusking, but battles for mating privileges sometimes involve fatal use of the tusks. Aggressive behaviors also include using the trunk to uproot and throw materials in the direction of an opponent, while submission is indicated by lowering the head, social affiliation by raising the head and lifting the ears, reassurance by "caressing" another with the trunk, and male sexual advertisement by urine dribbling; numerous other visual, tactile, and olfactory signals are used (Langbauer 2000; Poole and Granli 2011).

L. africana is well known for its loud trumpets and screams, when upset, but the species has a wide variety of sounds, including very low-frequency rumblings or growls, which communicate a broad range of complex messages involving agonistic, defensive, affiliative, parental care, mating, and social interaction (Poole, Kahumbu, and Whyte 2013). Even infants have a range of vocalizations, including trumpets associated with aggression or excitement and rumbles that seem to solicit continuation of nursing (Stoeger-Horwath,

African savannah elephant (*Loxodonta africana*) feeding. Photo by Barbara Smith DVM / Shutterstock.com.

Stoeger, and Schwammer 2007; Wesolek et al. 2009). Based on several studies, and eliminating overlapping accounts, Soltis (2010) found the vocal repertoire of *L. africana* to consist of perhaps nine acoustically distinct call types: grunts and barks, which appear limited to immature animals; revs, croaks, and chuffs, limited to adults; and rumbles, trumpets, snorts, and roars that occur in both age classes. There seems to be a difference in the calling patterns between the two sexes, the males giving far fewer types of calls than the females (Langbauer 2000).

According to Poole (2011), most elephant sounds are produced when air from the large lungs passes over the vocal chords or larynx, with extended resonation by the trunk and the honeycomb nasal passages in the skull, and possible modification also by the hyoid apparatus and pharyngeal pouch at the base of the tongue. Calls recorded in Kenya had frequencies ranging from 8 hertz in some rumbles to a maximum of around 10,000 hertz in some trumpets and snorts. Elephants, together with the Sirenia, are unique among modern mammals in having a reptilian-like cochlear structure that may facilitate auditory sensitivity to low-frequency sounds. When an elephant emits a low-frequency rumble, a corresponding seismic wave with similar characteristics is transmitted in the ground and can be discerned and interpreted by other elephants.

McComb, Reby, and Moss (2011) used experimental playback of recorded voices to show that elephants are capable of discriminating between individual voices of other elephants up to 2.5 km away. Most such long-distance communication is infrasonic, that is, with a fundamental frequency below the lower limit of human hearing, about 20 hertz. According to Wittemyer (2011), field work has demonstrated that infrasound is useful at distances exceeding 4 km and possibly up to 10 km under ideal atmospheric conditions. McComb et al. (2000) reported that females could distinguish the calls of female family and bond group members from those of other females, and could also discriminate between the calls of family units outside of their bond groups. Such investigation has shown the importance of old matriarchs with superior discriminatory abilities, which can use their contact calls to bring together or "bunch" a group. Garstang (2004) noted that very low-frequency calls, when emitted by females in or about to come into estrus, may attract potential mates up to several kilometers away. Leong et al. (2003) found elevated rates of female low-frequency vocalization in advance of estrus and suggested the calls may attract males prior to ovulation; once present in the female groups, the males might then switch to chemical and visual cues to detect potential mates. Langbauer (2000) cited playback experiments, during which bull elephants in musth condition have walked up to 1.2 km toward a loudspeaker playing an estrous call, though most non-musth males walked away. Wittemyer (2011) noted that males probably use infrasonic calls to advertise musth state and dominance status, thus allowing sexually active males to avoid one another.

Studies by Hollister-Smith et al. (2007, 2011), Moss (1983), Poole (1987a, 1987b, 1989a, 1989b), and Poole et al. (2011a) in the Amboseli ecosystem of Kenya, have provided remarkable insights into the mating system of *L. africana*. Males more than about 25 years old annu-

ally enter the condition known as musth, characterized by copious secretions from the temporal gland behind the eye, continuous discharge of urine, a great increase in aggressive behavior, and the seeking of and association with female groups. Of 533 groups observed to include an estrous female, 86.4 percent also contained a male in musth. This condition, accompanied by dramatic surges of circulating testosterone levels, lasts only a few days or weeks in young bulls, but length steadily increases from 4–8 weeks in males 36–40 years old to 10 weeks at 46–50 years, and then falls to less than 8 weeks in males over 50. There is a corresponding rise and fall in known paternity success. Musth does not occur synchronously, but it does come at about the same time each year in any given male, and it is especially frequent during and just after the rainy season.

A male in musth is dominant to other males and usually can defeat them in combat even if the latter are larger and normally higher-ranking. Two musth males interact aggressively upon meeting; the larger is dom-

African savannah elephant (*Loxodonta africana*), reaching up to feed. Photo by Jez Bennett / Shutterstock.com.

inant, but if they are closely matched in size and condition, there is likely to be a violent battle. At least four males are known to have died in fights during musth at Amboseli. Although a non-musth male is capable of successful mating, a female actively avoids most courting males, tending to choose a large male that is in musth, perhaps one with whom she has long been familiar. That male then guards the female from other suitors. Bulls 45 and over are much more successful in courtship than are younger males, and are estimated to father at least two offspring each year. While males as young as 26 sired young at Amboseli, males 45–50 years of age produced calves at an average annual rate six times that of 30-year-old males. Moreover, just three males were responsible for approximately 30 percent of calves that could be assigned a father. Of cases of known paternity and condition, 52 calves were fathered by males in musth and only 7 by non-musth males. Tragically, this remarkable selective process may be disappearing more rapidly than *Loxodonta* itself, because old males, which have the largest tusks, are the first victims of the ivory trade (see below).

Females are polyestrous and have an estrous cycle of 14–15 weeks (Allen 2006). Estrus lasts 2–6 days (Moss 1983). Births may occur at any time of year, but there evidently is a calving peak just before the height of the rainy season, so that the young have a cool environment with an abundance of good cover. In Zambia nearly 88 percent of all conceptions occur from November to April, when almost all the rain falls (Hanks 1979). In Uganda breeding also occurs primarily during the rainy season (Smith and Buss 1973). The Amboseli elephants of Kenya are not highly seasonal, as 80 percent of conceptions take place over the 9 months from February to September, but there are peaks in February and May, which occur 2 months after the rainfall peaks in December and March. Conceptions at Amboseli are clearly a function of females attaining sufficient condition to carry a calf to term, and few females enter estrus in years of lower rainfall when food and water are scarce (Lee, Lindsay, and Moss 2011). It appears that a change in diet from browse and dry grass with a low crude protein content at the end of the dry season to fresh green grass with a high protein content at the height of the rains may stimulate ovulation and fertile mating (Hanks 1979).

The average gestation period is 22 months, with a recorded range of 17–25 months. Long-term studies at Amboseli have confirmed a median length of 660 days, but indicate that calves born after less than 642 days of gestation were premature and that those born more than 679 days after the mother's observed estrus were likely conceived during a subsequent estrus (Poole et al. 2011a). According to Laursen and Bekoff (1978), there is normally a single young, but twins occur in 1–2 percent of births. The newborn weighs 90–120 kg, is about 90 cm tall, and can stand after 15 minutes. Other members of the group wait until the young has the strength to roam with them (usually about 2 days). The young depends on its mother's milk for 2 years, after which it learns to forage for itself, but most calves continue to suckle for as long as they are tolerated (Poole, Kahumbu, and Whyte 2013). Based on 30 years of study at Amboseli, Lee and Moss (2011) reported full weaning to take place at an average age of 55 months and that a calf will be 25–41 months old when its mother conceives her next calf. A calf under 24 months of age will almost always die if its mother is lost, and its chances of survival are still greatly reduced if it is orphaned when under 9 years old, despite support from other members of its family unit. Females of all ages in a family act as "allomothers," cooperating to protect and rear calves; grandmothers can enhance survival of their grand-calves, while at the same time continuing to bear their own young (Moss and Lee 2011a). Allomothers rarely suckle calves aside from their own, but they do maintain close contact with the young, providing reassurance, leading them away from predators and other dangers, and watching to be sure they are not left behind when the family moves on (Lee and Moss 2011). Notwithstanding the size and power of their parents, elephant calves are taken on occasion by lions (Loveridge et al. 2006).

Full adult height is reached at about 30 years for females, but males continue to grow well into their 50s (Poole, Kahumbu, and Whyte 2013). Males may begin producing sperm when 10–14 years old but, under natural conditions, cannot successfully compete for mating until they are more than 20 years old. Female elephants are also evidently capable of reproduction at a relatively early age for such a large mammal, though under natural conditions they usually wait until they are older. A study of 238 females that reached sexual maturity at Amboseli in Kenya determined average age to be 12.2 years at first conception and 14.2 years at first

African savannah elephant (*Loxodonta africana*), group in Amboseli. Photo by Oleg Znamenskiy / Shutterstock.com.

birth (Moss and Lee 2011a). In Zambia's South Luangwa National Park, prior to a severe period of poaching, the mean age of females at first birth was 16 years; following the poaching, in nearby North Luangwa National Park, mean age at first birth was 11.3 years, apparently as a result of social disruption and low population density (Owens and Owens 2009). In another heavily exploited population, in the Samburu National Reserve of central Kenya, the earliest age at first birth was 8.5 years and mean age was 11.34 years, thereby suggesting that reproductive effort increases in response to increased mortality (Wittemyer, Daballen, and Douglas-Hamilton 2013). Females remain fertile until they are 55–65 years old, producing calves at intervals of 2–9 (usually 4–5) years; the shorter interbirth intervals correspond to early death of the first calf. Long-term studies of natural mortality at Amboseli indicate an average life expectancy at birth of 37.4 years for males and 46.7 years for females; the oldest known male was 65 years, the oldest female 69 years (Lee, Lindsay, and Moss 2011).

A few individuals of *L. africana* have been employed in Botswana to carry tourists on safari (Iversen 1995), and the species may have been represented at least in part by the elephants trained in Ptolemaic Egypt for military purposes (see discussion in the above account

of *L. cyclotis*). However, the historical and modern utilization of *L. africana* by people certainly does not compare to that of *Elephas maximus*. Captive breeding of African elephants has been difficult. Dale (2010) noted that the first known birth in Europe was in 1943 at the Munich Zoo and that the first in North America was not until 1978 at the Knoxville Zoo. He recorded data on 67 live births between 1978 and 2008 at 28 zoos and other facilities in North America, Europe, and Japan, but referred to the greatest success having been at the Ramat Gan Zoo in Israel, where one female gave birth six times. Sukumar (2003) wrote that there were fewer than 1,000 African elephants in captivity worldwide and that the population was declining rapidly because of a low birth rate and high death rate. He did not distinguish between the two species, but Wiese and Willis (2006) stated that, while most of the North American captive population descends from *L. africana*, recent studies suggest the presence of some genetic material from *L. cyclotis*. Resolving that issue would be important in developing a conservation breeding program, but there are numerous other problems and general recognition that the overall captive population is not reproducing at a sustainable level. Captive elephant females, generally having been taken from the wild as infants and not fully exposed to allomothering and other natural sociality, often have poor success as mothers and lose their own young (Poole and Moss 2008). In addition to low calf survival, adult females experience a high rate of ovarian acyclicity (Brown et al. 2004; Proctor, Freeman, and Brown 2010), and the existing cohort of reproductively viable females is aging (Hutchins and Keele 2006). While it also has been suggested that longevity of zoo elephants actually is shorter than that of elephants living under fully natural conditions, Wiese and Willis (2004) concluded that the former have a life expectancy at least as great as that of the latter.

Although most of the captive elephant population is wild born, and its low reproductive rate might result in extinction within a few decades (Wiese and Willis 2006), restrictions imposed by CITES severely limit importation and are unlikely to be removed (Dale 2010). Because of the popularity of elephants, legal and political barriers to importation may even increase, thereby seemingly placing greater importance on developing viable captive breeding programs in zoos (Hutchins and Keele 2006). However, there has been a surprising degree of debate regarding the ethics of keeping elephants in zoos and other forms of captivity.

With their huge body, the largest (4.5–6.5 kg) brain among land mammals, remarkable intelligence, extended memory and communication capacity, prolonged rearing period, social intricacies, and emotional complexity, elephants may be capable of stress and suffering, in response to both physical pain and loss of family members, at levels bearing some approach to our own (Brown, Wielebnowski, and Cheeran 2008; Poole and Moss 2008). Opposition to subjecting elephants to what are considered the harsh conditions and social deprivation of captivity has even led to legal efforts to prevent importation of animals that otherwise would have been destroyed in culling operations (Hutchins and Keele 2006). Some zoos have agreed to forego keeping elephants, while others have built elaborate exhibits that provide a degree of comfort for a few individuals but require funding that might have helped protect entire wild populations (Garrison 2008). Responsible zoos carry out research and attempt to develop programs enabling elephant groups to function and breed in what is considered a semi-natural manner (Hutchins, Smith, and Keele 2008). But there remains considerable evidence that most cannot meet physical and social needs, and that some condone treatment

African savannah elephant (*Loxodonta africana*), interaction in Amboseli. Photo by Eduard Kyslynskyy/Shutterstock.com.

highly abusive to the animals and dangerous to humans; on average, one keeper per year is killed by an elephant (*Loxodonta* or *Elephas*) in a North American zoo or circus (Hancocks 2008).

In the wild, the African savannah elephant does not ordinarily threaten people, though conflict and competition occur, especially in areas of expanding agricultural development (Dublin and Hoare 2004; Hoare and du Toit 1999; O'Connell-Rodwell et al. 2000). It has been projected that by 2050, 63 percent of elephant range will be affected by infrastructure development, human population growth and rapid urban and agricultural expansion (Nellemann et al. 2013). Elephants raid human settlements, feed on and trample crops, and destroy farm installations. They occasionally kill or injure people trying to protect their property, force villagers to spend nights in the fields guarding crops, and have even caused schools to close because of danger to children (Kiiru 2008). In Kenya, reportedly, 108 persons were killed by elephants from 1990 to 1993

(Kiiru 1995) and over 200 in the 7 years up to 2006 (WWF International 2006). In perspective, it might be noted that in the year 2010 an estimated 8,484 persons in Kenya died in road traffic accidents (World Health Organization 2013).

Countermeasures against elephants in areas of agricultural development have been devastating. Such killing, together with exploitation for ivory and sport hunting, had eliminated *L. africana* from almost all of South Africa by about 1920 (Hall-Martin 1992) and wiped out elephants in most of the savannah zone of West Africa by about the same time (Barnes 1999; Roth and Douglas-Hamilton 1991). Kingdon (1979) reported that more than 500,000 elephants had been shot in East Africa during the previous 30 years, mostly along the frontiers of agricultural expansion. Large populations continued to exist, but as noted above, they often became unnaturally restricted in their distribution and ability to migrate. In some cases, legal protection allowed substantial increases of formerly depleted

African savannah elephant (*Loxodonta africana*), adult male in musth. Photo by Julian W / Shutterstock.com.

elephant populations in national parks and other reserves but did not provide for the space, food, and water the animals eventually would require (Cumming 1981; Laws 1981; Poché 1980). This process was especially prevalent in the savannah zone of eastern Africa, from Uganda to northeastern South Africa. On the grounds that constrained and often increasing populations were destroying their own habitat and that of other species, the governments of some countries have carried out large-scale programs to reduce elephant numbers. There has been considerable controversy about such culling, though many authorities have accepted its local necessity. Given the continued expansion of people and agriculture, this sort of management is likely to spread throughout the remaining range of the elephant. Unfortunately, therefore, even if some elephant populations are saved, the species may be lost with respect to its natural behavior and continent-wide ecological role.

Western (1989b) pointed out that many field studies of elephants have contributed to a negative image of the ecology of the species, but only because the investigations dealt with artificially compressed populations that were said to be damaging parks or commercial forests. He explained that the elephant has a key role throughout the savannah and tropical forest zones that it inhabits. When numbers and seasonal movements are natural, the elephant opens up dense woodland, prevents the spread of brush, and makes gaps through the jungle. It thereby allows the penetration of sunlight, promotes the growth of a great variety of plant species, and encourages a more abundant and more diverse fauna of smaller animals. If this role is applied across its entire original range, the elephant has been more important than any other species except people in shaping the ecology of Africa. Its loss will reduce biodiversity and increase extinction rates throughout the continent.

The most immediate factor jeopardizing African elephants is the demand for their tusks by the international ivory trade. This problem is not new, as ivory has been an important commodity throughout recorded history (Carruthers et al. 2008; Spinage 1973; Sukumar 2003; Whyte 2005). Egypt was importing from Somalia in 700 BC, and India from Ethiopia in the sixth century BC., and much ivory was still leaving East Africa around 100 AD. However, large-scale European exploitation within the range of *L. africana* probably

did not begin until the sixteenth century when the Portuguese were taking ivory from Angola and Mozambique. Intensive hunting subsequently started near the Cape of Good Hope in the 1600s and spread through what is now Namibia in the 1700s and Botswana and Zimbabwe in the 1800s. By the early 1900s, many local populations of *L. africana* had been extirpated, especially in the southern part of the continent, and there were suggestions that the species might already be nearing extinction. That view may have stemmed from observations in relatively well-known areas and did not give full consideration to the vast numbers of the species, the great extent and diversity of their habitat, and the capacity for a reduced population to recover, once commercial exploitation became unprofitable in its range.

In the early nineteenth century there may still have been as many as 27 million elephants (both *L. africana* and *L. cyclotis*) on 19 million sq km of suitable habitat in Africa. Subsequently there was a dramatic decline in association with habitat modification and hunting, with total numbers falling to fewer than 10 million by the mid–nineteenth century (Milner-Gulland and Beddington 1993). From 1860 to 1930, 25,000–100,000 elephants were killed annually for their ivory, mostly to supply material for the manufacture of piano keys in Europe and the United States (Conniff 1987; Ricciuti 1980; Sukumar 2003). Long after the Civil War, Americans unwittingly fostered a continuation of the slave trade in that the commercial pursuit of ivory was often associated with the capture of slaves, who then were forced to carry the tusks until both commodities were sold. The end of the slave trade and the development of wildlife protection regulations by the early 1900s may have facilitated the partial recovery of remaining elephant populations in eastern and southern Africa. Indeed, Spinage (1973) reported that surviving elephants had entered an exponential phase of population increase. However, Caughley, Dublin, and Parker (1990) suggested that although elephants did increase within national parks through the 1950s and 1960s, they declined over the much larger unprotected regions. Milner-Gulland and Beddington (1993) suggested that overall numbers held at several million for most of the first half of the twentieth century.

A reintensification of commercial exploitation had certainly developed by the 1970s, when the price of

African savannah elephant (*Loxodonta africana*), males sparring. Photo by Juliya Shangarey / Shutterstock.com.

ivory rose along with that of gold and other precious materials in conjunction with the energy crisis. During that decade the price of raw ivory increased from about US$5 to US$50 per kg. As a result, illegal killing became widespread, and many elephant populations, especially in East Africa, were devastated. The estimated number of elephants in Kenya, for example, fell from 167,000 in 1973 to 70,000 in 1977 (Kingdon 1979). There was also renewed conservation interest during the decade, leading to such measures as the placing of *L. africana* on appendix 2 of the CITES (it later was moved to appendix 1 but then partly returned to appendix 2), its classification as vulnerable by the IUCN (it was reclassified as endangered in 1996 but then changed back to vulnerable in 2004), and its listing as threatened by the USDI, along with issuance of regulations that limited but did not prohibit the importation of ivory.

The conservation efforts failed, and by the late 1980s it was generally realized that the world's most spectac- ular land animal was in danger of extinction. Nearly all the major populations that had been monitored since the 1970s were found to be in rapid decline (Douglas-Hamilton 1987). The price of ivory continued to rise, reaching more than US$100 per kg at several times in the 1980s (it would later go far higher), and poaching became more rampant than ever. Annual ivory exports from Africa, which had been about 200 metric tons in the early 1950s, were around 900–1,000 metric tons from 1979 to 1985 (Cobb and Western 1989; Cumming, Du Toit, and Stuart 1990). An estimated 100,000 elephants were being killed each year to supply this market, 80 percent of them illegally (Western 1989c). About the same number had been killed in the worst years of the late nineteenth century, but then the overall populations had been much larger. Most of the ivory went to the Far East, where it was carved into products for local use or export. After Japan, the United States was the largest importer of carved ivory. In 1986 the amount of ivory carvings brought into the United

States represented 32,000 dead elephants. In that year the United States also imported the raw tusks of several hundred elephants and skins representing 11,000 elephants. The annual retail value of elephant products sold in the United States was estimated at $100 million (Thomsen 1988).

Such commerce proved disastrous, with the estimated number of African elephants falling from at least 1.3 million in 1979 to 625,000 in 1989 (Cobb and Western 1989). Again the East African savannahs bore the brunt of the assault, with numerical declines during the period from 65,000 to 19,000 in Kenya, 134,000 to 40,000 in Sudan, 316,000 to 80,000 in Tanzania, and 150,000 to 41,000 in Zambia. Of particular concern was the devastation of the great, relatively undisturbed population in the savannah region of Democratic Republic of Congo; the estimated drop there, thought to have occurred mostly in the late 1980s, was from 170,000 to 21,000 (Alers et al. 1992). Just to the south, in Angola, large numbers of elephants were killed by

rebel military forces, which supported themselves through the sale of ivory (Chase and Griffin 2011).

The adverse effects of ivory exploitation have involved not only a numerical decline but also the upsetting of the age and social structure of populations (Cobb and Western 1989; Ottichilo 1986; Poole and Thomsen 1989; Western 1989c). Since old bulls generally have the largest tusks, they are the main initial target of ivory hunters. In some areas the number of males dropped to only 5 percent of the adult population, thus making it difficult for estrous females to attract a mate. The next victims are the old females, which the family units depend upon for leadership and which are the most successful mothers. Present populations in some areas consist mostly of juveniles less than 15 years old, but even those small-tusked animals are eventually taken. From 1979 to 1987 the average size of a tusk on the market fell from 9.8 kg to 4.7 kg; thus twice as many elephants must be killed to obtain a metric ton of ivory. Hunter, Martin, and Milliken

African savannah elephant (*Loxodonta africana*), old male accompanying group of females and young. Photo by David Steele / Shutterstock.com.

(2004) reported average size to have fallen still further, to just 3.68 kg.

Some authorities predicted near total extinction of *Loxodonta* in the wild, except in a few tightly protected reserves, if the trends of the 1980s continued (Caughley, Dublin, and Parker 1990). Any annual rate of kill in excess of 5 percent, the normal rate at which an elephant population can increase under favorable conditions, will eventually lead to extinction. Analyses by Basson, Beddington, and May (1991) indicated that the maximum sustainable ivory yield is only 1 to 5 percent of the ivory biomass of a pristine population and that even that yield might best be achieved by using only the tusks of elephants that die naturally. And yet in the 1980s ivory hunting was causing an overall loss of 10 percent to the living component of the species, and there were fears that this rate would rise exponentially (Poole and Thomsen 1989). Such concerns led to intensified conservation efforts, including international programs to raise funds and carry out research by the IUCN, the World Wildlife Fund, the New York Zoological Society, and other private groups. The United States Congress passed the African Elephant Conservation Act to control commerce and help support conservation work in Africa. Pursuant to that act, in June 1989, the United States completely banned the importation of ivory. Bans also were established by the European Community and, in part, by Japan. In October 1989 African elephants were transferred to appendix 1 of the CITES, a measure that largely ended legal international trade in ivory. In 1996 the IUCN formally reclassified the species as endangered.

On 18 March 1991, two years after being petitioned to also reclassify African elephants from threatened to endangered, the USDI proposed such a measure but excluded elephant populations in the countries of Zimbabwe, Botswana, and South Africa (Smith 1991). There reportedly had been no substantial declines in those three countries during the previous several decades, and management programs had been set up to cull excessive elephants, market the ivory, and use the profits for conservation. On 10 August 1992 the USDI announced that it would not even proceed with the reclassification in the rest of Africa (Short 1992). This decision may have involved a concession to interests seeking (1) continued interstate commerce in ivory already legally in the United States, (2) continued importation of sport-hunted trophies of *Loxodonta*, and (3) recognition of a potential benefit of regulated commerce in ivory. Through this shortsighted and exploitation-oriented decision the United States not only failed to contribute anything new to elephant conservation but may also have lost its last and greatest opportunity to save all the large wildlife of Africa. An enormous wave of public enthusiasm had developed for the elephant by 1990 and could have been used by

African savannah elephant (*Loxodonta africana*), matriarch with family group. Photo by 2630ben / Shutterstock.com.

American leadership to rally the levels of funding and effort needed for an international environmental campaign of unprecedented scope.

The USDI decision seemed to correspond to an official consensus that the ivory trade, if properly regulated, provides an economic value to the elephant that can encourage and fund its conservation. Starting in 1990 there were repeated proposals, primarily by countries of southern Africa, to move certain elephant populations back to appendix 2 of the CITES and thereby facilitate the export of ivory. Those efforts, opposed by such prominent field authorities as Douglas-Hamilton (1992), were regularly defeated, so that a general ban on international trade in ivory essentially remained in effect. However, at the Conference of the Parties to CITES in June 1997, it was decided to transfer the elephant populations of Botswana, Namibia, and Zimbabwe back to appendix 2 and to provide for resumption of the international trade in ivory under certain conditions. The population of South Africa was moved to appendix 2 in 2000. Proposals by Tanzania and Zambia to transfer their elephant populations to appendix 2 did not obtain necessary support from the CITES parties (Nellemann et al. 2013). Sales from stockpiles of ivory obtained legally in the appendix 2 countries were allowed in 1999 and 2008. Many authorities argue that such sales, and indeed any commerce in ivory, stimulate consumer demand and thus intensified poaching. Others, such as Stiles (2004) believe limited international trade, under strict supervision by governments and CITES, would get ivory away from criminal elements and thus help control poaching. Stiles (2004) noted, however, that improved monitoring and regulatory measures were necessary before commerce could be reopened. Later, Stiles (2014) presented a reasonable case that the 1999 and 2008 sales had not been the primary factors involved in the recent upswing in the ivory market and illegal hunting. He argued further that efforts to completely close markets and destroy ivory stockpiles actually create a perception of scarcity, which in turn leads to illicit activity and price speculation.

Poole et al. (2011b), a team of authorities among those most experienced in long-term studies of elephants, expressed opposition to commercial trade of ivory in any form. They observed that ivory hunting, as currently ongoing, has always been extremely damaging to the very fabric of elephant society, because it targets animals with the largest tusks and thus eliminates most of the older, more knowledgeable females and reproductively fit males. The result is devastating to the psychological state of remaining individuals and the normal structure of families. Females and young may form small units but with little cohesion. Groups of orphans wander in leaderless groups, with little chance of long-term survival. Any promotion of the use of ivory or the legitimization of its commercial value will inevitably stimulate demand and increased killing. While officials in some relatively stable countries claim that the sale of ivory derived from carefully controlled legal hunting, or from existing stockpiles, could benefit elephants, in reality such revenue is sporadic and contributes little to conservation needs. Meanwhile, in other countries beset by war, corruption, or limited resources, the slaughter will continue as long as there is a demand for ivory. Poole et al. (2011b) also went on to condemn trophy hunting, which destroys the primary breeding males and disrupts the psychological and genetic health of affected populations, while most licensing revenue goes not to conservation, as often claimed, but to commercial and political interests.

Dublin, Milliken, and Barnes (1995) indicated that the CITES ban had not halted illegal killing, that poaching may even have been increasing in some areas, and that the ability of the African countries to enforce protective measures was severely limited and actually deteriorating in effectiveness. However, there seems little doubt that the CITES ban, while not fully effective, did help prevent a total collapse of the last major populations of *L. africana*. Douglas-Hamilton (2009) observed that the price of ivory dropped dramatically following the ban and that in the following two decades there was a recovery of formerly decimated key elephant populations. This is shown to some extent by the numerical estimates resulting from periodic surveys carried out under the aegis of the African Elephant Specialist Group of the IUCN Species Survival Commission.

Those figures, which are broken down into the categories, "definite, probable, possible, and speculative" (Blanc et al. 2007), are sometimes misunderstood. It may not be correct that the best total estimate for an area is limited to the "definite" number or to "definite" plus "probable." Often, the only available figures fall into the "possible" and/or "speculative" categories. Moreover, the range areas where the estimates apply are categorized as "known, possible, and doubtful." In particular, "possible" refers to areas of suitable habitat in historical range, for which there are no data that would rule out elephants. A reasonable guess for such an area might be considered "speculative" but may be the best information available. The four categories of numbers are not overlapping; they must be combined to get a total for the maximum number that might be present. However, the totals presented for each country are minimum estimates; there are large areas of elephant range for which numbers have not been estimated. Considering the various caveats, it seems not unreasonable to use the combined figure of the four categories to provide an idea of overall trends. The publications do not separate numbers for *L. africana* and *L. cyclotis*, but do provide maps and tables of sites that allow figures for each species to be extracted. Also, because of calculation methodology, the published

IUCN totals do not always agree with totals derived by adding numbers for individual areas.

For three successive surveys, the estimated total numbers of *L. africana* throughout its range were: 394,765 (Said et al. 1995), 494,757 (Blanc et al. 2003), and 591,200 (Blanc et al. 2007). Those figures, while affected heavily by variations in technique and availability of data, do suggest the CITES ban was working and that recovery was underway, at least in some areas. The indicated numerical increases correspond well with an annual rate of 4 to 5 percent, which is considered normal for an elephant population under favorable conditions (see above).

It must be noted, however, that the increases were restricted entirely to parts of eastern and southern Africa. On the savannahs of western and central Africa, *L. africana* was undergoing a decline comparable to that in the nearby forests occupied by *L. cyclotis* (see account thereof). Bouché et al. (2011) reported a maximum of 12,533 elephants remaining in the entire

African savannah elephant (*Loxodonta africana*), mother and calf. Photo by Villiers Steyn/Shutterstock.com.

Sudanian/Sahelian region (habitat of *L. africana*) extending from Senegal to Chad and the northern Central African Republic. The animals were concentrated in nominally protected areas, but populations were thought to have declined there by 50 percent from four decades earlier. Surviving groups were mostly isolated, surrounded by agriculture, and hence subject to human animosity. Of the 23 surveyed elephant populations, half were estimated to number less than 200 individuals; historically, most groups of that size were extirpated within a few decades. To the south, the savannahs of Democratic Republic of Congo, which may have had 170,000 elephants in the 1970s, were reported by Blanc et al. (2007) to have not more than 10,845. Most of those were in and around Garamba National Park in the northeast; *L. africana* had been practically swept from the savannah zone in the southern half of the country. In neighboring Angola, only a few fragmented populations survived after many elephants, perhaps over 100,000, were killed during the civil war that began in 1975 (Chase and Griffin 2011). A maximum of 2,530 for Angola was reported by Blanc et al. (2007); nearly all were in the extreme southeastern corner of the country, within the Kavango-Zambezi Transfrontier Conservation Area (see above).

The situation has been no better in the northeastern part of the range of *L. africana*. Because of civil war, illegal hunting, and human population movements, the number of elephants in what is now the country of South Sudan fell from around 133,000 in the 1970s to a maximum of 40,000 by 1992 (Said et al. 1995). According to the IUCN African Elephant Specialist Group (2013), only about 13,000 were still present, while another 1,500 were in neighboring western Ethiopia, 100 in extreme western Eritrea, and perhaps 70 in extreme southern Somalia. The distinctive subspecies *L. a. orleansi*, which formerly occurred across much of northern Somalia and eastern Ethiopia (Yalden, Largen, and Kock 1986), has declined steadily in the past century and now is found only in and around Babile Elephant Sanctuary in east-central Ethiopia, where 324 individuals were counted by Demeke, Renfree, and Short (2012).

Therefore, the numerical increases, following the CITES ban on most international ivory trade, were restricted to the savannahs extending from Uganda and Kenya south to northeastern South Africa and then west through Botswana to Namibia. And much of that region would soon see disaster return. Analysis by Wittemyer et al. (2014), utilizing demographic data and reports from the MIKE (Monitoring the Illegal Killing of Elephants) sites that had been established across Africa in 2002 by CITES, showed that poaching for ivory began to increase markedly around 2008 and peaked in 2011, when approximately 40,000 elephants were killed. Many of those were *L. cyclotis* in central Africa, but most were *L. africana* in the east and south, where populations began to decline in 2010, when the rate of poaching reached unsustainable levels. Nellemann et al. (2013) reported that both poaching and illicit trade in ivory had increased sharply since 2007 and were comparable to the 1970s and 1980s. The MIKE sites of eastern Africa showed a three-fold increase in illegal kills from 2006 to 2011. The CITES Secretariat et al. (2013) stated that the MIKE analysis indicates that elephants were being illegally killed at a rate of 7.4 percent in 2012, exceeding natural population growth rates, which are usually no more than 5 percent. The overall weight and number of large-scale seizures (more than 500 kg) of illegal ivory by law-enforcement in 2013 exceeded those of any previous year. The CITES Secretariat (2014) noted that poaching numbers in 2013 were lower but remained at unsustainable levels, with a corresponding decline in elephant populations. Assessment of the most recent survey by the IUCN African Elephant Specialist Group (2013) indicates the total number of *L. africana* as 544,410, down about 7.9 percent from the previous survey (Blanc et al. 2007). Of particular concern is Tanzania, where the count fell from 167,003 to 117,456—a drop of 29.7 percent.

The intensified killing was precipitated by a rise in the price of ivory that was so overwhelming that even some of the foremost authorities on the subject (Stiles, Martin, and Vigne 2011) initially expressed doubts about the figures and cautioned that circulation of inflated values could further stimulate poaching. However, firsthand assessment showed that the average price paid by craftsmen or factory owners in Beijing for good quality tusks was US$750 per kg in 2010 and US$2,100 per kg in early 2014 (Gao and Clark 2014; Stiles 2014). Poachers themselves received as little as US$50 per kg, while illicit middlemen dealers in Africa sold tusks to Chinese importers for US$250–$300 per kg.

African savannah elephant (*Loxodonta africana*), adult male. Photo by Johan Swanepoel / Shutterstock.com.

Nellemann et al. (2013) noted that poaching is sometimes facilitated by armed conflicts that provide an abundance of weapons and optimal conditions for illegal activity. Farther along the trade chain, highly organized criminal networks operate with relative impunity, moving large shipments of ivory from Africa to markets in Asia.

The CITES Secretariat et al. (2013) stated that elephant poaching is ultimately driven and sustained by demand from consumers willing to pay for illegal ivory. In particular, recent commercial intensification seems associated with a rise in available household wealth in China, which has become the world's largest consumer of illegal ivory. The current situation is comparable to development of a prosperous Western middle class in the late nineteenth century, which saw pianos with ivory keys as a status symbol, hence amplifying elephant hunting of the period (Conniff 1987; Ricciuti 1980), and to the burgeoning wealth of Japan in the 1970s, which was connected to a demand for ivory

hanko (personal signature seals) and thus to a steep rise in ivory importation and to the enormous illegal slaughter of elephants at that time (Milner-Gulland and Mace 1991; Sukumar 2003). Japan was then the world's largest ivory market, and the United States was second. The Japanese market has since faded, but to this day the United States ivory market is the second largest in the world (Nellemann 2013). Restrictions by CITES and national law apply only to importation; it is still legal to work and sell African elephant ivory that entered the United States prior to 1989.

In contrast, the market in China was dormant for much of the twentieth century, but recently increased to the extent that the country is now the world's largest destination for illegal ivory. Nellemann (2013) explained a remarkable correlation between elephant poaching, the Chinese demand for ivory, and changes in wealth and consumer spending patterns. While China's economy had grown exponentially for 20 years, much of that new wealth was saved rather than spent, with savings rates increasing sharply from 1990 to 2006. In that last year, growth in savings stalled, while private consumption rose sharply. And it was the following year when the ominous rise in poaching and illicit trade in ivory was first recorded. According to Gao and Clark (2014), there had been a revival of the once nearly extinct Chinese ivory carving industry owing to a social movement focused on preserving China's cultural heritage. The previous decade had also seen a boom in investment in art. Owning or being able to present an ivory carving as a gift, especially a high-end piece, bestows a sense of prestige (or "face"). A 2013 National Geographic survey, polling 600 mid- to upper-income consumers in nine of China's largest cities found that 84 percent planned to buy ivory goods in the future (Bredar 2013). It is such a demand, and such a mindset, that would need to be modified if *Loxodonta* is to persist in anything approaching a natural state. As Stiles (2014) observed: "With 1.4 billion people and an economy that will soon be the largest in the world, coupled with a cultural desire for ivory, the Chinese control the fate of elephants."

PROBOSCIDEA; ELEPHANTIDAE; **Genus**
ELEPHAS Linnaeus, 1758

Asian Elephant, or Indian Elephant

The single living species, *E. maximus*, originally occurred in much of, and still is found in parts of, India, Sri Lanka, Nepal, Bhutan, Bangladesh, Burma, southern China, Laos, Thailand, Cambodia, Viet Nam, the Malay Peninsula, and Sumatra (Sukumar 2003). For information on the extinct pygmy species, *E. falconeri* or *E. tiliensis*, sometimes said to have survived into early historical time on the Greek island of Tilos, see account of family Elephantidae.

The historical limits of the range of *E. maximus* are not well known, and there is some question as to

Asian elephant (*Elephas maximus*). Photo by Dennis Jacobsen / Shutterstock.com.

whether records represent natural occurrence or introduction of animals or parts thereof by human agency. To the west, it is probable that the species occupied the Indus Valley of Pakistan around 4,000 to 5,000 years ago and that it was present along the southern coast of Iran (Sukumar 2003). There also are written accounts and skeletal remains showing the range included Iraq, Syria, and southeastern Turkey until about 3,000 years ago (Albayrak and Lister 2012; Becker 2005; Bishop 1921; Bökönyi 1985; Lister et al. 2013; Osborn and Osbornova 1998; Sukumar 2003). There have been suggestions that the elephants of ancient Southwest Asia actually represented *Loxodonta* or even *Mammuthus* (Corbet 1978; Sikes 1971), though now this does not appear likely. Tsahar et al. (2009) did not report the remains of any kind of elephant from the Holocene of the region now governed by Israel, Jordan, and the Palestinian National Authority. At the other end of the range of *E. maximus*, numerous accounts and specimens indicate the species was common in the Yellow River Valley of northern China until around 3,000 years ago and that it persisted in the Cháng Jiāng (Yangtze) Valley of central China until about 2,300 years ago (Bishop 1921; Tong and Patou-Mathis 2003; Wang 2011; Zhang, Ma, and Feng 2006).

There is less certainty regarding distribution on the islands of Southeast Asia, other than Sri Lanka and Sumatra. Corbet and Hill (1992) referred to the occurrence of *E. maximus* on Java as "prehistoric," but Cranbrook, Payne, and Leh (2007) believed the population there survived past 1350 A.D., possibly to the end of the eighteenth century. The species also is reported to have occurred on Borneo in the Pleistocene but to have disappeared, and a living population in the northeastern part of that island, primarily in Malaysia's Sabah state, is usually considered to have resulted from introduction of domesticated work animals several hundred years ago. However, based on a study of mitochondrial DNA, Fernando et al. (2003) indicated that the current population of Borneo diverged from other Asian elephants (including those of Sumatra) around 300,000 years ago and thus is a remnant native group. Cranbrook, Payne, and Leh (2007) agreed with the distinction but noted that paleontological and archeological evidence for Pleistocene to late Holocene occurrence was very limited and questionable. They provided information on the historical transport of elephants for use as work animals or gifts to royalty, which supported the view, first proposed by Shim (2003), that the living animals on Borneo actually had descended from the extinct Javan population. Furthermore, they suggested that elephants had first been brought from Java to the island of Sulu in the southwestern Philippines, where they bred and founded a feral population. Some were subsequently brought to Borneo, where they in turn founded the existing population, though the Sulu lineage itself was extirpated in the eighteenth century through conflict with local farmers. Another small feral population, descended from work animals released in 1962, still exists on Interview Island in the Andamans, south of Burma (Ali 2005).

Shoshani (2005b) and Shoshani and Eisenberg (1982) recognized three subspecies of *E. maximus*: *E. m. indicus* of mainland Asia, *E. m. maximus* of Sri Lanka, and *E. m. sumatranus* of Sumatra, the last being distinguished by having 20 pairs of ribs. Corbet and Hill (1992) wrote that *E. m. maximus* is distinguished by being tuskless but that all other Asian elephants could be assigned to a single subspecies, for which they used the name *E. m. bengalensis*. Kurt, Hartl, and Tiedemann (1995) noted that tuskless bulls are generally rare in *Elephas*, but that on Sri Lanka 93 percent of males lack tusks. They suggested the condition was the result of a long period of selective human hunting and capturing of tusked males in the isolated Sri Lankan population. Modeling by Sukumar (2003) indicated that tusklessness may have developed through genetic drift in the Pleistocene, though tusked males, which had been imported and then escaped, could have reintroduced the trait that now shows in a small proportion of wild Sri Lankan males. From ancient times until the early nineteenth century, there was an extensive trade in elephants between Sri Lanka, India, and other areas. Studies of mitochondrial DNA by Fernando et al. (2000) and Fleischer et al. (2001) indicate there are two major evolutionary clades of *E. maximus*, possibly originating from separate ancestral species, which now have been somewhat intermingled through human transport. Those studies support subspecific status for the population on Sumatra but not that on Sri Lanka. A more comprehensive mitochondrial DNA assessment by Vidya, Sukumar, and Melnick (2009) indicated both clades originated from a single ancestral species and that their present genetic structure has not been altered

Asian elephant (*Elephas maximus*), individual of Bornean population; note long tail. Photo by Tomáš Peš.

by human trade. That assessment also supported genetic distinction of the population on Borneo, for which the name *E. maximus borneensis* had once been proposed. However, both Fernando et al. (2003) and Shoshani (2005b) believed that formal subspecific designation should await morphological analysis. Actually, if the population of Borneo had originated from Javan stock, as suggested by Cranbrook, Payne, and Leh (2007), the correct name for the relevant subspecies might be *E. m. sondaicus*, which had once been proposed for the extinct wild population of Java (see above). An additional subspecific name, *E. m. asurus*, still is sometimes used, as by Becker (2006), for the population that in ancient times inhabited what is now Iraq, Syria, and Turkey.

In the species *E. maximus* head and body length is 550–640 cm, tail length is 120–150 cm, and shoulder height is 250–320 cm. Shoshani and Eisenberg (1982) reported that females weigh an average of 2,720 kg and that large bulls weigh 5,400 kg, though Wittemyer (2011) gave average weight of males as around 3,600 kg. One exceptionally huge circus animal reportedly weighed 6,700 kg (*Washington Post*, 31 March 1988, D-9). There recently have been reports of the existence of a population of exceptionally large animals in western Nepal, in which males have an estimated height of up to 343 cm (Lister and Blashford-Snell 2000), and of a "pygmy" form of *E. maximus*, with an adult height of just 150 cm, in Kerala State of southern India (Srivastava 2005). The elephant of Borneo is also sometimes referred to as a "pygmy" race, but available measurements show that, while up to a fifth smaller than the elephant of India, it is about the same size as those of the Malay Peninsula and Sumatra (Cranbrook, Payne, and Leh 2007).

The hair covering of *E. maximus* is scant; the hairs are long, stiff, and bristly (see discussion of hair in account of family Elephantidae). There is a tuft of hair at the tip of the tail. The coloration of the skin is dark gray to brown, often mottled about the forehead, ears, base of the trunk, and chest, with flesh-colored blotches that are perhaps caused by some skin disease. True coloration of the skin often is masked by the color of the soil on which the animal lives, as it constantly throws dirt over its back and wallows in the mud. *Elephas* has a heart with a double apex that often is confused with a "double heart."

Elephas is distinguished from *Loxodonta* by considerably smaller ears, 19 pairs of ribs (20 in *E. maximus sumatranus*), and 33 caudal vertebrae. Its forehead is flat, and the top of the head is the highest point of the animal. *Loxodonta* has large ears, 21 pairs of ribs, and a maximum of 26 caudal vertebrae. The forehead is more convex, and the back more sloping, so the shoulders are the highest point. In addition, the trunk of African elephants has two finger-like processes at its tip in contrast to the single tip of the trunk of *Elephas*. Both genera begin life with 5 hooves (nails) on each foot, though typically the hind foot of *Elephas* shows 4 by adulthood, as compared to as few as 3 in *L. africana* but 4–5 in *L. cyclotis*. Normally, tusks are absent in female *Elephas* but present in female *Loxodonta*.

The Asian elephant originally occurred in a wide variety of habitats, including grassy plains, thick tropical forest, semi-evergreen forest, moist deciduous forest, dry deciduous forest, dry thorn forest, and cultivated and secondary forests and scrublands. Across that range of habitats, elephants are seen from sea level to over 3,000 meters in elevation (Choudhury et al. 2008). According to Shoshani and Eisenberg (1982), populations of the species in southern India and Sri Lanka are now restricted to single-monsoon, dry, thorn-scrub forest. Overall distribution is patchy and limited mostly to forest-grassland ecotone. Seasonal migrations were formerly extensive but now have been reduced by human agricultural development. Seasonal movements of 30–40 km still occur in southern India and

Asian elephant (*Elephas maximus*), adult male and female. Photo by Rudy Umans / Shutterstock.com.

Sri Lanka. Elephants usually do not feed for more than a few days in any one place. There are two major feeding peaks every 24 hours. The animals feed and move about during the morning, evening, and night and rest during the middle of the day. Drinking takes place at least once a day, the animals never being far from a source of fresh water. Shade is essential during much of the day, and excess heat is radiated through the ears, which are incessantly moving and flapping. Mean daily movements of around 1 or 2 km were found by Alfred et al. (2012) for individual females in Borneo. Through analysis of video of captives urged on by mahouts, Hutchinson et al. (2003) determined that *Elephas* can move at speeds of 25 km/hour, using an unusual gait in which the vertical movements of the shoulder indicate walking, but hip motion indicates running. D. L. Johnson (1980) reported that the elephant is an excellent swimmer, covering up to 48 km at speeds of up to 2.7 km/hr. Chevalier-Skolnikoff and Liska (1993) cited a number of instances of "tool use" by *Elephas*, including the manipulation of sticks and branches with the trunk to scratch the body or repel insects and the accurate throwing of objects in an apparent effort to repel people or other animals. The senses of hearing, vision, and olfaction are thought to be well developed. Elephants have the best low-frequency hearing of any mammal yet tested, with hearing in the low frequencies 10–100 times better than that of humans (Langbauer 2000).

Asian elephant (*Elephas maximus*) feeding. Photo by Anan Kaewkhammul/Shutterstock.com.

McKay (1973) found that in Sri Lanka *Elephas* consumes a wide variety of grasses and also large amounts of bark, roots, leaves, and small stems. Cultivated crops such as bananas, paddy, and sugar cane are favored foods, and as a result the elephant often becomes a pest in agricultural areas. An adult Asian elephant appears to have a daily food intake of approximately 150 kg net weight. When feeding on long grasses, the animal uses its trunk as a "hand" to grasp a number of stems, which it pulls up and inserts directly into its mouth. When grasses are too short to be picked up in this way, the elephant scrapes the ground with its forefeet until a loose pile of grasses is formed, then sweeps the pile into its mouth with the "hand" of the trunk. When feeding on shrubs the elephant uses the end of the trunk to break off small twigs and branches. It removes bark from larger branches by inserting the branch into its mouth and then, with a turning motion of the trunk tip, rotating the branch against either the molars or a tusk, thus stripping the bark.

Shoshani and Eisenberg (1982) noted that overall population densities usually vary from 0.12/sq km to 1.00/sq km but can be greater than 7.00/sq km when temporary aggregations form during the wet season. In southern Sri Lanka, long-term densities of around 0.50 to 1.00/sq km have been reported for Ruhuna National Park (Katugaha, de Silva, and Santiapillai 1999) and close to 2.00/sq km in Uda Walawe National Park (de Silva, Ranjeewa, and Weerakoon 2011). Fernando et al. (2008) and Sukumar (2003) questioned some early studies that reported relatively small home ranges, noting that modern use of radio telemetry for extended periods generally, though not always, indicates larger areas. Overall range size, like that of *Loxodonta africana* differs widely, depending on habitat productivity, with a cited variation from 32 sq km in primary forests of Malaysia to around 3,000–4,000 sq km in dry fragmented forests of northern India. Sukumar (1989a, 1989b) found overlapping home ranges of 105–320 sq km in southern India. Based on satellite tracking in Borneo, Alfred et al. (2012) reported minimum annual home range of groups to be 250–400 sq km in nonfragmented forest but around 600 sq km in forests that had been fragmented by human activity. Fernando et al. (2008) considered the mean home range of 10 elephants radio-tracked in southern Sri Lanka, 115 sq km, to be relatively small, possibly in response to an abundance of successional vegetation and perennial water sources. However, ranges of males in musth were considerably larger, in one case 564 sq km during a 3-month period, and may signify a search for mates. Overlap of ranges was high, and, in line with all other reports on elephant behavior, territoriality was absent. The elephants showed high fidelity to their ranges and no geographically distinct seasonal movements or migratory behavior. In contrast, McKay (1973) had reported that each group he studied in Sri Lanka occupied an overall area that included a rainy-season range and a dry-season range.

Elephas is gregarious, and, while bulls sometimes live alone, cows are always members of a group (though they may often appear individually). Fernando and Lande (2000) considered social structure to differ substantially between the Asian and African savannah elephant, mainly because in the former they found no evidence of organization above independent family units, but, as they acknowledged, their sample (four

groups in Sri Lanka) was too small to be conclusive. Based on more comprehensive studies of groups in both Samburu National Reserve in Kenya and Uda Walawe National Park in Sri Lanka, de Silva and Wittemyer (2012) emphasized that the *multitiered* society of *Loxodonta africana* is different from the *multilevel* society of *Elephas*. The latter is found in smaller groups, does not maintain coherent core groups, demonstrates markedly less social connectivity at the population level, and is socially less influenced by seasonal differences in ecological conditions. The Asian elephant does, however, maintain a complex, well-networked society, in which an individual may have relationships with many others.

Observations by Sukumar (2003) in southern India suggest five possible levels in the social organization of *Elephas maximus*, thus indicating some degree of similarity with the tier structure reported for *Loxodonta africana* (see account thereof): family, consisting of a single adult cow and her offspring; joint family, with

Asian elephant (*Elephas maximus*), group. Photo by Guillermo Pis Gonzalez / Shutterstock.com.

two or more related adult cows and their young; bond group, with two or more related families; clan, with 50–125 animals of several families that may associate on occasion; and subpopulation, with sometimes over 200 elephants that seem to have a broadly coordinated movement pattern under certain environmental conditions. Those groupings apparently correspond to the mother-calf association, family unit, bond group, clan, and aggregation, as discussed above in the account of *L. africana*, though study of *Elephas* in this regard has been much less intensive. Moreover, Sukumar (2003) noted that elephants in the rainforests of peninsular Malaysia do not seem to have any higher social level than the family or bond group, and that group sizes there are smaller than in the South Asian dry forests. Vidya and Sukumar (2005a) suggested organizational similarity between Asian rainforest elephants and *L. cyclotis* of the African forests. Based on 23 years of study of *E. maximus* in Ruhuna National Park in southern Sri Lanka, Katugaha, de Silva, and Santiapillai

(1999) reported the mean size of all groups consisting of females and young to be 7.5 (range 3–21).

As with *L. africana*, social organization of *Elephas* is primarily matriarchal. Groups consist of related mothers, daughters, and sisters, and their young (Vidya and Sukumar 2005b). Movements are initiated by the oldest and usually the largest female (Shoshani and Eisenberg 1982). In Sri Lanka, McKay (1973) found groups to contain both "nursing units," consisting of lactating females and their infants, and "juvenile care units," containing females with juveniles. There was a degree of flexibility within these units in that the females might associate at a given time or divide into smaller subunits. The units and subunits of each group remained in close proximity but moved away from one another when foraging. All unit members occupying a common range were considered to compose a "herd," probably a term equivalent to "bond group" or "clan." Such herds tended to remain distinct from all other herds occurring in the same general area.

Asian elephant (*Elephas maximus*), mother and young. Photo by Petra Christen / Shutterstock.com.

McKay (1973) found that in Sri Lanka at least one adult male was present in 40 percent of the herds observed and that an adult male was associated with any one herd 25–30 percent of the time. However, for 23 years of observation reported by Katugaha, de Silva, and Santiapillai (1999), 80 percent of groups had no adult bull. Males stay with the group only when one or more females are in estrus. Males that are not associated with the female group usually remain solitary and disperse over relatively small, widely overlapping home ranges. Sometimes they gather in small but temporary bull units. In Sri Lanka, such groups with as many as seven animals have been observed (Kurt 1974). Males exhibit a high level of tolerance for one another except possibly when cows are in estrus. If there is more than one male in a group, a dominant individual is soon recognized, and serious fighting is rare (Shoshani and Eisenberg 1982). Young males appear to leave the matriarchal groups and become solitary at about the time of puberty, whereas young females remain in their natal units (Vidya and Sukumar 2005b).

De Silva (2010) identified 14 distinct vocalizations in a wild population of *Elephas*, most of which are comparable to those of *Loxodonta*. "Rumbles" and "growls," which may be inaudible to humans, are used most often and facilitate social contact and coordinated movements in dense vegetation. "Trumpets" always occur in response to disturbance and sometimes are accompanied by other threats, such as throwing objects with the trunk or charging. "Barks," evidently unique to *Elephas*, are short, spontaneous calls accompanying group movement or aggression. "Roars" and "longroars" are given primarily during movement but also disturbance and distress. "Croak-rumbles" and "chirp-rumbles" seem to be associated with contact between females and mothers and calves. "Bark-rumbles, roar-rumbles," and "longroar-rumbles" occur mainly during movement and searching. "Squeaks" and "squeals" appear unique to Asian elephants, particularly females, and are associated with disturbance and excitement. Squeaks usually are accompanied by blows (loud, rapid exhalations) and trunk-bounces, where the trunk is curled beneath the chin and then rapidly extended so that the leading edge hits the ground loudly. The "musth chirp-rumble" appears unique to male *Elephas* and is associated with courtship of and competition for females.

As with *Loxodonta*, communication in the Asian elephant involves much more than vocalization. Apparently, rumbles are transmitted not only as sound waves through air, but at slower speeds as seismic waves through the ground. Foot stomping may also function to send such messages. Anatomical features of the elephant's foot, bones, skull, and inner ear seem particularly suited to receive seismic transmission. Observations suggest *E. maximus* uses this ability to learn of the approach of conspecifics and even to detect earthquakes (O'Connell-Rodwell, Arnason, and Hart 2000; O'Connell-Rodwell, Hart, and Arnason 2001). Olfactory communication is also important. The Asian elephant emits large amounts of complex chemical mixtures through urination, exhalation, and secretions from the temporal gland, interdigital glands, and ears (Rasmussen and Krishnamurthy 2000). Cohesiveness and harmony of matriarchal groups are strengthened by female-to-female chemical signals (Rasmussen 1999). Females advertise a forthcoming ovulation through steadily increasing concentrations of certain urinary pheromones. Older males are apparently experienced at detecting the precise reproductive status of females. During musth, the males themselves secrete fluid copiously from their temporal glands, located on each side of the head between the eye and ear, and dribble strongly odoriferous urine; females can assess those fluids and are evidently drawn to them (Rasmussen, Krishnamurthy, and Sukumar 2005; Rasmussen and Schulte 1998).

While musth has been described only recently in *Loxodonta africana* (see account thereof), the phenomenon has been well known in *Elephas* since ancient times (Sukumar 2003). Captive bulls in musth become aggressive, fail to respond to commands from their keepers, and thus are kept restrained for the duration of the period. In southern India, males have been observed to first enter musth when 15 years old, though until age 25 intensity is low and duration only a few days to weeks. Vidya and Sukumar (2005a) noted that males in musth have significantly higher levels of testosterone and other androgens than males not in musth. Rasmussen and Greenwood (2003) found that the temporal secretions, urine, and exhalations of an Asian elephant in musth contain the pheromone frontalin, the concentrations of which become increasingly evident as the animal matures. Olfactory detection of frontalin appears to

Asian elephant (*Elephas maximus*), mother with young of two ages. Photo by Utopia_88/Shutterstock.com.

warn other males to keep away and thus avoid costly combat, but to attract females approaching ovulation.

Elephas is polyestrous, the estrous cycle is 14–16 weeks, and estrus lasts 3–7 days (Sukumar 2003). Eisenberg and Lockhart (1972) and McKay (1973) found no evidence of reproductive seasonality in Sri Lanka. However, Santiapillai, Chambers, and Ishwaran (1984) reported that in Ruhuna National Park, an area of relatively low rainfall in southern Sri Lanka, mating seems to coincide with the dry season (June–September) and parturition with the rainy season (October–January). Katugaha, de Silva, and Santiapillai (1999) noted that in 23 years of study in that area, 85 percent of the young were observed from December to May, when the park receives most of the annual rainfall. Reproduction is also seasonal in nearby Uda Walawe National Park, where most births occur during the long inter-monsoon dry season from April to July (de Silva et al. 2013). An investigation of demographic data on 2,350 semi-captive elephants in the logging camps of Burma showed that 41 percent of births occurred from December to March, the cool, dry period and the beginning of the hot season, with conceptions occurring during the resting, non-logging period from February to June (Mumby et al. 2013).

Dittrich (1966) reported that 32 births of male calves in European zoos and circuses indicated gestation periods of 615–668 days, with an average of 644 days, and that 17 births of female calves indicated gestations of 628–668 days, with an average of 648. There is normally a single calf, which at birth weighs 50–150 kg (the average is about 107 kg). Newborn baby elephants have a coat of widely spaced brown hairs that produces a halo effect as it stands out from the body. As the animal grows older the hairy coat becomes less noticeable (adult elephants may appear to be naked, but throughout life they have a scattered coat of hair). Shoshani and Eisenberg (1982) wrote that females may produce a calf every 2.5–4.0 years in favorable habitats and every 5–8 years in other areas. An infant can stand shortly after birth and can follow its mother in her daily routine after a few days. It suckles with its mouth (not the trunk) from its mother or another lactating cow. After several months it begins to eat grass and foliage. De Silva et al. (2013) reported that in the Uda Walawe population, calves are usually weaned as soon as a sibling is born, when they are around 4 years old, but that a few may nurse longer and that some females were observed nursing two calves simultaneously.

According to Shoshani and Eisenberg (1982), young males are capable of some independent movement at 4 years, and by 7–8 years they may form subgroups of their own or temporarily associate with older bulls. Growth slows in females at 10–12 years, in males at 15 years, and full size is attained at about 17 years. Males may become sexually fertile at as early as 9 years but usually do not reach maturity until 14–15 years, and even then they are not capable of the social domi-nance usually necessary for successful reproductive activity. Sukumar (1989a) indicated that bulls may not be able to mate before they are 20–25 years and that cows in India first give birth at an average age of 17–18 years. However, Sukumar (2003) noted that in Sri Lanka mean age of female maturity was likely to be 10–12 years. In a relatively undisturbed population in Uda Walawe National Park, Sri Lanka, average age at first birth was 13.4 years, median interbirth interval was approximately 6 years, and fecundity remained constant until around the age of 60; some females lived to over 65 years (de Silva et al. 2013). In contrast, semi-captive females working in the timber camps of Burma showed a decline in fecundity after the age of 19 (Hayward et al. 2014). One captive female in India last calved when she was 62 years old and then lived to 75, and another female died at an estimated age of 79 (Sukumar 2003; Sukumar et al. 1997). A male elephant in the Taipei Zoo died at the age of 86 years (Wiese and Willis 2004).

Asian elephant (*Elephas maximus*), adult male in musth. Photo by Dr. Ajay Kumar Singh / Shutterstock.com.

Historically, *Elephas* has bred more readily in captivity than has *Loxodonta*. Circus females in the United States bore surviving young as early as 1880 (Kreger 2008). In Western zoos, the first known live birth of an Asian elephant was at Buenos Aires in 1905, the first in Europe at Vienna in 1906, and the first in the United States at Salt Lake City in 1918 (Dale 2010). Although more births followed, and the North American population of *Elephas* increased steadily, most of the growth was the result of importation from the wild or other captive facilities. As importation slowed in the 1980s and 1990s and older animals died, overall numbers began a decline. To sustain the current population of about 150 elephants in North American zoos would require at least 8 births per year, a major increase over the present average of 3 per year (Faust, Thompson, and Earnhardt 2006). Internationally, at least 870 Asian elephants are kept in zoos and circuses outside of countries where they still occur naturally, but there are no self-sustaining captive populations (Kurt, Mar, and Garaï 2008). Efforts to reach such a level have been frustrated by imbalanced sex ratios, low female fecundity, high calf mortality, inadequate accommodation, and emergence of new diseases in zoo animals (Hedges et al. 2006; Rees 2003). Moreover, as discussed above in the account of *Loxodonta africana*, much controversy centers on the probity of even keeping elephants in captivity. Of particular concern regarding the Asian elephant, which is the species more commonly used in entertainment, is whether circus animals undergo abusive training techniques, poor maintenance, and careless transportation (Garrison 2008) or live comfortably within a benevolent performing community partnered with conservation organizations (Schmitt 2008). Recently, Ringling Bros. and Barnum & Bailey Circus, although having prevailed in a lawsuit alleging mistreatment of elephants (*Washington Post*, 17 May 2014, A-14), announced it would phase the animals out of acts (*Washington Post*, 6 March 2015, A-3). Subsequently, the Circus indicated it would close down entirely in May 2017, in large part because of a sharp drop in ticket sales after elephants were removed (*Washington Post*, 16 January 2017, A-3).

It is recorded that elephants performed in circuses in China during the reign of Emperor Wu Di, 141–87 BC, and that they had been used for labor and warfare during the earlier Shang dynasty, 1600–1046 BC

(Wang 2011). The history of human taming of *E. maximus* goes back much farther, possibly to 6000 BC, though the first substantive archeological records are from the Indus River civilization around 2500 BC, and people probably were riding elephants there by 1500 BC (Lahiri Choudhury 2008). An encompassing "elephant culture" subsequently spread across the Indian subcontinent and into Southeast Asia and saw both the incorporation of *E. maximus* into religious ritual and its use as a devastating machine of war (Sukumar 2003). The military role of elephants in the region was on a far greater scale than in the West and extended over a much longer period. In some battles, each side fielded thousands of the animals, which were commonly armored, directed by a mahout riding on the neck, and equipped with a fortified platform on the back for archers. The Mauryan Empire, which came to occupy most of the subcontinent, reputedly had 9,000 war elephants around 300 BC, and the Mughal Empire, which gained control of much the same territory, had 12,000 around 1600 AD. Both those states had rules and practices providing for the capture, care, and training of war elephants, and even for the protection of the wild populations from which the animals were taken. With the decline of the Mughals in the 1700s and the British occupation of India in the 1800s, the veneration and combat use of *E. maximus* declined. However, the British army did use the species extensively for transport and construction, roles that continued through World War II, and more recently elephants carried supplies for Vietnamese and Burmese insurgents (Kistler 2006). The animals, intelligent and docile when properly tamed and treated well, are still employed for various ceremonial, recreational, and labor purposes, notably the handling of timber.

Methods to capture wild elephants include lassoing, chasing to exhaustion, using pitfalls, and decoying a bull with a tame female; all were used into the twentieth century (Lahiri Choudhury 2008). Probably the most common procedure, one still in use today, is the *kheddah*, which involves building a strong stockade of high posts set close together in a circular form, with wings leading outward from the front to make a V-shaped entrance to the enclosure. The structure must be in an area with sufficient food, water, and cover, where a group of elephants might linger for a few days, during which time drivers slowly press the animals in

Asian elephant (*Elephas maximus*), mother and young. Photo by Alexander F. Meyer at Tierpark, Berlin.

the right direction. When the elephants near the long entrance, the drivers noisily force them through the V and into the corral. Then a gate is closed, and the process of selecting and taming is started. Elephants that have already been trained may be of great assistance in this process. Leimgruber et al. (2011) noted that elephant mortality during a *kheddah* was probably at least 30 percent and that the subsequent breaking in of the animals involved a combination of physical restraint, beatings, and food deprivation.

The treatment of elephants captured and maintained for service in South and Southeast Asia is controversial, with some authorities citing detailed protective laws, careful recordkeeping, conscientious handling, veterinary supervision, and restricted work hours (Aung and Nyunt 2002; Bist et al. 2002; Khawnual and Clarke 2002; Lehnhardt and Galloway 2008), and other observers pointing to inadequate regulation, decline in mahout quality, and abuse of the animals (Jayewardene 2002; Lohanan 2002; Mikota, Hammatt, and Fahrimal

2008; Suprayogi, Sugardjito, and Lilley 2002). Although those elephants are often referred to as "domestic," they are not in the same category as dogs (*Canis familiaris*) or cattle (*Bos taurus*), which have been maintained in fully controlled and selectively bred populations for thousands of years. Groups of captive *E. maximus* have never existed in sustained breeding populations but have been consistently supplemented with wild stock.

The elephants that work in timber camps in India and Burma are commonly kept in groups containing adult females, calves, and males of various ages, thus mimicking the social structure of wild populations. They are provided only part of the food they require and are released to forage in adjacent forests at night, where they may socialize with wild elephants. Many, if not most, of the calves born in captivity are sired by wild bulls. Notwithstanding the claims and counterclaims regarding the care of captive *E. maximus*, lumber camps with working elephants in India and Burma have been substantially more successful with respect to increasing fecundity, longevity, and calf survival than have Western zoos and circuses (Kurt, Mar, and Garaï 2008; Mumby et al. 2013; Sukumar 2003; Sukumar et al. 1997). Nonetheless, the Indian and Burmese captive populations are not self-sustaining and are slowly declining. Leimgruber et al. (2008) concluded that to maintain the then-estimated number in Burma, 6,000, would require bringing in 100 wild elephants annually. However, such removal from Burma's present wild population, thought to number barely 2,000, would result in its extinction within 31 years. Throughout much of the twentieth century, well over 100 wild elephants were captured each year to augment the captive Burmese population, and that is considered a major factor in the overall decline of the wild population. Recent removals from the wild have been lower, but the latest available estimate of the number in captivity in Burma is just 4,755 (Leimgruber et al. 2011).

A downward trend in captive *E. maximus* has been seen elsewhere as source populations decline, captive females age, regulations become more restrictive, and the timber industry mechanizes. Numbers in Thailand fell from perhaps 100,000 in 1900 to 13,000 in 1950 and 4,800 in 1982 (Dobias 1987; Lair 1988); the most recent estimate is around 2,500 (Kashio 2002). Other sharp declines have occurred in Cambodia, from 500 (Sukumar 1989a) to 93 (Maltby and Bourchier 2011); Sri Lanka, from 532 in 1970 (Jayewardene 2002) to 112 (Fernando et al. 2011); and Viet Nam, from 600 in 1980 (Cuong, Lien, and Giao 2002) to just 82 (Ly 2011). There are still approximately 3,567 captive elephants in India (Baskaran et al. 2011) and 500 in Laos (Khounboline 2011), but fewer than 1,000 in all the other countries that also have wild populations (Alfred et al. 2011; Azmi and Gunaryadi 2011; Islam et al. 2011; Jigme and Williams 2011; Pradhan, Williams and Dhakal 2011; Saaban et al. 2011; Zhang 2011).

The capture of elephants for labor and military purposes may have eliminated or greatly depleted wild populations in the Indus and Ganges basins by around 300 BC and in much of central India by 1600 AD (Sukumar 2003). By the late nineteenth century, expansion of agriculture and settlement had reduced distribution in India to the forests of the Himalayan foothills, Assam, the east-central region centered on the Chota Nagpur Plateau, and the far south; the current overall distribu-

Asian elephant (*Elephas maximus*), with mahout and tourists. Photo by Alexander F. Meyer at Bandhavgarh National Park, India.

tion in the country, though badly fragmented, remains about the same (Baskaran et al. 2011). Populations farther west, in Syria and Mesopotamia, may have been wiped out by ivory and recreational hunting in the first half of the first millennium BC (Lobban and de Liederkerke 2000). During the Shang Dynasty (1600–1046 BC), *E. maximus* was widespread in the Yellow River Valley of northern China, and at the end of that period could still be found in southern Shandong Province, but its habitat was contracting in association with climatic cooling, an increasing human population, and expanding agriculture (Wang 2011). The average rate of southward retreat has been calculated at 0.5 km per year (Zhang, Ma, and Feng 2006). The species was still in the Cháng Jiāng (Yangtze) Valley of central China until around 300 BC but by 1000 AD was largely restricted to the southern provinces of Yunnan, Guangxi, and Guangdong. It survived until the seventeenth century in Guangxi (Gao 1981), but today is confined to a small part of southern Yunnan (Zhang 2011).

Like *Loxodonta*, *Elephas* has been hunted ruthlessly through history for its valuable ivory. Martin and Vigne (1989) indicated that the ivory-carving industry of India dates back at least 4,000 years and that by the early nineteenth century this activity was being supported in large part by importation of tusks from the African *Loxodonta* (see accounts thereof) to replace stocks from the depleted *Elephas*. The current threat to the latter may not be quite as encompassing as in Africa, because the female Asian elephant usually lacks tusks and the male also does in Sri Lanka. However, selective hunting of males may result in skewed sex ratios, loss of genetic variation, and demographic and social changes. In a study of a heavily poached population in southern India, Sukumar, Ramakrishnan, and Santosh (1998) found the most disparate sex ratio ever recorded for an elephant population (Asian or African), with a count of 605 adult females and just 6 males. The CITES Secretariat (2014) noted difficulty in obtaining reliable data on the poaching problem in Asia, but indicated the illegal killing of Asian elephants may have increased in recent years.

Currently, *Elephas* is most seriously threatened by agricultural expansion, which inevitably results in conflicts with people, and by the loss, degradation, and fragmentation of habitat, which result from growing numbers of people. Chartier, Zimmermann, and Ladle (2011) reported that in eastern India conflict expanded significantly when forest cover was reduced to under 30–40 percent. Raiding and trampling of crops and destruction of storehouses and other property occur almost throughout the remaining range of the Asian elephant and constitute a major economic problem in some areas (Choudhury et al. 2008; Fernando et al. 2011; Islam et al. 2011; Kashio 2002; Khounboline 2011; Lahiri Choudhury 2008; Maltby and Bourchier 2011; Sukumar 1989a, 2003; Zhang 2011). People sometimes respond with lethal countermeasures, but

Asian elephant (*Elephas maximus*) at work camp in Thailand. Photo by Kikujungboy / Shutterstock.com.

Elephas has proved dangerous as well. From 1991 to 2005, 3,140 people were killed by elephants in India (Lahiri Choudhury 2008), and more recently the average annual death toll in the country has risen to over 400 (Rangarajan et al. 2010). It might be added in perspective, though, that in the one year of 2010 India had an estimated 231,027 road traffic deaths (World Health Organization 2013).

The severe fragmentation of elephant habitat in Asia has raised questions as to the viability of isolated populations and brought attention to the prospect of maintaining natural corridors between different groups (Choudhury et al. 2008). Sukumar (2003) calculated that to insure long-term conservation of a population would require at least 1,000 animals, with a normal sex ratio, in a protected area of at least 2,000 sq km. But fewer than 10 populations of *Elephas* are thought to have over 1,000 animals. Blake and Hedges (2004) questioned the reliability of most numerical estimates of wild elephants in Asia, pointing out that an overall figure of around 30,000–50,000 has been repeated by both the media and authorities since the 1970s without rigorous study and despite major losses of habitat over the period. At least partly within that context is the most recent available estimate for the total number of wild Asian elephants—46,321—as compiled from the following figures: mainland India, 27,694 (Baskaran et al. 2011); Sri Lanka, 5,879 (Fernando et al. 2011); Sumatra, 2,600 (Azmi and Gunaryadi 2011); Thailand, 2,250 (Kashio 2002); Borneo (possibly feral), 2,060 (Alfred et al. 2011); Burma, under 2,000 (Leimgruber et al. 2011); peninsular Malaysia, 1,450 (Saaban et al. 2011); Laos, 700 (Khounboline 2011); Bhutan, 513 (Jigme and Williams 2011); Cambodia, 400 (Maltby and Bourchier 2011); Bangladesh, 325 (Islam et al. 2011); Nepal, 125 (Pradhan, Williams, and Dhakal 2011); China, under 200 (Zhang 2011); Viet Nam, 85 (Ly 2011); Andaman Islands (feral), 40 (Baskaran et al. 2011). At least one of those figures, that for Viet Nam, may accurately reflect an extremely sharp decline in recent decades. Remarkably, the elephant and other large mammals appear to have persisted there in substantial numbers during the many years (1941–1975) when warfare curtailed agriculture and other human activity in forested areas (Nowak 1976). Shortly after the war ended, Viet Nam still had as many as 2,000 wild elephants, but subsequent hunting for ivory and conflict with farmers rapidly reduced numbers and distribution (Cuong, Lien, and Giao 2002).

In contrast, while *Elephas* now occupies just 3.5 percent of its original range in India, numerical estimates for the country have remained relatively stable or even increased in recent years. The largest numbers, about 10,000, are in the extreme south along the Western Ghats Mountains. The largest single population in that region, and indeed in the world, contains an estimated 5,839 elephants occurring from the Malnad Plateau on the north to the Palghat Gap on the south. While that population is reportedly increasing in some areas, it is subject to ivory poaching and may be jeopardized by agricultural, hydroelectric, and highway development, which could cut off north-south movements (Baskaran 2013). For nearby Sri Lanka, Santiapillai, Chambers, and Ishwaran (1984) had reported a decline from about 12,000 at the beginning of the nineteenth century to just 3,000. More recently there has been some evidence of numerical increase or stability, though Fernando et al. (2005) cautioned that historical respect and reverence for elephants was rapidly eroding and that irresponsible agricultural expansion on the island was creating problems and causing people to view the animals as a dangerous pest.

The Asian elephant is on appendix 1 of the CITES, and therefore international commerce in its ivory is totally banned among signatory nations. The USDI lists the species as endangered and also administers the Asian Elephant Conservation Act of 1997, which provides for funding of relevant research and protective measures (Stromayer 2002). While the IUCN classifies the overall species *E. maximus* as endangered (Choudhury et al. 2008), it treats the subspecies *E. m. sumatranus* of the island of Sumatra as critically endangered (Gopala et al. 2011). Until the late twentieth century, Sumatra was thought to have some of the largest populations in Asia, but conversion of forest to settlement and agriculture has fragmented habitat and created conflicts with people. Consequent killing and a large-scale government program to capture or relocate offending animals, accompanied by high mortality, have greatly reduced or even eliminated entire wild populations. Since 1985, when there were at least 4,800 elephants in Sumatra, 69 percent of potential habitat has been lost, and the species has disappeared in 23 of the 43 known ranges. The factors responsible seem certain to continue.

Woolly mammoth (*Mammuthus primigenius*). Painting by Helga Schulze, based on numerous descriptions and illustrations, including those found in Guthrie (2013), Haynes (1991), and Lister and Bahn (2007).

PROBOSCIDEA; ELEPHANTIDAE; **Genus** *MAMMUTHUS* Burnett, 1829

Woolly Mammoth

Among the most remarkable mammalogical discoveries of the past century is that a single species of mammoth, *M. primigenius*, evidently survived into historical time, until about 4,000 calendar years ago, on Wrangel Island in the Arctic Ocean off extreme northeastern Siberia (Vartanyan, Garutt, and Sher 1993). Any doubts about the circumstances were brushed aside when Vartanyan et al. (2008) issued a list of 124 specimens found on the island from 1989 to 2000—tusks, other teeth, and bones—all radiocarbon dated. Of the dates, 106 are Holocene, from 8,980 to 3,685 radiocarbon years ago. Unfortunately, radiocarbon years do not correspond exactly to familiar calendar years. Precise conversion is a laborious process; while some publications calibrate the difference and/or provide equivalencies, others do not. The calendar year dates given in the remainder of this ac-

count were taken or calculated from equivalencies shown in Haile et al. (2009), MacDonald et al. (2012), Nyström et al. (2012), and Thomas (2012). The age range of the Holocene Wrangel specimens corresponds approximately to 10,000 to 4,000 calendar years ago. Radiocarbon dating has recently been used, in conjunction with archeological materials, to establish that the First Dynasty of Egypt, a common indicator of the beginning of historical time, was founded almost exactly 5,100 calendar years ago (Dee et al. 2013). Of the mammoth specimens from Wrangel, 28 are less than 5,100 calendar years old. Subsequent to the discovery on Wrangel, remains of *M. primigenius*, dating as recently as 6,400 calendar years ago, were found on St. Paul Island in the Bering Sea off southwestern Alaska; estimates of Holocene sea level changes in the area suggest the island probably remained large enough to support a mammoth population until about 5,600 calendar years ago (Enk et al. 2009; Guthrie 2004).

M. primigenius, popularly known as the woolly mammoth, also once occurred over most of Europe and approximately the northern halves of Asia and

North America, but is thought to have disappeared rapidly from most of those regions in the late Pleistocene or early Holocene. Prior to the discovery on Wrangel Island, the final extinction of *M. primigenius* was usually dated at around 10,000 calendar years ago (Agenbroad 1984; Vereschagin and Baryshnikov 1984). Another species, *M. columbi*, survived about as long but was restricted to North America, mainly the western and southeastern United States and Mexico (Lister and Bahn 2007). Although the relationship of the two species is not precisely understood, Enk et al. (2011), using an analysis of mitochondrial DNA recovered from remains, suggested the two occasionally hybridized in a manner comparable to that reported between the two living species of *Loxodonta* (see account thereof). They also suggested that another named species of late Pleistocene North American mammoth, *M. jeffersonii*, might actually be based on specimens resulting from such hybridization. Specimens assigned to *M. jeffersonii* have been reported mostly from the Great Lakes region of the United States (Pasenko and Schubert 2004).

The Holocene mammoth population of Wrangel Island was initially termed a "dwarf" form of *M. primigenius* (Vartanyan, Garutt, and Sher 1993) and has often been referred to as such by authorities and the media. However, further assessment and collection of additional specimens have shown that the members of the population, while relatively small mammoths, were approximately the same size as those in the population of *M. primigenius*, which had occupied the nearby mainland in the late Pleistocene and earliest Holocene (Kuzmin and Orlova 2004; Reumer, Mol, and de Vos 2002; Tikhonov, Agenbroad, and Vartanyan 2003; Vartanyan et al. 2008). Mammoths had been undergoing an overall size reduction since the middle Pleistocene, and their average size seems to have been smaller to the north. The species *M. trogontherii*, of the early to middle Pleistocene of Eurasia, attained a shoulder height of 450 cm (Kurten 1968). That species may have weighed about 15,000–20,000 kg, twice as much as the largest modern *Loxodonta*, and would rival the extinct proboscidean *Deinotherium giganteum* and the extinct rhinoceros *Indricotherium* as the largest land mammal ever to exist (Christiansen 2004; Fortelius and Kappelman 1993). Apparently, *M. trogontherii* spread into North America and gave rise to *M. columbi*, which occupied the southern part of the continent from the

middle to late Pleistocene, and was up to 400 cm tall at the shoulder (Lister and Bahn 2007). Meanwhile, the smaller *M. primigenius* developed in Siberia and entered Europe and North America toward the end of the middle Pleistocene (Lister and Sher 2001; Lister et al. 2005). Certain populations of mammoths, especially those isolated on islands, became far smaller than most modern *Loxodonta* and *Elephas* (see account of family Elephantidae). *M. exilis*, a true dwarf descendant of *M. columbi*, found on the Channel Islands off southern California until near the end of the Pleistocene, had an estimated shoulder height of just 120–180 cm (Roth 1996).

The Holocene population of *M. primigenius* on Wrangel Island was described as a subspecies, *M. p. vrangeliensis*, based primarily on what were considered to be its relatively small molar teeth. That designation was challenged by Reumer, Mol, and de Vos (2002), who pointed out that the description covered only a few specimens, apparently female, and that equally small specimens of *M. primigenius* were known from the mainland of Eurasia. Tikhonov, Agenbroad, and Vartanyan (2003) agreed that the Wrangel animals should no longer be considered dwarfs or in the same category as *M. exilis* of the Channel Islands, which had been isolated from other mammoth populations for over 40,000 years. Based on newly collected specimens, the Holocene Wrangel mammoths were reported to vary in size, with a shoulder height from 180 to at least 230 cm, which probably corresponded to that of the last mammoths from the late Pleistocene of the northern Siberian mainland. Tusk and molar size of the two groups was also about the same, though the molars of the Wrangel population were considered to be relatively narrow and to have a frequency of enamel plates exceeding the average in the Siberian population. Vartanyan et al. (2008) continued to use the name *M. p. vrangeliensis*, but, based on the lack of any trend toward molar size reduction over time, suggested that the subspecies had originated on the mainland prior to reaching the island and becoming isolated in the Holocene.

According to Haynes (1991), typical *Mammuthus primigenius* of the late Pleistocene of Europe and North America was approximately the same size as modern *Loxodonta africana*, with a shoulder height estimated at 300–350 cm in males and 230–250 cm in females.

Woolly mammoth (*Mammuthus primigenius*), skeleton. © Beatrissa / Shutterstock.com.

Lister and Bahn (2007) listed an identical weight range for the two of about 3,600–5,400 kg. However, the two species were unlike one another in body profile, partly because of differences in the vertebral spines. In *Loxodonta* the longest spines are located at the front shoulder and decrease in length to the lumbar region, where they lengthen again, creating a dished appearance. In *Mammuthus* the longest spines are also near the front shoulder but are slanted backward, creating a hump that slopes anteriorly. This hump results from the vertebrae and also a mass of hair covering the shoulders; it is not a stored fat deposit (though an insulating layer of fat may have been distributed generally under the skin). In addition, the long axis of the skull is oriented more nearly vertically in *Mammuthus* than it is in *Loxodonta*, thereby creating a domed profile to the head, somewhat like that seen in *Elephas* but higher. The difference in the appearance of the two genera also results partly from the fact that the limb bones of *Mammuthus* were shorter but generally much broader than those of *Loxodonta*, and this suggests that the weight of *Mammuthus* was greater than that of a modern elephant of similar stature.

Paleolithic humans left drawings of the woolly mammoth on the walls of caves, and a number of car-casses have been recovered intact, complete with hide, hair, and internal organs, from permanently frozen ground in Siberia and Alaska. According to Haynes (1991), such evidence shows that the body, trunk, tail, and limbs of *M. primigenius* were covered with a heavy coat of hair right down to the toes, though not to the soles of the feet. That coat consisted of very dense, woolen underfur, 5–15 cm thick, and straight, coarse outer hairs up to 100 cm long. The coat apparently functioned to maintain a constant body temperature in the cold regions where the mammoth lived, in contrast to the warmer climates where the much more sparsely haired modern elephants are found. The original color of the pelage is questionable; the hair of most preserved specimens is black, brown, reddish, orange, or yellowish, but in some cases this may represent a modification after death. The ears of *M. primigenius* were small compared with those of modern *Loxodonta* and *Elephas*, probably an adaptation to prevent heat loss. Its trunk ended in two projections at the tip, the front one narrow and about 10 cm long, the back one broader and about 5 cm long; they may have acted like a finger and thumb to pick flowers, buds, and the short steppe grasses (Lister and Bahn 2007). Both sexes had tusks that were relatively larger than those of modern elephants and usually much more curved. The tusks, though quite variable in form, had a tendency to spiral, first downward and outward, then upward and inward. Lister and Bahn (2007) noted that typical male wooly mammoth tusks measured 240–270 cm along the curve and weighed about 45 kg, and that the largest on record was 420 cm and 85 kg. Female tusks were considerably smaller, thinner, and less curved. The grinding teeth or molars had numerous ridges of enamel. E. Anderson (1984) noted that the molar teeth of *M. primigenius* were the most complex of any mammoth's, with extremely thin, closely appressed plates.

The distribution of *M. primigenius* seems to have been closely associated with tundra, boreal forest, and cold steppe environments in regions that were near or had been subject to glaciation (Agenbroad 1984). For North America, Haynes (2009) estimated a population density of 0.1–0.3 mammoths per sq km. The remains of individuals and groups provide some clues about the natural history of the species (Haynes 1991; Lister and Bahn 2007). The woolly mammoth probably moved about extensively in search of food and may have undertaken regular long-distance migrations, especially southward during the winter, though clear evidence for this is lacking. The diet seems to have been variable but evidently was dominated by grasses and sedges and included less browse or woody plants than that of modern elephants. As in the latter, *M. primigenius* apparently lived in groups composed of related females and their young, which sometimes came together in large aggregations on the steppe. Adult males lived separately and seem to have competed for females during the mating season. Temporal glands were present, indicating that males seasonally entered musth, and some remains show signs of violent battles between bulls. The gestation period likely was similar to that of modern elephants, around 21–22 months, with mating probably concentrated in the summer and fall and births from spring to summer. Weight at birth was around 90 kg. Age of sexual maturity is estimated at 18–20 years but study of skeletal remains indicates growth continued until the age of about 25 in females and 40 in males. Life expectancy was about 60 years.

Mammoths persisted longer than most other large Pleistocene mammals. Even before the discovery on Wrangel Island, some reported radiocarbon dates indicated survival until 9,000 calendar years ago, possibly

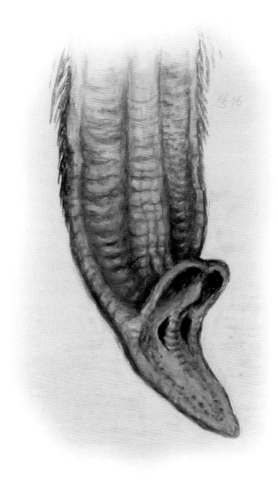

Woolly mammoth (*Mammuthus primigenius*), close-up of trunk. Painting by Helga Schulze, based on numerous descriptions and illustrations, including those found in Guthrie (2013), Haynes (1991), and Lister and Bahn (2007).

even less, in North America (Agenbroad 1984; Grayson 1989). Such recent dates may involve contamination of the involved materials and are no longer widely accepted (Fiedel 2009; Fisher 2009; Haynes 2002; Lister and Bahn 2007; P. S. Martin 2005; Martin and Steadman 1999). Nonetheless, a list of 189 reported dates in the last 50,000 years (the range of radiocarbon dating) for North American localities with *Mammuthus* (here considered to comprise *primigenius, columbi, jeffersonii,* and *exilis*) includes 20 dates approximately 11,500 to 8,000 calendar years old. Those recent dates, while questionable, are restricted to two regions, one around the Great Lakes and westward in extreme south-central Canada and the north-central United States, and the other in the southwestern United States, and suggest the possibility of refugial populations (Agenbroad 2005). That compilation of site distribution and dates, together with studies of numerous samples of mitochondrial DNA recovered from the remains of *M. primigenius* (Debruyne et al. 2008; Palkopoulou et al. 2013), show that the range of the species was dynamic, expanding and contracting across continents a number of times, with population clades developing in warmer intervals, when Eurasia and North America were isolated, and then crossing the Bering Land Bridge when sea levels fell during glacial advances.

The overall final retreat of the woolly mammoth was northward, toward Beringia and the Siberian coast, perhaps in association with loss of steppe habitat and expansion of forests (MacDonald et al. 2012). While reliable dates of around 11,700–11,000 calendar years ago have been reported for west-central Siberia, northeastern Russia, and Estonia, most comparable dates for mainland Eurasia are from sites on the Taymyr, Gydan,

and Yamal Peninsulas, which extend from northern Siberia into the Arctic Ocean (Kuzmin 2010; Kuzmin and Orlova 2004; Mol et al. 2006; Stuart et al. 2002). The very latest known date for Eurasia, aside from Wrangel Island, is about 10,700 calendar years ago and was taken from remains found on the New Siberian Islands, which at that time were connected to the mainland of northeastern Siberia (Nikolskiy, Sulerzhitsky, and Pitulko 2011). However, Vartanyan (2008) suggested the possibility that the species survived along the northern coast of the western Chukotka Peninsula and the adjacent exposed continental shelf as late as 9,000 calendar years ago. Mitochondrial DNA of *M. primigenius* has been recovered from a layer of sediment (not from visible remains of the animal) near Stevens Village in east-central Alaska, dated at 10,500–7,600 calendar years before the present (Haile et al. 2009).

The cause of extinction of *M. primigenius*, other mammoths, and the rest of the late Pleistocene megafauna remains open to question, with some authorities

Woolly mammoth (*Mammuthus primigenius*). Diorama at Royal British Columbia Museum in Victoria (Canada). The display is from 1979, and the fur is musk ox hair. Photo by Flying Puffin—MammutUploaded by FunkMonk.

suggesting climatic or associated changes (Grayson and Meltzer 2004; Guthrie 2006; Kuzmin 2010; Nikolskiy, Sulerzhitsky, and Pitulko 2011; Stuart 2005), others arguing that the primary agent was improved hunting ability by *Homo sapiens* (Alroy 1999; Fiedel and Haynes 2004; Haynes 2002, 2007; Koch and Barnosky 2006; Sandom et al. 2014; Surovell, Waguespack, and Brantingham 2005), and still others pointing to human hunting pressure as a final blow following a period of environmental deterioration (MacDonald et al. 2012; Nogués-Bravo et al. 2008; Solow, Roberts, and Robbirt 2006). Review of the dialogue indicates there were indeed severe climatic fluctuations at the end of the Pleistocene, which could have greatly constricted habitats capable of supporting mammoth populations. Expansion of forests and other unfavorable conditions during that period, along with decline of the steppe environment, do seem compelling factors in the northward withdrawal of *Mammuthus primigenius*. However, it is doubtful that such conditions could by themselves have suddenly obliterated the species (as well as the southerly *M. columbi*) throughout all habitats across three continents. Modern elephant populations have demonstrated repeated resilience and the ability to recover from all forms of environmental stress except overhunting by people (Haynes 1991). Likewise, the single factor that corresponded with the disappearance of mammoths from the mainland of Eurasia and North America was the spread of advanced human hunters (P. S. Martin 1984, 2005). Although the herds were vast and the animals formidable, determined bands of well-armed people, probably taking mainly female and young mammoths concentrated in remnant suitable habitat, could easily have been responsible for pushing *Mammuthus* beyond the point of recovery. The sequence of dates in the Old World suggests a geographic wave of extinction perhaps coinciding with modern human expansion, with *M. primigenius* disappearing from western and central Europe by about 13,500 calendar years ago (Stuart 2005) but not from northern Siberia until a few thousand years later (see above).

In addition to the 106 Holocene specimens of *M. primigenius* from Wrangel Island, dating around 10,000–4,000 calendar years ago, there are 18 from the late Pleistocene, dating from over 40,000 to about 14,000 calendar years ago (Vartanyan 2008). The reason for the 4,000 year gap in records is not clear, especially because early reports indicated the island and its mammoth population had become isolated about 14,000 calendar years ago. Alternatively, it was speculated that the late Pleistocene population became extinct and that the island was recolonized in the early Holocene. However, reassessment of hydrographic and morphometric data and analysis of both mitochondrial and nuclear DNA indicate the Wrangel mammoths were initially part of an overall north Siberian population that could move to the island when sea level was low, and even afterward across winter ice, and that the island did not completely separate from the mainland until about 10,000 calendar years ago (Nyström et al. 2010, 2012; Thomas 2012; Vartanyan 2008). Subsequent to isolation in the Holocene there was a substantial drop in genetic diversity, reflecting the much smaller size of the population, perhaps 500 individuals, now restricted to the island. But for the 6,000-calendar year duration of the population's existence there was no further decline of diversity or apparent numbers, thus suggesting that extinction, when it did come, resulted from a sudden event, such as the arrival of human hunters.

There is no direct evidence that people did hunt mammoths on Wrangel. The earliest human settlement there has been dated at about 3,600 calendar years ago. Guthrie (2004) noted that the time of extinction on Wrangel did coincide with the marine expansion of the northern Denbigh-Arctic small-tool-tradition peoples. He did not, however, think that humans were responsible for eliminating *M. primigenius* on St. Paul Island off southwestern Alaska, which evidently was not discovered by people until modern times. There it appears that as sea levels rose, available habitat became too small to sustain a mammoth population; such a situation would have developed around 5,600 calendar years ago. For Wrangel Island, Nyström et al. (2010) noted both that final extinction of the woolly mammoth likely happened later than the most recently dated specimen, and that the first arrival of humans probably predates the earliest known settlement. The proximity of the two events suggests that the disappearance of *M. primigenius* from Wrangel was the culmination of the wave of human killing that had begun thousands of years before. In any case, the species was now definitely extinct.

But even so, it was not to be allowed to rest in peace. Oddly enough, intensive human exploitation

Woolly mammoth (*Mammuthus primigenius*), tusk on Wrangel Island. Photo by Sergey Gorshkov / Minden Pictures.

continued into modern times and to this day remains an issue of serious conservation concern. The mythical elephant graveyards, long sought by ivory hunters in Africa, actually existed in Siberia. Untold numbers of woolly mammoths had perished there, perhaps by drowning or landslides during spring thaws, and subsequently become buried in ice or frozen ground (Haynes 1991). Their huge tusks were preserved, not by true fossilization, but by freezing, and they remained perfectly suitable for carving into decorative objects. Trade in mammoth ivory may have begun in Roman times and had become an enormous commercial enterprise by the eighteenth and nineteenth centuries. From 1809 to 1910, 6,000 tons of ivory, representing an estimated 46,000 *M. primigenius*, were taken from the Siberian tundra. Vast quantities are thought to remain, though availability depends on various environmental, economic, and political conditions. After the general importation of *Loxodonta* into the United States was banned in 1989, commercial interests in eastern Asia

attempted to send in ivory from that genus by claiming it was from *Mammuthus*. Such efforts posed a continued threat to the survival of *Loxodonta* and led to the development of microscopic techniques to distinguish the ivory of modern elephants from that of mammoths (Espinoza and Mann 1994).

Currently, international trade in mammoth ivory is legal, and data thereon is readily available (CITES Secretariat 2014). Overall, the total volume in trade went from 17.3 tons in 1997 to 95 tons in 2012, a more than five-fold increase. Nearly all the ivory originates on the Siberian tundra, and since 2007 China has accounted for virtually all global imports by weight. Hong Kong is the center of the carving industry and the largest marketplace (E. Martin and C. Martin 2011). The predominant buyers of mammoth ivory products, mostly figurines and beads for jewelry, are increasingly prosperous consumers from mainland China. Hence, the prime factors driving the exploitation of *Mammuthus* and *Loxodonta* are the same. Will their ultimate fates also be the same?

Order Sirenia

Dugong, Steller's Sea Cow, and Manatees

This order of aquatic mammals contains two Recent families: Dugongidae, with the genera *Dugong* (dugong, one species—*D. dugon*) and *Hydrodamalis* (Steller's sea cow, one species—*H. gigas*, extinct), and Trichechidae, with the single genus *Trichechus* (manatees, three species—*T. inunguis*, *T. manatus*, and *T. senegalensis*). The dugong inhabits coastal regions in tropical parts of the Old World, but some individuals go into the fresh water of estuaries and up rivers. Steller's sea cow occurred in the Bering Sea and was the only Recent member of this order adapted to cold waters. *T. manatus* lives along the coast and in connecting rivers in the southeastern United States, Central America, the West Indies, and northern South America; *T. inunguis* inhabits the Amazon Basin; and *T. senegalensis* is found along the coast and in the rivers of western Africa. McKenna and Bell (1997) listed Sirenia as an infraorder of suborder Tethytheria of order Uranotheria (see account of order Proboscidea), though most authorities have continued to treat Sirenia as an order (e.g., Domning, Zalmout, and Gingerich 2010; Marsh, O'Shea, and Reynolds 2012; Shoshani 2005c), with Tethytheria sometimes considered a superorder comprising Sirenia, Proboscidea, and the extinct order Desmostylia (Seiffert 2013). There has also been growing acceptance that Sirenia is part of the clade Afrotheria (see account of class Mammalia).

These massive, fusiform (spindle-shaped) animals have paddle-like forelimbs, no hind limbs or dorsal fin, and a tail in the form of a horizontally flattened fin. Adults of the living forms generally are 250–400 cm in length and weigh as much as 908 kg. The head is rounded, the mouth is small, and the muzzle is abruptly cut off. The nostrils, which are valvular and separate, are located on the upper surface of the muzzle. The eyelids, though small, are capable of contraction, and a well-developed nictitating membrane is present. There is no external ear flap. The neck is short. Females have two mammae, one on each side in the axilla under the flipper. The testes in males are abdominal (borne permanently within the abdomen).

The skin is thick, tough, and often wrinkled. Sirenians often are said to be nearly hairless, but they do have hairs sparsely distributed over the body, all of which are thought to be sensory sinus hairs. This is a unique condition, as no other mammal is known to have tactile hairs except in restricted areas, primarily the face. Reep, Marshall, and Stoll (2002) found specimens of *Trichechus manatus* to possess approximately 1,500 postcranial hairs on each side of the body, with density decreasing dorsal to ventral. External hairs were 2–9 mm long, mostly separated from one another by 20–40 mm, and connected internally to a capsule with an elongated circumferential blood sinus and innervation by 20–50 axons. Thus, a total of around 90,000 axons project centrally and provide an underwater tactile system capable of conveying detailed environmental information. In addition, sirenians have approximately 2,000 stiff, thickened vibrissae in the orofacial region around the lips. Those bristles, unique among mammals, function together with modified facial musculature as a

West Indian manatee (*Trichechus manatus*). Photo by R. K. Bonde, US Geological Survey, through Cathy A. Beck.

system to acquire, manipulate, and ingest vegetation (Marsh, O'Shea, and Reynolds 2012).

The heavy sirenian skeleton is characterized by both pachyostosis, or thickening of the bones, especially of the ribs, and osteosclerosis, in which cancellous or spongy bone is replaced with denser, compact bone, resulting in absence of pneumatic cavities. Those conditions, referred to in combination as "pachyosteosclerosis," produce greater specific gravity and are an adaptation to remaining submerged and maintaining horizontal trim in shallow waters (Domning and de Buffrénil 1991; Domning, Zalmout, and Gingerich 2010; Kaiser 1974). The skull is large in proportion to the size of the body, though the brain is relatively among the smallest found in mammals (O'Shea and Reep 1990). The nasal opening is located far back on the skull and directed dorsally. The lower jaws are heavy and united for a considerable distance. The forelimb has a well-developed skeletal support, but there is no trace of the skeletal elements of the hind limb. The pelvis consists of one or two pairs of bones suspended in muscle. The vertebrae are separate and distinct throughout the spinal column; one genus (*Trichechus*) has six neck vertebrae (nearly all other mammals have seven).

The dentition is highly modified and often reduced (functional teeth are lacking in *Hydrodamalis*). Most of the incisor teeth are reduced or absent, and canines are present only in certain fossil species. When the incisors are present, there is a space between them. The cheek teeth, which are arranged in a continuous series, number from 2 to 10 in each half of each jaw. The anterior part of the palate and the corresponding surface of the lower jaw are covered with rough, horny plates, used as an aid in chewing food. The tongue is small and somewhat roughened anteriorly.

The genera differ in a number of ways. The tail fin is deeply notched In *Dugong* and *Hydrodamalis* but more or less evenly rounded in *Trichechus*. The upper lip is more deeply cleft in manatees than in *Dugong* and *Hydrodamalis*. In *Dugong* there are 2/2 or 3/3 functional cheek

Dugong (*Dugong dugon*). Photo by Doug Perrine / npl / Minden Pictures.

teeth at any given time, and the males have one pair of tusklike incisors (usually unerupted in females). In *Trichechus* functional incisors are not present, but the cheek teeth are numerous and indefinite in number (up to 10 in each half of each jaw). The cheek teeth of *Trichechus* are replaced consecutively from the rear; an individual cheek tooth is worn down as it moves forward.

Sirenians differ conspicuously in size of rostrum (or snout) and, particularly, in the degree of its deflection relative to the palatal plane (Domning 1978; Marsh, O'Shea, and Reynolds 2012; Marshall et al. 2003). *Dugong dugon* has the most deflected snout (67–72°) of the historical species, which allows the perioral region to be placed almost flat against the substrate—an energetically advantageous position for a bottom feeder. *T. inunguis* and *T. senegalensis* have the least deflected snouts (25–41° and 15–40°, respectively), presumably an adaptation for feeding on natant and emerging vegetation. The snout deflection of *T. manatus* is intermediate (29–52°), reflecting its generalist niche, with feeding at any level from bottom to surface. Deflection in the extinct *Hydrodamalis gigas* (35–45°) probably expressed feeding at or near the surface.

Sirenians are solitary, travel in pairs, or associate in larger groups, sometimes in the hundreds. Generally slow and inoffensive, they spend all their life in the water. They are vegetarians and feed on various water plants. They are the only fully aquatic herbivorous mammals (Marsh, O'Shea, and Reynolds 2012) and the only living mammals that have evolved to exploit plant life in the sea margin (Anderson 1979). The ordinal name "Sirenia" is related to the supposed mermaid-like nursing of dugongs (thought to be the origin of the myths of the Sirens) and manatees. The only reliable observations of nursing in manatees, however, have revealed that the young suckle while the mother is underwater in a horizontal position, belly downward. Anderson (1984b) reported that suckling in the dugong is somewhat similar but that the calf usually is in an inverted position.

Order Sirenia is often classified together with orders Proboscidea and Hyracoidea in a mammalian superorder, Paenungulata, though there is also support for joining Sirenia with just Proboscidea (and several extinct orders) in another superorder, Tethytheria (see accounts of class Mammalia and order Proboscidea). Emergence of Sirenia evidently occurred in the warm bays and marshes lining the Tethys Seaway, which had formed when the northern and southern land masses of the world split apart in the Mesozoic (Domning 2013). However, based on available fossil material, the known geological range of order Sirenia is early or middle Eocene to Recent, with the earliest and most primitive known relatively complete specimens from Jamaica (Domning, Zalmout, and Gingerich 2010). The represented animals retained four legs and would have been fully capable of terrestrial locomotion, but also demonstrated typical sirenian adaptations, such as pachyosteosclerosis, and probably spent much time in the water feeding on aquatic vegetation (Marsh, O'Shea, and Reynolds 2012). Recently, a petrosal from the skull of an even more primitive sirenian found in Tunisia has been reported (Benoit et al. 2013).

By the middle Eocene the order was present in southeastern North America, the West Indies, southern Europe, northern and eastern Africa, and south-central Asia, and three distinct families—Dugongidae, Prorastomidae, and Protosirenidae—had evolved (Dawson and Krishtalka 1984; Domning, Morgan, and Ray 1982). On the basis of morphological studies, Domning (1994) argued that Trichechidae branched off shortly thereafter, in the late Eocene or early Oligocene, and he criticized molecular analyses suggesting that such divergence did not occur until the early Miocene. A comprehensive new molecular assessment has indicated that the lineage of *Trichechus* split from that of *Dugong* and *Hydrodamalis* about 42 million years ago in the Eocene and that the lineages of the latter two genera diverged about 29 million years ago in the Oligocene (Springer et al. 2015). Sirenians apparently were more abundant from the Oligocene to the Pliocene than they are now, with peak diversity of about a dozen known genera during the Miocene (Domning, Zalmout, and Gingerich 2010). Their comparative scarcity at the present time probably results from climatic changes in the Pliocene and Pleistocene and, more recently, exploitation by humans for food, hides, and oil. The number of individual sirenians remaining in the world, perhaps under 140,000, may be smaller than that of any other mammalian order (see also account of order Dermoptera).

SIRENIA; **Family DUGONGIDAE**

Dugong and Steller's Sea Cow

This family contains two Recent subfamiles: Dugonginae, with the single Recent genus *Dugong* (dugong), found in coastal regions of the tropics of the Old World; and Hydrodamalinae, with the single Recent genus *Hydrodamalis* (Steller's sea cow), formerly found in the Bering Sea (Rathbun 1984). A recent molecular analysis has confirmed the affinity of *Hydrodamalis* to *Dugong*, rather than to *Trichechus* (Springer et al. 2015).

Dugong is generally 250–350 cm in length, whereas *Hydrodamalis* measured up to 1,000 cm. The flippers in Steller's sea cow were curiously bent and were said to have been used to pull an individual along the bottom of the ocean as it foraged. Members of the family Dugongidae lack nails on their flippers. The deeply notched tail fin has two pointed, lateral lobes (the tail fin in manatees is more or less evenly rounded). The upper lip, which is more deeply cleft in the adult dugong than in the young, is not as deeply cleft as that of the manatees. The two nostrils are located on top of the anterior end of the snout. In the dugong (and apparently all sirenians) the eyelids contain a number of glands that produce an oily secretion to protect the eye against water.

The rostrum of the dugong skull is bent strongly downward, much more so than that of *Trichechus*, whereas the rostrum of the skull of the Steller's sea cow was only slightly inclined (see above account of order Sirenia). *Hydrodamalis* lacked functional teeth but did possess functional, rough oral plates. Male *Dugong* have one pair of tusklike incisors (usually unerupted in females), which are directed downward and forward and are partially covered with enamel; they project about 30 mm externally. A dugong has 6/6 functional cheek teeth (premolars and molars) during its lifetime, but only 2/2 or 3/3 are erupted and in wear simultaneously (Marsh and Dutton 2013). The simple, peglike cheek teeth are covered with cement, lack enamel in adulthood, and wear quickly; the last two molars are open-rooted and grow throughout life.

Left, dugong (*Dugong dugon*); photo by Alex Churilov/Shutterstock.com. *Right,* West Indian manatee (*Trichechus manatus*); photo by R. K. Bonde, US Geological Survey, through Cathy A. Beck These profile images show the much greater rostral deflection in *Dugong.*

The geological range of this family, aside from the modern distribution, is middle Eocene to late Pliocene in Europe and North Africa, middle Eocene to Miocene in Madagascar, late Oligocene to Pleistocene in Japan, Miocene in India and Sri Lanka, middle Eocene to Pliocene in North America, and early Miocene to early Pliocene in South America (Domning and Ray 1986; Domning, Zalmout, and Gingerich 2010; Marsh, O'Shea, and Reynolds 2012; Rathbun 1984). The two living subfamilies, Dugonginae and Hydrodamalinae, probably diverged from the extinct subfamily Halitheriinae in, respectively, the Oligocene and Miocene (Marsh, O'Shea, and Reynolds 2012).

SIRENIA; DUGONGIDAE; **Genus *DUGONG***
Lacépède, 1799

Dugong

The single species, *D. dugon,* originally occurred regularly in the Red Sea and Gulfs of Suez and Aqaba, along the eastern coast of Africa as far south as Mozambique, around Madagascar and nearby islands in the Indian Ocean, in the Arabian Gulf, on the western and southern coasts of India and around Sri Lanka and the Andaman and Nicobar Islands, off Southeast and East Asia from Burma to Singapore and to as far north as Taiwan and the Ryukyu Islands, throughout the East Indies to the Solomon Islands and around Palau, in the Southwest Pacific as far as Vanuatu and New Caledonia, and all along the coast of Australia as far south as Shark Bay on the west and Moreton Bay on the east (Dobbs, Lawler and Kwan 2012; Husar 1978a; Marsh et al. 2002; Nishiwaki et al. 1979). Occasional occurrences have been reported as far south as KwaZulu-Natal Province on the east coast of South Africa (Marsh and Dutton 2013), off Korea and Kyushu (Hirasaka 1934), around Guam and Yap in the Pacific (Eldredge 2003, Nishiwaki et al. 1979), at Eauripik Atoll on the western side and at Kosrae Island at the eastern edge of the Federated States of Micronesia (Buden and Haglelgam 2010), as far south as Albany on the southwest coast of Western Australia (Heinsohn 2008), and along the coast of New South Wales in eastern Australia, nearly to the border with Victoria and about 700 km south of current regularly occupied range (Allen, Marsh, and Hodgson 2004; Nishiwaki and Marsh 1985). Mid-Holocene remains from the coast of Victoria, near Melbourne, were reported by Fitzgerald (2005).

Although Kingdon (1971) indicated that *D. dugon* was present within early historical time in the eastern Mediterranean Sea, such occurrence does not now seem likely and was not referred to by Marsh and Dutton (2013) or Marsh, O'Shea, and Reynolds (2012). Prista et al. (2013) observed that the region has abundant sea grasses and other suitable conditions, but that no sirenians have been naturally present in the

Dugong (*Dugong dugon*). Photo by Mike Parry / npl / Minden Pictures.

Mediterranean since the end of the Pliocene. Rudolph and Smeenk (2008) noted that in modern times *D. dugon* has strayed into the Mediterranean, having traversed the Suez Canal (which opened in 1869). Allen (1942) cited a report of a female dugong killed in a sea cave on the coast of Palestine, sometime between 1869 and 1930, but suggested the possibility that the involved animal was a monk seal (*Monachus monachus*).

Adults usually have a total length of 240–270 cm and weigh 230–360 kg, but recorded maximums are 406 cm and 908 kg (Husar 1978a). There is little sexual dimorphism, though females may grow to a larger average size than do males (Marsh, Heinsohn, and Marsh 1984). The coloration is variable but is usually dull brownish gray above and somewhat lighter below. The skin is thick, tough, relatively smooth, and covered with widely scattered hairs. The forelimbs are modified into flippers about 35–45 cm in length; they are used for propulsion by the young, but adults propel themselves by means of the flukelike tail, using the flippers only for steering. When the dugong grazes on the ocean floor, it uses the flippers for "walking," and not to probe for food. Although captives have been said to use flippers to convey food to the mouth, Marsh, O'Shea, and Reynolds (2012) wrote that, while captives have used flippers to remove debris from their mouth parts, there is no evidence that the dugong, in contrast to the manatees, uses flippers to manipulate food. The dugong has a relatively small and simple stomach and also has unique adaptations for feeding on marine grasses. The upper lip protrudes considerably beyond the lower; it is deeply cleft, forming a large U-shaped, muscular pad that overhangs the small, downwardly opening mouth. On the sides of this facial pad are two ridges that bear short, sturdy, blunt bristles; the lower lip also has bristles. The distal portion of the palate and lower jaw have rough, horny pads that are used to grasp sea grasses during feeding. The whole upper lip is extended and curved around the base of a plant, which is then grasped with the mouth pads and pulled up by the roots. The bristles of the lips move out in a medial-to-lateral manner, like a swimmer's breaststroke, and then move back to sweep sea grasses into the sides of the mouth. The action is manipulative but not truly prehensile, in contrast to *T. manatus*, which moves its upper bristle fields in a lateral-to-medial grasping action (Marshall et al. 2003).

The adult dental formula is: i 1/0, c 0/0, pm 0/0, m 2–3/2–3 = 10–14. The upper incisors are rootless and straight, forming short, thick tusks in males more than 12–15 years old. In females the incisors usually do not pierce the gum, which implies that these teeth have a sex-related rather than a food-gathering function, and apparently are used in combat over mating rights with estrous females. Adult males invariably carry conspicuous scars that seem to have been made by competing males. The tusks also may in some way assist during mating and are used in exploratory activity (Marshall et al. 2003). The molars are circular in section, peglike,

and thick. The degenerate enameled crowns wear quickly, exposing the much softer and less wear-resistant dentine; the last two molars are open-rooted and grow throughout life, constituting the entire cheek dentition of most old adults (Marsh, O'Shea, and Reynolds 2012). Young animals have as many as six cheek teeth (including three premolars) on each side of each jaw; however, each row of cheek teeth migrates forward, with worn teeth dropping out anteriorly, until only two molars remain in each quadrant of old animals (Rathbun 1984).

The dugong occurs in the shallow waters of coastal regions of tropical seas, where there is an abundance of vegetation. It is more strictly marine than manatees and is seldom found in fresh water localities. Long-distance migration is unknown, but off some parts of Africa, southeastern Asia, and Australia seasonal changes in abundance are associated with the monsoons and may reflect movements in response to rough weather and availability of food; in some areas

Dugong (*Dugong dugon*). Photo by Mike Parry / npl / Minden Pictures.

there also seem to be regular daily movements between feeding areas and deeper waters (Husar 1978a). The dugong may rest in deep water during the day and move toward the shore to feed at night, but Nishiwaki and Marsh (1985) noted that there is diurnal inshore feeding in some areas. Anderson (1986) determined that a population in Shark Bay, off Western Australia, has cyclic movements, abandoning feeding sites when temperatures fall below 19° C and seeking warmer waters for the winter. Anderson (1984a) noted that the distance to those winter waters is more than 160 km. Hobbs et al. (2007) reported that a male dugong crossed at least 1,000 km of open sea to the Cocos Islands in the eastern Indian Ocean, the longest movement recorded for the species and demonstrating its capacity to colonize distant, unoccupied locations.

Average swimming speed is 10 km/hr, but animals can nearly double this rate if pressed (Husar 1978a). Dives seem to vary in length depending on such factors as water depth and forage species but have been re-

Dugong (*Dugong dugon*), close-up showing prehensile orofacial bristles. Photo by Mike Parry / npl / Minden Pictures.

ported normally to last around 1–3 minutes (Anderson 1982b; Nishiwaki and Marsh 1985). Feeding generally is at depths of 1–5 meters (Anderson 1984a). In a study of 15 individuals equipped with time-depth recorders off northern Australia, Chilvers et al. (2004) found 47 percent of daily activity within 1.5 meters of the sea surface and 72 percent less than 3 meters from the surface. Mean maximum dive depth was 4.8 meters, mean dive duration 2.7 minutes, and number of dives per hour averaged 11.8. Maximum dive depth was 20.5 meters, and maximum dive time was 12.3 minutes. Marsh, O'Shea, and Reynolds (2012) indicated that some dives reach depths of over 25 meters, evidently to reduce risk of shark attack from below.

Dugong is mostly herbivorous, and historical distribution broadly coincided with the tropical Indo-Pacific distribution of food plants, the phanerogamous sea grasses of the families Potamogetonaceae and Hydrocharitaceae. Although sea grasses are the primary food of the dugong, it has been reported to feed on brown algae following a cyclone that damaged the sea grass bed; green algae, marine algae, and some crabs also have been found in dugong stomachs (Husar 1978a). The dugong uproots whole plants when they are accessible, excavating long, serpentine, bare furrows, or "feeding trails," in sea grass meadows on the sea bottom; it feeds only on the leaves if a whole plant cannot be uprooted (Marsh and Dutton 2013; Marsh, O'Shea, and Reynolds 2012).

According to Anderson (1982a), population densities in Shark Bay range from 0.12 to 12.8 individuals per km of coastline. The dugong is essentially a gregarious animal, though it is frequently solitary. In the past, huge groups, sometimes containing thousands of individuals, were reported in various areas. Groups of 12 to about 300 have been reported in modern aerial surveys, and the largest aggregation on record, estimated to contain some 670 animals in two main groupings, was in the Arabian Gulf in winter at a presumed warm-water source (Marsh, O'Shea, and Reynolds 2012). Such groups are transitional and numbers usually are far smaller. Anderson (1982a) suggested that some aggregations do not form simply in response to favorable conditions but may involve deliberate assembly for protection against predators (especially sharks) and a process by which younger animals learn advantageous patterns of movement. Most dugongs in Shark Bay were found in the company of at least one other individual, and about 10–12 percent of the animals there were calves. Additional information, as summarized by Husar (1978a) and Lekagul and McNeely (1977), suggests there may be lasting pair bonds between mated dugongs and that family groups may form within the larger groups. Marsh and Dutton (2013) stated that the only defined long-lasting social unit is the cow and her calf, but also suggested that social behavior is variable.

Anderson (2002) discussed two apparently different mating systems in *Dugong*. The first, "lek polygyny," was observed over 4.5 months in a shallow cove in Shark Bay, Western Australia. Some 13–16 solitary adults, assumed to be males, continuously occupied, patrolled, and defended mutually exclusive territories, 2 km or less in maximum extent, frequently vocalizing during such activity. Confrontations with neighbors or intruders took place at territorial boundaries and frequently evolved into violent fighting that lasted less than a minute and terminated in pursuits extending for several hundred meters as the intruder fled. Known or assumed females entered a territory several times and engaged in apparent mating behavior with the resident. The cove had very little edible vegetation and no apparent attraction for females other than mating, and thus appeared to be a classic lek as described for other mammals, particularly ungulates. In contrast, "scramble promiscuity," comparable to but not as persistent as that in manatees (*Trichechus*), has been reported for *Dugong* in Moreton Bay in southeastern Queensland. There, clusters of up to 20 adults, presumably males, have been seen apparently following a focal individual, presumably a female. The followers sometimes engaged in violent combat, and some appeared to engage in mating behavior with the relatively passive focal animal. Marsh, O'Shea, and Reynolds (2012) referred to other such mating herds and also to observations of lone pairs of mating dugongs, but noted that the dugong is probably promiscuous, with multiple males mating with individual females during a single estrus.

Dugong (*Dugong dugon*) fin walking. Photo by Doug Perrine / npl / Minden Pictures.

Dugong has been reported to make whistling sounds when frightened; calves make a bleating, lamblike cry. Anderson and Barclay (1995) recorded a variety of "chirp-squeaks, barks, trills," and other sounds, some quite complex, which may be associated with identification and territorial advertisement and defense. Using playback of recorded sounds, Ichikawa et al. (2011) found that dugong chirps, but not trills, apparently functioned to signal how far one animal is from another.

Husar (1978a) wrote that breeding seems to occur throughout the year, with no well-defined season, though births apparently peak from July to September in Sri Lanka and parts of Australia. Marsh, Heinsohn, and Marsh (1984) reported that most births in northeastern Australia take place from September to December. Based on data from Australia, Anderson (2002) indicated that mating takes place from August to January, females are polyestrous, behavioral estrus may last just one day, the gestation period is 12–14 months, and parturition occurs from July to March. Normally, there is a single young; twins are reportedly very rare and have not been scientifically confirmed (Marsh, O'Shea, and Reynolds 2012). The newborn is 100–120 cm long and weighs 20–35 kg (Nishiwaki and Marsh 1985). It is born underwater and swims immediately to the surface for its first breath of air. The baby clings to its mother's back as she browses through the shoals of sea grass, submerging when the mother submerges and rising when she rises (Lekagul and McNeely 1977). The young have been seen to suckle while underwater in an inverted position behind the mother's axilla, and they are sometimes carried above the water when the mother rolls (Anderson 1984b). Lactation may last at least 18 months, though the calves begin to graze within 3 months of birth. Calves may sometimes separate from the herd to form nursery subgroups of their own. Males reach sexual maturity when 9–15 years old, females bear their first calf when 6–17 years old, calving interval is around 3–7 years, and maximum rate of population increase is thought to be about 5 percent (Marsh and Dutton 2013; Marsh, Heinsohn, and Marsh 1984). A captive male was still living at an age of 25 years (Weigl 2005), but examination of wild specimens indicates maximum longevity is about 73 years (Marsh 1995).

Sharks are probably the main natural enemy of the dugong, but individuals have been seen to "gang up" on sharks in shallow water and drive them off by butting them with the head (Lekagul and McNeely 1977). At times of tiger shark (*Galeocerdo cuvier*) abundance, the dugong has been found to feed more intensively on the edges of sea grass meadows, where escape routes into deep water refuges are available, even though forage quality in such habitat is lower than in the more productive but dangerous central portions of the meadows (Wirsing, Heithaus, and Dill 2007). A report of a devastating attack by about 10 killer whales on a tightly bunched group of approximately 40 dugongs was related by Anderson and Prince (1985).

People have had a far more serious long-term effect. Specialized human cultures based on dugong hunting have developed in several areas, such as the Torres Strait, between Australia and New Guinea (Nishiwaki and Marsh 1985). However, the dugong has been hunted for food practically everywhere within its range, its meat being likened to tender veal. It has been killed in large numbers, both historically and in recent years and both for subsistence and for commercial marketing. An average specimen yields 115 kg of meat and fat and 17 liters of oil (Dobbs, Lawler, and Kwan 2012). The hide has been used to make a good grade of leather and the bones and teeth used to make ivory artifacts and a good grade of charcoal for sugar refining. Several Asian cultures have prized dugong products for supposed medicinal and aphrodisiac properties. Large gill nets, harpooning, explosives, and even cyanide are used to capture and kill the dugong (Husar 1978a; Marsh et al. 2002). Accidental entanglement and drowning of the dugong in mesh nets and traps, set by persons seeking to catch fish, is a serious problem and often the most severe threat reported throughout the range of the species (Marsh, O'Shea, and Reynolds 2012). Furthermore, the coastal sea grass ecosystems, on which the dugong depends, are very sensitive to human pressures (Marsh et al. 2002). Sea grass beds may be destroyed directly by mining and trawling or lost through reduced light intensity and smothering by the sedimentation and turbidity resulting from dredging, land clearing, and other environmental disruption. Additional concerns include chemical pollution, oil spills, vessel strikes, and acoustic disturbance.

Marsh et al. (2002) considered the dugong to be represented throughout much of its range by relict populations separated by large areas where its numbers have

Dugong (*Dugong dugon*) feeding. Photo by Kristina Vackova / Shutterstock.com.

been greatly reduced or where it is already extirpated. Marsh, O'Shea, and Reynolds (2012) estimated the dugong's remaining extent of occurrence as 860,000 sq km spanning approximately 128,000 km of coastline across at least 38 countries. Marsh and Sobtzick (2015) compiled data indicating a sharp decline of the dugong along the east coast of Queensland since the 1960s; they suggested that if the magnitude of that decline was typical of the entire range of the species, most of which is currently experiencing the same and even more severe causative factors, the species would qualify for an IUCN classification of critically endangered. The information for the remainder of this account was derived from those three sources, as supplemented by the other references cited.

Regionally, the dugong may be in greatest jeopardy in the western part of its distribution, along the East African coast, around nearby islands, and in the Red Sea. There have been no extensive surveys in the Red Sea since 1987, when it was estimated that up to 4,000 individuals might be present there, but the area is increasingly subject to habitat disturbance and hunting pressure, especially on the western side (Preen et al. 2012). Surveys along the coast of Egypt in 2001–2003 found just 12–17 animals (Hanafy et al. 2006). Occasional reports come from the Gulf of Aqaba, as far north as Eilat, Israel. To the south, the dugong was formerly abundant, with herds of around 500 reported off Somalia and Kenya in the 1960s. Numbers declined precipitously there and in Tanzania in the 1970s and 1980s (Muir and Kiszka 2007). Most remaining groups are extremely small and fragmented. Currently the most significant population in the entire western Indian Ocean region, with an estimated 250 animals, is in the Bazaruto Archipelago off southern Mozambique (Muir and Kiszka 2012). Some hundreds may also survive around Madagascar, but probably just a few individuals still occur in the Comoros Archipelago and at Mayotte Island and Aldabra Atoll, and there are no recent records from the Seychelles. According to Haskins and Davis (2008), "huge herds were freely harvested" in the Mascarene Islands (Mauritius, Rodrigues, Réunion) in the seventeenth century; numbers had been reduced by the eighteenth century, and there are no definite records since the nineteenth century, though there was an unconfirmed sighting in 1999.

A remarkable contrast is presented by the situation in the Arabian Gulf (Preen et al. 2012). There is evidence that the dugong was hunted in Abu Dhabi about 7,000 years ago, and the species was still being marketed for food there at least into the 1980s. Moreover, the gulf is the most oil-polluted marine area in the world and in recent years has been beset by other environmental disruption and war. Until the 1980s there were suggestions that the population of the entire Gulf contained fewer than 100 individuals and was on the

brink of extinction. Since then, however, widespread aerial surveys, together with research and conservation efforts by some of the Gulf States, have provided a new perspective. The Arabian Gulf, particularly the waters off Saudi Arabia, Bahrain, Qatar, and the United Arab Emirates, is now thought to contain a stable dugong population of close to 6,000 animals, the largest outside Australia.

Eastward, the dugong's status becomes more typical. The species was common along the coasts of southern India and northern Sri Lanka in the nineteenth and early twentieth centuries, with groups of hundreds reported from the intervening Gulf of Mannar, but very few animals are thought to remain. There also are records from the west coast of India as far north as the Gulf of Kutch, where a small, isolated population may persist. It is not known if any animals exist on the coasts of Pakistan or Bangladesh. India's Andaman and Nicobar Islands once supported large dugong populations, but numbers have been declining since the 1950s, and

Dugong (*Dugong dugon*), mother and young. Photo by Geoff Taylor / Lochman LT.

only about 100 individuals are estimated to survive there. There have been repeated statements that the dugong is extinct in India's Maldive and Lakshadweep (Laccadive) island groups (Hines et al. 2012; Marsh, O'Shea, and Reynolds 2012; Marsh and Sobtzick 2015; Marsh et al. 2002; Nishiwaki and Marsh 1985). However, as a source for that information, authorities generally cite Husar (1975, 1978a), who in turn cites Snow (1970), who actually does not refer at all to the dugong's status in those islands, past or present.

Little seems known of the dugong on the coasts of Southeast and East Asia, though generally small, scattered groups remain from among once larger populations. One group was recently found to persist along the northern portion of the Burmese coast (Hines et al. 2012). One of the largest known populations, about 200 animals, is found on the western or Andaman coast of Thailand (Hines, Adulyanukosol, and Duffus 2005), while another 50 are on the eastern coast in the Gulf of Thailand. To the south, the dugong continues to exist around peninsular Malaysia and Singapore, including a possible breeding population in the Johor River estuary. The species once occurred all along the coast of Cambodia and Vietnam, at least as far north as Haiphong, but numbers have fallen precipitously since 1975. Presence of a remnant population off eastern Cambodia and around the Vietnamese islands of Phu Quoc and Con Son has been recently confirmed, though there is also an active market in that area for meat and dugong body parts used for medicinal purposes (Hines et al. 2012). Hunting for such products is thought to have greatly reduced dugong numbers in neighboring south China since the 1960s, though the species still exists around Hainan Island and along the nearby mainland coast. A population once inhabited the Pearl River estuary adjacent to Hong Kong, but the last records from that area, as well as Taiwan, date from the 1940s (Nishiwaki and Marsh 1985). A very small, permanent population still exists around Okinawa, though it is much reduced since the early twentieth century and is considered critically endangered, in part because landfill for construction of an American military airfield threatens important sea grass beds; individuals also occur on occasion in the rest of the Ryukyu Islands, from Amami Oshima in the north to Iriomote in the south (Ikeda and Mukai 2012; Shirakihara et al. 2007).

The East Indies, consisting mostly of Indonesia, insular Malaysia, and the Philippines, represent about 50 percent of the dugong's potential habitat, but information is limited. The species is thought to occur in the inshore waters of all island groups, but all evidence indicates that local populations are fragmented and very small, containing just around ten individuals each. Until the 1970s the dugong was believed fairly common throughout the Philippines, but now groups are sparse and scattered, and many places are no longer occupied. In the Sulu Sea, between the Philippines and Borneo, dugong numbers appear to be low and the species rare (Rajamani 2013). The total number of animals in Indonesia has been estimated variously at 1,000 or 10,000.

Populations farther east, in Papua New Guinea, the Solomon Islands, Vanuatu, and New Caledonia, were considered by Dobbs, Lawler, and Kwan (2012) to be mostly relictual and declining. The southern coast of Papua New Guinea is along the Torres Strait, which holds the world's largest dugong population (see below), though most islands in the strait belong to Australia. Survival of the dugong in the Solomons was considered uncertain by Bass (2010), because of subsistence hunting and degradation of coastal habitat, though she reported increasing numbers in some areas. The situation in Vanuatu, where suitable habitat is restricted, is not well understood, but individuals or small groups evidently occur throughout the archipelago. In contrast, recent surveys of the coastal waters of New Caledonia yielded a population estimate of about 1,800 animals, the highest in the world other than in Australia and the Arabian Gulf (Garrigue, Patenaude, and Marsh 2008).

The magnitude of the Australian populations was long not understood. Although early accounts suggested abundance, such as an 1893 report of a herd about 8 km long and 300 meters wide in Moreton Bay, southeastern Queensland, there was an overall impression of increasing rarity throughout the country by the early twentieth century (Allen 1942). Numbers in Australia were included by Nishiwaki et al. (1979) in an estimate of roughly 30,000 animals in the entire Indo-Pacific range of the species. Nishiwaki and Marsh (1985) indicated that extensive aerial surveys since 1974 were showing relatively large populations off Australia, but that derivation of numerical estimates was not yet possible. Reynolds and Odell (1991) did suggest the

Dugong (*Dugong dugon*). Photo by Doug Perrine / npl / Minden Pictures.

total number in Australian waters might exceed 80,000 and probably was more than half the world's total. Subsequently published totals have remained around that figure. A tabulation of recent estimates from aerial surveys (Marsh, O'Shea, and Reynolds 2012) provides the following approximate numbers for the various parts of the Australian range: Shark Bay, Western Australia, 12,000; remainder of the coast of Western Australia, 3,000; coast of Northern Territory west of Gulf of Carpentaria, 15,000; Gulf of Carpentaria 12,000; Torres Strait, 17,000; eastern coast of Cape York Peninsula, northern Queensland, 9,000; waters of Great Barrier Reef south of Cooktown, eastern Queensland, 2,000; Hervey Bay and Moreton Bay, southeastern Queensland, 3,000.

Notwithstanding those seemingly high numbers, at least some of the Australian populations are in jeopardy. In particular, off that part of eastern Queensland south of Cooktown, dugong numbers have fallen drastically since the early 1960s, when as many as 72,000 individuals may have been present (Marsh et al. 2005). The main factor is loss of sea grass habitat because of runoff from coastal development and agriculture. Remedial measures have not been fully effective, and the dugong is now regarded as critically endangered in that area. To the north, a longstanding conflict centers on traditional indigenous hunting of the dugong in the Torres Strait between Cape York and New Guinea. The hunt provides the native people with meat and oil, is an important aspect of their culture, and is legal pursuant to a treaty between Australia and Papua New Guinea. As many as 1,226 animals have been reported killed annually, and the actual number is probably considerably higher, but the sustainable take has been calculated at just 80–190 per year (Heinsohn et al. 2004; Marsh et al. 2004). Aerial surveys show the Torres population to fluctuate in size over time, apparently because of movements in and out of the area in response to sea grass conditions. However, it is thought that there has been a general decline of around 30 percent since the mid-twentieth century and, without restrictions on hunting, extinction is probable in the twenty-first century.

The estimated numbers given above for the different parts of Australia total 73,000. The information given above for other regions of the dugong's range suggests the following plausible totals: Red Sea, 2,000; the rest of the East African coast and nearby islands, 2,000; Arabian Gulf, 6,000; India, Sri Lanka, and Andaman and Nicobar Islands, 1,000; Southeast Asian coast, 2,000; East Indies, 10,000; and islands of the western Pacific (including Palau, see below), 3,000. The overall total is close to that suggested by Marsh and Dutton (2013), who thought there were likely more than 100,000 worldwide. While that figure may seem encouraging, the difficulties confronting even the large and relatively well-protected populations of Australia, as well as the multiplicity of problems elsewhere, fully justify the current classification of the dugong as vulnerable by the IUCN (Marsh and Sobtzick 2015) and its placement on appendix 1 of the CITES.

The official USDI List of Endangered and Threatened Wildlife indicates the dugong is classified as endangered throughout its range. As explained by Babij (2003), the species had that same classification when the U.S. Endangered Species Act of 1973 came into effect. However, in 1988 the USDI announced that, because certain required procedures had not been followed at the time of original listing, the classification did not apply within present or former territory of the United States. At that time, the Trust Territory of the Pacific, including Palau, was under the jurisdiction of the United States, and thus the dugong population there was not covered by the USDI listing. In August 1993 the USDI proposed to extend endangered status to that population, but further action was continually delayed. Throughout the bureaucratic process, data accumulated showing that the Palau population, among the most isolated in the world, numbered fewer than 200 individuals and was critically endangered by unsustainable subsistence hunting and habitat loss (Marine Mammal Commission 1994; Marsh et al. 1995; Marsh et al. 2002; Rathbun et al. 1988). Not until December 2003, 10 years after the proposal and 30 years

after the original error, was the final endangered listing rule issued (Babij 2003). Meanwhile, however, Palau had become an independent republic. And as the final rule referred only to Palau, the legal status of any dugong appearing off Guam or other Pacific islands still under U.S. jurisdiction seems questionable.

SIRENIA; DUGONGIDAE; **Genus** *HYDRODAMALIS* Retzius, 1794

Steller's Sea Cow, or Great Northern Sea Cow

The single Recent species, *H. gigas*, occurred in historical time around Bering and Copper Islands in the Commander Islands of the western Bering Sea (Rice 1977), and likely also to the southeast in the western Aleutian Islands (Domning, Thomason, and Corbett 2007) and to the northeast at St. Lawrence Island (Crerar et al. 2014). In the late Pleistocene, that species evidently had occurred all along the rim of the North Pacific from Japan to California, but subsequent hunting by primitive peoples probably restricted it to the Bering Sea region (Domning 1978).

The modern discovery of the species came in 1741, when a Russian expedition led by Captain Vitus Bering was stranded on the uninhabited island that now bears his name. The ship's crew, mostly sick with scurvy, eventually found the meat and fat of *H. gigas* agreeable and easily procured. Subsequent visitors to the Commander Islands also hunted the sea cow for food. The hide was used for making skin boat covers and shoe leather. Most of what we know of the appearance and habits of the living animal, as well as the internal anatomy of its soft parts, comes from the accounts of Georg Wilhelm Steller (1751, 1753, 1925), a German naturalist who accompanied the Bering expedition. Steller's writings were exhaustively studied and translated by Leonhard Stejneger (1936), a biologist with the United States National Museum, who spent much time collecting specimens on Bering Island in the late nineteenth century and assembling records and reports of voyagers who followed Steller and who also referred to the sea cow. A lengthy discussion of the work of Steller, Stejneger, and other observers,

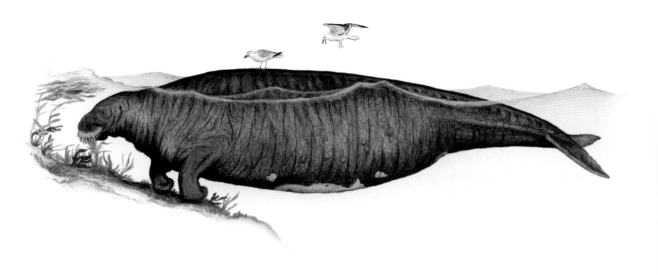

Steller's sea cow (*Hydrodamalis gigas*). The external appearance of the sea cow is controversial, and the numerous depictions thereof vary considerably, especially with respect to size of head, arrangement of orofacial bristles, structure of limbs, and shape of tail and flukes. This reconstructive painting by Helga Schulze is based primarily on her examination of photographs of skeletal material and review of the original descriptions by Steller (1751, 1753, 1925).

together with interpretations in relation to modern sirenian biology, was provided by Marsh, O'Shea, and Reynolds (2012). An assessment of the characteristics of *H. gigas*, based on examination of skeletal remains and accounts of persons who saw live specimens, was made by Domning (1978).

H. gigas was the largest Recent sirenian and possibly the largest non-cetacean mammal to exist in historical times. An adult female measured by Steller was 752 cm from the tip of the nose to the point of the tail flipper. Its greatest circumference was 620 cm, and its estimated weight was about 4,000 kg. Scheffer (1972) and Scheich et al. (1986) calculated that the largest specimens were at least 800 cm long and weighed 10,000 kg. A more refined analysis by Domning (1978) indicates an 800-cm specimen would have weighed just 6,300 kg but that an animal 10 meters long, a length probably reached by the fossil species *H. cuestae* of California, would have weighed 11,196 kg. Since *H. gigas* also was formerly present in warmer southerly waters, where growth potential was optimal, it would have been expected to attain the size of *H. cuestae*.

The head of *Hydrodamalis* was small in proportion to the body, much more so than in *Dugong* and *Trichechus*. The nostrils were at the forward tip of the head, the eye openings were barely 13 mm across and reportedly lacked lids, and there was no external ear. From the shoulders to the umbilical region the body grew rapidly wider, and from there to the anus it again grew slender. The tail had two pointed lobes forming a caudal flipper. The forelimbs were unlike the flippers of living sirenians, being small, somewhat clawlike, and truncated by absence of the phalanges. There were no hind limbs externally, but internally there was a stout pelvic bone resembling that of *Dugong*. The skin was naked and was covered with a very thick, barklike, extremely uneven-appearing epidermis, from which the German name *Borkèntier* (bark animal) was derived. The skin was dark brown to gray brown in color, occasionally spotted or streaked with white. The forelimbs were covered with short, brushlike hairs. There is a large dried sample of this skin at the Zoological Institute in St. Petersburg, but a reported specimen in the Hamburg Zoological Museum may actually represent a cetacean (Forstén and Youngman 1982).

According to Forstén and Youngman (1982), the condylobasal length of the skull was 638–722 mm, the zygomatic width was 324–373 mm, the nasal basin on top extended past the eye opening, and the braincase was relatively small. *Hydrodamalis* had no functional teeth; mastication was accomplished by keratinized pads, one each in the rostral areas on the upper and lower jaws. The upper and lower pads had corresponding V-shaped crests and valleys, between which the food was ground.

Following are excerpts from Steller's own account:

Usually entire families keep together, the male with the female, one grown offspring and a little, tender one. To me they appear to be monogamous. They bring forth their young at all seasons, generally however in autumn, judging from the many new-born seen at that time; from the fact that I observed them to mate preferably in the early spring, I conclude that the fetus remains in the uterus more than a year. That they bear not more than one calf I conclude from the shortness of the uterine cornua and the dual number of mammae, nor have I ever seen more than one calf about each cow.

These gluttonous animals eat incessantly, and because of their enormous voracity keep their heads always under water with but slight concern for their life and security, so that one may pass in the very midst of them in a boat even unarmed and safely single out from the herd the one he wishes to hook. All they do while feeding is to lift the nostrils every four or five minutes out of the water, blowing out air and a little water with a noise like that of a horse snorting. While browsing they move slowly forward, one foot after the other, and in this manner half swim, half walk like cattle or sheep grazing. Half the body is always out of water. . . . In winter these animals become so emaciated that not only the ridge of the backbone but every rib shows.

Their capture was effected by a large iron hook . . . the other end being fastened by means of an iron ring to a very long, stout rope, held by thirty men on shore. . . . The harpooner stood in the bow of the boat with the hook in his hand and struck as soon as he was near enough to do so, whereupon the men on shore grasping the other end of the rope pulled the desperately resisting animal laboriously towards them. Those in the boat, however, made the animal fast by means of another rope and wore it out with continual blows, until, tired and completely motionless, it was attacked with bayonets, knives and other weapons and pulled up on land. Immense slices were cut from the still living animal, but all it did was shake its tail furiously.

Hydrodamalis was the only Recent sirenian adapted to cold waters. It was quite numerous in the shallow bays and inlets around Bering Island when first discovered. It may have been restricted largely to a zone within a few hundred meters of the coast, where it could feed upon the kelp beds and extensive growths of various marine algae that grew in the shallower waters. It also was said to prefer areas near the mouths of streams, where apparently it could obtain fresh water. It swam slowly, usually against a current, propelling itself by undulations of its tail. It was never reported to dive, even when pursued by hunters, or even to completely submerge, as part of its back always was seen above the surface; gulls would alight thereon and feed on skin parasites. Domning (1978) stated that floating was the preferred posture for normal activity, unlike in any other known marine mammal, but that buoyancy may not have been fully obligate, as the ability for total submergence under adverse conditions probably would have been too valuable to lose. As noted by Steller, the small forelimbs sometimes were used for locomotion in shallow water, their clawlike structure providing traction on the sea bottom. Those appendages also served to brace and support the animal, to scrape seaweed

Steller's sea cow (*Hydrodamalis gigas*). This reconstruction features the relatively smaller head, simplified limbs, and extended tail shown in some depictions. © Martin Camm / Wildlife Art Co. / Minden Pictures.

from rocks, and perhaps to fight conspecifics and assist in mating.

Based on Steller's observations, extrapolation from living sirenians, and ecological conditions, Anderson (2002) developed a hypothesis regarding the social and reproductive aspects of *H. gigas*. Females may have established and defended linear shoreline territories, and males may have fought one another for mating rights. Stormy open coasts, a narrow habitat, and spacing of females would have favored limited group size and pair bonding. The species thus would have been monogamous, living in groups consisting of an adult pair and one or more offspring. The male courted and protected the female, even attempting to follow and assist her if she had been hooked by human hunters. Large animals, perhaps including the older offspring in a family, protected the young. Selection for a short breeding season became essential for the young to develop sufficiently to survive the rigors of winter. Steller had reported mating in early spring and births at any time of year, most frequently in the autumn, but he was in the area only from November 1741 to August 1742. Concentration of births at the onset of winter would seem unlikely; Steller may have seen many young in November that had been born the previous spring. Probably the species was polyestrous, with a February–March mating season, a gestation of 15–16 months, parturition in June–July, a lactation period of 36–48 months, and an interbirth interval of 5–10 years. Also extrapolating from living sirenians, Turvey and Risley (2006) assumed a life expectancy of 90 years for *H. gigas*.

The ancestral lineage of the subfamily Hydrodamalinae apparently entered the North Pacific in the early Miocene when the Central American land bridge was absent, and began to evolve away from dependence on tropical sea grasses and toward adaptations for a cooler climate and feeding on kelp and algae (Domning 1978; Domning and Furusawa 1994; Hitoshi and Naoki 1994). The earliest known species of *Hydrodamalis*, *H. cuestae*, has been recorded from the late Miocene to the late Pliocene of California and Baja California. Another fossil species, *H. spissa*, is known from the early Pliocene of Japan. *H. gigas*, which was present in historical time, also is represented by early to middle Pleistocene fossils from Japan and late Pleistocene remains from Amchitka Island in the Aleutians and Monterey Bay off California.

Until recently, definitive post-Pleistocene records and remains of *H. gigas* were known only from the Commander Islands, where many specimens were recovered subsequent to extinction of the species (Mat-

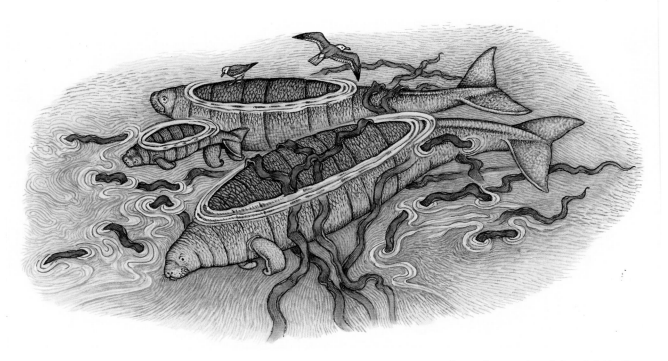

Steller's sea cow (*Hydrodamalis gigas*), family group. Painting by Patricia A. Wynne / American Museum of Natural History through Ross D. E. MacPhee.

tioli and Domning 2006). Some of the specimens have been radiocarbon dated from around 2,500 to 500 years before the present but most seem associated with the slaughter by Europeans that began when Bering's expedition arrived in 1741. There is no evidence that any humans visited the Commander Islands before then. However, the Aleutian Islands, which start about 400 km to the southeast, had been inhabited by native peoples for thousands of years. Specimens from archeological sites on Buldir, Kiska, and Adak islands indicate the sea cow may have been hunted there around 2,000–1,000 years ago, and a detailed account taken from people on Attu, at the western end of the Aleutians, suggests a substantial population of *H. gigas* survived there until the eighteenth century (Domning, Thomason, and Corbett 2007; Savinetsky, Kiseleva, and Khassanov 2004). Bone samples recently collected on St. Lawrence Island, approximately 1,600 km northeast of the Commanders, provide evidence that a sea cow population there became extinct about 1,100 years ago, because of climate change and/or hunting by the Inuit people (Crerar et al. 2014). A number of reports of historical occurrence at other localities, based on observations, specimens, and local tradition, were discussed by Domning (1978) and considered doubtful.

The sea cow population of the Commander Islands is the only one with a reasonably documented history. Stejneger thought that upon discovery in 1741 there probably were no more than 1,500 animals around Bering Island. Domning (1978) suggested a total of 2,000 for Bering and the smaller Copper Island. Both authorities considered that hunting quickly drove the population to extinction. However, another postulated factor was depletion of the sea cow's food source, kelp, by exploding numbers of sea urchins that had been held in check by sea otter (*Enhydra lutris*) predation (Anderson 1995). When Bering's crew returned to the mainland and spread word that the Commanders had an abundant population of *Enhydra*, which has an extremely valuable pelt, fur-hunting expeditions began to visit the islands regularly, where they killed both the sea otter for commerce and the sea cow for food. Turvey and Risley (2006) considered those factors in a model of sea cow extinction dynamics, and concluded that loss of the sea otter and increase of sea urchins did not contribute to the situation. Rather, the sea cow population, which they thought may originally have contained as many as 2,900 animals, was exploited by the fur hunters and other visitors seeking provisions in a highly wasteful manner and at a rate more than seven times above sustainability. The sea cow apparently was eliminated off Copper Island before 1754 and almost certainly around Bering Island by 1768, after which it was not reported by expeditions to the area. There were accounts of its later presence in various parts of the North Pacific, but such were dismissed by Domning (1978). *H. gigas* is classified as extinct by the IUCN (Domning, Anderson, and Turvey 2008).

SIRENIA; **Family TRICHECHIDAE; Genus** *TRICHECHUS* Linnaeus, 1758

Manatees

The single Recent genus, *Trichechus*, contains three species (Anderson 1997; Caldwell and Caldwell 1985; Domning 1994; Domning and Hayek 1986; Husar 1977, 1978b, 1978c; Keith Diagne 2015; Lefebvre et al. 2001; Marmontel, Weber Rosas, and Kendall 2012; Rice 1977; Self-Sullivan and Mignucci-Giannoni 2012):

T. inunguis (Amazonian manatee), throughout the Amazon Basin of northern South America, from Ilha de Marajó as far as northeastern Ecuador and eastern Peru and possibly into extreme northern Bolivia, also in lower reaches of Tocantins River;

T. manatus (West Indian manatee), coastal waters and some connecting rivers from Virginia to Florida and around the Gulf of Mexico and the Caribbean Sea to eastern Brazil, Orinoco Basin, Greater and formerly Lesser Antilles, sporadically in the Bahamas;

T. senegalensis (West African manatee), coastal waters and connecting rivers and lakes from the Senegal River and tributaries on the border of southern Mauritania to the Longa River of central Angola, occurs over 2,000 km up the Niger River and 1,600 km up the Benue River into southeastern Chad, isolated population formerly in Lake Chad and its southern tributaries (see below).

West Indian manatee (*Trichechus manatus*). Photo by Hans Leijnse / NiS / Minden Pictures.

Using morphological characters, Domning and Hayek (1986) recognized two living subspecies of *T. manatus*, *T. m. latirostris* in the southeastern United States (concentrating in Florida during winter) and *T. m. manatus* in the rest of the range of the species. Analysis of mitochondrial DNA by Garcia-Rodriguez et al. (1998) indicated three distinctive lineages of *T. manatus*, one centered in Florida and the West Indies, one from coastal Mexico to the rivers of South America that flow into the Caribbean, and one on the northeastern Atlantic coast of South America. The three groupings were reportedly separated from one another by the same level of genetic divergence as that between the species *T. manatus* and *T. inunguis*. Domning (2005) suggested the molecular similarity in Florida and the West Indies might indicate relatively recent dispersal from the latter area after an earlier population in Florida had disappeared during the cold of the late Pleistocene (see below). A new study, emphasizing nuclear DNA, showed substantial genetic divergence

between the manatee populations of Florida and Puerto Rico and supported the distinction of *T. m. latirostris* and *T. m. manatus* (Hunter et al. 2012). Meanwhile, another study of mitochondrial DNA (Cantanhede 2005) generally supported the work of Garcia-Rodriguez et al. (1998), but suggested the lineage of *T. manatus* in Florida and the West Indies (and also Colombia) is more closely related to *T. inunguis* than that lineage is to the other two lineages of *T. manatus*. Therefore, *T. manatus* would be paraphyletic, and it would become taxonomically necessary either to treat all New World *Trichechus* as a single species or to treat each lineage of *T. manatus*, as well as *T. inunguis*, as separate species.

Any need for such choice may have been forestalled by a more encompassing assessment of both mitochondrial and nuclear DNA (Vianna et al. 2006), which indicated the possibility of paraphyly but considered it more likely that *T. inunguis* represents an ancient lineage, separate from the three groups of *T. manatus*, which in turn are more closely related to *T. senegalen-*

West Indian manatee (*Trichechus manatus*), close-up showing prehensile orofacial bristles. Photo by Ethan Daniels/Shutterstock.com.

sis. Such a situation would correspond to the morphological assessment of Domning (1982b, 1994; see below). The three genetic groupings of *T. manatus* ascertained by Vianna et al. (2006) were considered "very distinct" but to have a somewhat heterogeneous geographic distribution: (1) Florida, Mexico, Greater Antilles, Central America, and Caribbean coast of South America; (2) Mexico, Central America, and Caribbean coast of South America; and (3) northeastern coast of South America (Brazil and the Guyanas). Moreover, there was evidence that gene flow was restricted by a barrier following a line along the Lesser Antilles to Trinidad, which may have been more pronounced when sea levels were lower during glacial periods in the Pleistocene, and that another barrier separated Puerto Rico and Hispaniola from the rest of the West Indies. Vianna et al. (2006) also reported hybridization, including evidence of backcrossing, between *T. manatus* and *T. inunguis* in an area at the mouth of the Amazon where the two species are sympatric. The hybrid zone was thought likely to extend west toward Guyana, perhaps as a result of ocean currents pushing

the Amazon's discharge in that direction, thereby making the area tolerable for the fresh water-adapted *T. inunguis*. The latter species was found not to have the distinct lineages shown by *T. manatus*. However, Satizábal et al. (2012) reported evidence of two genetically distinguishable populations of *T. inunguis*, one in the Peruvian portion of the Amazon Basin, the other in Colombia and Brazil.

A population in the Rio Arauazinho, a 120 km-long tributary of the Rio Aripuanã in central Brazil, was recently reported as having some affinity to *T. inunguis* but to be a "dwarf manatee" with the specific name *T. pygmaeus* (Van Roosmalen 2015). The report included description of the skull of one individual, said to represent an adult male, and an account of a living individual, said to be an adult male about 130 cm long and 60 kg in weight, which was maintained in a state of semi-captivity for 4 months. The species was considered critically endangered, with a total population of fewer than 100 individuals. The same species, first named *T. bernhardi* and then *T. pygmaeus*, was originally described by Van Roosmalen, Van Hooft, and de Iongh (2007). However, as that original account was online only and is no longer available, it apparently was invalid pursuant to Chapter 3, Article 8 of the International Code of Zoological Nomenclature (http://iczn.org/iczn/index.jsp, accessed on 28 February 2016). According to Hammer (2008), D. P. Domning (Laboratory of Evolutionary Biology, Department of Anatomy, Howard University, Washington, DC) believed on the basis of the original account that *T. pygmaeus* actually comprised immature individuals of *T. inunguis*. Domning (pers. comm.) subsequently reviewed Van Roosmalen (2015) and considered both the skull and live animal to represent juvenile specimens of *T. inunguis*.

The range of *T. inunguis* sometimes is said to include the Orinoco Basin, but such reports apparently result from misidentification of *T. manatus* in the early nineteenth century (Thornback and Jenkins 1982). It also is unlikely that *T. inunguis* reaches the Casiquiare Canal or other natural connections between the Amazon and Orinoco basins (Domning 1981). *T. manatus* once was present as far south as the state of Espirito Santo at about 20° S on the coast of Brazil (Domning 1982a; Lefebvre et al. 2001). In 1964 *T. manatus* was introduced in Lake Gatun, part of the Panama Canal, to aid in control of aquatic vegetation, and the population

Amazonian manatee (*Trichechus inunguis*). Photo by Nick Gordon / npl / Minden Pictures.

there subsequently increased and reportedly extended its range to the Pacific Ocean (Domning and Hayek 1986; Montgomery, Gale, and Murdoch 1982; Muschett 2009). However, neither Mou Sue et al. (1990) nor Lefebvre et al. (2001) could confirm the presence of the manatee on the Pacific coast of Panama. In the United States, *T. manatus* concentrates in Florida during winter but moves across a much larger region in the warm seasons. Williams and Domning (2004) reported discovery of a rib of the species, radiocarbon dated at 2,055 years before the present, near the Little Miami River in southwestern Ohio, and also a radius-ulna, probably somewhat older, on a gravel bar on the Mississippi River in eastern Arkansas. Both were thought to represent natural occurrences, though there had been no historical records of *T. manatus* in the Mississippi River drainage. Subsequently, in October 2006, a manatee appeared at Memphis, Tennessee, having moved about 1,400 km up the Mississippi River, and there have been other recent occurrences along the east coast as far north as Massachusetts (Deutsch and Reynolds 2012).

Although *T. senegalensis* has been reported from the southern river tributaries of Lake Chad, its presence in the Lake itself is not well documented (Allen 1942; Powell 2013). According to Keith Diagne (2015), the species was once abundant in the Chad Basin, including the southern tributary Chari, Bamingui, Bahr-Kieta, and Logone Rivers. It had become rare in the region by 1924, apparently disappeared from Lake Chad by 1929, and now reportedly is gone from the southern tributaries as well. Lake Chad is greatly reduced in size because of drying and desertification and no longer provides much suitable habitat. Currently in Chad, the manatee occurs only in Lakes Léré and Tréne, as well as the Mayo-Kebbi River, which feeds the lakes from the east. Those waters connect to the Benue and Niger Rivers, which ultimately reach the Atlantic and thus are not part of the isolated Chad Basin. However, the headwaters of the Mayo-Kebbi do come within a few kilometers of the Logone River, and periods of flood could have provided means for the manatee to move between the two systems.

Manatees have a rounded body, a small head, and a squarish snout. The upper lip is deeply split, and each half is capable of moving independently of the other. The nostrils are borne at the tip of the muzzle, the eyes are small, and there are no external ears. The flattened tail fin is more or less evenly rounded (not notched as in the dugong and Steller's sea cow) and is the sole means of propulsion in adults. The flexible flippers are used for aiding motion over the bottom, scratching, touching, and even embracing other manatees, and moving food into and cleaning the mouth (Caldwell and Caldwell 1985). Vestigial nails are present on the flippers, except in *T. inunguis*. The skin is 5.1 cm thick. Bristlelike short hairs are scattered singly over the body at intervals of about 1.25 cm and have a tactile func-

tion (see account of order Sirenia). Females have two mammae, one on each side in the axilla behind the flipper.

The skeleton is of extremely dense bone (osteosclerosis), a condition that increases specific gravity and may contribute to neutral buoyancy (Caldwell and Caldwell 1985). The stout ribs consist wholly of dense bone. Manatees have only six neck vertebrae, whereas nearly all other mammals have seven. Nasal bones are often present in the skull (these bones are absent or vestigial in the dugong and sea cow). Manatees have two upper incisors, which are concealed beneath horny plates and are lost before maturity. The cheek teeth number up to 10 in each half of each jaw, though they include unerupted, incomplete ones in the rear. Eight fully erupted cheek teeth are commonly present. They are low-crowned, enameled, and divided into cuspidate cross-crests. They lack cement and have closed roots. The cheek teeth are

West African manatee (*Trichechus senegalensis*). Photo from Toba Aquarium, Japan.

replaced horizontally; they form at the back of the jaw and wear down as they move forward. This may be an adaptation to eating grasses containing silica phytoliths. A somewhat comparable process of replacement occurs in the order Proboscidea and in the genus *Peradorcas* of the marsupial order Diprotodontia.

According to Reep and Bonde (2006), the facial region of manatees is fleshy and contains an abundance of hair, about 30 times denser than that on the rest of the body. The upper and lower lips contain long, stiff bristles. The face has an expanded region, the oral disk, between the upper lip and nostrils. This region contains bristlelike hairs, intermediate in stiffness to the body hairs and perioral bristles, and which are used as tactile feelers. When feeding, a manatee expands and flattens the snout region, causing the perioral bristles of the upper lip to protrude outward from their usual relaxed position inside folds of skin. Simultaneously, the lower jaw drops down to open the mouth. Then the upper lip bristles move from each side toward each other in a sweeping motion as they gather vegetation and bring it into the mouth. As the lower jaw closes, its perioral bristles evert and push the food deeper into the mouth. In contrast to the feeding method of *Dugong*, all trichechids use their upper bristle fields in a prehensile or grasping manner, moving them in a lateral-to-medial direction (Marshall et al. 2003).

Rathbun (1984) wrote that the known geological range of the family Trichechidae is early Pliocene to Recent in Atlantic North America, early Miocene to Recent in South America, and Recent in West Africa. Domning (1982b, 1994), however, indicated that Trichechidae arose in the late Eocene or early Oligocene and speculated that the family is descended from proto-sirenian stock that became isolated during the Eocene in South America. Until the late Miocene, trichechids may have been restricted to the coastal rivers and estuaries of that continent. By the Pliocene a branch of the family leading to modern *T. inunguis* had reached the interior of the Amazon Basin, where it adapted to feeding on the newly available floating meadows of nutrient-rich lakes. Another Pliocene branch, the forerunner of *T. manatus*, spread into the Caribbean, where its wear-resistant dentition proved superior to that of the dugongids, then present in the region, for dealing with a diet modified by the increased silt runoff of the period. During subsequent phases of climatic warming, Caribbean

T. manatus was able to disperse northward; periodic ecological barriers then developed, allowing evolution of a series of endemic subspecies in the United States, including *T. m. bakerorum* in the late Pleistocene, which became extinct, and the existing *T. m. latirostris* (Domning 2005). Meanwhile, in the Pliocene or Pleistocene, a closely related branch of *Trichechus* dispersed across the Atlantic to West Africa and gave rise to *T. senegalensis*. Additional information on each of the three living species is given separately below.

Trichechus inunguis (Amazonian Manatee)

Husar (1977) wrote that this species is smaller and more slenderly proportioned than either *T. manatus* or *T. senegalensis*; the largest recorded specimen was a 280-cm male. Timm, Albuja V., and Clauson (1986) cited a commercial hunter in Ecuador as saying that the largest manatee he had taken weighed 480 kg. Domning and Hayek (1986) stated that the skin is smooth, slick, and rubber-like, not wrinkled as in the other two species. The general coloration is gray, and most individuals have a distinct white or bright pink patch on the breast. Fine hairs are sparsely distributed over the body, and there are thick bristles on both the upper and lower lips. The flippers, unlike those of the other two species, are elongate and usually lack nails. The skull is relatively long and narrow.

The Amazonian manatee occupies a vast overall range but is patchily distributed, concentrating in areas of nutrient-rich flooded forest, which cover around 300,000 sq km (Marmontel 2008). It occurs exclusively in fresh water (Thornback and Jenkins 1982). It has been found all around Ilha de Marajó, at the mouth of the Amazon on the Atlantic coast, but even there salt water has little or no influence (Domning 1981). This species has been maintained successfully in waters with temperatures of 22–30° C (Husar 1977). According to Marmontel, Weber Rosas, and Kendall (2012), highly productive white water is the preferred habitat, but the species occurs throughout lowland forested areas below 300 meters and also occupies a great range of black and mixed water lakes, including expanses of *várzea* (floodplains and seasonally flooded forest), and *igapó* (permanently waterlogged swamp forest). It prefers shallow (1–4 meters deep), calm waters with easy access to patches of aquatic vegetation. It seems to re-

Amazonian manatee (*Trichechus inunguis*), mother and calf. Photo by Doug Perrine / npl / Minden Pictures.

quire connections to large rivers and abundant macrophytic plants such as grasses, bladderworts, and water lilies (Timm, Albuja V., and Clauson 1986). It usually surfaces once every couple of minutes to breathe, but the longest recorded submergence is 14 minutes (Husar 1977). A captive juvenile released and radio-tracked for 20 days was found to be equally active by day and by night, to move about 2.6 km per day, and to spend most of its time where food, especially floating vegetation, was most abundant (Montgomery, Best, and Yamakoshi 1981). The diet of *T. inunguis* consists mostly of vascular aquatic vegetation (Caldwell and Caldwell 1985). Captive adults consume 9–15 kg of leafy vegetables per day (Husar 1977).

Observations by Best (1983) and a radio-tracking study by Arraut et al. (2010) showed that manatee populations of the central Amazon Basin make annual movements and also undergo fasting for much of the year. From about February or March to late June, when water levels are high, the animals stay mainly in *várzea* lakes. In July and August, when water levels begin to fall, some individuals enter rivers, where they can continue to find food but must also expend energy to hold their positions in the current. However, most manatees become restricted to the deep parts of lakes, especially the long narrow *rias*, during the dry season from around September to March. Were the manatees to remain in the *várzeas*, particularly during the time of lowest water, October–November, the drastic reduction in aquatic space would lead to risk of habitat drying out and also to greatly increased vulnerability to predators, such as caimans, jaguars, and humans. However, in the *rias*, there are no obvious food sources until water levels rise 1–2 meters; available evidence suggests that the animals there fast for nearly 7 months, though possibly they find some vegetation. The manatee's large fat reserves and low metabolic rate—only 36 percent of the usual rate for eutherians—allow the species to survive at this time. Both studies were done primarily around Ria Amanã in the Brazilian state of

Amazonas, where Best (1983) reported 500–1,000 manatees to congregate during the dry season.

There have been other reports of large aggregations of manatees in the middle reaches of the Amazon, but such assemblies appear rare today. Loose groups of 4–8 individuals may be present in feeding areas. As with *T. manatus*, estrous females reportedly are pursued by groups of males (Marmontel, Odell, and Reynolds 1992). A study of captives (Sousa-Lima, Paglia, and da Fonseca 2002) found vocalizations to consist of one to four notes and to be mostly harmonic, with 1–12 frequency bands; females tended to have greater fundamental frequency and shorter note duration than males. A mother-calf pair showed similar patterns that may involve individual recognition.

Breeding has been reported to occur throughout the year, at least in some areas, but Timm, Albuja V., and Clauson (1986) received information that births occur mainly in January in one part of Amazonian Ecuador and in June in another. Evidence gathered by Best (1982) indicates that in the central Amazon Basin breeding is seasonal, with nearly all births taking place from December to July, mainly from February to May, the period of rising river levels. Data compiled by Husar (1977) and Marmontel, Odell, and Reynolds (1992) suggest gestation lasts approximately 1 year and that there normally is a single young weighing 10–15 kg. Cantanhede et al. (2005) noted that females first give birth when 5–7 years old, have a 2–3 year calving interval, and live to approximately 40 years in captivity. Sousa-Lima, Paglia, and da Fonseca (2002) reported nursing to last 25 months. The mother-calf bond seems to be long-lasting; the mother carries her calf on her back or clasped to her side. Da Silva (2003, 2004) reported that a 30-year-old female had been in captivity at the Aquatic Mammal Lab of Instituto Nacional de Pesquisa da Amazonia (INPA), Manaus, Brazil, since 1974 and first produced a calf in 1998. She subsequently had a stillborn calf but immediately began breastfeeding two orphaned young and continued to do so for two years. She then started ovulating, and a male was placed in her tank. She was receptive 4–14 February

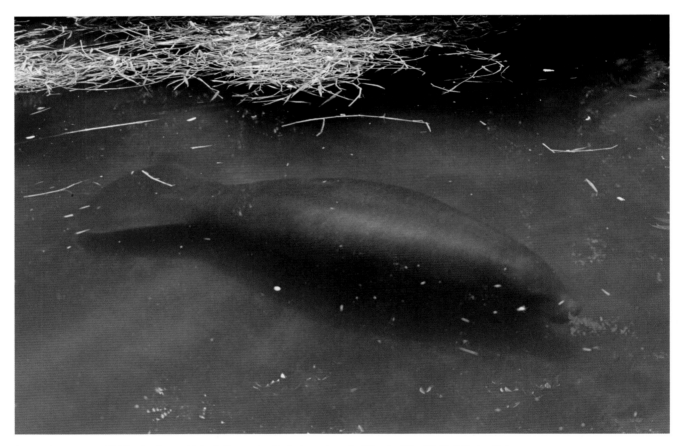

Amazonian manatee (*Trichechus inunguis*) in shallow water. Photo from above by Silvestre Silva / FLPA / Minden Pictures.

2003 and gave birth to a male calf (12 kg, 86.5 cm) on 3 February 2004.

Available records (Domning 1982a; Domning, Kendall, and Orozco 2004; Thornback and Jenkins 1982) indicate that in Brazil the Amazonian manatee was being exploited commercially for its meat as early as the 1600s and that even by the late eighteenth century there was some concern for its survival. Commercial hunting continued, however, with about 1,000–2,000 individuals being marketed each year from 1780 to 1925. Many other animals were taken for subsistence purposes. The main product during this period was *mixira*, fried meat packed in lard. The commercial kill rose sharply starting in 1935, when there was an increased demand for the manatee's hide to manufacture heavy-duty leather. In the peak year of 1940, some 6,300 manatee hides were exported from the state of Amazonas. That industry collapsed about 1954, when synthetic replacements became widely available, but then there was a renewed intensification of commercial meat hunting, which took about 3,000 manatees each year until the early 1960s. The annual take then fell to about 1,000–2,000, probably reflecting a severe decline of the species. The manatee was protected legally in Brazil in 1973, but subsistence and local commercial killing has continued.

Much the same situation applies throughout the range of the species, where it is now protected by law but subject to incessant illegal hunting and generally thought to be declining (Marmontel 2008; Marsh, O'Shea, and Reynolds 2012; Thornback and Jenkins 1982). The Siona people of Ecuador, who are traditional hunters of the manatee, were said to have had a self-imposed ban on taking the species because of its rarity, but such a ban was not fully effective and has not affected meat hunting by settlers and military personnel. Most reports indicate numbers now are lower than in the past in Ecuador, Peru, and Colombia. Traditional harpooning remains the most widespread technique for hunting manatees, but use of nets has been increasing. Periods of drought may restrict the manatee to small areas, making it easier to hunt. Other problems include accidental drowning in commercial fishing nets, degradation of food supplies by soil erosion caused by deforestation, and construction of hydroelectric dams that isolate populations and limit gene flow. The large-scale destruction of Amazonian rainforests also leads to a decline in rainfall, a drop in water levels, and increased mortality of the manatee (Best 1983). The indiscriminate release of mercury in mining activities has put the entire aquatic fauna of the Amazon Basin at risk (Rosas 1994).

An overall minimum estimate of 10,000 individuals for the species was cited by Husar (1977), but Marmontel, Weber Rosas, and Kendall (2012) questioned that figure, and some of the information in their account suggests populations could be larger. In particular, during a period of drought, the number of manatees reported killed for 10 towns in the Brazilian state of Amazonas was 443 in 1995, 648 in 1997, and 475 in 1998. Such a rate of known mortality in just part of the range would indicate that total numbers throughout the Amazon Basin would have been substantially higher than 10,000 in the 1990s. Also, using calculations of genetic diversity within the species, based on samples of mitochondrial DNA taken from 84 individuals, Cantanhede et al. (2005) suggested that *T. inunguis* might have experienced a remarkable demographic recovery since legal protection was established in 1973, in which the estimated number of reproductive females increased from 22,300 to 455,000. Those estimates, however, seem to assume maximum theoretical population growth and no premature mortality. They have not been repeated by other authorities (Marmontel 2008; Marmontel, Weber Rosas, and Kendall 2012; Marsh, O'Shea, and Reynolds 2012; Vianna et al. 2006), who indicate instead that the high genetic diversity of the species probably reflects an earlier period of population expansion, prior to exploitation, and that overall populations are now substantially decreasing. In any case, *T. inunguis* is classified as vulnerable by the IUCN (Marmontel 2008) and as endangered by the USDI and is on appendix 1 of the CITES.

Trichechus manatus (West Indian Manatee)

This species, on average, is the largest living sirenian, though there is some uncertainty regarding overall size. According to Husar (1978c), total length of adults varies from 250 to more than 450 cm, and weight is 200 to more than 600 kg, but average adults are 300–400 cm long and weigh less than 500 kg. Odell (2003) cited records of a 411-cm individual being the longest collected in recent years, and of a 375-cm female that weighed 1,620 kg. Sexual dimorphism in size has not been

West Indian manatee (*Trichechus manatus*). Photo by R. K. Bonde, US Geological Survey, through Cathy A. Beck.

bays, estuaries, lagoons, and rivers. It utilizes both salt and fresh water, though it consistently prefers the latter (Lefebvre et al. 1989). It also seems to prefer water above about 20° C but can endure water as cold as 13.5° C. According to Laist, Taylor, and Reynolds (2013), individuals in Florida, particularly juveniles, are vulnerable to death from cold stress when water temperatures fall below 18–20° C for lengthy periods, or to temperatures of 10–12° C or less for periods of a few days or less. Even in southernmost Florida, winter water temperatures can fall below 18° C for a week or more at a time and to 10° C for shorter periods. To survive, almost all manatees in Florida rely on warm-water refuges in the southern two-thirds of the peninsula (south of about latitude 29° S). Most such refuges are discharges from power plants and natural springs, or passive thermal basins that temporarily trap relatively warm water for a week or more. Statewide winter counts of manatees from 1999 to 2011 found 48.5 percent of manatees at power plant outfalls, 17.5 percent at natural springs, and 34.9 percent at passive thermal basins or sites with no known warm-water features. Deutsch and Reynolds (2012) noted that primary warm-water aggregation sites included seven power plants, four artesian springs, and two thermal basins.

documented, though females seem bulkier. The finely wrinkled skin is slate gray to brown in color. Fine, colorless hairs, 30–45 mm long, are sparsely distributed over the body, and there are stiff, stout bristles on the muscular, prehensile pads of the upper lip and on the lower lip (see above accounts of order Sirenia and family Trichechidae). Nails are present on the dorsal surface of the flippers. The broad skull has a relatively short and downturned snout. Reep and Bonde (2006) commented that the brain of *T. manatus* is perhaps the smallest, relative to body size, of any mammal, and also is smooth-surfaced, which is unusual for larger mammals, but that such factors do not necessarily imply low intelligence. The manatee brain has other unusual features, including large clusters of neurons in the cerebral cortex, which possibly process information from the prehensile facial bristles.

The information for the remainder of this account, if not cited otherwise, was taken from Hartman (1979) and refers to *T. manatus* in Florida. This species inhabits shallow (usually 1–4 meters deep) coastal waters,

West Indian manatee (*Trichechus manatus*); unusual display of open mouth and prehensile nature of upper bristle fields. Photo by Alex Mustard / npl / Minden Pictures.

Laist, Taylor, and Reynolds (2013) explained further that strong fidelity to one or more such warm-water refuges has created four relatively discrete, geographic subpopulations in Florida, with the following percentages of the overall population (then counted at 5,076 animals): (1) Atlantic Coast, 45.6 percent; (2) St. Johns River Region, 5 percent; (3) Southwest (around Tampa Bay and southward), 36.6 percent; and (4) Northwest (the northwestern peninsular portion of the state), 12.8 percent. The Atlantic Coast and Southwest Florida subpopulations were found to rely principally on power plants (66.6 percent and 47.4 percent, respectively, of the animals in those groups). However, the two subpopulations in more northerly parts of the peninsula, where winter water temperatures tend to be lowest, rely almost exclusively on natural springs. In the St. Johns River Region, 99.1 percent of manatees wintered at Blue Spring in Volusia County. In Northwest Florida, 88.6 percent were at springs, almost all of those at Kings Bay and Homosassa Springs, though some animals stayed just downstream in the Crystal River. Natural springs appear to offer the best protection against cold stress, though power plants have effectively extended the foraging grounds available to overwintering animals.

When air temperatures rise above 10° C in the spring and summer, Florida manatees commonly leave their refugia and wander all along the northern coast of the Gulf of Mexico and up the Atlantic coast, occasionally reaching the Carolinas and as far as Virginia (Lefebvre et al. 2001). One well-known individual entered Chesapeake Bay in 1994, where he was captured and taken again to Florida, but he moved back as far as Rhode Island in 1995, before returning to Florida on his own (Reid 1995). In 2006 a manatee appeared along the southern coast of Cape Cod, Massachusetts, and in 2008 a manatee, possibly the same individual, was seen in the same area but then appeared on the north side of Cape Cod, possibly having traversed the Cape Cod Canal (http://buzzardsbay.org /buzzards-bay-manatees.htm, accessed 18 April 2015).

However, in a 12-year radio-tracking study on the Atlantic coast, Deutsch et al. (2003) found most manatees that wintered in Florida to move no farther than Georgia in the warm season. Median waterway distance between the southerly winter range and northerly warm-season range was 280 km (range 11–831 km),

though 12 percent of individuals were resident year-round in less than 50 km. No study animals crossed to the Gulf coast. Southward autumn and northward spring migrations lasted an average of 10 and 15 days at respective mean rates of 33.5 and 27.3 km per day. During a season, individuals usually occupied only one or two core areas that encompassed about 90 percent of daily locations and moved an average of 2.5 km per day. Adult males traveled a significantly greater distance per day than did adult females for most of the warm season, which corresponded roughly to the breeding period, but there was no difference between sexes in daily travel rate during winter. Manatees sometimes left their warm-season ranges in response to cold fronts that dropped water temperatures. Most manatees returned faithfully to the same seasonal ranges year after year. Seasonal movements of four individuals tracked first as calves with their mothers, then as independent subadults after weaning, provided evidence for strong natal philopatry to specific warm-season and winter ranges, as well as to migratory patterns.

On the coast of the Gulf of Mexico, the manatee was reported to occur as far west as New Orleans in the nineteenth century, though there were very few precise records anywhere between Florida and Mexico before 1970. Subsequently, a number of close sightings and recoveries of dead individuals were made in Mississippi, Louisiana, the Galveston area, southern Texas, and Tamaulipas in northeastern Mexico (Fernandez and Jones 1990; Lazcano-Barrero and Packard 1989; Powell and Rathbun 1984). Indications were that most Gulf Coast records represented the subspecies *T. manatus latirostris* from Florida, but that records from southern Texas and Tamaulipas were of animals that had come from farther south in Mexico and that belonged to *T. m. manatus*, which is found in the remainder of the range of the species (Domning and Hayek 1986). Occurrences in Texas were considered to have declined in association with a drastic reduction of the population in Mexico (Powell and Rathbun 1984). Later, Fertl et al. (2005) compiled 377 sighting, capture, and carcass records from Alabama, Louisiana, Mississippi, and Texas, the great majority since 1990, which apparently corresponded to a demographic and geographic expansion of the Florida population. Genetic analysis of samples from an individual captured near Houston in 1995 and from a carcass found east of

West Indian manatee (*Trichechus manatus*). Photo by R. K. Bonde, US Geological Survey, through Cathy A. Beck.

Matagorda Bay, Texas, in 2001 indicated the specimens were referable to *T. m. latirostris*. Many recent reports came from the vicinity of New Orleans, though nearly all seem attributable to animals that approached the city through waterways leading from Lakes Borgne, Pontchartrain, and Maurepas, rather than having ascended the Mississippi River (but see above generic account for records farther north in the Mississippi Valley). A few individuals were seen near the mouth of the Mississippi and at New Orleans in the lock of the Inner Harbor Navigational (Industrial) Canal between the River and Lake Pontchartrain. Williams (2005) provided additional accounts of the manatee in the Lake. Pabody et al. (2009) compiled many records indicating a considerable increase of occurrence in Alabama waters from 1985 to 2007.

There is evidence of long-range, offshore migrations between manatee population centers. Individuals have been caught up to 15 km off the coast of Guyana. They have also been reported in rivers 230 km from the sea in Florida and 800 km from the coast in South America (Husar 1978c). From January to April 2007, a female and calf were observed in a power plant canal at Havana, Cuba, apparently having swum about 700 km from the mother's known winter habitat in Florida (Alvarez-Alemán, Beck, and Powell 2010). Several individuals are also known to have moved from Florida to the Bahamas and taken up residence there, though it is unlikely that there is deliberate or repeated crossing of the intervening deep waters and strong currents (Lefebvre et al. 2001). Many manatees appear to be nomadic and to move hundreds of kilometers, pausing for days, weeks, months, or seasons in estuaries and rivers that supply their needs. They follow established travel routes, preferring channels that are 2 meters or more in depth and shunning those that are less than 1 meter deep. Animals generally swim 1–3 meters below the surface of the water. They use their tails not only to propel themselves through the water but also as rudders by means of which they control roll, pitch, and yaw. The flippers are used in precise maneuvering and in minor corrective movements to stabilize, position, and orient the animal. Speed normally is 3–7 km/hr but can reach 25 km/hr when the animal is pressed. The manatee avoids fast currents, and animals migrating through the Intracoastal Canal were never seen swimming against currents that exceeded 6 km/hr. According to Husar (1978c), average submergence time is 259 seconds, but a dive of 980 seconds has been recorded.

Populations of *T. manatus* in large tropical rivers may have a lifestyle comparable to that reported for *T.*

inunguis (see account thereof). Castelblanco-Martínez et al. (2009) studied a population more than 1,100 km from the Caribbean coast in the Orinoco River on the border between Colombia and Venezuela. In the dry period (January–April), the animals have little vegetative food available and are isolated in *buceaderos*, zones retaining high water. They probably survive by feeding on dead material from the bottom or algae on rocks. As the river starts to rise (May–June) the manatees start moving to look for food. In the high water period (July–October) the manatees spread out in the *esteros*, the flooded plain of the Orinoco and its tributaries, and occupy fields of floating grass or inundated forests.

T. manatus may be active both by day and by night. During a 24-hour period adults have been observed to feed 6–8 hours in sessions that usually lasted 1–2 hours, to rest 6–10 hours, and to move as much as 12.5 km. The manatee rests by hanging suspended near the surface of the water or by lying prone on the bottom.

West Indian manatee (*Trichechus manatus*), aggregation at warm water discharge from Riviera Power Plant. Photo by John Reynolds and Florida Power & Light.

In both positions the animal lapses into a somnolent state, with eyes closed and body motionless. In the Drowned Cayes area of Belize, the manatee is commonly observed resting in depressions in the substrate; such holes are significantly deeper and have slower surface water velocity than areas without resting holes (Bacchus, Dunbar, and Self-Sullivan 2009). Sound appears to be the manatee's principal sensory mode, but in clear water the preferred method of environmental exploration is visual. In addition, the prevalence of mouthing in social interaction suggests that the species has a chemoreceptive sense by which it can recognize odors in the water. Comfort activities include stretching, using the flippers to scratch and clean the mouth, rubbing against logs and rocks, and rooting in the substrate. Rubbing also may involve deposition of scent, by which an estrous female signals receptivity to wandering males (Rathbun and O'Shea 1984).

The manatee is herbivorous, utilizing over 60 species of fresh water and marine plants, including those located on the bottom, floating at the surface, and growing on banks (Reep and Bonde 2006). Some invertebrates are ingested together with vegetation and may provide an important amount of protein. In Florida, individuals were observed pulling invertebrates, such as tunicates, off of dock structures and into their mouths and were heard crunching the shells (Courbis and Worthy 2003). Captives have deliberately eaten dead fish, and wild manatees off Jamaica have been seen to take fish entangled in nets (Powell 1978). The diet consists mainly of submerged vascular plants, but emergent and floating vegetation sometimes is eaten. Hydrilla (*Hydrilla*) and water hyacinth (*Eichhornia*) are staple foods in many rivers of Florida. For many years *T. manatus* has been deliberately employed for the control of water hyacinth and other aquatic weeds in waterways in Guyana (Haigh 1991). Sea grasses are taken in salt-water areas. Captives have consumed up to one-fourth of their body weight per day in wet greens, though 5–10 percent is normal in the wild (Reep and Bonde 2006). Feeding occurs from the surface to a depth of 4 meters. Individuals are drawn to sources of fresh water, which they appear to require for osmoregulation. However, such a need was questioned by Reynolds and Ferguson (1984), who observed two large manatees 61 km northeast of the Dry Tortugas Islands (about 110 km west of Key West, Florida), which

lack fresh water. Based on a number of cited studies, Marsh, O'Shea, and Reynolds (2012) concluded that *T. manatus* has a good tolerance for salt water but physiologically must drink fresh water periodically. Very limited fresh water is believed the main factor restricting manatee numbers in the Bahamas (Lefebvre et al. 2001).

In their search for receptive mates, males reportedly utilize a much larger home range than do females. Some information also suggests that individuals come together deliberately in order to learn the location of favorable habitat and migration routes (Thornback and Jenkins 1982). Generally, however, the West Indian manatee is considered a weakly social, essentially solitary species. The only lasting association seems to be that between a cow and her calf. Temporary, casual groups form in favorable areas for purposes of migration, feeding, resting, or cavorting. Such groups may be randomly made up of juveniles and adults of both sexes, though Reynolds (1981) suggested a tendency for subadult males to associate. A population wintering in the Crystal River of Florida one year contained 31 adults, 13 juveniles, and 6 calves and was divided about equally between males and females. Much larger aggregations have formed at warm-water refuges (see above), especially certain power plants where one-day winter counts have frequently exceeded 400 and sometimes approached 1,000 (Marsh, O'Shea, and Reynolds 2012).

There is no evidence of communal defense or mutual aid and little or no indication of a social hierarchy. However, wild females have been repeatedly observed to "adopt" and suckle orphaned calves, often for the full course of dependency (Marsh, O'Shea, and Reynolds 2012). Individuals indulge in what appears to be play. They exchange gentle nibbles, kisses, and embraces that are age- and sex-independent. In studies at several facilities holding captives, Harper and Schulte (2005) ascertained nearly all contact to be nonaggressive and suggested that the manatee may be more social than previously thought.

The manatee exhibits exceptional acoustic sensitivity; sound is doubtless the major directional determinant in social interactions. *T. manatus* can emit calls, which to the human ear sound like high-pitched squeals, chirp-squeals, screams, and grunts, under conditions of fear, aggravation, protest, internal con-

West Indian manatee (*Trichechus manatus*). Photo by Vladimir Wrangel/Shutterstock.com.

flict, male sexual arousal, and play. Vocalizations are probably non-navigational and lack ultrasonic signals, pulsed emissions, or directional sound fields. The most predictable vocal exchange between manatees involves screams of alarm, by which a cow calls her calf in case of danger, and the responding squeals of the young. In an interesting experiment, Phillips, Niezreckia, and Beusse (2004) placed hydrophones in a spring where manatees were present and used one to broadcast recordings of a frightened calf. A number of adult females then encircled the speaker in a protective manner, their tails inside the circle and their heads looking outward.

O'Shea and Poché (2006) made extensive recordings of wild *T. manatus* in Florida, noting that vocalizations are complex, single-note calls with multiple harmonics, frequency modulations, and non-harmonically related overtones. Structural characteristics appear to provide information on individual identity, basic motivational state, and perhaps size of the signaler; however, calls are used primarily to maintain contact between individuals, particularly females and juveniles. In 43 instances of a female being separated from her calf at distances of up to 40 meters, movement toward rejoining was preceded by calls made by either cow or calf, and calling by both continued at least until the pair made physical contact. In such cases, as many as 20 calls per minute occurred, and females and calves responded only to each others' vocalizations, strongly suggesting individual recognition by sound. Males were heard less frequently, but were recorded vocalizing while cavorting in groups of 6–10, when disturbed from sleep, and in response to calls of other manatees when resting, feeding, or traveling.

When a female is in estrus, she may be accompanied for a period of 1 week to more than a month by as many as 22 males, numbers of which join and leave the group each day (Rathbun et al. 1995). Such a mating system has been designated "scramble promiscuity," and the groups that form may move up to 160 km before breaking up (Anderson 2002). The courtship of the males is

West Indian manatee (*Trichechus manatus*), male pursuing estrous female. Photo by Fred Bavendam / Minden Pictures.

relentless, but the cow generally seeks to escape and to vigorously repulse their advances. Rathbun and O'Shea (1984) suggested mate selection is done by the female. She may eventually mate with several of the males (Reep and Bonde 2006). There is constant and sometimes violent pushing and shoving by the bulls as they compete for a position next to the estrous female; such activity seems to be the only display of aggression in the species.

In Florida, mating behavior appears to peak in April and May (Reep and Bonde 2006), and most calves are born in spring and early summer (Marmontel 1995); very few births occur in the winter months (Marsh, O'Shea, and Reynolds 2012). Calves have been seen throughout the year in the Dominican Republic and Puerto Rico (Belitsky and Belitsky 1980; Powell, Belitsky, and Rathbun 1981). Females are thought to be polyestrous and to have an estrous cycle of 28–42 days (Marsh, O'Shea, and Reynolds 2012). The interbirth interval probably is normally 2.5–3.0 years but may be

shorter if an infant is lost. The gestation period has never been precisely determined but probably is about 12–13 months. The cow seeks the shelter of a backwater to give birth. There normally is a single young but also a documented twinning incidence of 1.4 to 4.0 percent (Marsh, O'Shea, and Reynolds 2012). The newborn is about 120–130 cm long, weighs 28–36 kg, and is dark in color (Caldwell and Caldwell 1985; Marmontel 1995; Rathbun et al. 1995). Within half a day of birth it is capable of swimming and surfacing on its own, though it occasionally rides on the mother's back. While the calf begins to take some vegetation at about 1–3 months, it continues to suckle until leaving its mother at about 1–2 years. Males appear to reach full reproductive maturity at 9–10 years (Odell, Forrester, and Asper 1978). However, laboratory studies indicate that males may be physiologically capable of mating when they are only 2–3 years old (Hernandez et al. 1995). Females are sexually mature at 3–7 years, one was reproductively active in captivity for 35 years, and one was estimated to have reached an age of 59 years in the wild; a captive-born specimen of *T. manatus* was still living at a known age of 61 years (Marmontel 1995; Marsh, O'Shea, and Reynolds 2012).

Data compiled by Deutsch, Self-Sullivan, and Mignucci-Giannoni (2008), Lefebvre et al. (1989, 2001) and Thornback and Jenkins (1982) indicate that commercial exploitation of the subspecies *T. manatus manatus* began in the sixteenth century, and, combined with extensive subsistence hunting, resulted in severely reduced populations in most areas. The manatee had probably disappeared from the Lesser Antilles by the eighteenth century and is continuing to decline almost everywhere else despite legal protection in all countries. Threats include killing for meat, drainage of swamps and other habitat degradation, accidental drowning in fishing nets and flood control structures, vessel collisions, and pollution. Perhaps the one national population that still appears relatively dense and stable is that of Belize. O'Shea and Salisbury (1991)

reported that numbers there had been stable since 1977 and were greater than the number in any other country bordering the Caribbean, apparently because of a relatively low human population and maintenance of a high-quality habitat, including shallow waters sheltered by the second longest coral reef in the world. Such conditions still seem to prevail in Belize, and legal protection has been in effect there since 1935, though poaching remains a serious problem.

Self-Sullivan and Mignucci-Giannoni (2012) listed estimated numbers as 1,000 for Belize, but just 100–500 for each of the other Central and South American countries that border the Caribbean. Nonetheless, Lefebvre et al. (2001) indicated that the manatee still occurs in most waterways of the Orinoco Delta in Venezuela and throughout the middle Orinoco and its tributaries. And Castelblanco-Martínez et al. (2009) made 870 sightings of *T. manatus* in about 40 km of the Orinoco along the border of Colombia and Venezuela during 2001–2005, including one simultaneous observation of

West Indian manatee (*Trichechus manatus*), mother nursing calf. Photo by Todd Pusser / npl / Minden Pictures.

17 individuals. Self-Sullivan and Mignucci-Giannoni (2012) listed estimates of about 100 individuals each for the island countries of Cuba, Haiti, the Dominican Republic, and Trinidad and Tobago, and 50 for Jamaica, but 618 in the U.S. territory of Puerto Rico. Only vagrant individuals appear at other islands in the West Indies, including the Bahamas, where the manatee is naturally rare. Guyana reportedly had thousands as late as the 1970s, but now only about 100 are thought to be there, as well as in French Guiana, with an even smaller number in Surinam.

The total number for all of Brazil is estimated at 500, but that figure is based on surveys in the 1990s by Luna et al. (2008). The manatee formerly occurred as far south as the Brazilian state of Espírito Santo, but populations disappeared there by the nineteenth century, from Bahía in the 1960s, and subsequently from Sergipe. More recently, the species has been reported from several isolated stretches of the Brazilian coast, particularly in the state of Amapa near the mouth of the Amazon, at the Rio Mearim in Maranhao, from Piauí to western Ceará, from eastern Ceará to northern Pernambuco, including Barra de Mamanguape in Paraiba, and in Alagoas (Borobia and Lodi 1992; Domning 1981, 1982a; Lefebvre et al. 2001; Whitehead 1977). In 1994 the Brazilian government began a program to rehabilitate orphaned calves and return them to the wild in the states of Alagoas and Paraíba; 30 such animals had been released by 2011; most of them survived, with several of the females eventually producing calves of their own (Normande et al. 2015).

The manatee formerly occurred along the entire eastern coast of Mexico but by the 1980s had declined severely through accidental and deliberate killing by fishermen and no longer was commonly found north of the wetlands at Alvarado in southern Veracruz. It also still occurs in eastern Yucatan, especially at Chetumal Bay, and in the wetlands of Tabasco, Campeche, and northern Chiapas (Lefebvre et al. 2001). Self-Sullivan and Mignucci-Giannoni (2012) listed an estimate of 1,500 individuals in Mexico, as well as a total of 5,928 for the entire range of the subspecies *T. manatus manatus*.

The subspecies *T. manatus latirostris* of Florida originally contained an estimated several thousand animals but was so greatly reduced by hunters that in 1893 it was given legal protection by the state (Caldwell and Caldwell 1985). A slow recovery then began (Allen 1942), and by the 1970s about 1,000 individuals were thought present in Florida (Thornback and Jenkins 1982). Coverage by the United States Marine Mammal Protection Act of 1972 and the Endangered Species Act of 1973 apparently provided further help. Aerial surveys in 1992 showed at least 1,856 manatees remaining in Florida, with a roughly equal number on each coast (Ackerman 1995). Counts in early 1996 indicated the true number in Florida was close to 3,000. Although Domning (1996) cautioned that figure might reflect improved census techniques and ideal survey conditions, population analyses have indicated a modest increase over the years, except perhaps in the southwest (Deutsch and Reynolds 2012; Marsh, O'Shea, and Reynolds 2012). Subsequent range-wide counts at winter warm-water aggregation sites have yielded higher figures, including 5,076 (2,296 on the west coast, 2,780 on the east coast) in 2010 (Deutsch and Reynolds 2012) and 6,063 (2,730 west, 3,333 east) in February 2015 (http://myfwc .com/research/manatee/research/population -monitoring/synoptic-surveys/, accessed 1 May 2015); those figures do not include unobserved animals.

Notwithstanding the above, some statistical assessments suggest that current and projected mortality rates are not sustainable in the long term (Marsh, O'Shea, and Reynolds 2012). Many animals are killed and injured by vandals, entrapment in gates of locks and dams, collision with barges, and, especially, the propellers of power boats (O'Shea et al. 1985). Indeed it is a sad fact that most manatees in Florida bear scars from propellers or collisions with boats and that such marks are the chief means by which biologists identify the individuals they study (Reep and Bonde 2006; Reid, Rathbun, and Wilcox 1991). Manatee deaths, based on recovered remains and tabulated by the Florida Fish and Wildlife Conservation Commission, averaged 313 per year from 2010 to 2014 (http://myfwc.com/media /2970592/preliminary.pdf, accessed 29 April 2015). According to Deutsch and Reynolds (2012), nearly half (42.5 percent) of documented adult mortality in Florida is directly attributable to human activity. Watercraft collisions cause 25 percent of reported deaths and about 35 percent of reported deaths of known causes. That factor is likely to intensify as human population and boat usage continue to grow,

West Indian manatee (*Trichechus manatus*). Photo by Steven David Miller / npl / Minden Pictures.

and could potentially become severe enough to reduce manatee numbers to inviable levels. Another serious long-term threat is loss of warm-water sources, especially as aging power plants close down and remaining ones convert from cooling technology that produces large thermal plumes to more efficient methods. Elimination of a few key winter refuges could have catastrophic consequences. Also, many spring systems have been altered, so they are no longer accessible to manatees, and spring flows have declined as a growing human population demands more water. Natural deaths, aside from losses to cold stress, also can be disastrous. In 1996, 151 manatees were confirmed or suspected to have died in southwestern Florida from the toxic microorganism red tide; additional outbreaks occurred in later years (Deutsch, Self-Sullivan, and Mignucci-Giannoni 2008).

Marsh, O'Shea, and Reynolds (2012) cited information indicating that three and possibly all four of the Florida subpopulations have grown in recent years, but concluded that, because of threats from loss of warm-water habitat and increasing boat traffic in Florida, the subspecies *T. m. latirostris* is projected to decline by at least 10 percent over the next 40 years. They also projected a decline of at least 20 percent during that period for *T. m. manatus*. The IUCN classifies both subspecies individually as endangered but classifies the full species *T. manatus* as vulnerable (Deutsch 2008; Deutsch, Self-Sullivan, and Mignucci-Giannoni 2008; Self-Sullivan and Mignucci-Giannoni 2008). The USDI also lists *T. manatus* as endangered, but recently proposed reclassifying the entire species as threatened (Herrington and Muñiz 2016). The proposal was based primarily on the status of *T. m. latirostris* in Florida but provides a tabulation showing that, in the 20 countries in which *T. m. manatus* occurs, the subspecies is nationally listed as endangered or critically endangered in 11 (there is no information for the other 9) and that population trend is decreasing or unknown in 16 of the countries, stable in 3, and increasing only in the Bahamas, where the total number of individuals is estimated at 10. *T. manatus* is on appendix 1 of the CITES.

Trichechus senegalensis (West African Manatee)

According to Husar (1978b) and Kouadio (2012), adults average about 300 cm in length and 450–500 kg in weight, though some exceptional individuals reach 400 cm and over 1,000 kg. The skin is finely wrinkled and grayish brown in color. Fine, colorless hairs are distributed over the body, and there are stout bristles on the upper and lower lip pads. The flippers have nails on the dorsal surface. The skull is broad and has a shortened, slightly deflected snout. Domning and Hayek (1986) found *T. senegalensis* and *T. manatus* to be phenetically similar and noted that the reduced rostral deflection of the former probably is associated with its dependence on emergent or overhanging, rather than submerged, vegetation.

According to Keith Diagne (2015), the West African manatee occurs in coastal marine waters, brackish estuaries, and adjacent rivers. Optimal coastal habitats include lagoons and other protected areas with shallow, calm waters and abundant mangroves or emergent herbaceous growth. The manatee can be found 110 km offshore among the shallow coastal flats, mangrove creeks, and abundant sea grasses of the Bijagos Archipelago of Guinea-Bissau. It can travel freely from salt to fresh water and is found in marine habitats with a source of fresh water. It ascends most major rivers in its range until stopped by cataracts or shallow water. In riverine habitats that have major fluctuations in flow rates and water levels, it seems to prefer areas that have access to deep pools or connecting lakes for refuge during the dry season and that seasonally flood into swamps or forests with abundant grasses and sedges.

An individual may travel 30–40 km per day through lagoons and rivers (Reynolds and Odell 1991). During the rainy season in Guinea-Bissau, the manatee may as-

West African manatee (*Trichechus senegalensis*). Photo from Toba Aquarium, Japan.

cend the smaller rivers and inlets looking for new feeding areas and freshwater springs; when the water level in the rivers decreases at the beginning of the dry season, it may move to the mouth of the larger rivers and to coastal zones (Silva and Araújo 2001). The overall range apparently is limited to water with a temperature above 18° C (Husar 1978b). In some areas *T. senegalensis* feeds principally at night and travels in the late afternoon and at night, but in other places it can be seen traveling and feeding during all hours of day and night; it usually rests during the day in water 1–2 meters deep (Keith Diagne 2015). It may spend 4–6 hours feeding and sometimes travel several km between the feeding and resting areas (Powell 2013). The diet is believed to consist mostly of emergent grasses and plants, and also to include fruits that fall in the water (Kouadio 2012). A free-ranging adult might consume 8,000 kg of vegetation in a year (Husar 1978b). Populations in some rivers depend heavily on overhanging bank growth, and those in estuarine areas feed exclusively on mangroves (Domning and Hayek 1986). In Sierra Leone the manatee reportedly removes fish from nets and consumes rice in such quantities that it is considered a pest (Reeves, Tuboku-Metzger, and Kapindi 1988).

Roth and Waitkuwait (1986) reported that *T. senegalensis* lives singly or in family groups of four, or rarely six, animals. Husar (1978b) related a report of 15 individuals being seen together and indicated that social behavior probably is similar to that of *T. manatus*. Aggregations may form in association with breeding, with mating and births occurring just before the rainy season in some areas and separated by a gestation period of 12–13 months, but newborn calves have been reported throughout the year (Powell 2013). Parturition supposedly occurs in shallow lagoons; usually there is a single calf, and the newborn is about a meter long.

Husar (1978b) wrote that there had been no large-scale commercial exploitation of *T. senegalensis* comparable to that of other manatees but that populations had declined markedly in some areas because of local hunting for meat. Reeves, Tuboku-Metzger, and Kap-

indi (1988) pointed out additional problems, such as accidental capture in fishing nets and persecution for alleged depredations on netted fish and growing rice. Surveys in 1980–81 (Nishiwaki 1984; Nishiwaki et al. 1982) yielded a rough overall estimate of 9,000–15,000 individual *T. senegalensis*, exclusive of Angola and Congo. The largest numbers, perhaps several thousand in each area, were found in Gabon and the lower reaches of the Niger and Benue Rivers in Nigeria. Numbers were also considerable in Cameroon, Ghana, Ivory Coast, and the upper reaches of the Niger but were fewer from Senegal to Liberia. The species apparently has disappeared from the Lake Chad Basin (see above generic account).

According to an extensive IUCN account by Keith Diagne (2015), numbers still appear high in Gabon and Cameroon, but are depleted in Nigeria, estimated at just 750–800 in Ivory Coast, and declining alarmingly in Ghana. Substantial populations still occur to the west in Guinea Bissau, Gambia, and Senegal; the species persists in the inland river systems of Mali, Niger, and Burkina Faso; and there may be a sizeable population in the lower Congo River. Indeed, the West African manatee still evidently occurs in all the countries of its historical range from Mauritania to Angola, but the species is severely threatened, with total numbers now estimated at fewer than 10,000. Intensive subsistence and market hunting continues throughout the range, as does accidental drowning in fishing nets. Destruction of mangroves and other coastal habitat is rampant and some populations have been isolated by construction of dams along the Niger and Benue rivers. Marsh, O'Shea, and Reynolds (2012) expected declines to intensify and considered *T. senegalensis* the sirenian species at greatest risk of extinction. It is classified as vulnerable by the IUCN (Keith Diagne 2015) and as threatened by the USDI. In 2013 it was moved from appendix 2 to appendix 1 of the CITES, mainly because of concerns that excessive numbers were being killed for meat and potentially entering commercial trade (Keith Diagne 2013).

Order Cingulata

CINGULATA; **Family DASYPODIDAE**

Armadillos

This order contains the single Recent family Dasypodidae, which comprises 9 living genera and 21 species and is found from the central and southeastern United States to the Strait of Magellan. This order and the order Pilosa were sometimes considered infraorders of what had been designated the order Xenarthra, which is now often treated as a higher clade or magnorder (Gardner 2007e; McKenna and Bell 1997; see account of class Mammalia). Xenarthra was retained at ordinal level by Vizcaíno and Loughry (2008) and their inclusive authorities. One species of Cingulata, *Dasypus novemcinctus*, is the only xenarthran that occurs in the United States. A detailed account of that species, plus a succinct review of the phylogeny, diversity, and characteristics of order Cingulata was provided by Loughry and McDonough (2013).

The sequence of genera presented here is based primarily on the molecular analyses of Delsuc and Douzery (2008), Delsuc et al. (2012), and Möller-Krull et al. (2007), the latter two of whom recognized four subfamilies: Dasypodinae, with the single genus *Dasypus*; Euphractinae, with the genera *Euphractus*, *Chaetophractus*, and *Zaedyus*; Chlamyphorinae, with *Calyptophractus* and *Chlamyphorus*; and Tolypeutinae, with *Priodontes*, *Cabassous*, and *Tolypeutes*. Also considered in arranging the sequence was the morphological assessment of Gaudin and Wible (2006), which indicated support for most of the same groupings and for regarding *Tolypeutes* as the most derived genus, but which associated *Chlamyphorus* (including *Calyptophractus*) with the genera of Euphractinae. Likewise, Gardner (2005d) did not recognize Chlamyphorinae as a subfamily, and placed *Calyptophractus* and *Chlamyphorus* in Euphractinae. McKenna and Bell (1997) and Wetzel et al. (2007) put those two genera in a tribe, Chlamyphorini, within Euphractinae, and also divided Tolypeutinae into tribes, Tolypeutini for the genus *Tolypeutes*, and Priodontini for *Cabassous* and *Priodontes*. Earlier molecular analyses (Delsuc, Stanhope, and Douzery 2003; Delsuc, Vizcaíno, and Douzery 2004; Delsuc et al. 2002), while corroborating the content of Dasypodinae, Euphractinae, and Tolypeutinae, had been unable to assess the relationships of the rare *Calyptophractus* and *Chlamyphorus*. Remarkably, when appropriate specimens of those two genera did become available, analysis (Delsuc et al. 2012; Möller-Krull et al. 2007) confirmed their ancient lineage and supported treating them as members of a separate subfamily, as had been done long before by Cabrera (1957).

The word "armadillo" is of Spanish origin and refers to the armor-like covering or carapace of these animals. The skin is remarkably modified to provide a double-layered covering of horn and bone over most of the upper surface and sides of the animals and some protection to the underparts and limbs. The covering consists of bands or plates connected or surrounded by flexible skin. The scutes on the back and sides, which are quadrate or polygonal in shape, are usually ar-

Chacoan fairy armadillo (*Calyptophractus retusus*). Photo by Thomas and Sabine Vinke, Paraguay Salvaje.

ranged into shoulder and hip shields with movable rings between them. This body armor is flexible in most kinds of armadillos when they are alive but becomes rigid after they die; it often is sold to tourists as baskets. The skin between the bands and plates provides further flexibility. The limbs, which are covered on their outer or exposed margins with irregular horny shields, can be withdrawn into the space between the body and the shoulder and hip shields. The top of the head has a shield, and the tail is usually encased by bony rings or plates. The scutes are covered by a layer of horny epidermis. The undersurface of the body and the inner surfaces of the limbs are covered with soft, haired skin. In many species hairs also project through openings between the scutes on the back.

While it seems obvious that a major benefit of possessing a carapace is protection from predators, Superina and Loughry (2012) explained there are no data available to test that assumption, such as showing that armadillos have a lower risk of predation than similarly sized, non-armored mammals in the same habitats and active at the same times of day. In fact, the carapace could provide protection in other ways, such as shielding armadillos from thorny vegetation, minimizing injury during aggressive interaction with conspecifics, or limiting attachment sites for external parasites. Additionally, possession of armor is linked with relatively slow plantigrade locomotion, which in turn may have promoted what are among the lowest metabolic rates reported for any placental mammal. The high thermal conductance of the carapace, coupled with low metabolism and the absence of much insulating fat or fur, may benefit armadillos in maximizing heat loss and minimizing heat production in hot environments.

As a result of the armor-like covering, the body is heavy; the giant armadillo (*Priodontes*) weighs as much as 60 kg. Head and body length in the order is 85–1,000 mm and tail length is 25–500 mm. The scutes are brown to pinkish in color, and the pelage is grayish brown to white. Most species of armadillo have

Giant armadillo (*Priodontes maximus*). Photo by Alexander F. Meyer at Bioparque Los Ocarros, Colombia.

moderate-sized ears. The snout varies considerably in length and has a long, protrusible tongue. The forefeet have three, four, or five digits with powerful, curved claws, and the hind feet have five digits with claws. The lower bones of the legs are united proximally and distally. In females the mammae are usually located in the chest region, but an abdominal pair is occasionally present.

The skull is dorsoventrally flattened, and the lower jaw is elongate. The teeth are small, peglike, ever-growing, and numerous. There are usually 7–9 teeth in each half of the jaw, but *Priodontes* may have more than 40 small teeth in each jaw. The order is distinguished from all others, except Pilosa, by what are known as xenarthrous vertebrae, which have secondary and sometimes even further articulations between the vertebrae of the lumbar series. They lend support, particularly to the hips, which is especially valuable to

armadillos for digging (see account of class Mammalia, clade Xenarthra).

Armadillos generally inhabit open areas such as savannahs and pampas but also occur in forests. They travel singly, in pairs, or occasionally in small bands. They are terrestrial in habit, powerful diggers and scratchers, and either diurnal or nocturnal. When not active, they usually live in underground burrows. They generally walk on the tips of the foreclaws and the soles of their hind feet with the heels touching the ground. They can run fairly rapidly. Although they can defend themselves with their claws and sometimes by biting, the usual reaction to danger is to run or burrow rapidly into the ground and then anchor themselves in the burrow. If overtaken while running, or if they do not have the opportunity to burrow, some species draw in their feet so that the edges of the armor are in contact with the ground, and a few species roll themselves into a ball. Armadillos have relatively good senses of sight, smell, and hearing. They feed on insects and other invertebrates, some small vertebrates, plant material, and carrion.

The gestation period may be prolonged by delayed implantation. There may be as many as 12 young per birth, but usually 2–4 and often only 1. The young are covered with a soft, leathery skin that gradually hardens with age. Maturity commonly comes in 1 year, but life span is relatively long and is known to be more than 20 years in some genera.

The geological range of family Dasypodidae is late Paleocene to Recent in South America and late Pliocene to Recent in North America (McKenna and Bell 1997). A related family of order Cingulata, the heavily armored Glyptodontidae, appeared in South America in the middle Eocene and entered North America in the late Pliocene. Glyptodonts were characterized by an essentially immobile dorsal carapace, as opposed to the more flexible scutes of armadillos; some late Pleistocene species attained a weight of 1,000–2,000 kg (Fernicola, Vizcaíno and Fariña 2008). Al-

though generally assumed to have disappeared along with New World proboscideans and giant ground sloths at the end of the Pleistocene or in the early Holocene, some radiocarbon dates suggest the possibility of survival into very early historical time in South America. Fiedel (2009) discussed such reports but listed the latest credible date for a glyptodont as 9,100 years before the present. Barnosky and Lindsey (2010) accepted a date of just 7,663 calendar years ago for the survival of the large genus *Doedicurus* at a site in Argentina.

CINGULATA; DASYPODIDAE; **Genus** *DASYPUS* Linnaeus, 1758

Long-nosed Armadillos

There are three subgenera and seven species (Abba and Vizcaíno 2014; Redford and Eisenberg 1992; Vizcaíno 1995; Wetzel 1982, 1985b; Wetzel and Mondolfi 1979; Wetzel et al. 2007):

Long-nosed, or nine-banded, armadillo (*Dasypus novemcinctus*), individual in Florida. Photo by Svetlana Foote / Shutterstock.com.

subgenus *Dasypus* Linnaeus, 1758

D. novemcinctus, central and southeastern United States to Peru and Uruguay, Grenada in the Lesser Antilles, Trinidad and Tobago;

D. septemcinctus, eastern and southern Brazil, eastern Bolivia, Paraguay, extreme northern Argentina;

D. hybridus, southern Paraguay, northern and central Argentina, Uruguay, extreme southern Brazil;

D. yepesi, known only from Jujuy and Salta provinces in extreme northern Argentina, but possibly also occurs in adjacent parts of Bolivia and Paraguay;

D. sabanicola, *llanos* of east-central Colombia and central Venezuela;

subgenus *Hyperoambon* Peters, 1864

D. kappleri, east of the Andes and south of the Orinoco in Colombia, Venezuela, Ecuador, Peru, northern Bolivia, northern Brazil, and the Guianas;

subgenus *Cryptophractus* Fitzinger, 1856

D. pilosus, known only from the Andes Mountains of west-central Peru.

Gardner (2005d) did not refer to subgenera. Wetzel et al. (2007) considered *Hyperoambon* valid as a subgenus but did not specify the same for *Cryptophractus*.

Head and body length is 240–573 mm, tail length is 125–483 mm, and weight is 1–10 kg. In species other than *D. pilosus*, hair is almost lacking on the upper parts and is sparsely scattered on the underparts. The hairs are pale yellowish, and the remainder of the body is mottled brownish and yellowish white. *D. pilosus* has a thick dorsal covering of long reddish to reddish gray hair (Wetzel et al. 2007). Coloring is the same in both sexes, and there is no seasonal variation. The body of these armadillos is broad and depressed, the muzzle is obtusely pointed, and the legs are short. The forefoot has four toes, of which the middle two are the largest; the hind foot has five toes. The number of movable bands in this genus varies from 6 to 11 depending on the species and, within any given species, on geographic location. *Dasypus novemcinctus*, often called the "nine-banded armadillo," usually has eight bands

in the northern and southern parts of its range and nine in the central part (northern South America).

Long-nosed armadillos are partial to dense shady cover and limestone formations from sea level to 3,000 meters. *D. novemcinctus* requires access to fresh water and generally is found near creeks, rivers, and lakes, though not in swamps. It can swim, but not well, and is also capable of submerging and walking across the bottom of a body of water, though such activity seems rare and motivating factors are not known (Loughry and McDonough 2013). Burrows up to 7.5 meters long are constructed at a depth of 0.5–3.5 meters, and large nests of leaves and grass are assembled therein. Burrows of *D. novemcinctus* are usually about 1.0–2.0 meters long and 0.5–1.0 meter deep, have a single opening 18–22 cm across, and have two chambers, one near the entrance, presumably for observation, and one farther down for nesting; an individual generally has several burrows, using one for a series of days before switching to another (Loughry and McDonough 2013). Nests of *D. novemcinctus*, resembling miniature haystacks, also have been found above the ground in clumps of saw palmetto (Layne and Waggener 1984).

Long-nosed armadillos are mainly nocturnal but are frequently seen foraging during daylight. Layne and Glover (1985) reported that activity in south-central Florida peaked when temperatures were 20°–25° C and thus concluded that animals were more nocturnal in summer. When rooting about, they grunt almost constantly and do not appear to be alarmed by humans. Hunters report that if they stand still, armadillos will occasionally bump against their feet without noticing. When alarmed, however, they hurry away toward a burrow. If overtaken, they characteristically curl up as much as possible so that the armor protects their soft underparts. They forage in a jerky, nervous fashion, poking into many holes, crevices, and leaf piles in search of arthropods, small reptiles, and amphibians. According to studies cited by Lowery (1974) and Wetzel and Mondolfi (1979), the diet of *D. novemcinctus* consists predominantly of animal matter, especially insects such as beetles and ants. However, Loughry and McDonough (2013) noted that the percentage of insects consumed can vary dramatically with location and season, and that other invertebrates, plant matter, and occasionally small vertebrates are taken as well; there is no documented evidence for carrion feeding in the wild.

Greater long-nosed armadillo (*Dasypus kappleri*). Photo by Emilio Constantino through Mariella Superina.

D. novemcinctus is capable of widespread dispersal. In an experiment, all armadillos inhabiting a 1,600-ha. area of Florida were exterminated, with 451 animals (which had been previously marked) killed in 2004–2006 (McDonough, Lockhart, and Loughry 2007). And yet continuous monitoring showed total numbers in the area to remain stable during those years, as armadillos from surrounding areas immigrated to fill the ecological vacuum. The study supported previous reports of large numbers of transient armadillos moving extensively and may help explain how *D. novemcinctus* successfully invaded much of the United States since the nineteenth century (see below). Loughry et al. (2013) suggested that the invasion was achieved despite a relatively low natural population growth rate for the species and instead reflects its life-history strategy, in which populations consist of a core of long-term residents that move very little over time, but also an equal number of transients that are available to move into new areas. Notwithstanding the "clonal" nature of this species' reproduction, genetic assessment of populations indicates no association of related individuals following dispersal of littermates (Loughry and McDonough 2001, 2013).

Loughry and McDonough (2013) and McDonough and Loughry (2008) reviewed population characteristics and social behavior in *D. novemcinctus*. Reported densities for studies in the United States were 0.10–3.90, 0.15, 0.15–0.18, 0.18, 0.23, 0.25, 0.30, 0.35, 0.73, 1.45, 1.50, 2.80, and 3.75 individuals per ha. Home range estimates, 0.63–20.1 ha., varied widely between populations, possibly reflecting greater concentration of food resources in more mesic areas, but there was also variation within single populations. There was no general difference in range size between sexes, but older animals did tend to use larger areas. Females shared 66–73 percent of their ranges with neighboring females, while the ranges of males typically overlapped by 15–22 percent and those of breeding males by just 12 percent. Because of the multiplicity of overlap, an individual usually has no exclusive area within its home range. Although *D. novemcinctus* is not usually reported to be territorial, individuals, at least on occasion, do defend their ranges against others. In particular, breeding males are aggressive to other males in an apparent effort to obtain exclusive access to female mates. However, individuals usually ignore others foraging nearby. The armadillo is a relatively solitary and asocial animal, and no sustained grouping has been observed, though several adult females may frequent a common burrow. In the United States there is a distinct peak of reproductive pairing in June and July, which is recognizable by the close proximity and touching of two adults and by the female's tail wagging. All available evidence suggests *D. novemcinctus* is mildly polygynous. A male sometimes pairs, at different times during the same season, with two or three females and may sire more than one litter, but a female is never seen with more than one male in a season. Breeding male *D. novemcinctus* seem to mark the margins of their home range by rubbing the substrate with anal glands; such chemical secretions appear to be the primary means of communication in the species, while vocalization is limited to grunting.

In a study of a dense population of *D. novemcinctus* in southern Texas, McDonough (1994) observed considerable evidence of aggression, including chases and "boxing," in which animals balanced on their hind feet and tails and clawed at each other with their front feet. Fully adult males were commonly aggressive toward younger males, especially during the mating season, and near-term and lactating females were aggressive toward young born the previous year. In contrast, Layne and Glover (1977) observed no aggression in a study in Florida; average home range of 12 *D. novemcinctus* there was 5.7 (1.1–13.8) ha. Relatively little is known about the other species of *Dasypus*, but for *D. sabanicola* population densities of up to about 280 per sq km and home ranges of 1.7–11.6 ha. have been reported (Wetzel 1982; Wetzel et al. 2007).

In North America, *D. novemcinctus* commonly mates in July and August, but implantation is delayed until November. The period of delay usually is about 3.5 months, though it can last much longer. Captive females with no access to males have been known to give birth up to two years after removal from the wild (Loughry and McDonough 2013). Shortly after implantation, the single blastocyst divides to produce four embryos (Peppler 2008). Data compiled by Hayssen, Van Tienhoven, and Van Tienhoven (1993) indicate that mating may occur over a much longer period in Texas, from around June to November, and that implantation may be delayed up to 5 months. Depending on when a female implants her egg, the young of *D. novemcinctus* can be born anytime from early spring to early summer (Loughry and McDonough 2013). Births have been recorded in Texas during March and April and in Mexico from February to April. The period of actual developmental gestation is about 4.5 months. The normal litter consists of 4 genetically identical young, all of the same sex, from the single fertilized egg. Each weighs about 30–50 grams. They are weaned after 4–5 months and attain sexual maturity at about 1 year of age. Litters appear to remain intact for part of the summer, with siblings foraging together and sharing the same burrows, but seem to break up by early autumn (Loughry and McDonough 2001). In *D. hybridus*, of South America, implantation occurs about 1 June, and births take place in October (Barlow 1967). Litter size has been reported to be 4–12 young in *D. hybridus*, 2–12 in *D. kappleri*, 4 in *D. sabanicola*, and usually 4–8 in *D. septemcinctus* (Hayssen, Van Tienhoven, and Van Tienhoven 1993). Although the "clonal" nature of the young has been confirmed only in *D. novemcinctus*, females of all species of *Dasypus* are presumed to produce such genetically identical siblings (Boily 2008). Data from Tall Timbers Research Station in Florida suggest that longevity of *D. novemcinctus* in the wild might be 8–12 years (Loughry and McDonough 2013), but a captive of that species was still living after 22 years (McDonough 1994).

Although these armadillos are sometimes accused of stealing eggs, starting erosion, and undermining buildings, they are of economic importance due to their destruction of noxious insects. They also construct burrows in which other animals seek shelter. Some persons value armadillos as a source of food, and their armored hides are made into baskets that are sold to tourists. They are easily tamed but should not be

Llanos long-nosed armadillo (*Dasypus sabanicola*). Photo by Mariella Superina.

confined to small quarters. *D. novemcinctus* does not thrive under the conditions usually provided in zoos. This species has proved valuable to medical researchers engaged in studies of multiple births, organ transplants, birth defects, and such diseases as leprosy, typhus, and trichinosis (Yager and Frank 1972). Its production of four identical or "clonal" siblings provides a valuable research tool that allows partitioning of the genetic and environmental components of phenotypic variability (Boily 2008).

D. novemcinctus is particularly noted for its susceptibility to leprosy both in the laboratory and under natural conditions (Storrs and Burchfield 1985). Other than humans, it is the only natural host of *Mycobacterium leprae*, the bacterium that causes leprosy. The disease evidently was initially transmitted to armadillos from people, though details of the process are un-

known (Loughry and McDonough 2013). There is now a large concentration of infected armadillos in Louisiana and southern Texas. While there evidently is an increased risk of leprosy in people through exposure to armadillos in that area, *D. novemcinctus* actually has greatly benefited the effort to develop specific diagnostic antigens and vaccines against the disease. It has proven easy to maintain, and its long life span, cool body temperature, and unusual reproductive traits make it useful in a variety of studies (Truman 2008).

The IUCN classifies *D. sabanicola* and *D. hybridus* as near threatened. The former is declining because of intense subsistence hunting, conversion of its limited floodplain habitat to industrial agriculture and timber and palm oil plantations, and the use of agrochemicals that severely reduce availability of insects, its main food source; such impacts are expected to increase significantly over the next decade (Superina et al. 2014). *D. hybridus* was formerly common but is now rare or absent over much of its original range due to habitat loss through agriculture and urbanization, accidental mortality on roads, and direct hunting for use as food (Abba and Gonzalez 2014).

Conversely, the species *D. novemcinctus* is expanding at both ends of its vast range. In Argentina, it originally occurred only as far as Entre Rios Province (Wetzel et al. 2007), but a specimen was recently taken in southwestern Buenos Aires Province (Zamorano and Scillato-Yane 2008). In the north, this species is the only xenarthran that occurs as far as the United States, though apparently it was long restricted to the lower Rio Grande Valley of Texas. Increased human settlement and travel in Texas in the mid-nineteenth century may have resulted in initial translocations, and agricultural practices and reduction of prairie fires allowed more brush to invade the grasslands that had acted as a barrier to armadillo dispersal (Loughry and Mc-

Donough 2013). Around 1880 *D. novemcinctus* began moving to the north and east at an overall expansion rate of around 4–10 km per year (Fitch, Goodrum, and Newman 1952; Taulman and Robbins 1996). It first appeared in numbers in Louisiana not long before 1925 and subsequently spread to every parish in the state (Lowery 1974). It also moved into Oklahoma, Missouri, Arkansas, Mississippi, and Alabama. The species, evidently represented by animals from Texas (Loughry and McDonough 2013), was deliberately and/or accidentally introduced in Florida in the 1920s and 1930s and, mostly on its own, proceeded to occupy much of that state as well as Georgia and South Carolina (Cleveland 1970; Fitch, Goodrum, and Newman 1952). Expansion to the northwest has been slower, but Meaney, Bissell, and Slater (1987) reported a specimen from Colorado. Taulman and Robbins (1996) showed that established breeding populations were present in southern Kansas and western Tennessee, and that individual animals had been recorded as far north as

Long-nosed, or nine-banded, armadillo (*Dasypus novemcinctus*), individual near New Orleans. Photo by Henryk Sadura / Shutterstock.com.

Nebraska. They suggested that anthropogenic factors, including development of an extensive road and rail transportation network in the United States, had facilitated the armadillo's invasion.

Subsequently, Taulman and Robbins (2014) reported that the established breeding range extended into northern Kansas, central Illinois, southwestern Indiana, and western Kentucky. The northern limits of the range were now close to a line (−8° C isopleth) beyond which cold winter temperatures probably would preclude further expansion of permanent populations, though there were individual records as far as southern South Dakota, northern Iowa, and southern Wisconsin. Likewise, westward expansion of the breeding range had apparently stopped in western Texas, beyond which annual precipitation fell below 50 cm, though there were individual records from New Mexico (see also Frey and Stuart 2009). In contrast, in the northeast, the breeding range had advanced no farther than northern South Carolina, evidently because of factors other than temperature or precipitation, though there were individual records as far as northeastern Kentucky and southern Virginia. Loughry and McDonough (2013) cited and did not question a rough calculation, published in 1995, that the total number of armadillos in the United States was 30–50 million.

The movement of *D. novemcinctus* into the United States since the nineteenth century is probably only the most recent of several northward fluctuations of the range of *Dasypus*, resulting from varying climatic conditions. It was present intermittently in Florida since Blancan times (about 2 million years ago). During part of the late Pleistocene, a large (20–30 kg), extinct species, *D. bellus*, had much the same range in the United States as that now occupied by *D. novemcinctus*, reaching to Missouri, Illinois and Indiana, and to just past the northern border of North Carolina (Klippel and Parmalee 1984; Taulman and Robbins 1996, 2014).

CINGULATA; DASYPODIDAE; **Genus** *EUPHRACTUS* Wagler, 1830

Six-banded, or Yellow, Armadillo

The single species, *E. sexcinctus*, was reported by Wetzel et al. (2007) to occur in two disjunct habitats: a relatively small savannah region of southern Surinam, the extreme northern part of the adjacent Brazilian state of Pará, and eastward into the state of Amapá; and a larger *cerrado* region in eastern and southern Brazil, southeastern Peru, Bolivia, Paraguay, northern Argentina, and Uruguay. However, Abba, Lima, and Superina (2014) cited recent confirmation of the species in *cerrado* and Amazon savannah habitat of northern, northwestern, central, and eastern Pará, generally between the two previously reported regions of occurrence.

According to Redford and Wetzel (1985), head and body length is 401–495 mm, tail length is 119–241 mm, and weight is 3.2–6.5 kg. However, Eisenberg and Redford (1999) noted that *Euphractus* stores fat and in captivity can weigh up to 11 kg. Most individuals have a moderately hairy covering. The prevailing color is yellowish to reddish brown. Although it bears a general resemblance to certain other genera of armadillos, *Euphractus* is distinguished by its pointed and flattened head and its six to eight movable bands. The head shield is composed of rather large plates arranged in a fairly definite pattern. The plates of the well-armored tail are arranged in two to four distinctive bands at the base. Holes in a few plates above the base of the tail seem to be openings for scent glands, since the characteristic odor of the animal has been traced to this region. All five toes on the forefoot have claws, of which the second is the longest. Wetzel et al. (2007) wrote that *Euphractus* can be distinguished from other euphractines by the lack of a movable band of scutes at the anterior margin of the scapular shield and by the 9 upper and 10 lower teeth on each side.

According to Eisenberg and Redford (1999), the six-banded armadillo is most commonly found on savannahs, in other open vegetation formations, and at forest edges. It appears to use higher, drier habitat, though has been reported near streams. It is a good digger, making burrows with a single inverted U-shaped entrance. Unlike some other kinds of armadillos, it frequently reuses burrows. Moeller (1975) indicated that it continually digs new passageways in search of food, generally burrowing just one or two meters into the earth and then widening the underground area enough to turn around. Defecation always takes place outside the den. When digging, this animal does not throw the dirt to the side (as do moles) but scratches it up with the forefeet and then kicks it behind with the hind feet.

Six-banded, or yellow, armadillo (*Euphractus sexcinctus*). Photo by Rexford D. Lord.

Eisenberg and Redford (1999) stated that *Euphractus* is omnivorous, eating a broad range of animal and plant matter, including carrion, small vertebrates, insects, bromeliad fruits, tubers, and palm nuts; captives have killed and eaten large rats. Remains of rodents were found in the stomach of a wild individual (Wetzel et al. 2007). Sometimes this armadillo becomes abundant around plantations, where it causes damage by eating sprouting corn and other crops. Its burrows can cause dangerous accidents when horses stumble into them on the pampas. Moeller (1975) indicated that large numbers sometimes gather around the carcass of a dead animal, but that otherwise *Euphractus* is essentially solitary.

In Brazil, population density has been estimated at 0.14 individuals per ha. (Bonato et al. 2008), and minimum home range has been found to average 93.3 ha. (Carter 1983). At the Wroclaw Zoo in Poland young may be born at any time of the year after a gestation period of 60–65 days; litters contain one to three young and may include one or both sexes (Gucwinska 1971). A female taken in July in the Mato Grosso of Brazil was pregnant with two embryos, a male and a female, almost ready to be born. In the Pantanal wetlands of southern Brazil, Tomas et al. (2013) observed mating behavior only from July to November, suggesting a slightly defined breeding period from the middle of the dry season to the onset of the rainy season; this would be consistent with records of pregnant females in central Brazil in September and October and in Uruguay in January. Females build a nest before giving birth and lactate for about 55 days (Wetzel et al. 2007). The young weigh 95–115 grams at birth, open their eyes after 22–25 days, take solid food after a month, and mature by 9 months (Redford and Wetzel 1985). One specimen of this genus lived 22 years and 1 month in captivity (Weigl 2005).

Euphractus adapts well to habitat modifications and has been observed in timber plantations, sugar cane plantations, pasturelands, and areas with subsistence agriculture. There are no major threats to the genus, but it is hunted for food and medicinal use, as well as for its carapace to make handicrafts (Abba, Lima, and Superina 2014).

CINGULATA; DASYPODIDAE; **Genus** *CHAETOPHRACTUS* Fitzinger, 1871

Hairy Armadillos, or Peludos

Three species are commonly recognized (Wetzel 1985b; Wetzel et al. 2007):

C. *vellerosus*, extreme northern Chile, southern Bolivia, western Paraguay, northern and central Argentina;

C. *nationi*, western Bolivia, extreme northwestern Argentina;

C. *villosus*, northern Paraguay, extreme southern Bolivia, most of Argentina, parts of Chile from

Valparaiso to Magallanes Province in the extreme south.

C. *villosus* has been introduced to Tierra del Fuego, where it appears to be doing well (Loughry and Mc-Donough 2013). There has long been some question as to whether C. *nationi* might be only a high altitude subspecies of C. *vellerosus* (Gardner 2005d; Redford and Eisenberg 1992; Wetzel et al. 2007). Recent comprehensive analyses found no clear morphological or molecular distinction between the two and indicated that C. *nationi* should be considered a synonym of C. *vellerosus*, but also suggested that *Chaetophractus* is paraphyletic, with C. *vellerosus* being more closely related to *Zaedyus pichiy* than to C. *villosus* (Abba et al. 2015).

Head and body length is 220–400 mm, and tail length is 90–175 mm. Wetzel (1985b) listed average weights of about 0.84 kg for C. *vellerosus* and 2.02 kg for C. *villosus*. C. *nationi* weighs 1.5–3.0 kg (Peredo 1999). The armor of *Chaetophractus* consists of a shield on the head (which has a granular surface), a small

Andean hairy armadillo (*Chaetophractus nationi*). Photo by Alexander F. Meyer at Huachipa Zoo, Peru.

shield between the ears on the back of the neck, and the carapace (which protects the shoulders, back, sides, and rump). The head shield of *C. nationi* is about 60 mm long and 60 mm wide. The banded portion of the carapace has about 18 bands, of which usually 7–8 are movable. These animals have more hairs than most armadillos. Hairs project from the scales of the body armor, and the limbs and belly are covered with whitish or light brown hairs.

These armadillos usually inhabit open areas and seem best adapted to dry or semi-desert conditions. They are powerful diggers and live in burrows. On farms in the eastern Pampas of Argentina, *C. villosus* burrows in several kinds of grassland and wooded habitats, with a variety of soil conditions, while *C. vellerosus* is more specialized, using primarily native woodland with calcareous soil (Abba, Vizcaíno, and Cassini 2007). *C. nationi* is found only in grasslands at elevations of 2,400–4,000 meters (Perez Zubieta, Abba, and Superina 2014). The burrows of *C. vellerosus* are usually on sloping sand dunes and several meters long and more than a meter deep (Greegor 1985). In a study of 56 burrows of *C. villosus* on farmland in the Argentinian pampas, Abba, Udrizar Sauthier, and Vizcaíno (2005) found simple and complex structures, both located in high terrain that does not flood. Simple structures, made when animals are in search of food or temporary shelter, had a mean length of 70 cm and mean depth of 50 cm; they were found exclusively in soils with primarily organic material, where arthropods, grubs, and annelids were frequent. Complex structures, used as home burrows, were larger and branching, up to 485 cm long and 100 cm deep, and found in areas with a high proportion of calcareous material, little organic material, and low moisture, resulting in a hard soil. Seven of the 34 complex structures had enlarged chambers near the mouth and at the end of the galleries.

Activity may be largely nocturnal in summer, to avoid the desert heat, and diurnal in winter, but there seems to be considerable variation (Abba and Cassini 2008). When pursued, the species *C. villosus* at first attempts to run away, often emitting a characteristic snarling sound. If unable to find a hole, it tries to burrow into the ground. If overtaken while running or if it does not have chance to burrow, this species draws in its feet so that the edges of its armor are in contact with the ground and thus effectively protects itself against canid and avian predators. *C. villosus* anchors itself in its burrow by spreading its feet sideward and bending its body so that the free hind edges of the bands grasp the walls of the burrow. Greegor (1980b) found that during a 3-day period one individual *C. vellerosus* moved an average of 1,032 meters per night within an overall area of 3.4 ha. Abba (2011) reported an average home range of 2,670 sq meters for *C. vellerosus*. Peredo (1999) estimated population density of *C. nationi* in Bolivia as 10.6 individuals per sq km.

The members of this genus regularly burrow under animal carcasses to obtain maggots and other insects and are said sometimes to burrow into carcasses. Under certain conditions they obtain grubs and insects from a few centimeters below the surface of the ground by the unusual method of forcing a hole in the ground with the head and then turning the body in a circle so that a conical hole is formed without any digging. They have been observed to kill small snakes by throwing themselves upon the snakes and cutting them with the edges of the shell. According to Greegor (1980a, 1985), *C. vellerosus* feeds mostly on insects during the summer and takes a substantial number of rodents, lizards, and other small vertebrates. It also relies heavily on plant material, especially in the winter, when over half of its diet consists of vegetation.

In the Bolivian Chaco, *C. vellerosus* has a short, concentrated reproductive season from November to February, which is during the austral spring and summer (Cuéllar 2008). In Buenos Aires Province, eastern Argentina, pregnant female *C. villosus* have been observed in late winter (September) and litters found in late spring and early summer (December–January), while spermatogenesis in males continues through most of the year except from about mid-May to July (Luaces et al. 2013). Merrett (1983) stated that the gestation period of *C. villosus* is 60–75 days and that there is said to be more than one litter annually. Litters usually consist of two young, often one male and one female. The young weigh about 155 grams at birth, open their eyes after 16–30 days, are weaned at 50–60 days, and reach sexual maturity at 9 months. A captive specimen of *C. villosus* lived for 23 years and 6 months and another was still living after 25 years and 2 months (Weigl 2005).

Hairy armadillos are regularly hunted and eaten by people and are killed as agricultural pests. They also

Big hairy armadillo (*Chaetophractus villosus*). Photo by Alexander F. Meyer at Temaiken Zoo, Argentina.

are intensively harvested commercially, mainly for their carapace, which is used to make *charangos* (musical instruments) and handicrafts. Nonetheless, the widespread species *C. vellerosus* and *C. villosus* are considered common, with stable populations (Abba, Cuéllar, and Superina 2014; Abba, Poljak, and Superina 2014). The much more restricted *C. nationi* is being killed at an unsustainable rate and is also losing habitat to agriculture and excavation of sand for concrete production. During the 1990s an estimated 1,950 individual *C. nationi* were taken each year in Bolivia, from a total population of only 3,212–13,250 individuals, and extinction within a decade appeared possible (Peredo 1999). Apparently the species still survives, but it is classified by the IUCN as vulnerable (Perez Zubieta, Abba, and Superina 2014), because its population evidently experienced a decline exceeding 30 percent during the previous 10 years. It is now legally protected and is on appendix 2 of the CITES. Abba et al. (2015), while proposing that *C. nationi* be

regarded as a synonym of *C. vellerosus*, nonetheless argued that conservation measures should be implemented as soon as possible to save the involved high-altitude population from extinction.

CINGULATA; DASYPODIDAE; **Genus**
ZAEDYUS Ameghino, 1889

Pichi

The single species, *Z. pichiy*, occurs in central and southern Argentina and in the Andean grasslands of Chile, south to the Strait of Magellan (Wetzel 1985b). The species was originally restricted to Argentina and did not enter Chile until the nineteenth century (Superina and Abba 2014a).

Head and body length is 260–335 mm, and tail length is 100–140 mm. Weight of 100 specimens averaged 977 grams and ranged from 700 to 1,500 grams

Pichi (*Zaedyus pichiy*). Photo by Zixian/Shutterstock.com.

(Superina and Abba 2014a). The head shield and body carapace are dark brown with yellow or whitish lateral edges, and the tail shield is yellowish. The posterior edges of the dorsal plates are thickly set with fine blackish hairs interspersed with longer yellowish brown and whitish bristles. The underparts of the body are covered with coarse, yellowish white hairs. The digits on all limbs are distally separate and have well-developed claws. The ears are very small. This armadillo resembles *Euphractus* externally, but Jorge, Meritt, and Benirschke (1977) found that karyologically *Zaedyus* appears more closely related to *Chaetophractus* than to *Euphractus*. It differs from *Chaetophractus vellerosus*, which is about the same size, in having much shorter ears, a narrower head shield, and pointed marginal scutes. Females have two pectoral mammary glands (Superina and Abba 2014a).

The pichi occurs farther south than any other armadillo, mostly in arid to semiarid grassland habitat. Elevational range is sea level to 2,500 meters (Superina

and Abba 2014b). It digs and uses burrows, often in sandy soil, for shelter, thermoregulation, and giving birth to and nursing young; depth varies seasonally and by latitude and can be as great as 1.5 meters in winter (Superina and Abba 2014a). It can anchor itself in its burrow by wedging the serrated edges of the carapace into the surrounding dirt. Its usual defense reaction is to draw in its feet so that the edges of its armor are in contact with the ground. Captives are known to become torpid at low ambient temperatures (Wetzel et al. 2007), and long ago Hatcher (in Allen 1905) reported the pichi to hibernate in winter, at least in some localities. Recent studies have confirmed that observation (Superina 2008; Superina and Abba 2014a; Superina and Boily 2007). This species is the only extant xenarthran known to enter shallow daily torpor and prolonged deep hibernation, with minimal subcutaneous temperatures of 14.6° C. In Mendoza Province, west-central Argentina, hibernation usually lasts from mid-May to mid-August; considerable fat reserves are built up beforehand. In the wild, only subadults, which may lack the necessary energy reserves to enter torpor, have been observed outside their burrows during winter. Normal body temperature averages 35.1° C, but during daily torpor typically drops to below 30° C, sometimes as low as 24.5° C, during a period of 4–6 hours. Environmental stress, such as reduced food quality, leads to prolonged and deeper torpor. The diet includes insects, worms, small vertebrates, and any other animal food the pichi can find, including dead creatures. Redford and Eisenberg (1992) wrote that it also eats plant material, especially pods of the *Prosopis* tree.

Life history data were gathered by Superina (2008) and Superina and Abba (2014a). *Zaedyus* is predominantly solitary, with social interaction between wild individuals observed only during the breeding season. Also at that time, captive adult males aggressively defended areas against other males. Several vocalizations have been recorded, including a purring sound, when an animal is threatened, which can change to a grunt or scream. The reproductive season that lasts from spring to early summer, varies from 3 to 5 months depending on latitude, and is regulated by photoperiod. Mating in the wild has been seen in September and may extend through October, births occur from October to January, and lactating females have been observed from November to February. There is no evidence of delayed implantation, and gestation length is 58–60 days. There is an annual litter of one or two young, exceptionally three. They each weigh approximately 50 grams. They do not leave the burrow until about 40 days old and are partially or fully weaned. Sexual maturity is attained at 9–10 months, but some females do not reproduce until their second year. A captive pichi lived for 12 years and 6 months (Weigl 2005).

In some areas *Z. pichiy* is kept as a house pet, but its flesh is said to have an excellent flavor and is highly prized by some people. Although widespread, the species is heavily hunted for food and sport, especially in northern and eastern portions of its range, and it is also affected by overgrazing of its habitat by cattle. It has become less abundant in the provinces of Buenos Aires and Mendoza, with local extinctions recorded in some areas. The species is now classified as near threatened by the IUCN, but it may qualify for a more serious category due to ongoing exploitation levels, which have led to a decline in the order of 20 percent over the past 12 years (Superina and Abba 2014b).

CINGULATA; DASYPODIDAE; **Genus** ***CALYPTOPHRACTUS*** Fitzinger, 1871

Chacoan Fairy Armadillo, Greater Pichiciego

The single species, *C. retusus*, is found in the Gran Chaco region of south-central Bolivia, northwestern Paraguay, and extreme northern Argentina (Wetzel et al. 2007). The name *Burmeisteria* Gray, 1865 often has been used for this genus, but Wetzel (1985b) explained that if the species *C. retusus* is retained in its own genus, the correct name is *Calyptophractus*. *C. retusus* often has been assigned to the genus *Chlamyphorus*, together with *C. truncatus* (e.g., by Eisenberg and Redford 1999, Gaudin and Wible 2006, and McKenna and Bell 1997), but was considered to represent a separate genus by Gardner (2005d) and Wetzel et al. (2007). Molecular analysis by Delsuc et al. (2012) placed the divergence between *C. retusus* and *C. truncatus* around 17 million years ago and supported assigning each to a different genus.

Descriptive information was taken in part from Delsuc et al. (2012), Eisenberg and Redford (1999), Moeller (1990c), and Wetzel et al. (2007). Head and body length is 116–189 mm, tail length is about 35–43 mm, and weight is around 100 grams. The 24 dorsal plates are yellowish brown in color. The underparts, sides, and legs below the carapace are covered with long, silky, grayish white hair. The head is covered by a broad, rounded shield that extends laterally and ventrally to the level of the eye. The eyes are small, and the ears are reduced but visible. The dorsal carapace is thin, flexible, attached to the skin of the back for its entire extent, not underlain by fur, and separate from the round rump shield. The tail is rounded, is partly covered with plates, and has a pointed, not spatulate, tip. The forelimbs have five digits with proportionally very large claws. *Calyptophractus* resembles *Chlamyphorus* but is larger, has the entire carapace attached to the body, as in other armadillos, and has a less defined head shield that lacks a posterior row of large scutes. Unlike that of *Chlamyphorus*, the rear

Chacoan fairy armadillo, or greater pichiciego (*Calyptophractus retusus*). Photo by Thomas and Sabine Vinke, Paraguay Salvaje.

shield of *Calyptophractus* is discontinuous, consisting of small plates scattered all over the anal region, except in the center, and separated from each other by intervals of naked skin.

The Chacoan fairy armadillo is patchily distributed, being restricted to loose, sandy soils and absent from areas with clay soils; it can be found in disturbed habitat and may be encountered close to villages and other populated areas (Cuéllar et al. 2014). *Calyptophractus* is presumed to be largely fossorial (Wetzel et al. 2007), but is not as good a digger as *Chlamyphorus*; it probably presses itself against the ground when surprised in the open. Cries almost like those of human infants have been noted in *Calyptophractus*. A captive did well on boiled rice and grapefruit. An individual of this genus that had been in captivity in Bolivia for two months in a yard where it could burrow appeared to be in good health when received at the US National Zoo. It was given soil in which to burrow and a wide assortment

of foods, including mealworms, raw eggs, and meat, but it died after a few days. The immediate cause of death seems to have been a respiratory infection, perhaps caused by bacteria, to which the animal had no resistance and which were in the soil supplied to it.

Calyptophractus retusus has been variously classified by the IUCN—insufficiently known, vulnerable, near threatened—but is now designated data deficient because virtually nothing is known about the species except that its population is severely fragmented within microhabitats subject to further decline (Cuéllar et al. 2014). It may also be in jeopardy because of loss of habitat to agriculture and excessive collection (Thornback and Jenkins 1982). It is killed mercilessly by local people, not for food, but because it is considered an evil omen that must be quickly destroyed lest there be death in the family (Aguiar and da Fonseca 2008). Its population size probably has declined about 20–25 percent since 2000 (Delsuc et al. 2012).

CINGULATA; DASYPODIDAE; **Genus**
CHLAMYPHORUS Harlan, 1825

Pink Fairy Armadillo, Lesser Pichiciego

The single species, *C. truncatus*, is found in central Argentina, from western Mendoza to southern Buenos Aires Province, and from southern Catamarca to north-central Rio Negro Province (Borghi et al. 2011; Wetzel 1985b). *Calyptophractus* (see account thereof) sometimes has been included in *Chlamyphorus*.

This is the smallest armadillo. Redford and Eisenberg (1992) listed head and body lengths of 84–117 mm and tail lengths of 27–35 mm. Weight was given as 85 grams by Carter and Encarnaçao (1983) and 90 grams by Moeller (1990c). The armor is pale pink in color. The body under the shell, the underparts, and the legs are covered with fine, soft, white hairs. The head and body armor is anchored to two large, rough prominences on the bones above the eyes and only by a narrow ridge of flesh along the spine.

There is free movement between all of the approximately 24 bands across the neck and back. *Chlamyphorus* is the only armadillo whose dorsal shell is almost separated from the body. The head shield is well-defined and has a row of large osseous plates at the posterior margin, giving the appearance of a distinct "step" in the dorsal surface; the ears are not visible (Delsuc et al. 2012). The separate armored plate at the rear is securely attached to the pelvic bones. The spatula-shaped tail projects from a notch at the lower edge of the rear plate. The tail cannot be raised and is dragged on the ground. The forefeet are enlarged and molelike, each having five long stout claws. The lips are horny and stiff, the nose is pointed, and the nostrils are ducted downward. Females have two mammae.

The pink fairy armadillo inhabits dry grasslands as well as sandy plains with thornbushes and cacti. It occurs from sea level to an elevation of 1,500 meters (Superina, Abba, and Roig 2014). Although frequently described as nocturnal, Rood (1970) found a captive to

Pink fairy armadillo, or lesser pichiciego (*Chlamyphorus truncatus*). *Upper left*, photo by Mariella Superina. *Upper right*, photo by Guillermo Ferraris through Mariella Superina. *Lower left*, photo by Mariella Superina and Paul Vogt. *Lower right*, photo by Guillermo Ferraris through Mariella Superina.

have active periods spaced throughout the day and night. One observer reported seeing an individual of this genus while he was riding a horse, but, before he could dismount, the pichiciego had burrowed out of sight. To accomplish such rapid digging, the animal supports the rear end of the body with the rigid tail and uses the hind feet to kick back the ground loosened by the forefeet. Rood (1970) observed the animal to then pack the soil posteriorly with its rear armor plate, thereby moving freely beneath the surface without leaving a permanent tunnel. After entering the burrow, it may also use the rear plate effectively to close the burrow opening, much like corking a bottle. This small armadillo often burrows near an anthill. It seems to prefer dry soil that feels uncomfortably warm to the human hand. When infrequent rains moisten the soil, the animal may leave its burrow.

The diet apparently consists mostly of insects and their larvae. Several hundred ants were found in the digestive tract of one specimen of *Chlamyphorus*, and this genus also is known to feed on worms, snails, roots, and other plant material. *Chlamyphorus* is difficult to keep in captivity, but Rood (1970) maintained a specimen on a diet of bread, milk, whole oats, and occasional beetles and beetle larvae. That animal slept in an underground burrow and came aboveground only once each day to eat. It was captured on 15 July 1967 and was still living in January 1970. It was then given to the Brookfield Zoo, where it died on 1 December 1971 (Weigl 2005). Wetzel et al. (2007) indicated that litter size is one.

In contrast to *Calyptophractus*, the pink fairy armadillo is amiably tolerated by local people, who, however, say it has become much less common in the last several decades (Aguiar and da Fonseca 2008). It was formerly classified as endangered by the IUCN, then as near threatened, but now is designated data deficient (Superina, Abba, and Roig 2014). Habitat degradation is going on throughout its range, especially from plowing for agriculture and soil compaction from cattle and goat ranching. Other threats include predation by domestic cats and illegal collection of individuals to be kept as pets or sold; most animals removed from the wild die within 8 days. However, actual effect of such problems on populations is not well understood. *Chlamyphorus truncatus* also is listed as endangered by the USDI.

Giant Armadillo

The single species, *P. maximus*, is found throughout most of South America east of the Andes, from northwestern Venezuela to northeastern Argentina (Cabrera 1957; Handley 1976; Wetzel 1985b).

Head and body length is 75–100 cm, and tail length is about 50 cm. Adults weigh as much as 60 kg, but maximum recorded weights may apply to overstuffed zoo residents. Redford and Eisenberg (1992) listed weights of 18.7–32.3 kg. Silveira et al. (2009) recorded mean weights of 44.4 kg for five wild adult males and 28.0 kg for four females. The hair covering is scant; only a few hairs are scattered between the plates. The coloration is dark brown except on the head and tail and a band around the lower edge of the shell, which is whitish. There are 11–13 movable bands on the back and 3–4 bands on the back of the neck. The carapace is very flexible. The head shield is oval and is not expanded between the eyes. The plates on the tail are closely set and are not arranged in rows. The claws of the forefeet are powerful, and the claw on the third finger measures about 203 mm along the curve. The giant armadillo is so much larger than other armadillos that size alone serves to distinguish adults. It occasionally possesses as many as 100 small teeth, but these are shed as the animal ages. Females have two mammae.

The giant armadillo is a powerful and rapid digger and shelters in burrows of its own construction. It has been reported to occur primarily in unbroken, relatively undisturbed forest and usually near water. In a study in Brazil, however, Carter (1983) found 68 percent of its burrows in grassland, 28 percent in brushland, and only 3 percent in woodland. Nearly half of those burrows were located in active termite mounds. The average distance to the woods for burrows located in grassland or brushland was 192 meters. Burrows averaged 41 cm in width and 31 cm in height. Burrows of different ages were found in close proximity, indicating that *Priodontes* reuses certain areas. Carter and Encarnaçao (1983) found this genus also to remain longer in a burrow than did other armadillo genera investigated; one female stayed in one hole for 17 days. Activity is strictly nocturnal. Despite its

Giant armadillo (*Priodontes maximus*). Photo by Frederico Mosquera through Mariella Superina.

rigid appearance, the giant armadillo is fairly agile and often balances itself on its hind legs and tail with its forefeet off the ground. Unlike certain other armadillos, it cannot completely enclose itself in the carapace; if closely pursued, it may try to dig itself in. It also digs extensively to obtain food. The diet consists primarily of termites, but ants, other insects, spiders, worms, larvae, snakes, and carrion are also consumed. *Priodontes* frequently is accused of eating garden vegetables, but it probably digs in gardens in search of insects, not vegetables. Wallace and Painter (2013) showed that *Priodontes* does feed on figs in the wild and suggested that fruit may be seasonally important in its diet.

Carter (1985) reported mean nightly movement to be 2,765 meters and minimum home range to average 452.5 ha. In a more extensive study in Emas National Park, Brazil, Silveira et al. (2009) reported that five individuals, radio-tracked for an average of 27 days, had a mean minimum nightly movement of 1,800 meters and a mean home range of 10 sq km. Minimum density in the park was estimated at 3.36 animals per 100 sq km. The home range of two individuals showed very little overlap. Based on camera-trapping, Noss, Peña, and Rumiz (2004) suggested somewhat more extensive overlap between several males and between a male and a female in Kaa-Iya del Gran Chaco National Park, Bolivia, where they estimated population densities of 5.77–6.28 per 100 sq km. Merrett (1983) wrote that the gestation period is said to be 4 months and that there are one or two young. They weigh up to 113 grams each and have tough, leathery skin. They are weaned at 4–6 weeks and reach sexual maturity at 9–12 months. Weigl (2005) listed an individual born in the wild about 1972 that subsequently died in the San Antonio Zoo in January 1988, and another that died in the same zoo at an age of 11 years and 7 months. However, Superina, Miranda, and Plese (2008) observed that in the previous decade only a few zoos, all in the natural range of the giant armadillo, had been able to maintain the species and that there had been no reproductive success.

The IUCN now classifies the giant armadillo as vulnerable (Anacleto et al. 2014). It occurs over a vast region but is rare throughout its range and is very patchily distributed. It is avidly hunted for its meat and is heavily affected by loss of habitat to agriculture and deforestation. It is sometimes sought as a pet or to be sold as a "living fossil" on the black market, where the price for a grown specimen may reach into the tens of thousands of US dollars (Aguiar and da Fonseca 2008). However, the species usually does not survive long in captivity. Such problems have led to an estimated population decline of at least 30 percent in about the past 21 years. The species has disappeared from large parts of its southern range and may no longer be present in Uruguay. It also is listed as endangered by the USDI and is on appendix 1 of the CITES.

CINGULATA; DASYPODIDAE; **Genus** *CABASSOUS* McMurtie, 1831

Naked-tailed Armadillos

There are four species (Cuarón, March, and Rockstroh 1989; Hayssen 2014; Hayssen et al. 2013; McCarthy 1982; Smith et al. 2011; Wetzel 1980, 1985b; Wetzel et al. 2007):

C. unicinctus, east of the Andes from Venezuela to southern Brazil and eastern Paraguay;

C. centralis, extreme southern Mexico and Belize to northeastern Venezuela and the lower slopes of the Andes in western Colombia and northern Ecuador;

C. chacoensis, the Gran Chaco of western Paraguay and northern Argentina, possibly also the adjacent part of Brazil and southeastern Bolivia;

C. tatouay, southern Brazil, eastern Paraguay, Uruguay, northeastern Argentina in the provinces of Misiones and Buenos Aires.

Naked-tailed armadillos are closely related to *Priodontes* and resemble that genus closely except in their smaller size and lack of obvious and extensive covering of scutes on the tail (Wetzel et al. 2007). Head and body length is 300–490 mm, and tail length is 90–200 mm. Meritt (1985a) reported weights of 2.0–3.5 kg for *C. centralis* and 2.2–4.8 kg for *C. unicinctus*. Redford and Eisenberg (1992) listed weights of 3.4–6.4 kg for *C. tatouay*, the largest species, and Hayssen (2014) noted a weight of 1.2 kg for an adult female of *C. chacoensis*, the smallest species. The coloration above is dark brownish to almost black, the lateral edges of the carapace are

Naked-tailed armadillo (*Cabassous unicinctus*). Photo by Alexander F. Meyer at Huachipa Zoo, Peru.

yellowish, and the underparts are dull yellowish gray. There are five large claws on the forefeet; the middle claw is especially large and sickle-shaped. The snout is short and broad, the head is broad, and the ears are widely separated. The movable transverse bands across the middle of the back are the most numerous in the family, varying in number from 10 to 13. The tail is slender and shorter than the head and body and only slightly armored with small, thin, widely spaced plates. The scapular and pelvic shields are attached to the body almost to the base of the limbs. The lack of complete armor on the tail is unique among armadillos.

Naked-tailed armadillos live in a variety of habitats, including grasslands, semiarid and moist lowlands, upland areas, and riversides (Meritt 1985a). They occupy burrows, the entrances of which usually are in open ground or at the base of an embankment. Loughry and McDonough (2013) suggested that, because of their rapid excavation ability, they do not make permanent burrows but dig themselves into the ground wherever they happen to be when it is time to rest. Captives do not construct a nest even when provided with material. In some areas activity is nocturnal and begins shortly after sunset, though Bonato et al. (2008) found *C. unicinctus* to be almost entirely diurnal. The species *C. centralis* prefers dry to mesic forests and is one of the most fossorial of armadillos (Hayssen et al. 2013). It walks on the tips of the claws of its forefeet and on the soles of its hind feet. When danger threatens, it runs quite rapidly for short distances, burrows into the ground, or goes into water. Like that of the other species, its diet consists largely of ants and termites, which are apparently located by scent and dug out of litter and soil. The sickle-like claw is used to cut small roots as the armadillo digs after insect colonies in dead roots or stumps; the prey is extracted from the tunnels by the long, extensible tongue. *C. centralis* sometimes completely buries itself while digging for insects. It is slower than *Dasypus novemcinctus*.

Animals are usually found alone in the wild, but breeding groups of up to two males and four females have been established in captivity, where the individuals generally ignore one another. A variety of sounds has been reported, including a low buzz or growl, a loud grunt, and low gurgling squeals (Wetzel et al. 2007). A home range size of 64.9 ha. for *C. unicinctus* was listed by Loughry and McDonough (2013). In *cer-rado* habitat of southeastern Brazil, Bonato et al. (2008) found *C. unicinctus* to have a population density of 0.27 individuals per ha. In that area, reproduction apparently continues through both the warm-wet season from October to March and the cool-dry season from April to September. The single young weighs about 100 grams at birth. A specimen of *C. centralis*, born in the wild about 1979, was captured and lived in zoos until 11 March 1987 (Weigl 2005).

Cabassous chacoensis is classified as near threatened by the IUCN (Meritt, Superina, and Abba 2014). It is jeopardized by habitat degradation from agricultural activity, subsistence hunting for food by local people, and predation by domestic dogs. Its population may have declined by 20–25 percent over about the last 15 years. The status of *C. centralis* is not well known, but it is apparently rare, patchily distributed, and rapidly losing habitat, which may soon justify its classification as vulnerable by the IUCN (Tirira et al. 2014).

CINGULATA; DASYPODIDAE; **Genus**
TOLYPEUTES Illiger, 1811

Three-banded Armadillos

There are two species (Wetzel 1985b):

> *T. tricinctus*, east-central Brazil;
> *T. matacus*, central and eastern Bolivia, the Mato Grosso of central Brazil, the Chaco region of Paraguay, northern and central Argentina.

Head and body length is 218–277 mm, tail length is 38–80 mm, and weight is 1.00–1.59 kg (Redford and Eisenberg 1992; Wetzel 1985b; Wetzel et al. 2007). The overall coloration is blackish brown. Most individuals have three movable bands, but some have only two bands and others have four. These are the only armadillos that can completely enclose themselves by rolling into a sphere. Unlike most genera in the family, in *Tolypeutes* the sides of the two large shells are free from the skin; there is considerable space in the shell into which the head, legs, and tail can be fitted when the animal "rolls up." The second, third, and fourth toes of the hind foot are grown together, with nails almost like hoofs; the first and fifth toes are slightly separated from the others and have normal claws. The short, thick tail,

Three-banded armadillo (*Tolypeutes matacus*). Photo by Frederico Mosquera through Mariella Superina.

with prominent tubercles, is almost inflexible. *T. tricinctus* has five claws on the forefoot, but *T. matacus* has only four.

Jorge, Meritt, and Benirschke (1977) found a remarkable difference in karyotype between *T. matacus* and other armadillos. In this species there are no truly acrocentric autosomes, and the diploid number (2n=38) is the lowest found among the xenarthrans they examined. All other armadillos have diploid numbers from 50 to 64.

In the Mato Grosso *T. matacus* has been found in grassy or marshy areas between scattered forests. This species apparently does not dig holes but utilizes the abandoned burrows of anteaters (order Pilosa). It runs relatively rapidly with a peculiar gait, only the tips of the foreclaws touching the ground. When danger threatens, it leaves a small opening between the edges of the shell and the extremities; then, if it is touched on the chest or abdomen, it snaps the shells together like a steel trap. This behavior seems to be quite an effective defense against natural enemies. According to

Merritt (2008), *T. matacus* can often escape human hunters with speed and directional changes. It is encountered at night along dirt roads, but may also be active during warm sunny winter days.

In a study of the food habits of *T. matacus* in the Chaco of Argentina, Bolkovic, Caziani, and Protomastro (1995) found beetle larvae to be the most frequently consumed item throughout the year, ants and termites to be important in the dry season from July to November, and fruits to be especially significant during the summer rains. Ants and termites are obtained by probing into the ground, under bark, and into nests with the powerful forelegs and claws. In captivity, fruits, leaves, boiled rice, soft bread in milk or tea, ants, ant eggs or termite larvae, and mealworms are eaten.

Sources cited by Wetzel (1982) indicate that population densities of *T. matacus* reach 7 per sq km and that this species is primarily solitary, though groups of up to 12 individuals may be found together in a single shallow nest during the cold season. Meritt (2008) in-

Three-banded armadillo (*Tolypeutes matacus*). *Left,* shell almost closed, head above, tail below. *Right,* shell fully closed. Photos by Alexander F. Meyer at Cincinnati Zoo.

dicated that such congregations contain not more than six animals and are temporary, lasting 1–4 days. Population densities of 1.9 and 7.0 per sq km and home ranges of 2.7–14.0 ha. have been reported for *T. matacus* in the Bolivian Chaco (Cuéllar 2008). Population density of *T. tricinctus* was estimated at 1.2 animals per sq km in the *cerrado,* and mean home range was estimated at 122 hectares, with adult males having significantly larger home ranges (238 hectares) than females; male and female ranges overlapped one another but ranges of adult males did so only at the edges (Miranda et al. 2014b).

In the Bolivian Chaco, pregnancy rates in female *T. matacus* reportedly have two annual peaks, July–September (winter) and December–February (summer) (Cuéllar 2008). According to Meritt (1976), *Tolypeutes* is probably uniparous. From 1969 to 1974 at the Lincoln Park Zoo there were 18 live births of *T. matacus,* in January, February, March, May, June, August, October, November, and December. In each case there was only a single young. Meritt (1976) stated that in the Paraguayan Chaco most births take place from November to January but that in captivity the season seems to shift a little; the majority of births at Lincoln Park occurred from October to December. The only information on gestation was that an adult male and female produced a young 120 days after they were placed together. However, Loughry and McDonough (2013) listed a gestation period of 104–116 days for *T. matacus.* Meritt (1976) observed that a newborn *Tolypeutes* is a miniature adult. The claws are fully developed and hardened, and the eyes and ear pinnae are closed, but they open in the third or fourth week of life. The flexible carapace has a leathery texture, and the individual scute markings are already apparent. Right from birth the baby is capable of coordinated movements, including walking and rolling itself into a sphere. Birth weight is about 113 grams (Hayssen, Van Tienhoven, and Van Tienhoven 1993). Weaning is completed by 72 days, and sexual maturity comes at 9–12 months (Merrett 1983). A female *T. matacus* was living at the Lincoln Park Zoo at an age of approximately 36 years (Weigl 2005).

Wetzel (1982) indicated that *T. tricinctus* was known by not more than six museum specimens, was not located during several intensive field surveys, and probably had disappeared over much of the southeastern highlands of Brazil. Cardoso da Silva and Oren (1993) noted that the difficulty in locating *T. tricinctus* may be associated with its restricted habitat within patches of tropical deciduous forest on the more elevated sectors of the Brazilian plateau. Such habitat is naturally fragmented and highly threatened by agricultural development, cutting for charcoal, and mining for the underlying calcareous deposits. The species was so rare that some authorities believed it extinct, and it was listed as endangered by the IUCN. It was rediscovered in 1988 at a few isolated localities, but is continuing to decline because of heavy hunting and habitat loss, and is now classified as vulnerable by the IUCN (Miranda et al. 2014b). *T. matacus* is declining because of the same factors and is designated near threatened (Noss, Superina, and Abba 2014). It has been eliminated over a large part of central Argentina (Roig 1991).

Order Pilosa

Sloths and Anteaters

This order, the living components of which are 4 families, 5 genera, and 10 species, inhabits Mexico, Central America, and South America. At least 4 additional genera and 12 species are thought to have occurred in the West Indies until early historical time. This order and the order Cingulata were formerly considered infraorders of what had been designated the order Xenarthra (for further details, see account of class Mammalia, particularly the discussion of clade Xenarthra). Gardner (2005d, 2007a, 2007f, 2007g) recognized two living suborders of Pilosa: Vermilingua, with the families Cyclopedidae (silky anteater) and Myrmecophagidae (tamanduas and giant anteater), and Folivora, with the families Megalonychidae (West Indian sloths and two-toed sloths) and Bradypodidae (three-toed sloths). McKenna and Bell (1997) used the name "Phyllophaga" for the suborder containing the sloths. The sequence of families and genera presented here follows that suggested by the molecular analyses of Delsuc and Douzery (2008), Delsuc, Vizcaíno, and Douzery (2004), Delsuc et al. (2012), and Möller-Krull et al. (2007), which show the lineage of Cyclopedidae as having diverged the earliest. The sequence also considers the morphological analyses of Gaudin (2003, 2004), Gaudin and Branham (1998), and White and MacPhee (2001).

The two pilosan suborders are very different from one another and highly specialized. Living members vary in head and body length from about 150 mm in *Cyclopes* to 1,200 mm for the giant anteater, *Myr-*mecophaga. The forefoot usually has two or three digits that are much larger than the others, when present, and all digits have long, sharp, strong claws. The hind foot usually has five toes. The bones of the lower forelimb are separate, whereas the bones of the lower hind leg are either separate, united at the ankle, or joined at both ends. The mammae are located near the armpit, on the chest, or on the abdomen. Females have a common urinary and genital duct. The testes are located in the abdominal cavity between the rectum and the urinary bladder. Pilosans have a double posterior vena cava that returns blood from the posterior part of the body to the heart, whereas most other mammals have only a single vena cava. The skull is elongated in anteaters but short and rounded in sloths. The subordinal term "Vermilingua" (meaning "worm tongue") refers to the slender, elongated, sticky tongue of the anteaters (McDonald, Vizcaíno, and Bargo 2008). The neck vertebrae vary in number from six to nine, and the pelvis is narrow and elongate.

Although this order was at one time placed within another nominal order, Edentata, the name of which signifies toothlessness, only the anteaters actually have no teeth. None of the living folivorans has incisors or canines. The cheek teeth of the tree sloths lack enamel. They are open at the root, a condition incident to continuous growth throughout life. The premolars usually resemble the molars, so the cheek teeth present a uniform series, except in *Choloepus*, in which canine-like teeth are present. The order is distinguished from all others, except Cingulata, by what are known as xenar-

Left, lesser anteater or northern tamandua (*Tamandua mexicana*), mother and young; photo by Leonardo Mercon / Shutterstock.com. *Right*, two-toed tree sloth (*Choloepus didactylus*), mother and young; photo by Worldswildlifewonders / Shutterstock.com. These photos express the striking morphological differences of the order but the arboreal habits of most living species.

throus vertebrae (see account of class Mammalia, clade Xenarthra). From Cingulata, Pilosa is distinguished by having a primary body covering of hair and the absence of dermal armor (Gardner 2007f).

The giant anteater lives on the ground and is active by day or night. Other anteaters dwell mainly in trees and are usually nocturnal. Living sloths are restricted to trees; some are nocturnal and others may be active at any time. The species of Pilosa are chiefly solitary but may form small, loose associations. All have a good sense of smell, anteaters have poor eyesight and hearing, and sloths have good vision but poor hearing. Anteaters are basically insectivorous, while sloths are herbivorous.

Pilosa is an exclusively New World order; reports of fossils from the Eocene of Europe and Paleocene of Asia are now known to be incorrect (Gardner 2007e; Gaudin 1999; McDonald, Vizcaíno, and Bargo 2008). Most of the group's evolution occurred in South America,

where its fossil history ranges from the Eocene to Recent. Several genera have been reported from the late Miocene of North America, but it was not until the middle Pliocene, a few million years ago, that a major invasion of that continent began. Anteaters and tree sloths moved no farther north than Mexico but giant ground sloths occupied most of the continent.

PILOSA; **Family CYCLOPEDIDAE; Genus** *CYCLOPES* Gray, 1821

Silky Anteater

The single genus and species, *Cyclopes didactylus*, is found from Oaxaca and southern Veracruz in Mexico to northern and western Colombia and coastal Ecuador, and east of the Andes in southern Colombia, eastern Ecuador, eastern Peru, northern Bolivia, eastern

Venezuela, Trinidad, the Guianas, and western and northeastern Brazil, with an isolated population in coastal Brazil in the states of Rio Grande do Norte, Paraíba, Pernambuco, and Alagoas (Chacón, Racero-Casarrubia, and Rodríguez-Ortiz 2013; Gardner 2007g; Hall 1981; Miranda and Superina 2014). Cyclopedidae sometimes has been considered part of Myrmecophagidae, but was treated as a separate family by Gardner (2005e, 2007g), Gaudin and McDonald (2008), Hayssen, Miranda, and Pasch (2012), and McKenna and Bell (1997).

Head and body length is 153–230 mm, and tail length is 165–295 mm. Weight was listed as 300–500 grams by Moeller (1990c) and as 175–357 grams by Wetzel (1985a). The relatively long, prehensile tail is naked on the underside. The pelage is soft and silky with a buffy gray to golden yellow color; it is darker above with a dark line along the top of the head, neck, and back. The tip of the nose is pink, the soles of the feet are reddish, and the eyes are black. Two digits on the forefoot have large, curved, and sharp claws; two other digits are present but inconspicuous. The long hind feet have four claw-bearing digits and a heel pad and a peculiar "joint" in the sole that permits the claws to be turned back under the foot for grasping.

Cyclopedidae can be distinguished cranially from Myrmecophagidae by its proportionally shorter face and larger braincase. The basicranial–basifacial axis of its skull is strongly curved. In lateral view the skull is strongly tapered anteriorly, with the rostrum and mandible turning downward in the front. Ventrally, the glenoid, or depression where the mandible articulates with the cranium, is well separated from the porus acousticus, or the anterior opening of the tympanic around the inner ear, whereas in Myrmecophagidae the anterior margin of the tympanic forms the posterior edge of the glenoid. In Cyclopedidae the jugal has been completely lost. Cyclopedidae retains prominent

Silky anteater (*Cyclopes didactylus*). Photo by Alexander F. Meyer at Huachipa Zoo, Peru.

coronoid and angular processes on the mandible, but in Myrmecophagidae they are reduced (Gaudin and Branham 2008; Hayssen, Miranda, and Pasch 2012; McDonald, Vizcaíno, and Bargo 2008).

The silky anteater inhabits tropical forests and is reported to climb about treetops in search of ants, which it secures with its long, wormlike, sticky tongue. *Cyclopes* seems to feed exclusively on ants. Contrary to what has sometimes been reported, no termites have been identified in any dietary study (Miranda et al. 2009). Apparently it almost never descends to the ground, though it can walk well on flat surfaces by placing the sides of the forefeet on the surface and turning the claws inward. W. J. Schaldach, Jr., obtained a specimen in the city of Matias Romero, Oaxaca, which had walked into the kitchen of his house. In Brazil, C. T. Carvalho observed a *Cyclopes* crossing from one wooded area to another on the ground in the park at the Museo Goeldi.

Activity is entirely nocturnal. Sunquist and Montgomery (1973) reported that an individual followed by radiotelemetry was active almost continuously at night from within 15 minutes of sunset until 1.5–3.0 hours before dawn. Bouts of activity averaged about 4 hours in length. During the day this individual rested on or among shaded vines, in or below the crowns of trees; no two days were spent in the same tree. Like other xenarthrans, *Cyclopes* has a lower metabolic rate than would be expected for its size. A study of wild individuals in Panama found daily requirements for energy, food, and water to be only one- to two-thirds of those of typical eutherian mammals with the same body size (Nagy and Montgomery 2012). Such low intake results in a low body temperature that may vary with ambient temperature. Whereas normal body temperature for a mammal is usually 36–38° C, that of the silky anteater averages 33° C and has been recorded to drop as low as 16° C (Rodrigues et al. 2008).

It is said that the silky anteater frequents the silk-cotton tree (*Ceiba*), which has seed pods that are a massive ball of soft, silverish fibers. The sheen of this silky mass and that of the little anteater are so strikingly alike that when the animal is placed next to a freshly opened pod a person can scarcely tell the difference. This protective coloration helps the silky anteater escape the keen eyes of its chief predators, the harpy eagle, various eagle-hawks (*Spizaetus, Spizastur*), and the spectacled owl (*Pulsatrix*). When alarmed and taking a defensive

attitude, the silky anteater raises up on its hind legs, grasps opposite sides of a limb with its feet, and wraps its tail securely around a twig. The forefeet, with their powerful sharp claws, are held close to the face to strike swiftly and forcibly if the enemy comes within reach. This animal is, however, slow-moving and inoffensive and merely tries to defend itself in the only way possible.

Bhagratty et al. (2013) determined population density in a mangrove swamp on Trinidad to be 5.50 individuals per sq km in one year and 4.62 per sq km the following year. Home range was 0.8 ha. for one individual in the state of Tocantins, Brazil, and 10.0 ha. for another (Rodrigues et al. 2008). Montgomery (1985a, 1985b) reported a density of 0.77 per ha. on Barro Colorado Island. An adult male had a home range of about 11 ha. that overlapped the ranges of at least two adult females but not the range of an adjacent male. Female home ranges averaged 2.8 ha. and did not overlap one another. Most of the females were assumed to be carrying or accompanied by young. Depending on their age and sex, individuals ate 700–5,000 ants per day.

Silky anteater (*Cyclopes didactylus*). Photo by Flávia Miranda.

Silky anteater (*Cyclopes didactylus*). Photo by Alexander F. Meyer at Huachipa Zoo, Peru.

Being small, arboreal, and nocturnal, the silky anteater is naturally difficult to locate, and even the native people do not often see it. Little is known of its habits, and it has seldom been kept in captivity for more than short periods, though Merrett (1983) noted that one specimen was maintained for 2 years and 4 months. He suggested also that the gestation period is 120–150 days. According to Cuarón (2014a), there is usually a single young per litter, but sometimes two. Births in Mexico have occurred in March and April, and a lactating female with an infant was recorded in May. Miranda et al. (2014a) reported births in September or October. Moeller (1990c) reported the young are raised by both parents, which regurgitate semidigested insects to feed it. Sometimes the male anteater carries the baby on his back. Information summarized by Hayssen, Miranda, and Pasch (2012) indicated that a female may have two pregnancies in a year, that weaning occurs when the young is about one-half to two-thirds maternal mass, and that a foraging mother leaves her offspring for about 8 hours each night. There was an earlier observation of a young placed by the mother in a nest of dry leaves in a hole in a tree trunk. Cuarón (2014a) indicated that the mother changes such sites each day, carrying the young on her back.

Although *C. didactylus* has a large range and the overall species is not considered threatened, it is highly susceptible to disruption of forest habitat and several subspecies may warrant concern. The one population in greatest need of attention actually has not yet been taxonomically distinguished but likely will be when genetic analysis has been carried out. That is the pale-colored population isolated on the coast of extreme eastern Brazil (Aguiar and da Fonseca 2008). While formally classified as "data deficient" by the IUCN, the population is dependent on the Atlantic Forest, which is rapidly being destroyed, fragmented, and replaced by sugar cane plantations. Only 5 percent of the original suitable habitat remains intact (Miranda and Superina 2014).

Molecular assessment indicates the divergence of the lineage of Cyclopedidae from that of Myrmecophagidae occurred about 44 million years ago in the middle Eocene (Delsuc et al. 2012). A genus apparently closely related to *Cyclopes* is known from the late Miocene of Argentina (McDonald, Vizcaíno, and Bargo 2008; McKenna and Bell 1997). Otherwise, this family has no recorded geological history.

PILOSA; **Family MYRMECOPHAGIDAE**

Giant Anteater and Tamanduas

This family of two genera and three species occurs in Central America and South America as far south as Paraguay, Uruguay, and northern Argentina. Cyclopedidae (see account thereof) formerly was considered a subfamily of Myrmecophagidae.

Head and body length ranges from 470 mm in *Tamandua* to about 1,200 mm in *Myrmecophaga*. The hair is coarse. Anteaters have elongated, tapered snouts and tubular mouths. The long tongue has minute posteriorly directed spines (filiform papillae) and is covered with a sticky secretion from the large salivary glands when the animal is feeding. The tail is prehensile in *Tamandua* but not in *Myrmecophaga*. There are four claw-bearing digits on the forefoot and five digits on the hind foot. The foreclaws are long and sharp and can be used as powerful weapons of defense, though their usual function is the opening of ant and termite nests. The ears are short and rounded, and the eyes are small. Anteaters usually have no teeth. The skull is elongate and would appear to be fragile, but the bony walls are thick and unusually hard. The mammae of females are

Lesser anteater, or southern tamandua (*Tamandua tetradactyla*). Photo by Eric Isselee/Shutterstock.com.

located in the chest and abdominal regions. Characters distinguishing Myrmecophagidae from Cyclopedidae are described above in the account of the latter family.

Anteaters inhabit tropical forests and savannahs, sheltering in trees, hollow logs, or burrows constructed by other animals. They feed on ants, termites, and other insects. They use their strong foreclaws to rip open ant and termite nests, and they capture the insects with their long, sticky tongue. They have a good sense of smell, but sight and hearing are not well developed. Anteaters occur singly or in pairs, usually a female and her young.

The geological range of this family is late Oligocene or early Miocene to Recent in South America, early Pleistocene to Recent in Mexico, and Recent in Central America (McDonald, Vizcaíno, and Bargo 2008; McKenna and Bell 1997). The early Pleistocene record in Mexico is based on a fossil of *Myrmecophaga tridactyla* from a site in northwestern Sonora, more than 3,000 km north of the present range of the species (Shaw and McDonald 1987).

PILOSA; MYRMECOPHAGIDAE; **Genus** *TAMANDUA* Gray, 1825

Lesser Anteaters, or Tamanduas

There are two species (Coitiño et al. 2013; Cuarón 2014b; Gardner 2007g; Torres 2009; Wetzel 1975):

T. mexicana, Colima and southern Tamaulipas in Mexico to northwestern Venezuela and northwestern Peru west of the Cordillera Oriental;

T. tetradactyla, South America east of the Andes, from Venezuela and Trinidad to northern Argentina and Uruguay.

Head and body length is 470–770 mm, and tail length is 402–672 mm. Meritt (1975) reported that captive specimens weighed 2–7 kg. Eisenberg and Redford (1999) listed weights of 3.2–5.4 kg for *T. mexicana* and 3.8–7.0 kg for *T. tetradactyla*. The body is covered with short, dense, coarse hair, and coloration varies considerably. Wetzel (1975) observed that *T. mexicana* always has a vivid black area on the trunk, continuous from the shoulders to the rump and widening behind the shoulders to encircle the body. In *T. tetradactyla* this black "vest" is present only in specimens from the southeastern portion of the range of the species—this being the area most removed from the range of *T. mexicana*. The general background coloration in both species varies from blonde to tan or brown. The underside of the tail and its entire terminal portion are naked with irregular black markings. There are four clawed digits on the forefoot. The claw of the third digit is curved and is the largest; the claw of the first digit is the smallest. Each hind foot has five clawed digits. The snout is elongated. The mouth opening is about the diameter of a lead pencil. To compensate for the lack of teeth, a portion of the stomach is a muscular gizzard comparable to that of gallinaceous birds. Hayssen (2011b) cited a record of a female *T. tetradactyla* having two pairs of pectoral mammae.

Tamanduas inhabit tropical forests and savannahs. They exhibit both arboreal and terrestrial and both diurnal and nocturnal activity (Navarrete and Ortega 2011). They walk on the outside of the hand to avoid

Lesser anteater, or northern tamandua (*Tamandua mexicana*). Photo by Matthieu Gallet/Shutterstock.com.

forcing the tips of the large claws into their palms. An individual may have a daily period of activity that begins at any time of day or night and continues for about 8 hours (Montgomery 1985b). Tamanduas commonly shelter in hollow trees. Their movements on the ground appear rather clumsy, and unlike *Myrmecophaga*, they do not seem to be able to gallop. Nonetheless, Montgomery (1985b) reported *T. mexicana* to be scansorial, moving, feeding, and resting both on the ground and in trees. Esser, Brown, and Liefting (2010) observed an individual swimming, evidently comfortably, across a 400-meter-wide section of the Panama Canal. If attacked while in a tree, tamanduas defend themselves by assuming a tripod position formed by the hind feet and the tail, which leaves the arms free. They outstretch the arms and bare their claws until the enemy comes into reach. If, however, tamanduas are attacked while on the ground, they protect the back by leaning against a tree or rock, and they seize the opponent with their strong forearms. In both cases their protection

lies in the great strength of their arms and the tearing power of their foreclaws.

Tamanduas feed on ants, tree and ground termites, and bees. They visit 50–80 ant and termite colonies daily, feeding at each for less than a minute and rarely causing major damage to any one nest (Cuarón 2014b). During a study of *T. mexicana* on Barro Colorado Island, Panama, Brown (2011) observed several individuals consuming ripe fruits of the palm tree *Attalea butyracea* and suggested that tamanduas regularly seek out fruit as a supplement to their insect diet. Montgomery (1985a) found *T. mexicana* on Barro Colorado Island to consume about 9,000 ants per day. Population density in that area was 0.05 animals per ha., and home range was about 25 ha. In the Patanal wetlands of Brazil, Desbiez and Medri (2010) found density of *T. tetradactyla* to be 0.0034 per ha. Montgomery and Lubin (1977) reported home ranges of 350 ha. and 400 ha. for two *T. tetradactyla*. During 4 months of radio-tracking in the Brazilian *cerrado*, Tro-

Lesser anteater, or southern tamandua (*Tamandua tetradactyla*). Photo by Rexford D. Lord.

vati and de Brito (2009) reported an adult female *T. tetradactyla* established a home range of 1.06 sq km. In that same habitat, Rodrigues, Marinho-Filho, and dos Santos (2001) found a male *T. tetradactyla* radio-tracked for 9 months and a female tracked for 6 months to have respective home ranges of 1.0 and 3.4 sq km. Navarrete and Ortega (2011) wrote that adult home ranges of *T. mexicana* may overlap but that females tend to be spaced farther apart than males, and also that the species is territorial and has been observed drag-marking, apparently with anal glands. Adults rarely vocalize, but female *T. tetradactyla* reportedly emit low moans when in estrus (Rodrigues et al. 2008).

Cuarón (2014b) indicated that reproduction of *T. mexicana* appears not to be seasonal, with births, newborn individuals, and lactating females recorded in Mexico in March, May, and December. Genoways and Timm (2003) indicated that in Nicaragua neonates were found in June and July, partly grown young in March and June, and a lactating female in April. Ro-

drigues et al. (2008) wrote that female *T. tetradactyla* give birth twice a year, with one young per litter. Estrous cycles of one female *T. tetradactyla* averaged 42.5 days (Hayssen 2011b). Estrus is believed to last 2–3 days (Benirschke 2008). Pinto da Silveira (1968) suggested that females are polyestrous, with a gestation period of 130–150 days. Merrett (1983), however, cited reports of 160 and 190 days. He noted also that twins have been recorded, that the young are carried on the back or flanks of the mother, and that the two separate after about a year. Cuarón (2014b) wrote that initially the young is placed in a nest, usually within a hollow tree, and that only when bigger is it carried on the mother's back. The first vaginal bleeding in a female—evidence of possible sexual maturity—was observed at an age of 6 months. Captive specimens have lived as long as 19 years (Weigl 2005).

Although tamanduas are widely distributed and there is no immediate concern for the survival of either species, Aguiar and da Fonseca (2008) noted that they

607

are subject to loss and fragmentation of habitat by agricultural expansion, to predation by domestic dogs, and to hunting out of misplaced fear of the animals or for their skins. Several Central American subspecies of *T. mexicana* may warrant conservation concern and further systematic evaluation. The population of *T. tetradactyla* in the Brazilian state of Rio Grande do Sul is considered vulnerable because of severe degradation of its forest habitat.

PILOSA; MYRMECOPHAGIDAE; **Genus** *MYRMECOPHAGA* Linnaeus, 1758

Giant Anteater

The known, original, modern distribution of the single species, *M. tridactyla*, was from southern Belize to Ecuador west of the Andes, and throughout South America east of the Andes to as far south as northern Argentina and Uruguay (Cabrera 1957; Gardner 2007g; Hall 1981). The Central American range is not well understood, and the species is now apparently absent from Belize, Guatemala, and El Salvador. Its continued presence in northern Honduras and parts of Costa Rica was recently confirmed (Colindres and Días 2014; McCain 2001; Reyes, Matamoros, and Glowinski 2010). There had been no records from Nicaragua for over a century (Genoways and Timm 2003), but Koster (2008) recently showed the species to occur in the northern part of the country.

Head and body length is usually 1,000–1,200 mm, tail length is 600–900 mm, and weight is usually 18–39 kg. Morford and Meyers (2003) reported individuals in zoos weighing as much as 60–65 kg and that on average females slightly outweighed males. The body is narrow, the rostrum is much longer than the braincase, and the long, nonprehensile tail is heavily crested dorsally and ventrally with long hair (Gardner 2007g). The color is gray, and the diagonal stripe is black with white borders. The hair is coarse, stiff, and longest on the tail.

Giant anteater (*Myrmecophaga tridactyla*). Photo by Luiz Kagiyama / Shutterstock.com.

This animal is easily recognized by its large size, cylindrical snout, diagonal stripe, and bushy tail. There are three large claws and one small claw on each forefoot and five relatively small claws on each hind foot.

The powerful claws on the forefeet and the long, extensible tongue are the instruments for food gathering. Ant and termite mounds are ripped apart with the claws, and the eggs, cocoons, and adult insects are picked up with the saliva-coated tongue. Moeller (1990c) noted that *Myrmecophaga* may consume as many as 35,000 ants or termites in a single day. Many ant and termite nests may be raided during the course of a day, with feeding at each seeming to end when the soldier caste of the insects arrives at the breach to defend the nest, generally by noxious chemical secretions or by biting (Eisenberg and Redford 1999). Beetle larvae are also eaten in the wild, and fruit is taken on occasion, at least in captivity. Typical diets in zoos include eggs and milk beaten together, mealworms, ground beef, and dry dog and cat food. The salivary glands appear to secrete only when the animal is feeding. The tongue can be extended as much as 610 mm, more than the cranial length, but its diameter is only 10–15 mm at the widest point, just outside the mouth. Naples (1999) reported that a unique hyoid muscle arrangement enables *Myrmecophaga* to project the tongue with great speed and precise positional control. That arrangement, combined with an elongated secondary palate, accommodates the retracted tongue within the oropharynx without compromising the animal's ability to breathe.

The giant anteater is found in savannahs, grasslands, swampy areas, and humid forests. It has sometimes been reported to be active mainly during daylight in areas uninhabited by people and during the night in densely populated areas. However, a study in the Brazilian Patanal showed that a more substantive factor might involve temperature (Mourão and Medri 2007). The period of greatest activity was found to begin at 1800 hours on hot days, but whenever the daily minimum temperature declined, activity began earlier. The anteater used predominantly forest habitats for rest and open habitats for activity. Forest patches served as a temperature buffer, because they were cooler than open habitats during hot hours of the day and warmer during the cold hours. In the higher and cooler Serra da Canastra National Park in Brazil, Shaw, Machado-Neto, and Carter (1987) found activity to begin at

Giant anteater (*Myrmecophaga tridactyla*) with tongue extended. Photo of individual in Oklahoma City Zoo by Esdeem/Shutterstock.com.

1300–1400 hours, peak at 1800–1900 hours, and gradually diminish until 0200 hours. In contrast, on the warm lowland Venezuelan *llanos*, Montgomery and Lubin (1977) found a radio-tracked individual to be nocturnally active, from 1900 to 0800 hours.

Myrmecophaga walks with its nose close to the ground and with the side and knuckles of the forefeet on the ground. It takes to water readily and can swim across wide rivers. Unlike *Tamandua* and *Cyclopes*, the adult giant anteater does not regularly climb trees, but Widholzer and Voss (1978) indicated it was adept at climbing out of zoo enclosures. Young, Coelho, and Wieloch (2003) reported a number of cases in which wild and captive individuals climbed termite mounds, fences, and trees. Van Roosmalen (2015) referred to an arboreal form of *Myrmecophaga*, possibly a new species, in the Rio Aripuanã Basin of central Brazil. Although *Myrmecophaga* is a powerful digger, it does not construct burrows but merely seeks secluded spots in which to curl up, tucking its head between its forelegs and covering its head and body with the fanlike tail. Reported resting sites include forest patches and open grassland (Rodrigues et al. 2008). The giant anteater usually reacts to danger with a slow, clumsy gallop. It does not fight unless forced, and then it uses its strongly clawed forefeet to grasp and claw its adversary.

An average population density of 2.9 individuals per sq km, somewhat higher than reported for natural habitat, was found in a timber plantation with the tree *Acacia mangium*, which apparently supported a high concentration of insect prey (Kreutz, Fischer, and Linsenmair 2012). In the Patanal wetlands of Brazil, Desbiez and Medri (2010) determined density as 0.15 per sq km, and Medri and Mourão (2007) reported home ranges of 4.0–7.5 sq km for four males and 11.9 sq km for a female, with considerable overlap of those areas. Home ranges of at least 9 sq km (Pinto da Silveira 1969) and 25 sq km (Montgomery and Lubin 1977) have been reported, but extensive studies in Serra da Canastra National Park (Shaw, Carter, and Machado-Neto 1985; Shaw, Machado-Neto, and Carter 1987) suggest considerably smaller sizes. Ranges there averaged 3.67 sq km for adult females and 2.74 sq km for adult males. The ranges of females overlapped one another by an average of about 29 percent, and the male ranges by about 4 percent. Minimum population density in the area was calculated to be 1.3 per sq km. Animals usually were seen alone. Observations of agonistic interaction, generally between males, suggest that occupied space is defended, thus perhaps contradicting earlier reports that the genus is not territorial. Such interaction varied from slow circling to chases, serious fighting, and injuries. Other observations also suggest that the genus is usually solitary in the wild except for females with young, but Widholzer and Voss (1978) stated that a captive group of three adult males and two adult females did well together. Kreutz, Fischer, and Linsenmair (2009) observed a bloody 20-minute engagement, in which two wild individuals sought to damage one another with their foreclaws. For a time, the apparently dominant animal pranced aggressively on its forelegs and wagged its tail, while the other animal sat on its haunches and kept its tail flattened on the ground, "continuously screaming and roaring." Schmidt (2012) observed a captive group of three adult females for a year and recorded some agonistic behavior but relatively little violence and no serious injuries; the only vocalization was a "short low-frequency whistle," and that was considered unusual.

The following life history information was compiled from Benirschke (2008), Knott et al. (2013), Merrett (1983), Miranda, Bertassoni, and Abba (2014), and Rodrigues et al. (2008). *Myrmecophaga* is polyestrous, estrous cycles last 42–74 days, and females may return to estrus within 2 months after giving birth. Breeding occurs throughout the year in captivity and apparently also in the wild. The gestation period is about 170–190 days, but there are records of periods as short as 142

Giant anteater (*Myrmecophaga tridactyla*). Photo by Flávia Miranda.

days. A period of delayed implantation may be involved. The single young weighs 1–2 kg at birth, opens its eyes after 6 days, and is weaned at 4–6 weeks. It is carried on the mother's back for approximately 6 months, perhaps longer. In the wild, at least 3 years growth is required to reach full size. Age of sexual maturity has been variously reported as 2.0–4.0 years in the wild but females in zoos have reproduced for the first time at 18–22 months. The young remains with the mother until she again becomes pregnant. Some captive females have remained reproductively active until 20–24 years old. Weigl (2005) listed several captive specimens that lived about 30–31 years.

The giant anteater is classified as vulnerable by the IUCN (Miranda, Bertassoni, and Abba 2014) and is on appendix 2 of the CITES. The genus reportedly is now uncommon and localized in most regions, mainly because of habitat loss and expansion of the human population and agricultural activity. In Brazil, many individuals are killed by burning of sugar cane plantations prior to harvest. Although legally protected, *Myrmecophaga* is hunted in some areas for its skin, for use as food or pets, and because it is believed to be a pest or a danger to domestic dogs. The giant anteater is considered the most threatened mammal of Central America (see above). It has also been extirpated in Uruguay and in the Brazilian states of Espírito Santo, Rio de Janeiro, Santa Catarina, and Rio Grande do Sul, and is critically endangered in Paraná. One of the largest remaining populations is in Emas National Park in the *cerrado* of south-central Brazil. The park covers about 1,400 sq km but is surrounded by farms and cattle pasture, thereby largely isolating the anteater population, which now shows evidence of inbreeding (Collevatti et al. 2007).

PILOSA; Family MEGALONYCHIDAE

Two-toed Tree Sloths and West Indian Sloths

This family includes a single living genus, *Choloepus*, with 2 species of two-toed tree sloths of Central and South America, and 4 genera and 12 species of sloths that evidently survived in the West Indies until very early historical time, less than 5,000 calendar years ago (see account of *Mammuthus*, order Proboscidea, for comments on the definition of historical time and the distinction between calendar and radiocarbon dates). The West Indian sloths sometimes have been referred to as "ground sloths," such as those that occurred on the American mainland in the Pleistocene, some genera of which attained gigantic size (see below). However, the West Indian genera were comparatively small and appear to have been partly arboreal (MacPhee 2009).

White and MacPhee (2001) assigned the living and West Indian genera to two subfamilies: Choloepodinae, with *Neocnus*, *Choloepus*, and *Acratocnus*; and Megalocninae, with *Megalocnus* and *Parocnus*. Gaudin (2004) supported the affinity of *Megalocnus* and *Parocnus*, and of *Choloepus* and *Neocnus*, but indicated some question as to which of those two groupings *Acratocnus* is closer to. Gaudin (2011) found *Acratocnus* and

Two-toed tree sloth (*Choloepus didactylus*). Photo by Seubsai / Shutterstock.com.

Neocnus to exhibit more similarities to one another in the anatomy of their orbitotemporal bones than either does to *Choloepus*. Eight additional names for late Pleistocene to Holocene genera of West Indian sloths are considered synonyms of those mentioned above (White and MacPhee 2001; see also Alcover et al. 1998). Another described genus, *Paulocnus* Hooijer, 1962 of Curaçao, likely dates from the mid-Pleistocene, and two more, *Galerocnus* Arredondo and Rivero 1997, and *Paramiocnus* Arredondo and Arredondo, 2000, are based on extremely scanty material of uncertain age (MacPhee 2009).

Even for the better known late Quaternary genera of West Indian sloths, there has been much uncertainty regarding time of extinction and the role of humans therein. At least some genera reportedly were restricted to the Pleistocene (McKenna and Bell 1997; Morgan and Woods 1986), while others were thought to have persisted until after the arrival of European colonists (Allen 1942). The earliest evidence for Native American presence in the West Indies dates from about 6,000 years ago (MacPhee 2009), though the initial invasion of advanced peoples, who introduced agriculture and pottery, occurred around 2,000 years ago (Rouse 1989). P. S. Martin (1984) suggested that most West Indian sloth genera did not disappear until after the human invasion, a view supported by MacPhee and Flemming (1999). A comprehensive assessment by MacPhee (2009) indicates extinction did come after people first arrived on the islands, not rapidly in response to excessive hunting, but probably within one to two thousand years and not close to the time of European colonization. In any case, the West Indian sloths survived thousands of years beyond the extinction of the mainland ground sloths, which took place roughly 11,600 calendar years ago in North America and 11,000 calendar years ago in South America (Fiedel 2009). In determining which West Indian sloths may have lived into historical time, and thus warrant coverage herein, consideration has been given to the statement by MacPhee, White, and Woods (2000) that "it is certain that some and perhaps all Quaternary sloth extinctions occurred after the arrival of humans on the islands." Accordingly, unless a genus has been reliably dated only to the Pleistocene or earlier or is based on questionable material, it will be covered by the following accounts.

At one time, *Choloepus* was placed together with *Bradypus* in the family Bradypodidae, but available evidence now suggests those two tree sloth genera represent convergent surviving lines from separate ancestral stocks. Gaudin (2004) considered his cladistic analysis to imply "that the split between the two extant sloth genera is ancient, dating back perhaps as much as 40 [million years], and that the similarities between the two taxa, including their suspensory locomotor habits, present one of the most dramatic examples of convergent evolution known among mammals." Molecular estimates of the time of divergence center at around 27 million but range from about 21 to 36 million years ago (Delsuc et al. 2012).

Choloepus was also sometimes put in its own family, Choloepidae, though in recent years it has been recognized as belonging together with certain extinct ground sloths in Megalonychidae (Barlow 1984). *Bradypus*, in turn, has sometimes been associated with the extinct ground sloths of the families Megatheriidae and Nothrotheriidae, though, as indicated above, the lineage of Bradypodidae may have emerged from ancestral pilosan stock prior to evolution of the other sloth families (Gaudin 2004; Gaudin and McDonald 2008). Webb (1985) suggested that the early Miocene ancestors of Megalonychidae, Megatheriidae, and Nothrotheriidae were arboreal and that a fourth family of extinct sloths, the Mylodontidae, was always terrestrial. That position was supported by Pujos et al. (2012). However, Steadman et al. (2005) observed that, at the time of their extinction, all the mainland genera of extinct sloths were facultatively "ground" sloths rather than arboreal. P. S. Martin (2005) listed about 20 named genera that occurred in the late Pleistocene on the mainland. Those animals varied in size from about that of a medium dog to that of an elephant. The largest species was *Megatherium americanum* of South America, which was about 6 meters long and weighed over 6,000 kg, while the largest in North America was *Eremotherium rusconii*, which weighed a little over half as much.

Those genera of Megalonychidae that may have persisted into early historical time in the West Indies were not among the larger kinds. They weighed from less than 5 kg to around 200 kg. In appearance they may have seemed intermediate to living anteaters and tree sloths. Originally thought to have been ground

Extinct giant ground sloth (*Mylodon darwinii*). Painting by Helga Schulze, based in part on skeletal illustrations and descriptive information in Bargo, Toledo, and Vizcaíno (2006), Haro, Tauber, and Krapovickas (2016), and McAfee (2016).

dwellers, they likely were partly or largely arboreal. For the most part, they were probably inoffensive plant eaters, like the living two-toed tree sloth. Defining dental and skeletal characters that unite *Choloepus* and the West Indian sloths include: 5 upper and 4 lower teeth on each side of the jaw, modification of the anterior-most upper and lower teeth into either caniniforms or incisiforms that are separated by a long diastema from the molariforms, upper caniniform always positioned at anterior end of maxilla adjacent to premaxilla contact but the latter not contributing to its alveolus and not bearing any teeth, molars quadrangular or elliptical, mandibular symphyseal region usually elongated, condyle of jaw positioned above tooth row, limbs relatively gracile and pentadactyl, calcaneal tuberosity mediolaterally expanded, femoral third trochanter present, tibia and fibula separate, pes not rotated as in other sloths and more plantigrade (McDonald and De Iuliis 2008; White and MacPhee 2001). Additional characters of Megalonychidae, as expressed in the surviving *Choloepus*, are given below in the account of that genus.

The family Mylodontidae was sometimes included in accounts of Recent mammals (e.g., the 1964, 1968, and 1975 editions of this book), based on supposed survival of the genus *Mylodon* Owen, 1840, one of the giant ground sloths, into modern times. In April 1888 specimens now assigned to the species *M. darwinii* were found by Dr. Otto Nordenskjöld in Eberhardt Cave, now known as Cueva del Milodón. The cave is situated near Bahia Ultima Esperanza in southern Chile (51°35' S, 72°38' W). The specimens supposedly were found in direct association with human remains, tools, and bones and pieces of fur of the guanaco (*Lama guanicoe*). The sloth remains consisted of several pieces of hide with long, reddish hair and studded with oblong dermal ossicles similar to those previously known from other species of *Mylodon*. The find was reported by Einar Lonnberg in 1899 in a paper entitled "On some remains of '*Neomylodon listai*' Ameghino."

There long was speculation that those specimens represented a population that may still have been living at the time of discovery (Allen 1942; Heuvelmans 1958), but they and other more recently found material now have been radiocarbon dated to no later than about 11,800 calendar years before the present (Steadman et al. 2005). Remains of *Mylodon* from other sites in South America have been credibly dated as late as 11,000 calendar years ago, and it is not likely that the genus survived much longer than that (Fiedel 2009).

However, Barnosky and Lindsey (2010) accepted a date of 9,535 calendar years ago for another mylodontid genus, *Catonyx*, in Argentina, and also listed a reliable date of just 7,871 years ago for the genus *Megatherium* in Argentina. There have been additional reports of ground sloth persistence into the mid-Holocene (Cione, Figini, and Tonni 2001; Coltorti et al. 2012; Correal and Van der Hammen 2003; Ficcarelli et al. 2003), and there has been considerable debate as to the role of human hunting in the disappearance of the ground sloths and the rest of the Pleistocene megafauna (see account of *Mammuthus*). Also, Oren (1993) noted that native peoples throughout Brazilian Amazonia report the current or very recent existence of a creature, descriptions of which are surprisingly consistent with the characters expected of a relatively small (approximately 1.8 meters long), forest-dwelling, mylodontid ground sloth. Oren (2001) related additional reports of such animals, including accounts from seven hunters who claim to have killed one, and noted that the descriptions seemed more applicable to the family Megalonychidae.

The Megalonychidae may be represented by fossil material from the middle Eocene of the Seymour Peninsula of Antarctica and probably by early Oligocene specimens from Puerto Rico. The latter, along with likely late Pliocene specimens from Grenada and the well-defined genus *Imagocnus* MacPhee and Iturralde-Vinent, 1994, from the early Miocene of Cuba, demonstrate the very early megalonychid invasion of the West Indies. Otherwise, the known geological range of the family is late Miocene to Recent in South America, late Pleistocene to Recent in Central America, late Miocene to late Pleistocene in North America, and middle Pleistocene to Recent in the West Indies (MacPhee 2005; MacPhee, Singer, and Diamond 2000; McKenna and Bell 1997).

PILOSA; MEGALONYCHIDAE; Genus *NEOCNUS* Arredondo, 1961

Lesser West Indian Tree Sloths

There are five species (MacPhee, White, and Woods 2000; White and MacPhee 2001):

N. gliriformis, known only by skeletal remains from Cuba;

N. major, known only by skeletal remains from Cuba;

N. comes, known only by skeletal remains from Hispaniola (Haiti and Dominican Republic);

N. dousman, known only by skeletal remains from Hispaniola;

N. toupiti, known only by skeletal remains from Hispaniola.

Hall (1981) placed *N. comes* in a separate genus, *Synocnus* Paula Couto, 1967. White and MacPhee (2001) indicated that collection of additional material may show that *N. gliriformis* and *N. major* are conspecific.

Skeletal and dental characters include: cranium domed; postorbital constriction weak; rostral flare slight; pterygoid inflation absent; symphyseal spout present; first maxillary and mandibular tooth triangular in cross section; second maxillary molariform tooth anterolaterally concave; last maxillary molariform broadest on lingual side; mandibular caniniform grooved posterointernally; last mandibular molariform with deep lingual groove; humeral head spherical and skewed laterally; humerus short and slender but with prominent and squared supracondylar ridge; ulna gracile and anteriorly bowed; femoral shaft anteroposteriorly flat and medially bowed with prong on anterior aspect; tibial and fibular shafts bowed (White and MacPhee 2001). The overall body size was relatively small; weight of *N. major* has been estimated at 15–22 kg (White 1993), and weight of *N. comes* may have been about 23 kg (Hall 1981), but *N. toupiti* was the smallest known West Indian sloth, being significantly smaller in linear dimensions and long-bone cortical cross-sectional thickness than the existing genera *Choloepus* and *Bradypus* (MacPhee, White, and Woods 2000).

MacPhee (2009) stated that *Neocnus* was the most arboreal genus of West Indian sloths and probably did not spend considerable time on the ground. Nonetheless, Pujos et al. (2012) indicated that, while *Neocnus* was adapted for climbing and a semiarboreal lifestyle, it was not suspensory like the living tree sloths. MacPhee, White, and Woods (2000) referred to *N. toupiti* as a small, extremely gracile arborealist. The closely re-

Lesser West Indian tree sloth (*Neocnus comes*). Painting by Helga Schulze, somewhat hypothetical but based in part on skeletal illustrations and descriptive information in sources cited in this account.

lated *N. dousman* was also small, whereas *N. comes* was substantially larger and possibly not fully arboreal. Steadman et al. (2005) observed that *N. comes* did have skeletal features consistent with arboreality, but it is unknown whether that species was comparable to the existing small tree-dwelling sloths, which probably have survived because of their cryptic nature, living high in trees, being silent and relatively immobile, and having algally camouflaged fur. Prior to the arrival of humans, forays on the ground would have been relatively safe for a small sloth in the West Indian islands, which lack native placental carnivores. Woods and Ottenwalder (1992) referred to *N. comes* as "slightly arboreal" and noted that its apparent remains occur in cave and sinkhole deposits all over Haiti. Presumably,

to reach such sites, animals would have been moving on the ground.

The original description of *N. comes* indicated that its remains had been found intermixed with fragments of pottery and with the bones of humans and domestic pigs. Such association gave the impression that *N. comes* still lived when advanced Native American peoples had occupied Hispaniola, around 2,000 years ago, and even after European colonists had arrived and introduced pigs (*Sus*), about 500 years ago. Therefore, *N. comes*, as well as other West Indian ground sloths, were sometimes considered components of the historical fauna that had been extirpated by human agency (Allen 1942; E. Anderson 1984; Hall 1981; Woods and Ottenwalder 1992). However, MacPhee (2009) explained that,

615

while it long has been accepted that most late Quaternary extinctions in the West Indies occurred after the coming of Native Americans, thinking has changed as to how recent those losses occurred. He pointed out further that the likelihood of a real association of the remains of *N. comes* and those of the domestic pig always seemed doubtful. None of the other reports of association of West Indian sloth bones with human remains or artifacts has stood up to critical evaluation. Modified sloth bones are apparently absent at Native American archeological sites, indicating that the animals could not have been hunted very frequently, possibly because they were becoming rare.

Steadman et al. (2005) reported the most recent radiocarbon date for *N. comes*, as obtained by direct assessment of bone material, to be approximately 4,390 years before the present; pursuant calibration provided equivalent calendar dates ranging from 5,260 to 4,840 years ago. The range of Holocene dates for *N. comes* from other sites on Haiti was about 8,326–4,486 radiocarbon years and 9,490–4,970 calendar years ago. One specimen of *N. dousman* was dated at 9,897 radiocarbon years ago. Also, Morgan and Woods (1986) indicated that some sites where *N. comes* has been found date from 3,715 radiocarbon years ago. However, MacPhee, Iturralde-Vinent, and Jiménez Vázquez (2007) observed that the final collapse of megalonychids on both Hispaniola and Cuba may have occurred about 4,200 radiocarbon years ago, which would be just over 1,000 years after people first arrived on those islands. Timing of the West Indian extinction does suggest human agency as the cause, perhaps not by direct hunting but indirectly though effects on the environment (McDonald and De Iuliis 2008). Radiocarbon dates for *N. comes* demonstrate that the species survived the Pleistocene–Holocene transition, when Hispaniola became warmer and wetter, only to perish in the mid-Holocene during a relatively stable period. These findings would seem to counter the argument that climate change was a major factor in extinction. Perhaps more significantly, the West Indian sloths persisted 5,000–6,000 years longer than the continental ground sloths, and both groups of animals disappeared after the human invasion of their ranges. The situation is comparable to the respective extinctions of *Mammuthus* (see account thereof) on the mainland and much later on Wrangel Island, and implies the role of people in both cases.

PILOSA; MEGALONYCHIDAE; Genus *CHOLOEPUS* Illiger, 1811

Two-toed Tree Sloths

The single known genus, *Choloepus*, contains two species (Cabrera 1957; Gardner and Naples 2007; Hall 1981; Hayssen 2011a; McCarthy, Anderson, and Cruz D. 1999; Wetzel and Avila-Pires 1980):

> *C. didactylus*, east of the Andes and south of the Orinoco River in Colombia, northeastern Ecuador, Venezuela, the Guianas, eastern Peru, and northern Brazil;
>
> *C. hoffmanni*, eastern Honduras and Nicaragua to northwestern Venezuela and to Colombia and Ecuador west of the Andes, also east of the Andes from eastern Ecuador and southern Colombia to central Bolivia and western Brazil.

Head and body length is 540–740 mm, the tail is absent or vestigial, and weight is 3.0–9.0 kg. A study of *C. hoffmanni* in Costa Rica found males to average 5.9 kg, females 6.3 kg (Peery and Pauli 2012). The pelage consists of long guard hairs and short underfur. General body hair is blond, buff, tan, or light brown in adults (Hayssen 2011a). The shoulders and top of the head are the darkest, and the face is usually paler. According to Aiello (1985), the hairs of both *Choloepus* and *Bradypus* differ in form and structure from those of all other mammals, apparently being specialized to encourage colonization by algae. Although the sloths are usually said to benefit from this arrangement by being camouflaged, they actually may gain nutrients either by absorption through the skin or by licking the algae. The algal growth during wet seasons often gives the coat a greenish cast. The skull is rounded and short in the facial region. The teeth usually number 5/4 on a side, giving a total of 18 teeth. The teeth grow throughout life and have cupped grinding surfaces. The ears are inconspicuous, and the eyes are directed forward. The stomach is complex for the digestion of vegetation. Females have two mammae located in the chest region.

The limbs are long, but unlike in *Bradypus*, the forelegs are only slightly longer than the hind legs. All limbs terminate in narrow, curved feet. In *Choloepus* there are only two digits on each forefoot, and these are closely bound together with skin for their entire length;

Two-toed tree sloth (*Choloepus hoffmanni*), individual in suspended position. Photo by Worldswildlifewonders / Shutterstock.com.

each is armed with a large hooklike claw about 75 mm in length. The hind feet have three toes with hooklike claws. The claws and limbs are used for hanging from tree branches. According to Hayssen (2011a), *Choloepus* is usually stockier and larger than *Bradypus*, its face is much more elongated, with a more-prominent fleshy nose, and its neck is shorter. *Bradypus* has 8 or 9 cervical vertebrae compared with only 5–6 in *C. hoffmanni* and 6–8 in *C. didactylus*. *Choloepus* has an anterior caniniform tooth but *Bradypus* does not.

Two-toed tree sloths, like the related three-toed sloths, spend much of their life upside down, even eating, sleeping, mating, and giving birth in that position. They sleep with the head raised and placed between the forelegs on the chest. Often all four feet are placed so close together that the animal has the appearance of a bunch of dried leaves. Frequently they choose a

Two-toed tree sloth (*Choloepus hoffmanni*), individual in resting position. Photo by Worldswildlifewonders / Shutterstock.com.

position that permits them to rest their back on a lower limb or sit in a fork; the feet, however, are always hooked to a branch. The sitting resting posture is common and may have a role in allowing gravity to facilitate digestion (Clauss 2004; see also account of *Bradypus*). Almost all movements are extremely slow and by means of a hand-over-hand motion. They swim voluntarily, using a breast stroke and with the body right side up. When on the ground, where they do come about every 3–8 days to urinate and defecate, they have some difficulty standing, but they can rise on their palms and soles and crawl for short distances. Climbing is accomplished with seemingly little effort and some speed even when supports are vertical (Mendel 1981). Recorded average and maximum speed of *C. didactylus*, while suspended, is 0.5–0.6 km per hour and 1.6 km per hour, respectively; speed on the ground is 0.25 km per hour (Adam 1999).

Unlike that of most mammals, the body temperature of sloths varies considerably depending on the temperature of their surroundings. As a result, they are physiologically restricted to a limited equatorial habitat of constant temperature. *Choloepus* has the lowest and most variable body temperature of any mammal, with ranges typically reported at around 30–36° C (Hayssen 2011a). Temperature falls throughout the cooler hours of the night, during wet weather, and when the animals are inactive. Shortly after daybreak sloths show minimal activity and can be found high in the tree canopy warming themselves in the sun (Gilmore, Duarte, and Peres da Costa 2008).

It may seem strange that such slow-moving creatures have been able to persist in regions where there are large birds of prey and carnivorous mammals that can climb well. When sloths descend to the ground, they slowly drag their bodies along the surface, which makes them easy prey. Often they are killed by jaguars, ocelots, and other cats. They do defend themselves, however, and can strike fairly quickly with their long forefeet, which have large, hooked, sharp claws. They also use their

Two-toed tree sloth (*Choloepus didactylus*). Photo by Jonas Livet.

teeth effectively, and both means of defense can inflict severe wounds. If left alone, however, sloths are inoffensive. They have color vision, a good sense of smell, and a poorly developed sense of hearing. They probably owe their survival to such factors as protective coloration, remaining motionless during the day, nocturnal habits, heavy fur, thick skin, and an extreme tenacity to life.

Two-toed sloths inhabit tropical forests and are almost entirely arboreal and nocturnal. According to Sunquist and Montgomery (1973), *C. hoffmanni* in the Panama Canal Zone shows almost no crepuscular activity; most individuals remain inactive until the hour before sunset. In the study area *Choloepus* was much more active than *Bradypus* and was seldom found occupying the same tree on successive days. Because of its greater mobility, *Choloepus* is exposed to a wider variety of foods and tends to eat a wider variety than *Bradypus*. Its diet consists of leaves, tender twigs, and fruits. Animal matter is probably also taken (Chiarello 2008; Taube et al. 2001). The arms are often used to pull food within reach of the mouth. Sloths have a metabolic rate only 40–45 percent of what might be expected for their body weight, but, as they sleep or rest up to 20 hours per day, they probably perform only about 10 percent of the work of other, similarly sized mammals (Gilmore, Duarte, and Peres da Costa 2008).

Data from an extensive rescue operation after flooding in Surinam indicated a density of 56 *C. didactylus* per sq km within an area of 15 sq km (Adam 1999). Other reported densities for *Choloepus* have ranged from 0.13 to 2.70 individuals per ha. (Chiarello 2008). Meritt and Meritt (1976) reported *C. hoffmanni* to occur in Panama at a ratio of 11 females to 1 male and speculated that such a proportion would be advantageous to a species with unusually long gestation. On the other hand, they noted that in *C. hoffmanni* the females tend to gather in groups but males are loners, so the sampling techniques may have missed the latter and not adequately expressed the true sex ratio. The latter view was suggested by an extensive study of *C. hoffmanni* in Costa Rica (Peery and Pauli 2012), where sampled adults numbered 42 males and 53 females. Males had home ranges averaging 18.7 ha., a portion of which was used exclusively, though the average adult male's range modestly overlapped the ranges of 5.5 other adult males. Generally, adult males were segregated and appeared to defend their core areas; several

instances of aggression, presumably territorial disputes among males, were observed. Home ranges of adult males encompassed an average of 3.2 adult females, and most females occurred within the home range of more than one adult male. Supplementary genetic analyses indicated a mating system involving a mixture of polygyny and promiscuity. Five of 14 tested adult males sired offspring with more than one female during a single breeding season, but no male sired more than two offspring. Adult males in close proximity were more closely related than males separated by greater distances; natal dispersal was found to be substantially greater in females.

Vaughan et al. (2007), also studying *C. hoffmanni* in Costa Rica, found home range to average 21.5 ha. in males and just 1.69 ha. in females, but suggested that median size, 4.4 ha., provided a more accurate picture of sloth behavior, since animals occasionally moved

Two-toed tree sloth (*Choloepus hoffmanni*), mother and 2-month-old young. Photo by Suzi Eszterhas / Minden Pictures.

long distances to reach favored food trees. Home ranges generally contained several kinds of habitat—forest, plantation, pasture—but seemed to function more as defensible territories than as a means to access specific resources. Nonetheless, during at least part of the year several individuals were observed resting or feeding in the same tree. Merrett (1983) indicated that several individuals could be kept together in captivity. Adults rarely vocalize, but when in danger, they make a sound like a long shrill whistle; young have a bleating distress call (Gilmore, Duarte, and Peres da Costa 2008).

Breeding in *C. hoffmanni* in nature appears somewhat seasonal, with pregnancy proceeding during the rainy season and births occurring at the beginning of the dry season (Taube et al. 2001). Female *Choloepus* bear a single young; it hooks its tiny claws into the long hair on the mother's breast and abdomen, where it clings while she climbs or rests. Meritt (1985b) reported that young *C. hoffmanni* weigh 350–454 grams at birth, first hang upside down at 20–25 days, take some solid food when just a couple of weeks old, and regularly feed away from the mother at 5 months; they may, however, continue a regular association with the latter for at least 2 years. Eisenberg and Maliniak (1978), using a controlled breeding program and subjecting females to x-ray examination, estimated the true gestation period of *C. hoffmanni* at around 11.5 months. They found that when lactation and successful rearing take place, interbirth intervals may average 14–16 months. Sexual maturity may not occur until 3.5 years in females and occurs at 4–5 years in males. Females may live for 20 years with no decline in fecundity.

There is little evidence of breeding seasonality for *C. didactylus*, either in the wild or in captivity (Gilmore, Duarte, and Peres de Costa 2008). Troll et al. (2013) determined the ovarian cycle of that species to average between 31.4 and 32.5 days. In a study of captive *C. didactylus*, Eisenberg and Maliniak (1985) estimated gestation at 10–11 months. If an infant died, females were likely to give birth again after an average interval of 423 days. If the young survived to social weaning, average interbirth interval was 800 (472–1,093) days. Females could remain reproductively active until age 25; one male was sexually active over a period of about 16 years and lived to at least 21 years. Veselovsky (1966) reported that a female *C. didactylus* at the Prague Zoo gave birth to a single young 5 months

and 20 days after mating had been observed, but both Adam (1999) and Taube et al. (2001) considered Eisenberg and Maliniak's (1985) estimate to be the most reliable. Taube (2001) also reported sexual maturity in *C. didactylus* to come at the end of the third year of life for females and during the third year for males.

Choloepus readily adapts to and breeds in captivity, and there are numerous records of individuals living more than 10 years (McCrane 1966; Snyder and Moore 1968; Veselovsky 1966). A wild-born male of *C. hoffmanni* lived to an age of about 35 years at the Bronx Zoo, and a female *C. didactylus* was still living in the National Zoo at an age of 36 years and 10 months (Weigl 2005). There is an increasing capture of wild individuals, especially young, for sale as pets to tourists. That trade, along with severe habitat degradation and fragmentation, may be jeopardizing populations in Colombia and Central America (Moreno and Plese 2006; Plese and Chiarello 2014). Those populations, which extend into northwestern Venezuela and coastal Ecuador, are separated by the Andes from the rest of the species; Aguiar and da Fonseca (2008) wrote that those populations by themselves would almost certainly qualify for an IUCN threatened category, owing to rampant degradation of rainforest.

PILOSA; MEGALONYCHIDAE; **Genus**
ACRATOCNUS Anthony, 1916

Greater West Indian Tree Sloths

There are three possibly Holocene species (MacPhee, White, and Woods 2000; White and MacPhee 2001):

- *A. odontrigonus*, known only by skeletal remains from Puerto Rico;
- *A. ye*, known only by skeletal remains from Hispaniola;
- *A. antillensis*, known only by skeletal remains from Cuba.

An additional species, *A. simorhynchus*, has been reported only from the late Pleistocene of the Dominican Republic but may eventually prove to be a synonym of *A. ye* (MacPhee 2009).

Skeletal and dental characters of the genus include: cranium relatively tall and domed; prominent postor-

Greater West Indian tree sloth (*Acratocnus odontrigonus*). Painting by Helga Schulze, somewhat hypothetical but based in part on skeletal illustrations and descriptive information in sources cited in this account.

bital constriction; sagittal crest present; pronounced rostral mediolateral flare; pterygoid inflation absent; symphyseal spout pointed and short; first maxillary tooth spike shaped, trigonal, anteriorly projecting, and curved (caniniform); last maxillary molariform convex and narrowest lingually; first mandibular caniniform straight, trigonal, and lacking posterointernal groove; last mandibular molariform convex lingually; limbs slender; humeral head globular; humerus with large entepicondylar foramen; femur with lesser trochanter; terminal phalanges evidently provided with stout claws; caudal vertebrae wide; tail believed to have been short but rather stout (Allen 1942; Hall 1981; White and MacPhee 2001). Weight has been estimated at 17–32 kg for *A. odontrigonus* and 25–36 kg for *A. antillensis* (White 1993).

A study of limb morphology indicated that *Acratocnus* consistently falls within the range of arboreal mammals (White 1993), though it does not appear to have been suspensory like the living tree sloths (Pujos et al. 2012). Earlier accounts tended to refer to *Acratocnus* simply as a "ground sloth" (Allen 1942; Hall 1981), and E. Anderson (1984) stated that it was "possibly semiarboreal." However, more recent assessments have tended to view the genus as primarily arboreal (MacPhee 2009; McDonald and De Iuliis 2008; Steadman et al. 2005).

Nonetheless, most remains of sloths from Puerto Rico and Hispaniola are found in deep sinkholes and caves, which functioned as natural traps for animals moving on the ground (Allen 1942; MacPhee, White, and Woods 2000; Woods and Ottenwalder 1992).

Like *Neocnus*, *Acratocnus* was for a time considered to have lived until quite recently and to have been extirpated by human agency (Allen 1942; Hall 1981). However, there was little substantive evidence for such a position, and there is even some question as to whether *Acratocnus* survived the Pleistocene. McFarlane (1999) indicated the genus probably disappeared from Puerto Rico before the mid-Holocene, without trace of anthropogenic contact. Steadman et al. (2005) stated that no bones of *Acratocnus odontrigonus*, Puerto Rico's only species of sloth, had been successfully radiocarbon dated, though it was suspected that, if organically well-preserved bones were found, at least some of the resulting dates would fall within the Holocene. MacPhee (2009) noted that there was no positive evidence that *A. odontrigonus* had survived into the time of human presence on Puerto Rico, and that there were no radiometric dates or finds of cultural association for *A. antillensis* of Cuba, but that *A. ye* is known from Trou Wòch Sa Wo, an archeological site in southwestern Haiti. Sloth material found there, while not identified as to species, yielded a radiocarbon date of 3,755 years before the present, the youngest on record for West Indian megalonychids (MacPhee, White, and Woods 2000).

PILOSA; MEGALONYCHIDAE; Genus *MEGALOCNUS* Leidy, 1868

Greater West Indian Ground Sloths

Two species are recognized (White and MacPhee 2001):

M. zile, known only by skeletal remains from Hispaniola (Dominican Republic and Haiti) and nearby Île de la Tortue;

M. rodens, known only by skeletal remains from Cuba and nearby Isla de Pinos.

As explained by White and MacPhee (2001), *M. rodens* exhibits a substantial range of intraspecific variation and now incorporates a number of formerly accepted genera, species, and subspecies.

Skeletal and dental characters include: cranium long, of relatively uniform width, and flattened superiorly; postorbital constriction absent; jugal expanded; pterygoid inflation present; paroccipital process greatly enlarged and free standing; symphyseal spout absent; first maxillary tooth pseudorodentiform or incisiform rather than caniniform; last maxillary molariform medially narrow; last mandibular molariform convex; deltoid and pectoral crests of humerus confluent; ulnar shaft straight; femur with nonspherical and nonflat head and anteroposteriorly deep distal end; shaft of tibia and fibula straight; scapular spine divergent at vertebral border; prescapular fossa much larger than postscapular fossa (White and MacPhee 2001). *Megalocnus* is the only West Indian sloth genus other than *Parocnus* that can be justifiably called "megafaunal," with weight of *M. rodens* estimated to have been as great as 270 kg (MacPhee, White, and Woods 2000). However, there is much variation in that

Greater West Indian ground sloth (*Megalocnus rodens*). Painting by Helga Schulze, based on skeletal illustrations and descriptive information in sources cited in this account.

Two-toed tree sloth (*Choloepus didactylus*); individual walking on ground in manner possibly suggestive of extinct West Indian genera. Photo by Daniel Heuclin/npl/Minden Pictures.

species and calculations based on femoral head diameter for a number of different individuals by White (1993) produced a range of estimates from about 35 to 173 kg.

E. Anderson (1984) wrote that *Megalocnus* reached the size of a black bear (*Ursus americanus*) and displayed a drawing of a somewhat bearlike, terrestrial mammal. However, MacPhee (2009) observed that, while *Megalocnus* was likely to have spent a considerable amount of time on the ground, it displayed morphological indicators consistent with some amount of climbing or other arboreal activity. Analysis of the limb bones of *M. rodens* by White (1993) suggests terrestrial function in some aspects but also a degree of arboreality. Pujo et al. (2012) indicated *Megalocnus* was terrestrial. McDonald and De Iuliis (2008) cited a biochemical analysis of bone samples of *M. rodens*, which recovered the amino acid tyrosine, which is involved in hair pigmentation, allowing the inference that the

species was tawny or dark brown. Stearic acid was also recovered, indicating the animal was herbivorous.

Original accounts of *Megalocnus*, in contrast to those of the other West Indian sloths, suggested the genus disappeared before arrival of people in the West Indies (Allen 1942). Recently, however, a molariform tooth, referable in size and distinctive features to *M. rodens*, was discovered at a rock shelter near Havana, Cuba, and radiocarbon dated to about 4,190 years before the present, which was calibrated to a range of 4,840–4,580 calendar years ago (MacPhee, Iturralde-Vinent, and Jiménez Vázquez 2007). Respective dates for the oldest known archeological site in Cuba are 5,140 radiocarbon years and 6,280–5,590 calendar years ago. It is thus likely that West Indian sloths survived for more than a millennium after the arrival of people on the islands. Such a long period conflicts with the typical finding that prehistoric extinctions of endemic species occurred extremely rapidly after first contact with humans.

PILOSA; MEGALONYCHIDAE; Genus *PAROCNUS* Miller, 1929

Lesser West Indian Ground Sloths

Two species are recognized (White and MacPhee 2001):

- *P. serus*, known only by skeletal remains from Hispaniola (Dominican Republic and Haiti) and nearby Île de la Tortue and Île de la Gonave;
- *P. browni*, known only by skeletal remains from Cuba.

The distinguishing characters of this genus include: sagittal crest double; symphyseal spout greatly elongated and spatulate; anterior maxillary teeth small and triangular; mandibular caniniform teeth with deep inner groove; molariform teeth subquadrate; mandibular mental foramina very large; humeral head flattened; entepicondylar foramen of humerus absent; femoral head medially oriented, nonspherical, and flattened; femoral shaft wide, anteroposteriorly compressed for the upper two-thirds, and distally narrowed and rounded; lesser trochanter absent or inconspicuous; distal tibial articular surface with distinct separation; scapula with rounded borders and oval-shaped glenoid fossa; pre- and postscapular fossae approximately equal in size (White and MacPhee 2001). Weight has been estimated at over 70 kg (Allen 1942), though White's (1993) analysis of limb bones yielded a range of 23–41 kg.

Like *Megalocnus*, *Parocnus* probably spent considerable time on the ground, though its morphology suggests some arboreal capacity (MacPhee 2009). White's (1993) assessment indicated *Parocnus* to consistently show the most terrestrial adaptations of any of the West Indian sloths, a view supported by Pujos et al. (2012). Steadman et al. (2005) considered *Parocnus*, as well as *Megalocnus*, to have been entirely terrestrial or nearly so and thus to have been particularly vulnerable to hunting by human invaders. *P. serus* originally was de-

Three-toed tree sloth (*Bradypus variegatus*). Photo by Sharp / Shutterstock.com.

scribed from the same cave in Haiti where *Neocnus comes* was discovered, and its remains also were reportedly intermixed with fragments of pottery and the bones of humans and domestic pigs, thus suggesting the species still lived at the time of European colonization (Allen 1942). However, as discussed above in the account of *Neocnus*, the evidence for such late survival no longer seems credible, though *P. serus* is known to have occurred at a site on Haiti where sloth bones have been radiocarbon dated as recently as 3,715 years before the present (MacPhee 2009). Remains of *P. browni* from a limestone cave in central Cuba have a radiocarbon date of about 4,960 years before the present, which has been calibrated to a range of 6,350–4,950 calendar years ago (Steadman et al. 2005).

PILOSA; **Family BRADYPODIDAE;** *Genus* ***BRADYPUS*** Linnaeus, 1758

Three-toed Tree Sloths

The single known genus, *Bradypus*, contains two subgenera and four species (Anderson and Handley 2001; Gardner 2007f; Hirsch and Chiarello 2011; Moraes-Barros, Silva, and Morgante 2011; Moraes-Barros et al. 2010; Moreno and Plese 2006; Wetzel 1985a; Wetzel and Kock 1973):

subgenus *Bradypus* Linnaeus, 1758

B. variegatus, eastern Honduras, Nicaragua, Costa Rica, Panama, western and southeastern Colombia, Ecuador, western Venezuela, Peru east of the Andes, Brazil except east of Rio Branco and north of Amazon, northern and eastern Bolivia, Paraguay, extreme northern Argentina;

B. pygmaeus, Isla Escudo de Veraguas off northwestern Caribbean coast of Panama;

B. tridactylus, eastern Venezuela south of Orinoco, Guianas, Brazil west to Rio Branco and south to Amazon;

subgenus *Scaeopus* Peters, 1864

B. torquatus, coastal forests of eastern Brazil from state of Sergipe to state of Rio de Janeiro.

Although the range of *B. torquatus* is sometimes reported to have originally extended as far north as the state of Rio Grande do Norte, Hirsch and Chiarello (2011) presented compelling evidence that the northernmost verifiable record is from Sergipe and that the kind of forest habitat farther north is not suitable for the species. Chagas et al. (2009) confirmed continued presence in Sergipe. Paula Couto (1979) considered *Scaeopus* a full genus, but that view was not followed by Anderson and Handley (2001) or Gardner (2005e, 2007f).

For the subgenus *Bradypus*, head and body length is 413–750 mm, tail length is 20–110 mm, and adult weight is 2.25–6.30 kg. For 39 adult *B. torquatus*, Lara-Ruiz and Chiarello (2005) recorded head and body length as 590–752 mm and weight as 4.6–10.1 kg; females were on average larger than males, which is generally true for the genus. The pelage consists of long, thick overhairs that are grooved longitudinally and short underfur of fine texture. Most hairs point downward when the animal is hanging beneath a branch. Thus, most of the hairs are directed opposite to the hairs of most other mammals. The general coloration is grayish brown, slightly darker on the head and face, and on the shoulders there is usually a light area with brown markings. Frequently the coloration appears to be greenish because of algal growth on the coat. Male *B. tridactylus* have a dorsal orange-yellow patch with a broad, tapering, black central streak (Hayssen 2009). *B. torquatus*, the maned sloth, has long, dark hairs on the head and neck.

These sloths have some resemblance to *Choloepus* (see account thereof), but the forelegs are substantially longer than the hind legs, and there are three digits on both the forefeet and the hind feet. The digits are closely united, and each terminates in a long, hooklike claw. There is a tail in *Bradypus*, though it is short, stout, and blunt, having the appearance of the stump of an amputated limb. Whereas most mammals have seven neck vertebrae, *Bradypus* has eight or nine, thus allowing greater flexibility, a desirable feature in an animal that is so limited in its other movements. In this genus the head can be turned through an arc of 270°. The eighth and ninth neck vertebrae sometimes bear a pair of short, movable ribs. The head is small and round, and the eyes and ears are small. As in the Megalonychidae, there are 5 upper and 4 lower teeth on

Three-toed tree sloth (*Bradypus variegatus*). Photo by Alexander F. Meyer at Huachipa Zoo, Peru.

each side, thus a total of 18 teeth. Females have two mammae in the chest region.

Three-toed sloths inhabit forests and spend practically their entire life in trees, where they hang beneath the limbs or sit in a fork. Observations of *B. variegatus* in Venezuela showed that, contrary to what is sometimes said, sloths do not spend most of their time hanging from boughs, but, when stationary, sit for 77 percent of the time; the sitting-resting posture may assist digestion by facilitating stratification and sorting of food particles in the stomach (Clauss 2004; Urbani and Bosque 2007). Arboreal locomotion is usually by using the claws as hooks to move along beneath a branch or to climb a tree. In the Panama Canal Zone, Sunquist and Montgomery (1973) found *B. variegatus* to be active both by night and by day. *B. tridactylus* and *B. torquatus* also have been reported to exhibit both nocturnal and diurnal activity (Chiarello 2008). Sloths apparently descend to the ground about once or twice a week to urinate and defecate and on occasion to move from one tree to another. Normal terrestrial locomotion is by a slow crawl on the soles and forearms (Mendel 1985). They swim readily. If attacked on the ground, they endeavor to catch their adversary with their claws, with which they can inflict severe wounds. Although normally the movement of sloths in trees is slow and methodical, they can progress rapidly when pressed.

Sunquist and Montgomery (1973) reported that in *B. variegatus*, at least, individuals changed trees about four times less often from day to day than did individuals of *Choloepus hoffmanni*. Moreover, only 11 percent of the individuals of *B. variegatus* showed movements of 38 meters or more per day, whereas 54 percent of the individuals of *C. hoffmanni* had a daily movement of 38 meters or more. Chiarello (1998) found *B. torquatus* to travel an average of only about 24 meters per day, with averages of 17 meters during daylight and 5 meters at night, indicating a predominantly diurnal period of activity; animals rested for an average of 74 percent of daytime. Since *Bradypus* tends to stay in one tree for a long period, feeding is restricted to that tree and to other plants supported by the tree. *Bradypus* therefore tends to be a more specialized feeder than *Choloepus*, the members of which are exposed to many more species of plants from which to choose. Studies in which food items were quantified showed leaves to constitute 94–99 percent of the diet, with fruit and flowers together making up another 1 percent (Chiarello 2008). Urbani and Bosque (2007) found leaves, mostly those of Cecropiaceae, Clethraceae, and Clusiaceae, to form 99.4 percent of the diet of *B. variegatus*. Food is pulled to the mouth with slow movements of the forelegs. In part because of their more specialized diet in the wild,

Three-toed tree sloth (*Bradypus variegatus*). Photo by Seaphotoart/Shutterstock.com.

Three-toed tree sloth (*Bradypus variegatus*). Photo by Alexander F. Meyer at Dallas World Aquarium.

three-toed sloths are more difficult to breed in captivity than are two-toed sloths (Crandall 1964).

Montgomery and Sunquist (1978) reported population density of *B. variegatus* on Barro Colorado Island as 6–7 per ha. and individual home range as usually less than 2 ha. In Costa Rica, Vaughan et al. (2007) found *B. variegatus* to form home ranges of 4–7 ha., which contained fewer tree species than characteristic of the ranges of *Choloepus hoffmanni*. Reported home ranges for *B. torquatus*, have been 3–5 ha. for three individuals over 12–24 months (Cassano 2004), 0.5–6.0 ha. over a 14-month period (Chiarello 1998), and around 2–11 ha. for five individuals during periods of about 10–36 months (Chiarello et al. 2004). *Bradypus* is solitary; aggression has been reported between members of the same sex, which is predominantly by striking with claws and rarely by biting (Hayssen 2009, 2010). In contrast to what sometimes has been reported for *Cho-*

loepus hoffmanni, there seems to be a one-to-one sex ratio in *B. variegatus*. During a study of *B. variegatus* in Costa Rica from February 2010 through May 2012, Pauli and Peery (2012) found average home ranges of 5.4 ha. for males and 4.7 ha. for females. Male ranges intersected, but most males maintained exclusive use of a core area; male ranges overlapped more extensively with those of females, as did those of females with one another. The mating system there appeared strongly polygynous; just 16 percent of adult males (3 of 19 individuals genetically tested) sired 85 percent of the juveniles (17 of 20 individuals), and nearly three-quarters of adult males did not sire offspring. One male sired half of all sampled juveniles. In contrast, most females did produce offspring; if they did so more than once, in some cases the father was different.

In French Guiana, most births of *B. tridactyulus* were found to occur from February to August, with a peak in May, which generally corresponds with the long rainy season from April to July; no reproduction was observed from October to December (Taube et al. 2001). *B. variegatus* appears to have a definite, but somewhat irregular breeding season that varies according to climatic variations between different locations and between years (Taube et al. 2001). In Costa Rica, Pauli and Peery (2012) reported *B. variegatus* to generally breed before or during the rainy season, with young often born in November and December but also seen in April and May. Although Pinder (1993) reported *B. torquatus* to breed throughout the year, studies in the state of Espírito Santo and review of records from southern Bahia and northern Rio de Janeiro (Dias et al. 2009) indicate births occur predominantly at the end of the wet season and beginning of the dry season (February–April) and that mating is concentrated in the late dry and early wet seasons (August–October). Energy-demanding gestation (assumed to last 6 months) and lactation thus take place during the least stressful period of the year, when temperatures are higher and preferred food items are more plentiful. Female *B. torquatus* seem to have a shrill cry in association with the mating period. Bezerra et al. (2008) reported that a female *B. variegatus* also emitted loud, high frequency vocalizations while mating. The young of that species have a whistling call when separated from the mother (Hayssen 2010).

The gestation period of *B. tridactylus* is probably about 6 months and interbirth interval is approximately

12 months (Taube et al. 2001). Litter size in *Bradypus* normally is one, though Bezerra et al. (2008) reported a female *B. variegatus* carrying two young of equal size during a 5-day period. Infant *B. torquatus* were observed to ingest solid food (leaves) as early as 2 weeks of age, but nursing continued until they were 2–4 months old (Lara-Ruiz and Chiarello 2005). The young becomes independent at 8–11 months of age; the mother then abandons it, and the two are unlikely to encounter each other again, as their home ranges overlap very little, if at all (Gilmore, Duarte, and Peres da Costa 2008). Studies of *B. variegatus* on Barro Colorado Island (Montgomery and Sunquist 1978) indicate that females give birth in successive years, the gestation period is 5–6 months, and the young cease nursing at 3–4 weeks but depend on the mother for mobility for another 5 months. Merrett (1983) stated that the single young weighs 200–250 grams at birth and is carried on the mother's abdomen. Sexual maturity probably is attained at 3–4 years (Gilmore, Duarte, and Peres da Costa 2008).

Three-toed sloths can hardly be kept in captivity outside their natural habitat, because they do not adapt to substitute diets that seem to lack certain trace elements, they are very susceptible to cold, and they have no defense against even common diseases (Taube et al, 2001). Successful captive breeding has not been reported (Superina, Miranda, and Plese 2008). One *B. tridactylus* in the São Paulo Zoo lived to an age of 11 years and 8 months (Weigl 2005). A male *B. torquatus* was known to be living in the wild at an age of at least 12 years (Lara-Ruiz and Chiarello 2005).

All species of *Bradypus* may be jeopardized by habitat destruction and excessive hunting. In particular, *B. torquatus* is classified as vulnerable by the IUCN (Chiarello and Moraes-Barros 2014) and endangered by the USDI. Thornback and Jenkins (1982) had reported that the Atlantic coastal forest of eastern Brazil, which this sloth inhabits, was being rapidly cut over for lumber extraction and charcoal production and to make way for plantations and cattle pasture. Since that time, deforestation has decreased dramatically, but it has not stopped,

Three-toed tree sloth (*Bradypus variegatus*). Photo by Nacho Such / Shutterstock.com.

and populations and genetic diversity remain in jeopardy. Hirsch and Chiarello (2011) reported two major gaps in the natural distribution of the species, the larger between the states of Bahia and Espírito Santo and the smaller between Espírito Santo and Rio de Janeiro; these gaps probably result from discontinuities in the ombrophilous Atlantic Forest, the vegetation type preferred by *B. torquatus*. The three resulting populations of *B. torquatus* are considered genetically distinct, possibly at the subspecific level, hence compounding the need for conservation measures.

The subspecies *B. variegatus brasiliensis* occurs in the same region of eastern Brazil and is also threatened by habitat destruction and hunting pressure but has a more continuous distribution (Oliver and Santos 1991). It has been eliminated in the state of Paraná, and *B. variegatus* also has disappeared in northern Argentina. The species is declining in Colombia due to deforestation and because of the capture of young animals that are sold as pets to tourists (Moraes-Barros, Chiarello, and Plese 2014; Moreno and Plese 2006).

The newly discovered species *B. pygmaeus*, found only on Isla Escudo de Veraguas, a 4.3 sq km island about 17.6 km off the coast of Panama, is classified as critically endangered by the IUCN (Voirin et al. 2014). It originally was thought restricted to about 10.67 ha. of mangrove habitat and nearby areas, where Kaviar, Shockey, and Sundberg (2012) counted 79 individuals, which would amount to 7.4 individuals per ha. However, recent tracking studies have found the species in dense tropical rainforest in the interior of the island. If a density of 7.4 per ha. is consistent throughout the 430-ha. island, the total population size could be as high as 3,182 individuals, though 500–1,500 is more likely (Voirin 2015). In any case, numbers are declining because habitat is being logged to obtain firewood and construction materials. The species is legally protected but is hunted freely by fishermen and other visitors to the island (Aguiar and da Fonseca 2008). Both *B. pygmaeus* and *B. variegatus* are now on appendix 2 of the CITES.

Order Scandentia

Tree Shrews

This order of two families—Ptilocercidae, with the single genus and species *Ptilocercus lowii*, and Tupaiidae, with 5 Recent genera and 23 species—is found in forested areas of eastern Asia from India and southwestern China eastward through the Malay Peninsula to Borneo and the Philippines. Although Ptilocercidae is sometimes treated as only a subfamily, there now

seems general agreement that it is distinct from and basal to Tupaiidae (Emmons 2000; Helgen 2005; Janečka et al. 2007; Olson, Sargis, and Martin 2004; Roberts et al. 2011; Sargis 2001, 2002a, 2002b, 2002c, 2002d, 2004; Wible 2009, 2011).

In contrast, the inter-ordinal classification of Scandentia has long been contentious and remains unsettled. The various systematic arrangements proposed over the past century include: (1) grouping Scan-

Northern tree shrew (*Tupaia belangeri*). Photo by Alexei Abramov.

dentia with Macroscelidea to form Menotyphla, which was recognized either as a full order or as a suborder of the defunct order Insectivora (see account of order Afrosoricida); (2) placing the members of Scandentia within infraorder Lemuriformes, suborder Prosimii, order Primates; and (3) treating Scandentia as a separate order closely related to Dermoptera, Primates, and Chiroptera, which together would form the monophyletic supraordinal group Archonta. Molecular analyses have generally supported eliminating Chiroptera from that grouping and recognizing the clade Euarchonta, comprising only Scandentia, Dermoptera, and Primates (see account of class Mammalia, particularly under Euarchontoglires). Some morphological assessments, however, have continued to suggest validity of a clade comprising Dermoptera and Chiroptera (see Sargis 2002a, who also provides a review of the history of efforts to classify Scandentia and related groups). A recent study combining morphological, paleontological, and molecular data supported the affinity of Scandentia and Dermoptera, which would be placed in the supraordinal group Sundatheria, with Primates in a sister relationship to that group, and with Chiroptera not closely related (O'Leary et al. 2013). That last position is accepted herein.

In external appearance tree shrews resemble long-snouted squirrels, but, with the exception of *Ptilocercus*, they can readily be distinguished from squirrels by the absence of long whiskers. *Ptilocercus*, the pen-tailed tree shrew, is easily identified by its tail, which is naked except for a whitish feather-shaped arrangement of the hairs near the end. Tree shrews have a slender body, with most adults generally weighing less than 400 grams. Head and body length is about 100–230 mm, and tail length is approximately 90–225 mm. The pelage is of the usual mammalian kind, consisting of long, straight guard hairs and shorter, softer, more woolly underfur. Some forms have pale shoulder stripes, and others have facial markings. The ears are squirrel-like, that is, comparatively small and cartilaginous, except in *Ptilocercus*, in which they are larger and more membranous. The feet of tree shrews are naked beneath; the soles have tubercle-like pads. The long and supple digits bear sharp, moderately curved claws.

Some primate-like characters of tree shrews are the relatively large braincase, the remarkable resemblance of the carotid and subclavian arteries to those of humans, and the permanent sac, or scrotum, for the testes in males. The orbits are completely encircled by bone. The upper incisors are large, caninelike, and separated, and the small canines resemble the premolars. The upper molars are broad, with a W-shaped pattern of cusps. The dental formula is: i 2/3, c 1/1, pm 3/3, m 3/3 = 38.

In addition to their external resemblance to squirrels, tree shrews are generally diurnal, and some of their actions and movements are comparable. They are swift runners. *Ptilocercus* differs in being mainly nocturnal; when on the ground it progresses in a series of hops, but it is primarily arboreal. Within the family Tupaiidae there are arboreal, terrestrial, and scansorial species (Emmons 2000). Tree shrews have keen senses of smell and hearing and good vision. Their diet consists primarily of insects and fruits but occasionally includes other animal food and various kinds of plant material.

Most species of Scandentia are declining, mainly because of loss and fragmentation of habitat, and some are considered in danger of extinction. All species are on appendix 2 of the CITES. That designation may stem in part from the original general addition of the entire order Primates to the appendices of the CITES in 1975, when that order commonly was thought to include Tupaiidae. There is not believed to be any current large-scale trade in these animals, though some species are collected for medicinal purposes or laboratory use (see account of genus *Tupaia*).

There long were doubts concerning the identification of fossil specimens assigned to this order, but Jacobs (1980) described tupaiids from the middle Miocene of Pakistan and discussed other fossil material that seems to belong to this or to closely related groups. Butler (1980) stated that Scandentia appears to have been distinct from other mammals at least as far back as the beginning of the Tertiary and possibly in the Cretaceous. That view was supported by Janečka et al. (2007), though O'Leary et al. (2013) put the divergence of Scandentia at less than 50 million years ago. The known geological range of family Tupaiidae is middle Eocene to Recent in Asia, but Ptilocercidae is known only from the Recent (McKenna and Bell 1997).

Pen-tailed tree shrew (*Ptilocercus lowii*). Photo by Louise H. Emmons.

SCANDENTIA; **Family PTILOCERCIDAE;** Genus *PTILOCERCUS* Gray, 1848

Pen-tailed, or Feather-tailed, Tree Shrew

The single genus and species, *Ptilocercus lowii*, inhabits the Malay Peninsula (including extreme southern Thailand), Singapore, Sumatra and nearby Bangka and Siberut Islands, and northern and western Borneo and nearby Serasan, Riau, and Batu Islands. Ptilocercidae has at times been considered a subfamily of Tupaiidae (Corbet and Hill 1992) but now generally is regarded as a separate family (Helgen 2005).

Head and body length is 100–140 mm, tail length is 130–190 mm, and weight is about 25–75 grams. The fur is soft in texture, usually dark grayish brown above and yellowish gray beneath. The tail is naked and dark, except for the terminal part, which has whitish hairs on opposite sides, producing a feather-like form as all the hairs are in the same plane. The entire tail resembles an old-fashioned quill pen, hence the common names. The head is moderately tapering, and the whiskers are elongated and rather rigid. In contrast to

the small thick ears of *Tupaia*, the ears of *Ptilocercus* are rather large, thin, and membranous. The limbs are nearly equal in length. The forefeet, hind feet, and pads are relatively larger than in the other genera of tree shrews. Each foot has five digits, all of which bear short, sharp claws. Some of the skeletal features that distinguish Ptilocercidae from Tupaiidae are discussed below in the account of the latter family. Emmons (2000) wrote: "The stunningly brilliant white eyeshine of *P. lowii* is the brightest I have ever observed in a mammal."

The feather-tailed tree shrew lives in primary or secondary forests from sea level to about 1,000 meters; it is nocturnal and arboreal (Lekagul and McNeely 1977). In Borneo, Emmons (2000) found it to be active an average of 8.7 hours per night, with a well-defined rest period in the middle. Average distance traveled per night was 1,073 meters. *Ptilocercus* is an expert climber, using the tail for support and balance. Perhaps the tail also serves as an organ of touch, and it may assist in leaps. On the ground this mammal proceeds in a series of hops, with the tip of the tail inclined upward. It nests in holes in tree trunks or branches or among epiphytes about 12–20 meters above the ground. Nests consist of dried leaves, twigs, and fibers of soft wood. The diet consists of insects, small vertebrates such as geckos, and fruits (Gould 1978; Lekagul and McNeely 1977). In the wild, Emmons (2000) found the degree of frugivory unclear; *Ptilocercus* did eat fruit sporadically but, in contrast to *Tupaia*, could not be captured in traps baited with fruit.

Captives may be fierce at first but usually soon become tame. When annoyed, *Ptilocercus* emits a hoarse, snarling hiss with the mouth open. It also has soft to loud birdlike chirps and chirrups, which are raspy to musical (Emmons 2000). A pair observed in captivity carried their tails downward and outstretched, at the same time moving them back and forth like a pendulum. *Ptilocercus* sleeps with its body rolled up in a tight ball, and the tail may be curled so that the plume covers the face.

Although *Ptilocercus* usually occurs in pairs, as many as five individuals have been found in a single nest. A group of three females and a male in Borneo nested together in a hollow tree and shared a large home range where each individual had a largely separate foraging zone of about 2.0–6.4 ha. (Emmons

Pen-tailed tree shrew (*Ptilocercus lowii*). Photo by Lim Boo Liat.

2000). Available evidence indicates that young are raised in the same communal nest where the older animals live and that *Ptilocercus* lacks the "absentee" system of maternal care known in some species of *Tupaia*. In Borneo, Emmons (2000) observed a lactating female on 22 August, a very small young on 28 September, two young in November, and two presumed subadults on 9 May, thus suggesting a litter size of one or two. Young apparently share the den with the mother until they reach adult size, which would be after the birth of the

next litter. One specimen lived 2 years and 8 months in captivity (Lekagul and McNeely 1977).

Although not in an IUCN threatened category, *P. lowii* is on appendix 2 of the CITES and is considered rare and continuing to decline throughout most of its range. A major problem is loss of forest-canopy habitat due to agricultural expansion and conversion of land to non-tree crops; it survives in older mosaics of natural forest and tree plantations (Han and Stuebing 2008a).

Large tree shrew (*Tupaia tana*). Photo by Alexander F. Meyer at National Zoo, Washington, DC.

SCANDENTIA; Family TUPAIIDAE

Tree Shrews

This family of 4 Recent genera and 23 species is found in forested areas of eastern Asia from India and southwestern China eastward through the Malay Peninsula to Borneo and the Philippines. The sequence of genera presented herein considers the discussions of Olson, Sargis, and Martin (2004, 2005), suggesting *Dendrogale* as the most basal genus, followed by *Anathana*, with *Urogale* (see account thereof) either forming a clade with *Anathana* or being immediately related to and possibly subsumed by *Tupaia*. The family Tupaiidae is sometimes considered to include Ptilocercidae (see account thereof) as a subfamily, Ptilocercinae, but there now is general agreement the two families are separate.

Head and body length is 100–230 mm, tail length is 90–230 mm, and weight is 35–350 grams. Additional descriptive, natural history, and paleontolog-

ical information on family Tupaiidae in general is provided above in the account of order Scandentia. Regarding ecological and behavioral traits, Sargis (2002d) noted that Tupaiidae is diurnal and terrestrial (scansorial), has more extended limbs and a less abducted hallux, and is not capable of grasping, whereas Ptilocercidae is nocturnal and arboreal, has more flexed limbs and a more abducted hallux, and is capable of grasping.

A series of detailed morphological studies has also supported the distinction of Tupaiidae and Ptilocercidae. Wible (2011) pointed out that the skull of *Tupaia*, across the frontal and parietal bones, is much wider than that of *Ptilocercus*, both anterior to the postorbital process and at the postorbital constriction. The rostrum is relatively broader in *Tupaia* than in *Ptilocercus*. Also, in *Tupaia* the angulation of the postorbital bar is more vertical, and there is a large supraorbital foramen, which is lacking in *Ptilocercus*. Laterally, *Tupaia* has a relatively smaller temporal fossa, and the antorbital and supraorbital margins are defined by a distinct orbital crest; the margin is indistinct in *Ptilocercus.* The angle between the anterior root of the zygoma and the occlusal plane of the upper teeth is less than 20° in *Tupaia* but approaches 60° in *Ptilocercus.* In ventral view, the palate is relatively shorter in *Ptilocercus,* and the mesocranium is longer. The auditory bullae are more inflated in *Tupaia* and conceal the promontoria of the petrosals, part of which is visible in *Ptilocercus.* Regarding the middle ear, Wible (2009) explained that the anteromedial diverticula of the two genera are similar in position and entotympanic composition, but that *Tupaia* differs in having much more substantial septa demarcating the lateral and posteromedial limits. Laterally, its anterior septum forms a complete wall, filling the gap between tympanic roof and floor, the anterior pole of the promontorium, and the anterior internal aspect of the bulla; in contrast, the anterior septum in *Ptilocercus* is minute and prominent only on the anterior internal aspect of the bulla.

Postcranial skeletal characters reflect the basically terrestrial adaptations of Tupaiidae and arboreal specializations of Ptilocercidae. With respect to the axial skeleton, Sargis (2001) reported that in *Ptilocercus* the ribs are wide, the atlas is wide, the axis spinous process projects cranially, the thoracic spinous processes are short and wide, the lumbar spinous processes are very short, and the lumbar transverse processes are short and face laterally. In Tupaiidae, the ribs are relatively narrow, the atlas is narrow, the axis spinous process projects caudally, the thoracic spinous processes are long and narrow, the lumbar spinous processes are long to very long (except in *Dendrogale,* in which they are intermediate in length), and the lumbar transverse processes are long and face ventrally (except that in *Dendrogale* they also are short). The ribs and vertebrae of *Ptilocercus* provide a stable thorax enabling the animal to bridge gaps between tree branches. The vertebral columns of Tupaiidae are more mobile and allow more flexion and extension of the spine, thereby

Common tree shrew (*Tupaia glis*). Photo by Alexander F. Meyer at Bronx Zoo.

increasing stride length and speed for bounding or galloping.

The two most striking differences in the cervical vertebrae of Tupaiidae are found in the atlas and axis (Sargis 2001). The atlas of *Ptilocercus* is craniocaudally expanded on the dorsal side, while it is quite thin in Tupaiidae. The atlas of *Ptilocercus* actually is about the same length or longer than that of the species *Tupaia tana*, which is approximately six times larger than *Ptilocercus* in body size, while it is much longer than that of *Dendrogale*, which is about the same size as *Ptilocercus*. The craniocaudally long atlas would restrict vertebral mobility in *Ptilocercus*, while the short atlas of Tupaiidae would allow greater mobility. In addition, the cervical vertebrae of *Ptilocercus* articulate tightly, restricting mobility, while those of Tupaiidae exhibit more intervertebral space, allowing greater mobility.

Regarding the forelimb, Sargis (2002b) reported that in *Ptilocercus* the scapula is short and wide, the coracoid process is long and extends caudally, the lesser tuberosity is robust and has a strong medial protrusion, the capitulum is rounded (separated from trochlea), an entepicondylar foramen is present, the olecranon process is short, and the radial central fossa is circular. In Tupaiidae the scapula is long and narrow, the coracoid process is shorter and extends medially, the lesser tuberosity is gracile and has a weak medial protrusion, the capitulum is flatter (continuous with trochlea), an entepicondylar foramen is usually present (but absent in *Urogale*), the olecranon process is usually longer, and the radial central fossa is ovoid.

With respect to the hind limb, Sargis (2002c) reported that in *Ptilocercus* the ilio-pubic angle is large (e.g., above 145°), the pubic symphysis is short, the anterior inferior iliac spine is small, the ilium is very narrow, the acetabulum is elliptical, the greater trochanter is small, the third trochanter is small, the femoral condyle is shallow, and the patellar groove is short and wide. In Tupaiidae the ilio-pubic angle is smaller (e.g., below 130°), the pubic symphysis is long, the anterior inferior iliac spine is large, the ilium is wide, the acetabulum is circular, the greater trochanter is large, the third trochanter is large, the femoral condyle is deep, and the patellar groove is long and narrow.

The limbs of *Ptilocercus* are better adapted for arboreal locomotion, while those of Tupaiidae are better adapted for terrestrial or scansorial locomotion (Sargis 2002a, 2002b, 2002c). The limbs of *Ptilocercus* appear habitually flexed and exhibit more mobility in their joints—a necessity for movement on uneven, discontinuous arboreal supports. The tarsus of *Ptilocercus* facilitates inversion of the foot, and its grasping hallux is capable of a great range of abduction. Tupaiids, in contrast, are characterized by more extended limbs and less mobility in their joints. Such restriction limits movements more to the parasagittal plane, increasing efficiency of locomotion on a more even and continuous surface. Even the most arboreal tupaiids remain similar to their terrestrial relatives in their limb morphology, which probably reflects the terrestrial ancestry of Tupaiidae. The forelimb of *Urogale* is unique among tupaiids in that it exhibits adaptations for scratch-digging.

SCANDENTIA; TUPAIIDAE; **Genus *DENDROGALE***
Gray, 1848

Small Smooth-tailed Tree Shrews

There are two species (Han and Stuebing 2008b; Timmins et al. 2003):

> *D. murina*, southeastern Thailand, southern Laos, Cambodia, southern and probably northern Viet Nam;
> *D. melanura*, mountains of northern Borneo in the Malaysian states of Sarawak and Sabah.

Timmins et al. (2003) noted that the only records of *D. murina* from northern Viet Nam, two sightings of single animals at Tam Dao National Park, might possibly represent individuals that had been collected illegally elsewhere, confiscated by authorities, and released. However, on balance it seemed likely the occurrences were natural and indicative of the range of the species being much more extensive than originally thought.

Head and body length is 100–150 mm and tail length is 90–145 mm. Lekagul and McNeely (1977) gave the weight of *D. murina* as 35–55 grams. *D. murina* is light

Small smooth-tailed tree shrew (*Dendrogale murina*). Photo by Roland Seitre / Minden Pictures.

in color and has facial markings, whereas *D. melanura* is dark and lacks facial markings. In *D. murina* the upper parts are brownish or blackish, and buff, ochraceous, or tawny. The underparts and the inner sides of the legs are buffy. The side of the head has a blackish line from the base of the whiskers through the eye to the ear and light, usually buffy, lines above and below the black band. The tail is dark above and darker distally; it is ochraceous buff below with a dark line down the center. The claws of *D. murina* are small. In *D. melanura* the upper parts are mixed blackish and ochraceous buff or cinnamon rufous, the darker color predominating; the underparts and the inner sides of the legs are ochraceous. Short ochraceous lines are sometimes present immediately above and below the eye. The claws of *D. melanura* are long. Shoulder stripes are not present in either species. These are the only small members of the family Tupaiidae with round, uniformly even-haired tails.

During the day these tree shrews are quite active, running about on the lower branches of trees and shrubs looking for insects. *Dendrogale* is more arboreal than either *Tupaia* or *Anathana*. In captivity it will accept fruits and chopped meat. Timmins et al. (2003) reported *D. murina* to occur at elevations of 20–1,500 meters in a wide variety of habitats, from evergreen forest through deciduous forest to bushes in rocky savannahs, often in or close to bamboo, and almost exclusively at 30–300 cm above the ground. The species was seen in morning and afternoon and seemed oblivious to people; if disturbed, it fled through tangled vegetation, never along the ground. Most observations were of single animals, but pairs were seen in April and June. In one area, where the animals were at high density, calls were heard all morning and consisted of a series of 4–9 short, sharp, high-pitched notes. *D. melanura* is found in mountains from about 900 to 3,350 meters; one specimen was taken at the top of Mount Dulit, where it was living amid the moss-covered, stunted jungle. *D. melanura* is designated "data deficient" by the IUCN, but it is thought to be declining because of loss of forest habitat to agricultural expansion. There have been no records since the 1970s, despite several efforts to locate it (Han and Stuebing 2008b).

SCANDENTIA; TUPAIIDAE; **Genus**
ANATHANA Lyon, 1913

Indian Tree Shrew

The single species, *A. ellioti*, is found in peninsular India south of the Ganges River (Roonwal and Mohnot 1977). The range in the west extends north to the Satpura Hills in southwestern Madhya Pradesh, and in the east north to Bihar in West Bengal (Corbet and Hill 1992; Tiple and Talmale 2013). *Anathana* is considered very closely related to *Tupaia* (Corbet and Hill 1992; Helgen 2005) and has sometimes been placed in the latter genus. Some recent studies have indicated *Anathana* to be sister to a clade of *Tupaia* and *Urogale*, while other studies have suggested *Anathana* falls within the range of variation of *Tupaia*; in any case, there has been no definitive conclusion (Olson, Sargis, and Martin 2004, 2005; Roberts et al. 2011).

Head and body length is 175–200 mm, and tail length is 160–190 mm. R. D. Martin (1984) listed weight as 160 grams. The upper parts are usually speckled yellow and brown, with the middle of the back, the rump, and sometimes the upper tail tinged with reddish, producing a yellowish to reddish brown effect. Some individuals are blackish and orangish above. The underparts are whitish or buffy, and the feet are buff colored. A whitish or cream-colored shoulder stripe is present. This genus differs from *Tupaia* in its larger and thickly haired ears and in cranial and dental features, but the general appearance is close to that of *Tupaia*.

This tree shrew is little known compared with some species of *Tupaia*, but it probably lives in much the same way. It is usually found in forests, and unlike *Tupaia*, it does not frequent human dwellings (Roonwal and Mohnot 1977). Recently, Tiple and Talmale (2013) recorded 43 sightings of *Anathana* in the forests of central India. They considered the genus diurnal, both terrestrial and arboreal, and to move on the forest floor in search of food, which includes seeds, fruits, insects, earthworms, and small vertebrates. It was reported to make a rough nest on trees and to mark a territory by rubbing its chin and throat on rocks and leaving drops of urine on rocks in its path.

Chorazyna and Kurup (1975) observed several individuals on a slope covered with shrubs and stones at an elevation of 1,400 meters. At night the animals sheltered in holes among the rocks, sometimes using a system of corridors with two or three entrances. They usually left the shelters at dawn and returned 2 hours before sunset. Most of their active time was spent foraging—seeking insects on the ground, digging out worms, and sometimes leaping after flying insects. They also ate fruits. They climbed skillfully, even on vertical rocks, but rarely entered trees. They appeared to be largely solitary, always foraging alone, and there was usually one individual per hole, but occasionally groups of three or four were seen playing together for up to 20 minutes. A call described as a staccato squeak was heard twice. Roonwal and Mohnot (1977) reported that a pregnant female was taken in June. Hayssen, Van Tienhoven, and Van Tienhoven (1993)

Indian tree shrew (*Anathana ellioti*). Photo by Ashish Tirkey.

listed a record of five embryos being found in a female.

Tiple and Talmale (2013) indicated that Anathana was plentiful in central India in the early twentieth century but that populations there now are low because of habitat loss and fragmentation. Molur (2008) reported no major threats but that numbers are decreasing and that there is some trade for medicinal purposes. Like all scandentians, *Anathana* is on appendix 2 of the CITES.

SCANDENTIA; TUPAIIDAE; Genus *UROGALE*
Mearns, 1905

Philippine Tree Shrew

The single species, *U. everetti*, is widely distributed on Mindanao and has also been collected on Dinagat and Siargao islands in the Philippines (Heaney and Rabor 1982). Although *Urogale* was long treated as a distinctive genus by most authorities, its classification has become quite problematic. Several studies of morphology showed *Urogale* to have unique characters, which might warrant either retaining it as a full genus or at least leaving the question open (Olson, Sargis, and Martin 2004; Sargis 2002b, 2002d). An investigation combining DNA hybridization and cranial morphometrics suggested *Urogale* should be moved to the genus *Tupaia* (Han, Sheldon, and Stuebing 2000), but another DNA assessment was inconclusive on the issue (Olson, Sargis, and Martin 2005). More recent and more extensive analyses of both nuclear (Roberts, Sargis, and Olson 2009) and mitochondrial (Roberts et al. 2011) DNA have provided compelling evidence that *Urogale* is indeed nested within and should be synonymized with *Tupaia*, and that the latter would have to be considered paraphyletic without such inclusion. The precise position that *Urogale everetti* would have among the species of *Tupaia* is not clear, though it reportedly seems more aligned with *T. glis* and its relatives than with the other Philippine species, *T. palawanensis* and *T. moellendorffi*. It may have colonized the Philippines via overwater dispersal early in the radiation of *Tupaia*.

Notwithstanding their conclusion on the matter, Roberts, Sargis, and Olson (2009) stated that *Urogale* is "geographically as well as morphologically unique among tree shrews"; it is the farthest east and the only one on oceanic islands in the Philippines. Roberts et al. (2011) added that *Urogale* is "unquestionably morphologically (and probably behaviorally and ecologically . . .) distinct from all other extant tree shrews," though its distinctiveness almost certainly results mainly from derived characters. Retention of *Urogale* as a genus here is based on the reservations inferred from those statements, on the morphological characters that maintained generic status for a century, and on the continued recognition of such status by Heaney et al. (2010).

Head and body length of *Urogale everetti* is 170–220 mm, and tail length is 115–185 mm. Weight is 180–268 grams (Heaney et al. 2010). A blackish and tawny mixture on the upper parts produces a brownish color; the underparts are orangish to orangish red, being brightest on the chest region. There often is an indistinct orangish shoulder stripe. The specimens from Dinagat are relatively light in color with a metallic golden sheen dorsally; those from Siargao are much darker, with the dorsum nearly black. This tree shrew is distinguished externally by the elongated snout and the even-haired, rounded tail. The second pair of upper incisors is enlarged and caninelike.

The forelimb of *Urogale* is unique among tupaiids in that it exhibits adaptations for scratch-digging (Sargis 2002b, 2004). *Urogale* is the only tupaiid in which the entepicondylar foramen of the humerus is always absent—a feature that may strengthen and buttress the medial epicondyle, which anchors the flexor musculature that powerfully flexes the wrist and digits during scratch-digging. And in *Urogale* the medial epicondyle is especially large, which contributes to the flexion. Also almost completely unique to *Urogale* is the consistent and large perforation of the olecranon fossa, which allows powerful extension of the antebrachium. The particularly long claws and ungual phalanges are typical of digging mammals. Among tupaiids, *Urogale* has the deepest knee, enabling powerful extension thereof.

Philippine tree shrew (*Urogale everetti*). Photo by D. S. Balete through Lawrence R. Heaney and Kayleigh Kueffner.

According to Heaney et al. (2010), in the wild the Philippine tree shrew is common to scarce in primary and secondary forest at 750–2,500 meters, apparently most common in montane forest at 1,200–1,800 meters. It is diurnal, with an activity peak in early morning. It forages both on the ground and in low trees and shrubs, feeding primarily on insects, occasionally on earthworms and soft fruit, and rarely on small vertebrates. Observations in captivity (Wharton 1950) also show *Urogale* to be omnivorous, feeding on such items as meat, insects, fruits and vegetables, mice, lizards, and earthworms. It opens eggs with a skill suggesting it eats them in the wild. *Urogale* has a large appetite; it will consume several bananas or several two-ounce pieces of raw beef daily. Water is readily lapped up. Information from native people, as reported by Wharton (1950), indicated that *Urogale* nests in the ground and in cliffs and feeds mainly in the morning. When several animals were placed in a dirt-floored cage, they "exhibited tendencies to root and dig like miniature pigs." *Urogale* usually sleeps curled up in a tight ball, sometimes with the top of its head under its body. It is a good climber and swift runner. Several calls are uttered: whimpering, snorting, chirping squeals, and a protracted distress call. In captivity, males tend to be more inquisitive than females.

Urogale has bred in several zoos (Wharton 1950). The gestation period is probably 54–56 days, the number of young is one or two, and the female is receptive to the male soon after giving birth. On one occasion the female would not allow the male to come near the nest for several days before a birth. One young male had a complete body covering of hair and opened its eyes 19 days after it was born. It climbed about the cage with considerable agility and shared its mother's meals in the fifth week. It weighed 123 grams at 6 weeks of age, 145 grams at 7 weeks, and 185 grams at 12 weeks. Jones (1982) reported that a captive specimen lived 11 years and 6 months.

Although *Urogale* has some tolerance to light habitat disturbance, such as selective logging, it does not do well in heavily disturbed forest. Much of its habitat below an elevation of 1,000 meters has been lost. However, while it was formerly classified as vulnerable by the IUCN and is on appendix 2 of the CITES, populations now are considered widespread and stable (Tabaranza et al. 2008).

SCANDENTIA; TUPAIIDAE; **Genus** *TUPAIA* Raffles, 1821

Tree Shrews

The systematics of *Tupaia* and some related genera are in a state of flux but the following sequence of 19 species is based on the reviews of, and the phylogenetic relationships suggested by, Corbet and Hill (1992), Han, Sheldon, and Stuebing (2000), Helgen (2005), Roberts, Sargis, and Olson (2009), Roberts et al. (2011), Sargis, Campbell, and Olson (2014), and Sargis et al. (2013a, 2013b, 2014):

T. belangeri, eastern Nepal and Bangladesh to southeastern China, Indochina, and the Malay Peninsula above the Isthmus of Kra, Hainan;

T. glis, Malay Peninsula below the Isthmus of Kra and many small nearby islands, Singapore;

T. longipes, northwestern Borneo;

T. salatana, southern Borneo;

T. discolor, Bangka Island off southeastern Sumatra;

T. ferruginea, Sumatra and Tanahbala Island to the west;

T. chrysogaster, Siberut, North and South Pagai, and Sipora Islands off western Sumatra;

T. hypochrysa, Java;

T. gracilis, lowland Borneo except southeast, and the islands of Banggi, Karimata, Belitung, and Bangka to the west;

T. dorsalis, lowland Borneo;

T. nicobarica, Great and Little Nicobar Islands in the Bay of Bengal;

T. javanica, western Sumatra and nearby Nias Island, Java, Bali;

T. palawanensis, Palawan and Balabac Islands in the western Philippines;

Northern tree shrew (*Tupaia belangeri*). Photo by Alexander F. Meyer at Gladys Porter Zoo, Brownsville, Texas.

T. moellendorffi, Busuanga, Culion, and Cuyo
Islands in the western Philippines;

T. minor, Malay Peninsula including peninsular
Thailand, Sumatra, Borneo and the nearby islands
of Lingga, Banggi, Balambangan, and Laut;

T. picta, parts of northwestern and eastern
Borneo;

T. tana, Sumatra, Bangka and Belitung Islands,
Borneo and the nearby islands of Tana Balu,
Tana Mara, Tuangku, Lingga, Banggi, Serasan,
Big Tambelan, and Bunoa;

T. montana, mountains of northwestern Borneo;

T. splendidula, lowlands of southern Borneo and
the nearby islands of Bunguran, Laut, Karimata, and Riabu.

Much study has been devoted to whether the genera *Anathana* and, especially, *Urogale* (see accounts thereof) should be included within *Tupaia*. It is possible that, as constituted above, *Tupaia* is paraphyletic—that is, inclusive only of species descended from a common ancestor but not all the species descended from that ancestor. Conversely, certain of the above species have sometimes been placed in another named genus or subgenus, *Lyonogale* Conisbee, 1953, though there also has been much disagreement about its status. *Lyonogale* was considered a full genus by Butler (1980), Luckett (1980), and R. D. Martin (1984) but was included in the synonymy of *Tupaia* by Corbet and Hill (1991, 1992), Helgen (2005), Olson, Sargis, and Martin (2004, 2005), Sargis, Campbell, and Olson (2014), and Yates (1984). Moreover, R. D. Martin (1984) considered *Lyonogale* to comprise the species *T. tana* and *T. dorsalis*, but Dene, Goodman, and Prychodko (1978) recognized it as a subgenus consisting of *T. montana*, *T. minor*, *T. tana*, and *T. palawanensis*. Han, Sheldon, and Stuebing (2000) found those same four species to form a related group, though did not expressly refer to *Lyonogale* as a subgenus. Olson, Sargis, and Martin (2004) provided a review of the history of the name *Lyonogale*.

Dene, Goodman, and Prychodko (1978) regarded *T. chinensis* of Thailand and southern China to be a species distinct from *T. belangeri*. Helgen (2005) treated *chinensis* as a subspecies of *T. belangeri* but noted that the two might be better recognized as closely related parapatric species. Helgen (2005) did not list *T. salatana*, *T. discolor*, *T. ferruginea*, and *T. hypochrysa*

as separate species, though he noted that many insular forms are distinctive. Sargis, Campbell, and Olson (2014) considered there to be insufficient evidence to recognize *T. moellendorffi* as a species separate from *T. palawanensis*, yet indicated that populations assigned to the former on Culion and Cuyo Islands are morphologically distinct. They also suggested that the population on Balabac, assigned to *T. palawanensis*, might possibly represent a different species.

Head and body length in *Tupaia* is about 120–230 mm, and tail length may be somewhat more or less than head and body length. Weight is about 50–300 grams. Emmons (2000) reported averages of about 208 mm in head and body length and 177 mm in tail length for the large, terrestrial *T. tana*, and about 128 mm and 160 mm, respectively, in the small, more arboreal *T. minor*. There is no difference in size between the males and females of a species. Compared with other tree shrew genera, *Tupaia* is scantily haired. The upper parts are ochraceous, reddish, olive, or shades of brown and gray to almost black. The underparts are whitish or buff to dark brown. A light shoulder stripe may be present. According to R. D. Martin (1984), the species *T. tana* and *T. dorsalis* are distinguished by a conspicuous black dorsal stripe, well-developed canine teeth, robust claws, and large size.

The generic name is derived from *tupai*, a Malay word for squirrels. *Tupaia* resembles a squirrel externally but can easily be distinguished by the longer nose and the absence of long black whiskers. *Tupaia* is distinguished from the other genera of tree shrews by the following features: the lower lobe of the ear is smaller than the upper part; the naked area on top of the nose is cut squarely across instead of being slightly prolonged backward in the midline; and the tail is covered by long hairs. Studies by Dr. Heinrich Spranckel, of Frankfurt, Germany, on the skin glands of various mammals have demonstrated the presence of a throat gland in *T. glis* comparable to the glands in certain insectivores and different from the glands in other primates. Female *Tupaia* have been reported to have two, four, or six mammae, depending on species (Emmons 2000; Sargis et al. 2013b).

These tree shrews are diurnal. Emmons (2000) found five species in Borneo to be active about 11–12 hours per day, with activity ending at around 1700–1800 in the evening; there were rest periods concen-

Common tree shrew (*Tupaia glis*). Photo by Kajornyot Wildlife Photography / Shutterstock.com.

trated towards the middle of the day. Average distance traveled per day ranged from 871 meters in *T. minor* to 1,973 meters in *T. longipes*. A few species seem to be mainly arboreal and shelter in nests some distance above the ground, but most members of this genus actually spend most of their time on the ground and in low bushes. Nests are commonly spherical and composed of leaves or other plant material. *T. tana* usually builds its nests in holes in living or fallen trees, *T. minor* in hollows of tall trees, *T. gracilis* in exposed clumps of vines and undergrowth, *T. longipes* in simple underground burrows, and *T. montana* in natural crevices or cavities under the moss mat; all of those species have multiple sleeping sites and move from one to another almost daily (Emmons 2000). Tree shrews seem constantly on the move, searching for food in all possible crevices. The diet consists mainly of insects but also includes other animal food, fruits, seeds, and leaves. Observations by Emmons (1991, 2000) indicate that frugivory is more significant than usually thought and that tree shrews have a distinctive fruit feeding behavior. They often spend long periods next

to a fruit tree in season or make multiple visits to it each day. They may hold food between their forepaws and sit on their haunches while eating. However, like fruit bats, they masticate fruits, suck the juices, and spit out the fibers in a wad.

Natural population density has been reported to be about 6–12 individuals per ha. in Thailand, 2–5 per ha. in peninsular Malaysia, 0.1–1.2 per ha. in Borneo, and 1.6–3.2 per ha. on Palawan; territoriality has been demonstrated in every species studied and is probable in others (Dans 1993; Emmons 2000; Langham 1982; Lekagul and McNeely 1977; Sorenson 1974). In a field study in Borneo, home range size of five species averaged 1.5–10.5 ha., but was not well correlated with body size; the largest species, *T. tana*, had a range smaller than those of three smaller species (Emmons 2000). Also in that field study, the basic social unit was found to consist of an adult male-female pair. In all species, known resident adults of the same sex had nonoverlapping territories. However, there were some significant behavioral differences between species. One pair of *T. minor* used almost perfectly coincident home ranges, with the animals meeting three or four times a day and sometimes foraging together, though radio-tracking showed that male-female pairs never slept together in the same nest. Monogamous pairs of *T. montana* also spent much time together in congruent territories, but male and female *T. gracilis* foraged alone and were rarely seen together, and male *T. longipes* appeared to extend their territories into the ranges of more than one female. Pairs of *T. tana* seemed more loosely associated than those of other species, with territories of the male and female coinciding only approximately and also overlapping the territory of at least one more individual of the same sex. Hence, polygamy and even polyandry seemed plausible.

In further studies of *T. tana*, Munshi-South, Emmons, and Bernard (2007) referred to the social system as "dispersed pair living." The territories of both male and female were spatially concordant, but both overlapped the territories of one to three other individuals of the same sex. Males extended their territories during periods of abundant fruiting. Average territorial size over four years was 6.9, 4.0, 5.5, and 5.0 ha. for adult males and 3.4, 3.5, 4.1, and 4.2 ha. for adult females. Subadults used smaller ranges. It was hypothesized that a female was drawn to pair with a male that could control a large and high-quality foraging territory, so that she could avoid foraging competition.

Observations in captivity indicate the pair bond between male and female is pronounced (Martin 1968), though wild males of some species may associate with more than one female (Emmons 2000). In a study of *T. glis* on Singapore, Kawamichi and Kawamichi (1979, 1982) found males to pair with 1–3 females. In that area there was little overlap between the home ranges of adult residents of the same sex, but those of opposite sexes overlapped completely, and a male's range sometimes included the ranges of more than one female. Aggressive chases were directed only against individuals of the same sex. Home range size in the study area averaged 10,174 sq meters for males and 8,809 sq meters for

Pygmy tree shrew (*Tupaia minor*). Photo by Klaus Rudloff at London Zoo.

females. Groups changed composition mainly through loss of juveniles, with young males departing sooner than females. In a study of *T. tana* on Borneo, Munshi-South (2008) used genetic samples to determine that young females dispersed from their natal areas at a significantly greater rate than males.

During extensive field studies in Borneo, Emmons (2000) noted only a few cases of violent territorial behavior or pronounced dominance. Although not reflecting natural behavior, observations of captives indicate that males and females of some species establish linear dominance hierarchies based mainly on aggressive interaction (Hasler and Sorenson 1974; Sorenson 1970, 1974). In captive groups of *T. glis*, one male despot was seen to harass and dominate all other males, and he was the only one to mate with the females present. In a captive group of *T. montana*, however, there were two top-ranking, mutually tolerant males. Eight distinct vocalizations have been identified in *Tupaia* (Binz and Zimmermann 1989). Those include loud "squeals" of aggression, modulating "screams" to indicate immediate danger, "chatters" in response to disturbance, and rhythmic "clucking" and "whistles" associated with courtship and mating. Emmons (2000) added that each species can be identified in the wild by its distinctive alarm calls, and that each has at least two types of calls that seem to signal different levels of arousal. *T. minor*, for example, makes a soft birdlike peeping when slightly aroused and a loud squirrel-like staccato chatter when more strongly alarmed. *T. tana* chatters when surprised at close range but makes birdlike whistles when spotting an intruder at a distance.

Breeding can occur year-round in some wild populations, but in others is apparently restricted to certain months depending on food supplies (Emmons 2000; Langham 1982). In peninsular Malaysia, Langham (1982) found the main reproductive season to extend from February to June. In the Danum Valley of northern Borneo, several species showed a broad breeding season from August to November and a second one from March to May, and females sometimes gave birth twice in each season; since normal litter size was two, a female could thus have up to eight young annually (Emmons 2000). In Singapore, Kawamichi and Kawamichi (1982) reported that *T. glis* probably is reproductively inactive from August through November and then has estrus in December, births in February

followed by a postpartum estrus, and more births in April. Some species of tree shrews are relatively easy to keep in captivity, and extensive observations have been made on reproductive activity therein. Females of *T. montana* exhibit a 9- to 12-day estrous cycle, a 23- to 29-day pseudopregnancy cycle, and a 49- to 51-day gestation period (Sorenson and Conway 1968). In *T. belangeri* the estrous cycle is 8–39 days, reported gestation is 40–52 days, and litter size is 1–3 (Gensch 1963; Hasler and Sorenson 1974; Mallinson 1974). Birth weight is about 10–12 grams (Hayssen, Van Tienhoven, and Van Tienhoven 1993). In the wild, in Borneo, young seem to reach adult size at about 8 months of age, and females reach sexual maturity at about 1 year (Emmons 2000). Record longevity for *Tupaia* is held by a captive *T. glis* that lived 12 years and 5 months (Jones 1982).

In a laboratory investigation of *T. belangeri* Martin (1968) found that there was usually a fertile postpartum estrus and that there was strong evidence for delayed implantation, with the blastocyst implanting about halfway through the interbirth interval. He observed the young to be born and reared in a separate "juvenile nest," while the parents slept in a "parental nest." The young apparently were visited and suckled by the female only once every 48 hours. The young emerged and moved to the parental nest to sleep at about 36 days, by which time they were weaned. Sexual maturity was attained at about 3 months in both sexes, and females produced their first young at 4.5 months. Emmons (2000) and Emmons and Biun (1991) demonstrated basically the same "absentee" system of maternal care in wild *T. tana* on Borneo. The mother gave birth to a litter in a nest within a tree hole and then usually visited the young every other morning. The young emerged from the nest at about 34 days, and the mother subsequently spent much more time with them; the male had no contact with the young. Such a system may serve to protect the young from predation or possibly allow the mother to save energy by avoiding a long and frequent commute to the nestlings.

As indicated above, all species of *Tupaia*, as well as other genera of Scandentia, are on appendix 2 of the CITES, reflecting historical affinity of the group to Primates. Because tree shrews are primate-like and relatively easy to maintain in the laboratory, there is increasing interest in using them, particularly *Tupaia belangeri*, for medical and biological research (Cao

Large tree shrew (*Tupaia tana*). Photo by Klaus Rudloff at London Zoo.

et al. 2003). Because of their susceptibility to infection with human hepatitis B virus, tree shrews have been used to establish human hepatitis virus-induced hepatitis and human HBV- and aflatoxin B1-associated HCC models. As they are phylogenetically close to primates and have a well-developed visual system and color vision in some species, they have been utilized to establish myopia models. And because subordinate males undergo dramatic behavioral, physiological, and neuroendocrine changes similar to those observed in depressed human patients, tree shrews have been successfully employed to experimentally study psychosocial stress. However, their potential currently remains limited because they cannot be bred on a large scale in cages.

All species of *Tupaia* are found within a region of expanding agriculture, logging, and other human activity and consequent destruction of forest habitat; most are likely to be declining. The IUCN classifies *T. chrysogaster* and *T. nicobarica* as endangered, as both

Mountain tree shrew (*Tupaia montana*). Photo by Tomáš Peš.

are restricted to small islands where habitat loss and fragmentation are rampant (Meijaard and MacKinnon 2008; Saha and Bhatta 2008). Additional IUCN accounts state that the following species are now decreasing and rare in at least some major parts of their ranges: *T. dorsalis* (Han and Stuebing 2008c), *T. gracilis*

(Han and Stuebing 2008d), *T. javanica* (Han and Maharadatunkamsi 2008), *T. longipes* (Han and Stuebing 2008e), *T. minor* (Han and Stuebing 2008f), *T. picta* (Han and Stuebing 2008g), *T. splendidula* (Han and Maryanto 2008), and *T. tana* (Han and Stuebing 2008h).

Until recently, the range of *T. glis* was thought to include Sumatra, Java, and a number of other islands. Even so, the species was believed to be decreasing and to have become rare, especially on Java, where only a single specimen has been recorded in the past 100 years

(Han 2008). Conservation concern has now been compounded by recognition that the populations of Sumatra and Tanahbala Island, Bangka Island, and Java are separate species—respectively, *T. ferruginea*, *T. discolor*, and *T. hypochrysa*—the last of which may be immediately vulnerable to extinction (Sargis et al. 2013a, 2013b, 2014). *T. moellendorffi*, which occurs on small islands mostly converted to agriculture and which has not been collected since 1962, is also of particular concern (Heaney 2008).

Order Dermoptera

Colugos, or Flying Lemurs

This order of gliding mammals, containing the single Recent family Cynocephalidae, with the two genera and species *Galeopterus variegatus* and *Cynocephalus volans*, is found in southeastern Asia and the East Indies. McKenna and Bell (1997) used the familial name "Galeopithecidae" in place of "Cynocephalidae." Use of the latter and the specific sequence here is in accordance with Corbet and Hill (1992), who, however, did not list *Galeopterus* as a genus separate from *Cynocephalus*. Recognition here of such a distinction follows Stafford (2005) and Stafford and Szalay (2000). The systematic position of order Dermoptera has long been contentious. It has been treated as a suborder of the defunct order Insectivora (Van Valen 1967), conjoined with order Chiroptera in the supraordinal clade Volitantia (Simmons 1995), considered a suborder of Primates (McKenna and Bell 1997), placed with Primates in the supraordinal Primatomorpha (Beard 1993), and combined with order Scandentia in the supraordinal Sundatheria, a group considered closely related to Primates and unrelated to Chiroptera (O'Leary et al. 2013). That last position is accepted herein. For further information and references on the issue, see the accounts of class Mammalia and order Scandentia.

Head and body length in the living Dermoptera is 340–420 mm, tail length is 175–270 mm, and weight is usually 1.0–1.75 kg. There is great variation in color and pattern. Reportedly, the upper parts of males tend to be some shade of brown and those of females tend to be grayish, especially in *Galeopterus variegatus*, but that is not consistent (Lim 2007). Scattered white spots are present on the upper parts of *G. variegatus*. The underparts of both genera are paler and without spots. The shaded and mottled color pattern blends well with the bark of trees.

Colugos have a large gliding membrane, the patagium, attached to the neck and sides of the body. The development of this membrane is greater than that found in other volant mammals, such as flying squirrels (Rodentia) and gliding possums (Diprotodontia), the gliding surfaces of which are stretched only between the limbs, with the digits of the four feet, and the tail left free. In Dermoptera the membrane extends along the limbs to the tips of the digits and tail. The extensive patagium aids greatly in midair maneuverability, allowing change in glide direction of up to about 60°; this ability seems important in dense tropical rainforest, where a straight gliding path is not always possible (Lim 2007).

The four limbs and tail are long and slender. The limbs are equal in length, the feet are broad, and all five digits are tipped with strong, sharp, recurved claws for climbing. The head is broad and somewhat doglike, the ears are short, and the eyes are large. The rims of the eye orbits are projecting, so the eyes are well protected. Vision is thought to be somewhat binocular, giving the depth perception needed when gliding toward a tree. Colugos appear to lack a tapetum lucidum, the structure responsible for "eye shine"; the reddish to yellowish

Sunda colugo (*Galeopterus variegatus*). Photo by Norman Lim.

glow seen in some flash photographs of colugos might be a reflection from the highly vascularized choroid, resembling the phenomenon of human "red eye" in flash photos (Moritz et al. 2013). Females have a single pair of mammae located at the sides of the body, almost in the "armpits."

The skull has a broad and flattened cranium, well-developed zygomatic arches and postorbital processes, flattened tympanic bullae, a broad and flat palate, and large incisive foramina. The dental formula is: i 2/3, c 1/1, pm 2/2, m 3/3 = 34. The first upper incisors are reduced and separated by a wide gap. The second upper incisors are caniniform and double rooted, a unique condition for mammals. The first two lower incisors on each side are developed into peculiar "comb teeth" that vaguely resemble the teeth of true lemurs (Primates). The difference is that each of the first two lower incisors of Dermoptera may have as many as 20 prongs radiating from one root, whereas in true lemurs each prong of the comb is a single tooth. In Dermoptera the first two lower incisors occlude with the edentulous palate between the first upper incisors. The third lower incisors on each side are different, having five cusps that decrease in size posteriorly. The comb teeth of Dermoptera may act as food strainers or as scrapers, or they may be used to groom the fur. Aimi and Inagaki (1988) thought both those functions probable in *Galeopterus*, but Wischusen (1990) observed no specialized use by *Cynocephalus* of

Philippine colugo (*Cynocephalus volans*). Photo by Sergey Volkov.

the lower incisors during foraging or grooming. Examination of specimens by Stafford and Szalay (2000) tended to support such a behavioral difference in the two genera; the tines of the incisors of *Cynocephalus* seemed too fragile for the functions indicated.

Colugos seem relatively common and can even be found in forested preserves and adjacent areas within Singapore (Agoramoorthy, Sha, and Hsu 2006; Lim 2007; Lim et al. 2013), but surprisingly little is known about them. They are considered totally arboreal, seldom if ever descending to the ground, where they are nearly helpless, cannot stand erect, and try to climb any object encountered. They are skillful though slow climbers, ascending unbranched vertical trunks in a series of lurches with head up and limbs spread to grasp the tree. When moving about on branches or while feeding, colugos are in an upside-down position. At such times the gliding membrane is drawn down under the forelegs so that it will not catch on branches. There is no indication that the tail is prehensile. Although occasionally seen by day, colugos are generally nocturnal. They come out of their shelters around dusk, climb a short distance up a tree, and glide off to seek food.

They usually glide to the same spots on the same food trees night after night. Each glide may cover 100 meters or more. One individual lost only 10.5–12.0 meters of elevation during a measured glide of 136 meters between two trees. Colugos utter a rasping cry that probably is an alarm call. Usually a single young is born, rarely two. The mother may leave the young in a nest tree or carry it with her while foraging. In the latter case the young clings with its claws and milk teeth to the belly fur and nipples of the mother.

Although the diet sometimes has been reported to include fruits, buds, and flowers, as well as leaves, the stomach contents of about eight specimens revealed only green vegetable matter. Captive specimens, forced to feed on fruit, took it most reluctantly. Colugos pull food within reach of the mouth and then bite off leaves or parts of fruit. In the wild they obtain water by licking wet leaves, bark, or moss; they do not go down to a stream or pond to drink (Dzulhelmi and Abdullah 2009). Colugos are difficult to keep in captivity and usually die quickly because of improper diet or dampness of the underparts, probably resulting from unsuitable cage facilities. There is an old report of an individual kept as a pet for 17 years and 6 months, after which it escaped, but that account cannot now be verified, and the precise source is unknown (but see account of *Cynocephalus*).

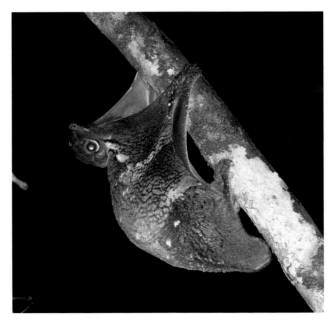

Sunda colugo (*Galeopterus variegatus*), tail turned down. Photo by Tomáš Peš.

The conservation status of Dermoptera is rather controversial, with some reports suggesting large and stable populations and others pointing to decreasing numbers in the face of massive deforestation (see generic accounts). Lim (2007) estimated around 100,000 animals left in the world and noted the possibility that the actual number could be lower due to habitat destruction and hunting. That estimate seems low in light of the status of populations in the Philippines (see account of *Cynocephalus*), but if it is accurate, the number of individuals of Dermoptera would probably be smaller than that of any other living order of mammals.

The family Cynocephalidae was long known only from the Recent of Asia, but Beard (1993) reported a fossil from the late Eocene of southern Thailand, and Marivaux et al. (2006) described additional material from the middle Eocene to late Oligocene of Thailand, Burma, and Pakistan. McKenna and Bell (1997) also listed the following fossil families as part of Dermoptera: Paromomydiae, early Paleocene to middle Eocene in North America, early to middle Eocene in Europe; Plagiomenidae, early Paleocene to late Oligocene in North America; and Mixodectidae, early Paleocene in North America. However, detailed morphological studies by MacPhee, Cartmill, and Rose (1989) did not confirm the affinity of Plagiomenidae, and Stafford and Szalay (2000) disagreed with inclusion of Paromomyidae in Dermoptera. There also has been disagreement about whether the Paromomyidae had a gliding ability (Krause 1991; Runestad and Ruff 1995).

DERMOPTERA; CYNOCEPHALIDAE; **Genus** *GALEOPTERUS* Thomas, 1908

Sunda Colugo

The single species usually recognized, *G. variegatus*, is found in Laos, Viet Nam, Cambodia, southern and peninsular Thailand, peninsular Burma, mainland Malaysia and many nearby islands including Langkawi to the west and Tioman and Aur to the east, Singapore, Sumatra and Banyak Islands to the west and the Lingga and Bangka islands to the east, Java, and Borneo and many nearby islands including some in the Riau, Natuna, and Anambas Archipelagoes (Boeadi and Steinmetz 2008; Corbet and Hill 1992; Francis 2008; Lekagul

and McNeely 1977; Lim 2007; Stafford and Szalay 2000). Presence of the species as far north as northern Viet Nam and Laos has only recently been confirmed, but reported occurrence in central Thailand is considered unlikely; records north of the Malay Peninsula are quite spotty (Lim 2007). A recent analysis of mitochondrial and nuclear DNA, as well as morphometric data, indicates *G. variegatus* should be divided into three species, with *G. peninsulae* of the Malay Peninsula and *G. borneanus* of Borneo showing closest affinity to one another, and *G. variegatus* of Java being somewhat farther removed (Janečka et al. 2008).

A number of "dwarfed forms" of *G. variegatus* have also sometimes been given specific names. Such

Sunda colugo (*Galeopterus variegatus*), mother holding young in patagium; the mother's tail is turned up to avoid soiling of interfemoral portion of the patagium. Photo by Norman Lim.

populations occur on Langkawi Island off the west coast and Aur Island off the east coast of the Malay Peninsula, Bakung and Sebangka Islands in the Lingga Archipelago off eastern Sumatra, Serasan Island in the Natuna Archipelago in the South China Sea, and Sebuku Island off the southeastern coast of Borneo (Stafford and Szalay 2000). Another dwarfed population was recently discovered in central Laos (Stalder et al. 2010). Specimens collected there are about 20 percent smaller than average *G. variegatus*, and it has been suggested they also represent a different species (Lim 2007). However, Stafford and Szalay (2000) concluded that all the dwarfed forms are not morphologically distinguishable from larger morphs of *G. variegatus*, other than in size, and do not warrant specific distinction.

Galeopterus was usually considered part of *Cynocephalus* until a detailed study of craniodental morphology by Stafford and Szalay (2000) showed the two to be highly distinctive at generic level. Notwithstanding its somewhat reduced incisors (see above ordinal

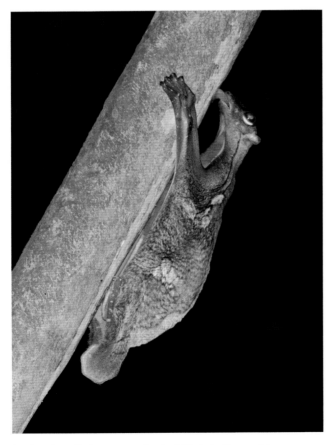

Sunda colugo (*Galeopterus variegatus*). Photo by Norman Lim.

account) and its smaller average body size, *Cynocephalus* has larger molar teeth than *Galeopterus*. It also has a broader and deeper rostrum, a greater degree of postorbital constriction, enhanced ectocranial ridges, and a deeper mandible, all of which are associated with more robust masticatory musculature. It appears adapted to a diet requiring a greater degree of shearing by the anterior (canine and premolar) dentition and crushing by the molariform dentition. The anterior shearing teeth are larger and more bladelike and have more robust roots than those of *Galeopterus*, and the bite force is more anteriorly directed. *Galeopterus* has a more gracile masticatory system and seems adapted to shredding or puncturing with its anterior dentition.

Head and body length is 330–420 mm, tail length is 175–270 mm, span of the gliding membrane is 700–1,200 mm, and weight is 925–1,750 grams; females are on average larger than males (Francis 2008; Kraft 1990; Lekagul and McNeely 1977). In one study of wild individuals in Singapore, females were found to be an average of 30 percent larger than males (Byrnes et al. 2011). The fur on the upper surface of the gliding membrane is mottled grayish brown, resembling the bark of a tree. The underside is paler and not spotted. There seem to be two color phases, to some extent expressing sexual dimorphism, with females tending to be gray and males tending to be brighter in color, usually some shade of brown or reddish brown. Gray individuals are said to be seen more often, possibly indicating a higher ratio of females. Lim (2007) examined 89 museum skins, finding all predominantly reddish-brown individuals to be males and all females to be grayish. However, 17 of the male skins also were grayish.

According to Lekagul and McNeely (1977), the Sunda colugo inhabits both lowland and mountainous areas and may be found in primary or secondary forests, coconut groves, and rubber plantations. In primary and secondary forest it spends the day in holes or hollows of trees about 25–50 meters above the ground, but in coconut plantations it curls up in a ball or hangs from a palm frond, with all four feet close together. Boeadi and Steinmetz (2008) noted that *Galeopterus* occurs in evergreen forest below 1,000 meters in elevation and has been reported from mangrove forest. While forest dependent, it can be found in secondary habitats close to human populations and may even breed on plantations in Thailand and Viet Nam. Kraft (1990) wrote that the

Sunda colugo (*Galeopterus variegatus*), mother gliding and carrying young. Photo by Norman Lim.

diet consists chiefly of the leaves of certain trees, but that buds, flowers, and fruit also are consumed, especially the flower buds of the coconut palm.

In a study of wild colugos in Singapore, Byrnes et al. (2011) found individuals of both sexes to be inactive for about 69 percent of a 24-hour day and to spend about 26 percent of the time foraging. Only 0.1 percent of the time involved gliding. Mean glide distance was 31.1 meters for males and 32.3 meters for females, notwithstanding the much heavier average weight of the latter; maximum glide distance was 145 meters. On average males glided 18.2 times and traveled 542.7 meters per night, while females glided only 8.9 times and moved 105.9 meters. Nonetheless, both sexes visited on average five food trees each night, and each foraged for equivalent periods of time. Females thus may concen-

trate their active time on foraging and acquiring resources, while males appear to spend additional time searching for mates or defending territories.

Population density in the forested preserves of Singapore was estimated at 0.535 individuals per ha. (Lim and Ng 2010). Several individuals may share the same shelter, but little is known about social life. MacKinnon (1984) noted that apart from mothers with young, animals move about singly, but several cover the same area and use the same feeding trees. Six independent animals were found in an area of less than 0.5 ha. on a coconut plantation in Java. Lim (2007) observed males fighting and chasing one another, and cited a report of five males fighting one another as they pursued a female. He suggested that a male might establish a territory large enough to encompass the home ranges of several females. He considered aggressive behavior among females nonexistent and observed up to three perching within a 5-meter radius without conflict. Males made a cracking or ripping sound during conflict. An animal caught and handled makes a wailing sound. Dzulhelmi and Abdullah (2009) distinguished several vocalizations, including a loud pitching noise made by a female in courtship and a squeaking call to other individuals, emitted at such times as before gliding.

Breeding appears to be nonseasonal, and there is normally a single young, but twins have been reported on occasion; a gestation period of 150–180 days has been inferred through field and laboratory study of *G. variegatus* (Lim 2007). Medway (1978) stated that lactating females with unweaned young have proved to be pregnant, and evidently births may follow in rapid succession. Lekagul and McNeely (1977) wrote that the young is born in a marsupial-like undeveloped state and that the gliding membrane of the mother can be folded into a soft, warm pouch to hold the young. When a female is pregnant or carrying her unweaned young on her underside (see above ordinal account), she experiences an increase in body mass of up to 300 grams, but this does not seem to impede her gliding capability; Lim (2007) witnessed glides of up to 20 meters in such circumstances. He also observed and photographed a reddish brown adult carrying a young, but doubted that males have an active role in parenting. He indicated a long-lasting bond between mother and young, though details on life history are not well understood.

Although *G. variegatus* is not in an IUCN threatened category, Boeadi and Steinmetz (2008) observed that overall populations are decreasing. The species is threatened by deforestation, especially in the northern part of its range, and by localized hunting in Java. It has been reported to occur now only in the western part of that island, though Lim (2007) mentioned a recent report from Lebakhajo in the east. He also reported that the species is hunted for meat and for parts used in Chinese medicine, and that it is a fairly easy target for poachers. Recent discovery that at least three populations of the Sunda colugo are genetically and morphologically divergent lineages and may represent distinct species (see above) indicates need for a reassessment of conservation priorities in tandem with revised taxonomy (Janečka et al. 2008).

DERMOPTERA; CYNOCEPHALIDAE; **Genus** *CYNOCEPHALUS* Boddaert, 1768

Philippine Colugo

The single species, *C. volans*, is found in the southern Philippines, where it has been specifically recorded from the islands of Basilan, Biliran, Bohol, Dinagat, Leyte, Maripipi, Mindanao, Samar, and Siargao (Heaney et al. 2010). The name *Galeopithecus* Pallas, 1783 has sometimes been used for this genus. *Galeopterus* (see account thereof) sometimes has been included in *Cynocephalus*.

Head and body length is 330–380 mm, tail length is 220–300 mm, span of the gliding membrane is about 700 mm, and weight is about 1,000–1,400 grams (Heaney et al. 2010; Kraft 1990). The upper parts are brown or gray-brown, males being darker than females, and the underparts are paler. Some individuals have a prominent yellowish blaze between the eyes and down the middle of the face (see photo in account of order Dermoptera). The fur is unusually soft. Although smaller, on average, than *Galeopterus*, the build of *Cynocephalus* is more robust, and it has a deeper mandible and larger teeth with more shearing surfaces and canine-like incisors (Lim 2007). More details on the differences between the two genera are found in the above accounts of Dermoptera and *Galeopterus*.

The Philippine colugo is common in primary and secondary forest, in mixed forest and orchards, and even in highly disturbed habitats; it occurs from sea level to about 500 meters elevation on small islands and up to about 1,250 meters on Mindanao (Heaney et al. 2010). It is nocturnal and crepuscular, is entirely arboreal, and commonly glides from tree to tree. In a radio-tracking study, Wischusen (1990) found most day roosts in thick foliage. He also observed that females carrying young usually transferred from tree to tree by moving through the branches rather than by gliding. Total nightly movements covered 1,011–1,764 meters, with the colugo feeding exclusively on young foliage. Wischusen and Richmond (1998) reported *C. volans* to typically have 12 foraging bouts per night, or 1 per hour during the period of activity, with a mean duration of 9.4 minutes. Animals were observed feeding on the young leaves of 35 of the 76 species of trees present in the study area and had a preference for larger trees.

Philippine colugo (*Cynocephalus volans*). Photos by Masahiro Iijima / ardea.com.

Wischusen (1990) determined monthly home range to measure 6.4–13.4 ha. and to overlap extensively. Allogrooming and other friendly interaction were observed between adults of opposite sexes and between adults and young, but adult males sometimes displayed hostility toward one another. Female *C. volans* were observed carrying young from December to June, and Wischusen (1990) suggested that breeding may continue throughout the year. Indeed, female *C. volans*, with young 223–254 mm long, have also been noted in March, April, and May. Gestation has been reported to last about 60 days, though observations by Wischusen (1990) suggest a period of approximately 150 days, perhaps involving delayed implantation. Litter size is one, rarely two. Birth weight is about 36 grams (Hayssen, Van Tienhoven, and Van Tienhoven 1993). Young *C. volans* are not weaned until about 6 months and may not attain full adult size until 3 years (Wischusen 1990; Wischusen, Ingle, and Richmond 1992). Although colugos are generally considered extremely difficult to keep in captivity, a female *C. volans* was obtained by the University of the Philippines, Los Baños, in September 1986 and was still living there in February 1991 after 4 years and 5 months (Weigl 2005).

Wischusen (1990) expressed concern about the very rapid loss of forested habitat within the range of *C. volans*, which is limited to a small and rapidly developing region, and believed it may be in danger of imminent extinction. He calculated its original numbers at 9,450,000 but estimated that only about 1,000,000 individuals survived, many of which were in isolated forest fragments with little chance of genetic exchange. Very slow rates of maturation and reproduction would compound the difficulty of recovery.

In contrast, Heaney et al. (2010) referred to *C. volans* as widespread and common, with populations generally stable. While somewhat vulnerable to habitat destruction, its ability to persist in heavily disturbed forest makes it more resilient than some other species. Although once classified as vulnerable by the IUCN, it is no longer in a threatened category. Gonzalez et al. (2008) did indicate that logging remains a threat and noted also that the fur of *C. volans* is used for making hats on Bohol and that the species is persecuted on Samar as a bad omen. Perhaps the greatest natural danger to *C. volans* is the so-called monkey-eating eagle (*Pithecophaga jefferyi*), the diet of which actually consists of 54–90 percent colugos (Lim 2007).

Literature Cited

Note: References with the same senior author and one or two coauthors are listed in alphabetical order. References with the same senior author and three or more coauthors (cited as "et al." in the text) are listed chronologically. References by the same single or senior author, regardless of whether or not that author used a middle name(s) or initial(s), are combined in one sequence.

Abba, A. M. 2011. Natural history of the screaming hairy armadillo *Chaetophractus vellerosus* (Mammalia: Xenarthra: Dasypodidae). Revista Chilena de Historia Natural 84:51–64.

Abba, A. M., and G. H. Cassini. 2008. Ecology and conservation of three species of armadillos in the Pampas region, Argentina. Pp. 300–305 in The biology of the Xenarthra (S. F. Vizcaíno and W. J. Loughry, eds.). University Press of Florida, Gainesville, Florida.

Abba, A. M., E. Cuéllar, and M. Superina. 2014. *Chaetophractus vellerosus*. The IUCN Red List of Threatened Species 2014: e.T4368A47438470. http://dx.doi.org/10.2305/IUCN.UK.2014-1.RLTS.T4368A47438470.en. Downloaded on 09 May 2015.

Abba, A. M., and E. Gonzalez. 2014. *Dasypus hybridus*. The IUCN Red List of Threatened Species 2014: e.T6288A47440329. http://dx.doi.org/10.2305/IUCN.UK.2014-1.RLTS.T6288A47440329.en. Downloaded on 09 May 2015.

Abba, A. M., E. Lima, and M. Superina. 2014. *Euphractus sexcinctus*. The IUCN Red List of Threatened Species 2014: e.T8306A47441708. http://dx.doi.org/10.2305/IUCN.UK.2014-1.RLTS.T8306A47441708.en. Downloaded on 07 May 2015.

Abba, A. M., S. Poljak, and M. Superina. 2014. *Chaetophractus villosus*. The IUCN Red List of Threatened Species 2014: e.T4369A47438745. http://dx.doi.org/10.2305/IUCN.UK.2014-1.RLTS.T4369A47438745.en. Downloaded on 09 May 2015.

Abba, A. M., D. E. Udrizar Sauthier, and S. F. Vizcaíno. 2005. Distribution and use of burrows and tunnels of *Chaetophractus villosus* (Mammalia, Xenarthra) in the eastern Argentinean pampas. Acta Theriologica 50:115–124.

Abba, A. M., and S. Vizcaíno. 2014. *Dasypus yepesi*. The IUCN Red List of Threatened Species 2014: e.T61924A47444043. http://dx.doi.org/10.2305/IUCN.UK.2014-1.RLTS.T61924A47444043.en. Downloaded on 06 May 2015.

Abba, A. M., S. Vizcaíno, and M. H. Cassini. 2007. Effects of land use on the distribution of three species of armadillos in the Argentinean Pampas. Journal of Mammalogy 88:502–507.

Abba, A. M., G. H. Cassini, G. Valverde, M. Tilak, S. F. Vizcaíno, M. Superina, and F. Delsuc. 2015. Systematics of hairy armadillos and the taxonomic status of the Andean hairy armadillo (*Chaetophractus nationi*). Journal of Mammalogy 96:673–689.

Abbott, I. 2001. The bilby *Macrotis lagotis* (Marsupialia: Peramelidae) in south-western Australia: original range limits, subsequent decline, and presumed regional extinction. Records of the Western Australian Museum 20:271–305.

Abbott, I. 2002. Origin and spread of the cat, *Felis catus*, on mainland Australia, with a discussion of the magnitude of its early impact on native fauna. Wildlife Research 29:51–74.

Abbott, I. 2006. Mammalian faunal collapse in Western Australia, 1875–1925: the hypothesised role of epizootic disease and a conceptual model of its origin, introduction, transmission, and spread. Australian Zoologist 33:530–561.

Abbott, I. 2008a. Historical perspectives of the ecology of some conspicuous vertebrate species in south-west Western Australia. Conservation Science Western Australia 6:1–214.

Abbott, I. 2008b. The spread of the cat, *Felis catus*, in Australia: re-examination of the current conceptual model with additional information. Conservation Science Western Australia 7:1–17.

Abbott, I. 2011. The importation, release, establishment, spread, and early impact on prey animals of the red fox *Vulpes vulpes* in Victoria and adjoining parts of south-eastern Australia. Australian Zoologist 35:463–533.

Abbott, I. 2012. Original distribution of *Trichosurus vulpecula* (Marsupialia: Phalangeridae) in Western Australia, with particular reference to occurrence outside the southwest. Journal of the Royal Society of Western Australia 95:83–93.

Abbott, I. 2013. Extending the application of Aboriginal names to Australian biota: *Dasyurus* (Marsupialia: Dasyuridae) species. Victorian Naturalist 130:109–126.

Abensperg-Traun, M. 1991a. A study of home range, movements, and shelter use in adult and juvenile echidnas, *Tachyglossus aculeatus* (Monotremata: Tachyglossidae), in Western Australian wheatbelt reserves. Australian Mammalogy 14:13–21.

Abensperg-Traun, M. 1991b. Survival strategies of the echidna *Tachyglossus aculeatus* Shaw 1792 (Monotremata: Tachyglossidae). Biological Conservation 58:317–328.

Abensperg-Traun, M., and E. S. De Boer. 1992. The foraging ecology of a termite- and ant-eating specialist, the echidna *Tachylossus aculeatus* (Monotremata: Tachyglossidae). Journal of Zoology 226:243–257.

Ackerman, B. B. 1995. Aerial surveys of manatees: a summary and progress report. Pp. 13–33 in Population biology of the Florida manatee (T. J. O'Shea, B. B. Ackerman, and H. F. Percival, eds.). National Biological Service, Information and Technology Report 1, United States Department of the Interior, Washington, DC.

Adam, P. J. 1999. *Choloepus didactylus*. Mammalian Species 621:1–8.

Agenbroad, L. D. 1984. New World mammoth distribution. Pp. 90–108 in Quaternary extinctions: a prehistoric revolution (P. S. Martin and R. G. Klein, eds.). University of Arizona Press, Tucson, Arizona.

Agenbroad, L. D. 1985. The distribution and chronology of mammoth in the New World. Acta Zoologica Fennica 170:221–224.

Agenbroad, L. D. 2005. North American proboscideans: mammoths: the state of knowledge, 2003. Quaternary International 126–128:73–92.

Aggundey, I. R. 1977. First record of *Potamogale velox* in Kenya. Mammalia 41:368.

Aggundey, I. R., and D. A. Schlitter. 1986. Annotated checklist of the mammals of Kenya. II. Insectivora and Macroscelidea. Annals of Carnegie Museum 55:325–347.

Agoramoorthy, G., C. M. Sha, and M. J. Hsu. 2006. Population, diet and conservation of Malayan flying lemurs in altered and fragmented habitats in Singapore. Biodiversity and Conservation 15:2177–2185.

Aguiar, J. M., and G. A. B. da Fonseca. 2008. Conservation status of the Xenarthra. Pp. 215–231 in The biology of the Xenarthra (S. F. Vizcaíno and W. J. Loughry, eds.). University Press of Florida, Gainesville, Florida.

Aiello, A. 1985. Sloth hair: unanswered questions. Pp. 213–218 in The evolution and ecology of armadillos, sloths, and vermilinguas (G. G. Montgomery, ed.). Smithsonian Institution Press, Washington, DC.

Aimi, M., and H. Inagaki. 1988. Grooved lower incisors in flying lemurs. Journal of Mammalogy 69:138–140.

Aitken, P. F. 1971a. Rediscovery of the large desert sminthopsis (*Sminthopsis psammophilus* Spencer) on Eyre Peninsula, South Australia. Victorian Naturalist 88:103–111.

Aitken, P. F. 1971b. The distribution of the hairy-nosed wombat [*Lasiorhinus latifrons* (Owen)]. Part I: Yorke Peninsula, Eyre Peninsula, the Gawler Ranges and Lake Harris. South Australian Naturalist 45:93–104.

Aitken, P. F. 1977. The little pygmy possum [*Cercartetus lepidus* (Thomas)] found living on the Australian mainland. South Australian Naturalist 51:63–66.

Aitken, P. F. 1979. The status of endangered Australian wombats, bandicoots, and the marsupial mole. Pp. 61–65 in The status of endangered Australian wildlife (M. J. Tyler, ed.). Proceedings of the Centenary Symposium of the Royal Zoological Society of South Australia, Adelaide, South Australia, Australia.

Aiyar, S., B. Barkbu, N. Batini, H. Berger, E. Detragiache, A. Dizioli, C. Ebeke, H. Lin, L. Kaltani, S. Sosa, A. Spilimbergo, and P. Topalova. 2016. The refugee surge in Europe: economic challenges. International Monetary Fund, Washington, DC.

Alacs, E. A., P. B. S. Spencer. P. J. de Tores, and S. L. Krauss. 2011. Population genetic structure of island and mainland populations of the quokka, *Setonix brachyurus* (Macropodidae): a comparison of AFLP and microsatellite markers. Conservation Genetics 12:297–309.

Albanese, M. S., G. M. Martin, P. Teta, and D. Flores. 2015. *Thylamys pallidior*. The IUCN Red List of Threatened Species 2015: e.T14888655A51222283. http://dx.doi.org/10.2305/IUCN.UK.2015-4.RLTS.T14888655A51222283.en. Downloaded on 27 December 2015.

Albayrak, E., and A. M. Lister. 2012. Dental remains of fossil elephants from Turkey. Quaternary International 276–277:198–211.

Albuja V., L., and B. D. Patterson. 1996. A new species of northern shrew-opossum (Paucituberculata: Caenolestidae) from the Cordillera del Condor, Ecuador. Journal of Mammalogy 77:41–53.

Alcover, J. A., X. Campillo, M. Macias, and A. Sans. 1998. Mammal species of the world: additional data on insular mammals. American Museum Novitates 3248:1–29.

Alers, M. P. T., A. Blom, C. Sikubwabo Kiyengo, T. Masunda, and R. F. W. Barnes. 1992. Preliminary assessment of the status of the forest elephant in Zaire. African Journal of Ecology 30:279–291.

Alfred, R., L. Ambu, S. K. S. S. Nathan, and B. Goossens. 2011. Current status of Asian elephants in Borneo. Gajah 35:29–35.

Alfred, R., A. H. Ahmad, J. Payne, C. Williams, L. N. Ambu, P. M. How, and B. Goossens. 2012. Home range and ranging behaviour of Bornean elephant (*Elephas maximus borneensis*) females. PLoS ONE 7(2): e31400. doi:10.1371/journal.pone.0031400.

Ali, R. 2005. An update on the elephants of Interview Island. Journal of the Bombay Natural History Society 102:221–223.

Allen, G. M. 1936. Zoological results of the George Vanderbilt African Expedition of 1934. Part II—the forest elephant of Africa. Proceedings of the Academy of Natural Sciences of Philadelphia 88:15–44.

Allen, G. M. 1939. A checklist of African mammals. Bulletin of the Museum of Comparative Zoology 83:1–763.

Allen, G. M. 1942. Extinct and vanishing mammals of the Western Hemisphere with the marine species of all the oceans. American Committee for International Wildlife Protection Special Publication 11:1–620.

Allen, J. A. 1905. Mammalia of southern Patagonia. Reports of the Princeton University Expeditions to Patagonia 3:1–210.

Allen, S., H. Marsh, and A. Hodgson. 2004. Occurrence and conservation of the dugong (Sirenia: Dugongidae) in New South Wales. Proceedings of the Linnean Society of New South Wales 125: 211–216.

Allen, W. R. 2006. Ovulation, pregnancy, placentation and husbandry in the African elephant (*Loxodonta africana*). Philosophical Transactions of the Royal Society B 361:821–834.

Alonso-Mejía, A., and R. A. Medellín. 1992. *Marmosa mexicana*. Mammalian Species 421:1–4.

Alroy, J. 1999. Putting North America's end-Pleistocene megafaunal extinction in context. Large-scale analyses of spatial patterns, extinction rates, and size distributions. Pp. 105–143 in Extinctions in near time. Causes, contexts, and consequences (R. D. E. MacPhee, ed.). Kluwer Academic/Plenum, New York, New York.

Alvarez-Alemán, A., C. A. Beck, and J. A. Powell. 2010. First report of a Florida manatee (*Trichechus manatus latirostris*) in Cuba. Aquatic Mammals 36:148–153.

Amrine-Madsen, H., M. Scally, M. Westerman, M. J. Stanhope, C. Krajewski, and M. S. Springer. 2003a. Nuclear gene sequences provide evidence for the monophyly of australidelphian marsupials. Molecular Phylogenetics and Evolution 28:186–196.

Amrine-Madsen, H., K.-P. Koepfli, R. K. Wayne, and M. S. Springer. 2003b. A new phylogenetic marker, apolipoprotein B, provides compelling evidence for eutherian relationships. Molecular Phylogenetics and Evolution 28:225–240.

Anacleto, T.C.S., F. Miranda, I. Medri, E. Cuellar, A. M. Abba, and M. Superina. 2014. *Priodontes maximus*. The IUCN Red List of Threatened Species 2014: e.T18144A47442343. http://dx.doi.org/10.2305/IUCN.UK.2014-1.RLTS.T18144A47442343.en. Downloaded on 20 January 2016.

Andanje, S., B. R. Agwanda, G. W. Ngaruiya, R. Amin, and G. B. Rathbun. 2010. Sengi (elephant-shrew) observations from northern coastal Kenya. Journal of East African Natural History 99:1–8.

Anderson, E. 1984. Who's who in the Pleistocene: a mammalian bestiary. Pp. 40–89 in Quaternary extinctions: a prehistoric revolution (P. S. Martin and R. G. Klein, eds.). University of Arizona Press, Tucson, Arizona.

Anderson, P. K. 1979. Dugong behavior: on being a marine mammalian grazer. Biologist 61:113–144.

Anderson, P. K. 1982a. Studies of dugongs at Shark Bay, Western Australia. I. Analysis of population size, composition, dispersion, and habitat use on the basis of aerial survey. Australian Wildlife Research 9:69–84.

Anderson, P. K. 1982b. Studies of dugongs at Shark Bay, Western Australia. II. Surface and subsurface observations. Australian Wildlife Research 9:85–99.

Anderson, P. K. 1984a. Dugong. Pp. 298–299 in The encyclopedia of mammals (D. Macdonald, ed.). Facts on File, New York, New York.

Anderson, P. K. 1984b. Suckling in *Dugong dugon*. Journal of Mammalogy 65:510–511.

Anderson, P. K. 1986. Dugongs of Shark Bay, Australia—seasonal migration, water temperature, and forage. National Geographic Research 2:473–490.

Anderson, P. K. 1995. Competition, predation, and the evolution and extinction of Steller's sea cow, *Hydrodamalis gigas*. Marine Mammal Science 11:391–394.

Anderson, P. K. 2002. Habitat, niche, and evolution of sirenian mating systems. Journal of Mammalian Evolution 9:55–98.

Anderson, P. K., and R. M. R. Barclay. 1995. Acoustic signals of solitary dugongs: physical characteristics and behavioral correlates. Journal of Mammalogy 76:1226–1237.

Anderson, P. K., and R. I. T. Prince. 1985. Predation on dugongs: attacks by killer whales. Journal of Mammalogy 66:554–556.

Anderson, R. P., and C. O. Handley, Jr. 2001. A new species of three-toed sloth (Mammalia: Xenarthra) from Panamá, with a review of the genus *Bradypus*. Proceedings of the Biological Society of Washington 114:1–33.

Anderson, S. 1997. Mammals of Bolivia, taxonomy and distribution. Bulletin of the American Museum of Natural History 231:1–652.

Anderson, T. J. C., A. J. Berry, J. N. Amos, and J. M. Cook. 1988. Spool-and-line tracking of the New Guinea bandicoot, *Echymipera kalubu* (Marsupialia, Peramelidae). Journal of Mammalogy 69:114–120.

Ansell, W. F. H. 1960. Mammals of Northern Rhodesia. Government Printer, Lusaka, Northern Rhodesia, Rhodesia.

Ansell, W. F. H. 1977. Order Proboscidea. Part 11 in The mammals of Africa: an identification manual (J. Meester and H. W. Setzer, eds.). Smithsonian Institution Press, Washington, DC.

Ansell, W. F. H., and P. D. H. Ansell. 1973. Mammals of the northeastern montane areas of Zambia. Puku 7:21–69.

Aplin, K. P. 1987. Basicranial anatomy of the Early Miocene diprotodontian *Wynyardia bassiana* (Marsupialia: Wynyardiidae) and its implications for wynyardiid phylogeny and classification. Pp. 369–391 in Possums and opossums (M. Archer, ed.). Surrey Beatty & Sons, Sydney, Australia.

Aplin, K. 2008. Suborder Phalangerida. Pp. 209–210, 222–224, 226, 237–238, 258, 261, and 265 in The mammals of Australia. 3rd edition (S. Van Dyck and R. Strahan, eds.). Reed New Holland, Sydney, Australia.

Aplin, K. 2015a. Family Notoryctidae (marsupial moles). Pp. 210–219 in Handbook of the mammals of the world, volume 5, Monotremes and marsupials (D. E. Wilson and R. A. Mittermeier, eds.). Lynx Edicions, Barcelona, Spain.

Aplin, K. 2015b. Family Acrobatidae (feather-tailed gliders and feather-tailed possum). Pp. 574–591 in Handbook of the mammals of the world, volume 5, Monotremes and marsupials (D. E. Wilson and R. A. Mittermeier, eds.). Lynx Edicions, Barcelona, Spain.

Aplin, K. P., and M. Archer. 1987. Recent advances in marsupial systematics with a new syncretic classification. Pp. xv–lxxii in Possums and opossums (M. Archer, ed.). Surrey Beatty & Sons, Sydney, Australia.

Aplin, K., C. Dickman, and L. Salas. 2008. *Dorcopsulus macleayi*. The IUCN Red List of Threatened Species 2008: e.T6801A12807686. http://dx.doi.org/10.2305/IUCN.UK.2008.RLTS.T6801 A12807686.en. Downloaded on 30 December 2015.

Aplin, K., and K. Helgen. 2008. *Spilocuscus wilsoni*. The IUCN Red List of Threatened Species 2008: e.T136443A4292460. http://dx.doi.org /10.2305/IUCN.UK.2008.RLTS.T136443A4292460.en. Downloaded on 30 December 2015.

Aplin, K. P, K. M. Helgen, and D. P. Lunde. 2010. A Review of *Peroryctes broadbenti*, the giant bandicoot of Papua New Guinea. American Museum Novitates 3696:1–41.

Aplin, K., and J. Pasveer. 2007. Mammals and other vertebrates from late Quaternary archaeological sites on Pulau Kobroor, Aru Islands, eastern Indonesia. Pp. 41–62 in The archaeology of the Aru Islands, eastern Indonesia, Terra Australis 22 (S. O'Connor, M. Spriggs, and P. Veth, eds.). Australian National University E Press, Canberra, Australia.

Aplin, K., J. Pasveer, and W. E. Boles. 1999. Late Quaternary vertebrates from the Bird's Head Peninsula, Irian Jaya, Indonesia, including descriptions of two previously unknown marsupial species. Records of the Western Australian Museum (Supplement) 57:351–387.

Aplin, K. P., and P. A. Woolley. 1993. Notes on the distribution and reproduction of the Papuan bandicoot *Microperoryctes papuensis* (Peroryctidae, Peramelemorphia). Science in New Guinea 19:109–112.

Aplin, K., C. Dickman, L. Salas, and K. Helgen. 2008a. *Tachyglossus aculeatus*. The IUCN Red List of Threatened Species 2008: e.T413 12A10439835. http://dx.doi.org/10.2305/IUCN.UK.2008.RLTS.T41 312A10439835.en. Downloaded on 06 July 2012.

Aplin, K., C. Dickman, L. Salas, J. Woinarski, and J. Winter. 2008b. *Macropus agilis*. The IUCN Red List of Threatened Species 2008: e.T40560A10318471. http://dx.doi.org/10.2305/IUCN.UK.2008 .RLTS.T40560A10318471.en. Downloaded on 05 January 2016.

Archer, M. 1974. New information about the Quaternary distribution of the thylacine (Marsupialia, Thylacinidae) in Australia. Journal of the Royal Society of Western Australia 57:43–50.

Archer, M. 1975. *Ningaui*, a new genus of tiny dasyurids (Marsupialia) and two species, *N. timealeyi* and *N. ridei*, from arid western Australia. Memoirs of the Queensland Museum 17:237–249.

Archer, M. 1976. Revision of the marsupial genus *Planigale* Troughton (Dasyuridae). Memoirs of the Queensland Museum 17:341–365.

Archer, M. 1977. Revision of the dasyurid marsupial genus *Antechinomys* Krefft. Memoirs of the Queensland Museum 18:17–29.

Archer, M. 1979a. The status of Australian dasyurids, thylacinids, and myrmecobiids. Pp. 29–43 in The status of endangered Australian wildlife (M. J. Tyler, ed.). Proceedings of the Centenary Symposium of the Royal Zoological Society of South Australia, Adelaide, South Australia, Australia.

Archer, M. 1979b. Two new species of *Sminthopsis* Thomas (Dasyuridae: Marsupialia) from northern Australia, *S. butleri* and *S. douglasi*. Australian Zoologist 20:327–345.

Archer, M. 1981. Results of the Archbold Expeditions. No. 104. Systematic revision of the marsupial dasyurid genus *Sminthopsis* Thomas. Bulletin of the American Museum of Natural History 168:61–224.

Archer, M. 1982a. Review of the dasyurid (Marsupialia) fossil record, integration of data bearing on phylogenetic interpretation, and suprageneric classification. Pp. 397–443 in Carnivorous marsupials (M. Archer, ed.). Royal Zoological Society of New South Wales, Mosman, New South Wales, Australia.

Archer, M. 1982b. A review of Miocene thylacinids (Thylacinidae, Marsupialia), the phylogenetic position of the Thylacinidae and the problem of apriorisms in character analysis. Pp. 445–476 in Carnivorous marsupials (M. Archer, ed.). Royal Zoological Society of New South Wales, Mosman, New South Wales, Australia.

Archer, M. 1984. The Australian marsupial radiation. Pp. 633–808 in Vertebrate zoogeography and evolution in Australasia (M. Archer and G. Clayton, eds.). Hesperian Press, Carlisle, Western Australia, Australia.

Archer, M. 1987. The relationships of the macropodoids (Marsupialia) and the polarity of some morphological features within the Phalangeriformes. Pp. 741–747 in Possums and opossums (M. Archer, ed.). Surrey Beatty & Sons, Sydney, Australia.

Archer, M. 1989. Emendations to the nomenclature of recently established palaeontological taxa. Australian Zoologist 25(3):67.

Archer, M., and S. J. Hand. 2006. The Australian marsupial radiation. Pp. 575–646 in Evolution and biogeography of Australasian vertebrates (J. R. Merrick, M. Archer, G. M. Hickey, and M. S. Y. Lee, eds.). Auscipub, Oatlands, New South Wales, Australia.

Archer, M., and J. A. W. Kirsch. 1977. The case for the Thylacomyidae and Myrmecobiidae (Gill, 1872), or why are marsupial families so extended? Proceedings of the Linnean Society of New South Wales 102:18–25.

Archer, M., and J. Kirsch. 2006. The evolution and classification of marsupials. Pp. 1–21 in Marsupials (P. J. Armati, C. R. Dickman, and I. D. Hume, eds.). Cambridge University Press, Cambridge, United Kingdom.

Archer, M., M. D. Plane, and N. S. Pledge. 1978. Additional evidence for interpreting the Miocene Obdurodon insignis (Woodburne and Tedford, 1975), to be a fossil platypus (Ornithorhynchidae: Monotremata) and a reconsideration of the status of Ornithorhynchus agilis (De Vis, 1885). Australian Zoologist 20:9–27.

Archer, M., T. F. Flannery, A. Ritchie, and R. E. Molnar. 1985. First Mesozoic mammal from Australia—an early Cretaceous monotreme. Nature 318:363–366.

Archer, M., F. A. Jenkins, Jr., S. J. Hand, P. Murray, and H. Godthelp. 1992. Description of the skull and non-vestigial dentition of a Miocene platypus (Obdurodon dicksoni n. sp.) from Riversleigh, Australia, and the problem of monotreme origins. Pp. 15–27 in Platypus and echidnas (M. L. Augee, ed.). Royal Zoological Society of New South Wales, Mosman, New South Wales, Australia.

Archer, M., P. Murray, S. Hand, and H. Godthelp. 1993. Reconsideration of monotreme relationships based on the skull and dentition of the Miocene Obdurodon dicksoni. Pp. 75–94 in Mammal phylogeny: Mesozoic differentiation, multituberculates, monotremes, early therians, and marsupials (F. S. Szalay, M. J. Novacek, and M. C. McKenna, eds.). Springer-Verlag, New York, New York.

Archer, M., H. Godthelp, M. Gott, Y. Wang, and A. Musser. 1999. The evolutionary history of notoryctids, yingabalanarids, yalkaparidontids and other enigmatic groups of Australian mammals. Australian Mammalogy 21:13–15.

Archer, M., R. Beck, M. Gott, S. Hand, H. Godthelp, and K. Black. 2011. Australia's first fossil marsupial mole (Notoryctemorphia) resolves controversies about their evolution and palaeoenvironmental origins. Proceedings of the Royal Society of London B Biological Sciences 278:1498–1506.

Archibald, J. D., and K. D. Rose. 2005. Womb with a view: the rise of placentals. Pp. 1–8 in The rise of placental mammals: origins and relationships of the major clades (K. D. Rose and J. D. Archibald, eds.). Johns Hopkins University Press, Baltimore, Maryland.

Archie, E. A., C. J. Moss, and S. C. Alberts. 2005. The ties that bind: genetic relatedness predicts the fission and fusion of social groups in wild African elephants. Proceedings of the Royal Society of London B Biological Sciences 273:513–522.

Archie, E. A., C. J. Moss, and S. C. Alberts. 2011. Friends and relations: kinship and the nature of female elephant social relationships. Pp. 238–245 in The Amboseli elephants: a long-term perspective on a long-lived mammal (C. J. Moss, H. Croze, and P. C. Lee, eds.). University of Chicago Press, Chicago, Illinois.

Archie, E. A., T. A. Morrison, C. A. H. Foley, C. J. Moss, and S. C. Alberts. 2006. Dominance rank relationships among wild female African elephants, Loxodonta africana. Animal Behaviour 71:117–127.

Archie, E. A., C. L. Fitzpatrick, C. J. Moss, and S. C. Alberts. 2011. The population genetics of the Amboseli and Kilimanjaro elephants. Pp. 38–47 in The Amboseli elephants: a long-term perspective on a long-lived mammal (C. J. Moss, H. Croze, and P. C. Lee, eds.). University of Chicago Press, Chicago, Illinois.

Ardente, N., D. Gettinger, R. Fonseca, H. de Godoy Bergallo, and F. Martins-Hatano. 2013. Mammalia, Didelphimorphia, Didelphidae, Glironia venusta (Thomas, 1912) and Chironectes minimus (Zimmermann, 1780): distribution extension for eastern Amazonia. Check List 9:1104–1107.

Argot, C. 2004. Evolution of South American mammalian predators (Borhyaenoidea): anatomical and palaeobiological implications. Zoological Journal of the Linnean Society 140:487–521.

Armbruster, P., and R. Lande. 1993. A population viability analysis for African elephant (Loxodonta africana): how big should reserves be? Conservation Biology 7:602–610.

Armstrong, D. M. 1972. Distribution of mammals in Colorado. Monograph of the Museum of Natural History, University of Kansas 3:1–415.

Armstrong, D. M., and J. K. Jones, Jr. 1971. Mammals from the Mexican state of Sinaloa. I. Marsupialia, Insectivora, Edentata, Lagomorpha. Journal of Mammalogy 52:747–757.

Armstrong, L. A., C. Krajewski, and M. Westerman. 1998. Phylogeny of the dasyurid marsupial genus Antechinus based on cytochrome-b, 12S-rRNA, and protamine-P1 genes. Journal of Mammalogy 79:1379–1389.

Arnason, U., and A. Janke. 2002. Mitogenomic analyses of eutherian relationships. Cytogenetic and Genome Research 96:20–32.

Arnason, U., J. A. Adegoke, A. Gullberg, E. H. Harley, A. Janke, and M. Kullberg. 2008. Mitogenomic relationships of placental mammals and molecular estimates of their divergences. Gene 421:37–51.

Arraut, E. M., M. Marmontel, J. E. Mantovani, E. M. L. M. Novo, D. W. Macdonald, and R. E. Kenward. 2010. The lesser of two evils: seasonal migrations of Amazonian manatees in the Western Amazon. Journal of Zoology 280:247–256.

Arrese, C., and P. B. Runham. 2002. Redefining the activity pattern of the honey possum (Tarsipes rostratus). Australian Mammalogy 23:169–172.

Arslanov, Kh. A., G. T. Cook, S. Gulliksen, D. D. Harkness, T. Kankainen, E. M. Scott, S. Vartanyan, and G. I. Zaitseva. 1998. Consensus dating of mammoth remains from Wrangel Island. Radiocarbon 40:289–294.

Arundel, J. H., I. K. Barker, and I. Beveridge. 1977. Diseases of marsupials. Pp. 141–154 in The biology of marsupials (B. Stonehouse and D. P. Gilmore, eds.). University Park Press, Baltimore, Maryland.

Asher, R. J. 1999. A morphological basis for assessing the phylogeny of the "Tenrecoidea" (Mammalia, Lipotyphla). Cladistics 15:231–252.

Asher, R. J. 2002. Phylogenetics of the Tenrecidae (Mammalia): a response to Douady et al., 2002. Molecular Phylogenetics and Evolution 26:328–330.

Asher, R. J. 2005. Insectivoran-grade placentals. Pp. 50–70 in The rise of placental mammals: origins and relationships of the major clades (K. D. Rose and J. D. Archibald, eds.). Johns Hopkins University Press, Baltimore, Maryland.

Asher, R. J. 2007. A web-database of mammalian morphology and a reanalysis of placental phylogeny. BMC Evolutionary Biology 7(108).

Asher, R. J. 2010. Tenrecoidea. Pp. 99–106 in Cenozoic mammals of Africa (L. Werdelin and W. J. Sanders, eds.). University of California Press, Berkeley, California.

Asher, R. J., N. Bennett, and T. Lehmann. 2009. The new framework for understanding placental mammal evolution. BioEssays 31:853–864.

Asher, R. J., and K. M. Helgen. 2010. Nomenclature and placental mammal phylogeny. BMC Evolutionary Biology 10:102.

Asher, R. J., and K. M. Helgen. 2011. High level mammalian taxonomy: a response to Hedges (2011). Zootaxa 3092:63–64.

Asher, R. J., and M. Hofreiter. 2006. Tenrec phylogeny and the noninvasive extraction of nuclear DNA. Systematic Biology 55:181–194.

Asher, R. J., I. Horovitz, and M. R. Sánchez-Villagra. 2004. First combined cladistic analysis of marsupial mammal interrelationships. Molecular Phylogenetics and Evolution 33:240–250.

Asher, R. J., M. J. Novacek, and J. H. Geisler. 2003. Relationships of endemic African mammals and their fossil relatives based on morphological and molecular evidence. Journal of Mammalian Evolution 10:131–194.

Asher, R. J., J. Meng, J. R. Wible, M. C. McKenna, G. W. Rougier, D. Dashzeveg, and M. J. Novacek. 2005. Stem lagomorphs and the antiquity of Glires. Science 307:1091–1094.

Asher, R. J., S. Maree, G. Bronner, N. C. Bennett, P. Bloomer, P. Czechowski, M. Meyer, and M. Hofreiter. 2010. A phylogenetic estimate for golden moles (Mammalia, Afrotheria, Chrysochloridae). BMC Evolutionary Biology (Supplement) 10(69):1–13.

Ashley, E., D. Lunney, J. Robertshaw, and R. Harden. 1990. Distribution and status of bandicoots in New South Wales. Pp. 43–50 in Bandicoots and bilbies (J. H. Seebeck, P. R. Brown, R. L. Wallis, and C. M. Kemper, eds.). Surrey Beatty & Sons, Chipping Norton, New South Wales, Australia.

Aslin, H. J. 1974. The behaviour of Dasyuroides byrnei (Marsupialia) in captivity. Zeitschrift für Tierpsychologie 35:187–208.

Aslin, H. J. 1975. Reproduction in Antechinus maculatus Gould (Dasyuridae). Australian Wildlife Research 2:77–80.

Aslin, H. J. 1980. Biology of a laboratory colony of Dasyuroides byrnei (Marsupialia: Dasyuridae). Australian Zoologist 20:457–471.

Aslin, H. J. 1983. Reproduction in Sminthopsis ooldea (Marsupialia: Dasyuridae). Australian Mammalogy 6:93–95.

Astúa, D. 2006. Range extension and first Brazilian record of the rare Hyladelphys kalinowskii (Hershkovitz, 1992) (Didelphimorphia, Didelphidae). Mammalia 70:174–176.

Astúa, D. 2015. Family Didelphidae (opossums). Pp. 70–186 in Handbook of the mammals of the world, volume 5, Monotremes and marsupials (D. E. Wilson and R. A. Mittermeier, eds.). Lynx Edicions, Barcelona, Spain.

Atherton, R. G., and A. T. Haffenden. 1982. Observations on the reproduction and growth of the long-tailed pygmy possum, Cercartetus caudatus (Marsupialia: Burramyidae), in captivity. Australian Mammalogy 5:253–259.

Atramentowicz, M. 1992. Optimal litter size: does it cost more to raise a large litter in Caluromys philander? Canadian Journal of Zoology 70:1511–1515.

Attard, M. R. G., U. Chamoli, T. L. Ferrara, T. L. Rogers, and S. Wroe. 2011. Skull mechanics and implications for feeding behaviour in a large marsupial carnivore guild: the thylacine, Tasmanian devil and spotted-tailed quoll. Journal of Zoology 285:292–300.

Augee, M. L. 1978. Monotremes and the evolution of homeothermy. Australian Zoologist 20:111–119.

Augee, M. L. 2008. Short-beaked echidna Tachyglossus aculeatus (Shaw, 1792). Pp. 37–39 in The mammals of Australia. 3rd edition (S. Van Dyck and R. Strahan, eds.). Reed New Holland, Sydney, Australia.

Augee, M. L., L. A. Beard, and G. C. Grigg. 1992. Home range of echidnas in the Snowy Mountains. Pp. 225–231 in Platypus and echidnas (M. L. Augee, ed.). Royal Zoological Society of New South Wales, Mosman, New South Wales, Australia.

Augee, M. L., T. J. Bergin, and C. Morris. 1978. Observations on behaviour of echidnas at Taronga Zoo. Australian Zoologist 20:121–129.

Augee, M. L., E. H. M. Ealey, and I. P. Price. 1975. Movements of echidnas, Tachyglossus aculeatus, determined by marking-recapture and radio-tracking. Australian Wildlife Research 2:93–101.

Augee, M. L., and B. A. Gooden. 1992. Evidence for electroreception from field studies of the echidna, Tachyglossus aculeatus. Pp. 211–215 in Platypus and echidnas (M. L. Augee, ed.). Royal Zoological Society of New South Wales, Mosman, New South Wales, Australia.

Aung, U. T., and U. T. Nyunt. 2002. The care and management of the domesticated Asian elephant in Myanmar. Pp. 89–102 in Giants on our hands. Proceedings of the International Workshop on the Domesticated Asian Elephant (I. Baker and M. Kashio, eds.). FAO Regional Office for Asia and the Pacific, Bangkok, Thailand.

Australasian Mammal Assessment Workshop. 2008a. Potorous platyops. The IUCN Red List of Threatened Species 2008: e.T18103A7659664. http://dx.doi.org/10.2305/IUCN.UK.2008.RLTS.T18103A7659664.en. Downloaded on 30 December 2015.

Australasian Mammal Assessment Workshop. 2008b. Caloprymnus campestris. The IUCN Red List of Threatened Species 2008: e.T362 6A9986323. http://dx.doi.org/10.2305/IUCN.UK.2008.RLTS.T362 6A9986323.en. Downloaded on 30 December 2015.

Australasian Mammal Assessment Workshop. 2008c. Lagorchestes leporides. The IUCN Red List of Threatened Species 2008: e. T11163A3258788. http://dx.doi.org/10.2305/IUCN.UK.2008.RLTS .T11163A3258788.en. Downloaded on 02 January 2016.

Australasian Mammal Assessment Workshop. 2008d. Macropus greyi. The IUCN Red List of Threatened Species 2008: e.T12625A3366597. http://dx.doi.org/10.2305/IUCN.UK.2008.RLTS.T12625A3366597.en. Downloaded on 05 January 2016.

Australian Department of Sustainability, Environment, Water, Population and Communities. 2011. Kangaroos and wallabies. Australian Department of Sustainability, Environment, Water, Population and Communities, Canberra, ACT, Australia. 1 April 2011. http://www.environment.gov.au/biodiversity/trade-use/wild-harvest/kangaroo/index.html.

Australian National Parks and Wildlife Service. 1977, 1978. Australian endangered species. Mammals, nos. 1–23.

Australian National Parks and Wildlife Service. 1988. Kangaroos in Australia: conservation status and management. Australian National Parks and Wildlife Service Occasional Papers 14:1–45.

Auty, J. 2004. Red plague grey plague: the kangaroo myths and legends. Australian Mammalogy 26:33–36.

Averianov, A. O., J. D. Archibald, and T. Martin. 2003. Placental nature of the alleged marsupial from the Cretaceous of Madagascar. Acta Palaeontologica Polonica 48:149–151.

Azmi, W., and D. Gunaryadi. 2011. Current status of Asian elephants in Indonesia. Gajah 35:55–61.

Babij, E. 2003 (Division of Scientific Authority, Fish and Wildlife Service, United States Department of the Interior). Endangered and Threatened Wildlife and Plants; determination of endangered status for the dugong (Dugong dugon) in the Republic of Palau. Federal Register 68:70185–70190.

Bacchus, M.-L. C., S. G. Dunbar, and C. Self-Sullivan. 2009. Characterization of resting holes and their use by the Antillean manatee (Trichechus manatus manatus) in the Drowned Cayes, Belize. Aquatic Mammals 35:62–71.

Bailey, P. 1992. A red kangaroo, Macropus rufus, recovered 25 years after marking in north-western New South Wales. Australian Mammalogy 15:141.

Baillie, J., and B. Groombridge. 1996. IUCN red list of threatened animals. IUCN, Gland, Switzerland.

Baillie, J. E. M., S. T. Turvey, and C. Waterman. 2009. Survival of Attenborough's long-beaked echidna Zaglossus attenboroughi in New Guinea. Oryx 43:146–148.

Bajpai, S., V. V. Kapur, J. G. M. Thewissen, B. N. Tiwari, and D. P. Das. 2005. First fossil marsupials from India: early Eocene Indodelphis n. gen. and Jaegeria n. gen. from Vastan Lignite Mine, District Surat, Gujarat, western India. Journal of the Palaeontological Society of India 50:147–151.

Baker, A. J., K. Lengel, K. McCafferty, and H. Hellmuth. 2005. Black-and-rufous sengi (Rhynchocyon petersi) at the Philadelphia Zoo. Afrotherian Conservation 3:6–7.

Baker, A. M. 2015. Family Dasyuridae (carnivorous marsupials). Pp. 232–348 in Handbook of the mammals of the world, volume 5, Monotremes and marsupials (D. E. Wilson and R. A. Mittermeier, eds.). Lynx Edicions, Barcelona, Spain.

Baker, A. M., T. Y. Mutton, and H. B. Hines. 2013. A new dasyurid marsupial from Kroombit Tops, south-east Queensland, Australia: the silver-headed antechinus, Antechinus argentus sp. nov. (Marsupialia: Dasyuridae). Zootaxa 3746:201–239.

Baker, A. M., T. Y. Mutton, and S. Van Dyck. 2012. A new dasyurid marsupial from eastern Queensland, Australia: the buff-footed antechinus, Antechinus mysticus sp. nov. (Marsupialia: Dasyuridae). Zootaxa 3515:1–37.

Baker, A. M., and S. Van Dyck. 2013a. Taxonomy and redescription of the fawn antechinus, Antechinus bellus (Thomas) (Marsupialia: Dasyuridae). Zootaxa 3613:201–228.

Baker, A. M., and S. Van Dyck. 2013b Taxonomy and redescription of the yellow-footed antechinus, Antechinus flavipes (Waterhouse) (Marsupialia: Dasyuridae). Zootaxa 3649:001–062.

Baker, A. M., and S. Van Dyck. 2013c. Taxonomy and redescription of the Atherton antechinus, Antechinus godmani (Thomas) (Marsupialia: Dasyuridae). Zootaxa 3670:401–439.

Baker, A. M., T. Y. Mutton, H. B. Hines, and S. Van Dyck. 2014. The black-tailed antechinus, Antechinus arktos sp. nov.: a new species of carnivorous marsupial from montane regions of the Tweed Volcano caldera, eastern Australia. Zootaxa 3765:101–133.

Baker, A. M., T. Y. Mutton, E. D. Mason, and E. L. Gray. 2015. A taxonomic assessment of the Australian dusky antechinus complex: a new species, the Tasman Peninsula dusky antechinus (Antechinus vandycki sp. nov.) and an elevation to species of the mainland dusky antechinus (Antechinus swainsonii mimetes (Thomas)). Memoirs of the Queensland Museum–Nature 59:75–126.

Baker, M. L., J. P. Wares, G. A. Harrison, and R. D. Miller. 2004. Relationships among the families and orders of marsupials and the major mammalian lineages based on recombination activating gene-1. Journal of Mammalian Evolution 11:1–16.

Baker, N. 2016. Papuan forest wallaby Dorcopsulus macleayi from Moro, Southern Highlands Province, Papua New Guinea. Southeast Asia Vertebrate Records 2016:16–18.

Banfield, A. W. F. 1974. The mammals of Canada. University of Toronto Press, Toronto, Ontario.

Banks, S. C., L. F. Skerratt, and A. C. Taylor. 2002. Female dispersal and relatedness structure in common wombats (Vombatus ursinus). Journal of Zoology 256:389–399.

Bannister, J. L., J. H. Calaby, L. J. Dawson, J. K. Ling, J. A. Mahoney, G. M. McKay, B. J. Richardson, W. D. L. Ride, and D. W. Walton. 1988. Zoological catalogue of Australia, volume 5, Mammalia. Australian Government Publication Service, Canberra, Australia.

Bargo, M. S., N. Toledo, and S. F. Vizcaíno. 2006. Muzzle of South American Pleistocene ground sloths (Xenarthra, Tardigrada). Journal of Morphology 267:248–263.

Barker, I. K., I. Beveridge, A. J. Bradley, and A. K. Lee. 1978. Observations on spontaneous stress-related mortality among males of the dasyurid marsupial Antechinus stuartii Macleay. Australian Journal of Zoology 26:435–447.

Barker, R. D., and G. Caughley. 1990. Distribution and abundance of kangaroos (Marsupialia: Macropodidae) at the time of European contact: Tasmania. Australian Mammalogy 13:157–166.

Barker, S. C. 1990. Behaviour and social organisation of the allied rock-wallaby Petrogale assimilis (Ramsay, 1877) (Marsupialia: Macropodoidea). Australian Wildlife Research 17:301–311.

Barkley, L. J. 2007. Genus Glironia O. (Thomas, 1912). Pp. 12–14 in Mammals of South America, volume 1, Marsupials, xenarthrans, shrews, and bats (A. L. Gardner, ed.). University of Chicago Press, Chicago, Illinois.

Barkley, L. J., and J. O. Whitaker, Jr. 1984. Confirmation of Caenolestes in Peru with information on diet. Journal of Mammalogy 65:328–330.

Barlow, J. C. 1967. Edentates and pholidotes. Pp. 178–191 in Recent mammals of the world: a synopsis of families (S. Anderson and J. K. Jones, Jr., eds.). Ronald Press, New York, New York.

Barlow, J. C. 1984. Xenarthrans and pholidotes. Pp. 219–239 in Orders and families of Recent mammals of the world (S. Anderson and J. K. Jones, Jr., eds.). John Wiley & Sons, New York, New York.

Barnes, R. F. W. 1996. The conflict between humans and elephants in the central African forests. Mammal Review 26:67–80.

Barnes, R. F. W. 1999. Is there a future for elephants in West Africa? Mammal Review 29:175–200.

Barnes, R. F. W., A. Blom, and M. P. T. Alers. 1995. A review of the status of forest elephants Loxodonta africana in central Africa. Biological Conservation 71:125–132.

Barnes, R. F. W., K. L. Barnes, M. P. T. Alers, and A. Blom. 1991. Man determines the distribution of elephants in the rain forests of northeastern Gabon. African Journal of Ecology 29:54–63.

Barnes, R. F. W., M. Agnagna, M. P. T. Alers, A. Blom, G. Doungoube, M. Fay, T. Masunda, J. C. Ndo Nkoumou, C. Sikubwabo Kiyengo, and M. Tchamba. 1993. Elephants and ivory poaching in the forests of equatorial Africa. Oryx 27:27–34.

Barnes, R. F. W., K. Beardsley, F. Michelmore, K. L. Barnes, M. P. T. Alers, and A. Blom. 1997. Estimating forest elephant numbers with dung counts and a geographic information system. Journal of Wildlife Management 61:1384–1393.

Barnett, A. 1991. Records of the grey-bellied shrew opossum, *Caenolestes caniventer*, and Tate's shrew-opossum, *Caenolestes tatei* (Caenolestidae, Marsupialia), from Ecuadorian montane forests. Mammalia 55:443–445.

Barnosky, A. D., and E. L. Lindsey. 2010. Timing of Quaternary megafaunal extinction in South America in relation to human arrival and climate change. Quaternary International 217:10–29.

Barriel, V., E. Thuet, and P. Tassy. 1999. Molecular phylogeny of Elephantidae. Extreme divergence of the extant forest elephant. Comptes Rendus, Life Sciences 322:447–454.

Barritt, M. K. 1978. Two further specimens of the little pigmy possum [*Cercartetus lepidus* (Thomas)] from the Australian mainland. South Australian Naturalist 53:12–13.

Barros, C. S., R. Crouzeilles, and F. A. S. Fernandez. 2008. Reproduction of the opossums *Micoureus paraguayanus* and *Philander frenata* in a fragmented Atlantic Forest landscape in Brazil: is seasonal reproduction a general rule for Neotropical marsupials? Mammalian Biology 73:463–467.

Barry, R. E., and P. J. Mundy. 1998. Population dynamics of two species of hyraxes in the Matobo National Park, Zimbabwe. African Journal of Ecology 36:221–233.

Barry, R. E., and P. J. Mundy. 2002. Seasonal variation in the degree of heterospecific association of two synoptic hyraxes (*Heterohyrax brucei* and *Procavia capensis*) exhibiting synchronous parturition. Behavioral Ecology and Sociobiology 52:177–181.

Barry, R. E., and J. Shoshani. 2000. *Heterohyrax brucei*. Mammalian Species 645:1–7.

Baskaran, N. 2013. An overview of Asian Elephants in the Western Ghats, southern India: implications for the conservation of Western Ghats ecology. Journal of Threatened Taxa 5:4854–4870.

Baskaran, N., S. Varma, C. K. Sar, and R. Sukumar. 2011. Current status of Asian elephants in India. Gajah 35:47–54.

Bass, D. K. 2010. Status of Dugong *Dugong dugon* and Australian snubfin dolphin *Orcaella heinsohni*, in the Solomon Islands. Pacific Conservation Biology 16:133–143.

Basson, M., J. R. Beddington, and R. M. May. 1991. An assessment of the maximum sustainable yield of ivory from African elephant populations. Mathematical Biosciences 104:73–95.

Baugh, T. 1987. Man and manatee: planning for the future. Endangered Species Technical Bulletin 12(9):7.

Baverstock, P. R. 1984. The molecular relationships of Australian possums and gliders. Pp. 1–8 in Possums and gliders (A. Smith and I. Hume, eds.). Surrey Beatty & Sons, Norton, New South Wales, Australia.

Baverstock, P. R., M. Adams, and M. Archer. 1984. Electrophoretic resolution of species boundaries in the *Sminthopsis murina* complex (Dasyuridae). Australian Journal of Zoology 32:823–832.

Baverstock, P. R., J. Birrell, and M. Krieg. 1987. Albumin immunologic relationships among Australian possums: a progress report. Pp. 229–334 in Possums and opossums (M. Archer, ed.). Surrey Beatty & Sons, Sydney, Australia

Baverstock, P. R., M. Archer, M. Adams, and B. J. Richardson. 1982. Genetic relationships among 32 species of Australian dasyurid marsupials. Pp. 641–650 in Carnivorous marsupials (M. Archer,

ed.). Royal Zoological Society of New South Wales, Mosman, New South Wales, Australia.

Baverstock, P. R., M. Adams, M. Archer, N. L. McKenzie, and R. How. 1983. An electrophoretic and chromosomal study of the dasyurid marsupial genus *Ningaui* Archer. Australian Journal of Zoology 31:381–392.

Baverstock, P. R., B. J. Richardson, J. Birrell, and M. Krieg. 1989. Albumin relationships of the Macropodidae (Marsupialia). Systematic Zoology 38:38–50.

Baverstock, P. R., T. Flannery, K. Aplin, J. Birrell, and M. Krieg. 1990. Albumin immunologic relationships of the bandicoots (Perameloidea: Marsupialia)—a preliminary report. Pp. 13–18 in Bandicoots and bilbies (J. H. Seebeck, P. R. Brown, R. L. Wallis, and C. M. Kemper, eds.). Surrey Beatty & Sons, Chipping Norton, New South Wales, Australia.

Baynes, A. 1982. Dasyurids (Marsupialia) in late Quaternary communities in southwestern Australia. Pp. 503–510 in Carnivorous marsupials (M. Archer, ed.). Royal Zoological Society of New South Wales, Mosman, New South Wales, Australia.

Baynes, A. 1984. Native mammal remains from Wilgie Mia aboriginal ochre mine: evidence on the pre-European fauna of the western arid zone. Records of the Western Australian Museum 11:297–310.

Beard, K. C. 1993. Phylogenetic systematics of the Primatomorpha, with special reference to Dermoptera. Pp. 129–150 in Mammal phylogeny: Mesozoic differentiation, multituberculates, monotremes, early therians, and marsupials (F. S. Szalay, M. J. Novacek, and M. C. McKenna, eds.). Springer-Verlag, New York, New York.

Beard, L. A., and G. C. Grigg. 2000. Reproduction in the short-beaked echidna, *Tachyglossus aculeatus*: field observations at an elevated site in south-east Queensland. Proceedings of the Linnean Society of New South Wales 122:89–99.

Beard, L. A., G. C. Grigg, and M. L. Augee. 1992. Reproduction by echidnas in a cold climate. Pp. 93–100 in Platypus and echidnas (M. L. Augee, ed.). Royal Zoological Society of New South Wales, Mosman, New South Wales, Australia.

Beaune, D., B. Fruth, L. Bollache, G. Hohmann, and F. Bretagnolle. 2013. Doom of the elephant-dependent trees in a Congo tropical forest. Forest Ecology and Management 295:109–117.

Beck, R. M. D. 2012. An "ameridelphian" marsupial from the early Eocene of Australia supports a complex model of Southern Hemisphere marsupial biogeography. Naturwissenschaften 99:715–729.

Beck, R. M. D., H. Godthelp, V. Weisbecker, M. Archer, and S. J. Hand. 2008. Australia's oldest marsupial fossils and their biogeographical implications. PLoS ONE 3(3): e1858. doi:10.1371/journal.pone.0001858.

Becker, C. 2005. Small numbers, large potential—new prehistoric finds of elephant and beaver from the Khabur River/Syria. Munibe (Antropologia-Arkeologia) 57:445–456.

Bedford, J. M., O. B. Mock, and S. M. Goodman. 2004. Novelties of conception in insectivorous mammals (Lipotyphla), particularly shrews. Biological Reviews 79:891–909.

Belcher, C. A. 1998. Susceptibility of the tiger quoll, *Dasyurus maculatus*, and the eastern quoll, *D. viverrinus*, to 1080-poisoned baits in control programmes for vertebrate pests in eastern Australia. Wildlife Research 25:33–40.

Belcher, C. A. 2003. Demographics of tiger quoll (*Dasyurus maculatus maculatus*) populations in south-eastern Australia. Australian Journal of Zoology 51:611–626.

Belcher, C., S. Burnett, and M. Jones. 2008. Spotted-tailed quoll, *Dasyurus maculatus*, Pp. 60–62 in The mammals of Australia. 3rd edition

(S. Van Dyck and R. Strahan, eds.). Reed New Holland, Sydney, Australia.

Belcher, C. A., and J. P. Darrant. 2004. Home range and spatial organization of the marsupial carnivore, *Dasyurus maculatus maculatus* (Marsupialia: Dasyuridae) in south-eastern Australia. Journal of Zoology 262:271–280.

Belcher, C. A., and J. P. Darrant. 2006. Den use by the spotted-tailed quoll, *Dasyurus maculatus*, in south-eastern Australia. Australian Mammalogy 28:59–64.

Belitsky, D. W., and C. L. Belitsky. 1980. Distribution and abundance of manatees *Trichechus manatus* in the Dominican Republic. Biological Conservation 17:313–319.

Belova, K., and L. Hellman. 2003. Immunoglobulin genetics of *Ornithorhynchus anatinus* (platypus) and *Tachyglossus aculeatus* (short-beaked echidna). Comparative Biochemistry and Physiology Part A 136:811–819.

Ben-Ami, D., K. Boom, L. Boronyak, C. Townend, D. Ramp, D. B. Croft, and M. Bekoff. 2014. The welfare ethics of the commercial killing of free-ranging kangaroos: an evaluation of the benefits and costs of the industry. Animal Welfare 23:1–10.

Bender, H. 2006. Structure and function of the eastern grey kangaroo (*Macropus giganteus*) foot thump. Journal of Zoology 268:415–422.

Benirschke, K. 2008. Reproductive parameters and placentation in anteaters and sloths. Pp. 160–171 in The biology of the Xenarthra (S. F. Vizcaíno and W. J. Loughry, eds.). University Press of Florida, Gainesville, Florida.

Bennett, M. B., and J. G. Garden. 2004. Locomotion and gaits of the northern brown bandicoot, *Isoodon macrourus* (Marsupialia: Peramelidae). Journal of Mammalogy 85:296–301.

Benoit, J., O. Maeva, and R. Tabuce. 2013. The petrosal of the earliest elephant-shrew *Chambius* (Macroscelidea: Afrotheria) from the Eocene of Djebel Chambi (Tunisia) and the evolution of middle and inner ear of elephant-shrews. Journal of Systematic Palaeontology 11:907–923.

Benoit, J., S. Adnet, E. El Mabrouk, H. Khayati, M. Ben Haj Ali, L. Marivaux, G. Merzeraud, S. Merigeaud, M. Vianey-Liaud, and R. Tabuce. 2013. Cranial remain from Tunisia provides new clues for the origin and evolution of Sirenia (Mammalia, Afrotheria) in Africa. PLoS ONE 8(1):e54307. doi:10.1371/journal.pone .0054307.

Benoit, J., N. Crumpton, S. Merigeaud, and R. Tabuce. 2014. Petrosal and bony labyrinth morphology supports paraphyly of *Elephantulus* within Macroscelididae (Mammalia, Afrotheria). Journal of Mammalian Evolution 21:173–193.

Benoit, J., T. Lehmann, M. Vatter, R. Lebrun, S. Merigeaud, L. Costeur, and R. Tabuce. 2015. Comparative anatomy and three-dimensional geometric-morphometric study of the bony labyrinth of Bibymalagasia (Mammalia, Afrotheria). Journal of Vertebrate Paleontology 35(3):e930043.

Benshemesh, J. 2004. Recovery plan for marsupial moles *Notoryctes typhlops* and *N. caurinus*, 2005–2010. Northern Territory Department of Infrastructure, Planning and Environment, Alice Springs, Australia.

Benshemesh, J. 2008. Itjaritjari *Notoryctes typhlops* (Stirling, 1889). Pp. 412–413 in The mammals of Australia. 3rd edition (S. Van Dyck and R. Strahan, eds.). Reed New Holland, Sydney, Australia.

Benshemesh, J. 2014. Backfilled tunnels provide a novel and efficient method of revealing an elusive Australian burrowing mammal. Journal of Mammalogy 95:1054–1063.

Benshemesh, J., and K. Aplin. 2008. Kakarratul *Notoryctes caurinus* (Thomas, 1920). Pp. 410–411 in The mammals of Australia. 3rd edition (S. Van Dyck and R. Strahan, eds.). Reed New Holland, Sydney, Australia.

Benshemesh, J., and A. Burbidge. 2008. *Notoryctes caurinus*. The IUCN Red List of Threatened Species 2008: e.T14878A4467686. http://dx .doi.org/10.2305/IUCN.UK.2008.RLTS.T14878A4467686.en. Downloaded on 03 January 2016.

Benshemesh, J., and K. Johnson. 2003. Biology and conservation of marsupial moles (*Notoryctes*). Pp. 464–474 in Predators with pouches: the biology of carnivorous marsupials (M. Jones, C. Dickman, and M. Archer, eds.). CSIRO Publishing, Collingwood, Victoria, Australia.

Benstead, J. P., K. H. Barnes, and C. M. Pringle. 2001. Diet, activity patterns, foraging movement and responses to deforestation of the aquatic tenrec *Limnogale mergulus* (Lipotyphla: Tenrecidae) in eastern Madagascar. Journal of Zoology 254:119–129.

Benstead, J. P., and L. E. Olson. 2003. *Limnogale mergulus*, web-footed tenrec or aquatic tenrec. Pp. 1267–1273 in The natural history of Madagascar (S. M. Goodman and J. P. Benstead, eds.). University of Chicago Press, Chicago, Illinois.

Bercovitch, F. B., J. R. Tobey, C. H. Andrus, and L. Doyle. 2006. Mating patterns and reproductive success in captive koalas (*Phascolarctos cinereus*). Journal of Zoology 270:512–516.

Best, R. C. 1982. Seasonal breeding in the Amazonian manatee, *Trichechus inunguis* (Mammalia: Sirenia). Biotrópica 14:76–78.

Best, R. C. 1983. Apparent dry-season fasting in Amazonian manatees (Mammalia: Sirenia). Biotrópica 15:61–64.

Bethge, P., S. Munks, H. Otley, and S. Nicol. 2003. Diving behaviour, dive cycles and aerobic dive limit in the platypus *Ornithorhynchus anatinus*. Comparative Biochemistry and Physiology Part A 136:799–809.

Bethge, P., S. Munks, H. Otley, and S. Nicol. 2009. Activity patterns and sharing of time and space of platypuses, *Ornithorhynchus anatinus*, in a subalpine Tasmanian lake. Journal of Mammalogy 90:1350–1356.

Beyers, R. L., J. A. Hart, A. R. E. Sinclair, F. Grossmann, B. Klinkenberg, and S. Dino. 2011. Resource wars and conflict ivory: the impact of civil conflict on elephants in the Democratic Republic of Congo—the case of the Okapi Reserve. PLoS ONE 6(11):e27129. doi:10.1371/journal.pone.0027129.

Bezerra, B. M., A. da Silva Souto, L. G. Halsey, and N. Schiel. 2008. Observation of brown-throated three-toed sloths: mating behaviour and the simultaneous nurturing of two young. Journal of Ethology 26:175–178.

Bhagratty, H., K. Taylor, A. Lawrence, E. S. Devenish-Nelson, and H. P. Nelson. 2013. Population density of silky anteaters (*Cyclopes didactylus* Xenarthra: Cyclopedidae) in a protected mangrove swamp on the island of Trinidad. Mammalia 77:447–450.

Bilton, A. D., and D. B. Croft. 2004. Lifetime reproductive success in a population of female red kangaroos *Macropus rufus* in the sheep rangelands of western New South Wales: environmental effects and population dynamics. Australian Mammalogy 26:45–60.

Bininda-Emonds, O. R. P., M. Cardillo, K. E. Jones, R. D. E. MacPhee, R. M. D. Beck, R. Grenyer, S. A. Price, R. A. Vos, J. L. Gittleman, and A. Purvis. 2007. The delayed rise of present-day mammals. Nature 446:507–512.

Bininda-Emonds, O. R. P., M. Cardillo, K. E. Jones, R. D. E. MacPhee, R. M. D. Beck, R. Grenyer, S. A. Price, R. A. Vos, J. L. Gittleman, and A. Purvis. 2008. Corrigendum: The delayed rise of present-day mammals. Nature 456:274. http://www.molekularesystematik .unioldenburg.de/download/Publications/mammalST_mainText _highlight.pdf. Accessed on 25 April 2009.

Binladen, J., M. T. P. Gilbert, and E. Willerslev. 2007. 800,000 year old mammoth DNA, modern elephant DNA or PCR artefact? Biology Letters 3:55–57.

Binz, H., and E. Zimmermann. 1989. The vocal repertoire of adult tree shrews (*Tupaia belangeri*). Behaviour 109:142–162.

Birney, E. C., J. A. Monjeau, C. J. Phillips, R. S. Sikes, and I. Kim. 1996. *Lestodelphys halli*: new information on a poorly known Argentine marsupial. Mastozoologia Neotropical 3:171–181.

Bishop, C. W. 1921. The elephant and its ivory in ancient China. Journal of the American Oriental Society 41:290–306.

Bist, S. S., J. V. Cheeran, S. Choudhury, P. Barua, and M. K. Misra. 2002. The domesticated Asian elephant in India. Pp. 129–148 in Giants on our hands. Proceedings of the International Workshop on the Domesticated Asian Elephant (I. Baker and M. Kashio, eds.). FAO Regional Office for Asia and the Pacific, Bangkok, Thailand.

Black, K. H., M. Archer, and S. J. Hand. 2012. New Tertiary koala (Marsupialia, Phascolarctidae) from Riversleigh, Australia, with a revision of phascolarctid phylogenetics, paleoecology, and paleobiodiversity. Journal of Vertebrate Paleontology 32:125–138.

Black, K. H., G. J. Price, M. Archer, and S. J. Hand. 2014. Bearing up well? Understanding the past, present and future of Australia's koalas. Gondwana Research 25:1186–1201.

Blacket, M. J., C. Kemper, and R. Brandle. 2008. Planigales (Marsupialia: Dasyuridae) of eastern Australia's interior: a comparison of morphology, distributions and habitat preferences, with particular emphasis on South Australia. Australian Journal of Zoology 56:195–205.

Blacket, M. J., C. Krajewski, A. Labrinidis, B. Cambron, S. Cooper, and M. Westerman. 1999. Systematic relationships within the dasyurid marsupial tribe Sminthopsini—a multigene approach. Molecular Phylogenetics and Evolution 12:140–155.

Blacket, M. J., M. Adams, C. Krajewski, and M. Westerman. 2000. Genetic variation within the dasyurid marsupial genus *Planigale*. Australian Journal of Zoology 48:443–459.

Blacket, M. J., M. Adams, S. J. B. Cooper, C. Krajewski, and M. Westerman. 2001. Systematics and evolution of the dasyurid marsupial genus *Sminthopsis*: I. The Macroura species group. Journal of Mammalian Evolution 8:149–170.

Blacket, M. J., S. J. B. Cooper, C. Krajewski, and M. Westerman. 2006. Systematics and evolution of the dasyurid marsupial genus *Sminthopsis*: II. The Murina species group. Journal of Mammalian Evolution 13:125–138.

Blake, S. 2002. The ecology of forest elephant distribution and its implications for conservation. Ph.D. thesis, University of Edinburgh, Edinburgh, United Kingdom.

Blake, S. 2004. Do leopards kill forest elephants? Evidence from northern Congo. Mammalia 68:225–227.

Blake, S., I. Douglas-Hamilton, and W. B. Karesh. 2001. GPS telemetry of forest elephants in central Africa: results of a preliminary study. African Journal of Ecology 39:178–186.

Blake, S., and S. Hedges. 2004. Sinking the flagship: the case of forest elephants in Asia and Africa. Conservation Biology 18:1191–1202.

Blake, S., and C. Inkamba-Nkulu. 2004. Fruit, minerals, and forest elephant trails: do all roads lead to Rome? Biotropica 36:392–401.

Blake, S., P. Bouché, H. Rasmussen, A. Orlando, and I. Douglas-Hamilton. 2003. The last Sahelian elephants. Ranging behavior, population status and recent history of the desert elephants of Mali. Save the Elephants, Nairobi, Kenya.

Blake, S., S. Strindberg, P. Boudjan, C. Makombo, I. Bila-Isia, O. Ilambu, F. Grossmann, L. Bene-Bene, B. De Semboli, V. Mbenzo, D. S'hwa, R. Bayogo, L. Williamson, M. Fay, J. Hart, and F. Maisels. 2007. Forest elephant crisis in the Congo Basin. PLoS Biology 5(4):e111. doi:10.1371/journal.pbio.0050111.

Blake, S., S. L. Deem, S. Strindberg, F. Maisels, L. Momont, I.-B. Isia, I. Douglas-Hamilton, W. B. Karesh, and M. D. Kock. 2008. Road-less wilderness area determines forest elephant movements in the Congo Basin. PLoS ONE 3(10):e3546. doi:10.1371/journal.pone .0003546.

Blake, S., S. L. Deem, E. Mossimbo, F. Maisels, and P. Walsh. 2009. Forest elephants: tree planters of the Congo. Biotropica 41:459–468.

Blanc, J. 2008. *Loxodonta africana*. The IUCN Red List of Threatened Species 2008: e.T12392A3339343. http://dx.doi.org/10 .2305/IUCN.UK.2008.RLTS.T12392A3339343.en. Downloaded on 18 July 2012.

Blanc, J. J., C. R. Thouless, J. A. Hart, H. T. Dublin, I. Douglas-Hamilton, G. C. Craig, and R. F. W. Barnes. 2003. African elephant status report 2002: an update from the African Elephant Database. IUCN, Gland, Switzerland.

Blanc, J. J., R. F. W. Barnes, G. C. Craig, H. T. Dublin, C. R. Thouless, I. Douglas-Hamilton, and J. A. Hart. 2007. African elephant status report 2007: an update from the African Elephant Database. IUCN, Gland, Switzerland.

Blanco, R. E., W. W. Jones, and N. Milne. 2013. Is the extant southern short-tailed opossum a pigmy sabretooth predator? Journal of Zoology 291:100–110.

Bloomer, P. 2009. Extant hyrax diversity is vastly underestimated. Afrotherian Conservation 7:11–16.

Bluff, L. A., L. Clausen, A. Hill, and M. D. Bramwel. 2011. A decade of monitoring the remnant Victorian population of the brush-tailed rock-wallaby (*Petrogale penicillata*). Australian Mammalogy 33:195–201.

Boeadi, and R. Steinmetz. 2008. *Galeopterus variegatus*. The IUCN Red List of Threatened Species 2008:e.T41502A10479343. http://dx.doi .org/10.2305/IUCN.UK.2008.RLTS.T41502A10479343.en. Downloaded on 01 June 2015.

Böer, M. 1998. Observations on reproduction in the common wombat *Vombatus ursinus* in captivity. Pp. 129–146 in Wombats (R. T. Wells, P. A. Pridmore, B. St. John, M. D. Gaughwin, and J. Ferris, eds.). Surrey Beatty & Sons, Chipping Norton, New South Wales, Australia.

Boily, P. 2008. The use of armadillo clones from the genus *Dasypus* as experimental models to investigate the source of physiological variability. Pp. 126–129 in The biology of the Xenarthra (S. F. Vizcaíno and W. J. Loughry, eds.). University Press of Florida, Gainesville, Florida.

Boisserie, J.-R., F. Lihoreau, M. Orliac, R. E. Fisher, E. M. Weston, and S. Ducrocq. 2010. Morphology and phylogenetic relationships of the earliest known hippopotamids (Cetartiodactyla, Hippopotamidae, Kenyapotaminae). Zoological Journal of the Linnean Society 158:325–366.

Bökönyi, S. 1985. Subfossil elephant remains from Southwestern Syria. Paléorient 11:161–163.

Bolkovic, M. L., S. M. Caziani, and J. J. Protomastro. 1995. Food habits of the three-banded armadillo (Xenarthra: Dasypodidae) in the dry Chaco, Argentina. Journal of Mammalogy 76:1199–1204.

Bollinger, A., and T. C. Backhouse. 1960. Blood studies on the echidna *Tachyglossus aculeatus*. Proceedings of the Zoological Society of London 135:91–97.

Bolton, B. L., and P. K. Latz. 1978. The western hare-wallaby, *Lagorchestes hirsutus* (Gould) (Macropodidae), in the Tanami Desert. Australian Wildlife Research 5:285–293.

Bonato, V., E. G. Martins, G. Machado, C. Q. da-Silva, and S. F. dos Reis. 2008. Ecology of the armadillos *Cabassous unicinctus* and *Euphractus sexcinctus* (Cingulata: Dasypodidae) in a Brazilian Cerrado. Journal of Mammalogy 89:168–174.

Borghi, C. E., C. M. Campos, S. M. Giannoni, V. E. Campos, and C. Sillero-Zubiri. 2011. Updated distribution of the pink fairy armadillo, *Chlamyphorus truncatus* (Xenarthra, Dasypodidae), the world's smallest armadillo. Edentata 12:14–19.

Borobia, M., and L. Lodi. 1992. Recent observations and records of the West Indian manatee *Trichechus manatus* in northeastern Brazil. Biological Conservation 59:37–43.

Bos, D. G., and S. M. Carthew. 2001. Population ecology of *Ningaui yvonneae* (Dasyuridae: Marsupialia) in the Middleback Ranges, Eyre Peninsula, South Australia. Wildlife Research 28:507–515.

Bos, D. G., and S. M. Carthew. 2003. The influence of behaviour and season on habitat selection by a small mammal. Ecography 26:810–820.

Bothma, J. du P. 1977. Order Hyracoidea. Part 12 in The mammals of Africa: an identification manual (J. Meester and H. W. Setzer, eds.). Smithsonian Institution Press, Washington, DC.

Bouché, P., I. Douglas-Hamilton, G. Wittemyer, A. J. Nianogo, J.-L. Doucet, P. Lejeune, C. Vermeulen. 2011. Will elephants soon disappear from West African savannahs? PLoS ONE 6(6):e20619. doi:10.1371/journal.pone.0020619.

Bourke, D. W. 1989. Observations on the behaviour of the grey dorcopsis wallaby, *Dorcopsis luctuosa* (Marsupialia: Macropodidae), in captivity. Pp. 633–640 in Kangaroos, wallabies and rat-kangaroos (G. C. Grigg, P. Jarman, and I. Hume, eds.). Surrey Beatty & Sons, Chipping Norton, New South Wales, Australia.

Bowen, M., and R. Goldingay. 2000. Distribution and status of the eastern pygmy possum (*Cercartetus nanus*) in New South Wales. Australian Mammalogy 21:153–164.

Bown, T. M., and E. L. Simons. 1984. First record of marsupials (Metatheria: Polyprotodontia) from the Oligocene in Africa. Nature 308:447–449.

Bowyer, J. C., G. R. Newell, C. J. Metcalfe, and M. B. D. Eldridge. 2003. Tree-kangaroos *Dendrolagus* in Australia: are *D. lumholtzi* and *D. bennettianus* sister taxa? Australian Zoologist 32:207–213.

Bozinovic, F., G. Ruiz, and M. Rosenmann. 2004. Energetics and torpor of a South American "living fossil," the microbiotheriid *Dromiciops gliroides*. Journal of Comparative Physiology B 174:293–297.

Bradley, A. J. 1997. Reproduction and life history in the red-tailed phascogale, *Phascogale calura* (Marsupialia: Dasyuridae): the adaptive-stress senescence hypothesis. Journal of Zoology 241:739–755.

Bradley, A. J. 2003. Stress, hormones and mortality in small carnivorous marsupials. Pp. 254–267 in Predators with pouches: the biology of carnivorous marsupials (M. Jones, C. Dickman, and M. Archer, eds.). CSIRO Publishing, Collingwood, Victoria, Australia.

Bradley, A. J., W. K. Foster, and D. A. Taggart. 2008. Red-tailed phascogale *Phascogale calura* (Gould, 1844). Pp. 101–102 in The mammals of Australia. 3rd edition (S. Van Dyck and R. Strahan, eds.). Reed New Holland, Sydney, Australia.

Bradshaw, C. J. A., and B. W. Brook. 2005. Disease and the devil: density-dependent epidemiological processes explain historical population fluctuations in the Tasmanian devil. Ecography 28:181–190.

Bradshaw, D., and F. Bradshaw. 2012. The physiology of the honey possum, *Tarsipes rostratus*, a small marsupial with a suite of highly specialised characters: a review. Journal of Comparative Physiology B 182:469–489.

Bradshaw, S. D., R. D. Phillips, S. Tomlinson, R. S. Holley, S. Jennings, and F. J. Bradshaw. 2007. Ecology of the honey possum, *Tarsipes rostratus*, in Scott National Park, Western Australia. Australian Mammalogy 29:25–38.

Brandt, A. L., Y. Hagos, Y. Yacob, V. A. David, N. J. Georgiadis, J. Shoshani, and A. L. Roca. 2014. The elephants of Gash-Barka, Eritrea: nuclear and mitochondrial genetic patterns. Journal of Heredity 105:82–90.

Brattstrom, B. H. 1973. Social and maintenance behavior of the echidna, *Tachyglossus aculeatus*. Journal of Mammalogy 54:50–70.

Braun, J. K., N. L. Pratt, and M. A. Mares. 2010. *Thylamys pallidior* (Didelphimorphia: Didelphidae). Mammalian Species 42(856):90–98.

Braun, J. K., R. A. Van Den Bussche, P. K. Morton, and M. A. Mares. 2005. Phylogenetic and biogeographic relationships of mouse opossums *Thylamys* (Didelphimorphia, Didelphidae) in southern South America. Journal of Mammalogy 86:147–159.

Bredar, J. 2013. The ivory trade: thinking like a businessman to stop the business. National Geographic Voices. http://voices.national geographic.com/2013/02/26/the-ivory-trade-thinking-like-a -businessman-to-stop-the-business/. Accessed on 20 March 2015.

Brito, D., and G. A. B. D. Fonseca. 2006. Evaluation of minimum viable population size and conservation status of the long-furred woolly mouse opossum *Micoureus paraguayanus*: an endemic marsupial of the Atlantic Forest. Biodiversity and Conservation 15:1713–1728.

Brito, D., and C. E. D. V. Grelle. 2004. Effectiveness of a reserve network for the conservation of the endemic marsupial *Micoureus travassosi* in Atlantic Forest remnants in southeastern Brazil. Biodiversity and Conservation 13: 2519–2536

Brito, D., D. Astua de Moraes, D. Lew, P. Soriano, and L. Emmons. 2008. *Cryptonanus unduaviensis*. The IUCN Red List of Threatened Species 2008: e.T136749A4335121. http://dx.doi.org/10.2305/IU CN.UK.2008.RLTS.T136749A4335121.en. Downloaded on 29 December 2015.

Brito, D., D. Astua de Moraes, D. Lew, P. Soriano, and L. Emmons. 2015. *Caluromys philander*. The IUCN Red List of Threatened Species 2015: e.T3649A22175720. http://dx.doi.org/10.2305/IUCN.UK.2015 -4.RLTS.T3649A22175720.en. Downloaded on 18 January 2016.

Bronner, G. N. 1991. Comparative hyoid morphology of nine chrysochlorid species (Mammalia: Chrysochloridae). Annals of the Transvaal Museum 35:295–311.

Bronner, G. N. 1995a. Cytogenetic properties of nine species of golden moles (Insectivora: Chrysochloridae). Journal of Mammalogy 76:957–971.

Bronner, G. N. 1995b. Systematic revision of the golden mole genera: *Amblysomus, Chlorotalpa* and *Calcochloris* (Insectivora: Chrysochloromorpha; Chrysochloridae). Ph.D. thesis, University of Natal, Durban, South Africa.

Bronner, G. N. 1996. Geographic patterns of morphometric variation in the Hottentot golden mole, *Amblysomus hottentotus* (Insectivora: Chrysochloridae). A multivariate analysis. Mammalia 60:729–751.

Bronner, G. N. 2000. New species and subspecies of golden mole (Chrysochloridae: *Amblysomus*) from Mpumalanga, South Africa. Mammalia 64:41–54.

Bronner, G. N. 2013a. *Amblysomus corriae* Fynbos golden-mole. Pp. 226–227 in Mammals of Africa, volume 1, Introductory chapters and Afrotheria (J. Kingdon, D. Happold, M. Hoffmann, T. Butynski, M. Happold, and J. Kalina, eds.). Bloomsbury, London, United Kingdom.

Bronner, G. N. 2013b. *Amblysomus septentrionalis* Highveld golden-mole. Pp. 232–233 in Mammals of Africa, volume 1, Introductory chapters and Afrotheria (J. Kingdon, D. Happold, M. Hoffmann, T. Butynski, M. Happold, and J. Kalina, eds.). Bloomsbury, London, United Kingdom.

Bronner, G. N. 2013c. *Chlorotalpa sclateri* Sclater's golden-mole. Pp. 240–241 in Mammals of Africa, volume 1, Introductory chapters and Afrotheria (J. Kingdon, D. Happold, M. Hoffmann, T. Butynski, M. Happold, and J. Kalina, eds.). Bloomsbury, London, United Kingdom.

Bronner, G. N. 2013d. *Neamblysomus julianae* Juliana's golden-mole. Pp. 256–257 in Mammals of Africa, volume 1, Introductory chapters and Afrotheria (J. Kingdon, D. Happold, M. Hoffmann, T. Butynski, M. Happold, and J. Kalina, eds.). Bloomsbury, London, United Kingdom.

Bronner, G. 2015a. *Chrysochloris stuhlmanni*. The IUCN Red List of Threatened Species 2015: e.T40601A21288271. http://dx.doi.org /10.2305/IUCN.UK.2015-2.RLTS.T40601A21288271.en. Downloaded on 19 January 2016.

Bronner, G. 2015b. *Chrysochloris visagiei*. The IUCN Red List of Threatened Species 2015: e.T4812A21287855. http://dx.doi.org/10 .2305/IUCN.UK.2015-2.RLTS.T4812A21287855.en. Downloaded on 19 January 2016.

Bronner, G. 2015c. *Cryptochloris wintoni*. The IUCN Red List of Threatened Species 2015: e.T5748A21287143. http://dx.doi.org/10 .2305/IUCN.UK.2015-2.RLTS.T5748A21287143.en. Downloaded on 19 January 2016.

Bronner, G. 2015d. *Cryptochloris zyli*. The IUCN Red List of Threatened Species 2015: e.T5749A21286235. http://dx.doi.org/10.2305 /IUCN.UK.2015-2.RLTS.T5749A21286235.en. Downloaded on 20 January 2016.

Bronner, G. 2015e. *Chrysospalax villosus*. The IUCN Red List of Threatened Species 2015: e.T4829A21290416. http://dx.doi.org/10.2305 /IUCN.UK.2015-2.RLTS.T4829A21290416.en. Downloaded on 20 January 2016.

Bronner, G. 2015f. *Chrysospalax trevelyani*. The IUCN Red List of Threatened Species 2015: e.T4828A21289898. http://dx.doi.org/10 .2305/IUCN.UK.2015-2.RLTS.T4828A21289898.en. Downloaded on 20 January 2016.

Bronner, G. 2015g. *Chlorotalpa duthieae*. The IUCN Red List of Threatened Species 2015: e.T4768A21285581. http://dx.doi.org/10.2305 /IUCN.UK.2015-2.RLTS.T4768A21285581.en. Downloaded on 20 January 2016.

Bronner, G. 2015h. *Chlorotalpa sclateri*. The IUCN Red List of Threatened Species 2015: e.T4766A21285759. http://dx.doi.org/10.2305 /IUCN.UK.2015-2.RLTS.T4766A21285759.en. Downloaded on 20 January 2016.

Bronner, G. 2015i. *Carpitalpa arendsi*. The IUCN Red List of Threatened Species 2015: e.T40596A21289173. http://dx.doi.org/10.2305 /IUCN.UK.2015-2.RLTS.T40596A21289173.en. Downloaded on 20 January 2016.

Bronner, G. N., and N. C. Bennett (subeds.). 2005. Order Afrosoricida. Pp. 1–21 in The mammals of the Southern African Subregion. 3rd edition (J. D. Skinner and C. T. Chimimba, eds.). Cambridge University Press, Cape Town, South Africa.

Bronner, G. N., and P. D. Jenkins. 2005. Order Afrosoricida. Pp. 71–81 in Mammal species of the world: a taxonomic and geographic reference. 3rd edition (D. E. Wilson and D. M. Reeder, eds.). Johns Hopkins University Press, Baltimore, Maryland.

Bronner, G., and S. Mynhardt. 2015a. *Amblysomus corriae*. The IUCN Red List of Threatened Species 2015: e.T62006A21284863. http://dx.doi.org/10.2305/IUCN.UK.2015-2.RLTS.T62006 A21284863.en. Downloaded on 20 January 2016.

Bronner, G., and S. Mynhardt. 2015b. *Amblysomus hottentotus*. The IUCN Red List of Threatened Species 2015: e.T41316A21286316. http://dx.doi.org/10.2305/IUCN.UK.2015-2.RLTS.T41316 A21286316.en. Downloaded on 20 January 2016.

Bronner, G., and S. Mynhardt. 2015c. *Amblysomus marleyi*. The IUCN Red List of Threatened Species 2015: e.T62007A21284544. http://dx.doi.org/10.2305/IUCN.UK.2015-2.RLTS.T62007 A21284544.en. Downloaded on 20 January 2016.

Broome, L. S. 2008. Mountain pygmy-possum *Burramys parvus* (Broom, 1896). Pp. 210–212 in The mammals of Australia. 3rd edition (S. Van Dyck and R. Strahan, eds.). Reed New Holland, Sydney, Australia.

Broome, L., M. Archer, H. Bates, H. Shi, F. Geiser, B. McAllan, D. Heinze, S. Hand, T. Evans, and S. Jackson. 2012. A brief review of the life history of, and threats to, *Burramys parvus* with a prehistory-based proposal for ensuring that it has a future. Pp 114–126 in Wildlife and climate change: towards robust conservation strategies for Australian fauna (D. Lunney and P. Hutchings, eds.). Royal Zoological Society of New South Wales, Mosman, New South Wales, Australia.

Broome, L., F. Ford., M. Dawson, K. Green, D. Little, and N. McElhinney. 2013. Re-assessment of mountain pygmy-possum *Burramys parvus* population size and distribution of habitat in Kosciuszko National Park. Australian Zoologist 36:381–403.

Brosset, A. 1989. Camouflage chez le yapock *Chironectes minimus*. Revue d'Écologie 44:279–281.

Broughton, S. K., and C. R. Dickman. 1991. The effect of supplementary food on home range of the southern brown bandicoot, *Isoodon obesulus* (Marsupialia: Peramelidae). Australian Journal of Ecology 16:71–78.

Brown, B. E. 2004. Atlas of New World marsupials. Fieldiana: Zoology (new series) 102:1–308.

Brown, D. D. 2011. Fruit-eating by an obligate insectivore: palm fruit consumption in wild northern tamanduas (*Tamandua mexicana*) in Panama. Edentata 12:63–65.

Brown, J. C. 1964. Observations on the elephant shrews (Macroscelididae) of equatorial Africa. Proceedings of the Zoological Society of London 143:103–119.

Brown, J. L., N. Wielebnowski, and J. V. Cheeran. 2008. Pain, stress, and suffering in elephants. What is the evidence and how can we measure it? Pp. 121–145 in Elephants and ethics, toward a morality of existence (C. Wemmer and C. A. Christen, eds.). Johns Hopkins University Press, Baltimore, Maryland.

Brown, J. L., D. Olson, M. Keele, and E. W. Freeman. 2004. Survey of the reproductive cyclicity status of Asian and African elephants in North America. Zoo Biology 23:309–321.

Brown, M., S. M. Carthew, and S. J. B. Cooper. 2007. Monogamy in an Australian arboreal marsupial, the yellow-bellied glider (*Petaurus australis*). Australian Journal of Zoology 55:185–195.

Brown, M., H. Cooksley, S. M. Carthew, and S. J. B. Cooper. 2006. Conservation units and phylogeographic structure of an arboreal marsupial, the yellow-bellied glider (*Petaurus australis*). Australian Journal of Zoology 54:305–317.

Brown, R. 1973. Has the thylacine really vanished? Animals 15:416–419.

Brugière, D., I. Badjinca, C. Silva, A. Serra, and M. Barry. 2006. On the road to extinction? The status of elephant *Loxodonta africana* in Guinea Bissau and western Guinea, West Africa. Oryx 40:442–446.

Bryant, S. L. 1989. Growth, development, and breeding pattern of the long-nosed potoroo, *Potorous tridactylus* (Kerr, 1792) in Tasmania. Pp. 449–456 in Kangaroos, wallabies and rat-kangaroos (G. Grigg, P. Jarman, and I. Hume, eds.). Surrey Beatty & Sons, Chipping Norton, New South Wales, Australia.

Bryden, M. M. 1989. Morphology and physiology of the Eutheria. Pp. 783–809 in Fauna of Australia: Mammalia, volume 1B (D. W. Walton and B. J. Richardson, eds.). Australian Government Publishing Service, Canberra, Australia.

Bublitz, J. 1987. Untersuchungen zur Systematik der Rezenten Caenolestidae (Trouessart, 1898). Bonner Zoologische Monographien 23:1–96

Bucher, J. E., and H. I. Fritz. 1977. Behavior and maintenance of the woolly opossum (*Caluromys*) in captivity. Laboratory Animal Science 27:1007–1012.

Buchmann, O. L. K., and E. R. Guiler. 1974. Locomotion in the potoroo. Journal of Mammalogy 55:203–206.

Buchmann, O. L. K., and E. R. Guiler. 1977. Behavior and ecology of the Tasmanian devil, *Sarcophilus harrisii*. Pp. 155–168 in The biology of marsupials (B. Stonehouse and D. P. Gilmore, eds.). University Park Press, Baltimore, Maryland.

Buckley, M. 2013. A molecular phylogeny of *Plesiorycteropus* reassigns the extinct mammalian order "Bibymalagasia." PLoS ONE 8(3): e59614. doi:10.1371/journal.pone.0059614.

Buden, D. W., and J. Haglelgam. 2010. Review of crocodile (Reptilia: Crocodilia) and dugong (Mammalia: Sirenia) sightings in the Federated States of Micronesia. Pacific Science 64:577–583.

Bulte, E. H., R. D. Horan, and J. F. Shogren. 2003. Is the Tasmanian tiger extinct? A biological-economic re-evaluation. Ecological Economics 45:271–279.

Burbidge, A. 2008a. *Dasykaluta rosamondae*. The IUCN Red List of Threatened Species 2008: e.T40527A10329507. http://dx.doi.org /10.2305/IUCN.UK.2008.RLTS.T40527A10329507.en. Downloaded on 26 August 2015.

Burbidge, A. 2008b. *Bettongia pusilla*. The IUCN Red List of Threatened Species 2008: e.T136805A4341945. http://dx.doi.org/10.2305 /IUCN.UK.2008.RLTS.T136805A4341945.en. Downloaded on 01 December 2015.

Burbidge, A. 2008c. Crescent nailtail wallaby *Onychogalea lunata* (Gould, 1841). Pp. 357–359 in The mammals of Australia. 3rd edition (S. Van Dyck and R. Strahan, eds.). Reed New Holland, Sydney, Australia.

Burbidge, A., N. Cooper, and K. Morris. 2008. *Pseudantechinus roryi*. The IUCN Red List of Threatened Species 2008: e.T136620A4319128. http://dx.doi.org/10.2305/IUCN.UK.2008.RLTS.T136620 A4319128.en. Downloaded on 05 January 2016.

Burbidge, A., C. Dickman, and K. Johnson. 2008. *Chaeropus ecaudatus*. The IUCN Red List of Threatened Species 2008: e.T4322A10787179. http://dx.doi.org/10.2305/IUCN.UK.2008.RLTS.T4322A10787179 .en. Downloaded on 05 January 2016.

Burbidge, A., and K. A. Johnson. 2008a. Spectacled hare-wallaby *Lagorchestes conspicillatus* (Gould, 1842). Pp. 314–316 in The mammals of Australia. 3rd edition (S. Van Dyck and R. Strahan, eds.). Reed New Holland, Sydney, Australia.

Burbidge, A., and K. Johnson. 2008b. *Lagorchestes asomatus*. The IUCN Red List of Threatened Species 2008: e.T11160A3258155. http://dx.doi.org/10.2305/IUCN.UK.2008.RLTS.T11160 A3258155.en. Downloaded on 02 January 2016.

Burbidge, A., and K. Johnson. 2008c. *Onychogalea lunata*. The IUCN Red List of Threatened Species 2008: e.T15331A4514401. http://dx. doi.org/10.2305/IUCN.UK.2008.RLTS.T15331A4514401.en. Downloaded on 02 January 2016.

Burbidge, A., K. Johnson, and K. Aplin. 2008. *Perameles eremiana*. The IUCN Red List of Threatened Species 2008: e.T16570A6095488. http://dx.doi.org/10.2305/IUCN.UK.2008.RLTS.T16570 A6095488.en. Downloaded on 06 January 2016

Burbidge, A., K. Johnson, and C. Dickman. 2008. *Macrotis leucura*. The IUCN Red List of Threatened Species 2008: e.T12651A3369111. http://dx.doi.org/10.2305/IUCN.UK.2008.RLTS.T12651A33 69111.en. Downloaded on 05 January 2016.

Burbidge, A., N. L. McKenzie, and P. J. Fuller. 2008. Long-tailed dunnart *Sminthopsis longicaudata* (Spencer, 1909). Pp. 148–150 in The mammals of Australia. 3rd edition (S. Van Dyck and R. Strahan, eds.). Reed New Holland, Sydney, Australia.

Burbidge, A., N. McKenzie, and T. Start. 2008. *Petrogale burbidgei*. The IUCN Red List of Threatened Species 2008: e.T16744A6348472. http://dx.doi.org/10.2305/IUCN.UK.2008.RLTS.T16744A63 48472.en. Downloaded on 22 December 2015.

Burbidge, A., and J. C. Short. 2008. Burrowing bettong *Bettongia lesueur* (Quoy and Gaimard, 1824). Pp. 288–290 in The mammals of Australia. 3rd edition (S. Van Dyck and R. Strahan, eds.). Reed New Holland, Sydney, Australia.

Burbidge, A., and M. J. Webb. 2008. Scaly-tailed possum *Wyulda squamicaudata* (Alexander, 1919). Pp. 277–278 in The mammals of Australia. 3rd edition (S. Van Dyck and R. Strahan, eds.). Reed New Holland, Sydney, Australia.

Burbidge, A., J. Woinarski, and K. Morris. 2008. *Isoodon auratus*. The IUCN Red List of Threatened Species 2008: e.T10863A3223718. http://dx.doi.org/10.2305/IUCN.UK.2008.RLTS.T10863A32 23718.en. Downloaded on 06 January 2016

Burbidge, A., K. A. Johnson, P. J. Fuller, and P. F. Aitken. 2008a. Central hare-wallaby *Lagorchestes asomatus* (Finlayson, 1943). Pp. 312–313 in The mammals of Australia. 3rd edition (S. Van Dyck and R. Strahan, eds.). Reed New Holland, Sydney, Australia.

Burbidge, A., J. Woinarski, J. Reed, J. van Weenen, K. E. Moseby, and K. Morris. 2008b. *Petrogale lateralis*. The IUCN Red List of Threatened Species 2008: e.T16751A6358423. http://dx.doi.org/10.2305 /IUCN.UK.2008.RLTS.T16751A6358423.en. Downloaded on 22 December 2015.

Burbidge, A., J. Woinarski, J. Winter, and M. Runcie. 2008c. *Petropseudes dahli*. The IUCN Red List of Threatened Species 2008: e.T40580A10322017. http://dx.doi.org/10.2305/IUCN.UK.2008 .RLTS.T40580A10322017.en. Downloaded on 30 December 2015.

Burbidge, A., P. Menkhorst, M. Ellis, and P. Copley. 2008d. *Macropus fuliginosus*. The IUCN Red List of Threatened Species 2008: e.T40563 A10319795. http://dx.doi.org/10.2305/IUCN.UK.2008.RLTS.T405 63A10319795.en. Downloaded on 05 January 2016.

Burk, A., and M. S. Springer. 2000. Intergeneric relationships among Macropodoidea (Metatheria: Diprotodontia) and the chronicle of kangaroo evolution. Journal of Mammalian Evolution 7:213–237.

Burk, A., M. Westerman, and M. S. Springer. 1998. The phylogenetic position of the musky rat-kangaroo and the evolution of bipedal hopping in kangaroos (Macropodidae: Diprotodontia). Systematic Biology 47:457–474.

Burk, A., M. Westerman, D. J. Kao, J. R. Kavanagh, and M. S. Springer. 1999. An analysis of marsupial interordinal relationships based on 12S rRNA, tRNA valine, 16S rRNA, and cytochrome *b* sequences. Journal of Mammalian Evolution 6:317–334.

Burnett, S. 2008a. Atherton antechinus *Antechinus godmani* (Thomas, 1923). Pp. 89–90 in The mammals of Australia. 3rd edition (S. Van Dyck and R. Strahan, eds.). Reed New Holland, Sydney, Australia.

Burnett, S. 2008b. Common planigale *Planigale maculata* (Gould, 1851). Pp. 112–113 in The mammals of Australia. 3rd edition (S. Van Dyck and R. Strahan, eds.). Reed New Holland, Sydney, Australia.

Burnett, S., and M. S. Crowther. 2008a. Rusty antechinus *Antechinus adustus* (Thomas, 1923). Pp. 81–82 in The mammals of Australia. 3rd edition (S. Van Dyck and R. Strahan, eds.). Reed New Holland, Sydney, Australia.

Burnett, S., and M. S. Crowther. 2008b. Subtropical antechinus *Antechinus subtropicus* (Van Dyck and Crowther, 2000). Pp. 97–98 in The mammals of Australia. 3rd edition (S. Van Dyck and R. Strahan, eds.). Reed New Holland, Sydney, Australia.

Burnett, S., and C. Dickman. 2008. *Dasyurus maculatus*. The IUCN Red List of Threatened Species 2008: e.T6300A12601070. http://dx.doi.org/10.2305/IUCN.UK.2008.RLTS.T6300A12601070.en. Downloaded on 05 January 2016.

Burnett, S., and M. Ellis. 2008. *Thylogale stigmatica*. The IUCN Red List of Threatened Species 2008: e.T40574A10335692. http://dx.doi.org/10.2305/IUCN.UK.2008.RLTS.T40574A10335692.en. Downloaded on 05 January 2016.

Burnett, S., and B. Mott. 2004. Spotted-tailed northern quolls, *Dasyurus hallucatus*: a precautionary tale. Memoirs of the Queensland Museum 49:760.

Burnett, S., and J. Winter. 2008a. *Sminthopsis douglasi*. The IUCN Red List of Threatened Species 2008: e.T20290A9182739. http://dx.doi.org/10.2305/IUCN.UK.2008.RLTS.T20290A9182739.en. Downloaded on 04 January 2016.

Burnett, S., and J. Winter. 2008b. *Antechinus godmani*. The IUCN Red List of Threatened Species 2008: e.T1583A5194035. http://dx.doi.org/10.2305/IUCN.UK.2008.RLTS.T1583A5194035.en. Downloaded on 24 December 2015.

Burnett, S., and J. Winter. 2008c. *Hemibelideus lemuroides*. The IUCN Red List of Threatened Species 2008: e.T9869A13023084. http://dx.doi.org/10.2305/IUCN.UK.2008.RLTS.T9869A13023084.en. Downloaded on 30 December 2015.

Burnett, S., and J. Winter. 2008d. *Bettongia tropica*. The IUCN Red List of Threatened Species 2008: e.T2787A9481109. http://dx.doi.org/10.2305/IUCN.UK.2008.RLTS.T2787A9481109.en. Downloaded on 01 December 2015.

Burnett, S., and J. Winter. 2008e. *Petrogale persephone*. The IUCN Red List of Threatened Species 2008: e.T16747A6353468. http://dx.doi.org/10.2305/IUCN.UK.2008.RLTS.T16747A6353468.en. Downloaded on 22 December 2015.

Burnett, S., J. Winter, and R. Martin. 2008. *Petaurus gracilis*. The IUCN Red List of Threatened Species 2008: e.T16727A6320516. http://dx.doi.org/10.2305/IUCN.UK.2008.RLTS.T16727A6320516.en. Downloaded on 30 December 2015.

Burney, D. A., H. F. James, F. V. Grady, J.-G. Rafamantanantsoaj, Ramilisonina, H. T. Wright, and J. B. Cowart. 1997. Environmental change, extinction and human activity: evidence from caves in NW Madagascar. Journal of Biogeography 24:755–767.

Burney, D. A., L. P. Burney, L. R. Godfrey, W. L. Jungers, S. M. Goodman, H. T. Wright, A. J. T. Jull. 2004. A chronology for late prehistoric Madagascar. Journal of Human Evolution 47:25–63.

Butler, P. M. 1980. The tupaiid dentition. Pp. 171–204 in Comparative biology and evolutionary relationships of tree shrews (W. P. Luckett, ed.). Plenum Press, New York, New York.

Butler, P. M. 1995. Fossil Macroscelidea. Mammal Review 25:3–14.

Butynski, T., F. Dowsett-Lemaire, and H. Hoeck. 2015. *Dendrohyrax dorsalis*. The IUCN Red List of Threatened Species 2015: e.T6410 A21282601. http://dx.doi.org/10.2305/IUCN.UK.2015-2.RLTS.T6410A21282601.en. Downloaded on 21 January 2016.

Butynski, T., H. Hoeck, and Y. A. de Jong. 2015a. *Dendrohyrax arboreus*. The IUCN Red List of Threatened Species 2015: e.T6409 A21282806. http://dx.doi.org/10.2305/IUCN.UK.2015-2.RLTS.T6409A21282806.en. Downloaded on 21 January 2016.

Butynski, T., H. Hoeck, and Y. A. de Jong. 2015b. *Heterohyrax brucei*. The IUCN Red List of Threatened Species 2015: e.T9997A21283287. http://dx.doi.org/10.2305/IUCN.UK.2015-2.RLTS.T9997A21283287.en. Downloaded on 21 January 2016.

Butynski, T., H. Hoeck, L. Koren, and Y. A. de Jong. 2015. *Procavia capensis*. The IUCN Red List of Threatened Species 2015: e.T41766 A21285876. http://dx.doi.org/10.2305/IUCN.UK.2015-2.RLTS.T41766A21285876.en. Downloaded on 21 January 2016.

Byrne, R. W., and L. A. Bates. 2011. Elephant cognition: what we know about what elephants know. Pp. 174–182 in The Amboseli elephants: a long-term perspective on a long-lived mammal (C. J. Moss, H. Croze, and P. C. Lee, eds.). University of Chicago Press, Chicago, Illinois.

Byrnes, G., N. T.-L. Lim, C. Yeong, and A. J. Spence. 2011. Sex differences in the locomotor ecology of a gliding mammal, the Malayan colugo (*Galeopterus variegatus*). Journal of Mammalogy 92:444–451.

Cabrera, A. 1957, 1961. Catalogo de los mamiferos de América del Sur. Revista del Museo. Argentino de Ciencias Naturales "Bernardo Rivadavia" 4:1–732.

Cáceres, N. C., and A. P. Carmignotto. 2006. *Caluromys lanatus*. Mammalian Species 803:1–6.

Cáceres, N. C., L. Ferreira, and A. P. Carmignotto. 2007. The occurrence of the mouse opossum *Marmosops ocellatus* (Marsupialia, Didelphidae) in western Brazil. Mammalian Biology 72:45–48.

Cáceres, N. C., I. R. Ghizoni-Jr, and M. E. Graipel. 2002. Diet of two marsupials, *Lutreolina* and *Micoureus demerarae*, in a coastal Atlantic Forest island of Brazil. Mammalia 66:331–339.

Cáceres, N. C., R. P. Napoli, W. H. Lopes, J. Casella, and G. S. Gazeta. 2007. Natural history of the marsupial *Thylamys macrurus* (Mammalia, Didelphidae) in fragments of savannah in southwestern Brazil. Journal of Natural History 41:1979–1988.

Calaby, J. H. 1966. Mammals of the upper Richmond and Clarence Rivers, New South Wales. CSIRO Division of Wildlife Research Technical Paper 10:1–55.

Calaby, J. H. 1971. The current status of Australian Macropodidae. Australian Zoologist 16:17–31.

Calaby, J. H., H. Dimpel, and I. M. Cowan. 1971. The mountain pygmy possum, *Burramys parvus* Broom (Marsupialia), in the Kosciuszko National Park, New South Wales. CSIRO Division of Wildlife Research Technical Paper 23:1–11.

Calaby, J. H., and G. C. Grigg. 1989. Changes in macropodoid communities and populations in the past 200 years, and the future. Pp. 813–820 in Kangaroos, wallabies and rat-kangaroos (G. Grigg,

P. Jarman, and I. Hume, eds.). Surrey Beatty & Sons, Chipping Norton, New South Wales, Australia.

Calaby, J. H., and B. J. Richardson. 1988a. Potoroidae. Pp. 53–59 in Zoological catalogue of Australia, volume 5, Mammalia (D. W. Walton, ed.). Australian Government Publishing Service, Canberra, Australia.

Calaby, J. H., and B. J. Richardson. 1988b. Macropodidae. Pp. 60–80 in Zoological catalogue of Australia, volume 5, Mammalia (D. W. Walton, ed.). Australian Government Publishing Service, Canberra, Australia.

Calaby, J. H., and C. White. 1967. The Tasmanian devil (*Sarcophilus harrisii*) in northern Australia in Recent times. Australian Journal of Science 29:473–475.

Calaby, J. H., L. K. Corbett, G. B. Sharman, and P. G. Johnston. 1974. The chromosomes and systematic position of the marsupial mole, *Notoryctes typhlops*. Australian Journal of Biological Sciences 27:529–532.

Caldwell, D. K., and M. C. Caldwell. 1985. Manatees—*Trichechus manatus*, *Trichechus senegalensis*, and *Trichechus inunguis*. Pp. 33–66 in Handbook of marine mammals, volume 3, The sirenians and baleen whales (S. H. Ridgway and R. Harrison, eds.). Academic Press, London, United Kingdom.

Calef, G. W. 1988. Maximum rate of increase in the African elephant. African Journal of Ecology 26:323–327.

Callister, D. J. 1991. A review of the Tasmanian brushtail possum industry. Traffic Bulletin 12(3):49–58.

Caloi, L., T. Kotsakis, M. R. Palombo, and C. Petronio. 1996. The Pleistocene dwarf elephants of the Mediterranean islands. Pp. 234–239 in The Proboscidea. Evolution and palaeoecology of elephants and their relatives (J. Shoshani and P. Tassy, eds.). Oxford University Press, Oxford, United Kingdom.

Calzada, J., M. Delibes, C. Keller, F. Palomares, and W. Magnusson. 2008. First record of the bushy-tailed opossum, *Glironia venusta* (Thomas, 1912) (Didelphimorphia), from Manaus, Amazonas, Brazil. Acta Amazonica 38:807–809.

Campeau-Péloquin, A., J. A. W. Kirsch, M. D. B. Eldridge, and F.-J. Lapointe. 2001. Phylogeny of the rock-wallabies, *Petrogale* (Marsupialia: Macropodidae) based on DNA/DNA hybridization. Australian Journal of Zoology 49:463–486.

Campos-Arceiz, A., and S. Blake. 2011. Megagardeners of the forest—the role of elephants in seed dispersal. Acta Oecologica 37:542–553.

Cantanhede, A. M., V. M. Ferreira Da Silva, I. P. Farias, T. Hrbek, S. M. Lazzarini, and J. Alves-Gomes. 2005. Phylogeography and population genetics of the endangered Amazonian manatee, *Trichechus inunguis* (Natterer, 1883) (Mammalia, Sirenia). Molecular Ecology 14:401–413.

Cao J., E.-B. Yang, J. J. Su, Y. Li, and P. Chow. 2003. The tree shrews: adjuncts and alternatives to primates as models for biomedical research. Journal of Medical Primatology 32:123–130.

Capelli, C., R. D. E. MacPhee, A. L. Roca, F. Brisighelli, N. Georgiadis, S. J. O'Brien, and A. D. Greenwood. 2006. A nuclear DNA phylogeny of the woolly mammoth (*Mammuthus primigenius*). Molecular Phylogenetics and Evolution 40:620–627.

Caramaschi, F. P., F. F. Nascimento, R. Cerqueira, and C. R. Bonvicino. 2011. Genetic diversity of wild populations of the grey short-tailed opossum, *Monodelphis domestica* (Didelphimorphia: Didelphidae), in Brazilian landscapes. Biological Journal of the Linnean Society 104:251–263.

Cardillo, M., O. R. P. Bininda-Emonds, E. Boakes, and A. Purvis. 2004. A species-level phylogenetic supertree of marsupials. Journal of Zoology 264:11–31.

Cardoso da Silva, J. M., and D. C. Oren. 1993. Observations on the habitat and distribution of the Brazilian three-banded armadillo *Tolypeutes tricinctus*, a threatened Caatinga endemic. Mammalia 57:149–152.

Carleton, M. D., and G. G. Musser. 2005. Order Rodentia. Pp. 745–1600 in Mammal species of the world: a taxonomic and geographic reference. 3rd edition (D. E. Wilson and D. M. Reeder, eds.). Johns Hopkins University Press, Baltimore, Maryland.

Carmignotto, A. P., and T. Monfort. 2006. Taxonomy and distribution of the Brazilian species of *Thylamys* (Didelphimorphia: Didelphidae). Mammalia 70:126–144.

Carmignotto, A. P., N. de la Sancha, D. Flores, and P. Teta. 2011. *Cryptonanus chacoensis*. The IUCN Red List of Threatened Species 2011: e.T136845A4346572. http://dx.doi.org/10.2305/IUCN.UK.2011-2.RLTS.T136845A4346572.en. Downloaded on 05 January 2016.

Carr, S. G., and A. C. Robinson. 1997. The present status and distribution of the desert rat-kangaroo *Caloprymnus campestris* (Marsupialia: Potoroidae). South Australian Naturalist 72(1/2):4–27.

Carrick, F. N., T. R. Grant, and P. D. Temple-Smith. 2008. Platypus *Ornithorhynchus anatinus* (Shaw, 1799). Pp. 32–35 in The mammals of Australia. 3rd edition (S. Van Dyck and R. Strahan, eds.). Reed New Holland, Sydney, Australia.

Carruthers, J., A. Boshoff, R. Slotow, H. C. Biggs, G. Avery, and W. Matthews. 2008. The elephant in South Africa: history and distribution. Pp. 23–83 in Elephant management: a scientific assessment for South Africa (R. J. Scholes and K. G. Mennell, eds.). Wits University Press, Johannesburg, South Africa.

Carter, T. S. 1983. The burrows of giant armadillos, *Priodontes maximus* (Edentata: Dasypodidae). Säugetierkundliche Mitteilungen 31:47–53.

Carter, T. S. 1985. Armadillos of Brazil. National Geographic Society Research Reports 20:101–107.

Carter, T. S., and C. D. Encarnaçao. 1983. Characteristics and use of burrows by four species of armadillos in Brazil. Journal of Mammalogy 64:103–108.

Carthew, S. M. 2004. Distribution and status of possums and gliders in South Australia. Pp. 63–70 in The biology of Australian possums and gliders (R. L. Goldingay and S. M. Jackson, eds.). Surrey Beatty & Sons, Chipping Norton, New South Wales, Australia.

Carthew, S. M., and D. G. Bos. 2008. Southern ningaui *Ningaui yvonneae* (Kitchener, Stoddart and Henry, 1983). Pp. 120–121 in The mammals of Australia. 3rd edition (S. Van Dyck and R. Strahan, eds.). Reed New Holland, Sydney, Australia.

Carthew, S. M., and B. R. Cadzow. 2008. Little pygmy-possum *Cercartetus lepidus* (Thomas, 1888). Pp. 217–219 in The mammals of Australia. 3rd edition (S. Van Dyck and R. Strahan, eds.). Reed New Holland, Sydney, Australia.

Carthew, S. M., B. R. Cadzow, and J. N. Foulkes. 2008. Western pygmy-possum *Cercartetus concinnus* (Gould, 1845). Pp. 215–216 in The mammals of Australia. 3rd edition (S. Van Dyck and R. Strahan, eds.). Reed New Holland, Sydney, Australia.

Carvalho, B. de A., L. F. B. Oliveira, and M. S. Mattevi. 2009. Phylogeny of *Thylamys* (Didelphimorphia, Didelphidae) species, with special reference to *Thylamys karimii*. Iheringia Serie Zoologia 99:419–425.

Carvalho, B. de A., L. F. B. Oliveira, A. Langguth, C. C. Freygang, R. S. Ferraz, and M. S. Mattevi. 2011. Phylogenetic relationships and phylogeographic patterns in *Monodelphis* (Didelphimorphia: Didelphidae). Journal of Mammalogy 92:121–133.

Cassano, C. 2004. Research on the maned sloth (*Bradypus torquatus*) in Bahia, Brazil. Edentata 6:56.

Castelblanco-Martínez, D. N., A. L. Bermúdez-Romero, I. V. Gómez-Camelo, F. C. Weber Rosas, F. Trujillo, and E. Zerda-Ordoñez. 2009. Seasonality of habitat use, mortality and reproduction of the vulnerable Antillean manatee *Trichechus manatus manatus* in the Orinoco River, Colombia: implications for conservation. Oryx 43:235–242.

Castro-Arellano, I., H. Zarza, and R. A. Medellín. 2000. *Philander opossum*. Mammalian Species 638:1–8.

Caughley, G. 1984. The grey kangaroo overlap zone. Australian Wildlife Research 11:1–10.

Caughley, G., H. Dublin, and I. Parker. 1990. Projected decline of the African elephant. Biological Conservation 54:157–164.

Caughley, G., G. C. Grigg, and L. Smith. 1985. The effect of drought on kangaroo populations. Journal of Wildlife Management 49:679–685.

Caughley, G., R. G. Sinclair, and G. R. Wilson. 1977. Numbers, distribution, and harvesting rate of kangaroos on the inland plains of New South Wales. Australian Wildlife Research 4:99–108.

Ceballos, G. 2014. Gray mouse opossum. Pp. 82–84 in Mammals of Mexico (G. Ceballos, ed.). Johns Hopkins University Press, Baltimore, Maryland.

Celis-Diez, J. L., J. Hetz, P. A. Marín-Vial, G. Fuster, P. Necochea, R. A. Vásquez, F. M. Jaksic, and J. J. Armesto. 2012. Population abundance, natural history, and habitat use by the arboreal marsupial *Dromiciops gliroides* in rural Chiloé Island, Chile. Journal of Mammalogy 93:134–148.

Cerqueira, R. 1985. The distribution of *Didelphis* in South America (Polyprotodontia, Didelphidae). Journal of Biogeography 12:135–145.

Cerqueira, R., and B. Lemos. 2000. Morphometric differentiation between Neotropical black-eared opossums, *Didelphis marsupialis* and *D. aurita* (Didelphimorphia, Didelphidae). Mammalia 64: 319–327.

Cerqueira, R., and C. J. Tribe. 2007. Genus *Didelphis* (Linnaeus, 1758). Pp. 17–25 in Mammals of South America, volume 1, Marsupials, xenarthrans, shrews, and bats (A. L. Gardner, ed.). University of Chicago Press, Chicago, Illinois.

Chacón, J., J. Racero-Casarrubia, and E. Rodríguez-Ortiz. 2013. Nuevos registros de *Cyclopes didactylus* (Linnaeus, 1758) para Colombia. Edentata 14:78–84.

Chagas, R. R. D., J. P. Souza-Alves, L. Jerusalinsky, and S. F. Ferrari. 2009. New records of *Bradypus torquatus* (Pilosa: Bradypodidae) from southern Sergipe, Brazil. Edentata 8–10:21–24.

Charif, R. A., R. R. Ramey, W. R. Langbauer, Jr., K. B. Payne, R. B. Martin, and L. M. Brown. 2005. Spatial relationships and matrilineal kinship in African savanna elephant (*Loxodonta africana*) clans. Behavioral Ecology and Sociobiology 57:327–338.

Chartier, L., A. Zimmermann, and R. J. Ladle. 2011. Habitat loss and human-elephant conflict in Assam, India: does a critical threshold exist? Oryx 45:528–533.

Chase, M. J., and C. R. Griffin. 2011. Elephants of south-east Angola in war and peace: their decline, re-colonization and recent status. African Journal of Ecology 49:353–361.

Chasen, F. N. 1940. A handlist of Malaysian mammals. Bulletin of the Raffles Museum, Singapore 15:1–209.

Chatterjee, S. 1997. The rise of birds. Johns Hopkins University Press, Baltimore, Maryland.

Chemisquy, M. A., and D. A. Flores. 2012. Taxonomy of the southernmost populations of *Philander* (Didelphimorphia, Didelphidae), with implications for the systematics of the genus. Zootaxa 3481:60–72.

Chevalier-Skolnikoff, S., and J. Liska. 1993. Tool use by wild and captive elephants. Animal Behaviour 46:209–219.

Chiarello, A. G. 1998. Activity budgets and ranging patterns of the Atlantic Forest maned sloth *Bradypus torquatus* (Xenarthra: Bradypodidae). Journal of Zoology 246:1–10.

Chiarello, A. G. 2008. Sloth ecology. An overview of field studies. Pp. 269–280 in The biology of the Xenarthra (S. F. Vizcaíno and W. J. Loughry, eds.). University Press of Florida, Gainesville, Florida.

Chiarello, A., and N. Moraes-Barros. 2014. *Bradypus torquatus*. The IUCN Red List of Threatened Species 2014: e.T3036A47436575. http://dx.doi.org/10.2305/IUCN.UK.2014-1.RLTS.T3036A47 436575.en. Downloaded on 21 May 2015.

Chiarello, A. G., D. J. Chivers, C. Bassi, M. A. F. Maciel, L. S. Moreira, and M. Bazzalo. 2004. A translocation experiment for the conservation of maned sloths, *Bradypus torquatus* (Xenarthra, Bradypodidae). Biological Conservation 118:421–430.

Chilvers, B. L., S. Delean, N. J. Gales, D. K. Holley, I. R. Lawler, H. Marsh, and A. R. Preen. 2004. Diving behaviour of dugongs, *Dugong dugon*. Journal of Experimental Marine Biology and Ecology 304:203–224.

Chorazyna, H., and G. U. Kurup. 1975. Observations on the ecology and behaviour of *Anathana ellioti* in the wild. Proceedings of the International Congress of Primatology 5:342–344.

Choudhury, A., D. K. Lahiri Choudhury, A. Desai, J. W. Duckworth, P. S. Easa, A. J. T. Johnsingh, P. Fernando, S. Hedges, M. Gunawardena, F. Kurt, U. Karanth, A. Lister, V. Menon, H. Riddle, A. Rübel, and E. Wikramanayake (IUCN SSC Asian Elephant Specialist Group). 2008. *Elephas maximus*. The IUCN Red List of Threatened Species 2008: e.T7140A12828813. http://dx.doi.org/10.2305/IUCN.UK .2008.RLTS.T7140A12828813.en. Downloaded on 28 March 2015.

Christensen, P. 1975. The breeding burrow of the banded ant-eater or numbat (*Myrmecobius fasciatus*). Western Australian Naturalist 13:32–34.

Christensen, P., K. Maisey, and D. H. Perry. 1984. Radiotracking the numbat, *Myrmecobius fasciatus*, in the Perup Forest of Western Australia. Australian Wildlife Research 11:275–288.

Christiansen, P. 2004. Body size in proboscideans, with notes on elephant metabolism. Zoological Journal of the Linnean Society 140:523–549.

Churchill, S. 1997. Habitat use, distribution and conservation status of the nabarlek, *Petrogale concinna*, and sympatric rock-dwelling mammals in the northern territory. Australian Mammalogy 19:297–308.

Churchill, S. 2001. Recovery plan for the sandhill dunnart (*Sminthopsis psammophila*). South Australia Department for Environment and Heritage, Adelaide, South Australia, Australia.

Cifelli, R. L. 1993. Theria of metatherian-eutherian grade and the origin of marsupials. Pp. 205–215 in Mammal phylogeny: Mesozoic differentiation, multituberculates, monotremes, early therians, and marsupials (F. S. Szalay, M. J. Novacek, and M. C. McKenna, eds.). Springer-Verlag, New York, New York.

Cione, A. L., A. J. Figini, and E. P. Tonni. 2001. Did the megafauna range to 4300 BP in South America? Radiocarbon 43:69–75.

Cione, A. L., E. P. Tonni, and L. Soibelzon. 2009. Did humans cause the late Pleistocene–early Holocene mammalian extinctions in South America in a context of shrinking open areas? Pp. 125–144 in American megafaunal extinctions at the end of the Pleistocene (G. Haynes, ed.). Springer, New York, New York.

CITES (Convention on International Trade in Endangered Species of Wild Fauna and Flora) Secretariat. 2012. Elephant conservation, illegal killing and ivory trade. CITES, Geneva, Switzerland.

CITES (Convention on International Trade in Endangered Species of Wild Fauna and Flora) Secretariat. 2014. Elephant conservation, illegal killing and ivory trade. CITES, Geneva, Switzerland.

CITES (Convention on International Trade in Endangered Species of Wild Fauna and Flora) Secretariat, IUCN/SSC African Elephant Specialist Group, and TRAFFIC International. 2013. Status of African elephant populations and levels of illegal killing and the illegal trade in ivory: a report to the African Elephant Summit. CITES, Geneva, Switzerland.

Clancy, T. F., and D. B. Croft. 1990. Home range of the common wallaroo, *Macropus robustus erubescens*, in far western New South Wales. Australian Wildlife Research 17:659–673.

Clancy, T. F., and D. B. Croft. 2008. Common wallaroo *Macropus robustus* (Gould, 1841). Pp. 346–348 in The mammals of Australia. 3rd edition (S. Van Dyck and R. Strahan, eds.). Reed New Holland, Sydney, Australia.

Claridge, A., J. Seebeck, and R. Rose. 2007. Bettongs, potoroos and the musky rat-kangaroo. CSIRO Publishing, Collingwood, Victoria, Australia.

Claridge, A. W., J. M. Trappe, and D. L. Claridge. 2001. Mycophagy by the swamp wallaby (*Wallabia bicolor*). Wildlife Research 28:643–645.

Claridge, A. W., and R. Van Der Ree. 2004. Recovering endangered populations in fragmented landscapes: the squirrel glider *Petaurus norfolcensis* on the south-west slopes of New South Wales. Pp. 678–687 in The Conservation of Australia's forest fauna. 2nd edition (D. Lunney, ed.). Royal Zoological Society of New South Wales, Mosman, New South Wales, Australia.

Clark, J. R. (Director, Fish and Wildlife Service, United States Department of the Interior). 2000. Endangered and Threatened Wildlife and Plants; final determination of threatened status for the koala. Federal Register 65:26762–26771.

Clark, M. J., and W. E. Poole. 1967. The reproductive system and embryonic diapause in the female grey kangaroo, *Macropus giganteus*. Australian Journal of Zoology 15:441–459.

Clark, T. W., N. Mazur, S. J. Cork, S. Dovers, and R. Harding. 2000. Koala conservation policy process: appraisal and recommendations. Conservation Biology 14:681–690.

Clarke, J. L., M. E. Jones, and P. J. Jarman. 1989. A day in the life of a kangaroo: activities and movements of eastern grey kangaroos *Macropus giganteus* at Wallaby Creek. Pp. 611–618 in Kangaroos, wallabies and rat-kangaroos (G. Grigg, P. Jarman, and I. Hume, eds.). Surrey Beatty & Sons, Chipping Norton, New South Wales, Australia.

Clauss, M. 2004. The potential interplay of posture, digestive anatomy, density of ingesta and gravity in mammalian herbivores: why sloths do not rest upside down. Mammal Review 34:241–245.

Clemens, W. A. 1989. Diagnosis of the class Mammalia. Pp. 401–406 in Fauna of Australia: Mammalia, volume 1B (D. W. Walton and B. J. Richardson, eds.). Australian Government Publishing Service, Canberra, Australia.

Clemens, W. A., B. J. Richardson, and P. R. Baverstock. 1989. Biogeography and phylogeny of the Metatheria. Pp. 527–559 in Fauna of Australia: Mammalia, volume 1B (D. W. Walton and B. J. Richardson, eds.). Australian Government Publishing Service, Canberra, Australia.

Cleveland, A. G. 1970. The current geographic distribution of the armadillo in the United States. Texas Journal of Science 22:87–92.

Close, R. L., and J. N. Bell. 1997. Fertile hybrids in two genera of wallabies: *Petrogale* and *Thylogale*. Journal of Heredity 88:393–397.

Close, R. L., J. D. Murray, and D. A. Briscoe. 1990. Electrophoretic and chromosome surveys of the taxa of short-nosed bandicoots within the genus *Isoodon*. Pp. 19–27 in Bandicoots and bilbies (J. H. Seebeck, P. R. Brown, R. L. Wallis, and C. M. Kemper, eds.). Surrey Beatty & Sons, Chipping Norton, New South Wales, Australia.

Cobb, S., and D. Western. 1989. The ivory trade and the future of the African elephant. Pachyderm 12:32–37.

Cockburn, A., and K. A. Lazenby-Cohen. 1992. Use of nest trees by *Antechinus stuartii*, a semelparous lekking marsupial. Journal of Zoology 226:657–680.

Coitiño, H. I., F. Montenegro, A. Fallabrino, E. M. González, and D. Hernández. 2013. Distribución actual y potencial de *Cabassous tatouay* y *Tamandua tetradactyla* en el límite sur de su distribución: implicancias para su conservación en Uruguay. Edentata 14:23–34.

Colby, S. L., and J. M. Ortman. 2015. Projections of the size and composition of the U.S. population: 2014 to 2060. United States Census Bureau, Report Number P25-1143.

Colgan, D., and T. F. Flannery. 1992. Biochemical systematic studies in the genus *Petaurus* (Marsupialia: Petauridae). Australian Journal of Zoology 40:245–256.

Colgan, D., T. F. Flannery, J. Trimble, and K. Aplin. 1993. Electrophoretic and morphological analysis of the systematics of the *Phalanger orientalis* (Marsupialia) species complex in Papua New Guinea and Solomon Islands. Australian Journal of Zoology 41:355–378.

Colindres, J. E. M, and G. A. C. Días 2014. Confirmación de la presencia del oso hormiguero gigante *Myrmecophaga tridactyla centralis* (Xenarthra: Myrmecophagidae) en la Reserva Biósfera Río Plátano, Departamento de Gracias a Dios, Honduras, con descripción y comentarios sobre su estatus taxonómico. Edentata 15:9–15.

Collevatti, R. G., K. C. E. Leite, G. H. B. de Miranda, and F. H. G. Rodrigues. 2007. Evidence of high inbreeding in a population of the endangered giant anteater, *Myrmecophaga tridactyla* (Myrmecophagidae), from Emas National Park, Brazil. Genetics and Molecular Biology 30:112–120.

Collins, L. R. 1973. Monotremes and marsupials: a reference for zoological institutions. Smithsonian Institution Press, Washington, DC.

Coltorti, M., J. D. Fazia, F. P. Rios, and G. Tito. 2012. Ñuagapua (Chaco, Bolivia): evidence for the latest occurrence of megafauna in association with human remains in South America. Journal of South American Earth Sciences 33:56–67.

Comstock, K. E., N. Georgiadis, J. Pecon-Slattery, A. L. Roca, E. A. Ostrander, S. J. O'Brien, and S. K. Wasser. 2002. Patterns of molecular genetic variation among African elephant populations. Molecular Ecology 11:2489–2498.

Conniff, R. 1987. When the music in our parlors brought death to darkest Africa. Audubon 89(4):76–93.

Connolly, J. H., and D. L. Obendorf. 1998. Distribution, captures and physical characteristics of the platypus (*Ornithorhynchus anatinus*) in Tasmania. Australian Mammalogy 20:231–237.

Cook, J., N. Oreskes, P. T. Doran, W. R. L. Anderegg, B. Verheggen, E. W. Maibach, J. S. Carlton, S. Lewandowsky, A. G. Skuce, S. A. Green, D. Nuccitelli, P. Jacobs, M. Richardson, B. Winkler,

R. Painting, and K. Rice. 2016. Consensus on consensus: a synthesis of consensus estimates on human-caused global warming. Environmental Research Letters 11:doi:10.1088/1748-9326/11/4/04800213.

Cooke, B. 2006. Kangaroos. Pp. 647–672 in Evolution and biogeography of Australasian vertebrates (J. R. Merrick, M. Archer, G. M. Hickey, and M. S. Y. Lee, eds.). Auscipub, Oatlands, New South Wales, Australia.

Cooke, B. D. 1998. Did introduced rabbits *Oryctolagus cuniculus* (L.) displace common wombats *Vombatus ursinus* (Shaw) from part of their range in South Australia? Pp. 263–270 in Wombats (R. T. Wells, P. A. Pridmore, B. St. John, M. D. Gaughwin, and J. Ferris, eds.). Surrey Beatty & Sons, Chipping Norton, New South Wales, Australia.

Cooper, C. E. 2011. *Myrmecobius fasciatus* (Dasyuromorphia: Myrmecobiidae). Mammalian Species 43(881):129–140.

Cooper, C. E., and P. C. Withers. 2004a. Influence of season and weather on activity patterns of the numbat (*Myrmecobius fasciatus*) in captivity. Australian Journal of Zoology 52:475–485.

Cooper, C. E., and P. C. Withers. 2004b. Patterns of body temperature variation and torpor in the numbat, *Myrmecobius fasciatus* (Marsupialia: Myrmecobiidae). Journal of Thermal Biology 29:277–284.

Cooper, N. K. 2008. Rory's pseudantechinus *Pseudantechinus roryi* (Cooper, Aplin, and Adams, 2000). Pp. 75–76 in The mammals of Australia. 3rd edition (S. Van Dyck and R. Strahan, eds.). Reed New Holland, Sydney, Australia.

Cooper, N. K., K. P. Aplin, and M. Adams. 2000 A new species of false antechinus (Marsupialia: Dasyuromorphia: Dasyuridae) from the Pilbara region, Western Australia. Records of the Western Australian Museum 20:115–136.

Cooper, S. J. B., M. Adams, and A. Labrinidis. 2000. Phylogeography of the Australian dunnart *Sminthopsis crassicaudata* (Marsupialia: Dasyuridae). Australian Journal of Zoology 48:461–473.

Copley, P. B. 1983. Studies on the yellow-footed rock-wallaby, *Petrogale xanthopus* Gray (Marsupialia: Macropodidae). I. Distribution in South Australia. Australian Wildlife Research 10:47–61.

Copley, P., M. Ellis, and J. van Weenen. 2008. *Petrogale xanthopus*. The IUCN Red List of Threatened Species 2008: e.T16750A6358247. http://dx.doi.org/10.2305/IUCN.UK.2008.RLTS.T16750A6358247.en. Downloaded on 22 December 2015.

Copley, P. B., V. T. Read, A. C. Robinson, and C. H. S. Watts. 1990. Preliminary studies of the Nuyts Archipelago bandicoot *Isoodon obesulus nauticus* on the Franklin Islands, South Australia. Pp. 345–356 in Bandicoots and bilbies (J. H. Seebeck, P. R. Brown, R. L. Wallis, and C. M. Kemper, eds.). Surrey Beatty & Sons, Chipping Norton, New South Wales, Australia.

Corbet, G. B. 1977a. Subfamily Potamogalinae. Part 1.2 in The mammals of Africa: an identification manual (J. Meester and H. W. Setzer, eds.). Smithsonian Institution Press, Washington, DC.

Corbet, G. B. 1977b. Family Macroscelididae. Part 1.5 in The mammals of Africa: an identification manual (J. Meester and H. W. Setzer, eds.). Smithsonian Institution Press, Washington, DC.

Corbet, G. B. 1978. The mammals of the Palaearctic Region: a taxonomic review. British Museum (Natural History), London, United Kingdom.

Corbet, G. B. 1979. The taxonomy of *Procavia capensis* in Ethiopia, with special reference to the aberrant tusks of *P. c. capillosa* Brauer (Mammalia, Hyracoidea). Bulletin of the British Museum (Natural History) Zoology 36:251–259.

Corbet, G. B. 1995. A cladistic look at classification within the subfamily Macroscelidinae based upon morphology. Mammal Review 25:15–17.

Corbet, G. B., and J. Hanks. 1968. A revision of the elephant-shrews, family Macroscelididae. Bulletin of the British Museum (Natural History) Zoology 16:1–111.

Corbet, G. B., and J. E. Hill. 1991. A world list of mammalian species. Natural History Museum Publications, London, and Oxford University Press, Oxford, United Kingdom.

Corbet, G. B., and J. E. Hill. 1992. The mammals of the Indomalayan region: a systematic review. Oxford University Press, Oxford, United Kingdom.

Corbett, L. K. 1975. Geographical distribution and habitat of the marsupial mole, *Notoryctes typhlops*. Australian Mammalogy 1:375–378.

Cork, S. J., T. W. Clark, and N. Mazur. 2000a. Introduction: an interdisciplinary effort for koala conservation. Conservation Biology 14:606–609.

Cork, S. J., T. W. Clark, and N. Mazur. 2000b. Conclusion and recommendations for koala conservation. Conservation Biology 14:702–704.

Correal U., G., and T. Van der Hammen. 2003. Supervivencia de mastodontes, megaterios y presencia del hombre en el Valle Del Magdalena (Colombia) entre 6000 Y 5000 ap. Revista Academia Colombiana de Ciencias 27:159–164.

Costa, L. P., Y. L. R. Leite, and J. L. Patton. 2003. Phylogeography and systematic notes on two species of gracile mouse opossums, genus *Gracilinanus* (Marsupialia: Didelphidae) from Brazil. Proceedings of the Biological Society of Washington 116:275–292.

Costa, L., D. Astua de Moraes, D. Brito, P. Soriano, D. Lew, and C. Delgado. 2008. *Cryptonanus guahybae*. The IUCN Red List of Threatened Species 2008: e.T136705A4329673. http://dx.doi.org/10.2305/IUCN.UK.2008.RLTS.T136705A4329673.en. Downloaded on 29 December 2015.

Costa, L. P., D. Astua de Moraes, D. Brito, D. Lew, and T. Tarifa. 2015. *Caluromys lanatus*. The IUCN Red List of Threatened Species 2015: e.T3648A22175609. http://dx.doi.org/10.2305/IUCN.UK.2015-4.RLTS.T3648A22175609.en. Downloaded on 18 January 2016.

Coulson, G. 2008a. Western grey kangaroo *Macropus fuliginosus* (Desmarest, 1817). Pp. 333–334 in The mammals of Australia. 3rd edition (S. Van Dyck and R. Strahan, eds.). Reed New Holland, Sydney, Australia.

Coulson, G. 2008b. Eastern grey kangaroo *Macropus giganteus* (Shaw, 1790). Pp. 335–338 in The mammals of Australia. 3rd edition (S. Van Dyck and R. Strahan, eds.). Reed New Holland, Sydney, Australia.

Courbis, S. S., and G. A. J. Worthy. 2003. Opportunistic carnivory by Florida manatees (*Trichechus manatus latirostris*). Aquatic Mammals 29:104–107.

Cousins, D. 1993. The puzzle of the pygmy elephant. International Zoo News 40(5):18–24.

Cousins, D. 1994. The African forest elephant and its status in captivity. Part I: Europe. International Zoo News 41(2):9–18.

Cowan, P. E. 2005. Brushtail possum *Trichosurus vulpecula* (Kerr, 1792). Pp. 56–80 in The Handbook of New Zealand mammals. 2nd edition (C. M. King, ed.). Oxford University Press, South Melbourne, Australia.

Craig, S. A. 1985. Social organization, reproduction, and feeding behaviour of a population of yellow-bellied gliders, *Petaurus australis* (Marsupialia: Petauridae). Australian Wildlife Research 12:1–18.

Craig, S. A. 1986. A record of twins in the yellow-bellied glider (Petaurus australis Shaw) (Marsupialia: Petauridae) with notes on the litter size and reproductive strategy of the species. Victorian Naturalist 103:72–75.

Cranbrook, Earl of, J. Payne, and C. M. U. Leh. 2007. Origin of the elephants Elephas maximus L. of Borneo. Sarawak Museum Journal, new series, 63(84):95–125.

Crandall, L. S. 1964. The management of wild animals in captivity. University of Chicago Press, Chicago, Illinois.

Crane, M. J., D. B. Lindenmayer, and R. B. Cunningham. 2010. The use of den trees by the squirrel glider (Petaurus norfolcensis) in temperate Australian woodlands. Australian Journal of Zoology 58:39–49.

Crawley, M. C. 1973. A live-trapping study of Australian brush-tailed possums, Trichosurus vulpecula (Kerr), in the Orongorongo Valley, Wellington, New Zealand. Australian Journal of Zoology 75–90.

Creighton, G. K. 1985. Phylogenetic inference, biogeographic interpretations, and the patterns of speciation in Marmosa (Marsupialia: Didelphidae). Acta Zoologica Fennica 170:121–124.

Creighton, G. K., and A. L. Gardner. 2007a. Genus Gracilinanus (Gardner and Creighton, 1989). Pp. 41–51 in Mammals of South America, volume 1, Marsupials, xenarthrans, shrews, and bats (A. L. Gardner, ed.). University of Chicago Press, Chicago, Illinois.

Creighton, G. K., and A. L. Gardner. 2007b. Genus Marmosa (Gray, 1821). Pp. 51–61 in Mammals of South America, volume 1, Marsupials, xenarthrans, shrews, and bats (A. L. Gardner, ed.). University of Chicago Press, Chicago, Illinois.

Creighton, G. K., and A. L. Gardner. 2007c. Genus Thylamys (Gray, 1843). Pp. 107–117 in Mammals of South America, volume 1, Marsupials, xenarthrans, shrews, and bats (A. L. Gardner, ed.). University of Chicago Press, Chicago, Illinois.

Crerar L. D., A. P. Crerar, D. P. Domning, and E. C. M. Parsons. 2014. Rewriting the history of an extinction—was a population of Steller's sea cows (Hydrodamalis gigas) at St. Lawrence Island also driven to extinction? Biology Letters 10:20140878. http://dx.doi.org/10.1098/rsbl.2014.0878. Downloaded on 21 April 2015.

Croft, D. B. 1980. Behaviour of red kangaroos, Macropus rufus (Desmarest, 1822), in northwestern New South Wales, Australia. Australian Mammalogy 4:5–58.

Croft, D. B. 1987. Socio-ecology of the antilopine wallaroo, Macropus antilopinus, in the Northern Territory, with observations on sympatric Macropus robustus woodwardii and Macropus agilis. Australian Wildlife Research 14:243–255.

Croft, D. B. 1989. Social organization of the Macropodoidea. Pp. 505–525 in Kangaroos, wallabies and rat-kangaroos (G. C. Grigg, P. Jarman, and I. Hume, eds.). Surrey Beatty & Sons, Chipping Norton, New South Wales, Australia.

Croft, D. B. 1991a. Home range of the euro, Macropus robustus erubescens. Journal of Arid Environments 20:99–111.

Croft, D. B. 1991b. Home range of the red kangaroo Macropus rufus. Journal of Arid Environments 20:83–98.

Croft, D. B. 2005. The future of kangaroos: going, going, gone? Pp. 223–243 in Kangaroos: myths and realities (M. Wilson and D. Croft, eds.). Australian Wildlife Protection Council, Melbourne, Victoria, Australia.

Croft, D. B., and T. F. Clancy. 2008. Red kangaroo Macropus rufus (Desmarest, 1822). Pp. 352–354 in The mammals of Australia. 3rd edition (S. Van Dyck and R. Strahan, eds.). Reed New Holland, Sydney, Australia.

Crosby, K., and C. A. Norris. 2003. Periotic morphology in the trichosurin possums Strigocuscus celebensis and Wyulda squamicaudata (Diprotodontia, Phalangeridae) and a revised diagnosis of the tribe Trichosurini. American Museum Novitates 3414:1–14.

Crosby, K., M. Bassarova, M. Archer, and K. Carbery. 2004. Fossil possums in Australasia: discovery, diversity and evolution. Pp. 161–176 in The biology of Australian possums and gliders (R. L. Goldingay and S. M. Jackson, eds.). Surrey Beatty & Sons, Chipping Norton, New South Wales, Australia.

Crowcroft, P. 1977. Breeding of wombats (Lasiorhinus latifrons) in captivity. Zoologische Garten 47:313–322.

Crowther, M. S. 2002a. Distributions of species of the Antechinus stuartii–A. flavipes complex as predicted by bioclimatic modeling. Australian Journal of Zoology 50:77–91.

Crowther, M. S. 2002b. Morphological variation within Antechinus agilis and Antechinus stuartii (Marsupialia: Dasyuridae). Australian Journal of Zoology 50:339–356.

Crowther, M. S. 2008. Yellow-footed antechinus Antechinus flavipes Waterhouse (1838). Pp. 83–84 in The mammals of Australia. 3rd edition (S. Van Dyck and R. Strahan, eds.). Reed New Holland, Sydney, Australia.

Crowther, M. S., and R. W. Braithwaite. 2008. Brown antechinus Antechinus stuartii (Macleay, 1841). Pp. 94–96 in The mammals of Australia. 3rd edition (S. Van Dyck and R. Strahan, eds.). Reed New Holland, Sydney, Australia.

Crowther, M. S., C. R. Dickman, and A. J. Lynam. 1999. Sminthopsis griseoventer boullangerensis (Marsupialia: Dasyuridae), a new subspecies in the S. murina complex from Boullanger Island, Western Australia. Australian Journal of Zoology 47:215–243.

Crowther, M. S., J. Sumner, and C. R. Dickman. 2003. Speciation of Antechinus stuartii and A. subtropicus (Marsupialia: Dasyuridae) in eastern Australia: molecular and morphological evidence. Australian Journal of Zoology 51:443–462.

Crowther, M. S., P. B. S. Spencer, D. Alpers, and C. R. Dickman. 2002. Taxonomic status of the mardo, Antechinus flavipes leucogaster (Marsupialia: Dasyuridae): a morphological, molecular, reproductive and bioclimatic approach. Australian Journal of Zoology 50:627–647.

Croze, H., and W. K. Lindsay. 2011. Amboseli ecosystem context: past and present. Pp. 11–28 in The Amboseli elephants: a long-term perspective on a long-lived mammal (C. J. Moss, H. Croze, and P. C. Lee, eds.). University of Chicago Press, Chicago, Illinois.

Croze, H., and C. J. Moss. 2011. Patterns of occupancy in time and space. Pp. 89–105 in The Amboseli elephants: a long-term perspective on a long-lived mammal (C. J. Moss, H. Croze, and P. C. Lee, eds.). University of Chicago Press, Chicago, Illinois.

Cuarón, A. D. 2014a. Silky anteater. Pp. 98–100 in Mammals of Mexico (G. Ceballos, ed.). Johns Hopkins University Press, Baltimore, Maryland.

Cuarón, A. D. 2014b. Mexican anteater. Pp. 100–101 in Mammals of Mexico (G. Ceballos, ed.). Johns Hopkins University Press, Baltimore, Maryland.

Cuarón, A. D., I. J. March, and P. M. Rockstroh. 1989. A second armadillo (Cabassous centralis) for the faunas of Guatemala and Mexico. Journal of Mammalogy 70:870–871.

Cuarón, A. D., L. Emmons, K. Helgen, F. Reid, D. Lew, B. Patterson, C. Delgado, and S. Solari. 2008. Chironectes minimus. The IUCN Red List of Threatened Species 2008: e.T4671A11076156. http://dx

.doi.org/10.2305/IUCN.UK.2008.RLTS.T4671A11076156.en. Downloaded on 12 September 2015.

Cuéllar, E. 2008. Biology and ecology of armadillos in the Bolivian Chaco. Pp. 306–312 in The biology of the Xenarthra (S. F. Vizcaíno and W. J. Loughry, eds.). University Press of Florida, Gainesville, Florida.

Cuéllar, E., D. A. Meritt, F. Delsuc, M. Superina, and A. M. Abba. 2014. *Calyptophractus retusus*. The IUCN Red List of Threatened Species 2014: e.T4703A47439036. http://dx.doi.org/10.2305/IUCN.UK .2014-1.RLTS.T4703A47439036.en. Downloaded on 06 May 2015.

Cumming, D. H. M. 1981. The management of elephant and other large mammals in Zimbabwe. Pp. 91–118 in Problems in management of locally abundant wild mammals (P. A. Jewell and S. Holt, eds.). Academic Press, New York, New York.

Cumming, D. H. M., and R. F. Du Toit. 1989. The African Elephant and Rhino Group Nyeri meeting. Pachyderm 11:4–6.

Cumming, D. H. M., R. F. Du Toit, and S. N. Stuart. 1990. African elephants and rhinos: status survey and conservation action plan. IUCN (World Conservation Union), Gland, Switzerland.

Cunha, A. A., and V. J. Vieira. 2002. Support diameter, incline, and vertical movements of four didelphid marsupials in the Atlantic Forest of Brazil. Journal of Zoology 258:419–426.

Cuong, T. V., T. T. Lien, and P. M. Giao. 2002. The present status and management of domesticated Asian elephants in Viet Nam. Pp. 111–128 in Giants on our hands. Proceedings of the International Workshop on the Domesticated Asian Elephant (I. Baker and M. Kashio, eds.). FAO (Food and Agriculture Organization of the United Nations) Regional Office for Asia and the Pacific, Bangkok, Thailand.

Cuthbert, R. J., and M. J. H. Denny. 2013. Aspects of the ecology of the kalubu bandicoot (*Echymipera kalubu*) and observations on Raffray's bandicoot (*Peroryctes raffrayanus*), Eastern Highlands Province, Papua New Guinea. Australian Mammalogy 36:21–28.

Cuttle, P. 1982. Life history of the dasyurid marsupial *Phascogale tapoatafa*. Pp. 13–22 in Carnivorous marsupials (M. Archer, ed.). Royal Zoological Society of New South Wales, Mosman, New South Wales, Australia.

Dale, R. H. 2010. Birth statistics for African (*Loxodonta africana*) and Asian (*Elephas maximus*) elephants in human care: history and implications for elephant welfare. Zoo Biology 29:87–103.

Danquah, E., and S. K. Oppong. 2014. Survey of forest elephants *Loxodonta cyclotis* (Matschie, 1900) (Mammalia: Proboscidea: Elephantidae) in the Bia Conservation Area, Ghana. Journal of Threatened Taxa 6:6399–6405.

Dans, A. T. L. 1993. Population estimate and behavior of Palawan tree shrew, *Tupaia palawanensis* (Scandentia, Tupaiidae). Asia Life Sciences 2:201–214.

Da Silva, V. M. F. 2003. Another captive pregnancy in an Amazonian manatee. Sirenews 40:5.

Da Silva, V. M. F. 2004. New Amazonian manatee captive birth. Sirenews 41:3.

Da Silveira, T. B., F. R. de Melo, and J. E. P. Lima. 2014. New field data on reproduction, diet, and activity of *Glironia venusta* (Thomas, 1912) (Didelphimorphia, Didelphidae) in northern Brazil. Mammalia 78:217–222.

Dawson, L. 1982a. Taxonomic status of fossil devils (*Sarcophilus*, Dasyuridae, Marsupialia) from late Quaternary eastern Australian localities. Pp. 517–25 in Carnivorous marsupials (M. Archer, ed.). Royal Zoological Society of New South Wales, Mosman, New South Wales, Australia.

Dawson, L. 1982b. Taxonomic status of fossil thylacines (*Thylacinus*, Thylacinidae, Marsupialia) from late Quaternary deposits in eastern Australia. Pp. 527–536 in Carnivorous marsupials (M. Archer, ed.). Royal Zoological Society of New South Wales, Mosman, New South Wales, Australia.

Dawson, L., and T. Flannery. 1985. Taxonomic and phylogenetic status of living and fossil kangaroos and wallabies of the genus *Macropus* Shaw (Macropodidae: Marsupialia), with a new subgeneric name for the larger wallabies. Australian Journal of Zoology 33:473–498.

Dawson, M. R., and L. Krishtalka. 1984. Fossil history of the families of Recent mammals. Pp. 11–58 in Orders and families of Recent mammals of the world (S. Anderson and J. K. Jones, Jr., eds.). John Wiley & Sons, New York, New York.

Dawson, T. J. 1995. Kangaroos: biology of the largest marsupials. Cornell University Press, Ithaca, New York.

Dawson, T. J., K. J. McTavish, and B. A. Ellis. 2004. Diets and foraging behaviour of red and eastern grey kangaroos in arid shrub land: is feeding behaviour involved in the range expansion of the eastern grey kangaroo into the arid zone? Australian Mammalogy 26:169–178.

Dawson, T. J., E. Finch, L. Freedman, I. D. Hume, M. B. Renfree, and P. D. Temple-Smith. 1989. Morphology and physiology of the Metatheria. Pp. 451–504 in Fauna of Australia: Mammalia, volume 1B (D. W. Walton and B. J. Richardson, eds.). Australian Government Publishing Service, Canberra, Australia.

Dawson, T. J., K. J. McTavish, A. J. Munn, and B. A. Ellis. 2006. Water use and the thermoregulatory behaviour of kangaroos in arid regions: insights into the colonisation of arid rangelands in Australia by the eastern grey kangaroo (*Macropus giganteus*). Journal of Comparative Physiology B 176:45–53.

Debruyne, R., 2001. New mitochondrial data demonstrating a close relationship between *Mammuthus primigenius* (Blumenbach, 1799) and *Loxodonta africana* (Blumenbach, 1797). Pp. 630–632 in Proceedings of the First International Congress of the World of Elephants (G. Cavarretta, P. Gioia, M. Mussi, and M. R. Palombo, eds.). Consiglio Nazionale delle Ricerche, Rome, Italy.

Debruyne, R. 2005. A case study of apparent conflict between molecular phylogenies: the interrelationships of African elephants. Cladistics 21:31–50.

Debruyne, R., V. Barriel, and P. Tassy. 2003. Mitochondrial cytochrome b of the Lyakhov mammoth (Proboscidea, Mammalia): new data and phylogenetic analyses of Elephantidae. Molecular Phylogenetics and Evolution 26:421–34.

Debruyne, R., A. Van Holt, V. Barriel, and P. Tassy. 2003. Status of the so-called African pygmy elephant (*Loxodonta pumilio* (Noack 1906)): phylogeny of cytochrome b and mitochondrial control region sequences. Comptes Rendus Biologies 326:687–697.

Debruyne, R., G. Chu, C. E. King, K. Bos, M. Kuch, C. Schwarz, P. Szpak, D. R. Gröcke, P. Matheus, G. Zazula, D. Guthrie, D. Froese, B. Buigues, C. de Marliave, C. Flemming, D. Poinar, D. Fisher, J. Southon, A. N. Tikhonov, R. D. E. MacPhee, and H. N. Poinar. 2008. Out of America: ancient DNA evidence for a New World origin of late Quaternary woolly mammoths. Current Biology 18:1320–1326.

Dee, M., D. Wengrow, A. Shortland, A. Stevenson, F. Brock, L. Girdland Flink, and C. Bronk Ramsey. 2013. An absolute chronology for early Egypt using radiocarbon dating and Bayesian statistical modelling. Proceedings of the Royal Society A 469:20130395.

http://dx.doi.org/10.1098/rspa.2013.0395. Downloaded on 29 March 2015.

Deems, E. F., Jr., and D. Pursley (eds.). 1978. North American furbearers. International Association of Fish and Wildlife Agencies, University of Maryland Press, College Park, Maryland.

Delaney, R. 1997. Reproductive ecology of the allied rock-wallaby, *Petrogale assimilis*. Australian Mammalogy 19:209–218.

De la Sancha, N. U., and G. D'Elía. 2015. Additions to the Paraguayan mammal fauna: the first records of two marsupials (Didelphimorphia, Didelphidae) with comments on the alpha taxonomy of *Cryptonanus* and *Philander*. Mammalia 79:343–356.

De la Sancha, N. U., G. D'Elía, and P. Teta. 2012. Systematics of the subgenus of mouse opossums *Marmosa* (*Micoureus*) (Didelphimorphia, Didelphidae) with noteworthy records from Paraguay. Mammalian Biology 77:229–236.

De la Sancha, N. U., S. Solari, and R. D. Owen. 2007. First records of *Monodelphis kunsi* Pine (Didelphimorphia, Didelphidae) from Paraguay, with an evaluation of its distribution. Mastozoologia Neotropical 14:241–247.

De la Sancha, N., and P. Teta. 2015. *Thylamys macrurus*. The IUCN Red List of Threatened Species 2015: e.T21867A22173324. http://dx.doi.org/10.2305/IUCN.UK.2015-4.RLTS.T21867A22173324.en. Downloaded on 27 December 2015.

De la Sancha, N., P. Teta, D. Flores, and M. S. Albanese. 2015. *Thylamys pusillus*. The IUCN Red List of Threatened Species 2015: e.T201936A22172657. http://dx.doi.org/10.2305/IUCN.UK.2015-4.RLTS.T201936A22172657.en. Downloaded on 27 December 2015.

Del Carmen Peralta, M., and V. Pacheco. 2014. Rediscovery of *Marmosops juninensis* (Tate, 1931) (Didelphimorphia: Didelphidae) in the Yungas of Peru. Check List 10:436–440.

D'Elía, G., N. Hurtado, and A. D'Anatro. 2016. Alpha taxonomy of *Dromiciops* (Microbiotheriidae) with the description of 2 new species of monito del monte. Journal of Mammalogy 97:1136–1152.

Delroy, L. B., J. Earl, I. Radbone, A. C. Robinson, and M. Hewett. 1986. The breeding and re-establishment of the brush-tailed bettong, *Bettongia penicillata*, in South Australia. Australian Wildlife Research 13:387–396.

Delsink, A. K., J. J. van Altena, D. Grobler, H. J. Bertschinger, J. F. Kirkpatrick, and R. Slotow. 2007. Implementing immunocontraception in free-ranging African elephants at Makalali conservancy. Journal of the South African Veterinary Association 78:25–30.

Delsuc, F., and E. J. P. Douzery. 2008. Recent advances and future prospects in xenarthran molecular phylogenetics. Pp. 11–23 in The biology of the Xenarthra (S. F. Vizcaíno and W. J. Loughry, eds.). University Press of Florida, Gainesville, Florida.

Delsuc, F., M. J. Stanhope, and E. J. P. Douzery. 2003. Molecular systematics of armadillos (Xenarthra, Dasypodidae): contribution of maximum likelihood and Bayesian analyses of mitochondrial and nuclear genes. Molecular Phylogenetics and Evolution 28:261–275.

Delsuc, F., S. F. Vizcaíno, and E. J. P. Douzery. 2004. Influence of Tertiary paleoenvironmental changes on the diversification of South American mammals: a relaxed molecular clock study within xenarthrans. BMC Evolutionary Biology 4:11. http://biomedcentral.com/1471-2148/4/11. Accessed on 4 May 2015.

Delsuc, F., M. Scally, O. Madsen, M. J. Stanhope, W. W. de Jong, F. M. Catzeflis, M. S. Springer, and E. J. P. Douzery. 2002. Molecular phylogeny of living xenarthrans and the impact of character and taxon sampling on the placental tree rooting. Molecular Biology and Evolution 19:1656–1671.

Delsuc, F., M. Superina, M.-K. Tilak, E. J. P. Douzery, and A. Hassanin. 2012. Molecular phylogenetics unveils the ancient evolutionary origins of the enigmatic fairy armadillos. Molecular Phylogenetics and Evolution 62:673–680.

Demeke, Y., M. B. Renfree, and R. V. Short. 2012. Historical range and movements of the elephants in Babile Elephant Sanctuary, Ethiopia. African Journal of Ecology 50:439–445.

Dene, H., M. Goodman, and W. Prychodko. 1978. An immunological examination of the systematics of Tupaioidea. Journal of Mammalogy 59:697–706.

Dennis, A. J. 2002. The diet of the musky rat-kangaroo, *Hypsiprymnodon moschatus*, a rainforest specialist. Wildlife Research 29:209–219.

Dennis, A. J. 2003. Scatter-hoarding by musky rat-kangaroos, *Hypsiprymnodon moschatus*, a tropical rain-forest marsupial from Australia: implications for seed dispersal. Journal of Tropical Ecology 19:619–627.

Dennis, A. J., and P. M. Johnson. 2008a. Musky rat-kangaroo *Hypsiprymnodon moschatus* (Ramsay, 1876). Pp. 281–283 in The mammals of Australia. 3rd edition (S. Van Dyck and R. Strahan, eds.). Reed New Holland, Sydney, Australia.

Dennis, A. J., and P. M. Johnson. 2008b. Rufous bettong *Aepyprymnus rufescens* (Gray, 1837). Pp. 285–286 in The mammals of Australia. 3rd edition (S. Van Dyck and R. Strahan, eds.). Reed New Holland, Sydney, Australia.

Dennis, A. J., and H. Marsh 1997. Seasonal reproduction in musky rat-kangaroos, *Hypsiprymnodon moschatus*: a response to changes in resource availability. Wildlife Research 24:561–578.

Desbiez, A. L. J., and I. M. Medri. 2010. Density and habitat use by giant anteaters (*Myrmecophaga tridactyla*) and southern tamanduas (*Tamandua tetradactyla*) in the Pantanal Wetland, Brazil. Edentata 11:4–10.

De Silva, S. 2010. Acoustic communication in the Asian elephant, *Elephas maximus maximus*. Behaviour 147:825–852.

De Silva, S., A. D. G. Ranjeewa, and D. Weerakoon. 2011. Demography of Asian elephants (*Elephas maximus*) at Uda Walawe National Park, Sri Lanka, based on identified individuals. Biological Conservation 144:1742–1752.

De Silva, S., and G. Wittemyer. 2012. a comparison of social organization in Asian elephants and African savannah elephants. International Journal of Primatology 33:1125–1141.

De Silva, S., C. E. Webber, U. S. Weerathunga, T. V. Pushpakumara, D. K. Weerakoon, and G. Wittemyer. 2013. Demographic variables for wild Asian elephants using longitudinal observations. PLoS ONE 8(12):e82788. doi:10.1371/journal.pone.0082788.

De Tores, P. J. 2008a. Western ringtail possum *Pseudocheirus occidentalis* (Thomas, 1888). Pp. 253–255 in The mammals of Australia. 3rd edition (S. Van Dyck and R. Strahan, eds.). Reed New Holland, Sydney, Australia.

De Tores, P. J. 2008b. Quokka *Setonix brachyurus* (Quoy and Gaimard, 1830). Pp. 402–404 in The mammals of Australia. 3rd edition (S. Van Dyck and R. Strahan, eds.). Reed New Holland, Sydney, Australia.

De Tores, P. J., and A. N. Start. 2008. Woylie *Bettongia penicillata* (Gray, 1837). Pp. 291–292 in The mammals of Australia. 3rd edition (S. Van Dyck and R. Strahan, eds.). Reed New Holland, Sydney, Australia.

De Tores, P. J., M. W. Hayward, M. J. Dillon, and R. I. Brazell. 2007. Review of the distribution, causes for the decline and recommendations for management of the quokka, *Setonix brachyurus* (Macropodidae: Marsupialia), an endemic macropodid marsupial from south-west Western Australia. Conservation Science Western Australia 6:13–73.

De Tores, P., A. Burbidge, K. Morris, and T. Friend. 2008. *Setonix brachyurus*. The IUCN Red List of Threatened Species 2008: e.T20165A9156418. http://dx.doi.org/10.2305/IUCN.UK.2008 .RLTS.T20165A9156418.en. Downloaded on 26 June 2017.

Deutsch, C. J. 2008. *Trichechus manatus* ssp. *latirostris*. The IUCN Red List of Threatened Species 2008: e.T22106A9359881. http://dx.doi .org/10.2305/IUCN.UK.2008.RLTS.T22106A9359881.en. Downloaded on 02 May 2015.

Deutsch, C. J., and J. E. Reynolds, III. 2012. Florida manatee status and conservation issues. A primer. Pp. 23–35 in Sirenian conservation. Issues and strategies in developing countries (E. M. Hines, J. E. Reynolds, III, L. V. Aragones, A. A. Mignucci-Giannoni, and M. Marmontel, eds.). University Press of Florida, Gainesville, Florida.

Deutsch, C. J., C. Self-Sullivan, and A. Mignucci-Giannoni. 2008. *Trichechus manatus*. The IUCN Red List of Threatened Species 2008: e.T22103A9356917. http://dx.doi.org/10.2305/IUCN.UK.2008 .RLTS.T22103A9356917.en. Downloaded on 02 May 2015.

Deutsch, C. J., J. P. Reid, R. K. Bonde, D. E. Easton, H. I. Kochman, and T. J. O'Shea. 2003. Seasonal movements, migratory behavior, and site fidelity of West Indian manatees along the Atlantic coast of the United States. Wildlife Monographs 151:1–77.

Diamond, J. 2008. The last giant kangaroo. Nature 454:835–836.

Dias, B. B., L. A. D. dos Santos, P. Lara-Ruiz, C. R. Cassano, L. Pinder, and A. G. Chiarello. 2009. First observation on mating and reproductive seasonality in maned sloths *Bradypus torquatus* (Pilosa: Bradypodidae). Journal of Ethology 27:97–103.

Dias, I. M. G., F. C. Almeida, G. Amato, R. DeSalle, and C. G. Fonseca. 2010. Delineating geographic boundaries of the woolly mouse opossums, *Micoureus demerarae* and *Micoureus paraguayanus* (Didelphimorphia: Didelphidae). Conservation Genetics 11:1579–1585.

Diaz, M., and R. Barquez. 2008a. *Cryptonanus ignitus*. The IUCN Red List of Threatened Species 2008: e.T41320A10442855. http://dx.doi .org/10.2305/IUCN.UK.2008.RLTS.T41320A10442855.en. Downloaded on 29 December 2015.

Diaz, M., and R. Barquez. 2008b. *Thylamys venustus*. The IUCN Red List of Threatened Species 2008: e.T136626A4319732. http://dx.doi .org/10.2305/IUCN.UK.2008.RLTS.T136626A4319732.en. Downloaded on 27 December 2015.

Diaz, M., and R. Barquez. 2008c. *Thylamys cinderella*. The IUCN Red List of Threatened Species 2008: e.T136617A4318831. http://dx.doi .org/10.2305/IUCN.UK.2008.RLTS.T136617A4318831.en. Downloaded on 27 December 2015.

Diaz, M., and R. Barquez. 2008d. *Thylamys sponsorius*. The IUCN Red List of Threatened Species 2008: e.T136213A4260624. http://dx.doi .org/10.2305/IUCN.UK.2008.RLTS.T136213A4260624.en. Downloaded on 27 December 2015.

Díaz, M., D. A. Flores, and R. Barquez. 2002. A new species of gracile mouse opossum, genus *Gracilinanus* (Didelphimorphia: Didelphidae), from Argentina. Journal of Mammalogy 83:824–833.

Diaz, M., and P. Teta. 2008. *Rhyncholestes raphanurus*. The IUCN Red List of Threatened Species 2008: e.T19710A9005805. http://dx.doi .org/10.2305/IUCN.UK.2008.RLTS.T19710A9005805.en Downloaded on 18 January 2016.

Díaz-N., J. F. 2012. New records of *Marmosops noctivagus* (Tschudi, 1845) (Didelphimorpia: Didelphidae) and first record of *Marmosops bishopi* (Pine, 1981) for Colombia. Check List 8:805–809.

Díaz-N., J. F., M. Gómez-Laverde, and C. Sánchez-Giraldoas. 2011. Rediscovery and redescription of *Marmosops handleyi* (Pine, 1981) (Didelphimorphia: Didelphidae), the least known Andean slender mouse opossum. Mastozoología Neotropical 18:45–61.

Díaz-N., J. F., and C. Sánchez-Giraldo. 2008. Notable altitudinal range extension of the northern naked-tailed armadillo *Cabassous centralis* (Cingulata: Dasypodidae) in Colombia. Brenesia 69:75–76.

Dickman, C. R. 2005. Marsupials of the world: an introduction. Pp. 1–67 in Walker's marsupials of the world (R. M. Nowak). Johns Hopkins University Press, Baltimore, Maryland.

Dickman, C. R. 2008a. Agile antechinus *Antechinus agilis* (Dickman, Parnaby, Crowther and King, 1998). Pp. 83–84 in The mammals of Australia. 3rd edition (S. Van Dyck and R. Strahan, eds.). Reed New Holland, Sydney, Australia.

Dickman, C. R. 2008b. Dusky antechinus *Antechinus swainsonii* (Waterhouse, 1840). Pp. 99–100 in The mammals of Australia. 3rd edition (S. Van Dyck and R. Strahan, eds.). Reed New Holland, Sydney, Australia.

Dickman, C. R. 2008c. Grey-bellied dunnart *Sminthopsis griseoventer* (Kitchener, Stoddart and Henry, 1984). Pp. 141–143 in The mammals of Australia. 3rd edition (S. Van Dyck and R. Strahan, eds.). Reed New Holland, Sydney, Australia.

Dickman, C. R. 2015. Family Peramelidae (bandicoots and echymiperas). Pp. 362–398 in Handbook of the mammals of the world, volume 5, Monotremes and marsupials (D. E. Wilson and R. A. Mittermeier, eds.). Lynx Edicions, Barcelona, Spain.

Dickman, C. R., and R. W. Braithwaite. 1992. Postmating mortality of males in the dasyurid marsupials, *Dasyurus* and *Parantechinus*. Journal of Mammalogy 73:143–147.

Dickman, C., S. Burnett, and N. McKenzie. 2008. *Sminthopsis murina*. The IUCN Red List of Threatened Species 2008: e.T40547A10332196. http://dx.doi.org/10.2305/IUCN.UK.2008.RLTS.T40547 A10332196.en. Downloaded on 04 January 2016.

Dickman, C., D. Lunney, and P. Menkhorst. 2008. *Cercartetus nanus*. The IUCN Red List of Threatened Species 2008: e.T40578A10321445. http://dx.doi.org/10.2305/IUCN.UK.2008.RLTS.T40578A10321445 .en. Downloaded on 09 December 2015.

Dickman, C. R., and E. Stodart. 2008. Long-nosed bandicoot *Perameles nasuta* (Geoffroy, 1804). Pp. 189–190 in The mammals of Australia. 3rd edition (S. Van Dyck and R. Strahan, eds.). Reed New Holland, Sydney, Australia.

Dickman, C. R., D. H. King, M. Adams, and P. R. Baverstock. 1988. Electrophoretic identification of a new species of *Antechinus* (Marsupialia: Dasyuridae) in south-eastern Australia. Australian Journal of Zoology 36:455–463.

Dickman C. R., H. E. Parnaby, M. S. Crowther, and D. H. King. 1998. *Antechinus agilis* (Marsupialia: Dasyuridae), a new species from the *A. stuartii* complex in south-eastern Australia. Australian Journal of Zoology 46:1–26.

Dickman, C., A. Burbidge, K. Aplin, and J. Benshemesh. 2008. *Notoryctes typhlops*. The IUCN Red List of Threatened Species 2008: e.T14879A4467905. http://dx.doi.org/10.2305/IUCN.UK.2008 .RLTS.T14879A4467905.en. Downloaded on 03 January 2016.

Dimpel, H., and J. H. Calaby. 1972. Further observations on the mountain pigmy possum (*Burramys parvus*). Victorian Naturalist 89:101–106.

Di Stefano, J., M. Swan, A. Greenfield, and G. Coulson. 2010. Effect of habitat type, sex and time of day on space use by the swamp wallaby. Pp. 187–196 in Macropods: the biology of kangaroos, wallabies and rat-kangaroos (G. Coulson and M. Eldridge, eds.). CSIRO Publishing, Collingwood, Victoria, Australia.

Dittrich, L. 1966. Breeding Indian elephants *Elephas maximus* at Hanover Zoo. International Zoo Yearbook 6:193–196.

Dixon, J. M. 1971. *Burramys parvus* Broom (Marsupialia) from Falls Creek area of the Bogong high plains, Victoria. Victorian Naturalist 88:133–138.

Dixon, J. M. 1978. The first Victorian and other records of the little pigmy possum *Cercartetus lepidus* (Thomas). Victorian Naturalist 95:4–7.

Dixon, J. M. 1988. Notes on the diet of three mammals presumed to be extinct: the pig-footed bandicoot, the lesser bilby and the desert rat kangaroo. Victorian Naturalist 105:208–211.

Dobbs, K., I. Lawler, and D. Kwan. 2012. Dugongs in Australia and the Pacific. Pp. 99–105 in Sirenian conservation. Issues and strategies in developing countries (E. M. Hines, J. E. Reynolds, III, L. V. Aragones, A. A. Mignucci-Giannoni, and M. Marmontel, eds.). University Press of Florida, Gainesville, Florida.

Dobias, R. J. 1987. Elephants in Thailand: an overview of their status and conservation. Tigerpaper 14(1):19–24.

Dobroruka, L. J. 1973. Yellow-spotted dassie *Heterohyrax brucei* (Gray, 1868) feeding on a poisonous plant. Säugetierkundliche Mitteilungen 21:365.

Dolgin, E. 2012. Rewriting evolution. Nature 486:460–462.

Domning, D. P. 1978. Sirenian evolution in the North Pacific Ocean. University of California Publications in Geological Sciences 118:1–176.

Domning, D. P. 1981. Distribution and status of manatees *Trichechus* spp. near the mouth of the Amazon River, Brazil. Biological Conservation 19:85–97.

Domning, D. P. 1982a. Commercial exploitation of manatees *Trichechus* in Brazil, c. 1785–1973. Biological Conservation 22:101–126.

Domning, D. P. 1982b. Evolution of manatees: a speculative history. Journal of Paleontology 56:599–619.

Domning, D. P. 1994. A phylogenetic analysis of the Sirenia. Proceedings of the San Diego Society of Natural History 29:177–189.

Domning, D. P. 1996. Bad news or good? Sirenews 25:1–2.

Domning, D. P. 2005. Fossil Sirenia of the West Atlantic and Caribbean region. VII. Pleistocene *Trichechus manatus* (Linnaeus, 1758). Journal of Vertebrate Paleontology 25:685–701.

Domning, D. P. 2013. Order Sirenia—Dugongs, Manatees. Pp. 201–202 in Mammals of Africa, volume 1, Introductory chapters and Afrotheria (J. Kingdon, D. Happold, M. Hoffmann, T. Butynski, M. Happold, and J. Kalina, eds.). Bloomsbury, London, United Kingdom.

Domning, D. P., P. K. Anderson, and S. Turvey. 2008. *Hydrodamalis gigas*. The IUCN Red List of Threatened Species 2008: e.T10303A3191997. http://dx.doi.org/10.2305/IUCN.UK.2008.RLTS.T10303A3191997.en. Downloaded on 20 April 2015.

Domning, D. P., and V. de Buffrénil. 1991. Hydrostasis in the Sirenia: quantitative data and functional interpretations. Marine Mammal Science 7:331–368.

Domning, D. P., and H. Furusawa. 1994. Summary of taxa and distribution of Sirenia in the North Pacific Ocean. The Island Arc 3:506–512.

Domning, D. P., and L.-A. C. Hayek. 1986. Interspecific and intraspecific morphological variation in manatees (Sirenia: *Trichechus*). Marine Mammal Science 2:87–144.

Domning, D. P., S. Kendall, and D. L. Orozco. 2004. How many manatees really were killed in Brazil by past commercial exploitation? Sirenews 41:3–5.

Domning, D. P., G. S. Morgan, and C. E. Ray. 1982. North American Eocene sea cows (Mammalia: Sirenia). Smithsonian Contributions in Paleobiology 52:1–69.

Domning, D. P., and C. E. Ray. 1986. The earliest Sirenian (Mammalia: Dugongidae) from the eastern Pacific Ocean. Marine Mammal Science 2:263–276.

Domning, D. P., J. Thomason, and D. G. Corbett. 2007. Steller's sea cow in the Aleutian Islands. Marine Mammal Science 23:976–983.

Domning, D. P., I. S. Zalmout, and P. D. Gingerich. 2010. Sirenia. Pp. 147–160 in Cenozoic mammals of Africa (L. Werdelin and W. J. Sanders, eds.). University of California Press, Berkeley, California.

Doody, J. S., D. Rhind, C. M. Castellano, and M. Bass. 2012. Rediscovery of the scaly-tailed possum (*Wyulda squamicaudata*) in the eastern Kimberley. Australian Mammalogy 34:260–262.

Dorst, J., and P. Dandelot. 1969. A field guide to the larger mammals of Africa. Houghton Mifflin, Boston, Massachusetts.

Douady, C. J., F. Catzeflis, J. Raman, M. S. Springer, and M. J. Stanhope. 2003. The Sahara as a vicariant agent, and the role of Miocene climatic events, in the diversification of the mammalian order Macroscelidea (elephant shrews). Proceedings of the National Academy of Sciences of the United States of America 100:8325–8330.

Douglas-Hamilton, I. 1973. On the ecology and behavior of the Lake Manyara elephants. East African Wildlife Journal 11:401–403.

Douglas-Hamilton, I. 1987. African elephants: population trends and their causes. Oryx 21:11–24.

Douglas-Hamilton, I. 1992. In defence of the ivory trade ban. Oryx 26:1–4.

Douglas-Hamilton, I. 2009. The current elephant poaching trend. Pachyderm 45:154–157.

Douglas-Hamilton, I., and F. Michelmore. 1996. *Loxodonta africana*: range and distribution, past and present. Pp. 321–326 in The Proboscidea. Evolution and palaeoecology of elephants and their relatives (J. Shoshani and P. Tassy, eds.). Oxford University Press, Oxford, United Kingdom.

Drake, N. A., R. M. Blench, S. J. Armitage, C. S. Bristow, and K. H. White. 2011. Ancient watercourses and biogeography of the Sahara explain the peopling of the desert. Proceedings of the National Academy of Sciences of the United States of America 108:458–462.

Dressen, W. 1993. On the behaviour and social organization of agile wallabies, *Macropus agilis* (Gould, 1842), in two habitats of northern Australia. Zeitschrift für Säugetierkunde 58:201–211.

Druce, H. C., R. L. Mackey, K. Pretorius, and R. Slotow. 2013. The intermediate-term effects of PZP immunocontraception: behavioural monitoring of the treated elephant females and associated family groups. Animal Conservation 16:180–187.

Dublin, H. T. 2011. African Elephant Specialist Group report. Pachyderm 50:1–6.

Dublin, H. T., and R. E. Hoare. 2004. Searching for solutions: the evolution of an integrated approach to understanding and mitigating human-elephant conflict in Africa. Human Dimensions of Wildlife 9:271–278.

Dublin, H. T., T. Milliken, and R. F. W. Barnes. 1995. Four years after the CITES ban: illegal killing of elephants, ivory trade and stockpiles. Report of the IUCN (World Conservation Union) Species Survival Commission African Elephant Specialist Group.

Ducrocq, S., E. Buffetaut, H. Buffetaut-Tong, J.-J. Jaeger, Y. Jongkanjanasoontorn, and V. Suteethorn. 1992. First fossil marsupial from South Asia. Journal of Vertebrate Paleontology 12:395–399.

Dumbacher, J. P., E. J. Carlen, and G. B. Rathbun. 2016. *Petrosaltator* gen. nov., a new genus replacement for the North African sengi *Elephantulus rozeti* (Macroscelidea; Macroscelididae). Zootaxa 4136:567–579.

Dumbacher, J. P., G. B. Rathbun, H. A. Smit, and S. J. Eiseb. 2012. Phylogeny and taxonomy of the round-eared sengis or elephant-shrews, genus *Macroscelides* (Mammalia, Afrotheria, Macroscelidea). PLoS ONE 7(3):e32410. doi:10.1371/journal.pone.0032410.

Dumbacher, J. P., G. B. Rathbun, T. O. Osborne, M. Griffin, and S. J. Eiseb. 2014. A new species of round-eared sengi (genus *Macroscelides*) from Namibia. Journal of Mammalogy 95:443–454.

Dunlop, J. N., N. K. Cooper, and R. J. Teale. 2008. Pilbara ningaui *Ningaui timealeyi* (Archer, 1973). Pp. 119–120 in The mammals of Australia. 3rd edition (S. Van Dyck and R. Strahan, eds.). Reed New Holland, Sydney, Australia.

Dwiyahreni, A. A., M. F. Kinnaird, T. G. O'Brien, J. Supriatna, and N. Andayani. 1999. Diet and activity of the bear cuscus, *Ailurops ursinus*, in north Sulawesi, Indonesia. Journal of Mammalogy 80:905–912.

Dwyer, P. D. 1977. Notes on *Antechinus* and *Cercartetus* (Marsupialia) in the New Guinea highlands. Proceedings of the Royal Society of Queensland 88:69–73.

Dwyer, P. D. 1983. An annotated list of mammals from Mt. Erimbari, Eastern Highlands Province, Papua New Guinea. Science in New Guinea 10:28–38.

Dzulhelmi, M. N., and M. T. Abdullah. 2009 An ethogram construction for the Malayan flying lemur (*Galeopterus variegatus*) in Bako National Park, Sarawak, Malaysia. Journal of Tropical Biology and Conservation 5:31–42.

Easter, T., N. Greenwald, B. Cummings, and S. Uhlmann. 2015. Petition to reclassify and uplist African elephants from threatened to endangered under the Endangered Species Act as two separate species: forest elephants (*Loxodonta cyclotis*) and savannah elephants (*Loxodonta africana*). Center for Biological Diversity, Portland, Oregon.

Eberhard, I. H. 1978. Ecology of the koala, *Phascolarctos cinereus* (Goldfuss) Marsupialia: Phascolarctidae, in Australia. Pp. 315–327 in The ecology of arboreal folivores (G. G. Montgomery, ed.). Smithsonian Institution Press, Washington, DC.

Edmunds, R. M., J. W. Goertz, and G. Linscombe. 1978. Age ratios, weights, and reproduction of the Virginia opossum in northern Louisiana. Journal of Mammalogy 59:884–885.

Edwards, D., and M. Westerman. 1992. DNA-DNA hybridisation and the position of Leadbeater's possum (*Gymnobelideus leadbeateri* McCoy) in the family Petauridae (Marsupialia: Diprotodontia). Australian Journal of Zoology 40:563–571.

Edwards, G. P. 1989. The interaction between macropodids and sheep: a review. Pp. 795–804 in Kangaroos, wallabies and rat-kangaroos (G. Grigg, P. Jarman and I. Hume, eds.). Surrey Beatty & Sons, Chipping Norton, New South Wales, Australia.

Edwards, G. P., and E. H. M. Ealey. 1975. Aspects of the ecology of the swamp wallaby, *Wallabia bicolor* (Marsupialia: Macropodidae). Australian Mammalogy 1:307–317.

Eggert, L. S., C. A. Rasner, and D. S. Woodruff. 2002. The evolution and phylogeography of the African elephant inferred from mitochondrial DNA sequence and nuclear microsatellite markers. Proceedings of the Royal Society of London B Biological Sciences 269:1993–2006.

Eisenberg, J. F. 1975. Tenrecs and solenodons in captivity. International Zoo Yearbook 15:6–12.

Eisenberg, J. F. 1981. The mammalian radiations. University of Chicago Press, Chicago, Illinois.

Eisenberg, J. F. 1989. Mammals of the neotropics: the northern neotropics. University of Chicago Press, Chicago, Illinois.

Eisenberg, J. F., and E. Gould. 1967. The maintenance of tenrecoid insectivores in captivity. International Zoo Yearbook 7:194–196.

Eisenberg, J. F., and E. Gould. 1970. The tenrecs: a study in mammalian behavior and evolution. Smithsonian Contributions to Zoology 27:1–138.

Eisenberg, J. F., and M. C. Lockhart. 1972. An ecological reconnaissance of Wilpattu National Park, Ceylon. Smithsonian Contributions to Zoology 101:1–118.

Eisenberg, J. F., and E. Maliniak. 1974. The reproduction of the genus *Microgale* in captivity. International Zoo Yearbook 14:108–110.

Eisenberg, J. F., and E. Maliniak. 1985. Maintenance and reproduction of the two-toed sloth *Choloepus didactylus* in captivity. Pp. 327–331 in The evolution and ecology of armadillos, sloths, and vermilinguas (G. G. Montgomery, ed.). Smithsonian Institution Press, Washington, DC.

Eisenberg, J. F., and E. Maliniak. 1978. Reproduction by the two-toed sloth *Choloepus hoffmanni* in captivity. American Society of Mammalogists, Technical Paper Abstract, Annual Meeting 58:41–42.

Eisenberg, J. F., and K. H. Redford. 1999. Mammals of the Neotropics, volume 3, The central Neotropics. Ecuador, Peru, Bolivia, Brazil. University of Chicago Press, Chicago, Illinois.

Eisentraut, M. 1970. Beitrag zur Fortpflanzungsbiologie der Zwerbeutelratte *Marmosa murina* (Didelphidae, Marsupialia). Zeitschrift für Säugetierkunde 35:159–172.

Eldredge, L. G. 2003. The marine reptiles and mammals of Guam. Micronesia 35–36:653–660.

Eldridge, M. D. B. 1997. Taxonomy of rock-wallabies, *Petrogale* (Marsupialia: Macropodidae). II. An historical review. Australian Mammalogy 19:113–122.

Eldridge, M. D. B. 2008a. Superfamily Macropodoidea. Pp. 279–281, 284, 305–307, and 406 in The mammals of Australia. 3rd edition (S. Van Dyck and R. Strahan, eds.). Reed New Holland, Sydney, Australia.

Eldridge, M. D. B. 2008b. Rock-wallabies: *Petrogale*. Pp. 361–362 in The mammals of Australia. 3rd edition (S. Van Dyck and R. Strahan, eds.). Reed New Holland, Sydney, Australia.

Eldridge, M. D. B. 2008c. Yellow-footed rock-wallaby *Petrogale xanthopus* (Gray, 1855). Pp. 392–394 in The mammals of Australia. 3rd edition (S. Van Dyck and R. Strahan, eds.). Reed New Holland, Sydney, Australia.

Eldridge, M. D. B., and T. L. Browning. 2002. Molecular genetic analysis of the naturalized Hawaiian population of the brush-tailed rock-wallaby, *Petrogale penicillata* (Marsupialia: Macropodidae). Journal of Mammalogy 83:437–444.

Eldridge, M. D. B., and R. L. Close. 1992. Taxonomy of rock wallabies, *Petrogale* (Marsupialia: Macropodidae). I. A revision of the eastern *Petrogale* with the description of three new species. Australian Journal of Zoology 40:605–625.

Eldridge, M. D. B., and R. L. Close. 1993. Radiation of chromosome shuffles. Current Opinion in Genetics and Development 3:915–922.

Eldridge, M. D. B., and R. L. Close. 1997. Chromosomes and evolution in rock-wallabies, *Petrogale* (Marsupialia: Macropodidae). Australian Mammalogy 19:123–135.

Eldridge, M. D. B., and R. L. Close. 2008a. Allied rock-wallaby *Petrogale assimilis* (Ramsay, 1877). Pp. 363–364 in The mammals of Australia. 3rd edition (S. Van Dyck and R. Strahan, eds.). Reed New Holland, Sydney, Australia.

Eldridge, M. D. B., and R. L. Close. 2008b. Godman's rock-wallaby *Petrogale godmani* (Thomas, 1923). Pp. 372–373 in The mammals of Australia. 3rd edition (S. Van Dyck and R. Strahan, eds.). Reed New Holland, Sydney, Australia.

Eldridge, M. D. B., and R. L. Close. 2008c. Herbert's rock-wallaby *Petrogale herberti* (Thomas, 1926). Pp. 373–374 in The mammals of Australia. 3rd edition (S. Van Dyck and R. Strahan, eds.). Reed New Holland, Sydney, Australia.

Eldridge, M. D. B., and R. L. Close. 2008d. Unadorned rock-wallaby *Petrogale inornata* (Gould, 1842). Pp. 375–376 in The mammals of Australia. 3rd edition (S. Van Dyck and R. Strahan, eds.). Reed New Holland, Sydney, Australia.

Eldridge, M. D. B., and R. L. Close. 2008e. Mareeba rock-wallaby *Petrogale mareeba* (Eldridge and Close, 1992). Pp. 381–382 in The mammals of Australia. 3rd edition (S. Van Dyck and R. Strahan, eds.). Reed New Holland, Sydney, Australia.

Eldridge, M. D. B., and R. L. Close. 2008f. Brush-tailed rock-wallaby *Petrogale penicillata* (Gray, 1825). Pp. 382–384 in The mammals of Australia. 3rd edition (S. Van Dyck and R. Strahan, eds.). Reed New Holland, Sydney, Australia.

Eldridge, M. D. B., and R. L. Close. 2008g. Sharman's rock-wallaby *Petrogale sharmani* (Eldridge and Close, 1992). Pp. 391–392 in The mammals of Australia. 3rd edition (S. Van Dyck and R. Strahan, eds.). Reed New Holland, Sydney, Australia.

Eldridge, M. D. B., and G. M. Coulson. 2015. Family Macropodidae (kangaroos and wallabies). Pp. 630–735 in Handbook of the mammals of the world, volume 5, Monotremes and marsupials (D. E. Wilson and R. A. Mittermeier, eds.). Lynx Edicions, Barcelona, Spain.

Eldridge, M. D. B., P. G. Johnston, and P. S. Lowry. 1992. Chromosomal rearrangements in rock wallabies, *Petrogale* (Marsupialia: Macropodidae). VII. G-banding analysis of *Petrogale brachyotis* and *P. concinna*: species with dramatically altered karyotypes. Cytogenetics and Cell Genetics 61:34–39.

Eldridge, M. D. B., L. A. Moore, and R. L. Close. 2008. Cape York rock-wallaby *Petrogale coenensis* (Eldridge and Close, 1992). Pp. 368–369 in The mammals of Australia. 3rd edition (S. Van Dyck and R. Strahan, eds.). Reed New Holland, Sydney, Australia.

Eldridge, M. D. B., and D. J. Pearson. 2008. Black-footed rock-wallaby *Petrogale lateralis* (Gould, 1842). Pp. 376–380 in The mammals of Australia. 3rd edition (S. Van Dyck and R. Strahan, eds.). Reed New Holland, Sydney, Australia.

Eldridge, M. D. B., and W. R. Telfer. 2008. Short-eared rock-wallaby *Petrogale brachyotis* (Gould, 1841). Pp. 365–366 in The mammals of Australia. 3rd edition (S. Van Dyck and R. Strahan, eds.). Reed New Holland, Sydney, Australia.

Eldridge, M. D. B., A. C. C. Wilson, C. J. Metcalfe, A. E. Dollin, J. N. Bell, P. M. Johnson, P. G. Johnston, and R. L. Close. 2001. Taxonomy of rock-wallabies, *Petrogale* (Marsupialia: Macropodidae). III. Molecular data confirms the species status of the purple-necked rock-wallaby (*Petrogale purpureicollis* Le Souef). Australian Journal of Zoology 49:323–343.

Eldridge M. D. B., P. M. Johnson, P. Hensler, J. K. Holden, and R. L. Close. 2008. The distribution of three parapatric, cryptic species of rock-wallaby (*Petrogale*) in north-east Queensland: *P. assimilis*, *P. mareeba* and *P. sharmani*. Australian Mammalogy 30:37–42.

Eldridge, M. D. B., K. Heckenberg, L. E. Neaves, C. J. Metcalfe, S. Hamilton, P. M. Johnson, and R. L. Close. 2011. Genetic differentiation and introgression amongst *Thylogale* (pademelons) taxa in eastern Australia. Australian Journal of Zoology 59:103–117.

Ellerman, J. R., and T. C. S. Morrison-Scott. 1966. Checklist of Palaearctic and Indian mammals. British Museum (Natural History), London, United Kingdom.

Ellis, M. 1992. The mulgara, *Dasycercus cristicauda* (Krefft, 1867): a new dasyurid record for New South Wales. Australian Zoologist 28:57–58.

Ellis, M., P. Wilson, and S. Hamilton. 1991. The golden bandicoot, *Isoodon auratus* Ramsay 1887, in western New South Wales during European times. Australian Zoologist 27:36–37.

Ellis, M., P. Menkhorst, J. van Weenen, A. Burbidge, P. Copley, M. Denny, J. Woinarski, P. Mawson, and K. Morris. 2008a. *Macropus robustus*. The IUCN Red List of Threatened Species 2008: e.T40565A10334447. http://dx.doi.org/10.2305/IUCN.UK.2008.RLTS.T40565A10334447.en. Downloaded on 02 January 2016.

Ellis, M., J. Van Weenen, P. Copley, C. Dickman, P. Mawson, and J. Woinarski. 2008b. *Macropus rufus*. The IUCN Red List of Threatened Species 2008: e.T40567A10334778. http://dx.doi.org/10.2305/IUCN.UK.2008.RLTS.T40567A10334778.en. Downloaded on 22 December 2015.

Ellis, W., F. Bercovitch, S. FitzGibbon, P. Roe, J. Wimmer, A. Melzer, and R. Wilson. 2011. Koala bellows and their association with the spatial dynamics of free-ranging koalas. Behavioral Ecology 22:372–377.

Emison, W. B., J. W. Porter, K. C. Norris, and G. J. Apps. 1975. Ecological distribution of the vertebrate animals of the volcanic plains—Otway Range area of Victoria. Victoria Fisheries and Wildlife Paper 6:1–93.

Emison, W. B., J. W. Porter, K. C. Norris, and G. J. Apps. 1978. Survey of the vertebrate fauna in the Grampians-Edenhope area of southwestern Victoria. Memoirs National Museum of Victoria 39:281–363.

Emmons, L. H. 1991. Frugivory in treeshrews (*Tupaia*). American Naturalist 138:642–649.

Emmons, L. H. 1997. Neotropical rainforest mammals: a field guide. 2nd edition. University of Chicago Press, Chicago, Illinois.

Emmons, L. H. 2000. Tupai. A field study of Bornean treeshrews. University of California Press, Berkeley, California

Emmons, L. H. 2007. Genus *Caluromysiops* (Sanborn, 1951). Pp. 11–12 in Mammals of South America, volume 1, Marsupials, xenarthrans, shrews, and bats (A. L. Gardner, ed.). University of Chicago Press, Chicago, Illinois.

Emmons, L. H., and A. Biun. 1991. Malaysian treeshrews. National Geographic Research and Exploration 7:70–81.

Enders, R. K. 1966. Attachment, nursing, and survival of young in some didelphids. Symposia of the Zoological Society of London 15:195–203.

Engelmann, G. F. 1985. The phylogeny of the Xenarthra. Pp. 51–64 in The evolution and ecology of armadillos, sloths, and vermilinguas (G. G. Montgomery, ed.). Smithsonian Institution Press, Washington, DC.

Enk, J. M., D. R. Yesner, K. J. Crossen, D. W. Veltre, and D. H. O'Rourke. 2009. Phylogeographic analysis of the mid-Holocene mammoth from Qagnax Cave, St. Paul Island, Alaska. Palaeogeography, Palaeoclimatology, Palaeoecology 273:184–190.

Enk, J., A. Devault, R. Debruyne, C. E King, T. Treangen, D. O'Rourke, S. L. Salzberg, D. Fisher, R. MacPhee, and H. Poinar. 2011. Complete Columbian mammoth mitogenome suggests interbreeding with woolly mammoths. Genome Biology 12:R51. http://genomebiology.com/2011/12/5/R51. Accessed on 29 March 2015.

Epstein, B., M. Jones, R. Hamede, S. Hendricks, H. McCallum, E. P. Murchison, B. Schönfeld, C. Wiench, P. Hohenlohe, and A. Storfer. 2016. Rapid evolutionary response to a transmissible cancer in Tasmanian devils. Nature Communications 7:12684.

Espinoza, E. O., and M.-J. Mann. 1994. Mammoth or elephant ivory? Forensics provides the key. Endangered Species Technical Bulletin 19(1):8–9.

Esser, H., D. Brown, and Y. Liefting. 2010. Swimming in the northern tamandua (Tamandua mexicana) in Panama. Edentata 11:70–72.

Evans, K., and S. Harris. 2008. Adolescence in male African elephants, Loxodonta africana, and the importance of sociality. Animal Behaviour 76:779–787.

Evans, M., and G. Gordon. 2008. Bridled nailtail wallaby Onychogalea fraenata (Gould, 1841). Pp. 355–357 in The mammals of Australia. 3rd edition (S. Van Dyck and R. Strahan, eds.). Reed New Holland, Sydney, Australia.

Evans, M. C. 2008. Home range, burrow-use and activity patterns in common wombats (Vombatus ursinus). Wildlife Research 35:455–462.

Ewer, R. F. 1968. A preliminary survey of the behaviour in captivity of the dasyurid marsupial, Sminthopsis crassicaudata (Gould). Zeitschrift für Tierpsychologie 25:319–365.

Fadem, B. H., and R. S. Rayve. 1985. Characteristics of the oestrous cycle and influence of social factors in grey short-tailed opossums (Monodelphis domestica). Journal of Reproduction and Fertility 73: 337–342.

Fancourt, B. A., C. E. Hawkins, and S. C. Nicol. 2013. Evidence of rapid population decline of the eastern quoll (Dasyurus viverrinus) in Tasmania. Australian Mammalogy 35:195–205.

Fanning, F. D. 1982. Reproduction, growth, and development in Ningaui sp. (Dasyuridae, Marsupialia) from the Northern Territory. Pp. 23–37 in Carnivorous marsupials (M. Archer, ed.). Royal Zoological Society of New South Wales, Mosman, New South Wales, Australia.

FAO (Food and Agricultural Organization of the United Nations). 2010. Global forest resources assessment 2010: main report. FAO, Rome, Italy.

Faria, M. B., J. A. de Oliveira, and C. R. Bonvicino. 2013. Filogeografia de populações brasileiras de Marmosa (Marmosa) murina (Didelphimorphia, Didelphidae). Revista Nordestina de Biologia 21:27–52.

Faria, M. B., F. F. Nascimento, J. A. de Oliveira, and C. R. Bonvicino. 2013. Biogeographic determinants of genetic diversification in the mouse opossum Gracilinanus agilis (Didelphimorphia: Didelphidae). Journal of Heredity 104:613–626.

Faust, L. J., S. D. Thompson, and J. M. Earnhardt. 2006. Is reversing the decline of Asian elephants in North American zoos possible? An individual-based modeling approach. Zoo Biology 25:201–218.

Fay, J. M., and M. Agnagna. 1991. A population survey of forest elephants (Loxodonta africana cyclotis) in northern Congo. African Journal of Ecology 29:177–187.

Fayrer-Hosken, R. A., D. Grobler, J. J. van Altena, H. J. Bertschinger, J. F. Kirkpatrick. 2001. African elephants and contraception. Fayrer-Hosken et al. reply. Nature 411:766.

Feiler, A. 1978a. Bemerkungen über Phalanger der "orientalis-Gruppe" nach Tate (1945). Zoologische Abhandlungen (Dresden) 34:385–395.

Feiler, A. 1978b. Über artliche Abgrenzung und innerartliche Ausformung bei Phalanger maculatus (Mammalia, Marsupialia, Phalangeridae). Zoologische Abhandlungen (Dresden) 35:1–30.

Feldhamer, G. A., L. C. Drickamer, S. H. Vessey, J. F. Merritt, and C. Krajewski. 2007. Mammalogy: adaptation, diversity, ecology. 3rd edition. Johns Hopkins University Press, Baltimore, Maryland.

Ferguson, A., and B. Turner. 2013. Reproductive parameters and behaviour of captive short-beaked echidna (Tachyglossus aculeatus acanthion) at Perth Zoo. Australian Mammalogy 35:84–92.

Fernandez, S., and S. C. Jones. 1990. Manatee stranding on the coast of Texas. Texas Journal of Science 42:103–104.

Fernando, P., and R. Lande. 2000. Molecular genetic and behavioral analysis of social organization in the Asian elephant (Elephas maximus). Behavioral Ecology and Sociobiology 48:84–91.

Fernando, P., M. E. Pfrender, S. E. Encalada, and R. Lande. 2000. Mitochondrial DNA variation, phylogeography and population structure of the Asian elephant. Heredity 84:362–372.

Fernando, P., T. N. C. Vidya, J. Payne, M. Stuewe, G. Davison, R. J. Alfred, P. Andau, E. Bosi, A. Kilbourn, and D. J. Melnick. 2003. DNA analysis indicates that Asian elephants are native to Borneo and are therefore a high priority for conservation. PLoS Biology 1(1):110–115.

Fernando, P., E. Wikramanayake, D. Weerakoon, L. K. A. Jayasinghea, M. Gunawardena, and H. K. Janaka. 2005. Perceptions and patterns of human-elephant conflict in old and new settlements in Sri Lanka: insights for mitigation and management. Biodiversity and Conservation 14:2465–2481.

Fernando, P., E. D. Wikramanayake, H. K. Janaka, L. K. A. Jayasinghe, M. Gunawardena, S. W. Kotagama, D. Weerakoon, and J. Pastorini. 2008. Ranging behavior of the Asian elephant in Sri Lanka. Zeitschrift für Säugetierkunde 73:2–13.

Fernando, P., J. Jayewardene, T. Prasad, W. Hendavitharana, and J. Pastorini. 2011. Current status of Asian elephants in Sri Lanka. Gajah 35:93–103.

Fernicola, J. C., S. F. Vizcaíno, and R. A. Fariña. 2008. The evolution of armored xenarthrans and a phylogeny of the glyptodonts. Pp. 79–85 in The biology of the Xenarthra (S. F. Vizcaíno and W. J. Loughry, eds.). University Press of Florida, Gainesville, Florida.

Ferretti, M. P., L. Rook, and D. Torre. 2003. Stegotetrabelodon (Proboscidea, Elephantidae) from the late Miocene of southern Italy. Journal of Vertebrate Paleontology 23:659–666.

Fertl, D., A. J. Schiro, G. T. Regan, C. A. Beck, N. Adimey, L. Price-May, A. Amos, G. A. J. Worthy, and R. Crossland. 2005. Manatee occurrence in the northern Gulf of Mexico, west of Florida. Gulf and Caribbean Research 17:69–94.

Ficcarelli, G., M. Coltorti, M. Moreno-Espinosa, P. L. Pieruccini, L. Rook, and D. Torre. 2003. A model for the Holocene extinction of the mammal megafauna in Ecuador. Journal of South American Earth Sciences 15:835–845.

Fiedel, S. 2009. Sudden deaths: the chronology of terminal Pleistocene megafaunal extinction. Pp. 21–37 in American megafaunal extinctions at the end of the Pleistocene (G. Haynes, ed.). Springer, New York, New York.

Fiedel, S., and G. Haynes. 2004. A premature burial: comments on Grayson and Meltzer's "Requiem for overkill." Journal of Archaeological Science 31:121–131.

Fielden, L. J. 1991. Home range and movements of the Namib Desert golden mole Eremitalpa granti namibensis (Chrysochloridae). Journal of Zoology 223:675–686.

Fielden, L. J., M. R. Perrin, and G. C. Hickman. 1990. Feeding ecology and foraging behaviour of the Namib Desert golden mole, Eremitalpa granti namibensis (Chrysochloridae). Journal of Zoology 220:367–389.

Figueirido, B., and C. M. Janis. 2011. The predatory behaviour of the thylacine: Tasmanian tiger or marsupial wolf? Biology Letters 7:937–940.

Fillios, M., C. Gordon, F. Koch, and M. Letnic. 2010. The effect of a top predator on kangaroo abundance in arid Australia and its implications for archaeological faunal assemblages. Journal of Archaeological Science 37:986–993.

Finlayson, G. R., S. T. Finlayson, and C. R. Dickman. 2010. Returning the rat-kangaroos: translocation attempts in the family Potoroidae (superfamily Macropodoidea) and recommendations for conservation. Pp. 245–262 in Macropods: the biology of kangaroos, wallabies and rat-kangaroos (G. Coulson and M. Eldridge, eds.). CSIRO Publishing, Collingwood, Victoria, Australia.

Finlayson, G. R., A. N. Diment, P. Mitrovski, G. G. Thompson, and S. A. Thompson. 2010. Estimating western ringtail possum (*Pseudocheirus occidentalis*) density using distance sampling. Australian Mammalogy 32:197–200.

Finlayson, H. H. 1943. The red centre. Man and beast in the heart of Australia. Angus and Robertson, Ltd., Sydney, Australia,

Firestone, K. B. 2000. Phylogenetic relationships among quolls revisited: the mtDNA control region as a useful tool. Journal of Mammalian Evolution 7:1–22.

Firestone, K. B., M. S. Elphinstone, W. B. Sherwin, and B. A. Houlden. 1999. Phylogeographical population structure of tiger quolls *Dasyurus maculatus* (Dasyuridae: Marsupialia), an endangered carnivorous marsupial. Molecular Ecology 8:1613–1625.

Fischer, M. S. 1989. Hyracoids, the sister-group of perissodactyls. Pp. 37–56 in The evolution of perissodactyls (D. R. Prothero and R. M. Schoch, eds.). Clarendon Press, New York, New York.

Fischer, M. S., and P. Tassy. 1993. The interrelation between Proboscidea, Sirenia, Hyracoidea, and Mesaxonia: the morphological evidence. Pp. 217–234 in Mammal phylogeny: Mesozoic differentiation, multituberculates, monotremes, early therians, and marsupials (F. S. Szalay, M. J. Novacek, and M. C. McKenna, eds.). Springer-Verlag, New York, New York.

Fish, F. E. 1993. Comparison of swimming kinematics between terrestrial and semiaquatic opossums. Journal of Mammalogy 74:275–284.

Fisher, A. 2008. Long-tailed planigale *Planigale ingrami* (Thomas, 1906). Pp. 110–111 in The mammals of Australia. 3rd edition (S. Van Dyck and R. Strahan, eds.). Reed New Holland, Sydney, Australia.

Fisher, D. C. 2009. Paleobiology and extinction of proboscideans in the Great Lakes region of North America. Pp. 55–75 in American megafaunal extinctions at the end of the Pleistocene (G. Haynes, ed.). Springer, New York, New York.

Fisher, D. O., and A. W. Goldizen. 2001. Maternal care and infant behaviour of the bridled nailtail wallaby (*Onychogalea fraenata*). Journal of Zoology 255:321–330.

Fisher, D. O., S. D. Hoyle, and S. P. Blomberg. 2000. Population dynamics and survival of an endangered wallaby: a comparison of four methods. Ecological Applications 10:901–910.

Fisher, D. O., and M. C. Lara. 1999. Effects of body size and home range on access to mates and paternity in male bridled nailtail wallabies. Animal Behaviour 58:121–130.

Fisher, D. O., S. Nuske, S. Green, J. M. Seddon, and B. McDonald. 2011. The evolution of sociality in small, carnivorous marsupials: the lek hypothesis revisited. Behavioral Ecology and Sociobiology 65:593–605.

Fishlock, V., and P. C. Lee. 2013. Forest elephants: fission-fusion and social arenas. Animal Behaviour 85:357–363.

Fishlock, V., P. C. Lee, and T. Breuer. 2008. Quantifying forest elephant social structure in Central African bai environments. Pachyderm 44:17–26.

Fitch, H. S., P. Goodrum, and C. Newman. 1952. The armadillo in the southeastern United States. Journal of Mammalogy 33:21–37.

Fitzgerald, A. E. 1976. Diet of the opossum, *Trichosurus vulpecula* (Kerr) in the Orongorongo Valley, Wellington, New Zealand, in relation to food-plant availability. New Zealand Journal of Zoology 3:399–419.

Fitzgerald, A. E. 1978. Aspects of the food and nutrition of the brush-tailed opossum, *Trichosurus vulpecula* (Kerr, 1792), Marsupialia: Phalangeridae, in New Zealand. Pp. 289–303 in The ecology of arboreal folivores (G. G. Montgomery, ed.). Smithsonian Institution Press, Washington, DC.

Fitzgerald, E. M. G. 2005. Holocene record of the dugong (*Dugong dugon*) from Victoria, southeast Australia. Marine Mammal Science 21:355–361.

FitzGibbon, C. D. 1997. The adaptive significance of monogamy in the golden-rumped elephant-shrew. Journal of Zoology 242:167–177.

FitzGibbon, C. D., and G. B. Rathbun. 1994. Surveying *Rhynchocyon* elephant-shrews in tropical forest. African Journal of Ecology 32:50–57.

FitzGibbon, C. D., and G. B. Rathbun. 2015. *Rhynchocyon chrysopygus*. The IUCN Red List of Threatened Species 2015: e.T19705 A21287265. http://dx.doi.org/10.2305/IUCN.UK.2015-2.RLTS.T19 705A21287265.en. Downloaded on 20 January 2016.

Fitzgibbon, S. I., R. S. Wilson, and A. W. Goldizen. 2011. The behavioural ecology and population dynamics of a cryptic ground-dwelling mammal in an urban Australian landscape. Austral Ecology 36:722–732.

Flannery, T. F. 1983. Revision in the macropodid subfamily Sthenurinae (Marsupialia: Macropodoidea) and the relationships of the species of *Troposodon* and *Lagostrophus*. Australian Mammalogy 6:15–28.

Flannery, T. F. 1987. A new species of *Phalanger* (Phalangeridae: Marsupialia) from montane western Papua New Guinea. Records of the Australian Museum 39:183–193.

Flannery, T. F. 1989. Phylogeny of the Macropodoidea: a study in convergence. Pp. 1–46 in Kangaroos, wallabies and rat-kangaroos (G. C. Grigg, P. Jarman, and I. Hume, eds.). Surrey Beatty & Sons, Chipping Norton, New South Wales, Australia.

Flannery, T. F. 1990. *Echymipera davidi*, a new species of Perameliformes (Marsupialia) from Kiriwina Island, Papua New Guinea, with notes on the systematics of the genus *Echymipera*. Pp. 29–35 in Bandicoots and bilbies (J. H. Seebeck, P. R. Brown, R. L. Wallis, and C. M. Kemper, eds.) Surrey Beatty & Sons, Chipping Norton, New South Wales, Australia.

Flannery, T. F. 1992. Taxonomic revision of the *Thylogale brunii* complex (Macropodidae: Marsupialia) in Melanesia, with description of a new species. Australian Mammalogy 15:7–23.

Flannery, T. F. 1993. Taxonomy of *Dendrolagus goodfellowi* (Macropodidae: Marsupialia) with description of a new subspecies. Records of the Australian Museum 45:33–42.

Flannery, T. 1995a. Mammals of New Guinea. 2nd edition. Reed Books, Chatswood, New South Wales, Australia.

Flannery, T. 1995b. Mammals of the south-west Pacific and Moluccan Islands. Reed Books, Chatswood, New South Wales, Australia.

Flannery, T. F., M. Archer, and G. Maynes. 1987. The phylogenetic relationships of living phalangerids (Phalangeroidea: Marsupialia) with a suggested new taxonomy. Pp. 477–506 in Possums and opossums (M. Archer, ed.). Surrey Beatty & Sons, Sydney, Australia.

Flannery, T. F., and B. Boeadi. 1995. Systematic revision within the *Phalanger ornatus* complex (Phalangeridae: Marsupialia), with description of a new species and subspecies. Australian Mammalogy 18:35–44.

Flannery, T. F., B. Boeadi, and A. L. Szalay. 1995. A new tree-kangaroo (*Dendrolagus*: Marsupialia) from Irian Jaya, Indonesia, with notes on ethnography and the evolution of tree-kangaroos. Mammalia 59:65–84.

Flannery, T. F., and J. H. Calaby. 1987. Notes on the species of *Spilocuscus* (Marsupialia: Phalangeridae) from northern New Guinea and the Admiralty and St. Matthias Island groups. Pp. 547–558 in Possums and opossums (M. Archer, ed.). Surrey Beatty & Sons, Sydney, Australia.

Flannery, T. F., and C. P. Groves. 1998. A revision of the genus *Zaglossus* (Monotremata, Tachyglossidae), with description of new species and subspecies. Mammalia 62:367–396.

Flannery, T., and K. Helgen. 2008. *Ailurops melanotis*. The IUCN Red List of Threatened Species 2008: e.T136218A4261150. http://dx.doi .org/10.2305/IUCN.UK.2008.RLTS.T136218A4261150.en. Downloaded on 30 December 2015.

Flannery, T. F., and P. Schouten. 1994. Possums of the world. GEO Productions, Chatswood, New South Wales, Australia.

Flannery, T. F., and L. Seri. 1990. *Dendrolagus scottae* n. sp. (Marsupialia: Macropodidae): a new tree-kangaroo from Papua New Guinea. Records of the Australian Museum 42:237–245.

Flannery, T. F., and F. Szalay. 1982. *Bohra paulae*, a new giant fossil tree kangaroo (Marsupialia: Macropodidae) from New South Wales, Australia. Australian Mammalogy 5:83–94.

Flannery, T. F., P. V. Kirch, J. Specht, and M. Spriggs. 1988. Holocene mammal faunas from archaeological sites in island Melanesia. Archaeology in Oceania 23:89–94.

Flannery, T. F., M. Archer, T. H. Rich, and R. Jones. 1995a. A new family of monotremes from the Cretaceous of Australia. Nature 377:418–420.

Flannery, T. F., P. Bellwood, P. White, A. Moore, B. Boeadi, and G. Nitihaminoto. 1995b. Fossil marsupials (Macropodidae, Peroryctidae) and other mammals of Holocene age from Halmahera, North Moluccas, Indonesia. Alcheringa 19:17–25.

Flannery, T. F., P. Bellwood, J. P. White, T. Ennis, G. Irwin, K. Schubert, and S. Balasubramaniam. 1998. Mammals from Holocene archaeological deposits on Gebe and Morotai Islands, northern Moluccas, Indonesia. Australian Mammalogy 20:391–400.

Flannery, T., T. Leary, R. Singadan, J. Menzies, D. Wright, K. Helgen, C. Dickman, and L. Salas. 2008. *Pseudochirops coronatus*. The IUCN Red List of Threatened Species 2008: e.T40582A10322910. http://dx.doi.org/10.2305/IUCN.UK.2008.RLTS.T40582 A10322910.en. Downloaded on 30 December 2015.

Fleck, D. W., and J. D. Harder. 1995. Ecology of marsupials in two Amazonian rain forests in northeastern Peru. Journal of Mammalogy 76:809–818.

Fleischer, R. C., E. A. Perry, K. Muralidharan, E. E. Stevens, C. M. Wemmer. 2001. Phylogeography of the Asian elephant (*Elephas maximus*) based on mitochondrial DNA. Evolution 55:1882–1892.

Fleming, M. R. 1980. Thermoregulation and torpor in the sugar glider, *Petaurus breviceps* (Marsupialia: Petauridae). Australian Journal of Zoology 28:521–534.

Fleming, M. R., and A. Cockburn. 1979. *Ningaui*: a new genus of dasyurid for Victoria. Victorian Naturalist 96:142–145.

Fleming, M. R., and H. Frey. 1984. Aspects of the natural history of feathertail gliders (*Acrobates pygmaeus*) in Victoria. Pp. 403–408 in Possums and gliders (A. Smith and I. Hume, eds.). Surrey Beatty & Sons, Norton, New South Wales, Australia.

Fleming, T. H. 1972. Aspects of the population dynamics of three species of opossums in the Panama Canal Zone. Journal of Mammalogy 53:619–23.

Fleming, T. H. 1973. The reproductive cycles of three species of opossums and other mammals in the Panama Canal Zone. Journal of Mammalogy 54:439–55.

Fletcher, M., C. J. Southwell, N. W. Sheppard, G. Caughley, D. Grice, G. C. Grigg, and L. A. Beard. 1990. Kangaroo population trends in the Australian rangelands, 1980–1987. Search 21:28–29.

Fletcher, T. P. 1985. Aspects of reproduction in the male eastern quoll, *Dasyurus viverrinus* (Shaw) (Marsupialia: Dasyuridae), with notes on polyoestry in the female. Australian Journal of Zoology 33:101–110.

Flores, D. A., R. M. Barquez, and M. M. Díaz. 2008. A new species of *Philander* (Brisson, 1762) (Didelphimorphia, Didelphidae). Mammalian Biology 73:14–24.

Flores, D. A., and M. M. Díaz. 2009. Postcranial skeleton of *Glironia venusta* (Didelphimorphia, Didelphidae, Caluromyinae): description and functional morphology. Zoosystematics and Evolution 85:311–339.

Flores, D. A., M. M. Díaz, and R. M. Barquez. 2000. Mouse opossums (Didelphimorphia, Didelphidae) of northwestern Argentina: systematics and distribution. Zeitschrift für Säugetierkunde 65:321–339.

Flores, D. A., M. M. Díaz, and R. M. Barquez. 2007. Systematics and distribution of marsupials in Argentina: a review. University of California Publications in Zoology 134:579–669.

Flores, D., and S. Solari. 2011. *Monodelphis handleyi*. The IUCN Red List of Threatened Species 2011: e.T199833A9129077. http://dx.doi .org/10.2305/IUCN.UK.2011-2.RLTS.T199833A9129077.en. Downloaded on 05 January 2016.

Flores, D., and P. Teta. 2011a. *Monodelphis unistriatus*. The IUCN Red List of Threatened Species 2011: e.T13703A4353243. http://dx.doi .org/10.2305/IUCN.UK.2011-2.RLTS.T13703A4353243.en. Downloaded on 05 January 2016.

Flores, D., and P. Teta. 2011b. *Thylamys citellus*. The IUCN Red List of Threatened Species 2011: e.T199835A9129311. http://dx.doi.org/10 .2305/IUCN.UK.2011-2.RLTS.T199835A9129311.en. Downloaded on 27 December 2015.

Flores, D., and P. Teta. 2011c. *Thylamys pulchellus*. The IUCN Red List of Threatened Species 2011: e.T199834A9129207. http://dx.doi.org /10.2305/IUCN.UK.2011-2.RLTS.T199834A9129207.en. Downloaded on 27 December 2015.

Flynn, J. J., and G. D. Wesley-Hunt. 2005. Carnivora. Pp. 175–198 in The rise of placental mammals: origins and relationships of the major clades (K. D. Rose and J. D. Archibald, eds.). Johns Hopkins University Press, Baltimore, Maryland.

Fons, R. 1990. Living insectivores. Pp. 425–519 in Grzimek's encyclopedia of mammals, volume 1 (B. Grzimek, ed.). McGraw-Hill, New York, New York.

Fontúrbel, F. E., A. B. Candia, and C. Botto-Mahan. 2014. Nocturnal activity patterns of the monito del monte (*Dromiciops gliroides*) in native and exotic habitats. Journal of Mammalogy 95:1199–1206.

Fontúrbel, F. E., and J. E. Jiménez. 2009. Underestimation of abundances of the monito del monte (*Dromiciops gliroides*) due to a sampling artifact. Journal of Mammalogy 90:1357–1362.

Fontúrbel, F. E., E. A. Silva-Rodríguez, N. H. Cárdenas, and J. E. Jiménez. 2010. Spatial ecology of monito del monte (*Dromiciops gliroides*) in a fragmented landscape of southern Chile. Mammalian Biology 75:1–9.

Fontúrbel, F. E., M. Franco, M. A. Rodríguez-Cabal, M. D. Rivarola, and G. C. Amico. 2012. Ecological consistency across space: a synthesis of the ecological aspects of *Dromiciops gliroides* in Argentina and Chile. Naturwissenschaften 99:873–881.

Formoso, A. E., D. E. Udrizar Sauthier, P. Teta, and U. F. J. Pardiñas. 2011. Dense-sampling reveals a complex distributional pattern between the southernmost marsupials *Lestodelphys* and *Thylamys* in Patagonia, Argentina. Mammalia 75:371–379.

Forstén, A., and P. M. Youngman. 1982. *Hydrodamalis gigas*. Mammalian Species 1651–3.

Fortelius, M., and J. Kappelman. 1993. The largest land mammal ever imagined. Zoological Journal of the Linnean Society 107:85–101.

Foster, W. K., W. Caton, J. Thomas, S. Cox, and D. A. Taggart. 2008. Timing of births and reproductive success in captive red-tailed phascogales, *Phascogale calura*. Journal of Mammalogy 89:1136–1144.

Foulkes, J. N. 2008. Ooldea dunnart *Sminthopsis ooldea* (Troughton, 1963). Pp. 152–154 in The mammals of Australia. 3rd edition (S. Van Dyck and R. Strahan, eds.). Reed New Holland, Sydney, Australia.

Fourie, P. B. 1977. Acoustic communication in the rock hyrax. Zeitschrift für Tierpsychologie 44:194–219.

Fox, B. J. 2008. Common dunnart *Sminthopsis murina* (Waterhouse, 1838). Pp. 150–152 in The mammals of Australia. 3rd edition (S. Van Dyck and R. Strahan, eds.). Reed New Holland, Sydney, Australia.

Francis, C. M. 2008. A field guide to the mammals of south-east Asia. New Holland Publishers, London, United Kingdom.

Franco, M., A. Quijano, and M. Soto-Gamboa. 2011. Communal nesting, activity patterns, and population characteristics in the near-threatened monito del monte, *Dromiciops gliroides*. Journal of Mammalogy 92:994–1004.

Frankham, G. J., K. A. Handasyde, and M. D. B. Eldridge. 2012. Novel insights into the phylogenetic relationships of the endangered marsupial genus *Potorous*. Molecular Phylogenetics and Evolution 64:592–602.

Frederick, H., and C. N. Johnson. 1996. Social organisation in the rufous bettong, *Aepyprymnus rufescens*. Australian Journal of Zoology 44:9–17.

Frey, J. K., and J. N. Stuart. 2009. Nine-banded armadillo (*Dasypus novemcinctus*) records in New Mexico, USA. Edentata 8–10:54–55.

Friend, G. R. 1990. Breeding and population dynamics of *Isoodon macrourus* (Marsupialia: Peramelidae): studies from the wet-dry tropics of northern Australia. Pp. 357–365 in Bandicoots and bilbies (J. H. Seebeck, P. R. Brown, R. L. Wallis, and C. M. Kemper, eds.). Surrey Beatty & Sons, Chipping Norton, New South Wales, Australia.

Friend, G. R., and D. J. Pearson. 2008. Little long-tailed dunnart *Sminthopsis dolichura* (Kitchener, Stoddart and Henry, 1984). Pp. 134–135 in The mammals of Australia. 3rd edition (S. Van Dyck and R. Strahan, eds.). Reed New Holland, Sydney, Australia.

Friend, J. A. 1989. Myrmecobiidae. Pp. 583–590 in Fauna of Australia: Mammalia, volume 1B (D. W. Walton and B. J. Richardson, eds.). Australian Government Publishing Service, Canberra, Australia.

Friend, J. A. 1990a. The numbat *Myrmecobius fasciatus* (Myrmecobiidae): history of decline and potential for recovery. Proceedings of the Ecological Society of Australia 16:369–77.

Friend, J. A. 1990b. Status of bandicoots in Western Australia. Pp. 73–84 in Bandicoots and bilbies (J. H. Seebeck, P. R. Brown, R. L. Wallis, and C. M. Kemper, eds). Surrey Beatty & Sons, Chipping Norton, New South Wales, Australia.

Friend, J. A. 2008a. Numbat *Myrmecobius fasciatus* (Waterhouse, 1836). Pp. 163–165 in The mammals of Australia. 3rd edition (S. Van Dyck and R. Strahan, eds.). Reed New Holland, Sydney, Australia.

Friend, J. A. 2008b. Western barred bandicoot *Perameles bougainville* (Quoy and Gaimard, 1824). Pp. 182–184 in The mammals of Australia. 3rd edition (S. Van Dyck and R. Strahan, eds.). Reed New Holland, Sydney, Australia.

Friend, J. A. 2008c. Gilbert's potoroo *Potorous gilbertii* (Gould, 1841). Pp. 297–298 in The mammals of Australia. 3rd edition (S. Van Dyck and R. Strahan, eds.). Reed New Holland, Sydney, Australia.

Friend, J. A. 2015. Family Myrmecobiidae (numbat). Pp. 222–231 in Handbook of the mammals of the world, volume 5, Monotremes and marsupials (D. E. Wilson and R. A. Mittermeier, eds.). Lynx Edicions, Barcelona, Spain.

Friend, J. A., and N. D. Thomas. 2003. Conservation of the numbat (*Myrmecobius fasciatus*). Pp. 452–463 in Predators with pouches: the biology of carnivorous marsupials (M. Jones, C. Dickman, and M. Archer, eds.). CSIRO Publishing, Collingwood, Victoria, Australia.

Friend, T. 2003. Dibbler (*Parantechinus apicalis*) recovery plan. Western Australian Threatened Species and Communities Unit, Department of Conservation and Land Management, Wanneroo, Australia.

Friend, T. 2008. *Phascogale calura*. The IUCN Red List of Threatened Species 2008: e.T16888A6544803. http://dx.doi.org/10.2305/IUCN.UK.2008.RLTS.T16888A6544803.en. Downloaded on 04 January 2016.

Friend, T., and A. Burbidge. 2008a. *Myrmecobius fasciatus*. The IUCN Red List of Threatened Species 2008: e.T14222A4424357. http://dx.doi.org/10.2305/IUCN.UK.2008.RLTS.T14222A4424357.en. Downloaded on 21 November 2015.

Friend, T., and A. Burbidge. 2008b. *Potorous gilbertii*. The IUCN Red List of Threatened Species 2008: e.T18107A7660937. http://dx.doi.org/10.2305/IUCN.UK.2008.RLTS.T18107A7660937.en. Downloaded on 30 December 2015.

Friend, T., A. Burbidge, and K. Morris. 2008. *Parantechinus apicalis*. The IUCN Red List of Threatened Species 2008: e.T16138A5410795. http://dx.doi.org/10.2305/IUCN.UK.2008.RLTS.T16138A5410795.en. Downloaded on 04 January 2016.

Friend, T., K. Morris, and J. van Weenen. 2008. *Macrotis lagotis*. The IUCN Red List of Threatened Species 2008: e.T12650A3368711. http://dx.doi.org/10.2305/IUCN.UK.2008.RLTS.T12650A3368711.en. Downloaded on 05 January 2016.

Friend, T., and J. Richards. 2008. *Perameles bougainville*. The IUCN Red List of Threatened Species 2008: e.T16569A6092834. http://dx.doi.org/10.2305/IUCN.UK.2008.RLTS.T16569A6092834.en. Downloaded on 06 January 2016.

Friend, T., K. Morris, A. Burbidge, and N. McKenzie. 2008. *Tarsipes rostratus*. The IUCN Red List of Threatened Species 2008: e.T40583A10323458. http://dx.doi.org/10.2305/IUCN.UK.2008.RLTS.T40583A10323458.en. Downloaded on 30 December 2015.

Fry, E. 1971. The scaly-tailed possum *Wyulda squamicaudata* in captivity. International Zoo Yearbook 11:44–45.

Furlan, E., P. A. Umina, P. J. Mitrovski, N. Gust, J. Griffiths, and A. R. Weeks. 2010. High levels of genetic divergence between Tasmanian and Victorian platypuses, *Ornithorhynchus anatinus*, as revealed by microsatellite loci. Conservation Genetics 11:319–323.

Furlan, E. M., J. Griffiths, N. Gust, K. A. Handasyde, T. R. Grant, B. Gruber, and A. R. Weeks. 2013. Dispersal patterns and population structuring among platypuses, *Ornithorhynchus anatinus*, throughout south-eastern Australia. Conservation Genetics 14:837–853.

Gabuniya, L. K., N. S. Shevyreva, and V. D. Gabuniya. 1985. On the first find of fossil Marsupialia in Asia. Doklady Akademii Nauk SSSR 281:684–685.

Gaeth, A. P., R. V. Short, and M. B. Renfree. 1999. The developing renal, reproductive, and respiratory systems of the African elephant suggest an aquatic ancestry. Proceedings of the National Academy of Sciences of the United States of America 96:5555–5558.

Galliez, M., M. De Souza Leite, T. L. Queiroz, F. A. Dos Santos Fernandez. 2009. Ecology of the water opossum *Chironectes minimus* in Atlantic Forest streams of southeastern Brazil. Journal of Mammalogy 90:93–103.

Gambaryan, P. P., and A. N. Kuznetsov. 2013. An evolutionary perspective on the walking gait of the long-beaked echidna. Journal of Zoology 290:58–67.

Ganslosser, U. 1984. On the occurrence of female coalitions in tree kangaroos (Marsupialia, Macropodidae, *Dendrolagus*). Australian Mammalogy 7:219–221.

Ganslosser, U. 1990. True kangaroos. Pp. 360–379 in Grzimek's encyclopedia of mammals, volume 1 (B. Grzimek, ed.). McGraw-Hill, New York, New York.

Ganslosser, U. 1992. Behavioural data support the currently proposed phylogeny of the Macropodoidea (Marsupialia). Australian Mammalogy 15:89–104.

Ganslosser, U. 2003. Musky rat-kangaroos (Hypsiprymnodontidae). Pp. 69–72 in Grzimek's animal life encyclopedia, 2nd edition, volume 13, Mammals II (M. Hutchins, D. G. Kleiman, V. Geist, and M. C. McDade, eds.). Gale Group, Farmington Hills, Michigan.

Gao, Y. 1981. On the present status, historical distribution, and conservation of wild elephant in China. Acta Theriologica Sinica 1:19–26.

Gao, Y., and S. G. Clark. 2014. Elephant ivory trade in China: trends and drivers. Biological Conservation 180:23–30.

Garbutt, N. 2007. Mammals of Madagascar: a complete guide. Yale University Press, New Haven, Connecticut.

Garcia-Rodriguez, A. I., B. W. Bowen, D. Domning, A. A. Mignucci-Giannoni, A. Marmontel, R. A. Montoya-Ospina, B. Morales-Vela, M. Rudin, R. K. Bonde, and P. M. McGuire. 1998. Phylogeography of the West Indian manatee (*Trichechus manatus*): how many populations and how many taxa? Molecular Ecology 7:1137–1149.

Gardner, A. L. 1973. The systematics of the genus *Didelphis* (Marsupialia: Didelphidae) in North and Middle America. Special Publications of the Museum of Texas Tech University 4:1–81.

Gardner, A. L. 1981. Review of The mammals of Suriname, by A. M. Husson. Journal of Mammalogy 62:445–448.

Gardner, A. L. 2005a. Order Didelphimorphia. Pp. 3–18 in Mammal species of the world: a taxonomic and geographic reference, 3rd edition (D. E. Wilson and D. M. Reeder, eds.). Johns Hopkins University Press, Baltimore, Maryland.

Gardner, A. L. 2005b. Order Paucituberculata. Pp. 19–20 in Mammal species of the world: a taxonomic and geographic reference. 3rd edi-

tion (D. E. Wilson and D. M. Reeder, eds.). Johns Hopkins University Press, Baltimore, Maryland.

Gardner, A. L. 2005c. Order Microbiotheria. P. 21 in Mammal species of the world: a taxonomic and geographic reference. 3rd edition (D. E. Wilson and D. M. Reeder, eds.). Johns Hopkins University Press, Baltimore, Maryland.

Gardner, A. L. 2005d. Order Cingulata. Pp. 94–99 in Mammal species of the world: a taxonomic and geographic reference. 3rd edition (D. E. Wilson and D. M. Reeder, eds.). Johns Hopkins University Press, Baltimore, Maryland.

Gardner, A. L. 2005e. Order Pilosa. Pp. 100–103 in Mammal species of the world: a taxonomic and geographic reference. 3rd edition (D. E. Wilson and D. M. Reeder, eds.). Johns Hopkins University Press, Baltimore, Maryland.

Gardner, A. L. (ed.). 2007a. Mammals of South America, volume 1, Marsupials, xenarthrans, shrews, and bats. University of Chicago Press, Chicago, Illinois.

Gardner, A. L. 2007b. Cohort Marsupialia. Pp. 1–11 in Mammals of South America, volume 1, Marsupials, xenarthrans, shrews, and bats (A. L. Gardner, ed.). University of Chicago Press, Chicago, Illinois.

Gardner, A. L. 2007c. Tribe Monodelphini. Pp. 39–43 in Mammals of South America, volume 1, Marsupials, xenarthrans, shrews, and bats (A. L. Gardner, ed.). University of Chicago Press, Chicago, Illinois.

Gardner, A. L. 2007d. Genus *Hyladelphys* (Voss, Lunde, and Simmons, 2001). P. 50 in Mammals of South America, volume 1, Marsupials, xenarthrans, shrews, and bats (A. L. Gardner, ed.). University of Chicago Press, Chicago, Illinois.

Gardner, A. L. 2007e. Magnorder Xenarthra. Pp. 127–128 in Mammals of South America, volume 1, Marsupials, xenarthrans, shrews, and bats (A. L. Gardner, ed.). University of Chicago Press, Chicago, Illinois.

Gardner, A. L. 2007f. Order Pilosa (Flower, 1883). Pp. 157–164 in Mammals of South America, volume 1, Marsupials, xenarthrans, shrews, and bats (A. L. Gardner, ed.). University of Chicago Press, Chicago, Illinois.

Gardner, A. L. 2007g. Suborder Vermilingua (Illiger, 1811). Pp. 168–177 in Mammals of South America, volume 1, Marsupials, xenarthrans, shrews, and bats (A. L. Gardner, ed.). University of Chicago Press, Chicago, Illinois.

Gardner, A. L., and G. K. Creighton. 1989. A new generic name for Tate's (1933) *Microtarsus* group of South American mouse opossums (Marsupialia: Didelphidae). Proceedings of the Biological Society of Washington 102:3–7.

Gardner, A. L., and G. K. Creighton. 2007a. Genus *Marmosops* (Matschie, 1916). Pp. 61–74 in Mammals of South America, volume 1, Marsupials, xenarthrans, shrews, and bats (A. L. Gardner, ed.). University of Chicago Press, Chicago, Illinois.

Gardner, A. L., and G. K. Creighton. 2007b. Genus *Micoureus* (Lesson, 1842). Pp. 74–82 in Mammals of South America, volume 1, Marsupials, xenarthrans, shrews, and bats (A. L. Gardner, ed.). University of Chicago Press, Chicago, Illinois.

Gardner, A. L., and M. Dagosto. 2007. Tribe Metachirini (Reig, Kirsch, and Marshall, 1985). Pp. 35–39 in Mammals of South America, volume 1, Marsupials, xenarthrans, shrews, and bats (A. L. Gardner, ed.). University of Chicago Press, Chicago, Illinois.

Gardner, A. L., and V. L. Naples. 2007. Family Megalonychidae P. (Gervais, 1855). Pp. 165–168 in Mammals of South America, volume 1,

Marsupials, xenarthrans, shrews, and bats (A. L. Gardner, ed.). University of Chicago Press, Chicago, Illinois.

Gardner, A. L., and J. L. Patton. 1972. New species of *Philander* (Marsupialia: Didelphidae) and *Mimon* (Chiroptera: Phyllostomatidae) from Peru. Occasional Papers of the Museum of Zoology of Louisiana State University 43:1–12.

Gardner, A. L., and M. E. Sunquist. 2003. Opossum *Didelphis virginiana*. Pp. 3–29 in Wild mammals of North America, biology, management, and conservation (G. A. Feldhamer, B. C. Thompson, and J. A. Chapman, eds.). Johns Hopkins University Press, Baltimore, Maryland.

Gardner, J. L., and M. Serena. 1995. Spatial organisation and movement patterns of adult male platypus, *Ornithorhynchus anatinus* (Monotremata: Ornithorhynchidae). Australian Journal of Zoology 43:91–103.

Garrigue, C., N. Patenaude, and H. Marsh. 2008. Distribution and abundance of the dugong in New Caledonia, southwest Pacific. Marine Mammal Science 24:81–90.

Garrison, J. 2008. The challenges of meeting the needs of captive elephants. Pp. 237–256 in Elephants and ethics, toward a morality of existence (C. Wemmer and C. A. Christen, eds.). Johns Hopkins University Press, Baltimore, Maryland.

Garstang, M. 2004. Long-distance, low-frequency elephant communication. Journal of Comparative Physiology A 190:791–805.

Gaski, A. L. 1988. 'Roo update: a short history of U.S. kangaroo skin and product imports, 1984–1987. Traffic (U.S.A.) 8(2):1–5.

Gates, G. R. 1978. Vision in the monotreme echidna (*Tachyglossus aculeatus*). Australian Zoologist 20:147–169.

Gaudin, T. J. 1999. The morphology of xenarthrous vertebrae (Mammalia: Xenarthra). Fieldiana: Geology (new series) 41:1–38.

Gaudin, T. J. 2003. Phylogeny of the Xenarthra. Senckenbergiana Biologica 83:27–40.

Gaudin, T. J. 2004. Phylogenetic relationships among sloths (Mammalia, Xenarthra, Tardigrada): the craniodental evidence. Zoological Journal of the Linnean Society 140:255–305.

Gaudin, T. J. 2011. On the osteology of the auditory region and orbital wall in the extinct West Indian sloth genus *Neocnus* (Arredondo, 1961) (Placentalia, Xenarthra, Megalonychidae). Annals of Carnegie Museum 80:5–28.

Gaudin, T. J., and D. G. Branham. 1998. The phylogeny of the Myrmecophagidae (Mammalia, Xenarthra, Vermilingua) and the relationship of *Eurotamandua* to the Vermilingua. Journal of Mammalian Evolution 5:237–265.

Gaudin, T. J., and G. McDonald. 2008. Morphology-based investigations of the phylogenetic relationships among extant and fossil xenarthrans. Pp. 24–36 in The biology of the Xenarthra (S. F. Vizcaíno and W. J. Loughry, eds.). University Press of Florida, Gainesville, Florida.

Gaudin, T. J., and J. R. Wible. 2006. The phylogeny of living and extinct armadillos (Mammalia, Xenarthra, Cingulata): a craniodental analysis. Pp. 153–198 in Amniote paleobiology. Perspectives on the evolution of mammals, birds, and reptiles. A volume honoring James Allen Hopson (M. T. Carrano, T. J. Gaudin, R. W. Blob, and J. R. Wible, eds.). University of Chicago Press, Chicago, Illinois.

Gautier, A., R. Schild, F. Wendorf, and T. W. Stafford, Jr. 1994. One elephant doesn't make a savanna. Palaeoecological significance of *Loxodonta africana* in the Holocene Sahara. Sahara 6:7–20.

Gaylard, A. 1992. The plight of the elusive tree dassie. Naturalist (South Africa) 36(1):12–15.

Gaylard, A. (subed.). 2005. Order Hyracoidea. Pp. 41–50 in The mammals of the Southern African Subregion. 3rd edition (J. D. Skinner and C. T. Chimimba, eds.). Cambridge University Press, Cape Town, South Africa.

Geiser, F. 1986. Thermoregulation and torpor in the kultarr, *Antechinomys laniger* (Marsupialia: Dasyuridae). Journal of Comparative Physiology B 156:751–757.

Geiser, F. 1993. Hibernation in the eastern pygmy possum, *Cercartetus nanus* (Marsupialia: Burramyidae). Australian Journal of Zoology 41:67–75.

Geiser, F. 2003. Thermal biology and energetics of carnivorous marsupials. Pp. 238–253 in Predators with pouches: the biology of carnivorous marsupials (M. Jones, C. Dickman, and M. Archer, eds.). CSIRO Publishing, Collingwood, Victoria, Australia.

Geiser, F. 2007. Yearlong hibernation in a marsupial mammal. Naturwissenschaften 94:941–944.

Geiser, F., and R. V. Baudinette. 1988. Daily torpor and thermoregulation in the small dasyurid marsupials *Planigale gilesi* and *Ningaui yvonneae*. Australian Journal of Zoology 36:473–481.

Geiser, F., and L. S. Broome. 1991. Hibernation in the mountain pygmy possum *Burramys parvus* (Marsupialia). Journal of Zoology 223:593–602.

Geiser, F., and G. Körtner. 2000. Torpor and activity patterns in freeranging sugar gliders *Petaurus breviceps* (Marsupialia). Oecologia 123:350–357.

Geiser, F., and G. Körtner. 2004. Thermal biology, energetics, and torpor in the possums and gliders. Pp. 186–198 in The biology of Australian possums and gliders (R. L. Goldingay and S. M. Jackson, eds.). Surrey Beatty & Sons, Chipping Norton, New South Wales, Australia.

Geiser, F., and G. M. Martin. 2013. Torpor in the Patagonian opossum (*Lestodelphys halli*): implications for the evolution of daily torpor and hibernation. Naturwissenschaften 100:975–981.

Geiser, F., and P. Masters. 1994. Torpor in relation to reproduction in the mulgara, *Dasycercus cristicauda* (Dasyuridae: Marsupialia). Journal of Thermal Biology 19:33–40.

Geiser, F., L. Matwiejczyk, and R. V. Baudinette. 1986. From ectothermy to heterothermy: the energetics of the kowari, *Dasyuroides byrnei* (Marsupialia: Dasyuridae). Physiological Zoology 59:220–229.

Geiser, F., M. L. Augee, H. C. K. McCarron, and J. Raison. 1984. Correlates of torpor in the insectivorous marsupial *Sminthopsis murina*. Australian Mammalogy 7:185–191.

Geisler, J. H., and J. M. Theodor. 2009. Hippopotamus and whale phylogeny. Nature 458. doi:10.1038/nature07776.

Geisler, J. H., J. M. Theodor, M. D. Uhen, and S. E. Foss. 2007. Phylogenetic relationships of cetaceans to terrestrial artiodactyls. Pp. 19–31 in The evolution of artiodactyls (D. R. Prothero and S. E. Foss, eds.). Johns Hopkins University Press, Baltimore, Maryland.

Gemmell, N. J., T. R. Grant, P. S. Western, J. M. Watson, N. D. Murray, and J. A. M. Graves. 1992. Preliminary molecular studies of platypus family and population structure. Pp. 277–284 in Platypus and echidnas (M. L. Augee, ed.). Royal Zoological Society of New South Wales, Mosman, New South Wales, Australia.

Gemmell, R. T. 1988. The oestrous cycle length of the bandicoot *Isoodon macrourus*. Australian Wildlife Research 15:633–635.

Genest, H., and F. Petter. 1977. Subfamilies Tenrecinae and Oryzorictinae. Part 1.1 in The mammals of Africa: an identification manual (J. Meester and H. W. Setzer, eds.). Smithsonian Institution Press, Washington, DC.

Genoways, H. H., and R. M. Timm. 2003. The xenarthrans of Nicaragua. Mastozoología Neotropical 10:231–253.

Gensch, W. 1963. Breeding tupaias. International Zoo Yearbook 4:75–76.

George, G. G. 1979. The status of endangered Papua New Guinea mammals. Pp. 93–100 in The status of endangered Australian wildlife (M. J. Tyler, ed.). Royal Zoological Society of South Australia, Adelaide, South Australia, Australia.

George, G. G. 1987. Characterisation of the living species of cuscus (Marsupialia: Phalangeridae). Pp. 507–526 in Possums and opossums (M. Archer, ed.), Surrey Beatty & Sons, Sydney, Australia.

George, G. G., and G. M. Maynes. 1990. Status of New Guinea bandicoots. Pp. 93–105 in Bandicoots and bilbies (J. H. Seebeck, P. R. Brown, R. L. Wallis, and C. M. Kemper, eds). Surrey Beatty & Sons, Chipping Norton, New South Wales, Australia.

George, G. G., and U. Schürer. 1978. Some notes on macropods commonly misidentified in zoos. International Zoo Yearbook 18:152–156.

Gettinger, D., T. C. Modesto, H. de Godoy Bergallo, and F. Martins-Hatano. 2011. Mammalia, Didelphimorphia, Didelphidae, *Monodelphis kunsi* (Pine, 1975): distribution extension and first record for eastern Amazonia. Check List 7:585–588.

Gewalt, W. 1990. Other opossums. Pp. 239–252 in Grzimek's encyclopedia of mammals, volume 1 (B. Grzimek, ed.). McGraw-Hill, New York, New York.

Gheerbrant, E. 1998. The oldest known proboscidean and the role of Africa in the radiation of modern orders of placentals. Bulletin of the Geological Society of Denmark 44:181–185.

Gheerbrant, E. 2009. Paleocene emergence of elephant relatives and the rapid radiation of African ungulates. Proceedings of the National Academy of Sciences of the United States of America 106:10717–10721.

Gheerbrant, E., D. P. Domning, and P. Tassy. 2005. Paenungulata (Sirenia, Proboscidea, Hyracoidea, and relatives). Pp. 84–105 in The rise of placental mammals: origins and relationships of the major clades (K. D. Rose and J. D. Archibald, eds.). Johns Hopkins University Press, Baltimore, Maryland.

Gheerbrant, E., and J.-C. Rage. 2006. Paleobiogeography of Africa: how distinct from Gondwana and Laurasia? Palaeogeography, Palaeoclimatology, Palaeoecology 241:224–246.

Gheerbrant, E., J. Sudre, and H. Cappetta. 1996. A Palaeocene proboscidean from Morocco. Nature 383:68–71.

Gheerbrant, E., J. Sudre, H. Cappetta, and G. Bignot. 1998. *Phosphatherium escuilliei* du Thanétien du Bassin des Ouled Abdoun (Maroc), plus ancien proboscidien (Mammalia) d'Afrique. GEOBIOS 30:247–269.

Gheerbrant, E., J. Sudre, H. Cappetta, M. Iarochène, M. Amaghzaz, and B. Bouya. 2002. A new large mammal from the Ypresian of Morocco: evidence of surprising diversity of early proboscideans. Acta Palaeontologica Polonica 47:493–506.

Gheerbrant, E., J. Sudre, P. Tassy, M. Amaghzaz, B. Bouya, and M. Iarochène. 2005. Nouvelles données sur *Phosphatherium escuilliei* (Mammalia, Proboscidea) de l'Éocène inférieur du Maroc, apports à la phylogénie des Proboscidea et des ongulés lophodontes. Geodiversitas 27:239–333.

Giannini, N., F. Abdala, and D. A. Flores. 2004. Comparative postnatal ontogeny of the skull in *Dromiciops gliroides* (Marsupialia: Microbiotheriidae). American Museum Novitates 3460:1–17.

Giarla, T. C., and S. A. Jansa. 2014. The role of physical geography and habitat type in shaping the biogeographical history of a recent radiation of Neotropical marsupials (*Thylamys*: Didelphidae). Journal of Biogeography 41:1547–1558.

Giarla, T. C., R. S. Voss, and S. A. Jansa. 2010. Species limits and phylogenetic relationships in the didelphid marsupial genus *Thylamys* based on mitochondrial DNA sequences and morphology. Bulletin of the American Museum of Natural History 346:1–67.

Gilbert, C., S. Maree, and T. J. Robinson. 2008. Chromosomal evolution and distribution of telomeric repeats in golden moles (Chrysochloridae, Mammalia). Cytogenetic and Genome Research 121:110–119.

Gilbert, C., P. C. O'Brien, G. Bronner, F. Yang, A. Hassanin, M. A. Ferguson-Smith, and T. J. Robinson. 2006. Chromosome painting and molecular dating indicate a low rate of chromosomal evolution in golden moles (Mammalia, Chrysochloridae). Chromosome-Research 14:793–803.

Gilmore, D., D. F. Duarte, and C. Peres da Costa. 2008. The physiology of two- and three-toed sloths. Pp. 130–142 in The biology of the Xenarthra (S. F. Vizcaíno and W. J. Loughry, eds.). University Press of Florida, Gainesville, Florida.

Gilmore, D. P. 1977. The success of marsupials as introduced species. Pp. 169–178 in The biology of marsupials (B. Stonehouse and D. P. Gilmore, eds.). University Park Press, Baltimore, Maryland.

Gingerich, P. D. 2005. Cetacea. Pp. 234–252 in The rise of placental mammals: origins and relationships of the major clades (K. D. Rose and J. D. Archibald, eds.). Johns Hopkins University Press, Baltimore, Maryland.

Glass, B. P. 1985. History of classification and nomenclature in Xenarthra (Edentata). Pp.1–3 in The evolution and ecology of armadillos, sloths, and vermilinguas (G. G. Montgomery, ed.). Smithsonian Institution Press, Washington, DC.

Glen, A. S. 2008. Population attributes of the spotted-tailed quoll (*Dasyurus maculatus*) in north-eastern New South Wales. Australian Journal of Zoology 56:137–142.

Glen, A. S., and C. R. Dickman. 2006. Home range, denning behaviour and microhabitat use of the carnivorous marsupial *Dasyurus maculatus* in eastern Australia. Journal of Zoology 268:347–354.

Glen, A. S., and C. R. Dickman. 2008. Niche overlap between marsupial and eutherian carnivores: does competition threaten the endangered spotted-tailed quoll? Journal of Applied Ecology 45:700–707.

Glen, A. S., and C. R. Dickman. 2011. Why are there so many spotted-tailed quolls *Dasyurus maculatus* in parts of north-eastern New South Wales? Australian Zoologist 35:711–718.

Glen, A. S., P. J. de Tores, D. R. Sutherland, and K. D. Morris. 2009. Interactions between chuditch (*Dasyurus geoffroii*) and introduced predators: a review. Australian Journal of Zoology 57:347–356.

Godfrey, G. K. 1969. Reproduction in a laboratory colony of the marsupial mouse *Sminthopsis larapinta* (Marsupialia: Dasyuridae). Australian Journal of Zoology 17:637–654.

Godfrey, G. K. 1975. A study of oestrus and fecundity in a laboratory colony of mouse opossums (*Marmosa robinsoni*). Journal of Zoology 175:541–555.

Godfrey, G. K., and P. Crowcroft. 1971. Breeding the fat-tailed marsupial mouse *Sminthopsis crassicaudata* in captivity. International Zoo Yearbook 11:33–38.

Godfrey, G. K., and W. L. R. Oliver. 1978. The reproduction and development of the pigmy hedgehog. Dodo 15:38–52.

Godin, A. J. 1977. Wild mammals of New England. Johns Hopkins University Press, Baltimore, Maryland.

Goin, F. J., J. A. Case, M. O. Woodburne, S. F. Vizcaíno, and M. A. Reguero. 1999. New discoveries of "opposum-like" marsupials from Antarctica (Seymour Island, Medial Eocene). Journal of Mammalian Evolution 6:335–365.

Goin, F. J., J. Francisco, M. A. Reguero, R. Pascual, H. Von Koenigswald, M. O. Woodburne, J. A. Case, S. A. Marenssi, C. Vieytes, S. F. Vizcaíno. 2006. First gondwanatherian mammal from Antarctica. Pp. 135–144 in Cretaceous–Tertiary high-latitude palaeoenvironments, James Ross Basin, Antarctica (J. E. Francis, D. Pirrie, and J. A. Crame, eds.). Geological Society, London, Special Publication 258.

Goin, F. J., N. Zimicz, M. A. Reguero, S. N. Santillana, S. A. Marenssi, and J. S. Moly. 2007. New marsupial (Mammalia) from the Eocene of Antarctica, and the origins and affinities of the Microbiotheria. Revista de la Asociación Geológica Argentina 62:597–603.

Goldenberg, S. Z., S. de Silva, H. B. Rasmussen, I. Douglas-Hamilton, and G. Wittemyer. 2014. Controlling for behavioural state reveals social dynamics among male African elephants, *Loxodonta Africana*. Animal Behaviour 95:111–119.

Goldingay, R. L. 1990. The foraging behaviour of a nectar feeding marsupial *Petaurus australis*. Oecologia 85:191–199.

Goldingay, R. L. 1992. Socioecology of the yellow-bellied glider (*Petaurus australis*) in a coastal forest. Australian Journal of Zoology 40:267–278.

Goldingay, R. L. 1994. Loud calls of the yellow-bellied glider, *Petaurus australis*: territorial behaviour by an arboreal marsupial? Australian Journal of Zoology 42:279–293.

Goldingay, R. L. 2008. Yellow-bellied glider *Petaurus australis* (Shaw, 1791). Pp. 228–230 in The mammals of Australia. 3rd edition (S. Van Dyck and R. Strahan, eds.). Reed New Holland, Sydney, Australia.

Goldingay, R. L. 2010. Direct male parental care observed in wild sugar gliders. Australian Mammalogy 32:177–178.

Goldingay, R. L., and S. M. Jackson. 2004. A review of the ecology of the Australian Petauridae. Pp. 376–400 in The biology of Australian possums and gliders (R. L. Goldingay and S. M. Jackson, eds.). Surrey Beatty & Sons, Chipping Norton, New South Wales, Australia.

Goldingay, R. L., and R. P. Kavanagh. 1990. Socioecology of the yellow-bellied glider, *Petaurus australis*, at Waratah Creek, NSW. Australian Journal of Zoology 38:327–41.

Goldingay, R. L., and R. P. Kavanagh. 1993. Home-range estimates and habitat of the yellow-bellied glider (*Petaurus australis*) at Waratah Creek, New South Wales. Wildlife Research 20:387–404.

Goldingay, R., and H. Possingham. 1995. Area requirements for viable populations of the Australian gliding marsupial *Petaurus australis*. Biological Conservation 73:161–167.

Goldingay, R. L., D. G. Quin, and S. Churchill. 2001. Spatial variability in the social organization of the yellow-bellied glider (*Petaurus australis*) near Ravenshoe, north Queensland. Australian Journal of Zoology 49:397–409.

Gombe, S. 1983. Reproductive cycle of the rock hyrax (*Procavia capensis*). African Journal of Ecology 21:129–133.

Gongora, J., A. B. Swan, A. Y. Chong, S. Y. W. Ho, C. S. Damayanti, S. Kolomyjec, T. Grant, E. Miller, D. Blair, E. Furlan, and N. Gust. 2012. Genetic structure and phylogeography of platypuses revealed by mitochondrial DNA. Journal of Zoology 286:110–119.

Gonzalez, J. C., C. Custodio, P. Carino, and R. Pamaong-Jose. 2008. *Cynocephalus volans*. The IUCN Red List of Threatened Species 2008: e.T6081A12410826. http://dx.doi.org/10.2305/IUCN.UK .2008.RLTS.T6081A12410826.en. Downloaded on 01 June 2015.

Goodman, S. M. 2003. *Oryzorictes*, mole tenrec or rice tenrec. Pp. 1278–1281 in The natural history of Madagascar (S. M. Goodman and J. P. Benstead, eds.). University of Chicago Press, Chicago, Illinois.

Goodman, S. M., J. U. Ganzhorn, and D. Rakotondravony. 2003. Introduction to the mammals. Pp. 1159–1186 in The natural history of Madagascar (S. M. Goodman and J. P. Benstead, eds.). University of Chicago Press, Chicago, Illinois.

Goodman, S. M., and P. D. Jenkins. 1998. The insectivores of the Réserve Spéciale D'Anjanaharibe-Sud, Madagascar. Fieldiana: Zoology (new series) 90:139–161.

Goodman, S. M., P. D. Jenkins, and M. Pidgeon. 1999. Lipotyphla (Tenrecidae and Soricidae) of the Reserve Naturelle Integrale d'Andohahela, Madagascar. Fieldiana: Zoology (new series) 94:187–216.

Goodman, S. M., and V. Soarimalala. 2004. A new species of *Microgale* (Lipotyphla: Tenrecidae: Oryzorictinae) from the Foret Des Mikea of southwestern Madagascar. Proceedings of the Biological Society of Washington 117:251–265.

Goodman, S., V. Soarimalala, and P. J. Stephenson. 2015a. *Microgale brevicaudata*. The IUCN Red List of Threatened Species 2015: e.T54007828A54008079. http://dx.doi.org/10.2305/IUCN.UK .2015-4.RLTS.T54007828A54008079.en. Downloaded on 18 January 2016.

Goodman, S., V. Soarimalala, and P. J. Stephenson. 2015b. *Microgale dobsoni*. The IUCN Red List of Threatened Species 2015: e.T405 87A21290278. http://dx.doi.org/10.2305/IUCN.UK.2015-4.RLTS .T40587A21290278.en. Downloaded on 18 January 2016.

Goodman, S., V. Soarimalala, and P. J. Stephenson. 2015c. *Microgale gymnorhyncha*. The IUCN Red List of Threatened Species 2015: e.T62014A21283447. http://dx.doi.org/10.2305/IUCN.UK.2015-4 .RLTS.T62014A21283447.en. Downloaded on 18 January 2016.

Goodman, S., V. Soarimalala, and P. J. Stephenson. 2015d. *Microgale talazaci*. The IUCN Red List of Threatened Species 2015: e.T413 15A21286699. http://dx.doi.org/10.2305/IUCN.UK.2015-4.RLTS .T41315A21286699.en. Downloaded on 18 January 2016.

Goodman, S., V. Soarimalala, and P. J. Stephenson. 2015e. *Microgale jenkinsae*. The IUCN Red List of Threatened Species 2015: e.T620 15A21283004. http://dx.doi.org/10.2305/IUCN.UK.2015-4.RLTS .T62015A21283004.en. Downloaded on 18 January 2016.

Goodman, S., V. Soarimalala, and P. J. Stephenson. 2015f. *Microgale dryas*. The IUCN Red List of Threatened Species 2015: e.T133 56A21287924. http://dx.doi.org/10.2305/IUCN.UK.2015-4.RLTS .T13356A21287924.en. Downloaded on 18 January 2016.

Goodman, S., V. Soarimalala, and P. J. Stephenson. 2015g. *Limnogale mergulus*. The IUCN Red List of Threatened Species 2015: e.T119 79A21287536. http://dx.doi.org/10.2305/IUCN.UK.2015-4.RLTS .T11979A21287536.en. Downloaded on 18 January 2016.

Goodman, S., P. J. Stephenson, and V. Soarimalala. 2015a. *Hemicentetes semispinosus*. The IUCN Red List of Threatened Species 2015: e.T40593A21289685. http://dx.doi.org/10.2305/IUCN.UK.2015-4 .RLTS.T40593A21289685.en. Downloaded on 18 January 2016.

Goodman, S., P. J. Stephenson, and V. Soarimalala. 2015b. *Hemicentetes nigriceps*. The IUCN Red List of Threatened Species 2015: e.T620 11A21283860. http://dx.doi.org/10.2305/IUCN.UK.2015-4.RLTS .T62011A21283860.en. Downloaded on 18 January 2016.

Goodman, S., P. J. Stephenson, and V. Soarimalala. 2015c. *Setifer setosus*. The IUCN Red List of Threatened Species 2015: e.T405 94A21289581. http://dx.doi.org/10.2305/IUCN.UK.2015-4.RLTS .T40594A21289581.en. Downloaded on 18 January 2016.

Goodman, S., P. J. Stephenson, and V. Soarimalala. 2015d. *Geogale aurita.* The IUCN Red List of Threatened Species 2015: e.T9048A21282443. http://dx.doi.org/10.2305/IUCN.UK.2015-4.RLTS.T9048A21282443.en. Downloaded on 18 January 2016.

Goodman, S., P. J. Stephenson, and V. Soarimalala. 2015e. *Oryzorictes hova.* The IUCN Red List of Threatened Species 2015: e.T40589A21290193. http://dx.doi.org/10.2305/IUCN.UK.2015-4.RLTS.T40589A21290193.en. Downloaded on 18 January 2016.

Goodman, S., P. J. Stephenson, and V. Soarimalala. 2015f. *Oryzorictes tetradactylus.* The IUCN Red List of Threatened Species 2015: e.T40591A21290044. http://dx.doi.org/10.2305/IUCN.UK.2015-4.RLTS.T40591A21290044.en. Downloaded on 18 January 2016.

Goodman, S., P. J. Stephenson, and V. Soarimalala. 2015g. *Microgale pusilla.* The IUCN Red List of Threatened Species 2015: e.T41314A21286156. http://dx.doi.org/10.2305/IUCN.UK.2015-4.RLTS.T41314A21286156.en. Downloaded on 18 January 2016.

Goodman, S., P. J. Stephenson, and V. Soarimalala. 2015h. *Microgale drouhardi.* The IUCN Red List of Threatened Species 2015: e.T62012A21283974. http://dx.doi.org/10.2305/IUCN.UK.2015-4.RLTS.T62012A21283974.en. Downloaded on 18 January 2016.

Goodman, S., P. J. Stephenson, and V. Soarimalala. 2015i. *Microgale longicaudata.* The IUCN Red List of Threatened Species 2015: e.T13344A21284348. http://dx.doi.org/10.2305/IUCN.UK.2015-4.RLTS.T13344A21284348.en. Downloaded on 18 January 2016.

Goodman, S., P. J. Stephenson, and V. Soarimalala. 2015j. *Microgale majori.* The IUCN Red List of Threatened Species 2015: e.T62016A21283078. http://dx.doi.org/10.2305/IUCN.UK.2015-4.RLTS.T62016A21283078.en. Downloaded on 18 January 2016.

Goodman, S., P. J. Stephenson, and V. Soarimalala. 2015k. *Microgale parvula.* The IUCN Red List of Threatened Species 2015: e.T13349A21288988. http://dx.doi.org/10.2305/IUCN.UK.2015-4.RLTS.T13349A21288988.en. Downloaded on 18 January 2016.

Goodman, S., P. J. Stephenson, and V. Soarimalala. 2015l. *Microgale principula.* The IUCN Red List of Threatened Species 2015: e.T13350A21288804. http://dx.doi.org/10.2305/IUCN.UK.2015-4.RLTS.T13350A21288804.en. Downloaded on 18 January 2016.

Goodman, S., P. J. Stephenson, and V. Soarimalala. 2015m. *Microgale taiva.* The IUCN Red List of Threatened Species 2015: e.T62019A21285499. http://dx.doi.org/10.2305/IUCN.UK.2015-4.RLTS.T62019A21285499.en. Downloaded on 18 January 2016.

Goodman, S., P. J. Stephenson, and V. Soarimalala. 2015n. *Microgale monticola.* The IUCN Red List of Threatened Species 2015: e.T29462A21288204. http://dx.doi.org/10.2305/IUCN.UK.2015-4.RLTS.T29462A21288204.en. Downloaded on 18 January 2016.

Goodman, S., P. J. Stephenson, and V. Soarimalala. 2015o. *Microgale nasoloi.* The IUCN Red List of Threatened Species 2015: e.T62017A21285248. http://dx.doi.org/10.2305/IUCN.UK.2015-4.RLTS.T62017A21285248.en. Downloaded on 18 January 2016.

Goodman, S. M., N. Vasey, and A. Burney. 2007. Description of a new species of subfossil shrew tenrec (Afrosoricida: Tenrecidae: *Microgale*) from cave deposits in southeastern Madagascar. Proceedings of the Biological Society of Washington 120:367–376.

Goodman, S. M., D. Rakotondravony, V. Soarimalada, J. B. Duchemin, and M. Duplantier. 2000. Syntopic occurrence of *Hemicentetes semispinosus* and *H. nigriceps* (Lipotyphla: Tenrecidae) on the central highlands of Madagascar. Mammalia 64:113–116.

Goodman, S. M., C. J. Raxworthy, C. P. Maminirina, and L. E. Olson. 2006. A new species of shrew tenrec (*Microgale jobihely*) from northern Madagascar. Journal of Zoology 270:384–398.

Goodwin, H. A., and J. M. Goodwin. 1973. List of mammals which have become extinct or are possibly extinct since 1600. International Union for the Conservation of Nature Occasional Papers 8:1–20.

Gopala, A., O. Hadian, Sunarto, A. Sitompul, A. Williams, P. Leimgruber, S. E. Chambliss, and D. Gunaryadi. 2011. *Elephas maximus* ssp. *sumatranus.* The IUCN Red List of Threatened Species 2011: e.T199856A9129626. http://dx.doi.org/10.2305/IUCN.UK.2011-2.RLTS.T199856A9129626.en. Downloaded on 29 March 2015.

Gordon, G. 2008a. Long-nosed echymipera *Echymipera rufescens* (Peters and Doria, 1875). Pp. 174–175 in The mammals of Australia. 3rd edition (S. Van Dyck and R. Strahan, eds.). Reed New Holland, Sydney, Australia.

Gordon, G. 2008b. Northern brown bandicoot *Isoodon macrourus* (Gould, 1842). Pp. 178–180 in The mammals of Australia. 3rd edition (S. Van Dyck and R. Strahan, eds.). Reed New Holland, Sydney, Australia.

Gordon, G. 2008c. Desert bandicoot *Perameles eremiana* (Spencer, 1897). Pp. 185–186 in The mammals of Australia. 3rd edition (S. Van Dyck and R. Strahan, eds.). Reed New Holland, Sydney, Australia.

Gordon, G., and L. S. Hall. 1995. Tail fat storage in arid zone bandicoots. Australian Mammalogy 18:87–90.

Gordon, G., L. S. Hall, and R. G. Atherton. 1990. Status of bandicoots in Queensland. Pp. 37–42 in Bandicoots and bilbies (J. H. Seebeck, P. R. Brown, R. L. Wallis, and C. M. Kemper, eds). Surrey Beatty & Sons, Chipping Norton, New South Wales, Australia.

Gordon, G., and F. Hrdina. 2005. Koala and possum populations in Queensland during the harvest period, 1906–1936. Australian Zoologist 33:69–99.

Gordon, G., F. Hrdina, and R. Patterson. 2005. Decline in the distribution of the koala *Phascolorctos cinereus* in Queensland. Australian Zoologist 33:345–358.

Gordon, G., and A. J. Hulbert. 1989. Peramelidae. Pp. 603–624 in Fauna of Australia, volume 1B, Mammalia (D. W. Walton and B. J. Richardson, eds.). Australian Government Publishing Service, Canberra, Australia.

Gordon, G., and B. C. Lawrie. 1977. The rufescent bandicoot, *Echymipera rufescens* (Peters and Doria) on Cape York Peninsula. Australian Wildlife Research 5:41–45.

Gordon, G., and B. C. Lawrie. 1980. The rediscovery of the bridled nail-tailed wallaby, *Onychogalea fraenata* (Gould) (Marsupialia: Macropodidae) in Queensland. Australian Wildlife Research 7:339–345.

Gordon, G., D. G. McGreevy, and B. C. Lawrie. 1978. The yellow-footed rock-wallaby, *Petrogale xanthopus* Gray (Macropodidae), in Queensland. Australian Wildlife Research 5:295–297.

Gordon, G., D. G. McGreevy, and B. C. Lawrie. 1990. Koala population turnover and male social organization. Pp. 189–192 in Biology of the koala (A. K. Lee, K. A. Handasyde, and G. D. Sanson, eds.). Surrey, Beatty & Sons, Sydney, Australia.

Gordon, G., P. McRae, L. Lim, D. Reimer, and G. Porter. 1993. The conservation status of the yellow-footed rock-wallaby in Queensland. Oryx 27:159–168.

Gordon, G., P. Menkhorst, T. Robinson, D. Lunney, R. Martin, and M. Ellis. 2008. *Phascolarctos cinereus.* The IUCN Red List of Threatened Species 2008: e.T16892A6549393. http://dx.doi.org/10.2305/IUCN.UK.2008.RLTS.T16892A6549393.en. Downloaded on 27 November 2015.

Gould, E. 1965. Evidence for echolocation in the Tenrecidae of Madagascar. Proceedings of the American Philosophical Society 109:352–360.

Gould, E. 1978. The behavior of the moonrat, *Echinosorex gymnurus* (Erinaceidae) and the pentail shrew, *Ptilocercus lowi* (Tupaiidae) with comments on the behavior of other insectivora. Zeitschrift für Tierpsychologie 48:1–27.

Gould, E., and J. F. Eisenberg. 1966. Notes on the biology of the Tenrecidae. Journal of Mammalogy 47:660–686.

Grant, T. R. 1973. Dominance and association among members of a captive and free-ranging group of grey kangaroos (*Macropus giganteus*). Animal Behaviour 21:449–456.

Grant, T. R. 1983. Behavioral ecology of monotremes. American Society of Mammalogists Special Publication 7:360–394.

Grant, T. R. 1992. The historical and current distribution of the platypus (*Ornithorhynchus anatinus*) in Australia. Pp. 232–254 in Platypus and echidnas (M. L. Augee, ed.). Royal Zoological Society of New South Wales, Mosman, New South Wales, Australia.

Grant, T. R. 1993. The past and present freshwater fishery in New South Wales and the distribution and status of platypus *Ornithorhynchus anatinus*. Australian Zoologist 29:105–113.

Grant, T. R. 2004a. Captures, capture mortality, age and sex ratios of platypuses, *Ornithorhynchus anatinus*, during studies over 30 years in the upper Shoalhaven River in New South Wales. Proceedings of the Linnean Society of New South Wales 125:217–226.

Grant, T. R. 2004b. Depth and substrate selection by platypuses, *Ornithorhynchus anatinus*, in the lower Hastings River, New South Wales. Proceedings of the Linnean Society of New South Wales 125:235–241.

Grant, T. R., and F. N. Carrick. 1978. Some aspects of the ecology of the platypus, *Ornithorhynchus anatinus*, in the upper Shoalhaven River, New South Wales. Australian Zoologist 20:181–199.

Grant, T. R., and M. Griffiths. 1992. Aspects of lactation and determination of sex ratios and longevity in a free-ranging population of platypuses, *Ornithorhynchus anatinus*, in the Shoalhaven River, NSW. Pp. 80–89 in Platypus and echidnas (M. L. Augee, ed.). Royal Zoological Society of New South Wales, Mosman, New South Wales, Australia.

Grant, T. R., M. Griffiths, and P. D. Temple-Smith. 2004. Breeding in a free-ranging population of platypuses, *Ornithorhynchus anatinus*, in the upper Shoalhaven River, New South Wales—a 27 year study. Proceedings of the Linnean Society of New South Wales 125:227–234.

Grant, T. R., and P. D. Temple-Smith. 1987. Observations on torpor in the small marsupial *Dromiciops australis* (Marsupialia: Microbiotheriidae) from southern Chile. Pp. 273–277 in Possums and opossums (M. Archer, ed.). Surrey Beatty & Sons, Sydney, Australia.

Grant, T. R., and P. D. Temple-Smith. 1998. Field biology of the platypus (*Ornithorhynchus anatinus*): historical and current perspectives. Philosophical Transactions of the Royal Society of London B 353:1081–1091.

Grant, T. R., and P. D. Temple-Smith. 2003. Conservation of the platypus, *Ornithorhynchus anatinus*: threats and challenges. Aquatic Ecosystem Health & Management 6:5–18.

Grant, T. R., G. C. Grigg, L. A. Beard, and M. L. Augee. 1992. Movements and burrow use by platypuses, *Ornithorhynchus anatinus*, in the Thredbo River, New South Wales. Pp. 263–267 in Platypus and echidnas (M. L. Augee, ed.). Royal Zoological Society of New South Wales, Mosman, New South Wales, Australia.

Graur, D. 1993. Towards a molecular resolution of the ordinal phylogeny of the eutherian mammals. FEBS (Federation of European Biochemical Societies) Letters 325:152–159.

Grayson, D. K. 1989. The chronology of North American late Pleistocene extinctions. Journal of Archaeological Science 16:153–165.

Grayson, D. K., and D. J. Meltzer. 2004. North American overkill continued? Journal of Archaeological Science 31:133–136.

Greegor, D. H., Jr. 1980a. Diet of the little hairy armadillo, *Chaetophractus vellerosus*, of northwestern Argentina. Journal of Mammalogy 61:331–334.

Greegor, D. H., Jr. 1980b. Preliminary study of movements and home range of the armadillo, *Chaetophractus vellerosus*. Journal of Mammalogy 61:334–335.

Greegor, D. H., Jr. 1985. Ecology of the little hairy armadillo *Chaetophractus vellerosus*. Pp. 397–405 in The evolution and ecology of armadillos, sloths, and vermilinguas (G. G. Montgomery, ed.). Smithsonian Institution Press, Washington, DC.

Green, K., and A. T. Mitchell. 1997. Breeding of the long-footed potoroo, *Potorous longipes* (Marsupialia: Potoroidae), in the wild: behaviour, births and juvenile independence. Australian Mammalogy 20:1–7.

Green, K., A. T. Mitchell, and P. Tennant. 1998. Home range and microhabitat use by the long-footed potoroo (*Potorous longipes*). Wildlife Research 25:357–372.

Green, K., M. K. Troy, A. T. Mitchell, P. Tennant, and T. W. May. 1999. The diet of the long-footed potoroo (*Potorous longipes*). Australian Journal of Ecology 24:151–156.

Green, R. H. 1967. Notes on the devil (*Sarcophilus harrisii*) and the quoll (*Dasyurus viverrinus*) in north-eastern Tasmania. Records of the Queen Victoria Museum 27:1–13.

Green, R. H. 1972. The murids and small dasyurids in Tasmania. Parts 5, 6, and 7. Records of the Queen Victoria Museum 46:1–34.

Green, R. H. 1987. The common wombat *Vombatus ursinus* (Shaw, 1800) in northern Tasmania—Part 1. Breeding, growth, and development. Records of the Queen Victoria Museum 91:1–20.

Green, R. H., and J. L. Rainbird. 1987. The common wombat *Vombatus ursinus* (Shaw, 1800) in northern Tasmania—Part 1. Breeding, growth, and development. Records of the Queen Victoria Museum 91:1–20.

Greenwell, R. 1993. Little elephant stories. BBC Wildlife, May, p. 53.

Greenwood, A. D., C. Capelli, G. Possnert, and S. Pääbo. 1999. Nuclear DNA sequences from late Pleistocene megafauna. Molecular Biology and Evolution 16:1466–1473.

Griffiths, M. 1968. Echidnas. Pergamon Press, Oxford, United Kingdom.

Griffiths, M. 1978. The biology of monotremes. Academic Press, New York, New York.

Griffiths, M., R. T. Wells, and D. J. Barrie. 1991. Observations on the skulls of fossil and extant echidnas (Monotremata: Tachyglossidae). Australian Mammalogy 14:87–101.

Grigg, G. 2002. Conservation benefit from harvesting kangaroos: status report at the start of a new millennium: a paper to stimulate discussion and research. Pp. 53–76 in A zoological revolution: using native fauna to assist in its own survival (D. Lunney and C. Dickman, eds.). Royal Zoological Society of New South Wales, Mosman, New South Wales, Australia.

Grigg, G. C., M. L. Augee, and L. A. Beard. 1992. Thermal relations of free-living echidnas during activity and in hibernation in a cold climate. Pp. 160–173 in Platypus and echidnas (M. L. Augee, ed.). Royal Zoological Society of New South Wales, Mosman, New South Wales, Australia.

Grigg, G. C., L. A. Beard, and M. L. Augee. 1989. Hibernation in a monotreme, the echidna (*Tachyglossus aculeatus*). Comparative Biochemistry and Physiology 92A:609–612.

Grigg, G. C., L. A. Beard, G. Caughley, D. Grice, J. A. Caughley, N. Sheppard, M. Fletcher, and C. Southwell. 1985. The Australian kangaroo populations, 1984. Search 16:277–279.

Grigg, G. C., L. A. Beard, J. A. Barnes, L. I. Perry, G. J. Fry, and M. Hawkins. 2003. Body temperature in captive long-beaked echidnas (*Zaglossus bartoni*). Comparative Biochemistry and Physiology Part A 136(4):911–916.

Grimwood, I. R. 1969. Notes on the distribution and status of some Peruvian mammals. Special Publication of the American Committee for International Wild Life Protection 21:1–86.

Groom, C. 2010. Justification for continued conservation efforts following the delisting of a threatened species: a case study of the woylie, *Bettongia penicillata ogilbyi* (Marsupialia : Potoroidae). Wildlife Research 37:183–193.

Groves, C. P. 1976. The origin of the mammalian fauna of Sulawesi (Celebes). Zeitschrift für Säugetierkunde 41:201–216.

Groves, C. P. 1982. The systematics of tree kangaroos (*Dendrolagus*; Marsupialia, Macropodidae). Australian Mammalogy 5:157–186.

Groves, C. P. 1987a. On the highland cuscuses (Marsupialia: Phalangeridae) of New Guinea. Pp. 559–567 in Possums and opossums (M. Archer, ed.). Surrey Beatty & Sons, Sydney, Australia.

Groves, C. P. 1987b. On the cuscuses (Marsupialia: Phalangeridae) of the Phalanger orientalis group from Indonesian territory. Pp. 569–579 in Possums and opossums (M. Archer, ed.). Surrey Beatty & Sons, Sydney, Australia.

Groves, C. P. 2000. What are the elephants of West Africa? Elephant 2(4)7–8.

Groves, C. P. 2005a. Order Monotremata. Pp. 1–2 in Mammal species of the world: a taxonomic and geographic reference. 3rd edition (D. E. Wilson and D. M. Reeder, eds.). Johns Hopkins University Press, Baltimore, Maryland.

Groves, C. P. 2005b. Order Notoryctemorphia. P. 22 in Mammal species of the world: a taxonomic and geographic reference. 3rd edition (D. E. Wilson and D. M. Reeder, eds.). Johns Hopkins University Press, Baltimore, Maryland.

Groves, C. P. 2005c. Order Dasyuromorphia. Pp. 23–37 in Mammal species of the world: a taxonomic and geographic reference. 3rd edition (D. E. Wilson and D. M. Reeder, eds.). Johns Hopkins University Press, Baltimore, Maryland.

Groves, C. P. 2005d. Order Peramelemorphia. Pp. 38–42 in Mammal species of the world: a taxonomic and geographic reference. 3rd edition (D. E. Wilson and D. M. Reeder, eds.). Johns Hopkins University Press, Baltimore, Maryland.

Groves, C. P. 2005e. Order Diprotodontia. Pp. 43–70 in Mammal species of the world: a taxonomic and geographic reference. 3rd edition (D. E. Wilson and D. M. Reeder, eds.). Johns Hopkins University Press, Baltimore, Maryland.

Groves, C. P., and T. Flannery. 1989. Revision of the genus Dorcopsis (Macropodidae: Marsupialia). Pp. 117–128 in Kangaroos, wallabies and rat-kangaroos (G. C. Grigg, P. Jarman, and I. Hume, eds.). Surrey Beatty & Sons, Chipping Norton, New South Wales, Australia.

Groves, C. P., and T. Flannery. 1990. Revision of the families and genera of bandicoots. Pp. 1–11 in Bandicoots and bilbies (J. H. Seebeck, P. R. Brown, R. L. Wallis, and C. M. Kemper, eds). Surrey Beatty & Sons, Chipping Norton, New South Wales, Australia.

Groves, C. P., and P. Grubb. 2000a. Do *Loxodonta cyclotis* and *L. africana* interbreed? Elephant 2(4):4–7.

Groves, C. P., and P. Grubb. 2000b. Are there pygmy elephants? Elephant 2(4):8–10.

Grubb, P., C. P. Groves, J. P. Dudley, and J. Shoshani. 2000. Living African elephants belong to two species: *Loxodonta africana* (Blumenbach, 1797) and *Loxodonta cyclotis* (Matschie, 1900). Elephant 2(4):1–4.

Gruber, K. F., R. S. Voss, and S. A. Jansa. 2007. Base-compositional heterogeneity in the RAG1 locus among didelphid marsupials: implications for phylogenetic inference and the evolution of GC content. Systematic Biology 56:1–14.

Grzimek, B. 1990. African elephant. Pp. 502–520 in Grzimek's encyclopedia of mammals, volume 4 (B. Grzimek, ed.). McGraw-Hill, New York, New York.

Grzimek, B., and U. Ganslosser. 1990. Ringtails and gliders. Pp. 312–324 in Grzimek's encyclopedia of mammals, volume 1 (B. Grzimek, ed.). McGraw-Hill, New York, New York.

Grzimek, B., and D. Heinemann. 1975. Subfamily true kangaroos (Macropodinae). Pp. 162–173 in Grzimek's animal life encyclopedia (B. Grzimek, ed.). Van Nostrand Reinhold, New York, New York.

Gucwinska, H. 1971. Development of six-banded armadillos *Euphractus sexcinctus* at Wroclaw Zoo. International Zoo Yearbook 11:88–89.

Guiler, E. R. 1961. Breeding season of the thylacine. Journal of Mammalogy 42:396–397.

Guiler, E. R. 1970a. Observations on the Tasmanian devil, *Sarcophilus harrisii* (Marsupialia: Dasyuridae). I. Numbers, home range, movements, and food in two populations. Australian Journal of Zoology 18:49–62.

Guiler, E. R. 1970b. Observations on the Tasmanian devil, *Sarcophilus harrisii* (Marsupialia: Dasyuridae). II. Reproduction, breeding, and growth of pouch young. Australian Journal of Zoology 18:63–70.

Guiler, E. R. 1971a. Food of the potoroo (Marsupialia, Macropodidae). Journal of Mammalogy 52:232–234.

Guiler, E. R. 1971b. The husbandry of the potoroo *Potorous tridactylus*. International Zoo Yearbook 11:21–22.

Guiler, E. R. 1971c. The Tasmanian devil, *Sarcophilus harrisii*, in captivity. International Zoo Yearbook 11:32–33.

Guiler, E. R. 1982. Temporal and spatial distribution of the Tasmanian devil, *Sarcophilus harrisii* (Dasyuridae: Marsupialia). Papers and Proceedings of the Royal Society of Tasmania 116:153–163.

Guiler, E. 1983. Tasmanian devil *Sarcophilus harrisii*. Pp. 27–28 in The Australian Museum complete book of Australian mammals (R. Strahan, ed.). Angus & Robertson, London, United Kingdom.

Guiler, E. R., and D. A. Kitchener. 1967. Further observations on longevity in the wild potoroo, *Potorous tridactylus*. Australian Journal of Science 30:105–106.

Guldemond, R., and R. Van Aarde. 2008. A meta-analysis of the impact of African elephants on savanna vegetation. Journal of Wildlife Management 72:892–899.

Guldemond, R., and R. Van Aarde. 2008. A meta-analysis of the impact of African elephants on savanna vegetation. Journal of Wildlife Management 72:892–899.

Gurovich, Y., and R. Beck. 2009. The phylogenetic affinities of the enigmatic mammalian clade Gondwanatheria. Journal of Mammalian Evolution 16:25–49.

Gust, N., and K. Handasyde. 1995. Seasonal variation in the ranging behavior of the platypus (*Ornithorhynchus anatinus*) on the Goulburn River, Victoria. Australian Journal of Zoology 43:193–208.

Guthrie, R. D. 2004. Radiocarbon evidence of mid-Holocene mammoths stranded on an Alaskan Bering Sea island. Nature 429:746–749.

Guthrie, R. D. 2006. New carbon dates link climatic change with human colonization and Pleistocene extinctions. Nature 441:207–209.

Guthrie, R. D. 2013. Frozen fauna of the mammoth steppe. The story of Blue Babe. University of Chicago Press. Chicago, Illinois.

Gutiérrez, E. E., S. A. Jansa, and R. S. Voss. 2010. Molecular systematics of mouse opossums (Didelphidae: *Marmosa*): assessing species limits using mitochondrial DNA sequences, with comments on phylogenetic relationships and biogeography. American Museum Novitates 3692:1–22.

Gutiérrez, E. E., P. J. Soriano, R. V. Rossi, J. J. Murillo, J. Ochoa-G., and M. Aguilera. 2011. Occurrence of *Marmosa waterhousei* in the Venezuelan Andes, with comments on its biogeographic significance. Mammalia 75:381–386.

Gutiérrez, E. E., R. P. Anderson, R. S. Voss, J. Ochoa-G., M. Aguilera, and S. A. Jansa. 2014. Phylogeography of *Marmosa robinsoni*: insights into the biogeography of dry forests in northern South America. Journal of Mammalogy 95:1175–1188.

Gwinn, R. N., G. H. Palmer, and J. L. Koprowski. 2011. Virginia opossum (*Didelphis virginiana virginiana*) from Yavapai County, Arizona. Western North American Naturalist 71:113–114.

Haffenden, A. T., and R. G. Atherton. 2008. Long-tailed pygmy-possum *Cercartetus caudatus* (Milne-Edwards, 1877). Pp. 213–214 in The mammals of Australia. 3rd edition (S. Van Dyck and R. Strahan, eds.). Reed New Holland, Sydney, Australia.

Haigh, M. D. 1991. The use of manatees for the control of aquatic weeds in Guyana. Irrigation and Drainage Systems 5:339–349.

Haile, J., D. G. Froese, R. D. E. MacPhee, R. G. Roberts, L. J. Arnold, A. V. Reyes, M. Rasmussen, R. Nielsen, B. W. Brook, S. Robinson, M. Demuro, M. T. P. Gilbert, K. Munch, J. J. Austin, A. Cooper, I. Barnes, P. Möller, and E. Willerslev. 2009. Ancient DNA reveals late survival of mammoth and horse in interior Alaska. Proceedings of the National Academy of Sciences of the United States of America 106:22352–22357.

Hall, E. R. 1955. Handbook of mammals of Kansas. University of Kansas Museum of Natural History Miscellaneous Publications 7:1–303.

Hall, E. R. 1981. The mammals of North America. John Wiley & Sons, New York, New York.

Hall, L. S. 1987. Syndactyly in marsupials—problems and prophecies. Pp. 245–255 in Possums and opossums (M. Archer, ed.). Surrey Beatty & Sons, Sydney, Australia.

Hall-Martin, A. J. 1992. Distribution and status of the African elephant *Loxodonta africana* in South Africa, 1652–1992. Koedoe 35:65–88.

Hallström, B. M., M. Kullberg, M. A. Nilsson, and A. Janke. 2007. Phylogenomic data analyses provide evidence that Xenarthra and Afrotheria are sister groups. Molecular Biology and Evolution 24:2059–2068.

Hamilton, A. T., and Springer, M. S. 1999. DNA sequence evidence for placement of the ground cuscus, *Phalanger gymnotis*, in the tribe Phalangerini (Marsupialia: Phalangeridae). Journal of Mammalian Evolution 6:1–17.

Hammer, J. 2008. Trials of a primatologist. Smithsonian 38(11):81–82, 84–86, 88, 90–95.

Han, K. H. 2008. *Tupaia glis*. The IUCN Red List of Threatened Species 2008: e.T41494A10468880. http://dx.doi.org/10.2305/IUCN.UK.2008.RLTS.T41494A10468880.en. Downloaded on 20 January 2016.

Han, K. H., and D. Maharadatunkamsi. 2008. *Tupaia javanica*. The IUCN Red List of Threatened Species 2008: e.T41496A10469813. http://dx.doi.org/10.2305/IUCN.UK.2008.RLTS.T41496A10469813.en. Downloaded on 31 May 2015.

Han, K. H., and I. Maryanto. 2008. *Tupaia splendidula*. The IUCN Red List of Threatened Species 2008: e.T41500A10478563. http://dx.doi.org/10.2305/IUCN.UK.2008.RLTS.T41500A10478563.en. Downloaded on 31 May 2015.

Han, K. H., F. H. Sheldon, and R. Stuebing. 2000. Interspecific relationships and biogeography of some Bornean tree shrews (Tupaiidae: *Tupaia*), based on DNA hybridization and morphometric comparisons. Biological Journal of the Linnean Society 70:1–14.

Han, K. H., and R. Stuebing. 2008a. *Ptilocercus lowii*. The IUCN Red List of Threatened Species 2008: e.T41491A10467786. http://dx.doi.org/10.2305/IUCN.UK.2008.RLTS.T41491A10467786.en. Downloaded on 31 May 2015.

Han, K. H., and R. Stuebing. 2008b. *Dendrogale melanura*. The IUCN Red List of Threatened Species 2008: e.T6405A12739829. http://dx.doi.org/10.2305/IUCN.UK.2008.RLTS.T6405A12739829.en. Downloaded on 30 May 2015.

Han, K. H., and R. Stuebing. 2008c. *Tupaia dorsalis*. The IUCN Red List of Threatened Species 2008: e.T41493A10468506. http://dx.doi.org/10.2305/IUCN.UK.2008.RLTS.T41493A10468506.en. Downloaded on 31 May 2015.

Han, K. H., and R. Stuebing. 2008d. *Tupaia gracilis*. The IUCN Red List of Threatened Species 2008: e.T41495A10469254. http://dx.doi.org/10.2305/IUCN.UK.2008.RLTS.T41495A10469254.en. Downloaded on 31 May 2015.

Han, K. H., and R. Stuebing. 2008e. *Tupaia longipes*. The IUCN Red List of Threatened Species 2008: e.T22451A9373564. http://dx.doi.org/10.2305/IUCN.UK.2008.RLTS.T22451A9373564.en. Downloaded on 31 May 2015.

Han, K. H., and R. Stuebing. 2008f. *Tupaia minor*. The IUCN Red List of Threatened Species 2008: e.T41497A10470177. http://dx.doi.org/10.2305/IUCN.UK.2008.RLTS.T41497A10470177.en. Downloaded on 31 May 2015.

Han, K. H., and R. Stuebing. 2008g. *Tupaia picta*. The IUCN Red List of Threatened Species 2008: e.T41499A10470900. http://dx.doi.org/10.2305/IUCN.UK.2008.RLTS.T41499A10470900.en. Downloaded on 31 May 2015.

Han, K. H., and R. Stuebing. 2008h. *Tupaia tana*. The IUCN Red List of Threatened Species 2008: e.T41501A10478912. http://dx.doi.org/10.2305/IUCN.UK.2008.RLTS.T41501A10478912.en. Downloaded on 31 May 2015.

Hanafy, M., M. A. Gheny, A. B. Rouphael, A. Salam, and M. Fouda. 2006. The dugong, *Dugong dugon*, in Egyptian waters: distribution, relative abundance and threats. Zoology in the Middle East 39:17–24.

Hancocks, D. 2008. Most zoos do not deserve elephants. Pp. 259–283 in Elephants and ethics, toward a morality of existence (C. Wemmer and C. A. Christen, eds.). Johns Hopkins University Press, Baltimore, Maryland.

Handasyde, K. 2000. Striped possums: the bold and the beautiful. Nature Australia 26(9):27–31.

Handasyde, K. A. 2008. Striped possum *Dactylopsila trivirgata* (Gray, 1858). Pp. 224–225 in The mammals of Australia. 3rd edition (S. Van Dyck and R. Strahan, eds.). Reed New Holland, Sydney, Australia.

Handasyde, K. A., I. R. McDonald, K. A. Than, J. Michaelides, and R. W. Martin. 1990. Reproductive hormones and reproduction in the koala. Pp. 203–210 in Biology of the koala (A. K. Lee, K. A. Handasyde, and G. D. Sanson, eds.). Surrey, Beatty & Sons, Sydney, Australia.

Handley, C. O., Jr. 1976. Mammals of the Smithsonian Venezuelan Project. Brigham Young University Science Bulletin Biological Series 20(5):1–89.

Handley, C. O., Jr., and L. K. Gordon. 1979. New species of mammals from northern South America: mouse possums, genus Marmosa Gray. Pp. 65–72 in Vertebrate ecology in the northern neotropics (J. F. Eisenberg, ed.). Smithsonian Institution Press, Washington, DC.

Hanks, J. 1979. The struggle for survival: the elephant problem. Mayflower Books, New York, New York.

Hansen, B. D., D. K. P. Harley, D. B. Lindenmayer, and A. C. Taylor. 2009. Population genetic analysis reveals a long-term decline of a threatened endemic Australian marsupial. Molecular Ecology 18:3346–3362.

Happold, D. C. D. 1987. The mammals of Nigeria. Clarendon Press, Oxford, United Kingdom.

Harding, H. R., F. N. Carrick, and C. D. Shorey. 1987. The affinities of the koala Phascolarctos cinereus (Marsupialia: Phascolarctidae) on the basis of sperm ultrastructure and development. Pp. 353–364 in Possums and opossums (M. Archer, ed.). Surrey Beatty & Sons, Sydney, Australia.

Harley, D. K. P. 2004. A review of recent records of Leadbeater's possum Gymnobelideus leadbeateri. Pp. 330–338 in The biology of Australian possums and gliders (R. L. Goldingay and S. M. Jackson, eds.). Surrey Beatty & Sons, Chipping Norton, New South Wales, Australia.

Harley, D. K. P., and A. Lill. 2007. Reproduction in a population of the endangered Leadbeater's possum inhabiting lowland swamp forest. Journal of Zoology 272:451–457.

Haro, J. A., A. A. Tauber, and J. M. Krapovickas. 2016. The manus of Mylodon darwinii Owen (Tardigrada, Mylodontidae) and its phylogenetic implications. Journal of Vertebrate Paleontology 36(5):e1188824.

Harper, F. 1945. Extinct and vanishing mammals of the Old World. American Committee for International Wildlife Protection Special Publication 12:1–850.

Harper, J. Y., and B. A. Schulte. 2005. Social interactions in captive female Florida manatees. Zoo Biology 24:135–144.

Harris, J. M. 2006. The discovery and early natural history of the eastern pygmy-possum, Cercartetus nanus. Proceedings of the Linnean Society of New South Wales 127:107–124.

Harris, J. M. 2009a. Cercartetus concinnus (Diprotodontia: Burramyidae). Mammalian Species 831:1–11.

Harris, J. M. 2009b. Cercartetus lepidus (Diprotodontia: Burramyidae). Mammalian Species 842:1–8.

Harris, J. M. 2015. Acrobates pygmaeus (Diprotodontia: Acrobatidae). Mammalian Species 47(920):32–44.

Harris, J. M., and R. L. Goldingay. 2005. Distribution, habitat and conservation status of the eastern pygmy-possum Cercartetus nanus in Victoria. Australian Mammalogy 27:185–210.

Harris, J. M., and K. S. Maloney. 2010. Petauroides volans (Diprotodontia: Pseudocheiridae). Mammalian Species 42(866):207–219.

Harris, J. M., R. L. Goldingay, L. Broome, and K. S. Maloney. 2007. Aspects of the ecology of the eastern pygmy-possum Cercartetus nanus at Jervis Bay, New South Wales. Australian Mammalogy 29:39–46.

Harrison, D. L., and P. J. Bates. 1991. The mammals of Arabia. Harrison Zoological Museum, Sevenoaks, Kent, United Kingdom.

Hartenberger, J.-L. 1986. Hypothèse paléontologique sur l'origine des Macroscelidea (Mammalia). Comptes Rendus de l'Académie des Sciences ser. 2 302:247–249.

Hartman, D. S. 1979. Ecology and behavior of the manatee (Trichechus manatus) in Florida. American Society of Mammalogists Special Publication 5:1–153.

Haskins, G., and P. Davis. 2008. Has the dugong gone the way of the dodo? Sirenews 49:16–17.

Hasler, J. F., and M. W. Sorenson. 1974. Behavior of the tree shrew, Tupaia chinensis, in captivity. American Midland Naturalist 91:294–314.

Hawkins, C. E., C. Baars, H. Hesterman, G. J. Hocking, M. E. Jones, B. Lazenby, D. Mann, N. Mooney, D. Pemberton, S. Pyecroft, M. Restani, and J. Wiersma. 2006. Emerging disease and population decline of an island endemic, the Tasmanian devil Sarcophilus harrisii. Biological Conservation 131:307–324.

Hawkins, C. E., H. McCallum, N. Mooney, M. Jones, and M. Holdsworth. 2008. Sarcophilus harrisii. The IUCN Red List of Threatened Species 2008: e.T40540A10331066. http://dx.doi.org/10.2305/IUCN.UK.2008.RLTS.T40540A10331066.en. Downloaded on 20 December 2015.

Hawkins, M. R. 1998. Time and space sharing between platypuses (Ornithorhynchus anatinus) in captivity. Australian Mammalogy 20:195–205.

Hawkins, M., and A. Battaglia. 2009. Breeding behaviour of the platypus (Ornithorhynchus anatinus) in captivity. Australian Journal of Zoology 57:283–293.

Haynes, G. 1991. Mammoths, mastodonts, and elephants: biology, behavior, and the fossil record. Cambridge University Press, Cambridge, United Kingdom.

Haynes, G. 2002. The catastrophic extinction of North American mammoths and mastodonts. World Archaeology 33:391–416.

Haynes, G. 2007. A review of some attacks on the overkill hypothesis, with special attention to misrepresentations and doubletalk. Quaternary International 169–170:84–94.

Haynes, G. 2009. Estimates of Clovis-Era megafaunal populations and their extinction risks. Pp. 39–54 in American megafaunal extinctions at the end of the Pleistocene (G. Haynes, ed.). Springer, New York, New York.

Hayssen, V. 2009. Bradypus tridactylus (Pilosa: Bradypodidae). Mammalian Species 839:1–9.

Hayssen, V. 2010. Bradypus variegatus (Pilosa: Bradypodidae). Mammalian Species 42(850):19–32.

Hayssen, V. 2011a. Choloepus hoffmanni (Pilosa: Megalonychidae). Mammalian Species 43(873):37–55.

Hayssen, V. 2011b. Tamandua tetradactyla (Pilosa: Myrmecophagidae). Mammalian Species 43(875):64–74.

Hayssen, V. 2014. Cabassous chacoensis (Cingulata: Dasypodidae). Mammalian Species 46(908):24–27.

Hayssen, V., F. Miranda, and B. Pasch. 2012. Cyclopes didactylus (Pilosa: Cyclopedidae). Mammalian Species 44(895):51–58.

Hayssen, V., A. Van Tienhoven, and A. Van Tienhoven. 1993. Asdell's patterns of mammalian reproduction: a compendium of species-specific data. Cornell University Press, Ithaca, New York.

Hayssen, V., J. Ortega, A. Morales-Leyva, and N. Martínez-Mendez. 2013. Cabassous centralis (Cingulata: Dasypodidae). Mammalian Species 45(898):12–17.

Haythornthwaite, A. S., and C. R. Dickman. 2006. Long-distance movements by a small carnivorous marsupial: how Sminthopsis youngsoni (Marsupialia: Dasyuridae) uses habitat in an Australian sandridge desert. Journal of Zoology 270:543–549.

Hayward, A. D., K. U. Mar, M. Lahdenpara, and V. Lummaa. 2014. Early reproductive investment, senescence and lifetime reproductive success in female Asian elephants. Journal of Evolutionary Biology 27:772–783.

Hayward, M. W. 2008. Home range overlap of the quokka *Setonix brachyurus* (Macropodidae: Marsupialia) suggests a polygynous mating system. Conservation Science Western Australia 7:57–64.

Hayward, M. W., P. J. de Tores, M. J. Dillon, and B. J. Fox. 2003. Local population structure of a naturally occurring metapopulation of the quokka (*Setonix brachyurus* Macropodidae: Marsupialia). Biological Conservation 110:343–355.

Hayward, M. W., P. J. de Tores, M. L. Augee, B. J. Fox, and P. B. Banks. 2004. Home range and movements of the quokka *Setonix brachyurus* (Macropodidae: Marsupialia), and its impact on the viability of the metapopulation on the Australian mainland. Journal of Zoology 263:219–228.

Hazlitt, S. L., M. D. B. Eldridge, and A. W. Goldizen. 2010. Strong matrilineal structuring in the brush-tailed rock-wallaby conformed by spatial patterns of mitochondrial DNA. Pp. 87–95 in Macropods: the biology of kangaroos, wallabies and rat-kangaroos (G. Coulson and M. Eldridge, eds.). CSIRO Publishing, Collingwood, Victoria, Australia.

Heaney, L. 2008. *Tupaia moellendorffi*. The IUCN Red List of Threatened Species 2008: e.T136406A4287364. http://dx.doi.org/10.2305/IUCN.UK.2008.RLTS.T136406A4287364.en. Downloaded on 31 May 2015.

Heaney, L. R., and D. S. Rabor. 1982. Mammals of Dinagat and Siargao Islands, Philippines. Occasional Papers of the Museum of Zoology, University of Michigan 699:1–30.

Heaney, L. R., M. L. Dolar, D. S. Balete, J. A. Esselstyn, E. A. Rickart, and J. L. Sedlock. 2010. Synopsis of Philippine mammals. Field Museum of Natural History, Chicago, Illinois. http://archive.fieldmuseum.org/philippine_mammals/. Accessed on 29 May 2015.

Hedges, S., M. J. Tyson, A. F. Sitompul, and H. Hammatt. 2006. Why inter-country loans will not help Sumatra's elephants. Zoo Biology 25:235–246.

Hedges, S. B. 2001. Afrotheria: plate tectonics meets genomics. Proceedings of the National Academy of Sciences of the United States of America 98:1–2.

Hedges, S. B. 2011. On the use of high-level taxonomic names. Zootaxa 2867:67–68.

Heinsohn, G. E. 2008. Dugong *Dugong dugon* (Müller, 1776). Pp. 711–713 in The mammals of Australia. 3rd edition (S. Van Dyck and R. Strahan, eds.). Reed New Holland, Sydney, Australia.

Heinsohn, R., R. C. Lacy, D. B. Lindenmayer, H. Marsh, D. Kwan, and I. R. Lawler. 2004. Unsustainable harvest of dugongs in Torres Strait and Cape York (Australia) waters: two case studies using population viability analysis. Animal Conservation 7:417–425.

Heinsohn, T. 2001. A giant among possums. Nature Australia 26(12):25–31.

Heinsohn, T. E. 2002a. Status of the common spotted cuscus *Spilocuscus maculatus* and other wild mammals on Selayar Island, Indonesia, with notes on Quaternary faunal turnover. Australian Mammalogy 24:199–207.

Heinsohn, T. E. 2002b. Possum extinctions at the marsupial frontier: the status of the northern common cuscus *Phalanger orientalis* on Santa Ana Island, Makira Province, Solomon Islands. Australian Mammalogy 24:247–248.

Heinsohn, T. E. 2003. Animal translocation: long-term human influences on the vertebrate zoogeography of Australasia (natural dispersal versus ethnophoresy). Australian Zoologist 32:351–376.

Heinsohn, T. E. 2004a. Phalangeroids as ethnotramps: a brief history of possums and gliders as introduced species. Pp. 506–526 in The biology of Australian possums and gliders (R. L. Goldingay and S. M. Jackson, eds.). Surrey Beatty & Sons, Chipping Norton, New South Wales, Australia.

Heinsohn, T. E. 2004b. Ecological variability in the common spotted cuscus *Spilocuscus maculatus* in the Australasian Archipelago—a review. Pp. 527–538 in The biology of Australian possums and gliders (R. L. Goldingay and S. M. Jackson, eds.). Surrey Beatty & Sons, Chipping Norton, New South Wales, Australia.

Heinsohn, T. 2005a. The cuscus that fooled science. Nature Australia 28(5):27–33.

Heinsohn, T. E. 2005b. Wallaby extinctions at the macropodid frontier: the changing status of the northern pademelon *Thylogale browni* (Marsupialia: Macropodidae) in New Ireland Province, Papua New Guinea. Australian Mammalogy 27:175–183.

Heinze, D., L. Broome, and I. Mansergh. 2004. A review of the ecology and conservation of the mountain pygmy-possum *Burramys parvus*. Pp. 254–267 in The biology of Australian possums and gliders (R. L. Goldingay and S. M. Jackson, eds.). Surrey Beatty & Sons, Chipping Norton, New South Wales, Australia.

Heinze, D. A., and A. M. Olejniczak. 2000. First observations of the mountain pygmy-possum *Burramys parvus* nesting in the wild. Australian Mammalogy 22:65–67.

Heise-Pavlov, S. R., S. L. Jackrel, and S. Meeks. 2011. Conservation of a rare arboreal mammal: habitat preferences of the Lumholtz's tree-kangaroo, *Dendrolagus lumholtzi*. Australian Mammalogy 33:5–12.

Helgen, K. M. 2003a. Major mammalian clades: a review under consideration of molecular and palaeontological evidence. Mammalian Biology 68:1–15.

Helgen, K. M. 2003b. The feather-tailed glider (*Acrobates pygmaeus*) in New Guinea. Treubia 33:107–111.

Helgen, K. M. 2005. Order Scandentia. Pp. 104–109 in Mammal species of the world: a taxonomic and geographic reference. 3rd edition (D. E. Wilson and D. M. Reeder, eds.). Johns Hopkins University Press, Baltimore, Maryland.

Helgen, K. M. 2007. A taxonomic and geographic overview of the mammals of Papua. Pp. 689–749 in The ecology of Papua (A. J. Marshall and B. M. Beehler, eds.). Periplus Editions, Singapore.

Helgen, K., K. Aplin, and C. Dickman. 2008. *Spilocuscus papuensis*. The IUCN Red List of Threatened Species 2008: e.T20638A9216739. http://dx.doi.org/10.2305/IUCN.UK.2008.RLTS.T20638A9216739.en. Downloaded on 30 December 2015.

Helgen, K., C. Dickman, and L. Salas. 2008a. *Pseudochirops albertisii*. The IUCN Red List of Threatened Species 2008: e.T18503A8354282. http://dx.doi.org/10.2305/IUCN.UK.2008.RLTS.T18503A8354282.en. Downloaded on 30 December 2015.

Helgen, K., C. Dickman, and L. Salas. 2008b. *Pseudochirulus schlegeli*. The IUCN Red List of Threatened Species 2008: e.T40641A10347153. http://dx.doi.org/10.2305/IUCN.UK.2008.RLTS.T40641A10347153.en. Downloaded on 30 December 2015.

Helgen, K. M., and T. F. Flannery. 2003. Taxonomy and historical distribution of the wallaby genus *Lagostrophus*. Australian Journal of Zoology 51:199–212.

Helgen, K. M., and T. F. Flannery. 2004a. A new species of bandicoot, *Microperoryctes aplini*, from western New Guinea Journal of Zoology 264:117–124.

Helgen, K. M., and T. F. Flannery. 2004b. Notes on the phalangerid marsupial genus *Spilocuscus*, with description of a new species from Papua. Journal of Mammalogy 85:825–833.

Helgen, K. M., and S. M. Jackson. 2015. Family Phalangeridae (cuscuses, brush-tailed possums and scaly-tailed possum). Pp. 456–497 in Handbook of the mammals of the world, volume 5, Monotremes and marsupials (D. E. Wilson and R. A. Mittermeier, eds.). Lynx Edicions, Barcelona, Spain.

Helgen, K. M., and E. G. Veatch. 2015. Recently extinct Australian marsupials and monotremes. Pp. 17–31 in Handbook of the mammals of the world, volume 5, Monotremes and marsupials (D. E. Wilson and R. A. Mittermeier, eds.). Lynx Edicions, Barcelona, Spain.

Helgen. K. M., and D. E. Wilson. 2001. Additional material of the enigmatic golden mole *Cryptochloris zyli*, with notes on the genus *Cryptochloris* (Mammalia: Chrysochloridae). African Zoology 36:110–112.

Helgen, K. M., R. T. Wells, B. P. Kear, W. R. Gerdtz, and T. F. Flannery. 2006. Ecological and evolutionary significance of sizes of giant extinct kangaroos. Australian Journal of Zoology 54:293–303.

Helgen, K., K. Aplin, C. Dickman, and L. Salas. 2008a. *Strigocuscus celebensis*. The IUCN Red List of Threatened Species 2008: e.T208 90A9234928. http://dx.doi.org/10.2305/IUCN.UK.2008.RLTS .T20890A9234928.en. Downloaded on 30 December 2015.

Helgen, K., T. Leary, R. Singadan, J. Menzies, and D. Wright. 2008b. *Spilocuscus kraemeri*. The IUCN Red List of Threatened Species 2008: e.T20637A9216570. http://dx.doi.org/10.2305/IUCN.UK .2008.RLTS.T20637A9216570.en. Downloaded on 30 December 2015.

Helgen, K. M., R. P. Miguez, J. L. Kohen, and L. E. Helgen. 2012. Twentieth-century occurrence of the long-beaked echidna *Zaglossus bruijnii* in the Kimberley region of Australia. ZooKeys 255:103–132.

Henry, S. 1984. Social organisation of the greater glider in Victoria. Pp. 221–228 in Possums and gliders (A. Smith and I. Hume, eds.). Surrey Beatty & Sons, Norton, New South Wales, Australia.

Henry, S., and S. A. Craig. 1984. Diet, ranging behaviour, and social organization of the yellow-bellied glider (*Petaurus australis* Shaw) in Victoria. Pp. 331–341 in Possums and gliders (A. Smith and I. Hume, eds.). Surrey Beatty & Sons, Norton, New South Wales, Australia.

Hercock, M. 2005. The wild kangaroo industry: developing the potential for sustainability. Environmentalist 24:73–86.

Hernandez, P., J. E. Reynolds, III, H. Marsh, and M. Marmontel. 1995. Age and seasonality in spermatogenesis of Florida manatees. Pp. 84–95 in Population biology of the Florida manatee (T. J. O'Shea, B. B. Ackerman, and H. F. Percival, eds.). National Biological Service Information and Technology Report 1:1–289.

Herridge, V. L., and A. M. Lister. 2012. Extreme insular dwarfism evolved in a mammoth. Proceedings of the Royal Society B 279:3193–3200.

Herrington, J., and E. Muñiz (Fish and Wildlife Service, United States Department of the Interior). 2016. Endangered and Threatened Wildlife and Plants; 12-month finding on a petition to downlist the West Indian manatee, and proposed rule to reclassify the West Indian manatee as threatened. Federal Register 81:1000–1026.

Hershkovitz, P. 1976. Comments on generic names of four-eyed opossums (family Didelphidae). Proceedings of the Biological Society of Washington 89:295–304.

Hershkovitz, P. 1981. *Philander* and four-eyed opossums once again. Proceedings of the Biological Society of Washington 93:943–946.

Hershkovitz, P. 1992a. Ankle bones: the Chilean opossum *Dromiciops gliroides* Thomas, and marsupial phylogeny. Bonner Zoologische Beiträge 43:181–213.

Hershkovitz, P. 1992b. The South American gracile opossums, genus *Gracilinanus* (Gardner and Creighton, 1989) (Marmosidae, Marsupialia): a taxonomic review with notes on general morphology and relationships. Fieldiana: Zoology (new series) 70:1–56

Hershkovitz, P. 1995. The staggered marsupial third lower incisor: hallmark of cohort Didelphimorphia, and description of a new genus and species with staggered i3 from the Albian (lower Cretaceous) of Texas. Bonner Zoologische Beiträge 45:153–169.

Hershkovitz, P. 1997. Composition of the family Didelphidae (Gray, 1821) (Didelphoidea: Marsupialia), with a review of the morphology and behavior of the included four-eyed opossums of the genus *Philander* (Tiedemann, 1808). Fieldiana: Zoology (new series) 86:1–103.

Hershkovitz, P. 1999. *Dromiciops gliroides* (Thomas, 1894), last of the Microbiotheria, with a review of the family Microbiotheriidae. Fieldiana: Zoology (new series) 93:1–60.

Heuvelmans, B. 1958. On the track of unknown animals. Hill & Wang, New York, New York.

Hickman, G. C. 1986. Swimming of *Amblysomus hottentotus* (Insectivora: Chrysochloridae), with notes on *Chrysospalax* and *Eremitalpa*. Cimbebasia, ser. A, 8:55–61.

Hickman, V. V., and J. L. Hickman. 1960. Notes on the habits of the Tasmanian dormouse phalangers *Cercartetus nanus* (Desmarest) and *Eudromicia lepida* (Thomas). Proceedings of the Zoological Society of London 135:365–374.

Hinds, L. A. 2008. Tammar wallaby *Macropus eugenii* (Desmarest, 1817). Pp. 330–332 in The mammals of Australia. 3rd edition (S. Van Dyck and R. Strahan, eds.). Reed New Holland, Sydney, Australia.

Hines, E. M., K. Andulyanukosol, and D. A. Duffus. 2005. Dugong (*Dugong dugon*) abundance along the Andaman coast of Thailand. Marine Mammal Science 21:536–549.

Hines, E. M., K. Andulyanukosol, S. Poochaviranon, P. Somany, L. S. Ath, N. Cox, K. Symington, T. Tun, A. Ilangakoon, H. H. de Iongh, L. V. Aragones, S. Lu, X. Jiang, X. Jing, E. D'Souza, V. Patankar, D. Sutaria, B. Jethva, and P. Solanki. 2012. Dugongs in Asia. Pp. 58–76 in Sirenian conservation. Issues and strategies in developing countries (E. M. Hines, J. E. Reynolds, III, L. V. Aragones, A. A. Mignucci-Giannoni, and M. Marmontel, eds.). University Press of Florida, Gainesville, Florida.

Hingst, E., P. S. D'Andrea, R. Santori, and R. Cerqueira. 1998. Breeding of *Philander frenata* (Didelphimorphia, Didelphidae) in captivity. Laboratory Animals 32:434–438.

Hirasaka, K. 1934. On the distribution of sirenians in the Pacific. Proceedings of the 5th Pacific Science Congress 5:4221–4222.

Hirsch, A., and A. G. Chiarello. 2011. The endangered maned sloth *Bradypus torquatus* of the Brazilian Atlantic Forest: a review and update of geographical distribution and habitat preferences. Mammal Review 42:35–54.

Hitchcock, G. 1997. First record of the spectacled hare-wallaby, *Lagorchestes conspicillatus* (Marsupialia: Macropodidae), in New Guinea. Science in New Guinea 23:47–51.

Hitoshi, F., and K. Naoki. 1994. Steller's sea-cow (Sirenia: *Hydrodamalis gigas*) from the middle Pleistocene Mandano Formation of the Boso Peninsula, central Japan. Fossils 56:26–32.

695

Hoare, R.E., and J. T. du Toit. 1999. Coexistence between people and elephants in African savannas. Conservation Biology 13:633–639.

Hobbs, J.-P. A., A. J. Frisch, J. Hender, and J. J. Gilligan. 2007. Long-distance oceanic movement of a solitary dugong (*Dugong dugon*) to the Cocos (Keeling) Islands. Aquatic Mammals 33:175–178.

Hocknull, S. A. 2005. Late Pleistocene-Holocene occurrence of *Chaeropus* (Peramelidae) and *Macrotis* (Thylacomyidae) from Queensland. Memoirs of the Queensland Museum 51:38.

Hoeck, H. N. 1975. Differential feeding behaviour of the sympatric hyrax *Procavia johnstoni* and *Heterohyrax brucei*. Oecologia 22:15–47.

Hoeck, H. N. 1978. Systematics of the Hyracoidea: toward a clarification. Bulletin of Carnegie Museum of Natural History 6:146–151.

Hoeck, H. N. 1982. Population dynamics, dispersal, and genetic isolation in two species of hyrax (*Heterohyrax brucei* and *Procavia johnstoni*) on habitat islands in the Serengeti. Zeitschrift für Tierpsychologie 59:177–210.

Hoeck, H. N. 1984. Hyraxes. Pp. 462–465 in The encyclopedia of mammals (D. Macdonald, ed.). Facts on File, New York, New York.

Hoeck, H. N. 1989. Demography and competition in hyrax. Oecologia 79:353–360.

Hoeck, H. N. 1990. Modern hyraxes. Pp. 539–546 in Grzimek's encyclopedia of mammals, volume 4 (B. Grzimek, ed.). McGraw-Hill, New York, New York.

Hoeck, H. N. 2011. Family Procaviidae (hyraxes). Pp. 28–47 in Handbook of the mammals of the world, volume 2, Hoofed mammals (D. E. Wilson and R. A. Mittermeier, eds.). Lynx Edicions, Barcelona, Spain.

Hoeck, H. N., H. Klein, and P. Hoeck. 1982. Flexible social organization in hyrax. Zeitschrift für Tierpsychologie 9:265–298.

Hoeck, H., F. Rovero, N. Cordeiro, T. Butynski, A. Perkin, and T. Jones. 2015. *Dendrohyrax validus*. The IUCN Red List of Threatened Species 2015: e.T136599A21288090. http://dx.doi.org/10.2305/IUCN.UK.2015-2.RLTS.T136599A21288090.en. Downloaded on 21 January 2016.

Hoffmann, M. 2008. *Microgale jobihely*. The IUCN Red List of Threatened Species 2008: e.T136628A4319967. http://dx.doi.org/10.2305/IUCN.UK.2008.RLTS.T136628A4319967.en. Downloaded on 18 January 2016.

Hoffmann, M., H. N. Hoeck, and K. De Smet. 2008. Does the bush hyrax occur in North Africa? Afrotherian Conservation 6:6–8.

Holland, N., and S. M. Jackson. 2002. Reproductive behaviour and food consumption associated with the captive breeding of platypus (*Ornithorhynchus anatinus*). Journal of Zoology 256:279–288.

Holleley, C. E., C. R. Dickman, M. S. Crowther, and B. P. Oldroyd. 2006. Size breeds success: multiple paternity, multivariate selection and male semelparity in a small marsupial, *Antechinus stuartii*. Molecular Ecology 15:3439–3448.

Hollister-Smith, J. A., J. H. Poole, E. A. Archie, E. A. Vance, N. J. Georgiadis, C. J. Moss, and S. C. Alberts. 2007. Age, musth and paternity success in wild male African elephants, *Loxodonta africana*. Animal Behaviour 74:287–296.

Hollister-Smith, J. A., J. H. Poole, C. J. Moss, and S. C. Alberts. 2011. Genetic paternity analysis of the Amboseli elephant population. Pp. 274–275 in The Amboseli elephants: a long-term perspective on a long-lived mammal (C. J. Moss, H. Croze, and P. C. Lee, eds.). University of Chicago Press, Chicago, Illinois.

Holmes, D. J. 1991. Social behavior in captive Virginia opossums, *Didelphis virginianus*. Journal of Mammalogy 72:402–410.

Holroyd, P. A. 2010. Macroscelidea. Pp. 89–98 in Cenozoic mammals of Africa (L. Werdelin and W. J. Sanders, eds.). University of California Press, Berkeley, California.

Holroyd, P. A., and J. C. Mussell. 2005. Macroscelidea and Tubulidentata. Pp. 71–83 in The rise of placental mammals: origins and relationships of the major clades (K. D. Rose and J. D. Archibald, eds.). Johns Hopkins University Press, Baltimore, Maryland.

Hooker, J. J. 2005. Perissodactyla. Pp. 199–214 in Mammal species of the world: a taxonomic and geographic reference. 3rd edition (D. E. Wilson and D. M. Reeder, eds.). Johns Hopkins University Press, Baltimore, Maryland.

Hoolihan, D. W., and A. W. Goldizen. 1998. The grouping dynamics of the black-striped wallaby. Wildlife Research 25:467–473.

Hope, J. H. 1981. A new species of *Thylogale* (Marsupialia: Macropodidae) from Mapala Rock Shelter, Jaya (Carstensz) Mountains, Irian Jaya (western New Guinea), Indonesia. Records of the Australian Museum 33:369–387.

Horovitz, I., and M. R. Sánchez-Villagra. 2003. A morphological analysis of marsupial mammal higher-level phylogenetic relationships. Cladistics 19:181–212.

Horovitz, I., T. Martin, J. Bloch, S. Ladevèze, C. Kurz, M. R. Sánchez-Villagra. 2009. Cranial anatomy of the earliest marsupials and the origin of opossums. PLoS ONE 4(12):e8278. doi:10.1371/journal.pone.0008278.

Horsup, A. 1998. A trapping survey of the northern hair-nosed wombat *Lasiorhinus krefftii*. Pp. 147–155 in Wombats (R. T. Wells, P. A. Pridmore, B. St. John, M. D. Gaughwin, and J. Ferris, eds.). Surrey Beatty & Sons, Chipping Norton, New South Wales, Australia.

Horsup, A. 2004. Recovery plan for the northern hairy-nosed wombat *Lasiorhinus krefftii* 2004–2008. Environmental Protection Agency, Queensland Parks and Wildlife Service, Brisbane, New South Wales, Australia.

Horsup, A., and C. N. Johnson. 2008. Northern hairy-nosed wombat *Lasiorhinus krefftii* (Owen, 1872). Pp. 202–204 in The mammals of Australia. 3rd edition (S. Van Dyck and R. Strahan, eds.). Reed New Holland, Sydney, Australia.

Hou, Z.-C., R. Romero, and D. E. Wildman. 2009. Phylogeny of the Ferungulata (Mammalia: Laurasiatheria) as determined from phylogenomic data. Molecular Phylogenetics and Evolution 52:660–664.

How, R. A. 1976. Reproduction, growth, and survival of young in the mountain possum, *Trichosurus caninus* (Marsupialia). Australian Journal of Zoology 24:189–199.

How, R. A. 1978. Population strategies of four species of Australian "possums." Pp. 305–313 in The ecology of arboreal folivores (G. G. Montgomery, ed.). Smithsonian Institution Press, Washington, DC.

How, R. A. 2008. Short-eared brushtail possum *Trichosurus caninus* (Ogilby, 1836). Pp. 270–272 in The mammals of Australia. 3rd edition (S. Van Dyck and R. Strahan, eds.). Reed New Holland, Sydney, Australia.

How, R. A., P. B. S. Spencer, and L. H. Schmitt. 2009. Island populations have high conservation value for northern Australia's top marsupial predator ahead of a threatening process. Journal of Zoology 278:206–217.

How, R. A., J. L. Barnett, A. J. Bradley, W. J. Humphreys, and R. W. Martin. 1984. The population biology of *Pseudocheirus peregrinus* in a *Leptospermum laevigatum* thicket. Pp. 261–268 in Possums and gliders (A. Smith and I. Hume, eds.). Surrey Beatty & Sons, Norton, New South Wales, Australia.

Howe, D. 1975. Observations on a captive marsupial mole, *Notoryctes typhlops*. Australian Mammalogy 1:361–365.

Hrdina, F., and G. Gordon. 2004. The koala and possum trade in Queensland, 1906–1936. Australian Zoologist 32:543–584.

Hu, Y., J. Meng, Y. Wang, and C. Li. 2005. Large Mesozoic mammals fed on young dinosaurs. Nature 433:149–152.

Hughes, R. L., L. S. Hall, K. P. Aplin, and M. Archer. 1987. Organogenesis and fetal membranes in the New Guinea pen-tailed possum, *Distoechurus pennatus* (Acrobatidae: Marsupialia). Pp. 715–724 in Possums and opossums (M. Archer, ed.). Surrey Beatty & Sons, Sydney, Australia.

Hulbert, A. J., G. Gordon, and T. J. Dawson. 1971. Rediscovery of the marsupial *Echymipera rufescens* in Australia. Nature 231:330–331.

Hume, I. D., P. J. Jarman, M. B. Renfree, and P. D. Temple-Smith. 1989. Macropodidae. Pp. 679–715 in Fauna of Australia: Mammalia, volume 1B (D. W. Walton and B. J. Richardson, eds.). Australian Government Publishing Service, Canberra, Australia.

Humphreys, W. F., R. A. How, A. J. Bradley, C. M. Kemper, and D. J. Kitchener. 1984. The biology of *Wyulda squamicaudata* Alexander 1919. Pp. 162–169 in Possums and gliders (A. Smith and I. Hume, eds.). Surrey Beatty & Sons, Norton, New South Wales, Australia.

Hunsaker, D., II. 1977. Ecology of New World marsupials. Pp. 95–156 in The biology of marsupials (D. Hunsaker, II, ed.). Academic Press, New York, New York.

Hunsaker, D., II, and D. Shupe. 1977. Behavior of New World marsupials. Pp. 279–347 in The biology of marsupials (D. Hunsaker, II, ed.). Academic Press, New York, New York.

Hunter, J. P., and C. M. Janis. 2006a. "Garden of Eden" or "Fool's Paradise"? Phylogeny, dispersal, and the southern continent hypothesis of placental mammal origins. Paleobiology 32:339–344.

Hunter, J. P., and C. M. Janis. 2006b. Spiny Norman in the Garden of Eden? Dispersal and early biogeography of Placentalia. Journal of Mammalian Evolution 13:89–123.

Hunter, M. E., A. A. Mignucci-Giannoni, K. P. Tucker, T. L. King, R. K. Bonde, B. A. Gray, and P. M. McGuire. 2012. Puerto Rico and Florida manatees represent genetically distinct groups. Conservation Genetics 13:1623–1635.

Hunter, N., E. Martin, and T. Milliken. 2004. Determining the number of elephants required to supply current unregulated ivory markets in Africa and Asia. Pachyderm 36:116–128.

Husar, S. L. 1975. A review of the literature of the dugong. United States Fish and Wildlife Service Wildlife Research Report 4:1–30.

Husar, S. L. 1977. *Trichechus inunguis*. Mammalian Species 72:1–4.

Husar, S. L. 1978a. *Dugong dugon*. Mammalian Species 88:1–7.

Husar, S. L. 1978b. *Trichechus senegalensis*. Mammalian Species 89:1–3.

Husar, S. L. 1978c. *Trichechus manatus*. Mammalian Species 93:1–5.

Husson, A. M. 1978. The mammals of Suriname. E. J. Brill, Leiden, Netherlands.

Hutchins, M., and Mike Keele. 2006. Elephant importation from range countries: ethical and practical considerations for accredited zoos. Zoo Biology 25:219–233.

Hutchins, M., B. Smith, and M. Keele. 2008. Zoos as responsible stewards of elephants. Pp. 285–305 in Elephants and ethics, toward a morality of existence (C. Wemmer and C. A. Christen, eds.). Johns Hopkins University Press, Baltimore, Maryland.

Hutchins, M., and R. Smith. 1990. Biology and status of wild tree kangaroos. Pp. 1–6 in The biology and management of tree kangaroos (M. Roberts and M. Hutchins, eds.). American Association of Zoo-

logical Parks and Aquariums Marsupial and Monotreme Advisory Group, Bethesda, Maryland.

Hutchinson, J. R., D. Famini, R. Lair, and Rodger Kram. 2003. Are fast-moving elephants really running? Nature 422:493–494.

Huttley, G. 2009. Do genomic datasets resolve the correct relationship among the placental, marsupial and monotreme lineages? Australian Journal of Zoology 57:167–174.

Ichikawa, K., T. Akamatsu, T. Shinke, K. Adulyanukosol, and N. Arai. 2011. Callback response of dugongs to conspecific chirp playbacks. Journal of the Acoustical Society of America 129:3623–3629.

Ikeda, K., and H. Mukai. 2012. Dugongs in Japan. Pp. 77–83 in Sirenian conservation. Issues and strategies in developing countries (E. M. Hines, J. E. Reynolds, III, L. V. Aragones, A. A. Mignucci-Giannoni, and M. Marmontel, eds.). University Press of Florida, Gainesville, Florida.

Ingleby, S. 1991a. Distribution and status of the spectacled hare-wallaby, *Lagorchestes conspicillatus*. Wildlife Research 18:501–519.

Ingleby, S. 1991b. Distribution and abundance of the northern nailtail wallaby, *Onychogalea unguifera* (Gould, 1841). Wildlife Research 18:655–676.

Ingleby, S., and G. Gordon. 2008. Northern nailtail wallaby *Onychogalea unguifera* (Gould, 1841). Pp. 359–361 in The mammals of Australia. 3rd edition (S. Van Dyck and R. Strahan, eds.). Reed New Holland, Sydney, Australia.

Isaac, J. L. 2005. Life history and demographics of an island possum. Australian Journal of Zoology 53:195–203.

Ishida, Y., Y. Demeke, P. J. Van Coeverden De Groot, N. J. Georgiadis, K. E. A. Leggett, V. E. Fox, and A. L. Roca. 2011a. Distinguishing forest and savanna African elephants using short nuclear DNA sequences. Journal of Heredity 102:610–616.

Ishida, Y., T. K. Oleksyk, N. J. Georgiadis, V. A. David, K. Zhao, R. M. Stephens, S.-O. Kolokotronis, and A. L. Roca. 2011b. Reconciling apparent conflicts between mitochondrial and nuclear phylogenies in African elephants. PLoS ONE 6(6):e20642. doi:10.1371/journal.pone.0020642.

Islam, M. A., S. Mohsanin, G. W. Chowdhury, S. U. Chowdhury, M. A. Aziz, M. Uddin, S. Saif, S. Chakma, R. Akter, I. Jahan, and I. Azam. 2011. Current status of Asian elephants in Bangladesh. Gajah 35:21–24.

IUCN. 2008. 2008 IUCN Red List of Threatened Species. IUCN, Gland, Switzerland. http://www.iucnredlist.org. Accessed on 06 March 2009.

IUCN African Elephant Specialist Group. 2013. 2013 provisional African elephant status report. IUCN, Gland, Switzerland. http://www.elephantdatabase.org/preview report/2013_africa/Loxodonta_africana/2012/Africa. Accessed on 20 March 2015.

Iversen, E. 1995. The domestication of the African elephant. Pachyderm 20:65–68.

Izor, R. J., and R. H. Pine. 1987. Notes on the black-shouldered opossum, *Caluromysiops irrupta*. Fieldiana: Zoology (new series) 39:117–124.

Jackson, C. R., N. R. Lubbe, M. P. Robertson, T. H. Setsaas, J. van der Waals, and N. C. Bennett. 2008. Soil properties and the distribution of the endangered Juliana's golden mole. Journal of Zoology 274:13–17.

Jackson, C. R., T. H. Setsaas, M. P. Robertson, M. Scantlebury, and N. C. Bennett. 2009. Insights into torpor and behavioural thermoregulation of the endangered Juliana's golden mole. Journal of Zoology 278:299–307.

Jackson, H. H. T. 1961. Mammals of Wisconsin. University of Wisconsin Press, Madison, Wisconsin.

Jackson, S. 2003. Australian mammals, biology and captive management. CSIRO Publishing, Collingwood, Victoria, Australia.

Jackson, S. M. 1999. Glide angle in the genus *Petaurus* and a review of gliding in mammals. Mammal Review 30:9–30.

Jackson, S. M. 2000a. Population dynamics and life history of the mahogany glider, *Petaurus gracilis*, and the sugar glider, *Petaurus breviceps*, in north Queensland. Wildlife Research 27:21–37.

Jackson, S. M. 2000b. Home-range and den use of the mahogany glider, *Petaurus gracilis*. Wildlife Research 27:49–60.

Jackson, S. M. 2008. Mahogany glider *Petaurus gracilis* (de Vis, 1883). Pp. 233–234 in The mammals of Australia. 3rd edition (S. Van Dyck and R. Strahan, eds.). Reed New Holland, Sydney, Australia.

Jackson, S. M. 2011. *Petaurus gracilis* (Diprotodontia: Petauridae). Mammalian Species 43(882):141–148.

Jackson, S. M. 2015a. Family Pseudocheiridae (ring-tailed possums and greater gliders). Pp. 498–530 in Handbook of the mammals of the world, volume 5, Monotremes and marsupials (D. E. Wilson and R. A. Mittermeier, eds.). Lynx Edicions, Barcelona, Spain.

Jackson, S. M. 2015b. Family Petauridae (striped possums, Leadbeater's possum and lesser gliders). Pp. 532–565 in Handbook of the mammals of the world, volume 5, Monotremes and marsupials (D. E. Wilson and R. A. Mittermeier, eds.). Lynx Edicions, Barcelona, Spain.

Jackson, S. M., and R. W. Thorington, Jr. 2012. Gliding mammals: taxonomy of living and extinct species. Smithsonian Contributions to Zoology 638:1–117.

Jacobs, L. L. 1980. Siwalik fossil tree shrews. Pp. 205–216 in Comparative biology and evolutionary relationships of tree shrews (W. P. Luckett, ed.). Plenum Press, New York, New York.

Janečka, J. E., W. Miller, T. H. Pringle, F. Wiens, A. Zitzmann, K. M. Helgen, M. S. Springer, and W. J. Murphy. 2007. Molecular and genomic data identify the closest living relative of primates. Science 318:792–794.

Janečka, J. E., K. M. Helgen, N. T.-L. Lim, M. Baba, M. Izawa, Boeadi, W. J. Murphy. 2008. Evidence for multiple species of Sunda colugo. Current Biology 18:R1001–R1002.

Janis, C. M., K. Buttrill, and B. Figueirido. 2014. Locomotion in extinct giant kangaroos: were sthenurines hop-less monsters? PLoS ONE 9(10):e109888. doi:10.1371/journal.pone.0109888.

Jansa, S. A., J. F. Forsman, and R. S. Voss. 2006. Different patterns of selection on the nuclear genes IRBP and DMP-1 affect the efficiency but not the outcome of phylogeny estimation for didelphid Marsupials. Molecular Phylogenetics and Evolution 38:363–380.

Jansa, S., and R. S. Voss. 2000. Phylogenetic studies on didelphid marsupials. 1. Introduction and preliminary results from nuclear IRBP gene sequences. Journal of Mammalian Evolution 7:43–77.

Jansa, S. A., and R. S. Voss. 2005. Phylogenetic relationships of the marsupial genus *Hyladelphys* based on nuclear gene sequences and morphology. Journal of Mammalogy 86:853–865.

Janssens, P. A., and A. M. T. Rogers. 1989. Metabolic changes during pouch vacation and weaning in macropodoids. Pp. 367–376 in Kangaroos, wallabies and rat-kangaroos (G. Grigg, P. Jarman, and I. Hume, eds.). Surrey Beatty & Sons, Chipping Norton, New South Wales, Australia.

Jarman, P. J. 1989. Sexual dimorphism in the Macropodoidea. Pp. 433–447 in Kangaroos, wallabies and rat-kangaroos (G. Grigg, P. Jarman, and I. Hume, eds.). Surrey Beatty & Sons, Chipping Norton, New South Wales, Australia.

Jarman, P. J., and P. Bayne. 1997. Behavioural ecology of *Petrogale penicillata* in relation to conservation. Australian Mammalogy 19:219–228.

Jarman, P. J., and J. H. Calaby. 2008. Red-necked wallaby *Macropus rufogriseus* (Desmarest, 1817). Pp. 349–351 in The mammals of Australia. 3rd edition (S. Van Dyck and R. Strahan, eds.). Reed New Holland, Sydney, Australia.

Jarman, P. J., and G. Coulson. 1989. Dynamics and adaptiveness of grouping in macropods. Pp. 527–547 in Kangaroos, wallabies and rat-kangaroos (G. Grigg, P. Jarman, and I. Hume, eds.). Surrey Beatty & Sons, Chipping Norton, New South Wales, Australia.

Jarman, P. J., A. K. Lee, and L. S. Hall. 1989. Natural history of the Eutheria. Pp. 810–828 in Fauna of Australia: Mammalia, volume 1B (D. W. Walton and B. J. Richardson, eds.). Australian Government Publishing Service, Canberra, Australia.

Jayewardene, J. 2002. The care and management of domesticated Asian elephants in Sri Lanka. Pp. 43–58 in Giants on our hands. Proceedings of the International Workshop on the Domesticated Asian Elephant (I. Baker and M. Kashio, eds.). FAO Regional Office for Asia and the Pacific, Bangkok, Thailand.

Jenkins, P. D. 1988. A new species of *Microgale* (Insectivora: Tenrecidae) from northeastern Madagascar. American Museum Novitates 2910:1–7.

Jenkins, P. D. 1992. Description of a new species of *Microgale* (Insectivora: Tenrecidae) from eastern Madagascar. Bulletin of the British Museum (Natural History) Zoology 58:53–59.

Jenkins, P. D. 1993. A new species of *Microgale* (Insectivora: Tenrecidae) from eastern Madagascar with an unusual dentition. American Museum Novitates 3067:1–11.

Jenkins, P. D. 2003. *Microgale*, shrew tenrecs. Pp. 1273–1278 in The natural history of Madagascar (S. M. Goodman and J. P. Benstead, eds.). University of Chicago Press, Chicago, Illinois.

Jenkins, P. D., and S. M. Goodman. 1999. A new species of *Microgale* (Lipotyphla, Tenrecidae) from isolated forest in southwestern Madagascar. Bulletin of the Natural History Museum, London (Zoology) 65:155–164.

Jenkins, P. D., S. M. Goodman, and C. J. Raxworthy. 1996. The shrew tenrecs (*Microgale*) (Insectivora: Tenrecidae) of the Réserve Naturelle Intégrale d'Andringitra, Madagascar. Fieldiana: Zoology (new series) 85:191–217.

Jenkins, P. D., C. J. Raxworthy, and R. A. Nussbaum. 1997. A new species of *Microgale* (Insectivora, Tenrecidae), with comments on the status of four other taxa of shrew tenrecs. Bulletin of the Natural History Museum, London (Zoology) 63:1–12.

Jennings, M. R., and G. B. Rathbun. 2001. *Petrodromus tetradactylus*. Mammalian Species 682:1–6.

Ji, Q., Z.-X. Luo, C.-X. Yuan, J. R. Wible, J. P. Zhang, and J. A. Georgi. 2002. The earliest known eutherian mammal. Nature 416:816–822.

Ji, Q., Z.-X. Luo, C.-X. Yuan, and A. R. Tabrum. 2006. A swimming mammaliaform from the middle Jurassic and ecomorphological diversification of early mammals. Science 311:1123–1127.

Jigme, K., and A. C. Williams. 2011. Current status of Asian elephants in Bhutan. Gajah 35:25–28.

Johnson, C. N. 1998. The evolutionary ecology of wombats. Pp. 34–41 in Wombats (R. T. Wells, P. A. Pridmore, B. St. John, M. D. Gaughwin, and J. Ferris, eds.). Surrey Beatty & Sons, Chipping Norton, New South Wales, Australia.

Johnson, C. N. 2005. What can the data on late survival of Australian megafauna tell us about the cause of their extinction? Quaternary Science Reviews 24:2167–2172.

Johnson, C. N., and D. G. Crossman. 1991. Dispersal and social organization of the northern hairy-nosed wombat *Lasiorhinus krefftii*. Journal of Zoology 225:605–613.

Johnson, C. N., and K. A. Johnson. 1983. Behaviour of the bilby, *Macrotis lagotis* (Reid) (Marsupialia: Thylacomyidae), in captivity. Australian Wildlife Research 10:77–87.

Johnson, C. N., and A. P. McIlwee. 1997. Ecology of the northern bettong, *Bettongia tropica*, a tropical mycophagist. Wildlife Research 24:549–559.

Johnson, C. N., and G. J. Prideaux. 2004. Extinctions of herbivorous mammals in the late Pleistocene of Australia in relation to their feeding ecology: no evidence for environmental change as cause of extinction. Austral Ecology 29:553–557.

Johnson, C. N., and S. Wroe. 2003. Causes of extinction of vertebrates during the Holocene of mainland Australia: arrival of the dingo, or human impact? Holocene 13:941–948.

Johnson, D. L. 1980. Problems in the land vertebrate zoogeography of certain islands and the swimming powers of elephants. Journal of Biogeography 7:383–398.

Johnson, K. A. 1989. Thylacomyidae. Pp. 625–635 in Fauna of Australia: Mammalia, volume 1B (D. W. Walton and B. J. Richardson, eds.). Australian Government Publishing Service, Canberra, Australia.

Johnson, K. A. 2008a. Bilby *Macrotis lagotis* (Reid, 1837). Pp. 191–193 in The mammals of Australia. 3rd edition (S. Van Dyck and R. Strahan, eds.). Reed New Holland, Sydney, Australia.

Johnson, K. A. 2008b. Lesser bilby *Macrotis leucura* (Thomas, 1887). Pp. 194–195 in The mammals of Australia. 3rd edition (S. Van Dyck and R. Strahan, eds.). Reed New Holland, Sydney, Australia.

Johnson, K. A. 2008c. Red-necked pademelon *Thylogale thetis* (Lesson, 1827). Pp. 400–401 in The mammals of Australia. 3rd edition (S. Van Dyck and R. Strahan, eds.). Reed New Holland, Sydney, Australia.

Johnson, K. A., and A. A. Burbidge. 2008a. Pig-footed bandicoot *Chaeropus ecaudatus* (Ogilby, 1838). Pp. 172–173 in The mammals of Australia. 3rd edition (S. Van Dyck and R. Strahan, eds.). Reed New Holland, Sydney, Australia.

Johnson, K. A., and A. A. Burbidge. 2008b. Rufous hare-wallaby *Lagorchestes hirsutus* (Gould, 1844). Pp. 317–319 in The mammals of Australia. 3rd edition (S. Van Dyck and R. Strahan, eds.). Reed New Holland, Sydney, Australia.

Johnson, K. A., and R. W. Rose. 2008. Tasmanian pademelon *Thylogale billardieri* (Desmarest, 1822). Pp. 395–397 in The mammals of Australia. 3rd edition (S. Van Dyck and R. Strahan, eds.). Reed New Holland, Sydney, Australia.

Johnson, K. A., and R. I. Southgate. 1990. Present and former status of bandicoots in the Northern Territory. Pp. 85–92 in Bandicoots and bilbies (J. H. Seebeck, P. R. Brown, R. L. Wallis, and C. M. Kemper, eds). Surrey Beatty & Sons, Chipping Norton, New South Wales, Australia.

Johnson, K. A., and D. W. Walton. 1989. Notoryctidae. Pp. 591–602 in Fauna of Australia, volume 1B, Mammalia (D. W. Walton and B. J. Richardson, eds.). Australian Government Publishing Service, Canberra, Australia.

Johnson, K. A., J. C. Z. Woinarski, and D. G. Langford. 2008. Carpentarian pseudantechinus *Pseudantechinus mimulus* (Thomas, 1906).

Pp. 71–72 in The mammals of Australia. 3rd edition (S. Van Dyck and R. Strahan, eds.). Reed New Holland, Sydney, Australia.

Johnson, M. B., S. L. Clifford, B. Goossens, S. Nyakaana, B. Curran, L. J. T. White, E. J. Wickings, and M. W. Bruford. 2007. Complex phylogeographic history of central African forest elephants and its implications for taxonomy. BMC Evolutionary Biology 7:244. doi:10.1186/1471-2148-7-244.

Johnson, P. M. 1978a. Reproduction of the rufous rat-kangaroo (*Aepyprymnus rufescens* (Gray)) in captivity with age estimation of pouch young. Queensland Journal of Agricultural and Animal Sciences 35:69–72.

Johnson, P. M. 1978b. Husbandry of the rufous rat-kangaroo *Aepyprymnus rufescens* and brush-tailed rock wallaby *Petrogale penicillata* in captivity. International Zoo Yearbook 18:156–157.

Johnson, P. M. 1979. Reproduction in the plain rock-wallaby, *Petrogale penicillata inornata* Gould, in captivity, with age estimation of the pouch young. Australian Wildlife Research 6:1–4.

Johnson, P. M. 1980. Observations of the behaviour of the rufous rat-kangaroo, *Aepyprymnus rufescens* (Gray), in captivity. Australian Wildlife Research 7:347–357.

Johnson, P. M. 1993. Reproduction of the spectacled hare-wallaby, *Lagorchestes conspicillatus* Gould (Marsupialia: Macropodidae), in captivity, with age estimation of the pouch young. Wildlife Research 20:97–101.

Johnson, P. M. 1997. Reproduction in the bridled nailtail wallaby, *Onychogalea fraenata* Gould (Marsupialia: Macropodidae), in captivity. Wildlife Research 24:411–415.

Johnson, P. M. 1998. Reproduction of the whiptail wallaby, *Macropus parryi* Bennett (Marsupialia: Macropodidae), in captivity with age estimation of the pouch-young. Wildlife Research 25:635–641.

Johnson, P. M. 2008a. Black-striped wallaby *Macropus dorsalis* (Gray, 1837). Pp. 329–330 in The mammals of Australia. 3rd edition (S. Van Dyck and R. Strahan, eds.). Reed New Holland, Sydney, Australia.

Johnson, P. M. 2008b. Whiptail wallaby *Macropus parryi* (Bennett, 1835). Pp. 343–345 in The mammals of Australia. 3rd edition (S. Van Dyck and R. Strahan, eds.). Reed New Holland, Sydney, Australia.

Johnson, P. M., and S. C. Delean. 1999. Reproduction in the Proserpine rock-wallaby, *Petrogale persephone* Maynes (Marsupialia: Macropodidae), in captivity, with age estimation and development of pouch young. Wildlife Research 26:631–639.

Johnson, P. M., and S. Delean. 2001. Reproduction in the northern bettong, *Bettongia tropica* Wakefield (Marsupialia: Potoroidae), in captivity, with age estimation and development of pouch young. Wildlife Research 28:79–85.

Johnson, P. M., and S. Delean. 2002a. Reproduction in the Proserpine rock-wallaby, *Petrogale persephone* Maynes (Marsupialia: Macropodidae), in captivity, with age estimation and development of pouch young. Wildlife Research 26:631–639.

Johnson, P. M., and S. Delean. 2002b. Development and age estimation of the pouch young of the black-striped wallaby *Macropus dorsalis*, with notes on reproduction. Australian Mammalogy 24:193–198.

Johnson, P. M., and S. Delean. 2002c. Reproduction of the purple-necked rock-wallaby, *Petrogale purpureicollis* Le Souef (Marsupialia: Macropodidae) in captivity, with age estimation and development of pouch young. Wildlife Research 29:463–468.

Johnson, P. M., and S. Delean. 2003. Reproduction of Lumholtz's tree-kangaroo, *Dendrolagus lumholtzi* (Marsupialia: Macropodidae) in captivity, with age estimation and development of pouch young. Wildlife Research 30:505–512.

Johnson, P. M., and M. D. B. Eldridge. 2008a. Purple-necked rock-wallaby *Petrogale purpureicollis* (Le Souef, 1924). Pp. 387–388 in The mammals of Australia. 3rd edition (S. Van Dyck and R. Strahan, eds.). Reed New Holland, Sydney, Australia.

Johnson, P. M., and M. D. B. Eldridge. 2008b. Proserpine rock-wallaby *Petrogale persephone* (Maynes, 1982). Pp. 385–386 in The mammals of Australia. 3rd edition (S. Van Dyck and R. Strahan, eds.). Reed New Holland, Sydney, Australia.

Johnson, P. M., and G. R. Newell. 2008. Lumholtz's tree kangaroo *Dendrolagus lumholtzi* (Collett, 1884). Pp. 310–311 in The mammals of Australia. 3rd edition (S. Van Dyck and R. Strahan, eds.). Reed New Holland, Sydney, Australia.

Johnson, P. M., and R. Strahan. 1982. A further description of the musky rat-kangaroo, *Hypsiprymnodon moschatus* (Ramsay, 1876) (Marsupialia, Potoroidae), with notes on its biology. Australian Zoologist 21:27–46.

Johnson, P. M., and K. Vernes. 1994. Reproduction in the red-legged pademelon, *Thylogale stigmatica* Gould (Marsupialia: Macropodidae), and age estimation and development of pouch young. Wildlife Research 21:553–558.

Johnson, P. M., and K. Vernes. 2008. Red-legged pademelon *Thylogale stigmatica* (Gould, 1860). Pp. 397–400 in The mammals of Australia. 3rd edition (S. Van Dyck and R. Strahan, eds.). Reed New Holland, Sydney, Australia.

Johnson, P. M., M. D. B. Eldridge, V. Kiernan, and R. J. Cupitt. 2001. A significant range extension of the purple-necked rock-wallaby *Petrogale purpureicollis*. Australian Mammalogy 23:71–73.

Johnson, P. M., S. Lloyd, T. Vallance, and M. D. B. Eldridge. 2005. First record of quadruplets in the musky rat-kangaroo *Hypsiprymnodon moschatus*. Australian Mammalogy 27:95–97.

Johnson-Murray, J. L. 1987. The comparative myology of the gliding membranes of *Acrobates*, *Petauroides*, and *Petaurus* contrasted with the cutaneous myology of *Hemibelideus* and *Pseudocheirus* (Marsupialia: Phalangeridae) and with selected gliding Rodentia (Sciuridae and Anomaluridae). Australian Wildlife Research 35:101–113.

Johnston, P. G. 2008. Long-nosed potoroo *Potorous tridactylus* (Kerr, 1792). Pp. 302–304 in The mammals of Australia. 3rd edition (S. Van Dyck and R. Strahan, eds.). Reed New Holland, Sydney, Australia.

Johnston, P. G., and G. B. Sharman. 1976. Studies on populations of *Potorous* Desmarest (Marsupialia). I. Morphological variation. Australian Journal of Zoology 24:573–588.

Johnston, P. G., and G. B. Sharman. 1977. Studies on populations of *Potorous* Desmarest (Marsupialia). II. Electrophoretic, chromosomal, and breeding studies. Australian Journal of Zoology 25:733–747.

Johnston, S. D., M. R. McGowan, P. O'Callaghan, R. Cox, and V. Nicolson. 2000. Studies of the oestrous cycle, oestrus and pregnancy in the koala (*Phascolarctos cinereus*). Journal of Reproduction and Fertility 120:49–57.

Jones, B. 2004. The possum fauna of Western Australia: decline, persistence and status. Pp. 49–60 in The biology of Australian possums and gliders (R. L. Goldingay and S. M. Jackson, eds.). Surrey Beatty & Sons, Chipping Norton, New South Wales, Australia.

Jones, B. A., R. A. How, and D. J. Kitchener. 1994. A field study of *Pseudocheirus occidentalis* (Marsupialia: Petauridae). II. Population studies. Wildlife Research 21:189–201.

Jones, C. 1978. *Dendrohyrax dorsalis*. Mammalian Species 113:1–4.

Jones, C. 1984. Tubulidentates, proboscideans, and hyracoideans. Pp. 523–536 in Orders and families of recent mammals of the world

(S. Anderson and J. K. Jones, Jr., eds.). John Wiley & Sons, New York, New York.

Jones, C. J., and F. Geiser. 1992. Prolonged and daily torpor in the feathertail glider, *Acrobates pygmaeus* (Marsupialia: Acrobatidae). Journal of Zoology 227:101–108.

Jones, F. W. 1923–1925. The mammals of South Australia. A. B. Jones, Government Printer, Adelaide, South Australia, Australia.

Jones, J. K., Jr. 1964. Distribution and taxonomy of mammals of Nebraska. University of Kansas Publications, Museum of Natural History 16:1–356.

Jones, J. K., Jr., H. H. Genoways, and J. D. Smith. 1974. Annotated checklist of mammals of the Yucatan Peninsula, Mexico. III. Marsupialia, Insectivora, Primates, Edentata, Lagomorpha. Occasional Papers of the Museum of Texas Tech University 23:1–12.

Jones, M. 2008a. Eastern quoll *Dasyurus viverrinus* (Shaw, 1800). Pp. 62–64 in The mammals of Australia. 3rd edition (S. Van Dyck and R. Strahan, eds.). Reed New Holland, Sydney, Australia.

Jones, M. 2008b. Tasmanian devil *Sarcophilus harrisii* (Boitard, 1841). Pp. 78–80 in The mammals of Australia. 3rd edition (S. Van Dyck and R. Strahan, eds.). Reed New Holland, Sydney, Australia.

Jones, M. E., G. C. Grigg, and L. A. Beard. 1997. Body temperatures and activity patterns of Tasmanian devils (*Sarcophilus harrisii*) and eastern quolls (*Dasyurus viverrinus*) through a subalpine winter. Physiological Zoology 70:53–60.

Jones M. E., and R. K. Rose. 2001. *Dasyurus viverrinus*. Mammalian Species 677:1–9.

Jones, M. E., R. K. Rose, and S. Burnett. 2001. *Dasyurus maculatus*. Mammalian Species 676:1–9.

Jones, M. E., and M. Stoddart. 1998. Reconstruction of the predatory behaviour of the extinct marsupial thylacine (*Thylacinus cynocephalus*). Journal of Zoology 246:239–246.

Jones, M. E., D. Paetkau, E. Geffen, and C. Moritz. 2004. Genetic diversity and population structure of Tasmanian devils, the largest marsupial carnivore. Molecular Ecology 13:2197–2209.

Jones, M. E., P. J. Jarman, C. M. Lees, H. Hesterman, R. K. Hamede, N. J. Mooney, D. Mann, C. E. Pukk, J. Bergfeld, and H. McCallum. 2007. Conservation management of Tasmanian devils in the context of an emerging, extinction-threatening disease: devil facial tumor disease. Ecohealth 4:326–337.

Jones, M. L. 1982. Longevity of captive mammals. Zoologische Garten 52:113–128.

Jones Lennon, M., D. A. Taggart, P. D. Temple-Smith, and M. D. B. Eldridge. 2011. The impact of isolation and bottlenecks on genetic diversity in the Pearson Island population of the black-footed rock-wallaby (*Petrogale lateralis pearsoni*; Marsupialia: Macropodidae). Australian Mammalogy 33:152–161.

Jonzén, N., T. Pople, J. Knape, and M. Sköld. 2010. Stochastic demography and population dynamics in the red kangaroo *Macropus rufus*. Journal of Animal Ecology 79:109–116.

Jorge, W., D. A. Meritt, Jr., and K. Benirschke. 1977. Chromosome studies in Edentata. Cytobios 18:157–172.

Kaiser, H. E. 1974. Morphology of the Sirenia. S. Karger, Basel, Switzerland.

Kashio, M. 2002. Summary of the International Workshop on the Domesticated Asian Elephant. Pp. 17–21 in Giants on our hands. Proceedings of the International Workshop on the Domesticated Asian Elephant (I. Baker and M. Kashio, eds.). FAO Regional Office for Asia and the Pacific, Bangkok, Thailand.

Katugaha, H. I. E., M., de Silva, and C. Santiapillai. 1999. A long-term study on the dynamics of the elephant (*Elephas maximus*) population in Ruhuna National Park, Sri Lanka. Biological Conservation 89:51–59.

Kaufmann, J. H. 1974. Social ethology of the whiptail wallaby, *Macropus parryi*, in northeastern New South Wales. Animal Behaviour 22:281–369.

Kaufmann, J. H. 1975. Field observations of the social behaviour of the eastern grey kangaroo, *Macropus giganteus*. Animal Behaviour 23:214–221.

Kavanagh, J. R., A. Burk-Herrick, M. Westerman, and M. S. Springer. 2004. Relationships among families of Diprotodontia (Marsupialia) and the phylogenetic position of the autapomorphic honey possum (*Tarsipes rostratus*). Journal of Mammalian Evolution 11:207–222.

Kavanagh, R. P., and R. J. Wheeler. 2004. Home-range of the greater glider *Petauroides volans* in tall montane forest of southeastern New South Wales, and changes following logging. Pp. 413–425 in The biology of Australian possums and gliders (R. L. Goldingay and S. M. Jackson, eds.). Surrey Beatty & Sons, Chipping Norton, New South Wales, Australia.

Kaviar, S., J. Shockey, and P. Sundberg 2012. Observations on the endemic pygmy three-toed sloth of Isla Escudo de Veraguas, Panama. PLoS ONE 7(11):e49854. doi:10.1371/journal.pone.0049854.

Kawamichi, T., and M. Kawamichi. 1979. Spatial organization and territory of tree shrews (*Tupaia glis*). Animal Behaviour 27:381–393.

Kawamichi, T., and M. Kawamichi. 1982. Social system and independence of offspring in tree shrews. Primates 23:189–205.

Kear, B. P., and B. N. Cooke. 2001. A review of macropodoid (Marsupialia) systematics with the inclusion of a new family. Memoirs of the Association of Australasian Palaeontologists 25:83–101.

Keast, A. 1982. The thylacine (Thylacinidae, Marsupialia): how good a pursuit carnivore? Pp. 675–684 in Carnivorous marsupials (M. Archer, ed.). Royal Zoological Society of New South Wales, Mosman, New South Wales, Australia.

Keeley, T., J. K. O'Brien, B. G. Fanson, K. Masters, and P. D. McGreevy. 2012. The reproductive cycle of the Tasmanian devil (*Sarcophilus harrisii*) and factors associated with reproductive success in captivity. General and Comparative Endocrinology 176:182–191.

Keith Diagne, L. 2013. African manatees raised to CITES appendix I. Sirenews 59:4–6.

Keith Diagne, L. 2015. *Trichechus senegalensis*. The IUCN Red List of Threatened Species 2015: e.T22104A81904980. http://dx.doi.org /10.2305/IUCN.UK.2015-4.RLTS.T22104A81904980.en. Downloaded on 21 January 2016.

Kellogg, M. E., S. Burkett, T. R. Dennis, G. Stone, B. A. Gray, P. M. McGuire, R. T. Zori, and R. Stanyon. 2007. Chromosome painting in the manatee supports Afrotheria and Paenungulata. BMC Evolutionary Biology 7:6–12.

Kelt, D. A., and D. R. Martínez. 1989. Notes on the distribution and ecology of two marsupials endemic to the Valdivian forests of southern South America. Journal of Mammalogy 70:220–224.

Kemp, T. S. 1982. Mammal-like reptiles and the origin of mammals. Academic Press, London, United Kingdom.

Kemp, T. S. 1983. The relationships of mammals. Zoological Journal of the Linnean Society 77:353–384.

Kemp, T. S. 2005. The origin and evolution of mammals. Oxford University Press, Oxford, United Kingdom.

Kemper, C. 1990. Status of bandicoots in South Australia. Pp. 67–72 in Bandicoots and bilbies (J. H. Seebeck, P. R. Brown, R. L. Wallis, and C. M. Kemper, eds). Surrey Beatty & Sons, Chipping Norton, New South Wales, Australia.

Kemper, C. M., S. J. B. Cooper, G. C. Medlin, M. Adams, D. Stemmer, K. M. Saint, M. C. McDowell, and J. J. Austin. 2011. Cryptic grey-bellied dunnart (*Sminthopsis griseoventer*) discovered in South Australia: genetic, morphological and subfossil analyses show the value of collecting voucher material. Australian Journal of Zoology 59:127–144.

Kennedy, M. 1992. Australasian marsupials and monotremes: an action plan for their conservation. IUCN (World Conservation Union), Gland, Switzerland.

Kerle, J. A., and R. A. How. 2008. Common brushtail possum *Trichosurus vulpecula* (Kerr, 1792). Pp. 274–276 in The mammals of Australia. 3rd edition (S. Van Dyck and R. Strahan, eds.). Reed New Holland, Sydney, Australia.

Kerle, J. A., and C. J. Howe. 1992. The breeding biology of a tropical possum, *Trichosurus vulpecula arnhemensis* (Phalangeridae: Marsupialia). Australian Journal of Zoology 40:653–665.

Kerle, J. A., G. M. McKay, and G. B. Sharman. 1991. A systematic analysis of the brushtail possum, *Trichosurus vulpecula* (Kerr, 1792) (Marsupialia: Phalangeridae). Australian Journal of Zoology 39:313–331.

Kerle, J. C., and A. Borsboom. 1984. Home range, den tree use, and activity patterns in the greater glider, *Petauroides volans*. Pp. 229–236 in Possums and gliders (A. Smith and I. Hume, eds.). Surrey Beatty & Sons, Norton, New South Wales, Australia.

Kerley, G. I. H. 1995. The round-eared elephant shrew *Macroscelides proboscideus* (Macroscelidea) as an omnivore. Mammal Review 25:39–44.

Kershenbaum, A., A. Ilany, L. Blaustein, and E. Geffen. 2012. Syntactic structure and geographical dialects in the songs of male rock hyraxes. Proceedings of the Royal Society B 279:2974–2981.

Khawnual, P., and B. Clarke. 2002. General care and reproductive management of pregnant and infant elephants at the Ayutthaya Elephant Camp. Pp. 249–258 in Giants on our hands. Proceedings of the International Workshop on the Domesticated Asian Elephant (I. Baker and M. Kashio, eds.). FAO Regional Office for Asia and the Pacific, Bangkok, Thailand.

Khounboline, K. 2011. Current status of Asian elephants in Lao PDR. Gajah 35:62–66.

Kielan-Jaworowska, Z., R. L. Cifelli, and Z.-X. Luo. 2004. Mammals from the age of dinosaurs: origins, evolution, and structure. Columbia University Press, New York, New York.

Kiiru, W. 1995. The current status of human-elephant conflict in Kenya. Pachyderm 19:15–20.

Kiiru, W. 2008. Human-elephant conflicts in Africa. Who has the right of way? Pp. 382–396 in Elephants and ethics, toward a morality of existence (C. Wemmer and C. A. Christen, eds.). Johns Hopkins University Press, Baltimore, Maryland.

Kingdon, J. 1971. East African mammals: an atlas of evolution in Africa. I. Academic Press, London, United Kingdom.

Kingdon, J. 1974. East African mammals: an atlas of evolution in Africa. II(A). Insectivores and bats. Academic Press, London, United Kingdom.

Kingdon, J. 1979. East African mammals: an atlas of evolution in Africa. III(B). Large mammals. Academic Press, London, United Kingdom.

Kingdon, J., D. Happold, M. Hoffmann, T. Butynski, M. Happold, and J. Kalina (eds.). 2013. Mammals of Africa, volume 1, Introductory chapters and Afrotheria. Bloomsbury, London, United Kingdom.

Kirkby, R. J. 1977. Learning and problem-solving in marsupials. Pp. 193–208 in The biology of marsupials (B. Stonehouse and D. P. Gilmore, eds.). University Park Press, Baltimore, Maryland.

Kirkpatrick, T. H. 1965. Studies of Macropodidae in Queensland. 2. Age estimation in the grey kangaroo, the red kangaroo, the eastern wallaroo and the red-necked wallaby, with notes on dental abnormalities. Queensland Journal of Agricultural and Animal Sciences 22:301–317.

Kirkpatrick, T. H. 1967. The grey kangaroo in Queensland. Queensland Agricultural Journal 93:550–552.

Kirkpatrick, T. H. 1968. Studies on the wallaroo. Queensland Agricultural Journal 94:362–365.

Kirkpatrick, T. H. 1970a. The agile wallaby. Queensland Agricultural Journal 96:169–170.

Kirkpatrick, T. H. 1970b. The swamp wallaby. Queensland Agricultural Journal 96:335–336.

Kirsch, J. A. W. 1968. Burrowing by the quenda, *Isoodon obesulus*. Western Australian Naturalist 10:178–180.

Kirsch, J. A. W. 1977a. The classification of marsupials. Pp. 1–50 in The biology of marsupials (D. Hunsaker, II, ed.). Academic Press, New York, New York.

Kirsch, J. A. W. 1977b. The comparative serology of Marsupialia, and a classification of marsupials. Australian Journal of Zoology Supplementary Series 52:1–152.

Kirsch, J. A. W. 1977c. The six-percent solution: second thoughts on the adaptedness of the Marsupialia. American Scientist 65:276–288.

Kirsch, J. A. W., and J. H. Calaby. 1977. The species of living marsupials: an annotated list. Pp. 9–26 in The biology of marsupials (B. Stonehouse and D. P. Gilmore, eds.). University Park Press, Baltimore, Maryland.

Kirsch, J. A. W., F.-J. Lapointe, and M. S. Springer. 1997. DNA-hybridisation studies of marsupials and their implications for metatherian classification. Australian Journal of Zoology 45:211–280.

Kirsch, J. A. W., and W. E. Poole. 1972. Taxonomy and distribution of the grey kangaroos, *Macropus giganteus* Shaw and *Macropus fuliginosus* (Desmarest), and their subspecies (Marsupialia: Macropodidae). Australian Journal of Zoology 20:315–339.

Kirsch, J. A. W., and P. F. Waller. 1979. Notes on the trapping and behavior of the Caenolestidae (Marsupialia). Journal of Mammalogy 60:390–395.

Kirsch, J. A. W., and M. A. Wolman. 2001. Molecular relationships of the bear cuscus, *Ailurops ursinus* (Marsupialia: Phalangeridae). Australian Mammalogy 23:23–30.

Kirsch, J. A. W., M. S. Springer, C. Krajewski, M. Archer, K. Aplin, and A. W. Dickerman. 1990. DNA / DNA hybridization studies of the carnivorous marsupials. I: The intergeneric relationships of bandicoots (Marsupialia: Perameloidea). Journal of Molecular Evolution 30:434–448.

Kirsch, J. A. W., O. Gauthier, A. Campeau-Peloquin, M. D. B. Eldridge, and F.-J. Lapointe. 2010. Phylogeny of the rock wallabies, *Petrogale* (Marsupialia: Macropodidae). Part II: Detection of hybridisation among macropodines. Australian Mammalogy 32:67–75.

Kistler, J. M. 2006. War elephants. Praeger, Westport, Connecticut.

Kitazoe Y, H. Kishino, P. J. Waddell, N. Nakajima, T. Okabayashi, T. Watabe, and Y. Okuhara. 2007. Robust time estimation reconciles views of the antiquity of placental mammals. PLoS ONE 2: e384.

Kitchener, A. C., T. Clegg, N. M. J. Thompson, H. Wiik, and A. A. Macdonald. 1993. First records of the Malay civet, *Viverra tangalunga* (Gray, 1832), on Seram with notes on the Seram bandicoot *Rhynchomeles prattorum* (Thomas, 1920). Zeitschrift für Säugetierkunde 58:378–380.

Kitchener, A. C., F. Maisels, L. Pearson, and P. Aczel. 2008. A golden mole (family Chrysochloridae) from savanna woodland in the Batéké Plateau, Gabon. Afrotherian Conservation 6:5–6.

Kitchener, D. J. 1972. The importance of shelter to the quokka, *Setonix brachyurus* (Marsupialia), on Rottnest Island. Australian Journal of Zoology 20:281–299.

Kitchener, D. J. 1981. Breeding, diet, and habitat preference of *Phascogale calura* (Gould, 1844) (Marsupialia: Dasyuridae) in the southern wheatbelt, Western Australia. Records of the Western Australian Museum 9:173–186.

Kitchener, D. J. 1988. A new species of false antechinus (Marsupialia: Dasyuridae) from the Kimberley, Western Australia. Records of the Western Australian Museum 14:61–71.

Kitchener, D. J. 1991. *Pseudantechinus mimulus* (Thomas, 1906) (Marsupialia: Dasyuridae): rediscovery and redescription. Records of the Western Australian Museum 15:191–202.

Kitchener, D. J., and N. Caputi. 1988. A new species of false antechinus (Marsupialia: Dasyuridae) from Western Australia, with remarks on the generic classification within the Parantechini. Records of the Western Australian Museum 14:35–59.

Kitchener, D. J., and J. A. Friend. 2008. Broad-faced potoroo *Potorous platyops* (Gould, 1844). Pp. 301–302 in The mammals of Australia. 3rd edition (S. Van Dyck and R. Strahan, eds.). Reed New Holland, Sydney, Australia.

Kitchener, D. J., and G. Sanson. 1978. *Petrogale burbidgei* (Marsupialia, Macropodidae), a new rock wallaby from Kimberley, Western Australia. Records of the Western Australian Museum 6:269–285.

Kitchener, D. J., J. Stoddart, and J. Henry. 1983. A taxonomic appraisal of the genus *Ningaui* Archer (Marsupialia: Dasyuridae), including description of a new species. Australian Journal of Zoology 31:361–379.

Kitchener, D. J., J. Stoddart, and J. Henry. 1984. A taxonomic revision of the *Sminthopsis murina* complex (Marsupialia: Dasyuridae) in Australia, including descriptions of four new species. Records of the Western Australian Museum 11:201–248.

Kjer, K. M., and R. L. Honeycutt. 2007. Site specific rates of mitochondrial genomes and the phylogeny of eutheria. BMC Evolutionary Biology 7(8).

Klein, R. G. 1984. Mammalian extinctions and Stone Age people in Africa. Pp. 553–573 in Quaternary extinctions: a prehistoric revolution (P. S. Martin and R. G. Klein, eds.). University of Arizona Press, Tucson, Arizona.

Klippel, W. E., and P. W. Parmalee. 1984. Armadillos in North American late Pleistocene contexts. Carnegie Museum of Natural History Special Publications 8:149–160.

Knott, K. K., B. M. Roberts, M. A. Maly, C. K. Vance, J. DeBeachaump, J. Majors, P. Riger, H. DeCaluwe, and A. J. Kouba. 2013. Fecal estrogen, progestagen and glucocorticoid metabolites during the estrous cycle and pregnancy in the giant anteater (*Myrmecophaga tridactyla*): evidence for delayed implantation. Reproductive Biology and Endocrinology 11:83. http://www.rbej.com/content/11/1/83.

Koch, P. L., and A. D. Barnosky. 2006. Late Quaternary extinctions: state of the debate. Annual Review of Ecology, Evolution, and Systematics 37:215–50.

Kolomyjec, S. H., J. Y. T. Chong, D. Blair, J. Gongora, T. R. Grant, C. N. Johnson, and C. Moran. 2009. Population genetics of the platypus (*Ornithorhynchus anatinus*): a fine-scale look at adjacent river systems. Australian Journal of Zoology 57:225–234.

Koontz, F. W., and N. J. Roeper. 1983. *Elephantulus rufescens*. Mammalian Species 204:1–5.

Körtner, G., and F. Geiser. 1996. Hibernation of mountain pygmy-possums (*Burramys parvus*) in the Australian alps. Pp. 31–38 in Adaptations to the cold, Tenth International Hibernation Symposium (F. Geiser, A. J. Hulbert, and S. C. Nicol, eds.). University of New England Press, Armidale, New South Wales, Australia.

Körtner, G., and F. Geiser. 1998. Ecology of natural hibernation in the mountain pygmy-possum (*Burramys parvus*). Oecologia 113:170–178.

Körtner, G., and F. Geiser. 2000. Torpor and activity patterns in free-ranging sugar gliders *Petaurus breviceps* (Marsupialia). Oecologia 123:350–357.

Körtner, G., C. R. Pavey, and F. Geiser. 2007. Spatial ecology of the mulgara in arid Australia: impact of fire history on home range size and burrow use. Journal of Zoology 273:350–357.

Koster, J. M. 2008. Giant anteaters (*Myrmecophaga tridactyla*) killed by hunters with dogs in the Bosawas Biosphere Reserve, Nicaragua. Southwestern Naturalist 53:414–416.

Kouadio, A. 2012. The West African manatee. Pp. 54–57 in Sirenian conservation. Issues and strategies in developing countries (E. M. Hines, J. E. Reynolds, III, L. V. Aragones, A. A. Mignucci-Giannoni, and M. Marmontel, eds.). University Press of Florida, Gainesville, Florida.

Kraft, R. 1990. Modern flying lemurs. Pp. 632–639 in Grzimek's encyclopedia of mammals, volume 1 (B. Grzimek, ed.). McGraw-Hill, New York, New York.

Krajewski, C., M. J. Blacket, and M. Westerman. 2000. DNA Sequence analysis of familial relationships among dasyuromorphian marsupials. Journal of Mammalian Evolution 7:95–108.

Krajewski, C., L. Buckley, and M. Westerman. 1997. DNA phylogeny of the marsupial wolf resolved. Proceedings of the Royal Society of London B Biological Sciences 264:911–917.

Krajewski, C., and M. Westerman. 2003. Molecular systematics of Dasyuromorphia. Pp. 3–20 in Predators with pouches: the biology of carnivorous marsupials (M. Jones, C. Dickman, and M. Archer, eds.). CSIRO Publishing, Collingwood, Victoria, Australia.

Krajewski, C., P. A. Woolley, and M. Westerman. 2000. The evolution of reproductive strategies in dasyurid marsupials: implications of molecular phylogeny. Biological Journal of the Linnean Society 71:417–435.

Krajewski, C., S. Wroe, and M. Westerman. 2000. Molecular evidence for the pattern and timing of cladogenesis in dasyurid marsupials. Zoological Journal of the Linnean Society 130:375–404.

Krajewski, C., A. C. Driskell, P. R. Baverstock, and M. J. Braun. 1992. Phylogenetic relationships of the thylacine (Mammalia: Thylacinidae) among dasyuroid marsupials: evidence from cytochrome b DNA sequences. Proceedings of the Royal Society of London B 250:19–27.

Krajewski, C., J. Painter, L. Buckley, and M. Westerman. 1994. Phylogenetic structure of the marsupial family Dasyuridae based on cytochrome b DNA sequences. Journal of Mammalian Evolution. 2:25–35.

Krajewski, C., L. Buckley, P. A. Woolley, and M. Westerman. 1996. Phylogenetic analysis of cytochrome b sequences in the dasyurid marsupial subfamily Phascogalinae: systematics and the evolution of reproductive strategies. Journal of Mammalian Evolution 3:81–91.

Krajewski, C., M. Blacket, L. Buckley, and M. Westerman. 1997. A multigene assessment of phylogenetic relationships within the dasyurid marsupial subfamily Sminthopsinae. Molecular Phylogenetics and Evolution 8:236–248.

Krajewski, C., G. R. Moyer, J. T. Sipiorski, M. G. Fain, and M. Westerman. 2004. Molecular systematics of the enigmatic "phascolosoricine" marsupials of New Guinea. Australian Journal of Zoology 52:389–415.

Krajewski, C., R. Torunsky, J. T. Sipiorski, and M. Westerman. 2007. Phylogenetic relationships of the dasyurid marsupial genus *Murexia*. Journal of Mammalogy 88:696–705.

Krajewski, C., F. E. Anderson, P. A. Woolley, and M. Westerman. 2012. Molecular evidence for a deep clade of dunnarts (Marsupialia: Dasyuridae: *Sminthopsis*). Journal of Mammalian Evolution 19:265–276.

Krause, D. W. 1991. Were paromomyids gliders? Maybe, maybe not. Journal of Human Evolution 21:177–188.

Krause, D. W. 2001. Fossil molar from a Madagascan marsupial. Nature 412:497–498.

Kreger, M. D. 2008. Canvas to concrete. Elephants and the circus-zoo relationship. Pp. 185–203 in Elephants and ethics, toward a morality of existence (C. Wemmer and C. A. Christen, eds.). Johns Hopkins University Press, Baltimore, Maryland.

Kreutz, K., F. Fischer, and K. E. Linsenmair. 2009. Observations of intraspecific aggression in giant anteaters (*Myrmecophaga tridactyla*). Edentata 8–10:6–7.

Kreutz, K., F. Fischer, and K. E. Linsenmair. 2012. Timber plantations as favourite habitat for giant anteaters. Mammalia 76:137–142.

Kröpelin, S., D. Verschuren, A.-M. Lézine, H. Eggermont, C. Cocquyt, P. Francus, J.-P. Cazet, M. Fagot, B. Rumes, J. M. Russell, F. Darius, D. J. Conley, M. Schuster, H. Von Suchodoletz, D. R. Engstrom. 2008. Climate-driven ecosystem succession in the Sahara: the past 6000 years. Science 320:765–768.

Kuhn, H. 1971. An adult female *Micropotamogale lamottei*. Journal of Mammalogy 52:477–478.

Kullberg, M., B. M. Hallström, U. Arnason, and A. Janke. 2008. Phylogenetic analysis of 1.5 Mbp and platypus EST data refute the Marsupionta hypothesis and unequivocally support Monotremata as sister group to Marsupialia/Placentalia. Zoologica Scripta 37:115–127.

Kumordzi, B. B., W. Oduro, S. K. Oppong, E. Danquah, A. Lister, M. K. Sam. 2008. An elephant survey in Digya National Park, Ghana, and implications for conservation and management. Pachyderm 44:27–34.

Kuntner, M., L. J. May-Collado, and I. Agnarsson. 2010. Phylogeny and conservation priorities of afrotherian mammals (Afrotheria, Mammalia). Zoologica Scripta 40:1–15.

Kurt, F. 1974. Remarks on the social structure and ecology of the Ceylon elephant in the Yala National Park. Pp. 618–634 in The behaviour of ungulates and its relation to management (V. Geist and F. Walther, eds.). IUCN, Gland, Switzerland.

Kurt, F., G. B. Hartl, and R. Tiedemann. 1995. Tuskless bulls in Asian elephant *Elephas maximus*: history and population genetics of a man-made phenomenon. Acta Theriologica, Supplement 3:125–143.

Kurt, F., K. U. Mar, and M. E. Garaï. 2008. Giants in chains: history, biology, and preservation of Asian elephants in captivity. Pp. 327–345 in Elephants and ethics, toward a morality of existence (C. Wemmer and C. A. Christen, eds.). Johns Hopkins University Press, Baltimore, Maryland.

Kurten, B. 1968. Pleistocene mammals of Europe. Aldine, Chicago, Illinois.

Kutt, A. S., S. Van Dyck, and S. J. Christie. 2005. A significant range extension for the chestnut dunnart *Sminthopsis archeri*

(Mammalia: Dasyuridae) in north Queensland. Australian Zoologist 33:265–268.

Kuyper, M. A. 1985. The ecology of the golden mole *Amblysomus hottentotus*. Mammal Review 15:3–11.

Kuzmin, Y. V. 2010. Extinction of the woolly mammoth (*Mammuthus primigenius*) and woolly rhinoceros (*Coelodonta antiquitatis*) in Eurasia: review of chronological and environmental issues. Boreas 39:247–261.

Kuzmin, Y. V., and L. A. Orlova. 2004. Radiocarbon chronology and environment of woolly mammoth (*Mammuthus primigenius* Blum.) in northern Asia: results and perspectives. Earth-Science Reviews 68:133–169.

Lada, H., and R. Mac Nally. 2008. Decline and potential recovery of yellow-footed antechinus in parts of south-eastern Australia: a perspective with implications for management. Ecological Management and Restoration 9:120–125.

Lada, H., R. Mac Nally, and A. C. Taylor. 2008. Distinguishing past from present gene flow along and across a river: the case of the carnivorous marsupial (*Antechinus flavipes*) on southern Australian floodplains. Conservation Genetics 9:569–580.

Lahiri Choudhury, D. K. 2008. Elephants and people in India. Historical patterns of capture and management. Pp. 149–164 in Elephants and ethics, toward a morality of existence (C. Wemmer and C. A. Christen, eds.). Johns Hopkins University Press, Baltimore, Maryland.

Laidlaw, W. S., S. Hutchings, and G. R. Newell. 1996. Home range and movement patterns of *Sminthopsis leucopus* (Marsupialia: Dasyuridae) in coastal dry heathland, Anglesea, Victoria. Australian Mammalogy 19:1–9.

Laidlaw, W. S., and B. A. Wilson. 1996. The home range and habitat utilisation of *Cercartetus nanus* (Marsupialia: Burramyidae) in coastal heathland, Anglesea, Victoria. Australian Mammalogy 19:63–68.

Lair, R. C. 1988. The number and distribution of domesticated elephants in Thailand. Natural History Bulletin of the Siam Society 36:143–160.

Laist, D. W., C. Taylor, and J. E. Reynolds, III. 2013. Winter habitat preferences for Florida manatees and vulnerability to cold. PLoS ONE 8(3):e58978. doi:10.1371/journal.pone.0058978.

Lambert, T. D., M. K. Halsey, J. W. Dittel, S. A. Mangan, E. Delfosse, G. H. Adler, and S. A. Schnitzer. 2011. First record of Alston's woolly mouse opossum (*Micoureus alstoni*) from the canal area of Central Panama. Mammalia 75:107–109.

Lamotte, M., and F. Petter. 1981. Une taupe dorée nouvelle do Cameroun (Mt Oku, 6°15'N, 10°26'E): *Chrysochloris stuhlmanni balsaci* ssp. nov. Mammalia 45:43–48.

Langbauer, W. R., Jr. 2000. Elephant communication. Zoo Biology 19:425–445.

Langham, N. P. E. 1982. The ecology of the common tree shrew, *Tupaia glis*, in peninsular Malaysia. Journal of Zoology 197:323–344.

Lara-Ruiz, P., and A. G. Chiarello. 2005. Life-history traits and sexual dimorphism of the Atlantic Forest maned sloth *Bradypus torquatus* (Xenarthra: Bradypodidae). Journal of Zoology 267:63–73.

Largen, M. J., and D. W. Yalden. 1987. The decline of elephant and black rhinoceros in Ethiopia. Oryx 21:103–106.

Larivière, S. 2001. *Ursus americanus*. Mammalian Species 647:1–11.

Laurance, W. F. 1990. Comparative responses of five arboreal marsupials to tropical forest fragmentation. Journal of Mammalogy 71:641–653.

Laurie, E. M. O., and J. E. Hill. 1954. List of land mammals of New Guinea, Celebes, and adjacent islands, 1758–1952. British Museum (Natural History), London, United Kingdom.

Laursen, L., and M. Bekoff. 1978. *Loxodonta africana*. Mammalian Species 92:1–8.

Lavergne, A., E. Douzery, T. Stichler, F. M. Catzeflis, and M. S. Springer. 1996. Interordinal mammalian relationships: evidence for paenungulate monophyly is provided by complete mitochondrial 12S rRNA sequences. Molecular Phylogenetics and Evolution 6:245–258.

Laws, R. J., and A. W. Goldizen. 2003. Nocturnal home ranges and social interactions of the brush-tailed rock-wallaby *Petrogale penicillata* at Hurdle Creek, Queensland. Australian Mammalogy 25:169–176.

Laws, R. M. 1981. Large mammal feeding strategies and related overabundance problems. Pp. 217–232 in Problems in management of locally abundant wild mammals (P. A. Jewell and S. Holt, eds.). Academic Press, New York, New York.

Laws, R. M., I. S. C. Parker, and R. C. B. Johnstone. 1975. Elephants and their habitats: the ecology of elephants in north Bunyoro, Uganda. Oxford University Press, Oxford, United Kingdom.

Lawson, L. P., C. Vernesi, S. Ricci, and F. Rovero. 2013. Evolutionary history of the grey-faced sengi, *Rhynchocyon udzungwensis*, from Tanzania: a molecular and species distribution modelling approach. PLoS ONE 8(8):e72506. doi:10.1371/journal.pone.0072506.

Layne, J. N., and D. Glover. 1977. Home range of the armadillo in Florida. Journal of Mammalogy 58:411–413.

Layne, J. N., and D. Glover. 1985. Activity patterns of the common long-nosed armadillo *Dasypus novemcinctus* in south-central Florida. Pp. 407–417 in The evolution and ecology of armadillos, sloths, and vermilinguas (G. G. Montgomery, ed.). Smithsonian Institution Press, Washington, DC.

Layne, J. N., and A. M. Waggener, Jr. 1984. Above-ground nests of the nine-banded armadillo in Florida. Florida Field Naturalist 12:58–61.

Layser, T. R., and I. O. Buss. 1985. Observations on morphological characteristics of elephant tusks. Mammalia 49:407–414.

Lazcano-Barrero, M. A., and J. M. Packard. 1989. The occurrence of manatees (*Trichechus manatus*) in Tamaulipas, Mexico. Marine Mammal Science 5:202–205.

Lazell, J. D., Jr., T. W. Sutterfield, and W. D. Giezentanner. 1984. The population of rock wallabies (genus *Petrogale*) on Oahu, Hawaii. Biological Conservation 30:99–108.

Lazenby-Cohen, K. A. 1991. Communal nesting in *Antechinus stuartii* (Marsupialia: Dasyuridae). Australian Journal of Zoology 39:273–283.

Lazenby-Cohen, K. A., and A. Cockburn. 1988. Lek promiscuity in a semelparous mammal, *Antechinus stuartii* (Marsupialia: Dasyuridae)? Behavioral Ecology and Sociobiology 22:195–202.

Leary, T., L. Seri, T. Flannery, D. Wright, S. Hamilton, K. Helgen, R. Singadan, J. Menzies, A. Allison, R. James, K. Aplin, L. Salas, and C. Dickman. 2008a. *Zaglossus bartoni*. The IUCN Red List of Threatened Species 2008: e.T136552A4309582. http://dx.doi.org/10.2305/IUCN.UK.2008.RLTS.T136552A4309582.en. Downloaded on 19 June 2015.

Leary, T., L. Seri, T. Flannery, D. Wright, S. Hamilton, K. Helgen, R. Singadan, J. Menzies, A. Allison, R. James, K. Aplin, L. Salas, and C. Dickman. 2008b. *Zaglossus bruijnii*. The IUCN Red List of

Threatened Species 2008: e.T23179A9426076. http://dx.doi.org/10.2305/IUCN.UK.2008.RLTS.T23179A9426076.en. Downloaded on 03 January 2016.

Leary, T., L. Seri, T. Flannery, D. Wright, S. Hamilton, K. Helgen, R. Singadan, J. Menzies, A. Allison, R. James, K. Aplin, L. Salas, and C. Dickman. 2008c. *Zaglossus attenboroughi*. The IUCN Red List of Threatened Species 2008: e.T136322A4274381. http://dx.doi.org/10.2305/IUCN.UK.2008.RLTS.T136322A4274381.en. Downloaded on 03 January 2016.

Leary, T., L. Seri, D. Wright, S. Hamilton, K. Helgen, R. Singadan, J. Menzies, A. Allison, R. James, C. Dickman, D. Lunde, K. Aplin, and P. Woolley. 2008d. *Murexia rothschildi*. The IUCN Red List of Threatened Species 2008: e.T13931A4366514. http://dx.doi.org/10.2305/IUCN.UK.2008.RLTS.T13931A4366514.en. Downloaded on 04 January 2016.

Leary, T., L. Seri, T. Flannery, D. Wright, S. Hamilton, K. Helgen, R. Singadan, J. Menzies, A. Allison, R. James, and P. Woolley. 2008e. *Dasyurus spartacus*. The IUCN Red List of Threatened Species 2008: e.T6301A12601528. http://dx.doi.org/10.2305/IUCN.UK.2008.RLTS.T6301A12601528.en. Downloaded on 05 January 2016.

Leary, T., D. Wright, S. Hamilton, R. Singadan, J. Menzies, F. Bonaccorso, K. Helgen, L. Seri, A. Allison, K. Aplin, C. Dickman, and L. Salas. 2008f. *Peroryctes broadbenti*. The IUCN Red List of Threatened Species 2008: e.T16710A6302300. http://dx.doi.org/10.2305/IUCN.UK.2008.RLTS.T16710A6302300.en. Downloaded on 21 November 2015.

Leary, T., D. Wright, S. Hamilton, R. Singadan, J. Menzies, F. Bonaccorso, K. Helgen, L. Seri, A. Allison, K. Aplin, C. Dickman, and L. Salas. 2008g. *Peroryctes raffrayana*. The IUCN Red List of Threatened Species 2008: e.T16711A6303217. http://dx.doi.org/10.2305/IUCN.UK.2008.RLTS.T16711A6303217.en. Downloaded on 21 November 2015.

Leary, T., D. Wright, S. Hamilton, R. Singadan, J. Menzies, F. Bonaccorso, K. Helgen, L. Seri, A. Allison, J. Winter, K. Aplin, C. Dickman, and L. Salas. 2008h. *Echymipera rufescens*. The IUCN Red List of Threatened Species 2008: e.T7019A12822745. http://dx.doi.org/10.2305/IUCN.UK.2008.RLTS.T7019A12822745.en. Downloaded on 06 January 2016.

Leary, T., D. Wright, S. Hamilton, R. Singadan, J. Menzies, F. Bonaccorso, K. Helgen, L. Seri, K. Aplin, C. Dickman, and L. Salas. 2008i. *Echymipera echinista*. The IUCN Red List of Threatened Species 2008: e.T7016A12822320. http://dx.doi.org/10.2305/IUCN.UK.2008.RLTS.T7016A12822320.en. Downloaded on 06 January 2016.

Leary, T., D. Wright, S. Hamilton, R. Singadan, J. Menzies, F. Bonaccorso, K. Helgen, and L. Seri. 2008j. *Echymipera davidi*. The IUCN Red List of Threatened Species 2008: e.T7017A12822440. http://dx.doi.org/10.2305/IUCN.UK.2008.RLTS.T7017A12822440.en. Downloaded on 06 January 2016.

Leary, T., D. Wright, S. Hamilton, R. Singadan, J. Menzies, F. Bonaccorso, K. Helgen, L. Seri, A. Allison, K. Aplin, C. Dickman, and L. Salas. 2008k. *Rhynchomeles prattorum*. The IUCN Red List of Threatened Species 2008: e.T19711A9006016. http://dx.doi.org/10.2305/IUCN.UK.2008.RLTS.T19711A9006016.en. Downloaded on 21 November 2015.

Leary, T., D. Wright, S. Hamilton, R. Singadan, J. Menzies, F. Bonaccorso, K. Helgen, L. Seri, K. Aplin, C. Dickman, and L. Salas. 2008l. *Microperoryctes longicauda*. The IUCN Red List of Threat-

ened Species 2008: e.T13388A3879271. http://dx.doi.org/10.2305/IUCN.UK.2008.RLTS.T13388A3879271.en. Downloaded on 21 November 2015.

Leary, T., R. Singadan, J. Menzies, K. Helgen, A. Allison, R. James, T. Flannery, K. Aplin, C. Dickman, and L. Salas. 2008m. *Strigocuscus pelengensis*. The IUCN Red List of Threatened Species 2008: e.T20892A9235085. http://dx.doi.org/10.2305/IUCN.UK.2008.RLTS.T20892A9235085.en. Downloaded on 30 December 2015.

Leary, T., R. Singadan, J. Menzies, K. Helgen, A. Allison, R. James, T. Flannery, K. Aplin, C. Dickman, and L. Salas. 2008n. *Spilocuscus rufoniger*. The IUCN Red List of Threatened Species 2008: e.T20639A9216946. http://dx.doi.org/10.2305/IUCN.UK.2008.RLTS.T20639A9216946.en. Downloaded on 30 December 2015.

Leary, T., L. Seri, T. Flannery, D. Wright, S. Hamilton, K. Helgen, R. Singadan, J. Menzies, A. Allison, R. James, L. Salas, and C. Dickman. 2008o. *Phalanger matanim*. The IUCN Red List of Threatened Species 2008: e.T16851A6505272. http://dx.doi.org/10.2305/IUCN.UK.2008.RLTS.T16851A6505272.en. Downloaded on 30 December 2015

Leary, T., R. Singadan, J. Menzies, K. Helgen, D. Wright, A. Allison, and S. Hamilton. 2008p. *Phalanger lullulae*. The IUCN Red List of Threatened Species 2008: e.T16846A6496309. http://dx.doi.org/10.2305/IUCN.UK.2008.RLTS.T16846A6496309.en. Downloaded on 30 December 2015.

Leary, T., R. Singadan, J. Menzies, K. Helgen, D. Wright, A. Allison, T. Flannery, L. Salas, and C. Dickman. 2008q. *Phalanger alexandrae*. The IUCN Red List of Threatened Species 2008: e.T16858A6518981. http://dx.doi.org/10.2305/IUCN.UK.2008.RLTS.T16858A6518981.en. Downloaded on 30 December 2015.

Leary, T., R. Singadan, J. Menzies, K. Helgen, D. Wright, A. Allison, T. Flannery, L. Salas, and C. Dickman. 2008r. *Phalanger matabiru*. The IUCN Red List of Threatened Species 2008: e.T136200A4258928. http://dx.doi.org/10.2305/IUCN.UK.2008.RLTS.T136200A4258928.en. Downloaded on 30 December 2015.

Leary, T., R. Singadan, J. Menzies, K. Helgen, D. Wright, A. Allison, L. Salas, and C. Dickman. 2008s. *Phalanger gymnotis*. The IUCN Red List of Threatened Species 2008: e.T16856A6514775. http://dx.doi.org/10.2305/IUCN.UK.2008.RLTS.T16856A6514775.en. Downloaded on 30 December 2015.

Leary, T., R. Singadan, J. Menzies, K. Helgen, D. Wright, A. Allison, S. Hamilton, L. Salas, and C. Dickman. 2008t. *Phalanger intercastellanus*. The IUCN Red List of Threatened Species 2008: e.T16857A6516977. http://dx.doi.org/10.2305/IUCN.UK.2008.RLTS.T16857A6516977.en. Downloaded on 30 December 2015.

Leary, T., R. Singadan, J. Menzies, K. Helgen, D. Wright, A. Allison, K. Aplin, L. Salas, and C. Dickman. 2008u. *Phalanger vestitus*. The IUCN Red List of Threatened Species 2008: e.T16850A6503469. http://dx.doi.org/10.2305/IUCN.UK.2008.RLTS.T16850A6503469.en. Downloaded on 30 December 2015.

Leary, T., R. Singadan, J. Menzies, D. Wright, K. Helgen, C. Dickman, and L. Salas. 2008v. *Pseudochirops corinnae*. The IUCN Red List of Threatened Species 2008: e.T18504A8356425. http://dx.doi.org/10.2305/IUCN.UK.2008.RLTS.T18504A8356425.en. Downloaded on 30 December 2015.

Leary, T., D. Wright, S. Hamilton, R. Singadan, J. Menzies, F. Bonaccorso, L. Salas, C. Dickman, and K. Helgen. 2008w. *Petaurus abidi*. The IUCN Red List of Threatened Species 2008: e.T16726A6319161. http://dx.doi.org/10.2305/IUCN.UK.2008.RLTS.T16726A6319161.en. Downloaded on 30 December 2015.

Leary, T., D. Wright, S. Hamilton, R. Singadan, J. Menzies, F. Bonaccorso, K. Helgen, L. Seri, and A. Allison. 2008x. *Dactylopsila tatei.*
The IUCN Red List of Threatened Species 2008: e.T6224A12583705.
http://dx.doi.org/10.2305/IUCN.UK.2008.RLTS.T6224A12583705
.en. Downloaded on 30 December 2015.

Leary, T., D. Wright, S. Hamilton, R. Singadan, J. Menzies, F. Bonaccorso, K. Helgen, L. Seri, A. Allison, and R. James. 2008y. *Dorcopsis atrata.* The IUCN Red List of Threatened Species 2008:
e.T6794A12806981. http://dx.doi.org/10.2305/IUCN.UK.2008
.RLTS.T6794A12806981.en. Downloaded on 30 December 2015.

Leary, T., D. Wright, R. Singadan, L. Seri, A. Allison, K. Aplin, R. James,
T. Flannery, C. Dickman, and L. Salas. 2008z. *Dorcopsis luctuosa.*
The IUCN Red List of Threatened Species 2008: e.T6799A12807319.
http://dx.doi.org/10.2305/IUCN.UK.2008.RLTS.T6799A12807319
.en. Downloaded on 30 December 2015.

Leary, T., R. Singadan, J. Menzies, K. Helgen, A. Allison, R. James,
T. Flannery, K. Aplin, C. Dickman, and L. Salas. 2008aa. *Dorcopsulus vanheurni.* The IUCN Red List of Threatened Species 2008:
e.T6802A12807878. http://dx.doi.org/10.2305/IUCN.UK.2008
.RLTS.T6802A12807878.en. Downloaded on 30 December 2015.

Leary, T., L. Seri, T. Flannery, D. Wright, S. Hamilton, K. Helgen, R.
Singadan, J. Menzies, A. Allison, and R. James. 2008bb. *Dendrolagus mayri.* The IUCN Red List of Threatened Species 2008:
e.T136668A4325261. http://dx.doi.org/10.2305/IUCN.UK.2008
.RLTS.T136668A4325261.en. Downloaded on 31 December 2015.

Leary, T., D. Wright, S. Hamilton, K. Helgen, R. Singadan, K. Aplin,
C. Dickman, L. Salas, T. Flannery, R. Martin, and L. Seri. 2008cc.
Dendrolagus pulcherrimus. The IUCN Red List of Threatened Spe-
cies 2008: e.T136696A4328700. http://dx.doi.org/10.2305/IUCN
.UK.2008.RLTS.T136696A4328700.en. Downloaded on 31 De-
cember 2015.

Leary, T., D. Wright, S. Hamilton, K. Helgen, R. Singadan, K. Aplin,
C. Dickman, L. Salas, T. Flannery, R. Martin, and L. Seri.
2008dd. *Dendrolagus scottae.* The IUCN Red List of Threatened
Species 2008: e.T6435A12773127. http://dx.doi.org/10.2305
/IUCN.UK.2008.RLTS.T6435A12773127.en. Downloaded on 31
December 2015.

Leary, T., L. Seri, D. Wright, S. Hamilton, K. Helgen, R. Singadan,
J. Menzies, A. Allison, R. James, C. Dickman, K. Aplin, T. Flannery,
R. Martin, and L. Salas. 2008ee. *Dendrolagus matschiei.* The IUCN
Red List of Threatened Species 2008: e.T6433A12770851. http://dx
.doi.org/10.2305/IUCN.UK.2008.RLTS.T6433A12770851.en.
Downloaded on 31 December 2015.

Leary, T., L. Seri, D. Wright, S. Hamilton, K. Helgen, R. Singadan,
J. Menzies, A. Allison, R. James, C. Dickman, K. Aplin, T. Flannery,
R. Martin, and L. Salas. 2008ff. *Dendrolagus goodfellowi.* The IUCN
Red List of Threatened Species 2008: e.T6429A12754290. http://dx
.doi.org/10.2305/IUCN.UK.2008.RLTS.T6429A12754290.en.
Downloaded on 02 January 2016.

Leary, T., L. Seri, T. Flannery, D. Wright, S. Hamilton, K. Helgen,
R. Singadan, J. Menzies, A. Allison, and R. James. 2008gg. *Dendrolagus notatus.* The IUCN Red List of Threatened Species 2008:
e.T136732A4333248. http://dx.doi.org/10.2305/IUCN.UK.2008
.RLTS.T136732A4333248.en. Downloaded on 02 January 2016.

Leary, T., L. Seri, D. Wright, S. Hamilton, K. Helgen, R. Singadan,
J. Menzies, A. Allison, R. James, C. Dickman, K. Aplin, T. Flannery,
R. Martin, and L. Salas. 2008hh. *Dendrolagus inustus.* The IUCN
Red List of Threatened Species 2008: e.T6431A12762670. http://dx

.doi.org/10.2305/IUCN.UK.2008.RLTS.T6431A12762670.en.
Downloaded on 02 January 2016.

Leary, T., L. Seri, D. Wright, S. Hamilton, K. Helgen, R. Singadan,
J. Menzies, A. Allison, R. James, C. Dickman, K. Aplin, L. Salas,
T. Flannery, and F. Bonaccorso. 2008ii. *Dendrolagus ursinus.* The
IUCN Red List of Threatened Species 2008: e.T6434A12771953.
http://dx.doi.org/10.2305/IUCN.UK.2008.RLTS.T6434
A12771953.en. Downloaded on 02 January 2016.

Leary, T., L. Seri, T. Flannery, D. Wright, S. Hamilton, K. Helgen, R.
Singadan, J. Menzies, A. Allison, and R. James. 2008jj. *Dendrolagus dorianus.* The IUCN Red List of Threatened Species 2008:
e.T6427A12759868. http://dx.doi.org/10.2305/IUCN.UK.2008
.RLTS.T6427A12759868.en. Downloaded on 02 January 2016.

Leary, T., L. Seri, T. Flannery, D. Wright, S. Hamilton, K. Helgen,
R. Singadan, J. Menzies, A. Allison, R. James, K. Aplin, L. Salas,
and C. Dickman. 2008kk. *Dendrolagus stellarum.* The IUCN Red
List of Threatened Species 2008: e.T136812A4342630. http://dx
.doi.org/10.2305/IUCN.UK.2008.RLTS.T136812A4342630.en.
Downloaded on 02 January 2016.

Leary, T., L. Seri, D. Wright, S. Hamilton, K. Helgen, R. Singadan,
J. Menzies, A. Allison, R. James, C. Dickman, K. Aplin, L. Salas,
T. Flannery, and F. Bonaccorso. 2008ll. *Dendrolagus spadix.* The
IUCN Red List of Threatened Species 2008: e.T6436A12760953.
http://dx.doi.org/10.2305/IUCN.UK.2008.RLTS.T6436
A12760953.en. Downloaded on 02 January 2016.

Leary, T., L. Seri, T. Flannery, D. Wright, S. Hamilton, K. Helgen,
R. Singadan, J. Menzies, A. Allison, and R. James. 2008mm. *Thylogale brunii.* The IUCN Red List of Threatened Species 2008: e.T218
70A9333216. http://dx.doi.org/10.2305/IUCN.UK.2008.RLTS.T218
70A9333216.en. Downloaded on 05 January 2016.

Leary, T., L. Seri, T. Flannery, D. Wright, S. Hamilton, K. Helgen,
R. Singadan, J. Menzies, A. Allison, and R. James. 2008nn. *Thylogale
browni.* The IUCN Red List of Threatened Species 2008: e.T218
74A9333558. http://dx.doi.org/10.2305/IUCN.UK.2008.RLTS.T218
74A9333558.en. Downloaded on 05 January 2016.

Leary, T., L. Seri, T. Flannery, D. Wright, S. Hamilton, K. Helgen,
R. Singadan, J. Menzies, A. Allison, and R. James. 2008oo. *Thylogale
lanatus.* The IUCN Red List of Threatened Species 2008: e.T136
255A4266028. http://dx.doi.org/10.2305/IUCN.UK.2008.RLTS
.T136255A4266028.en. Downloaded on 05 January 2016.

Leary, T., L. Seri, T. Flannery, D. Wright, S. Hamilton, K. Helgen,
R. Singadan, J. Menzies, A. Allison, R. James, K. Aplin, L. Salas, and
C. Dickman. 2008pp. *Thylogale calabyi.* The IUCN Red List of
Threatened Species 2008: e.T21873A9333404. http://dx.doi.org/10
.2305/IUCN.UK.2008.RLTS.T21873A9333404.en. Downloaded
on 05 January 2016.

Leary, T., L. Seri, D. Wright, S. Hamilton, K. Helgen, R. Singadan, J.
Menzies, A. Allison, R. James, C. Dickman, K. Aplin, T. Flannery,
R. Martin, and L. Salas. 2010. *Dendrolagus mbaiso.* The IUCN Red
List of Threatened Species 2010: e.T6437A12774651. http://dx.doi
.org/10.2305/IUCN.UK.2010-2.RLTS.T6437A12774651.en. Down-
loaded on 02 January 2016.

Lee, A. K., A. J. Bradley, and R. W. Braithwaite. 1977. Corticosteroid
levels and male mortality in *Antechinus stuartii.* Pp. 209–222 in The
biology of marsupials (B. Stonehouse and D. P. Gilmore, eds.). University Park Press, Baltimore, Maryland.

Lee, A. K., and F. N. Carrick. 1989. Phascolarctidae. Pp. 740–754 in
Fauna of Australia: Mammalia, volume 1B (D. W. Walton and B. J.

Richardson, eds.). Australian Government Publishing Service, Canberra, Australia.

Lee, K. E., J. M. Seddon, S. W. Corley, W. A. H. Ellis, S. D. Johnston, D. L. de Villiers, H. J. Preece, and F. N. Carrick. 2010. Genetic variation and structuring in the threatened koala populations of Southeast Queensland. Conservation Genetics 11:2091–2103.

Lee, P. C. 2011. Dominance in female elephants. Pp. 190–191 in The Amboseli elephants: a long-term perspective on a long-lived mammal (C. J. Moss, H. Croze, and P. C. Lee, eds.). University of Chicago Press, Chicago, Illinois.

Lee, P. C., W. K. Lindsay, and C. J. Moss. 2011. Ecological patterns of variability in demographic rates. Pp. 74–88 in The Amboseli elephants: a long-term perspective on a long-lived mammal (C. J. Moss, H. Croze, and P. C. Lee, eds.). University of Chicago Press, Chicago, Illinois.

Lee, P. C., and C. J. Moss. 2011. Calf development and maternal rearing strategies. Pp. 224–237 in The Amboseli elephants: a long-term perspective on a long-lived mammal (C. J. Moss, H. Croze, and P. C. Lee, eds.). University of Chicago Press, Chicago, Illinois.

Lee, P. C., J. H. Poole, N. Njiraini, C. N. Sayialel, and C. J. Moss. 2011. Male social dynamics: independence and beyond. Pp. 260–271 in The Amboseli elephants: a long-term perspective on a long-lived mammal (C. J. Moss, H. Croze, and P. C. Lee, eds.). University of Chicago Press, Chicago, Illinois.

Lefebvre, L. W., T. J. O'Shea, G. B. Rathbun, and R. C. Best. 1989. Distribution, status, and biogeography of the West Indian manatee. Pp. 567–610 in Biogeography of the West Indies: past, present, and future (C. A. Woods, ed.). Sandhill Crane Press, Gainesville, Florida.

Lefebvre, L. W., M. Marmontel, J. P. Reid, G. B. Rathbun, and D. P. Domning. 2001. Status and biogeography of the West Indian manatee. Pp. 425–474 in Biogeography of the West Indies: patterns and perspectives. 2nd edition (C. A. Woods and F. E. Sergile, eds.). CRC Press, Boca Raton, Florida.

Lehmann, T. 2007. Amended taxonomy of the order Tubulidentata (Mammalia, Eutheria). Annals of the Transvaal Museum 44:179–196.

Lehnhardt, J., and M. Galloway. 2008. Carrots and sticks, people and elephants. Pp. 167–182 in Elephants and ethics, toward a morality of existence (C. Wemmer and C. A. Christen, eds.). Johns Hopkins University Press, Baltimore, Maryland.

Leimgruber, P., B. Senior, Uga, M. Aung, M. A. Songer, T. Mueller, C. Wemmer, and J. D. Ballou. 2008. Modeling population viability of captive elephants in Myanmar (Burma): implications for wild populations. Animal Conservation 11:198–205.

Leimgruber, P., Z. M. Oo, M. Aung, D. S. Kelly, C. Wemmer, B. Senior, and M. Songer. 2011. Current status of Asian elephants in Myanmar. Gajah 35:76–86.

Leiner, N. O., E. Z. F. Setz, and W. R. Silva. 2008. Semelparity and factors affecting the reproductive activity of the Brazilian slender opossum (Marmosops paulensis) in southeastern Brazil. Journal of Mammalogy 89:153–158.

Leiner, N. O., and W. R. Silva. 2009. Territoriality in females of the slender opossum (Marmosops paulensis) in the Atlantic Forest of Brazil. Journal of Tropical Ecology 25:671–675.

Leirs, H., R. Verhagen, W. Verheyen, and M. R. Perrin. 1995. The biology of Elephantulus brachyrhynchus in natural miombo woodland in Tanzania. Mammal Review 25:45–49.

Lekagul, B., and J. A. McNeely. 1977. Mammals of Thailand. Sahakarnbhat, Bangkok, Thailand.

Lemke, T. O., A. Cadena, R. H. Pine, and J. Hernandez-Camacho. 1982. Notes on opossums, bats, and rodents new to the fauna of Colombia. Mammalia 46:225–234.

Lemos, B., and R. Cerqueira. 2002. Morphological differentiation in the white-eared opossum group (Didelphidae: Didelphis). Journal of Mammalogy 83:354–369.

Lemos, B., M. Weksler, and C. R. Bonvicino. 2000. The taxonomic status of Monodelphis umbristriata (Didelphimorphia: Didelphidae). Mammalia 64:329–337.

Lensing, J. E. 1983. Feeding strategy of the rock hyrax and its relation to the rock hyrax problem in southern South West Africa. Madoqua 13:177–196.

Leong, K. M., A. Ortolani, L. H. Graham, and A. Savagea. 2003. The use of low-frequency vocalizations in African elephant (Loxodonta africana) reproductive strategies. Hormones and Behavior 43:433–443.

Leopold, A. S. 1959. Wildlife of Mexico. University of California Press, Berkeley, California.

Le Quellec, J.-L. 1999. Répartition de la grande faune sauvage dans le nord de l'Afrique durant l'Holocène. L'Anthropologie 103:161–176.

Lerebours, B., H. Magnin, and J. Reynolds. 2013. Proposed reintroduction of the Antillean manatee (Trichechus manatus manatus) to Guadeloupe, French West Indies. Sirenews 59:12–15.

Letnic, M., M. Fillios, M. S. Crowther. 2012. Could direct killing by larger dingoes have caused the extinction of the thylacine from mainland Australia? PLoS ONE 7(5):e34877. doi:10.1371/journal.pone.0034877.

Leung, L. K.-P. 2008. Cinnamon antechinus Antechinus leo (Van Dyck, 1980). Pp. 91–92 in The mammals of Australia. 3rd edition (S. Van Dyck and R. Strahan, eds.). Reed New Holland, Sydney, Australia.

Leuthold, W., and J. B. Sale. 1973. Movements and patterns of habitat utilization of elephants in Tsavo National Park, Kenya. East African Wildlife Journal 11:369–384.

Levesque, D. L., D. Rakotondravony, and B. G. Lovegrove. 2012. Home range and shelter site selection in the greater hedgehog tenrec in the dry deciduous forest of western Madagascar. Journal of Zoology 287:161–168.

Levesque, D. L., O. M. A. Lovasoa, D. Rakotondravony, and B. G. Lovegrove. 2013. High mortality and annual fecundity in a free-ranging basal placental mammal, Setifer setosus (Tenrecidae: Afrosoricida). Journal of Zoology 291:205–212.

Lew, D., R. Pérez-Hernández, and J. Ventura. 2006. Two new species of Philander (Didelphimorphia, Didelphidae) from northern South America. Journal of Mammalogy 87:224–237.

Lew, D., M. López Fuster, J. Ventura, R. Pérez-Hernandez, and E. Gutiérrez. 2011a. Monodelphis reigi. The IUCN Red List of Threatened Species 2011: e.T136392A4285661. http://dx.doi.org/10.2305/IUCN.UK.2011-2.RLTS.T136392A4285661.en. Downloaded on 05 January 2016.

Lew, D., E. Gutiérrez, J. Ventura, M. López Fuster, and R. Pérez-Hernandez. 2011b. Marmosa xerophila. The IUCN Red List of Threatened Species 2011: e.T12815A3384469. http://dx.doi.org/10.2305/IUCN.UK.2011-2.RLTS.T12815A3384469.en. Downloaded on 07 November 2015.

Lew, D., R. Pérez-Hernandez, N. de la Sancha, D. Flores, and P. Teta. 2011c. Lutreolina crassicaudata. The IUCN Red List of Threatened

Species 2011: e.T40503A10319982. http://dx.doi.org/10.2305/IUCN.UK.2011-2.RLTS.T40503A10319982.en. Downloaded on 09 November 2015

Lewis, E. R., P. M. Nairns, J. U. M. Jarvis, G. Bronner, and M. J. Mason. 2006. Preliminary evidence for the use of microseismic cues for navigation by the Namib golden mole. Journal of the Acoustical Society of America 119:1260–1268.

Li, J., Y. Hou, Y. Li, and J. Zhang. 2012. The latest straight-tusked elephants (*Palaeoloxodon*)? "Wild elephants" lived 3000 years ago in North China. Quaternary International 281:84–88.

Lidicker, W. Z., Jr., and B. J. Marlow. 1970. A review of the dasyurid marsupial genus *Antechinomys* Krefft. Mammalia 34:212–227.

Lidicker, W. Z., Jr., and A. C. Ziegler. 1968. Report on a collection of mammals from eastern New Guinea including species keys for fourteen genera. University of California Publications in Zoology 87:1–64.

Ligabue-Braun, R., H. Verli, and C. R. Carlini. 2012. Venomous mammals: a review. Toxicon 59:680–695.

Lim, B. K., M. D. Engstrom, J. C. Patton, and J. W. Bickham. 2010. Molecular phylogenetics of Reig's short-tailed opossum (*Monodelphis reigi*) and its distributional range extension into Guyana. Mammalian Biology 75:287–293.

Lim, N. 2007. Colugo. The flying lemur of South-east Asia. Draco Publishing and Distribution and Raffles Museum of Biodiversity Research, National University of Singapore, Kent Ridge, Singapore.

Lim, N. T.-L., and P. K. L. Ng. 2010. Population assessment methods for the Sunda colugo *Galeopterus variegatus* (Mammalia: Dermoptera) in tropical forests and their viability in Singapore. Raffles Bulletin of Zoology 58:157–164.

Lim, N. T.-L., X. Giam, G. Byrnes, and G. R. Clements. 2013. Occurrence of the Sunda colugo (*Galeopterus variegatus*) in the tropical forests of Singapore: a Bayesian approach. Mammalian Biology 78:63–67.

Lim, T. L. 2008. Kowari *Dasyuroides byrnei* (Spencer, 1896). Family Dasyuridae, dasyurids. Pp. 52–54 in The mammals of Australia. 3rd edition (S. Van Dyck and R. Strahan, eds.). Reed New Holland, Sydney, Australia.

Lima, M., N. C. Stenseth, N. G. Yoccoz, and F. M. Jaksic. 2001. Demography and population dynamics of the mouse opossum (*Thylamys elegans*) in semi-arid Chile: seasonality, feedback structure and climate. Proceedings of the Royal Society of London B Biological Sciences 268:2053–2064.

Lima-Ribeiro, M. S., D. Nogués-Bravo, L. C. Terribile, P. Batra, and J. A. F. Diniz-Filho. 2013. Climate and humans set the place and time of Proboscidean extinction in late Quaternary of South America. Palaeogeography, Palaeoclimatology, Palaeoecology 392:546–556.

Lindenmayer, D. B., and J. M. Dixon. 1992. An additional historical record of Leadbeater's possum, *Gymnobelideus leadbeateri* McCoy, prior to the 1961 rediscovery of the species. Victorian Naturalist 109:217–218.

Lindenmayer, D. B., J. Dubach, and K. L. Viggers. 2002. Geographic dimorphism in the mountain brushtail possum (*Trichosurus caninus*): the case for a new species. Australian Journal of Zoology 50:369–393.

Lindenmayer, D. B., and R. A. Meggs. 1996. Use of den trees by Leadbeater's possum (*Gymnobelideus leadbeateri*). Australian Journal of Zoology 44:625–638.

Lindenmayer, D. B., H. A. Nix, J. P. McMahon, M. F. Hutchinson, and M. T. Tanton. 1991. The conservation of Leadbeater's possum, *Gymnobelideus leadbeateri* (McCoy): a case study of the use of bioclimatic modelling. Journal of Biogeography 18:371–383.

Lindsay, W. K. 2011. Habitat use, diet choice, and nutritional status in female and male Amboseli elephants. Pp. 51–73 in The Amboseli elephants: a long-term perspective on a long-lived mammal (C. J. Moss, H. Croze, and P. C. Lee, eds.). University of Chicago Press, Chicago, Illinois.

Linscombe, G. 1994. U.S. fur harvest (1970–1992) and fur value (1974–1992) statistics by state and region. Louisiana Department of Wildlife and Fisheries, Baton Rouge, Louisiana.

Lira, P. K., and F. A. S. Fernandez. 2009. A comparison of trapping- and radiotelemetry-based estimates of home range of the neotropical opossum *Philander frenatus*. Mammalian Biology 74:1–8.

Lister, A., and P. Bahn. 2007. Mammoths: giants of the Ice Age. University of California Press, Berkeley, California.

Lister, A. M., and J. Blashford-Snell. 2000. Exceptional size and form of Asian elephants in western Nepal. Elephant 2(4)33–36.

Lister, A. M., and A. V. Sher. 2001. The origin and evolution of the woolly mammoth. Science 294:1094–1097.

Lister, A. M., A. V. Sher, H. van Essen, and G. Wei. 2005. The pattern and process of mammoth evolution in Eurasia. Quaternary International 126–128:49–64.

Lister, A. M., W. Dirks, A. Assaf, M. Chazan, P. Goldberg, Y. H. Applbaum, N. Greenbaum, and L. K. Horwitz. 2013. New fossil remains of *Elephas* from the southern Levant: implications for the evolutionary history of the Asian elephant. Palaeogeography, Palaeoclimatology, Palaeoecology 386:119–130.

Lloyd, S. 2001. Oestrous cycle and gestation length in the musky rat-kangaroo, *Hypsiprymnodon moschatus* (Potoroidae: Marsupialia). Australian Journal of Zoology 49:37–44.

Lobban, R. A., and V. de Liedekerke. 2000. Elephants in ancient Egypt and Nubia. Anthrozoös 13:232–244.

Lobert, B., and A. K. Lee. 1990. Reproduction and life history of *Isoodon obesulus* in Victorian heathland. Pp. 311–318 in Bandicoots and bilbies (J. H. Seebeck, P. R. Brown, R. L. Wallis, and C. M. Kemper, eds). Surrey Beatty & Sons, Chipping Norton, New South Wales, Australia.

Lobos, G., A. Charrier, G. Carrasco, and R.E. Palma. 2005. Presence of *Dromiciops gliroides* (Microbiotheria: Microbiotheriidae) in the deciduous forests of central Chile. Mammalian Biology 70:376–380.

Lohanan, R. 2002. The elephant situation in Thailand and a plea for cooperation. Pp. 231–238 in Giants on our hands. Proceedings of the International Workshop on the Domesticated Asian Elephant (I. Baker and M. Kashio, eds.). FAO Regional Office for Asia and the Pacific, Bangkok, Thailand.

Long, J., M. Archer, T. Flannery, and S. Hand. 2002. Prehistoric mammals of Australia and New Guinea, one hundred million years of evolution. Johns Hopkins University Press, Baltimore, Maryland.

Long, J. L. 2003. Introduced mammals of the world, their history, distribution and influence. CSIRO Publishing, Collingwood, Victoria, Australia.

Lopes, G. P., and N. O. Leiner. 2015. Semelparity in a population of *Gracilinanus agilis* (Didelphimorphia: Didelphidae) inhabiting the Brazilian cerrado. Mammalian Biology 80:1–6.

López-Fuster, M. J., R. Pérez-Hernández, and J. Ventura. 2008. Morphometrics of genus *Caluromys* (Didelphimorphia: Didelphidae) in northern South America. ORSIS 23:97–114.

Loretto, D., E. Ramalho, and M. V. Vieira. 2005. Defense behavior and nest architecture of *Metachirus nudicaudatus* (Desmarest, 1817) (Marsupialia, Didelphidae). Mammalia 69:417–419.

Loss, S., L. P. Costa, and Y. L. R. Leite. 2011. Geographic variation, phylogeny and systematic status of *Gracilinanus microtarsus* (Mammalia: Didelphimorphia: Didelphidae). Zootaxa 2761:1–33.

Loughry, W. J., and C. M. McDonough. 2001. Natal recruitment and adult retention in a population of nine-banded armadillos. Acta Theriologica 46:393–406.

Loughry, W. J., and C. M. McDonough. 2013. The nine-banded armadillo. A natural history. University of Oklahoma Press, Norman, Oklahoma.

Loughry, W. J., C. Perez-Heydrich, C. M. McDonough, and M. K. Oli. 2013. Population dynamics and range expansion in nine-banded armadillos. PLoS ONE 8(7):e68311. doi:10.1371/journal.pone.0068311.

Louwman, J. W. W. 1973. Breeding the tailless tenrec *Tenrec ecaudatus* at Wassenaar Zoo. International Zoo Yearbook 11:125–126.

Louys, J., D. Curnoe, and H. Tong. 2007. Characteristics of Pleistocene megafauna extinctions in Southeast Asia. Palaeogeography, Palaeoclimatology, Palaeoecology 243:152–173.

Lovegrove, B. G., M. J. Lawes, and L. Roxburgh. 1999. Confirmation of pleisiomorphic daily torpor in mammals: the round-eared elephant shrew *Macroscelides proboscideus* (Macroscelidea). Journal of Comparative Physiology B 169:453–460.

Lovegrove, B. G., J. Raman, and M. R. Perrin. 2001. Daily torpor in elephant shrews (Macroscelidea: *Elephantulus* spp.) in response to food deprivation. Journal of Comparative Physiology B 171:11–21.

Loveridge A. J., J. E. Hunt, F. Murindagomo, and D. W. Macdonald. 2006. Influence of drought on predation of elephant (*Loxodonta africana*) calves by lions (*Panthera leo*) in an African wooded savannah. Journal of Zoology 270:523–530.

Lowery, G. H., Jr. 1974. The mammals of Louisiana and its adjacent waters. Louisiana State University Press, Baton Rouge, Louisiana.

Luaces, J. P., L. F. Rossi, V. Merico, M. Zuccotti, C. A. Redi, A. J. Solari, M. S. Merani, and S. Garagna. 2013. Spermatogenesis is seasonal in the large hairy armadillo, *Chaetophractus villosus* (Dasypodidae, Xenarthra, Mammalia). Reproduction, Fertility and Development 25:547–557.

Lucas, S. G., and Z. Luo. 1993. *Adelobasileus* from the upper Triassic of west Texas: the oldest mammal. Journal of Vertebrate Paleontology 13:309–334.

Luckett, W. P. 1980. The suggested evolutionary relationships and classification of tree shrews. Pp. 3–31 in Comparative biology and evolutionary relationships of tree shrews (W. P. Luckett, ed.). Plenum Press, New York, New York.

Luna, F. O., R. P. Lima, J. P. Araújo, and J. Z. O. Passavante. 2008. Conservation status of the Antillean manatee (*Trichechus manatus manatus* Linnaeus, 1758) in Brazil. Revista Brasileira de Zoociencias 10:145–153.

Lundie-Jenkins, G., L. K. Corbett, and C. M. Phillips. 1993. Ecology of the rufous hare-wallaby, *Lagorchestes hirsutus* Gould (Marsupialia: Macropodidae), in the Tanami Desert, Northern Territory. III. Interactions with introduced mammal species. Wildlife Research 20:495–511.

Lundie-Jenkins, G. and A. Payne. 2000. Recovery plan for the Julia Creek dunnart (*Sminthopsis douglasi*) 2000–2004. Queensland Parks and Wildlife Service, Brisbane, Australia.

Lunney, D. 2010. A history of the debate (1948–2009) on the commercial harvesting of kangaroos, with particular reference to New South Wales and the role of Gordon Grigg. Australian Zoologist 35:383–430.

Lunney, D., B. Law, and C. Rummery. 1997. An ecological interpretation of the historical decline of the brush-tailed rock-wallaby *Petrogale penicillata* in New South Wales. Australian Mammalogy 19:281–296.

Lunney, D., and N. McKenzie. 2008. *Macropus parma*. The IUCN Red List of Threatened Species 2008: e.T12627A3366970. http://dx.doi.org/10.2305/IUCN.UK.2008.RLTS.T12627A3366970.en. Downloaded on 05 January 2016.

Lunney, D., P. Menkhorst, and S. Burnett. 2008. *Sminthopsis leucopus*. The IUCN Red List of Threatened Species 2008: e.T20297A9183853. http://dx.doi.org/10.2305/IUCN.UK.2008.RLTS.T20297 A9183853.en. Downloaded on 04 January 2016.

Lunney, D., C. Dickman, P. Copley, T. Grant, S. Munks, F. Carrick, M. Serena, and M. Ellis. 2008a. *Ornithorhynchus anatinus*. The IUCN Red List of Threatened Species 2008: e.T40488A10317618. http://dx.doi.org/10.2305/IUCN.UK.2008.RLTS.T40488A10 317618.en. Downloaded on 03 January 2016.

Lunney, D., P. Menkhorst, J. Winter, M. Ellis, R. Strahan, M. Oakwood, S. Burnett, M. Denny, and R. Martin. 2008b. *Petauroides volans*. The IUCN Red List of Threatened Species 2008: e.T40579A10321730. http://dx.doi.org/10.2305/IUCN.UK.2008.RLTS.T40579A10 321730.en. Downloaded on 30 December 2015.

Lunney, D., E. Stalenberg, T. Santika, and J. R. Rhodes. 2014. Extinction in Eden: identifying the role of climate change in the decline of the koala in south-eastern NSW. Wildlife Research 41:22–34.

Luo, Z.-X. 2007. Transformation and diversification in early mammal evolution. Nature 450:1011–1019.

Luo, Z.-X., Q. Ji, J. R. Wible, and C.-X. Yuan. 2003. An early Cretaceous tribosphenic mammal and metatherian evolution. Science 302:1934–1940.

Luo, Z.-X., C.-X. Yuan, Q.-J. Meng, and Q. Ji. 2011. A Jurassic eutherian mammal and divergence of marsupials and placentals. Nature 476:442–445.

Ly, C. T. 2011. Current status of Asian elephants in Vietnam. Gajah 35:104–109.

Lyne, A. G. 1974. Gestation period and birth in the marsupial *Isoodon macrourus*. Australian Journal of Zoology 22:303–309.

Lyne, A. G. 1976. Observations on oestrus and the oestrous cycle in the marsupials *Isoodon macrourus* and *Perameles nasuta*. Australian Journal of Zoology 24:513–521.

Lyne, A. G. 1990. A brief review of bandicoot studies. Pp. xxiii–xxix in Bandicoots and bilbies (J. H. Seebeck, P. R. Brown, R. L. Wallis, and C. M. Kemper, eds). Surrey Beatty & Sons, Chipping Norton, New South Wales, Australia.

Lyne, A. G., and P. A. Mort. 1981. Comparison of skull morphology in the marsupial bandicoot genus *Isoodon*: its taxonomic implications and notes on a new species, *Isoodon arnhemensis*. Australian Mammalogy 4:107–133.

MacDonald, G. M., D. W. Beilman, Y. V. Kuzmin, L. A. Orlova, K. V. Kremenetski, B. Shapiro, R. K. Wayne, and B. Van Valkenburgh. 2012. Pattern of extinction of the woolly mammoth in Beringia. Nature Communications 3:893. doi:10.1038/ncomms1881. Downloaded on 02 February 2015.

MacFarlane, A. M., and G. Coulson. 2009. Boys will be boys: social affinity among males drives social segregation in western grey kangaroos. Journal of Zoology 277:37–44.

MacKinnon, K. 1984. Flying lemurs. Pp. 446–447 in The encyclopedia of mammals (D. Macdonald, ed.). Facts on File, New York, New York.

Mackowski, C. M. 1986. Distribution, habitat, and status of the yellow-bellied glider, *Petaurus australis* Shaw (Marsupialia: Petauridae), in northeastern New South Wales. Australian Mammalogy 9:141–144.

MacPhee, R. D. E. 1987a. The shrew tenrecs of Madagascar: systematic revision and Holocene distribution of *Microgale* (Tenrecidae, Insectivora). American Museum Novitates 2889:1–45.

MacPhee, R. D. E. 1987b. Systematic status of *Dasogale fontoynonti* (Tenrecidae, Insectivora). Journal of Mammalogy 68:133–135.

MacPhee, R. D. E. 1994. Morphology, adaptations, and relationships of *Plesiorycteropus*, and a diagnosis of a new order of eutherian mammals. Bulletin of the American Museum of Natural History 220:1–214.

MacPhee, R. D. E. 2005. 'First' appearances in the Cenozoic land-mammal record of the Greater Antilles: significance and comparison with South American and Antarctic records. Journal of Biogeography 32:551–564.

MacPhee, R. D. E. 2009. *Insulae infortunatae*: establishing a chronology for late Quaternary mammal extinctions in the West Indies. Pp. 169–193 in American megafaunal extinctions at the end of the Pleistocene (G. Haynes, ed.). Springer, New York, New York.

MacPhee, R. D. E., M. Cartmill, and K. D. Rose. 1989. Craniodental morphology and relationships of the supposed Eocene dermopteran *Plagiomene* (Mammalia). Journal of Vertebrate Paleontology 9:329–349.

MacPhee, R. D. E., and C. Flemming. 1999. *Requiem Æternam*. The last five hundred years of mammalian species extinctions. Pp. 333–371 in Extinctions in near time. Causes, contexts, and consequences (R. D. E. MacPhee, ed.). Kluwer Academic/Plenum, New York, New York.

MacPhee, R. D. E., M. A. Iturralde-Vinent, and O. Jiménez Vázquez. 2007. Prehistoric sloth extinctions in Cuba: implications of a new "last" appearance date. Caribbean Journal of Science 43:94–98.

MacPhee, R. D. E., and M. J. Novacek. 1993. Definition and relationships of Lipotyphla. Pp. 13–31 in Mammal phylogeny: Mesozoic differentiation, multituberculates, monotremes, early therians, and marsupials (F. S. Szalay, M. J. Novacek, and M. C. McKenna, eds.). Springer-Verlag, New York, New York.

MacPhee, R. D. E., R. Singer, and M. Diamond. 2000. Late Cenozoic land mammals from Grenada, Lesser Antilles Island-Arc. American Museum Novitates 3302:1–20.

MacPhee, R. D. E., J. L. White, and C. A. Woods. 2000. New megalonychid sloths (Phyllophaga, Xenarthra) from the Quaternary of Hispaniola. American Museum Novitates 3303:1–32.

Macqueen, P., J. M. Seddon, J. J. Austin, S. Hamilton, and A. W. Goldizen. 2010. Phylogenetics of the pademelons (Macropodidae: *Thylogale*) and historical biogeography of the Australo-Papuan region. Molecular Phylogenetics and Evolution 57:1134–1148.

Macqueen, P., A. W. Goldizen, J. J. Austin, and J. M. Seddon. 2011. Phylogeography of the pademelons (Marsupialia: Macropodidae: *Thylogale*) in New Guinea reflects both geological and climatic events during the Plio-Pleistocene. Journal of Biogeography 38:1732–1747.

Macrini, T. E. 2004. *Monodelphis domestica*. Mammalian Species 760:1–8.

Maddock, A. H. 1986. *Chrysospalax trevelyani*: an unknown and rare mammal endemic to southern Africa. Cimbebasia, ser. A, 8:88–90.

Madsen, O., P. M. Deen, G. Pesole, C. Saccone, and W. W. de Jong. 1997. Molecular evolution of mammalian aquaporin-2: further evidence that elephant shrew and aardvark join the paenungulate clade. Molecular Biology and Evolution 14:363–371.

Madsen, O., M. Scally, C. J. Douady, D. J. Kao, R. W. DeBryk, R. Adkins, H. M. Amrine, M. J. Stanhope, W. W. de Jong, and M. S. Springer. 2001. Parallel adaptive radiations in two major clades of placental mammals. Nature 409:610–614.

Mahoney, J. A. 1981. The specific name of the honey possum (Marsupialia: Tarsipedidae: *Tarsipes rostratus* Gervais and Verreaux, 1842). Australian Mammalogy 4:135–138.

Mahoney, J. A., and W. D. L. Ride. 1988. Dasyuridae. Pp. 14–33 in Zoological catalogue of Australia, volume 5, Mammalia (D. W. Walton, ed.). Australian Government Publishing Service, Canberra, Australia.

Main, A. R., and M. Yadav. 1971. Conservation of macropods in reserves in Western Australia. Biological Conservation 3:123–133.

Maisels, F., S. Strindberg, S. Blake, G. Wittemyer, J. Hart, E. A. Williamson, R. Aba, G. Abitsi, R. D. Ambahe, F. Amsini, P. C. Bakabana, T. C. Hicks, R. E. Bayogo, M. Bechem, R. L. Beyers, A. N. Bezangoye, P. Boundja, N. Bout, M. E. Akou, L. B. Bene, B. Fosso, E. Greengrass, F. Grossmann, C. Ikamba-Nkulu, O. Ilambu, B.-I. Inogwabini, F. Iyenguet, F. Kiminou, M. Kokangoye, D. Kujirakwinja, S. Latour, I. Liengola, Q. Mackaya, J. Madidi, B. Madzoke, C. Makoumbou, G.-A. Malanda, R. Malonga, O. Mbani, V. A. Mbendzo, E. Ambassa, A. Ekinde, Y. Mihindou, B. J. Morgan, P. Motsaba, G. Moukala, A. Mounguengui, B. S. Mowawa, C. Ndzai, S. Nixon, P. Nkumu, F. Nzolani, L. Pintea, A. Plumptre, H. Rainey, B. Bokoto de Semboli, A. Serckx, E. Stokes, A. Turkalo, H. Vanleeuwe, A. Vosper, and Y. Warren. 2013. Devastating decline of forest elephants in Central Africa. PLoS ONE 8(3):e59469. doi:10.1371/journal .pone.0059469.

Malekian, M., S. J. B. Cooper, and S. M. Carthew. 2010. Phylogeography of the Australian sugar glider (*Petaurus breviceps*): evidence for a new divergent lineage in eastern Australia. Australian Journal of Zoology 58:165–181.

Malekian, M., S. J. B. Cooper, J. A. Norman, L. Christidis, and S. M. Carthew. 2010. Molecular systematics and evolutionary origins of the genus *Petaurus* (Marsupialia: Petauridae) in Australia and New Guinea. Molecular Phylogenetics and Evolution 54:122–135.

Malia, M. J., Jr., R. M. Adkins, and M. W. Allard. 2002. Molecular support for Afrotheria and the polyphyly of Lipotyphla based on analyses of the growth hormone receptor gene. Molecular Phylogenetics and Evolution 24:91–101.

Mallick, S., M. M. Driessen, and G. J. Hocking. 1998a. Biology of the southern brown bandicoot (*Isoodon obesulus*) in south-eastern Tasmania. I. Diet. Australian Mammalogy 20:331–338.

Mallick, S., M. M. Driessen, and G. J. Hocking. 1998b. Biology of the southern brown bandicoot (*Isoodon obesulus*) in south-eastern Tasmania. II. Demography. Australian Mammalogy 20:339–347.

Mallick, S., M. M. Driessen, and G. J. Hocking. 2000. Demography and home range of the eastern barred bandicoot (*Perameles gunnii*) in south-eastern Tasmania. Wildlife Research 27:103–115.

Mallick, S., M. Haseler, G. J. Hocking, and M. M. Driessen. 1997. Past and present distribution of the eastern barred bandicoot (*Perameles gunnii*) in the Midlands, Tasmania. Pacific Conservation Biology 3:397–402.

Mallinson, J. J. C. 1974. Establishing mammal gestation at the Jersey Zoological Park. International Zoo Yearbook 14:184–187.

Maltby, M., and G. Bourchier. 2011. Current status of Asian elephants in Cambodia. Gajah 35:36–42.

Mansergh, I., and L. Broome. 1994. The mountain pygmy-possum of the Australian Alps. New South Wales University Press, Kensington, New South Wales, Australia.

Mansergh, I., and D. Scotts. 1990. Aspects of the life history and breeding of the mountain pygmy-possum, *Burramys parvus* (Marsupialia: Burramyidae), in alpine Victoria. Australian Mammalogy 13:179–191.

Maréchal, C., C. Maurois, and C. Chamberlan. 1998. Size (and structure) of forest elephant groups (*Loxodonta africana cyclotis* Matschie,1900) in the Odzala National Park, Republic of Congo. Mammalia 62:297–300.

Maree, S. 2015a. *Huetia leucorhina*. The IUCN Red List of Threatened Species 2015: e.T40597A21288887. http://dx.doi.org/10.2305/IUCN.UK.2015-2.RLTS.T40597A21288887.en. Downloaded on 20 January 2016.

Maree, S. 2015b. *Calcochloris tytonis*. The IUCN Red List of Threatened Species 2015: e.T4767A21285700. http://dx.doi.org/10.2305/IUCN.UK.2015-2.RLTS.T4767A21285700.en. Downloaded on 20 January 2016.

Maree, S. 2015c. *Eremitalpa granti*. The IUCN Red List of Threatened Species 2015: e.T7994A21283661. http://dx.doi.org/10.2305/IUCN.UK.2015-2.RLTS.T7994A21283661.en. Downloaded on 20 January 2016.

Maree, S. 2015d. *Calcochloris obtusirostris*. The IUCN Red List of Threatened Species 2015: e.T3519A21284422. http://dx.doi.org/10.2305/IUCN.UK.2015-2.RLTS.T3519A21284422.en. Downloaded on 20 January 2016.

Maree, S. 2015e. *Neamblysomus gunningi*. The IUCN Red List of Threatened Species 2015: e.T1087A21283546. http://dx.doi.org/10.2305/IUCN.UK.2015-2.RLTS.T1087A21283546.en. Downloaded on 20 January 2016.

Maree, S. 2015f. *Neamblysomus julianae*. The IUCN Red List of Threatened Species 2015: e.T1089A21285354. http://dx.doi.org/10.2305/IUCN.UK.2015-2.RLTS.T1089A21285354.en. Downloaded on 20 January 2016.

Maree, S. 2015g. *Neamblysomus julianae* (Bronberg Ridge subpopulation). The IUCN Red List of Threatened Species 2015: e.T62010A21284251. http://dx.doi.org/10.2305/IUCN.UK.2015-2.RLTS.T62010A21284251.en. Downloaded on 20 January 2016.

Maree, S., N. C. Bennett, and G. N. Bronner. 2005. The rough-haired golden mole *Chrysospalax villosus*. Afrotherian Conservation 3:2–3.

Margot, J. D. 2007. Molecular phylogeny of terrestrial artiodactyls: conflict and resolution. Pp. 4–18 in The evolution of artiodactyls (D. R. Prothero and S. E. Foss, eds.). Johns Hopkins University Press, Baltimore, Maryland.

Marine Mammal Commission. 1994. Annual report to Congress 1993. Marine Mammal Commission, Washington, DC.

Marivaux, L., L. Bocat, Y. Chaimanee, J.-J. Jaeger, B. Marandat, P. Srisuk, P. Tafforeau, C. Yamee, and J.-L. Welcomme. 2006. Cynocephalid dermopterans from the Palaeogene of South Asia (Thailand, Myanmar and Pakistan): systematic, evolutionary and palaeobiogeographic implications. Zoologica Scripta 35:395–420.

Marlow, B. J. 1961. Reproductive behavior of the marsupial mouse, *Antechinus flavipes* (Waterhouse) (Marsupialia), and the development of pouch young. Australian Journal of Zoology 9:203–218.

Marmontel, M. 1995. Age and reproduction in female Florida manatees. Pp. 98–119 in Population biology of the Florida manatee (T. J. O'Shea, B. B. Ackerman, and H. F. Percival, eds.). National Biological Service Information and Technology Report 1:1–289.

Marmontel, M. 2008. *Trichechus inunguis*. The IUCN Red List of Threatened Species 2008: e.T22102A9356406. http://dx.doi.org/10.2305/IUCN.UK.2008.RLTS.T22102A9356406.en. Downloaded on 29 April 2015.

Marmontel, M., D. K. Odell, and J. E. Reynolds, III. 1992. Reproductive biology of South American manatees. Pp. 295–312 in Reproductive biology of South American vertebrates (W. C. Hamlett, ed.). Springer-Verlag, New York, New York.

Marmontel, M., F. C. Weber Rosas, and S. Kendall. 2012. The Amazonian manatee. Pp. 47–53 in Sirenian conservation. Issues and strategies in developing countries (E. M. Hines, J. E. Reynolds, III, L. V. Aragones, A. A. Mignucci-Giannoni, and M. Marmontel, eds.). University Press of Florida, Gainesville, Florida.

Marples, T. G. 1973. Studies on the marsupial glider, *Schoinobates volans* (Kerr). IV. Feeding biology. Australian Journal of Zoology 21:213–216.

Marsh, H. 1995. The life history, pattern of breeding, and population dynamics of the dugong. Pp. 75–83 in Population biology of the Florida manatee (T. J. O'Shea, B. B. Ackerman, and H. F. Percival, eds.). National Biological Service Information and Technology Report 1:1–289.

Marsh, H., and P. Dutton. 2013. *Dugong dugon* dugong. Pp. 204–206 in Mammals of Africa, volume 1, Introductory chapters and Afrotheria (J. Kingdon, D. Happold, M. Hoffmann, T. Butynski, M. Happold, and J. Kalina, eds.). Bloomsbury, London, United Kingdom.

Marsh, H., G. E. Heinsohn, and L. M. Marsh. 1984. Breeding cycle, life history, and population dynamics of the dugong, *Dugong dugon* (Sirenia: Dugongidae). Australian Journal of Zoology 32:767–788.

Marsh, H., T. J. O'Shea, and J. E. Reynolds, III. 2012. Ecology and conservation of the Sirenia. Dugongs and manatees. Cambridge University Press, Cambridge, United Kingdom.

Marsh, H., and S. Sobtzick. 2015. *Dugong dugon*. The IUCN Red List of Threatened Species 2015: e.T6909A43792211. http://dx.doi.org/10.2305/IUCN.UK.2015-4.RLTS.T6909A43792211.en. Downloaded on 21 January 2016.

Marsh, H., G. B. Rathbun, T. J. O'Shea, and A. R. Preen. 1995. Can dugongs survive in Palau? Biological Conservation 72:85–89.

Marsh, H., H. Penrose, C. Eros, and J. Hughes. 2002. Dugong status report and action plans for countries and territories. Early Warning and Assessment Report Series, United Nations Environment Programme, Nairobi, Kenya.

Marsh, H., I. R. Lawler, D. Kwan, S. Delean, K. Pollock, and M. Alldredge. 2004. Aerial surveys and the potential biological removal technique indicate that the Torres Strait dugong fishery is unsustainable. Animal Conservation 7:435–443.

Marsh, H., G. De'ath, N. Gribble, and B. Lane. 2005. Historical marine population estimates: triggers or targets for conservation? The dugong case study. Ecological Applications 15:481–492.

Marshall, C. D., and J. F. Eisenberg. 1996. *Hemicentetes semispinosus*. Mammalian Species 541:1–4.

Marshall, C. D., H. Maeda, M. Iwata, M. Furuta, S. Asano, F. Rosas, and R. L. Reep. 2003. Orofacial morphology and feeding behaviour of the dugong, Amazonian, West African and Antillean manatees (Mammalia: Sirenia): functional morphology of the muscular-vibrissal complex. Journal of Zoology 259:245–260.

Marshall, L. G. 1977. *Lestodelphys halli*. Mammalian Species 81:1–3.

Marshall, L. G. 1978a. *Lutreolina crassicaudata*. Mammalian Species 91:1–4.

Marshall, L. G. 1978b. *Dromiciops australis*. Mammalian Species 99:1–5.

Marshall, L. G. 1978c. *Glironia venusta*. Mammalian Species 107:1–3.

Marshall, L. G. 1978d. *Chironectes minimus*. Mammalian Species 109:1–6.

Marshall, L. G. 1980. Systematics of the South American marsupial family Caenolestidae. Fieldiana: Zoology (new series) 5:1–145.

Marshall, L. G. 1982. Systematics of the South American marsupial family Microbiotheriidae. Fieldiana: Geology (new series) 10:1–75.

Marshall, L. G. 1984. Monotremes and marsupials. Pp. 59–116 in Orders and families of Recent mammals of the world (S. Anderson and J. K. Jones, Jr., eds.). John Wiley & Sons, New York, New York.

Marshall, L. G. 1987. Systematics of Itaboraian (middle Paleocene) age "opossum-like" marsupials from the limestone quarry at Sao Jose de Itaborai, Brazil. Pp. 91–160 in Possums and opossums (M. Archer, ed.). Surrey Beatty & Sons, Sydney, Australia.

Marshall, L. G., J. A. Case, and M. O. Woodburne. 1990. Phylogenetic relationships of the families of marsupials. Current Mammalogy 2:433–506.

Martin, E., and C. Martin. 2011. Large and mostly legitimate: Hong Kong's mammoth and elephant ivory trade. Pachyderm 50:37–49.

Martin, E. B., and L. Vigne. 1989. The decline and fall of India's ivory industry. Pachyderm 12:4–21.

Martin, G. M. 2009. On the identity of *Thylamys* (Marsupialia, Didelphidae) from the western pampas and south-central espinal, Argentina. Mastozoologia Neotropical 16:333–346.

Martin, G. M. 2010. Geographic distribution and historical occurrence of *Dromiciops gliroides* Thomas (Metatheria: Microbiotheria). Journal of Mammalogy 91:1025–1035.

Martin, G. M. 2011. Geographic distribution of *Rhyncholestes raphanurus* (Osgood, 1924) (Paucituberculata: Caenolestidae), an endemic marsupial of the Valdivian temperate rainforest. Australian Journal of Zoology 59:118–126.

Martin, G. M. 2013. Intraspecific variability in *Lestoros inca* (Paucituberculata, Caenolestidae), with reports on dental anomalies and eruption pattern. Journal of Mammalogy 94:601–617.

Martin, G.M. 2015. *Rhyncholestes raphanurus*. The IUCN Red List of Threatened Species 2015: e.T19710A22179691. http://dx.doi.org/10.2305/IUCN.UK.2015-4.RLTS.T19710A22179691.en. Downloaded on 18 January 2016.

Martin, G. M., L. J. M. De Santis, and G. J. Moreira. 2008. Southernmost record for a living marsupial. Mammalia 72:131–134.

Martin, G.M., D. Flores, and P. Teta. 2015a. *Lestodelphys halli*. The IUCN Red List of Threatened Species 2015: e.T11856A22175270. http://dx.doi.org/10.2305/IUCN.UK.2015-4.RLTS.T11856A22175270.en. Downloaded on 05 January 2016.

Martin, G. M., D. Flores, and P. Teta. 2015b. *Dromiciops gliroides*. The IUCN Red List of Threatened Species 2015: e.T6834A22180239. http://dx.doi.org/10.2305/IUCN.UK.2015-4.RLTS.T6834A22180239.en. Downloaded on 19 December 2015.

Martin, G. M., and D. E. Udrizar Sauthier. 2011. Observations on the captive behavior of the rare Patagonian opossum *Lestodelphys halli* (Thomas, 1921) (Marsupialia, Didelphimorphia, Didelphidae). Mammalia 75:281–286.

Martin, J. E., J. A. Case, J. W. M. Jagt, A. S. Schulp, and E. W. A. Mulder. 2005. A new European marsupial indicates a Late Cretaceous high-latitude transatlantic dispersal route. Journal of Mammalian Evolution 12:495–511.

Martin, J. K. 2008. Mountain brushtail possum *Trichosurus cunninghami* (Lindenmayer, Dubach, and Viggers, 2002). Pp. 272–274 in The mammals of Australia. 3rd edition (S. Van Dyck and R. Strahan, eds.). Reed New Holland, Sydney, Australia.

Martin, P. S. 1984. Prehistoric overkill: the global model. Pp. 354–403 in Quaternary extinctions: a prehistoric revolution (P. S. Martin and R. G. Klein, eds.). University of Arizona Press, Tucson, Arizona.

Martin, P. S. 2005. Twilight of the mammoths: Ice Age extinctions and the rewilding of America. University of California Press, Berkeley, California.

Martin, P. S., and J. E. Guilday. 1967. A bestiary for Pleistocene biologists. Pp. 1–62 in Pleistocene extinctions. The search for a cause (P. S. Martin and H. E. Wright, Jr., eds.). Yale University Press, New Haven, Connecticut.

Martin, P. S., and D. W. Steadman. 1999. Prehistoric extinctions on islands and continents. Pp. 17–55 in Extinctions in near time. Causes, contexts, and consequences (R. D. E. MacPhee, ed.). Kluwer Academic/Plenum, New York, New York.

Martin, R. D. 1968. Reproduction and ontogeny in tree-shrews (*Tupaia belangeri*), with reference to their general behaviour and taxonomic relationships. Zeitschrift für Tierpsychologie 25:409–495, 505–532.

Martin, R. D. 1984. Tree shrews. Pp. 441–445 in The encyclopedia of mammals (D. Macdonald, ed.). Facts on File, New York, New York.

Martin, R. W. 2005. Tree-kangaroos of Australia and New Guinea. CSIRO Publishing, Collingwood, Victoria, Australia.

Martin, R. W., and K. A. Handasyde. 1990. Population dynamics of the koala (*Phascolarctos cinereus*) in southeastern Australia. Pp. 75–84 in Biology of the koala (A. K. Lee, K. A. Handasyde, and G. D. Sanson, eds.). Surrey, Beatty & Sons, Sydney, Australia.

Martin, R. W., and K. A. Handasyde. 1999. The koala, natural history, conservation and management. 2nd edition. Krieger Publishing Company, Malabar, Florida.

Martin, R. W., K. A. Handasyde, and A. Krockenberger. 2008. Koala *Phascolarctos cinereus* (Goldfuss, 1817). Pp. 198–201 in The mammals of Australia. 3rd edition (S. Van Dyck and R. Strahan, eds.). Reed New Holland, Sydney, Australia.

Martin, R. W., and P. M. Johnson. 2008. Bennett's tree kangaroo *Dendrolagus bennettianus* (De Vis, 1887). Pp. 308–309 in The mammals of Australia. 3rd edition (S. Van Dyck and R. Strahan, eds.). Reed New Holland, Sydney, Australia.

Martin, R. W., and A. Lee. 1984. The koala, *Phascolarctos cinereus*, the largest marsupial folivore. Pp. 463–467 in Possums and gliders (A. Smith and I. Hume, eds.). Surrey Beatty & Sons, Norton, New South Wales, Australia.

Martinelli, A. G., M. L. F. Ferraz, and V. P. A. Teixeira. 2011. Range extension and first record of *Cryptonanus chacoensis* (Mammalia, Didelphimorphia, Didelphidae) in west Minas Gerais State, Brazil. Historia Natural (Corrientes) 1:113–118.

Martínez-Lanfranco, J. A., D. Flores, J. P. Jayat, and G. D'Elía. 2014. A new species of lutrine opossum, genus *Lutreolina* Thomas (Didelphidae), from the South American yungas. Journal of Mammalogy 95:225–240.

Martins, E. G., V. C. Bonato, C. Q. da Silva, and S. F. dos Reis. 2006. Partial semelparity in the neotropical didelphid marsupial *Gracilinanus microtarsus*. Journal of Mammalogy 87:915–920.

Maseko, B. C., M. A. Spocter, M. Haagensen, and P. R. Manger. 2012. Elephants have relatively the largest cerebellum size of mammals. Anatomical Record 295:661–672.

Masseti, M. 2001. Did endemic dwarf elephants survive on Mediterranean islands up to protohistorical times? Pp. 402–406 in Proceedings of the First International Congress of the World of Elephants (G. Cavarretta, P. Gioia, M. Mussi, and M. R. Palombo, eds.). Consiglio Nazionale delle Ricerche, Rome, Italy.

Masters, J. C., M. J. de Wit, and R. J. Asher. 2006. Reconciling the origins of Africa, India and Madagascar with vertebrate dispersal scenarios. Folia Primatologica 77:399–418.

Masters, P. 2003. Movement patterns and spatial organisation of the mulgara, *Dasycercus cristicauda* (Marsupialia: Dasyuridae), in central Australia. Wildlife Research 30:339–344.

Masters, P. 2008. Crest-tailed mulgara *Dasycercus critsticauda* (Krefft, 1867). Pp. 49–50 in The mammals of Australia. 3rd edition (S. Van Dyck and R. Strahan, eds.). Reed New Holland, Sydney, Australia.

Masters, P., and C. R. Dickman. 2012. Population dynamics of *Dasycercus blythi* (Marsupialia: Dasyuridae) in central Australia: how does the mulgara persist? Wildlife Research 39:419–428.

Matthee, C. A., G. Eick, S. Willows-Munro, C. Montgelard, A. T. Pardini, and T. J. Robinson. 2007. Indel evolution of mammalian introns and the utility of non-coding nuclear markers in eutherian phylogenetics. Molecular Phylogenetics and Evolution 42:827–837.

Mattioli, S., and D. P. Domning. 2006. An annotated list of extant skeletal material of Steller's sea cow (*Hydrodamalis gigas*) (Sirenia: Dugongidae) from the Commander Islands. Aquatic Mammals 32:273–288.

Maxwell, S., A. A. Burbidge, and K. D. Morris. 1996. The 1996 action plan for Australian marsupials and monotremes. Wildlife Australia, Department of the Environment, Water, Heritage and the Arts, Canberra, Australia.

Mayberry, C., S. K. Maloney, P. Mawson, and R. Bencini. 2010. Seasonal anoestrus in western grey kangaroos (*Macropus fuliginosus ocydromus*) in south-western Australia. Australian Mammalogy 32:189–196.

Maynes, G. M. 1973. Reproduction in the parma wallaby, *Macropus parma* Waterhouse. Australian Journal of Zoology 21:331–351.

Maynes, G. M. 1974. Occurrence and field recognition of *Macropus parma*. Australian Zoologist 18:72–87.

Maynes, G. M. 1977a. Breeding and age structure of the population of *Macropus parma* on Kawau Island, New Zealand. Australian Journal of Ecology 2:207–214.

Maynes, G. M. 1977b. Distribution and aspects of the biology of the parma wallaby, *Macropus parma*, in New South Wales. Australian Wildlife Research 4:109–125.

Maynes, G. M. 1982. A new species of rock wallaby, *Petrogale persephone* (Marsupialia: Macropodidae), from Proserpine, central Queensland. Australian Mammalogy 5:47–58.

Maynes, G. M. 1989. Zoogeography of the Macropodoidea. Pp. 47–66 in Kangaroos, wallabies and rat-kangaroos (G. C. Grigg, P. Jarman, and I. Hume, eds.). Surrey Beatty & Sons, Chipping Norton, New South Wales, Australia.

Maynes, G. M. 2008. Parma wallaby *Macropus parma* (Waterhouse, 1845). Pp. 341–343 in The mammals of Australia. 3rd edition (S. Van Dyck and R. Strahan, eds.). Reed New Holland, Sydney, Australia.

McAfee, R. K. 2016. Description of new postcranial elements of *Mylodon darwinii* Owen 1839 (Mammalia: Pilosa: Mylodontinae), and functional morphology of the forelimb. Ameghiniana 53:418–433.

McAllan, B. 2003. Timing of reproduction in carnivorous marsupials. Pp. 147–167 in Predators with pouches: the biology of carnivorous marsupials (M. Jones, C. Dickman, and M. Archer, eds.). CSIRO Publishing, Collingwood, Victoria, Australia.

McAllan, B. M. 2006. Dasyurid marsupials as models for the physiology of ageing in humans. Australian Journal of Zoology 54:159–172.

McAllan, B. M., C. R. Dickman, and M. S. Crowther. 2006. Photoperiod as a reproductive cue in the marsupial genus *Antechinus*: ecological and evolutionary consequences. Biological Journal of the Linnean Society 87:365–379.

McAllan, B. M., and F. Geiser. 2006. Photoperiod and the timing of reproduction in *Antechinus flavipes* (Dasyuridae: Marsupialia). Mammalian Biology 71:129–138.

McCain, C. M. 2001. First evidence of the giant anteater (*Myrmecophaga tridactyla*) in Honduras. Southwestern Naturalist 46:252–254.

McCallum, H., D. M. Tompkins, M. Jones, S. Lachish, S. Marvanek, B. Lazenby, G. Hocking, J. Wiersma, and C. E. Hawkins. 2007. Distribution and impacts of Tasmanian devil facial tumor disease. Ecohealth 4:318–325.

McCallum, H., M. Jones, C. Hawkins, R. Hamede, S. Lachish, D. L. Sinn, N. Beeton, and B. Lazenby. 2009. Transmission dynamics of Tasmanian devil facial tumor disease may lead to disease-induced extinction. Ecology 90:3379–3392.

McCarthy, M. A., and D. B. Lindenmayer. 1999. Conservation of the greater glider (*Petauroides volans*) in remnant native vegetation within exotic plantation forest. Animal Conservation 2:203–209.

McCarthy, T. J. 1982. *Chironectes*, *Cyclopes*, *Cabassous*, and probably *Cebus* in southern Belize. Mammalia 46:397–400.

McCarthy, T. J., D. L. Anderson, and G. A. Cruz D. 1999. Tree sloths (Mammalia: Xenarthra) in Nicaragua and Honduras, Central America. Southwestern Naturalist 44:410–414.

McComb, K., D. Reby, and C. J. Moss. 2011. Vocal communication and social knowledge in African elephants. Pp. 162–173 in The Amboseli elephants: a long-term perspective on a long-lived mammal (C. J. Moss, H. Croze, and P. C. Lee, eds.). University of Chicago Press, Chicago, Illinois.

McComb, K., C. Moss, S. Sayialel, and L. Baker. 2000. Unusually extensive networks of vocal recognition in African elephants. Animal Behaviour 59:1103–1109.

McCracken, H. E. 1990. Reproduction in the greater bilby, *Macrotis lagotis* (Reid)—a comparison with other perameloids. Pp. 199–204 in Bandicoots and bilbies (J. H. Seebeck, P. R. Brown, R. L. Wallis, and C. M. Kemper, eds.). Surrey Beatty & Sons, Chipping Norton, New South Wales, Australia.

McCrane, M. P. 1966. Birth, behaviour, and development of a hand-reared two-toed sloth. International Zoo Yearbook 6:153–163.

McCullough, D. R., and Y. McCullough. 2000. Kangaroos in outback Australia: comparative ecology and behavior of three coexisting species. Columbia University Press, New York, New York.

McDonald, H. G., and G. De Iuliis. 2008. Fossil history of sloths. Pp. 39–55 in The biology of the Xenarthra (S. F. Vizcaíno and W. J. Loughry, eds.). University Press of Florida, Gainesville, Florida.

McDonald, H. G., S. F. Vizcaíno, and S. Bargo. 2008. Skeletal anatomy and the fossil history of the Vermilingua. Pp. 64–78 in The biology of the Xenarthra (S. F. Vizcaíno and W. J. Loughry, eds.). University Press of Florida, Gainesville, Florida.

McDonough, C. M. 1994. Determinants of aggression in nine-banded armadillos. Journal of Mammalogy 75:189–198.

McDonough, C. M., J. M. Lockhart, and W. J. Loughry. 2007. Population dynamics of nine-banded armadillos: insights from a removal experiment. Southeastern Naturalist 6:381–392.

McDonough, C. M., and W. J. Loughry. 2008. Behavioral ecology of armadillos. Pp. 281–293 in The biology of the Xenarthra (S. F. Vizcaíno and W. J. Loughry, eds.). University Press of Florida, Gainesville, Florida.

McDowell, M. C., D. Haouchar, K. P. Aplin, M. Bunce, A. Baynes, and G. J. Prideaux. 2015. Morphological and molecular evidence supports specific recognition of the recently extinct *Bettongia anhydra* (Marsupialia: Macropodidae). Journal of Mammalogy 96:287–296.

McDowell, S. B., Jr. 1958. The Greater Antillean insectivores. Bulletin of the American Museum of Natural History 115:113–214.

McEvoy, J. S. 1970. Red-necked wallaby in Queensland. Queensland Division of Plant Industry Advisory Leaflet 1050:1–4.

McFarlane, D. A. 1999. Late Quaternary fossil mammals and last occurrence dates from caves at Barahona, Puerto Rico. Caribbean Journal of Science 35:238–248.

McGreevy, T. J., Jr., L. Dabek, and T. P. Husband. 2012. Tree kangaroo molecular systematics based on partial cytochrome b sequences: are Matschie's tree kangaroo (*Dendrolagus matschiei*) and Goodfellow's tree kangaroo (*D. goodfellowi buergersi*) sister taxa? Australian Mammalogy 34:18–28.

McGregor, F., and R. T. Wells. 1998. Population status of the southern hairy-nosed wombat *Lasiorhinus latifrons* in the Murraylands of South Australia. Pp. 218–227 in Wombats (R. T. Wells, P. A. Pridmore, B. St. John, M. D. Gaughwin, and J. Ferris, eds.). Surrey Beatty & Sons, Chipping Norton, New South Wales, Australia.

McIlroy, J. C. 2008. Common wombat *Vombatus ursinus* (Shaw, 1800). Pp. 206–208 in The mammals of Australia. 3rd edition (S. Van Dyck and R. Strahan, eds.). Reed New Holland, Sydney, Australia.

McKay, G. M. 1973. Behavior and ecology of the Asiatic elephant in southeastern Ceylon. Smithsonian Contributions to Zoology 125:1–113.

McKay, G. M. 1982. Nomenclature of the gliding possum genera *Petaurus* and *Petauroides* (Marsupialia: Petauridae). Australian Mammalogy 5:37–39.

McKay, G. M. 1988. Petauridae. Pp. 87–97 in Zoological catalogue of Australia, volume 5, Mammalia (D. W. Walton, ed.). Australian Government Publishing Service, Canberra, Australia.

McKay, G. M. 1989. Family Petauridae. Pp. 665–678 in Fauna of Australia: Mammalia, volum 1B (D. W. Walton and B. J. Richardson, eds.). Australian Government Publishing Service, Canberra, Australia.

McKay, G. M. 2008. Greater glider *Petauroides volans* (Kerr, 1792). Pp. 240–242 in The mammals of Australia. 3rd edition (S. Van Dyck and R. Strahan, eds.). Reed New Holland, Sydney, Australia.

McKay, G. M., and P. Ong. 2008. Common ringtail possum *Pseudocheirus peregrinus* (Boddaert, 1785). Pp. 255–257 in The mammals of Australia. 3rd edition (S. Van Dyck and R. Strahan, eds.). Reed New Holland, Sydney, Australia.

McKay, G. M., and J. W. Winter. 1989. Phalangeridae. Pp. 636–651 in Fauna of Australia: Mammalia, volume 1B (D. W. Walton and B. J. Richardson, eds.). Australian Government Publishing Service, Canberra, Australia.

McKenna, A. 2005. Husbandry and breeding of the striped possum *Dactylopsila trivirgata* at London ZSL. International Zoo Yearbook 39:169–176.

McKenna, M. C. 1975. Toward a phylogenetic classification of the Mammalia. Pp. 21–46 in Phylogeny of the primates: a multidisciplinary approach (W. P. Luckett and F. S. Szalay, eds.). Plenum Press, New York, New York.

McKenna, M. C., and S. K. Bell. 1997. Classification of mammals above the species level. Columbia University Press, New York, New York.

McKenzie, L. M., and D. W. Cooper. 1997. Hybridization between tammar wallaby (*Macropus eugenii*) populations from Western and South Australia. Journal of Heredity 88:398–400.

McKenzie, N. L., and M. Archer. 1982. *Sminthopsis youngsoni* (Marsupialia: Dasyuridae), the lesser hairy-footed dunnart, a new species from arid Australia. Australian Mammalogy 5:267–279.

McKenzie, N. L., and C. R. Dickman. 2008. Wongai ningaui *Ningaui ridei* (Archer, 1975). Pp. 117–118 in The mammals of Australia. 3rd edition (S. Van Dyck and R. Strahan, eds.). Reed New Holland, Sydney, Australia.

McKenzie, N., and C. Kemper. 2008. *Sminthopsis griseoventer*. The IUCN Red List of Threatened Species 2008: e.T41510A10482492. http://dx.doi.org/10.2305/IUCN.UK.2008.RLTS.T41510A10482492.en. Downloaded on 04 January 2016.

McKenzie, N. L., K. D. Morris, and C. R. Dickman. 2008. Golden bandicoot *Isoodon auratus* (Ramsay, 1887). Pp. 176–178 in The mammals of Australia. 3rd edition (S. Van Dyck and R. Strahan, eds.). Reed New Holland, Sydney, Australia.

McKnight, M. 2008a. *Sminthopsis butleri*. The IUCN Red List of Threatened Species 2008: e.T20295A9183520. http://dx.doi.org/10.2305/IUCN.UK.2008.RLTS.T20295A9183520.en. Downloaded on 04 January 2016.

McKnight, M. 2008b. *Dasyurus viverrinus*. The IUCN Red List of Threatened Species 2008: e.T6296A12600445. http://dx.doi.org/10.2305/IUCN.UK.2008.RLTS.T6296A12600445.en. Downloaded on 05 January 2016.

McKnight, M. 2008c. *Thylacinus cynocephalus*. The IUCN Red List of Threatened Species 2008: e.T21866A9332383. http://dx.doi.org/10.2305/IUCN.UK.2008.RLTS.T21866A9332383.en. Downloaded on 20 December 2015.

McKnight, M. 2008d. *Wyulda squamicaudata*. The IUCN Red List of Threatened Species 2008: e.T23091A9416462. http://dx.doi.org/10.2305/IUCN.UK.2008.RLTS.T23091A9416462.en. Downloaded on 27 November 2015.

McKnight, M. 2008e. *Potorous longipes*. The IUCN Red List of Threatened Species 2008: e.T18102A7658714. http://dx.doi.org/10.2305/IUCN.UK.2008.RLTS.T18102A7658714.en. Downloaded on 30 December 2015.

McKnight, M. 2008f. *Onychogalea fraenata*. The IUCN Red List of Threatened Species 2008: e.T15330A4514082. http://dx.doi.org/10.2305/IUCN.UK.2008.RLTS.T15330A4514082.en. Downloaded on 02 January 2016.

McKnight, M., P. Canty, R. Brandle, T. Robinson, and M. Watson. 2008. *Dasyuroides byrnei*. The IUCN Red List of Threatened Species 2008: e.T6265A12592863. http://dx.doi.org/10.2305/IUCN.UK.2008.RLTS.T6265A12592863.en. Downloaded on 05 January 2016.

McLean, I. G., E. Z. Cameron, W. L. Linklater, N. T. Schmitt, and K. S. M. Pulskamp. 2009. Partnerships in the social system of a small macropod marsupial, the quokka (*Setonix brachyurus*). Behaviour 146:89–112.

McLean, R. G., and S. R. Ubico. 1993. A first record of the water opossum (*Chironectes minimus*) from Guatemala. Southwestern Naturalist 38:402–404.

McManus, J. J. 1970. Behavior of captive opossums, *Didelphis marsupialis virginiana*. American Midland Naturalist 84:144–169.

McManus, J. J. 1974. *Didelphis virginiana*. Mammalian Species 40:1–6.

McNamara, J. A. 1997. Some smaller macropod fossils of South Australia. Proceedings of the Linnean Society of New South Wales 117:97–106.

McNee, A., and A. Cockburn. 1992. Specific identity is not correlated with behavioural and life-history diversity in *Antechinus stuartii* sensu lato. Australian Journal of Zoology 40:127–133.

McQuade, L. R. 1984. Taxonomic relationship of the greater glider *Petauroides volans* and lemur-like possum *Hemibelideus lemuroides*. Pp. 303–310 in Possums and gliders (A. Smith and I. Hume, eds.). Surrey Beatty & Sons, Norton, New South Wales, Australia.

Meaney, C. A., S. J. Bissell, and J. S. Slater. 1987. A nine-banded armadillo, *Dasypus novemcinctus* (Dasypodidae), in Colorado. Southwestern Naturalist 32:507–508

Medellín, R. A., G. Cancino Z., A. Clemente M., and R. O. Guerrero V. 1992. Noteworthy records of three mammals from Mexico. Southwestern Naturalist 37:427–428.

Medger, K., C. T. Chimimba, and N. C. Bennett. 2012. Seasonal reproduction in the eastern rock elephant-shrew: influenced by rainfall and ambient temperature? Journal of Zoology 288:283–293.

Medri, Í. M., and G. Mourão. 2007. Home range of giant anteaters (*Myrmecophaga tridactyla*) in the Pantanal wetland, Brazil. Journal of Zoology 266:365–375.

Medway, Lord. 1978. The wild mammals of Malaya (peninsular Malaysia) and Singapore. Oxford University Press, Kuala Lumpur, Malaysia.

Meester, J. 1972. A new golden mole from the Transvaal (Mammalia: Chrysochloridae). Annals of the Transvaal Museum 28:37–46.

Meester, J. 1977a. Family Chrysochloridae. Part 1.2 in The mammals of Africa: an identification manual (J. Meester and H. W. Setzer, eds.). Smithsonian Institution Press, Washington, DC.

Meester, J. 1977b. Order Tubulidentata. Part 10 in The mammals of Africa: an identification manual (J. Meester and H. W. Setzer, eds.). Smithsonian Institution Press, Washington, DC.

Meester, J., and H. W. Setzer. 1977. The mammals of Africa: an identification manual. Smithsonian Institution Press, Washington, DC.

Meester, J., I. L. Rautenbach, N. J. Dippenaar, and C. M. Baker. 1986. Classification of southern African mammals. Transvaal Museum Monograph 5:1–359.

Meijaard, E., and J. MacKinnon. 2008. *Tupaia chrysogaster*. The IUCN Red List of Threatened Species 2008: e.T22446A9373383. http://dx.doi.org/10.2305/IUCN.UK.2008.RLTS.T22446A9373383.en. Downloaded on 20 January 2016.

Melzer, A., F. Carrick, P. Menkhorst, D. Lunney, and B. St. John. 2000. Overview, critical assessment, and conservation implications of koala distribution and abundance. Conservation Biology 14:619–628.

Mendel, F. C. 1981. Use of hands and feet of two-toed sloths (*Choloepus hoffmanni*) during climbing and terrestrial locomotion. Journal of Mammalogy 62:413–21.

Mendel, F. C. 1985. Use of hands and feet of three-toed sloths (*Bradypus variegatus*) during climbing and terrestrial locomotion. Journal of Mammalogy 66:359–366.

Mendelssohn, H. 1965. Breeding the Syrian hyrax *Procavia capensis siriaca* (Schreber, 1784). International Zoo Yearbook 5:116–125.

Meng, J., and A. R. Wyss. 2005. Glires (Lagomorpha, Rodentia). Pp. 145–158 in The rise of placental mammals: origins and relationships of the major clades (K. D. Rose and J. D. Archibald, eds.). Johns Hopkins University Press, Baltimore, Maryland.

Menkhorst, P. 2008a. Hunted, marooned, re-introduced, contracepted: a history of koala management in Victoria. Pp. 73–92 in Too close for comfort: contentious issues in human-wildlife encounters (D. Lunney, A. Munn, and W. Meikle, eds.). Royal Zoological Society of New South Wales, Mosman, New South Wales, Australia.

Menkhorst, P. 2008b. *Gymnobelideus leadbeateri*. The IUCN Red List of Threatened Species 2008: e.T9564A13001448. http://dx.doi.org/10.2305/IUCN.UK.2008.RLTS.T9564A13001448.en. Downloaded on 01 December 2015.

Menkhorst, P. 2008c. *Bettongia gaimardi*. The IUCN Red List of Threatened Species 2008: e.T2783A9480319. http://dx.doi.org/10.2305/IUCN.UK.2008.RLTS.T2783A9480319.en. Downloaded on 01 December 2015.

Menkhorst, P., L. Broome, and M. Driessen. 2008. *Burramys parvus*. The IUCN Red List of Threatened Species 2008: http://dx.doi.org/10.2305/IUCN.UK.2008.RLTS.T3339A9775825.en. Downloaded on 07 December 2015.

Menkhorst, P., and M. Denny. 2008. *Thylogale billardierii*. The IUCN Red List of Threatened Species 2008: e.T40571A10335376. http://dx.doi.org/10.2305/IUCN.UK.2008.RLTS.T40571A10335376.en. Downloaded on 05 January 2016.

Menkhorst, P., and D. Lunney. 2008. *Potorous tridactylus*. The IUCN Red List of Threatened Species 2008: e.T41511A10482761. http://dx.doi.org/10.2305/IUCN.UK.2008.RLTS.T41511A10482761.en. Downloaded on 30 December 2015.

Menkhorst, P., S. Rhind, and M. Ellis. 2008. *Phascogale tapoatafa*. The IUCN Red List of Threatened Species 2008: e.T16890A6546705. http://dx.doi.org/10.2305/IUCN.UK.2008.RLTS.T16890A6546705.en. Downloaded on 04 January 2016.

Menkhorst, P., and J. Richards. 2008. *Perameles gunnii*. The IUCN Red List of Threatened Species 2008: e.T16572A6101370. http://dx.doi.org/10.2305/IUCN.UK.2008.RLTS.T16572A6101370.en. Downloaded on 06 January 2016.

Menkhorst, P. W., and J. H. Seebeck. 1990. Distribution and conservation status of bandicoots in Victoria. Pp. 51–60 in Bandicoots and bilbies (J. H. Seebeck, P. R. Brown, R. L. Wallis, and C. M. Kemper, eds.). Surrey Beatty & Sons, Chipping Norton, New South Wales, Australia.

Menkhorst, P. W., and J. H. Seebeck. 2008. Long-footed potoroo *Potorous longipes* (Seebeck and Johnston, 1980). Pp. 299–300 in The mammals of Australia. 3rd edition (S. Van Dyck and R. Strahan, eds.). Reed New Holland, Sydney, Australia.

Menkhorst, P., J. Winter, M. Ellis, M. Denny, S. Burnett, and D. Lunney. 2008. *Petaurus australis*. The IUCN Red List of Threatened Species 2008: e.T16730A6325601. http://dx.doi.org/10.2305/IUCN.UK.2008.RLTS.T16730A6325601.en. Downloaded on 30 December 2015.

Menzies, J. I. 1990. Notes on spiny bandicoots, *Echymipera* spp. (Marsupialia: Peramelidae) from New Guinea and description of a new species. Science in New Guinea 16:86–98.

Menzies, J. I., and J. C. Pernetta. 1986. A taxonomic revision of cuscuses allied to *Phalanger orientalis* (Marsupialia: Phalangeridae). Journal of Zoology series B 1:551–618.

Menzies, J. I., and R. K. Singadan. 2005. An improved diagnosis for *Dactylopsila megalura* (Marsupialia: Petauridae) of New Guinea. Australian Mammalogy 27:129–135.

Merchant, J. C. 1976. Breeding biology of the agile wallaby, *Macropus agilis* (Gould) (Marsupialia: Macropodidae), in captivity. Australian Wildlife Research 3:93–103.

Merchant, J. C. 1989. Lactation in macropodoid marsupials. Pp. 355–366 in Kangaroos, wallabies and rat-kangaroos (G. C. Grigg, P. Jarman, and I. Hume, eds.). Surrey Beatty & Sons, Chipping Norton, New South Wales, Australia.

Merchant, J. C. 2008a. Agile wallaby *Macropus agilis* (Gould, 1842). Pp. 323–324 in The mammals of Australia. 3rd edition (S. Van Dyck and R. Strahan, eds.). Reed New Holland, Sydney, Australia.

Merchant, J. C. 2008b. Swamp wallaby *Wallabia bicolor* (Desmarest, 1804). Pp. 404–406 in The mammals of Australia. 3rd edition (S. Van Dyck and R. Strahan, eds.). Reed New Holland, Sydney, Australia.

Meredith, R. W., M. Westerman, and M. S. Springer. 2008. A timescale and phylogeny for "bandicoots" (Peramelemorphia: Marsupialia) based on sequences for five nuclear genes. Molecular Phylogenetics and Evolution 47:1–20.

Meredith, R. W., M. Westerman, and M. S. Springer. 2009a. A phylogeny and timescale for the living genera of kangaroos and kin (Macropodiformes: Marsupialia) based on nuclear DNA sequences. Australian Journal of Zoology 56:395–410.

Meredith, R. W., M. Westerman, and M. S. Springer. 2009b. A phylogeny of Diprotodontia (Marsupialia) based on sequences for five nuclear genes. Molecular Phylogenetics and Evolution 51:554–571.

Meredith, R. W., M. Westerman, J. A. Case, and M. S. Springer. 2008. A phylogeny and timescale for marsupial evolution based on sequences for five nuclear genes. Journal of Mammalian Evolution 15:1–36.

Meredith, R. W., M. A. Mendoza, K. K. Roberts, M. Westerman, and M. S. Springer. 2010. A phylogeny and timescale for the evolution of Pseudocheiridae (Marsupialia: Diprotodontia) in Australia and New Guinea. Journal of Mammalian Evolution 17:75–99.

Meritt, D. A., Jr. 1975. The lesser anteater *Tamandua tetradactyla* in captivity. International Zoo Yearbook 15:41–44.

Meritt, D. A., Jr. 1976. The La Plata three-banded armadillo *Tolypeutes matacus* in captivity. International Zoo Yearbook 16:153–155.

Meritt, D. A., Jr. 1985a. Naked-tailed armadillos *Cabassous* sp. Pp. 389–391 in The evolution and ecology of armadillos, sloths, and vermilinguas (G. G. Montgomery, ed.). Smithsonian Institution Press, Washington, DC.

Meritt, D. A., Jr. 1985b. The two-toed Hoffman's sloth, *Choloepus hoffmanni* Peters. Pp. 333–341 in The evolution and ecology of armadillos, sloths, and vermilinguas (G. G. Montgomery, ed.). Smithsonian Institution Press, Washington, DC.

Meritt, D. A., Jr. 2008. Xenarthrans of the Paraguayan Chaco. Pp. 294–299 in The biology of the Xenarthra (S. F. Vizcaíno and W. J. Loughry, eds.). University Press of Florida, Gainesville, Florida.

Meritt, D. A., Jr., and G. F. Meritt. 1976. Sex ratios of Hoffmann's sloth, *Choloepus hoffmanni* Peters, and three-toed sloth, *Bradypus infuscatus* Wagler, in Panama. American Midland Naturalist 96:472–473.

Meritt, D. A., M. Superina, and A. M. Abba. 2014. *Cabassous chacoensis*. The IUCN Red List of Threatened Species 2014: e.T341 3A47437534. http://dx.doi.org/10.2305/IUCN.UK.2014-1.RLTS .T3413A47437534.en. Downloaded on 09 May 2015.

Merrett, P. K. 1983. Edentates. Zoological Trust of Guernsey, Guernsey, Channel Islands, United Kingdom.

Merz, G. 1986. Movement patterns and group size of the African forest elephant *Loxodonta africana cyclotis* in the Tai National Park, Ivory Coast. African Journal of Ecology 24:133–136.

Messer, M., A. S. Weiss, D. C. Shaw, and M. Westerman. 1998. Evolution of the monotremes: phylogenetic relationship to marsupials and eutherians, and estimation of divergence dates based on α-lactalbumin amino acid sequences. Journal of Mammalian Evolution 5:95–105.

Meyer, M., E. Palkopoulou, S. Baleka, M. Stiller, K. E. H. Penkman, K. W. Alt, Y. Ishida, D. Mania, S. Mallick, T. Meijer, H. Meller, S. Nagel, B. Nickel, S. Ostritz, N. Rohland, K. Schauer, T. Schüler, A. L. Roca, D. Reich, B. Shapiro, and M. Hofreiter. 2017. Palaeogenomes of Eurasian straight-tusked elephants challenge the current view of elephant evolution. eLife 2017 6:e25413. doi:10.7554/eLife.25413.

Miao, D. 1993. Cranial morphology and multituberculate relationships. Pp. 63–74 in Mammal phylogeny: Mesozoic differentiation, multituberculates, monotremes, early therians, and marsupials (F. S. Szalay, M. J. Novacek, and M. C. McKenna, eds.). Springer-Verlag, New York, New York.

Michelmore, F., K. Beardsley, R. F. W. Barnes, and I. Douglas-Hamilton. 1994. A model illustrating the changes in forest elephant numbers caused by poaching. African Journal of Ecology 32:89–99.

Mikkelsen, T. S., M. J. Wakefield, B. Aken, C. T. Amemiya, J. L. Chang, S. Duke, M. Garber, A. J. Gentles, L. Goodstadt, A. Heger, J. Jurka, M. Kamal, E. Mauceli, S. M. J. Searle, T. Sharpe, M. L. Baker, M. A. Batzer, P. V. Benos, K. Belov, M. Clamp, A. Cook, J. Cuff, R. Das, L. Davidow, J. E. Deakin, M. J. Fazzari, J. L. Glass, M. Grabherr, J. M. Greally, W. Gu, T. A. Hore, G. A. Huttley, M. Kleber, R. L. Jirtle, E. Koina, J. T. Lee, S. Mahony, M. A. Marra, R. D. Miller, R. D. Nicholls, M. Oda, A. T. Papenfuss, Z. E. Parra, D. D. Pollock, D. A. Ray, J. E. Schein, T. P. Speed, K. Thompson, J. L. VandeBerg, C. M. Wade, J. A. Walker, P. D. Waters, C. Webber, J. R. Weidman, X. Xie, M. C. Zody, Broad Institute Genome Sequencing Platform, Broad Institute Whole Genome Assembly Team, J. A. M. Graves, C. P. Ponting, M. Breen, P. B. Samollow, E. S. Lander, and K. Lindblad-Toh. 2007. Genome of the marsupial *Monodelphis domestica* reveals innovation in non-coding sequences. Nature 447:167–178.

Mikota, S. K., H. Hammatt, and Y. Fahrimal. 2008. Sumatran elephant crisis. Time for a change. Pp. 361–380 in Elephants and ethics, toward a morality of existence (C. Wemmer and C. A. Christen, eds.). Johns Hopkins University Press, Baltimore, Maryland.

Millar, R. P. 1971. Reproduction in the rock hyrax (*Procavia capensis*). Zoologica Africana 6:243–261.

Miller, G. S., Jr., and N. Hollister. 1922. A new phalanger from Celebes. Proceedings of the Biological Society of Washington 35:115–116.

Miller, S. D., J. Rottmann, K. J. Raedeke, and R. D. Taber. 1983. Endangered mammals of Chile: status and conservation. Biological Conservation 25:335–352.

Miller, W., D. I. Drautz, A. Ratan, B. Pusey, J. Qi, A. M. Lesk, L. P. Tomsho, M. D. Packard, F. Zhao, A. Sher, A. Tikhonov, B. Raney, N. Patterson, K. Lindblad-Toh, E. S. Lander, J. R. Knight, G. P. Irzyk, K. M. Fredrikson, T. T. Harkins, S. Sheridan, T. Pringle, and S. C. Schuster. 2008. Sequencing the nuclear genome of the extinct woolly mammoth. Nature 456:387–392.

Miller, W., D. I. Drautz, J. E. Janecka, A. M. Lesk, A. Ratan, L. P. Tomsho, M. Packard, Y. Zhang, L. R. McClellan, J. Qi, F. Zhao, M. T. P. Gilbert, L. Dalén, J. L. Arsuaga, P. G. P. Ericson, D. H. Huson, K. M. Helgen, W. J. Murphy, A. Göherström, and S. C. Schuster. 2009. The mitochondrial genome sequence of the Tasmanian tiger (*Thylacinus cynocephalus*). Genome Research 19:213–220.

Millis, A. L., and A. J. Bradley. 2001. Reproduction in the squirrel glider, *Petaurus norfolcensis* (Petauridae) in south-east Queensland. Australian Journal of Zoology 49:139–154.

Millis, A. L., D. A. Taggart, A. J. Bradley, J. Phelan, and P. D. Temple-Smith. 1999. Reproductive biology of the brush-tailed phascogale, *Phascogale tapoatafa* (Marsupialia: Dasyuridae). Journal of Zoology 248:325–335.

Mills, H. R., F. J. Bradshaw, C. Lambert, S. D. Bradshaw, and R. Bencini. 2012. Reproduction in the marsupial dibbler, *Parantechinus apicalis*; differences between island and mainland populations. General and Comparative Endocrinology 178:347–354.

Milner, J. M., and A. Gaylard. 2013. *Dendrohyrax arboreus* Southern tree hyrax (southern tree dassie). Pp. 152–155 in Mammals of Africa, volume 1, Introductory chapters and Afrotheria (J. Kingdon, D. Happold, M. Hoffmann, T. Butynski, M. Happold, and J. Kalina, eds.). Bloomsbury, London, United Kingdom.

Milner, J. M., and S. Harris. 1999. Habitat use and ranging behaviour of tree hyrax, *Dendrohyrax arboreus*, in the Virunga Volcanoes, Rwanda. African Journal of Ecology 37:281–294.

Milner-Gulland, E. J., and J. R. Beddington. 1993. The exploitation of elephants for the ivory trade: an historical perspective. Proceedings of the Royal Society of London B Biological Sciences 252:29–37.

Milner-Gulland, E. J., and R. Mace. 1991. The impact of the ivory trade on the African elephant *Loxodonta africana* population as assessed by data from the trade. Biological Conservation 55:215–229.

Minta, S. C., T. W. Clark, and P. Goldstraw. 1989. Population estimates and characteristics of the eastern barred bandicoot in Victoria, with recommendations for population monitoring. Pp. 47–75 in Management and conservation of small populations (T. W. Clark and J. H. Seebeck, eds.). Chicago Zoological Society, Chicago, Illinois.

Miranda, F., A. Bertassoni, and A. M. Abba. 2014. *Myrmecophaga tridactyla*. The IUCN Red List of Threatened Species 2014: e.T142 24A47441961. http://dx.doi.org/10.2305/IUCN.UK.2014-1.RLTS .T14224A47441961.en. Downloaded on 24 May 2015.

Miranda, F., and M. Superina. 2014. *Cyclopes didactylus* (Northeastern Brazil subpopulation). The IUCN Red List of Threatened Species 2014: e.T173393A47444393. http://dx.doi.org/10.2305/IUCN.UK .2014-1.RLTS.T173393A47444393.en. Downloaded on 21 May 2015.

Miranda, F., R. Veloso, M. Superina and F. J. Zara 2009. Food habits of wild silky anteaters (*Cyclopes didactylus*) of São Luis do Maranhão, Brazil. Edentata 8–10:1–5.

Miranda, F., D. A. Meritt, D. G. Tirira, and M. Arteaga. 2014a. *Cyclopes didactylus*. The IUCN Red List of Threatened Species 2014: e.T6019A47440020. http://dx.doi.org/10.2305/IUCN.UK.2014-1 .RLTS.T6019A47440020.en. Downloaded on 22 May 2015.

Miranda, F., N. Moraes-Barros, M. Superina, and A. M. Abba. 2014b. *Tolypeutes tricinctus*. The IUCN Red List of Threatened Species 2014: e.T21975A47443455. http://dx.doi.org/10.2305/IUCN.UK .2014-1.RLTS.T21975A47443455.en. Downloaded on 09 May 2015.

Mitchell, P. 1990a. The home ranges and social activity of koalas—a quantitative analysis. Pp. 171–187 in Biology of the koala (A. K. Lee, K. A. Handasyde, and G. D. Sanson, eds.). Surrey, Beatty & Sons, Sydney, Australia.

Mitchell, P. 1990b. Social behaviour and communication of koalas. Pp. 151–170 in Biology of the koala (A. K. Lee, K. A. Handasyde, and G. D. Sanson, eds.). Surrey, Beatty & Sons, Sydney, Australia.

Mitchell, P., and R. Martin. 1990. The structure and dynamics of koala populations—French Island in perspective. Pp. 97–108 in Biology of the koala (A. K. Lee, K. A. Handasyde, and G. D. Sanson, eds.). Surrey, Beatty & Sons, Sydney, Australia.

Mitrovski, P., A. A Hoffmann, D. A Heinze, and A. R Weeks. 2008. Rapid loss of genetic variation in an endangered possum. Biology Letters 4:134–138.

Moeller, E. 1975. Present-day edentates. Pp. 154–181 in Grzimek's animal life encyclopedia, volume 11 (B. Grzimek, ed.). Van Nostrand Reinhold, New York, New York.

Moeller, H. F. 1990a. Tasmanian wolf. Pp. 286–293 in Grzimek's encyclopedia of mammals, volume 1 (B. Grzimek, ed.). McGraw-Hill, New York, New York.

Moeller, H. F. 1990b. Tree kangaroos. Pp. 387–393 in Grzimek's encyclopedia of mammals, volume 1 (B. Grzimek, ed.). McGraw-Hill, New York, New York.

Moeller, H. F. 1990c. Modern xenarthrans. Pp. 583–626 in Grzimek's encyclopedia of mammals, volume 2 (B. Grzimek, ed.). McGraw-Hill, New York, New York.

Mol, D., A. Tikhonov, J. Van Der Plicht, R.-D. Kahlke, R. Debruyne, B. Van Geel, G. Van Reenen, J. P. Pals, C. De Marliave, and J. W. F. Reumer. 2006. Results of the CERPOLEX/*Mammuthus* Expeditions on the Taimyr Peninsula, arctic Siberia, Russian Federation. Quaternary International 142–143:186–202.

Möller-Krull, M., F. Delsuc, G. Churakov, C. Marker, M. Superina, J. Brosius, E. J. P. Douzery, and J. Schmitz. 2007. Retroposed elements and their flanking regions resolve the evolutionary history of xenarthran mammals (armadillos, anteaters, and sloths). Molecular Biology and Evolution 24:2573–2582.

Molsher, R. 2002. Kultarr (*Antechinomys laniger*) recovery plan. NSW National Parks and Wildlife Service, Hurstville, New South Wales, Australia.

Molur, S. 2008. *Anathana ellioti*. The IUCN Red List of Threatened Species 2008: e.T39593A10244083. http://dx.doi.org/10.2305 /IUCN.UK.2008.RLTS.T39593A10244083.en. Downloaded on 30 May 2015.

Montgomery, G. G. 1985a. Impact of vermilinguas (*Cyclopes, Tamandua*: Xenarthra = Edentata) on arboreal ant populations. Pp. 351–363 in The evolution and ecology of armadillos, sloths, and vermilinguas (G. G. Montgomery, ed.). Smithsonian Institution Press, Washington, DC.

Montgomery, G. G. 1985b. Movements, foraging, and food habits of the four extant species of neotropical vermilinguas (Mammalia; Myrmecophagidae). Pp. 365–377 in The evolution and ecology of armadillos, sloths, and vermilinguas (G. G. Montgomery, ed.). Smithsonian Institution Press, Washington, DC.

Montgomery, G. G., R. C. Best, and M. Yamakoshi. 1981. A radio-tracking study of the Amazonian manatee *Trichechus inunguis*. Biotrópica 13:81–85.

Montgomery, G. G., N. B. Gale, and W. P. Murdoch, Jr. 1982. Have manatee entered the eastern Pacific Ocean? Mammalia 46:257–258.

Montgomery, G. G., and Y. D. Lubin. 1977. Prey influences on movements of neotropical anteaters. Pp. 103–131 in Proceedings of the 1975 Predator Symposium (R. L. Phillips and C. Jonkel, eds.). Montana Forest and Conservation Experiment Station, University of Montana, Missoula, Montana.

Montgomery, G. G., and M. E. Sunquist. 1978. Habitat selection and use by two-toed and three-toed sloths. Pp. 329–359 in The ecology of arboreal folivores (G. G. Montgomery, ed.). Smithsonian Institution Press, Washington, DC.

Mooney, N., and D. E. Rounsevell. 2008. Thylacine *Thylacinus cynocephalus* (Harris, 1808). Pp. 167–168 in The mammals of Australia. 3rd edition (S. Van Dyck and R. Strahan, eds.). Reed New Holland, Sydney, Australia.

Moors, P. J. 1975. The urogenital system and notes on the reproductive biology of the female rufous rat-kangaroo, *Aepyprymnus rufescens* (Gray) (Macropodidae). Australian Journal of Zoology 23:355–361.

Moraes-Barros, N., A. Chiarello, and T. Plese. 2014. *Bradypus variegatus*. The IUCN Red List of Threatened Species 2014: e.T303 8A47437046. http://dx.doi.org/10.2305/IUCN.UK.2014-1.RLTS .T3038A47437046.en. Downloaded on 21 May 2015.

Moraes-Barros, N. D., J. A. B. Silva, and J. S. Morgante. 2011. Morphology, molecular phylogeny, and taxonomic inconsistencies in the study of *Bradypus* sloths (Pilosa: Bradypodidae). Journal of Mammalogy 92:86–100.

Moraes-Barros, N. D., A. P. Giorgi, S. Silva, and J. S. Morgante. 2010. Reevaluation of the geographical distribution of *Bradypus tridactylus* (Linnaeus, 1758) and *B. variegatus* (Schinz, 1825). Edentata 11:53–61.

Moraes Junior, E. A. 2004. Radio tracking of one *Metachirus nudicaudatus* (Desmarest, 1817) individual in Atlantic Forest of southeastern Brazil. Boletim do Museu de Biologia Mello Leitão 17:57–64.

Moraes Junior, E. A., and A. G. Chiarello. 2005. A radio tracking study of home range and movements of the marsupial *Micoureus demerarae* (Thomas) (Mammalia, Didelphidae) in the Atlantic Forest of south-eastern Brazil. Revista Brasileira de Zoologia 22:85–91.

Morcombe, M. K. 1967. The rediscovery after eighty-three years of the dibbler *Antechinus apicalis* (Marsupialia, Dasyuridae). Western Australian Naturalist 10:103–111.

Moreno, S., and T. Plese. 2006. The illegal traffic in sloths and threats to their survival in Colombia. Edentata 7:10–18.

Morford, S., and M. A. Meyers. 2003. Giant anteater (*Myrmecophaga tridactyla*) health care survey. Edentata 5:5–20.

Morgan, B. J. 2007. Group size, density and biomass of large mammals in the Réserve de Faune du Petit Loango, Gabon. African Journal of Ecology 45:508–518.

Morgan, B. J., and P. C. Lee. 2007. Forest elephant group composition, frugivory and coastal use in the Réserve de Faune du Petit Loango, Gabon. African Journal of Ecology 45:519–526.

Morgan, C. C., P. G. Foster, A. E. Webb, D. Pisani, J. O. McInerney, and M. J. O'Connell. 2013. Heterogeneous models place the root of the placental mammal phylogeny. Molecular Biology and Evolution 30:2145–2156.

Morgan, G. S., and C. A. Woods. 1986. Extinction and the zoogeography of West Indian mammals. Biological Journal of the Linnean Society 28:167–203.

Moritz, G. L., N. T.-L. Lim, M. Neitz, L. Peichl, N. J. Dominy. 2013. Expression and evolution of short wavelength sensitive opsins in colugos: a nocturnal lineage that informs debate on primate origins. Evolutionary Biology 40:542–553.

Morrant, D. S., and S. Petit. 2012. Strategies of a small nectarivorous marsupial, the western pygmy-possum, in response to seasonal variation in food availability. Journal of Mammalogy 93:1525–1535.

Morris, K., A. Burbidge, and T. Friend. 2008. *Pseudocheirus occidentalis*. The IUCN Red List of Threatened Species 2008: e.T18492A8336432. http://dx.doi.org/10.2305/IUCN.UK.2008.RLTS.T18492A8336432 .en. Downloaded on 30 December 2015.

Morris, K., A. Burbidge, and S. Hamilton. 2008. *Dasyurus geoffroii*. The IUCN Red List of Threatened Species 2008: e.T6294A12599937. http://dx.doi.org/10.2305/IUCN.UK.2008.RLTS.T6294A12599937 .en. Downloaded on 29 December 2015.

Morris, K. D., and P. Christensen. 2008. Western brush wallaby *Macropus irma* (Jourdan, 1837). Pp. 340–341 in The mammals of Australia. 3rd edition (S. Van Dyck and R. Strahan, eds.). Reed New Holland, Sydney, Australia.

Morris, K., T. Friend, and A. Burbidge. 2008. *Macropus irma*. The IUCN Red List of Threatened Species 2008: e.T12626A3366762. http://dx.doi.org/10.2305/IUCN.UK.2008.RLTS.T12626A3366762 .en. Downloaded on 05 January 2016.

Morris, K., B. Johnson, P. Orell, G. Gaikhorst, A. Wayne, and D. Moro. 2003. Recovery of the threatened chuditch (*Dasyurus geoffroii*): a case study. Pp. 435–451 in Predators with pouches: the biology of carnivorous marsupials (M. Jones, C. Dickman, and M. Archer, eds.). CSIRO Publishing, Collingwood, Victoria, Australia.

Morris, K., J. Woinarski, M. Ellis, T. Robinson, and P. Copley. 2008a. *Antechinomys laniger*. The IUCN Red List of Threatened Species 2008: e.T1581A5183739. http://dx.doi.org/10.2305/IUCN.UK .2008.RLTS.T1581A5183739.en. Downloaded on 04 January 2016.

Morris, K., T. Friend, A. Burbidge, and J. van Weenen. 2008b. *Macropus eugenii*. The IUCN Red List of Threatened Species 2008: e.T415 12A10483066. http://dx.doi.org/10.2305/IUCN.UK.2008.RLTS .T41512A10483066.en. Downloaded on 05 January 2016.

Morrow, G., N. A. Andersen, and S. C. Nicol. 2009. Reproductive strategies of the short-beaked echidna—a review with new data from a long-term study on the Tasmanian subspecies (*Tachyglossus aculeatus setosus*). Australian Journal of Zoology 57:275–282.

Morrow, G. and S. C. Nicol. 2009. Cool sex? Hibernation and reproduction overlap in the echidna. PLoS ONE 4(6):e6070. doi:10.1371/ journal.pone.0006070.

Morton, S. R. 1978a. An ecological study of *Sminthopsis crassicaudata* (Marsupialia: Dasyuridae). I. Distribution, study areas, and methods. Australian Wildlife Research 5:151–162.

Morton, S. R. 1978b. An ecological study of *Sminthopsis crassicaudata* (Marsupialia: Dasyuridae). II. Behaviour and social organization. Australian Wildlife Research 5:163–182.

Morton, S. R. 1978c. An ecological study of *Sminthopsis crassicaudata* (Marsupialia: Dasyuridae). III. Reproduction and life history. Australian Wildlife Research 5:183–211.

Morton, S. R. 1978d. Torpor and nest-sharing in free-living *Sminthopsis crassicaudata* (Marsupialia) and *Mus musculus* (Rodentia). Journal of Mammalogy 59:569–575.

Morton, S. R., and T. C. Burton. 1973. Observations on the behaviour of the macropodid marsupial *Thylogale billardieri* (Desmarest) in captivity. Australian Zoologist 18:1–14.

Morton, S. R., and C. R. Dickman. 2008a. Fat-tailed dunnart *Sminthopsis crassicaudata* (Gould, 1844). Pp. 150–152 in The mammals of Australia. 3rd edition (S. Van Dyck and R. Strahan, eds.). Reed New Holland, Sydney, Australia.

Morton, S. R., and C. R. Dickman. 2008b. Stripe-faced dunnart *Sminthopsis macroura* (Gould, 1845). Pp. 150–152 in The mammals of Australia. 3rd edition (S. Van Dyck and R. Strahan, eds.). Reed New Holland, Sydney, Australia.

Morton, S. R., C. R. Dickman, and T. P. Fletcher. 1989. Dasyuridae. Pp. 560–582 in Fauna of Australia: Mammalia, volume 1B (D. W. Walton and B. J. Richardson, eds.). Australian Government Publishing Service, Canberra, Australia.

Morton, S. R., and A. K. Lee. 1978. Thermoregulation and metabolism in *Planigale maculata* (Marsupialia: Dasyuridae). Journal of Thermal Biology 3:117–120.

Morton, S. R., J. W. Wainer, and T. P. Thwaites. 1980. Distributions and habitats of *Sminthopsis leucopus* and *S. murina* (Marsupialia: Dasyuridae) in southeastern Australia. Australian Mammalogy 3:19–30.

Morwood, M. J., R. P. Soejono, R. G. Roberts, T. Sutikna, C. S. M. Turney, K. E. Westaway, W. J. Rink, J.- X. Zhao, G. D. Van Den Bergh, R. A. Due, D. R. Hobbs, M. W. Moore, M. I. Bird, and L. K. Fifield. 2004. Archaeology and age of a new hominin from Flores in eastern Indonesia. Nature 431:1087–1091.

Moss, C. J. 1983. Oestrous behaviour and female choice in the African elephant. Behaviour 86:167–196.

Moss, C. J., H. Croze, and P. C. Lee. 2011. The Amboseli elephants: introduction. Pp. 1–7 in The Amboseli elephants: a long-term perspective on a long-lived mammal (C. J. Moss, H. Croze, and P. C. Lee, eds.). University of Chicago Press, Chicago, Illinois.

Moss, C. J., and P. C. Lee. 2011a. Female reproductive strategies: individual life histories. Pp. 187–204 in The Amboseli elephants: a long-term perspective on a long-lived mammal (C. J. Moss, H. Croze, and P. C. Lee, eds.). University of Chicago Press, Chicago, Illinois.

Moss, C. J., and P. C. Lee. 2011b. Female social dynamics: fidelity and flexibility. Pp. 205–223 in The Amboseli elephants: a long-term perspective on a long-lived mammal (C. J. Moss, H. Croze, and P. C. Lee, eds.). University of Chicago Press, Chicago, Illinois.

Mothé, D., L. S. Avilla, and M. A. Cozzuol. 2012. The South American gomphotheres (Mammalia, Proboscidea, Gomphotheriidae): taxonomy, phylogeny, and biogeography. Journal of Mammalian Evolution 20:23–32.

Mouchaty, S. K., A. Gullberg, A. Janke, and U. Arnason. 2000. Phylogenetic position of the tenrecs (Mammalia: Tenrecidae) of Madagascar based on analysis of the complete mitochondrial genome sequence of Echinops telfairi. Zoologica Scripta 29:307–317.

Mourão, G., and Í. M. Medri. 2007. Activity of a specialized insectivorous mammal (Myrmecophaga tridactyla) in the Pantanal of Brazil. Journal of Zoology 271:187–192.

Mou Sue, L. L., D. H. Chen, K. Bonde, and T. J. O'Shea. 1990. Distribution and status of manatees (Trichechus manatus) in Panama. Marine Mammal Science 6:234–241.

Moyal, A. 2004. Platypus, the extraordinary story of how a curious creature baffled the world. Johns Hopkins University Press, Baltimore, Maryland.

Muir, C. E., and J. J. Kiszka. 2007. Dugongs in eastern Africa: balancing on the brink. Sirenews 48:7–9.

Muir, C. E., and J. J. Kiszka. 2012. Eastern African dugongs. Pp. 84–90 in Sirenian conservation. Issues and strategies in developing countries (E. M. Hines, J. E. Reynolds, III, L. V. Aragones, A. A. Mignucci-Giannoni, and M. Marmontel, eds.). University Press of Florida, Gainesville, Florida.

Muldoon, K. M., D. D. De Blieux, E. L. Simons, and P. S. Chatrath. 2009. The subfossil occurrence and paleoecological significance of small mammals at Ankilitelo Cave, southwestern Madagascar. Journal of Mammalogy 90:1111–1131.

Mumby, H., A. Courtiol, K. U. Mar, and V. Lummaa. 2013. Birth seasonality and calf mortality in a large population of Asian elephants. Ecology and Evolution 3:3794–3803.

Munks, S. A., N. Mooney, D. Pemberton, and R. Gales. 2004. An update on the distribution and status of possums and gliders in Tasmania, including off-shore islands. Pp. 111–129 in The biology of Australian possums and gliders (R. L. Goldingay and S. M. Jackson, eds.). Surrey Beatty & Sons, Chipping Norton, New South Wales, Australia.

Munn, A. J., T. J. Dawson, S. R. McLeod, D. B. Croft, M. B. Thompson, and C. R. Dickman. 2009. Field metabolic rate and water turnover of red kangaroos and sheep in an arid rangeland: an empirically derived dry-sheep-equivalent for kangaroos. Australian Journal of Zoology 57:23–28.

Munny, P., P. Menkhorst, and J. Winter. 2008. Macropus giganteus. The IUCN Red List of Threatened Species 2008: e.T41513A10483378. http://dx.doi.org/10.2305/IUCN.UK.2008.RLTS.T41513A10483378 .en. Downloaded on 05 January 2016.

Munshi-South, J. 2008. Female-biased dispersal and gene flow in a behaviorally monogamous mammal, the large treeshrew (Tupaia tana). PLoS ONE 3(9):e3228. doi:10.1371/journal.pone.0003228.

Munshi-South, J. 2011. Relatedness and demography of African forest elephants: inferences from noninvasive fecal DNA analyses. Journal of Heredity 102:391–398.

Munshi-South, J., L. H. Emmons, and H. Bernard. 2007. Behavioral monogamy and fruit availability in the large treeshrew (Tupaia tana) in Sabah, Malaysia. Journal of Mammalogy 88:1427–1438.

Murata, Y., M. Nikaido, T. Sasaki, Y. Cao, Y. Fukumoto, M. Hasegawa, and N. Okada. 2003. Afrotherian phylogeny as inferred from complete mitochondrial genomes. Molecular Phylogenetics and Evolution 28:253–260.

Murata, Y., T. Yonezawa, I. Kihara, T. Kashiwamura, Y. Sugihara, M. Nikaido, N. Okada, H. Endo, and M. Hasegawa. 2009. Chronology of the extant African elephant species and case study of the species identification of the small African elephant with the molecular phylogenetic method. Gene 441:176–186.

Murphy, W. J., E. Eizirik, W. E. Johnson, Y. P. Zhang, O. A. Ryder, and S. J. O'Brien. 2001a. Molecular phylogenetics and the origins of placental mammals. Nature 409:614–618.

Murphy, W. J., E. Eizirik, S. J. O'Brien, O. Madsen, M. Scally, C. J. Douady, E. Teeling, O. A. Ryder, M. J. Stanhope, W. W. de Jong, and M. S. Springer. 2001b. Resolution of the early placental mammal radiation using bayesian phylogenetics. Science 294:2348–2351.

Murphy, W. J., T. H. Pringle, T. A. Crider, M. S. Springer, and W. Miller. 2007. Using genomic data to unravel the root of the placental mammal phylogeny. Genome Research 17:413–421.

Murray, P. F. 1998. Palaeontology and palaeobiology of wombats. Pp. 1–33 in Wombats (R. T. Wells, P. A. Pridmore, B. St. John, M. D. Gaughwin, and J. Ferris, eds.). Surrey Beatty & Sons, Chipping Norton, New South Wales, Australia.

Muschett, G. 2009. The manatees of the Panama Canal watershed. Ruford Small Grant for Nature Conservation Final Report, RSG 05.05.07.

Muschetto, E., G. R. Cueto, and O. V. Suarez. 2011. New data on the natural history and morphometrics of Lutreolina crassicaudata (Didelphimorphia) from central-eastern Argentina. Mastozoologia Neotropical 18:73–79.

Musser, A. M. 1998. Evolution, biogeography and palaeoecology of the Ornithorhynchidae. Australian Mammalogy 20:147–162.

Musser, A. M. 2003. Review of the monotreme fossil record and comparison of palaeontological and molecular data. Comparative Biochemistry and Physiology Part A 136:927–942.

Musser, A. M. 2006. Furry egg-layers: monotreme relationships and radiations. Pp. 523–550 in Evolution and biogeography of Australasian vertebrates (J. R. Merrick, M. Archer, G. M. Hickey, and M. S. Y. Lee, eds.). Auscipub, Oatlands, New South Wales, Australia.

Musser, G. G., and H. G. Sommer. 1992. Taxonomic notes on specimens of the marsupials Pseudocheirus schlegelii and P. forbesi (Diprotodontia, Pseudocheiridae) in the American Museum of Natural History. American Museum Novitates 3044:1–16.

Mustrangi, M. A., and J. L. Patton. 1997. Phylogeography and systematics of the slender mouse opossum Marmosops (Marsupialia, Didelphidae). University of California Publications in Zoology 130:1–86.

Mutinda, H., J. H. Poole, and C. J. Moss. 2011. Decision making and leadership in using the ecosystem. Pp. 246–259 in The Amboseli elephants: a long-term perspective on a long-lived mammal (C. J. Moss, H. Croze, and P. C. Lee, eds.). University of Chicago Press, Chicago, Illinois.

Myers, P., and J. L. Patton. 2007. Genus *Lestoros* (Oehser, 1934). Pp. 124–126 in Mammals of South America, volume 1, Marsupials, xenarthrans, shrews, and bats (A. L. Gardner, ed.). University of Chicago Press, Chicago, Illinois.

Myhrvold, C. L., H. A. Stone, and E. Bou-Zeid. 2012. What is the use of elephant hair? PLoS ONE 7(10):e47018. doi:10.1371/journal.pone.0047018.

Mzilikazi, N., and B. G. Lovegrove. 2004. Daily torpor in free-ranging rock elephant shrews, *Elephantulus myurus*: a year-long study. Physiological and Biochemical Zoology 77:285–296.

Nagy, K. A., and G. G. Montgomery. 2012. Field metabolic rate, water flux and food consumption by free-living silky anteaters (*Cyclopes didactylus*) in Panama. Edentata 13:61–65.

Nagy, K. A., R. S. Seymour, A. K. Lee, and R. Braithwaite. 1978. Energy and water budgets in free-living *Antechinus stuartii* (Marsupialia: Dasyuridae). Journal of Mammalogy 59:60–68.

Naples, V. L. 1999. Morphology, evolution and function of feeding in the giant anteater (*Myrmecophaga tridactyla*). Journal of Zoology 249:19–41.

Naughton-Treves, L., and W. Weber. 2001. Human dimensions of the African rain forest. Pp. 30–43 in African rain forest ecology and conservation (W. Weber, L. J. T. White, A. Vedder, and L. Naughton-Treves, eds.). Yale University Press, New Haven, Connecticut.

Navarrete, D., and J. Ortega. 2011. *Tamandua mexicana* (Pilosa: Myrmecophagidae). Mammalian Species 43(874):56–63.

Naylor, R., S. J. Richardson, and B. M. McAllan. 2008. Boom and bust: a review of the physiology of the marsupial genus *Antechinus*. Journal of Comparative Physiology 178:545–562.

Neal, B. R. 1995. The ecology and reproduction of the short-snouted elephant-shrew, *Elephantulus brachyrhynchus*, in Zimbabwe with a review of the reproductive ecology of the genus *Elephantulus*. Mammal Review 25:51–60.

Neaves, L. E., K. R. Zenger, D. W. Cooper, and M. D. B. Eldridge. 2010. Molecular detection of hybridization between sympatric kangaroo species in south-eastern Australia. Heredity 104:502–512.

Nellemann, C., R. K. Formo, J. Blanc, D. Skinner, T. Milliken, and T. De Meulenaer (eds.). 2013. Elephants in the dust. The African elephant crisis. A rapid response assessment. Produced as an inter-agency collaboration between UNEP, CITES, IUCN, and TRAFFIC. http://www.unep.org/pdf/RRAivory_draft7.pdf. Downloaded on 16 March 2015.

Nelson, J. E., and A. Goldstone. 1986. Reproduction in *Peradorcas concinna* (Marsupialia: Macropodidae). Australian Wildlife Research 13:501–505.

Nespolo, R. F., C. Verdugo, P. A. Cortés, and L. D. Bacigalupe. 2010. Bioenergetics of torpor in the microbiotherid marsupial, monito del monte (*Dromiciops gliroides*): the role of temperature and food availability. Journal of Comparative Physiology B 180:767–773.

Newsome, A. E. 1971a. Competition between wildlife and domestic livestock. Australian Veterinary Journal 47:577–586.

Newsome, A. E. 1971b. The ecology of red kangaroos. Australian Zoologist 16:32–50.

Newsome, A. E. 1975. An ecological comparison of the two arid-zone kangaroos of Australia, and their anomalous prosperity since the introduction of ruminant stock to their environment. Quarterly Review of Biology 50:389–424.

Newsome, A. E., P. C. Catling, B. D. Cooke, and R. Smyth. 2001. Two ecological universes separated by the dingo barrier fence in semi-arid Australia: interactions between landscapes, herbivory and carnivory, with and without dingoes. Rangeland Journal 23:71–98.

Nicholls, D. G. 1971. Daily and seasonal movements of the quokka, *Setonix brachyurus* (Marsupialia), on Rottnest Island. Australian Journal of Zoology 19:215–226.

Nicol, S. 2003. Monotreme biology. Comparative Biochemistry and Physiology Part A 136:795–798.

Nicol, S. C. 2015. Family Tachyglossidae (echidnas). Pp. 34–56 in Handbook of the mammals of the world, volume 5, Monotremes and marsupials (D. E. Wilson and R. A. Mittermeier, eds.). Lynx Edicions, Barcelona, Spain.

Nicol, S., and N. A. Andersen. 2007. The life history of an egg-laying mammal, the echidna (*Tachyglossus aculeatus*). Ecoscience 14:275–285.

Nicol, S. C., and G. E. Morrow. 2012. Sex and seasonality: reproduction in the echidna (*Tachyglossus aculeatus*). Pp. 143–153 in Living in a seasonal world: thermoregulatory and metabolic adaptations (T. Ruf, C. Bieber, W. Arnold, and E. Millesi, eds.). Springer, New York, New York.

Nicol, S. C., C. Vanpé, J. Sprent, G. Morrow, and N. A. Andersen. 2011. Spatial ecology of a ubiquitous Australian anteater, the short-beaked echidna (*Tachyglossus aculeatus*). Journal of Mammalogy 92:101–110.

Nicoll, M. E. 1985. The biology of the giant otter shrew *Potamogale velox*. National Geographic Society Research Reports 21:331–337.

Nicoll, M. E. 2003. *Tenrec ecaudatus*, tenrec, *tandraka, trandraka*. Pp. 1283–1287 in The natural history of Madagascar (S. M. Goodman and J. P. Benstead, eds.). University of Chicago Press, Chicago, Illinois.

Nicoll, M. E., and G. B. Rathbun. 1990. African Insectivora and elephant-shrews: an action plan for their conservation. IUCN (World Conservation Union), Gland, Switzerland.

Nikolaev, S., J. I. Montoya-Burgos, E. H. Margulies, NISC Comparative Sequencing Program, J. Rougemont, B. Nyffeler, and S. E. Antonarakis. 2007. Early history of mammals is elucidated with the ENCODE multiple species sequencing data. PLoS Genet 3(1):e2. https://doi.org/10.1371/journal.pgen.0030002.

Nikolskiy, P. A., L. D. Sulerzhitsky, and V. V. Pitulko. 2011. Last straw versus blitzkrieg overkill: climate-driven changes in the Arctic Siberian mammoth population and the late Pleistocene extinction problem. Quaternary Science Reviews 30:2309–2328.

Nilsson, M. A., A. Gullberg, A. E. Spotorno, U. Arnason, and A. Janke. 2003. Radiation of extant marsupials after the K/T boundary: evidence from complete mitochondrial genomes. Journal of Molecular Evolution 57:S3–S12.

Nilsson, M. A., U. Arnason, P. B. Spencer, and A. Janke. 2004. Marsupial relationships and a timeline for marsupial radiation in South Gondwana. Gene 340:189–196.

Nilsson, M. A., G. Churakov, M. Sommer, N. V. Tran, A. Zemann, J. Brosius, and J. Schmitz. 2010. Tracking marsupial evolution using archaic genomic retroposon insertions. PLOS Biol 8(7):e1000436. doi:10.1371/journal.pbio.1000436.

Nishihara, H., S. Maruyama, and N. Okada. 2009. Retroposon analysis and recent geological data suggest near-simultaneous divergence of the three superorders of mammals. Proceedings of the National Academy of Sciences of the United States of America 106:5235–5240.

Nishihara, H., N. Okada, and M. Hasegawa. 2007. Rooting the eutherian tree: the power and pitfalls of phylogenomics. Genome Biology 8(R199).

Nishihara, H., Y. Satta, M. Nikaido, J. G. M. Thewissen, M. J. Stanhope, and N. Okada. 2005. A retroposon analysis of afrotherian phylogeny. Molecular Biology and Evolution 22:1823–1833.

Nishihara, T. 2012. Demand for forest elephant ivory in Japan. Pachyderm 52:55–65.

Nishiwaki, M. 1984. Current status of the African manatee. Acta Zoologica Fennica 172:135–136.

Nishiwaki, M., and H. Marsh. 1985. Dugong—*Dugong dugon*. Pp. 1–31 in Handbook of marine mammals, volume 3, The sirenians and baleen whales (S. H. Ridgway and R. Harrison, eds.). Academic Press, London, United Kingdom.

Nishiwaki, M., T. Kasuya, N. Miyazoki, T. Tobayama, and T. Kataoka. 1979. Present distribution of the dugong in the world. Scientific Reports of the Whales Research Institute 31:133–141.

Nishiwaki, M., M. Yamaguchi, S. Shokita, S. Uchida, and T. Kataoka. 1982. Recent survey on the distribution of the African manatee. Scientific Reports of the Whales Research Institute 34:137–147.

Nogueira, J. C., A. C. S. Castro, E. V. C. Câmara, and B. G. O. Câmara. 2004. Morphology of the male genital system of *Chironectes minimus* and comparison to other didelphid marsupials. Journal of Mammalogy 85:834–841.

Nogués-Bravo, D., J. Rodríguez, J. Hortal, P. Batra, and M. B. Araújo. 2008. Climate change, humans, and the extinction of the woolly mammoth. PLoS Biology 6(4):e79.

Normande, I. C., F. de Oliveira Luna, A. C. Mendes Malhado, J. C. Gomes Borges, P. C. Viana Junior, F. L. Niemeyer Attademo, and R. J. Ladle. 2015. Eighteen years of Antillean manatee *Trichechus manatus manatus* releases in Brazil: lessons learnt. Oryx 49:338–344.

Norris, C. A. 1999. *Phalanger lullulae*. Mammalian Species 620:1–4.

Norris, C. A., and G. G. Musser. 2001. Systematic revision within the *Phalanger orientalis* complex (Diprotodontia, Phalangeridae): a third species of lowland gray cuscus from New Guinea and Australia. American Museum Novitates 3356:1–20.

Noss, A., R. Peña, and D. I. Rumiz. 2004. Camera trapping *Priodontes maximus* in the dry forests of Santa Cruz, Bolivia. Endangered Species Update 21(2):43–52.

Noss, A., M. Superina, and A. M. Abba. 2014. *Tolypeutes matacus*. The IUCN Red List of Threatened Species 2014: e.T21974A47443233. http://dx.doi.org/10.2305/IUCN.UK.2014-1.RLTS.T21974A47443233.en. Downloaded on 09 May 2015.

Notz, D., and J. Stroeve. 2016. Observed Arctic sea-ice loss directly follows anthropogenic CO2 emission. Science 354:747–750.

Novacek, M. J. 1986. The skull of leptictid insectivorans and the higher-level classification of eutherian mammals. Bulletin of the American Museum of Natural History 183:1–112.

Novacek, M. J. 1992. Mammalian phylogeny: shaking the tree. Nature 356:121–125.

Novacek, M. J. 1993. Reflections on higher mammalian phylogenetics. Journal of Mammalian Evolution 1:3–30.

Novak, M., M. E. Obbard, J. G. Jones, R. Newman, A. Booth, A. J. Satterthwaite, and G. Linscombe. 1987. Furbearer harvests in North America, 1600–1984. Ontario Ministry of Natural Resources, Toronto, Ontario.

Nowak, R. M. 1976. Wildlife of Indochina: tragedy or opportunity? National Parks and Conservation 50(6):13–18.

Nyári, A. S., A. T. Peterson, and G. B. Rathbun. 2010. Late Pleistocene potential distribution of the North African sengi or elephant-shrew *Elephantulus rozeti* (Mammalia:Macroscelidea). African Zoology 45:330–339.

Nyström, V., L. Dalén, S. Vartanyan, K. Lidén, N. Ryman, and A. Angerbjörn. 2010. Temporal genetic change in the last remaining population of woolly mammoth. Proceedings of the Royal Society B. doi:10.1098/rspb.2010.0301. Downloaded on 30 May 2012.

Nyström, V., J. Humphrey, P. Skoglund, N. J. McKeown, S. Vartanyan, P. W. Shaw, K. Lidén, M. Jakobsson, I. Barnes, A. Angerbjörn, A. Lister, and L. Dalén. 2012. Microsatellite genotyping reveals end-Pleistocene decline in mammoth autosomal genetic variation. Molecular Ecology 21:3391–3402.

Oakwood, M. 2000. Reproduction and demography of the northern quoll, *Dasyurus hallucatus*, in the lowland savanna of northern Australia. Australian Journal of Zoology 48:519–539.

Oakwood, M. 2002. Spatial and social organization of a carnivorous marsupial *Dasyurus hallucatus* (Marsupialia: Dasyuridae). Journal of Zoology 257:237–248.

Oakwood, M. 2008. Northern quoll *Dasyurus hallucatus* (Gould, 1842). Pp. 57–59 in The mammals of Australia. 3rd edition (S. Van Dyck and R. Strahan, eds.). Reed New Holland, Sydney, Australia.

Oakwood, M., J. Woinarski, and S. Burnett. 2008. *Dasyurus hallucatus*. The IUCN Red List of Threatened Species 2008: e.T6295A12600197. http://dx.doi.org/10.2305/IUCN.UK.2008.RLTS.T6295A12600197.en. Downloaded on 05 January 2016.

O'Brien, T. G., and M. F. Kinnaird. 1996. Changing populations of birds and mammals in north Sulawesi. Oryx 30:150–156.

O'Connell, M. A. 1983. *Marmosa robinsoni*. Mammalian Species 203:1–6.

O'Connell-Rodwell, C. E., B. T. Arnason, and L. A. Hart. 2000. Seismic properties of Asian elephant (*Elephas maximus*) vocalizations and locomotion. Journal of the Acoustical Society of America 108:3066–3072.

O'Connell-Rodwell, C. E., L. A. Hart, and B. T. Arnason. 2001. Exploring the potential use of seismic waves as a communication channel by elephants and other large mammals. American Zoologist 41:1157–1170.

O'Connell-Rodwell, C. E., T. Rodwell, M. Rice, and L. A. Hart. 2000. Living with the modern conservation paradigm: can agricultural communities co-exist with elephants? A five-year case study in East Caprivi, Namibia. Biological Conservation 93:381–391.

Odell, D. K. 2003. West Indian manatee *Trichechus manatus*. Pp. 855–864 in Wild mammals of North America: biology, management, and conservation (G. A. Feldhamer, B. C. Thompson, and J. A. Chapman, eds.). Johns Hopkins University Press, Baltimore, Maryland.

Odell, D. K., D. Forrester, and E. Asper. 1978. Growth and sexual maturation in the West Indian manatee. American Society of Mammalogists, Technical Paper Abstract, Annual Meeting 58:7–8.

Ojala-Barbour, R., C. M. Pinto, J. Brito M., L. Albuja V., T. E. Lee, Jr., and B. D. Patterson. 2013. A new species of shrew-opossum (Paucituberculata: Caenolestidae) with a phylogeny of extant caenolestids. Journal of Mammalogy 94:967–982.

Olbricht, G., C. Kern, and G. Vakhrusheva. 2006. Einige Aspekte der Fortpflanzungsbiologie von Kurzohr-Rüsselspringern (*Macroscelides proboscideus* A. Smith, 1829) in Zoologischen Gärten unter besonderer Berücksichtigung von Drillingswürfen. Zoologische Garten 75:304–316.

Olds, N., and J. Shoshani. 1982. *Procavia capensis*. Mammalian Species 171:1–7.

O'Leary, M. A., J. I. Bloch, J. J. Flynn, T. J. Gaudin, A. Giallombardo, N. P. Giannini, S. L. Goldberg, B. P. Kraatz, Z.-X. Luo, J. Meng, X. Ni, M. J. Novacek, F. A. Perini, Z. S. Randall, G. W. Rougier, E. J.

Sargis, M. T. Silcox, N. B. Simmons, M. Spaulding, P. M. Velazco, M. Weksler, J. R. Wible, and A. L. Cirranello. 2013. The placental mammal ancestor and the post–K-Pg radiation of placentals. Science 339:662–667.

Oliveira, E. V., P. Villa Nova, F. J. Goin, and L. dos Santos Avilla. 2011. A new hyladelphine marsupial (Didelphimorphia, Didelphidae) from cave deposits of northern Brazil. Zootaxa 3041:51–62.

Oliver, W. L. R., and I. B. Santos. 1991. Threatened endemic mammals of the Atlantic Forest region of south-east Brazil. Wildlife Preservation Trust Special Scientific Report 4:1–126.

Olivier, R. 1978. Distribution and status of the Asian elephant. Oryx 14:380–424.

Olson, L. E., and S. M. Goodman. 2003. Phylogeny and biogeography of tenrecs. Pp. 1235–1242 in The natural history of Madagascar (S. M. Goodman and J. P. Benstead, eds.). University of Chicago Press, Chicago, Illinois.

Olson, L. E., S. M. Goodman, and A. D. Yoder. 2004. Illumination of cryptic species boundaries in long-tailed shrew tenrecs (Mammalia: Tenrecidae; *Microgale*), with new insights into geographic variation and distributional constraints. Biological Journal of the Linnean Society 83:1–22.

Olson, L. E., E. J. Sargis, and R. D. Martin. 2004. Phylogenetic relationships among treeshrews (Scandentia): a review and critique of the morphological evidence. Journal of Mammalian Evolution 11:49–71.

Olson, L. E., E. J. Sargis, and R. D. Martin. 2005. Intraordinal phylogenetics of treeshrews (Mammalia: Scandentia) based on evidence from the mitochondrial 12S rRNA gene. Molecular Phylogenetics and Evolution 35:656–673.

Olson, L. E., Z. Rakotomalala, K. B. P. Hildebrandt, H. C. Lanier, C. J. Raxworthy, and S. M. Goodman. 2009. Phylogeography of *Microgale brevicaudata* (Tenrecidae) and description of a new species from western Madagascar. Journal of Mammalogy 90:1095–1110.

Opiang, M. D. 2009. Home ranges, movement, and den use in long-beaked echidnas, *Zaglossus bartoni*, from Papua New Guinea. Journal of Mammalogy 90:340–346.

Orell, P., and K. Morris. 1994. Chuditch recovery plan. Western Australian Department of Conservation and Land Management, Como, Western Australia, Australia.

Oren, D. C. 1993. Did ground sloths survive in recent times in the Amazonian region? Goeldiana Zoologia 19:1–11.

Oren, D. C. 2001. Does the endangered xenarthran fauna of Amazonia include remnant ground sloths? Edentata 4:2–5.

Osborn, D. J., and I. Helmy. 1980. The contemporary land mammals of Egypt (including Sinai). Fieldiana: Zoology (new series) 5:1–579.

Osborn, D. J., and J. Osbornová. 1998. The mammals of ancient Egypt. Aris & Phillips Ltd, Warminster, United Kingdom.

Osborne, M. J., and L. Christidis. 2001. Molecular phylogenetics of Australo–Papuan possums and gliders (Family Petauridae). Molecular Phylogenetics and Evolution 20:211–224.

Osborne, M. J., and L. Christidis. 2002a. Systematics and biogeography of pygmy possums (Burramyidae: *Cercartetus*). Australian Journal of Zoology 50:25–37.

Osborne, M. J., and L. Christidis. 2002b. Molecular relationships of the cuscuses, brushtail and scaly-tailed possums (Phalangerinae). Australian Journal of Zoology 50:135–149.

Osborne, M. J., L. Christidis, and J. A. Norman. 2002. Molecular phylogenetics of the Diprotodontia (kangaroos, wombats, koala, possums, and allies). Molecular Phylogenetics and Evolution 25:219–228.

O'Shea, T. J., and L. B. Poché, Jr. 2006. Aspects of underwater sound communication in Florida manatees (*Trichechus manatus latirostris*). Journal of Mammalogy 87:1061–1071.

O'Shea, T. J., and R. L. Reep. 1990. Encephalization quotients and life-history traits in the Sirenia. Journal of Mammalogy 71:534–543.

O'Shea, T. J., and C. A. L. Salisbury. 1991. Belize—a last stronghold for manatees in the Caribbean. Oryx 25:156–164.

O'Shea, T. J., C. A. Beck, R. K. Bonde, H. I. Kochman, and D. K. Odell. 1985. An analysis of manatee mortality patterns in Florida, 1976–81. Journal of Wildlife Management 49:1–11.

Ottichilo, W. K. 1986. Age structure of elephants in Tsavo National Park, Kenya. African Journal of Ecology 24:69–75.

Owen, D. 2004. Tasmanian tiger: the tragic tale of how the world lost its most mysterious predator. Johns Hopkins University Press, Baltimore, Maryland.

Owen, D., and D. Pemberton. 2005. Tasmanian devil: a unique and threatened animal. Allen & Unwin, Crows Nest, New South Wales, Australia.

Owens, M. J., and D. Owens. 2009. Early age reproduction in female savanna elephants (*Loxodonta africana*) after severe poaching. African Journal of Ecology 47:214–222.

Oxenham, H. H., and M. R. Perrin. 2009. The spatial organization of the four-toed elephant-shrew (*Petrodromus tetradactylus*) in Tembe Elephant Park, KwaZulu-Natal, South Africa. African Zoology 44:171–180.

Ozawa, T., S. Hayashi, and V. M. Mikhelson. 1997. Phylogenetic position of mammoth and Steller's sea cow within Tethytheria demonstrated by mitochondrial DNA sequences. Journal of Molecular Evolution 44:406–413.

Pabody, C. M., R. H. Carmichael, L. Rice, and M. Ross. 2009. A new sighting network adds to 20 years of historical data on fringe West Indian manatee (*Trichechus manatus*) populations in Alabama waters. Gulf of Mexico Science 27:52–61.

Pacheco, V., S. Solari, and B. Patterson. 2008. *Marmosops juninensis*. The IUCN Red List of Threatened Species 2008: e.T136364A4281010. http://dx.doi.org/10.2305/IUCN.UK.2008.RLTS.T136364A4281010.en. Downloaded on 10 November 2015.

Paddle, R. N. 1993. Thylacines associated with the Royal Zoological Society of New South Wales. Australian Zoologist 29:97–101.

Paddle, R. 2008. The most photographed of thylacines: Mary Roberts' Tyenna male—including a response to Freeman (2005) and a farewell to Laird (1968). Australian Zoologist 34:459–470.

Paddle, R. 2012. The thylacine's last straw: epidemic disease in a recent mammalian extinction. Australian Zoologist 36:75–92.

Painter, J., C. Krajewski, and M. Westerman. 1995. Molecular phylogeny of the marsupial genus *Planigale* (Dasyuridae). Journal of Mammalogy 76:406–413.

Palkopoulou, E., L. Dalé, A. M. Lister, S. Vartanyan, M. Sablin, A. Sher, V. N. Edmark, M. D. Brandström, M. Germonpré, I. Barnes, and J. A. Thomas. 2013. Holarctic genetic structure and range dynamics in the woolly mammoth. Proceedings of the Royal Society B 280:20131910. http://dx.doi.org/10.1098/rspb.2013.1910. Downloaded on 30 March 2015.

Palma, R. E. 1997. *Thylamys elegans*. Mammalian Species 572:1–4.

Palma, R. E. 2003. Evolution of American marsupials and their phylogenetic relationships with Australian metatherians. Pp. 21–29 in Predators with pouches: the biology of carnivorous marsupials (M. Jones, C. Dickman, and M. Archer, eds.). CSIRO Publishing, Collingwood, Victoria, Australia.

Palma, R. E., and A. E. Spotorno. 1999. Molecular systematics of marsupials based on the rRNA 12S mitochondrial gene: the phylogeny of Didelphimorphia and of the living fossil microbiotheriid *Dromiciops gliroides* Thomas. Molecular Phylogenetics and Evolution 13:525–535.

Palma, R. E., E. Rivera-Milla, T. L. Yates, P. A. Marquet and A. P. Meynard. 2002. Phylogenetic and biogeographic relationships of the mouse opossum *Thylamys* (Didelphimorphia, Didelphidae) in southern South America. Molecular Phylogenetics and Evolution 25:245–253.

Palma, R. E., D. Boric-Bargetto, J. P. Jayat, D. A. Flores, H. Zeballos, V. Pacheco, R. A. Cancino, F. D. Alfaro, E. Rodríguez-Serrano, and U. F. J. Pardiñas. 2014. Molecular phylogenetics of mouse opossums: new findings on the phylogeny of *Thylamys* (Didelphimorphia, Didelphidae). Zoologica Scripta 43:217–234.

Palmer, C., R. Taylor, and A. Burbidge. 2003. Recovery plan for the golden bandicoot *Isoodon auratus* and golden-backed tree-rat *Mesembriomys macrurus* 2004–2009. Northern Territory Department of Infrastructure Planning and Environment, Darwin, Northern Territory, Australia.

Panchetti, F., M. Scalici, G. M. Carpaneto, and G. Gibertini. 2008. Shape and size variations in the cranium of elephant-shrews: a morphometric contribution to a phylogenetic debate. Zoomorphology 127:69–82.

Paplinska, J. Z., R. L. C. Moyle, P. D. M. Temple-Smith, and M. B. Renfree. 2006. Reproduction in female swamp wallabies, *Wallabia bicolor*. Reproduction, Fertility and Development 18:735-743.

Paplinska, J. Z., M. D. B. Eldridge, D. W. Cooper, P. D. M. Temple-Smith, and M. B. Renfree. 2009. Use of genetic methods to establish male-biased dispersal in a cryptic mammal, the swamp wallaby (*Wallabia bicolor*). Australian Journal of Zoology 57:65–72.

Pardini, A. T., P. C. M. O'Brien, B. Fu, R. K. Bonde, F. F. B. Elder, M. A. Ferguson-Smith, F. Yang, and T. J. Robinson. 2007. Chromosome painting among Proboscidea, Hyracoidea and Sirenia: support for Paenungulata (Afrotheria, Mammalia) but not Tethytheria. Proceedings of the Royal Society of London B Biological Sciences 274:1333–1340.

Parker, P. 1977. An ecological comparison of marsupial and placental patterns of reproduction. Pp. 273–286 in The biology of marsupials (B. Stonehouse and D. P. Gilmore, eds.). University Park Press, Baltimore, Maryland.

Parker, S. A. 1971. Notes on the small black wallaroo *Macropus bernardus* (Rothschild, 1904) of Arnhem Land. Victorian Naturalist 88:41–43.

Parrott, M. L., S. J. Ward, and D. A. Taggart. 2005. Multiple paternity and communal maternal care in the feathertail glider (*Acrobates pygmaeus*). Australian Journal of Zoology 53:79–85.

Partridge, J. 1967. A 3,300 year old thylacine (Marsupialia: Thylacinidae) from the Nullarbor Plain, Western Australia. Journal of the Royal Society of Western Australia 50:57–59.

Pascual, R., M. Archer, E. O. Jaureguizar, J. L. Prado, H. Godthelp, and S. J. Hand. 1992. The first non-Australian monotreme: an early Paleocene South American platypus (Monotremata, Ornithorhynchidae). Pp. 1–14 in Platypus and echidnas (M. L. Augee, ed.). Royal Zoological Society of New South Wales, Mosman, New South Wales, Australia.

Pascual, R., F. J. Goin, L. Balarino, and D. E. Udrizar Sauthier. 2002. New data on the Paleocene monotreme *Monotrematum sudamericanum*, and the convergent evolution of triangulate molars. Acta Palaeontologica Polonica 47: 487–492.

Pasenko, M. R., and B. W. Schubert. 2004. *Mammuthus jeffersonii* (Proboscidea, Mammalia) from northern Illinois. PaleoBios 24(3):19–24.

Pasitschniak-Arts, M., and L. Marinelli.1998. *Ornithorhynchus anatinus*. Mammalian Species 585:1–9.

Patterson, B. D. 2007a. Order Paucituberculata (Ameghino, 1894). Pp. 119–120 in Mammals of South America, volume 1, Marsupials, xenarthrans, shrews, and bats (A. L. Gardner, ed.). University of Chicago Press, Chicago, Illinois.

Patterson, B. D. 2007b. Genus *Rhyncholestes* (Osgood, 1924). Pp. 126–127 in Mammals of South America, volume 1, Marsupials, xenarthrans, shrews, and bats (A. L. Gardner, ed.). University of Chicago Press, Chicago, Illinois.

Patterson, B. D. 2015. Family Caenolestidae (shrew-opossums). Pp. 188–197 in Handbook of the mammals of the world, volume 5, Monotremes and marsupials (D. E. Wilson and R. A. Mittermeier, eds.). Lynx Edicions, Barcelona, Spain.

Patterson, B. D., and M. H. Gallardo. 1987. *Rhyncholestes raphanurus*. Mammalian Species 286:1–5.

Patterson, B., and M. Gomez-Laverde. 2008. *Caenolestes convelatus*. The IUCN Red List of Threatened Species 2008: e.T40522A10328655. http://dx.doi.org/10.2305/IUCN.UK.2008.RLTS.T40522A10328655 .en. Downloaded on 18 November 2015.

Patterson, B. D., P. L. Meserve, and B. K. Lang. 1990. Quantitative habitat associations of small mammals along an elevational transect in temperate rainforests of Chile. Journal of Mammalogy 71:620–633.

Patterson, B. D., and M. A. Rogers. 2007. Order Microbiotheria (Ameghino, 1889). Pp. 117–119 in Mammals of South America, volume 1, Marsupials, xenarthrans, shrews, and bats (A. L. Gardner, ed.). University of Chicago Press, Chicago, Illinois.

Patterson, B., and S. Solari. 2008. *Glironia venusta*. IUCN Red List of Threatened Species 2008: e.T9245A12971485. http://dx.doi.org/10 .2305/IUCN.UK.2008.RLTS.T9245A12971485.en. Downloaded on 01 November 2015.

Patton, J. L., and L. P. Costa. 2003. Molecular phylogeography and species limits in rainforest didelphid marsupials of South America. Pp. 63–81 in Predators with pouches: the biology of carnivorous marsupials (M. Jones, C. Dickman, and M. Archer, eds.). CSIRO Publishing, Collingwood, Victoria, Australia.

Patton, J. L., and M. N. F. Da Silva. 1997. Definition of species of pouched four-eyed opossums (Didelphidae, *Philander*). Journal of Mammalogy 78:90–102.

Patton, J. L., and M. N. F. Da Silva. 2007. Genus *Philander* (Brisson, 1762). Pp. 27–33 in Mammals of South America, volume 1, Marsupials, xenarthrans, shrews, and bats (A. L. Gardner, ed.). University of Chicago Press, Chicago, Illinois.

Paula Couto, C. 1979. Tratado de Paleomastozoologia. Academia Brasileira de Ciências, Rio de Janeiro, Brazil.

Pauli, J. N., and M. Z. Perry. 2012. Unexpected strong polygyny in the brown-throated three-toed sloth. PLoS ONE 7(12):e51389. doi:10.1371/journal.pone.0051389.

Paull, D. J. 2008a. Superfamily Perameloidea. Pp. 169–171, 172, 174, 176, and 191 in The mammals of Australia. 3rd edition (S. Van Dyck and R. Strahan, eds.). Reed New Holland, Sydney, Australia.

Paull, D. J. 2008b. Southern brown bandicoot *Isoodon obesulus* (Shaw, 1797). Pp. 180–182 in The mammals of Australia. 3rd edition (S. Van Dyck and R. Strahan, eds.). Reed New Holland, Sydney, Australia.

Pauza, M., J. Richley, S. Robinson, and S. Fearn. 2013. Surviving in the south: a recent incursion of the agile wallaby (*Macropus agilis*) in Tasmania. Australian Mammalogy 36:95–98.

Pavan, S. E., R. V. Rossi, and H. Schneider. 2012. Species diversity in the *Monodelphis brevicaudata* complex (Didelphimorphia: Didelphidae) inferred from molecular and morphological data, with the description of a new species. Zoological Journal of the Linnean Society 165:190–223.

Pavey, C. R., C. J. Burwell, and J. Benshemesh. 2012. Diet and prey selection of the southern marsupial mole: an enigma from Australia's sand deserts. Journal of Zoology 287:115–123.

Pavey, C. R., and F. Geiser. 2008. Basking and diurnal foraging in the dasyurid marsupial *Pseudantechinus macdonnellensis*. Australian Journal of Zoology 56:129–135.

Payne, O. 1995. Koalas out on a limb. National Geographic 187(4):36–59.

Peace Parks Foundation. 2014. KAZA TFCA univisa now in effect. Peace Parks Foundation, Stellenbosch, South Africa. http://www.peaceparks.org/tfca.php?pid=19&mid=1008. Accessed on 20 March 2015.

Peacock, D., and I. Abbott. 2014. When the "native cat" would "plague": historical hyperabundance in the quoll (Marsupialia: Dasyuridae) and an assessment of the role of disease, cats and foxes in its curtailment. Australian Journal of Zoology 62:294–344.

Pearson, D. J. 1992. Past and present distribution and abundance of the black-footed rock-wallaby in the Warburton region of Western Australia. Wildlife Research 19:605–622.

Pearson, D. J., and M. D. B. Eldridge. 2008. Rothschild's rock-wallaby *Petrogale rothschildi* (Thomas, 1904). Pp. 389–390 in The mammals of Australia. 3rd edition (S. Van Dyck and R. Strahan, eds.). Reed New Holland, Sydney, Australia.

Pearson, D. J., and J. E. Kinnear. 1997. A review of the distribution, status and conservation of rock-wallabies in Western Australia. Australian Mammalogy 19:137–152.

Pearson, D. J., and N. L. McKenzie. 2008. Hairy-footed dunnart *Sminthopsis hirtipes* (Thomas, 1898). Pp. 143–145 in The mammals of Australia. 3rd edition (S. Van Dyck and R. Strahan, eds.). Reed New Holland, Sydney, Australia.

Pearson, D. J., and A. C. Robinson. 1990. New records of the sandhill dunnart, *Sminthopsis psammophila* (Marsupialia: Dasyuridae) in South and Western Australia. Australian Mammalogy 13:57–59.

Pearson, D., and J. Turner. 2000. Marsupial moles pop up in the Great Victoria and Gibson Deserts. Australian Mammalogy 22:115–119.

Pearson, D. J., A. A. Burbidge, J. Lochman, and A. N. Start. 2008. Monjon *Petrogale burbidgei* (Kitchener and Sanson, 1978). Pp. 367–368 in The mammals of Australia. 3rd edition (S. Van Dyck and R. Strahan, eds.). Reed New Holland, Sydney, Australia.

Pearson, O. P. 1983. Characteristics of a mammalian fauna from forests in Patagonia, southern Argentina. Journal of Mammalogy 64:476–492.

Pearson, O. P. 2007. Genus *Lestodelphys* (Tate, 1934). Pp. 50–51 in Mammals of South America, volume 1, Marsupials, xenarthrans, shrews, and bats (A. L. Gardner, ed.). University of Chicago Press, Chicago, Illinois.

Peery, M. Z., and J. N. Pauli. 2012. The mating system of a "lazy" mammal, Hoffmann's two-toed sloth. Animal Behaviour 84:555–562.

Pemberton, D., and D. Renouf. 1993. A field study of communication and social behaviour of the Tasmanian devil at feeding sites. Australian Journal of Zoology 41:507–526.

Pemberton, D., S. Gales, B. Bauer, R. Gales, B. Lazenby and K. Medlock. 2008. The diet of the Tasmanian devil, *Sarcophilus harrisii*, as determined from analysis of scat and stomach contents. Papers and Proceedings of the Royal Society of Tasmania 142:13–21.

Peppler, R. D. 2008. Reproductive biology of the nine-banded armadillo. Pp. 151–159 in The biology of the Xenarthra (S. F. Vizcaíno and W. J. Loughry, eds.). University Press of Florida, Gainesville, Florida.

Peredo, B. 1999. Bolivia's trade in hairy armadillos. Traffic Bulletin 18(1):41–45.

Peres, C. A. 1999. The structure of nonvolant mammal communities in different Amazonian forest types. Pp. 564–581 in Mammals of the neotropics, the central neotropics, volume 3, Ecuador, Peru, Bolivia, Brazil (J. F. Eisenberg and K. H. Redford). University of Chicago Press, Chicago, Illinois.

Pérez-Hernandez, R., D. Lew, E. Gutiérrez, and J. Ventura. 2011. *Gracilinanus dryas*. The IUCN Red List of Threatened Species 2011: e.T9418A12985010. http://dx.doi.org/10.2305/IUCN.UK.2011-2.RLTS.T9418A12985010.en. Downloaded on 05 January 2016.

Perez Zubieta, J., A. M. Abba, and M. Superina. 2014. *Chaetophractus nationi*. The IUCN Red List of Threatened Species 2014: e.T4367A47438187. http://dx.doi.org/10.2305/IUCN.UK.2014-1.RLTS.T4367A47438187.en. Downloaded on 09 May 2015.

Perrers, C. 1965. Notes on a pigmy possum, *Cercartetus nanus* Desmarest. Australian Zoologist 13:126.

Perrin, M. R., and L. J. Fielden. 1999. *Eremitalpa granti*. Mammalian Species 629:1–4.

Perrin, M., and G. B. Rathbun. 2013a. Order Macroscelidea—sengis (elephant-shrews). Pp. 258–260 in Mammals of Africa, volume 1, Introductory chapters and Afrotheria (J. Kingdon, D. Happold, M. Hoffmann, T. Butynski, M. Happold, and J. Kalina, eds.). Bloomsbury, London, United Kingdom.

Perrin, M., and G. B. Rathbun. 2013b. *Elephantulus rozeti* North African sengi (North African elephant-shrew). Pp. 272–273 in Mammals of Africa, volume 1, Introductory chapters and Afrotheria (J. Kingdon, D. Happold, M. Hoffmann, T. Butynski, M. Happold, and J. Kalina, eds.). Bloomsbury, London, United Kingdom.

Pestell, A. J. L., S. J. B. Cooper, K. M. Saint, and S. Petit. 2008. Genetic structure of the western pygmy possum *Cercartetus concinnus* Gould (Marsupialia: Burramyidae) based on mitochondrial DNA. Australian Mammalogy 29:191–200.

Petter, F. 1981. Remarques sur la systématique des chrysochlorides. Mammalia 45:49–53.

Pettigrew, J. D. 1999. Electroreception in monotremes. Journal of Experimental Biology 202:1447–1454.

Pettigrew, J. D., P. R. Manger, and S. L. B. Fine. 1998. The sensory world of the platypus. Philosophical Transactions of the Royal Society of London B 353:1199–1210.

Pettigrew, J. D., and L. Wilkens. 2003. Paddlefish and platypus: parallel evolution of passive electroreception in a rostral bill organ. Pp. 420–433 in Sensory processing in aquatic environments (S. P. Collin and N. J. Marshall, eds.). Springer, New York, New York.

Phillips, B. 1990. Koalas: the little Australians we'd all hate to lose. Australian National Parks and Wildlife Service, Canberra, Australia.

Phillips, B. T., and S. M. Jackson. 2003. Growth and development of the Tasmanian devil (*Sarcophilus harrisii*) at Healesville Sanctuary, Victoria, Australia. Zoo Biology 22:497–505.

Phillips, C. J., and J. K. Jones, Jr. 1968. Additional comments on reproduction in the woolly opossum (*Caluromys derbianus*) in Nicaragua. Journal of Mammalogy 49:320–321.

Phillips, C. J., and J. K. Jones, Jr. 1969. Notes on reproduction and development in the four-eyed opossum, *Philander opossum*, in Nicaragua. Journal of Mammalogy 50:345–348.

Phillips, M. J., T. H. Bennett, and M. S. Y. Lee. 2009. Molecules, morphology, and ecology indicate a recent, amphibious ancestry for echidnas. Proceedings of the National Academy of Sciences of the United States of America 106:17089–17094.

Phillips, M. J., and D. Penny. 2003. The root of the mammalian tree inferred from whole mitochondrial genomes. Molecular Phylogenetics and Evolution 28:171–185.

Phillips, M. J., and R. C. Pratt. 2008. Family-level relationships among the Australasian marsupial "herbivores" (Diprotodontia: koala, wombats, kangaroos and possums). Molecular Phylogenetics and Evolution 46:594–605.

Phillips, M. J., Y.-H. Lin, G. L. Harrison, and D. Penny. 2001. Mitochondrial genomes of a bandicoot and a brushtail possum confirm the monophyly of australidelphian marsupials. Proceedings of the Royal Society of London B Biological Sciences 268:1533–1538.

Phillips, M. J., P. A. McLenachan, C. Down, G. C. Gibb, and D. Penny. 2006. Combined mitochondrial and nuclear DNA sequences resolve the interrelations of the major Australasian marsupial radiations. Systematic Biology 55:122–137.

Phillips, R., C. Niezreckia, and D. O. Beusse. 2004. Determination of West Indian manatee vocalization levels and rate. Journal of the Accoustical Society of America 115:422–428.

Phillips, S. S. 2000. Population trends and the koala conservation debate. Conservation Biology 14:650–659.

Pickford, M. 2004. Revision of the early Miocene Hyracoidea (Mammalia) of East Africa. Comptes Rendus Palevol 3:675–690.

Pickford, M. 2005. Fossil hyraxes (Hyracoidea: Mammalia) from the late Miocene and Plio-Pleistocene of Africa, and the phylogeny of the Procaviidae. Palaeontologia Africana 141–161.

Pinder, L. 1993. Body measurements, karyotype, and birth frequencies of maned sloth (*Bradypus torquatus*). Mammalia 57:43–48.

Pine, R. H. 1972. A new subgenus and species of murine opossum (genus *Marmosa*) from Peru. Journal of Mammalogy 53:279–282.

Pine, R. H. 1973. Anatomical and nomenclatural notes on opossums. Proceedings of the Biological Society of Washington 86:391–402.

Pine, R. H. 1981. Reviews of the mouse opossums *Marmosa parvidens* Tate and *Marmosa invicta* Goldman (Mammalia: Marsupialia: Didelphidae) with description of a new species. Mammalia 45: 56–70.

Pine, R. H., P. L. Dalby, Jr., and J. O. Matson. 1985. Ecology, postnatal development, morphometrics, and taxonomic status of the short-tailed opossum, *Monodelphis dimidiata*, an apparently semelparous annual marsupial. Annals of Carnegie Museum 54:195–231.

Pine, R. H., D. A. Flores, and K. Bauer. 2013. The second known specimen of *Monodelphis unistriata* (Wagner) (Mammalia: Didelphimorphia), with redescription of the species and phylogenetic analysis. Zootaxa 3640:425–441.

Pine, R. H., and C. O. Handley, Jr. 1984. A review of the Amazonian short-tailed opossum *Monodelphis emiliae*. Mammalia 48:239–245.

Pine, R. H., and C. O. Handley, Jr. 2007. Genus *Monodelphis* (Burnett, 1830). Pp. 82–107 in Mammals of South America, volume 1, Marsupials, xenarthrans, shrews, and bats (A. L. Gardner, ed.). University of Chicago Press, Chicago, Illinois.

Pine, R. H., S. D. Miller, and M. L. Schamberger. 1979. Contributions to the mammalogy of Chile. Mammalia 43:339–376.

Pinto da Silveira, E. K. 1968. Notas sôbre a historia natural do tamanduá mirim (*Tamandua tetradactyla chiriquensis* J. A. Allen, 1904, Myrmecophagidae), com referências a fauna do Istmo do Panamá. Vellozia, Rio de Janeiro 6:9–31.

Pinto da Silveira, E. K. 1969. História natural do tamanduá-bandeira *Myrmecophaga tridactyla* (Linn, 1758), Myrmecophagidae. Vellozia, Rio de Janeiro 7:1–20.

Pires, A. D. S., and F. A. D. S. Fernandez. 1999. Use of space by the marsupial *Micoureus demerarae* in small Atlantic Forest fragments in south-eastern Brazil. Journal of Tropical Ecology 15:279–290.

Pires, M. M., E. G. Martins, M. N. F. Silva, and S. F. dos Reis. 2010. *Gracilinanus microtarsus* (Didelphimorphia: Didelphidae). Mammalian Species 42(851):33–40.

Pires Costa, A., and B. Patterson. 2008a. *Cryptonanus agricolai*. The IUCN Red List of Threatened Species 2008: e.T136545A4308227. http://dx.doi.org/10.2305/IUCN.UK.2008.RLTS.T136545A4308227.en. Downloaded on 29 December 2015.

Pires Costa, A., and B. Patterson. 2008b. *Thylamys karimii*. The IUCN Red List of Threatened Species 2008: e.T136653A4323798. http://dx.doi.org/10.2305/IUCN.UK.2008.RLTS.T136653A4323798.en. Downloaded on 27 December 2015.

Pires Costa, L. P., Y. L. R. Leite, and J. L. Patton. 2003. Phylogeography and systematic notes on two species of gracile mouse opossums, genus *Gracilinanus* (Marsupialia: Didelphidae) from Brazil. Proceedings of the Biological Society of Washington 116:275–292.

Plese, T., and A. Chiarello. 2014. *Choloepus hoffmanni*. The IUCN Red List of Threatened Species 2014: e.T4778A47439751. http://dx.doi.org/10.2305/IUCN.UK.2014-1.RLTS.T4778A47439751.en. Downloaded on 21 May 2015.

Poché, R. M.. 1980. Elephant management in Africa. Wildlife Society Bulletin 8:199–207.

Poduschka, W. 1980. Notes on the giant golden mole *Chrysospalax trevelyani* (Gunther, 1875) (Mammalia: Insectivora) and its survival chances. Zeitschrift für Säugetierkunde 45:193–206.

Poole, J. H. 1987a. Elephants in musth, lust. Natural History 96(11):46–55.

Poole, J. H. 1987b. Rutting behaviour in African elephants: the phenomenon of musth. Behaviour 102:283–316.

Poole, J. H. 1989a. Announcing intent: the aggressive state of musth in African elephants. Animal Behaviour 37:140–152.

Poole, J. H. 1989b. Mate guarding, reproductive success, and female choice in African elephants. Animal Behaviour 37:842–849.

Poole, J. H. 2011. Behavioral contexts of elephant acoustic communication. Pp. 125–161 in The Amboseli elephants: a long-term perspective on a long-lived mammal (C. J. Moss, H. Croze, and P. C. Lee, eds.). University of Chicago Press, Chicago, Illinois.

Poole, J. H., and P. Granli. 2011. Signals, gestures, and behavior of African elephants. Pp. 109–124 in The Amboseli elephants: a long-term perspective on a long-lived mammal (C. J. Moss, H. Croze, and P. C. Lee, eds.). University of Chicago Press, Chicago, Illinois.

Poole, J., P. Kahumbu, and I. Whyte. 2013. *Loxodonta africana* Savanna elephant. Pp. 182–194 in Mammals of Africa, volume 1, Introductory chapters and Afrotheria (J. Kingdon, D. Happold, M. Hoffmann, T. Butynski, M. Happold, and J. Kalina, eds.). Bloomsbury, London, United Kingdom.

Poole, J. H., and C. J. Moss. 2008. Elephant sociality and complexity. The scientific evidence. Pp. 69–98 in Elephants and ethics, toward a morality of existence (C. Wemmer and C. A. Christen, eds.). Johns Hopkins University Press, Baltimore, Maryland.

Poole, J. H., and J. B. Thomsen. 1989. Elephants are not beetles: implications of the ivory trade for the survival of the African elephant. Oryx 23:188–198.

Poole, J. H., P. C. Lee, N. Njiraini, and C. J. Moss. 2011a. Longevity, competition, and musth: male reproductive strategies. Pp. 272–286 in The Amboseli elephants: a long-term perspective on a long-lived mammal (C. J. Moss, H. Croze, and P. C. Lee, eds.). University of Chicago Press, Chicago, Illinois.

Poole, J. H., W. K. Lindsay, P. C. Lee, and C. J. Moss. 2011b. Ethical approaches to elephant conservation. Pp. 318–326 in The Amboseli elephants: a long-term perspective on a long-lived mammal (C. J. Moss, H. Croze, and P. C. Lee, eds.). University of Chicago Press, Chicago, Illinois.

Poole, W. E. 1973. A study of breeding in grey kangaroos, Macropus giganteus Shaw and M. fuliginosus (Desmarest), in central New South Wales. Australian Journal of Zoology 21:183–212.

Poole, W. E. 1975. Reproduction in the two species of grey kangaroos, Macropus giganteus Shaw and M. fuliginosus (Desmarest). II. Gestation, parturition, and pouch life. Australian Journal of Zoology 23:333–353.

Poole, W. E. 1976. Breeding biology and current status of the grey kangaroo, Macropus fuliginosus fuliginosus, of Kangaroo Island, South Australia. Australian Journal of Zoology 24:169–187.

Poole, W. E. 1977. The eastern grey kangaroo, Macropus giganteus, in south-east South Australia: its limited distribution and need of conservation. CSIRO Division of Wildlife Research Technical Paper 31:1–15.

Poole, W. E. 1978. Management of kangaroo harvesting in Australia. Australian National Parks and Wildlife Service Occasional Papers 2:1–28.

Poole, W. E. 1979. The status of the Australian Macropodidae. Pp. 13–27 in The status of endangered Australian wildlife (M. J. Tyler, ed.). Royal Zoological Society of South Australia, Adelaide, South Australia, Australia.

Poole, W. E. 1982. Macropus giganteus. Mammalian Species 187:1–8.

Poole, W. E. 1983a. Eastern grey kangaroo Macropus giganteus. Pp. 244–247 in The Australian Museum complete book of Australian mammals (R. Strahan, ed.). Angus & Robertson, London, United Kingdom.

Poole, W. E. 1983b. Western grey kangaroo Macropus fuliginosus. Pp. 248–249 in The Australian Museum complete book of Australian mammals (R. Strahan, ed.). Angus & Robertson, London, United Kingdom.

Poole, W. E. 1983c. Common wallaroo Macropus robustus. Pp. 250–251 in The Australian Museum complete book of Australian mammals (R. Strahan, ed.). Angus & Robertson, London, United Kingdom.

Poole, W. E. 1984. Management of kangaroo harvesting in Australia (1984). Australian National Parks and Wildlife Service Occasional Papers 9:1–25.

Poole, W. E. 1995a. Western grey kangaroo Macropus fuliginosus. Pp. 332–334 in Mammals of Australia (R. Strahan, ed.). Smithsonian Institution Press, Washington, DC.

Poole, W. E. 1995b. Eastern grey kangaroo Macropus giganteus. Pp. 335–338 in Mammals of Australia (R. Strahan, ed.). Smithsonian Institution Press, Washington, DC.

Poole, W. E., S. M. Carpenter, and N. G. Simms. 1990. Subspecific separation in the western grey kangaroo, Macropus fuliginosus: a morphometric study. Australian Wildlife Research 17:159–168.

Poole, W. E., and P. C. Catling. 1974. Reproduction in the two species of grey kangaroos, Macropus giganteus Shaw and M. fuliginosus (Desmarest). I. Sexual maturity and oestrus. Australian Journal of Zoology 22:277–302.

Poole, W. E., J. T. Wood, and N. G. Simms. 1991. Distribution of the tammar, Macropus eugenii, and the relationships of populations as determined by cranial morphometrics. Wildlife Research 18:625–639.

Pope, L. C., D. Blair, and C. N. Johnson. 2005. Dispersal and population structure of the rufous bettong, Aepyprymnus rufescens (Marsupialia: Potoroidae). Austral Ecology 30:572–580.

Pope, L., D. Storch, M. Adams, C. Moritz, and G. Gordon. 2001. A phylogeny for the genus Isoodon and a range extension for I. obesulus peninsulae based on mtDNA control region and morphology. Australian Journal of Zoology 49:411–434.

Pope, M. L., D. B. Lindenmayer, and R. B. Cunningham. 2004. Patch use by the greater glider (Petauroides volans) in a fragmented forest ecosystem. I. Home range size and movements. Wildlife Research 31:559–568.

Pople, A. R., S. C. Cairns, N. Menke, and N. Payne. 2006. Estimating the abundance of eastern grey kangaroos (Macropus giganteus) in south-eastern New South Wales, Australia. Wildlife Research 33:93–102.

Pople, A. R., G. C. Grigg, S. R. Phinn, N. Menke, C. McAlpine, and H. P. Possingham. 2010. Reassessing the spatial and temporal dynamics of kangaroo populations. Pp. 197–210 in Macropods: the biology of kangaroos, wallabies and rat-kangaroos (G. Coulson and M. Eldridge, eds.). CSIRO Publishing, Collingwood, Victoria, Australia.

Pople, T., and G. Grigg. 1999. Commercial harvesting of kangaroos in Australia. Australian Department of the Environment, Water, Heritage and the Arts, Canberra, ACT, Australia. Web 1 April 2011 http://www.environment.gov.au/biodiversity/trade-use/wild-harvest/kangaroo/harvesting/roobg-01.html.

Potter, S., S. J. B. Cooper, C. J. Metcalfe, D. A. Taggart, and M. D. B. Eldridge. 2012a. Phylogenetic relationships of rock-wallabies, Petrogale (Marsupialia: Macropodidae) and their biogeographic history within Australia. Molecular Phylogenetics and Evolution 62:640–652.

Potter, S., M. D. B. Eldridge, D. A. Taggart, and S. J. B. Cooper. 2012b. Multiple biogeographical barriers identified across the monsoon tropics of northern Australia: phylogeographic analysis of the brachyotis group of rock-wallabies. Molecular Ecology 21:2254–2269.

Potter, S., D. Rosauer, J. S. Doody, M. J. Webb and M. D. B. Eldridge. 2014a. Persistence of a potentially rare mammalian genus (Wyulda) provides evidence for areas of evolutionary refugia within the Kimberley, Australia. Conservation Genetics 15:1085–1094.

Potter, S., R. L. Close, D. A. Taggart, S. J. B. Cooper, and M. D. B. Eldridge. 2014b. Taxonomy of rock-wallabies, Petrogale (Marsupialia: Macropodidae). IV. Multifaceted study of the brachyotis group identifies additional taxa. Australian Journal of Zoology 62:401–414.

Poulakakis, N., G. E. Theodorou, E. Zouros, and M. Mylonas. 2002. Molecular phylogeny of the extinct Pleistocene dwarf elephant Palaeoloxodon antiquus falconeri from Tilos Island, Dodekanisa, Greece. Journal of Molecular Evolution 55:364–374.

Poulakakis, N., A. Parmakelis, P. Lymberakis, M. Mylonas, E. Zouros, D. S. Reese, S. Glaberman, and A. Caccone. 2006. Ancient DNA forces reconsideration of evolutionary history of Mediterranean pygmy elephantids. Biology Letters 2:451–454.

Poux, C., O. Madsen, J. Glos, W. W. de Jong, and M. Vences. 2008. Molecular phylogeny and divergence times of Malagasy tenrecs: influence of data partitioning and taxon sampling on dating analyses. BMC Evolutionary Biology 8:102. doi:10.1186/1471-2148-8-102.

Powell, J. A., Jr. 1978. Evidence of carnivory in manatees (*Trichechus manatus*). Journal of Mammalogy 59:442.

Powell, J. A. 2013. *Trichechus senegalensis* West African manatee. Pp. 210–212 in Mammals of Africa, volume 1, Introductory chapters and Afrotheria (J. Kingdon, D. Happold, M. Hoffmann, T. Butynski, M. Happold, and J. Kalina, eds.). Bloomsbury, London, United Kingdom.

Powell, J. A., Jr., D. W. Belitsky, and G. B. Rathbun. 1981. Status of the West Indian manatee (*Trichechus manatus*) in Puerto Rico. Journal of Mammalogy 62:642–646.

Powell, J. A., Jr., and G. B. Rathbun. 1984. Distribution and abundance of manatees along the northern coast of the Gulf of Mexico. Northeast Gulf Science 7:1–28.

Pradhan, N. M. B., A. C. Williams, and M. Dhakal. 2011. Current status of Asian elephants in Nepal. Gajah 35:87–92.

Preen, A., H. Das, M. Al-Rumaidh, and A. Hodgson. 2012. Dugongs in Arabia. Pp. 91–98 in Sirenian conservation. Issues and strategies in developing countries (E. M. Hines, J. E. Reynolds, III, L. V. Aragones, A. A. Mignucci-Giannoni, and M. Marmontel, eds.). University Press of Florida, Gainesville, Florida.

Price, G. J. 2008a. Taxonomy and palaeobiology of the largest-ever marsupial, *Diprotodon* (Owen, 1838) (Diprotodontidae, Marsupialia). Zoological Journal of the Linnean Society 153:369–397.

Price, G. J. 2008b. Is the modern koala (*Phascolarctos cinereus*) a derived dwarf of a Pleistocene giant? Implications for testing megafauna extinction hypotheses. Quaternary Science Reviews 27:2516–2521.

Price, G. J., and S. A. Hocknull. 2011. *Invictokoala monticola* gen. et sp. nov. (Phascolarctidae, Marsupialia), a Pleistocene plesiomorphic koala holdover from Oligocene ancestors. Journal of Systematic Palaeontology 9:327–335.

Prideaux, G. J. 2004. Systematics and evolution of the sthenurine kangaroos. University of California Publications in Geological Sciences 146:1–623.

Prideaux, G. J., and N. M. Warburton. 2010. An osteology-based appraisal of the phylogeny and evolution of kangaroos and wallabies (Macropodidae: Marsupialia). Zoological Journal of the Linnean Society 159:954–987.

Prince, R. I. T., and J. D. Richards. 2008. Banded hare-wallaby *Lagostrophus fasciatus* (Péron and Lesueur, 1807). Pp. 407–408 in The mammals of Australia. 3rd edition (S. Van Dyck and R. Strahan, eds.). Reed New Holland, Sydney, Australia.

Prista, G., M. Estevens, R. Agostinho, and M. Cachão. 2013. The disappearance of the European/North African Sirenia (Mammalia). Palaeogeography, Palaeoclimatology, Palaeoecology 387:1–5.

Proctor, C. M., W. Freeman, and J. L. Brown. 2010. Results of a second survey to assess the reproductive status of female Asian and African elephants in North America. Zoo Biology 29:127–139.

Proske, U., and E. Gregory. 2003. Electrolocation in the platypus—some speculations. Comparative Biochemistry and Physiology Part A 136:821–825.

Proske, U., J. E. Gregory, and A. Iggo. 1992. Activity in the platypus brain evoked by weak electrical stimulation of the bill. Pp. 204–210 in Platypus and echidnas (M. L. Augee, ed.). Royal Zoological Society of New South Wales, Mosman, New South Wales, Australia.

Proske, U., J. E. Gregory, and A. Iggo. 1998. A review of recent developments in the study of electroreception in the platypus. Australian Mammalogy 20:163–170.

Prothero, D. R. 1993. Ungulate phylogeny: molecular vs. morphological evidence. Pp. 173–181 in Mammal phylogeny: placentals (F. S. Szalay, M. J. Novacek, and M. C. McKenna, eds.). Springer-Verlag, New York, New York.

Prothero, D. R., and R. M. Schoch. 1989a. Origin and evolution of the Perissodactyla: summary and synthesis. Pp. 504–529 in The evolution of perissodactyls (D. R. Prothero and R. M. Schoch, eds.). Clarendon Press, New York, New York.

Prothero, D. R., and R. M. Schoch. 1989b. Classification of the Perissodactyla. Pp. 530–537 in The evolution of perissodactyls (D. R. Prothero and R. M. Schoch, eds.). Clarendon Press, New York, New York.

Prothero, D. R., and R. M. Schoch. 2002. Horns, tusks, and flippers: the evolution of hoofed mammals. Johns Hopkins University Press, Baltimore, Maryland.

Prowse, T. A. A., C. N. Johnson, R. C. Lacy, C. J. A. Bradshaw, J. P. Pollak, M. J. Watts, and B. W. Brook. 2013. No need for disease: testing extinction hypotheses for the thylacine using multi-species metamodels. Journal of Animal Ecology 82:355–364.

Pujos, F., T. J. Gaudin, G. De Iuliis, and C. Cartelle. 2012. Recent advances on variability, morpho-functional adaptations, dental terminology, and evolution of sloths. Mammalian Evolution 19:159–169.

Qiu, Z., B. Wang, L. Hong, D. Tao, and S. Yan. 2007. First discovery of deinothere in China. Vertebrata Palasiatica 45:261–277.

Querouil, S., F. Magliocca, and A. Gautier-Hion. 1999. Structure of population, grouping patterns and density of forest elephants in north-west Congo. African Journal of Ecology 37:161–167.

Quin, D. G., A. P. Smith, S. W. Green, and H. B. Hines. 1992. Estimating the home ranges of sugar gliders (*Petaurus breviceps*) (Marsupialia: Petauridae), from grid-trapping and radiotelemetry. Wildlife Research 19:471–487.

Quintela, F. M., M. B. Santos, A. Gava, and A. U. Christoff. 2011. Notes on morphology, geographic distribution, natural history, and cytogenetics of *Cryptonanus guahybae* (Didelphimorphia: Didelphidae). Mastozoologia Neotropical 18:247–257.

Rademaker, V., and R. Cerqueira. 2006. Variation in the latitudinal reproductive patterns of the genus *Didelphis* (Didelphimorphia: Didelphidae). Austral Ecology 31:337–342.

Rageot, R. 1978. Observaciones sobre el monito del monte. Departamento de Tecnología, Corporación Nacional Forestal, Ministerio de Agricultura, Temuco, Chile.

Rahm, U. 1990. Aardvark. Pp. 452–458 in Grzimek's encyclopedia of mammals, volume 4 (B. Grzimek, ed.). McGraw-Hill, New York, New York.

Raia, P., C. Barbera, and M. Conte. 2003. The fast life of a dwarfed giant. Evolutionary Ecology 17:293–312.

Rajamani, L. 2013. Using community knowledge in data-deficient regions: conserving the vulnerable dugong *Dugong dugon* in the Sulu Sea, Malaysia. Oryx 47:173–176.

Rampartab, C. 2015a. *Amblysomus septentrionalis*. The IUCN Red List of Threatened Species 2015: e.T62009A21284057. http://dx.doi.org/10.2305/IUCN.UK.2015-2.RLTS.T62009A21284057.en. Downloaded on 20 January 2016.

Rampartab, C. 2015b. *Amblysomus robustus*. The IUCN Red List of Threatened Species 2015: e.T62008A21284697. http://dx.doi.org/10.2305/IUCN.UK.2015-2.RLTS.T62008A21284697.en. Downloaded on 20 January 2016.

Rangarajan, M., A. Desai, R. Sukumar, P. S. Easa, V. Menon, S. Vincent, S. Ganguly, B. K. Talukdar, B. Singh, D. Mudappa, S. Chowdhary, A. N. Prasad. 2010. Gajah. Securing the future for elephants in India. Report of the Elephant Task Force. Ministry of Environment and Forests, Delhi, India.

Rasmussen, D. T. 1989. The evolution of the Hyracoidea: a review of the fossil evidence. Pp. 57–78 in The evolution of perissodactyls (D. R. Prothero and R. M. Schoch, eds.). Clarendon Press, New York, New York.

Rasmussen, D. T., M. Pickford, P. Mein, B. Senut, and G. C. Conroy. 1996. Earliest known procaviid hyracoid from the late Miocene of Namibia. Journal of Mammalogy 77:745–754.

Rasmussen L. E. L. 1999. Evolution of chemical signals in the Asian elephant, *Elephas maximus*: behavioural and ecological influences. Journal of Biosciences 24:241–251.

Rasmussen, L. E. L., and D. R. Greenwood. 2003. Frontalin: a chemical message of musth in Asian elephants (*Elephas maximus*). Chemical Senses 28:433–446.

Rasmussen, L. E. L., and V. Krishnamurthy. 2000. How chemical signals integrate Asian elephant society: the known and the unknown. Zoo Biology 19:405–423.

Rasmussen, L. E. L., V. Krishnamurthy, and R. Sukumar. 2005. Behavioural and chemical confirmation of the preovulatory pheromone, (Z)-7-dodecenyl acetate, in wild Asian elephants: its relationship to musth. Behaviour 142:351–396.

Rasmussen, L. E. L., and B. A. Schulte. 1998. Chemical signals in the reproduction of Asian (*Elephas maximus*) and African (*Loxodonta africana*) elephants. Animal Reproduction Science 53:19–34.

Raterman, D., R. W. Meredith, L. A. Ruedas, and M. S. Springer. 2006. Phylogenetic relationships of the cuscuses and brushtail possums (Marsupialia: Phalangeridae) using the nuclear gene BRCA1. Australian Journal of Zoology 54:353–361.

Rathbun, G. B. 1973. Territoriality in the golden-rumped elephant shrew. East African Wildlife Journal 11:405.

Rathbun, G. B. 1979. *Rhynchocyon chrysopygus*. Mammalian Species 117:1–4.

Rathbun, G. B. 1984. Sirenians. Pp. 537–548 in Orders and families of Recent mammals of the world (S. Anderson and J. K. Jones, Jr., eds.). John Wiley & Sons, New York, New York.

Rathbun, G. B. (subed.). 2005. Order Macroscelidea. Pp. 22–34 in The mammals of the Southern African Subregion. 3rd edition (J. D. Skinner and C. T. Chimimba, eds.). Cambridge University Press, Cape Town, South Africa.

Rathbun, G. B. 2008. *Rhynchocyon cirnei*. The IUCN Red List of Threatened Species 2008: e.T19709A9005441. http://dx.doi.org/10.2305/IUCN.UK.2008.RLTS.T19709A9005441.en. Downloaded on 20 January 2016.

Rathbun, G. B. 2009. Why is there discordant diversity in sengi (Mammalia: Afrotheria: Macroscelidea) taxonomy and ecology? African Journal of Ecology 47:1–13.

Rathbun, G. B. 2013a. *Petrodromus tetradactylus* four-toed sengi. Pp. 279–281 in Mammals of Africa, volume 1, Introductory chapters and Afrotheria (J. Kingdon, D. Happold, M. Hoffmann, T. Butynski, M. Happold, and J. Kalina, eds.). Bloomsbury, London, United Kingdom.

Rathbun, G. B. 2013b. Genus *Rhynchocyon* giant sengis. Pp. 282–283 in Mammals of Africa, volume 1, Introductory chapters and Afrotheria (J. Kingdon, D. Happold, M. Hoffmann, T. Butynski, M. Happold, and J. Kalina, eds.). Bloomsbury, London, United Kingdom.

Rathbun, G. B. 2013c. *Rhynchocyon chrysopygus* golden-rumped giant sengi. Pp. 283–284 in Mammals of Africa, volume 1, Introductory chapters and Afrotheria (J. Kingdon, D. Happold, M. Hoffmann, T. Butynski, M. Happold, and J. Kalina, eds.). Bloomsbury, London, United Kingdom.

Rathbun, G. B. 2015a. *Elephantulus rozeti*. The IUCN Red List of Threatened Species 2015: e.T42663A21289287. http://dx.doi.org/10.2305/IUCN.UK.2015-2.RLTS.T42663A21289287.en. Downloaded on 20 January 2016.

Rathbun, G. B. 2015b. *Elephantulus revoilii*. The IUCN Red List of Threatened Species 2015: e.T7137A21290721. http://dx.doi.org/10.2305/IUCN.UK.2015-2.RLTS.T7137A21290721.en. Downloaded on 20 January 2016.

Rathbun, G. B. 2015c. *Elephantulus rufescens*. The IUCN Red List of Threatened Species 2015: e.T42664A21289073. http://dx.doi.org/10.2305/IUCN.UK.2015-2.RLTS.T42664A21289073.en. Downloaded on 20 January 2016.

Rathbun, G. B., P. Agnelli, and G. Innocenti. 2014. Distribution of sengis in the Horn of Africa. Afrotherian Conservation 10:2–4.

Rathbun, G. B., and T. M. Butynski. 2008. *Rhynchocyon petersi*. The IUCN Red List of Threatened Species 2008: e.T19708A9004669. http://dx.doi.org/10.2305/IUCN.UK.2008.RLTS.T19708A9004669.en. Downloaded on 20 January 2016.

Rathbun, G. B., and J. Dumbacher. 2015. *Macroscelides micus*. The IUCN Red List of Threatened Species 2015: e.T45434566A45436004. http://dx.doi.org/10.2305/IUCN.UK.2015-2.RLTS.T45434566A45436004.en. Downloaded on 20 January 2016.

Rathbun, G. B., and S. Eiseb. 2015. *Macroscelides flavicaudatus*. The IUCN Red List of Threatened Species 2015: e.T45369877A45435876. http://dx.doi.org/10.2305/IUCN.UK.2015-2.RLTS.T45369877A45435876.en. Downloaded on 20 January 2016.

Rathbun, G. B., and C. FitzGibbon. 2015. *Petrodromus tetradactylus*. The IUCN Red List of Threatened Species 2015: e.T42679A21290893. http://dx.doi.org/10.2305/IUCN.UK.2015-2.RLTS.T42679A21290893.en. Downloaded on 20 January 2016.

Rathbun, G. B., and T. J. O'Shea. 1984. The manatee's simple social life. Pp. 300–301 in The encyclopedia of mammals (D. Macdonald, ed.). Facts on File, New York, New York.

Rathbun, G. B., and C. D. Rathbun. 2006a. Habitat use by radio-tagged Namib Desert golden moles (*Eremitalpa granti namibensis*). African Journal of Ecology 45:196–201.

Rathbun, G. B., and C. D. Rathbun. 2006b. Social structure of the bushveld sengi (*Elephantulus intufi*) in Namibia and the evolution of monogamy in the Macroscelidea. Journal of Zoology 269:391–399.

Rathbun, G. B., and H. Smit-Robinson. 2015a. *Macroscelides proboscideus*. The IUCN Red List of Threatened Species 2015: e.T45369602A45435551. http://dx.doi.org/10.2305/IUCN.UK.2015-2.RLTS.T45369602A45435551.en. Downloaded on 20 January 2016.

Rathbun, G. B., and H. Smit-Robinson. 2015b. *Elephantulus edwardii*. The IUCN Red List of Threatened Species 2015: e.T7136A21290344. http://dx.doi.org/10.2305/IUCN.UK.2015-2.RLTS.T7136A21290344.en. Downloaded on 20 January 2016.

Rathbun, G. B., and H. Smit-Robinson. 2015c. *Elephantulus rupestris*. The IUCN Red List of Threatened Species 2015: e.T7138A21290631.

http://dx.doi.org/10.2305/IUCN.UK.2015-2.RLTS.T7138
A21290631.en. Downloaded on 20 January 2016.

Rathbun, G. B., R. L. Brownell, Jr., K. Ralls, and J. Engbring. 1988. Status of dugongs in waters around Palau. Marine Mammal Science 4:265–270.

Rathbun, G. B., J. P. Reid, R. K. Bonde, and J. A. Powell. 1995. Reproduction in free-ranging Florida manatees. Pp. 135–156 in Population biology of the Florida manatee (T. J. O'Shea, B. B. Ackerman, and H. F. Percival, eds.). National Biological Service, Information and Technology Report 1, United States Department of the Interior, Washington, DC.

Rauhut, O. W. M., T. Martin, E. Ortiz-Jaureguizar, and P. Puerta. 2002. A Jurassic mammal from South America. Nature 416:165–168.

Rawlins, D. R., and K. A. Handasyde. 2002. The feeding ecology of the striped possum Dactylopsila trivirgata (Marsupialia: Petauridae) in far north Queensland. Australian Journal of Zoology 257:195–206.

Read, D. G. 1984. Reproduction and breeding season of Planigale gilesi and P. tenuirostris (Marsupialia: Dasyuridae). Australian mammalogy 7:161–173.

Read, D. G. 2008a. Giles' planigale Planigale gilesi (Aitken, 1972). Pp. 107–109 in The mammals of Australia. 3rd edition (S. Van Dyck and R. Strahan, eds.). Reed New Holland, Sydney, Australia.

Read, D. G. 2008b. Narrow-nosed planigale Planigale tenuirostris (Troughton, 1928). Pp. 114–116 in The mammals of Australia. 3rd edition (S. Van Dyck and R. Strahan, eds.). Reed New Holland, Sydney, Australia.

Reading, R. P., P. Myronuik, G. Backhouse, and T. W. Clark. 1992. Eastern barred bandicoot reintroductions in Victoria, Australia. Species 19:29–31.

Reason, R., D. Gierhahn, and M. Schollhamer. 2005. Gestation in aardvarks Orycteropus afer at Brookfield Zoo, Illinois. International Zoo Yearbook 39:222–225.

Redford, K. H., and J. F. Eisenberg.1992. Mammals of the neotropics: the southern cone. University of Chicago Press, Chicago, Illinois.

Redford, K. H., and R. M. Wetzel. 1985. Euphractus sexcinctus. Mammalian Species 252:1–4.

Redi, C. A., S. Garagna, M. Zuccotti, and E. Capanna. 2007. Genome size: a novel genomic signature in support of Afrotheria. Journal of Molecular Evolution 64:484–487.

Reep, R. L., and R. K. Bonde. 2006. The Florida manatee. Biology and conservation. University Press of Florida, Gainesville, Florida.

Reep, R. L., C. D. Marshall, and M. L. Stoll. 2002. Tactile hairs on the postcranial body in Florida manatees: a mammalian lateral line? Brain Behavior and Evolution 59:141–154.

Rees, P. A. 2003. Asian elephants in zoos face global extinction: should zoos accept the inevitable? Oryx 37:20–22.

Reeves, R. R., D. Tuboku-Metzger, and R. A. Kapindi. 1988. Distribution and exploitation of manatees in Sierra Leone. Oryx 22:75–84.

Regidor, H. A., M. Gorostiague, and S. Suhring. 1999. Reproduction and dental age classes of the little opossum (Lutreolina crassicaudata) in Buenos Aires, Argentina. Revista de Biologia Tropical 47:271–272.

Reid, F. A. 1997. A field guide to the mammals of Central America and southeast Mexico. Oxford University Press, New York, New York.

Reid, J. 1995. Chessie's most excellent adventure: the 1995 east coast tour. Sirenews 24:9–11.

Reid, J. P., G. B. Rathbun, and J. R. Wilcox. 1991. Distribution patterns of individually identifiable West Indian manatees (Trichechus manatus) in Florida. Marine Mammal Science 7:180–190.

Reig, O. A., J. A. W. Kirsch, and L. G. Marshall. 1987. Systematic relationships of the living and neocenozoic American "opossum-like" marsupials (suborder Didelphimorphia), with comments on the classification of these and of the Cretaceous and Paleogene New World and European metatherians. Pp. 1–89 in Possums and opossums (M. Archer, ed.). Surrey Beatty & Sons, Sydney, Australia.

Renfree, M. B. 1980. Embryonic diapause in the honey possum Tarsipes spencerae. Search 11:81.

Renfree, M. B. 1983. Marsupial reproduction: the choice between placentation and lactation. Oxford Reviews of Reproductive Biology 5:1–29.

Renfree, M. B. 1993. Ontogeny, genetic control, and phylogeny of female reproduction in monotreme and therian mammals. Pp. 4–20 in Mammal phylogeny: Mesozoic differentiation, multituberculates, monotremes, early therians, and marsupials (F. S. Szalay, M. J. Novacek, and M. C. McKenna, eds.). Springer-Verlag, New York, New York.

Renfree, M. B. 2008. Honey possum Tarsipes rostratus (Gervais and Verreaux, 1842). Pp. 258–260 in The mammals of Australia. 3rd edition (S. Van Dyck and R. Strahan, eds.). Reed New Holland, Sydney, Australia.

Renfree, M. B., E. M. Russell, and R. D. Wooller. 1984. Reproduction and life history of the honey possum, Tarsipes rostratus. Pp. 427–437 in Possums and gliders (A. Smith and I. Hume, eds.). Surrey Beatty & Sons, Norton, New South Wales, Australia.

Retief, T. A., N. C. Bennett, A. A. Kinahana, and P. W. Bateman. 2013. Sexual selection and genital allometry in the Hottentot golden mole (Amblysomus hottentotus). Mammalian Biology 78:356–360.

Reumer, J. W. F., D. Mol, and J. de Vos. 2002. The Wrangel dwarf mammoths were no island endemics. British Archaeological Reports, International Series 1095:415–419.

Reyes, H. O. P., W. A. Matamoros, and S. L. Glowinski. 2010. Distribution and conservation status of the giant anteater (Myrmecophaga tridactyla) in Honduras. Southwestern Naturalist 55:118–120.

Reynolds, J. E., III. 1981. Aspects of the social behaviour and herd structure of a semi-isolated colony of West Indian manatees, Trichechus manatus. Mammalia 45:431–451.

Reynolds, J. E., III, and J. C. Ferguson. 1984. Implications of the presence of manatees (Trichechus manatus) near the Dry Tortugas Islands. Florida Scientist 47:187–189.

Reynolds, J. S., III, and D. K. Odell. 1991. Manatees and dugongs. Facts on File, New York, New York.

Rhind, S. G. 2002. Reproductive demographics among brush-tailed phascogales (Phascogale tapoatafa) in south-western Australia. Wildlife Research 29:247–257.

Rhind, S. G. 2003. Communal nesting in the usually solitary marsupial, Phascogale tapoatafa. Journal of Zoology 261:345–351.

Rhind, S. G., J. S. Bradley, and N. K. Cooper. 2001. Morphometric variation and taxonomic status of brush-tailed phascogales, Phascogale tapoatafa (Meyer, 1793) (Marsupialia: Dasyuridae). Australian Journal of Zoology 49:345–368.

Rhind, S., J. Woinarski, and K. P. Aplin. 2008. Northern brush-tailed phascogale Phascogale pirata (Thomas, 1904). Pp. 103–104 in The mammals of Australia. 3rd edition (S. Van Dyck and R. Strahan, eds.). Reed New Holland, Sydney, Australia.

Ribble, D. O., and M. R. Perrin. 2005. Social organization of the eastern rock elephant-shrew (Elephantulus myurus): the evidence for mate guarding. Belgian Journal of Zoology 135 (Supplement):167–173.

Ricciuti, E. R. 1980. The ivory wars. Animal Kingdom 83(1):6–58.

Rice, D. W. 1977. A list of the marine mammals of the world. 3rd edition. United States National Marine Fisheries Service, National Oceanic and Atmospheric Administration Technical Report NMFS SSRF-711.

Rich, T. H. V., M. Fortellius, P. V. Rich, and D. A. Hooijer. 1987. The supposed *Zygomaturus* from New Caledonia is a rhinoceros: a second solution to an enigma and its palaeogeographic consequences. Pp. 769–778 in Possums and opossums (M. Archer, ed.). Surrey Beatty & Sons, Sydney, Australia.

Richards, J. 2007. Western barred bandicoot, burrowing bettong and banded hare-wallaby recovery plan 2007–2011. Western Australia Department of Conservation and Land Management, Wanneroo, Western Australia, Australia.

Richards, J., K. Morris, and A. Burbidge. 2008. *Bettongia lesueur.* The IUCN Red List of Threatened Species 2008: e.T2784A9480530. http://dx.doi.org/10.2305/IUCN.UK.2008.RLTS.T2784A9480530 .en. Downloaded on 01 December 2015.

Richards, J., K. Morris, A. Burbidge, and T. Friend. 2008a. *Lagostrophus fasciatus.* The IUCN Red List of Threatened Species 2008: e.T11171A3259511. http://dx.doi.org/10.2305/IUCN.UK.2008 .RLTS.T11171A3259511.en. Downloaded on 30 December 2015.

Richards, J., K. Morris, T. Friend, and A. Burbidge. 2008b. *Lagorchestes hirsutus.* The IUCN Red List of Threatened Species 2008: e.T11162A3258493. http://dx.doi.org/10.2305/IUCN.UK.2008. RLTS.T11162A3258493.en. Downloaded on 02 January 2016.

Richards, J. D., and J. Short. 2003. Reintroduction and establishment of the western barred bandicoot *Perameles bougainville* (Marsupialia: Peramelidae) at Shark Bay, Western Australia. Biological Conservation 109:181–195.

Richardson, B. J., and G. B. Sharman. 1976. Biochemical and morphological observations on the wallaroos (Macropodidae: Marsupialia) with a suggested new taxonomy. Journal of Zoology 179:499–513.

Ride, W. D. L. 1964. A review of Australian fossil marsupials. Journal of the Royal Society of Western Australia 47:97–131.

Ride, W. D. L. 1970. A guide to the native mammals of Australia. Oxford University Press, Melbourne, Victoria, Australia.

Rieck, B., et al. 2016. Endangered and Threatened Wildlife and Plants; removal of the Louisiana black bear from the Federal List of Endangered and Threatened Wildlife and removal of similarity-of-appearance protections for the American black bear. Federal Register 81:13124–13171.

Rigby, R. G. 1972. A study of the behaviour of caged *Antechinus stuartii.* Zeitschrift für Tierpsychologie 31:15–25.

Riley, J. 2002. Mammals on the Sangihe and Talaud Islands, Indonesia, and the impact of hunting and habitat loss. Oryx 36:288–296.

Rismiller, P. D., and M. W. McKelvey. 2000. Frequency of breeding and recruitment in the short-beaked echidna, *Tachyglossus aculeatus.* Journal of Mammalogy 81:1–17.

Rismiller, P. D., and M. W. McKelvey. 2003. Body mass, age and sexual maturity in short-beaked echidnas, *Tachyglossus aculeatus.* Comparative Biochemistry and Physiology Part A 136:851–865.

Rismiller, P. D., and M. W. McKelvey. 2009. Activity and behaviour of lactating echidnas (*Tachyglossus aculeatus multiaculeatus*) from hatching of egg to weaning of young. Australian Journal of Zoology 57:265–273.

Ritchie, E. G. 2005. An extension to the known range of the eastern grey kangaroo *Macropus giganteus* on Cape York Peninsula. Australian Mammalogy 27:225–226.

Ritchie, E. G. 2008. Antilopine wallaroo *Macropus antilopinus* (Gould, 1842). Pp. 325–326 in The mammals of Australia. 3rd edition (S.

Van Dyck and R. Strahan, eds.). Reed New Holland, Sydney, Australia.

Ritchie, E. G. 2010. Ecology and conservation of the antilopine wallaroo: an overview of current knowledge. Pp. 179–186 in Macropods: the biology of kangaroos, wallabies and rat-kangaroos (G. Coulson and M. Eldridge, eds.). CSIRO Publishing, Collingwood, Victoria, Australia.

Roberts, T. E., E. J. Sargis, and L. E. Olson. 2009. Networks, trees, and treeshrews: assessing support and identifying conflict with multiple loci and a problematic root. Systematic Biology 58:257–270.

Roberts, T. E., H. C. Lanier, E. J. Sargis, and L. E. Olson. 2011. Molecular phylogeny of treeshrews (Mammalia: Scandentia) and the timescale of diversification in Southeast Asia. Molecular Phylogenetics and Evolution 60:358–372.

Robertshaw, J. D., and R. H. Harden. 1989. Predation on Macropodoidea: a review. Pp. 735–753 in Kangaroos, wallabies and rat-kangaroos (G. Grigg, P. Jarman, and I. Hume, eds.). Surrey Beatty & Sons, Chipping Norton, New South Wales, Australia.

Robinson, N. A. 1995. Implications from mitochondrial DNA for management to conserve the eastern barred bandicoot (*Perameles gunnii*). Conservation Biology 9:114.

Robinson, N. A., N. D. Murray, and W. B. Sherwin. 1993. VNTR loci reveal differentiation between and structure within populations of the eastern barred bandicoot *Perameles gunnii.* Molecular Biology 2:195–207.

Robinson, N. A., W. B. Sherwin, and P. R. Brown. 1991. A note on the status of the eastern barred bandicoot, *Perameles gunnii*, in Tasmania. Wildlife Research 18:451–457.

Robinson, T., G. Gaikhorst, D. Pearson, and P. Copley. 2008. *Sminthopsis psammophila.* The IUCN Red List of Threatened Species 2008: e.T20293A9183034. http://dx.doi.org/10.2305/IUCN.UK.2008 .RLTS.T20293A9183034.en. Downloaded on 04 January 2016.

Robinson, T. J., and E. R. Seiffert. 2004. Afrotherian origins and interrelationships: new views and future prospects. Current Topics in Developmental Biology 63:37–60.

Roca, A. L. 2007. The mastodon mitochondrial genome: a mammoth accomplishment. Trends in Genetics 24:49–52.

Roca, A. L., N. Georgiadis, and S. J. O'Brien. 2005. Cytonuclear genomic dissociation in African elephant species. Nature Genetics 37:96–100.

Roca, A. L., N. Georgiadis, and S. J. O'Brien. 2007. Cyto-nuclear genomic dissociation and the African elephant species question. Quaternary International 169–170:4–16.

Roca, A. L., N. Georgiadis, J. Pecon-Slattery, and S. J. O'Brien. 2001. Genetic evidence for two species of elephant in Africa. Science 293:1473–1477.

Roca, A. L., Y. Ishida, A. L. Brandt, N. R. Benjamin, K. Zhao, and N. J. Georgiadis. 2015. Elephant natural history: a genomic perspective. Annual Review of Animal Biosciences 3:139–167.

Rocha, P. A., J. Ruiz-Esparza, R. Beltrão-Mendes, M. Alves da Cunha, J. A. Feijó, and S. F. Ferrari. 2012. Expansion of the known range of *Marmosops incanus* (Mammalia, Didelphimorphia, Didelphinae) to the right bank of the São Francisco River in north-east Brazil. Mammalia 76:441–445.

Roche, J. 1972. Systématique du genre *Procavia* et des damans en général. Mammalia 36:22–49.

Rodgers, D. H., and W. H. Elder. 1977. Movements of elephants in Luangwa Valley, Zambia. Journal of Wildlife Management 41:56–62.

Rodrigues, F. H. G., J. Marinho-Filho, and H. G. dos Santos. 2001. Home ranges of translocated lesser anteaters *Tamandua tetradactyla* in the cerrado of Brazil. Oryx 35:166–169.

Rodrigues, F. H. G., I. M. Medri, G. H. B. de Miranda, C. Camilo-Alves, and G. Mourão. 2008. Anteater behavior and ecology. Pp. 257–268 in The biology of the Xenarthra (S. F. Vizcaíno and W. J. Loughry, eds.). University Press of Florida, Gainesville, Florida.

Rodríguez-Cabal, M. A., G. C. Amico, A. J. Novaro, and M. A. Aizen. 2008. Population characteristics of Dromiciops gliroides (Philippi, 1893), an endemic marsupial of the temperate forest of Patagonia. Mammalian Biology 73:74–76.

Rogaev, E. I., Y. K. Moliaka, B. A. Malyarchuk, F. A. Kondrashov, M. V. Derenko, I. Chumakov, and A. P. Grigorenko. 2006. Complete mitochondrial genome and phylogeny of Pleistocene mammoth Mammuthus primigenius. PLoS Biology 4(3):e73.

Roger, E., S. W. Laffan, and D. Ramp. 2007. Habitat selection by the common wombat (Vombatus ursinus) in disturbed environments: implications for the conservation of a "common" species. Biological Conservation 137:437–449.

Rohland N., A.-S. Malaspinas, J. L. Pollack, M. Slatkin, P. Matheus, and M. Hofreiter. 2007. Proboscidean mitogenomics: chronology and mode of elephant evolution using mastodon as outgroup. PLoS Biology 5(8):e207.

Rohland, N., D. Reich, S. Mallick, M. Meyer, R. E. Green, N. J. Georgiadis, A. L. Roca, and M. Hofreiter. 2010. Genomic DNA sequences from mastodon and woolly mammoth reveal deep speciation of forest and savanna elephants. PLoS Biology 8(12):e1000564.

Roig, V. G. 1991. Desertification and distribution of mammals in the southern cone of South America. Pp. 239–279 in Latin American mammalogy: history, biodiversity, and conservation (M. A. Mares and D. J. Schmidly, eds.). University of Oklahoma Press, Norman, Oklahoma.

Romiguier, J., V. Ranwez, E. J. P. Douzery, and N. Galtier. 2012. Genomic evidence for large, long-lived ancestors to placental mammals. Molecular Biology and Evolution 30:5–13.

Romiguier, J., V. Ranwez, F. Delsuc, N. Galtier, E. J. P. Douzery. 2013. Less is more in mammalian phylogenomics: at-rich genes minimize tree conflicts and unravel the root of placental mammals. Molecular Biology and Evolution 30:2134–2144.

Rood, J. P. 1970. Notes on the behavior of the pygmy armadillo. Journal of Mammalogy 51:179.

Roonwal, M. L., and S. M. Mohnot. 1977. Primates of South Asia. Harvard University Press, Cambridge, Massachusetts.

Rosas, F. C. W. 1994. Biology, conservation, and status of the Amazonian manatee Trichechus inunguis. Mammal Review 24:49–59.

Rose, K. D. 1975. Elpidophorus, the earliest dermopteran (Dermoptera, Plagiomenidae). Journal of Mammalogy 56:675–679.

Rose, K. D., and R. J. Emry. 1993. Relationships of Xenarthra, Pholidota, and fossil "edentates": the morphological evidence. Pp. 81–102 in Mammal phylogeny: Mesozoic differentiation, multituberculates, monotremes, early therians, and marsupials (F. S. Szalay, M. J. Novacek, and M. C. McKenna, eds.). Springer-Verlag, New York, New York.

Rose, K. D., R. J. Emry, T. J. Gaudin, and G. Storch. 2005. Xenarthra and Pholidota. Pp. 106–126 in The rise of placental mammals: origins and relationships of the major clades (K. D. Rose and J. D. Archibald, eds.). Johns Hopkins University Press, Baltimore, Maryland.

Rose, R. W. 1978. Reproduction and evolution in female Macropodidae. Australian Mammalogy 2:65–72.

Rose, R. W. 1986. The habitat, distribution, and conservation status of the Tasmanian bettong, Bettongia gaimardi (Desmarest). Australian Wildlife Research 13:1–6.

Rose, R. W. 1987. Reproductive biology of the Tasmanian bettong (Bettongia gaimardi: Macropodidae). Journal of Zoology 212:59–67.

Rose, R. W. 1989. Reproductive biology of the rat-kangaroos. Pp. 307–315 in Kangaroos, wallabies and rat-kangaroos (G. Grigg, P. Jarman, and I. Hume, eds.). Surrey Beatty & Sons, Chipping Norton, New South Wales, Australia.

Rose, R. W., and K. A. Johnson. 2008. Tasmanian bettong Bettongia gaimardi (Desmarest, 1822). Pp. 287–288 in The mammals of Australia. 3rd edition (S. Van Dyck and R. Strahan, eds.). Reed New Holland, Sydney, Australia.

Rose, R. W., and D. J. McCartney. 1982. Reproduction of the red-bellied pademelon, Thylogale billardierii (Marsupialia). Australian Wildlife Research 9:27–32.

Rose, R. W., and R. K. Rose. 1998. Bettongia gaimardi. Mammalian Species 584:1–6.

Rose, T. A., A. Munn, D. Ramp, and P. B. Banks. 2006. Foot-thumping as an alarm signal in macropodoid marsupials: prevalence and hypotheses of function. Mammal Review 36:281–298.

Rosen, B. 1994. Mammoths in ancient Egypt? Nature 369:364.

Rosenberg, H. L., and R. Rose. 1999. Volar adhesive pads of the feathertail glider, Acrobates pygmaeus (Marsupialia; Acrobatidae). Canadian Journal of Zoology 77:233–248.

Rosenthal, M. A. 1975a. The management, behavior, and reproduction of the short-eared elephant shrew Macroscelides proboscideus (Shaw). M.A. thesis, Northeastern Illinois University, Chicago, Illinois.

Rosenthal, M. A. 1975b. Observations on the water opossum or yapok Chironectes minimus in captivity. International Zoo Yearbook 15:4–6.

Rossi, R. V., R. S. Voss, and D. P. Lunde. 2010. A revision of the didelphid marsupial genus Marmosa. Part 1. The species in Tate's "mexicana" and "mitis" sections and other closely related forms. Bulletin of the American Museum of Natural History 334:1–83.

Rossi, R. V., C. L. Miranda, T. S. Santos Júnior, and T. B. F. Semedo. 2010. New records and geographic distribution of the rare Glironia venusta (Didelphimorphia, Didelphidae). Mammalia 74:445–447.

Roth, H. H., and I. Douglas-Hamilton. 1991. Distribution and status of elephants in West Africa. Mammalia 55:489–527.

Roth, H. H., and E. Waitkuwait. 1986. Répartition et statut des grandes espèces de mammifères en Côte-d'Ivoire. III. Lamantins. Mammalia 50:227–242.

Roth, V. L. 1992. Inferences from allometry and fossils: dwarfing of elephants on islands. Oxford Surveys in Evolutionary Biology 8:259–288.

Roth, V. L. 1993. Dwarfism and variability in the Santa Rosa Island mammoth (Mammuthus exilis): an interspecific comparison of limb-bone sizes and shapes in elephants. Pp. 433–442 in Recent advances in California Islands research: Proceedings of the Third California Islands Symposium (F. G. Hochberg, ed.). Santa Barbara Museum of Natural History, Santa Barbara, California.

Roth, V. L. 1996. Pleistocene dwarf elephants of the California islands. Pp. 249–253 in The Proboscidea. Evolution and palaeoecology of elephants and their relatives (J. Shoshani and P. Tassy, eds.). Oxford University Press, Oxford, United Kingdom.

Rougier, G. W., A. G. Martinelli, A. M. Forasiepi, and M. J. Novacek. 2007. New Jurassic mammals from Patagonia, Argentina: a reappraisal of australosphenidan morphology and interrelationships. American Museum Novitates 3566:1–54.

Rounsevell, D. E., and S. J. Smith. 1982. Recent alleged sightings of the thylacine (Marsupialia, Thylacinidae) in Tasmania. Pp. 233–236 in

Carnivorous marsupials (M. Archer, ed.). Royal Zoological Society of New South Wales, Mosman, New South Wales, Australia.

Rouse, I. 1989. Peopling and repeopling of the West Indies. Pp. 119–136 in Biogeography of the West Indies: past, present, and future (C. A. Woods, ed.). Sandhill Crane Press, Gainesville, Florida.

Rovero, F., and G. B. Rathbun. 2015. *Rhynchocyon udzungwensis*. The IUCN Red List of Threatened Species 2015: e.T136309A21287423. http://dx.doi.org/10.2305/IUCN.UK.2015-2.RLTS.T136309A21287423.en. Downloaded on 20 January 2016.

Rovero, F., G. B. Rathbun, A. Perkin, T. Jones, D. O. Ribble, C. Leonard, R. R. Mwakisoma, and N. Doggart. 2008. A new species of giant sengi or elephant-shrew (genus *Rhynchocyon*) highlights the exceptional biodiversity of the Udzungwa Mountains of Tanzania. Journal of Zoology 274:126–133.

Rowe, T. 1988. Definition, diagnosis, and origin of Mammalia. Journal of Vertebrate Paleontology 8:241–264.

Rowe, T. 1993. Phylogenetic systematics and the early history of mammals. Pp. 129–145 in Mammal phylogeny: Mesozoic differentiation, multituberculates, monotremes, early therians, and marsupials (F. S. Szalay, M. J. Novacek, and M. C. McKenna, eds.). Springer-Verlag, New York, New York.

Rowe, T., and J. Gauthier. 1992. Ancestry, paleontology, and definition of the name Mammalia. Systematic Biology 41:372–378.

Rowe, T., T. H. Rich, P. Vickers-Rich, M. Springer, and M. O. Woodburne. 2008. The oldest platypus and its bearing on divergence timing of the platypus and echidna clades. Proceedings of the National Academy of Sciences of the United States of America 105:1238–1242.

Rudnai, J. A. 1984. Activity cycle and space utilization in captive *Dendrohyrax arboreus*. South African Journal of Zoology 19:124–128.

Rudolph, P., and C. Smeenk. 2008. Indo-Pacific marine mammals. Pp. 608–616 in Encyclopedia of marine mammals. 2nd edition (W. F. Perrin, B. Würsig, and J. G. M. Thewissen, eds.). Academic Press, New York, New York.

Ruedas, L. A., and J. C. Morales. 2005. Evolutionary relationships among genera of Phalangeridae (Metatheria: Diprotodontia) inferred from mitochondrial DNA. Journal of Mammalogy 86:353–365.

Runcie, M. J. 1999. Movements, dens and feeding behaviour of the tropical scaly-tailed possum (*Wyulda squamicaudata*). Wildlife Research 26:367–373.

Runcie, M. J. 2000. Biparental care and obligate monogamy in the rock-haunting possum, *Petropseudes dahli*, from tropical Australia. Animal Behaviour 59:1001–1008.

Runcie, M. J. 2004. Scent-marking and vocal communication in the rock-haunting possum *Petropseudes dahli*. Pp. 401–412 in The biology of Australian possums and gliders (R. L. Goldingay and S. M. Jackson, eds.). Surrey Beatty & Sons, Chipping Norton, New South Wales, Australia.

Runestad, J. A., and C. B. Ruff. 1995. Structural adaptations for gliding in mammals with implications for locomotor behavior in paromomyids. American Journal of Physical Anthropology 98:101–119.

Russell, E. M. 1974a. The biology of kangaroos (Marsupialia-Macropodidae). Mammal Review 4:1–59.

Russell, E. M. 1974b. Recent ecological studies on Australian marsupials. Australian Mammalogy 189–211.

Russell, E. M. 1979. The size and composition of groups in the red kangaroo, *Macropus rufus*. Australian Wildlife Research 6:237–244.

Russell, E. M., A. K. Lee, and G. R. Wilson. 1989. Natural history of the Metatheria. Pp. 505–526 in Fauna of Australia: Mammalia, volume 1B (D. W. Walton and B. J. Richardson, eds.). Australian Government Publishing Service, Canberra, Australia.

Russell, E. M., and M. B. Renfree. 1989. Tarsipedidae. Pp. 769–782 in Fauna of Australia: Mammalia, volume 1B (D. W. Walton and B. J. Richardson, eds.). Australian Government Publishing Service, Canberra, Australia.

Saaban, S., N. Bin Othman, M. N. Bin Yasak, B. M. Nor, A. Zafir, and A. Campos-Arceiz. 2011. Current status of Asian elephants in Peninsular Malaysia. Gajah 35:67–75.

Sadler, L. M., and S. J. Ward. 1999. Coalitions in male sugar gliders: are they natural? Journal of Zoology 248:91–96.

Saegusa, H. 2001. Comparisons of stegodon and elephantid abundances in the late Pleistocene of southern China. Pp. 345–349 in Proceedings of the First International Congress of the World of Elephants (G. Cavarretta, P. Gioia, M. Mussi, and M. R. Palombo, eds.). Consiglio Nazionale delle Ricerche, Rome, Italy.

Saha, S.S., and T. Bhatta. 2008. *Tupaia nicobarica*. The IUCN Red List of Threatened Species 2008:e.T22454A9373847. http://dx.doi.org/10.2305/IUCN.UK.2008.RLTS.T22454A9373847.en. Downloaded on 20 January 2016.

Said, M. Y., R. N. Chunge, G. C. Craig, C. R. Thouless, R. F. W. Barnes, and H. T. Dublin. 1995. African elephant database 1995. Occasional Papers, IUCN (World Conservation Union) Species Survival Commission 11:1–225.

Salas, L., C. Dickman, and K. Helgen. 2008. *Phalanger mimicus*. The IUCN Red List of Threatened Species 2008: e.T136450A4293490. http://dx.doi.org/10.2305/IUCN.UK.2008.RLTS.T136450A4293490.en. Downloaded on 30 December 2015.

Salas, L., C. Dickman, K. Helgen, and T. Flannery. 2008a. *Ailurops ursinus*. The IUCN Red List of Threatened Species 2008: e.T40637A10346312. http://dx.doi.org/10.2305/IUCN.UK.2008.RLTS.T40637A10346312.en. Downloaded on 30 December 2015.

Salas, L., C. Dickman, K. Helgen, S. Burnett, and R. Martin. 2008b. *Dactylopsila trivirgata*. The IUCN Red List of Threatened Species 2008: e.T6226A12585182. http://dx.doi.org/10.2305/IUCN.UK.2008.RLTS.T6226A12585182.en. Downloaded on 30 December 2015.

Salton, J. A., and F. S. Szalay. 2004. The tarsal complex of Afro-Malagasy Tenrecoidea: a search for phylogenetically meaningful characters. Journal of Mammalian Evolution 11:73–104.

Sánchez-Villagra, M.R., Y. Narita, and S. Kuratani. 2007. Thoracolumbar vertebral number: the first skeletal synapomorphy for afrotherian mammals. Systematics and Biodiversity 5:1–7.

Sanders, W. J., E. Gheerbrant, J. M. Harris, H. Saegusa, and C. Delmer. 2010. Proboscidea. Pp. 161–251 in Cenozoic mammals of Africa (L. Werdelin and W. J. Sanders, eds.). University of California Press, Berkeley, California,

Sanderson, K. J. 2004. A review of brain studies of possums and gliders. Pp. 177–185 in The biology of Australian possums and gliders (R. L. Goldingay and S. M. Jackson, eds.). Surrey Beatty & Sons, Chipping Norton, New South Wales, Australia.

Sandom, C., S. Faurby, B. Sandel, and J.-C. Svenning. 2014. Global late Quaternary megafauna extinctions linked to humans, not climate change. Proceedings of the Royal Society B 281: 20133254. http://dx.doi.org/10.1098/rspb.2013.3254. Downloaded on 12 February 2015.

Sanson, G. D., and S. K. Churchill. 2008. Nabarlek *Petrogale concinna* (Gould, 1842). Pp. 370–371 in The mammals of Australia. 3rd edition (S. Van Dyck and R. Strahan, eds.). Reed New Holland, Sydney, Australia.

Santiapillai, C., M. R. Chambers, and N. Ishwaran. 1984. Aspects of the ecology of the Asian elephant *Elephas maximus* L. in the Ruhuna National Park, Sri Lanka. Biological Conservation 29:47–61.

Santiapillai, C., and P. Jackson. 1990. The Asian elephant: an action plan for its conservation. IUCN (World Conservation Union), Gland, Switzerland.

Santori, R. T., O. Rocha-Barbosa, M. V. Vieira, J. A. Magnan-Neto, and M. F. C. Loguercio. 2005. Locomotion in aquatic, terrestrial, and arboreal habitat of thick-tailed opossum, *Lutreolina crassicaudata* (Desmarest, 1804). Journal of Mammalogy 86:902–908.

Santos-Filho, M., M. N. F. da Silva, B. A. Costa, C. G. Bantel, C. L. G. Vieira, D. J. Silva, and A. M. R. Franco. 2007. New records of *Glironia venusta* (Thomas, 1912) (Mammalia, Didelphidae), from the Amazon and Paraguay Basins, Brazil. Mastozoologia Neotropical 14:103–105.

Sargis, E. J. 2001. A preliminary qualitative analysis of the axial skeleton of tupaiids (Mammalia, Scandentia): functional morphology and phylogenetic implications. Journal of Zoology 253:473–483.

Sargis, E. J. 2002a. The postcranial morphology of *Ptilocercus lowii* (Scandentia, Tupaiidae): an analysis of primatomorphan and volitantian characters. Journal of Mammalian Evolution 9:137–160.

Sargis, E. J. 2002b. Functional morphology of the forelimb of tupaiids (Mammalia, Scandentia) and its phylogenetic implications. Journal of Morphology 253:10–42.

Sargis, E. J. 2002c. Functional morphology of the hindlimb of tupaiids (Mammalia, Scandentia) and its phylogenetic implications. Journal of Morphology 254:149–185.

Sargis, E. J. 2002d. A multivariate analysis of the postcranium of tree shrews (Scandentia, Tupaiidae) and its taxonomic implications. Mammalia 66:579–598.

Sargis, E. J. 2004. New views on tree shrews: the role of tupaiids in primate supraordinal relationships. Evolutionary Anthropology 13:56–66.

Sargis, E. J., K. K. Campbell, and L. E. Olson. 2014. Taxonomic boundaries and craniometric variation in the treeshrews (Scandentia, Tupaiidae) from the Palawan Faunal Region. Journal of Mammalian Evolution 21:111–123.

Sargis, E. J., N. Woodman, N. C. Morningstar, A. T. Reese, and L. E. Olson. 2013a. Morphological distinctiveness of Javan *Tupaia hypochrysa* (Scandentia, Tupaiidae). Journal of Mammalogy 94:938–947.

Sargis, E. J., N. Woodman, A. T. Reese, and L. E. Olson. 2013b. Using hand proportions to test taxonomic boundaries within the *Tupaia glis* species complex (Scandentia, Tupaiidae). Journal of Mammalogy 94:183–201.

Sargis, E. J., N. Woodman, N. C. Morningstar, A. T. Reese, and L. E. Olson. 2014. Island history affects faunal composition: the treeshrews (Mammalia: Scandentia: Tupaiidae) from the Mentawai and Batu Islands, Indonesia. Biological Journal of the Linnean Society 111:290–304.

Sarich, V. M. 1985. Xenarthran systematics: albumin immunological evidence. Pp. 77–81 in The evolution and ecology of armadillos, sloths, and vermilinguas (G. G. Montgomery, ed.). Smithsonian Institution Press, Washington, DC.

Sarich, V. M., J. M. Lowenstein, and B. J. Richardson. 1982. Phylogenetic relationships of *Thylacinus cynocephalus* (Marsupialia) as reflected in comparative serology. Pp. 703–709 in Carnivorous marsupials (M. Archer, ed.). Royal Zoological Society of New South Wales, Mosman, New South Wales, Australia.

Satizábal, P., A. A. Mignucci-Giannoni, S. Duchêne, D. Caicedo-Herrera, C. M. Perea-Sicchar, C. R. García-Dávila, F. Trujillo, and S. J. Caballero. 2012. Phylogeography and sex-biased dispersal across riverine manatee populations (*Trichechus inunguis* and *Trichechus manatus*) in South America. PLoS ONE 7(12):e52468. doi:10.1371/journal.pone.0052468.

Sauer, E. G. F. 1973. Zum Sozialverhalten der kurzohrigen Elefantenspitzmaus, *Macroscelides proboscideus*. Zeitschrift für Säugetierkunde 38:65–97.

Sauthier, D. E. U., M. Carrera, and U. F. J. Pardinas. 2007. Mammalia, Marsupialia, Didelphidae, *Lestodelphys halli*: new records, distribution extension and filling gaps. Check List 3:137–140.

Savinetsky, A. B., N. K. Kiseleva, and B. F. Khassanov. 2004. Dynamics of sea mammal and bird populations of the Bering Sea region over the last several millennia. Palaeogeography, Palaeoclimatology, Palaeoecology 209:335–352.

Scalici, M., and F. Panchetti. 2011. Morphological cranial diversity contributes to phylogeny in soft-furred sengis (Afrotheria, Macroscelidea). Zoology (Jena) 114:85–94.

Scally, M. O. Madsen, C. J. Douady, W. W. de Jong, M. J. Stanhope, and M. S. Springer. 2001. Molecular evidence for the major clades of placental mammals. Journal of Mammalian Evolution 8:239–277.

Scarlett, N. 1969. The bilby, *Thylacomys lagotis*, in Victoria. Victorian Naturalist 86:292–294.

Scheffer, V. B. 1972. The weight of the Steller sea cow. Journal of Mammalogy 53:912–914.

Scheich, H., G. Langner, C. Tidemann, R. B. Coles, and A. Guppy. 1986. Electroreception and electrolocation in platypus. Nature 319:401–402.

Schliebe, S., and K. Johnson. 2008. Endangered and Threatened Wildlife and Plants; determination of threatened status for the polar bear (*Ursus maritimus*) throughout its range. Federal Register 73:28212–28303.

Schlitter, D. A. 2005a. Order Macroscelidea. Pp. 82–85 in Mammal species of the world: a taxonomic and geographic reference. 3rd edition (D. E. Wilson and D. M. Reeder, eds.). Johns Hopkins University Press, Baltimore, Maryland.

Schlitter, D. A. 2005b. Order Tubulidentata. P. 86 in Mammal species of the world: a taxonomic and geographic reference. 3rd edition (D. E. Wilson and D. M. Reeder, eds.). Johns Hopkins University Press, Baltimore, Maryland.

Schmidt, T. L. 2012. Ethogram of the giant anteater (*Myrmecophaga tridactyla*) in captivity: an experience in the Temaikèn Foundation. Edentata 13:38–48.

Schmitt, D. 2008. View from the big top. Why elephants belong in North American circuses. Pp. 227–234 in Elephants and ethics, toward a morality of existence (C. Wemmer and C. A. Christen, eds.). Johns Hopkins University Press, Baltimore, Maryland.

Schmitt, L. H., A. J. Bradley, C. M. Kemper, D. J. Kitchener, W. F. Humphreys, and R. A. How. 1989. Ecology and physiology of the northern quoll, *Dasyurus hallucatus* (Marsupialia, Dasyuridae), at Mitchell Plateau, Kimberley, Western Australia. Journal of Zoology 217:539–558.

Schoeman, S., N. C. Bennett, M. Van Der Merwe, and A. S. Schoeman. 2004. Aseasonal reproduction in the Hottentot golden mole, *Amblysomus hottentotus* (Afrosoricida: Chrysochloridae) from KwaZulu-Natal, South Africa. African Zoology 39:41–46.

Schubert, M. 2011. A summary of the social system of the round-eared sengi. Afrotherian Conservation 8:12–13.

Schubert, M., N. Pillay, D. O. Ribble, and C. Schradin. 2009a. The round-eared sengi and the evolution of social monogamy: factors that constrain males to live with a single female. Ethology 115:972–985.

Schubert, M., C. Schradin, H. G. Rödel, N. Pillay, and D. O. Ribble. 2009b. Male mate guarding in a socially monogamous mammal, the round-eared sengi: on costs and trade-offs. Behavioral Ecology and Sociobiology 64:257–264.

Schulz, M., G. Wilks, and L. Broome. 2012. An uncharacteristic new population of the mountain pygmy-possum *Burramys parvus* in New South Wales. Australian Zoologist 36:22–28.

Schuttler, S. G., S. Blake, and L. S. Eggert. 2012. Movement patterns and spatial relationships among African forest elephants. Biotropica 44:445–448.

Schuttler S. G., J. A. Philbrick, K. J. Jeffery, and L. S. Eggert. 2014a. Fine-scale genetic structure and cryptic associations reveal evidence of kin-based sociality in the African forest elephant. PLoS ONE 9(2):e88074. doi:10.1371/journal.pone.0088074.

Schuttler, S. G., A. Whittaker, K. J. Jeffery, and L. S. Eggert. 2014b. African forest elephant social networks: fission-fusion dynamics, but fewer associations. Endangered Species Research 25:165–173.

Schwartz, C. W., and E. R. Schwartz. 1959. The wild mammals of Missouri. University of Missouri Press, Columbia, Missouri.

Schwartz, G. T., D. T. Rasmussen, and R. J. Smith. 1995. Body-size diversity and community structure and fossil hyracoids. Journal of Mammalogy 76:1088–1099.

Sclater, W. L. 1900. The mammals of South Africa. R. H. Porter, London, United Kingdom.

Scott, L. K., I. D. Hume, and C. R. Dickman. 1999. Ecology and population biology of long-nosed bandicoots (*Perameles nasuta*) at North Head, Sydney Harbour National Park. Wildlife Research 26:805–821.

Scott, M. P. 1987. The effect of mating and agonistic experience on adrenal function and mortality of male *Antechinus stuartii* (Marsupialia). Journal of Mammalogy 68:479–486.

Seebeck, J. H. 1981. *Potorous tridactylus* (Kerr) (Marsupialia: Macropodidae): its distribution, status, and habitat preferences in Victoria. Australian Wildlife Research 8:285–306.

Seebeck, J. H. 1992a. Breeding, growth, and development of captive *Potorous longipes* (Marsupialia: Potoroidae); and a comparison with *P. tridactylus*. Australian Mammalogy 15:37–45.

Seebeck, J. H. 1992b. Sub-fossil potoroos in south-eastern Australia; with a record of *Potorous longipes* from New South Wales. Victorian Naturalist 109:173–176.

Seebeck, J. H. 2001. *Perameles gunnii*. Mammalian Species 654:1–8.

Seebeck, J. H., A. F. Bennett, and D. J. Scotts. 1989. Ecology of the Potoroidae—a review. Pp. 67–88 in Kangaroos, wallabies and rat-kangaroos (G. C. Grigg, P. Jarman, and I. Hume, eds.). Surrey Beatty & Sons, Chipping Norton, New South Wales, Australia.

Seebeck, J. H., and P. G. Johnston. 1980. *Potorous longipes* (Marsupialia: Macropodidae); a new species from eastern Victoria. Australian Journal of Zoology 28:119–134.

Seebeck, J. H., and P. W. Menkhorst. 2008. Eastern barred bandicoot *Perameles gunnii* (Gray, 1838). Pp. 186–188 in The mammals of Australia. 3rd edition (S. Van Dyck and R. Strahan, eds.). Reed New Holland, Sydney, Australia.

Seebeck, J. H., and R. W. Rose. 1989. Potoroidae. Pp. 716–739 in Fauna of Australia: Mammalia, volume 1B (D. W. Walton and B. J. Richardson, eds.). Australian Government Publishing Service, Canberra, Australia.

Seebeck, J. H., P. R. Brown, R. L. Wallis, and C. M. Kemper (eds). 1990. Bandicoots and bilbies. Surrey Beatty & Sons, Chipping Norton, New South Wales, Australia.

Séguignes, M. 1989. Contribution à l'étude de la reproduction d'*Elephantulus rozeti* (Insectivora, Macroscelididae). Mammalia 53:377–384.

Seiffert, E. R. 2007. A new estimate of afrotherian phylogeny based on simultaneous analysis of genomic, morphological, and fossil evidence. BMC Evolutionary Biology 7(224).

Seiffert, E. R. 2010. The oldest and youngest records of afrosoricid placentals from the Fayum Depression of northern Egypt. Acta Palaeontologica Polonica 55:599–616.

Seiffert, E. R. 2013. Superorder Tethytheria. P. 172 in Mammals of Africa, volume 1, Introductory chapters and Afrotheria (J. Kingdon, D. Happold, M. Hoffmann, T. Butynski, M. Happold, and J. Kalina, eds.). Bloomsbury, London, United Kingdom.

Self-Sullivan, C., and A. Mignucci-Giannoni. 2008. *Trichechus manatus* ssp. *manatus*. 2014. The IUCN Red List of Threatened Species 2008: e.T22105A9359161. http://dx.doi.org/10.2305/IUCN.UK.2008.RLTS.T22105A9359161.en. Downloaded on 02 May 2015.

Self-Sullivan, C., and A. A. Mignucci-Giannoni. 2012. West Indian manatees (*Trichechus manatus*) in the wider Caribbean region. Pp. 36–46 in Sirenian conservation. Issues and strategies in developing countries (E. M. Hines, J. E. Reynolds, III, L. V. Aragones, A. A. Mignucci-Giannoni, and M. Marmontel, eds.). University Press of Florida, Gainesville, Florida.

Serena, M., and T. R. Soderquist. 1988. Growth and development of pouch young of wild and captive *Dasyurus geoffroii* (Marsupialia: Dasyuridae). Australian Journal of Zoology 36:533–543.

Serena, M., and T. R. Soderquist. 1989a. Nursery dens of *Dasyurus geoffroii* (Marsupialia: Dasyuridae), with notes on nest building behaviour. Australian Mammalogy 12:35–36.

Serena, M., and T. R. Soderquist. 1989b. Spatial organization of a riparian population of the carnivorous marsupial *Dasyurus geoffroii*. Journal of Zoology 219:373–383.

Serena, M., and T. R. Soderquist. 1990. Occurrence and outcome of polyoestry in wild western quolls, *Dasyurus geoffroii* (Marsupialia: Dasyuridae). Australian Mammalogy 13:205–208.

Serena, M., and T. Soderquist. 2008. Western quoll *Dasyurus geoffroii* (Gould, 1841). Pp. 54–56 in The mammals of Australia. 3rd edition (S. Van Dyck and R. Strahan, eds.). Reed New Holland, Sydney, Australia.

Serena, M., and G. Williams. 2010. Factors contributing to platypus mortality in Victoria. Victorian Naturalist 127:178–183.

Serena, M., and G. Williams. 2013. Movements and cumulative range size of the platypus (*Ornithorhynchus anatinus*) inferred from mark-recapture studies. Australian Journal of Zoology 60:352–359.

Seymour, R. S., P. C. Withers, and W. W. Weathers. 1998. Energetics of burrowing, running, and free-living in the Namib Desert golden mole (*Eremitalpa namibensis*). Journal of Zoology 244:107–117.

Shalmon, B. 2000. The ancient world of mammals in rock carvings from the southern Negev, eastern Sinai, and southern Jordan. Israel Journal of Zoology 46:172–173.

Sharman, G. B. 1970. Reproductive physiology of marsupials. Science 167:1221–1228.

Sharman, G. B., R. L. Close, and G. M. Maynes. 1990. Chromosome evolution, phylogeny, and speciation of rock wallabies (*Petrogale*: Macropodidae). Australian Journal of Zoology 37:351–363.

Sharman, G. B., and P. E. Pilton. 1964. The life history and reproduction of the red kangaroo (*Megaleia rufa*). Proceedings of the Zoological Society of London 142:29–48.

Sharman, G. B., C. E. Murtagh, P. M. Johnson, and C. M. Weaver. 1980. The chromosomes of a rat-kangaroo attributable to *Bettongia tropica* (Marsupialia: Macropodidae). Australian Journal of Zoology 28:59–63.

Sharpe, D. J., and R. L. Goldingay. 2007. Home range of the Australian squirrel glider, *Petaurus norfolcensis* (Diprotodontia). Journal of Mammalogy 88:1515–1522.

Sharpe D. J., and R. L. Goldingay. 2009. Vocal behaviour of the squirrel glider (*Petaurus norfolcensis*). Australian Journal of Zoology 57:55–64.

Shaw, C. A., and H. G. McDonald. 1987. First record of giant anteater (Xenarthra, Myrmecophagidae) in North America. Science 236:186–188.

Shaw, G. 2006. Reproduction. Pp. 83–107 in Marsupials (P. J. Armati, C. R. Dickman, and I. D. Hume, eds.). Cambridge University Press, Cambridge, United Kingdom.

Shaw, J. S., T. S. Carter, and J. C. Machado-Neto. 1985. Ecology of the giant anteater *Myrmecophaga tridactyla* in Serra da Canastra, Minas Gerais, Brazil: a pilot study. Pp. 379–384 in The evolution and ecology of armadillos, sloths, and vermilinguas (G. G. Montgomery, ed.). Smithsonian Institution Press, Washington, DC.

Shaw, J. S., J. C. Machado-Neto, and T. S. Carter. 1987. Behavior of free-living giant anteaters (*Myrmecophaga tridactyla*). Biotrópica 19:255–259.

Sheppe, W. A. 1973. Notes on Zambian rodents and shrews. Puku 7:167–190.

Sherwin, W. B., and P. R. Brown. 1990. Problems in the estimation of the effective population size of the eastern barred bandicoot *Perameles gunnii* at Hamilton, Victoria. Pp. 367–374 in Bandicoots and bilbies (J. H. Seebeck, P. R. Brown, R. L. Wallis, and C. M. Kemper, eds.). Surrey Beatty & Sons, Chipping Norton, New South Wales, Australia.

Shevill, D. I., and C. N. Johnson. 2008. Diet and breeding of the rufous spiny bandicoot *Echymipera rufescens australis*, Iron Range, Cape York Peninsula. Australian Mammalogy 29:169–175.

Shield, J. 1968. Reproduction of the quokka, *Setonix brachyurus*, in captivity. Journal of Zoology 155:427–444.

Shim, P. S. 2003. Another look at the Borneo elephant. Sabah Society Journal 20:7–14.

Shimmin, G.A., J. Skinner, and R. V. Baudinette. 2002. The warren architecture and environment of the southern hairy-nosed wombat (*Lasiorhinus latifrons*). Journal of Zoology 258:469–477.

Shimmin, G. A., D. A. Taggart, and P. D. Temple-Smith. 2002. Mating behaviour in the agile antechinus *Antechinus agilis* (Marsupialia: Dasyuridae). Journal of Zoology 258:39–48.

Shirakihara, M., H. Yoshida, H. Yokochi, H. Ogawa, T. Hosokawa, N. Higashi, and T. Kasuya. 2007. Current status and conservation needs of dugongs in southern Japan. Marine Mammal Science 23:694–706.

Short, H. L. (Office of Scientific Authority, Fish and Wildlife Service, United States Department of the Interior). 1992. Endangered and Threatened Wildlife and Plants; retention of threatened status for the continental population of the African elephant. Federal Register 57:35473–35485.

Short, H. L. (Office of Scientific Authority, Fish and Wildlife Service, United States Department of the Interior). 1995. Endangered and Threatened Wildlife and Plants; removal of three kangaroos from the List of Endangered and Threatened Wildlife. Federal Register 60:12887–12906.

Short, J., and G. Milkovits. 1990. Distribution and status of the brush-tailed rock-wallaby in south-eastern Australia. Australian Wildlife Research 17:169–179.

Short, J., and B. Turner. 1991. Distribution and abundance of spectacled hare-wallabies and euros on Barrow Island, Western Australia. Wildlife Research 18:421–429.

Short, J., and B. Turner. 1992. The distribution and abundance of the banded and rufous hare-wallabies, *Lagostrophus fasciatus* and *Lagorchestes hirsutus*. Biological Conservation 60:157–166.

Short, J., and B. Turner. 1993. The distribution and abundance of the burrowing bettong (Marsupialia: Macropodidae). Wildlife Research 20:525–534.

Short, J., S. D. Bradshaw, J. Giles, R. I. T. Prince, and G. R. Wilson. 1992. Reintroduction of macropods (Marsupialia: Macropodidae) in Australia—a review. Biological Conservation 62:189–204.

Shoshani, J. 1992a. The controversy continues: an overview of evidence for Hyracoidea-Tethytheria affinity. Israel Journal of Zoology 38:233–244.

Shoshani, J. 1992b. Cuvier vis-à-vis Huxley on the relationship of Hyracoidea and an update on an old controversy. Pp. 103–112 in "Ongulés / Ungulates 91" (F. Spitz, G. Janeau, G. Gonzalez, and S. Aulagnier, eds.). S.F.E.P.M.-I.R.G.M., Paris, France.

Shoshani, J. 1993. Hyracoidea-Tethytheria affinity based on myological data. Pp. 235–256 in Mammal phylogeny: Mesozoic differentiation, multituberculates, monotremes, early therians, and marsupials (F. S. Szalay, M. J. Novacek, and M. C. McKenna, eds.). Springer-Verlag, New York, New York.

Shoshani, J. 1998. Understanding proboscidean evolution: a formidable task. Tree 13:480–487.

Shoshani, J. 2005a. Order Hyracoidea. Pp. 87–89 in Mammal species of the world: a taxonomic and geographic reference. 3rd edition (D. E. Wilson and D. M. Reeder, eds.). Johns Hopkins University Press, Baltimore, Maryland.

Shoshani, J. 2005b. Order Proboscidea. Pp. 90–91 in Mammal species of the world: a taxonomic and geographic reference. 3rd edition (D. E. Wilson and D. M. Reeder, eds.). Johns Hopkins University Press, Baltimore, Maryland.

Shoshani, J. 2005c. Order Sirenia. Pp. 92–93 in Mammal species of the world: a taxonomic and geographic reference. 3rd edition (D. E. Wilson and D. M. Reeder, eds.). Johns Hopkins University Press, Baltimore, Maryland.

Shoshani, J., and J. F. Eisenberg. 1982. *Elephas maximus*. Mammalian Species 182:1–8.

Shoshani, J., C. A. Goldman, and J. G. M. Thewissen. 1988. *Orycteropus afer*. Mammalian Species 300:1–8.

Shoshani, J., and M. C. McKenna. 1998. Higher taxonomic relationships among extant mammals based on morphology, with selected comparisons of results from molecular data. Molecular Phylogenetics and Evolution 9:572–584.

Shoshani, J., and P. Tassy (eds.). 1996. The Proboscidea. Evolution and palaeoecology of elephants and their relatives. Oxford University Press, Oxford, United Kingdom.

Shoshani, J., and P. Tassy. 2005. Advances in proboscidean taxonomy & classification, anatomy & physiology, and ecology & behavior. Quaternary International 126–128:5–20.

Shoshani, J., R. C. Walter, M. Abraha, S. Berhe, P. Tassy, W. J. Sanders, G. H. Marchant, Y. Libsekal, T. Ghirmai, and D. Zinner. 2006.

A proboscidean from the late Oligocene of Eritrea, a "missing link" between early Elephantiformes and Elephantimorpha, and biogeographic implications. Proceedings of the National Academy of Sciences of the United States of America 103:17296–17301.

Shoshani, J., M. P. Ferretti, A. M. Lister, L. D. Agenbroad, H. Saegusa, D. Mol, and K. Takahashi. 2007. Relationships within the Elephantinae using hyoid characters. Quaternary International 169–170:174–185.

Shultz, S., and D. Roberts. 2013. *Dendrohyrax dorsalis* Western tree hyrax. Pp. 155–157 in Mammals of Africa, volume 1, Introductory chapters and Afrotheria (J. Kingdon, D. Happold, M. Hoffmann, T. Butynski, M. Happold, and J. Kalina, eds.). Bloomsbury, London, United Kingdom.

Sigg, D. P., and A. W. Goldizen. 2006. Male reproductive tactics and female choice in the solitary, promiscuous bridled nailtail wallaby (*Onychogalea fraenata*). Journal of Mammalogy 87:461–469.

Sikes, S. K. 1971. The natural history of the African elephant. American Elsevier, New York, New York.

Silcox, M. T., J. I. Bloch, E. J. Sargis, and D. M. Boyer. 2005. Euarchonta (Dermoptera, Scandentia, Primates). Pp. 127–144 in The rise of placental mammals: origins and relationships of the major clades (K. D. Rose and J. D. Archibald, eds.). Johns Hopkins University Press, Baltimore, Maryland.

Silva, M., and A. Araújo. 2001. Distribution and current status of the West African manatee (*Trichechus senegalensis*) in Guinea-Bissau. Marine Mammal Science 17:418–424.

Silveira, L., A. T. de Almeida Jácomo, N. M. Torres, R. Sollmann, and C. Vynne. 2009. Ecology of the giant armadillo (*Priodontes maximus*) in the grasslands of central Brazil. Edentata 8–10:25–34.

Simmons, N. B. 1993. Phylogeny of Multituberculata. Pp. 146–164 in Mammal phylogeny: Mesozoic differentiation, multituberculates, monotremes, early therians, and marsupials (F. S. Szalay, M. J. Novacek, and M. C. McKenna, eds.). Springer-Verlag, New York, New York.

Simmons, N. B. 1995. Bat relationships and the origin of flight. Symposia of the Zoological Society of London 67:27–43.

Simmons, N. B. 2005. Chiroptera. Pp. 159–174 in The rise of placental mammals: origins and relationships of the major clades (K. D. Rose and J. D. Archibald, eds.). Johns Hopkins University Press, Baltimore, Maryland.

Simonetta, A. M. 1968. A new golden mole from Somalia with an appendix on the taxonomy of the family Chrysochloridae (Mammalia, Insectivora). Italian Journal of Zoology, Supplement, new series 2:27–55.

Simonetta, A. M. 1979. First record of *Caluromysiops* from Colombia. Mammalia 43:247–248.

Simons, E. L., and T. M. Bown. 1984. A new species of *Peratherium* (Didelphidae: Polyprotodonta): the first African marsupial. Journal of Mammalogy 65:539–548.

Simons, E. L., P. A. Holroyd, and T. M. Bown. 1991. Early Tertiary elephant-shrews from Egypt and the origin of the Macroscelidea. Proceedings of the National Academy of Sciences of the United States of America 88:9734–9737.

Simpson, G. G. 1945. The principles of classification and a classification of the mammals. Bulletin of the American Museum of Natural History 85:1–350.

Sinclair, E. A., and M. Westerman. 1997. Phylogenetic relationships within the genus *Potorous* (Marsupialia: Potoroidae) based on allozyme electrophoresis and sequence analysis. Journal of Mammalian Evolution 4:147–161.

Skarpe, C., P. A. Aarrestad, H. P. Andreassen, S. S. Dhillion, T. Dimakatso, J. T. du Toit, D. J. Halley, H. Hytteborn, S. Makhabu, M. Mari, W. Marokane, G. Masunga, D. Modise, S. R. Moe, R. Mojaphoko, D. Mosugelo, S. Motsumi, G. Neo-Mahupeleng, M. Ramotadima, L. Rutina, L. Sechele, T. B. Sejoe, S. Stokke, J. E. Swenson, C. Taolo, M. Vandewalle, and P. Wegge. 2004. The return of the giants: ecological effects of an increasing elephant population. Ambio 33:276–282.

Sleightholme, S. R., and C. R. Campbell. 2014. A retrospective review of the breeding season of the thylacine; Guiler revisited. Australian Zoologist 37:238–244.

Slotow, R., G. van Dyk, J. Poole, B. Page, and A. Klocke. 2000. Older bull elephants control young males. Nature 408:425–426.

Smit, H. A., T. J. Robinson, J. Watson, and B. Jansen van Vuuren. 2008. A new species of elephant-shrew (Afrotheria: Macroscelidea: *Elephantulus*) from South Africa. Journal of Mammalogy 89:1257–1268.

Smit, H. A., B. Jansen van Vuuren, P. C. M. O'Brien, M. Ferguson-Smith, F. Yang, and T. J. Robinson. 2011. Phylogenetic relationships of elephant-shrews (Afrotheria, Macroscelididae). Journal of Zoology 284:133–143.

Smith, A. 1982. Is the striped possum (*Dactylopsila trivirgata*; Marsupialia, Petauridae) an arboreal anteater? Australian Mammalogy 5:229–234.

Smith, A. 1984a. Demographic consequences of reproduction, dispersal, and social interaction in a population of Leadbeater's possum. Pp. 359–373 in Possums and gliders (A. Smith and I. Hume, eds.). Surrey Beatty & Sons, Norton, New South Wales, Australia.

Smith, A. 1984b. Diet of Leadbeater's possum, *Gymnobelideus leadbeateri* (Marsupialia). Australian Wildlife Research 11:265–273.

Smith, A. P., and D. K. P. Harley. 2008. Leadbeater's possum *Gymnobelideus leadbeateri* (McCoy, 1867). Pp. 226–228 in The mammals of Australia. 3rd edition (S. Van Dyck and R. Strahan, eds.). Reed New Holland, Sydney, Australia.

Smith, K. M., and D. C. Fisher. 2013. Sexual dimorphism and intergeneric variation in proboscidean tusks: multivariate assessment of American mastodons (*Mammut americanum*) and extant African elephants. Journal of Mammalian Evolution 20:337–355.

Smith, M. 1982. Review of the thylacine (Marsupialia, Thylacinidae). Pp. 237–253 in Carnivorous marsupials (M. Archer, ed.). Royal Zoological Society of New South Wales, Mosman, New South Wales, Australia.

Smith, M. J. 1971. Breeding the sugar-glider *Petaurus breviceps* in captivity; and growth of pouch young. International Zoo Yearbook 11:26–28.

Smith, M. J. 1973. *Petaurus breviceps*. Mammalian Species 30:1–5.

Smith, M. J., B. K. Brown, and H. J. Frith. 1969. Breeding of the brush-tailed possum, *Trichosurus vulpecula* (Kerr), in New South Wales. CSIRO Wildlife Research 14:181–193.

Smith, M. J., and R. A. How. 1973. Reproduction in the mountain possum, *Trichosurus caninus* (Ogilby), in captivity. Australian Journal of Zoology 21:321–329.

Smith, M. J., and P. M. Johnson. 2008. Desert rat-kangaroo *Caloprymnus campestris* (Gould, 1843). Pp. 293–294 in The mammals of Australia. 3rd edition (S. Van Dyck and R. Strahan, eds.). Reed New Holland, Sydney, Australia.

Smith, M. J., and G. C. Medlin. 1982. Dasyurids of the northern Flinders Ranges before pastoral development. Pp. 563–572 in Carnivorous marsupials (M. Archer, ed.). Royal Zoological Society of New South Wales, Mosman, New South Wales, Australia.

Smith, M. J., and A. C. Robinson. 2008. Toolache wallaby *Macropus greyi* (Waterhouse, 1843). Pp. 338–339 in The mammals of Australia. 3rd edition (S. Van Dyck and R. Strahan, eds.). Reed New Holland, Sydney, Australia.

Smith, N. S., and I. O. Buss. 1973. Reproductive ecology of the female African elephant. Journal of Wildlife Management 37:524–534.

Smith, P. R., D. Owen, K. Atkinson, H. del Castillo, and E. Northcote-Smith. 2011. First records of the southern naked-tailed armadillo *Cabassous unicinctus* (Cingulata: Dasypodidae) in Paraguay. Edentata 12:53–57.

Smith, R. F. C. 1969. Studies on the marsupial glider, *Schoinobates volans* (Kerr). Australian Journal of Zoology 17:625–636.

Smith, R. N. (Director, Fish and Wildlife Service, United States Department of the Interior). 1991. Endangered and Threatened Wildlife and Plants; proposed endangered status for certain populations of the African elephant and revision of special rule. Federal Register 56:11392–11401.

Smithers, R. H. N. 1971. The mammals of Botswana. Trustees of the National Museums and Monuments of Rhodesia Museum Memoir 4:1–340.

Smithers, R. H. N. 1986. South African red data book—terrestrial mammals. South Africa National Scientific Programmes Report 125:1–216.

Snow, D. W. 1970. The eastern Indian Ocean islands. A summary of their geography, fauna and flora. IUCN Publications (new series) 17:212–223.

Snyder, H. K., R. Maia, L. D'Alba, A. J. Shultz, K. M. C. Rowe, K. C. Rowe, and M. D. Shawkey. 2012. Iridescent colour production in hairs of blind golden moles (Chrysochloridae). Biology Letters 8:393–396.

Snyder, R. L., and S. C. Moore. 1968. Longevity of captive mammals in Philadelphia Zoo. International Zoo Yearbook 8:175–183.

Soarimalala, V., and S. M. Goodman. 2003. The food habits of Lipotyphla. Pp. 1203–1205 in The natural history of Madagascar (S. M. Goodman and J. P. Benstead, eds.). University of Chicago Press, Chicago, Illinois.

Soarimalala, V., and S. M. Goodman. 2008. New distributional records of the recently described and endangered shrew tenrec *Microgale nasoloi* (Tenrecidae: Afrosoricida) from central western Madagascar. Mammalian Biology 73:468–471.

Soarimalala, V., S. Goodman, and P. J. Stephenson. 2015. *Microgale cowani*. The IUCN Red List of Threatened Species 2015: e.T40586A21290559. http://dx.doi.org/10.2305/IUCN.UK.2015-4.RLTS.T40586A21290559.en. Downloaded on 18 January 2016.

Soarimalala, V., M. Raheriarisena, and S. M. Goodman. 2010. New distributional records from central-eastern Madagascar and patterns of morphological variation in the endangered shrew tenrec *Microgale jobihely* (Afrosoricida: Tenrecidae). Mammalia 74:187–198.

Soarimalala, V., P. J. Stephenson, and S. Goodman. 2015. *Microgale grandidieri*. The IUCN Red List of Threatened Species 2015: e.T54008309A54008345. http://dx.doi.org/10.2305/IUCN.UK.2015-4.RLTS.T54008309A54008345.en. Downloaded on 18 January 2016

Soderquist, T. R. 1993. Maternal strategies of *Phascogale tapoatafa* (Marsupialia: Dasyuridae). I. Breeding seasonality and maternal investment. Australian Journal of Zoology 41:549–566.

Soderquist, T. R. 1994. Anti-predator behaviour of the brush-tailed phascogale (*Phascogale tapoatafa*). Victorian Naturalist 111:22–24.

Soderquist, T. R. 1995. Spatial organization of the arboreal carnivorous marsupial *Phascogale tapoatafa*. Journal of Zoology 237:385–398.

Soderquist, T. R., and L. Ealey. 1994. Social interactions and mating strategies of a solitary carnivorous marsupial, *Phascogale tapoatafa*, in the wild. Wildlife Research 21:527–542.

Soderquist, T. R., and A. Lill. 1995. Natal dispersal and philopatry in the carnivorous marsupial *Phascogale tapoatafa* (Dasyuridae). Ethology 99:297–312.

Soderquist, T., and S. Rhind. 2008. Brush-tailed phascogale *Phascogale tapoatafa* (Meyer, 1793). Pp. 105–107 in The mammals of Australia. 3rd edition (S. Van Dyck and R. Strahan, eds.). Reed New Holland, Sydney, Australia.

Solari, S. 2003. Diversity and distribution of *Thylamys* (Didelphidae) in South America, with emphasis on species from the western side of the Andes. Pp. 82–101 in Predators with pouches: the biology of carnivorous marsupials (M. Jones, C. Dickman, and M. Archer, eds.). CSIRO Publishing, Collingwood, Victoria, Australia.

Solari, S. 2004. A new species of *Monodelphis* (Didelphimorphia : Didelphidae) from southeastern Perú. Mammalian Biology 69:145–152.

Solari, S. 2007. New species of *Monodelphis* (Didelphimorphia: Didelphidae) from Peru, with notes on *M. adusta* (Thomas, 1897). Journal of Mammalogy 88:319–329.

Solari, S. 2015a. *Marmosa andersoni*. The IUCN Red List of Threatened Species 2015: e.T12812A22174790. http://dx.doi.org/10.2305/IUCN.UK.2015-4.RLTS.T12812A22174790.en. Downloaded on 18 January 2016.

Solari, S. 2015b. *Thylamys tatei*. The IUCN Red List of Threatened Species 2015: e.T136243A22173132. http://dx.doi.org/10.2305/IUCN.UK.2015-4.RLTS.T136243A22173132.en. Downloaded on 27 December 2015.

Solari, S., and N. Cáceres. 2015. *Caluromysiops irrupta*. The IUCN Red List of Threatened Species 2015: e.T3651A22172207. http://dx.doi.org/10.2305/IUCN.UK.2015-4.RLTS.T3651A22172207.en. Downloaded on 05 January 2016.

Solari, S., and D. Lew. 2015. *Caluromys derbianus*. The IUCN Red List of Threatened Species 2015: e.T3650A22175821. http://dx.doi.org/10.2305/IUCN.UK.2015-4.RLTS.T3650A22175821.en. Downloaded on 18 January 2016.

Solari, S., and J. Martínez-Cerón. 2015a. *Caenolestes condorensis*. The IUCN Red List of Threatened Species 2015: e.T136743A22180165. http://dx.doi.org/10.2305/IUCN.UK.2015-4.RLTS.T136743A22180165.en. Downloaded on 18 January 2016.

Solari, S., and J. Martínez-Cerón. 2015b. *Caenolestes caniventer*. The IUCN Red List of Threatened Species 2015: e.T40521A22180055. http://dx.doi.org/10.2305/IUCN.UK.2015-4.RLTS.T40521A22180055.en. Downloaded on 18 January 2016.

Solari, S., and B. Patterson. 2011. *Marmosa phaea*. The IUCN Red List of Threatened Species 2011: e.T136244A4264648. http://dx.doi.org/10.2305/IUCN.UK.2011-2.RLTS.T136244A4264648.en. Downloaded on 08 November 2015.

Solari, S., and R. H. Pine. 2008. Rediscovery and redescription of *Marmosa* (*Stegomarmosa*) *andersoni* Pine (Mammalia: Didelphimorphia: Didelphidae), an endemic Peruvian mouse opossum, with a reassessment of its affinities. Zootaxa 1756:49–61.

Solari, S., and P. Teta. 2008. *Thylamys elegans*. The IUCN Red List of Threatened Species 2008: e.T40517A10326191. http://dx.doi.org/10.2305/IUCN.UK.2008.RLTS.T40517A10326191.en. Downloaded on 27 December 2015.

Solari, S., V. Pacheco, F. Vivar, and L. H. Emmons. 2012. A new species of *Monodelphis* (Mammalia: Didelphimorphia: Didelphidae) from the montane forests of central Perú. Proceedings of the Biological Society of Washington 125:295–307.

Solow, A. R., D. L. Roberts, and K. M. Robbirt. 2006. On the Pleistocene extinctions of Alaskan mammoths and horses. Proceedings of the National Academy of Sciences of the United States of America 103:7351–7353.

Soltis, J. 2010. Vocal communication in African elephants (*Loxodonta africana*). Zoo Biology 29:192–209.

Sorenson, M. W. 1970. Behavior of tree shrews. Pp. 141–194 in Primate Behavior, volume 1 (L. A. Rosenblum, ed.). Academic Press, New York, New York.

Sorenson, M. W. 1974. A review of aggressive behavior in the tree shrews. Pp. 13–30 in Primate aggression, territoriality, and xenophobia: a comparative perspective (R. L. Holloway, ed.). Academic Press, New York, New York.

Sorenson, M. W., and C. H. Conway. 1968. The social and reproductive behavior of *Tupaia montana* in captivity. Journal of Mammalogy 49:502–512.

Sousa-Lima, R., A. P. Paglia, G. A. B. da Fonseca. 2002. Signature information and individual recognition in the isolation calls of Amazonian manatees, *Trichechus inunguis* (Mammalia: Sirenia). Animal Behaviour 63:301–310.

Southgate, R. I. 1990. Distribution and abundance of the greater bilby *Macrotis lagotis* Reid (Marsupialia: Peramelidae). Pp. 293–302 in Bandicoots and bilbies (J. H. Seebeck, P. R. Brown, R. L. Wallis, and C. M. Kemper, eds.). Surrey Beatty & Sons, Chipping Norton, New South Wales, Australia.

Southgate, R. I. 2015. Family Thylacomyidae (greater bilby). Pp. 350–361 in Handbook of the mammals of the world, volume 5, Monotremes and marsupials (D. E. Wilson and R. A. Mittermeier, eds.). Lynx Edicions, Barcelona, Spain.

Southgate, R., and S. M. Carthew. 2006. Diet of the bilby (*Macrotis lagotis*) in relation to substrate, fire and rainfall characteristics in the Tanami Desert. Wildlife Research 33:507–519.

Southgate, R., C. Palmer, M. Adams, P. Masters, B. Triggs, and J. Woinarski. 1996. Population and habitat characteristics of the golden bandicoot (*Isoodon auratus*) on Marchinbar Island, Northern Territory. Wildlife Research 23:647–664.

Spencer, P. B. S., and H. Marsh. 1997. Microsatellite DNA fingerprinting confirms dizygotic twinning and paternity in the allied rock wallaby, *Petrogale assimilis* (Marsupialia: Macropodidae). Australian Mammalogy 19:279–280.

Spencer, P. B. S., S. G. Rhind, and M. D. B. Eldridge. 2001. Phylogeographic structure within *Phascogale* (Marsupialia: Dasyuridae) based on partial cytochrome *b* sequence. Australian Journal of Zoology 49:369–377.

Spinage, C. A. 1973. A review of ivory exploitation and elephant population trends in Africa. East African Wildlife Journal 11:281–289.

Spotorno, A. E., J. C. Marin, M. Yévenes, L. I. Walker, R. Fernandez-Donoso, J. Pincheira, M. S. Berrios, and R. E. Palma. 1997. Chromosome divergences among American marsupials and the Australian affinities of the American *Dromiciops*. Journal of Mammalian Evolution 4:259–269.

Springer, M. S., and W. J. Murphy. 2007. Mammalian evolution and biomedicine: new views from phylogeny. Biological Reviews 82:375–392.

Springer, M. S., J. A. W. Kirsch, K. Aplin, and T. Flannery. 1990. DNA hybridization, cladistics, and the phylogeny of phalangerid marsupials. Journal of Molecular Evolution 30:298–311.

Springer, M. S., G. C. Cleven, O. Madsen, W. W. de Jong, V. G. Waddell, H. M. Amrine, and M. J. Stanhope. 1997. Endemic African mammals shake the phylogenetic tree. Nature 388:61–64.

Springer, M. S., M. Westerman, J. R. Kavanagh, A. Burk, M. O. Woodburne, D. J. Kao, and Carey Krajewski. 1998. The origin of the Australasian marsupial fauna and the phylogenetic affinities of the enigmatic monito del monte and marsupial mole. Proceedings of the Royal Society of London B Biological Sciences 265:2381–2386.

Springer, M. S., W. J. Murphy, E. Eizirik, and S. J. O'Brien. 2005. Molecular evidence for major placental clades. Pp. 37–49 in The rise of placental mammals: origins and relationships of the major clades (K. D. Rose and J. D. Archibald, eds.). Johns Hopkins University Press, Baltimore, Maryland.

Springer, M. S., A. V. Signore, J. L. A. Paijmans, J. Vélez-Juarbe, D. P. Domning, C. E. Bauer, K. Heb, L. Crerar, P. F. Campos, W. J. Murphy, R. W. Meredith, J. Gatesy, E. Willerslev, R. D. E. MacPhee, M. Hofreiter, and K. L. Campbell. 2015. Interordinal gene capture, the phylogenetic position of Steller's sea cow based on molecular and morphological data, and the macroevolutionary history of Sirenia. Molecular Phylogenetics and Evolution 91:178–193.

Srivastava, R. K. 2005. An encounter with Kalyanai (short statured elephant). Their behaviour and habitat: a new race. Indian Forester 131:598.

Stafford, B. J. 2005. Order Dermoptera. P. 110 in Mammal species of the world: a taxonomic and geographic reference. 3rd edition (D. E. Wilson and D. M. Reeder, eds.). Johns Hopkins University Press, Baltimore, Maryland.

Stafford, B. J., and F. S. Szalay. 2000. Craniodental functional morphology and taxonomy of dermopterans. Journal of Mammalogy 81:360–385.

Stalder, G. L., N. Broadis, A. M. Adi, and U. Streicher. 2010. Field observation, capture and anesthesia of a colugo (*Galeopterus variegatus*) in Laos. Zoologische Garten 79:105–108.

Stanhope, M. J., O. Madsen, V. G. Waddell, G. C. Cleven, W. W. de Jong, and M. S. Springer. 1998a. Highly congruent molecular support for a diverse superordinal clade of endemic African mammals. Molecular Phylogenetics and Evolution 9:501–508.

Stanhope, M. J., V. G. Waddell, O. Madsen, W. de Jong, S. B. Hedges, G. C. Cleven, D. Kao, and M. S. Springer. 1998b. Molecular evidence for multiple origins of Insectivora and for a new order of endemic African insectivore mammals. Proceedings of the National Academy of Sciences of the United States of America 95:9967–9972.

Stannard, H. J., and J. M. Old. 2010. Observation of reproductive strategies of captive kultarrs (*Antechinomys laniger*). Australian Mammalogy 32:179–182.

Stannard, H. J., C. R. Borthwick, O. Ong, and J. M. Old. 2013. Longevity and breeding in captive red-tailed phascogales (*Phascogale calura*). Australian Mammalogy 35:217–219.

Start, A. N., D. Moro, M. Adams, and R. Bencini. 2006. Dunnarts from Boullanger Island: new evidence and reassessment of a taxonomic issue with resource implications. Australian Mammalogy 28:51–58.

Steadman, D. W., P. S. Martin, R. D. E. MacPhee, A. J. T. Jull, H. G. McDonald, C. A. Woods, M. Iturralde-Vinent, and G. W. L. Hodgins. 2005. Asynchronous extinction of late Quaternary sloths on continents and islands. Proceedings of the National Academy of Sciences of the United States of America 102:11763–11768.

Stein, B. R., and J. L. Patton. 2007. Genus *Lutreolina* O. (Thomas, 1910). Pp. 25–27 in Mammals of South America, volume 1, Marsupials, xenarthrans, shrews, and bats (A. L. Gardner, ed.). University of Chicago Press, Chicago, Illinois.

Steiner, C., and F. M. Catzeflis. 2003. Mitochondrial diversity and morphological variation of *Marmosa murina* (Didelphidae) in French Guiana. Journal of Mammalogy 84:822–831.

Stejneger, L. 1936. Georg Wilhelm Steller. Harvard University Press, Cambridge, Massachusetts.

Steller, G. W. 1751. De bestiis marinis (The beasts of the sea). Petersburg Imperial Academy of Science, St. Petersburg, Russia.

Steller, G. W. 1753. Ausführliche Beschreibung von sonderbaren Meerthieren, mit Erläuterungen und nöthigen Kupfern versehen. In Verlag Karl Christian Kümmel, Halle, Germany.

Steller, G. W. 1925. Bering's Voyages, volume II, Journal of the sea voyage from Kamchatka to America and return on the second expedition 1741–1742. American Geographical Society, New York, New York.

Stephenson, P. J. 1993. Reproductive biology of the large-eared tenrec, *Geogale aurita* (Insectivora: Tenrecidae). Mammalia 57:553–563.

Stephenson, P. J. 1994. Notes on the biology of the fossorial tenrec, *Oryzorictes hova* (Insectivora: Tenrecidae). Mammalia 58:312–315.

Stephenson, P. J. 1995. Taxonomy of shrew-tenrecs (*Microgale* spp.) from eastern and central Madagascar. Journal of Zoology 235:339–350.

Stephenson, P. J. 2003a. Lipotyphla (ex Insectivora: *Geogale aurita*, large-eared tenrec). Pp. 1265–1267 in The natural history of Madagascar (S. M. Goodman and J. P. Benstead, eds.). University of Chicago Press, Chicago, Illinois.

Stephenson, P. J. 2003b. *Hemicentetes*, streaked tenrecs, sora, tsora. Pp. 1281–1283 in The natural history of Madagascar (S. M. Goodman and J. P. Benstead, eds.). University of Chicago Press, Chicago, Illinois.

Stephenson, P. J. 2015. *Potamogale velox*. The IUCN Red List of Threatened Species 2015: e.T18095A21285012. http://dx.doi.org /10.2305/IUCN.UK.2015-4.RLTS.T18095A21285012.en. Downloaded on 18 January 2016

Stephenson, P. J., S. Goodman, and V. Soarimalala. 2015a. *Microgale thomasi*. The IUCN Red List of Threatened Species 2015: e.T133 55A21288000. http://dx.doi.org/10.2305/IUCN.UK.2015-4.RLTS .T13355A21288000.en. Downloaded on 18 January 2016.

Stephenson, P. J., S. Goodman, and V. Soarimalala. 2015b. *Microgale fotsifotsy*. The IUCN Red List of Threatened Species 2015: e.T620 13A21283193. http://dx.doi.org/10.2305/IUCN.UK.2015-4.RLTS .T62013A21283193.en. Downloaded on 18 January 2016.

Stephenson, P. J., S. Goodman, and V. Soarimalala. 2015c. *Microgale gracilis*. The IUCN Red List of Threatened Species 2015: e.T133 43A21284192. http://dx.doi.org/10.2305/IUCN.UK.2015-4.RLTS .T13343A21284192.en. Downloaded on 18 January 2016.

Stephenson, P. J., S. Goodman, and V. Soarimalala. 2015d. *Microgale soricoides*. The IUCN Red List of Threatened Species 2015: e.T620 18A21285156. http://dx.doi.org/10.2305/IUCN.UK.2015-4.RLTS .T62018A21285156.en. Downloaded on 18 January 2016.

Stephenson, P. J., and P. A. Racey. 1994. Seasonal variation in resting metabolic rate and body temperature of streaked tenrecs, *Hemicentetes nigriceps* and *H. semispinosus* (Insectivora: Tenrecidae). Journal of Zoology 232:285–294.

Stiles, D. 2004. The ivory trade and elephant conservation. Environmental Conservation 31:309–321.

Stiles, D. 2014. Can elephants survive a continued ivory trade ban? National Geographic Voices. http://voices.nationalgeographic.com /2014/09/15/opinion-can-elephants-survive-a-continued-ivory -trade-ban/. Accessed on 18 March 2014.

Stiles, S., E. Martin, and L. Vigne. 2011. Exaggerated ivory prices can be harmful to elephants. Swara, East African Wildlife Society, October–December, pp. 16–20.

St. John, B. J. 1998. Management of southern hairy-nosed wombats *Lasiorhinus latifrons* in South Australia. Pp. 229–242 in Wombats

(R. T. Wells, P. A. Pridmore, B. St. John, M. D. Gaughwin, and J. Ferris, eds.). Surrey Beatty & Sons, Chipping Norton, New South Wales, Australia.

Stodart, E. 1977. Breeding and behaviour of Australian bandicoots. Pp. 179–191 in The biology of marsupials (B. Stonehouse and D. P. Gilmore, eds.). University Park Press, Baltimore, Maryland.

Stoeger-Horwath, A. S., S. Stoeger, and H. M. Schwammer. 2007. Call repertoire of infant African elephants: first insights into the early vocal ontogeny. Journal of the Accoustical Society of America 121:3922–3931.

Storrs, E. E., and H. P. Burchfield. 1985. Leprosy in wild common long-nosed armadillos *Dasypus novemcinctus*. Pp. 265–268 in The evolution and ecology of armadillos, sloths, and vermilinguas (G. G. Montgomery, ed.). Smithsonian Institution Press, Washington, DC.

Strahan, R. 2008. Eastern hare-wallaby *Lagorchestes leporides* (Gould, 1841). Pp. 320–321 in The mammals of Australia. 3rd edition (S. Van Dyck and R. Strahan, eds.). Reed New Holland, Sydney, Australia.

Streilein, K. E. 1982a. Ecology of small mammals in the semiarid Brazilian Caatinga. I. Climate and faunal composition. Annals of Carnegie Museum 51:79–106.

Streilein, K. E. 1982b. The ecology of small mammals in the semiarid Brazilian Caatinga. V. Agonistic behavior and overview. Annals of Carnegie Museum 51:345–369.

Stromayer, K. A. K. 2002. The Asian Elephant Conservation Act, the Asian Elephant Conservation Fund, and the conservation of the wild and the domesticated Asian elephant. Pp. 241–248 in Giants on our hands. Proceedings of the International Workshop on the Domesticated Asian Elephant (I. Baker and M. Kashio, eds.). FAO Regional Office for Asia and the Pacific, Bangkok, Thailand.

Stuart, A. J. 2005. The extinction of woolly mammoth (*Mammuthus primigenius*) and straight-tusked elephant (*Palaeoloxodon antiquus*) in Europe. Quaternary International 126–128:171–177.

Stuart, A. J., L. D. Sulerzhitsky, L. A. Orlova, Y. V. Kuzmin, and A. M. Lister. 2002. The latest woolly mammoths (*Mammuthus primigenius* Blumenbach) in Europe and Asia: a review of the current evidence. Quaternary Science Reviews 21:1559–1569.

Stuart, C., T. Stuart, and V. Pereboom. 2003. Aspects of the biology of the Cape sengi, *Elephantulus edwardii*, from the Western Escarpment, South Africa. Afrotherian Conservation 2:2–4.

Suckling, G. C. 1984. Population ecology of the sugar glider, *Petaurus breviceps*, in a system of fragmented habitats. Australian Wildlife Research 11:49–75.

Suckling, G. C. 2008. Sugar glider *Petaurus breviceps* (Waterhouse, 1839). Pp. 230–232 in The mammals of Australia. 3rd edition (S. Van Dyck and R. Strahan, eds.). Reed New Holland, Sydney, Australia.

Sukumar, R. 1989a. The Asian elephant: ecology and management. Cambridge University Press, Cambridge, United Kingdom.

Sukumar, R. 1989b. Ecology of the Asian elephant in southern India. I. Movement and habitat utilization patterns. Journal of Tropical Ecology 5:1–18.

Sukumar, R. 2003. The living elephants: evolutionary ecology, behavior, and conservation. Oxford University Press, New York, New York.

Sukumar, R., U. Ramakrishnan, and J. A. Santosh. 1998. Impact of poaching on an Asian elephant population in Periyar, southern India: a model of demography and tusk harvest. Animal Conservation 1:281–291.

Sukumar, R., V. Krishnamurthy, C. Wemmer, and M. Rodden. 1997. Demography of captive Asian elephants (*Elephas maximus*) in southern India. Zoo Biology 16:263–272.

Sumner, P., C. A. Arrese, and J. C. Partridge. 2005. The ecology of visual pigment tuning in an Australian marsupial: the honey possum *Tarsipes rostratus*. Journal of Experimental Biology 208:1803–1815.

Sunquist, M. E., S. N. Austad, and F. Sunquist. 1987. Movement patterns and home range in the common opossum (*Didelphis marsupialis*). Journal of Mammalogy 68:173–176.

Sunquist, M. E., and G. G. Montgomery. 1973. Activity patterns and rates of movement of two-toed and three-toed sloths (*Choloepus hoffmanni* and *Bradypus infuscatus*). Journal of Mammalogy 54:946–954.

Superina, M. 2008. The natural history of the pichi, *Zaedyus pichiy*, in western Argentina. Pp. 313–318 in The biology of the Xenarthra (S. F. Vizcaíno and W. J. Loughry, eds.). University Press of Florida, Gainesville, Florida.

Superina, M., and A. M. Abba. 2014a. *Zaedyus pichiy* (Cingulata: Dasypodidae). Mammalian Species 46(905):1–10.

Superina, M., and A. M. Abba. 2014b. *Zaedyus pichiy*. The IUCN Red List of Threatened Species 2014: e.T23178A47443734. http://dx.doi.org/10.2305/IUCN.UK.2014-1.RLTS.T23178A47443734.en. Downloaded on 08 May 2015.

Superina, M., A. M. Abba, and V. G. Roig. 2014. *Chlamyphorus truncatus*. The IUCN Red List of Threatened Species 2014: e.T4704A47439264. http://dx.doi.org/10.2305/IUCN.UK.2014-1.RLTS.T4704A47439264.en. Downloaded on 06 May 2015.

Superina, M., and P. Boily. 2007. Hibernation and daily torpor in an armadillo, the pichi (*Zaedyus pichiy*). Comparative Biochemistry and Physiology, Part A 148:893–898.

Superina, M., and W. J. Loughry. 2012. Life on the half-shell: consequences of a carapace in the evolution of armadillos (Xenarthra: Cingulata). Journal of Mammalian Evolution 19:217–224.

Superina, M., F. Miranda, and T. Plese. 2008. Maintenance of Xenarthra in captivity. Pp. 232–243 in The biology of the Xenarthra (S. F. Vizcaíno and W. J. Loughry, eds.). University Press of Florida, Gainesville, Florida.

Superina, M., F. Trujillo, M. Arteaga, and A. M. Abba. 2014. *Dasypus sabanicola*. The IUCN Red List of Threatened Species 2014: e.T6292A47441316. http://dx.doi.org/10.2305/IUCN.UK.2014-1.RLTS.T6292A47441316.en. Downloaded on 09 May 2015.

Suprayogi, B., J. Sugardjito, and R. P. H. Lilley. 2002. Management of Sumatran elephants in Indonesia: problems and challenges. Pp. 183–194 in Giants on our hands. Proceedings of the International Workshop on the Domesticated Asian Elephant (I. Baker and M. Kashio, eds.). FAO Regional Office for Asia and the Pacific, Bangkok, Thailand.

Surovell, T., N. Waguespack, and P. J. Brantingham. 2005. Global archaeological evidence for proboscidean overkill. Proceedings of the National Academy of Sciences of the United States of America 102:6231–6236.

Szalay, F. S. 1982. A new appraisal of marsupial phylogeny and classification. Pp. 621–640 in Carnivorous marsupials (M. Archer, ed.). Royal Zoological Society of New South Wales, Mosman, New South Wales, Australia.

Szalay, F. S. 1993. Metatherian taxon phylogeny: evidence and interpretation from the cranioskeletal system. Pp. 216–242 in Mammal phylogeny: Mesozoic differentiation, multituberculates, mono-

tremes, early therians, and marsupials (F. S. Szalay, M. J. Novacek, an M. C. McKenna, eds.). Springer-Verlag, New York, New York.

Szalay, F. S. 1994. Evolutionary history of the marsupials and an analysis of osteological characters. Cambridge University Press, Cambridge, United Kingdom.

Szalay, F. S. 1999. [Review of] Classification of mammals above the species level (M. C. McKenna and S. K. Bell, 1997, Columbia University Press, New York). Journal of Vertebrate Paleontology 19:191–195.

Tabaranza, B., J. C. Gonzalez, G. Rosell-Ambal, and L. Heaney. 2008. *Urogale everetti*. The IUCN Red List of Threatened Species 2008: e.T22784A9387093. http://dx.doi.org/10.2305/IUCN.UK.2008.RLTS.T22784A9387093.en. Downloaded on 30 May 2015.

Tabuce, R., R. J. Asher, and T. Lehmann. 2008. Afrotherian mammals: a review of current data. Mammalia 72:2–14.

Tabuce, R., B. Coiffait, P.-E. Coiffait, M. Mahboubi, and J.-J. Jaeger. 2001. A new genus of Macroscelidea (Mammalia) from the Eocene of Algeria: a possible origin for elephant-shrews. Journal of Vertebrate Paleontology 21:535–546.

Tabuce, R., L. Marivaux, M. Adaci, M. Bensalah, J.-L. Hartenberger, M. Mahboubi, F. Mebrouk, P. Tafforeau, and J.-J. Jaeger. 2007. Early Tertiary mammals from North Africa reinforce the molecular Afrotheria clade. Proceedings of the Royal Society of London B Biological Sciences 274:1159–1166.

Taggart, D., R. Martin, and A. Horsup. 2008. *Lasiorhinus krefftii*. The IUCN Red List of Threatened Species 2008: e.T11343A3269797. http://dx.doi.org/10.2305/IUCN.UK.2008.RLTS.T11343A3269797.en. Downloaded on 30 December 2015.

Taggart, D., R. Martin, and P. Menkhorst. 2008. *Vombatus ursinus*. The IUCN Red List of Threatened Species 2008: e.T40556A10334014. http://dx.doi.org/10.2305/IUCN.UK.2008.RLTS.T40556A10334014.en. Downloaded on 02 December 2015.

Taggart, D., P. Menkhorst, and D. Lunney. 2008. *Petrogale penicillata*. The IUCN Red List of Threatened Species 2008: e.T16746A6351546. http://dx.doi.org/10.2305/IUCN.UK.2008.RLTS.T16746A6351546.en. Downloaded on 22 December 2015.

Taggart, D., and T. Robinson. 2008. *Lasiorhinus latifrons*. The IUCN Red List of Threatened Species 2008: e.T40555A10333839. http://dx.doi.org/10.2305/IUCN.UK.2008.RLTS.T40555A10333839.en. Downloaded on 30 December 2015

Taggart, D. A., and P. D. Temple-Smith. 2008. Southern hairy-nosed wombat *Lasiorhinus latifrons* (Owen, 1845). Pp. 204–206 in The mammals of Australia. 3rd edition (S. Van Dyck and R. Strahan, eds.). Reed New Holland, Sydney, Australia.

Taggart, D. A., G. A. Shimmin, P. McCloud, and P. D. Temple-Smith. 1999. Timing of mating, sperm dynamics, and ovulation in a wild population of agile antechinus (Marsupialia: Dasyuridae). Biology of Reproduction 60:283–289.

Tangley, L. 1997. In search of Africa's forgotten forest elephant. Science 275:1417–1419.

Tate, G. H. H. 1933. A systematic revision of the marsupial genus *Marmosa*, with a discussion of the adaptive radiation of the murine opossums (*Marmosa*). Bulletin of the American Museum of Natural History 66:1–250.

Tate, G. H. H. 1945. Results of the Archbold Expeditions. No. 52. The marsupial genus *Phalanger*. American Museum Novitates 1283:1–30.

Taube, E., J. Keravec, J.-C. Vié, and J.-M. Duplantier. 2001. Reproductive biology and postnatal development in sloths, *Bradypus* and *Choloepus*: review with original data from the field (French Guiana) and from captivity. Mammal Review 31:173–188.

Taulman, J. F., and L. W. Robbins. 1996. Recent range expansion and distributional limits of the nine-banded armadillo (*Dasypus novemcinctus*) in the United States. Journal of Biogeography 23:635–648.

Taulman, J. F., and L. W. Robbins. 2014. Range expansion and distributional limits of the nine-banded armadillo in the United States: an update of Taulman & Robbins (1996). Journal of Biogeography 41:1626–1630.

Taylor, A., and T. Lehmann. 2015. *Orycteropus afer*. The IUCN Red List of Threatened Species 2015: e.T41504A21286437. http://dx.doi.org /10.2305/IUCN.UK.2015-2.RLTS.T41504A21286437.en. Downloaded on 21 January 2016.

Taylor, A. C., and J. Foulkes. 2004. Molecules and morphology: a taxonomic analysis of the common brushtail possum *Trichosurus vulpecula* with an emphasis on the central Australian form. Pp. 455–470 in The biology of Australian possums and gliders (R. L. Goldingay and S. M. Jackson, eds.). Surrey Beatty & Sons, Chipping Norton, New South Wales, Australia.

Taylor, A. C., W. B. Sherwin, and R. K. Wayne. 1994. Genetic variation of microsatellite loci in a bottlenecked species: the northern hairy-nosed wombat *Lasiorhinus krefftii*. Molecular Ecology 3:277–290.

Taylor, E. H. 1934. Philippine land mammals. Philippine Bureau of Science Monograph 30:1–548.

Taylor, J. M., J. H. Calaby, and T. D. Redhead. 1982. Breeding in wild populations of the marsupial-mouse *Planigale maculata sinualis* (Dasyuridae, Marsupialia). Pp. 83–87 in Carnivorous marsupials (M. Archer, ed.). Royal Zoological Society of New South Wales, Mosman, New South Wales, Australia.

Taylor, R. J. 1992. Seasonal changes in the diet of the Tasmanian bettong (*Bettongia gaimardi*), a mycophagous marsupial. Journal of Mammalogy 73:408–414.

Taylor, R. J. 1993. Habitat requirements of the Tasmanian bettong (*Bettongia gaimardi*), a mycophagous marsupial. Wildlife Research 20:699–710.

Taylor, W. A. (subed.). 2005. Order Tubulidentata. Pp. 35–40 in The mammals of the Southern African Subregion. 3rd edition (J. D. Skinner and C. T. Chimimba, eds.). Cambridge University Press, Cape Town, South Africa.

Taylor, W. A. 2011. Family Orycteropodidae (aardvark). Pp. 18–25 in Handbook of the mammals of the world, volume 2, Hoofed mammals (D. E. Wilson and R. A. Mittermeier, eds.). Lynx Edicions, Barcelona, Spain.

Taylor, W. A., and J. D. Skinner. 2000. Associative feeding between aardwolves (*Proteles cristatus*) and aardvarks (*Orycteropus afer*). Mammal Review 30:141–143.

Taylor, W. A., and J. D. Skinner. 2003. Activity patterns, home ranges and burrow use of aardvarks (*Orycteropus afer*) in the Karoo. Journal of Zoology 261:291–297.

Teeling, E. C., and S. B. Hedges. 2013. Making the impossible possible: rooting the tree of placental mammals. Molecular Biology and Evolution 30:1999–2000.

Telfer, W. R., and J. H. Calaby. 2008. Black wallaroo, *Macropus bernardus* (Rothschild, 1904). Pp. 327–328 in The mammals of Australia. 3rd edition (S. Van Dyck and R. Strahan, eds.). Reed New Holland, Sydney, Australia.

Telfer, W. R., and A. D. Griffiths. 2006. Dry-season use of space, habitats and shelters by the short-eared rock-wallaby (*Petrogale brachyotis*) in the monsoon tropics. Wildlife Research 33:207–214.

Temple-Smith, P., and T. Grant. 2001. Uncertain breeding: a short history of reproduction in monotremes. Reproduction, Fertility and Development 13:487–497.

Teta, P., and N. de la Sancha. 2008. *Chacodelphys formosa*. The IUCN Red List of Threatened Species 2008: e.T136547A4308404. http://dx .doi.org/10.2305/IUCN.UK.2008.RLTS.T136547A4308404 .en. Downloaded on 29 December 2015.

Teta, P., and U. F. J. Pardiñas. 2007. Mammalia, Didelphimorphia, Didelphidae, *Chacodelphys formosa* (Shamel, 1930): range extension. Check List 3:333–335.

Teta, P., U. F. J. Pardiñas, and G. D'Elía. 2006. Rediscovery of *Chacodelphys*: a South American marsupial genus previously known from a single specimen. Mammalian Biology 71:309–314.

Teta, P., E. Muschetto, S. Maidana, C. Bellomo, and P. Padula. 2007. *Gracilinanus microtarsus* (Didelphimorphia, Didelphidae) en la provincia de Misiones, Argentina. Mastozoologia Neotropical 14:113–115.

Teta, P., G. D'Elía, D. Flores, and N. de la Sancha. 2009. Diversity and distribution of the mouse opossums of the genus *Thylamys* (Didelphimorphia, Didelphidae) in northeastern and central Argentina. Gayana 73:180–199.

Theodor, J. M., K. D. Rose, and J. Erfurt. 2005. Artiodactyla. Pp. 215–233 in The rise of placental mammals: origins and relationships of the major clades (K. D. Rose and J. D. Archibald, eds.). Johns Hopkins University Press, Baltimore, Maryland.

Theodorou, G., N. Symeonidis, and E. Stathopoulou. 2007. *Elephas tiliensis* n. sp. from Tilos island (Dodecanese, Greece). Hellenic Journal of Geosciences 42:19–32.

Theuerkauf, J., H. Ellenberg, and Y. Guiro. 2000. Group structure of forest elephants in the Bossematié Forest Reserve, Ivory Coast. African Journal of Ecology 38:262–264.

Thewissen, J. G. M. 1985. Cephalic evidence for the affinities of Tubulidentata. Mammalia 49:257–284.

Thewissen, J. G. M., L. N. Cooper, M. T. Clementz, S. Bajpai, and B. N. Tiwari. 2007. Whales originated from aquatic artiodactyls in the Eocene epoch of India. Nature 450:1190–1194.

Thewissen, J. G. M., L. N. Cooper, M. T. Clementz, S. Bajpai, and B. N. Tiwari. 2009. Thewissen et al. reply. Nature 458, doi:10.1038/ nature07776.

Thomas, M. G. 2012. The flickering genes of the last mammoths. Molecular Ecology 21:3379–3381.

Thomas, M. G., E. Hagelberg, H. B. Jones, Z. Yang, and A. L. Lister. 2000. Molecular and morphological evidence on the phylogeny of the Elephantidae. Proceedings of the Royal Society of London B Biological Sciences 267:2493–2500.

Thomas, O. 1921. A new genus of opossum from southern Patagonia. Annals and Magazine of Natural History, series 9, 8:136–139.

Thompson, M. E., S. J. Schwager, and K. B. Payne. 2009. Heard but not seen: an acoustic survey of the African forest elephant population at Kakum Conservation Area, Ghana. African Journal of Ecology 48:224–231.

Thompson, V. D. 1995. Queensland koala (*Phascolarctos adustus*) and Victorian koala (*Phascolarctos cinereus victor*): North American regional studbook. Zoological Society of San Diego, San Diego, California.

Thomsen, J. B. 1988. Recent U.S. imports of certain products from the African elephant. Pachyderm 10:1–5, 21.

Thornback, J., and M. Jenkins. 1982. The IUCN mammal red data book. Part 1: Threatened mammalian taxa of the Americas and the

Australasian zoogeographic region (excluding Cetacea). International Union for Conservation of Nature, Gland, Switzerland.

Tikhonov, A., L. Agenbroad, and S. Vartanyan. 2003. Comparative analysis of the mammoth populations on Wrangel Island and the Channel Islands. Deinsea 9:415–420.

Timm, R. M., L. Albuja V., and B. L. Clauson. 1986. Ecology, distribution, harvest, and conservation of the Amazonian manatee *Trichechus inunguis* in Ecuador. Biotrópica 18:150–156.

Timm, R. M., and B. D. Patterson. 2007. Genus *Caenolestes* O. (Thomas, 1895). Pp. 120–124 in Mammals of South America, volume 1, Marsupials, xenarthrans, shrews, and bats (A. L. Gardner, ed.). University of Chicago Press, Chicago, Illinois.

Timmins, R. J., J. W. Duckworth, C. R. Robson, and J. L. Walston. 2003. Distribution, status and ecology of the mainland slender-tailed treeshrew *Dendrogale murina*. Mammal Review 33:272–283.

Tiple, A. D., and S. S. Talmale. 2013. Occurrence of the Indian tree shrew *Anathana ellioti* (Waterhouse) (Mammalia: Scandentia) in central India. Tigerpaper 40(2):14–18.

Tirira, D. G., J. Díaz-N., M. Superina, and A. M. Abba. 2014. *Cabassous centralis*. The IUCN Red List of Threatened Species 2014: e.T3412A47437304. http://dx.doi.org/10.2305/IUCN.UK.2014-1.RLTS.T3412A47437304.en. Downloaded on 09 May 2015.

Tisdell, C., and H. S. Nantha. 2007. Comparison of funding and demand for the conservation of the charismatic koala with those for the critically endangered wombat *Lasiorhinus krefftii*. Biodiversity and Conservation 16:1261–1281.

Todd, N. E. 2010. New phylogenetic analysis of the family Elephantidae based on cranial-dental morphology. Anatomical Record 293:74–90.

Tomas, W. M., Z. Campos, A. L. J. Desbiez, D. Kluyber, P. A. L. Borges, and G. Mourão. 2013. Mating behavior of the six-banded armadillo *Euphractus sexcinctus* in the Pantanal wetland, Brazil. Edentata 14:87–89.

Tong, H., and J. Liu. 2004. The Pleistocene-Holocene extinctions of mammals in China. Proceedings of the Annual Symposium of the Chinese Society of Vertebrate Paleontology 9:111–119.

Tong, H., and M. Patou-Mathis. 2003. Mammoth and other proboscideans in China during the late Pleistocene. Deinsea 9:421–428.

Torres, R. 2009. Ampliacion del limite austral de la distribucion del oso melero (*Tamandua tetradactyla*) en la Argentina. Notulas Faunisticas 39:1–5.

Travouillon, K. J., S. J. Hand, M. Archer, and K. H. Black. 2014. Earliest modern bandicoot and bilby (Marsupialia, Peramelidae, and Thylacomyidae) from the Miocene of the Riversleigh World Heritage Area, Northwestern Queensland, Australia. Journal of Vertebrate Paleontology 34:375–382.

Troll, S., J. Gottschalk, J. Seeburger, E. Ziemssen, M. Häfner, J. Thielebein, and A. Einspanier. 2013. Characterization of the ovarian cycle in the two-toed sloths (*Choloepus didactylus*): an innovative, reliable, and noninvasive method using fecal hormone analyses. Theriogenology 80:275–283.

Trovati, R. G., and B. A. de Brito. 2009. Nota sobre deslocamento e área de uso de tamanduá-mirim (*Tamandua tetradactyla*) translocado no Cerrado brasileiro. Neotropical Biology and Conservation 4:144–149.

Troy, S., and G. Coulson. 1993. Home range of the swamp wallaby, *Wallabia bicolor*. Wildlife Research 20:571–577.

Truman, R. W. 2008. Leprosy. Pp. 111–119 in The biology of the Xenarthra (S. F. Vizcaíno and W. J. Loughry, eds.). University Press of Florida, Gainesville, Florida.

Tsahar, E., I. Izhaki, S. Lev-Yadun, and G. Bar-Oz. 2009. Distribution and extinction of ungulates during the Holocene of the southern Levant. PLoS ONE 4(4):e5316. doi:10.1371/journal.pone.0005316.

Turkalo, A. K. 1996. Studying forest elephants by direct observation in the Dzanga clearing: an update. Pachyderm 22:59–60.

Turkalo, A., and R. Barnes. 2013. *Loxodonta cyclotis* forest elephant. Pp. 195–200 in Mammals of Africa, volume 1, Introductory chapters and Afrotheria (J. Kingdon, D. Happold, M. Hoffmann, T. Butynski, M. Happold, and J. Kalina, eds.). Bloomsbury, London, United Kingdom.

Turkalo, A. K., and J. M. Fay. 1996. Studying forest elephants by direct observation: preliminary results from the Dzanga clearing, Central African Republic. Pachyderm 21:45–54.

Turkalo, A. K., and J. M. Fay. 2001. Forest elephant behavior and ecology: observations from the Dzanga Saline. Pp. 207–213 in African rain forest ecology and conservation (W. Weber, L. J. T. White, A. Vedder, and L. Naughton-Treves, eds.). Yale University Press, New Haven, Connecticut.

Turner, J. M., L. Warnecke, G. Körtner, and F. Geiser. 2012. Opportunistic hibernation by a free-ranging marsupial. Journal of Zoology 286:277–284.

Turner, K. 1970. Breeding Tasmanian devils, *Sarcophilus harrisii*, at Westbury Zoo. International Zoo Yearbook 10:65.

Turney, C. S. M., T. F. Flannery, R. G. Roberts, C. Reide, L. K. Fifield, T. F. G. Higham, Z. Jacobs, N. Kemp, E. A. Colhouni, R. M. Kalin, and N. Oglek. 2008. Late-surviving megafauna in Tasmania, Australia, implicate human involvement in their extinction. Proceedings of the National Academy of Sciences of the United States of America 105:12150–12153.

Turvey, S. T. 2009. In the shadow of the megafauna: prehistoric mammal and bird extinctions across the Holocene. Pp. 17–39 in Holocene extinctions (S. T. Turvey, ed.). Oxford University Press, New York, New York.

Turvey, S. T., and C. L. Risley. 2006. Modeling the extinction of Steller's sea cow. Biology Letters 2:94–97.

Turvey, S. T., H. Tong, A. J. Stuart, and A. M. Lister. 2013. Holocene survival of late Pleistocene megafauna in China: a critical review of the evidence. Quaternary Science Reviews 76:156–166.

Tyndale-Biscoe, C. H. 1968. Reproduction and post-natal development in the marsupial *Bettongia lesueur* (Quoy and Gaimard). Australian Journal of Zoology 16:577–602.

Tyndale-Biscoe, C. H. 1973. Life of marsupials. American Elsevier, New York, New York.

Tyndale-Biscoe, C. H. 1989. The adaptiveness of reproductive processes. Pp. 277–285 in Kangaroos, wallabies and rat-kangaroos (G. Grigg, P. Jarman, and I. Hume, eds.). Surrey Beatty & Sons, Chipping Norton, New South Wales, Australia.

Tyndale-Biscoe, C. H., and R. B. MacKenzie. 1976. Reproduction in *Didelphis marsupialis* and *D. albiventris* in Colombia. Journal of Mammalogy 57:249–265.

Tyndale-Biscoe, C. H., and R. F. C. Smith. 1969. Studies on the marsupial glider, *Schoinobates volans* (Kerr). II. Population structure and regulatory mechanisms. Journal of Animal Ecology 38:637–649.

UNEP (United Nations Environmental Program). 2011. Garamba National Park. http://www.unep-wcmc.org/medialibrary/2011/06/24/e4384acd/Garamba%20National%20Park.pdf.

United Nations. 2015. World population prospects. The 2015 revision. Key findings and advance tables. Department of Economic and

Social Affairs, Population Division, United Nations, New York, New York.

Urban, M. C. 2015. Accelerating extinction risk from climate change. Science 348:571–573.

Urbani, B., and C. Bosque. 2007. Feeding ecology and postural behaviour of the three-toed sloth (*Bradypus variegatus flaccidus*) in northern Venezuela. Zeitschrift für Säugetierkunde 72:321–329.

USGCRP. 2017. Climate Science Special Report: Fourth National Climate Assessment, Volume I (D. J. Wuebbles, D. W. Fahey, K. A. Hibbard, D. J. Dokken, B. C. Stewart, and T. K. Maycock, eds.). United States Global Change Research Program, Washington, DC. doi: 10.7930/J0J964J6. https://science2017.globalchange.gov/.

Valeix, M., H. Fritz, V. Canévet, S. Le Bel, H. Madzikanda. 2009. Do elephants prevent other African herbivores from using waterholes in the dry season? Biodiversity and Conservation 18:569–576.

Valente, A. 2008. Kultarr *Antechinomys laniger* (Gould, 1856). Pp. 122–124 in The mammals of Australia. 3rd edition (S. Van Dyck and R. Strahan, eds.). Reed New Holland, Sydney, Australia.

Van Aarde, R. J., and T. P. Jackson. 2007. Megaparks for metapopulations: addressing the causes of locally high elephant numbers in southern Africa. Biological Conservation 134:289–297.

VandeBerg, J. L., and E. S. Robinson. 1997. The laboratory opossum (*Monodelphis domestica*) in laboratory research. ILAR Journal 38:4–12.

Van Den Bergh, G. D., R. Due Awe, M. J. Morwood, T. Sutikna, Jatmiko, E. W. Saptomo. 2008. The youngest stegodon remains in Southeast Asia from the late Pleistocene archaeological site Liang Bua, Flores, Indonesia. Quaternary International 182:16–48.

Van Den Bergh. G. D., H. J. M. Meijer, R. Due Awe, M. J. Morwood, K. Szabóc, L. W. Van Den Hoek Ostende, T. Sutikna, E. W. Saptomo, P. J. Piper, K. M. Dobney. 2009. The Liang Bua faunal remains: a 95 k. yr. sequence from Flores, East Indonesia. Journal of Human Evolution 57:527–537.

Van Der Ree, R, M. J. Harper, and M. Crane. 2006. Longevity in wild populations of the squirrel glider *Petaurus norfolcensis*. Australian Mammalogy 28:239–242.

Van Der Ree, R., and G. C. Suckling. 2008. Squirrel glider *Petaurus norfolcensis* (Kerr, 1792). Pp. 237–238 in The mammals of Australia. 3rd edition (S. Van Dyck and R. Strahan, eds.). Reed New Holland, Sydney, Australia.

Van Deusen, H. M., and J. K. Jones, Jr. 1967. Marsupials. Pp. 61–86 in Recent mammals of the world: a synopsis of families (S. Anderson and J. K. Jones, Jr., eds.). Ronald Press, New York, New York.

Van Deusen, H. M., and K. Keith. 1966. Range and habitat of the bandicoot, *Echymipera clara*, in New Guinea. Journal of Mammalogy 47:721–723.

Van Dijk, M. A. M., O. Madsen, F. Catzeflis, M. J. Stanhope, W. W. de Jong, and M. Pagel. 2001. Protein sequence signatures support the African clade of mammals. Proceedings of the National Academy of Sciences of the United States of America 98:188–193.

Van Dyck, S. 1980. The cinnamon antechinus, *Antechinus leo* (Marsupialia: Dasyuridae), a new species from the vine-forests of Cape York Peninsula. Australian Mammalogy 3:5–17.

Van Dyck, S. 1982a. The relationships of *Antechinus stuartii* and *A. flavipes* (Dasyuridae, Marsupialia) with special reference to Queensland. Pp. 723–766 in Carnivorous marsupials (M. Archer, ed.). Royal Zoological Society of New South Wales, Mosman, New South Wales, Australia.

Van Dyck, S. 1982b. The status and relationships of the Atherton antechinus, *Antechinus godmani* (Marsupialia: Dasyuridae). Australian Mammalogy 5:195–210.

Van Dyck, S. 1985. *Sminthopsis leucopus* (Marsupialia: Dasyuridae) in north Queensland rainforest. Australian Mammalogy 8:53–60.

Van Dyck, S. 1986. The chestnut dunnart, *Sminthopsis archeri* (Marsupialia: Dasyuridae), a new species from the savannahs of Papua New Guinea and Cape York Peninsula, Australia. Australian Mammalogy 9:111–124.

Van Dyck, S. 1987. The bronze quoll, *Dasyurus spartacus* (Marsupialia: Dasyuridae), a new species from the savannahs of Papua New Guinea. Australian Mammalogy 11:145–156.

Van Dyck, S. 2002. Morphology-based revision of *Murexia* and *Antechinus* (Marsupialia: Dasyuridae). Memoirs of the Queensland Museum 48:239–330.

Van Dyck, S, and M. S. Crowther. 2000. Reassessment of northern representatives of the *Antechinus stuartii* complex (Marsupialia: Dasyuridae): *A. subtropicus* sp. nov. & *A. adustus* new status. Memoirs of the Queensland Museum 45:451–475.

Van Dyck, S., and R. Strahan (eds.). 2008. The mammals of Australia. 3rd edition. Reed New Holland, Sydney, Australia.

Van Dyck, S., J. C. Z. Woinarski, and A. J. Press. 1994. The Kakadu dunnart, *Sminthopsis bindi* (Marsupialia: Dasyuridae), a new species from the stony woodlands of the Northern Territory. Memoirs of the Queensland Museum 37:311–323.

Van Gelder, R. G. 1977. Mammalian hybrids and generic limits. American Museum Novitates 2635:1–25.

Vanleeuwe, H., and A. Gautier-Hion. 1998. Forest elephant paths and movements at the Odzala National Park, Congo: the role of clearings and Marantaceae forests. African Journal of Ecology 36:174–182.

Van Rheede, T., T. Bastiaans, D. N. Boone, S. B. Hedges, W. W. de Jong, and O. Madsen. 2006. The platypus is in its place: nuclear genes and indels confirm the sister group relation of monotremes and therians. Molecular Biology and Evolution 23:587–597.

Van Roosmalen, M. G. M. 2015. Hotspot of new megafauna found in the Central Amazon (Brazil): the lower Rio Aripuanã Basin. Biodiversity Journal 6:219–244.

Van Roosmalen, M. G. M., P. Van Hooft, and H. H. de Iongh. 2007. A new species of living manatee from the Amazon. Shallow clearwater adapted dwarf manatee is already on the verge of extinction. Amazon Association for the Preservation of Nature, Manaus, Brazil. http://www.marcvanroosmalen.org/dwarfmanatee.htm.

Van Valen, L. 1967. New Paleocene insectivores and insectivore classification. Bulletin of the American Museum of Natural History 135:217–284.

Van Weenen, J. 2008. *Sminthopsis aitkeni*. The IUCN Red List of Threatened Species 2008: e.T20294A9183297. http://dx.doi.org/10.2305/IUCN.UK.2008.RLTS.T20294A9183297.en. Downloaded on 04 January 2016.

Van Weenen, J., and P. Menkhorst. 2008. *Antechinus minimus*. The IUCN Red List of Threatened Species 2008: e.T40525A10329188. http://dx.doi.org/10.2305/IUCN.UK.2008.RLTS.T40525A10329188.en. Downloaded on 24 December 2015.

Vartanyan, S. L., V. E. Garutt, and A. V. Sher. 1993. Holocene dwarf mammoths from Wrangel Island in the Siberian Arctic. Nature 362:337–340.

Vartanyan, S. L., K. A. Arslanov, J. A. Karhu, G. Possnert, and L. D. Sulerzhitsky. 2008. Collection of radiocarbon dates on the mammoths (*Mammuthus primigenius*) and other genera of Wrangel Island, northeast Siberia, Russia. Quaternary Research 70:51–59.

Vaughan, C., O. Ramírez, G. Herrera, and R. Guries. 2007. Spatial ecology and conservation of two sloth species in a cacao landscape in Limón, Costa Rica. Biodiversity and Conservation 16:2293–2310.

Ventura, J., R. Pérez-Hernández, and M. J. Lopez-Fuster. 1998. Morphometric assessment of the *Monodelphis brevicaudata* group (Didelphimorphia: Didelphidae) in Venezuela. Journal of Mammalogy 79:104–117.

Ventura, J., M. Salazar, R. Pérez-Hernández, M. López-Fuster. 2002. Morphometrics of the genus *Didelphis* (Didelphimorphia: Didelphidae) in Venezuela. Journal of Mammalogy 83:1087–1096.

Ventura, J., D. Lew, R. Pérez Hernández, and M. J. Lopez Fuster. 2005. Skull size and shape relationships between Venezuelan *Monodelphis* taxa (Didelphimorphia Didelphidae), including the recently described species *M. reigi* Lew & Pérez Hernández 2004. Tropical Zoology 18:227–235.

Vereschagin, N. K., and G. F. Baryshnikov. 1984. Quaternary mammalian extinctions in northern Eurasia. Pp. 483–516 in Quaternary extinctions: a prehistoric revolution (P. S. Martin and R. G. Klein, eds.). University of Arizona Press, Tucson, Arizona.

Vernes, K., and L. C. Pope. 2002. Fecundity, pouch young survivorship and breeding season of the northern bettong (*Bettongia tropica*) in the wild. Australian Mammalogy 23:95–100.

Vernes, K., and L. C. Pope. 2009. Reproduction in the northern brown bandicoot (*Isoodon macrourus*) in the Australian wet tropics. Australian Journal of Zoology 57:105–109.

Veselovsky, Z. 1966. A contribution to the knowledge of the reproduction and growth of the two-toed sloth *Choloepus didactylus* at Prague Zoo. International Zoo Yearbook 6:147–153.

Vianna J. A., R. K. Bonde, S. Caballero, J. P. Giraldo, R. P. Lima, A. Clark, M. Marmontel, B. Morales-Vela, M. J. De Souza, L. Parr, M. A. Rodríguez-Lopez, A. A. Mignucci-Giannoni, J. A. Powell, and F. R. Santos. 2006. Phylogeography, phylogeny and hybridization in trichechid sirenians: implications for manatee conservation. Molecular Ecology 15:433–447.

Vidya, T. N. C., and R. Sukumar. 2005a. Social and reproductive behaviour in elephants. Current Science 89:1200–1207.

Vidya, T. N. C., and R. Sukumar. 2005b. Social organization of the Asian elephant (*Elephas maximus*) in southern India inferred from microsatellite DNA. Journal of Ethology 23:205–210.

Vidya, T. N. C., R. Sukumar, and D. J. Melnick. 2009. Range-wide mtDNA phylogeography yields insights into the origins of Asian elephants. Proceedings of the Royal Society of London B Biological Sciences 276:893–202.

Vieira, E., D. Astua de Moraes, and D. Brito. 2008. *Thylamys velutinus*. The IUCN Red List of Threatened Species 2008: e.T40520A10327781. http://dx.doi.org/10.2305/IUCN.UK.2008.RLTS.T40520A10327781.en. Downloaded on 12 November 2015

Viggers, K. L., and D. B. Lindenmayer. 2004. A review of the biology of the short-eared possum *Trichosurus caninus* and the mountain brushtail possum *Trichosurus cunninghami*. Pp. 490–505 in The biology of Australian possums and gliders (R. L. Goldingay and S. M. Jackson, eds.). Surrey Beatty & Sons, Chipping Norton, New South Wales, Australia.

Vilela, J. F., C. A. D. M. Russo, and J. Alves de Oliveria. 2010. An assessment of morphometric and molecular variation in *Monodelphis dimidiata* (Wagner, 1847) (Didelphimorphia: Didelphidae). Zootaxa 2646:26–42.

Vizcaíno, S. F. 1995. Identificacion especifica de las "mulitas," genero *Dasypus* L. (Mammalia, Dasypodidae), del noroeste Argentino. Descripcion de una nueva especie. Mastozoología Neotropical 2:5–13.

Vizcaíno, S. F., and W. J. Loughry (eds.). 2008. The biology of the Xenarthra. University Press of Florida, Gainesville, Florida.

Vogel, P. (IUCN SSC Afrotheria Specialist Group). 2008a. *Micropotamogale lamottei*. The IUCN Red List of Threatened Species 2008: e.T13393A3880038. http://dx.doi.org/10.2305/IUCN.UK.2008.RLTS.T13393A3880038.en. Downloaded on 21 January 2016.

Vogel, P. (IUCN SSC Afrotheria Specialist Group). 2008b. *Micropotamogale ruwenzorii*. The IUCN Red List of Threatened Species 2008: e.T13394A3880319. http://dx.doi.org/10.2305/IUCN.UK.2008.RLTS.T13394A3880319.en. Downloaded on 21 January 2016.

Vogel, P. 2013. *Potamogale velox* Giant otter-shrew. Pp. 220–222 in Mammals of Africa, volume 1, Introductory chapters and Afrotheria (J. Kingdon, D. Happold, M. Hoffmann, T. Butynski, M. Happold, and J. Kalina, eds.). Bloomsbury, London, United Kingdom.

Voirin, B. 2015. Biology and conservation of the pygmy sloth, *Bradypus pygmaeus*. Journal of Mammalogy 96:703–707.

Voirin, B., D. Smith, A. Chiarello, and N. Moraes-Barros. 2014. *Bradypus pygmaeus*. The IUCN Red List of Threatened Species 2014: e.T61925A47444229. http://dx.doi.org/10.2305/IUCN.UK.2014-1.RLTS.T61925A47444229.en. Downloaded on 16 January 2016.

Vose, H. M. 1973. Feeding habits of the western Australian honey possum, *Tarsipes spenserae*. Journal of Mammalogy 54:245–247.

Voss, R. S., D. W. Fleck, and S. A. Jansa. 2009. On the diagnostic characters, ecogeographic distribution, and phylogenetic relationships of *Gracilinanus emiliae* (Didelphimorphia: Didelphidae: Thylamyini). Mastozoologia Neotropical 16:433–443.

Voss, R. S., A. L. Gardner, and S. A. Jansa. 2004. On the relationships of "*Marmosa*" *formosa* (Shamel, 1930) (Marsupialia: Didelphidae), a phylogenetic puzzle from the Chaco of northern Argentina. American Museum Novitates 3442:1–18.

Voss, R. S., and S. A. Jansa. 2003. Phylogenetic studies on didelphid marsupials II. Nonmolecular data and new IRBP sequences: separate and combined analyses of didelphine relationships with denser taxon sampling. Bulletin of the American Museum of Natural History 276:1–82.

Voss, R. S., and S. A. Jansa. 2009. Phylogenetic relationships and classification of didelphid marsupials, an extant radiation of New World metatherian mammals. Bulletin of the American Museum of Natural History 322:1–177.

Voss, R. S., D. P. Lunde, and S. A. Jansa. 2005. On the contents of *Gracilinanus* (Gardner and Creighton, 1989), with the description of a previously unrecognized clade of small didelphid Marsupials. American Museum Novitates 3482:1–34.

Voss, R. S., D. P. Lunde, and N. B. Simmons. 2001. Mammals of Paracou, French Guiana: a neotropical lowland rainforest fauna. Part 2. Nonvolant species. Bulletin of the American Museum of Natural History 263:1–236.

Voss, R. S., R. H. Pine, and S. Solari. 2012. A new species of the didelphid marsupial genus *Monodelphis* from eastern Bolivia. American Museum Novitates 3740:1–14.

Voss, R. S., T. Tarifa, and E. Yensen. 2004. An introduction to *Marmosops* (Marsupialia: Didelphidae), with the description of a new species and notes on the taxonomy and distribution of other Bolivian forms. American Museum Novitates 3466:1–40.

Voss, R. S., P. Myers, F. Catzeflis, A. P. Carmignotto, and J. Barreiro. 2009. The six opossums of Félix de Azara: identification, taxonomic history, neotype designations, and nomenclatural recommendations. Bulletin of the American Museum of Natural History 331:406–433.

Voss, R. S., B. K. Lim, J. F. Díaz-Nieto, and S. A. Jansa. 2013. A New species of *Marmosops* (Marsupialia: Didelphidae) from the Pakaraima Highlands of Guyana, with remarks on the origin of the endemic Pantepui mammal fauna. American Museum Novitates 3778:1–27.

Voss, R. S., E. E. Gutiérrez, S. Solari, R. V. Rossi, and S. A. Jansa. 2014. Phylogenetic relationships of mouse opossums (Didelphidae, *Marmosa*) with a revised subgeneric classification and notes on sympatric diversity. American Museum Novitates 3817:1–27.

Wainer, J. W. 1976. Studies of an island population of *Antechinus minimus* (Marsupialia, Dasyuridae). Australian Zoologist 19:1–7.

Wakefield, N. A. 1970a. Notes on Australian pigmy-possums (*Cercartetus*, Phalangeridae, Marsupialia). Victorian Naturalist 87:11–18.

Wakefield, N. A. 1970b. Notes on the glider-possum, *Petaurus australis* (Phalangeridae, Marsupialia). Victorian Naturalist 87:221–236.

Wakefield, N. A. 1971. The brush-tailed rock-wallaby (*Petrogale penicillata*) in western Victoria. Victorian Naturalist 88:92–102.

Wallace, R. E., and L. E. Painter. 2013. Observations on the diet of the giant armadillo (*Priodontes maximus* Kerr, 1792). Edentata 14:85–86.

Wallis, I. R., P. J. Jarman, B. E. Johnson, and R. W. Liddle. 1989. Nest sites and use of nests by rufous bettongs *Aepyprymnus rufescens*. Pp. 619–623 in Kangaroos, wallabies and rat-kangaroos (G. C. Grigg, P. Jarman, and I. Hume, eds.). Surrey Beatty & Sons, Chipping Norton, New South Wales, Australia.

Walton, D. W. 1988. Notoryctidae. Pp. 46–47 in Zoological catalogue of Australia, volume 5, Mammalia (D. W. Walton, ed.). Australian Government Publishing Service, Canberra, Australia.

Wang, Y. 2011. The elephant in cross-cultural perspective: from Han to Tang dynasties in Chinese history. Global History Review (Global History Center of Capital Normal University, Beijing, China) 3:851–867.

Warburton, B. 2005a. Dama wallaby. Pp. 32–39 in The handbook of New Zealand mammals. 2nd edition (C. M. King, ed.). Oxford University Press, South Melbourne, Australia.

Warburton, B. 2005b. Bennett's wallaby. Pp. 39–45 in The handbook of New Zealand mammals. 2nd edition (C. M. King, ed.). Oxford University Press, South Melbourne, Australia.

Warburton, B. 2005c. Parma wallaby. Pp. 45–49 in The handbook of New Zealand mammals. 2nd edition (C. M. King, ed.). Oxford University Press, South Melbourne, Australia.

Warburton, B. 2005d. Black-striped wallaby. P. 49 in The handbook of New Zealand mammals. 2nd edition (C. M. King, ed.). Oxford University Press, South Melbourne, Australia.

Warburton, B. 2005e. Brushtailed rock wallaby. Pp. 50–53 in The handbook of New Zealand mammals. 2nd edition (C. M. King, ed.). Oxford University Press, South Melbourne, Australia.

Warburton, B. 2005f. Swamp wallaby. Pp. 53–55 in The handbook of New Zealand mammals. 2nd edition (C. M. King, ed.). Oxford University Press, South Melbourne, Australia.

Warburton, N. M. 2006. Functional morphology of marsupial moles (Marsupialia, Notoryctidae). Verhandlungen des Naturwissenschaftlichen Vereins in Hamburg 42:39–149.

Ward, G. D. 1978. Habitat use and home range of radio-tagged opossums *Trichosurus vulpecula* (Kerr) in New Zealand lowland forest. Pp. 267–287 in The ecology of arboreal folivores (G. G. Montgomery, ed.). Smithsonian Institution Press, Washington, DC.

Ward, S. J. 1990a. Life history of the eastern pygmy possum, *Cercartetus nanus* (Burramyidae: Marsupialia), in south-eastern Australia. Australian Journal of Zoology 38:287–304.

Ward, S. J. 1990b. Life history of the feathertail glider, *Acrobates pygmaeus* (Acrobatidae: Marsupialia) in south-eastern Australia. Australian Journal of Zoology 38:503–517.

Ward, S. J. 1990c. Reproduction in the western pygmy-possum, *Cercartetus concinnus* (Marsupialia: Burramyidae), with notes on reproduction of some other small possum species. Australian Journal of Zoology 38:423–438.

Ward, S. J. 2004. Patterns of movement and nesting in feathertail gliders. Pp. 285–289 in The biology of Australian possums and gliders (R. L. Goldingay and S. M. Jackson, eds.). Surrey Beatty & Sons, Chipping Norton, New South Wales, Australia.

Ward, S. J., and M. B. Renfree. 1988a. Reproduction in females of the feathertail glider, *Acrobates pygmaeus* (Marsupialia). Journal of Zoology 216:225–239.

Ward, S. J., and M. B. Renfree. 1988b. Reproduction in males of the feathertail glider, *Acrobates pygmaeus* (Marsupialia). Journal of Zoology 216:241–251.

Ward, S. J., and V. Turner. 2008. Eastern pygmy-possum *Cercartetus nanus* (Geoffroy and Desmarest, 1817). Pp. 219–221 in The mammals of Australia. 3rd edition (S. Van Dyck and R. Strahan, eds.). Reed New Holland, Sydney, Australia.

Ward, S. J., and D. P. Woodside. 2008. Feathertail glider *Acrobates pygmaeus* (Shaw, 1794). Pp. 261–264 in The mammals of Australia. 3rd edition (S. Van Dyck and R. Strahan, eds.). Reed New Holland, Sydney, Australia.

Warren, W. C., L. W. Hillier, J. A. M. Graves, E. Birney, C. P. Ponting, F. Grützner, K. Belov, W. Miller, L. Clarke, A. T. Chinwalla, S.-P. Yang, A. Heger, D. P. Locke, P. Miethke, P. D. Waters, F. Veyrunes, L. Fulton, B. Fulton, T. Graves, J. Wallis, X. S. Puente, C. López-Otín, G. R. Ordóñez, E. E. Eichler, L. Chen, Z. Cheng, J. E. Deakin, A. Alsop, K. Thompson, P. Kirby, A. T. Papenfuss, M. J. Wakefield, T. Olender, D. Lancet, G. A. Huttley, A. F. A. Smit, A. Pask, P. Temple-Smith, M. A. Batzer, J. A. Walker, M. K. Konkel, R. S. Harris, C. M. Whittington, E. S. W. Wong, N. J. Gemmell, E. Buschiazzo, I. M. V. Jentzsch, A. Merkel, J. Schmitz, A. Zemann, G. Churakov, J. O. Kriegs, J. Brosius, E. P. Murchison, R. Sachidanandam, C. Smith, G. J. Hannon, E. Tsend-Ayush, D. McMillan, R. Attenborough, W. Rens, M. Ferguson-Smith, C. M. Lefèvre, J. A. Sharp, K. R. Nicholas, D. A. Ray, M. Kube, R. Reinhardt, T. H. Pringle, J. Taylor, R. C. Jones, B. Nixon, J.-L. Dacheux, H. Niwa, Y. Sekita, X. Huang, A. Stark, P. Kheradpour, M. Kellis, P. Flicek, Y. Chen, C. Webber, R. Hardison, J. Nelson, K. Hallsworth-Pepin, K. Delehaunty, C. Markovic, P. Minx, Y. Feng, C. Kremitzki, M. Mitreva1, J. Glasscock, T. Wylie, P. Wohldmann, P. Thiru, M. N. Nhan, C. S. Pohl, S. M. Smith, S. Hou, M. Nefedov, P. J. De Jong, M. B. Renfree, E. R. Mardis, and R. K. Wilson. 2008. Genome analysis of the platypus reveals unique signatures of evolution. Nature 453:175–183.

Wasser, S. K., A. M. Shedlock, K. Comstock, E. A. Ostrander, B. Mutayoba, and M. Stephens. 2004. Assigning African elephant DNA to geographic region of origin: applications to the ivory trade. Proceedings of the National Academy of Sciences of the United States of America 101:14847–14852.

Waters, P. D., G. Dobigny, P. J. Waddell, and T. J. Robinson. 2007. Evolutionary history of LINE-1 in the major clades of placental mammals. PLoS ONE 2:e158.

Watson, M., and M. Halley. 1999. Recovery Plan for the eastern barred bandicoot *Perameles gunnii* (mainland subspecies). Victoria Department of Natural Resources and Environment, East Melbourne, Australia.

Watson, M. L., and J. H. Calaby. 2008. Fawn antechinus *Antechinus bellus* (Thomas, 1904). Pp. 85–86 in The mammals of Australia. 3rd edition (S. Van Dyck and R. Strahan, eds.). Reed New Holland, Sydney, Australia.

Wayne, A., T. Friend, A. Burbidge, K. Morris, and J. van Weenen. 2008. *Bettongia penicillata*. The IUCN Red List of Threatened Species 2008: e.T2785A9480872. http://dx.doi.org/10.2305/IUCN.UK.2008 .RLTS.T2785A9480872.en. Downloaded on 30 December 2015.

Wayne, A. F., C. G. Ward, J. F. Rooney, C. V. Vellios, and D. B. Lindenmayer. 2005a. The life history of *Trichosurus vulpecula hypoleucus* (Phalangeridae) in the jarrah forest of south-western Australia. Australian Journal of Zoology 53:265–278.

Wayne, A. F., J. F. Rooney, C. G. Ward, C. V. Vellios, and D. B. Lindenmayer. 2005b. The life history of *Pseudocheirus occidentalis* (Pseudocheiridae) in the jarrah forest of south-western Australia. Australian Journal of Zoology 53:325–337.

Wayne, A. F., A. Cowling, D. B. Lindenmayer C. G. Ward, C. V. Vellios, C. F. Donnelly, and M. C. Calver. 2006. The abundance of a threatened arboreal marsupial in relation to anthropogenic disturbances at local and landscape scales in Mediterranean-type forests in south-western Australia. Biological Conservation 127:463–476.

Webb, J. K., G. P. Brown, T. Child, M. J. Greenlees, B. L. Phillips, and R. Shine. 2008. A native dasyurid predator (common planigale, *Planigale maculata*) rapidly learns to avoid a toxic invader. Austral Ecology 33:821–829.

Webb, M. J., J. A. Kerle, and J. W. Winter. 2008. Rock ringtail possum *Petropseudes dahli* (Collett, 1895). Pp. 243–244 in The mammals of Australia. 3rd edition (S. Van Dyck and R. Strahan, eds.). Reed New Holland, Sydney, Australia.

Webb, S. D. 1985. The interrelationships of tree sloths and ground sloths. Pp. 105–112 in The evolution and ecology of armadillos, sloths, and vermilinguas (G. G. Montgomery, ed.). Smithsonian Institution Press, Washington, DC.

Weigl, R. 2005. Longevity of mammals in captivity; from the living collections of the world. E. Schweizerbart'sche, Stuttgart, Germany.

Weksler, M., and C. Bonvicino. 2008. *Monodelphis umbristriatus*. The IUCN Red List of Threatened Species 2008: e.T136241A4264209. http://dx.doi.org/10.2305/IUCN.UK.2008.RLTS.T136241 A4264209.en. Downloaded on 05 January 2016.

Weksler, M., B. Patterson, and C. Bonvicino. 2008. *Marmosops handleyi*. The IUCN Red List of Threatened Species 2008: e.T12820A3385702. http://dx.doi.org/10.2305/IUCN.UK.2008.RLTS.T12820A3385702 .en. Downloaded on 10 November 2015.

Wells, R. T. 1978. Field observations of the hairy-nosed wombat, *Lasiorhinus latifrons* (Owen). Australian Wildlife Research 5:299–303.

Wells, R. T. 1989. Vombatidae. Pp. 755–768 in Fauna of Australia: Mammalia, volume 1B (D. W. Walton and B. J. Richardson, eds.). Australian Government Publishing Service, Canberra, Australia.

Wells, R. 1998. Preface. Pp. v–vi in Wombats (R. T. Wells, P. A. Pridmore, B. St. John, M. D. Gaughwin, and J. Ferris, eds.). Surrey Beatty & Sons, Chipping Norton, New South Wales, Australia.

Wells, R. T., and B. Green. 1998. Aspects of water metabolism in the southern hairy-nosed wombat *Lasiorhinus latifrons*. Pp. 61–66 in Wombats (R. T. Wells, P. A. Pridmore, B. St. John, M. D. Gaugh-win, and J. Ferris, eds.). Surrey Beatty & Sons, Chipping Norton, New South Wales, Australia.

Wesolek, C. M., J. Soltis, K. A. Leighty, and A. Savage. 2009. Infant African elephant rumble vocalizations vary according to social interactions with adult females. Bioacoustics 18:227–239.

West, M., D. Galloway, J. Shaw, A. Trouson, and M. C. J. Paris. 2004. Oestrous cycle of the common wombat, *Vombatus ursinus*, in Victoria, Australia. Reproduction, Fertility and Development 16: 339–346.

Westerman, M. 1991. Phylogenetic relationships of the marsupial mole, *Notoryctes typhlops* (Marsupialia: Notoryctidae). Australian Journal of Zoology 39:529–537.

Westerman, M., and C. Krajewski. 2000. Molecular relationships of the Australian bandicoot genera *Isoodon* and *Perameles* (Marsupialia: Peramelina). Australian Mammalogy 22:1–8.

Westerman, M., S. Loke, and M. S. Springer. 2004. Molecular phylogenetic relationships of two extinct potoroid marsupials, *Potorous platyops* and *Caloprymnus campestris* (Potoroinae: Marsupialia). Molecular Phylogenetics and Evolution 31:476–485.

Westerman, M., A. H. Sinclair, and P. A. Woolley. 1984. Cytology of the feathertail possum *Distoechurus pennatus*. Pp. 423–425 in Possums and gliders (A. Smith and I. Hume, eds.). Surrey Beatty & Sons, Norton, New South Wales, Australia.

Westerman, M., M. S. Springer, and C. Krajewski. 2001. Molecular relationships of the New Guinean bandicoot genera *Microperoryctes* and *Echymipera* (Marsupialia: Peramelina). Journal of Mammalian Evolution 8:93–105.

Westerman, M., J. Young, and C. Krajewski. 2008. Molecular relationships of species of *Pseudantechinus*, *Parantechinus* and *Dasykaluta* (Marsupialia: Dasyuridae). Australian Mammalogy 29:201–212.

Westerman, M., M. S. Springer, J. Dixon, and C. Krajewski. 1999. Molecular relationships of the extinct pig-footed bandicoot *Chaeropus ecaudatus* (Marsupialia: Perameloidea) using 12S rRNA sequences. Journal of Mammalian Evolution 6:271–287.

Westerman, M., A. Burk, H. M. Amrine-Madsen, G. J. Prideaux, J. A. Case, and M. S. Springer. 2002. Molecular evidence for the last survivor of an ancient kangaroo lineage. Journal of Mammalian Evolution 9:209–223.

Westerman, M., J. Young, S. Donnellan, P. A. Woolley, and C. Krajewski. 2006. Molecular relationships of New Guinean three-striped dasyures (*Myoictis*, Marsupialia: Dasyuridae). Journal of Mammalian Evolution 13:211–222.

Westerman, M., B. P. Kear, K. Aplin, R. W. Meredith, C. Emerling, and M. S. Springer. 2012. Phylogenetic relationships of living and recently extinct bandicoots based on nuclear and mitochondrial DNA sequences. Molecular Phylogenetics and Evolution 62:97–108.

Western, D. 1989a. The ecological role of elephants in Africa. Pachyderm 12:42–45.

Western, D. 1989b. Ivory trade under scrutiny. Pachyderm 12:2–3.

Western, D. 1989c. The undetected trade in rhino horn. Pachyderm 11:26–28.

Western, D., and W. K. Lindsay. 1984. Seasonal herd dynamics of a savanna elephant population. African Journal of Ecology 22:229–244.

Wetzel, R. M. 1975. The species of *Tamandua* Gray (Edentata, Myrmecophagidae). Proceedings of the Biological Society of Washington 88:95–112.

Wetzel, R. M. 1980. Revision of the naked-tailed armadillos, genus *Cabassous* McMurtie. Annals of Carnegie Museum 49:323–357.

Wetzel, R. M. 1982. Systematics, distribution, ecology, and conservation of South American edentates. Pymatuning Laboratory of Ecology Special Publication 6:345–375.

Wetzel, R. M. 1985a. The identification and distribution of Recent Xenarthra (= Edentata). Pp. 5–21 in The evolution and ecology of armadillos, sloths, and vermilinguas (G. G. Montgomery, ed.). Smithsonian Institution Press, Washington, DC.

Wetzel, R. M. 1985b. Taxonomy and distribution of armadillos, Dasypodidae. Pp. 23–46 in The evolution and ecology of armadillos, sloths, and vermilinguas (G. G. Montgomery, ed.). Smithsonian Institution Press, Washington, DC.

Wetzel, R. M., and F. D. de Avila-Pires. 1980. Identification and distribution of the Recent sloths of Brazil (Edentata). Revista Brasileira de Biologia 40:831–836.

Wetzel, R. M., and D. Kock. 1973. The identity of *Bradypus variegatus* Schinz (Mammalia, Edentata). Proceedings of the Biological Society of Washington 86:25–34.

Wetzel, R. M., and E. Mondolfi. 1979. The subgenera and species of long-nosed armadillos, genus *Dasypus* L. Pp. 43–63 in Vertebrate ecology in the northern neotropics (J. F. Eisenberg, ed.). Smithsonian Institution Press, Washington, DC.

Wetzel, R. M., A. L. Gardner, K. H. Redford, and J. F. Eisenberg. 2007. Order Cingulata (Illiger, 1811). Pp. 128–157 in Mammals of South America, volume 1, Marsupials, xenarthrans, shrews, and bats (A. L. Gardner, ed.). University of Chicago Press, Chicago, Illinois.

Wharton, C. H. 1950. Notes on the Philippine tree shrew, *Urogale everetti* Thomas. Journal of Mammalogy 31:352–354.

Whidden, H. P. 2002. Extrinsic snout musculature in Afrotheria and Lipotyphla. Journal of Mammalian Evolution 9:161–184.

White, J. L. 1993. Indicators of locomotor habits in xenarthrans: evidence for locomotor heterogeneity among fossil sloths. Journal of Vertebrate Paleontology 13:230–242.

White, J. L., and R. D. E. MacPhee. 2001. The sloths of the West Indies: a systematic and phylogenetic review. Pp. 201–235 in Biogeography of the West Indies. Patterns and perspectives. 2nd edition (C. A. Woods and F. E. Sergile, eds.). CRC Press, Boca Raton, Florida.

White, L. C., and J. J. Austin. 2017. Relict or reintroduction? Genetic population assignment of three Tasmanian devils (*Sarcophilus harrisii*) recovered on mainland Australia. Royal Society Open Science 4:170053. http://dx.doi.org/10.1098/rsos.170053.

Whitehead, P. J. P. 1977. The former southern distribution of New World manatees (*Trichechus* spp.). Biological Journal of the Linnean Society 9:165–189.

Whitehouse, A. M. 2002. Tusklessness in the elephant population of the Addo Elephant National Park, South Africa. Journal of Zoology 257:249–254.

Whyte, I., and R. Fayrer-Hosken. 2008. Playing elephant god. Ethics of managing wild African elephant populations. Pp. 399–417 in Elephants and ethics, toward a morality of existence (C. Wemmer and C. A. Christen, eds.). Johns Hopkins University Press, Baltimore, Maryland.

Whyte, I. J. (subed.). 2005. Order Proboscidea. Pp. 51–59 in The mammals of the Southern African Subregion. 3rd edition (J. D. Skinner and C. T. Chimimba, eds.). Cambridge University Press, Cape Town, South Africa.

Wible, J. R. 2009. The ear region of the pen-tailed treeshrew, *Ptilocercus lowii* (Gray, 1848) (Placentalia, Scandentia, Ptilocercidae). Journal of Mammalian Evolution 16:199–233.

Wible, J. R. 2011. On the treeshrew skull (Mammalia, Placentalia, Scandentia). Annals of Carnegie Museum 79:149–230.

Wible, J. R., and J. A. Hopson. 1993. Basicranial evidence for early mammal phylogeny. Pp. 45–62 in Mammal phylogeny: Mesozoic differentiation, multituberculates, monotremes, early therians, and marsupials (F. S. Szalay, M. J. Novacek, and M. C. McKenna, eds.). Springer-Verlag, New York, New York.

Wible, J. R., G. W. Rougier, and M. J. Novacek. 2005. Anatomical evidence for superordinal/ordinal eutherian taxa in the Cretaceous. Pp. 15–36 in The rise of placental mammals: origins and relationships of the major clades (K. D. Rose and J. D. Archibald, eds.). Johns Hopkins University Press, Baltimore, Maryland.

Wible, J. R., G. W. Rougier, M. J. Novacek, and R. J. Asher. 2007. Cretaceous eutherians and Laurasian origin for placental mammals near the K/T boundary. Nature 447:1003–1006.

Widholzer, F. L., and W. A. Voss. 1978. Breeding the giant anteater *Myrmecophaga tridactyla* at Sao Leopoldo Zoo. International Zoo Yearbook 18:122–123.

Wiese, R. J., and K. Willis. 2004. Calculation of longevity and life expectancy in captive elephants. Zoo Biology 23:365–373.

Wiese, R. J., and K. Willis. 2006. Population management of zoo elephants. International Zoo Yearbook 40:80–87.

Wildlife Conservation Society. 2012. Elephants safe in Congo Park amidst slaughter in surrounding forests. http://www.wcs.org/press/press-releases/elephants-safe-in-congo-park.aspx. Downloaded on 18 July 2012

Wildman, D. E., M. Uddin, J. C. Opazo, G. Liu, V. Lefort, S. Guindon, O. Gascuel, L. I. Grossman, R. Romero, and M. Goodman. 2007. Genomics, biogeography, and the diversification of placental mammals. Proceedings of the National Academy of Sciences of the United States of America 104:14395–14400.

Wilkinson, D. A., G. C. Grigg, and L. A. Beard. 1998. Shelter selection and home range of echidnas, *Tachyglossus aculeatus*, in the highlands of south-east Queensland. Wildlife Research 25:219–232.

Williams, L. 2005. Manatees appear to approve of the improved quality of the water in Lake Pontchartrain. Sirenews 44:11–12.

Williams, M. E., and D. P. Domning. 2004. Pleistocene or post-Pleistocene manatees in the Mississippi and Ohio River Valleys. Marine Mammal Science 20:167–176.

Wilson, B. A., and M. R. Bachmann. 2008. Swamp antechinus *Antechinus minimus* (Geoffroy, 1803). Pp. 93–94 in The mammals of Australia. 3rd edition (S. Van Dyck and R. Strahan, eds.). Reed New Holland, Sydney, Australia.

Wilson, B. A., and A. R. Bourne. 1984. Reproduction in the male dasyurid *Antechinus minimus maritimus* (Marsupialia: Dasyuridae). Australian Journal of Zoology 32:311–318.

Wilson, D. E. 2009. Class Mammalia. Pp. 17–47 in Handbook of the mammals of the world, volume 1, Carnivores (D. E. Wilson and R. A. Mittermeier, eds.). Lynx Edicions, Barcelona, Spain.

Wilson, D. E., and D. M. Reeder (eds.). 2005. Mammal species of the world: a taxonomic and geographic reference. 3rd edition. Johns Hopkins University Press, Baltimore, Maryland.

Wilson, V. J. 1975. Mammals of the Wankie National Park. Trustees of the National Museums and Monuments of Rhodesia Museum Memoir 5:1–147.

Winkel, K., and I. Humphrey-Smith. 1987. Diet of the marsupial mole, *Notoryctes typhlops* (Stirling, 1889) (Marsupialia: Notoryctidae). Australian Mammalogy 11:159–161.

Winter, J., S. Burnett, and R. Martin. 2008a. *Dendrolagus bennettianus*. The IUCN Red List of Threatened Species 2008: e.T6426A12759345. http://dx.doi.org/10.2305/IUCN.UK.2008.RLTS.T6426A12759345.en. Downloaded on 02 January 2016.

Winter, J., S. Burnett, and R. Martin. 2008b. *Petrogale sharmani*. The IUCN Red List of Threatened Species 2008: e.T16753A6358729. http://dx.doi.org/10.2305/IUCN.UK.2008.RLTS.T16753A6358729.en. Downloaded on 03 January 2016.

Winter, J., S. Burnett, and R. Martin. 2008c. *Petrogale coenensis*. The IUCN Red List of Threatened Species 2008: e.T16752A6358582. http://dx.doi.org/10.2305/IUCN.UK.2008.RLTS.T16752A6358582.en. Downloaded on 03 January 2016.

Winter, J., S. Burnett, and P. Menkhorst. 2008. *Macropus dorsalis*. The IUCN Red List of Threatened Species 2008: e.T40562A10318966. http://dx.doi.org/10.2305/IUCN.UK.2008.RLTS.T40562A10318966.en. Downloaded on 05 January 2016.

Winter, J., J. Woinarski, and A. Burbidge. 2008. *Lagorchestes conspicillatus*. The IUCN Red List of Threatened Species 2008: e.T11161A3258306. http://dx.doi.org/10.2305/IUCN.UK.2008.RLTS.T11161A3258306.en. Downloaded on 02 January 2016.

Winter, J., D. Lunney, M. Denny, S. Burnett, and P. Menkhorst. 2008. *Petaurus norfolcensis*. The IUCN Red List of Threatened Species 2008: e.T16728A6322169. http://dx.doi.org/10.2305/IUCN.UK.2008.RLTS.T16728A6322169.en. Downloaded on 30 December 2015.

Winter, J. W. 1979. The status of endangered Australian Phalangeridae, Petauridae, Burramyidae, Tarsipedidae, and the Koala. Pp. 45–59 in The status of endangered Australian wildlife (M. J. Tyler, ed.). Proceedings of the Centenary Symposium of the Royal Zoological Society of South Australia, Adelaide, South Australia, Australia.

Winter, J. W. 1984. Conservation studies of tropical rainforest possums. Pp. 469–481 in Possums and gliders (A. Smith and I. Hume, eds.). Surrey Beatty & Sons, Norton, New South Wales, Australia.

Winter, J. W., P. M. Johnson, and K. Vernes. 2008. Northern bettong *Bettongia tropica* (Wakefield, 1967). Pp. 293–294 in The mammals of Australia. 3rd edition (S. Van Dyck and R. Strahan, eds.). Reed New Holland, Sydney, Australia.

Winter, J. W., A. K. Krockenberger, and N. J. Moore. 2008. Green ringtail possum *Pseudochirops archeri* (Collen, 1884). Pp. 245–247 in The mammals of Australia. 3rd edition (S. Van Dyck and R. Strahan, eds.). Reed New Holland, Sydney, Australia.

Winter, J. W., and L. K.-P. Leung. 2008a. Common spotted cuscus *Spilocuscus maculatus* (Desmarest, 1818). Pp. 266–268 in The mammals of Australia. 3rd edition (S. Van Dyck and R. Strahan, eds.). Reed New Holland, Sydney, Australia.

Winter, J. W., and L. K.-P. Leung. 2008b. Southern common cuscus *Phalanger mimicus* (Thomas, 1922). Pp. 268–279 in The mammals of Australia. 3rd edition (S. Van Dyck and R. Strahan, eds.). Reed New Holland, Sydney, Australia.

Winter, J. W., and N. J. Moore. 2008. Herbert River ringtail possum *Pseudochirulus herbertensis* (Collett, 1884). Pp. 250–252 in The mammals of Australia. 3rd edition (S. Van Dyck and R. Strahan, eds.). Reed New Holland, Sydney, Australia.

Winter, J. W., N. J. Moore, and R. F. Wilson. 2008. Lemuroid ringtail possum *Hemibelideus lemuroides* (Collett, 1884). Pp. 238–240 in The mammals of Australia. 3rd edition (S. Van Dyck and R. Strahan, eds.). Reed New Holland, Sydney, Australia.

Winter, J. W., and M. Trenerry. 2008. Daintree River ringtail possum *Pseudochirulus cinereus* (Tate, 1945). Pp. 248–250 in The mammals of Australia. 3rd edition (S. Van Dyck and R. Strahan, eds.). Reed New Holland, Sydney, Australia.

Winter, J. W., H. A. Dillewaard, S. E. Williams, and E. E. Bolitho. 2004. Possums and gliders of north Queensland: distribution and conservation status. Pp. 26–50 in The biology of Australian possums and gliders (R. L. Goldingay and S. M. Jackson, eds.). Surrey Beatty & Sons, Chipping Norton, New South Wales, Australia.

Wirsing, A. J., M. R. Heithaus, and L. M. Dill. 2007. Living on the edge: dugongs prefer to forage in microhabitats that allow escape from rather than avoidance of predators. Animal Behaviour 74:93–101.

Wischusen, E. W. 1990. The foraging ecology and natural history of the Philippine flying lemur (*Cynocephalus volans*). Ph.D. dissertation, Cornell University, Ithaca, New York.

Wischusen, E. W., N. R. Ingle, and M. E. Richmond. 1992. Observations on the reproductive biology and social behaviour of the Philippine flying lemur (*Cynocephalus volans*). Malayan Nature Journal 46:65–71.

Wischusen, E. W., and M. E. Richmond. 1998. Foraging ecology of the Philippine flying lemur (*Cynocephalus volans*). Journal of Mammalogy 79:1288–1295.

Withers, P. C., and C. E. Cooper. 2009. Thermal, metabolic, and hygric physiology of the little red kaluta, *Dasykaluta rosamondae* (Dasyuromorphia: Dasyuridae). Journal of Mammalogy 90:752–760.

Withers, P. C., G. G. Thompson, and R. S. Seymour. 2000. Metabolic physiology of the north-western marsupial mole, *Notoryctes caurinus* (Marsupialia: Notoryctidae). Australian Journal of Zoology 48:241–258.

Witte, I. 2005. Kangaroos: misunderstood and maligned reproductive miracle workers. Pp. 188–207 in Kangaroos: myths and realities (M. Wilson and D. Croft, eds.). Australian Wildlife Protection Council, Melbourne, Victoria, Australia. Proceedings of the Royal Society of London B Biological Sciences 276:3513–3521.

Wittemyer, G. 2011. Family Elephantidae (elephants). Pp. 50–79 in Handbook of the mammals of the world, volume 2, Hoofed mammals (D. E. Wilson and R. A. Mittermeier, eds.). Lynx Edicions, Barcelona, Spain.

Wittemyer G., D. Daballen, and I. Douglas-Hamilton. 2013. Comparative demography of an at-risk African elephant population. PLoS ONE 8(1):e53726. doi:10.1371/journal.pone.0053726.

Wittemyer, G., I. Douglas-Hamilton, and W. M. Getz. 2005. The socioecology of elephants: analysis of the processes creating multitiered social structures. Animal Behaviour 69:1357–1371.

Wittemyer, G., and W. M. Getz. 2007. Hierarchical dominance structure and social organization in African elephants, *Loxodonta africana*. Animal Behaviour 73:671–681.

Wittemyer, G., W. M. Getz, F. Vollrath, and I. Douglas-Hamilton. 2007. Social dominance, seasonal movements, and spatial segregation in African elephants: a contribution to conservation behavior. Behavioral Ecology and Sociobiology 61:1919–1931

Wittemyer, G., J. B. A. Okello, H. B. Rasmussen, P. Arctander, S. Nyakaana, I. Douglas-Hamilton, and H. R. Siegismund. 2009. Where sociality and relatedness diverge: the genetic basis for hierarchical social organization in African elephants. Proceedings of the Royal Society of London B Biological Sciences 276:3513–3521.

Wittemyer, G., J. M. Northrupa, J. Blanc, I. Douglas-Hamilton, P. Omondif, and K. P. Burnhama. 2014. Illegal killing for ivory drives global decline in African elephants. Proceedings of the National

Academy of Sciences of the United States of America 111: 13117–13121.

Wodzicki, K., and J. E. C. Flux. 1971. The parma wallaby and its future. Oryx 11:40–47.

Woinarski, J. C. Z. 2004. In a land with few possums, even the common are rare: ecology, conservation and management of possums in the Northern Territory. Pp. 51–62 in The biology of Australian possums and gliders (R. L. Goldingay and S. M. Jackson, eds.). Surrey Beatty & Sons, Chipping Norton, New South Wales, Australia.

Woinarski, J. 2008a. Kakadu dunnart Sminthopsis bindi (Van Dyck, Woinarski, and Press, 1994). Pp. 129–130 in The mammals of Australia. 3rd edition (S. Van Dyck and R. Strahan, eds.). Reed New Holland, Sydney, Australia.

Woinarski, J. 2008b. Pseudantechinus bilarni. The IUCN Red List of Threatened Species 2008: e.T40636A10346155. http://dx.doi.org/10.2305/IUCN.UK.2008.RLTS.T40636A10346155.en. Downloaded on 05 January 2016.

Woinarski, J. 2008c. Macropus bernardus. The IUCN Red List of Threatened Species 2008: e.T12620A3366407. http://dx.doi.org/10.2305/IUCN.UK.2008.RLTS.T12620A3366407.en. Downloaded on 05 January 2016.

Woinarski, J. C. Z., A. A. Burbidge, and P. L. Harrison. 2014. The Action Plan for Australian Mammals 2012. CSIRO Publishing, Collingwood, Victoria, Australia.

Woinarski, J., and C. Dickman. 2008. Pseudantechinus mimulus. The IUCN Red List of Threatened Species 2008: e.T18447A8280035. http://dx.doi.org/10.2305/IUCN.UK.2008.RLTS.T18447A8280035.en. Downloaded on 05 January 2016.

Woinarski, J., S. Rhind, M. Oakwood. 2008. Phascogale pirata. The IUCN Red List of Threatened Species 2008: e.T16889A6546305. http://dx.doi.org/10.2305/IUCN.UK.2008.RLTS.T16889A6546305.en. Downloaded on 04 January 2016.

Woinarski, E. Ritchie, and J. Winter. 2008. Macropus antilopinus. The IUCN Red List of Threatened Species 2008: e.T40561A10319426. http://dx.doi.org/10.2305/IUCN.UK.2008.RLTS.T40561A10319426.en. Downloaded on 05 January 2016.

Woinarski, J., A. Burbidge, W. Telfer, N. McKenzie, and T. Start. 2008. Petrogale concinna. The IUCN Red List of Threatened Species 2008: e.T16761A6359479. http://dx.doi.org/10.2305/IUCN.UK.2008.RLTS.T16761A6359479.en. Downloaded on 03 January 2016.

Wood, D. H. 1970. An ecological study of Antechinus stuartii (Marsupialia) in a southeast Queensland rain forest. Australian Journal of Zoology 18:185–207.

Woodall, P. F. 1995. The male reproductive system and the phylogeny of elephant-shrews (Macroscelidea). Mammal Review 25:87–93.

Woodburne, M. O. 2003. Monotremes as pretribosphenic mammals. Journal of Mammalian Evolution 10:195–248.

Woodburne, M. O., R. H. Tedford, and M. Archer. 1987. New Miocene ringtail possums (Marsupialia: Pseudocheiridae) from South Australia. Pp. 639–679 in Possums and opossums (M. Archer, ed.). Surrey Beatty & Sons, Sydney, Australia.

Woodburne, M. O., and W. J. Zinsmeister. 1982. Fossil land mammal from Antarctica. Science 218:284–286.

Woodburne, M. O., and W. J. Zinsmeister. 1984. The first land mammal from Antarctica and its biogeographic implications. Journal of Paleontology 58:913–948.

Woodman, N., and N. B. Athfield. 2009. Post-Clovis survival of American mastodon in the southern Great Lakes region of North America. Quaternary Research 72:359–363.

Woods, C. A., and J. A. Ottenwalder. 1992. The natural history of southern Haiti. Florida Museum of Natural History, Gainesville, Florida.

Wooller, R. D., K. C. Richardson, C. A. M. Garavanta, V. M. Saffer, and K. A. Bryant. 2000. Opportunistic breeding in the polyandrous honey possum, Tarsipes rostratus. Australian Journal of Zoology 48:669–680.

Wooller, R. D., K. C. Richardson, V. M. Saffer, C. A. M. Garavanta, K. A. Bryant, A. N. Everaardt, and S. J. Wooller. 2004. The honey possum Tarsipes rostratus: an update. Pp. 312–317 in The biology of Australian possums and gliders (R. L. Goldingay and S. M. Jackson, eds.). Surrey Beatty & Sons, Chipping Norton, New South Wales, Australia.

Woolley, P. A. 1971a. Maintenance and breeding of laboratory colonies of Dasyuroides byrnei and Dasycercus cristicauda. International Zoo Yearbook 11:351–354.

Woolley, P. A. 1971b. Observations on the reproductive biology of the dibbler, Antechinus apicalis (Marsupialia: Dasyuridae). Journal of the Royal Society of Western Australia 54:99–102.

Woolley, P. A. 1982a. Observations on the feeding and reproductive status of captive feather-tailed possums, Distoechurus pennatus (Marsupialia: Burramyidae). Australian Mammalogy 5:285–287.

Woolley, P. A. 1982b. Phallic morphology of the Australian species of Antechinus (Dasyuridae, Marsupialia): a new taxonomic tool? Pp. 767–781 in Carnivorous marsupials (M. Archer, ed.). Royal Zoological Society of New South Wales, Mosman, New South Wales, Australia.

Woolley, P. A. 1983. Dibbler Parantechinus apicalis. P. 29 in The Australian Museum complete book of Australian mammals (R. Strahan, ed.). Angus & Robertson, London, United Kingdom.

Woolley, P. A. 1984. Reproduction in Antechinomys laniger ("spenceri" form) (Marsupialia: Dasyuridae): field and laboratory investigations. Australian Wildlife Research 11:481–489.

Woolley, P. A. 1988. Reproduction in the Ningbing antechinus (Marsupialia: Dasyuridae): field and laboratory observations. Australian Wildlife Research 15:149–156.

Woolley, P. A. 1989. Nest location by spool-and-line tracking of dasyurid marsupials in New Guinea. Journal of Zoology 218:689–700.

Woolley, P. A. 1990. Mulgaras, Dasycercus cristicauda (Marsupialia: Dasyuridae): their burrows, and records of attempts to collect live animals between 1966 and 1979. Australian Mammalogy 13:61–64.

Woolley, P. A. 1991a. Reproduction in Dasykaluta rosamondae (Marsupialia: Dasyuridae): field and laboratory observations. Australian Journal of Zoology 39:549–568.

Woolley, P. A. 1991b. Reproduction in Pseudantechinus macdonnellensis (Marsupialia: Dasyuridae): field and laboratory observations. Wildlife Research 18:13–25.

Woolley, P. A. 1991c. Reproductive pattern of captive Boullanger Island dibblers, Parantechinus apicalis (Marsupialia: Dasyuridae). Wildlife Research 18:157–163.

Woolley, P. A. 1992. New records of the Julia Creek dunnart, Sminthopsis douglasi (Marsupialia: Dasyuridae). Wildlife Research 19:779–784.

Woolley, P. A. 1994. The dasyurid marsupials of New Guinea: use of museum specimens to assess seasonality of breeding. Science in New Guinea 20:49–55.

Woolley, P. A. 2001. Observations on the reproductive biology of Myoictis wallacei, Neophascogale lorentzi, Dasyurus albopunctatus and Dasyurus spartacus, dasyurid marsupials endemic to New Guinea. Australian Mammalogy 23:63–66.

Woolley, P. A. 2003. Reproductive biology of some dasyurid marsupials of New Guinea. Pp. 169–182 in Predators with pouches: the biology of carnivorous marsupials (M. Jones, C. Dickman, and M. Archer, eds.). CSIRO Publishing, Collingwood, Victoria, Australia.

Woolley, P. A. 2005a. Revision of the three-striped dasyures, genus Myoictis (Marsupialia: Dasyuridae), of New Guinea, with description of a new species. Records of the Australian Museum 57:321–340.

Woolley, P. A. 2005b. The species of Dasycercus (Peters, 1875) (Marsupialia: Dasyuridae). Memoirs of Museum Victoria 62:213–221.

Woolley, P. A. 2006. Studies on the crest-tailed mulgara Dasycercus cristicauda and the brush-tailed mulgara Dasycercus blythi (Marsupialia: Dasyuridae). Australian Mammalogy 28:117–120.

Woolley, P. A. 2008a. Family Dasyuridae dasyurids. Pp. 45–46, 81, 107, and 122 in The mammals of Australia. 3rd edition (S. Van Dyck and R. Strahan, eds.). Reed New Holland, Sydney, Australia.

Woolley, P. A. 2008b. Brush-tailed mulgara Dasycercus blythi (Waite, 1904). Pp. 47–48 in The mammals of Australia. 3rd edition (S. Van Dyck and R. Strahan, eds.). Reed New Holland, Sydney, Australia.

Woolley, P. A. 2008c. Kaluta Dasykaluta rosamondae (Ride, 1964). Pp. 51–52 in The mammals of Australia. 3rd edition (S. Van Dyck and R. Strahan, eds.). Reed New Holland, Sydney, Australia.

Woolley, P. A. 2008d. Dibbler Parantechinus apicalis (Gray, 1842). Pp. 65–66 in The mammals of Australia. 3rd edition (S. Van Dyck and R. Strahan, eds.). Reed New Holland, Sydney, Australia.

Woolley, P. A. 2008e. Sandstone pseudantechinus Pseudantechinus bilarni (Johnson, 1954). Pp. 67–68 in The mammals of Australia. 3rd edition (S. Van Dyck and R. Strahan, eds.). Reed New Holland, Sydney, Australia.

Woolley, P. A. 2008f. Ningbing pseudantechinus Pseudantechinus ningbing (Kitchener, 1988). Pp. 73–74 in The mammals of Australia. 3rd edition (S. Van Dyck and R. Strahan, eds.). Reed New Holland, Sydney, Australia.

Woolley, P. A. 2008g. Woolley's pseudantechinus Pseudantechinus woolleyae (Kitchener and Caputi, 1988). Pp. 76–77 in The mammals of Australia. 3rd edition (S. Van Dyck and R. Strahan, eds.). Reed New Holland, Sydney, Australia.

Woolley, P. A. 2008h. Butler's dunnart Sminthopsis butleri (Archer, 1979). Pp. 130–131 in The mammals of Australia. 3rd edition (S. Van Dyck and R. Strahan, eds.). Reed New Holland, Sydney, Australia.

Woolley, P. A 2008i. Julia Creek dunnart Sminthopsis douglasi (Archer, 1979). Pp. 136–137 in The mammals of Australia. 3rd edition (S. Van Dyck and R. Strahan, eds.). Reed New Holland, Sydney, Australia.

Woolley, P. A. 2008j. Red-cheeked dunnart Sminthopsis virginiae (Tarragon, 1847). Pp. 158–159 in The mammals of Australia. 3rd edition (S. Van Dyck and R. Strahan, eds.). Reed New Holland, Sydney, Australia.

Woolley, P. 2008k. Dasycercus cristicauda. The IUCN Red List of Threatened Species 2008: e.T6266A12593048. http://dx.doi.org/10.2305/IUCN.UK.2008.RLTS.T6266A12593048.en. Downloaded on 05 January 2016.

Woolley, P. 2008l. Dasycercus blythi. The IUCN Red List of Threatened Species 2008: e.T6267A12593219. http://dx.doi.org/10.2305/IUCN.UK.2008.RLTS.T6267A12593219.en. Downloaded on 05 January 2016.

Woolley, P.A. 2011. Pseudantechinus mimulus: a little known dasyurid marsupial. Australian Mammalogy 33:57–67.

Woolley, P. A., C. Krajewski, and M. Westerman. 2015. Phylogenetic relationships within Dasyurus (Dasyuromorphia: Dasyuridae): quoll systematics based on molecular evidence and male characteristics. Journal of Mammalogy 96:37–46.

Woolley, P. A., M. Westerman, and C. Krajewski. 2007. Interspecific affinities within the genus Sminthopsis (Dasyuromorphia: Dasyuridae) based on morphology of the penis: congruence with other anatomical and molecular data. Journal of Mammalogy 88:1381–1392.

Woolley, P. A., S. A. Raftopoulos, G. J. Coleman, and S. M. Armstrong. 1991. A comparative study of circadian activity patterns of two New Guinean dasyurid marsupials, Phascolosorex dorsalis and Antechinus habbema. Australian Journal of Zoology 39:661–671.

Woolley, P., T. Leary, L. Seri, T. Flannery, D. Wright, S. Hamilton, K. Helgen, R. Singadan, J. Menzies, A. Allison, and R. James. 2008. Dasyurus albopunctatus. The IUCN Red List of Threatened Species 2008: e.T6299A12600801. http://dx.doi.org/10.2305/IUCN.UK.2008.RLTS.T6299A12600801.en. Downloaded on 05 January 2016.

Woolnough, A. P., and V. R. Steele. 2001. The palaeoecology of the Vombatidae: did giant wombats burrow? Mammal Review 31:33–45.

World Health Organization. 2013. Global status report on road safety 2013: supporting a decade of action. World Health Organization, Geneva, Switzerland.

Wrege, P. H., E. D. Rowland, N. Bout, and M. Doukaga. 2011. Opening a larger window onto forest elephant ecology. African Journal of Ecology 50:176–183.

Wright, P. G. 1984. Why do elephants flap their ears? South African Journal of Zoology 19:266–269.

Wroe, S. 1997. A reexamination of proposed morphology-based synapomorphies for the families of Dasyuromorphia (Marsupialia). 1. Dasyuridae. Journal of Mammalian Evolution 4:19–52.

Wroe, S. 1999. The geologically oldest dasyurid, from the Miocene of Riversleigh, north-west Queensland. Palaeontology 42:501–527.

Wroe, S. 2003. Australian marsupial carnivores: recent advances in palaeontology. Pp. 102–123 in Predators with pouches: the biology of carnivorous marsupials (M. Jones, C. Dickman, and M. Archer, eds.). CSIRO Publishing, Collingwood, Victoria, Australia.

Wroe, S. 2008. Family Thylacinidae thylacine. P. 166 in The mammals of Australia. 3rd edition (S. Van Dyck and R. Strahan, eds.). Reed New Holland, Sydney, Australia.

Wroe, S., C. McHenry, and J. Thomason. 2005. Bite club: comparative bite force in big biting mammals and the prediction of predatory behaviour in fossil taxa. Proceedings of the Royal Society of London B Biological Sciences 272:619–625.

Wroe, S., and J. Muirhead. 1999. Evolution of Australia's marsupicarnivores: Dasyuridae, Thylacinidae, Myrmecobiidae, Dasyuromorphia incertae sedis and Marsupialia incertae sedis. Australian Mammalogy 21:10–11.

Wroe, S., M. Ebach, S. Ahyong, C. De Muizon, and J. Muirhead. 2000. Cladistic analysis of dasyuromorphian (Marsupialia) phylogeny using cranial and dental characters. Journal of Mammalogy 81:1008–1024.

Wroe, S., T. Myers, F. Seebacher, B. Kear, A. Gillespie, M. Crowther, and S. Salisbury. 2003. An alternative method for predicting body mass: the case of the Pleistocene marsupial lion. Paleobiology 29:403–411.

Wroe, S., M. Crowther, J. Dortch, and J. Chong. 2004. The size of the largest marsupial and why it matters. Proceedings of the Royal Society of London B Biological Sciences (Supplement) 271:S34–S36.

WWF International. 2006. Human-animal conflict. WWF International, Gland, Switzerland. http://assets.wwf.ch/downloads/conflictw.pdf. Downloaded on 12 March 2015.

Yager, R. H., and C. B. Frank. 1972. The nine-banded armadillo for medical research. Institute for Laboratory Animal Research News 15(2):4–5.

Yalden, D. W., M. J. Largen, and D. Kock. 1986. Catalogue of the mammals of Ethiopia. 6. Perissodactyla, Proboscidea, Hyracoidea, Lagomorpha, Tubulidentata, Sirenia, and Cetacea. Italian Journal of Zoology, Supplement, new series, 21:31–103.

Yang, H., E. M. Golenberg, and J. Shoshani. 1996. Phylogenetic resolution within the Elephantidae using fossil DNA sequence from the American mastodon (Mammut americanum) as an outgroup. Proceedings of the National Academy of Sciences of the United States of America 93:1190–1194.

Yates, T. L. 1984. Insectivores, elephant shrews, tree shrews, and dermopterans. Pp. 117–144 in Orders and families of Recent mammals of the world (S. Anderson and J. K. Jones, Jr., eds.). John Wiley & Sons, New York, New York.

Young, R. J., C. M. Coelho, and D. R. Wieloch. 2003. A note on the climbing abilities of giant anteaters, Myrmecophaga tridactyla (Xenarthra, Myrmecophagidae). Boletim do Museu de Biologia Mello Leitão (nova série) 15:41–46.

Zack, S. P., T. A. Penkrot, J. I. Bloch, and K. D. Rose. 2005. Affinities of "hyopsodontids" to elephant shrews and a Holarctic origin of Afrotheria. Nature 434:497–501.

Zamorano, M., and G. J. Scillato-Yane. 2008. Registro de Dasypus (Dasypus) novemcinctus (Mammalia, Dasypodidae) en el sudoeste de la provincia de Buenos Aires, Argentina. BioScriba 1:17–26.

Zapata, S. C., D. Procopio, A. Travaini, and A. Rodríguez. 2013. Summer food habits of the Patagonian opossum, Lestodelphys halli (Thomas, 1921), in southern arid Patagonian shrub-steppes. Gayana 77:64–67.

Zarza, H., G. Ceballos, and M. A. Steele. 2003. Marmosa canescens. Mammalian Species 725:1–4.

Zenger, K. R., M. D. B. Eldridge, and P. G. Johnston. 2005. Phylogenetics, population structure and genetic diversity of the endangered southern brown bandicoot (Isoodon obesulus) in south-eastern Australia. Conservation Genetics 6:193–204.

Zhang, L. 2011. Current status of Asian elephants in China. Gajah 35:43–46.

Zhang, L., L. Ma, and L. Feng. 2006. New challenges facing traditional nature reserves: Asian elephant (Elephas maximus) conservation in China. Integrative Zoology 1:179–187.

Ziegler, A. C. 1972. Additional specimens of Planigale novaeguineae (Dasyuridae: Marsupialia) from Territory of Papua. Australian Mammalogy 1:43–45.

Ziegler, A. C. 1977. Evolution of New Guinea's marsupial fauna in response to a forested environment. Pp. 117–138 in The biology of marsupials (B. Stonehouse and D. P. Gilmore, eds.). University Park Press, Baltimore, Maryland.

Ziegler, A. C. 1981. Petaurus abidi, a new species of glider (Marsupialia: Petauridae) from Papua New Guinea. Australian Mammalogy 4:81–88.

Ziegler, R. 1999. Amphiperatherium, the last European opossum. Pp. 49–52 in The Miocene land mammals of Europe (G. E. Rössner and K. Heissig, eds.). Dr. Friedrich Pfeil, Munich, Germany.

Index

The scientific names of orders, families, and genera, which have titled accounts in the text, are in boldface type. The page numbers on which such accounts begin are also in boldface type. Other scientific names, and vernacular names, appear in ordinary type. If the scientific and common names of a mammal are identical, they usually are indexed separately (for example, **Antechinus** and antechinus). Information on an indexed subject sometimes is found in more than one account on the same page. Names in legends of illustrations usually are not indexed independently unless an illustration appears separately from the main text on the genus (or species) shown; usually in such a case only the scientific name is indexed. If no illustration of a genus appears on the page indexed for that genus, the reader should look one or two pages either before or after the account. However, not every account has an accompanying illustration (see preface).

GEOLOGICAL TIME

millions of years ago	epoch	period
280		**PERMIAN**
225		**TRIASSIC**
190		**JURASSIC**
135		**CRETACEOUS**
100		
65	Paleocene	**TERTIARY**
55	Eocene	
50		
38	Oligocene	
26	Miocene	
10		
7	Pliocene	
3	Pleistocene	**QUATERNARY**
1		
0.01	Holocene	
0.005	Historical Time	

WEIGHT
scales for comparison of metric and U.S. units of measurement

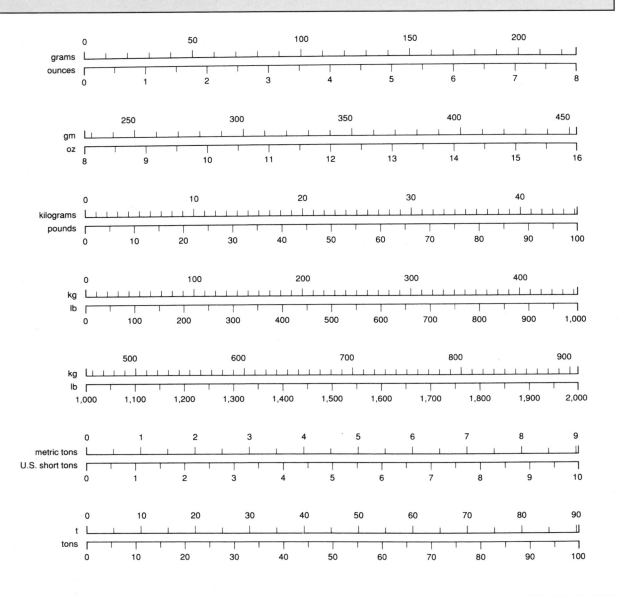

TEMPERATURE
scales for comparison of metric and U.S. units of measurement

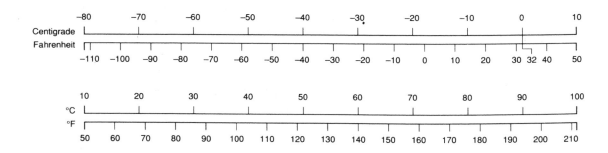